FOURTH EDITION

Matter & Interactions

VOLUME I
Modern Mechanics

VOLUME II
Electric and Magnetic Interactions

RUTH W. CHABAY
BRUCE A. SHERWOOD

North Carolina State University

WILEY

VICE PRESIDENT & PUBLISHER Petra Recter
EXECUTIVE EDITOR Jessica Fiorillo
ASSOCIATE EDITOR Aly Rentrop
EDITORIAL ASSISTANT Amanda Rillo
SENIOR MARKETING MANAGER Kristy Ruff
MARKETING ASSISTANT Claudine Scrivanich
SENIOR CONTENT MANAGER Kevin Holm
SENIOR PRODUCTION EDITOR Elizabeth Swain
DESIGN DIRECTOR Harry Nolan
TEXT/COVER DESIGN Maureen Eide
COVER IMAGE Ruth Chabay

Cover Description: The cover image is a snapshot from a VPython program that models the motion of a mass-spring system in 3D (see Computational Problem P70 in Chapter 4).

This book was set in 10/12 Times Ten Roman in LaTex by MPS and printed and bound by Quad Graphics. The cover was printed by Quad Graphics.

This book is printed on acid-free paper.

Complete text ISBN 978-1-118-87586-5
Complete text binder version ISBN 978-1-118-91451-9

Volume 1 ISBN 978-1-118-91449-6
Volume 1 Binder version ISBN 978-1-118-91452-6

Volume 2 ISBN 978-1-118-91450-2
Volume 2 Binder version ISBN 978-1-118-91453-3

Printed in the United States of America
SKY10035815_082022

Brief Contents

VOLUME I Modern Mechanics

VOLUME II Electric and Magnetic Interactions

The Supplements can be found at the web site, www.wiley.com/college/chabay

Contents

VOLUME I
Modern Mechanics

CHAPTER 1 *Interactions and Motion* 1

CHAPTER 2 *The Momentum Principle* 45

CHAPTER 3 *The Fundamental Interactions* 88

CHAPTER 4 *Contact Interactions* 130

The Supplements can be found at the web site, www.wiley.com/college/chabay

Preface

TO THE STUDENT

This textbook emphasizes a 20th-century perspective on introductory physics. Contemporary physicists build models of the natural world that are based on a small set of fundamental physics principles and on an understanding of the microscopic structure of matter, and they apply these models to explain and predict a very broad range of physical phenomena. In order to involve students of introductory physics in the contemporary physics enterprise, this textbook emphasizes:

- Reasoning directly from a small number of fundamental physics principles, rather than from a large set of special-case equations.
- Integrating contemporary insights, such as atomic models of matter, quantized energy, and relativistic dynamics, throughout the curriculum.
- Engaging in the full process of creating and refining physical models (idealizing, making approximations, explicitly stating assumptions, and estimating quantities).
- Reasoning iteratively about the time-evolution of system behavior, both on paper and through the construction and application of computational models.

Because the physical world is 3-dimensional, we work in 3D throughout the text. Many students find the approach to 3D vectors used in this book easier than standard treatments of 2D vectors.

Textbook and Supplemental Resources

Modern Mechanics (Volume 1, Chapters 1–12) focuses on the atomic structure of matter and interactions between material objects. It emphasizes the wide applicability and utility of a small number of fundamental principles: the Momentum Principle, the Energy Principle, and the Angular Momentum Principle, and the Fundamental Assumption of Statistical Mechanics. We study how to explain and predict the behavior of systems as different as elementary particles, molecules, solid metals, and galaxies.

Electric and Magnetic Interactions (Volume 2, Chapters 13–23) emphasizes the somewhat more abstract concepts of electric and magnetic fields and extends the study of the atomic structure of matter to include the role of electrons. The principles of electricity and magnetism are the foundation for much of today's technology, from cell phones to medical imaging.

Additional resources for students are freely available at this site:

www.wiley.com/college/chabay

The web resources include several supplements. A copy of Chapter 1 is provided for students who are currently using Volume 2 but whose previous physics course did not use Volume 1. This chapter introduces 3D vectors and vector algebra, and includes an introduction to computational modeling in VPython, which is used throughout the textbook.

Supplement S1 treats the kinetic theory of gases and heat engines, and can be used by students who have completed Chapter 12 on Entropy. Supplement S2 explains the basic principles of PN junctions in semiconductor devices, and can be used by students who have completed Chapter 21: Patterns of Field in Space. Supplement S3 includes a more mathematically sophisticated treatment of mechanical and electromagnetic waves and wave phenomena, and can be used by students who have completed Chapter 23 on Electromagnetic Radiation.

Answers to odd-numbered problems may be found at the end of the book.

The new Student Solutions Manual is available for purchase as a printed supplement and contains fully worked solutions for a subset of end of chapter problems.

Prerequisites

This book is intended for introductory calculus-based college physics courses taken by science and engineering students. It requires a basic knowledge of derivatives and integrals, which can be obtained by studying calculus concurrently.

Modeling

Matter & Interactions places a major emphasis on constructing and using physical models. A central aspect of science is the modeling of complex real-world phenomena. A physical model is based on what we believe to be fundamental principles; its intent is to predict or explain the most important aspects of an actual situation. Modeling necessarily involves making approximations and simplifying assumptions that make it possible to analyze a system in detail.

Computational Modeling

Computational modeling is now as important as theory and experiment in contemporary science and engineering. We introduce you to serious computer modeling right away to help you build a strong foundation in the use of this important tool.

In this course you will construct simple computational models based on fundamental physics principles. You do not need any prior programming experience–this course will teach you the small number of computational concepts you will need. Using VPython, a computational environment based on the Python programming language, you will find that after less than an hour you can write a simple computational model that produces a navigable 3D animation as a side effect of your physics code.

Computational modeling allows us to analyze complex systems that would otherwise require very sophisticated mathematics or that could not be analyzed at all without a computer. Numerical calculations based on the Momentum Principle give us the opportunity to watch the dynamical evolution of the behavior of a system. Simple models frequently need to be refined and extended. This can be done straightforwardly with a computer model but is often impossible with a purely analytical (non-numerical) model.

VPython is free, and runs on Windows, MacOS, and Linux. Instructions in Chapter 1 tell you how to install it on your own computer, and how to find a set of instructional videos that will help you learn to use VPython.

Questions

As you read the text, you will frequently come to a question that looks like this:

> QUESTION What should I do when I encounter a question in the text?

A question invites you to stop and think, to make a prediction, to carry out a step in a derivation or analysis, or to apply a principle. These questions are answered in the following paragraphs, but it is important that you make a serious effort to answer the questions on your own before reading further. Be honest in comparing your answers to those in the text. Paying attention to surprising or counterintuitive results can be a useful learning strategy.

Checkpoints

Checkpoints at the end of some sections ask you to apply new concepts or techniques. These may involve qualitative reasoning or simple calculations. You should complete these checkpoints when you come to them, before reading further. The goal of a checkpoint is to help you consolidate your understanding of the material you have just read, and to make sure you are ready to continue reading. Answers to checkpoints are found at the end of each chapter.

Conventions Used in Diagrams

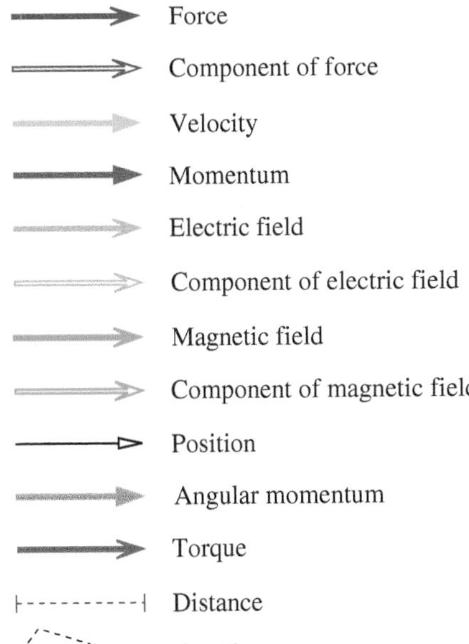

The conventions most commonly used to represent vectors and scalars in diagrams in this text are shown in the margin. In equations and text, a vector will be written with an arrow above it: $\vec{p}$.

TO THE INSTRUCTOR

The approach to introductory physics in this textbook differs significantly from that in most textbooks. Key emphases of the approach include:

- Starting from fundamental principles rather than secondary formulas
- Atomic-level description and analysis
- Modeling the real world through idealizations and approximations
- Computational modeling of physical systems
- Unification of mechanics and thermal physics
- Unification of electrostatics and circuits
- The use of 3D vectors throughout

Web Resources for Instructors

Instructor resources are available at this web site:

www.wiley.com/college/chabay

Resources on this site include lecture-demo software, textbook figures, clicker questions, test questions, lab activities including experiments and computational modeling, a computational modeling guide, and a full solutions manual. Contact your Wiley representative for information about this site.

Electronic versions of the homework problems are available in WebAssign:

www.webassign.net

Some instructor resources are available through WebAssign as well.

Other information may be found on the authors' *Matter & Interactions* web site:

matterandinteractions.org

Also on the authors' website are reprints of published articles about *Matter & Interactions*, including these:

- Chabay, R. & Sherwood, B. (1999). Bringing atoms into first-year physics. *American Journal of Physics* 67, 1045–1050.
- Chabay, R. W. & Sherwood, B. (2004). Modern mechanics. *American Journal of Physics* 72, 439–445.
- Chabay, R. W. & Sherwood, B. (2006). Restructuring Introductory E&M. *American Journal of Physics* 74, 329–336.
- Chabay, R. & Sherwood, B. (2008) Computational physics in the introductory calculus-based course. *American Journal of Physics* 76(4&5), 307–313.
- Beichner, R., Chabay, R., & Sherwood, B. (2010) Labs for the Matter & Interactions curriculum. *American Journal of Physics* 78(5), 456–460.

Computational Homework Problems

Some important homework problems require the student to write a simple computer program. The textbook and associated instructional videos teach VPython, which is based on the Python programming language, and which generates real-time 3D animations as a side effect of simple physics code written by students. Such animations provide powerfully motivating and instructive visualizations of fields and motions. VPython supports true vector computations, which encourages students to begin thinking about vectors as much more than mere components. VPython can be obtained at no cost for Windows, Macintosh, and Linux at vpython.org.

In the instructor resources section of matterandinteractions.org is "A Brief Guide to Computational Modeling in Matter & Interactions" which explains how to incorporate computation into the curriculum in a way that is easy for instructors to manage and which is entirely accessible to students with no prior programming experience. There you will also find a growing list of advanced computational physics textbooks that use VPython, which means that introducing students to Python and VPython in the introductory physics course can be of direct utility in later courses. Python itself is now widely used in technical fields.

Desktop Experiment Kit for Volume 2

On the authors' web site mentioned above is information about a desktop experiment kit for E&M that is distributed by PASCO. The simple equipment in this kit allows students to make key observations of electrostatic, circuit, and magnetic phenomena, tightly integrated with the theory (www.pasco.com, search for EM-8675). Several chapters contain optional experiments that can be done with this kit. This does not preclude having other, more complex laboratory experiences associated with the curriculum. For example, one such lab that we use deals with Faraday's law and requires signal generators, large coils, and oscilloscopes. You may have lab experiments already in place that will go well with this textbook.

What's New in the 4th Edition

The 4th edition of this text includes the following major new features:

- Increased support for computational modeling throughout, including sample code.
- Discussion throughout the text contrasting iterative and analytical problem solutions.
- Many new computational modeling problems (small and large).
- Improved discussion throughout the text of the contrast between models of a system as a point particle and as an extended system.
- An improved discussion of the Momentum Principle throughout Volume 1, emphasizing that the future momentum depends on two elements: the momentum now, and the impulse applied.
- Improved treatment of polarization surface charge in electrostatics (Chapter 14) and circuits (Chapter 18) based on the results of detailed 3D computational models.
- A more extensive set of problems at the end of each chapter, with improved indication of difficulty level.

In order to reduce cost and weight, some materials that have seen little use by instructors have been moved to the Wiley web site (www.wiley.com/college/chabay) where they are freely available. These materials include Supplement S1 (Chapter 13 in the 3rd Edition: kinetic theory of gases, thermal processes, and heat engines), Supplement S2 on PN junctions (formerly an optional section in Chapter 22 in the 3rd Edition), and Supplement S3 (a significantly extended version of Chapter 25 in the 3rd Edition: electromagnetic interference and diffraction, wave-particle duality, and a new section on mechanical waves and the wave equation).

Additional changes in the 4th Edition include:

- In Chapter 5, improved treatment of curving motion and an added section on the dynamics of multiobject systems.
- An improved sequence of topics in Chapter 6, with an explicit discussion of the role of energy in computational models, and an improved treatment of path independence, highlighting its limitation to point particles.
- A new section in Chapter 7 on the effect of the choice of reference frame on the form of the Energy Principle, and explicit instruction on how to model several kinds of friction in a computational model.
- In Chapter 8, discussion of the lifetime of excited states and on the probabilistic nature of energy transitions.
- In Chapter 9, now renamed "Translational, Rotational, and Vibrational Energy," improved treatment of the energetics of deformable systems.
- In Chapter 11, analysis of a physical pendulum.
- A detailed discussion in Chapter 16 of how to calculate potential difference by numerical path integration.
- An improved treatment of motional emf in the case of a bar dragged along rails (Chapter 20).

Suggestions for Condensed Courses

In a large course for engineering and science students with three 50-minute lectures and one 110-minute small-group studio lab per week, or in a studio format with five 50-minute sessions per week, it is possible to complete most but not all of the mechanics and E&M material in two 15-week semesters. In an honors course, or a course for physics majors, it is possible to do almost everything. You may be able to go further or deeper if your course has a weekly recitation session in addition to lecture and lab.

What can be omitted if there is not enough time to do everything? In mechanics, the one thing we feel should not be omitted is the introduction to entropy in terms of the statistical mechanics of the Einstein solid (Chapter 12). This is a climax of the integration of mechanics and thermal physics. One approach to deciding what mechanics topics can be omitted is to be guided by what foundation is required for Chapter 12. See other detailed suggestions below.

In E&M, one should not omit electromagnetic radiation and its effects on matter (Chapter 23). This is the climax of the whole E&M enterprise. One way to decide what E&M topics can be omitted is to be guided by what foundation is required for Chapter 23. See other detailed suggestions below.

Any starred section (*) can safely be omitted. Material in these sections is not referenced in later work. In addition, the following sections may be omitted:

Chapter 3 (The Fundamental Interactions): The section on determinism may be omitted.

Chapter 4 (Contact Interactions): Buoyancy and pressure may be omitted (one can return to these topics by using Supplement S1 on gases).

Chapter 7 (Internal Energy): If you are pressed for time, you might choose to omit the second half of the chapter on energy dissipation, beginning with Section 7.10.

Chapter 9 (Translational, Rotational, and Vibrational Energy): The formalism of finding the center of mass may be skipped, because the important applications have obvious locations of the center of mass. Although they are very instructive, it is possible to omit the sections contrasting point-particle with extended system models; you may also omit the analysis of sliding friction.

Chapter 10 (Collisions): A good candidate for omission is the analysis of collisions in the center-of-mass frame. Since there is a basic introduction to collisions in Chapter 3 (before energy is introduced), one could omit all of Chapter 10. On the other hand, the combined use of the Momentum Principle and the Energy Principle can illuminate both fundamental principles.

Chapter 11 (Angular Momentum): The main content of this chapter should not be omitted, as it introduces the third fundamental principle of mechanics, the Angular Momentum Principle. One might choose to omit most applications involving nonzero torque.

Chapter 12 (Entropy: Limits on the Possible): The second half of this chapter, on the Boltzmann distribution, may be omitted if necessary.

Chapter 15 (Electric Field of Distributed Charges): It is important that students acquire a good working knowledge of the patterns of electric field around some standard charged objects (rod, ring, disk, capacitor, sphere). If however they themselves are to acquire significant expertise in setting up physical integrals, they need extensive practice, and you might decide that the amount of time necessary for acquiring this expertise is not an appropriate use of the available course time.

Chapter 16 (Electric Potential): The section on dielectric constant can be omitted if necessary.

Chapter 17 (Magnetic Field): In the sections on the atomic structure of magnets, you might choose to discuss only the first part, in which one finds that the magnetic moment of a bar magnet is consistent with an atomic model. Omitting the remaining sections on spin and domains will not cause significant difficulties later.

Chapter 19 (Circuit Elements): The sections on series and parallel resistors and on internal resistance, meters, quantitative analysis of RC circuits, and multiloop circuits can be omitted. Physics and engineering students who need to analyze complex multiloop circuits will later take specialized courses on the topic; in the introductory physics course the emphasis should be on giving all students a good grounding in the fundamental mechanisms underlying circuit behavior.

Chapter 20 (Magnetic Force): We recommend discussing Alice and Bob and Einstein, but it is safe to omit the sections on relativistic field transformations. However, students often express high interest in the relationship between electric fields and magnetic fields, and here is an opportunity to satisfy some aspects of their curiosity. Motors and generators may be omitted or downplayed. The case study on sparks in air can be omitted, because nothing later depends critically on this topic, though it provides an introductory-level example of a phenomenon where an intuitively appealing model fails utterly, while a different model predicts several key features of the phenomenon. Another possibility is to discuss sparks near the end of the course, because it can be a useful review of many aspects of E&M.

Chapter 22 (Faraday's Law): Though it can safely be omitted, we recommend retaining the section on superconductors, because students are curious about this topic. The section on inductance may be omitted.

Chapter 23 (Electromagnetic Radiation): The treatment of geometrical optics may be omitted.

Acknowledgments

We owe much to the unusual working environment provided by the Department of Physics and the former Center for Innovation in Learning at Carnegie Mellon, which made it possible during the 1990s to carry out the research and development leading to the first edition of this textbook in 2002. We are grateful for the open-minded attitude of our colleagues in the Carnegie Mellon physics department toward curriculum innovations.

We are grateful to the support of our colleagues Robert Beichner and John Risley in the Physics Education Research and Development group at North Carolina State University, and to other colleagues in the NCSU physics department.

We thank Fred Reif for emphasizing the role of the three fundamental principles of mechanics, and for his view on the reciprocity of electric and gravitational forces. We thank Robert Bauman, Gregg Franklin, and Curtis Meyer for helping us think deeply about energy.

Much of Chapter 12 on quantum statistical mechanics is based on an article by Thomas A. Moore and Daniel V. Schroeder, "A different approach to introducing statistical mechanics," *American Journal of Physics*, vol. 65, pp. 26–36 (January 1997). We have benefited from many stimulating conversations with Thomas Moore, author of another introductory textbook that takes a contemporary view of physics, *Six Ideas that Shaped Physics*. Michael Weissman and Robert Swendsen provided particularly helpful critiques on some aspects of our implementation of Chapter 12.

We thank Hermann Haertel for opening our eyes to the fundamental mechanisms of electric circuits. Robert Morse, Priscilla Laws, and Mel Steinberg stimulated our thinking about desktop experiments. Bat-Sheva Eylon offered important guidance at an early stage. Ray Sorensen provided deep analytical critiques that influenced our thinking in several important areas. Randall Feenstra taught us about semiconductor junctions. Thomas Moore showed us a useful way to present the differential form of Maxwell's equations. Fred Reif helped us devise an assessment of student learning of basic E&M concepts. Uri Ganiel suggested the high-voltage circuit used to demonstrate the reality of surface charge. The unusual light bulb circuits at the end of Chapter 22 are based on an article by P. C. Peters, "The role of induced emf's in simple circuits," *American Journal of Physics* 52, 1984, 208–211. Thomas Ferguson gave us unusually detailed and useful feedback on the E&M chapters. Discussions with John Jewett about energy transfers were helpful. We thank Seth Chabay for help with Latin.

We thank David Andersen, David Scherer, and Jonathan Brandmeyer for the development of tools that enabled us and our students to write associated software.

The research of Matthew Kohlmyer, Sean Weatherford, and Brandon Lunk on student engagement with computational modeling has made major contributions to our instruction on computational modeling. Lin Ding developed an energy assessment instrument congruent with the goals of this curriculum.

We thank our colleagues David Brown, Krishna Chowdary, Laura Clarke, John Denker, Norman Derby, Ernst-Ludwig Florin, Thomas Foster, Jon D.H. Gaffney, Chris Gould, Mark Haugan, Joe Heafner, Robert Hilborn, Eric Hill, Andrew Hirsch, Leonardo Hsu, Barry Luokkala, Sara Majetich, Jonathan Mitschele, Arjendu Pattanayak, Jeff Polak, Prabha Ramakrishnan, Vidhya Ramachandran, Richard Roth, Michael Schatz, Robert Swendsen, Aaron Titus, Michael Weissman, and Hugh Young.

We thank a group of reviewers assembled by the publisher, who gave us useful critiques on the second edition of this textbook: Kelvin Chu, Michael Dubson, Tom Furtak, David Goldberg, Javed Iqbal, Shawn Jackson, Craig Ogilvie, Michael Politano, Norris Preyer, Rex Ramsier, Tycho Sleator, Robert Swendsen, Larry Weinstein, and Michael Weissman. We also thank the group who offered useful critiques on the third edition: Alex Small, Bereket Behane, Craig Wiegert, Galen Pickett, Ian Affleck, Jeffrey Bindel, Jeremy King, Paula Heron, and Surenda Singh.

We have benefited greatly from the support and advice of Stuart Johnson and Jessica Fiorillo of John Wiley & Sons. Elizabeth Swain of John Wiley & Sons was exceptionally skilled in managing the project. Helen Walden did a superb job of copyediting; any remaining errors are ours.

This project was supported, in part, by the National Science Foundation (grants MDR-8953367, USE-9156105, DUE-9554843, DUE-9972420, DUE-0320608, DUE-0237132, and DUE-0618504). We are grateful to the National Science Foundation and its reviewers for their long-term support of this challenging project. Opinions expressed are those of the authors, and not necessarily those of the Foundation.

How the Figures Were Made

Almost all of the figures in this book were produced by us (for the third edition the Aptara studio created the figures that show human figures, and the studio added full color to our two-color versions from the second edition). Our main tool was Adobe Illustrator. The many 3D computer-generated images were made using VPython, with optional processing in POV-Ray using a module written by Ruth Chabay to generate a POV-Ray scene description file corresponding to a VPython scene, followed by editing in Adobe Photoshop before exporting to Illustrator. We used TeXstudio for editing LaTeX, with a package due in part to the work of Aptara. All the computer work was done on Windows computers.

Ruth Chabay and Bruce Sherwood
Santa Fe, New Mexico, July 2014

Biographical Background

Ruth Chabay earned a Ph.D in physical chemistry from the University of Illinois at Urbana-Champaign; her undergraduate degree was in chemistry from the University of Chicago. She is Professor Emerita in the Department of Physics at North Carolina State University and was Weston Visiting Professor, Department of Science Teaching, at the Weizmann Institute of Science in Rehovot, Israel. She has also taught at the University of Illinois at Urbana-Champaign and Carnegie Mellon University. She is a Fellow of the American Physical Society.

Bruce Sherwood's Ph.D is in experimental particle physics from the University of Chicago; his undergraduate degree was in engineering science from Purdue University, after which he studied physics for one year at the University of Padua, Italy. He is Professor Emeritus in the Department of Physics at North Carolina State University. He has also taught at Caltech, the University of Illinois at Urbana-Champaign, and Carnegie Mellon University. He is a Fellow of the American Physical Society and of the American Association for the Advancement of Science.

Chabay and Sherwood have been joint recipients of several educational awards. At Carnegie Mellon University they received the Ashkin Award for Teaching in the Mellon College of Science in 1999 and the Teaching Award of the National Society of Collegiate Scholars in 2001. At North Carolina State University they received the Margaret Cox Award for excellence in teaching and learning with technology in 2005. In 2014 the American Association of Physics Teachers presented them with the David Halliday and Robert Resnick Award for Excellence in Undergraduate Physics Teaching.

CHAPTER

1

Interactions and Motion

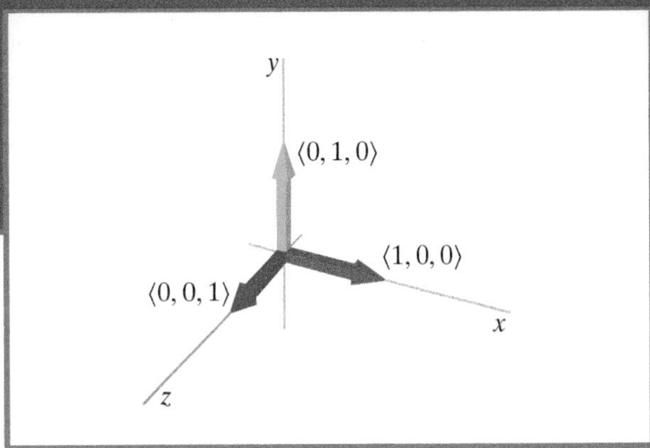

This textbook deals with the nature of matter and its interactions. The main goal of this textbook is to have you engage in a process central to science: constructing and applying physical models based on a small set of powerful fundamental physical principles and the atomic structure of matter. The variety of phenomena that we will be able to model, explain, and predict is very wide, including the orbit of stars around a black hole, nuclear fusion, the formation of sparks in air, and the speed of sound in a solid. This first chapter deals with the physical idea of interactions.

OBJECTIVES

After studying this chapter you should be able to

- Deduce from observations of an object's motion whether or not it has interacted with its surroundings.
- Mathematically describe position and motion in three dimensions.
- Mathematically describe momentum and change of momentum in three dimensions.
- Read and modify a simple computational model of motion at constant velocity.

1.1 KINDS OF MATTER

We will deal with material objects of many sizes, from subatomic particles to galaxies. All of these objects have certain things in common.

Atoms and Nuclei

Ordinary matter is made up of tiny atoms. An atom isn't the smallest type of matter, for it is composed of even smaller objects (electrons, protons, and neutrons), but many of the ordinary everyday properties of ordinary matter can be understood in terms of atomic properties and interactions. As you probably know from studying chemistry, atoms have a very small, very dense core, called the nucleus, around which is found a cloud of electrons. The nucleus contains protons and neutrons, collectively called nucleons. Electrons are kept close to the nucleus by electric attraction to the protons (the neutrons hardly interact with the electrons).

> QUESTION Recall your previous studies of chemistry. How many protons and electrons are there in a hydrogen atom? In a helium or carbon atom?

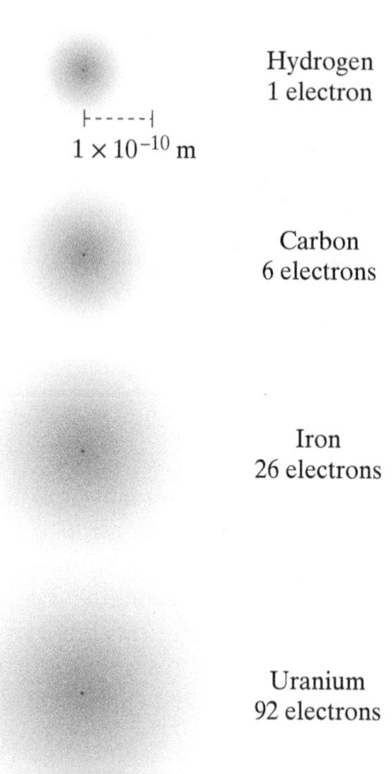

Figure 1.1 Atoms of hydrogen, carbon, iron, and uranium. The gray blur represents the electron cloud surrounding the nucleus. The black dot shows the location of the nucleus. On this scale, however, the nucleus would be much too small to see.

When you encounter a question in the text, you should think for a moment before reading on. Active reading contributes to significantly greater understanding. In the case of the questions posed above, if you don't remember the properties of these atoms, it may help to refer to the periodic table on the inside front cover of this textbook.

Hydrogen is the simplest atom, with just one proton and one electron. A helium atom has two protons and two electrons. A carbon atom has six protons and six electrons. Near the other end of the chemical periodic table, a uranium atom has 92 protons and 92 electrons. Figure 1.1 shows the relative sizes of the electron clouds in atoms of several elements but cannot show the nucleus to the same scale; the tiny dot marking the nucleus in the figure is much larger than the actual nucleus.

The radius of the electron cloud for a typical atom is about 1×10^{-10} meter. The reason for this size can be understood using the principles of quantum mechanics, a major development in physics in the early 20th century. The radius of a proton is about 1×10^{-15} meter, very much smaller than the radius of the electron cloud.

Nuclei contain neutrons as well as protons (Figure 1.2). The most common form or "isotope" of hydrogen has no neutrons in the nucleus. However, there exist isotopes of hydrogen with one or two neutrons in the nucleus (in addition to the proton). Hydrogen atoms containing one or two neutrons are called deuterium or tritium. The most common isotope of helium has two neutrons (and two protons) in its nucleus, but a rare isotope has only one neutron; this is called helium-3.

The most common isotope of carbon has six neutrons together with the six protons in the nucleus (carbon-12), whereas carbon-14 with eight neutrons is an isotope that plays an important role in dating archaeological objects.

Near the other end of the periodic table, uranium-235, which can undergo a fission chain reaction, has 92 protons and 143 neutrons, whereas uranium-238, which does not undergo a fission chain reaction, has 92 protons and 146 neutrons.

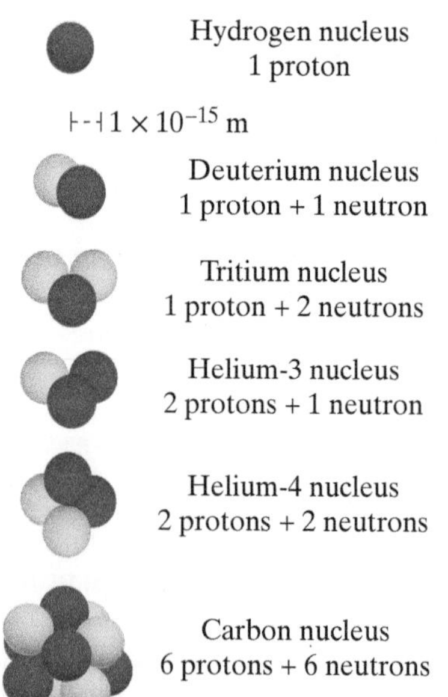

Figure 1.2 Nuclei of hydrogen, helium, and carbon. Note the *very* much smaller scale than in Figure 1.1!

Molecules and Solids

When atoms come in contact with each other, they may stick to each other ("bond" to each other). Several atoms bonded together can form a molecule—a substance whose physical and chemical properties differ from those of the constituent atoms. For example, water molecules (H_2O) have properties quite different from the properties of hydrogen atoms or oxygen atoms.

An ordinary-sized rigid object made of bound-together atoms and big enough to see and handle is called a solid, such as a bar of aluminum. A new kind of microscope, the scanning tunneling microscope (STM), is able to map the locations of atoms on the surface of a solid, which has provided new techniques for investigating matter at the atomic level. Two such images appear in Figure 1.3. You can see that atoms in a crystalline solid are arranged in a regular three-dimensional array. The arrangement of atoms on the surface depends on the direction along which the crystal is cut. The irregularities in the bottom image reflect "defects," such as missing atoms, in the crystal structure.

Liquids and Gases

When a solid is heated to a higher temperature, the atoms in the solid vibrate more vigorously about their normal positions. If the temperature is raised high enough, this thermal agitation may destroy the rigid structure of the solid. The atoms may become able to slide over each other, in which case the substance is a liquid.

At even higher temperatures the thermal motion of the atoms or molecules may be so large as to break the interatomic or intermolecular bonds completely, and the liquid turns into a gas. In a gas the atoms or molecules are quite free to move around, only occasionally colliding with each other or the walls of their container.

We will learn how to analyze many aspects of the behavior of solids and gases. We won't have much to say about liquids, because their properties are much harder to analyze. Solids are simpler to analyze than liquids because the atoms stay in one place (though with thermal vibration about their usual positions). Gases are simpler to analyze than liquids because between collisions the gas molecules are approximately unaffected by the other molecules. Liquids are the awkward intermediate state, where the atoms move around rather freely but are always in contact with other atoms. This makes the analysis of liquids very complex.

Figure 1.3 Two different surfaces of a crystal of pure silicon. The images were made with a scanning tunneling microscope. (Images courtesy of Randall Feenstra, IBM Corp.)

Planets, Stars, Solar Systems, and Galaxies

In our brief survey of the kinds of matter that we will study, we make a giant leap in scale from atoms all the way up to planets and stars, such as our Earth and Sun. We will see that many of the same principles that apply to atoms apply to planets and stars. By making this leap we bypass an important physical science, geology, whose domain of interest includes the formation of mountains and continents. We will study objects that are much bigger than mountains, and we will study objects that are much smaller than mountains, but we don't have time to apply the principles of physics to every important kind of matter.

Our Sun and its accompanying planets constitute our Solar System. It is located in the Milky Way galaxy, a giant rotating disk-shaped system of stars. On a clear dark night you can see a band of light (the Milky Way) coming from the huge number of stars lying in this disk, which you are looking at from a position in the disk, about two-thirds of the way out from the center of the disk. Our galaxy is a member of a cluster of galaxies that move around each other much as the planets of our Solar System move around the Sun (Figure 1.4). The Universe contains many such clusters of galaxies.

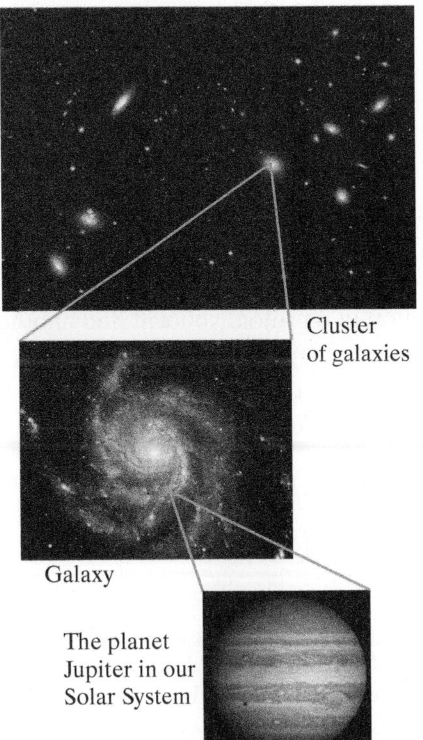

Figure 1.4 Our Solar System exists inside a galaxy, which itself is a member of a cluster of galaxies. (Photos courtesy NASA/JPL-Caltech)

Point Particles

It is common in physics to talk about the motion of a "point particle." What we mean by a particle is an object whose size, shape, and internal structure are not important to us in the current context, and which we can consider to be located at a single point in space. In modeling the motion of a real object (whether it is a galaxy or a proton), we often choose to make the simplifying assumption that it is a point particle, as if Superman or a giant space alien had come along and squeezed the object until it was compressed into a very tiny, structureless microscopic speck with the full mass of the original object.

Of course, there are many situations in which it would be absurd to use this approximation. The Earth, for example, is a large, complex object, with a core of turbulent molten rock, huge moving continents, and massive sloshing oceans. Radioactivity keeps its core hot; electromagnetic radiation from the Sun warms its surface; and thermal energy is also radiated away into space. If we are interested in energy flows or continental motion or earthquakes we need to consider the detailed structure and composition of the Earth. However, if what we want to do is model the motion of the Earth as it interacts with other objects in our Solar System, it works quite well to ignore this complexity, and to treat the Earth, the Sun, the Moon, and the other planets as if they were point particles.

Even most very tiny objects, such as atoms, protons, and neutrons, are not truly point particles—they do have finite size, and they have internal structure, which can influence their interactions with other objects. By contrast, electrons may really be point particles—they appear to have no internal structure, and attempts to measure the radius of an electron have not produced a definite number (recent experiments indicate only that the radius of an electron is less than 2×10^{-20} m, much smaller than a proton).

As we consider various aspects of matter and its interactions, it will be important for us to state explicitly whether or not we are modeling material objects as point particles, or as extended and perhaps deformable macroscopic chunks of matter. In Chapters 1–3 we will emphasize systems that can usefully be modeled as particles. In Chapter 4 we will begin to consider the detailed internal structure of material objects.

1.2 DETECTING INTERACTIONS

Objects made of different kinds of matter interact with each other in various ways: gravitationally, electrically, magnetically, and through nuclear interactions. How can we detect that an interaction has occurred? In this section we consider various kinds of observations that indicate the presence of interactions.

> QUESTION Before you read further, take a moment to think about your own ideas of interactions. How can you tell that two objects are interacting with each other?

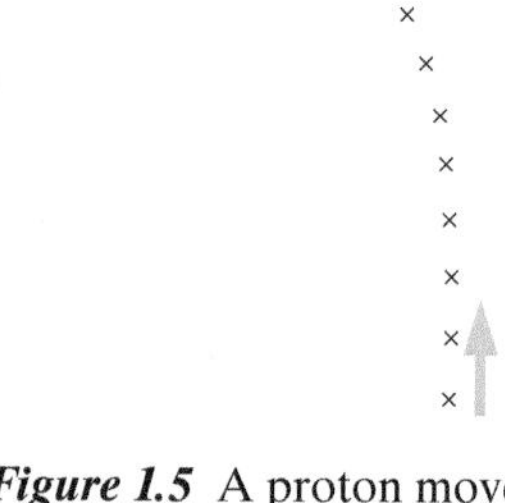

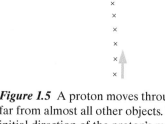

Figure 1.5 A proton moves through space, far from almost all other objects. The initial direction of the proton's motion is upward, as indicated by the arrow. The ×'s represent the position of the proton at equal time intervals.

Change of Direction of Motion

Suppose that you observe a proton moving through a region of outer space, far from almost all other objects. The proton moves along a path like the one shown in Figure 1.5. The arrow indicates the initial direction of the proton's motion, and the ×'s in the diagram indicate the position of the proton at equal time intervals.

> QUESTION Do you see evidence in Figure 1.5 that the proton is interacting with another object?

Evidently a change in direction is a vivid indicator of interactions. If you observe a change in direction of the motion of a proton, you will find another object somewhere that has interacted with this proton.

> QUESTION Suppose that the only other object nearby was another proton. What was the approximate initial location of this second proton?

Since two protons repel each other electrically, the second proton must have been located to the right of the bend in the first proton's path.

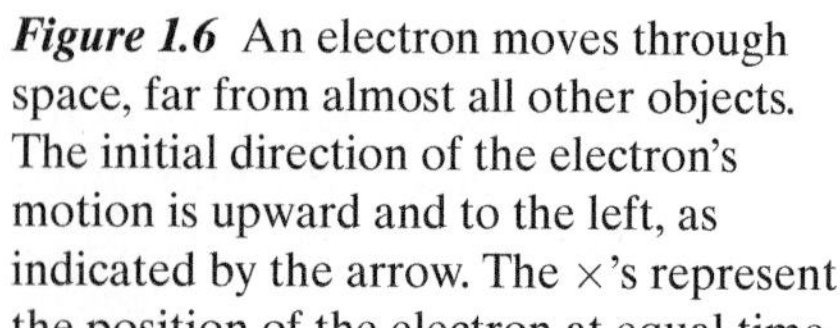

Figure 1.6 An electron moves through space, far from almost all other objects. The initial direction of the electron's motion is upward and to the left, as indicated by the arrow. The ×'s represent the position of the electron at equal time intervals.

Change of Speed

Suppose that you observe an electron traveling in a straight line through outer space far from almost all other objects (Figure 1.6). The path of the electron is shown as though a camera had taken multiple exposures at equal time intervals.

> QUESTION Where is the electron's speed largest? Where is the electron's speed smallest?

The speed is largest at the upper left, where the ×'s are farther apart, which means that the electron has moved farthest during the time interval between exposures. The speed is smallest at the bottom right, where the $\times$'s are closer together, which means that the electron has moved the least distance during the time interval between exposures.

> QUESTION Suppose that the only other object nearby was another electron. What was the approximate initial location of this other electron?

The other electron must have been located directly just below and to the right of the starting location, since electrons repel each other electrically.

Evidently a change in speed is an indicator of interactions. If you observe a change in speed of an electron, you will find another object somewhere that has interacted with the electron.

Velocity Includes Both Speed and Direction

In physics, the word "velocity" has a special technical meaning that is different from its meaning in everyday speech. In physics, the quantity called "velocity" denotes a combination of speed and direction. Even if the speed or direction of motion is changing, the velocity has a precise value (speed and direction) at any instant. In contrast, in everyday speech, "speed" and "velocity" are often used as synonyms. In physics and other sciences, however, words have rather precise meanings and there are few synonyms.

For example, consider an airplane that at a particular moment is flying with a speed of 1000 kilometers/hour in a direction that is due east. We say the velocity is 1000 km/h, east, where we specify both speed and direction. An airplane flying west with a speed of 1000 km/h would have the same speed but a different velocity.

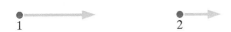

Figure 1.7 Two successive positions of a particle (indicated by a dot), with arrows indicating the velocity of the particle at each location. The shorter arrow indicates that the speed of the particle at location 2 is less than its speed at location 1.

We have seen that a change in an object's speed, or a change in the direction of its motion, indicates that the object has interacted with at least one other object. The two indicators of interaction, change of speed and change of direction, can be combined into one compact statement:

> A change of velocity (speed or direction or both) indicates the existence of an interaction.

In physics diagrams, the velocity of an object is represented by an arrow: a line with an arrowhead. The tail of the arrow is placed at the location of the object at a particular instant, and the arrow points in the direction of the motion of the object at that instant. The length of the arrow is proportional to the speed of the object. Figure 1.7 shows two successive positions of a particle at two different times, with velocity arrows indicating a change in speed of the particle (it's slowing down). Figure 1.8 shows three successive positions of a different particle at three different times, with velocity arrows indicating a change in direction but no change in speed. Note that the arrows themselves are straight; even if the path of the particle curves over time, at any instant the particle may be considered to be traveling in a specific direction.

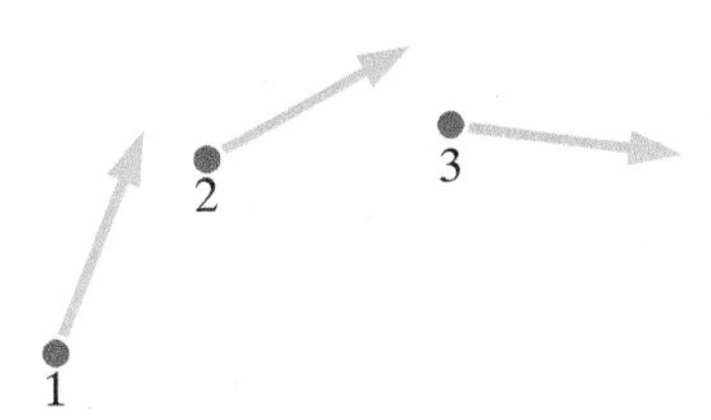

Figure 1.8 Three successive positions of a particle (indicated by a dot), with arrows indicating the velocity of the particle at each location. The arrows are the same length, indicating the same speed, but they point in different directions, indicating a change in direction and therefore a change in velocity.

We will see a little later that velocity is only one example of a physical quantity that has a "magnitude" (an amount or a size) and a direction. Other examples of such quantities are position relative to an origin in 3D space, changes in position or velocity, and force. In Section 1.4 we will see how to represent such quantities as vectors: single mathematical entities that combine information about magnitude and direction.

Uniform Motion

Suppose that you observe a rock moving along in outer space far from all other objects. We don't know what made it start moving in the first place; presumably a long time ago an interaction gave it some velocity and it has been coasting through the vacuum of space ever since.

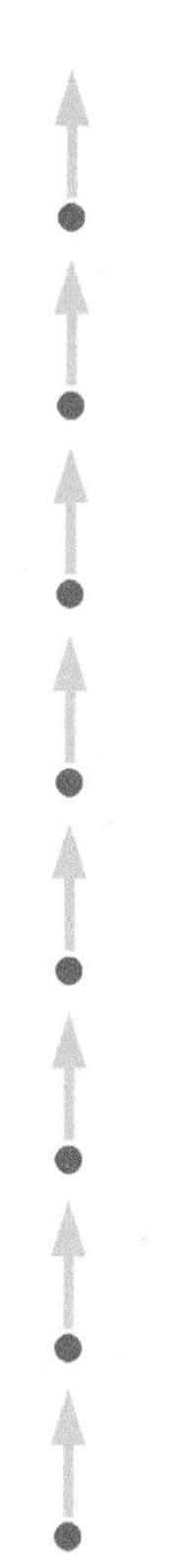
Figure 1.9 "Uniform motion"—no change in speed or direction.

It is an observational fact that such an isolated object moves at constant, unchanging speed, in a straight line. Its velocity does not change (neither its direction nor its speed changes). We call motion with unchanging velocity "uniform motion" (Figure 1.9). Other terms for uniform motion include "uniform velocity" and "constant velocity," since velocity refers to both speed and direction.

QUESTION Is an object at rest in uniform motion?

If an object remains at rest, then neither the speed nor direction of the object's velocity changes. This is a special case of uniform motion: the object's speed is constant (zero is a valid value of speed) and the direction of motion, while undefined, is not changing.

QUESTION If we observe an object in uniform motion, can we conclude that it has no interactions with its surroundings?

When we observe an object in uniform motion, one possibility is that it has no interactions at all with its surroundings. However, there is another possibility: the object may be experiencing multiple interactions that cancel each other out. In either case, we can correctly deduce that the "net" (total) interaction of the object with its surroundings is zero.

Checkpoint 1 **(a)** Which of the following do you see moving with constant velocity? (1) A ship sailing northeast at a speed of 5 meters per second (2) The Moon orbiting the Earth (3) A tennis ball traveling across the court after having been hit by a tennis racket (4) A can of soda sitting on a table (5) A person riding on a Ferris wheel that is turning at a constant rate. **(b)** In which of the following situations is there observational evidence for significant interaction between two objects? How can you tell? (1) A ball bounces off a wall with no change in speed. (2) A baseball that was hit by a batter flies toward the outfield. (3) A communications satellite orbits the Earth. (4) A space probe travels at constant speed toward a distant star. (5) A charged particle leaves a curving track in a particle detector.

1.3 NEWTON'S FIRST LAW OF MOTION

The basic relationship between change of velocity and interaction is summarized qualitatively by what is known as Newton's "first law of motion," though it was originally discovered by Galileo. In his original Latin, Newton said, "Corpus omne perseverare in statu suo quiescendi vel movendi uniformiter in directum, nisi quatenus a viribus impressis cogitur statum illum mutare." A literal translation is "Every body persists in its state of resting or of moving uniformly in a direction, except to the extent that it is compelled to change that state by forces pressed upon it." Expressing this in more modern language, we have this:

NEWTON'S FIRST LAW OF MOTION

Every body persists in its state of rest or of moving with constant speed in a constant direction, except to the extent that it is compelled to change that state by forces acting on it.

"Force" is the way in which the amount of interaction is quantified, and we'll discuss force in detail in Chapter 2. The words "except to the extent" imply that the stronger the interaction, the more change there will be in direction and/or speed. The weaker the interaction, the less change. If there is no net (total) interaction at all, the object's motion will be uniform (constant speed and direction); this could happen either because there are no interactions or because there are interactions that cancel each other, such as equally strong pushes to the left and right. It is important to remember that if an object is not moving at all, its velocity is not changing, so it too may be considered to be in uniform motion.

Newton's first law of motion is only qualitative, because it doesn't give us a way to calculate quantitatively how much change in speed or direction will be produced by a certain amount of interaction, a subject we will take up in the next chapter. Nevertheless, Newton's first law of motion is important in providing a conceptual framework for thinking about the relationship between interaction and motion.

This law represented a major break with ancient tradition, which assumed that constant pushing was required to keep something moving. This law says something radically different: no interactions at all are needed to keep something moving!

QUESTION To move a box across a table at constant speed in a straight line, you must keep pushing it. Does this contradict Newton's first law?

Since a constant interaction is required to keep the box moving, we might be tempted to conclude that Newton's first law of motion does not apply in many everyday situations. However, what matters is the *net* interaction of the box with its surroundings, which could be zero if there are multiple interactions that cancel each other out.

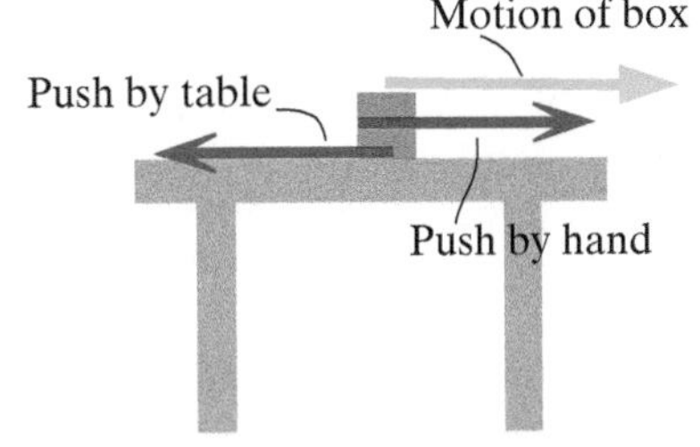

Figure 1.10 The red arrows represent the magnitude and direction of the pushes the box gets from your hand and from the friction with the table. If these pushes add up to zero, the box moves with constant speed in a straight line, indicated by the green arrow.

QUESTION In addition to your hand, what other objects in the surroundings interact with the box?

The table also interacts with the box, in a way that we call friction. If you push just hard enough to compensate exactly for the table friction, the sum of all the interactions is zero, and the box moves at constant speed as predicted by Newton's first law (Figure 1.10). (If you push harder than the table does, the box's speed steadily increases.)

It is difficult to observe motion without friction in everyday life, because objects almost always interact with many other objects, including air, flat surfaces, and so on. You may be able to think of situations in which you have seen an object keep moving at constant (or nearly constant) velocity, without being pushed or pulled. One example of a nearly friction-free situation is a hockey puck sliding on ice. The puck slides a long way at nearly constant speed in a straight line (constant velocity) because there is little friction with the ice. An even better example is the uniform motion of an object in outer space, far from all other objects.

QUESTION Is a change of position an indicator of an interaction?

Not necessarily. If the change of position occurs simply because a particle is moving at constant speed and direction, then a mere change of position is not an indicator of an interaction, since uniform motion is an indicator of zero net interaction. We need to know the object's velocity at each observation to be able to make further deductions.

QUESTION If you observe an object at rest in one location, and later you observe it again at rest but in a different location, can you conclude that an interaction took place?

Yes. You can infer that there must have been an interaction to give the object some velocity to move the object toward the new position, and another interaction to slow the object to a stop in its new position.

QUESTION Is it possible to deduce the existence of an interaction even though you do not observe a change?

As we saw when we considered pushing a box across a table at constant speed, sometimes we may find indirect evidence for an additional interaction.

When something doesn't change although we would normally expect a change due to a known interaction, we can logically deduce that an additional interaction must be occurring. For example, consider a helium-filled balloon that hovers motionless in the air despite the downward gravitational pull of the Earth. Evidently there is some additional kind of interaction that opposes the gravitational interaction. In this case, interactions with air molecules have the net effect of pushing up on the balloon ("buoyancy"). The lack of change implies that the effect of the air molecules exactly compensates for the gravitational interaction with the Earth.

The stability of the nucleus of an atom is another example of indirect evidence for an additional interaction. The nucleus contains positively charged protons that repel each other electrically, yet the nucleus remains intact. We conclude that there must be some other kind of interaction present, a nonelectric attractive interaction that overcomes the electric repulsion. This is evidence for a nonelectric interaction called the "strong interaction," which as we will see acts among protons and neutrons to hold the nucleus together. We will discuss the strong interaction in Chapter 3.

Other Indicators of Interaction

Change of velocity is not the only indication that an object has interacted with its surroundings, but it is the only change possible for a single object that is modeled as a point particle, which has neither shape nor internal structure. In later chapters we will examine other kinds of changes, such as change of temperature, change of shape or configuration, and change of identity (for example, in nuclear reactions). In Chapters 1–3, however, we will concentrate on how interactions change motion.

Checkpoint 2 (a) Apply Newton's first law to each of the following situations. In which situations can you conclude that the object is undergoing a net interaction with one or more other objects? (1) A book slides across the table and comes to a stop. (2) A proton in a particle accelerator moves faster and faster. (3) A car travels at constant speed around a circular race track. (4) A spacecraft travels at a constant speed toward a distant star. (5) A hydrogen atom remains at rest in outer space. **(b)** A spaceship far from all other objects uses its rockets to attain a speed of 1×10^4 m/s. The crew then shuts off the power. According to Newton's first law, which of the following statements about the motion of the spaceship after the power is shut off are correct? (Choose all statements that are correct.) (1) The spaceship will move in a straight line. (2) The spaceship will travel on a curving path. (3) The spaceship will enter a circular orbit. (4) The speed of the spaceship will not change. (5) The spaceship will gradually slow down. (6) The spaceship will stop suddenly.

1.4 DESCRIBING THE 3D WORLD: VECTORS

Physical phenomena take place in the 3D world around us. In order to be able to make quantitative predictions and give detailed, quantitative explanations, we need tools for describing precisely the positions and velocities of objects in 3D, and the changes in position and velocity due to interactions. These tools are mathematical entities called 3D "vectors." A symbol denoting a vector is written with an arrow over it:

$$\vec{r} \text{ is a vector}$$

In three dimensions a vector is a triple of numbers $\langle x,y,z\rangle$. Quantities like the position or velocity of an object can be represented as vectors:

$$\vec{r}_1 = \langle 3.2, -9.2, 66.3\rangle \text{ m} \quad \text{(a position vector)}$$
$$\vec{v}_1 = \langle -22.3, 0.4, -19.5\rangle \text{ m/s} \quad \text{(a velocity vector)}$$

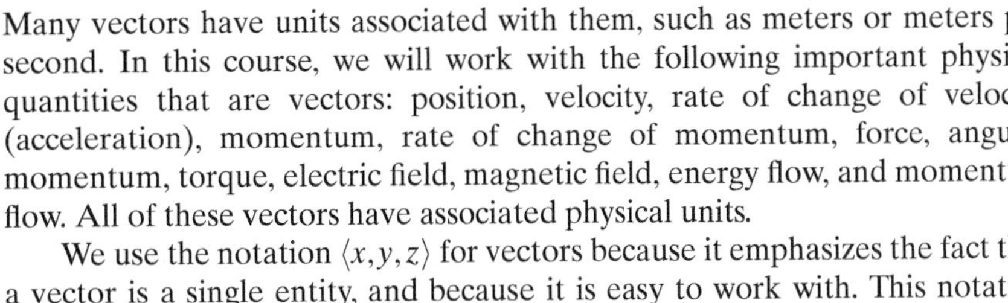

Many vectors have units associated with them, such as meters or meters per second. In this course, we will work with the following important physical quantities that are vectors: position, velocity, rate of change of velocity (acceleration), momentum, rate of change of momentum, force, angular momentum, torque, electric field, magnetic field, energy flow, and momentum flow. All of these vectors have associated physical units.

We use the notation $\langle x,y,z \rangle$ for vectors because it emphasizes the fact that a vector is a single entity, and because it is easy to work with. This notation appears in many calculus textbooks; you will probably encounter other ways of expressing vectors mathematically as well.

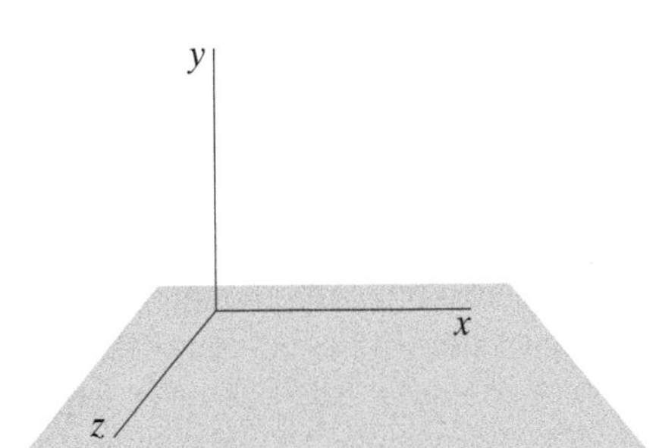

Figure 1.11 Right-handed 3D coordinate system. The xy plane is in the plane of the page, and the z axis projects out of the page, toward you.

Position Vectors

A position vector is a simple example of a physical vector quantity. We will use a 3D Cartesian coordinate system to specify positions in space and other vector quantities. Usually we will orient the axes of the coordinate system as shown in Figure 1.11: $+x$ axis to the right, $+y$ axis upward, and $+z$ axis coming out of the page, toward you. This is a "right-handed" coordinate system: if you hold the thumb, first, and second fingers of your right hand perpendicular to each other, and align your thumb with the x axis and your first finger with the y axis, your second finger points along the z axis. In some math textbook discussions of 3D coordinate systems, the x axis points out, the y axis points to the right, and the z axis points up. This is the same right-handed coordinate system, viewed from a different "camera position." Since we will sometimes consider motion in a single plane, it makes sense to orient the xy plane in the plane of a vertical page or computer display, so we will use the viewpoint in which the y axis points up.

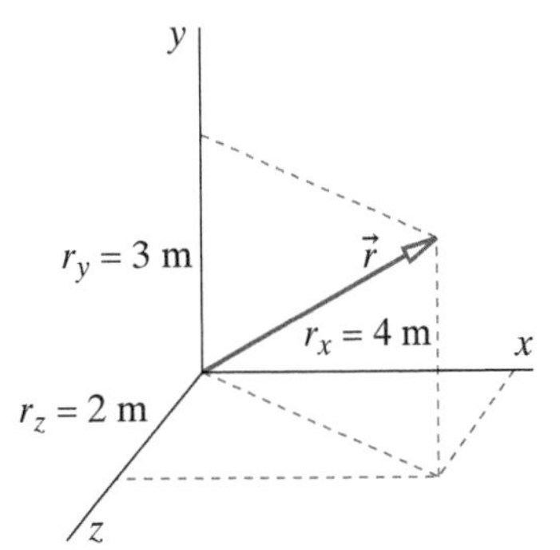

Figure 1.12 A position vector $\vec{r} = \langle 4,3,2 \rangle$ m and its x, y, and z components.

A position in 3D space can be considered to be a vector, called a *position vector*, pointing from an origin to that location. Figure 1.12 shows a position vector, represented by an arrow with its tail at the origin, that might represent your final position if you started at the origin and walked 4 meters along the x axis, then 2 meters parallel to the z axis, then climbed a ladder so you were 3 meters above the ground. Your new position relative to the origin is a vector that can be written like this:

$$\vec{r} = \langle 4,3,2 \rangle \text{ m}$$

Each of the numbers in the triple is called a "component" of the vector, and is associated with a particular axis. Usually the components of a vector are denoted symbolically by the subscripts x, y, and z:

$$\vec{v} = \langle v_x, v_y, v_z \rangle \quad \text{(a velocity vector)}$$
$$\vec{r} = \langle r_x, r_y, r_z \rangle \quad \text{(a position vector)}$$
$$\vec{r} = \langle x, y, z \rangle \quad \text{(alternative notation for a position vector)}$$

The components of the position vector $\vec{r} = \langle 4,3,2 \rangle$ m are:

$$r_x = 4\,\text{m} \quad \text{(the } x \text{ component)}$$
$$r_y = 3\,\text{m} \quad \text{(the } y \text{ component)}$$
$$r_z = 2\,\text{m} \quad \text{(the } z \text{ component)}$$

The x component of the vector $\vec{v}$ is the number v_x. The z component of the vector $\vec{v}_1 = \langle -22.3, 0.4, -19.5 \rangle$ m/s is -19.5 m/s. A component such as v_x is not a vector, since it is only one number.

QUESTION Can a vector be zero?

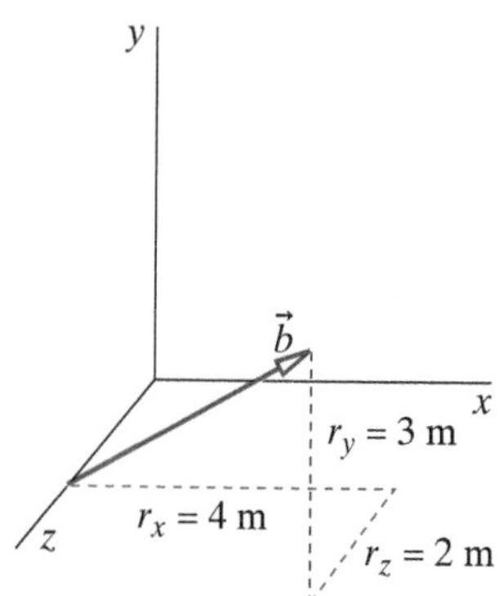

Figure 1.13 The arrow represents the vector $\vec{b} = \langle 4,3,2 \rangle$ m, drawn with its tail at location $\langle 0,0,2 \rangle$.

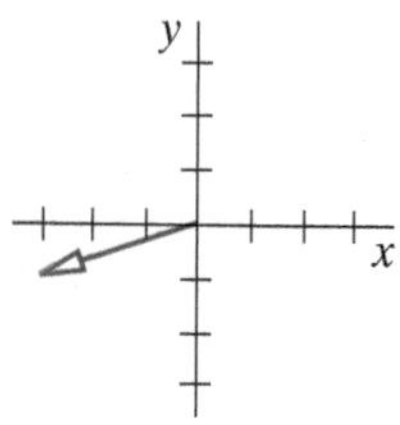

Figure 1.14 The position vector $\langle -3,-1,0 \rangle$, drawn at the origin, in the xy plane. The components of the vector specify the displacement from the tail to the tip. The z axis, which is not shown, comes out of the page, toward you.

The zero vector $\langle 0,0,0 \rangle$ is a legal vector, which we will sometimes write as $\vec{0}$. A zero position vector describes the position of an object located at the origin. A zero velocity vector describes the velocity of an object that is at rest at a particular instant.

Drawing Vectors

A position vector is special in that its tail is always at the origin of a coordinate system, but this is not the case for other vectors. It is important to note that the x component of a vector specifies the difference between the x coordinate of the tail of the vector and the x coordinate of the tip of the vector. It does not give any information about the location of the tail of the vector (compare Figures 1.12 and 1.13). By convention, arrows representing vector quantities such as velocity are usually drawn with the tail of the arrow at the location of the object.

In Figure 1.12 we represented your position vector relative to the origin graphically by an arrow whose tail is at the origin and whose arrowhead is at your position. The length of the arrow represents the distance from the origin, and the direction of the arrow represents the direction of the vector, which is the direction of a direct path from the initial position to the final position (the "displacement"; by walking and climbing you "displaced" yourself from the origin to your final position).

Since it is difficult to draw a 3D diagram on paper, when working on paper you will usually be asked to draw vectors that all lie in a single plane. Figure 1.14 shows an arrow in the xy plane representing the vector $\langle -3,-1,0 \rangle$.

Scalars

A quantity that is represented by a single number is called a *scalar.* A scalar quantity does not have a direction. Examples include the mass of an object, such as 50 kg, or the temperature, such as -20 °C. Vectors and scalars are very different entities; a vector can never be equal to a scalar, and a scalar cannot be added to a vector. Scalars can be positive, negative, or zero:

$$m = 50\,\text{kg}$$
$$T = -20\,°\text{C}$$

Vector Operations

Vectors are mathematical entities, and have their own mathematical operations. Some of these operations are the same as those you already know for scalars. Others, such as multiplication, are quite different, and division by a vector is not legal. Here are the vector operations that we will discuss and use in this textbook:

VECTOR OPERATIONS

Mathematical operations that are defined for vectors:

- Multiply or divide a vector by a scalar: $2\vec{a}$, $\vec{v}/5$
- Find the magnitude of a vector: $|\vec{a}|$
- Find a unit vector giving direction: $\hat{a}$
- Add one vector to another: $\vec{a}+\vec{b}$
- Subtract one vector from another: $\vec{a}-\vec{b}$
- Differentiate a vector: $d\vec{r}/dt$
- Dot product of two vectors (result is a scalar): $\vec{a} \bullet \vec{b}$
- Cross product of two vectors (result is a vector): $\vec{a} \times \vec{b}$

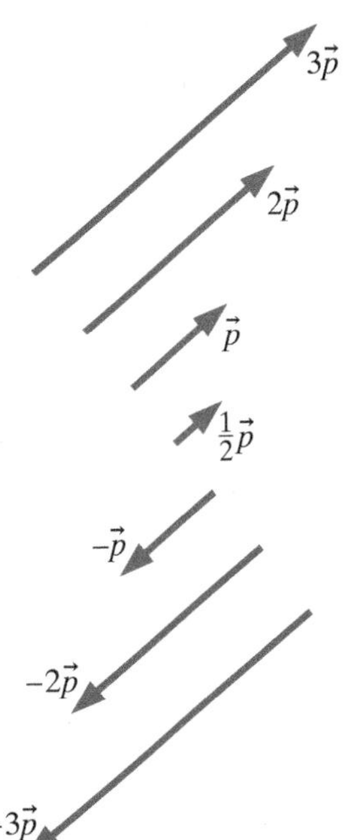

Figure 1.15 Multiplying a vector by a scalar changes the magnitude of the vector. Multiplying by a negative scalar reverses the direction of the vector.

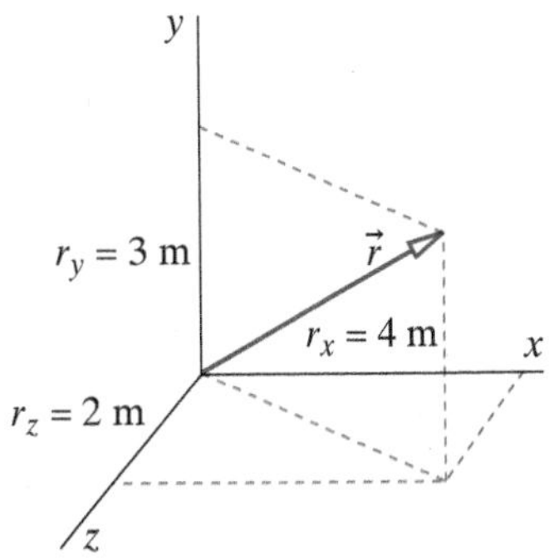

Figure 1.16 A vector representing a displacement from the origin.

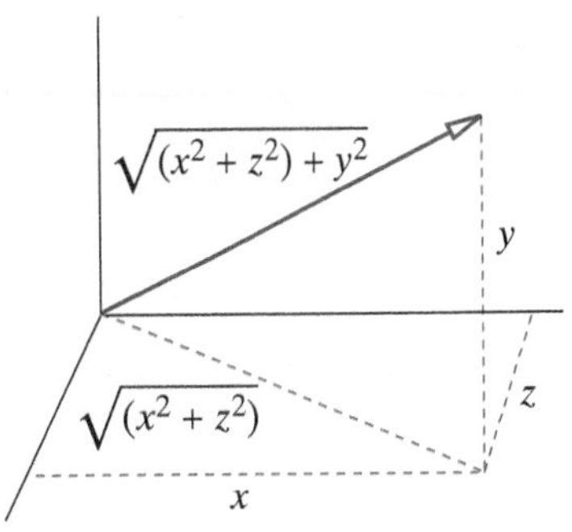

Figure 1.17 The magnitude of a vector is the square root of the sum of the squares of its components (3D version of the Pythagorean theorem).

The dot product will be introduced in Chapter 5, and the cross product in Chapter 11.

There are certain operations that are neither legal nor meaningful for vectors:

- A vector cannot be set equal to a scalar.
- A vector cannot be added to or subtracted from a scalar.
- A vector cannot occur in the denominator of an expression. (Although you can't divide by a vector, note that you can legally divide by the *magnitude* of a vector, which is a scalar.)
- As with scalars, you can't add or subtract vectors that have different units.

Multiplying a Vector by a Scalar

A vector can be multiplied (or divided) by a scalar. If a vector is multiplied by a scalar, each of the components of the vector is multiplied by the scalar:

$$\text{If } \vec{r} = \langle x,y,z\rangle, \text{ then } a\vec{r} = \langle ax,ay,az\rangle$$

$$\text{If } \vec{v} = \langle v_x,v_y,v_z\rangle, \text{ then } \frac{\vec{v}}{b} = \left\langle \frac{v_x}{b},\frac{v_y}{b},\frac{v_z}{b}\right\rangle$$

$$\frac{1}{2}\langle 6,-20,9\rangle = \langle 3,-10,4.5\rangle$$

Multiplication by a scalar "scales" a vector, keeping its direction the same but making its magnitude larger or smaller (Figure 1.15). Multiplying by a negative scalar reverses the direction of a vector.

$$(-1)\langle 0,0,4\rangle = \langle 0,0,-4\rangle$$

Checkpoint 3 You stand at location $\vec{r} = \langle 2,-3,5\rangle$ m. Your friend stands at location $\vec{r}/2$. What is your friend's position vector?

Magnitude

Figure 1.16 shows a vector representing a displacement of $\langle 4,3,2\rangle$ m from the origin. What is the distance from the tip of this vector to the origin? Using a 3D extension of the Pythagorean theorem for right triangles (Figure 1.17), we find that

$$\sqrt{(4\,\text{m})^2+(3\,\text{m})^2+(2\,\text{m})^2} = \sqrt{29}\,\text{m} = 5.39\,\text{m}$$

We say that the *magnitude* $|\vec{r}|$ of the position vector $\vec{r}$ is

$$|\vec{r}| = 5.39\,\text{m}$$

The magnitude of a vector is written either with absolute-value bars around the vector as $|\vec{r}|$, or simply by writing the symbol for the vector without the little arrow above it, r.

MAGNITUDE OF A VECTOR

$$\text{If the vector } \vec{r} = \langle r_x,r_y,r_z\rangle \text{ then } |\vec{r}| = \sqrt{r_x^2+r_y^2+r_z^2} \text{ (a scalar).}$$

The magnitude of a vector is always a positive number. The magnitude of a vector is a single number, not a triple of numbers, and it is a scalar, not a vector.

You may wonder how to find the magnitude of a quantity like $-3\vec{r}$, which involves the product of a scalar and a vector. This expression can be factored:

$$|-3\vec{r}| = |-3| \cdot |\vec{r}|$$

The magnitude of a scalar is its absolute value, so:

$$|-3\vec{r}| = |-3| \cdot |\vec{r}| = 3\sqrt{r_x^2 + r_y^2 + r_z^2}$$

Checkpoint 4 If $\vec{v} = \langle 2, -3, 5 \rangle$ m/s, what is $\left|-\frac{1}{2}\vec{v}\right|$?

Unit Vectors

One way to describe the direction of a vector is by specifying a *unit vector*. A unit vector is a vector of magnitude 1, pointing in some direction. A unit vector is written with a "hat" (caret) over it instead of an arrow. The unit vector $\hat{a}$ is called "a-hat."

QUESTION Is the vector $\langle 1,1,1 \rangle$ a unit vector?

The magnitude of $\langle 1,1,1 \rangle$ is $\sqrt{1^2 + 1^2 + 1^2} = 1.73$, so this is not a unit vector.

The vector $\langle 1/\sqrt{3}, 1/\sqrt{3}, 1/\sqrt{3} \rangle$ is a unit vector, since its magnitude is 1:

$$\sqrt{\left(\frac{1}{\sqrt{3}}\right)^2 + \left(\frac{1}{\sqrt{3}}\right)^2 + \left(\frac{1}{\sqrt{3}}\right)^2} = 1$$

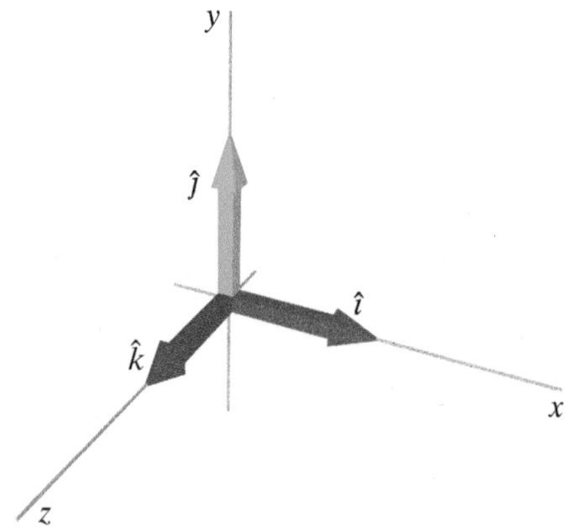

Figure 1.18 The unit vectors $\hat{\imath}, \hat{\jmath}, \hat{k}$.

Note that every component of a unit vector must be less than or equal to 1.

In our 3D Cartesian coordinate system, there are three special unit vectors, oriented along the three axes. They are called i-hat, j-hat, and k-hat, and they point along the x, y, and z axes, respectively (Figure 1.18):

$$\hat{\imath} = \langle 1,0,0 \rangle$$
$$\hat{\jmath} = \langle 0,1,0 \rangle$$
$$\hat{k} = \langle 0,0,1 \rangle$$

One way to express a vector is in terms of these special unit vectors:

$$\langle 0.02, -1.7, 30.0 \rangle = 0.02\hat{\imath} + (-1.7)\hat{\jmath} + 30.0\hat{k}$$

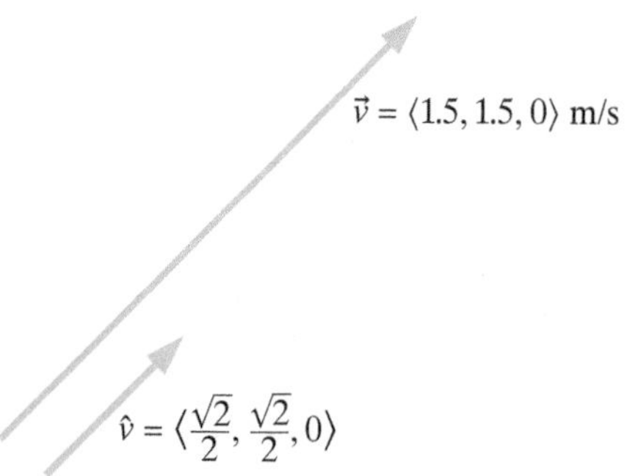

Figure 1.19 The unit vector $\hat{v}$ has the same direction as the vector $\vec{v}$, but its magnitude is 1, and it has no physical units.

Not all unit vectors point along an axis, as shown in Figure 1.19. For example, the vectors

$$\hat{g} = \langle 0.5774, 0.5774, 0.5774 \rangle \quad \text{and} \quad \hat{r} = \langle 0.424, 0.566, 0.707 \rangle$$

are both approximately unit vectors, since the magnitude of each is approximately equal to 1. Again, note that every component of a unit vector is less than or equal to 1.

Any vector may be factored into the product of a unit vector in the direction of the vector, multiplied by a scalar equal to the magnitude of the vector.

$$\vec{w} = |\vec{w}| \cdot \hat{w}$$

For example, a vector of magnitude 5, aligned with the y axis, could be written as:

$$\langle 0,5,0 \rangle = 5\langle 0,1,0 \rangle$$

Therefore, to find a unit vector in the direction of a particular vector, we just divide the vector by its magnitude:

FINDING A UNIT VECTOR

$$\hat{r} = \frac{\vec{r}}{|\vec{r}|} = \frac{\langle x,y,z\rangle}{\sqrt{(x^2+y^2+z^2)}}$$

$$\hat{r} = \left\langle \frac{x}{\sqrt{(x^2+y^2+z^2)}}, \frac{y}{\sqrt{(x^2+y^2+z^2)}}, \frac{z}{\sqrt{(x^2+y^2+z^2)}} \right\rangle$$

EXAMPLE **Magnitude and Direction**

Factor the vector $\vec{v} = \langle -22.3, 0.4, -19.5\rangle$ m/s into a magnitude times a unit vector.

Solution

$$|\vec{v}| = \sqrt{(-22.3)^2 + (0.4)^2 + (-19.5)^2}\ \text{m/s} = 29.6\ \text{m/s}$$

$$\hat{v} = \frac{\vec{v}}{|\vec{v}|} = \frac{\langle -22.3, 0.4, -19.5\rangle\ \text{m/s}}{29.6\ \text{m/s}} = \langle -0.753, 0.0135, -0.658\rangle$$

$$\vec{v} = (29.6\ \text{m/s})\langle -0.753, 0.0135, -0.658\rangle$$

We can now explain algebraically why multiplying a vector by a scalar changes the magnitude but not the direction of a vector. If we write the original vector as the product of a magnitude and a unit vector, after multiplying by a scalar the unit vector is unchanged, but the magnitude is increased or decreased:

$$\begin{aligned} \vec{a} &= \langle 3,-2,4\rangle = (5.385)\langle 0.577,-0.371,0.743\rangle \\ 2\cdot\vec{a} &= (2)(5.385)\langle 0.577,-0.371,0.743\rangle \\ &= 10.770\langle 0.577,-0.371,0.743\rangle \\ &= \langle 6,-4,8\rangle \end{aligned}$$

EQUALITY OF VECTORS

A vector is equal to another vector if and only if all the components of the vectors are equal.

$$\vec{w} = \vec{r} \quad \text{means that}$$

$$w_x = r_x \quad \text{and} \quad w_y = r_y \quad \text{and} \quad w_z = r_z$$

The magnitudes and directions of two equal vectors are the same:

$$|\vec{w}| = |\vec{r}| \quad \text{and} \quad \hat{w} = \hat{r}$$

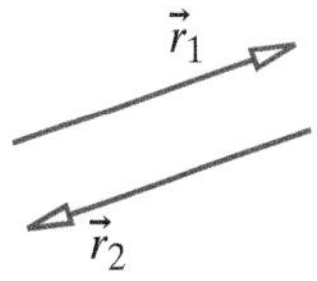

$\vec{r}_1$

$\vec{r}_2$

Figure 1.20 Are these two vectors equal? (Checkpoint 5)

Checkpoint 5 **(a)** Consider the vectors $\vec{r}_1$ and $\vec{r}_2$ represented by arrows in Figure 1.20. Are these two vectors equal? **(b)** If $\vec{a} = \langle 400, 200, -100\rangle$ m/s^2, and $\vec{c} = \vec{a}$, what is the unit vector $\hat{c}$ in the direction of $\vec{c}$?

Vector Addition and Subtraction

Vectors may be added, and one vector may be subtracted from another vector. However, a scalar cannot be added to or subtracted from a vector.

ADDING AND SUBTRACTING VECTORS

The sum or difference of two vectors is another vector, obtained by adding or subtracting the components of the vectors. Given two vectors $\vec{A} = \langle A_x, A_y, A_z\rangle$ and $\vec{B} = \langle B_x, B_y, B_z\rangle$, then

$$\begin{aligned} \vec{A} + \vec{B} &= \langle (A_x+B_x), (A_y+B_y), (A_z+B_z)\rangle \\ \langle 1,2,3\rangle + \langle -4,5,6\rangle &= \langle -3,7,9\rangle \\ \vec{A} - \vec{B} &= \langle (A_x-B_x), (A_y-B_y), (A_z-B_z)\rangle \\ \langle 1,2,3\rangle - \langle -4,5,6\rangle &= \langle 5,-3,-3\rangle \end{aligned}$$

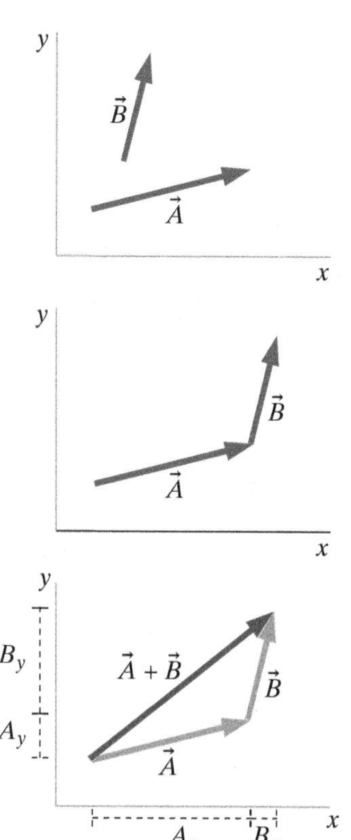

Figure 1.21 The procedure for adding two vectors graphically: To add $\vec{A}+\vec{B}$ graphically, move $\vec{B}$ so the tail of $\vec{B}$ is at the tip of $\vec{A}$, then draw a new arrow starting at the tail of $\vec{A}$ and ending at the tip of $\vec{B}$.

If $\vec{C}=\vec{A}+\vec{B}$, then $\vec{C}-\vec{B}=\vec{A}$ and so on, just as in scalar addition and subtraction. Note also that $\vec{A}-\vec{B}=\vec{A}+(-\vec{B})$, which is sometimes useful in the context of graphical subtraction (see below).

QUESTION Is adding the magnitudes of two vectors equivalent to adding two vectors, then taking the magnitude?

No. The magnitude of a vector is *not* in general equal to the sum of the magnitudes of the two original vectors! For example, the magnitude of the vector $\langle 3,0,0\rangle$ is 3, and the magnitude of the vector $\langle -2,0,0\rangle$ is 2, but the magnitude of the vector $(\langle 3,0,0\rangle+\langle -2,0,0\rangle)$ is 1, not 5!

Checkpoint 6 If $\vec{F}_1=\langle 300,0,-200\rangle$ and $\vec{F}_2=\langle 150,-300,0\rangle$, calculate the following quantities and make the requested comparisons: **(a)** $\vec{F}_1+\vec{F}_2$ **(b)** $|\vec{F}_1+\vec{F}_2|$ **(c)** $|\vec{F}_1|+|\vec{F}_2|$ **(d)** Is $|\vec{F}_1+\vec{F}_2|=|\vec{F}_1|+|\vec{F}_2|$? **(e)** $\vec{F}_1-\vec{F}_2$ **(f)** $|\vec{F}_1-\vec{F}_2|$ **(g)** $|\vec{F}_1|-|\vec{F}_2|$ **(h)** Is $|\vec{F}_1-\vec{F}_2|=|\vec{F}_1|-|\vec{F}_2|$?

The sum of two vectors has a geometric interpretation. In Figure 1.21 you first walk along displacement vector $\vec{A}$, followed by walking along displacement vector $\vec{B}$. What is your net displacement vector $\vec{C}=\vec{A}+\vec{B}$? The x component C_x of your net displacement is the sum of A_x and B_x. Similarly, the y component C_y of your net displacement is the sum of A_y and B_y.

GRAPHICAL ADDITION OF VECTORS

To add two vectors $\vec{A}$ and $\vec{B}$ graphically (Figure 1.21):

- Draw the first vector $\vec{A}$.
- Move the second vector $\vec{B}$ (without rotating it) so its tail is located at the *tip* of the first vector.
- Draw a new vector from the tail of vector $\vec{A}$ to the tip of vector $\vec{B}$.

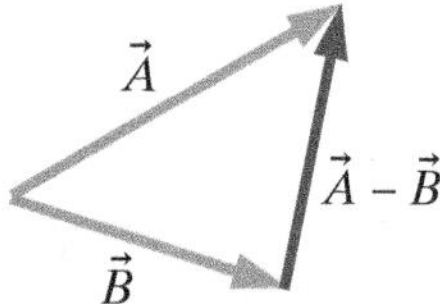

Figure 1.22 The procedure for subtracting vectors graphically: Draw vectors tail to tail; draw a new vector from the tip of the second vector to the tip of the first vector.

GRAPHICAL SUBTRACTION OF VECTORS

To subtract one vector $\vec{B}$ from another vector $\vec{A}$ graphically (Figure 1.22):

- Draw the first vector $\vec{A}$.
- Move the second vector $\vec{B}$ (without rotating it) so its tail is located at the *tail* of the first vector.
- Draw a new vector from the tip of vector $\vec{B}$ to the tip of vector $\vec{A}$.

Note that you can check this algebraically and graphically. As shown in Figure 1.22, since the tail of $\vec{A}-\vec{B}$ is located at the tip of $\vec{B}$, then the vector $\vec{A}$ should be the sum of $\vec{B}$ and $\vec{A}-\vec{B}$, as indeed it is:

$$\vec{B}+(\vec{A}-\vec{B})=\vec{A}$$

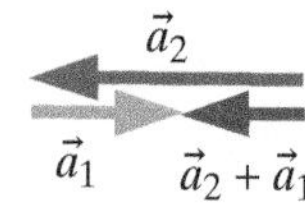

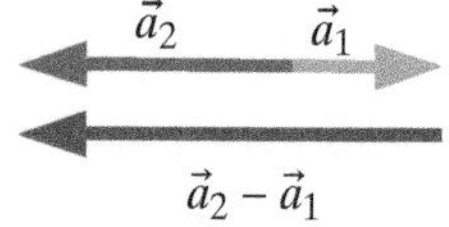

Figure 1.23 To add (top diagram) and subtract (bottom diagram) collinear vectors graphically, we offset the arrows slightly for clarity.

Graphical addition and subtraction of collinear vectors would be messy and difficult to interpret if we actually drew the arrows on top of each other. To make diagrams easier to interpret, we typically offset arrows slightly so we can see the results (Figure 1.23).

Checkpoint 7 Which of the following statements about the three vectors in Figure 1.24 are correct?
(a) $\vec{s}=\vec{t}-\vec{r}$ **(b)** $\vec{r}=\vec{t}-\vec{s}$ **(c)** $\vec{r}+\vec{t}=\vec{s}$ **(d)** $\vec{s}+\vec{t}=\vec{r}$ **(e)** $\vec{r}+\vec{s}=\vec{t}$

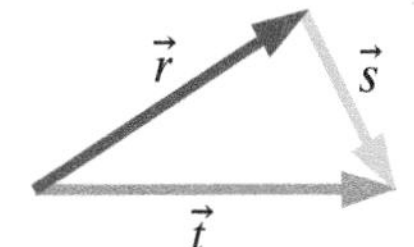

Figure 1.24 Checkpoint 7.

Commutativity and Associativity

Vector addition is commutative:

$$\vec{A}+\vec{B}=\vec{B}+\vec{A}$$

Vector subtraction is *not* commutative:

$$\vec{A}-\vec{B}\neq\vec{B}-\vec{A}$$

The associative property holds for vector addition and subtraction:

$$(\vec{A}+\vec{B})-\vec{C}=\vec{A}+(\vec{B}-\vec{C})$$

Applications of Vector Subtraction

Since we are interested in changes caused by interactions, we will frequently need to calculate the change in a vector quantity. For example, we may want to know the change in a moving object's position or the change in its velocity during some time interval. Finding such changes requires vector subtraction.

The Greek letter Δ (capital delta suggesting "D for Difference") is traditionally used to denote the change in a quantity (either a scalar or a vector). We use the subscript i to denote an *initial* value of a quantity, and the subscript f to denote the *final* value of a quantity.

Δ (DELTA) IS THE SYMBOL FOR A CHANGE

The symbol Δ (delta) means "final minus initial." If a vector $\vec{r}_i$ denotes the initial position of an object relative to the origin (its position at the beginning of a time interval), and $\vec{r}_f$ denotes the final position of the object, then

$$\Delta\vec{r}=\vec{r}_f-\vec{r}_i$$

$\Delta\vec{r}$ means "change of $\vec{r}$" or $\vec{r}_f-\vec{r}_i$ (displacement).
Δt means "change of t" or t_f-t_i (time interval).

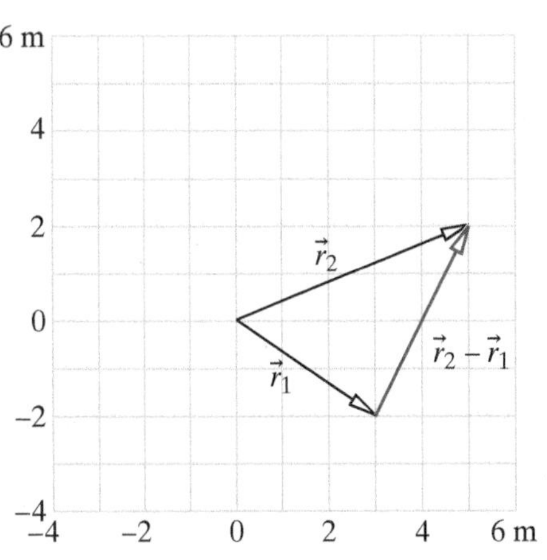

Figure 1.25 Relative position vector.

Since subtraction is not commutative, the order of the quantities matters: the symbol Δ (delta) always means "final minus initial," not "initial minus final." For example, when a child's height changes from 1.1 m to 1.2 m, the change is $\Delta y=+0.1$ m, a positive number. If your bank account dropped from \$150 to \$130, what was the change in your balance? Δ (bank account) = −20 dollars.

Another important application of vector subtraction is the calculation of relative position vectors, vectors that represent the position of one object relative to another object.

RELATIVE POSITION VECTOR

If object 1 is at location $\vec{r}_1$ and object 2 is at location $\vec{r}_2$ (Figure 1.25), the position of 2 relative to 1 is:

$$\vec{r}_{2\text{ relative to }1}=\vec{r}_2-\vec{r}_1$$

Checkpoint 8 At 10:00 AM you are at location $\langle -3,2,5\rangle$ m. By 10:02 AM you have walked to location $\langle 6,4,25\rangle$ m. **(a)** What is $\Delta\vec{r}$, the change in your position? **(b)** What is Δt, the time interval during which your position changed?

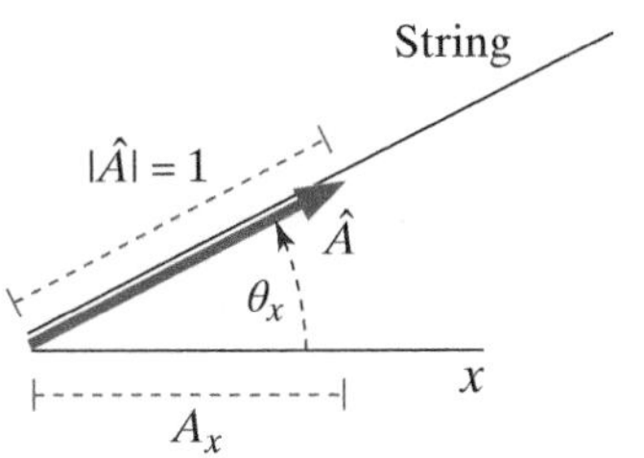

Figure 1.26 A unit vector whose direction is at a known angle from the $+x$ axis.

Unit Vectors and Angles

Suppose that a taut string is at an angle θ_x to the $+x$ axis, and we need a unit vector in the direction of the string. Figure 1.26 shows a unit vector $\hat{A}$ pointing

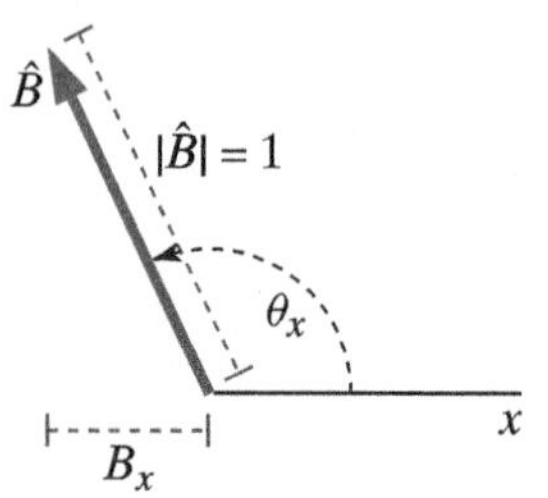

Figure 1.27 A unit vector in the second quadrant from the $+x$ axis.

along the string. What is the x component of this unit vector? Consider the triangle whose base is A_x and whose hypotenuse is $|\hat{A}| = 1$. From the definition of the cosine of an angle we have this:

$$\cos\theta = \frac{\text{adjacent}}{\text{hypotenuse}} = \frac{A_x}{1}, \text{ so } A_x = \cos\theta_x$$

In Figure 1.26 the angle θ_x is shown in the first quadrant (θ_x less than 90°), but this works for larger angles as well. For example, in Figure 1.27 the angle from the $+x$ axis to a unit vector $\hat{B}$ is in the second quadrant (θ_x greater than 90°) and $\cos\theta_x$ is negative, which corresponds to a negative value of B_x.

What is true for x is also true for y and z. Figure 1.28 shows a 3D unit vector $\hat{r}$ and indicates the angles between the unit vector and the x, y, and z axes. Evidently we can write

$$\hat{r} = \langle\cos\theta_x, \cos\theta_y, \cos\theta_z\rangle$$

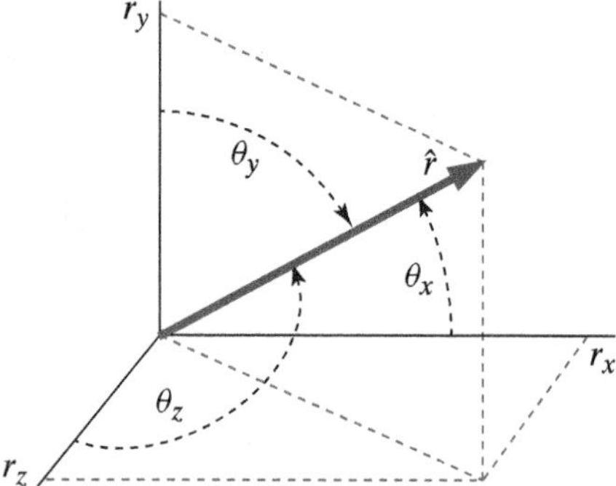

Figure 1.28 A 3D unit vector and its angles to the x, y, and z axes.

These three cosines of the angles between a vector (or unit vector) and the coordinate axes are called the "direction cosines" of the vector. The cosine function is never greater than 1, just as no component of a unit vector can be greater than 1.

A common special case is that of a unit vector lying in the xy plane, with zero z component (Figure 1.29). In this case $\theta_x + \theta_y = 90°$, so that $\cos\theta_y = \cos(90° - \theta_x) = \sin\theta_x$, therefore you can express the cosine of θ_y as the sine of θ_x, which is often convenient. However, in the general 3D case shown in Figure 1.28 there is no such simple relationship among the direction angles or among their cosines.

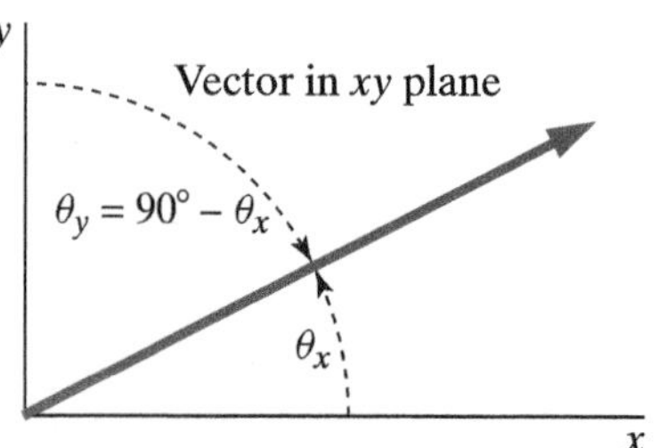

Figure 1.29 If a vector lies in the xy plane, $\cos\theta_y = \sin\theta_x$.

FINDING A UNIT VECTOR FROM ANGLES

To find a unit vector if angles are given:

- Redraw the vector of interest with its tail at the origin, and determine the angles between this vector and the axes.
- Imagine the vector $\langle 1,0,0\rangle$, which lies on the $+x$ axis. θ_x is the angle through which you would rotate the vector $\langle 1,0,0\rangle$ until its direction matched that of your vector. θ_x is positive, and $\theta_x \le 180°$.
- θ_y is the angle through which you would rotate the vector $\langle 0,1,0\rangle$ until its direction matched that of your vector. θ_y is positive, and $\theta_y \le 180°$.
- θ_z is the angle through which you would rotate the vector $\langle 0,0,1\rangle$ until its direction matched that of your vector. θ_z is positive, and $\theta_z \le 180°$.

EXAMPLE **From Unit Vector to Angles**

A vector $\vec{r}$ points from the origin to the location $\langle -600,0,300\rangle$ m. What is the angle that this vector makes to the x axis? To the y axis? To the z axis?

Solution

$$\hat{r} = \frac{\langle -600,0,300\rangle\ \text{m}}{\sqrt{(-600)^2 + (0)^2 + (300)^2}\ \text{m}} = \langle -0.894, 0, 0.447\rangle$$

But we also know that $\hat{r} = \langle\cos\theta_x, \cos\theta_y, \cos\theta_z\rangle$, so $\cos\theta_x = -0.894$, and the arccosine gives $\theta_x = 153.4°$. Similarly,

$$\cos\theta_y = 0, \text{ so } \theta_y = 90° \quad \text{(which checks; no } y \text{ component)}$$
$$\cos\theta_z = 0.447, \text{ so } \theta_z = 63.4°$$

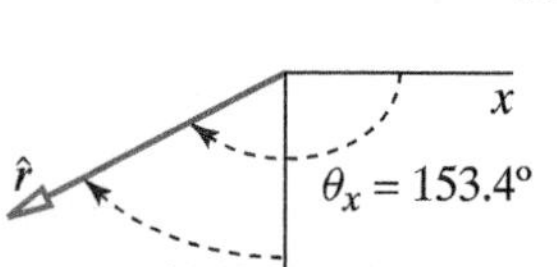

Figure 1.30 Look down on the xz plane. The difference in the two angles is 90°, as it should be.

Looking down on the xz plane in Figure 1.30, you can see that the difference between $\theta_x = 153.4°$ and $\theta_z = 63.4°$ is 90°, as it should be.

EXAMPLE **From Angle to Unit Vector**

A rope lying in the xy plane, pointing up and to the right, supports a climber at an angle of 20° to the vertical (Figure 1.31). What is the unit vector pointing up along the rope?

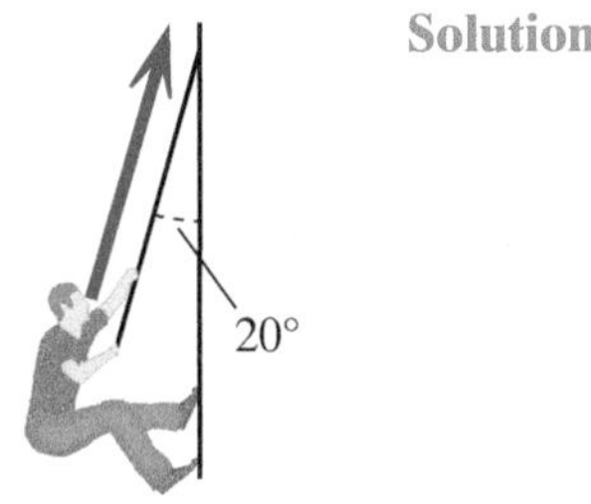

Figure 1.31 A climber supported by a rope.

Solution Follow the procedure given above for finding a unit vector from angles. In Figure 1.32 we redraw the vector with its tail at the origin, and we determine the angles between the vector and the axes.

If we rotate the unit vector $\langle 1,0,0\rangle$ from along the $+x$ axis to the vector of interest, we see that we have to rotate through an angle $\theta_x = 70°$. To rotate the unit vector $\langle 0,1,0\rangle$ from along the $+y$ axis to the vector of interest, we have to rotate through an angle of $\theta_y = 20°$. The angle from the $+z$ axis to our vector is $\theta_z = 90°$. Therefore the unit vector that points along the rope is:

$$\langle \cos 70°, \cos 20°, \cos 90°\rangle = \langle 0.342, 0.940, 0\rangle$$

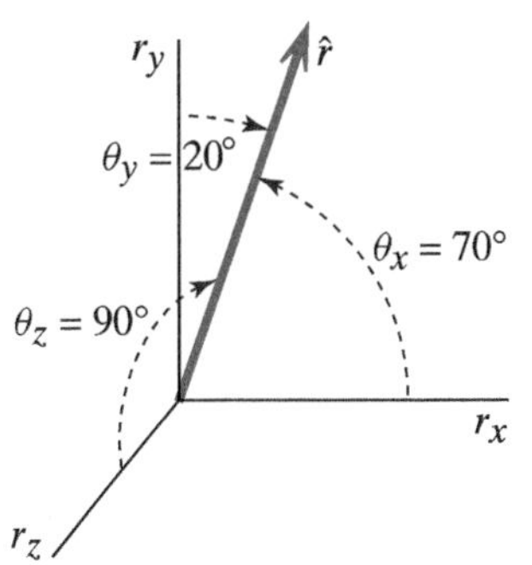

Figure 1.32 Redraw the vector with its tail at the origin. Identify the angles between the positive axes and the vector. In this example the vector lies in the xy plane.

You may have noticed that the y component of the unit vector can also be calculated as $\sin 70° = 0.940$, and it can be useful to recognize that a vector component can be obtained using sine instead of cosine. There is, however, an advantage in consistently calculating in terms of direction cosines. This is a method that always works, especially in 3D, and that helps avoid errors due to choosing the wrong trig function.

Checkpoint 9 (a) A unit vector lies in the xy plane, at an angle of 160° from the $+x$ axis, with a positive y component. What is the unit vector? (It helps to draw a diagram.) **(b)** A string runs up and to the left in the xy plane, making an angle of 40° to the vertical. Determine the unit vector that points along the string.

Reorienting the Coordinate Axes

In order to describe position and displacement we had to choose an origin and a set of axes. What if we had made different choices? Certain quantities related to vectors change when a different orientation is chosen for the coordinate axes, but others remain the same. Scalar quantities such as mass and temperature do not change. The magnitude of a vector remains the same when axes are oriented differently, even though the components of the vector do change—the x component of velocity will have a different value if the x axis is chosen to have a different orientation. (Because of this, a vector component is not considered mathematically to be a true scalar, even though it is a single number. However, this distinction is not going to be important in the context of the physics we will study in this course.)

Quantity	Unit	Symbol
mass	kilogram	kg
distance	meter	m
time	second	s
charge	coulomb	C

Figure 1.33 Basic SI units.

1×10^{3}	kilo	1×10^{-3}	milli
1×10^{6}	mega	1×10^{-6}	micro
1×10^{9}	giga	1×10^{-9}	nano
1×10^{12}	tera	1×10^{-12}	pico

Figure 1.34 Common metric prefixes.

1.5 SI UNITS

In this book we use the SI (Système Internationale) unit system, as is customary in technical work. The SI unit of mass is the kilogram (kg), the unit of distance is the meter (m), and the unit of time is the second (s) (Figure 1.33). In later chapters we will encounter other SI units, such as the newton (N), which is a unit of force. Many quantities combine SI units (for example, velocity, which has units of m/s).

It is essential to use SI units in physics equations; this may require that you convert from some other unit system to SI units. Common metric prefixes are shown in Figure 1.34. If mass is known in grams, you need to divide by 1000 and use the mass in kilograms. If a distance is given in centimeters, you need to divide by 100 to convert the distance to meters. If the time is measured in minutes, you need to multiply by 60 to use a time in seconds. A convenient way to do such conversions is to multiply by factors that are equal to 1, such as (1 min)/(60 s) or (100 cm)/(1 m). As an example, consider converting 60 miles per hour to SI units, meters per second. Start with the 60 mi/h and multiply by factors of 1:

$$\left(60\frac{\text{mi}}{\text{h}}\right)\left(\frac{1\,\text{h}}{60\,\text{min}}\right)\left(\frac{1\,\text{min}}{60\,\text{s}}\right)\left(\frac{5280\,\text{ft}}{1\,\text{mi}}\right)\left(\frac{12\,\text{in}}{1\,\text{ft}}\right)\left(\frac{2.54\,\text{cm}}{1\,\text{in}}\right)\left(\frac{1\,\text{m}}{100\,\text{cm}}\right)=26.8\frac{\text{m}}{\text{s}}$$

Observe how most of the units cancel, leaving final units of m/s.

Checkpoint 10 A snail moved 80 cm (80 centimeters) in 5 min. What was its average speed in SI units? Write out the factors as was done above.

1.6 SPEED AND VELOCITY

Because it has a magnitude and a direction, velocity is a vector quantity. Velocity is unusual among vector quantities because its magnitude has a special name: "speed." In everyday conversation the words "velocity" and "speed" are often used interchangeably, but because the vector quantity velocity (represented by $\vec{v}$) and the scalar quantity speed ($|\vec{v}|$) have quite different meanings in physics, we will need to use these terms precisely.

Motion is important in everyday life, and it may be somewhat surprising to reflect on the fact that neither velocity nor speed can be measured directly in a single measurement. To determine the speed of a moving object it is necessary to measure two times and two positions, and to calculate the speed by dividing the distance traveled by the time elapsed, as reflected in the familiar expression $v = d/t$. Even the radar guns commonly used to measure the speed of cars or baseballs do not make instantaneous measurements. The wavelength of radar waves reflected from a moving object is different from that of the original waves; the original and reflected waves must be compared over a period of time to determine the speed of the moving object.

Since the measurement interval is finite, it is of course possible that the moving object is speeding up or slowing down during that interval. So what speed do we actually measure, if we are not measuring the exact speed at one instant? We refer to the results of our measurements as "average speed" or, if we include direction, "average velocity."

EXAMPLE **A Sprinter's Average Speed**

At the 2008 summer Olympics in Beijing, the Jamaican sprinter Usain Bolt won the gold medal in the 100 m race, finishing in a time of 9.69 s, and setting a new world record for the event. What was his average speed? From videos of the event one can determine that he reached the 60 meter mark at a time 5.73 s after the start of the race. Was his speed constant?

Solution Bolt's average speed over the entire 100 m was

$$|\vec{v}_{\text{avg}}| = \frac{100\,\text{m}}{9.69\,\text{s}} = 10.32\,\text{m/s}$$

However, his average speed over the first 60 m was

$$|\vec{v}_{\text{avg}}| = \frac{60\,\text{m}}{5.73\,\text{s}} = 10.47\,\text{m/s}$$

Bolt clearly slowed down at the end of the race, beginning to celebrate his win even before reaching the finish line.

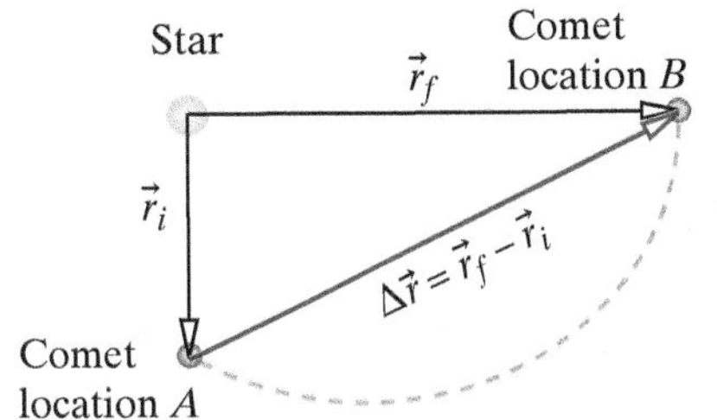

Figure 1.35 A comet orbiting a star moves from location A to location B in 5 years. The vector $\Delta\vec{r}$ is the displacement of the comet during this time period.

If we wanted to know Bolt's speed as accurately as possible at the instant he reached the 60 m mark, we would need to know his position at two times: just before he reached the mark, and just afterward. The smaller the time interval, the closer the average speed we calculate is to the actual instantaneous speed.

Vector Velocity

As with speed, to determine velocity we need two measurements separated in time. With this data we can find the "displacement," or change of position $\Delta\vec{r} = \vec{r}_f - \vec{r}_i$ of an object during a time interval, where $\vec{r}_i$ is the initial 3D position and $\vec{r}_f$ is the final 3D position (note that as with relative position vectors, we always calculate "final minus initial"). For example, Figure 1.35 shows the displacement of a comet over a 5 month period. Dividing the (vector) displacement by the (scalar) time interval $t_f - t_i$ (final time minus initial time) gives the average (vector) velocity of the object:

DEFINITION: AVERAGE VELOCITY

$$\vec{v}_{\text{avg}} = \frac{\vec{r}_f - \vec{r}_i}{t_f - t_i}$$

A more compact way of writing this expression, using the "Δ" symbol (capital Greek delta, defined in Section 1.4), is

$$\vec{v}_{\text{avg}} = \frac{\Delta\vec{r}}{\Delta t}$$

Note that we are using vector subtraction here to find the displacement of a single object from one time to another, while earlier we used vector subtraction to find the relative position of one object with respect to a second object at a single time. The mathematical operation is the same, but the physical meaning is different.

EXAMPLE **Average Velocity of a Bee**

Consider a bee in flight. At time $t_i = 15.0$ s after 9:00 AM, the bee's position vector was $\vec{r}_i = \langle 2,4,0\rangle$ m. At time $t_f = 15.1$ s after 9:00 AM, the bee's position vector was $\vec{r}_f = \langle 3,3.5,0\rangle$ m. What was the average velocity of the bee during this interval? Express this vector as the product of the magnitude of the velocity (speed) and a unit vector in the direction of the velocity.

Solution On the diagram shown in Figure 1.36, we draw and label three arrows representing the vectors $\vec{r}_i$, $\vec{r}_f$, and $\vec{r}_f - \vec{r}_i$. The tail of the latter arrow is placed at the bee's initial position. The vector $\vec{r}_f - \vec{r}_i$, which points in the direction of the bee's motion, is the displacement of the bee during this time interval.

We calculate the bee's displacement vector numerically by taking the difference of the two vectors, final minus initial:

$$\vec{r}_f - \vec{r}_i = \langle 3,3.5,0\rangle\,\text{m} - \langle 2,4,0\rangle\,\text{m} = \langle 1,-0.5,0\rangle\,\text{m}$$

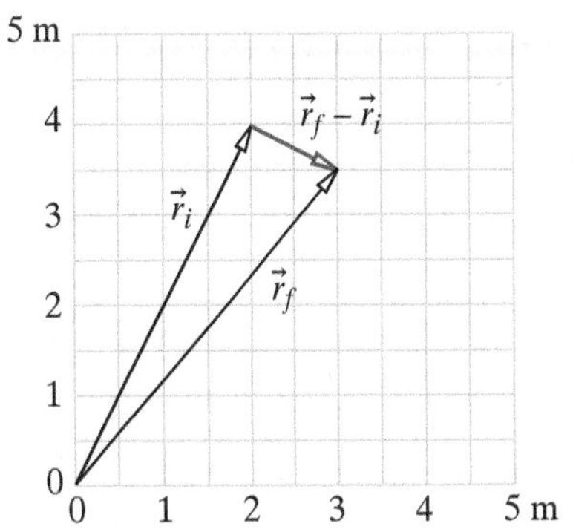

Figure 1.36 The displacement vector points from initial position to final position.

This numerical result should be consistent with our graphical construction. Look at the components of $\vec{r}_f - \vec{r}_i$ in Figure 1.36. Do you see that this vector has an x component of +1 and a y component of −0.5 m?

The average velocity of the bee, a vector quantity, is the (vector) displacement $\vec{r}_f - \vec{r}_i$ divided by the (scalar) time interval $t_f - t_i$.

$$\vec{v}_{\text{avg}} = \frac{\vec{r}_f - \vec{r}_i}{t_f - t_i} = \frac{\langle 1,-0.5,0\rangle\,\text{m}}{(15.1-15.0)\,\text{s}} = \frac{\langle 1,-0.5,0\rangle\,\text{m}}{0.1\,\text{s}} = \langle 10,-5,0\rangle\,\text{m/s}$$

Since we divided $\vec{r}_f - \vec{r}_i$ by a scalar $(t_f - t_i)$, the average velocity $\vec{v}_{\text{avg}}$ points in the direction of the bee's motion, assuming that the bee flew in a straight line.

The average speed of the bee is the magnitude of its velocity:

$$|\vec{v}_{\text{avg}}| = \sqrt{10^2 + (-5)^2 + 0^2}\ \text{m/s} = 11.18\ \text{m/s}$$

The direction of the bee's motion, expressed as a unit vector, is:

$$\hat{v}_{\text{avg}} = \frac{\vec{v}_{\text{avg}}}{|\vec{v}_{\text{avg}}|} = \frac{\langle 10, -5, 0\rangle\ \text{m/s}}{11.18\ \text{m/s}} = \langle 0.894, -0.447, 0\rangle$$

Note that the "m/s" units cancel; the result is dimensionless. We can check that this really is a unit vector:

$$\sqrt{0.894^2 + (-0.447)^2 + 0^2} = 0.9995$$

This is not quite 1.0 due to rounding the velocity coordinates and speed to three significant figures. To check, we can put the pieces back together and see what we get. The original vector factors into the product of the magnitude times the unit vector:

$$|\vec{v}|\hat{v} = (11.18\ \text{m/s})\langle 0.894, -0.447, 0\rangle = \langle 10, -5, 0\rangle\ \text{m/s}$$

This is the same as the original vector $\vec{v}$.

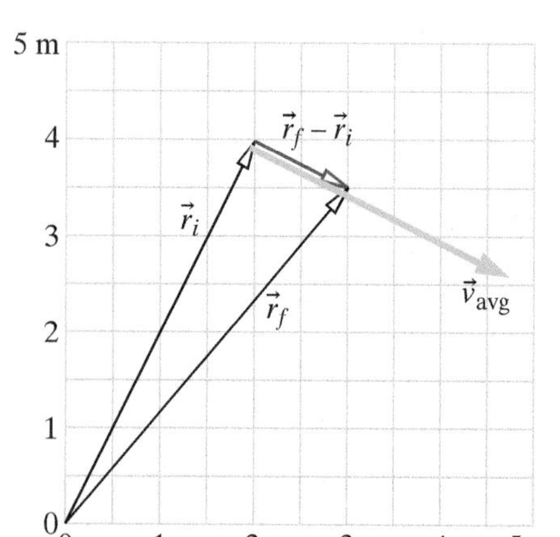

Figure 1.37 Average velocity vector: displacement divided by time interval.

Scaling a Vector to Fit on a Graph

We can plot the average velocity vector on the same graph that we use for showing the vector positions of the bee (Figure 1.37). However, note that velocity has units of meters per second whereas positions have units of meters, so we are mixing quantities on this diagram.

Moreover, the magnitude of the vector, 11.18 m/s, doesn't fit on a graph that is only 5 units wide (in meters). It is standard practice in such situations to scale down the arrow representing the vector to fit on the graph, preserving the correct direction. In Figure 1.37 we've scaled down the arrow representing the velocity vector by about a factor of 3 to make the arrow fit on the graph. Of course if there is more than one velocity vector we use the same scale factor for all the velocity vectors. The same kind of scaling is used with other physical quantities that are vectors, such as force and momentum, which we will encounter later.

Checkpoint 11 At a time 0.2 s after it has been hit by a tennis racket, a tennis ball is located at $\langle 5, 7, 2\rangle$ m, relative to an origin in one corner of a tennis court. At a time 0.7 s after being hit, the ball is located at $\langle 9, 2, 8\rangle$ m. **(a)** What is the average velocity of the tennis ball? **(b)** What is the average speed of the tennis ball? **(c)** What is the unit vector in the direction of the ball's average velocity?

1.7 PREDICTING A NEW POSITION

We can rewrite the velocity relationship in the form

$$(\vec{r}_f - \vec{r}_i) = \vec{v}_{\text{avg}}(t_f - t_i)$$

That is, the (vector) displacement of an object is its average (vector) velocity times the time interval. This is just the vector version of the simple notion that if you run at a speed of 7 m/s for 5 s you move a distance of $(7\,\text{m/s})(5\,\text{s}) = 35\,\text{m}$, or that a car going 50 mi/h for 2 h goes $(50\,\text{mi/h})(2\,\text{h}) = 100$ mi.

Is $(\vec{r}_f - \vec{r}_i) = \vec{v}_{\text{avg}}(t_f - t_i)$ a valid vector relation? Yes, multiplying a vector $\vec{v}_{\text{avg}}$ times a scalar $(t_f - t_i)$ yields a vector. We make a further rearrangement to obtain a relation for updating the position when we know the velocity:

THE POSITION UPDATE EQUATION

$$\vec{r}_f = \vec{r}_i + \vec{v}_{\text{avg}}(t_f - t_i)$$

or

$$\vec{r}_f = \vec{r}_i + \vec{v}_{\text{avg}}\Delta t$$

This equation says that if we know the starting position, the average velocity, and the time interval, we can predict the final position. This equation will be important throughout our work.

The position update equation $\vec{r}_f = \vec{r}_i + \vec{v}_{\text{avg}}\,\Delta t$ is a vector equation, so we can write out its full component form:

$$\langle x_f, y_f, z_f \rangle = \langle x_i, y_i, z_i \rangle + \langle v_{\text{avg},x}, v_{\text{avg},y}, v_{\text{avg},z} \rangle\,\Delta t$$

Because the x component on the left of the equation must equal the x component on the right (and similarly for the y and z components), this compact vector equation represents three separate component equations:

$$x_f = x_i + v_{\text{avg},x}\,\Delta t$$
$$y_f = y_i + v_{\text{avg},y}\,\Delta t$$
$$z_f = z_i + v_{\text{avg},z}\,\Delta t$$

EXAMPLE **Predicting the Position of a Ball**

At time $t_i = 12.18$ s after 1:30 PM a ball's position vector is $\vec{r}_i = \langle 20, 8, -12 \rangle$ m. The ball's velocity at that moment is $\vec{v} = \langle 9, -4, 6 \rangle$ m/s. At time $t_f = 12.21$ s after 1:30 PM, where will the ball be, assuming that its velocity hardly changes during this short time interval?

Solution

$$\vec{r}_f = \langle 20, 8, -12 \rangle\ \text{m} + (\langle 9, -4, 6 \rangle\ \text{m/s})(12.21 - 12.18)\ \text{s}$$
$$\vec{r}_f = \langle 20, 8, -12 \rangle\ \text{m} + \langle 0.27, -0.12, 0.18 \rangle\ \text{m}$$
$$\vec{r}_f = \langle 20.27, 7.88, -11.82 \rangle\ \text{m}$$

Checkpoint 12 A proton traveling with a velocity of $\langle 3 \times 10^5, 2 \times 10^5, -4 \times 10^5 \rangle$ m/s passes the origin at a time 9.0 s after a proton detector is turned on. Assuming that the velocity of the proton does not change, what will be its position at time 9.7 s?

Instantaneous Velocity

Figure 1.39 shows the path of a ball, with positions marked at 1 s intervals, and the table in Figure 1.38 lists the position information. While the ball is in the air, its velocity is constantly changing, due to interactions with the Earth (gravity) and with the air (air resistance).

Loc.	t (s)	Position (m)
A	0.0	$\langle 0,0,0 \rangle$
B	1.0	$\langle 22.3,26.1,0 \rangle$
C	2.0	$\langle 40.1,38.1,0 \rangle$
D	3.0	$\langle 55.5,39.2,0 \rangle$
E	4.0	$\langle 69.1,31.0,0 \rangle$
F	5.0	$\langle 80.8,14.8,0 \rangle$

Figure 1.38 Table showing elapsed time and position of the ball at each location marked by a dot in Figure 1.39.

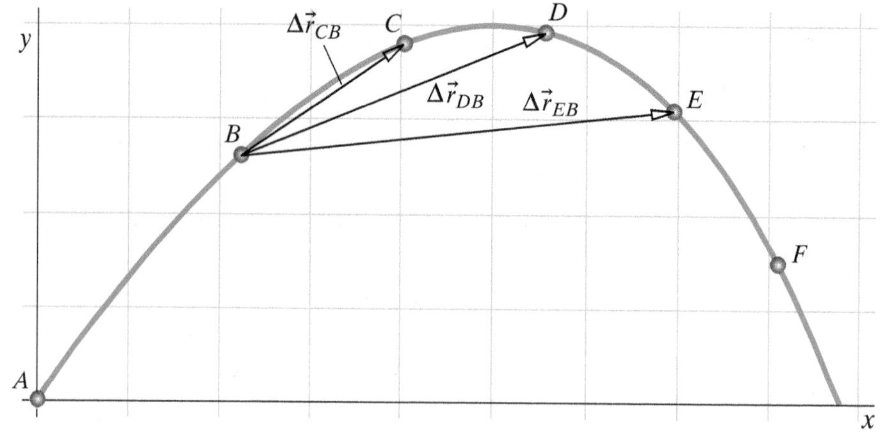

Figure 1.39 The trajectory of a ball through air. The axes represent the x and y distance from the ball's initial location; each square on the grid corresponds to 10 meters. The position of the ball at intervals of 1 s is represented by the colored dots. Three different displacements, corresponding to three different time intervals, are indicated by arrows on the diagram.

Suppose we ask: What is the velocity of the ball at the precise instant that it reaches location B? This quantity would be called the "instantaneous velocity" of the ball. We can start by approximating the instantaneous velocity of the ball by finding its average velocity over some larger time interval.

We can use the position and time data in Figure 1.38 to calculate the average velocity of the ball over three different intervals, by finding the ball's displacement during each interval, and dividing by the appropriate Δt for that interval:

$$\vec{v}_{EB} = \frac{\Delta\vec{r}_{EB}}{\Delta t} = \frac{\vec{r}_E - \vec{r}_B}{t_E - t_B} = \frac{(\langle 69.1,31.0,0\rangle - \langle 22.3,26.1,0\rangle)\,\text{m}}{(4.0-1.0)\,\text{s}}$$
$$= \langle 15.6,1.6,0\rangle \frac{\text{m}}{\text{s}}$$

$$\vec{v}_{DB} = \frac{\Delta\vec{r}_{DB}}{\Delta t} = \frac{\vec{r}_D - \vec{r}_B}{t_D - t_B} = \frac{(\langle 55.5,39.2,0\rangle - \langle 22.3,26.1,0\rangle)\,\text{m}}{(3.0-1.0)\,\text{s}}$$
$$= \langle 16.6,6.55,0\rangle \frac{\text{m}}{\text{s}}$$

$$\vec{v}_{CB} = \frac{\Delta\vec{r}_{CB}}{\Delta t} = \frac{\vec{r}_C - \vec{r}_B}{t_C - t_B} = \frac{(\langle 40.1,38.1,0\rangle - \langle 22.3,26.1,0\rangle)\,\text{m}}{(2.0-1.0)\,\text{s}}$$
$$= \langle 17.8,12.0,0\rangle \frac{\text{m}}{\text{s}}$$

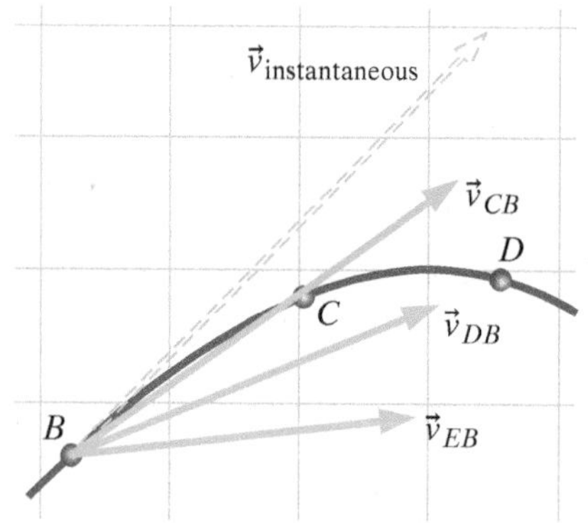

Figure 1.40 A segment of the trajectory shown in Figure 1.39. The three different average velocity vectors calculated above are shown by three arrows, each with its tail at location B. The three arrows representing average velocities are drawn with their tails at the location of interest. The dashed arrow represents the actual instantaneous velocity of the ball at location B. Note that since the units of velocity are m/s, these arrows use a different scale from the distance scale used for the path of the ball.

Not surprisingly, the average velocities over these different time intervals are not the same, because both the direction of the ball's motion and the speed of the ball were changing continuously during its flight. The three average velocity vectors that we calculated are shown in Figure 1.40.

> **QUESTION** Which of the three average velocity vectors depicted in Figure 1.40 best approximates the instantaneous velocity of the ball at location B?

Simply by looking at the diagram, we can tell that $\vec{v}_{CB}$ is closest to the actual instantaneous velocity of the ball at location B, because its direction is closest to the direction in which the ball is actually traveling. Because the direction of the instantaneous velocity is the direction in which the ball is moving at a particular instant, the instantaneous velocity is tangent to the ball's path. Of the three average velocity vectors we calculated, $\vec{v}_{CB}$ best approximates a tangent to the path of the ball. Evidently $\vec{v}_{CB}$, the velocity calculated with the shortest time interval, $t_C - t_B$, is the best approximation to the instantaneous velocity at location B. If we used even smaller values of Δt in our calculation of average velocity, such as 0.1 s, or 0.01 s, or 0.001 s, we would presumably have better and better estimates of the actual instantaneous velocity of the object at the instant when it passes location B.

We can draw two important conclusions about instantaneous velocity:

- The direction of the instantaneous velocity of an object is tangent to the path of the object's motion.
- Smaller time intervals yield more accurate estimates of instantaneous velocity.

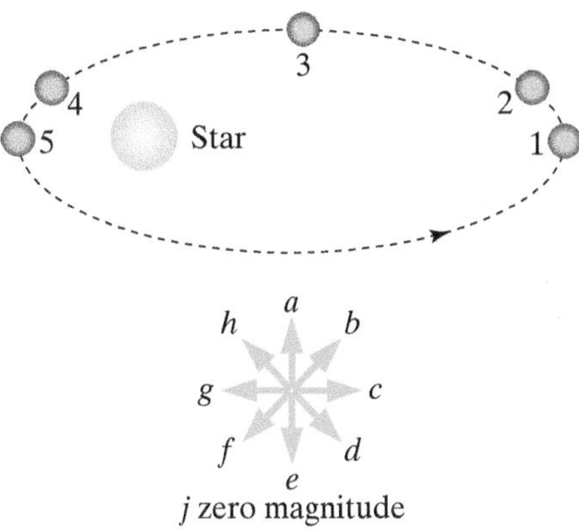

Figure 1.41 A comet goes around a star.

Checkpoint 13 A comet travels in an elliptical path around a star, in the direction shown in Figure 1.41. Which arrow best indicates the direction of the comet's instantaneous velocity vector at each of the numbered locations in the orbit?

Velocity Is a Rate of Change

You may already have learned about derivatives in calculus. Because instantaneous velocity is the rate of change of position, it is a derivative, the limit of $\Delta\vec{r}/\Delta t$ as the time interval Δt used in the calculation gets closer and closer to zero:

$$\vec{v} = \lim_{\Delta t \to 0} \frac{\Delta\vec{r}}{\Delta t}, \text{ which is written as } \vec{v} = \frac{d\vec{r}}{dt}$$

In Figure 1.40, the process of taking the limit is illustrated graphically. As smaller values of Δt are used in the calculation, the average velocity vectors approach the limiting value: the actual instantaneous velocity.

The rate of change (derivative) of a vector is also a vector. Differentiating a vector simply requires differentiating each component:

$$\vec{v} = \frac{d\vec{r}}{dt} = \frac{d}{dt}\langle x, y, z\rangle = \left\langle \frac{dx}{dt}, \frac{dy}{dt}, \frac{dz}{dt} \right\rangle = \langle v_x, v_y, v_z\rangle$$

The derivative of the position vector $\vec{r}$ gives components that are the components of the velocity, as we should expect.

Informally, you can think of $d\vec{r}$ as a very small ("infinitesimal") displacement, and dt as a very small ("infinitesimal") time interval. It is as though we had continued the process illustrated in Figure 1.40 to smaller and smaller time intervals, down to an extremely tiny time interval dt with a correspondingly tiny displacement $d\vec{r}$. The ratio of these tiny quantities is the instantaneous velocity.

The ratio of these two tiny quantities need not be small. For example, suppose that an object moves in the x direction a tiny distance of 1×10^{-15} m, the radius of a proton, in a very short time interval of 1×10^{-23} s:

$$\vec{v} = \frac{\langle 1 \times 10^{-15}, 0, 0\rangle \text{ m}}{1 \times 10^{-23} \text{ s}} = \langle 1 \times 10^{8}, 0, 0\rangle \text{ m/s}$$

which is one-third the speed of light (3×10^8 m/s)!

Acceleration

Velocity is the time rate of change of position: $\vec{v} = d\vec{r}/dt$. Similarly, we define "acceleration" as the time rate of change of velocity: $\vec{a} = d\vec{v}/dt$. Acceleration, which is itself a vector quantity, has units of meters per second per second, written as m/s/s or m/s^2.

DEFINITION: ACCELERATION

Instantaneous acceleration is the time rate of change of velocity:

$$\vec{a} = \frac{d\vec{v}}{dt}$$

Average acceleration can be calculated from a change in velocity:

$$\vec{a}_{\text{avg}} = \frac{\Delta\vec{v}}{\Delta t}$$

The units of acceleration are m/s^2.

EXAMPLE **Acceleration of a Car**

A car traveling in the $+x$ direction speeds up from 20 m/s to 26 m/s in 3 s. What is its average acceleration?

Solution

$$\vec{a}_{\text{avg}} = \frac{\Delta\vec{v}}{\Delta t} = \frac{(\langle 26,0,0\rangle - \langle 20,0,0\rangle)\ \text{m/s}}{3\ \text{s}} = \langle 2,0,0\rangle\ \text{m/s/s}$$

For another example, if you drop a rock near the surface of the Earth, its speed increases 9.8 m/s every second, so its acceleration is 9.8 m/s/s, as long as air resistance is negligible.

Because velocity is a vector, there are two parts to its time derivative, acceleration:

$$\vec{v} = |\vec{v}|\hat{v}$$

$$\vec{a} = \frac{d\vec{v}}{dt} = \frac{d|\vec{v}|}{dt}\hat{v} + |\vec{v}|\frac{d\hat{v}}{dt}$$

As we'll see in later chapters, these two parts of the acceleration are associated with pushing or pulling parallel to the motion (changing the speed) or perpendicular to the motion (changing the direction).

Checkpoint 14 **(a)** Powerful sports cars can go from zero to 25 m/s (about 60 mi/h) in 5 s. (1) What is the magnitude of the average acceleration? (2) How does this compare with the acceleration of a rock falling near the Earth's surface? **(b)** Suppose the position of an object at time t is $\langle 3+5t, 4t^2, 2t-6t^3\rangle$. (1) What is the instantaneous velocity at time t? (2) What is the instantaneous acceleration at time t? (3) What is the instantaneous velocity at time $t=0$? (4) What is the instantaneous acceleration at time $t=0$?

1.8 MOMENTUM

In trying to model the real world, physicists look for powerful ideas that are very general—that is, that apply to a very large range of systems and phenomena. Some of the most powerful and general principles involve "hidden" quantities—things we do not perceive directly. Momentum is such a quantity.

We have discussed velocity, a vector quantity that describes motion and can be determined from measurements of position and time; position, time, speed, and now 3D velocity are all familiar quantities. However, velocity is not the whole story. Consider the following thought experiment:

Suppose you gently toss a tennis ball to a friend, in such a way that just before the ball reaches her hands, its velocity is $\langle 0.3, -0.2, 0\rangle$ m/s. When your friend catches the ball, she must interact with the ball to stop its motion, changing its velocity from $\langle 0.3, -0.2, 0\rangle$ m/s to $\vec{0}$ m/s.

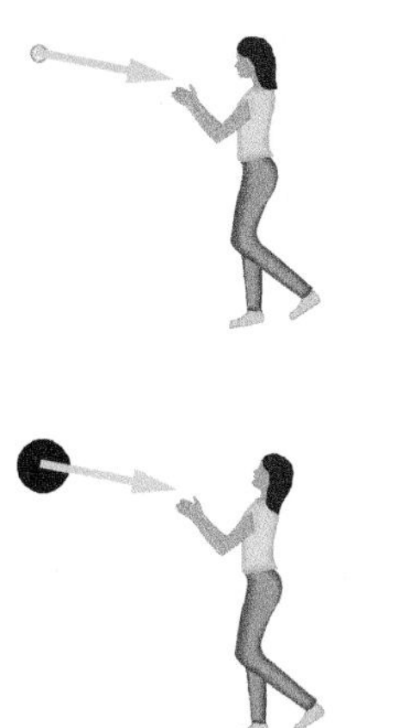

Figure 1.42 Catching a bowling ball requires a larger interaction than catching a tennis ball with the same velocity.

Now suppose that you again toss a ball to your friend with the same velocity; this time, however, the ball is a bowling ball! When the ball reaches her hands with velocity $\langle 0.3, -0.2, 0\rangle$ m/s, your friend must interact much more strongly with the ball to change its velocity to $\vec{0}$ m/s (Figure 1.42). Even though the change in the velocities of the two balls is identical, the amount of interaction needed to cause this change is very different. Evidently the mass of the moving object must explicitly be taken into account.

The larger the mass of the object, the stronger the interaction required to change its motion. Since the same is true for the velocity of the object (your friend would have had to interact more strongly to stop a tennis ball with velocity $\langle 40, 0, 0\rangle$ m/s), we will surmise that it is the product of mass and velocity that is important. This quantity is called "momentum"; since it is the product of a scalar and a vector, momentum is a vector. For reasons lost in the mists of history, the symbol used to represent momentum is $\vec{p}$. Momentum is of fundamental importance not only in classical (prequantum) mechanics but also in relativity and quantum mechanics. In Chapter 3 we will discuss the fact that momentum is a "conserved" quantity; the total momentum of the universe is constant.

APPROXIMATE DEFINITION OF MOMENTUM

$$\vec{p} \approx m\vec{v}$$

The units of momentum are kg · m/s. We will see in Section 1.10 that this expression is a good approximation for the momentum of objects traveling at speeds less than about one-tenth of the speed of light.

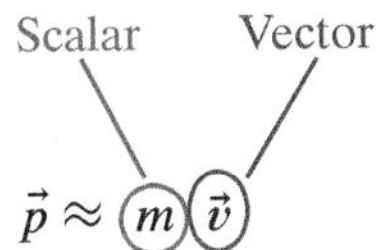

Figure 1.43 The approximate expression for momentum is the product of a scalar times a vector. The scalar factor, mass, must be positive, so the direction of an object's momentum is the same as the direction of its velocity.

Although we can see and compare velocities, momentum is a quantity that we can't see directly. We will encounter other important quantities that aren't directly visible, such as energy, angular momentum, and electric and magnetic fields. Like velocity, momentum is a vector quantity, so it has a magnitude and a direction. Since the mass m must be a positive number, this scalar factor cannot change the direction of the vector (Figure 1.43). Therefore the direction of an object's momentum is the same as the direction of its velocity.

Change of Momentum

Looking back at Newton's first law of motion, we can see that the idea that a body "persists in its state of rest or of moving with constant speed in a constant direction..." can be stated compactly as "the momentum of a body remains constant..." In Chapter 2 we will relate momentum change to interaction mathematically, using the concept of "force" to quantify interactions. This will allow us to predict quantitatively the motion of objects whose momentum is changed by interactions with their surroundings.

Change in momentum, therefore, is an important quantity. We have just noted that it was harder for the person to change the momentum of a bowling ball than to change the momentum of a tennis ball with the same velocity. Calculating a change of momentum requires vector subtraction.

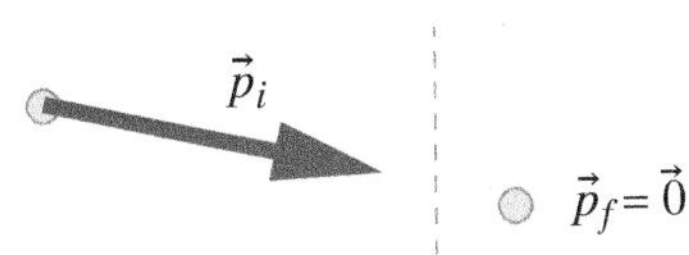

Figure 1.44 The system is the ball. The initial state is just before touching your friend's hands, and the final state is just after the ball has come to a stop in her hands.

EXAMPLE **Magnitude of Momentum Change**

Consider the tennis ball and bowling ball discussed above. What is the magnitude of the change in the momentum of each ball when your friend catches it?

Solution The mass of a regulation tennis ball is about 58 g, or 0.058 kg in S.I. units. The momentum of the tennis ball just before it reaches your friend's hands (Figure 1.44) is:

$$\vec{p}_i = (0.058\,\text{kg})\langle 0.3, -0.2, 0\rangle\ \text{m/s} = \langle 0.0174, -0.0116, 0\rangle\ \text{kg}\cdot\text{m/s}$$

The final momentum of the tennis ball is $\vec{0}$ kg · m/s, and the change in the tennis ball's momentum is:

$$\begin{aligned}\Delta\vec{p} &= \vec{p}_f - \vec{p}_i = \langle 0,0,0\rangle\ \text{kg}\cdot\text{m/s} - \langle 0.0174,-0.0116,0\rangle\ \text{kg}\cdot\text{m/s}\\ &= \langle -0.0174, 0.0116, 0\rangle\ \text{kg}\cdot\text{m/s}\end{aligned}$$

The magnitude of the tennis ball's change in momentum is:

$$|\Delta\vec{p}| = \sqrt{-0.0174^2 + 0.0116^2}\ \text{kg}\cdot\text{m/s} = 0.0209\ \text{kg}\cdot\text{m/s}$$

For a bowling ball of mass 5.8 kg (about 13 lb):

$$\begin{aligned}\vec{p}_i &= (5.8\ \text{kg})\langle 0.3,-0.2,0\rangle\ \text{m/s} = \langle 1.74,-1.16,0\rangle\ \text{kg}\cdot\text{m/s}\\ \Delta\vec{p} &= \langle 0,0,0\rangle\ \text{kg}\cdot\text{m/s} - \langle 1.74,-1.16,0\rangle\ \text{kg}\cdot\text{m/s}\\ &= \langle -1.74, 1.16, 0\rangle\ \text{kg}\cdot\text{m/s}\\ |\Delta\vec{p}| &= \sqrt{-1.74^2 + 1.16^2}\ \text{kg}\cdot\text{m/s} = 2.09\ \text{kg}\cdot\text{m/s}\end{aligned}$$

The change in velocity of each ball was the same, but the magnitude of the change in momentum of the bowling ball was 100 times larger than the change of momentum of the tennis ball.

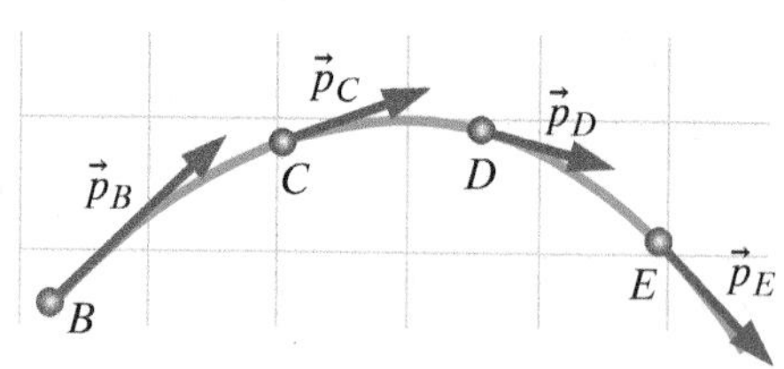

Figure 1.45 A portion of the trajectory of a ball moving through air, subject to gravity and air resistance. The arrows represent the momentum of the ball at the locations indicated by letters.

Momentum is a vector quantity and proportional to the velocity, so just as was the case with velocity, there are two aspects of momentum that can change: magnitude and direction. A mathematical description of change of momentum must include either a change in the magnitude of the momentum, or a change in the direction of the momentum, or both.

EXAMPLE **Change in Magnitude and Direction of Momentum**

Figure 1.45 shows a portion of the trajectory of a ball in air, subject to gravity and air resistance. At location B, the ball's momentum is $\vec{p}_B = \langle 3.03, 2.83, 0\rangle$ kg·m/s. At location C, the ball's momentum is $\vec{p}_C = \langle 2.55, 0.97, 0\rangle$ kg·m/s. Find the change in the ball's momentum between these locations, and show it on the diagram. What changed: the direction of the ball's momentum, the magnitude of the ball's momentum, or both?

Solution

$$\begin{aligned}\Delta\vec{p} &= \vec{p}_C - \vec{p}_B = \langle 2.55, 0.97, 0\rangle\ \text{kg}\cdot\text{m/s} - \langle 3.03, 2.83, 0\rangle\ \text{kg}\cdot\text{m/s}\\ &= \langle -0.48, -1.86, 0\rangle\ \text{kg}\cdot\text{m/s}\end{aligned}$$

Both the x and y components of the ball's momentum decreased, so $\Delta\vec{p}$ has negative x and y components. This is consistent with the graphical subtraction shown in Figure 1.46.

It is clear from the diagram that both the magnitude and direction of the ball's momentum changed. The arrow representing $\vec{p}_B$ is longer than the arrow representing $\vec{p}_C$, and the directions of the arrows are different.

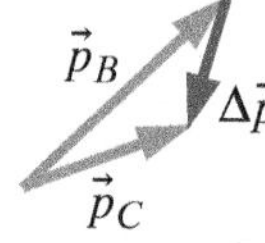

Figure 1.46 Graphical calculation of $\Delta\vec{p}$.

EXAMPLE **A Ball Bounces Off a Wall**

A tennis ball of mass 58 g travels with velocity $\langle 50, 0, 0\rangle$ m/s toward a wall. After bouncing off the wall, the tennis ball is observed to be moving at nearly the same speed, in the opposite direction. **(a)** Draw a diagram showing the initial and final momentum of the tennis ball. **(b)** What is the change in the momentum of the tennis ball? **(c)** Compare the change in the magnitude of the tennis ball's momentum to the magnitude of the change of the ball's momentum.

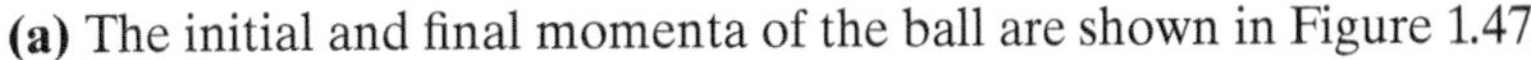

Solution **(a)** The initial and final momenta of the ball are shown in Figure 1.47.

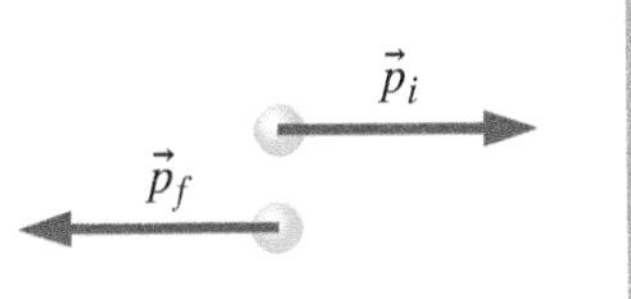

Figure 1.47 The initial and final momentum of the tennis ball.

(b)

$$\vec{p}_i = (0.058\,\text{kg})\langle 50,0,0\rangle\ \text{m/s} = \langle 2.9,0,0\rangle\ \text{kg}\cdot\text{m/s}$$
$$\vec{p}_f = (0.058\,\text{kg})\langle -50,0,0\rangle\ \text{m/s} = \langle -2.9,0,0\rangle\ \text{kg}\cdot\text{m/s}$$
$$\Delta\vec{p} = \langle -2.9,0,0\rangle\ \text{kg}\cdot\text{m/s} - \langle 2.9,0,0\rangle\ \text{kg}\cdot\text{m/s}$$
$$\Delta\vec{p} = \langle -5.80,0,0\rangle\ \text{kg}\cdot\text{m/s}$$

(c) The change in the magnitude of the ball's momentum was:

$$\Delta|\vec{p}| = |\vec{p}_f| - |\vec{p}_i|$$
$$\Delta|\vec{p}| = \sqrt{-2.9^2+0^2+0^2}\,\text{kg}\cdot\text{m/s} - \sqrt{2.9^2+0^2+0^2}\,\text{kg}\cdot\text{m/s}$$
$$\Delta|\vec{p}| = 0\,\text{kg}\cdot\text{m/s}$$

The magnitude of the change in the ball's momentum was:

$$|\Delta\vec{p}| = \sqrt{-5.8^2+0^2+0^2} = 5.8\,\text{kg}\cdot\text{m/s}$$

How do we make sense of this difference? The interaction with the wall made a large change in the (vector) momentum of the ball; the magnitude of this change is twice as large as the magnitude of the ball's original momentum. However, because the change in the ball's speed was negligible, the change in the magnitude of its momentum was also negligible. We will see in Chapter 2 that this distinction is important, because it is the change in the vector momentum that is proportional to the strength of an interaction with the surroundings. In discussing momentum change we will almost always be interested in $\Delta\vec{p}$ and its magnitude ($|\Delta\vec{p}|$), rather than in the change in the magnitude ($\Delta|\vec{p}|$).

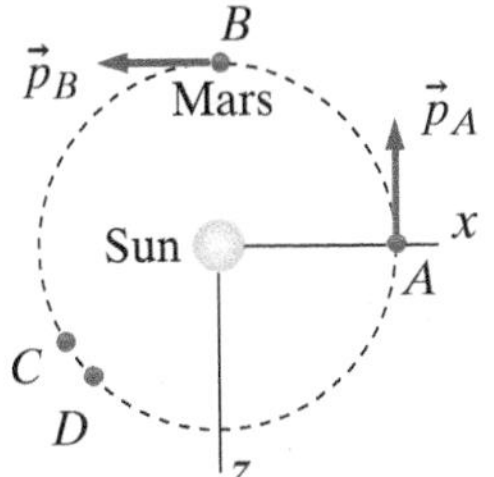

Figure 1.48 The nearly circular orbit of Mars around the Sun, viewed from above the orbital plane ($+x$ to the right, $+z$ down the page). Not to scale: The sizes of the Sun and Mars are exaggerated.

Checkpoint 15 The planet Mars has a mass of 6.4×10^{23} kg, and travels in a nearly circular orbit around the Sun, as shown in Figure 1.48. When it is at location A, the velocity of Mars is $\langle 0,0,-2.5\times 10^4\rangle$ m/s. When it reaches location B, the planet's velocity is $\langle -2.5\times 10^4,0,0\rangle$ m/s. We're looking down on the orbit from above the north poles of the Sun and Mars, with $+x$ to the right and $+z$ down the page. **(a)** What is $\Delta\vec{p}$, the change in the momentum of Mars between locations A and B? **(b)** On a copy of the diagram in Figure 1.48, draw two arrows representing the momentum of Mars at locations C and D, paying attention to both the length and direction of each arrow. **(c)** What is the direction of the change in the momentum of Mars between locations C and D? Draw the vector $\Delta\vec{p}$ on your diagram.

1.9 USING MOMENTUM TO UPDATE POSITION

If you know the momentum of an object, you can calculate the change in position of the object over a given time interval. This is straightforward if the object is traveling at a speed low enough that the approximate expression for momentum can be used, since in this case $\vec{v} \approx \vec{p}/m$.

EXAMPLE **Displacement of an Ice Skater**

An ice skater whose mass is 50 kg moves with constant momentum $\langle 400,0,300\rangle$ kg·m/s. At a particular instant in her skating program she passes location $\langle 0,0,3\rangle$ m. What was her location at a time 3 s earlier?

Solution

$$\vec{v} \approx \frac{\vec{p}}{m} = \frac{\langle 400,0,300\rangle\,\text{kg}\cdot\text{m/s}}{50\,\text{kg}} = \langle 8,0,6\rangle\,\text{m/s}$$

$$(\vec{r}_f - \vec{r}_i) = \vec{v}\Delta t$$

$$\begin{aligned}\vec{r}_i &= \vec{r}_f - \vec{v}\Delta t\\ &= \langle 0,0,3\rangle\,\text{m} - (\langle 8,0,6\rangle\,\text{m/s})(3\,\text{s})\\ &= \langle -24,0,-15\rangle\,\text{m}\end{aligned}$$

Checkpoint 16 At time $t_1 = 12$ s, a car with mass 1300 kg is located at $\langle 94,0,30\rangle$ m and has momentum $\langle 4500,0,-3000\rangle$ kg·m/s. The car's momentum is not changing. At time $t_2 = 17$ s, what is the position of the car?

1.10 MOMENTUM AT HIGH SPEEDS

■ The largest particle accelerator currently in operation is the Large Hadron Collider at CERN, in Geneva, Switzerland (http://cern.ch/), where the Higgs boson was recently found. There are many other accelerators in both the United States and other countries.

Although most of the objects we encounter in our daily lives move at speeds that are much less than the speed of light, motion at very high speeds is not unusual. Every day many particles enter the Earth's atmosphere traveling at speeds near the speed of light. Some of these particles are protons, which react with atomic nuclei in the atmosphere to produce showers of high-speed particles that rain down on the Earth. To study high-speed particles in a controlled way, scientists use machines called "particle accelerators."

Experiments on particles moving at very high speeds, close to the speed of light $c = 3\times 10^8$ m/s, show that changes in $m\vec{v}$ (the approximate momentum) are not really proportional to the strength of the interactions. As we keep applying a force to a particle near the speed of light, the speed of the particle barely increases, and it is not possible to increase a particle's speed beyond the speed of light.

Because of these experiments we can define momentum in a precise (not approximate) way. We observe that changes in the following quantity are truly proportional to the amount of interaction:

DEFINITION OF MOMENTUM

For a particle of mass m, momentum is defined as the product of mass times velocity, multiplied by a proportionality factor gamma:

$$\vec{p} = \gamma m \vec{v}$$

The proportionality factor γ (lowercase Greek gamma) is defined as

$$\gamma = \frac{1}{\sqrt{1-\left(\frac{|\vec{v}|}{c}\right)^2}}$$

In these equations $\vec{p}$ represents momentum, m is the mass of the object, $\vec{v}$ is the velocity of the object, and c is the speed of light (3×10^8 m/s). Momentum has units of kg·m/s. The factor γ is a positive number that is always greater than or equal to one, and it has no units.

This is the "relativistic" definition of momentum. Albert Einstein in 1905 in his Special Theory of Relativity predicted that this would be the appropriate definition for momentum at high speeds, a prediction that has been abundantly verified in a wide range of experiments. In Chapter 6 we will see that the factor γ is also important in expressions for the energy of objects in motion.

EXAMPLE

Momentum of a Fast-Moving Proton

Suppose that a proton (mass 1.7×10^{-27} kg) in an accelerator at CERN (the large particle physics laboratory in Geneva, Switzerland) is traveling with a velocity of $\langle 2\times10^7, 1\times10^7, -3\times10^7\rangle$ m/s. **(a)** What is the momentum of the proton? **(b)** What is the magnitude of the momentum of the proton?

Solution

(a)

$$|\vec{v}| = \sqrt{(2\times10^7)^2+(1\times10^7)^2+(-3\times10^7)^2}\text{ m/s} = 3.7\times10^7\text{ m/s}$$

$$\frac{|\vec{v}|}{c} = \frac{3.7\times10^7\text{ m/s}}{3\times10^8\text{ m/s}} = 0.12$$

$$\gamma = \frac{1}{\sqrt{1-(0.12)^2}} = 1.007$$

$$\vec{p} = (1.007)(1.7\times10^{-27}\text{ kg})\langle 2\times10^7, 1\times10^7, -3\times10^7\rangle\text{ m/s}$$

$$\vec{p} = \langle 3.4\times10^{-20}, 1.7\times10^{-20}, -5.1\times10^{-20}\rangle\text{ kg}\cdot\text{m/s}$$

(b)

$$|\vec{p}| = \sqrt{(3.4\times10^{-20})^2+(1.7\times10^{-20})^2+(-5.1\times10^{-20})^2}\text{ kg}\cdot\text{m/s}$$

$$|\vec{p}| = 6.4\times10^{-20}\text{ kg}\cdot\text{m/s}$$

$\lvert\vec{v}\rvert$ m/s	$\lvert\vec{v}\rvert/c$	γ
0	0	1.0000
3	1×10^{-8}	1.0000
300	1×10^{-6}	1.0000
3×10^6	0.01	1.0001
3×10^7	0.1	1.0050
1.5×10^8	0.5	1.1547
2.997×10^8	0.999	22.3663
2.9997×10^8	0.9999	70.7124
3×10^8	1	Infinite! Impossible!

Figure 1.49 Values of γ calculated for some speeds. γ is shown to four decimal places, which is more accuracy than we will usually need.

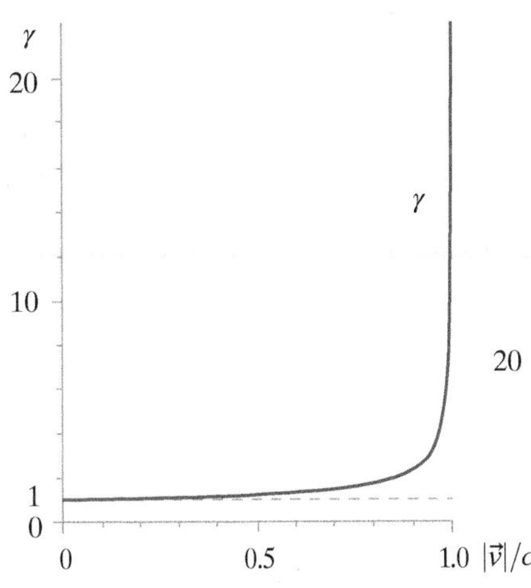

Figure 1.50 γ as a function of $|\vec{v}|/c$. At low speeds, γ is approximately equal to 1. At very high speeds, $|\vec{v}|\approx c$, γ increases rapidly.

Approximate Expression for Momentum

In the example above, we found that $\gamma = 1.007$. Since in that calculation we used only two significant figures, we could have used the approximation that $\gamma \approx 1.0$ without affecting our answer. Let's examine the expression for γ to see whether we can come up with a guideline for when it is reasonable to use the approximate expression

$$\vec{p} \approx 1\cdot m\vec{v}$$

Looking at the expression for γ,

$$\gamma = \frac{1}{\sqrt{1-\left(\frac{|\vec{v}|}{c}\right)^2}}$$

we see that it depends only on the ratio of the speed of the object to the speed of light (the object's mass doesn't appear in this expression).

If $|\vec{v}|/c$ is a very small number, then $1-(|\vec{v}|/c)^2 \approx 1-0 \approx 1$, so $\gamma \approx 1$.

APPROXIMATION: MOMENTUM AT LOW SPEEDS

When $|\vec{v}| \ll c$ then $\gamma \approx 1$, and

$$\vec{p} \approx 1\cdot m\vec{v} \approx m\vec{v}$$

Some values of $(|\vec{v}|/c)$ and γ are displayed in Figure 1.49. From this table you can see that even at the very high speed where $|\vec{v}|/c = 0.1$, which means that $|\vec{v}| = 3\times10^7$ m/s, the relativistic factor γ is only slightly different from 1.0. For large-scale objects such as a space rocket, whose speed is typically only about 1×10^4 m/s, we can ignore the factor γ, and momentum is to a good approximation $\vec{p}\approx m\vec{v}$. It is only for high-speed cosmic rays or particles produced in high-speed particle accelerators that we need to use the full relativistic definition for momentum, $\vec{p} = \gamma m\vec{v}$.

Figure 1.50 displays graphically the data shown in Figure 1.49. For speeds up to about half the speed of light, γ is very nearly equal to 1, but at very high speeds, approaching the speed of light, γ increases rapidly. From examining Figure 1.50, you can see why it is not possible to exceed the speed of light. As you make a particle go faster and faster, approaching the speed of light, additional increases in the speed become increasingly difficult, because a tiny increase in speed means a huge increase in γ and a corresponding huge increase in momentum, requiring huge amounts of interaction. In fact, for the speed to equal the speed of light, the momentum would have to increase to be infinite! There is a cosmic speed limit in the Universe, 3×10^8 m/s.

The Role of Approximations and Models in Physics

> QUESTION Earlier in this chapter we used the approximate expression $\vec{p} \approx m\vec{v}$ for momentum. Is this legitimate? Shouldn't we always use the exact equation?

The table of values of γ in Figure 1.49 shows that for many real-world situations in which objects are traveling at speeds small compared to the speed of light, our calculators will give the same result whether we actually calculate γ or simply use the approximation that $\gamma \approx 1$. Our measurements and calculations simply aren't precise enough to make the distinction useful.

You may have heard physics described as an "exact science." This is not a good description of what physics is actually about. Physics and physicists try to describe and understand what happens in the real world, which is a messy, complicated place. For example, we'll see in the next chapter that it is reasonably easy to predict the motion of a thrown ball in the highly simplified case where there is no air resistance or deflection by wind. Even if we can really get rid of air resistance by throwing the ball in a big vacuum chamber, it makes sense to simplify the analysis by neglecting the tiny but real effects of gravitational attraction by nearby mountains, the Moon, and Mars, and also to ignore the fact that the gravitational attraction changes very slightly as the ball gets closer to or farther from the center of the Earth. Moreover, to predict *exactly* the future motion of the ball, we would need to know *exactly* its initial position and velocity. This is impossible, because meter sticks and clocks aren't exact.

> QUESTION If physics isn't exact, what is the point?

We can often learn a great deal by carrying out an approximate analysis of a complex situation. In some cases the differences between an approximate analysis and a hypothetical exact analysis may be negligible. For example, it certainly makes sense to ignore the gravitational attraction of Mars when predicting the motion of a ball thrown on Earth. In other cases, a simplified model may allow us to identify the most important interactions and effects in a complex situation. A comparison of the predictions of such a model with data from real-world observations can suggest refinements to our model that would make its predictions more accurate.

A good way of describing what physics can (and can't) do is that with physics we construct, analyze, and refine simplified, idealized, approximate "models" of real-world phenomena, in the hope that such analyses will give us useful but necessarily approximate understanding of the real world.

Updating Position at High Speeds

For speeds near the speed of light, we cannot use the approximation $\vec{v} \approx \vec{p}/m$. Instead, we need to start with the definition of momentum $\vec{p} = \gamma m\vec{v}$, and solve for $\vec{v}$. The detailed derivation is given at the end of the chapter. The result is:

$$\vec{v} = \frac{\vec{p}/m}{\sqrt{1 + \left(\frac{|\vec{p}|}{mc}\right)^2}}$$

EXAMPLE **Displacement of a Fast Proton**

A proton with constant momentum $\langle 0,0,2.72\times10^{-19}\rangle$ kg · m/s leaves the origin 10.0 s after an accelerator experiment is started. What is the location of the proton 2 ns later? (ns = nanosecond = 1×10^{-9} s)

Solution

$$\vec{v} = \frac{\vec{p}/m}{\sqrt{1+\left(\frac{|\vec{p}|}{mc}\right)^2}} = \frac{\langle 0,0,2.72\times10^{-19}\rangle\,\text{kg}\cdot\text{m/s}}{(1.7\times10^{-27}\,\text{kg})\sqrt{1+\left(\frac{2.72\times10^{-19}\,\text{kg}\cdot\text{m/s}}{(1.7\times10^{-27}\,\text{kg})(3\times10^{8}\,\text{m/s})}\right)^2}}$$

$$= \langle 0,0,1.4\times10^{8}\rangle\,\text{m/s}$$

$$\vec{r}_f = \vec{r}_i + \vec{v}_{\text{avg}}\,\Delta t$$

$$= \langle 0,0,0\rangle\,\text{m} + \langle 0,0,1.4\times10^{8}\,\text{m/s}\rangle(2\times10^{-9}\,\text{s})$$

$$= \langle 0,0,0.28\rangle\,\text{m}$$

The proton traveled 28 cm in 2 ns.

Checkpoint 17 What is the momentum of an electron traveling at a velocity of $\langle 0,0,-2\times10^{8}\rangle$ m/s? (Masses of particles are given on the inside back cover of this textbook.) What is the magnitude of the momentum of the electron?

1.11 COMPUTATIONAL MODELING

Computational modeling plays an important role not only in physics theory and experiment, but in nearly all other scientific and engineering fields. Creating simple computational models based on fundamental physics principles can allow us to see more clearly how these principles govern the real-time behavior of all physical systems. Such models can allow us to apply physics principles to complicated systems. Appropriate computational tools can let us visualize the time evolution of the behavior of 3D physical systems, and can also help us visualize vector quantities such as velocity, momentum, and other quantities we will encounter in later chapters. For these reasons, computational modeling is included as an integral part of this textbook.

QUESTION But what if I don't know how to program?

Don't worry! No prior experience is required. You will be able to learn all you need to know to build simple but powerful computational models as we go along in the course.

VPython

We have chosen to use VPython to build our computational models. VPython is an extension of the widely used Python programming language. VPython supports 3D vector algebra, and allows you to create real-time dynamic 3D animations as a side effect of physics calculations. VPython is free and open source, and runs on Windows, MacOS, and Linux.

Here is an example of a small VPython program. The display generated by the program is shown in Figure 1.51.

```
from visual import *
ball = sphere(pos = vector(-3,-3,0),
        color = color.orange,
        radius = 0.5)
velocity = vector(3,1.5,4)
arrow(pos = ball.pos,
      axis = velocity,
      color = color.green)
```

Figure 1.51 This is the display generated when you run the VPython program shown at the right.

QUESTION In the code above, the sphere named "ball" is created at the position $\langle -3,-3,0\rangle$. Considering this, look at the display in Figure 1.51, and decide where you think the origin of the coordinate system is located.

■ To install VPython on your computer, go to http://vpython.org and follow the directions for your operating system.
See homework problems P68–P77 at the end of this chapter to find the instructional videos for VPython.

The origin of the 3D Cartesian coordinate system, $\langle 0,0,0\rangle$, is in the center of the display. The orientation of the axes is the same as the one we have been using, with $+x$ to the right, $+y$ up, and $+z$ out of the plane of the display, toward you. The location of the ball, below and to the left of the origin, is consistent with the position $\langle -3,-3,0\rangle$ assigned to the sphere in the program.

QUESTION What could we change in the code to make the ball larger?

The sphere named "ball" is created with a radius of 0.5. To make the ball larger, we could specify a larger radius.

QUESTION How could the program be extended to animate the motion of the ball?

In the example above a velocity has already been given. To use this velocity to update the position of the ball, we need to instruct the computer to do exactly the same calculation you have been doing by hand. A translation of these calculations into VPython syntax would look like this:

$\Delta t = 0.1$	`delta_t = 0.1`
$\vec{r}_f = \vec{r}_i + \vec{v}\Delta t$	`ball.pos = ball.pos + velocity * delta_t`

A statement such as `x = x + 1` is not wrong in a computer language. Such a statement instructs the computer to pick up the current value of the variable x, add 1 to it, and replace the value of x with this new value. Subscripts identifying "current" and "new" values are not needed, since anything to the right of the equal sign refers to the current value, and anything to the left of the equal sign is the new value. Therefore the statement

```
ball.pos = ball.pos + velocity * delta_t
```

tells the computer to pick up the current position of the ball (`ball.pos`), add the quantity `velocity * delta_t` to it, and update `ball.pos` to this new value. To animate the motion of the ball over a period of 4 seconds, we can instruct the computer to repeat this calculation over and over again, using a "while loop":

```
delta_t = 0.1
t = 0
while t < 4:
    rate(100)
    ball.pos = ball.pos + velocity * delta_t
    t = t + delta_t
```

To get an animation of the object's motion as a side effect of our calculations, we had to add the statement `rate(100)` inside the while loop. This statement tells VPython to allow only 100 repetitions of the while loop to be done each second of real time, and allows it to render a new image of the scene many times per second, producing a movie-like animation as a side effect of our physics calculations.

For clarity, we can refer to the time step Δt as "virtual time" and to the time required to do a calculation as "real time." If you were doing the calculation by hand, it might take you 60 real seconds to calculate the new position of an object at a time 0.1 virtual second in the future. In contrast, it takes a computer a very short time—about 5 microseconds (5×10^{-6} s). In neither case is the time taken to do the calculation equal to the "virtual" elapsed time of 0.1 second.

It is much easier to learn to write simple programs in VPython by watching videos than by reading text. There is a series of YouTube videos that were developed to help physics students learn VPython. Homework problems P68–P77 at the end of this chapter ask you to watch these videos and do the challenge tasks at the end of each video. Problem P78 includes a complete program to study and modify. Your instructor may also assign more extended computational activities both to help you get started and to accompany future topics. In future chapters we will assume that you have watched these videos and done these simple introductory exercises.

1.12 *THE PRINCIPLE OF RELATIVITY

Sections marked with an asterisk ("*") are optional. They provide additional information and context, but later sections of the textbook don't depend critically on them. This optional section deals with some deep issues about the "reference frame" from which you observe motion. Newton's first law of motion only applies in an "inertial reference frame," which we will discuss here in the context of the principle of relativity.

A great variety of experimental observations has led to the establishment of the following principle, first recognized by Galileo:

THE PRINCIPLE OF RELATIVITY

Physical laws work in the same way for observers in uniform motion as for observers at rest.

This principle is called "the principle of relativity." (Einstein's extensions of this principle are known as "special relativity" and "general relativity.") Phenomena observed in a room in uniform motion (for example, on a train moving with constant speed on a smooth straight track) obey the same physical laws in the same way as experiments done in a room that is not moving. According to this principle, Newton's first law of motion should be true both for an observer moving at constant velocity and for an observer at rest.

For example, suppose that you're riding in a car moving with constant velocity, and you're looking at a map lying on the dashboard. As far as you're concerned, the map isn't moving, and no interactions are required to hold it still on the dashboard. Someone standing at the side of the road sees the car go by, sees the map moving at a high speed in a straight line, and can see that no interactions are required to hold the map still on the dashboard. Both you and the bystander agree that Newton's first law of motion is obeyed: the bystander sees the map moving with constant velocity in the absence of interactions, and you see the map not moving at all (a zero constant velocity) in the absence of interactions.

On the other hand, if the car suddenly speeds up, it moves out from under the map, which ends up in your lap. To you it looks like "the map sped up in the backwards direction" without any interactions to cause this to happen, which looks like a violation of Newton's first law of motion. The problem is that you're strapped to the car, which is an accelerated reference frame, and Newton's first law of motion applies only to nonaccelerated reference frames, called "inertial" reference frames. Similarly, if the car suddenly turns to the right, moving out from under the map, the map tends to keep going in its original direction, and to you it looks like "the map moved to the left" without any interactions. So a change of speed or a change of direction of the car (your reference frame) leads you to see the map behave in a strange way.

The bystander, who is in an inertial (nonaccelerating) reference frame, doesn't see any violation of Newton's first law of motion. The bystander's reference frame is an inertial frame, and the map behaves in an understandable way, tending to keep moving with unchanged speed and direction when the car changes speed or direction.

The Cosmic Microwave Background

The principle of relativity, and Newton's first law of motion, apply only to observers who have a constant speed and direction (or zero speed) relative to the "cosmic microwave background," which provides the only backdrop and frame of reference with an absolute, universal character. It used to be that the basic reference frame was loosely called "the fixed stars," but stars and galaxies have their own individual motions within the Universe and do not constitute an adequate reference frame with respect to which to measure motion.

The cosmic microwave background is low-intensity electromagnetic radiation with wavelengths in the microwave region, which pervades the Universe, radiating in all directions. Measurements show that our galaxy is moving through this microwave radiation with a large, essentially constant velocity, toward a cluster of a large number of other galaxies. The way we detect our motion relative to the microwave background is through the "Doppler shift" of the frequencies of the microwave radiation, toward higher frequencies in front of us and lower frequencies behind. This is essentially the same phenomenon as that responsible for a fire engine siren sounding at a higher frequency when it is approaching us and a lower frequency when it is moving away from us.

The discovery of the cosmic microwave background provided major support for the "Big Bang" theory of the formation of the Universe. According to the Big Bang theory, the early Universe must have been an extremely hot mixture of charged particles and high-energy, short-wavelength electromagnetic radiation (visible light, x-rays, gamma rays, etc.). Electromagnetic radiation interacts strongly with charged particles, so light could not travel very far without interacting, making the Universe essentially opaque. Also, the Universe was so hot that electrically neutral atoms could not form without the electrons immediately being stripped away again by collisions with other fast-moving particles.

As the Universe expanded, the temperature dropped. Eventually the temperature was low enough for neutral atoms to form. The interaction of electromagnetic radiation with neutral atoms is much weaker than with individual charged particles, so the radiation was now essentially free, dissociated from the matter, and the Universe became transparent. As the Universe continued to expand (the actual space between clumps of matter got bigger), the wavelengths of the electromagnetic radiation got longer, until today this fossil radiation has wavelengths in the relatively low-energy, long-wavelength microwave portion of the electromagnetic spectrum.

Inertial Frames of Reference

It is an observational fact that in reference frames that are in uniform motion with respect to the cosmic microwave background, far from other objects (so that interactions are negligible), an object maintains uniform motion. Such frames are called "inertial frames" and are reference frames in which

Newton's first law of motion is valid. All of these reference frames are equally valid; the cosmic microwave background simply provides a concrete example of such a reference frame.

QUESTION Is the surface of the Earth an inertial frame?

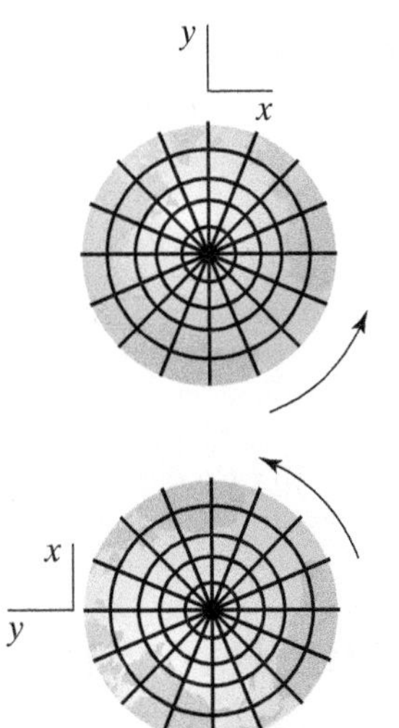

Figure 1.52 Axes tied to the Earth rotate through 90° in a quarter of a day (6 h).

No! The Earth is rotating on its axis, so the velocity of an object sitting on the surface of the Earth is constantly changing direction, as is a coordinate frame tied to the Earth (Figure 1.52). Moreover, the Earth is orbiting the Sun, and the Solar System itself is orbiting the center of our Milky Way galaxy, and our galaxy is moving toward other galaxies. So the motion of an object sitting on the Earth is actually quite complicated and definitely not uniform with respect to the cosmic microwave background.

However, for many purposes the surface of the Earth can be considered to be (approximately) an inertial frame. For example, it takes 6 hours for the rotation of the Earth on its axis to make a 90° change in the direction of the velocity of a "fixed" point. When a hockey puck slides across an ice rink with negligible friction, during these few seconds the puck moves in nearly a straight line at constant speed, and if the puck is struck its velocity change is much larger than the very small velocity change of the approximate inertial frame of the Earth's surface.

Similarly, although the Earth is in orbit around the Sun, it takes 365 days to go around once, so for a period of a few days or even weeks the Earth's orbital motion is nearly in a straight line at constant speed. For our purposes we will consider the Earth's surface to represent an approximately inertial reference frame.

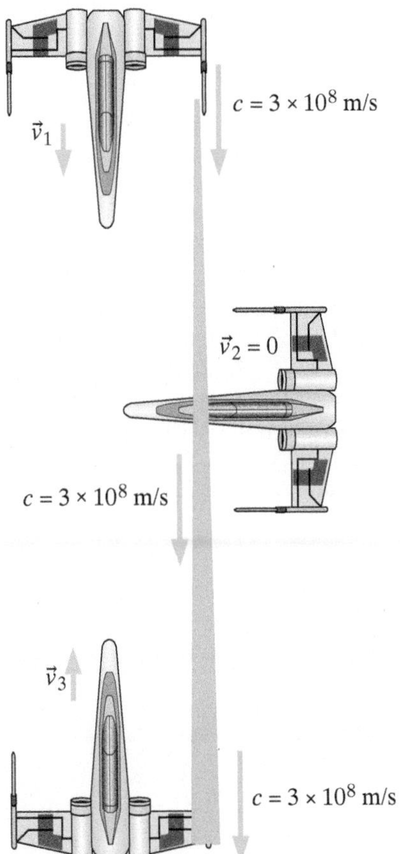

Figure 1.53 Light emitted by the top spaceship is measured to have the same speed by observers in all three ships.

The Special Theory of Relativity

Einstein's Special Theory of Relativity (published in 1905) built on the basic principle of relativity introduced by Galileo but added the conjecture that the speed of a beam of light must be the same as measured by observers in different frames of reference in uniform motion with respect to each other. In Figure 1.53, observers on each spaceship measure the speed of the light c emitted by the ship at the top to be the same ($c = 3 \times 10^8$) m/s, despite the fact that they are moving at different velocities.

This additional condition seems peculiar and has far-reaching consequences. After all, the map on the dashboard of your car has different speeds relative to different observers, depending on the motion of the observer. Yet a wide range of experiments has confirmed Einstein's conjecture: all observers measure the same speed for the same beam of light, $c = 3 \times 10^8$ m/s. (The color of the light is different for the different observers, but the speed is the same.)

On the other hand, if someone on the ship at the top throws a ball or a proton or some other piece of matter, the speed of the object will be different for observers on the three ships; it is only light whose speed is independent of the observer.

Einstein's theory has interesting consequences. For example, it predicts that time will be measured to run at different rates in different frames of reference. These predictions have been confirmed by many experiments. These unusual effects are large only at very high speeds (a sizable fraction of the speed of light), which is why we don't normally observe these effects in everyday life, and why we can use nonrelativistic calculations for low-speed phenomena.

However, for the Global Positioning System (GPS) to give adequate accuracy it is necessary to take Einstein's Special Theory of Relativity into account. The atomic clocks on the satellites run slower than our clocks, due to the speed of the satellites, and the difference in clock rate depends on γ. Although γ for the GPS satellites orbiting the Earth is nearly 1, it differs just enough that making the approximation $\gamma \approx 1$ would make the GPS hopelessly inaccurate and useless. It is even necessary to apply corrections based on Einstein's General Theory of Relativity, which correctly predicts an additional change in clock rate due to the gravity of the Earth being weaker at the high altitude of the GPS satellites.

Checkpoint 18 A spaceship at rest with respect to the cosmic microwave background emits a beam of red light. A different spaceship, moving at a speed of 2.5×10^8 m/s toward the first ship, detects the light. Which of the following statements are true for observers on the second ship? (More than one statement may be correct.) **(a)** They observe that the light travels at 3×10^8 m/s. **(b)** The light is not red. **(c)** They observe that the light travels at 5.5×10^8 m/s. **(d)** They observe that the light travels at 2.5×10^8 m/s.

1.13 *UPDATING POSITION AT HIGH SPEED

If $v \ll c$, $\vec{p} \approx m\vec{v}$ and $\vec{v} \approx \vec{p}/m$. But at high speed it is more complicated to determine the velocity from the (relativistic) momentum. Here is a way to solve for $\vec{v}$ in terms of $\vec{p}$:

$$|\vec{p}| = \frac{1}{\sqrt{1-(|\vec{v}|/c)^2}} m|\vec{v}|$$

Divide by m and square: $\dfrac{|\vec{p}|^2}{m^2} = \dfrac{|\vec{v}|^2}{1-(|\vec{v}|/c)^2}$

Multiply by $(1-(|\vec{v}|/c)^2)$: $\dfrac{|\vec{p}|^2}{m^2} - \left(\dfrac{|\vec{p}|^2}{m^2c^2}\right)|\vec{v}|^2 = |\vec{v}|^2$

Collect terms: $\left(1+\dfrac{|\vec{p}|^2}{m^2c^2}\right)|\vec{v}|^2 = \dfrac{|\vec{p}|^2}{m^2}$

$$|\vec{v}| = \frac{|\vec{p}|/m}{\sqrt{1+\left(\frac{|\vec{p}|}{mc}\right)^2}}$$

The expression above gives the magnitude of $\vec{v}$, in terms of the magnitude of $\vec{p}$. To get an expression for the vector $\vec{v}$, recall that any vector can be factored into its magnitude times a unit vector in the direction of the vector, so

$$\vec{p} = |\vec{p}|\hat{p} \quad \text{and} \quad \vec{v} = |\vec{v}|\hat{v}$$

But since $\vec{p}$ and $\vec{v}$ are in the same direction, $\hat{v} = \hat{p}$, so

$$\vec{v} = |\vec{v}|\hat{p} = \frac{|\vec{p}|/m}{\sqrt{1+\left(\frac{|\vec{p}|}{mc}\right)^2}}\hat{p} = \frac{(|\vec{p}|\hat{p})/m}{\sqrt{1+\left(\frac{|\vec{p}|}{mc}\right)^2}}$$

$$\vec{v} = \frac{\vec{p}/m}{\sqrt{1+\left(\frac{|\vec{p}|}{mc}\right)^2}}$$

THE RELATIVISTIC POSITION UPDATE EQUATION

$$\vec{r}_f = \vec{r}_i + \frac{1}{\sqrt{1+\left(\frac{|\vec{p}|}{mc}\right)^2}}\left(\frac{\vec{p}}{m}\right)\Delta t \quad \text{(for small } \Delta t\text{)}$$

Note that at low speeds $|\vec{p}| \approx m|\vec{v}|$, and the denominator is

$$\sqrt{1+\left(\frac{|\vec{v}|}{c}\right)^2} \approx 1$$

so the equation becomes the familiar $\vec{r}_f = \vec{r}_i + (\vec{p}/m)\Delta t$.

SUMMARY

Interactions
Interactions are indicated by

- Change of velocity (change of direction and/or change of speed)
- Change of identity
- Change of shape of multiparticle system
- Change of temperature of multiparticle system
- Indirect evidence for interactions

Newton's first law of motion
Every body persists in its state of rest or of moving with constant speed in a constant direction, except to the extent that it is compelled to change that state by forces acting on it.

Vectors
A 3D *vector* is a quantity with magnitude and a direction, which can be expressed as a triple $\langle x, y, z \rangle$. A vector is indicated by an arrow: $\vec{r}$.
A *scalar* is a single number.
Legal mathematical operations involving vectors include:

- Adding one vector to another vector
- Subtracting one vector from another vector
- Multiplying or dividing a vector by a scalar
- Finding the magnitude of a vector
- Taking the derivative of a vector

Operations that are *not* legal with vectors include:

- A vector cannot be added to a scalar.
- A vector cannot be set equal to a scalar.
- A vector cannot appear in the denominator (you can't divide by a vector).

A unit vector $\hat{r} = \vec{r}/|\vec{r}|$ has magnitude 1.
A vector can be factored using a unit vector: $\vec{F} = |\vec{F}|\hat{F}$.
Direction cosines: $\hat{r} = \langle \cos\theta_x, \cos\theta_y, \cos\theta_z \rangle$

The symbol Δ
The symbol Δ (delta) means "change of": $\Delta t = t_f - t_i$, $\Delta\vec{r} = \vec{r}_f - \vec{r}_i$.
Δ always means "final minus initial."

Velocity and change of position
Definition of average velocity

$$\vec{v}_{\text{avg}} = \frac{\Delta\vec{r}}{\Delta t} = \frac{\vec{r}_f - \vec{r}_i}{t_f - t_i}$$

Velocity is a vector. $\vec{r}$ is the position of an object (a vector). t is the time. Average velocity is equal to the change in position divided by the time elapsed. SI units of velocity are meters per second (m/s).
The position update equation

$$\vec{r}_f = \vec{r}_i + \vec{v}_{\text{avg}}\Delta t$$

The final position (vector) is the vector sum of the initial position plus the product of the average velocity and the elapsed time.
Definition of instantaneous velocity

$$\vec{v} = \lim_{\Delta t \to 0} \frac{\Delta\vec{r}}{\Delta t} = \frac{d\vec{r}}{dt}$$

The instantaneous velocity is the limiting value of the average velocity as the time elapsed becomes very small.
Velocity in terms of momentum

$$\vec{v} = \frac{\vec{p}/m}{\sqrt{1+\left(\frac{|\vec{p}|}{mc}\right)^2}} \text{ or } \vec{v} \approx \vec{p}/m \text{ at low speeds}$$

Acceleration
Acceleration is the time rate of change of velocity: $\vec{a} = d\vec{v}/dt$

Momentum

Definition of momentum

$$\vec{p} = \gamma m\vec{v}$$

where $\gamma = \dfrac{1}{\sqrt{1-(|\vec{v}|/c)^2}}$ (lowercase Greek gamma)

Momentum (a vector) is the product of the relativistic factor "gamma" (a scalar), mass, and velocity.

Combined into one equation: $\vec{p} = \dfrac{1}{\sqrt{1-(|\vec{v}|/c)^2}}m\vec{v}$.

Approximation for momentum at low speeds

$$\vec{p} \approx m\vec{v} \text{ at speeds such that } |\vec{v}| \ll c$$

Useful numbers:

Radius of a typical atom $\approx 1 \times 10^{-10}$ meter
Radius of a proton or neutron $\approx 1 \times 10^{-15}$ meter
Speed of light: 3×10^8 m/s

These and other useful data and conversion factors are given on the inside back cover of the textbook.

QUESTIONS

Q1 Why do we use a spaceship in outer space, far from other objects, to illustrate Newton's first law? Why not a car or a train? (More than one of the following statements may be correct.) (1) A car or train touches other objects, and interacts with them. (2) A car or train can't travel fast enough. (3) The spaceship has negligible interactions with other objects. (4) A car or train interacts gravitationally with the Earth. (5) A spaceship can never experience a gravitational force.

Q2 In the periodic table on the inside front cover of this book (or one you find on the internet), for each element there is given the "atomic number," the number of protons or electrons in an atom, and the "atomic mass," which is essentially the number of nucleons, protons plus neutrons, in the nucleus, averaged over the various isotopes of the element, which differ in the number of neutrons. Make a graph of the number of neutrons vs. the number of protons in the elements. You needn't graph every element, just enough to see the trend. What do you observe about the data? (This reflects the need for more neutrons in proton-rich nuclei in order to prevent the electric repulsion of the protons of each other from destroying the nucleus.)

Q3 Which of the following observers might observe something that appears to violate Newton's first law of motion? Explain why. (1) A person standing still on a street corner (2) A person riding on a roller coaster (3) A passenger on a starship traveling at $0.75c$ toward the nearby star Alpha Centauri (4) An airplane pilot doing aerobatic loops (5) A hockey player coasting across the ice.

Q4 Place a ball on a book and walk with the book in uniform motion. Note that you don't really have to do anything to the ball to keep the ball moving with constant velocity (relative to the ground) or to keep the ball at rest (relative to you). Then stop suddenly, or abruptly change your direction or speed. What does Newton's first law of motion predict for the motion of the ball (assuming that the interaction between the ball and the book is small)? Does the ball behave as predicted? It may help to take the point of view of a friend who is standing still, watching you.

Q5 Which of the following statements about the velocity and momentum of an object are correct? (1) The momentum of an object is always in the same direction as its velocity. (2) The momentum of an object can be either in the same direction as its velocity or in the opposite direction. (3) The momentum of an object is perpendicular to its velocity. (4) The direction of an object's momentum is not related to the direction of its velocity. (5) The direction of an object's momentum is tangent to its path.

Q6 Answer the following questions about the factor γ (gamma) in the full relativistic equation for momentum: **(a)** Is γ a scalar or a vector quantity? **(b)** What is the minimum possible value of γ? **(c)** Does γ reach its minimum value when an object's speed is high or low? **(d)** Is there a maximum possible value for γ? **(e)** Does γ become large when an object's speed is high or low? **(f)** Does the approximation $\gamma \approx 1$ apply when an object's speed is low or when it is high?

Q7 In which of these situations is it reasonable to use the approximate equation for the momentum of an object, instead of the full relativistically correct equation? (1) A car traveling on an interstate highway (2) A commercial jet airliner flying between New York and Seattle (3) A neutron traveling at 2700 meters per second (4) A proton in outer space traveling at 2×10^8 m/s (5) An electron in a television tube traveling 3×10^6 m/s

Q8 Moving objects left the traces labeled *A–F* in Figure 1.54. The dots were deposited at equal time intervals (for example, one dot each second). In each case the object starts from the square. Which trajectories show evidence that the moving object was interacting with another object somewhere? If there is evidence of an interaction, what is the evidence?

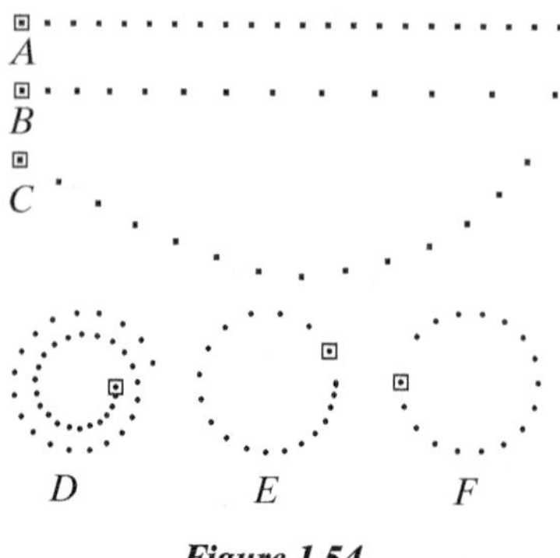

Figure 1.54

Q9 A car moves along a straight road. It moves at a speed of 50 km/h for 4 minutes, then during 4 minutes it gradually speeds up to 100 km/h, continues at this speed for 4 minutes, then during

4 minutes gradually slows to a stop. Make a sketch like the figures in Section 1.2, marking dots for the position along the road every minute.

Q10 A spaceship far from all other objects uses its thrusters to attain a speed of 1×10^4 m/s. The crew then shuts off the power. According to Newton's first law, what will happen to the motion of the spaceship from then on?

Q11 Which of the following are vectors? **(a)** $\vec{r}/2$ **(b)** $|\vec{r}|/2$ **(c)** $\langle r_x, r_y, r_z \rangle$ **(d)** $5 \cdot \vec{r}$

PROBLEMS

The difficulty of a problem is represented by the number of dots preceding the problem number.

Section 1.4

•P12 Figure 1.55 shows several arrows representing vectors in the xy plane. **(a)** Which vectors have magnitudes equal to the magnitude of $\vec{a}$? **(b)** Which vectors are equal to $\vec{a}$?

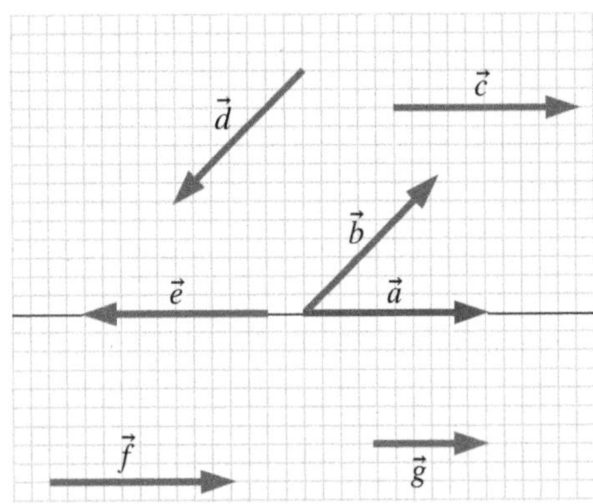

Figure 1.55

•P13 What is the magnitude of the vector $\vec{v}$, where $\vec{v} = \langle 8 \times 10^6, 0, -2 \times 10^7 \rangle$ m/s?

•P14 In Figure 1.56 three vectors are represented by arrows in the xy plane. Each square in the grid represents one meter. For each vector, write out the components of the vector, and calculate the magnitude of the vector.

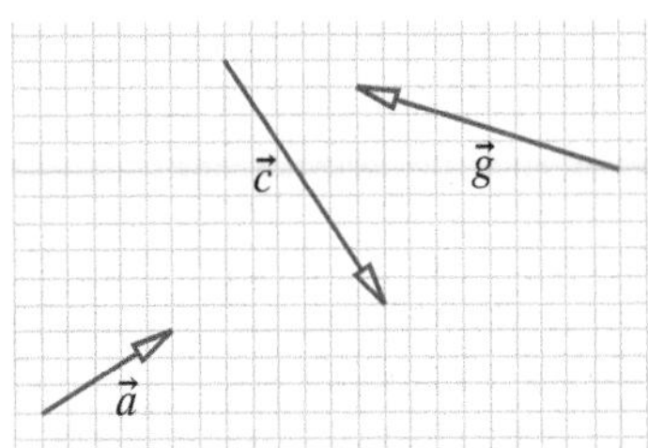

Figure 1.56

•P15 The following questions refer to the vectors depicted by arrows in Figure 1.57. **(a)** What are the components of the vector $\vec{a}$? (Note that since the vector lies in the xy plane, its z component is zero.) **(b)** What are the components of the vector $\vec{b}$? **(c)** Is this statement true or false? $\vec{a} = \vec{b}$ **(d)** What are the components of the vector $\vec{c}$? **(e)** Is this statement true or false? $\vec{c} = -\vec{a}$ **(f)** What are the components of the vector $\vec{d}$? **(g)** Is this statement true or false? $\vec{d} = -\vec{c}$

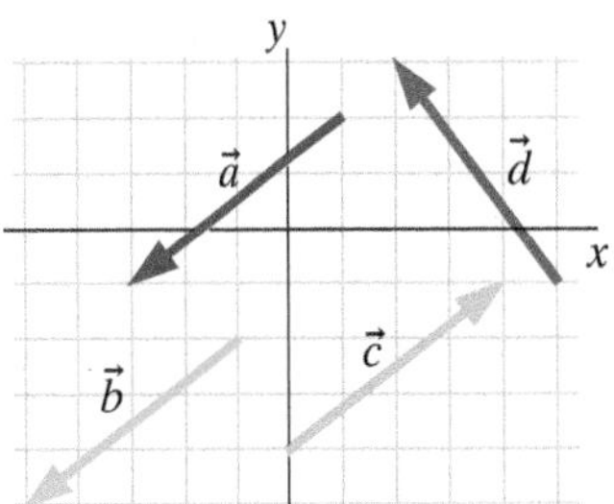

Figure 1.57

•P16 On a piece of graph paper, draw arrows representing the following vectors. Make sure the tip and tail of each arrow you draw are clearly distinguishable. **(a)** Placing the tail of the vector at $\langle 5,2,0 \rangle$, draw an arrow representing the vector $\vec{p} = \langle -7,3,0 \rangle$. Label it $\vec{p}$. **(b)** Placing the tail of the vector at $\langle -5,8,0 \rangle$, draw an arrow representing the vector $-\vec{p}$. Label it $-\vec{p}$.

•P17 What is the result of multiplying the vector $\vec{a}$ by the scalar f, where $\vec{a} = \langle 0.02, -1.7, 30.0 \rangle$ and $f = 2.0$?

•P18 **(a)** In Figure 1.58, what are the components of the vector $\vec{d}$? **(b)** If $\vec{e} = -\vec{d}$, what are the components of $\vec{e}$? **(c)** If the tail of vector $\vec{d}$ were moved to location $\langle -5,-2,4 \rangle$ m, where would the tip of the vector be located? **(d)** If the tail of vector $-\vec{d}$ were placed at location $\langle -1,-1,-1 \rangle$ m, where would the tip of the vector be located?

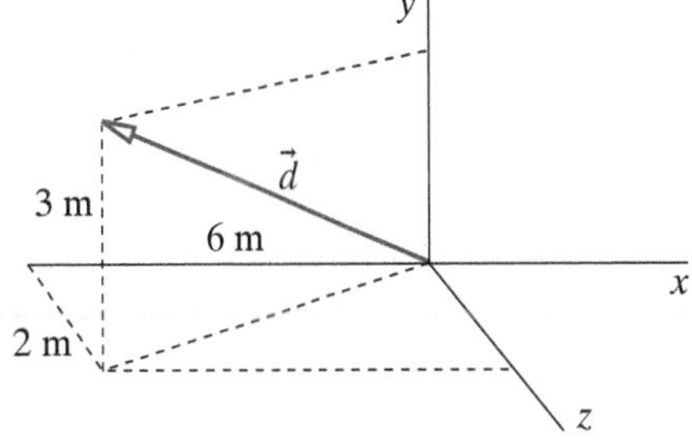

Figure 1.58

•P19 What is the unit vector in the direction of $\langle 2,2,2 \rangle$? What is the unit vector in the direction of $\langle 3,3,3 \rangle$?

•P20 **(a)** On a piece of graph paper, draw the vector $\vec{f} = \langle -2,4,0 \rangle$, putting the tail of the vector at $\langle -3,0,0 \rangle$. Label the vector $\vec{f}$. **(b)** Calculate the vector $2\vec{f}$, and draw this vector on the graph, putting its tail at $\langle -3,-3,0 \rangle$, so you can compare it to the original vector. Label the vector $2\vec{f}$. **(c)** How does the magnitude of $2\vec{f}$ compare to the magnitude of $\vec{f}$? **(d)** How does the direction of $2\vec{f}$ compare to the direction of $\vec{f}$? **(e)** Calculate the vector $\vec{f}/2$, and draw this vector on the graph, putting its tail at $\langle -3,-6,0\rangle$, so you can compare it to the other vectors. Label the vector $\vec{f}/2$. **(f)** How does the magnitude of $\vec{f}/2$ compare to the magnitude of $\vec{f}$? **(g)** How does the direction of $\vec{f}/2$ compare to the direction of $\vec{f}$? **(h)** Does multiplying a vector by a scalar change the magnitude of the vector? **(i)** The vector $a(\vec{f})$ has a magnitude three times as great as that of $\vec{f}$, and its direction is opposite to the direction of $\vec{f}$. What is the value of the scalar factor a?

•**P21** Write the vector $\vec{a} = \langle 400, 200, -100\rangle$ m/s^2 as the product $|\vec{a}| \cdot \hat{a}$.

•**P22 (a)** On a piece of graph paper, draw the vector $\vec{g} = \langle 4,7,0\rangle$ m. Put the tail of the vector at the origin. **(b)** Calculate the magnitude of $\vec{g}$. **(c)** Calculate $\hat{g}$, the unit vector pointing in the direction of $\vec{g}$. **(d)** On the graph, draw $\hat{g}$. Put the tail of the vector at $\langle 1,0,0\rangle$ m so you can compare $\hat{g}$ and $\vec{g}$. **(e)** Calculate the product of the magnitude $|\vec{g}|$ times the unit vector $\hat{g}$, $(|\vec{g}|)(\hat{g})$.

•**P23** A proton is located at $\langle 3\times 10^{-10}, -3\times 10^{-10}, 8\times 10^{-10}\rangle$ m. **(a)** What is $\vec{r}$, the vector from the origin to the location of the proton? **(b)** What is $|\vec{r}|$? **(c)** What is $\hat{r}$, the unit vector in the direction of $\vec{r}$?

•**P24** In Figure 1.59, the vector $\vec{r}_1$ points to the location of object 1 and $\vec{r}_2$ points to the location of object 2. Both vectors lie in the xy plane. **(a)** Calculate the position of object 2 relative to object 1, as a relative position vector. **(b)** Calculate the position of object 1 relative to object 2, as a relative position vector.

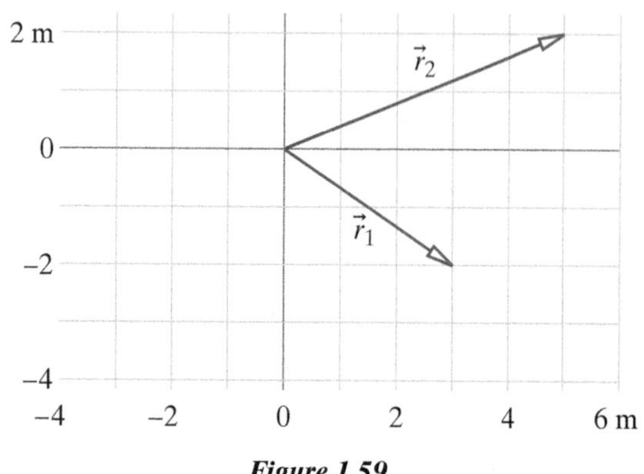

Figure 1.59

•**P25 (a)** What is the vector whose tail is at $\langle 9.5,7,0\rangle$ m and whose head is at $\langle 4,-13,0\rangle$ m? **(b)** What is the magnitude of this vector?

•**P26** A man is standing on the roof of a building with his head at the position $\langle 12,30,13\rangle$ m. He sees the top of a tree, which is at the position $\langle -25,35,43\rangle$ m. **(a)** What is the relative position vector that points from the man's head to the top of the tree? **(b)** What is the distance from the man's head to the top of the tree?

•**P27** A star is located at $\langle 6\times 10^{10}, 8\times 10^{10}, 6\times 10^{10}\rangle$ m. A planet is located at $\langle -4\times 10^{10}, -9\times 10^{10}, 6\times 10^{10}\rangle$ m. **(a)** What is the vector pointing from the star to the planet? **(b)** What is the vector pointing from the planet to the star?

•**P28** A planet is located at $\langle -1\times 10^{10}, 8\times 10^{10}, -3\times 10^{10}\rangle$. A star is located at $\langle 6\times 10^{10}, -5\times 10^{10}, 1\times 10^{10}\rangle$. **(a)** What is $\vec{r}$, the vector from the star to the planet? **(b)** What is the magnitude of $\vec{r}$? **(c)** What is $\hat{r}$, the unit vector (vector with magnitude 1) in the direction of $\vec{r}$?

•**P29** A proton is located at $\langle x_p, y_p, z_p\rangle$. An electron is located at $\langle x_e, y_e, z_e\rangle$. What is the vector pointing from the electron to the proton? What is the vector pointing from the proton to the electron?

••**P30** A cube is 3 cm on a side, with one corner at the origin. What is the unit vector pointing from the origin to the diagonally opposite corner at location $\langle 3,3,3\rangle$ cm? What is the angle from this diagonal to one of the adjacent edges of the cube?

Section 1.6

•**P31** A "slow" neutron produced in a nuclear reactor travels from location $\langle 0.2,-0.05,0.1\rangle$ m to location $\langle -0.202,0.054,0.098\rangle$ m in 2 microseconds ($1\mu s = 1\times 10^{-6}$ s). **(a)** What is the average velocity of the neutron? **(b)** What is the average speed of the neutron?

•**P32** The position of a baseball relative to home plate changes from $\langle 15,8,-3\rangle$ m to $\langle 20,6,-1\rangle$ m in 0.1 s. As a vector, write the average velocity of the baseball during this time interval.

•**P33** You jog at a steady speed of 2 m/s. You start from the location $\langle 0,0,0\rangle$ and for the first 200 s your direction is given by the unit vector $\langle 1,0,0\rangle$. Next you jog for 300 s in the direction given by the unit vector $\langle \cos 45°, 0, \cos 45°\rangle$. Finally you jog for 150 s in the direction given by the unit vector $\langle \cos 60°, 0, \cos 30°\rangle$. **(a)** Now what is your position? **(b)** What was your average velocity?

•**P34** The position of a golf ball relative to the tee changes from $\langle 50,20,30\rangle$ m to $\langle 53,18,31\rangle$ m in 0.1 second. As a vector, write the velocity of the golf ball during this short time interval.

•**P35** The crew of a stationary spacecraft observe an asteroid whose mass is 4×10^{17} kg. Taking the location of the spacecraft as the origin, the asteroid is observed to be at location $\langle -3\times 10^3, -4\times 10^3, 8\times 10^3\rangle$ m at a time 18.4 s after lunchtime. At a time 21.4 s after lunchtime, the asteroid is observed to be at location $\langle -1.4\times 10^3, -6.2\times 10^3, 9.7\times 10^3\rangle$ m. Assuming that the velocity of the asteroid does not change during this time interval, calculate the vector velocity $\vec{v}$ of the asteroid.

••**P36** A spacecraft traveling at a velocity of $\langle -20,-90,40\rangle$ m/s is observed to be at a location $\langle 200,300,-500\rangle$ m relative to an origin located on a nearby asteroid. At a later time the spacecraft is at location $\langle -380,-2310,660\rangle$ m. **(a)** How long did it take the spacecraft to travel between these locations? **(b)** How far did the spacecraft travel? **(c)** What is the speed of the spacecraft? **(d)** What is the unit vector in the direction of the spacecraft's velocity?

••**P37** Here are the positions at three different times for a bee in flight (a bee's top speed is about 7 m/s).

Time	6.3 s	6.8 s	7.3 s
Position	$\langle -3.5,9.4,0\rangle$ m	$\langle -1.3,6.2,0\rangle$ m	$\langle 0.5,1.7,0\rangle$ m

(a) Between 6.3 s and 6.8 s, what was the bee's average velocity? Be careful with signs. **(b)** Between 6.3 s and 7.3 s, what was the bee's average velocity? Be careful with signs. **(c)** Of the two average velocities you calculated, which is the best estimate of the bee's instantaneous velocity at time 6.3 s? **(d)** Using the best information available, what was the displacement of the bee during the time interval from 6.3 s to 6.33 s?

Section 1.7

•**P38** At time $t_1 = 12$ s, a car is located at $\langle 84,78,24\rangle$ m and has velocity $\langle 4,0,-3\rangle$ m/s. At time $t_2 = 18$ s, what is the position of the car? (The velocity is constant in magnitude and direction during this time interval.)

•P39 An electron passes location $\langle 0.02, 0.04, -0.06\rangle$ m, and 2 μs later is detected at location $\langle 0.02, 1.84, -0.86\rangle$ m (1 microsecond is 1×10^{-6} s). **(a)** What is the average velocity of the electron? **(b)** If the electron continues to travel at this average velocity, where will it be in another 5 μs?

•P40 After World War II the U.S. Air Force carried out experiments on the amount of acceleration a human can survive. These experiments, led by John Stapp, were the first to use crash dummies as well as human subjects, especially Stapp himself, who became an effective advocate for automobile safety belts. In one of the experiments Stapp rode a rocket sled that decelerated from 140 m/s (about 310 mi/h) to 70 m/s in just 0.6 s. **(a)** What was the absolute value of the (negative) average acceleration? **(b)** The acceleration of a falling object if air resistance is negligible is 9.8 m/s/s, called "one g." What was the absolute value of the average acceleration in g's? (Stapp eventually survived a test at 46 g's!)

••P41 At a certain instant a ball passes location $\langle 7, 21, -17\rangle$ m. In the next 3 s, the ball's average velocity is $\langle -11, 42, 11\rangle$ m/s. At the end of this 3 s time interval, what is the height y of the ball?

••P42 You throw a ball. Assume that the origin is on the ground, with the $+y$ axis pointing upward. Just after the ball leaves your hand its position is $\langle 0.06, 1.03, 0\rangle$ m. The average velocity of the ball over the next 0.7 s is $\langle 17, 4, 6\rangle$ m/s. At time 0.7 s after the ball leaves your hand, what is the height of the ball above the ground?

••P43 Figure 1.60 shows the trajectory of a ball traveling through the air, affected by both gravity and air resistance. Here are the positions of the ball at several successive times:

Location	t(s)	Position (m)
A	0.0	$\langle 0,0,0\rangle$
B	1.0	$\langle 22.3, 26.1, 0\rangle$
C	2.0	$\langle 40.1, 38.1, 0\rangle$

(a) What is the average velocity of the ball as it travels between location A and location B? **(b)** If the ball continued to travel at the same average velocity during the next second, where would it be at the end of that second? (That is, where would it be at time t=2 s?) **(c)** How does your prediction from part (b) compare to the actual position of the ball at t=2 s (location C)? If the predicted and observed locations of the ball are different, explain why.

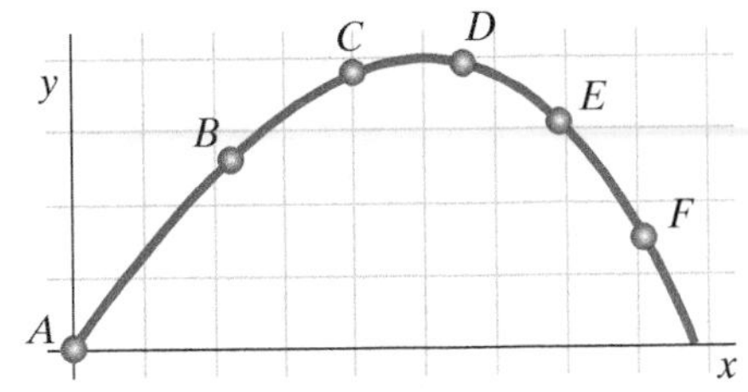

Figure 1.60

••P44 At 6 s after 3:00, a butterfly is observed leaving a flower whose location is $\langle 6, -3, 10\rangle$ m relative to an origin on top of a nearby tree. The butterfly flies until 10 s after 3:00, when it alights on a different flower whose location is $\langle 6.8, -4.2, 11.2\rangle$ m relative to the same origin. What was the location of the butterfly at a time 8.5 s after 3:00? What assumption did you have to make in calculating this location?

Section 1.8

•P45 A baseball has a mass of 0.155 kg. A professional pitcher throws a baseball 90 mi/h, which is 40 m/s. What is the magnitude of the momentum of the pitched baseball?

•P46 A hockey puck with a mass of 0.4 kg has a velocity of $\langle 38, 0, -27\rangle$ m/s. What is the magnitude of its momentum, $|\vec{p}|$?

•P47 What is the magnitude (in kg·m/s) of the momentum of a 1000 kg airplane traveling at a speed of 500 mi/h? (Note that you need to convert speed to meters per second.)

•P48 A baseball has a mass of about 155 g. What is the magnitude of the momentum of a baseball thrown at a speed of 100 miles per hour? (Note that you need to convert mass to kilograms and speed to meters per second. See the inside back cover of the textbook for conversion factors.)

•P49 If a particle has momentum $\vec{p} = \langle 4, -5, 2\rangle$kg·m/s, what is the magnitude $|\vec{p}|$ of its momentum?

•P50 An object with mass 1.6 kg has momentum $\langle 0, 0, 4\rangle$ kg·m/s. **(a)** What is the magnitude of the momentum? **(b)** What is the unit vector corresponding to the momentum? **(c)** What is the speed of the object?

•P51 A tennis ball of mass m traveling with velocity $\langle v_x, 0, 0\rangle$ hits a wall and rebounds with velocity $\langle -v_x, 0, 0\rangle$. **(a)** What was the change in momentum of the tennis ball? **(b)** What was the change in the magnitude of the momentum of the tennis ball?

•P52 A basketball has a mass of 570 g. Heading straight downward, in the $-y$ direction, it hits the floor with a speed of 5 m/s and rebounds straight up with nearly the same speed. What was the momentum change $\Delta\vec{p}$?

••P53 A basketball has a mass of 570 g. Moving to the right and heading downward at an angle of 30° to the vertical, it hits the floor with a speed of 5 m/s and bounces up with nearly the same speed, again moving to the right at an angle of 30° to the vertical. What was the momentum change $\Delta\vec{p}$?

•P54 The first stage of the giant Saturn V rocket reached a speed of 2300 m/s at 170 s after liftoff. **(a)** What was the average acceleration in m/s/s? **(b)** The acceleration of a falling object if air resistance is negligible is 9.8 m/s/s, called "one g." What was the average acceleration in g's?

••P55 A 50 kg child is riding on a carousel (merry-go-round) at a constant speed of 5 m/s. What is the magnitude of the change in the child's momentum $|\Delta\vec{p}|$ in going all the way around (360°)? In going halfway around (180°)? It is very helpful to draw a diagram, and to do the vector subtraction graphically.

••P56 Figure 1.61 shows a portion of the trajectory of a ball traveling through the air. Arrows indicate its momentum at several locations.

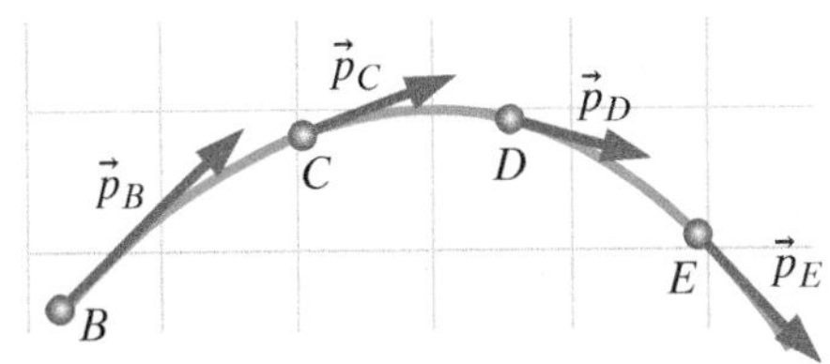

Figure 1.61

At various locations, the ball's momentum is:
$\vec{p}_B = \langle 3.03, 2.83, 0\rangle$ kg·m/s
$\vec{p}_C = \langle 2.55, 0.97, 0\rangle$ kg·m/s
$\vec{p}_D = \langle 2.24, -0.57, 0\rangle$ kg·m/s

$\vec{p}_E = \langle 1.97, -1.93, 0 \rangle$ kg · m/s
$\vec{p}_F = \langle 1.68, -3.04, 0 \rangle$ kg · m/s
(a) Calculate the change in the ball's momentum between each pair of adjacent locations. **(b)** On a copy of the diagram, draw arrows representing each $\Delta\vec{p}$ you calculated in part (a). **(c)** Between which two locations is the magnitude of the change in momentum greatest?

Section 1.9

•P57 What is the velocity of a 3 kg object when its momentum is $\langle 60, 150, -30 \rangle$ kg · m/s?

•P58 A 1500 kg car located at $\langle 300, 0, 0 \rangle$ m has a momentum of $\langle 45000, 0, 0 \rangle$ kg · m/s. What is its location 10 s later?

••P59 An ice hockey puck of mass 170 g enters the goal with a momentum of $\langle 0, 0, -6.3 \rangle$ kg · m/s, crossing the goal line at location $\langle 0, 0, -26 \rangle$ m relative to an origin in the center of the rink. The puck had been hit by a player 0.4 s before reaching the goal. What was the location of the puck when it was hit by the player, assuming negligible friction between the puck and the ice? (Note that the ice surface lies in the xz plane.)

••P60 A space probe of mass 400 kg drifts past location $\langle 0, 3 \times 10^4, -6 \times 10^4 \rangle$ m with momentum $\langle 6 \times 10^3, 0, -3.6 \times 10^3 \rangle$ kg · m/s. Assuming the momentum of the probe does not change, what will be its position 2 minutes later?

Section 1.10

•P61 A proton in an accelerator attains a speed of 0.88c. What is the magnitude of the momentum of the proton?

•P62 An electron with a speed of 0.95c is emitted by a supernova, where c is the speed of light. What is the magnitude of the momentum of this electron?

•P63 A "cosmic-ray" proton hits the upper atmosphere with a speed 0.9999c, where c is the speed of light. What is the magnitude of the momentum of this proton? Note that $|\vec{v}|/c = 0.9999$; you don't actually need to calculate the speed $|\vec{v}|$.

•P64 A proton in a particle accelerator is traveling at a speed of 0.99c. (Masses of particles are given on the inside back cover of this textbook.) **(a)** If you use the approximate nonrelativistic equation for the magnitude of momentum of the proton, what answer do you get? **(b)** What is the magnitude of the correct relativistic momentum of the proton? **(c)** The approximate value (the answer to part a) is significantly too low. What is the ratio of magnitudes you calculated (correct/approximate)?

•P65 When the speed of a particle is close to the speed of light, the factor γ, the ratio of the correct relativistic momentum $\gamma m\vec{v}$ to the approximate nonrelativistic momentum $m\vec{v}$, is quite large. Such speeds are attained in particle accelerators, and at these speeds the approximate nonrelativistic equation for momentum is a very poor approximation. Calculate γ for the case where $|\vec{v}|/c = 0.9996$.

•P66 An electron travels at speed $|\vec{v}| = 0.996c$, where $c = 3 \times 10^8$ m/s is the speed of light. The electron travels in the direction given by the unit vector $\hat{v} = \langle 0.655, -0.492, -0.573 \rangle$. The mass of an electron is 9×10^{-31} kg. **(a)** What is the value of γ? You can simplify the calculation if you notice that $(|\vec{v}|/c)^2 = (0.996)^2$. **(b)** What is the speed of the electron? **(c)** What is the magnitude of the electron's momentum? **(d)** What is the vector momentum of the electron? Remember that any vector can be "factored" into its magnitude times its unit vector, so that $\vec{v} = |\vec{v}|\hat{v}$.

•P67 If $|\vec{p}|/m$ is 0.85c, what is $|\vec{v}|$ in terms of c?

COMPUTATIONAL PROBLEMS

These problems are intended to introduce you to using a computer to model matter, interactions, and motion in 3D. You do not need to know how to program; you will learn what you need to know by doing these problems. In later chapters you will build on these small calculations to build models of physical systems.

To install the free 3D programming environment VPython, go to http://vpython.org and (carefully) follow the instructions for your operating system (Windows, MacOS, or Linux). Note the instructions given there on how to zoom and rotate the "camera" when viewing a 3D scene you have created.

More detailed and extended versions of some of these computational modeling problems may be found in the lab activities included in the *Matter & Interactions 4th Edition* resources for instructors.

•P68 Watch the first introductory VPython video, *VPython Instructional Videos 1: 3D Objects,* at vpython.org/video01.html and complete the challenge activity at the end of the video.

•P69 **(a)** Write a VPython program that creates eight spheres, each placed at one corner of a cube centered on the origin. The length of a side of the cube should be 6 units, and the radius of each sphere should be 0.5. Use at least two different colors for the spheres. **(b)** Add to the program an arrow whose tail is at one corner of the cube and whose tip is at the corner diagonally opposite. Figure 1.62 shows the display from one possible solution to this problem.

Figure 1.62

•P70 Write a VPython program that represents the x, y, and z axes by three cylinders of different colors. The display from one possible solution is shown in Figure 1.63.

Figure 1.63

•**P71** Write a VPython program that represents the x, y, and z axes by three boxes (rectangular solids) of different colors. The display from one possible solution is shown in Figure 1.64.

Figure 1.64

•**P72** Watch the second introductory VPython video, *VPython Instructional Videos 2: Variable Assignment*, at vpython.org/video02.html and complete the challenge activity at the end of the video.

•**P73** Based on Problem P72, write a program that shows an arrow pointing from one small box to another in such a way that when you change only the position of the first box, making no other changes, the arrow and the other box move too, so that the two boxes remain linked by the arrow.

•**P74** Watch the third introductory VPython video, *VPython Instructional Videos 3: Beginning Loops*, at vpython.org/video03.html and complete the challenge activity at the end of the video.

••**P75** **(a)** Write a VPython program that uses three while loops to create a display in which each of the axes (x, y, z) is represented by a linear array of boxes, with spaces between the boxes. Figure 1.65 shows a possible example. **(b)** (Optional) Rewrite your program to produce the same display using only one while loop.

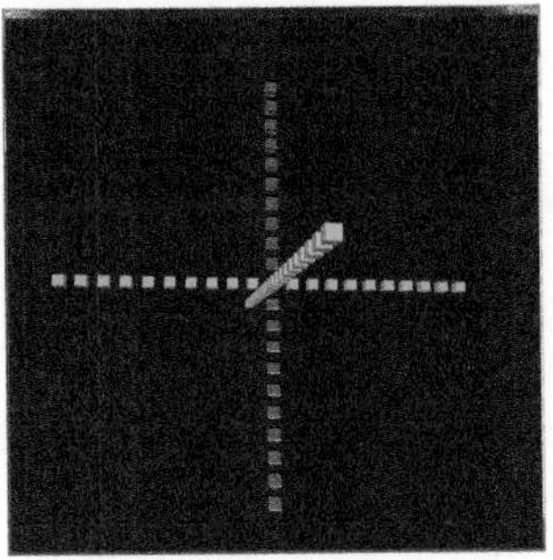

Figure 1.65

••**P76** Write a VPython program that creates three circles of spheres: one in the xy plane, one in the yz plane, and one in the xz plane. Each ring should be centered on the origin. Use either three while loops or one while loop. A possible solution is shown in Figure 1.66.

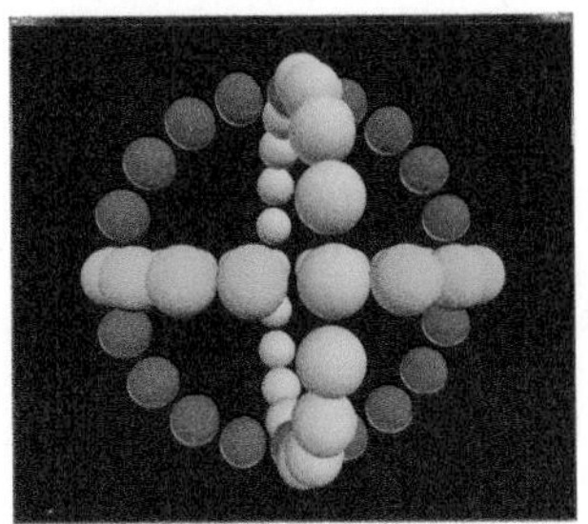

Figure 1.66

•**P77** Watch the fourth introductory VPython video, *VPython Instructional Videos 4: Loops and Animation*, at vpython.org/video04.html and complete the challenge activity at the end of the video.

••**P78** Consider the VPython program shown below. **(a)** An important skill is being able to read and understand an existing program, in order to be able to make useful modifications. *Before running the program*, study the program carefully line by line, then answer the following questions: (1) What is the initial velocity of the particle? (2) Is the particle initially located in front of the box or behind it? (3) In which line of code is the position of the particle updated? (4) What is the value of the time step Δt? (5) Will the particle bounce off of the red box, or travel through it? **(b)** Now run the program, and see if your answers were correct. **(c)** Modify the program to start the particle from an initial position on the $+x$ axis, to the right of and in front of the red box. Give the particle a velocity that will make it travel to the left, along the x axis, passing in front of the box.

```
from visual import *
box(pos=vector(0,0,-1),
    size=(5,5,0.5),
    color=color.red,
    opacity = 0.4)
particle = sphere(pos=vector(-5,0,-5),
                  radius=0.3,
                  color=color.cyan,
                  make_trail = True)
v = vector(0.5,0,0.5)
delta_t = 0.05
t = 0
while t < 20:
    rate(100)
    particle.pos = particle.pos + v * delta_t
    t = t + delta_t
```

•••**P79** Modify the program shown above to make the particle bounce off the red box instead of passing through it. On the Help menu available in IDLE (the VPython editor), choose "Python docs" or search the web for "Python if" to find out how to use an `if` statement. You may also find it helpful to look at the example program `bounce2.py`, included in the examples installed with VPython.

ANSWERS TO CHECKPOINTS

Numerical answers are given to three significant figures.

1 (a) 1, 4, **(b)** 1: change of direction, 2: change of speed and direction, 3: change of direction, 5: change of direction

2 (a) 1: change of speed, 2: change of speed, 3: change of direction, **(b)** 1, 4

3 $\langle 1, -1.5, 2.5 \rangle$ m

4 3.08 m/s

5 (a) no, **(b)** $\langle 0.873, 0.436, -0.218 \rangle$

6 (a) $\langle 450, -300, -300 \rangle$, **(b)** 577, **(c)** 696, **(d)** No, **(e)** $\langle 150, 300, -200 \rangle$, **(f)** 391, **(g)** 25.1, **(h)** No

7 a, b, e

8 (a) $\langle 9, 2, 20 \rangle$ m, **(b)** 120 s

9 (a) $\langle -0.949, 0.342, 0 \rangle$, **(b)** $\langle -0.643, 0.766, 0 \rangle$

10 2.67×10^{-3} m/s

11 (a) $\langle 8, -10, 12 \rangle$ m/s, **(b)** 17.5 m/s, **(c)** $\langle 0.456, -0.570, 0.684 \rangle$

12 $\langle 2.1 \times 10^5, 1.4 \times 10^5, -2.8 \times 10^5 \rangle$ m

13 (1) *a*, (2) *h*, (3) *g*, (4) *f*, (5) *e*

14 (a) (1) 5 m/s/s, (2) about half as big **(b)** (1) $\langle 5, 8t, 2 - 18t^2 \rangle$, (2) $\langle 0, 8, -36t \rangle$, (3) $\langle 5, 0, 2 \rangle$, (4) $\langle 0, 8, 0 \rangle$

15 (a) $\langle -1.6 \times 10^{28}, 0, -1.6 \times 10^{28} \rangle$ kg·m/s, **(b and c)** If you make a careful diagram, with the two arrows tail to tail, you'll see that the arrow representing $\Delta\vec{p}$ from *C* to *D* points up and to the right on the page (toward the Sun).

16 $\langle 111.31, 0, 18.46 \rangle$ m

17 $\langle 0, 0, -2.415 \times 10^{-22} \rangle$ kg·m/s, 2.415×10^{-22} kg·m/s

18 a, b

CHAPTER

2

The Momentum Principle

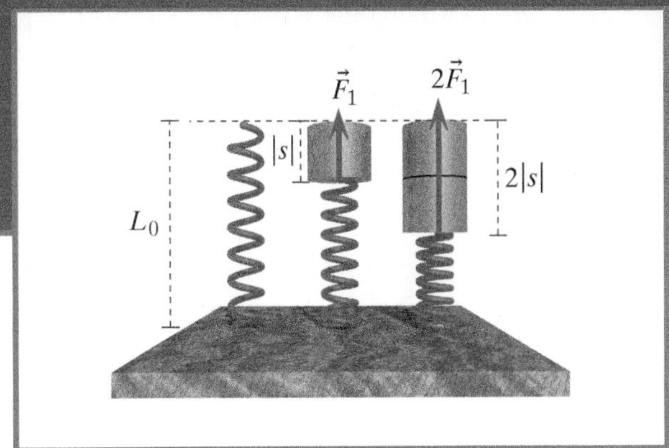

OBJECTIVES

After studying this chapter, you should be able to

- Use both iterative and analytical techniques to predict the future motion of a system that is subjected to a constant net force.
- Use an iterative approach to predict the future momentum of an object that is subjected to a varying net force.
- Draw and interpret graphs of the components of position or velocity *vs.* time.
- Calculate the force exerted by a spring on an object in contact with it.
- Calculate the approximate gravitational force on an object near the Earth's surface.

2.1 THE MOMENTUM PRINCIPLE

In many real situations, the momentum (and therefore the velocity) of a moving object are continually changing due to interactions of the object with its surroundings. In order to predict the motion of such a system, whether it is a meteor, a tennis ball, or an electron, we need to be able to express mathematically the relationship between interaction and momentum change. The Momentum Principle makes a quantitative connection between amount of interaction and change of momentum.

The Momentum Principle is the first of three fundamental principles of mechanics that together make it possible to predict and explain a very broad range of real-world phenomena. The other two fundamental principles we will encounter are the Energy Principle (Chapter 6) and the Angular Momentum Principle (Chapter 11). A fundamental principle is powerful because it applies in absolutely every situation. In its most general form, a fundamental principle expresses the idea that certain important quantities are *conserved*; that is, that the total value of the quantity in the Universe does not change. However, in order to predict the details of what will happen to a system we will often write the principle in a form relating the change in a quantity to the details of the interaction causing the change.

System and Surroundings

The Momentum Principle relates an interaction quantitatively to a change in momentum. To analyze a change in momentum, we need to specify clearly the objects whose momentum change we wish to know. One or more objects can be considered to be a "system." Everything that is not included in the system is part of the "surroundings." For example, the red dashed line in Figure 2.1 indicates that we have chosen the skater as the system, and that therefore all other objects, including the wall, the ice, the air, and the Earth, are part of the surroundings. The Momentum Principle relates the change in momentum of a system to the amount of interaction with its surroundings. No matter what system we choose, the Momentum Principle will correctly predict the behavior of the system.

In this chapter we will always consider a single object to be the system, and everything else in the Universe to be the surroundings. However, in later chapters we will see that it is possible, and often useful, to choose a system comprising several objects. A fundamental principle such as the Momentum Principle applies to any system, no matter how complex.

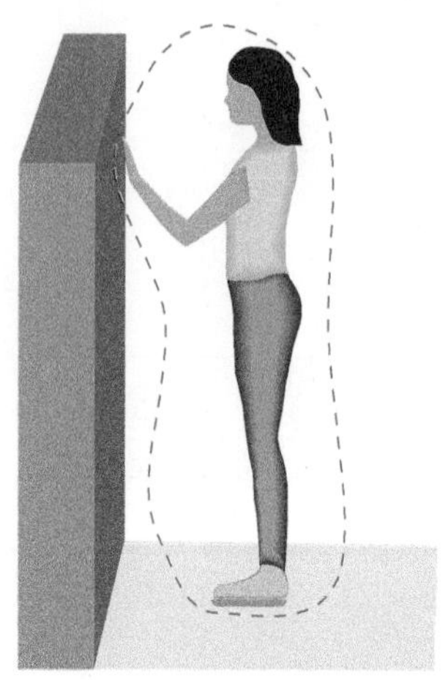

Figure 2.1 A skater pushes off from a wall. If we choose the skater as the system (indicated by the dashed line), the wall, the ice, the air, and the Earth are included in the surroundings.

Cause and Effect

The Momentum Principle is a fundamental principle that is also known as Newton's second law. It restates and extends Newton's first law of motion in a quantitative, causal form that can be used to predict the behavior of objects. The validity of the Momentum Principle has been verified through a very wide variety of observations and experiments, involving large and small objects, moving slowly or at speeds near the speed of light. It is a summary of the way interactions affect motion in the real world.

THE MOMENTUM PRINCIPLE

$$\Delta\vec{p} = \vec{F}_{\text{net}}\,\Delta t$$

The change of momentum of a system (the effect) is equal to the net force acting on the system times the duration of the interaction (the cause). The time interval Δt must be small enough that the net force is nearly constant during this time interval.

The Momentum Principle is a fundamental principle because: it applies to every possible system, no matter how large or small (from clusters of galaxies to subatomic particles), and no matter how fast it is moving; it is true for every kind of interaction (electric, gravitational, etc.); it relates an effect (change in momentum) to a cause (an interaction).

To begin to understand this equation, we will consider each quantity involved.

$$\Delta\vec{p} = \vec{F}_{\text{net}}\,\Delta t$$

Change of Momentum $\Delta\vec{p}$

As we saw in the previous chapter, the change in momentum $\Delta\vec{p} = \vec{p}_f - \vec{p}_i$ of a system can involve a change in the magnitude of momentum, the direction of momentum, or both magnitude and direction.

$$\Delta\vec{p} = \vec{F}_{\text{net}}\,\Delta t$$

Force $\vec{F}$

Scientists and engineers employ the concept of "force" to quantify interactions between two objects. Force is a vector quantity because a force has a magnitude and is exerted in a particular direction. Examples of forces include the following:

- The repulsive electric force a proton exerts on another proton
- The attractive gravitational force the Earth exerts on you
- The force that a compressed spring exerts on your hand
- The force on a spacecraft by expanding gases in a rocket engine
- The force of the surrounding air on the propeller of an airplane

Measuring a Force

A simple way to measure force is to use the stretch or compression of a spring. In Figure 2.2 we hang a block from a spring, and note that the spring is stretched a distance s. Then we hang two such blocks from the spring, and we see that the spring is stretched twice as much. By experimentation, we find that any spring made of the same material and produced to the same specifications behaves in the same way.

Similarly, we can observe how much the spring compresses when the same blocks are supported by it (Figure 2.3). We find that one block compresses the spring by the same distance $|s|$, and two blocks compress it by $2|s|$. (Compression can be considered negative stretch, because the length of the spring decreases.)

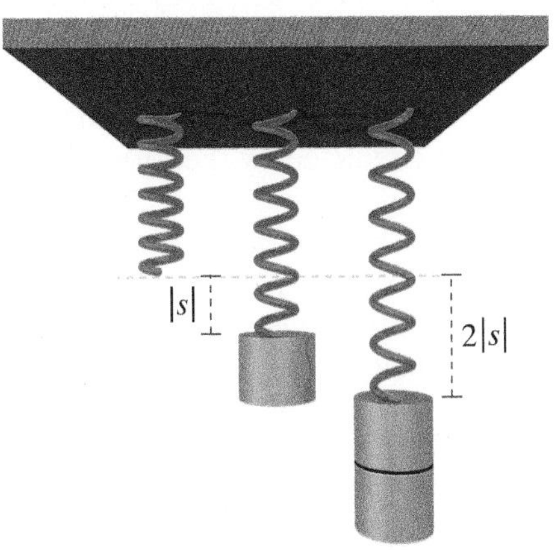

Figure 2.2 Stretching of a spring is a measure of force.

Unit of Force

We can use a spring to make a scale for measuring forces, calibrating it in terms of how much force is required to produce a given stretch. The SI unit of force is the newton, abbreviated as N. One newton is a rather small force. A newton is approximately the downward gravitational force of the Earth on a small apple, or about a quarter of a pound. If you hold a small apple at rest in your hand, you apply an upward force of about one newton, compensating for the downward pull of the Earth.

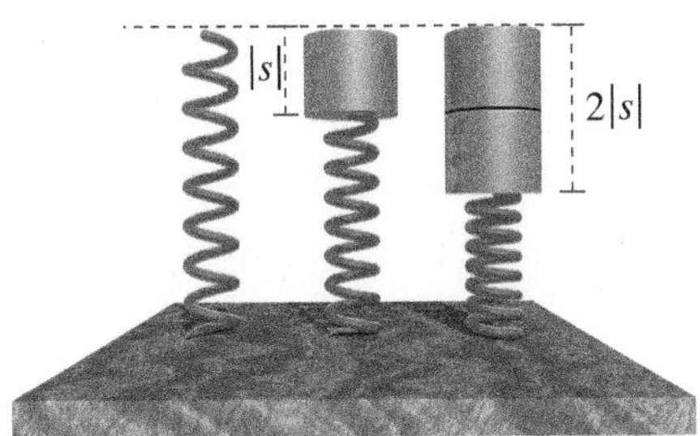

Figure 2.3 Compression of a spring is also a measure of force.

The Meaning of *net*

In physics the word *net* has a precise meaning: it is the sum of all contributions to a quantity, both pluses and minuses.

Because a force has a magnitude and a direction, it is a vector, so the *net* force on a system is the vector sum of all the forces exerted on the system by all the objects in the surroundings. It is the net force $\vec{F}_{\text{net}}$ on a system, acting for some time Δt, that causes a change of momentum.

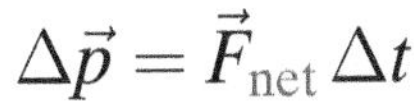

DEFINITION OF NET FORCE

$$\vec{F}_{\text{net}} \equiv \vec{F}_1 + \vec{F}_2 + \cdots$$

The "net" force acting on a system at an instant is the vector sum of *all* the forces exerted on the system by all the objects in the surroundings at that instant.

The net force acting on a system at an instant is the vector sum of all of the forces exerted on the system by all the objects in the surroundings, which are called "external" forces (Figure 2.4). There may be forces internal to the system, exerted by one object in the system on another object in the system, but such internal forces cannot change the momentum of the system. We will see in detail why this is so in Chapter 3, but the basic reason is that internal forces cancel each other's effects. A force that object 1 in a system exerts on object 2 in the system does change the momentum of object 2, but object 2 exerts a force in the opposite direction on object 1 that changes the momentum of object 1 in the opposite way, so that the changes in momenta of the two objects add up to zero.

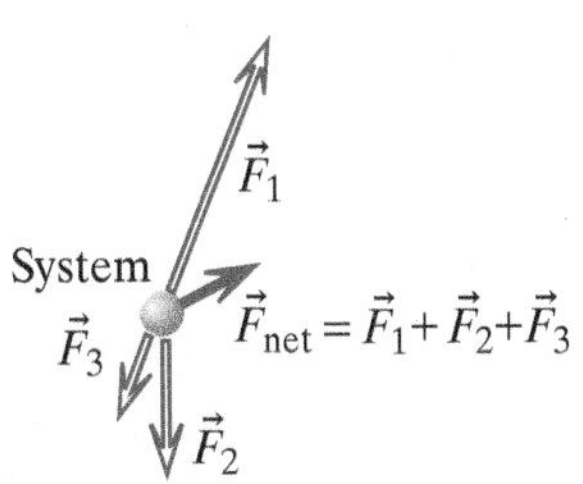

Figure 2.4 The net force on the system is the vector sum of all the individual forces acting on the system.

QUESTION A ball falling toward the Earth consists of many atoms. Each atom in the ball exerts forces on neighboring atoms in the ball, the Earth exerts forces on each atom in the ball, and the air exerts forces on atoms at the surface of the ball. Taking the ball as the system, and the Earth and air as the surroundings, which of these forces are external forces and which are internal forces?

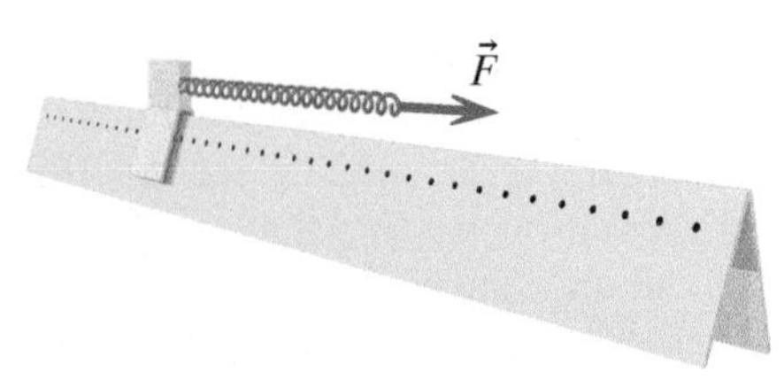

Figure 2.5 A block mounted on a nearly frictionless air track, pulled by a spring.

The Earth and the air are part of the surroundings, so the force exerted by the Earth and the force exerted by the air are both external forces. They are forces exerted on the chosen system (the ball) by objects in the surroundings (the Earth and the air). Forces that atoms in the ball exert on neighboring atoms in the ball are internal forces. The internal forces don't contribute to the net force $\vec{F}_{net}$ that changes the ball's momentum as predicted by the Momentum Principle, $\Delta\vec{p} = \vec{F}_{net}\,\Delta t$.

We find experimentally that the magnitude of the net force acting on an object affects the magnitude of the change in its momentum. Many introductory physics laboratories have air tracks like the one illustrated in Figure 2.5. The long triangular base has many small holes in it, and air under pressure is blown out through these holes. The air forms a cushion under the glider, allowing it to coast smoothly with very little friction.

Suppose that we place a block on a glider on a long air track, and attach a calibrated spring to it (Figure 2.5). We pull on the spring so that it is stretched a distance s, so that it exerts a force F on the block, and we pull for a short time. Choose the block and glider as the system, so that the net force on the system is just the spring, since friction with the track is very small due to the air cushion.

We observe that the momentum of the system of block and glider increases from zero to an amount mv. If instead we pull on the spring so that it stretches a distance $2s$, the spring exerts a force $2F$ on the system, and we observe that the system's momentum increases to $2mv$ in the same amount of time. Apparently the magnitude of the change in momentum is proportional to the magnitude of the net force applied to the object.

$$\Delta\vec{p} = \vec{F}_{net}\,\Delta t$$

Duration of Interaction Δt

Another experiment will show us that not only the magnitude of the force, but also the length of time during which it acts on an object, affects the change of momentum of the object. Suppose that we again place a block on an air track and attach a spring to it. We pull on the spring for a time Δt with a force F, and we observe that the momentum of the system of block plus glider increases from zero to an amount mv. However, if we repeat the experiment but pull for a time twice as long with the same force F, the system's momentum increases to $2mv$; the change in its momentum is twice as great. We observe that the magnitude of the change in momentum is directly proportional to the length of time $\Delta t = t_2 - t_1$ during which the force acts on the object.

$$\Delta\vec{p} = \vec{F}_{net}\,\Delta t$$

Impulse $\vec{F}_{net}\,\Delta t$

The amount of interaction affecting an object includes both the strength of the interaction (expressed as the net force $\vec{F}_{net}$) and the duration Δt of the interaction. Either a bigger force or applying the force for a longer time will cause more change of momentum. The product of a force and a time interval is called "impulse."

DEFINITION OF IMPULSE

$$\text{Impulse} = \vec{F}\Delta t$$

Impulse has units of N · s (newton-seconds)

The time interval Δt must be small enough that the net force is nearly constant during this time interval.

Now we can state the Momentum Principle this way:

The change of momentum of a system is equal to the net impulse applied to the system.

QUESTION Are the units of impulse and momentum the same? If not, how are they related?

The units of impulse are N · s (a force of 1 N acting during a time of 1 second gives an impulse of 1 N · s). The units of momentum are kg · m/s, so evidently 1 N · s = 1 kg · m/s, and therefore 1 N = 1 kg · m/s^2.

EXAMPLE **Impulse Applied to a Baseball**

Just before it touches the bat, the momentum of a baseball is $\langle 0,0,3\rangle$ kg · m/s. After the batter has hit the ball, and the ball is no longer in contact with the bat, the momentum of the baseball is $\langle -2,2,-1\rangle$ kg · m/s. What was the impulse applied to the ball during its contact with the bat?

Solution

$$\Delta\vec{p} = \vec{F}_{\text{net}}\,\Delta t$$
$$\langle -2,2,-1\rangle \text{ kg}\cdot\text{m/s} - \langle 0,0,3\rangle \text{ kg}\cdot\text{m/s} = \vec{F}_{\text{net}}\,\Delta t$$
$$\langle -2,2,-4\rangle \text{ kg}\cdot\text{m/s} = \vec{F}_{\text{net}}\,\Delta t$$

Without more information we cannot find the values of $\vec{F}_{\text{net}}$ or Δt separately.

EXAMPLE **Collision with a Wall**

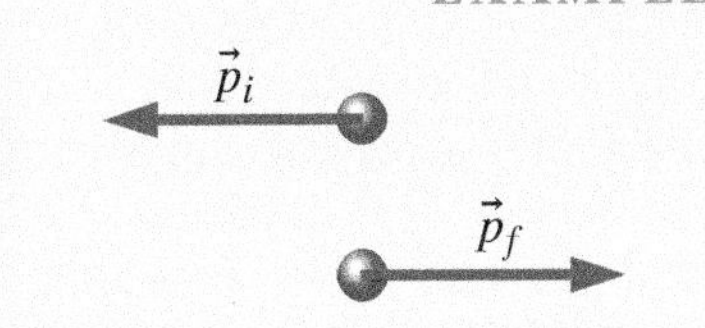

Figure 2.6 A ball hits a wall and rebounds.

A tennis ball traveling in the $-x$ direction with magnitude of momentum 2.8 kg · m/s hits a wall and rebounds (Figure 2.6). After the collision the magnitude of the ball's momentum is nearly the same, but the ball is traveling in the $+x$ direction. What was the impulse applied to the ball by the wall?

Solution Although the magnitude of the ball's momentum did not change significantly, its direction did change, so the impulse is not zero.

$$\Delta\vec{p} = \vec{F}_{\text{net}}\,\Delta t$$
$$(\langle 2.8,0,0\rangle \text{ kg}\cdot\text{m/s} - \langle -2.8,0,0\rangle \text{ kg}\cdot\text{m/s}) = \vec{F}_{\text{net}}\,\Delta t$$
$$\langle 5.6,0,0\rangle \text{ kg}\cdot\text{m/s} = \vec{F}_{\text{net}}\,\Delta t$$

Note that in this collision the magnitude of the change in the ball's momentum (which is equal to the magnitude of the impulse) was about twice as large as the ball's original momentum.

EXAMPLE **Finding Direction of Impulse Graphically**

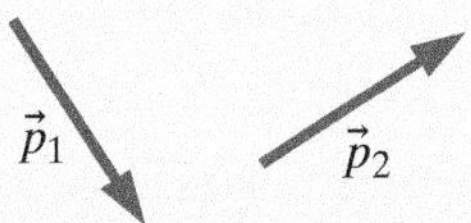

Figure 2.7 Initial and final momenta of a hockey puck. Both momenta lie in the xz plane; this is a top view, looking down on the ice from above.

A hockey puck is sliding across the ice with momentum $\vec{p}_1$ (shown in Figure 2.7) when a hockey player hits it sharply with a hockey stick. Afterward, the puck has momentum $\vec{p}_2$. Both $\vec{p}_1$ and $\vec{p}_2$ lie in the xz plane. What was the direction of the impulse applied to the puck by the stick? What was the direction of the net force exerted on the puck during its short contact with the hockey stick?

Solution We can find the direction of $\Delta\vec{p}$ graphically, as shown in Figure 2.8. From the vector equation

$$\Delta\vec{p} = \vec{F}_{\text{net}}\,\Delta t$$

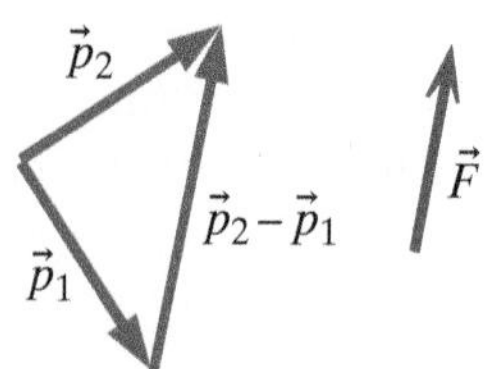

Figure 2.8 To find $\Delta\vec{p}$, the vectors $\vec{p}_1$ and $\vec{p}_2$ are placed tail to tail. The net force on the puck was in the direction of the observed change in momentum $\Delta\vec{p}$.

we see that $\Delta\vec{p}$ and $\vec{F}_{net}\,\Delta t$ must have the same directions. Since Δt is a positive scalar, the direction of the net force is the same as the direction of the impulse, because multiplying the vector $\vec{F}_{net}$ by the positive scalar Δt can't change the direction of the vector.

Checkpoint 1 (1) Two external forces, $\langle 40, -70, 0\rangle$ N and $\langle 20, 10, 0\rangle$ N, act on a system. What is the net force acting on the system? (2) A hockey puck initially has momentum $\langle 0, 2, 0\rangle$ kg·m/s. It slides along the ice, gradually slowing down, until it comes to a stop. **(a)** What was the impulse applied by the ice and the air to the hockey puck? **(b)** It took 3 seconds for the puck to come to a stop. During this time interval, what was the net force on the puck by the ice and the air (assuming that this force was constant)?

2.2 LARGE FORCES AND SHORT TIMES

The Momentum Principle can be a powerful tool for analyzing real events. We'll consider two situations in which there is a sudden impact: a very large force, applied to a system for a very short time, that produces a dramatic change in momentum. In one case we will deduce the approximate magnitude of the force, and in the other case we will infer the duration of the interaction. In both cases we will need to make the simplifying assumption that the net force on the system is constant during the short interaction time. This is probably not really true in either case, but by considering this constant value to be the average force exerted on the system during the short interaction, we can extract useful information about the impact.

By making these simplifying assumptions and approximations, we are making a "model" of each physical system. Making and using models is an activity central to physics, and the criteria for a "good" physical model will depend on how we intend to use the model. In fact, one of the most important problems a scientist or engineer faces is deciding what interactions must be included in a model of a real physical, chemical, or biological system, and what interactions can reasonably be ignored.

EXAMPLE

Colliding Students

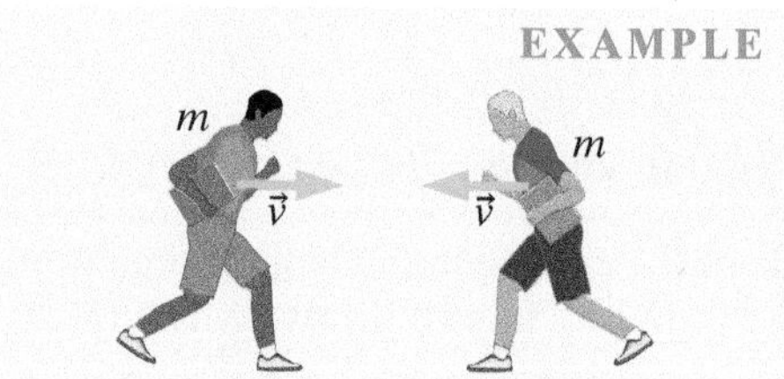

Figure 2.9 Two students running fast collide head-on.

Two students who are late for tests are running to classes in opposite directions as fast as they can. They turn a corner, run into each other head-on, and come to a complete stop (Figure 2.9). Using physics principles, estimate the force that one student exerts on the other during the collision. You will need to estimate some quantities; give reasons for your choices and provide checks showing that your estimates are physically reasonable.

This problem is rather ill-defined and doesn't seem much like a "textbook" problem. No numbers have been given, yet you're asked to estimate the force of the collision.

QUESTION How can our answer possibly be meaningful if we have to estimate quantities?

Since we have no idea how big such a force might be, even an approximate answer will be informative. This kind of problem is typical of the kinds of problems engineers and scientists encounter in their professional work.

To begin, we'll make a diagram of the forces due to objects in the environment acting on a system consisting of the student on the left. Such a diagram is called a "free-body diagram." If we are modeling the system as a point particle, we usually represent the system as a dot.

Solution

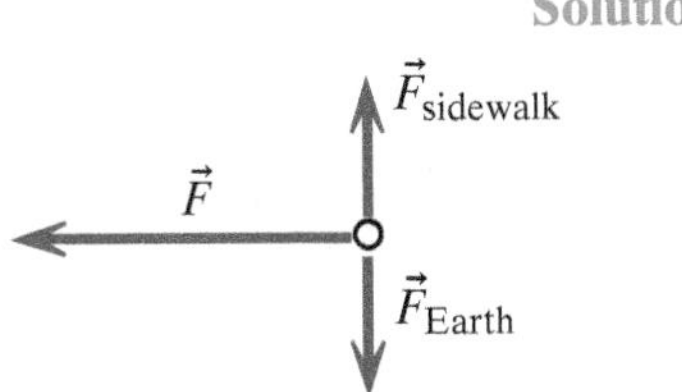

Figure 2.10 Free-body diagram showing forces on the left student. Forces that we assume to be negligible have been omitted.

System: The student on the left (we could choose either student)
Surroundings: Other student, Earth, sidewalk, air
Free-body diagram: Figure 2.10
Initial time: Just before impact
Final time: When speeds become zero
Simplifying assumptions:

- The students have the same mass m and same initial speed $|\vec{v}_i|$.
- Friction and air resistance are negligible during this short Δt.

QUESTION We don't know the upward force exerted on the student by the sidewalk, so how can we solve this problem?

We know that the y component of the student's momentum does not change during the collision. From this we can deduce that the y component of the net force on the student is zero, and therefore the downward gravitational force exerted by the Earth and the upward force exerted by the sidewalk are equal in magnitude: $\vec{F}_{\text{sidewalk}} + \vec{F}_{\text{Earth}} = \vec{0}$. Therefore, the only force we need to consider is the unknown force $\vec{F}$ exerted by the other student.

Applying the Momentum Principle to the system of the student on the left, we get:

$$\Delta\vec{p} = \vec{F}_{\text{net}}\,\Delta t$$
$$(\langle 0,0,0\rangle - \langle mv_i,0,0\rangle) = \vec{F}_{\text{net}}\,\Delta t$$

To go further we must estimate some quantities. We need the initial momentum of the student on the left, $\langle mv_i,0,0\rangle$, and we need the duration of the collision, Δt. Let's choose these values:

$$m \approx 60\text{ kg (1 kg is about 2.2 lb)}$$
$$|\vec{v}_i| \approx 6\text{ m/s (elite sprinters run 100 m in about 10 s)}$$
$$\Delta t \approx ???$$

We know the collision occurs during a short time, but how short? One way to estimate a time is to estimate the distance traveled during the collision. In this case, this distance is the amount by which the student's body is compressed by the impact. Let's guess that the student's body is compressed by about 5 cm during the collision, which will be painful. During this collision, the student's speed drops from 6 m/s to 0, so we'll estimate the average speed as the "arithmetic average" $(6\text{ m/s} + 0\text{ m/s})/2 = 3\text{ m/s}$ (more about this later in the chapter).

All motion is in the x direction, so:

$$\Delta x = v_{\text{avg},x}\,\Delta t$$
$$0.05\text{ m} \approx 3\text{ m/s}\,\Delta t$$
$$\Delta t \approx 0.017\text{ s}$$

The whole interaction lasts only about 0.02 s! The force exerted by one student on the other must be very large. Let's see about how large it is:

$$(\langle 0,0,0\rangle\text{ kg}\cdot\text{m/s} - \langle mv_i,0,0\rangle\text{ kg}\cdot\text{m/s}) = \vec{F}_{\text{net}}\,\Delta t$$
$$(\langle 0,0,0\rangle\text{ kg}\cdot\text{m/s} - \langle 60\cdot 6,0,0\rangle\text{ kg}\cdot\text{m/s}) \approx \vec{F}_{\text{net}} \times 0.017\text{ s}$$
$$\vec{F}_{\text{net}} \approx \langle -21100,0,0\rangle\text{ N}$$

Check:

- The units check ($\text{N}\cdot\text{s} = \text{kg}\cdot\text{m/s}$, so $(\text{kg}\cdot\text{m/s})/\text{s} = \text{N}$).
- Is the result reasonable?

21,000 N is a very large force. As we will see in Section 2.4, the gravitational force on a 60 kg student (the "weight") is only about 600 N. The force of the impact is about 35 times the weight of the student! It's like having a stack of 35 students sitting on you. If the students hit heads instead of stomachs, the squeeze might be less than 1 cm, and the force would be over 5 times as large! This is why heads can break in such a collision. Our result of 21,000 N shows why collisions are so dangerous. Collisions involve very large forces acting for very short times, giving impulses of ordinary magnitude.

> **QUESTION** Estimating masses or distances is usually not difficult, but most people find it more challenging to estimate time durations, especially if the times are very short. Why would a guess of 1 second for Δt have been a poor guess?

At an average speed of 3 m/s, each student would have traveled 3 meters during a one-second impact. This is clearly not physically possible—they would have had to pass through each other to do this!

> **QUESTION** We estimated various quantities to get this result. Is the result still useful?

A useful model should omit extraneous detail but retain the most important features of the real-world situation. When we make many estimates, our goal is often to determine the order of magnitude of the answer: is the force on a colliding student closer to 0.0001 newton or 10,000 newtons? Knowing the order of magnitude to expect in an answer can be critically important in solving a real problem, designing an experiment, or designing a crash helmet. The designer of a crash helmet needs an approximate value for the maximum force it must withstand. Actual collisions will vary, so there is no "right" or simple answer.

> **QUESTION** What simplifications did we make in our model of the colliding students?

Our model of the colliding student situation is good enough for many purposes. However, we left out some aspects of the actual motion. For example, we mostly ignored the flexible structure of the students, how much their shoes slip on the ground during the collision, and so on. We have also quite sensibly neglected the gravitational force of Mars on the students, because it is so tiny compared to the force of one student on the other.

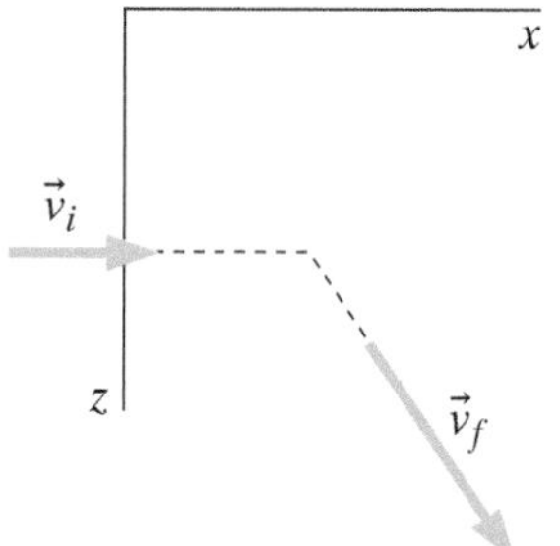

Figure 2.11 Top view of the path of a hockey puck that is hit by a hockey stick. (Assuming our usual coordinate system with y up, the ice surface lies in the xz plane.) The impact occurs at the location where the trajectory changes direction.

EXAMPLE **Striking a Hockey Puck**

In a video of a hockey game, we see a 0.16 kg hockey puck sliding across the ice, as shown in Figure 2.11. A player hits the puck sharply with a hockey stick; the stick breaks, and the direction of motion of the puck changes. From examining the video frame by frame, we find that before it was hit the velocity of the puck was $\langle 20,0,0 \rangle$ m/s, and after the impact the velocity of the puck was $\langle 20,0,31.7 \rangle$ m/s. When we pile weights on the side of a similar hockey stick we find that the stick breaks under a force of about 1000 N (this is roughly 225 lb). How long was the puck in contact with the hockey stick? Explain what assumptions were made in the solution.

Solution This problem can't be solved simply by plugging numbers into an equation. Instead, as will often be the case, we need to start from a fundamental principle, and work forward until the solution emerges. We will start by writing down the Momentum Principle, which relates change in momentum of a system to net impulse applied. If we can find both the change in momentum of the puck and the net force applied to it, we will be able to deduce the contact time Δt.

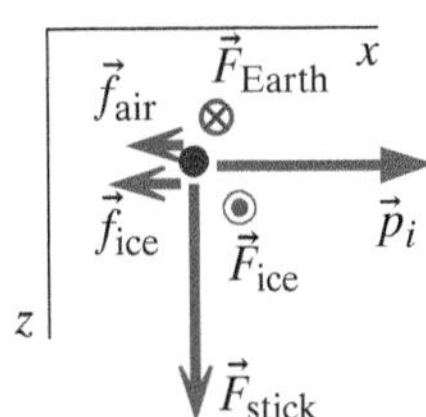

Figure 2.12 Free-body diagram for the hockey puck during the impact. The viewpoint is above the ice, looking down at the xz plane.

System: The hockey puck
Surroundings: Earth, ice, hockey stick, air (free-body diagram: Figure 2.12)
Initial time: When stick first makes contact with puck
Final time: When stick breaks

$$\vec{p}_f - \vec{p}_i = \vec{F}_{\text{net}}\,\Delta t$$

$$0.16\,\text{kg} \cdot \langle 20, 0, 31.7\rangle\ \text{m/s} - 0.16\,\text{kg} \cdot \langle 20, 0, 0\rangle\ \text{m/s} = \vec{F}_{\text{net}}\,\Delta t$$

$$\langle 0, 0, 5.072\rangle\ \text{m/s} = \vec{F}_{\text{net}}\,\Delta t$$

Only the z component of the puck's momentum changed, so the net impulse must have been in the z direction. (This implies that the downward gravitational force by the Earth and the upward force by the ice are equal in magnitude and opposite in direction. In Chapter 4 we will see how the solid ice can exert an upward force.) Assuming that the forces by the air and the ice are small compared to the force on the puck by the hockey stick, we can now simplify the free-body diagram, as shown in Figure 2.13.

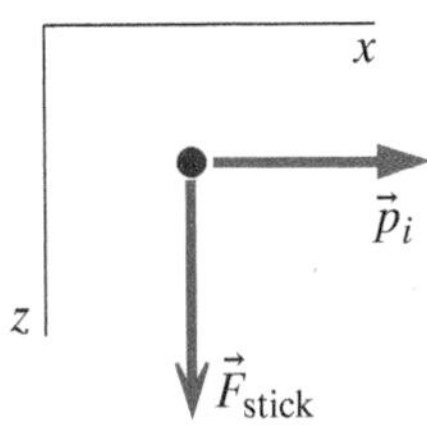

Figure 2.13 Simplified free-body diagram, taking into account assumptions and deductions. The viewpoint is above the ice, looking down at the xz plane.

The direction of the force is the unit vector $\hat{F}$:

$$\hat{F} = \langle 0, 0, 1\rangle$$

Because the stick broke, the magnitude of the force must have been about 1000 N. Therefore

$$\vec{F}_{\text{net}} = \left|\vec{F}\right|\hat{F} = \langle 0, 0, 1000\rangle\ \text{N}$$

From the Momentum Principle:

$$\langle 0, 0, 5.072\rangle\ \text{m/s} = \langle 0, 0, 1000\rangle\ \text{N}\ \Delta t$$

QUESTION Since we can't divide by a vector, how do we solve for Δt?

Recall that a vector equation can be written as three separate equations, one for each component. To solve for Δt we must consider the z component separately:

$$p_{fz} - p_{iz} = F_{\text{net},z}\,\Delta t$$

$$\Delta t = \frac{p_{fz} - p_{iz}}{F_{\text{net},z}} = \frac{5.072\ \text{kg}\cdot\text{m/s}}{1000\ \text{N}}$$

$$\Delta t = 0.005\ \text{s}$$

We assumed that:

- The force by the hockey stick on the puck was constant during the contact time.
- The friction force due to the ice surface was negligible during the contact time.
- Air resistance was negligible during the contact time.

QUESTION How good were our assumptions?

The neglect of sliding friction and air resistance is probably pretty good, since a hockey puck slides for long distances on ice with nearly constant speed.

We know the hockey stick exerts a maximum force of $F_{\text{stick}} = 1000$ N, because we observe that the stick breaks. We approximate the force as nearly constant during contact. Actually, this force grows quickly from zero at first contact to 1000 N, then abruptly drops to zero when the stick breaks.

Our analysis of the stick contact time (0.005 s) isn't exact. However, it is adequate to get a reasonably good determination of the short interaction time, something that we wouldn't know without using the Momentum Principle.

> QUESTION How can we decide if our result is reasonable? Could the actual contact time have been much longer?

The very short duration of the impact is consistent with what we hear: a sharp, short crack. For further verification, let's consider how far the puck would move during contacts of differing times.

During 0.005 s the puck moves (20 m/s)(0.005 s) = 0.1 m in the x direction, sliding along the blade of the stick. Also during this time v_z increases from 0 to 31.7 m/s, with an average value of about 15.8 m/s, so the z displacement is about (15.8 m/s)(0.005 s) = 0.08 m during contact. A more accurate sketch of the path of the puck might show a bend as in Figure 2.14.

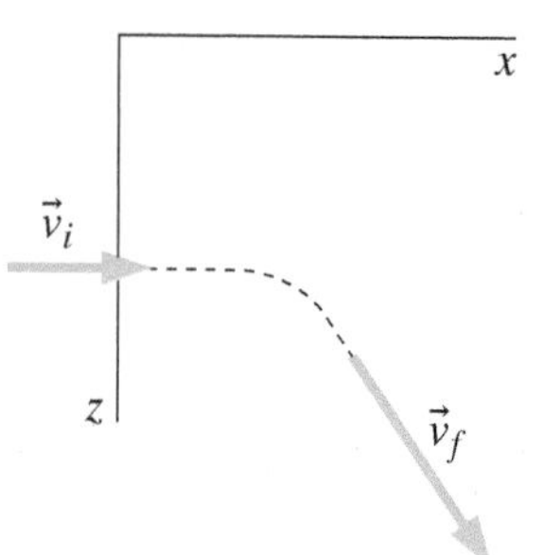

Figure 2.14 A more accurate overhead view of the path of the hockey puck, showing the bend during impact.

> QUESTION Suppose we were to guess that the contact time between the stick and the puck was about one second. We know that the original speed of the puck was 20 m/s. At this speed, how far would the puck have traveled in 1 s?

$$(20 \text{ m/s})(1 \text{ s}) = 20 \text{ m}$$

It is clear that the puck did not move 20 m during the impact. We must conclude that the impact took much less than 1 s.

> QUESTION If we guess that the puck may have slid 20 cm along the stick during the impact, what would the contact time be?

$$\Delta t \approx \frac{(0.20\,\text{m})}{(20 \text{ m/s})} = 0.01\,\text{s}$$

This estimate of the contact time differs from our result by only a factor of around 2, instead of a factor of 100, so it would be a much better estimate. Using an object's speed and estimating the distance traveled during an interval allows us to come up with a much better estimate of interaction times than we get simply by guessing.

Idealized Models

In the previous two examples, we made "idealized" models of real physical situations. By idealized models, we mean models that involve simple, clean, stripped-down situations, free of messy complexities (Figure 2.15). Ideally, a hockey puck will slide forever on a smooth ice surface, but a real puck sliding on real ice eventually comes to a stop. An ideal gas is a fictitious gas in which the molecules don't interact at all with each other. In contrast, the molecules of a real gas do interact, but only when they come close to each other. A model of a single Earth orbiting a Sun is an idealized model because it leaves out the effects of the other bodies in the Solar System.

The behavior of idealized models allows us to investigate simple patterns of motion, and learn what factors are important in determining these patterns. Once we understand these factors, we can revise and extend our models, including more interactions and complexities, to see what effects these have.

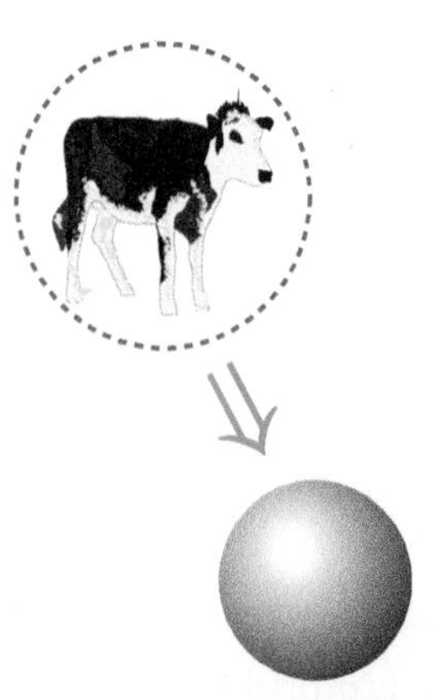

Figure 2.15 An old physics joke begins, "Consider a spherical cow ..." Sometimes this degree of idealization is actually appropriate.

An important aspect of physical modeling is that we make appropriate approximations and assumptions to simplify the messy, real-world situation enough to permit us to apply fundamental principles such as the Momentum Principle, the Energy Principle (see Chapter 6), and the Angular Momentum Principle (see Chapter 11).

Checkpoint 2 (a) In the colliding students example, how was Δt, the duration of the collision, estimated? **(b)** In the hockey puck example, why were only the z components of $\Delta\vec{p}$ and $\vec{F}_{\text{net}}$ used to get Δt?

2.3 PREDICTING THE FUTURE

Since $\Delta\vec{p} = \vec{p}_f - \vec{p}_i$ ("final minus initial"), we can rearrange the Momentum Principle algebraically to the update form:

THE MOMENTUM PRINCIPLE (UPDATE FORM)

$$\vec{p}_f = \vec{p}_i + \vec{F}_{\text{net}}\,\Delta t$$

or, choosing "now" as the initial time, and "a short time in the future" as the final time, we can write:

$$\vec{p}_{\text{future}} = \vec{p}_{\text{now}} + \vec{F}_{\text{net,now}}\,\Delta t$$

The time interval Δt must be small enough that the net force is nearly constant during this time interval.

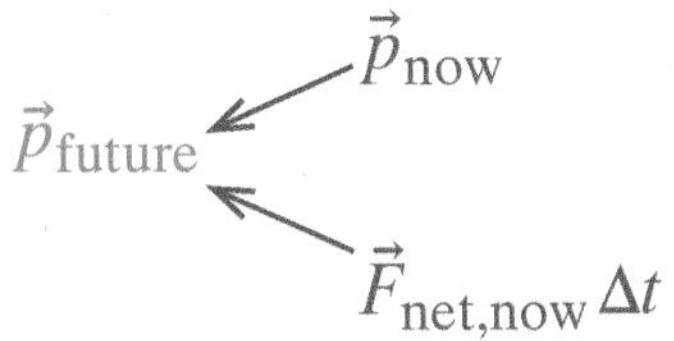

Figure 2.16 The future momentum of a system depends on two things: its momentum now and the net force acting on it now.

This way of writing the Momentum Principle emphasizes the important idea that the momentum that an object will have a short time in the future depends on *two* things: its momentum right now and the total force acting on the object right now. Figure 2.16 summarizes this graphically.

Even without numerical values, we can use the update form of the Momentum Principle to reason qualitatively, but rigorously, about physical situations.

A Skydiver

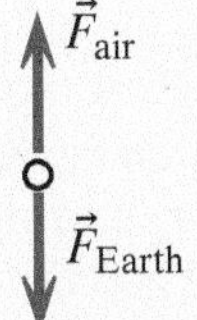

Figure 2.17 At a particular instant the upward force of the air becomes equal in magnitude to the downward force of the Earth on a skydiver. At this instant the net force on the skydiver becomes zero.

A skydiver jumps out of an airplane and her parachute opens. For a while she falls faster and faster. Then at a particular instant, which we will choose to be *now*, the upward air resistance force acting on her and her parachute (this force increases as her speed increases) gets large enough to be exactly equal in magnitude and opposite in direction to the downward gravitational force due to the Earth (Figure 2.17). Therefore, at this instant the net force on the skydiver becomes zero. Which of the following answers correctly predicts the skydiver's motion a short time in the future from now? **(a)** She will float without moving up or down. **(b)** She will fall faster and faster. **(c)** She will fall at a constant speed.

Solution Even though this problem does not involve numbers, we must use the Momentum Principle to reason formally about the future momentum of the skydiver. This principle tells us that $\vec{p}_{\text{future}}$ depends on two things: her current momentum $\vec{p}_{\text{now}}$, and the net impulse $\vec{F}_{\text{net,now}}\Delta t$ acting on her during the next short time interval, as indicated in Figure 2.18. The net impulse will be zero because the net force on her is zero. Therefore, her momentum a short time in the future will be exactly equal to her momentum right now. Because her momentum does not change, her speed does not change, and she will fall at constant speed. Answer (c) is correct.

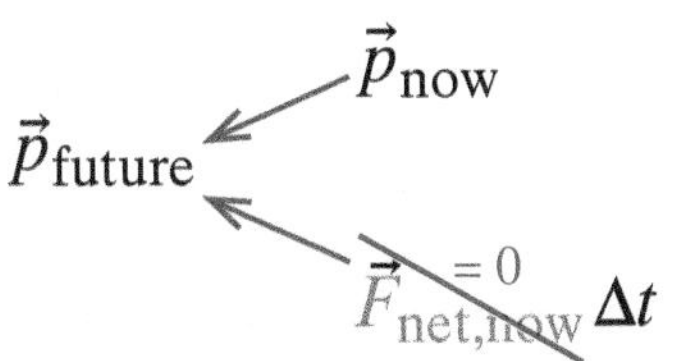

Figure 2.18 At the moment of interest, the momentum of the skydiver now is the only nonzero contribution to the future momentum of the skydiver.

Almost everyone who is learning physics occasionally makes mistakes when thinking about problems like this one. Many of these mistakes are due to forgetting about one of the two things that determine an object's future momentum. For example, if you picked answer (a) above, perhaps you considered only the net force acting on the skydiver, and forgot to take her current momentum into account. You can avoid many mistakes by starting from the Momentum Principle, and by making sure you have taken both current momentum and net impulse into account.

How Do We Take the Past into Account?

The momentum of an object at any particular instant depends on its history—the interactions it underwent in the past. However, the Momentum Principle refers only to the momentum of an object *now* and in the *future*—it does not mention the past.

For example, suppose you hit a ball with a tennis racket. At a time one-tenth of a second after the ball has left the racket, it is no longer interacting with the racket, so at that moment the net force on the ball includes forces by the Earth and by the air, but not by the racket. This may seem counterintuitive: the fact that the racket interacted with the ball in the past is important!

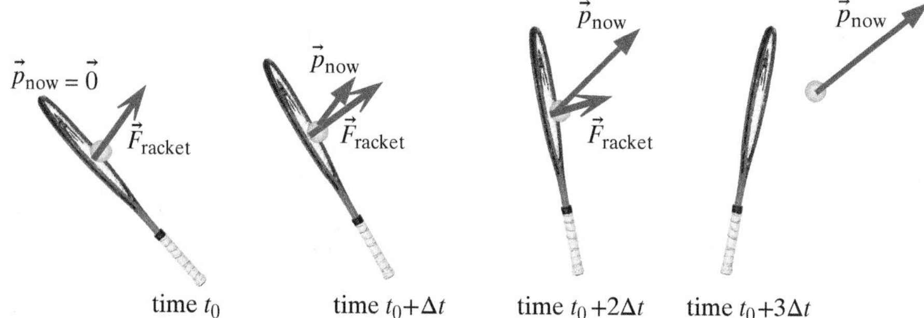

Figure 2.19 Several steps in the process of hitting a tennis ball. At each time, the momentum of the ball $\vec{p}_{now}$ includes the effects of all prior impulses delivered by the racket. Only the force by the racket is shown; forces on the ball by the Earth and the air are not shown.

If we reflect for a moment, though, we see that the effects of all past interactions have already been added up into the object's momentum at the current instant, $\vec{p}_{now}$. Imagine what happened a few seconds ago (Figure 2.19). Suppose that you tossed the ball straight up in the air, and that your racket first made contact with the ball at time t_0 when the ball had reached the top of its trajectory and was momentarily motionless. At that time (the first image in Figure 2.19) the ball's momentum was zero, but because at that instant the racket exerted a force on the ball, after a fraction of a second at time $t_0 + \Delta t$ (the second image in Figure 2.19) the ball's momentum was zero plus the net impulse—a nonzero quantity. Considering the next time step, we see that the ball's initial momentum was now not zero, and it increased even more because of the impulse delivered by your racket in the next small time step. This continued, step by step, until the ball lost contact with your racket. At that instant, the momentum of the ball was the sum of all the impulses delivered by the racket during the time it was in contact with the ball, and the racket no longer exerted a force on the ball.

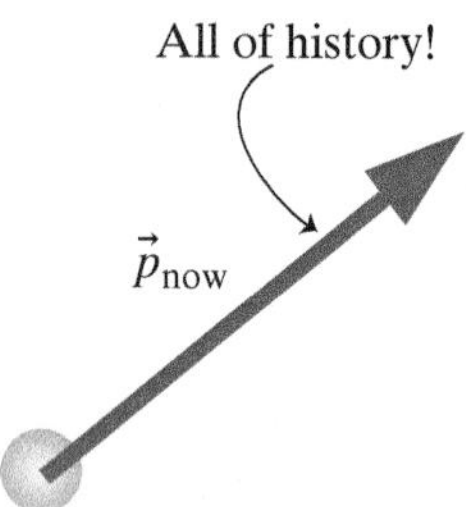

Figure 2.20 The entire history of an object—all impulses that have ever been applied to it—has been added up into the current momentum of the object.

The fact that all of the past history of an object and its interactions has been incorporated into its current momentum, $\vec{p}_{now}$, is extremely important (Figure 2.20). Because this is true, we are free to pick any instant of time as the initial time of interest, and consider only the interactions of the object with its surroundings at this instant, in order to predict its future momentum. As long as we know an object's current momentum, we do not need to know anything about how the object came to be moving this way. If we had to know all of the details of all of an object's past interactions, predicting the future motion of an object would almost never be possible!

In the following section we will work through a process like this quantitatively, dividing a time interval into very small steps and adding up all the impulses exerted on an object, to predict its motion over an extended time period.

Checkpoint 3 (1) You drop a piece of paper, and observe that it eventually falls at a constant speed. Which of the following statements about this situation is based on a fundamental physics principle? **(a)** Because the paper is moving downward, we know that it experiences a nonzero net downward force. **(b)** Since the momentum of the paper does not change from one instant to the next, $\vec{p}_{\text{future}} = \vec{p}_{\text{now}}$, and therefore the net force on the paper must be zero. (2) You give a push to a toy car, which rolls away smoothly on a wooden floor. Why does the car keep moving after your hand is no longer touching it? **(a)** The momentum of the car just after it leaves your hand reflects the total impulse given to the car by your hand. **(b)** Your hand continues to exert a force on the car even after the car and hand no longer touch.

2.4 ITERATIVE PREDICTION: CONSTANT NET FORCE

The Momentum Principle states that if the momentum of a system changes, a nonzero net force must have been applied to the system. The simplest case of a nonzero net force is that of a constant force: a force whose magnitude and direction do not change even though the position and momentum of the system change. In the real world, there are very few situations in which the net force on a system is truly constant. However, this can be a useful approximate model for some real phenomena. One example of an approximately constant force is the gravitational force exerted by the Earth on an object near the Earth's surface.

Approximate Gravitational Force

In general, the gravitational interaction between two objects depends on the distance between the objects, and its magnitude and direction vary as the positions of the objects vary. In Chapter 3 we will discuss the general form of the force law that describes the gravitational attraction that all objects in the Universe have for each other. However, anywhere near the Earth's surface it is approximately true that a 1 kg object is attracted to the Earth by a gravitational force of magnitude 9.8 N, and a 2 kg object is attracted by a force of magnitude 2×9.8 N (Figure 2.21). We define $g = +9.8\,\text{N/kg}$ to be the strength of this interaction in newtons per kilogram, so we can say that a mass m near the surface of the Earth is pulled toward the Earth by a force whose magnitude is mg. We can treat this force as approximately constant in magnitude and direction, even if the object moves.

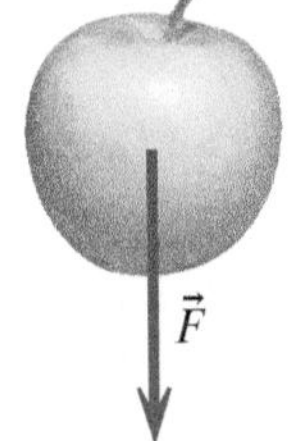

Figure 2.21 The magnitude of the gravitational force by the Earth on a small apple is about 1 newton.

APPROXIMATE GRAVITATIONAL FORCE

$$|\vec{F}_{\text{grav}}| \approx mg$$

where $g = +9.8\,\text{N/kg}$

for an object of mass m near the surface of the Earth.

Iterative Prediction of Motion

The most general method for predicting the motion of an object that interacts with its surroundings is to use an iterative approach. To "iterate" means "to repeat" or "to perform again"; the word "iteration" implies repeating the same process many times. The basic idea we will use in predicting the motion of interacting systems is to divide up the time interval of interest into many very small time intervals, and iteratively apply the Momentum Principle over successive intervals. The time intervals must be small enough that neither the velocity of the object nor the net force acting on it changes significantly in one time step. We will update momentum and position iteratively like this:

ITERATIVE PREDICTION OF MOTION

Repeat:

- Calculate the net force $\vec{F}_{net,now}$ acting on the system.
- Update momentum: $\vec{p}_{future} = \vec{p}_{now} + \vec{F}_{net,now}\Delta t$.
- Update position: $\vec{r}_{future} = \vec{r}_{now} + \vec{v}_{avg}\,\Delta t$.

In the final step we use the approximation $\vec{v}_{avg} \approx \vec{p}_{future}/m$.

This general procedure can be used to predict the entire path along which an object will travel. As we will see in Section 2.6, it can be used even if the net force is not constant. The order of the steps turns out to be important; to get good accuracy it is necessary to update the momentum of the system before updating its position.

EXAMPLE **A Ball Without Air Resistance**

Let's apply this procedure to calculate the trajectory of a ball moving without air resistance. It is necessary to neglect air resistance in order to assume a constant force, because the air resistance force depends on the velocity of the object. At low speeds, this can be a reasonable approximation.

Solution In outline, here is the calculation we will do. We need two initial values:

- The initial position of the ball
- The initial momentum of the ball

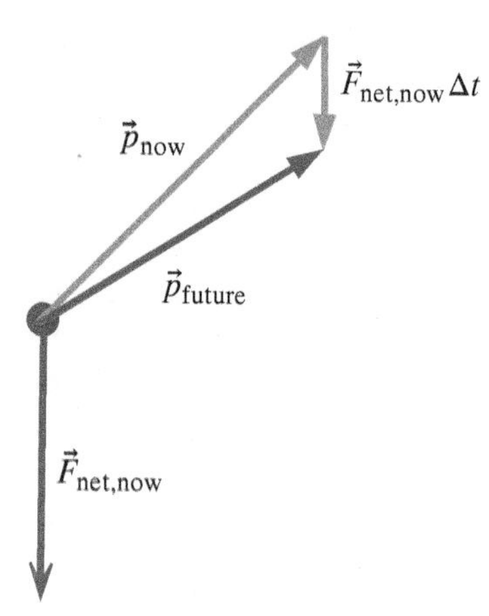

Figure 2.22 Time step 1 in the calculation of the motion of a ball without air resistance.

We will do the following calculation iteratively (repeatedly):

- Find the net force $\vec{F}_{net,now}$ on the ball.
- Apply the Momentum Principle to find $\vec{p}_{future}$.
- Use $(\vec{p}_{future}/m)$ to approximate the average velocity, and use this to find the new position of the ball.

We'll consider a tennis ball of mass 0.06 kg that is hit by a racket so its initial velocity just after losing contact with the racket is $\langle 10,10,0\rangle$ m/s, so its initial momentum is $\langle 0.6,0.6,0\rangle$ kg · m/s. The initial position of the ball is $\langle 0,2,0\rangle$ m. We'll use a time step of 0.4 s. In the absence of air resistance, the net force on the ball is just the downward gravitational force exerted by the Earth. The result of the first step in our calculation is shown graphically in Figure 2.22, and the actual calculations are shown below.

Time Step 1:

$$\begin{aligned}\vec{F}_{net,now} &= \langle 0,-9.8\times 0.06,0\rangle \text{ N/kg} = \langle 0,-0.588,0\rangle \text{ N}\\ \vec{p}_{future} &= \langle 0.6,0.6,0\rangle \text{ kg}\cdot\text{m/s} + \langle 0,-0.588,0\rangle \text{ N}\times 0.4\text{ s}\\ &= \langle 0.6,0.365,0\rangle \text{ kg}\cdot\text{m/s}\\ \vec{r}_{future} &= \langle 0,2,0\rangle \text{ m} + \left(\frac{\langle 0.6,0.365,0\rangle \text{ kg}\cdot\text{m/s}}{0.06\text{ kg}}\right)\times 0.4\text{ s}\\ &= \langle 4,4.32,0\rangle \text{ m}\end{aligned}$$

QUESTION The y component of the ball's momentum changed, but the x and z components did not. Why not?

The y component of the net force is nonzero, but the other components are zero. The y component of a force can only affect the y component of an object's momentum. Therefore the x and z components of the ball's momentum (and hence the x and z components of its velocity) are not changed by the gravitational force.

> **QUESTION** Why did we use $\vec{p}_{\text{future}}$ to obtain a value for the average velocity of the ball, instead of an average of $\vec{p}_{\text{future}}$ and $\vec{p}_{\text{now}}$?

We will see later that although in the special case of a constant force an average of $\vec{p}_{\text{future}}$ and $\vec{p}_{\text{now}}$ would provide a more accurate estimate of $\vec{v}_{\text{avg}}$, in the general case of a changing force this is often not true. As long as we take small enough time steps, so that the changes in momentum are small, $\vec{p}_{\text{future}}/m$ provides an adequate estimate of $\vec{v}_{\text{avg}}$.

> **QUESTION** What value should we use for $\vec{p}_{\text{now}}$ in the second time step?

To take the next time step, we advance the "clock." If our initial time was t_0, the time is now $t_0+\Delta t$. The value of $\vec{p}_{\text{now}}$ at this time is the value we previously called $\vec{p}_{\text{future}}$: $\langle 0.6, 0.365, 0\rangle$ kg·m/s, and the value of $\vec{r}_{\text{now}}$ at this time is the value we previously called $\vec{r}_{\text{future}}$: $\langle 4, 4.32, 0\rangle$ m.

Time Step 2:

$$\begin{aligned}
\vec{F}_{\text{net,now}} &= \langle 0, -9.8\times 0.06, 0\rangle \text{ N/kg} = \langle 0, -0.588, 0\rangle \text{ N}\\
\vec{p}_{\text{future}} &= \langle 0.6, 0.365, 0\rangle \text{ kg}\cdot\text{m/s} + \langle 0, -0.588, 0\rangle \text{ N}\times 0.4\text{ s}\\
&= \langle 0.6, 0.130, 0\rangle \text{ kg}\cdot\text{m/s}\\
\vec{r}_{\text{future}} &= \langle 4, 4.32, 0\rangle \text{ m} + \left(\frac{\langle 0.6, 0.365, 0\rangle \text{ kg}\cdot\text{m/s}}{0.06\text{ kg}}\right)\times 0.4\text{ s}\\
&= \langle 8, 5.30, 0\rangle \text{ m}
\end{aligned}$$

Figure 2.23 displays the results of four steps in the calculation of the path of the ball. Even with only four steps the trajectory looks roughly like what is observed in the real world. If we used smaller time steps, the straight line segments representing displacement would be shorter, and the trajectory would look more and more like the actual curving path of the ball.

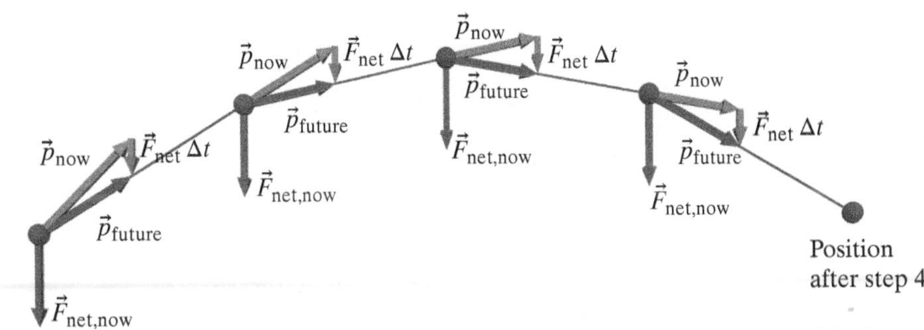

Figure 2.23 Four steps in the iterative prediction of the path of a ball in the absence of air resistance. In each step the future momentum of the ball is calculated, then this new momentum is used to update the position of the ball.

Checkpoint 4 (1) For the third time step in the iterative solution above, what value should you use for $\vec{p}_{\text{now}}$? **(a)** $\langle 0.6, 0.6, 0\rangle$ kg·m/s **(b)** $\langle 0.6, 0.365, 0\rangle$ kg·m/s **(c)** $\langle 0.6, 0.130, 0\rangle$ kg·m/s **(d)** A value that has not yet been calculated. (2) What value should you use for $\vec{r}_{\text{now}}$? **(a)** $\langle 0, 2, 0\rangle$ m **(b)** $\langle 4, 4.32, 0\rangle$ m **(c)** $\langle 8, 5.30, 0\rangle$ m **(d)** A value that has not yet been calculated.

2.5 ANALYTICAL PREDICTION: CONSTANT NET FORCE

In contrast to an iterative solution, an analytical solution to a problem yields a mathematical function that gives the value of a quantity of interest as a function of time. If such a function exists, one can use it to find the position or velocity of a system at a given time in one step, instead of taking many small steps to get this information. It is not always possible to get an analytical solution, so iterative methods are of more general applicability. One case in which an iterative solution is possible is the case of constant net force. We will derive this analytical solution to illustrate the approach and the nature of the results.

Approximations for Average Velocity

In the previous iterative solution for the motion of a tennis ball, we used the approximation

$$\frac{\vec{p}_{\text{future}}}{m} \approx \vec{v}_{\text{future}} \approx \vec{v}_{\text{avg}}$$

for the average velocity over each time interval Δt.

> **QUESTION** Why didn't we calculate the average velocity directly from the definition?

We can't calculate the average velocity directly from the definition

$$\vec{v}_{\text{avg}} \equiv \frac{\Delta\vec{r}}{\Delta t} = \frac{\vec{r}_f - \vec{r}_i}{\Delta t}$$

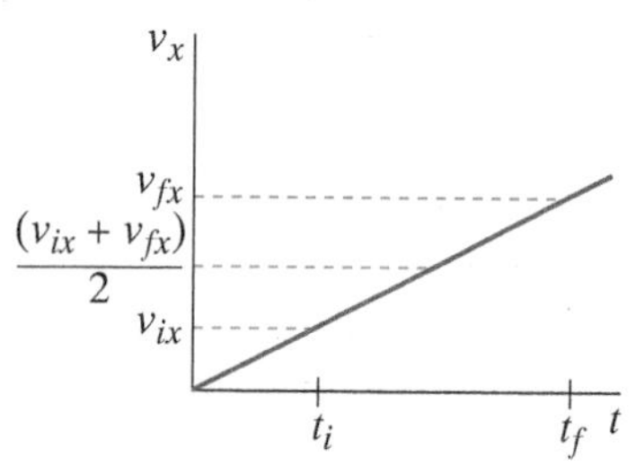

Figure 2.24 v_x is changing at a constant rate (linearly with time), so the arithmetic average is equal to $v_{\text{avg},x}$.

because the final location $\vec{r}_f$ of the moving ball is unknown. Since we do not have any other information that would allow us to determine the average velocity, we need to find an approximate value for the average velocity in this situation. There are two possible approaches:

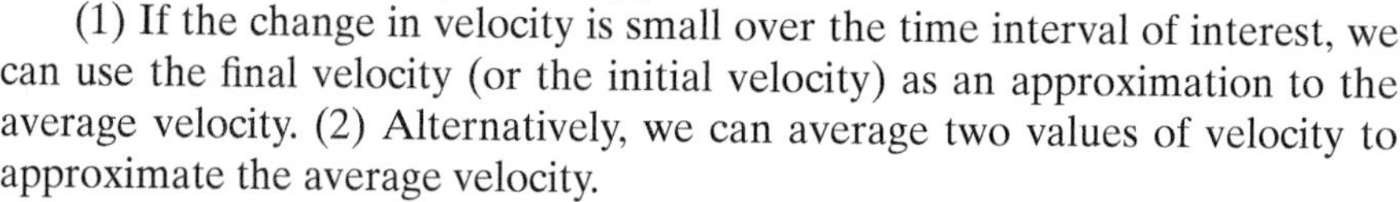

(1) If the change in velocity is small over the time interval of interest, we can use the final velocity (or the initial velocity) as an approximation to the average velocity. (2) Alternatively, we can average two values of velocity to approximate the average velocity.

> **QUESTION** What might be a better approximation for $\vec{v}_{\text{avg}}$ if the net force on a system is constant?

Let's consider using the arithmetic average:

$$\vec{v}_{\text{avg}} \approx \frac{\vec{v}_{\text{future}} + \vec{v}_{\text{now}}}{2}$$

The arithmetic average lies between the two values. For example, the arithmetic average of 6 and 8 is $(6+8)/2 = 14/2 = 7$, halfway between 6 and 8. The arithmetic average is often a good approximation, but it is not necessarily equal to the true average $\vec{v}_{\text{avg}} = \Delta\vec{r}/\Delta t$. The arithmetic average does not give the true average unless each component of $\vec{v}$ is changing at a constant rate, as in Figure 2.24, which is the case only if the net force is constant. The detailed proof that $\vec{v}_{\text{avg}} = (\vec{v}_i + \vec{v}_f)/2$ when $\vec{v}$ changes at a constant rate is more complicated than one might expect, and is given in optional Section 2.8.

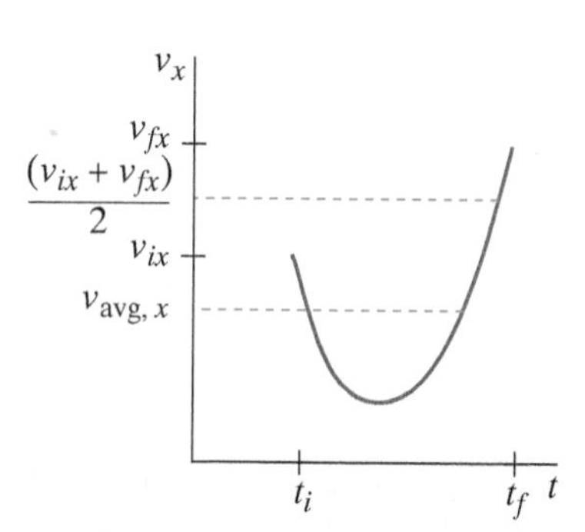

Figure 2.25 When v_x does not change linearly with time, the arithmetic average may be significantly different from the true average value of $v_{\text{avg},x}$.

> **QUESTION** When is the arithmetic average a poor approximation for $\vec{v}_{\text{avg}}$?

The arithmetic average is a poor approximation for $\vec{v}_{\text{avg}}$ if the rate of change of velocity (the acceleration $\vec{a} = d\vec{v}/dt$) is not constant. Figure 2.25 shows an example of such a situation. At low speeds ($v \ll c$), the acceleration is directly proportional to the rate of change of momentum because $\vec{F}_{\text{net}} = \Delta\vec{p}/\Delta t \approx m(d\vec{v}/dt)$, so we can conclude that $d\vec{v}/dt$ will be constant if the net force on the system is constant.

EXAMPLE

The Arithmetic Average Can Be a Poor Approximation

If you drive 50 mi/h in the $+x$ direction for four hours, and then 20 mi/h in the $+x$ direction for an hour, how far have you traveled (Figure 2.26)?

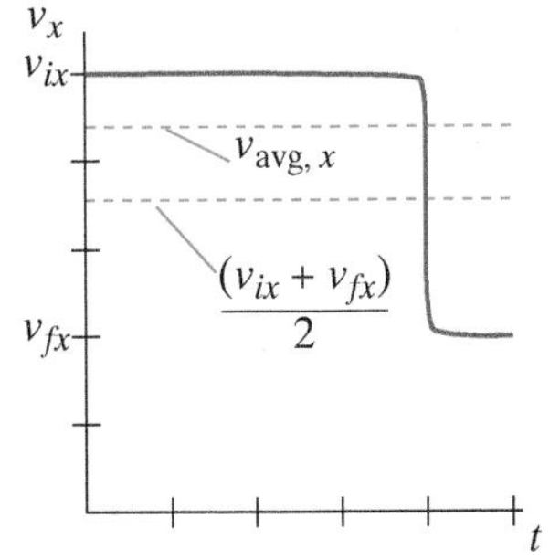

Solution

If we use v_x at the beginning and at the end of this 5 h time interval to determine the arithmetic average velocity, you would calculate that you have traveled 175 mi:

$$v_{\text{avg},x} \approx \frac{50+20}{2}\text{ mi/h} = 35\text{ mi/h}$$

$$x_f - x_i \approx (35\text{ mi/h})(5\text{ h}) = 175\text{ mi}$$

However, you have actually traveled significantly farther:

$$x_f - x_i = (50\text{ mi/h})(4\text{ h}) + (20\text{ mi/h})(1\text{ h}) = 220\text{ mi}$$

The x component of your actual average velocity was:

$$v_{\text{avg},x} = \frac{220\text{ mi}}{5\text{ h}} = 44\text{ mi/h}$$

v_x, v_{ix}, $v_{\text{avg},x}$, $\frac{(v_{ix}+v_{fx})}{2}$, v_{fx}, t

Figure 2.26 Because v_x of the car does not change linearly with time, $v_{\text{avg},x}$ is significantly different from the arithmetic average of $v_{\text{i,x}} + v_{\text{f,x}}$.

APPROXIMATE AVERAGE VELOCITY

$$\vec{v}_{\text{avg}} \approx \frac{\vec{v}_{\text{future}} + \vec{v}_{\text{now}}}{2}$$

This is exactly correct if all components of $\vec{v}$ change linearly with time (acceleration is constant). However, it may be a poor approximation if the net force is not constant.

Analytical Solution for Constant Net Force

The term *analytical* describes a solution that can be written symbolically (algebraically). In many cases it is not possible to find an analytical solution for the motion of a system. However, if the net force on a system is constant, we can easily find an analytical solution of the form $\vec{r}(t) = \ldots$.

Consider a system on which there is a constant net force in the $+x$ direction during the time Δt. For ordinary speeds $\gamma \approx 1$, so $v_x \approx p_x/m$. Using (a) the Momentum Principle in its update form, (b) the position update equation, and (c) the special result for average velocity in the case of constant net force, we can easily derive equations for the position and velocity as a function of time.

For convenience we'll start measuring the time t from 0, and write t to represent Δt:

$$p_{xf} = p_{xi} + F_{\text{net},x}t$$

$$v_{xf} = v_{xi} + \frac{F_{\text{net},x}}{m}t$$

$$v_{\text{avg},x} = \frac{\left(v_{xi} + \frac{F_{\text{net},x}}{m}t\right) + v_{xi}}{2}$$

$$v_{\text{avg},x} = v_{xi} + \frac{1}{2}\frac{F_{\text{net},x}}{m}t$$

$$x_f = x_i + v_{\text{avg},x}t$$
$$x_f = x_i + v_{xi}t + \frac{1}{2}\frac{F_{\text{net},x}}{m}t^2$$

To emphasize that we now have an *analytical solution,* we will write equations that express position and velocity as a function of time, given that the net force is constant:

ANALYTICAL SOLUTION: CONSTANT NET FORCE

$$x(t) = x_i + v_{xi}t + \frac{1}{2}\frac{F_{\text{net},x}}{m}t^2$$
$$v_x(t) = v_{xi} + \frac{F_{\text{net},x}}{m}t$$

If the net force had acted in the y or z direction the solution would be the same, except for the subscripts indicating direction. The time t can be arbitrarily large as long as the net force on the system is constant during the time interval 0 to t.

The time derivative of v_x is the x component of acceleration, which is constant since $F_{\text{net},x}$ is constant:

$$\frac{dv_x}{dt} = \frac{F_{\text{net},x}}{m} = a_x$$

EXAMPLE

Constant Net Force on a Fan Cart

Figure 2.27 A fan cart on a track. (Ruth Chabay/Bruce Sherwood)

An easy way to apply a nearly constant force is to mount a battery-powered fan on a cart (Figure 2.27). If the fan directs air backwards, the interaction with the air pushes the cart forward with a nearly constant force. (Airboats used in the very shallow Florida Everglades are built in a similar way, with large fans on top of the boats propelling them through the swamp.) Suppose that you have a fan cart whose mass is 0.5 kg. You turn on the fan, then give the cart a shove so that when it is at location $\langle 0.2, 0, 0\rangle$ m it is traveling with velocity $\langle 0.3, 0, 0\rangle$ m/s. At a time 0.9 seconds later, you observe that the cart is at location $\langle 0.146, 0, 0\rangle$ m. What was the force exerted by the air on the fan cart?

Solution

Since both net force and motion are parallel to the track, we only need to consider the x components of position, velocity, and force. Using the equations derived above for the motion of a system that experiences a constant net force:

$$x_f = x_i + v_{xi}t + \frac{1}{2}\frac{F_{\text{net},x}}{m}t^2$$
$$0.0.146\text{ m} = 0.2\text{ m} + (0.3\text{ m/s})(0.9\text{ s}) + \frac{F_{\text{net},x}}{2(0.5\text{ kg})}(0.9\text{ s})^2$$
$$F_{\text{net},x} = -0.4\text{ N}$$

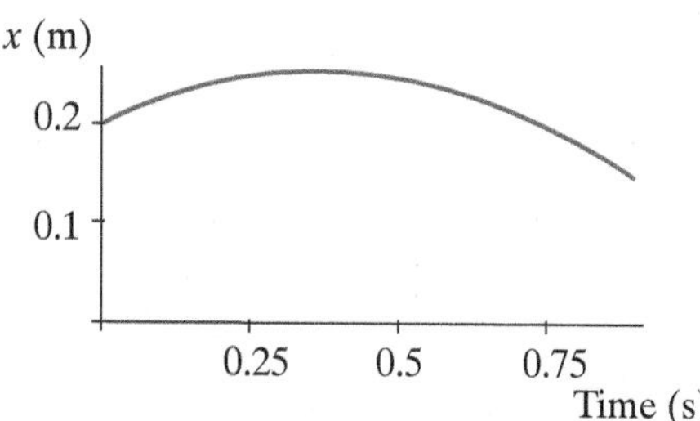

Figure 2.28 A plot of x vs. t for the fan cart. Note that $v_x(t)$ is the derivative of this function. Where the slope is zero, v_x is zero; this occurs at the time when v_x changes sign, from positive to negative.

We see that the net force was in the $-x$ direction, opposite to the initial direction of motion of the cart. We conclude that the cart initially moved in the $+x$ direction, slowing down, coming to a stop, then moving at increasing speed in the $-x$ direction. A graph of x vs. t for the cart is shown in Figure 2.28. (We used a computer to calculate all the (x,t) values needed for this graph.)

EXAMPLE **Kicking a Ball on the Moon**

A soccer ball, whose mass is 0.45 kg, is initially at rest on the surface of the Moon, at location $\langle 0,0,0\rangle$ m. On the surface of the Moon, the constant $g \approx 1.6$N/kg. An astronaut kicks the ball, giving it an initial velocity of $\langle 15,15,0\rangle$ m/s; this speed of $\sqrt{15^2+15^2} = 21$ m/s can be compared to the top speed of about 35 m/s achieved by professional soccer players. What are the coordinates of the highest position the ball reaches? How long is the ball in the air before hitting the ground? How far does the ball travel before hitting the ground?

Solution It is not immediately obvious how to find the maximum height of the ball, because this is not a quantity that appears in the equations we derived above. If we think about the ball's motion, we note that at the top of the ball's trajectory, v_y is instantaneously zero. This is a useful observation because we can solve for the time at which this occurs, and then we can calculate the x and y coordinates of the ball at that time. Because the Moon is airless, there is truly no air resistance, and the net force on the ball is the downward gravitational force $\langle 0,-mg,0\rangle$, which does not affect the x component of the ball's momentum. Again setting the initial time $t_0 = 0$, and substituting t for Δt, we get:

$$v_{yf} = v_{yi} + \frac{(-mg)}{m}t$$

$$0 = 15 \text{ m/s} + (-1.6\text{N/kg})t$$

$$t = 9.375 \text{ s}$$

$$y_f = y_i + v_{yi}t + \frac{1}{2}\frac{(-mg)}{m}t^2$$

$$y_f = 0 + (15 \text{ m/s})(9.375 \text{ s}) + (0.5)(-1.6\text{N/kg})(9.375 \text{ s})^2$$

$$y_f = 70.3 \text{ m}$$

$$x_f = x_i + v_{xi}t$$

$$x_f = 0 + (15 \text{ m/s})(9.375 \text{ s})$$

$$x_f = 140.6 \text{ m}$$

The coordinates of the highest position of the ball are $\langle 140.6,70.3,0\rangle$ m, assuming we choose our coordinate axes so that the motion is in the $z = 0$ plane. Note that the maximum height of 70 m is about the height of a 20-story building.

Next we find the time of flight and the distance the ball goes. One way to find the flight time is to note that on level ground $y_f = 0$ when the ball hits:

$$y_f = y_i + v_{yi}t + \frac{1}{2}\frac{(-mg)}{m}t^2$$

$$0 = 0 + (15 \text{ m/s})t + (0.5)(-1.6\text{N/kg})t^2$$

$$t = 18.75 \text{ s}$$

$$x_f = x_i + v_{xi}t$$

$$x_f = 0 + (15 \text{ m/s})(18.75 \text{ s})$$

$$x_f = 281.3 \text{ m}$$

On Earth, where $g = 9.8$ N/kg, even if there were no atmosphere this soccer ball would go only about 46 m, less than half the length of a soccer field. The "hang time" of 18.75 s is very long; on Earth (in the absence of air resistance) it would be only 3 s.

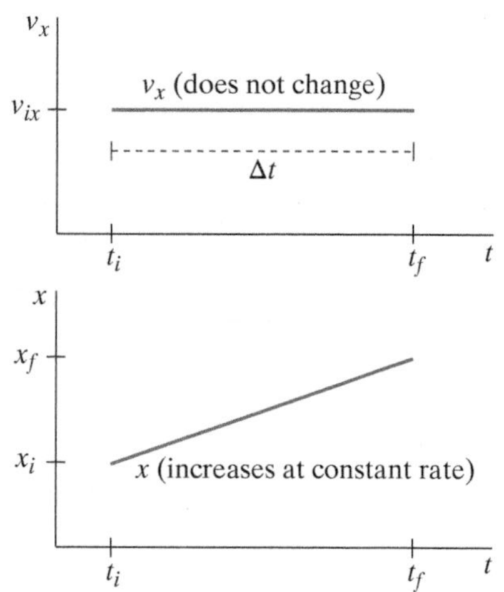

Figure 2.29 Motion graphs for the kicked ball. Top: v_x vs. t; bottom: x vs. t. v_x does not change because the net force acted only in the y direction.

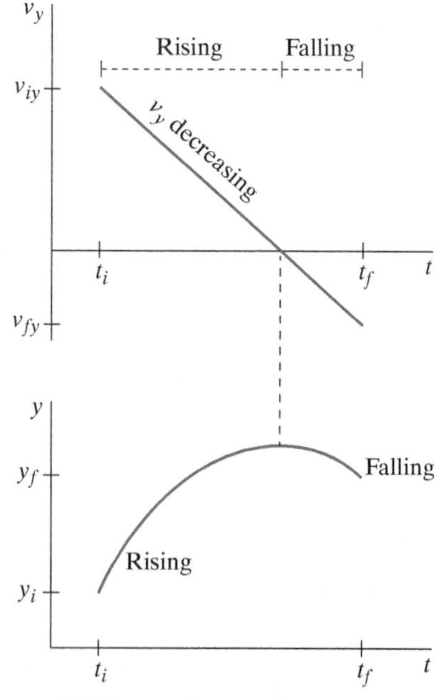

Figure 2.30 Motion graphs for the kicked ball. Top: v_y vs. t; bottom: y vs. t.

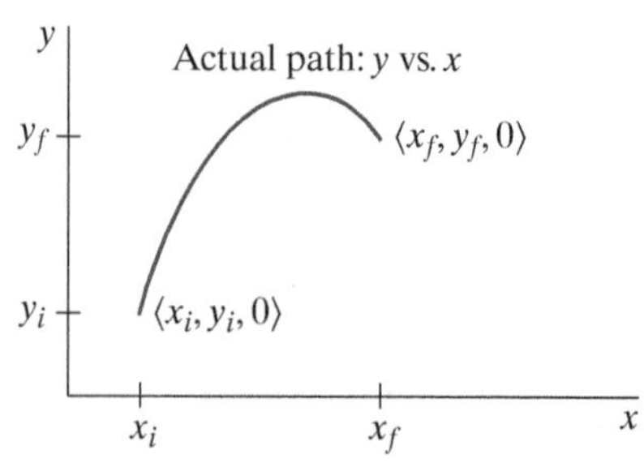

Figure 2.31 The actual trajectory of the thrown ball, with negligible air resistance (y vs. x).

The approach we took above is equivalent to a more formal calculus-based approach. We could differentiate the equation giving y as a function of t and find its maximum by setting $dy/dt = 0$. Either approach will yield a correct answer.

Graphs of Motion

Graphs can be a compact way of representing particular aspects of motion. It is difficult to graph all components of a system's position, velocity, or momentum at once, so typically we make one graph for each component. For example, Figures 2.29 and 2.30 show graphs of position and velocity components vs. time for the soccer ball in the preceding example.

QUESTION In Figure 2.29, why is the graph of v_x vs. t a horizontal line?

Since the x component of the net force on the ball is zero, v_x doesn't change, and the graph of v_x vs. t is simply a horizontal line with a slope of zero. (The slope of v_x vs. t is the x component of the acceleration: $a_x = dv_x/dt$. In this case $a_x = 0$.)

QUESTION How is the graph of v_x vs. time related to the graph of x vs. time?

The graph of x vs. time is a straight line, rising because the ball is moving in the $+x$ direction. The slope of the x vs. t graph is equal to v_x, since v_x is the rate of change of x ($v_x = dx/dt$). Similarly, the slope of the y vs. t graph at any time is equal to v_y at that time.

QUESTION What is the significance of the point at which $v_y = 0$ in Figure 2.30?

When v_y is equal to zero, the slope of y vs. t is momentarily zero, and y is a maximum. At this instant the ball is neither rising nor falling. Before that point the slope is positive, corresponding to $v_y > 0$, and after that point the slope is negative, corresponding to $v_y < 0$. In Figure 2.30 the graph of v_y has a constant negative slope (the top graph), because the y component of the force has the constant value $-mg$, which makes the y component of momentum decrease at a constant rate. In the next section we will see that the equation for y is a quadratic function in the time, which is consistent with the parabolic form of the graph of y vs. time t.

QUESTION How would a plot of p_y vs. time compare to the plot of v_y vs. time (Figure 2.30)?

Since at speeds that are low compared to the speed of light $\vec{p} \approx m\vec{v}$, $p_y \approx mv_y$. A graph of p_y vs. time would look very similar to the graph of v_y vs. time, but the units of the y axis would be different.

The actual path of the ball, the graph of y vs. x (Figure 2.31), is also an inverted parabola. Since x increases linearly with t, whether we plot y vs. t or y vs. x we'll see a similar curve. The units of the horizontal axis are different, of course (meters instead of seconds).

Checkpoint 5 A ball is kicked on Earth from a location $\langle 9, 0, -5\rangle$ m (on the ground) with initial velocity $\langle -10, 13, -5\rangle$ m/s. Neglecting air resistance: **(a)** What is the velocity of the ball 0.6 s after being kicked? **(b)** What is the location of the ball 0.6 s after being kicked? **(c)** What is the maximum height reached by the ball?

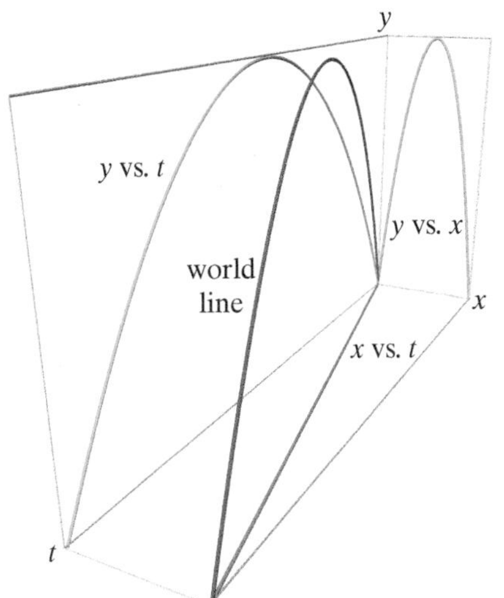

Figure 2.32 A "world line" and its projections.

Projections of a "World Line"

Figure 2.32 illustrates an interesting way to think of the graphs of y vs. x, y vs. t, and x vs. t. The red curve in Figure 2.32 is a graphical description of the motion in x, y, and t of a ball that moves in the xy plane. Imagine shining a light on the red curve to project a shadow onto the xy plane, or the yt plane, or the xt plane. These "projections" are shown in cyan in the figure. Compare these projections with the individual graphs shown in Figures 2.29, 2.30, and 2.31.

The red curve, a function of x, y, and t, is called a "world line." World lines play an important role in textbooks on special relativity. In reality, a world line should be a function of x, y, z, and t, but it's very difficult even with a computer to draw a four-dimensional curve!

2.6 ITERATIVE PREDICTION: VARYING NET FORCE

A familiar example of a force that is not constant is the force exerted by a spring that is stretched or compressed. The magnitude of this force depends on the amount of stretch or compression of the spring. The force is directed along the line of the spring.

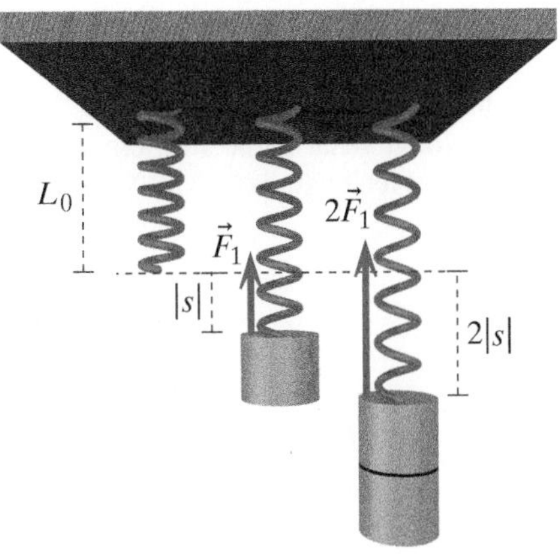

Figure 2.33 The magnitude of the force exerted by a spring is proportional to the absolute value of the stretch of the spring. For an elongated spring, stretch is positive.

The Spring Force

A "force law" describes mathematically how a force depends on the situation. For a spring, it is determined experimentally that the force exerted by a spring on an object attached to the spring is given by the following equation:

MAGNITUDE OF THE SPRING FORCE

$$|\vec{F}_{\text{spring}}| = k_s|s|$$

$|s|$ is the absolute value of the stretch: $s = L - L_0$.

L_0 is the length of the relaxed spring.

L is the length of the spring when stretched or compressed.

k_s is the "spring stiffness" (also called "spring constant").

The force acts in a direction to restore the spring to its relaxed length.

The constant k_s is a positive number, and is a property of the particular spring: the stiffer the spring, the larger the spring stiffness, and the larger the force needed to stretch the spring. Note that s is positive if the spring is stretched ($L > L_0$, Figure 2.33) and negative if the spring is compressed ($L < L_0$, Figure 2.34). This equation is sometimes called "Hooke's law." It is valid as long as the spring is not stretched or compressed too much.

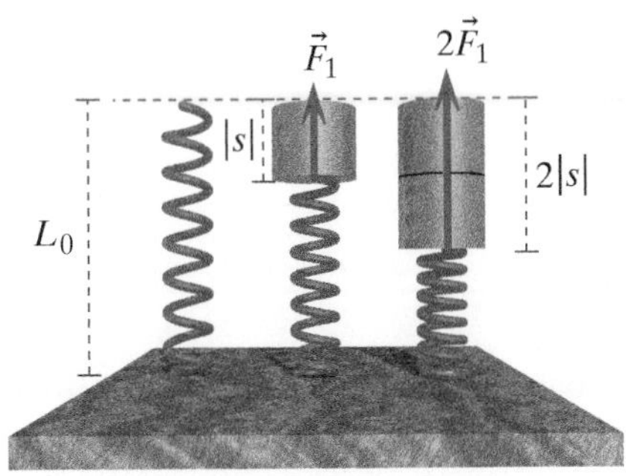

Figure 2.34 The magnitude of the force exerted by a spring is proportional to the absolute value of the stretch of the spring. For a compressed spring, stretch is negative.

> QUESTION Suppose a certain spring has been calibrated so that we know that its spring stiffness k_s is 500 N/m. You pull on the spring and observe that it is 0.01 m (1 cm) longer than it was when relaxed. What is the magnitude of the force exerted by the spring on your hand?

The force law gives $|\vec{F}_{\text{spring}}| = (500\,\text{N/m})(|+0.01\,\text{m}|) = 5\,\text{N}$. Note that the total length of the spring doesn't matter; it's just the amount of stretch or compression that matters. Because of the "reciprocity" of the electric forces between the protons and electrons in the spring and those in your hand, as we'll study in Chapter 3, the force you exert on the spring is equal in magnitude and opposite in direction to the force the spring exerts on your hand.

QUESTION Suppose that instead of pulling on the spring, you push on it, so the spring becomes shorter than its relaxed length. If the relaxed length of the spring is 10 cm, and you compress the spring to a length of 9 cm, what is the magnitude of the force exerted by the spring on your hand?

The stretch of the spring in SI units is

$$s = L - L_0 = (0.09\,\text{m} - 0.10\,\text{m}) = -0.01\,\text{m}$$

The force law gives

$$|\vec{F}_{\text{spring}}| = (500\,\text{N/m})(|-0.01\,\text{m}|) = 5\,\text{N}$$

The magnitude of the force is the same as in the previous case. Of course the direction of the force exerted by the spring on your hand is now different, so we will need to write a full vector equation to incorporate this information.

Checkpoint 6 (1) You push on a spring whose stiffness is 11 N/m, compressing it until it is 2.5 cm shorter than its relaxed length. What is the magnitude of the force the spring now exerts on your hand? (2) A different spring is 0.17 m long when it is relaxed. **(a)** When a force of magnitude 250 N is applied, the spring becomes 0.24 m long. What is the stiffness of this spring? **(b)** This spring is compressed so that its length is 0.15 m. What magnitude of force is required to do this?

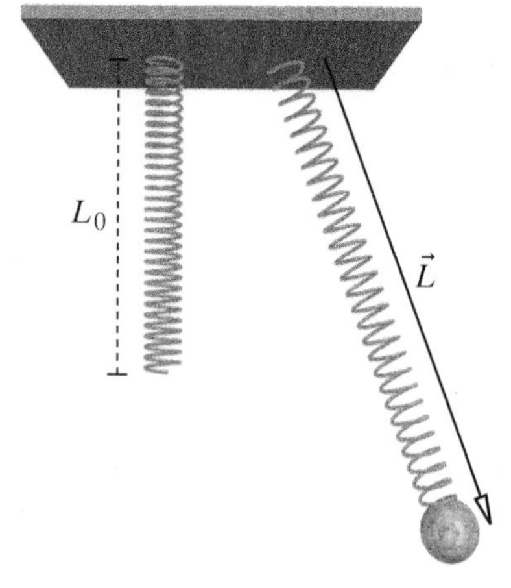

Figure 2.35 The vector $\vec{L}$ points from the location where the spring is attached to the ceiling to the location of the mass. The arrow is offset for clarity in the diagram. L_0 is the length of the relaxed spring.

The Spring Force as a Vector

To write an equation for the spring force as a vector, combining magnitude and direction into one expression, we need to consider the general case, in which the spring may be stretched both vertically and horizontally. Since the force exerted by the spring will be directed along the axis of the spring, we need a unit vector in that direction. We will get this unit vector from the vector $\vec{L}$, shown in Figure 2.35, which is a relative position vector: it specifies the position of the movable end of the spring relative to the fixed end of the spring. $\vec{L}$ extends from the point at which the spring is attached to a support to the mass at the other end of the spring. We can factor the relative position vector $\vec{L}$ into the product of a magnitude (the length of the spring) and a unit vector:

$$\vec{L} = |\vec{L}|\hat{L}$$

The stretch of the spring can now be written as

$$s = |\vec{L}| - L_0$$

- When the spring is longer than its relaxed length, the value of s will be positive, but the direction of the force will be given by $-\hat{L}$.
- When the spring is shorter than its relaxed length, the value of s will be negative, but the direction of the force will be given by $+\hat{L}$.

The general expression for the vector spring force is therefore:

THE VECTOR SPRING FORCE

$$\vec{F} = -k_s s\hat{L}$$

The stretch $s = |\vec{L}| - L_0$ and may be positive or negative.
The scalar L_0 is the length of the relaxed spring.
The vector $\vec{L}$ extends from the point of attachment of the spring to the movable end.
k_s is the "spring stiffness" (also called "spring constant").

QUESTION Consider the two stretched springs in Figure 2.33. What is the unit vector $\hat{L}$ for each stretched spring (assuming the usual coordinate system)? Does $\hat{L}$ depend on how much the spring is stretched?

In each case $\hat{L} = \langle 0, -1, 0\rangle$. Although the magnitude of $\vec{L}$ depends on the stretch, the unit vector $\hat{L}$, which has magnitude 1, is the same in both cases. Similarly, for both compressed springs in Figure 2.34, $\hat{L} = \langle 0, 1, 0\rangle$.

EXAMPLE **Force Due to a Stretched Spring**

Suppose that the stiffness of the rightmost spring shown in Figure 2.35 is 9 N/m, and its relaxed length is 21 cm. At the instant shown, the location of the green mass is $\langle 0.07, -0.33, 0\rangle$ m relative to an origin at the point of attachment of the spring. What is the force exerted by the spring on the green mass at this instant?

Solution

$$\vec{L} = \langle 0.07, -0.33, 0\rangle\ \text{m} - \langle 0, 0, 0\rangle\ \text{m} = \langle 0.07, -0.33, 0\rangle\ \text{m}$$

$$|\vec{L}| = \sqrt{(0.07\ \text{m})^2 + (-.33\ \text{m})^2} = 0.337\ \text{m}$$

$$\hat{L} = \frac{\langle 0.07, -0.33, 0\rangle\ \text{m}}{0.337\ \text{m}} = \langle 0.208, -.979, 0\rangle$$

$$s = 0.337\ \text{m} - 0.21\ \text{m} = 0.127\ \text{m}$$

$$\vec{F} = -(9\ \text{N/m})(0.127\ \text{m})\langle 0.208, -.979, 0\rangle = \langle -0.238, 1.12, 0\rangle\ \text{N/m}$$

Check: The x component of the force is negative, and the y component is positive, as they should be.

Checkpoint 7 (1) A spring of stiffness 13 N/m, with relaxed length 20 cm, stands vertically on a table as shown in Figure 2.36. Use the usual coordinate system, with $+x$ to the right, $+y$ up, and $+z$ out of the page, toward you. **(a)** When the spring is compressed to a length of 13 cm, what is the unit vector $\hat{L}$? **(b)** When the spring is stretched to a length of 24 cm, what is the unit vector $\hat{L}$? (2) A different spring of stiffness 95 N/m, and with relaxed length 15 cm, stands vertically on a table, as shown in Figure 2.36. With your hand you push straight down on the spring until your hand is only 11 cm above the table. Find **(a)** the vector $\vec{L}$, **(b)** the magnitude of $\vec{L}$, **(c)** the unit vector $\hat{L}$, **(d)** the stretch s, **(e)** the force $\vec{F}$ exerted on your hand by the spring.

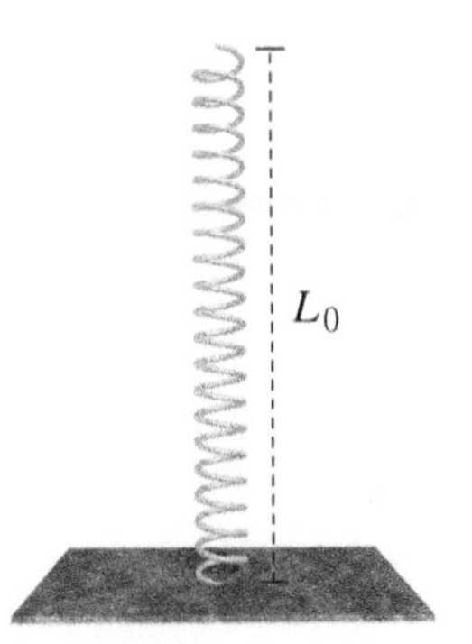

Figure 2.36 A relaxed vertical spring. The tabletop lies in the xz plane, and y is up, as usual.

Motion of a Block–Spring System

If we attach a block to the top of a spring, push down on the block, and then release it, the block will oscillate up and down. (This repetitive motion is described as "periodic.") As the spring stretches and compresses, the force exerted on the block by the spring changes in magnitude and direction. There is also a constant gravitational force on the block. Because the net force on the block is continually changing, we can't use a one-step calculation to predict its motion. We need to apply the Momentum Principle iteratively to predict the location and velocity of the block at any instant.

> QUESTION Can we use the equations we derived for the constant force situation?

No, those equations would give us the wrong answer here, because the net force on the block is not constant. We need to solve the problem iteratively, using the same procedure we used to find the path of a ball.

We need two initial values:

- The initial position of the block
- The initial momentum of the block

We will do the following calculation iteratively (repeatedly):

- Find the net force $\vec{F}_{\text{net,now}}$ on the block.
- Apply the Momentum Principle to find $\vec{p}_{\text{future}}$.
- Use $(\vec{p}_{\text{future}}/m)$ to approximate the average velocity, and use this to find the new position of the block.

> QUESTION Why is it important to calculate the net force for each iteration? Couldn't we just calculate it once and use this value in each iteration?

As the block moves up and down, the length of the spring changes, so the force exerted by the spring on the block is different each time we take a step.

EXAMPLE **Block on Spring: 1D, Nonconstant Net Force**

A spring has a relaxed length of 20 cm (0.2 m) and its spring stiffness is 8 N/m (Figure 2.36). You glue a 60 g block (0.06 kg) to the top of the spring, and push the block down, compressing the spring so its total length is 10 cm (Figure 2.37). You make sure the block is at rest, then you quickly move your hand away. The block begins to move upward, because the upward force on the block by the spring is greater than the downward force on the block by the Earth. Make a graph of y vs. time for the block during a 0.3 s interval after you release the block.

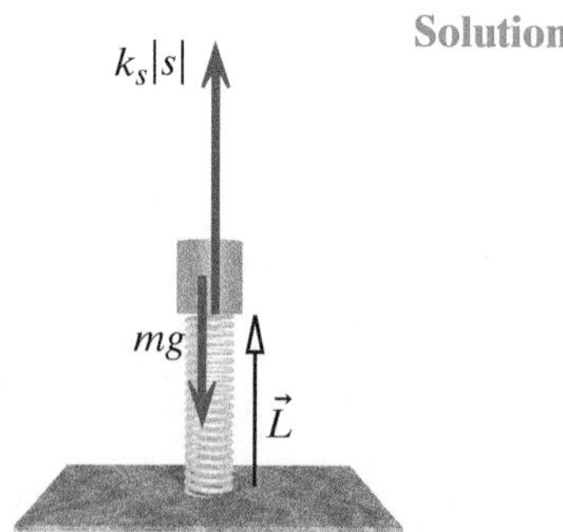

Figure 2.37 You compress the spring, make sure the block is at rest, then release the block. The red arrows show the forces on the block, due to the surroundings, at the instant just after you release the block.

Solution System: Block
Surroundings: Spring, Earth
Diagram: Figure 2.37

For convenience we place the origin at the base of the spring. We will use the shorter notation $\vec{p}_i$ and $\vec{p}_f$ rather than $\vec{p}_{\text{now}}$ and $\vec{p}_{\text{future}}$ to refer to the momentum at the beginning and end of each time step.

$$\vec{L} = \langle 0,0.1,0\rangle - \langle 0,0,0\rangle = \langle 0,0.1,0\rangle \text{ m}$$

$$|\vec{L}| = 0.1$$

$$\hat{L} = \langle 0,1,0\rangle$$

$$\vec{F}_{\text{spring}} = -k_s(|\vec{L}| - L_0)\langle 0,1,0\rangle = \langle 0,-k_s(|\vec{L}| - L_0),0\rangle$$

$$\vec{F}_{\text{Earth}} = \langle 0,-mg,0\rangle$$

The initial momentum of the block is zero, since it is at rest when you release it:

$$\vec{p}_i = \langle 0,0,0 \rangle$$

The net force on the block will be the sum of the forces on the block by the Earth and by the spring. Because both the force by the spring and the gravitational force by the Earth act in the $\pm y$ direction, and the initial x and z components of the block's momentum are zero, we could consider only the y components of force, momentum, and position in our solution. However, writing out all the vectors helps considerably in avoiding sign errors in forces and momenta.

To get an approximate answer, let's divide the 0.3 s time interval into three intervals each 0.1 s long. (It would be better to use even shorter intervals, but this would be unduly tedious if done by hand.)

First time step (Figure 2.38):

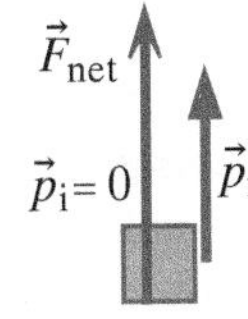

Figure 2.38 At the beginning of time step 1, the net force on the block is in the $+y$ direction, so the y component of the block's momentum will increase.

$$|\vec{L}| = 0.1\,\text{m}$$
$$s = 0.1\,\text{m} - 0.2\,\text{m} = -0.1\,\text{m}$$
$$\vec{F}_{\text{spring}} = -8\,\text{N/m}(-0.1\,\text{m})\langle 0,1,0 \rangle = \langle 0,+0.8,0 \rangle\,\text{N}$$
$$\vec{F}_{\text{Earth}} = \langle 0,-0.06\,\text{kg}\cdot 9.8\,\text{N/kg},0 \rangle = \langle 0,-0.588,0 \rangle\,\text{N}$$
$$\vec{F}_{\text{net}} = \langle 0,+0.212,0 \rangle\,\text{N}$$
$$\vec{p}_f = \langle 0,0,0 \rangle + \langle 0,+0.212,0 \rangle\,\text{N}\,(0.1\,\text{s}) \quad \text{(Momentum Principle)}$$
$$\vec{p}_f = \langle 0,+0.0212,0 \rangle\,\text{kg}\cdot\text{m/s}$$
$$\vec{v}_{\text{avg}} \approx \vec{v}_f$$
$$\vec{v}_f = \frac{\vec{p}_f}{m} = \frac{\langle 0,+0.0212,0 \rangle\,\text{kg}\cdot\text{m/s}}{0.06\,\text{kg}} = \langle 0,+0.353,0 \rangle\,\text{m/s}$$
$$\vec{r}_f = \langle 0,0.1,0 \rangle\,\text{m} + \langle 0,+0.353,0 \rangle\,\text{m/s}\,(0.1\,\text{s}) \quad \text{(position update)}$$
$$\vec{r}_f = \langle 0,0.135,0 \rangle\,\text{m}$$

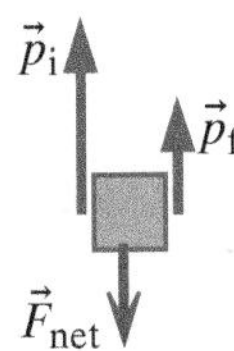

Figure 2.39 At the beginning of time step 2, the net force on the block is now in the $-y$ direction. The block will move upward, but its upward momentum will decrease.

Second time step (Figure 2.39):
Now we advance the clock. At the beginning of time step 2, we need to recalculate $F_{\text{spring},y}$, because the length of the spring has changed. We find that both the magnitude and the direction of the net force are different from the values we found at the beginning of step 1. The momentum of the block at this time ("now") reflects the forces that the block experienced during step 1.

$$|\vec{L}| = 0.135\,\text{m}$$
$$s = -0.0647\,\text{m}$$
$$\vec{F}_{\text{spring}} = \langle 0,+0.520,0 \rangle\,\text{N}$$
$$\vec{F}_{\text{net}} = \langle 0,-0.0707,0 \rangle\,\text{N}$$
$$\vec{p}_f = \langle 0,0.0141,0 \rangle\,\text{kg}\cdot\text{m/s}$$
$$\vec{r}_f = \langle 0,0.159,0 \rangle\,\text{m}$$

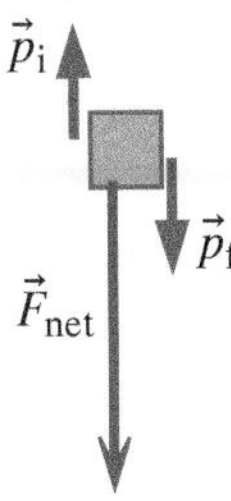

Figure 2.40 At the beginning of time step 3, the net force on the block is in the $-y$ direction, and the final momentum of the block will be downward.

QUESTION What value did we use for $\vec{p}_i$ in time step 2?

$\vec{p}_f$ from step 1 became $\vec{p}_i$ in step 2, as we advanced the clock.

Third time step (Figure 2.40):
At the beginning of the third time step we find that the net force on the block has changed again. The momentum of the block reflects the impulses it has experienced during the first two steps:

$$|\vec{L}| = 0.159\,\text{m}$$
$$s = -0.0411\,\text{m}$$
$$\vec{F}_{\text{net}} = \langle 0, -0.259, 0\rangle\ \text{N}$$
$$\vec{p}_f = \langle 0, -0.0118, 0\rangle\ \text{kg}\cdot\text{m/s}$$
$$\vec{r}_f = \langle 0, 0.139, 0\rangle\ \text{m}$$

The graph of y vs. time for our three-step calculation is shown in Figure 2.41.

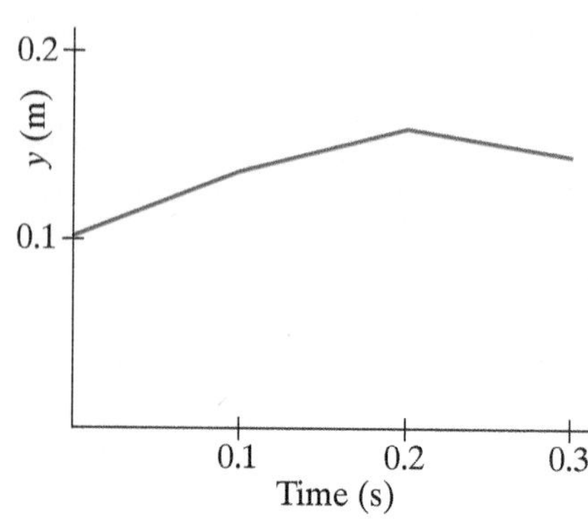

Figure 2.41 A graph of the y component of the block's position vs. time for the three-step iterative calculation.

QUESTION Is this result reasonable?

Yes, because we expect the block to oscillate up and down on the spring.

QUESTION What approximations were made in this calculation?

We made the approximation that the net force did not change significantly over each small time step. During each time step, we also used the final velocity as an approximation for the average velocity. (This is simpler than computing the arithmetic average, and not necessarily worse in a situation where the force is changing. In this case, it turns out that using the arithmetic average would actually have given a less accurate answer.)

Figure 2.42 shows the graph of y vs. t produced when the iterative calculation above is carried out for 10 time steps of 0.1 s.

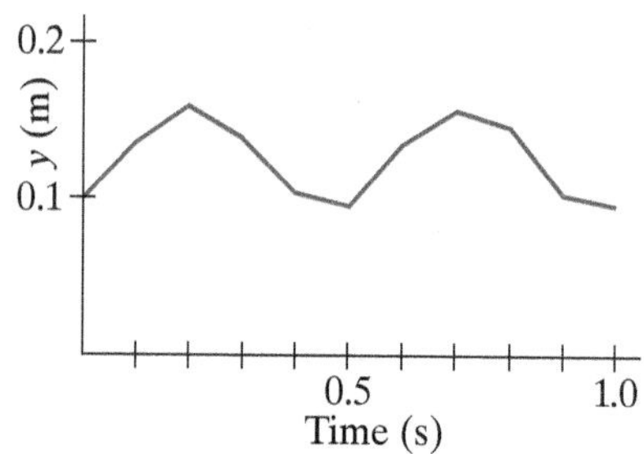

Figure 2.42 A graph of the y component of the block's position vs. time for an iterative calculation carried out for 10 steps of 0.1 s.

QUESTION What features of the graph in Figure 2.42 reflect inaccuracies in the calculations?

Although the graph does show the oscillatory motion of the block, its jagged lines reflect the fact that 0.1 s is too large a step size to produce an accurate result. There are two issues:

- The graph is jagged because we're connecting temporally distant computed points by long straight lines.
- The computed points themselves are inaccurate because we used a rather long time step of 0.1 s, which is clearly large compared to the time scale of the changes occurring in the motion.

Since the motion of the block–spring system repeats over and over, this motion is called *periodic* motion. The time interval between maxima in the plot of y vs. t is called the *period* of the motion, and is usually represented by the symbol T. From the graph in Figure 2.42 we can see that for this particular mass–spring system the period T appears to be around half a second. To get a better estimate of the period, we would need a better graph.

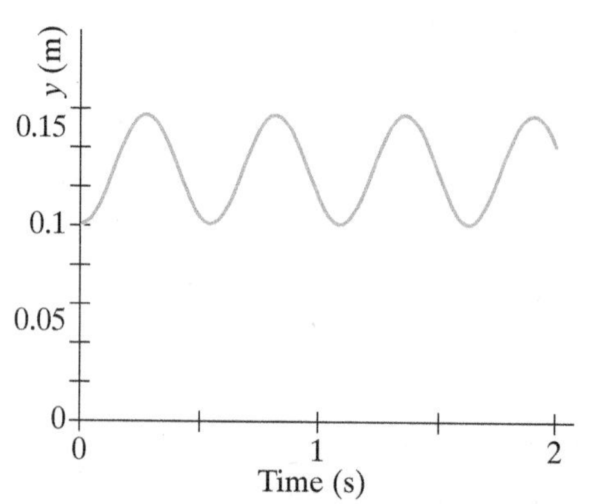

Figure 2.43 A graph of the y component of the block's position vs. time for an iterative calculation using a step size of 0.02 s.

Checkpoint 8 **(a)** In step 2 of the mass–spring example above, the net force on the block was downward, but the block moved upward. Explain why this was possible, and what the effect of the net force was during this time step. **(b)** In step 3 of the mass–spring example above, only the results of the calculations are given. Carry out the calculations for this step yourself, to be sure you understand in detail the procedure used. Compare your results to the values given above.

Improving Accuracy

QUESTION How could we improve the accuracy of the preceding calculations?

In general, decreasing the time step size will produce more accurate results, because the assumption that $\vec{F}_{net}$ is constant during the time Δt will be more valid. Using a step size of 0.02 s produces a smoothly oscillating graph of y vs. t, as shown by the curve in Figure 2.43. Now the straight lines connecting the computed points are short, making the graph look more smooth, and the computed points themselves are more accurate thanks to using a shorter time step, so that the force and velocity are more nearly constant during each time interval. This prediction is in agreement with experiments with masses oscillating on springs.

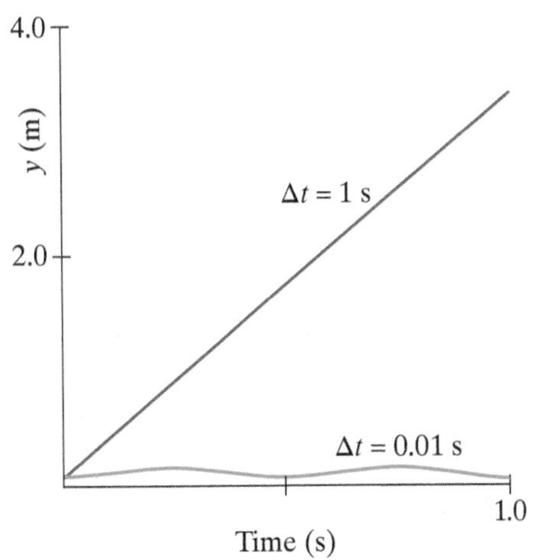

Figure 2.44 If we use a time step of 1 s to predict the motion of the mass–spring system, we predict that the mass will be more than 3 m above the floor! Note that the vertical scale of this graph is very different from the scale of the preceding graphs.

QUESTION Suppose that we had chosen to use a Δt of 1 s. What effect would this have had on our prediction of the block's motion?

The periodicity of the system's motion is due to the particular way in which the net force on the block is changing with time. Since 1 s is actually longer than the period of the motion, assuming that the net force is constant over this interval will lead to an extremely inaccurate result. The resulting graph of y vs. t, shown in Figure 2.44, shows that the predicted y position of the mass after 1 s would be more than 3 m above the floor! A 1 s time step is clearly too large for this system; if the motion we want to predict is periodic, we need a time step that is much smaller than the period of the motion.

There is no single rule for picking an appropriate time step, and sometimes we will need to use a guess-and-check strategy to refine our choice. One approach to deciding if we have chosen an appropriate value for Δt is to try reducing the step size significantly, and see if the predicted motion changes. However, it will be much less tedious to do calculations for many small time steps if we can instruct a computer to do the calculation for us.

QUESTION Would you expect the iterative calculation to be more accurate if we updated the position using $(\vec{v}_i + \vec{v}_f)/2$ instead of using the velocity at the end of the time interval?

For the blue curve in Figure 2.45 the velocity at the end of the time interval, $\vec{v}_f$, was used to update the position, whereas for the red curve the arithmetic average, $(\vec{v}_i + \vec{v}_f)/2$, was used. Using the arithmetic average (red curve) clearly produces a very inaccurate prediction: the oscillation will not grow larger and larger in the real world! Here we see an example of the fact that the arithmetic average isn't necessarily the best way to estimate the average velocity in an iterative calculation. In more advanced courses on computational modeling one learns special numerical techniques for optimizing speed and accuracy. For our purposes in this introductory course, we get good results by calculating the net force, then updating the momentum, then updating the position using the final velocity.

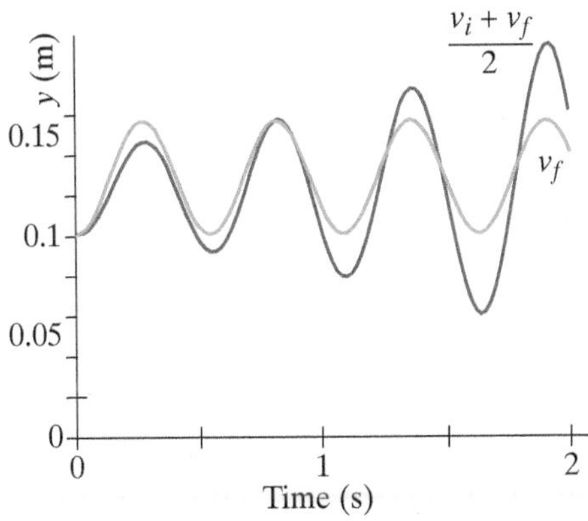

Figure 2.45 Two graphs of the y component of the block's position vs. time for an iterative calculation using a step size of 0.02 s. Blue curve: the velocity at the end of the time interval, $\vec{v}_f$, was used to update the position. Red curve: the arithmetic average, $(\vec{v}_i + \vec{v}_f)/2$, was used. In this calculation the arithmetic average gives significantly worse results.

Why Not Just Use Calculus?

You might wonder why we don't simply use calculus to predict the motion of physical systems. There are two answers to this question:

First: In fact, we actually are using calculus, in its most fundamental form. Step by step, we add up a large number of small increments of the momentum of an object, and a large number of small displacements of the object, to calculate a large change in its momentum and position over a long time, and this corresponds to a numerical evaluation of an integral:

$$\Delta\vec{p}_1 + \Delta\vec{p}_2 + \cdots \approx \int_i^f d\vec{p} = \int_i^f \vec{F}_{net}\, dt$$

$$\Delta\vec{r}_1 + \Delta\vec{r}_2 + \cdots \approx \int_i^f d\vec{r} = \int_i^f \vec{v}_{avg}\, dt$$

When we calculate these summations with finite time steps, we call the process *numerical integration* (Figures 2.46 and 2.47). The integral sign used in calculus is a distorted "S" meaning "sum" of an infinite number of infinitesimal quantities, from the initial time i to the final time f. Evaluating a definite integral in calculus corresponds to taking an infinite number of infinitesimally small time steps in a summation.

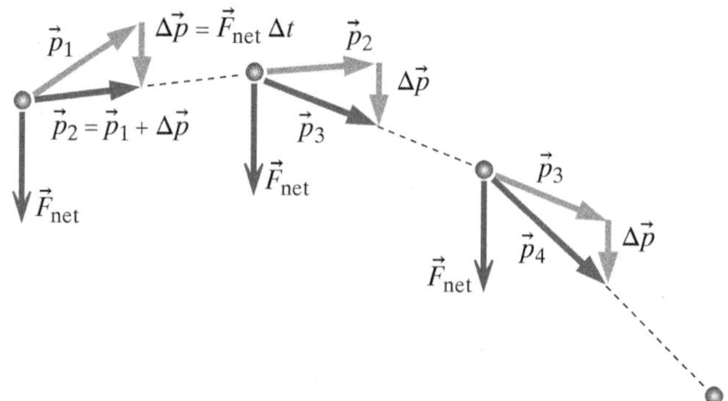

Figure 2.46 When we predict motion iteratively, we are numerically adding a large number of small increments ($\Delta\vec{p}_i$) to the starting momentum to get the momentum at some future time, as shown in Figure 2.47. This is called "numerical integration."

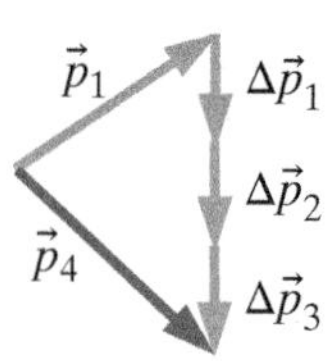

Figure 2.47 Numerical integration: adding all the $\Delta\vec{p}_i$s to $\vec{p}_1$ (Figure 2.46) gives the future momentum $\vec{p}_4$.

Second: A more interesting answer is that the motion of most physical systems actually *cannot* be predicted using calculus in any way other than by numerical integration. In a few special cases calculus does give a general result without carrying out a numerical integration. For example, we have seen that an object subjected to a constant force has a constant rate of change of momentum and velocity, and calculus can be used to obtain a prediction for the position as a function of time. An analytical solution can be derived for the motion in one dimension of a mass attached to a very low mass spring, in the absence of air resistance or friction, as we will see in Chapter 4. The elliptical orbits of two stars around each other can be predicted mathematically without an iterative approach, although the math is quite challenging.

However, in only slightly more complicated situations, an analytical solution may not be possible at all. For example, the general motion of three stars around each other has never been successfully analyzed in this way. The basic problem is that it is usually relatively easy to take the derivative of a known function, but it is often impossible to determine in algebraic form the integral of a known function, which is what would be involved in long-term prediction (adding up a large number of small momentum increments due to known forces).

In contrast, a step-by-step procedure of the kind we carried out for the mass–spring system can easily be extended to three or more bodies in three dimensions. It is also possible to include the effects of various kinds of friction and damping in an iterative calculation. This is why we learn to use the step-by-step prediction method: because it is a powerful technique of increasing importance in modern science and engineering, thanks to the availability of powerful computers to do the repetitive work for us.

2.7 ITERATIVE CALCULATIONS ON A COMPUTER

While the iterative scheme is very general, doing it by hand is tedious. It is not difficult to program a computer to do these calculations repetitively. Computers are now fast enough that it is possible to get high accuracy simply by taking very short time steps, so that during each step the net force and velocity do not change much.

A computer program is simply a sequence of instructions that specify how to perform a calculation. There are many programming languages that can be

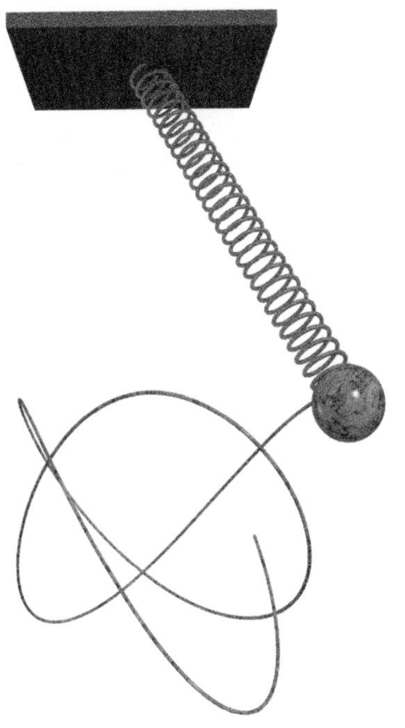

Figure 2.48 A snapshot from a computer program to calculate and animate the motion of a mass–spring system in 3D. This program was written in the language VPython (http://vpython.org), which generates real-time 3D visualizations of motion.

used to do this, but the basic organization of the program will be the same in almost all cases:

STRUCTURING ITERATIVE CALCULATIONS ON A COMPUTER

- Define the values of constants such as g or k_s to use in the program.
- Specify the masses, initial positions, and initial momenta of the interacting objects.
- Specify an appropriate value for Δt, small enough that the objects don't move very far during one update.
- Create a "loop" structure for repetitive calculations:

Repeat
- Calculate the net (vector) force $\vec{F}_{\text{net}}$ acting on the system.
- Update the momentum of the system: $\vec{p}_f = \vec{p}_i + \vec{F}_{\text{net}}\,\Delta t$.
- Update the position: $\vec{r}_f = \vec{r}_i + \vec{v}_{\text{avg}}\,\Delta t$.

As before, we use the approximation $\vec{v}_{\text{avg}} \approx \vec{p}_f/m$.

Figure 2.48 shows a display created by such a computer program.

QUESTION Do computer calculations give answers that are exactly correct?

Neither a computer nor your own calculator give "exact" answers, for several reasons. First, when we take a finite time step Δt, we are making the approximation that $\vec{F}_{\text{net}}$ is constant over this time interval. In a situation in which the net force is varying, this approximation may never be exactly true. However, if we take small enough time steps, this approximation can give very good results.

Second, real numbers (typically called "floating-point numbers" in a computational context) cannot be represented infinitely precisely inside a calculator or computer. When arithmetic operations are done with floating-point numbers there is always a small round-off error. For example, if you start with zero and repeatedly add to it the number 1×10^{-5}, doing this 1×10^{5} times, you may get the result 0.99999999998 instead of 1.0.

Although round-off error does accumulate as the number of steps taken increases, it turns out that taking more, smaller steps does make calculations more accurate. The increased accuracy of a single smaller step more than compensates for the accumulation of round-off error in many steps.

Time Step Size in a Computer Calculation

When doing a calculation by hand, there is a trade-off between accuracy and time required to do many iterations (the smaller the time step, the more calculations must be done). Since computers are quite fast, it is reasonable to use much smaller time steps in a computer calculation than one would use by hand. However, even a fast computer can take a very long time to do calculations that use an unnecessarily tiny time step—it would not be reasonable to use a time step of 1×10^{-20} s in a prediction of the Earth's motion around the Sun.

If the motion is periodic, as in the case of an oscillating mass–spring system or a planet orbiting a star, it is important to use a time step that is much shorter than the period of the motion (the repetition time).

A standard method of checking the accuracy of a computer calculation is to decrease the time step and repeat the calculation. If the results do not change significantly, the original time step was adequately small.

Checkpoint 9 Jupiter goes around the Sun in 4333 Earth days. Which of the following would be a reasonable value to try for Δt in a *computer* calculation of the orbit? **(a)** 1 day **(b)** 4333 days **(c)** 0.01 second **(d)** 800 days

The Euler–Cromer Method of Numerical Integration

If you have progressed far enough in your study of calculus, you may have recognized that the iterative procedures we have been using to predict motion involve the Euler method of numerical integration. Actually we have been using a variation on this method, called the Euler–Cromer method, which much improves the accuracy of these iterative predictions. The key feature of this method is the order of the steps:

- First calculate the net force using current positions
- Second, use this force to update the momentum of the system
- Third, use this new momentum to update the position

Doing these calculations in a different order gives much less accurate predictions. (The proof of this is beyond the scope of this textbook; if you are interested you can find more information online.)

Iterative Calculations in VPython

■ The computational homework problems at the end of Chapter 1, and the instructional videos referenced there, introduce you to the key concepts needed to understand VPython code.

The most important part of an iterative computer model is the calculation loop, which contains all the instructions that need to be repeated for each time step. Consider the case of a fan cart moving under the influence of the nearly constant force of the air on the fan. Our computational loop might look like this code, which is an excerpt from a complete program:

```
while True:
    rate(100)
    F_fan = vector(-0.4,0,0)
    F_net = F_fan
    p_cart = p_cart + F_net * deltat
    cart.pos = cart.pos + (p_cart/m_cart) * deltat
```

In contrast, the computational loop in a program modeling a block hanging from a vertical spring might look like this:

```
while True:
    rate(100)
    F_grav = vector(0, -g*m_block, 0)
    L = block.pos - spring.pos
    Lhat = L/mag(L)
    s = mag(L) - L0
    F_spring = -ks * s * Lhat
    F_net = F_grav + F_spring
    p_block = p_block + F_net * deltat
    block.pos = block.pos + (p_block/m_block) * deltat
```

QUESTION What code is essentially the same in both loops?

The last two lines of code are essentially the same. In the next-to-last line, although the objects involved have different names (one is a block, one is a cart), the Momentum Principle is used to update the momentum of each object. In the last line, the position update equation is used to update the position of each object by approximating the average velocity by $\vec{p}_f/m$.

QUESTION What is the major difference between these two computations?

The forces on the two objects are very different. The fan cart is subject only to a constant force, while the net force on the block is the sum of two forces: a constant gravitational force and a spring force that varies with the stretch of the spring.

The instructions that specify how to calculate the spring force mirror what you would do on paper, using your calculator. If $\vec{r}_0$ is the location of the fixed end of the spring, then here is a translation of the algebraic expressions to VPython expressions:

$\vec{L} = \vec{r} - \vec{r}_0$	`L = block.pos - spring.pos`
$\hat{L} = \vec{L}/\lvert\vec{L}\rvert$	`Lhat = L/mag(L)`
$s = \lvert\vec{L}\rvert - L0$	`s = mag(L) - L0`
$\vec{F} = -k_s \cdot s \cdot \hat{L}$	`F_spring = -ks * s * Lhat`

QUESTION Where is $|\vec{L}|$ calculated?

VPython provides a function for calculating the magnitude of a vector. In the code above, the `mag()` function is used to get the magnitude of $\vec{L}$. We could have calculated it separately instead, giving it a name such as Lmag, as is done below:

```
Lmag = sqrt(L.x**2 + L.y**2 + L.z**2)
Lhat = L/Lmag
```

Checkpoint 10 Some code would need to be added in front of each computational loop discussed above in order to make a runnable program. What things would you need to instruct the computer to do before beginning the loop? **(a)** Create objects **(b)** Define constants **(c)** Set initial positions **(d)** Set initial momentum **(e)** Calculate the net force

2.8 *DERIVATION: SPECIAL-CASE AVERAGE VELOCITY

Here are two proofs, one geometric and one algebraic (using calculus), for the following special-case result concerning average velocity:

$$v_{\text{avg},x} = \frac{(v_{ix} + v_{fx})}{2} \quad \text{if } v_x \text{ changes at a constant rate.}$$

The results are similar for $v_{\text{avg},y}$ and $v_{\text{avg},z}$.

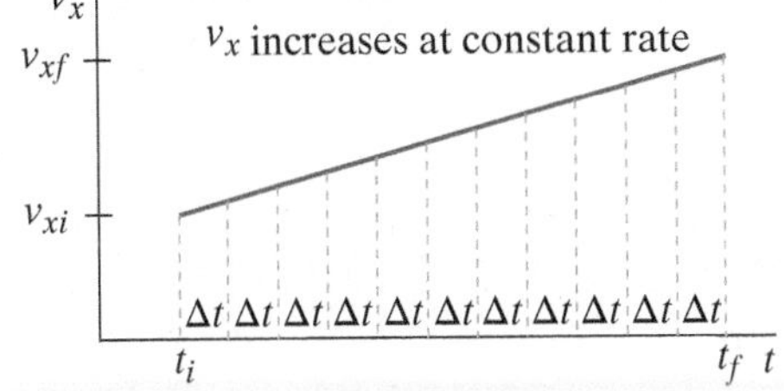

Figure 2.49 Graph of v_x vs. t (constant force), divided into narrow vertical slices each of height v_x and width Δt.

Geometric Proof

If $F_{\text{net},x}$ is constant, $p_{fx} = p_{ix} + F_{\text{net},x}\Delta t$ implies that p_x changes at a constant rate. At speeds small compared to the speed of light, $v_x \approx p_x/m$, so a graph of v_x vs. time is a straight line, as in Figure 2.49. Using this graph, we form narrow vertical slices, each of height v_x and narrow width Δt.

Within each narrow slice v_x changes very little, so the change in position during the brief time Δt is approximately $\Delta x = v_x \Delta t$. Therefore the change in x is approximately equal to the area of the slice of height v_x and width Δt (Figure 2.50).

If we add up the areas of all these slices, we get approximately the area under the line in Figure 2.49, and this is also equal to the total displacement $\Delta x_1 + \Delta x_2 + \Delta x_3 + \cdots = x_f - x_i$. If we go to the limit of an infinite number of slices, each with infinitesimal width, the sum of slices really *is* the area, and this area we have shown to be equal to the change in position. This kind of sum of an infinite number of infinitesimal pieces is called an "integral" in calculus.

The area under the line is a trapezoid, and from geometry we know that the area of a trapezoid is the average of the two bases times the altitude:

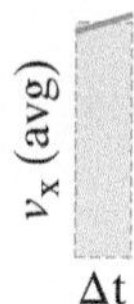

Figure 2.50 One narrow slice has an area given approximately by $v_x\Delta t$. This is equal to Δx, the displacement of the object.

$$\text{area} = \frac{(\text{top} + \text{bottom})}{2}(\text{altitude})$$

Turn Figure 2.50 on its side, as in Figure 2.51, and you see that the top and bottom have lengths v_{ix} and v_{fx}, while the altitude of the trapezoid is the total time $(t_f - t_i)$. Therefore we have the following result:

$$\text{trapezoid area } = x_f - x_i = \frac{(v_{ix} + v_{fx})}{2}(t_f - t_i)$$

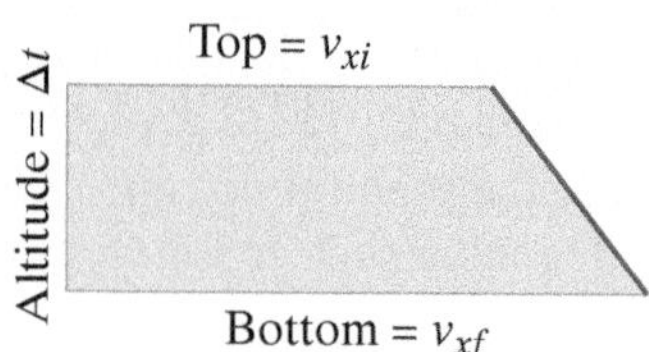

Figure 2.51 The area of the whole trapezoid is equal to the total displacement. The figure in Figure 2.50 has been rotated 90 degrees clockwise.

Dividing by $(t_f - t_i)$, we have this:

$$\frac{x_f - x_i}{t_f - t_i} = \frac{(v_{ix} + v_{fx})}{2}$$

By definition, however, the x component of average velocity is the change in x divided by the total time, so we have proved that

$$v_{\text{avg},x} = \frac{(v_{ix} + v_{fx})}{2} \quad \text{if } v_x \text{ changes linearly with time}$$

The proof depended critically on the straight-line ("linear") change in velocity, which occurs if $F_{\text{net},x}$ is constant (and $v \ll c$). Otherwise we wouldn't have a trapezoidal area. That's why the result isn't true in general; it's an important but special case.

Algebraic Proof Using Calculus

An algebraic proof using calculus can also be given. We will use the x component of the derivative version of the Momentum Principle (more about this in a later chapter):

$$\Delta\vec{p} = \vec{F}_{\text{net}}\,\Delta t \quad \text{implies that} \quad \frac{\Delta\vec{p}}{\Delta t} = \vec{F}_{\text{net}}$$

In the limit we have

$$\lim_{\Delta t \to 0}\frac{\Delta\vec{p}}{\Delta t} = \frac{d\vec{p}}{dt} = \vec{F}_{\text{net}} \quad \text{and} \quad \frac{dp_x}{dt} = F_{\text{net},x}$$

If $F_{\text{net},x}$ is a constant, the time derivative of p_x is a constant, so we have

$$p_x = F_{\text{net},x}t + p_{ix} \quad \text{since } p_x = p_{ix} \text{ when } t = 0$$

You can check this by taking the derivative with respect to time t, which gives the original equation $dp_x/dt = F_{\text{net},x}$. At speeds small compared to the speed of light, $v_x \approx p_x/m$, so we can write

$$v_x = \frac{F_{\text{net},x}}{m}t + v_{ix} \quad \text{since } v_x = v_{ix} \text{ when } t = 0$$

However, the x component of velocity is the rate at which x is changing:

$$v_x = \frac{dx}{dt} = \frac{F_{\text{net},x}}{m}t + v_{ix}$$

Now the question is, can you think of a function of x that has this time derivative? Since the time derivative of t^2 is $2t$, the following equation for x has the appropriate derivative:

$$x = \frac{1}{2}\frac{F_{\text{net},x}}{m}t^2 + v_{ix}t + x_i \quad \text{since } x = x_i \text{ when } t = 0$$

You can check this by taking the derivative with respect to t, which gives the equation for v_x, since $d(\frac{1}{2}t^2)/dt = t$ and $d(t)/dt = 1$.

The average velocity that we seek is the change in position divided by the total time:

$$v_{\text{avg},x} = \frac{x - x_i}{t} = \frac{1}{2}\frac{F_{\text{net},x}}{m}t + v_{ix} = \frac{1}{2}(v_{fx} - v_{ix}) + v_{ix}$$

where we have used the equation we previously derived for the velocity:

$$v_{fx} = v_x = \frac{F_{\text{net},x}}{m}t + v_{ix}$$

Simplifying the expression for $v_{\text{avg},x}$ we have the proof:

$$v_{\text{avg},x} = \frac{(v_{ix} + v_{fx})}{2} \quad \text{if } v_x \text{ changes at a constant rate}$$

2.9 *RELATIVISTIC MOTION

Inertial Frames of Reference

Newton's first law is valid only in an "inertial frame" of reference, one in uniform motion (or at rest) with respect to the pervasive "cosmic microwave background" (see the optional discussion at the end of Chapter 1). Since the Momentum Principle is a quantitative version of Newton's first law, we expect the Momentum Principle to be valid in an inertial reference frame, but not in a reference frame that is not in uniform motion. Let's check that this is true.

If you view some objects from a space ship that is moving uniformly with velocity $\vec{v}_s$ with respect to the cosmic microwave background, all of the velocities of those objects have the constant $\vec{v}_s$ subtracted from them, as far as you are concerned. For example, a rock moving at the same velocity as your spacecraft would have $\vec{v}_{\text{rock}} = (\vec{v} - \vec{v}_s) = \vec{0}$ in your reference frame: it would appear to be stationary as it coasted along beside your spacecraft.

With a constant space ship velocity, we have $\Delta\vec{v}_s = \vec{0}$, and the change of momentum of the moving object reduces to the following (for speeds small compared to c):

$$\Delta[m(\vec{v} - \vec{v}_s)] = \Delta(m\vec{v}) - \Delta(m\vec{v}_s) = \Delta(m\vec{v}) \quad \text{since } \Delta(m\vec{v}_s) = \vec{0}$$

Therefore,

$$\Delta[m(\vec{v} - \vec{v}_s)] = \Delta(m\vec{v}) = \vec{F}_{\text{net}}\,\Delta t$$

If the velocity $\vec{v}_s$ of the space ship doesn't change (it represents an inertial frame of reference), the form (and validity) of the Momentum Principle is unaffected by the motion of the space ship.

However, if your space ship increases its speed, or changes direction, $\Delta\vec{v}_s \neq \vec{0}$, an object's motion relative to you changes without any force acting on it. In that case the Momentum Principle is not valid for the object, because you are not in an inertial frame.

Fast-Moving Objects

Applying the Momentum Principle to an object moving at a speed near the speed of light is a bit more complex algebraically, but conceptually exactly the same as what we have done previously. The derivation of the equation for speed in terms of momentum for a fast-moving particle may be found in Chapter 1.

EXAMPLE **Fast Proton: 1D, Constant Net Force, Relativistic**

A proton in a particle accelerator is moving with velocity $\langle 0.960c, 0, 0\rangle$, so the speed is $0.960 \times 3 \times 10^8$ m/s $= 2.88 \times 10^8$ m/s. A constant electric force is applied to the proton to speed it up, $\vec{F}_{\text{net}} = \langle 5 \times 10^{-12}, 0, 0\rangle$ N. What is the proton's speed as a fraction of the speed of light after 20 nanoseconds (1 ns $= 1 \times 10^{-9}$ s)?

Solution System: The proton
Surroundings: Electric charges in the accelerator
Free-body diagram: Figure 2.52
Momentum Principle:

Proton

Figure 2.52 The free-body diagram for the proton.

$$\vec{p}_f = \vec{p}_i + \vec{F}_{\text{net}}\,\Delta t$$

$$\langle p_{fx}, 0, 0\rangle = \langle \gamma_i m v_{ix}, 0, 0\rangle + \langle F_{\text{net},x}, 0, 0\rangle\,\Delta t$$

Find the x component of the initial momentum:

$$\gamma = \frac{1}{\sqrt{1 - \left(\frac{0.960c}{c}\right)^2}} = 3.57$$

$$\begin{aligned} p_{ix} &= \gamma_i m v_{ix} = 3.57 \cdot (1.7 \times 10^{-27}\text{ kg})(0.96 \cdot 3 \times 10^8\text{ m/s}) \\ &= 1.75 \times 10^{-18}\text{ kg}\cdot\text{m/s} \end{aligned}$$

Find the x component of the impulse:

$$F_{\text{net},x}\,\Delta t = 5 \times 10^{-12}\text{ N} \cdot (20 \times 10^{-9}\text{ s}) = 1 \times 10^{-19}\text{ N}\cdot\text{s}$$

Apply the Momentum Principle:

$$\begin{aligned} p_{fx} &= (1.75 \times 10^{-18}\text{ kg}\cdot\text{m/s}) + (1 \times 10^{-19}\text{ N}\cdot\text{s}) \\ &= 1.85 \times 10^{-18}\text{ kg}\cdot\text{m/s} \end{aligned}$$

Find the final speed:

$$\frac{v_{fx}}{c} = \frac{\frac{p_{fx}}{mc}}{\sqrt{1+\left(\frac{p_{fx}}{mc}\right)^2}}$$

$$\frac{p_{fx}}{mc} = \frac{1.85 \times 10^{-18}\ \text{kg}\cdot\text{m/s}}{(1.7 \times 10^{-27}\ \text{kg})(3 \times 10^{8}\ \text{m/s})} = 3.62 \quad \text{(no units)}$$

$$\frac{v_{fx}}{c} = \frac{3.62}{\sqrt{1+3.62^2}} = 0.964$$

$$v_{fx} = 0.964c$$

Although the magnitude of the momentum increased from 1.75×10^{-18} kg·m/s to 1.85×10^{-18} kg·m/s, the speed didn't increase very much, because the proton's initial speed, $0.960c$, was already close to the cosmic speed limit, c. Because the speed hardly changed, the distance the proton moved during the 20 ns was approximately equal to

$$(0.964 \times 3 \times 10^{8}\ \text{m/s})(20 \times 10^{-9}\ \text{s}) = 5.8\ \text{m}$$

2.10 *MEASUREMENTS AND UNITS

Using the Momentum Principle requires a consistent way to measure length, time, mass, and force, and a consistent set of units. We state the definitions of the standard Système Internationale (SI) units, and we briefly discuss some subtle issues underlying this choice of units.

Units: Meters, Seconds, Kilograms, Coulombs, and Newtons

Originally the meter was defined as the distance between two scratches on a platinum bar in a vault in Paris, and a second was 1/86,400th of a "mean solar day." Now, however, the second is defined in terms of the frequency of light emitted by a cesium atom, and the meter is defined as the distance light travels in 1/299,792,458th of a second, or about 3.3×10^{-9} s (3.3 ns). The speed of light is defined to be exactly 299,792,458 m/s (very close to 3×10^{8} m/s). As a result of these modern redefinitions, it is really speed (of light) and time that are the internationally agreed-upon basic units, not length and time.

By international agreement, one kilogram is the mass of a platinum block kept in that same vault in Paris. As a practical matter, other masses are compared to this standard kilogram by using a balance-beam or spring weighing scale (more about this in a moment). The newton, the unit of force, is defined as that force which acting for 1 s imparts to 1 kg a velocity change of 1 m/s. We could make a scale for force by calibrating the amount of stretch of a spring in terms of newtons.

The coulomb, the SI unit of electric charge, is defined in terms of electric currents. The charge of a proton is 1.6×10^{-19} C.

Some Subtle Issues

What we have just said about SI units is sufficient for practical purposes to predict the motion of objects, but here are some questions that might bother you. Is it legitimate to measure the mass that appears in the Momentum Principle by seeing how that mass is affected by gravity on a balance-beam scale? Is it legitimate to use the Momentum Principle to define the units of force, when the concept of force is itself associated with the same law? Is this all circular reasoning, and the Momentum Principle merely a definition with little content? Here is a chain of reasoning that addresses these issues.

Measuring Inertial Mass

When we use balance-beam or spring weighing scales to measure mass, what we're really measuring is the "gravitational mass"—that is, the mass that appears in the law of gravitation and is a measure of how much this object is affected by the gravity of the Earth. In principle, this "gravitational mass" could be different from the so-called "inertial mass"—the mass that appears in the definition of momentum. It is possible to compare the inertial masses of two objects, and we find experimentally that inertial and gravitational mass seem to be entirely equivalent.

Here is a way to compare two inertial masses directly, without involving gravity. Starting from rest, pull on the first object with a spring stretched by some amount s for an amount of time Δt, and measure the increase of speed Δv_1. Then, starting from rest, pull on the second object with the same spring stretched by the same amount s for the same amount of time Δt, and measure the increase of speed Δv_2. We *define* the ratio of the inertial masses as $m_1/m_2 = \Delta v_2/\Delta v_1$. Since one of these masses could be the standard kilogram kept in Paris, we now have a way of measuring inertial mass in kilograms. Having defined inertial mass this way, we find experimentally that the Momentum Principle is obeyed by both of these objects in all situations, not just in the one special experiment we used to compare the two masses.

Moreover, we find to extremely high precision that the inertial mass in kilograms measured by this comparison experiment is exactly the same as the gravitational mass in kilograms obtained by comparing with a standard kilogram on a balance-beam scale (or using a calibrated spring scale), and that it doesn't matter what the objects are made of (wood, copper, glass, etc.). This justifies the convenient use of ordinary weighing scales to determine inertial mass.

Is This Circular Reasoning?

The definitions of force and mass may sound like circular reasoning, and the Momentum Principle may sound like just a kind of definition, with no real content, but there is real power in the Momentum Principle. Forget for a moment the definition of force in newtons and mass in kilograms. The experimental fact remains that any object if subjected to a single force by a spring with constant stretch experiences a change of momentum (and velocity) proportional to the duration of the interaction. Note that it is not a change of *position* proportional to the time (that would be a constant speed), but a change of *velocity*. That's real content. Moreover, we find that the change of velocity is proportional to the amount of stretch of the spring. That too is real content.

Then we find that a different object undergoes a different rate of change of velocity with the same spring stretch, but after we've made one single comparison experiment to determine the mass relative to the standard kilogram, the Momentum Principle works in all situations. That's real content.

Finally come the details of setting standards for measuring force in newtons and mass in kilograms, and we use the Momentum Principle in helping set these standards. Logically, however, this comes after having established the law itself.

SUMMARY

System is a portion of the Universe acted on by the *surroundings.*

Force is a quantitative measure of interactions; units are newtons (N).

Net force $\vec{F}_{\text{net}}$ is the vector sum of all the forces acting on a system.

Impulse is the product of force times time $\vec{F}\Delta t$; momentum change equals net impulse (the impulse due to the net force).

Physical models are tractable approximations/idealizations of the real world.

The Momentum Principle

$$\Delta\vec{p} = \vec{F}_{\text{net}}\,\Delta t$$

$$\vec{p}_f = \vec{p}_i + \vec{F}_{\text{net}}\,\Delta t \quad \text{(update form)}$$

or, choosing "now" as the initial time, and "a short time in the future" as the final time, we can write:

$$\vec{p}_{\text{future}} = \vec{p}_{\text{now}} + \vec{F}_{\text{net,now}}\,\Delta t$$

The time interval Δt must be small enough that the net force is nearly constant during this time interval.

ITERATIVE PREDICTION OF MOTION

Repeat

- Calculate the net (vector) force $\vec{F}_{\text{net}}$ acting on the system.
- Update the momentum of the system: $\vec{p}_f = \vec{p}_i + \vec{F}_{\text{net}}\,\Delta t$.
- Update the position: $\vec{r}_f = \vec{r}_i + \vec{v}_{\text{avg}}\,\Delta t$.

Use $\vec{p}_f/m$ to approximate the $\vec{v}_{\text{avg}}$ in each step.

Special-case result for average velocity if net force is constant

$$v_{\text{avg},x} = \frac{(v_{ix} + v_{fx})}{2} \quad \text{if } v_x \text{ changes linearly with time}$$

Results are similar for $v_{\text{avg},y}$ and $v_{\text{avg},z}$.

Analytical solution for motion with constant net force

$$x(t) = x_i + v_{xi}t + \frac{1}{2}\frac{F_{\text{net},x}}{m}t^2$$

$$v_x(t) = v_{xi} + \frac{F_{\text{net},x}}{m}t$$

Specific forces

Force by a spring

$$\vec{F} = -k_s s\hat{L}$$

The stretch $s = |\vec{L}| - L_0$ and may be positive or negative.
The scalar L_0 is the length of the relaxed spring.
The vector $\vec{L}$ extends from the point of attachment of the spring to the movable end.
k_s is the "spring stiffness" (also called "spring constant").

Approximate gravitational force near the Earth's surface

$$|\vec{F}_{\text{grav}}| \approx mg \quad \text{where } g = +9.8\,\text{N/kg}$$

for an object of mass m near the surface of the Earth.

QUESTIONS

Q1 Because the *change* of the momentum is equal to the net impulse, the relationship of momentum itself to the net force is somewhat indirect, as can be seen in this question. An object is initially moving in the $+x$ direction with magnitude of momentum p, with a net force of magnitude F acting on the object in either the $+x$ or $-x$ direction. After a very short time, say whether the magnitude of the momentum increases, decreases, or stays the same in each of the following situations: **(a)** The net force acts in the $+x$ direction and F is constant. **(b)** The net force acts in the $+x$ direction and F is increasing. **(c)** The net force acts in the $+x$ direction and F is decreasing. **(d)** The net force acts in the $-x$ direction and F is constant. **(e)** The net force acts in the $-x$ direction and F is increasing. **(f)** The net force acts in the $-x$ direction and F is decreasing.

Q2 An object is moving in the $+y$ direction. Which, if any, of the following statements might be true? Check all that apply. **(a)** The net force on the object is zero. **(b)** The net force on the object is in the $-y$ direction. **(c)** The net force on the object is in the $+y$ direction.

Q3 You observe three carts moving to the left. Cart A moves to the left at nearly constant speed. Cart B moves to the left, gradually speeding up. Cart C moves to the left, gradually slowing down. Which cart or carts, if any, experience a net force to the left?

Q4 In order to pull a sled across a level field at constant velocity you have to exert a constant force. Doesn't this violate Newton's first and second laws of motion, which imply that no force is required to maintain a constant velocity? Explain this seeming contradiction.

Q5 In a lab experiment you observe that a pendulum swings with a "period" (time for one round trip) of 2 s. In an iterative calculation of the motion, which of the following would NOT be a reasonable choice for Δt, for either hand or computer iterative calculations? **(a)** 1 s **(b)** 0.1 s **(c)** 0.05 s **(d)** 0.01 s

Q6 A comet passes near the Sun. When the comet is closest to the Sun, it is 9×10^{10} m from the Sun. You need to choose a time step to use in predicting the comet's motion. Which of the following would be a reasonable distance for the comet to move in one time step, doing an iterative calculation *by hand*? **(a)** 1×10^2 m **(b)** 1×10^{10} m **(c)** 1×10^{11} m **(d)** 1×10^9 m

Q7 A ball moves in the direction of the arrow labeled *c* in Figure 2.53. The ball is struck by a stick that briefly exerts a force on the ball in the direction of the arrow labeled *e*. Which arrow best describes the direction of $\Delta\vec{p}$, the change in the ball's momentum?

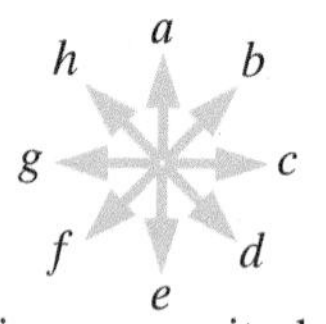

Figure 2.53

PROBLEMS

Section 2.1

•P8 A system is acted upon by two forces, $\langle 18,47,-23\rangle$ N, and $\langle -20,-13,41\rangle$ N. What is the net force acting on the system?

•P9 A truck driver slams on the brakes and the momentum of the truck changes from $\langle 65{,}000,0,0\rangle$ kg·m/s to $\langle 26{,}000,0,0\rangle$ kg·m/s in 4.1 s due to a constant force of the road on the wheels of the truck. As a vector, write the net force exerted on the truck by the surroundings.

•P10 At a certain instant a particle is moving in the $+x$ direction with momentum +8 kg·m/s. During the next 0.13 s a constant force acts on the particle, with $F_x = -7$ N and $F_y = +5$ N. What is the *magnitude* of the momentum of the particle at the end of this 0.13 s interval?

•P11 At $t = 16.0$ s an object with mass 4 kg was observed to have a velocity of $\langle 9,29,-10\rangle$ m/s. At $t = 16.2$ s its velocity was $\langle 18,20,25\rangle$ m/s. What was the average net force acting on the object?

••P12 A proton (mass 1.7×10^{-27} kg) interacts electrically with a neutral HCl molecule located at the origin. At a certain time t, the proton's position is $\langle 1.6 \times 10^{-9},0,0\rangle$ m and the proton's velocity is $\langle 3600,600,0\rangle$ m/s. The force exerted on the proton by the HCl molecule is $\langle -1.12 \times 10^{-11},0,0\rangle$ N. At a time $t + 3.4 \times 10^{-14}$ s, what is the approximate velocity of the proton? (You may assume that the force was approximately constant during this interval.)

••P13 A Ping-Pong ball is acted upon by the Earth, air resistance, and a strong wind. Here are the positions of the ball at several times.

Early time interval:

- At $t = 12.35$ s, the position was $\langle 3.17,2.54,-9.38\rangle$ m.
- At $t = 12.37$ s, the position was $\langle 3.25,2.50,-9.40\rangle$ m.

Late time interval:

- At $t = 14.35$ s, the position was $\langle 11.25,-1.50,-11.40\rangle$ m.
- At $t = 14.37$ s, the position was $\langle 11.27,-1.86,-11.42\rangle$ m.

(a) In the early time interval, from $t = 12.35$ s to $t = 12.37$ s, what was the average momentum of the ball? The mass of the Ping-Pong ball is 2.7 grams (2.7×10^{-3} kg). Express your result as a vector. **(b)** In the late time interval, from $t = 14.35$ s to $t = 14.37$ s, what was the average momentum of the ball? Express your result as a vector. **(c)** In the time interval from $t = 12.35$ s (the start of the early time interval) to $t = 14.35$ s (the start of the late time interval), what was the average net force acting on the ball? Express your result as a vector.

••P14 A 0.7 kg block of ice is sliding by you on a very slippery floor at 2.5 m/s. As it goes by, you give it a kick perpendicular to its path. Your foot is in contact with the ice block for 0.003 s. The block eventually slides at an angle of 22 degrees from its original direction. The overhead view shown in Figure 2.54 is approximately to scale. The arrow represents the average force your toe applies briefly to the block of ice.
(a) Which of the possible paths shown in the diagram corresponds to the correct overhead view of the block's path? **(b)** Which components of the block's momentum are changed by the impulse applied by your foot? (Check all that apply. The diagram shows a top view, looking down on the xz plane.) **(c)** What is the unit vector $\hat{p}$ in the direction of the block's momentum after the kick? **(d)** What is the x component of the block's momentum after the kick? **(e)** Remember that $\vec{p} = |\vec{p}|\hat{p}$. What is the magnitude of the block's momentum after the kick? **(f)** Use your answers to the preceding questions to find the z component of the block's momentum after the kick (drawing a diagram is helpful). **(g)** What was the magnitude of the average force you applied to the block?

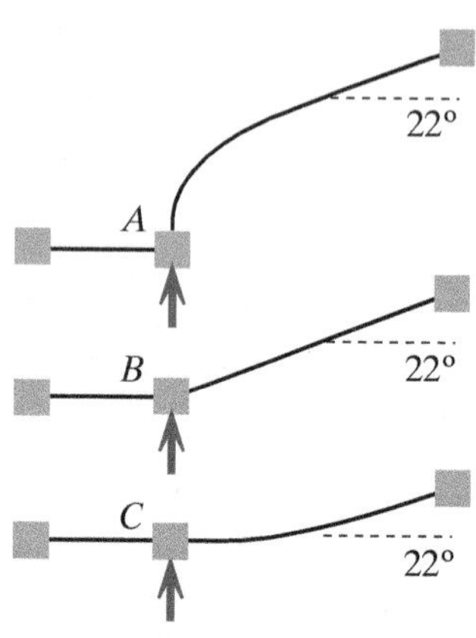

Figure 2.54

••P15 In outer space a rock of mass 5 kg is acted on by a constant net force $\langle 29,-15,40\rangle$ N during a 4 s time interval. At the end of this time interval the rock has a velocity of

$\langle 114, 94, 112 \rangle$ m/s. What was the rock's velocity at the beginning of the time interval?

Section 2.2

••P16 A steel safe with mass 2200 kg falls onto concrete. Just before hitting the concrete its speed is 40 m/s, and it smashes without rebounding and ends up being 0.06 m shorter than before. What is the approximate magnitude of the force exerted on the safe by the concrete? How does this compare with the gravitational force of the Earth on the safe? Explain your analysis carefully, and justify your estimates on physical grounds.

••P17 In a crash test, a truck with mass 2500 kg traveling at 24 m/s smashes head-on into a concrete wall without rebounding. The front end crumples so much that the truck is 0.72 m shorter than before. **(a)** What is the average speed of the truck during the collision (that is, during the interval between first contact with the wall and coming to a stop)? **(b)** About how long does the collision last? (That is, how long is the interval between first contact with the wall and coming to a stop?) **(c)** What is the magnitude of the average force exerted by the wall on the truck during the collision? **(d)** It is interesting to compare this force to the weight of the truck. Calculate the ratio of the force of the wall to the gravitational force mg on the truck. This large ratio shows why a collision is so damaging. **(e)** What approximations did you make in your analysis?

••P18 A tennis ball has a mass of 0.057 kg. A professional tennis player hits the ball hard enough to give it a speed of 50 m/s (about 120 mi/h). The ball hits a wall and bounces back with almost the same speed (50 m/s). As indicated in Figure 2.55, high-speed photography shows that the ball is crushed 2 cm (0.02 m) at the instant when its speed is momentarily zero, before rebounding.

Making the very rough approximation that the large force that the wall exerts on the ball is approximately constant during contact, determine the approximate magnitude of this force. *Hint:* Think about the approximate amount of time it takes for the ball to come momentarily to rest. (For comparison note that the gravitational force on the ball is quite small, only about (0.057 kg)(9.8 N/ kg) $\approx$ 0.6 N. A force of 5 N is approximately the same as a force of one pound.)

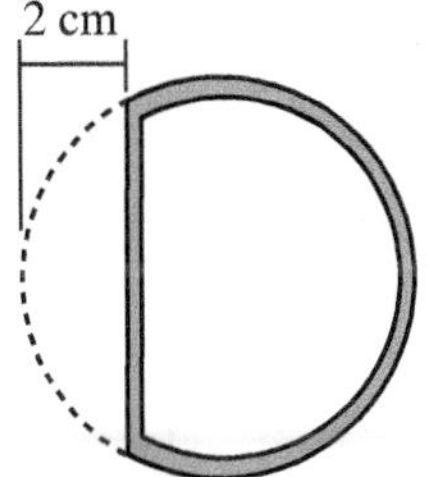

Figure 2.55

••P19 An object is on a collision course with the Earth and is predicted to hit in the center of the target, coming in vertically. The object is roughly spherical, with an approximate diameter of 100 m (the meteor that damaged Chelyabinsk, Russia, in February 2013 had a diameter of about 20 m; the object that killed off the dinosaurs 65 million years ago is thought to have had a diameter of about 10 km). If we start immediately, we can rendezvous with the object when it is 2.5×10^{11} m from the Sun, traveling toward the Sun at a speed of 30 km/s. There is a proposal to implant a rocket engine on the surface of the object to deflect the object enough to miss the Earth. Perform an approximate feasibility analysis of this critical mission. Give details of your design, including the estimates, assumptions, and idealizations you make. Note that the first stage of the most powerful rocket ever used, the Saturn V, had a thrust of 3.4×10^7 N and burned for 170 s. Can the Earth be saved?

Section 2.4

For all problems in this section, use the approximation that $\vec{v}_{avg} \approx \vec{p}_f/m$ for each time step.

••P20 Suppose that you are navigating a spacecraft far from other objects. The mass of the spacecraft is 1.5×10^5 kg (about 150 tons). The rocket engines are shut off, and you're coasting along with a constant velocity of $\langle 0, 20, 0 \rangle$ km/s. As you pass the location $\langle 12, 15, 0 \rangle$ km you briefly fire side-thruster rockets, so that your spacecraft experiences a net force of $\langle 6 \times 10^4, 0, 0 \rangle$ N for 20 s. After turning off the thrusters, you then continue coasting with the rocket engines turned off. (The ejected gases have a mass that is very small compared to the mass of the spacecraft.) **(a)** Using a step size of 20 seconds, predict where you will be 40 s after you began firing the thrusters. (Note that for the second step the thrusters are off.) **(b)** Where would you have been if you had not fired the thrusters?

••P21 You throw a metal block of mass 0.25 kg into the air, and it leaves your hand at time $t = 0$ at location $\langle 0, 2, 0 \rangle$ m with velocity $\langle 3, 4, 0 \rangle$ m/s. At this low velocity air resistance is negligible. Using the iterative method shown in Section 2.4 with a time step of 0.05 s, calculate step by step the position and velocity of the block at $t = 0.05$ s, $t = 0.10$ s, and $t = 0.15$ s.

••P22 A small space probe, of mass 240 kg, is launched from a spacecraft near Mars. It travels toward the surface of Mars, where it will land. At a time 20.7 s after it is launched, the probe is at the location $\langle 4.30 \times 10^3, 8.70 \times 10^2, 0 \rangle$ m, and at this same time its momentum is $\langle 4.40 \times 10^4, -7.60 \times 10^3, 0 \rangle$ kg·m/s. At this instant, the net force on the probe due to the gravitational pull of Mars plus the air resistance acting on the probe is $\langle -7 \times 10^3, -9.2 \times 10^2, 0 \rangle$ N. Assuming that the net force on the probe is approximately constant over this time interval, what are the momentum and position of the probe 20.9 s after it is launched? Divide the interval into two time steps, and use the approximation $\vec{v}_{avg} \approx \vec{p}_f/m$.

••P23 A soccer ball of mass 0.43 kg is rolling with velocity $\langle 0, 0, 2.2 \rangle$ m/s, when you kick it. Your kick delivers an impulse of magnitude 1.3 N·s in the $-x$ direction. The net force on the rolling ball, due to the air and the grass, is 0.25 N in the direction opposite to the direction of the ball's momentum. Using a time step of 0.5 s, find the position of the ball at a time 1.5 s after you kick it, assuming that the ball is at the origin at the moment it is kicked. Use the approximation $\vec{v}_{avg} \approx \vec{p}_f/m$.

••P24 As your spaceship coasts toward Mars, you need to move a heavy load of 1200 kg along a hallway of the spacecraft that has a 90° right turn, without touching the walls, floor, or ceiling, by working remotely, using devices attached to the load that can be programmed to fire blasts of compressed air for up to 1.0 s in any desired direction. During a blast the load is subjected to a force of 20 N. The center of the load must move 3 m along the first section of the hallway, starting from rest, then 4 m along the second section, ending at rest. Let the starting point be $\langle 0, 0, 0 \rangle$ m, with the first section ending at $\langle 0, 3, 0 \rangle$ m and the second section ending at $\langle 4, 3, 0 \rangle$ m. Using just three blasts of compressed air, choose the times when these blasts should be scheduled, their durations, and their directions. How long does it take to complete the entire move?

Section 2.5

•P25 A runner starts from rest and in 3 s reaches a speed of 8 m/s. Assume that her speed changed at a constant rate (constant net force). **(a)** What was her average speed during this 3 s interval? **(b)** How far did she go in this 3 s interval?

•P26 The driver of a car traveling at a speed of 18 m/s slams on the brakes and comes to a stop in 4 s. If we assume that the car's speed changed at a constant rate (constant net force): **(a)** What was the car's average speed during this 4 s interval? **(b)** How far did the car go in this 4 s interval?

•P27 On a straight road with the $+x$ axis chosen to point in the direction of motion, you drive for 3 h at a constant 30 mi/h, then in a few seconds you speed up to 60 mi/h and drive at this speed for 1 h. **(a)** What was the x component of average velocity for the 4 h period, using the fundamental definition of average velocity, which is the displacement divided by the time interval? **(b)** Suppose that instead you use the equation $v_{avg,x} = (v_{ix} + v_{fx})/2$. What do you calculate for the x component of average velocity? **(c)** Why does the equation used in part (b) give the wrong answer?

•P28 A ball of mass 0.4 kg flies through the air at low speed, so that air resistance is negligible. **(a)** What is the net force acting on the ball while it is in motion? **(b)** Which components of the ball's momentum will be changed by this force? **(c)** What happens to the x component of the ball's momentum during its flight? **(d)** What happens to the y component of the ball's momentum during its flight? **(e)** What happens to the z component of the ball's momentum during its flight? **(f)** In this situation, why is it legitimate to use the expression for average y component of velocity, $v_{avg,y} = (v_{iy} + v_{fy})/2$, to update the y component of position?

•P29 For each graph of v_x vs. t numbered 1–6 in Figure 2.56, choose the letter (a–i) corresponding to the appropriate description of motion of a fan cart moving along a track. Not all descriptions will be used. Assume the usual coordinate system ($+x$ to the right, $+y$ up, $+z$ out of the page).

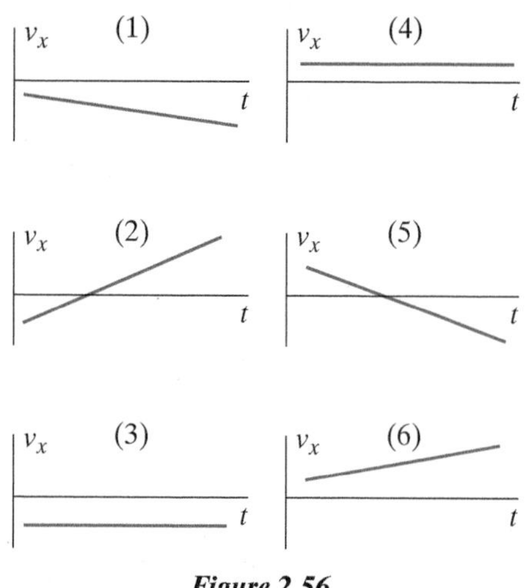

Figure 2.56

(a) A cart moves to the left, gradually slowing down. **(b)** A cart moves to the right, gradually speeding up. **(c)** A cart moves to the left at constant speed. **(d)** A cart moves to the left, gradually slowing down, stops, and moves to the right, speeding up. **(e)** A cart remains stationary and does not move. **(f)** A cart moves to the right, gradually slowing down. **(g)** A cart moves to the right, gradually slowing down, stops, and moves to the left, speeding up. **(h)** A cart moves to the left, gradually speeding up. **(i)** A cart moves to the right at constant speed.

•P30 A cart rolls with low friction on a track. A fan is mounted on the cart, and when the fan is turned on, there is a constant force acting on the cart. Three different experiments are performed: **(a)** Fan off: The cart is originally at rest. You give it a brief push, and it coasts a long distance along the track in the $+x$ direction, slowly coming to a stop. **(b)** Fan forward: The fan is turned on, and you hold the cart stationary. You then take your hand away, and the cart moves forward, in the $+x$ direction. After traveling a long distance along the track, you quickly stop and hold the cart. **(c)** Fan backward: The fan is turned on facing the "wrong" way, and you hold the cart stationary. You give it a brief push, and the cart moves forward, in the $+x$ direction, slowing down and then turning around, returning to the starting position, where you quickly stop and hold the cart. Figure 2.57 displays four graphs of p_x (numbered 1–4), the x component of momentum, vs. time. The graphs start when the cart is at rest, and end when the cart is again at rest. Match the experiment with the correct graph.

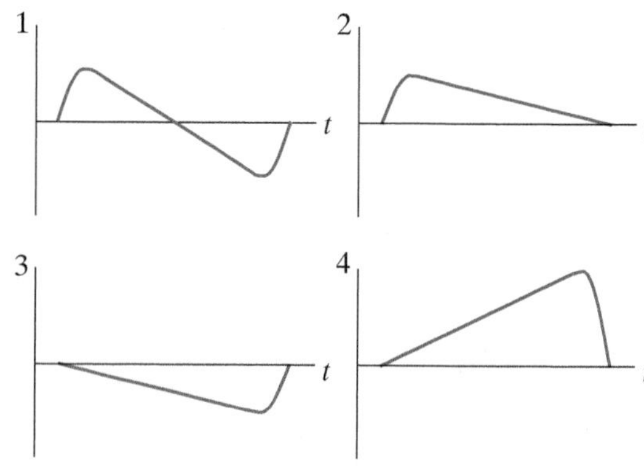

Figure 2.57

•P31 Consider the three experiments described in Problem 30. Figure 2.58 displays four graphs of $F_{net,x}$, the x component of the net force acting on the cart, vs. time. The graphs start when the cart is at rest, and end when the cart is again at rest. Match the experiment with the graph.

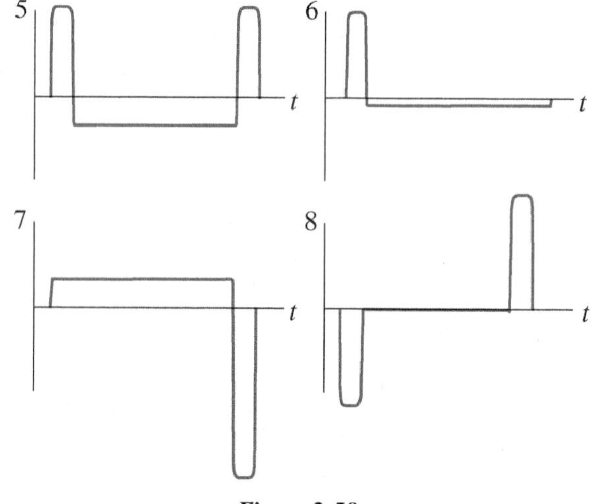

Figure 2.58

•P32 Consider the three experiments described in Problem 30. Figure 2.59 displays four graphs of x, positioned along the track, vs. time. The graphs start when the cart is at rest, and end when the cart is again at rest. Match the experiment with the graph.

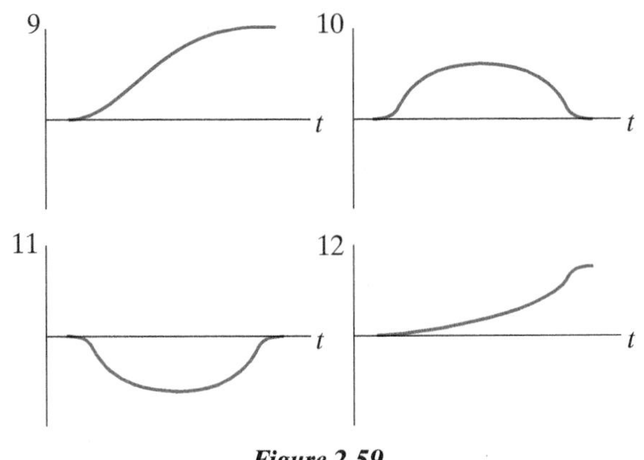

Figure 2.59

••**P33** You are a detective investigating why someone was hit on the head by a falling flowerpot. One piece of evidence is a home video taken in a 4th-floor apartment, which happens to show the flowerpot falling past a tall window. Inspection of individual frames of the video shows that in a span of 6 frames the flowerpot falls a distance that corresponds to 0.85 of the window height seen in the video. (*Note:* Standard video runs at a rate of 30 frames per second.) You visit the apartment and measure the window to be 2.2 m high. What can you conclude? Under what assumptions? Give as much detail as you can.

••**P34** A soccer ball is kicked at an angle of 59° to the horizontal with an initial speed of 20 m/s. Assume for the moment that we can neglect air resistance. **(a)** For how much time is the ball in the air? **(b)** How far does it go (horizontal distance along the field)? **(c)** How high does it go?

••**P35** A ball is kicked from a location $\langle 9,0,-6\rangle$ (on the ground) with initial velocity $\langle -11,16,-6\rangle$ m/s. The ball's speed is low enough that air resistance is negligible. **(a)** What is the velocity of the ball 0.5 s after being kicked? (Use the Momentum Principle!) **(b)** In this situation (constant force), which velocity will give the most accurate value for the location of the ball 0.5 s after it is kicked: the arithmetic average of the initial and final velocities, the final velocity of the ball, or the initial velocity of the ball? **(c)** What is the average velocity of the ball over this time interval (a vector)? **(d)** Use the average velocity to find the location of the ball 0.5 s after being kicked.

Now consider a different time interval: the interval between the initial kick and the moment when the ball reaches its highest point. We want to find how long it takes for the ball to reach this point, and how high the ball goes. **(e)** What is the y component of the ball's velocity at the instant when the ball reaches its highest point (the end of this time interval)? **(f)** Fill in the known quantities in the update form of the Momentum Principle, $mv_{yf} = mv_{yi} + F_{\text{net},y}\,\Delta t$, leaving as symbols anything that is unknown. **(g)** How long does it take for the ball to reach its highest point? **(h)** Knowing this time, first find the y component of the average velocity during *this* time interval, then use it to find the maximum height attained by the ball.

Now take a moment to reflect on the reasoning used to solve this problem. You should be able to do a similar problem on your own, without prompting. Note that the only equations needed were the Momentum Principle and the expression for the arithmetic average velocity.

••**P36** A small dense ball with mass 1.5 kg is thrown with initial velocity $\langle 5,8,0\rangle$ m/s at time $t = 0$ at a location we choose to call the origin ($\langle 0,0,0\rangle$). Air resistance is negligible. **(a)** When the ball reaches its maximum height, what is its velocity (a vector)? It may help to make a simple diagram. **(b)** When the ball reaches its maximum height, what is t? You know how v_y depends on t, and you know the initial and final velocities. **(c)** Between the launch at $t = 0$ and the time when the ball reaches its maximum height, what is the average velocity (a vector)? You know how to determine average velocity when velocity changes at a constant rate. **(d)** When the ball reaches its maximum height, what is its location (a vector)? You know how average velocity and displacement are related. **(e)** At a later time the ball's height y has returned to zero, which means that the average value of v_y from $t = 0$ to this time is zero. At this instant, what is the time t? **(f)** At the time calculated in part (e), when the ball's height y returns to zero, what is x? (This is called the "range" of the trajectory.) **(g)** At the time calculated in part (e), when the ball's height y returns to zero, what is v_y? **(h)** What was the angle to the x axis of the initial velocity? **(i)** What was the angle to the x axis of the velocity at the time calculated in part (e), when the ball's height y returned to zero?

••**P37** Apply the general results obtained in the full analysis of motion under the influence of a constant force in Section 2.5 to answer the following questions. You hold a small metal ball of mass m a height h above the floor. You let go, and the ball falls to the floor. Choose the origin of the coordinate system to be on the floor where the ball hits, with y up as usual. Just after release, what are y_i and v_{iy}? Just before hitting the floor, what is y_f? How much time Δt does it take for the ball to fall? What is v_{fy} just before hitting the floor? Express all results in terms of m, g, and h. How would your results change if the ball had twice the mass?

••**P38** In a cathode ray tube (CRT) used in older television sets, a beam of electrons is steered to different places on a phosphor screen, which glows at locations hit by electrons. The CRT is evacuated, so there are few gas molecules present for the electrons to run into. Electric forces are used to accelerate electrons of mass m to a speed $v_0 \ll c$, after which they pass between positively and negatively charged metal plates that deflect the electron in the vertical direction (upward in Figure 2.60, or downward if the sign of the charges on the plates is reversed).

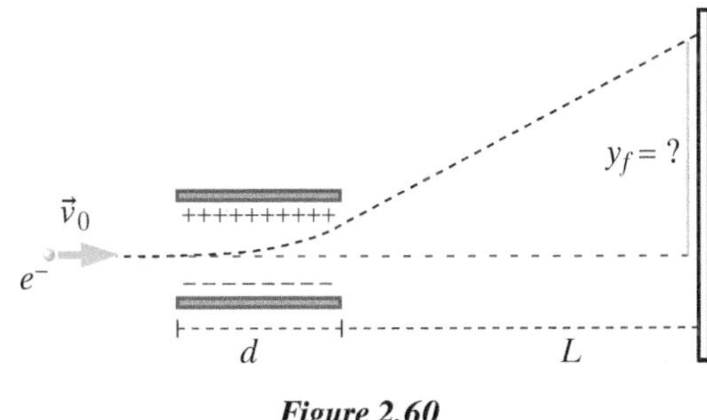

Figure 2.60

While an electron is between the plates, it experiences a uniform vertical force F, but when the electron is outside the plates there is negligible force on it. The gravitational force on the electron is negligibly small compared to the electric force in this situation. The length of the metal plates is d, and the phosphor screen is a distance L from the metal plates. Where does the electron hit the screen? (That is, what is y_f?)

••P39 The performance of two different cars, car 1 and car 2, was measured on a long horizontal test track. Car 1 started from rest and ran with constant acceleration until it was halfway down the track and then stopped accelerating, continuing to run at the attained speed to the end of the track. Car 2 started from rest and ran with a constant acceleration for the entire distance. It was observed that both cars covered the test distance in the same amount of time. **(a)** What was the ratio of the average speed of car 1 to that of car 2? **(b)** What was the ratio of the initial acceleration of car 1 to that of car 2? **(c)** What was the ratio of the final speed of car 1 to that of car 2?

••P40 A driver starts from rest on a straight test track that has markers every 0.1 km. The driver presses on the accelerator and for the entire period of the test holds the car at constant acceleration. The car passes the 0.1 km post at 8 s after starting the test. **(a)** What was the car's acceleration? **(b)** What was the car's speed as it passed the 0.1 km post? **(c)** What does the speedometer read at the 0.2 km post? **(d)** When does the car pass the 0.2 km post?

Section 2.6

For all problems in this section, use the approximation that $\vec{v}_{\text{avg}} \approx \vec{p}_f/m$ for each time step.

•P41 The stiffness of a particular spring is 40 N/m. One end of the spring is attached to a wall. When you pull on the other end of the spring with a steady force of 2 N, the spring elongates to a total length of 18 cm. What was the relaxed length of the spring? (Remember to convert to SI units.)

••P42 A spring with a relaxed length of 25 cm and a stiffness of 11 N/m stands vertically on a table. A block of mass 70 g is attached to the top of the spring. You pull the block upward, stretching the spring until its length is now 28 cm, hold the block at rest for a moment, and then release it. Using a time step of 0.1 s, predict the position and momentum of the block at a time 0.2 s after you release the block.

••P43 A block is attached to the top of a spring that stands vertically on a table. The spring stiffness is 55 N/m, its relaxed length is 23 cm, and the mass of the block is 350 g. The block is oscillating up and down as the spring stretches and compresses. At a particular time you observe that the velocity of the block is $\langle 0, 0.0877, 0\rangle$ m/s, and the position of the block is $\langle 0, 0.0798, 0\rangle$ m, relative to an origin at the base of the spring. Using a time step of 0.1 s, determine the position of the block 0.2 s later.

••P44 A paddle ball toy consists of a flat wooden paddle and a small rubber ball that are attached to each other by an elastic band (Figure 2.61). You have a paddle ball toy for which the mass of the ball is 0.015 kg, the stiffness of the elastic band is 0.9 N/m, and the relaxed length of the elastic band is 0.30 m. You are holding the paddle so the ball hangs suspended under it, when your cat comes along and bats the ball around, setting it in motion. At a particular instant the the momentum of the ball is $\langle -0.02, -0.01, -0.02\rangle$ kg · m/s, and the moving ball is at location $\langle -0.2, -0.61, 0\rangle$ m relative to an origin located at the point where the elastic band is attached to the paddle. **(a)** Determine the position of the ball 0.1 s later, using a Δt of 0.1 s. **(b)** Starting with the same initial position ($\langle -0.2, -0.61, 0\rangle$ m) and momentum ($\langle -0.02, -0.01, -0.02\rangle$ kg · m/s), determine the position of the ball 0.1 s later, using a Δt of 0.05 s. **(c)** If your answers are different, explain why.

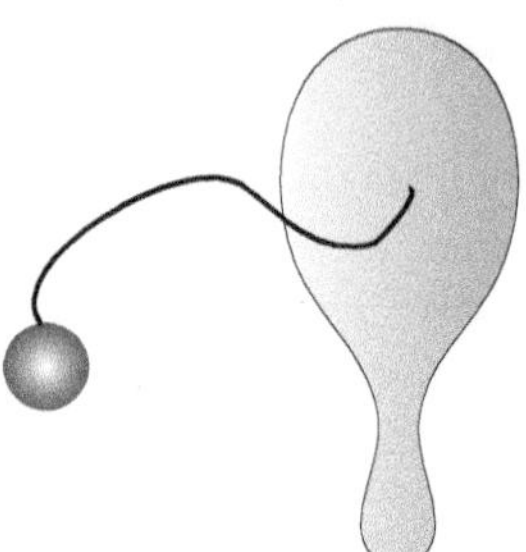

Figure 2.61

••P45 A block of mass 0.5 kg is placed at rest on a relaxed vertical spring-like device of length 0.1 m which exerts an upward force of $F_y = -2 \times 10^5 \times s^3$, where s is the stretch. With a time step of 0.04 s, calculate the length L of the device and the y component of the velocity of the block at $t = 0.04s$ and $t = 0.08s$.

Section 2.9

•P46 A proton has mass 1.7×10^{-27} kg. What is the magnitude of the impulse required to increase its speed from $0.990c$ to $0.994c$?

••P47 SLAC, the Stanford Linear Accelerator Center, located at Stanford University in Palo Alto, California, accelerates electrons through a vacuum tube 2 mi long (it can be seen from an overpass of the Junipero Serra freeway that goes right over the accelerator). Electrons that are initially at rest are subjected to a continuous force of 2×10^{-12} newton along the entire length of 2 mi (1 mi is 1.6 km) and reach speeds very near the speed of light. **(a)** Determine how much time is required to increase the electrons' speed from $0.93c$ to $0.99c$. (That is, the quantity $|\vec{v}|/c$ increases from 0.93 to 0.99.) **(b)** Approximately how far does the electron go in this time? What is approximate about your result?

COMPUTATIONAL PROBLEMS

You should do the introductory computational problems at the end of Chapter 1 before doing these problems.

More detailed and extended versions of some of these computational modeling problems may be found in the lab activities included in the *Matter & Interactions, 4th Edition*, resources for instructors.

A note on graphs: Graphs in VPython appear in a separate window. A graph is dynamic; it appears point by point as the program runs, and scales itself automatically. To make a graph in VPython requires three things, as shown in the program shown below:

(1) At the beginning of the program, type these two lines:
```
from visual import *
from visual.graph import *
```
(2) Create one or more `gcurve` objects, such as the one called `speed` in the program shown below.

(3) Inside the computational loop, add one or more `plot` operations specifying the pair of values to plot, such as the statement `speed.plot( pos=(t, v) )` in the program shown below.

Consult the VPython help for full details on graphing.

```
from visual import *
# Import graphing:
from visual.graph import *
...
# Create a graphing curve:
speed = gcurve(color=color.yellow)
...
while ship.pos.x < L:
    rate(20000)
    # update position and momentum:
    ....
    # Calculate speed v, and
    # add to the graphing curve:
    speed.plot( pos=(t, v) )
    # Update the time:
    t = t + dt
```

••**P48** Write an iterative computational model that predicts and displays (as a real-time animation) the motion of a fan cart on a low-friction track. A typical track is 2 m long, and a typical fan cart (which can be represented by a box object) is about 10 cm long and has a mass of about 0.8 kg (including the fan). Use a time step $\Delta t = 0.01$ s. Run your program often while writing it. It is much easier to find and fix errors when creating the program incrementally. **(a)** Start the cart at the left edge of the track, and give it an initial velocity of $\langle 0.5, 0, 0 \rangle$ m/s. **(b)** With the fan off ($\vec{F}_{\text{net}} = \vec{0}$) determine how long it takes for the cart to reach the right end of the track. **(c)** To model the behavior of the cart with the fan on, add a constant force to your model. By experimenting with your program, find values for the force and the initial velocity that lead to this behavior: The cart starts at the right end of the track and moves to the left, gradually slowing down. It stops at the left end of the track, and moves back to the right, gradually speeding up. **(d)** What happens when the initial velocity of the cart has a nonzero y component? Make the cart leave a trail so you can see its path clearly. Explain why your model predicts this behavior.

••**P49** Starting with the program you wrote in Problem P48, **(a)** Add commands to create a graph of the x component of the cart's position (`cart.pos.x`) vs. time as the cart moves. Explain the shape of the graph. **(b)** Change your graph to plot the x component of the cart's momentum vs. time. Explain the shape of the graph.

••**P50** Write an iterative computational model that predicts and displays the 3D motion of a 55 g tennis ball that leaves a tennis racket with a speed of 55 m/s. Neglect air resistance (at this speed, this approximation is actually not realistic). Experiment by changing the initial velocity of the ball to see what kinds of different trajectories can be produced.

•••**P51** The nearest stars are a group of three stars orbiting each other, Alpha Centauri A, Alpha Centauri B, and Proxima Centauri, located about 4.3 light years from Earth (one light year is the distance light travels in one Earth year). Suppose a space tug is able to pull a cargo ship with a constant force of $g = 9.8$ N/kg times the mass of the cargo ship, for many years. Starting from rest, speed up the cargo ship until you're halfway to the nearest stars, then pull back with the same force to slow the cargo ship back to rest when you reach the nearest stars. Determine the maximum speed attained (which occurs at the halfway point) and how many Earth years the trip takes. You must use the relativistic position update equation, because the speed approaches the speed of light. It is interesting to make a graph of the speed as a function of time, which does not consist of straight lines as it would for nonrelativistic motion. Time passes slowly on board the cargo ship, a relativistic effect that we'll discuss in Chapter 20.

ANSWERS TO CHECKPOINTS

1 (1) $\langle 60, -60, 0 \rangle$ N; **(2) (a)** $\langle 0, -2, 0 \rangle$ kg·m/s **(b)** $\langle 0, -0.667, 0 \rangle$ N

2 (a) By estimating the distance one student "traveled" during the collision–that is, how much the student's body was compressed. Using the approximation that $v_{\text{avg}} \approx (v_i + 0)/2$ we divided distance by average speed to estimate contact time. **(b)** Only the z component of $\vec{p}$ changed, so we were able to deduce that $\vec{F}_{\text{net}}$ was in the z direction.

3 (1) b **(2)** a

4 (1) c **(2)** c

5 (a) $\langle -10, 7.12, -5 \rangle$ m/s **(b)** $\langle 3, 6.04, -8 \rangle$ m **(c)** 8.62 m

6 (1) 0.275 N **(2) (a)** 3571 N/m **(b)** 71.4 N

7 (1) **(a)** $\langle 0, 1, 0 \rangle$, **(b)** $\langle 0, 1, 0 \rangle$; (2) **(a)** $\langle 0, 0.11, 0 \rangle$ m, **(b)** 0.11 m, **(c)** $\langle 0, 1, 0 \rangle$, **(d)** -0.04 m, **(e)** $\langle 0, 3.8, 0 \rangle$ N

8 (a) The momentum of the block a short time in the future depends on two things: its momentum now and the net force acting on it now. At the beginning of time step 2 the net force is downward but the momentum is upward. The sum of the momentum now plus the downward impulse over the next Δt gives a new momentum that is smaller in magnitude, but still upward, so the block moves upward.

9 a

10 a, b, c, d

CHAPTER

3

The Fundamental Interactions

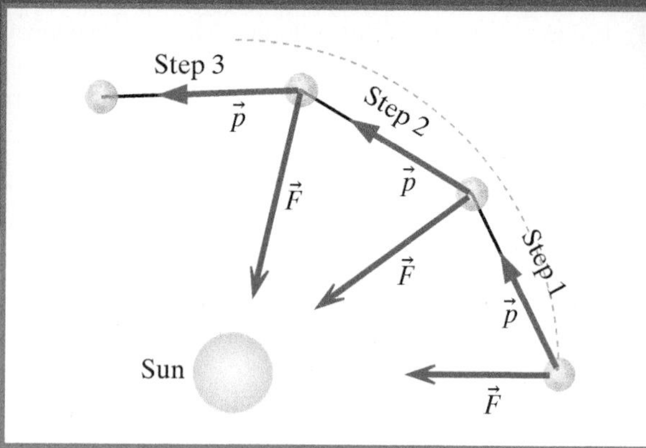

OBJECTIVES

After studying this chapter you should be able to

- Calculate the 3D gravitational or electric force exerted on a system by objects in its surroundings.
- Iteratively predict the motion of an object that interacts gravitationally or electrically with its surroundings, by hand or with a computer.
- Analyze simple collisions by applying the Momentum Principle to a system of more than one particle.

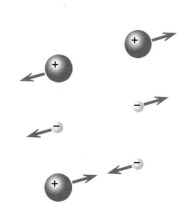

Figure 3.1 The electric interaction: protons repel each other; electrons repel each other; protons and electrons attract each other.

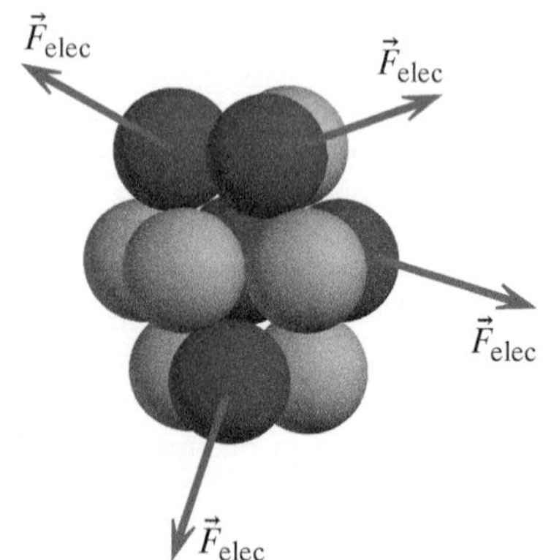

Figure 3.2 The strong interaction: the protons in the nucleus of an atom exert repulsive electric forces on each other, but the strong interaction (which involves neutrons as well as protons) holds the nucleus together despite this electric repulsion.

3.1 THE FUNDAMENTAL INTERACTIONS

In the world we see what seem to be many different kinds of interactions. Planets orbit stars. A falling leaf whirls in the wind. You bend a metal rod. Carbon and oxygen react to form carbon dioxide. In a nuclear power station, uranium nuclei split, making water boil, which drives electric generators. Despite the variety of effects we observe, it became clear in the 20th century that all the changes we see are due to just four different kinds of fundamental interactions: gravitational, electromagnetic, "strong" (also referred to as the nuclear interaction), and "weak."

- The *gravitational* interaction is responsible for an attraction between objects that have mass. For example, the Earth exerts a gravitational force on the Moon, and the Moon exerts a gravitational force on the Earth.
- The *electromagnetic* interaction is responsible for attraction or repulsion between objects that have electric charge. Electric forces are responsible for sparks, static cling, and the behavior of electronic circuits, and magnetic forces are responsible for the operation of motors driven by electric current. Protons repel each other electrically, as do electrons, whereas protons and electrons attract each other (Figure 3.1). Electric forces bind protons and electrons to each other in atoms, and are responsible for the chemical bonds between atoms in molecules. The force of a stretched or compressed spring is due to electric forces between the atoms that make up the spring.
- The *strong* or nuclear interaction occurs between objects made of quarks, such as protons and neutrons, which are held together in the nucleus of an atom despite the large mutual electric repulsion of the protons (Figure 3.2). (The neutrons are not electrically charged and don't exert electric forces.)
- The *weak* interaction affects all kinds of elementary particles but is much weaker than the strong and electromagnetic interactions. An example of its effects is seen in the instability of a neutron. If a neutron is removed from a nucleus, with an average lifetime of about 15 minutes the neutron decays

into a proton, an electron, and a ghostly particle called the antineutrino. This change is brought about by the weak interaction.

In this chapter we will be concerned primarily with gravitational and electric interactions. In later chapters dealing with energy we will encounter situations in which the strong interaction plays an important role. An example of the weak interaction appears in the section on conservation of momentum later in this chapter. The weak interaction is not important in most everyday interactions, so we will not discuss it extensively. We will defer a discussion of magnetic interaction, the other part of the electromagnetic interaction, until later chapters.

Although it continues to be fruitful to classify interactions into four types, it was found in the second half of the 20th century that the electromagnetic interaction and the weak interaction can be considered to be different manifestations of one type of interaction, now called the "electroweak" interaction. Soon after this discovery, it became possible to unify the strong interaction and the electroweak interaction within one powerful theory, the "Standard Model," which also explains the nature of subatomic particles such as the proton and neutron. At present there seem to be really only two fundamental categories of interactions: those explained by the Standard Model and those explained by the gravitational interaction. Physicists are aggressively searching for ways to unify the Standard Model with gravity; this is one of the major scientific quests of the present era.

3.2 THE GRAVITATIONAL FORCE

The motion of stars and planets is in important ways simpler than other mechanical phenomena, because there is no friction to worry about. These massive bodies interact through the gravitational force, which is always an attractive force. Studying how to predict the motion of stars and planets is one of the most direct ways to understand in general how the Momentum Principle determines the behavior of objects in the real world. The basic ideas used to predict the motion of stars and planets can be applied to a wide range of everyday and atomic phenomena.

In the 1600s Isaac Newton deduced that there must be an attractive force associated with a gravitational interaction between any pair of objects. The gravitational force acts along a line connecting the two objects (Figure 3.3), is proportional to the mass of one object and to the mass of the other object, and is inversely proportional to the square of the distance between the *centers* of the two objects (not the gap between their surfaces). The gravitational force exerted on object 2 by object 1 is expressed by this compact vector equation:

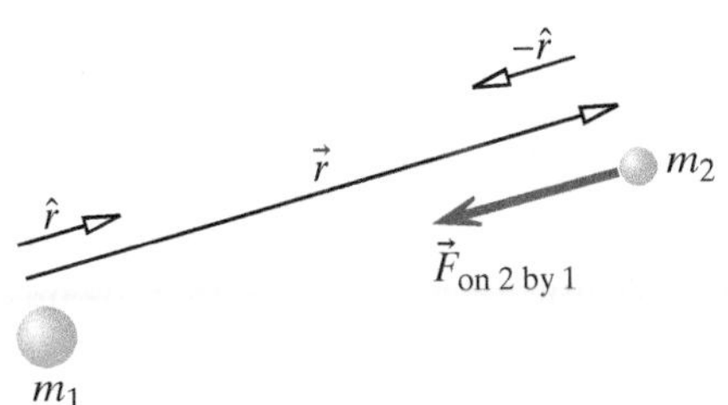

Figure 3.3 The gravitational force exerted on object 2 by object 1.

THE GRAVITATIONAL FORCE

$$\vec{F}_{\text{grav on 2 by 1}} = -G\frac{m_1 m_2}{|\vec{r}|^2}\hat{r}$$

m_1 and m_2 are the masses of the objects.

$\vec{r} = \vec{r}_2 - \vec{r}_1$ extends from the center of object 1 to the center of object 2.

The universal constant $G = 6.7 \times 10^{-11}\dfrac{\text{N}\cdot\text{m}^2}{\text{kg}^2}$.

This is often called Newton's "universal law of gravitation," reflecting its central importance in the development of human understanding of the Universe. The word "universal" is used because all massive objects in the Universe attract all other massive objects in the Universe.

The gravitational force involves a lot of different symbols and may look pretty intimidating at first. Let's take the expression apart and look at the individual pieces to try to make sense of the expression.

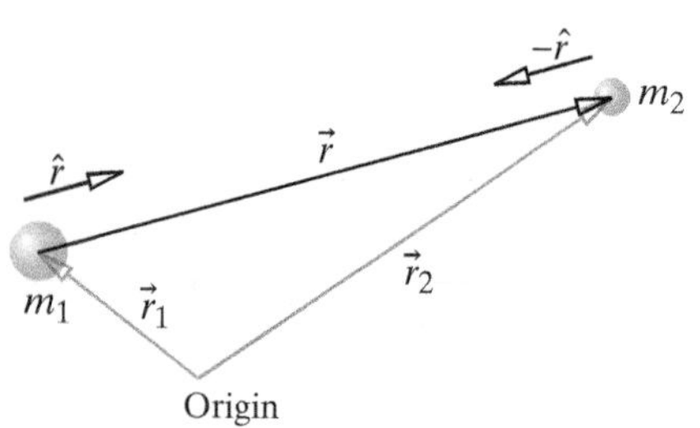

Figure 3.4 $\vec{r} = \vec{r}_2 - \vec{r}_1$, the location of object 2 relative to object 1: "final minus initial." The gravitational force on object 2 is in the direction of $-\hat{r}$.

The Relative Position Vector

The relative position vector $\vec{r}$ extends from the center of object 1 to the center of object 2, as shown in Figure 3.4. (This vector can also be written as $\vec{r}_{2-1}$; we will use the notation $\vec{r}$, which is more compact, but requires that you remember that object 1 is the initial location and object 2 is the final location.) The unit vector $\hat{r}$ points in the same direction as $\vec{r}$, but has magnitude 1. The magnitude of $\vec{r}$ is the distance between the centers of the two objects.

The Direction of the Gravitational Force

The direction of the gravitational force on object 2 by object 1 is specified by $-\hat{r}$ (Figure 3.4), which is the unit vector $\hat{r}$, in combination with the minus sign, as indicated in Figure 3.5.

$$\vec{F}_{\text{on 2 by 1}} = -G\frac{m_1 m_2}{|\vec{r}|^2}\hat{r}$$

Figure 3.5 The direction of the gravitational force is opposite to the direction of the unit vector $\hat{r}$, which points from object 1 to object 2.

QUESTION Why is the minus sign necessary?

The minus sign is necessary because the force on object 2 due to object 1 is in the direction opposite to $\hat{r}$ (Figure 3.4).

Mass and Magnitude of the Gravitational Force

As highlighted in Figure 3.6, the gravitational force is proportional to the product of the two masses, $m_1 m_2$. If you double either of these masses, keeping the other one the same, the force will be twice as big.

$$\vec{F}_{\text{on 2 by 1}} = -G\frac{m_1 m_2}{|\vec{r}|^2}\hat{r}$$

Figure 3.6 The magnitude of the gravitational force depends on the masses of both interacting objects.

QUESTION If both masses are doubled, how much larger will the force be?

If you double both of the masses, the force will be four times as big.

Note that since $m_2 m_1 = m_1 m_2$, the magnitude of the force exerted on object 1 by object 2 is exactly the same as the magnitude of the force exerted on object 2 by object 1, but the direction is opposite. This important property of the gravitational force is called "reciprocity" and is discussed in more detail in Section 3.4.

Distance and Magnitude of the Gravitational Force

$$\vec{F}_{\text{on 2 by 1}} = -G\frac{m_1 m_2}{|\vec{r}|^2}\hat{r}$$

Figure 3.7 The gravitational force is an "inverse square" force.

The gravitational force is an "inverse square" force. As highlighted in Figure 3.7, the square of the center-to-center distance appears in the denominator. This means that the gravitational force depends very strongly on the distance between the objects. For example, if you double the distance between them, the only thing that changes is the denominator, which gets four times bigger (2 squared is 4), so the force is only 1/4 as big as before:

$$|\vec{F}_d| = \frac{Gm_1 m_2}{|\vec{r}|^2}$$

$$|\vec{F}_{2d}| = \frac{Gm_1 m_2}{|2\vec{r}|^2} = \frac{1}{4}\frac{Gm_1 m_2}{|\vec{r}|^2} = \frac{1}{4}|\vec{F}_d|$$

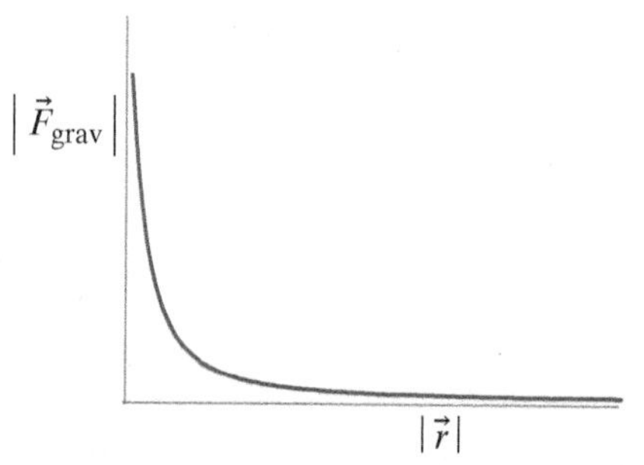

Figure 3.8 The magnitude of an inverse square force is large at small distances, and very small at large distances.

QUESTION If you move the masses 10 times farther apart than they were originally, how does the gravitational force change?

The force goes down by a factor of 100. Evidently when two objects are very far apart, the gravitational forces they exert on each other will be vanishingly small: big denominator, small force (Figure 3.8).

The Gravitational Constant G

$$G = 6.7 \times 10^{-11} \frac{\text{N} \cdot \text{m}^2}{\text{kg}^2}$$

We say that the gravitational constant G is "universal" because it is the same for any pair of interacting masses, no matter how big or small they are, or where they are located in the Universe. Because G is universal, it can be measured for a particular pair of objects and then used with other pairs of objects. As we describe in Section 3.17, Cavendish was the first person to make such a measurement.

Checkpoint 1 A star exerts a gravitational force of magnitude 4×10^{25} N on a planet. **(a)** What is the magnitude of the gravitational force that the planet exerts on the star? **(b)** If the mass of the planet were twice as large, what would be the magnitude of the gravitational force on the planet? **(c)** If the distance between the star and planet (with their original masses) were three times larger, what would be the magnitude of this force?

Calculating Gravitational Force

A useful way to think about the gravitational force is to factor it into a scalar part and a direction, like this:

A vector is a magnitude times a direction: $\vec{F}_{\text{grav}} = |\vec{F}_{\text{grav}}|\hat{F}_{\text{grav}}$

- Magnitude: $|\vec{F}_{\text{grav}}| = G\dfrac{m_1 m_2}{|\vec{r}|^2}$
- Direction (unit vector): $\hat{F}_{\text{grav}} = -\hat{r}$

It is often simplest to calculate the magnitude and direction separately, then combine them to get the vector force. That way you can focus on one thing at a time rather than getting confused (or intimidated!) by the full complexity of the vector force.

Calculating the gravitational force on an object due to another object requires several steps. Use the equation for gravitational force as a guide that tells you to carry out these steps:

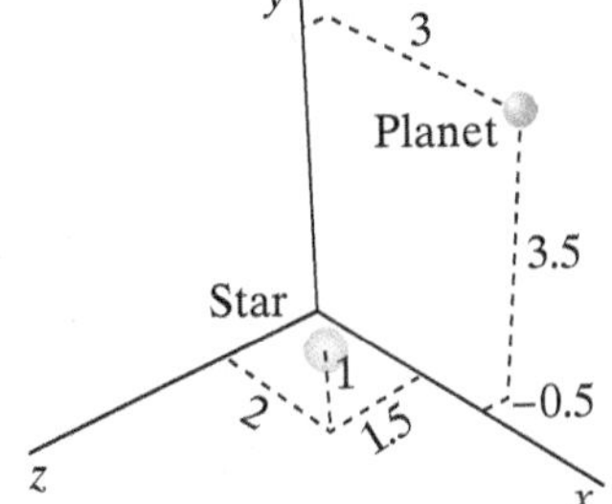

Figure 3.9 A star and a planet interact gravitationally. Distances shown should be multiplied by 1×10^{11} m.

CALCULATING GRAVITATIONAL FORCE

- Calculate $\vec{r} = \vec{r}_2 - \vec{r}_1$, the position of the center of object 2 relative to the center of object 1.
- Calculate $|\vec{r}|$, the center-to-center distance between the objects.
- Calculate $Gm_1m_2/|\vec{r}|^2$, the magnitude of the force.
- Calculate $-\hat{r} = -\vec{r}/|\vec{r}|$, the direction of the force.
- Multiply the magnitude times the direction to get the vector force.

EXAMPLE **The Force on a Planet by a Star**

Figure 3.9 shows a star of mass 4×10^{30} kg located at this moment at position $\langle 2 \times 10^{11}, 1 \times 10^{11}, 1.5 \times 10^{11} \rangle$ m and a planet of mass 3×10^{24} kg located at position $\langle 3 \times 10^{11}, 3.5 \times 10^{11}, -0.5 \times 10^{11} \rangle$ m. (These are typical values for stars and planets. Notice that the mass of the star is much greater than that of the planet.) In Figure 3.9 we show the x, y, and z components of the positions, to be multiplied by 1×10^{11} m. Make sure you understand how the numbers on the diagram correspond to the positions given as vectors. **(a)** Calculate the gravitational force exerted on the planet by the star. **(b)** Calculate the gravitational force exerted on the star by the planet.

Solution

QUESTION Which object is object 1, and which is object 2?

In this case **(a)** the planet is object 2 (we want the force on the planet), and the star is object 1.

Relative position vector:

$$\vec{r} = \vec{r}_2 - \vec{r}_1$$
$$\vec{r} = \langle 3\times10^{11}, 3.5\times10^{11}, -0.5\times10^{11}\rangle\ \text{m} - \langle 2\times10^{11}, 1\times10^{11}, 1.5\times10^{11}\rangle\ \text{m}$$
$$\vec{r} = \langle 1\times10^{11}, 2.5\times10^{11}, -2\times10^{11}\rangle\ \text{m}$$

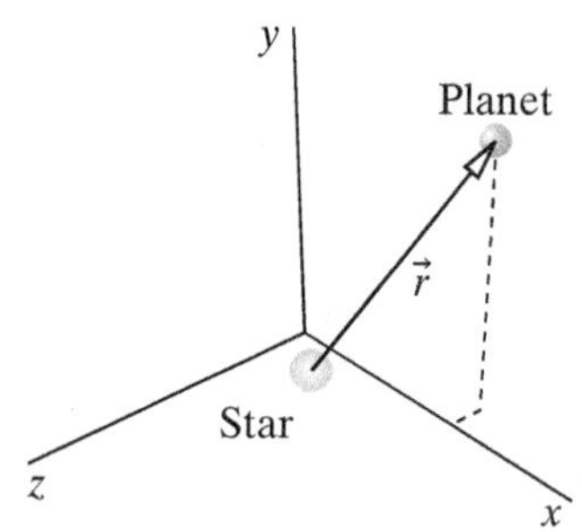

Figure 3.10 The position vector of the planet relative to the star. Initial location: center of star. Final location: center of planet.

Check: Do the signs of the components make sense? Yes, see Figure 3.10.
Distance:

$$|\vec{r}| = \sqrt{(1\times10^{11})^2 + (2.5\times10^{11})^2 + (-2\times10^{11})^2}\ \text{m}$$
$$|\vec{r}| = 3.35\times10^{11}\ \text{m}$$

Magnitude of force:

$$|\vec{F}_{\text{on } P \text{ by } S}| = G\frac{m_1 m_2}{|\vec{r}|^2}$$
$$|\vec{F}_{\text{on } P \text{ by } S}| = \left(6.7\times10^{-11}\frac{\text{N}\cdot\text{m}^2}{\text{kg}^2}\right)\frac{(3\times10^{24}\ \text{kg})(4\times10^{30}\ \text{kg})}{(3.35\times10^{11}\ \text{m})^2}$$
$$|\vec{F}_{\text{on } P \text{ by } S}| = 7.15\times10^{21}\ \text{N}$$

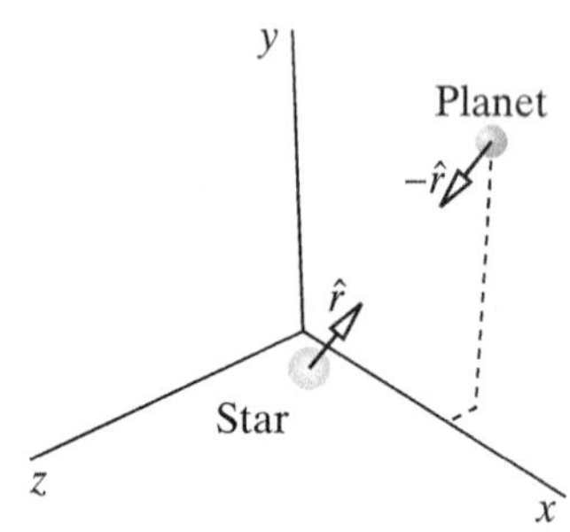

Figure 3.11 Unit vectors.

Direction of force on planet:

$$\hat{F}_{\text{on } P \text{ by } S} = -\hat{r} = -\frac{\vec{r}}{|\vec{r}|}$$
$$\hat{F}_{\text{on } P \text{ by } S} = -\frac{\langle 1\times10^{11}, 2.5\times10^{11}, -2\times10^{11}\rangle\ \text{m}}{3.35\times10^{11}\ \text{m}}$$
$$\hat{F}_{\text{on } P \text{ by } S} = \langle -0.298, -0.745, 0.596\rangle$$

Check: Units? Okay, unit vector has no units. Direction? Figure 3.11 shows that $\hat{F}_{\text{on } P \text{ by } S}$ points from the planet toward the star, which is correct for the direction of the force. Magnitude of the unit vector? $\sqrt{(-0.298)^2 + (-0.745)^2 + (0.596)^2} = 0.9995$, okay to three significant figures.

Calculate force as a vector:

$$\vec{F}_{\text{on } P \text{ by } S} = |\vec{F}_{\text{on } P \text{ by } S}|\hat{F}_{\text{on } P \text{ by } S}$$
$$\vec{F}_{\text{on } P \text{ by } S} = (7.15\times10^{21}\ \text{N})\langle -0.298, -0.745, 0.596\rangle$$
$$\vec{F}_{\text{on } P \text{ by } S} = \langle -2.13\times10^{21}, -5.33\times10^{21}, 4.26\times10^{21}\rangle\ \text{N}$$

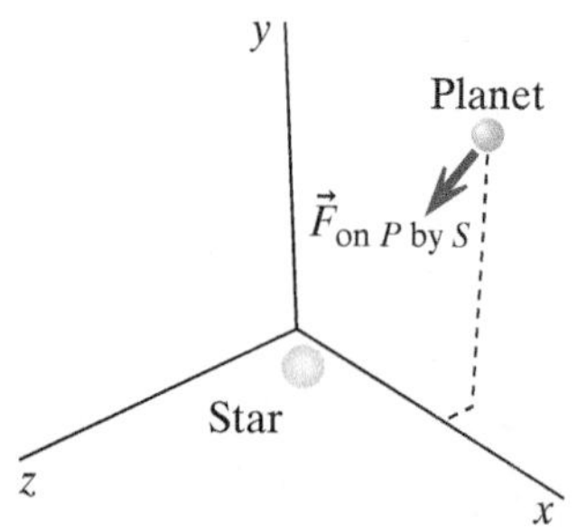

Figure 3.12 Gravitational force on the planet by the star.

Check: Direction? Figure 3.12 shows the force in the correct direction. Magnitude? Do a rough "order of magnitude" check: the distance between star and planet is *very* roughly 1×10^{11} m, the mass of the star is *very* roughly 1×10^{30} kg, the mass of the planet is *very* roughly 1×10^{24} kg, and the gravitational constant is *very* roughly $1\times10^{-11}\ \text{N}\cdot\text{m}^2/\text{kg}^2$. Therefore we expect the magnitude of the force to be *very* roughly this:

$$|\vec{F}_{\text{on } P \text{ by } S}| \approx \left(1\times10^{-11}\frac{\text{N}\cdot\text{m}^2}{\text{kg}^2}\right)\frac{(1\times10^{24}\ \text{kg})(1\times10^{30}\ \text{kg})}{(1\times10^{11}\ \text{m})^2} = 1\times10^{21}\ \text{N}$$

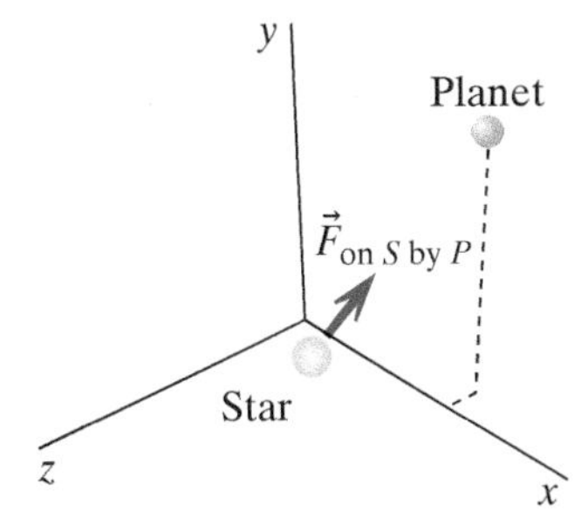

Figure 3.13 The gravitational force exerted on the star by the planet.

The fact that we get within an order of magnitude of the result 7.15×10^{21} N is evidence that we haven't made any huge mistakes.

The magnitude of the force on the star by the planet (part **b**) will be the same. The only change is in the direction, because $\vec{r}_{S-P} = -\vec{r}_{P-S}$.

$$\vec{F}_{\text{on } S \text{ by } P} = \langle 2.13 \times 10^{21}, 5.33 \times 10^{21}, -4.26 \times 10^{21} \rangle \text{ N}$$

Note that the signs of the force components for the force on the star are consistent with Figure 3.13. You might be puzzled that the planet pulls just as hard on the star as the star pulls on the planet. We'll discuss this in more detail later in this chapter.

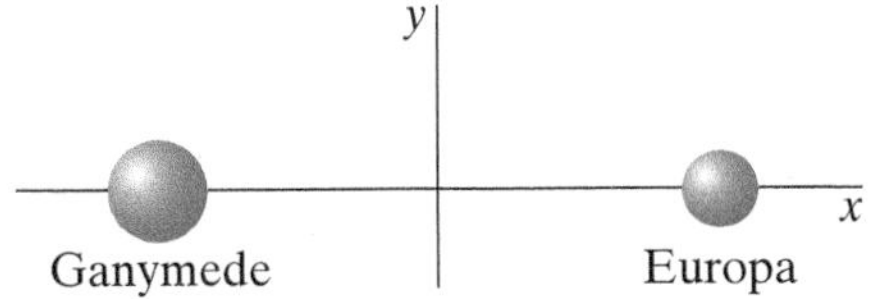

Figure 3.14 Two of Jupiter's moons, aligned along the x axis.

Checkpoint 2 At a particular instant Ganymede and Europa, two of the moons of Jupiter, are aligned as shown in Figure 3.14. Coordinate axes are shown in the diagram. **(a)** In calculating the gravitational force on Ganymede by Europa: (1) What is the direction of $\vec{r}$? (2) What is the direction of $-\hat{r}$? (3) What is the direction of the gravitational force? **(b)** In calculating the gravitational force on Europa by Ganymede: (1) What is the direction of $\vec{r}$? (2) What is the direction of $-\hat{r}$? (3) What is the direction of the gravitational force?

3.3 APPROXIMATE GRAVITATIONAL FORCE NEAR THE EARTH'S SURFACE

Earlier we used the expression mg to represent the magnitude of the gravitational force on an object near the Earth's surface. This is an approximation to the actual force, but it is a good one. The magnitude of the gravitational force that the Earth exerts on an object of mass m near the Earth's surface (Figure 3.15) is

$$F_g = G\frac{M_E m}{(R_E + y)^2}$$

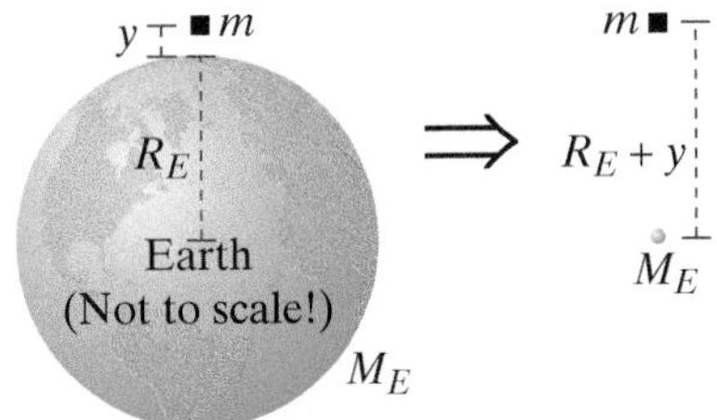

Figure 3.15 Determining the gravitational force by the Earth on an object a height y above the surface. The nearly spherical Earth acts as if all its mass were concentrated at its center.

where y is the distance of the object above the surface of the Earth, R_E is the radius of the Earth, and M_E is the mass of the Earth.

A spherical object of uniform density can be treated as if all its mass were concentrated at its center (see Section 3.16 for a discussion of this). As a result, we can treat the effect of the Earth on the object as though the Earth were a tiny, very dense ball a distance $R_E + y$ away.

Consider the difference in the gravitational force exerted by the Earth on an object when it is one meter above the surface compared with when it is two meters above the surface. The radius of the Earth is about 6.4×10^6 m. Then

$$F_{1\text{ m}} = G\frac{M_E m}{(6.400001 \times 10^6 \text{ m})^2} \quad \text{whereas} \quad F_{2\text{ m}} = G\frac{M_E m}{(6.400002 \times 10^6 \text{ m})^2}$$

For most purposes this difference is not significant. In fact, for all interactions of objects near the surface of the Earth, it makes sense to use the same approximate value, $R_E = 6.4 \times 10^6$ m, for the distance from the center of the Earth to the object. This simplifies calculation of gravitational forces by allowing us to combine all the constants into a single lumped constant, g:

$$F_g \approx \left(G\frac{M_E}{R_E^2}\right) m = gm$$

$$g = G\frac{M_E}{R_E^2}$$

QUESTION The mass of the Earth is 6×10^{24} kg and the radius of the Earth is 6.4×10^6 m. What is the value of g?

$$g = \left(6.7 \times 10^{-11} \frac{\text{N} \cdot \text{m}^2}{\text{kg}^2}\right) \frac{(6 \times 10^{24}\,\text{kg})}{(6.4 \times 10^6\,\text{m})^2}$$
$$= 9.8 \frac{\text{N}}{\text{kg}}$$

The constant g, called "the magnitude of the gravitational field," has the value $g = +9.8$ newtons/kilogram near the Earth's surface. Note that g is a positive number.

APPROXIMATE GRAVITATIONAL FORCE NEAR EARTH'S SURFACE

$$F_g \approx mg$$

where $g = G\frac{M_E}{R_E^2} = +9.8\frac{\text{N}}{\text{kg}}$.

M_E is the mass of the Earth, and R_E is the radius of the Earth. g is the magnitude of the gravitational field near the Earth's surface. The units of g are newtons/kg.

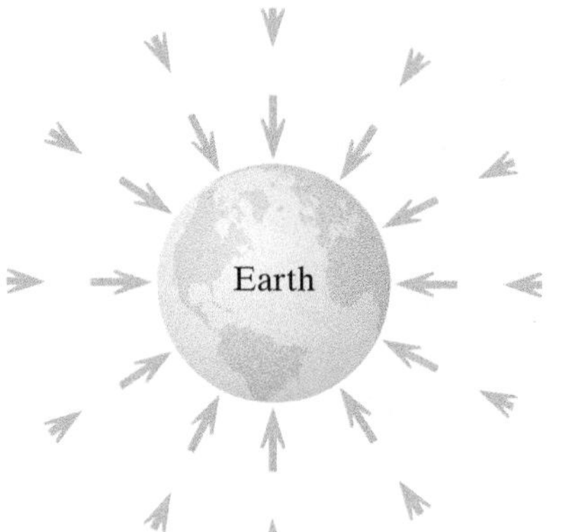

Figure 3.16 The "gravitational field" around the Earth. At a position where the field is $\vec{g}$, an object of mass m experiences a force $m\vec{g}$. The magnitude g of the field decreases with distance away from the Earth.

The "gravitational field" $\vec{g}$ at a location in space is defined to be the vector gravitational force per kg, so the force on a 1 kg mass placed at that location would be $(1\,\text{kg})\vec{g}$. A 2 kg mass would experience a force $(2\,\text{kg})\vec{g}$ at that location. In general, a mass m will experience a force $m\vec{g}$ of magnitude mg. The gravitational field around the Earth is shown in Figure 3.16.

The gravitational field is a vector quantity. The Earth's gravitational field at any location points toward the center of the Earth. Assuming the usual coordinate system, if you stand at the North Pole, which is located on the $+y$ axis, you would experience a gravitational field $\vec{g} = \langle 0, -g, 0 \rangle$. Note that g is a *positive* number, the magnitude of the gravitational field $\vec{g}$.

If the Earth consisted of layers of uniform-density spherical shells, the Earth's gravitational field would be perfectly spherically symmetric, and g would be truly constant all over the surface of the Earth. However, the Earth due to its spin bulges out at the equator, so g is smaller at the equator than at the poles because you're farther from the center of the Earth. Also, regions in the Earth of higher or lower density make g above these regions vary slightly from the average value (and the direction may also be affected), and geologists use these small variations to find valuable minerals and oil. The fact that a reference frame tied to the spinning Earth is not an inertial frame further complicates measurements of $\vec{g}$, though the effect is negligibly small for the purposes of this course.

We will use the concept of field extensively when we study electric and magnetic fields in later chapters. Until then, we will refer to it only in the context of the gravitational field near the Earth's surface ($g = +9.8\,\text{N/kg}$).

EXAMPLE **A Falling Object**

Drop a rock with mass m near the surface of the Earth. What is its (vector) acceleration?

Solution Use the Momentum Principle: $\Delta\vec{p} = \vec{F}_{\text{net}}\,\Delta t$. Since the speed is small compared to the speed of light, and the mass isn't changing, we can write $\vec{p} \approx m\vec{v}$. If we neglect air resistance, the only force is the gravitational force, so $\Delta(m\vec{v}) = m\vec{g}\,\Delta t$. Divide through by m and by Δt and we get this:

$$\frac{\Delta\vec{v}}{\Delta t} = \vec{g}$$

However, the acceleration of the rock is the rate of change of the velocity, $\frac{\Delta\vec{v}}{\Delta t}$, so we have

$$\vec{a} = \vec{g} = \langle 0, -g, 0 \rangle$$

The rock accelerates toward the ground (the $-y$ direction) at a rate of 9.8 m/s/s. Evidently the units of N/kg are equivalent to the units of m/s/s.

The reason that the mass dropped out of the calculation is that both the momentum and the gravitational force are proportional to m, which is not true for other kinds of force. This has the interesting effect that if the gravitational force is the only force acting, a heavy rock and a light rock will fall in exactly the same way. However, the air resistance force depends on how wide the rock is, and how blunt it is, and how fast the rock is moving, and is not proportional to m, so if the air resistance force is significant, different masses will fall differently. An obvious example is that if you drop a metal ball and a partially crumpled sheet of paper, the metal ball will hit the ground before the paper does.

Checkpoint 3 A roughly spherical asteroid has a mass of 3.1×10^{20} kg and a radius of 270 km. **(a)** What is the value of the constant g at a location on the surface of the asteroid? **(b)** What would be the magnitude of the gravitational force exerted by the asteroid on a 70 kg astronaut standing on the asteroid's surface? **(c)** How does this compare to the gravitational force on the same astronaut when standing on the surface of the Earth?

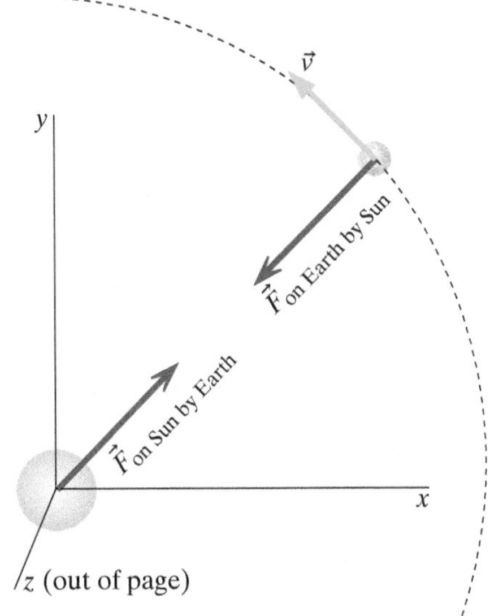

Figure 3.17 The Sun and the Earth exert equal and opposite forces on each other.

3.4 RECIPROCITY

An important aspect of the gravitational interaction is that the force that object 1 exerts on object 2 is equal and opposite to the force that object 2 exerts on object 1 (Figure 3.17). That the magnitudes must be equal is clear from the algebraic form of the gravitational force, because $m_1 m_2 = m_2 m_1$. The directions of the forces are along the line connecting the centers, and in opposite directions.

This property of gravitational interactions is called "reciprocity" or "Newton's third law of motion" and as we will see later in this chapter it is also a property of electric interactions, including the electric interactions between neighboring atoms.

RECIPROCITY

$$\vec{F}_{\text{on 1 by 2}} = -\vec{F}_{\text{on 2 by 1}}$$

(gravitational and electric forces)

The force that the Earth exerts on the massive Sun in Figure 3.17 is just as big as the force that the Sun exerts on the Earth, so in the same time interval the momentum changes are equal in magnitude and opposite in direction ("equal and opposite" for short):

$$\Delta\vec{p}_1 = \vec{F}_{\text{on 1 by 2}}\,\Delta t = -\Delta\vec{p}_2$$

However, the velocity change $\Delta\vec{v} = \Delta\vec{p}/m$ of the Sun is extremely small compared to the velocity change of the Earth, because the mass of the Sun is enormous in comparison with the mass of the Earth. The mass of our Sun, which is a rather ordinary star, is 2×10^{30} kg. This is enormous compared to the mass of the Earth (6×10^{24} kg) or even to the mass of the largest planet in our Solar System, Jupiter (2×10^{27} kg). Nevertheless, very accurate measurements of small velocity changes of distant stars have been used to infer the presence of unseen planets orbiting those stars.

2 grams

Magnetic forces do not have the property of reciprocity. Two electrically charged particles that are both moving can interact magnetically as well as electrically, and the magnetic forces that these two particles exert on each other need not be equal in magnitude or opposite in direction. Reciprocity applies to gravitational and electric forces, but not in general to magnetic forces acting between individual moving charges.

As we will see in Chapter 4, the forces associated with contact between two objects are actually electric forces at the microscopic level. Thus contact forces also have the property of reciprocity.

Why Reciprocity?

The algebraic form of the expression for the gravitational force indicates that reciprocity should hold. A diagram may be helpful in explaining *why* the forces behave this way.

3 grams

Figure 3.18 Reciprocity: There are six forces acting on the 3 g object and six forces acting on the 2 g object.

Consider two small objects, a 2 g object made up of two 1 g balls, and a 3 g object made up of three 1 g balls (Figure 3.18). The distance between centers of the two objects is large compared to the size of either object, so the distances between pairs of 1 g balls are about the same for all pairs. You can see that in the 2 g object at the top of Figure 3.18 each ball has three gravitational forces exerted on it by the distant balls in the 3 g object.

> QUESTION How many of these individual forces act on the 2 g object?

There is a net force of $2 \times 3 = 6$ times the force associated with one pair of balls.

> QUESTION Similarly, consider the forces acting on the 3 g object at the bottom of Figure 3.18. How many of these individual forces act on the 3 g object?

There is again a net force of $3 \times 2 = 6$ times the force associated with one pair of balls.

The effect is that the force exerted by the 3 g object on the 2 g object has the same magnitude as the force exerted by the 2 g object on the 3 g object. The same reciprocity holds, for similar reasons, for the electric forces between a lithium nucleus containing 3 protons and a helium nucleus containing 2 protons.

> **Checkpoint 4** A 60 kg person stands on the Earth's surface. **(a)** What is the approximate magnitude of the gravitational force on the person by the Earth? **(b)** What is the magnitude of the gravitational force on the Earth by the person?

3.5 PREDICTING MOTION OF GRAVITATIONALLY INTERACTING OBJECTS

In Chapter 2 we used an iterative approach to predict the motion of a system that is interacting with its surroundings:

Repeat:

- Calculate the net (vector) force $\vec{F}_{net}$ acting on the system.
- Update the momentum of the system: $\vec{p}_f = \vec{p}_i + \vec{F}_{net}\,\Delta t$.
- Update the position: $\vec{r}_f = \vec{r}_i + \vec{v}_{avg}\,\Delta t$.

The same iterative approach can be used to predict the motion of objects that interact gravitationally. The only difference is in the force we use: the full $1/r^2$ gravitational force. The gravitational force changes magnitude and direction as the objects move, so at the beginning of every time step we need to recalculate the gravitational force, both magnitude and direction.

EXAMPLE

Earth and Sun

Let's try to predict approximately the motion of the Earth around the Sun. Choose a coordinate system whose origin is at the center of the Sun, and in which the Earth orbits the Sun in the xy plane, as shown in Figure 3.19. The mass of the Sun is 2×10^{30} kg; the mass of the Earth is 6×10^{24} kg. Suppose that at a particular instant the Earth has these "initial conditions": position $\langle 1.5\times10^{11},0,0\rangle$ m and velocity $\langle 0,3\times10^{4},0\rangle$ m/s (Figure 3.19). (This is close to the actual speed of the Earth in its nearly circular orbit; to calculate this speed divide the circumference of the orbit by the orbital period.) What does the Momentum Principle predict will be the location of the Earth 3 months later? How well does the prediction match the actual orbit?

Figure 3.19 Initial location and momentum of the Earth. The origin of the coordinate system is at the center of the Sun. Sizes of the Sun and Earth are exaggerated.

Solution

System: Earth
Surroundings: Sun
We will make the approximation that the effect of the Earth on the Sun is negligible, because the Sun's mass is so much larger than the mass of the Earth. Therefore we can neglect changes in the position of the Sun. We will also neglect the interactions of the Earth with all other planets and stars.

To get an approximate answer, we'll divide the 3 mo time interval into three intervals each 1 mo long. This will give us only a rough approximation to the motion, because the gravitational force acting on the Earth changes quite a lot in one month, which is one-twelfth of the way around the circular orbit. We need to convert one month into SI units (seconds):

$$\Delta t = (1\text{ mo})\left(30\frac{\text{days}}{\text{mo}}\right)\left(24\frac{\text{hr}}{\text{day}}\right)\left(60\frac{\text{min}}{\text{hr}}\right)\left(60\frac{\text{sec}}{\text{min}}\right) = 2.6\times10^{6}\text{ s}$$

Initial conditions (position and momentum of the Earth):

$$\vec{r}_E = \langle 1.5\times10^{11},0,0\rangle\text{ m}$$
$$\vec{p}_i = (6\times10^{24}\text{ kg})\langle 0,3\times10^{4},0\rangle\text{ m/s} = \langle 0,1.8\times10^{29},0\rangle\text{ kg}\cdot\text{m/s}$$

A note on significant figures: To make the example easier to read, we show all numbers to only two significant figures. In the actual calculations, however, many significant figures were used, because this is how a computer would do the calculations. In a computer calculation, rounding is typically done only on the final results.

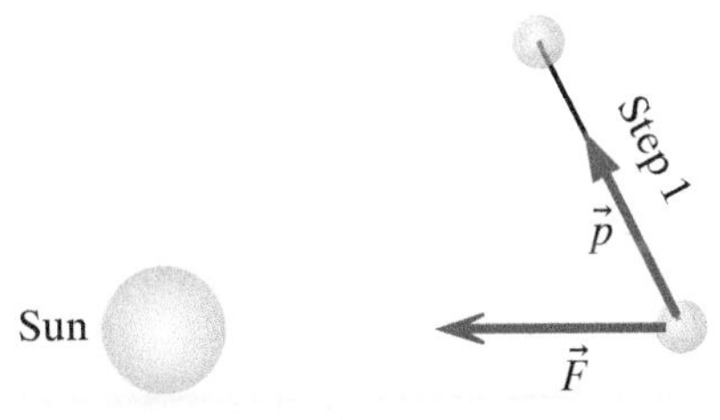

Figure 3.20 Step 1 of three-step iterative calculation. Gravitational force on Earth at beginning of step and final momentum of Earth are shown.

Time step 1 (Figure 3.20):

$$\vec{r} = \langle 1.5\times10^{11},0,0\rangle\text{ m} - \langle 0,0,0\rangle\text{ m} = \langle 1.5\times10^{11},0,0\rangle\text{ m}$$
$$|\vec{r}| = 1.5\times10^{11}\text{ m}$$
$$\hat{r} = \langle 1,0,0\rangle$$

$$\vec{F}_{\text{on }E\text{ by }S} = (6.7\times10^{-11}\text{ N/ kg})\frac{(6\times10^{24}\text{ kg})(2\times10^{30}\text{ kg})}{(1.5\times10^{11}\text{ m})^2}\langle -1,0,0\rangle$$
$$= \langle -3.6\times10^{22},0,0\rangle\text{ N}$$
$$\vec{p}_f = \langle 0,1.8\times10^{29},0\rangle\text{ kg}\cdot\text{m/s} + \langle -3.6\times10^{22},0,0\rangle\text{ N}\cdot(2.6\times10^{6}\text{ s})$$
$$= \langle -9.3\times10^{28},1.8\times10^{29},0\rangle\text{ kg}\cdot\text{m/s}$$

$$\vec{v}_{\text{avg}} \approx \vec{v}_f \approx \frac{\vec{p}_f}{m} = \frac{\langle -9.3\times 10^{28}, 1.8\times 10^{29}, 0\rangle \text{ kg}\cdot\text{m/s}}{6\times 10^{24}\text{ kg}}$$
$$= \langle -1.6\times 10^4, 3\times 10^4, 0\rangle \text{ m/s}$$
$$\vec{r}_{E,f} = \langle 1.5\times 10^{11}, 0, 0\rangle \text{ m} + \langle -1.6\times 10^4, 3\times 10^4, 0\rangle \text{ m/s}\cdot(2.6\times 10^6\text{ s})$$
$$= \langle 1.1\times 10^{11}, 7.8\times 10^{10}, 0\rangle \text{ m}$$

Time Step 2 (Figure 3.21):

$$\vec{r} = \langle 1.1\times 10^{11}, 7.8\times 10^{10}, 0\rangle \text{ m}$$
$$|\vec{r}| = 1.348\times 10^{11}\text{ m}$$
$$\hat{r} = \langle 0.816, 0.578, 0\rangle$$
$$\vec{F}_{\text{on } E \text{ by } S} = \langle -3.6\times 10^{22}, -2.6\times 10^{22}, 0\rangle \text{ N}$$
$$\vec{p}_f = \langle -1.9\times 10^{29}, 1.1\times 10^{29}, 0\rangle \text{ kg}\cdot\text{m/s}$$
$$\vec{r}_{E,f} = \langle 2.9\times 10^{10}, 1.3\times 10^{11}, 0\rangle \text{ m}$$

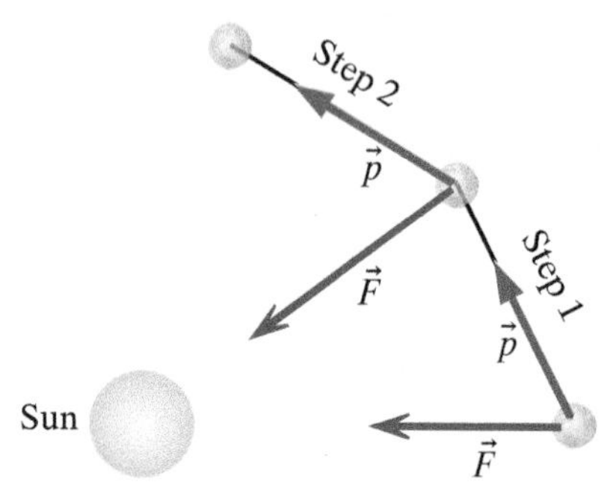

Figure 3.21 Results of first two steps of the iterative calculation. The gravitational force on the Earth is different in the two steps.

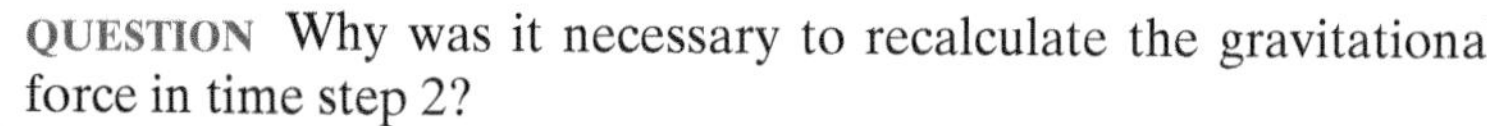

QUESTION Why was it necessary to recalculate the gravitational force in time step 2?

The position of the Earth changed, so the direction (and magnitude) of the gravitational force were not the same as in step 1.

Time Step 3 (Figure 3.22):

Following the same procedure, we must again calculate $\vec{r}$ and $\vec{F}_{\text{on } E \text{ by } S}$, then update the momentum and the position. The results of the third iteration are shown in Figure 3.22.

$$\vec{r}_{E,f} = \langle -6.3\times 10^{10}, 1.2\times 10^{11}, 0\rangle \text{ m}$$

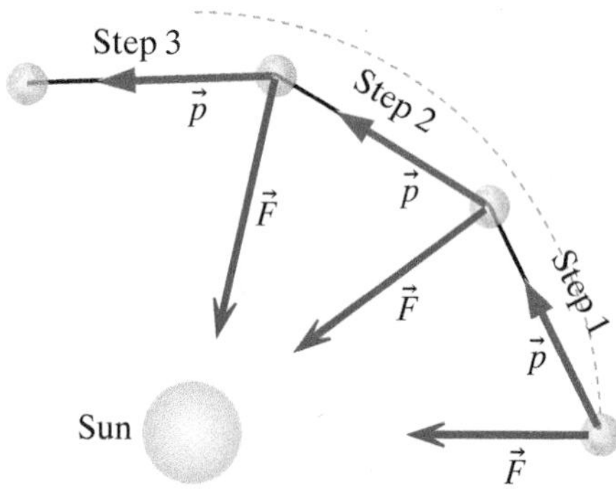

Figure 3.22 The results of all three steps in the iterative calculation of the motion of the Earth under the influence of the Sun. The dashed arc shows the actual orbit of the Earth in three months.

Check: Is the path reasonable? Yes. Although this was a rough calculation, it does predict that the Earth will orbit around the Sun in a roughly circular orbit. However, one indication of the sizable inaccuracy in our calculation is that in 3 mo the Earth should have gone only one-quarter of the way around the circle, but it has gone significantly farther than this in Figure 3.22.

QUESTION How could we improve the accuracy of this calculation?

When we use the Momentum Principle in the form $\vec{p}_f = \vec{p}_i + \vec{F}_{\text{net}}\Delta t$, we're assuming that $\vec{F}_{\text{net}}$ doesn't change much during the interval Δt. This isn't true for a gravitational force if the position changes a lot. If we use smaller time steps, the displacement of the Earth will be smaller in each step, and the gravitational force will change much less. Using smaller time steps should make our predicted path much closer to the actual orbit.

You might wonder about the effect of the approximation that $\vec{v}_{\text{avg}} \approx \vec{p}_f/m$. It turns out in this case, as in many others, that this approximation gives a significantly more accurate result than we would get using the arithmetic average velocity.

Clearly, the procedure used above is very tedious if done by hand. The natural way to carry out an iterative prediction of motion is to program a computer to do the repetitive calculations. In a computer program one would choose a step size much smaller than 1 mo, to ensure that the change in $\vec{F}_{\text{net}}$ is small over the interval Δt.

QUESTION How would you have to change the calculation described above to include in the prediction how the Sun will move?

You would need to update the momentum and position of the Sun each time you update the momentum and position of the Earth. Fortunately, the force acting on the Sun is just the opposite of the force acting on the Earth: $\vec{F}_{\text{on } S \text{ by } E} = -\vec{F}_{\text{on } E \text{ by } S}$ thanks to the reciprocity of gravitational forces (Section 3.4). So it is unnecessary to repeat all of the usual steps in calculating the gravitational force that acts on the Sun; just reverse the force vector.

Because its mass is so large, you would find that the Sun moves very little. Nevertheless, in recent years many planets have been discovered orbiting distant stars by observing the very small wobbles of the stars caused by the gravitational forces of the planets acting on the stars.

EXAMPLE

If the Step Size Is Too Big

By taking a step size of 1 mo, we predicted that the Earth would orbit the Sun, though our prediction was rather imprecise, because in 1 mo the gravitational force and the momentum change a lot. It is instructive to see what happens if we take an even bigger step size, making the results even more imprecise. Taking a step size of 3 mo instead of 1 mo, what will we predict the location of the Earth to be 3 mo later?

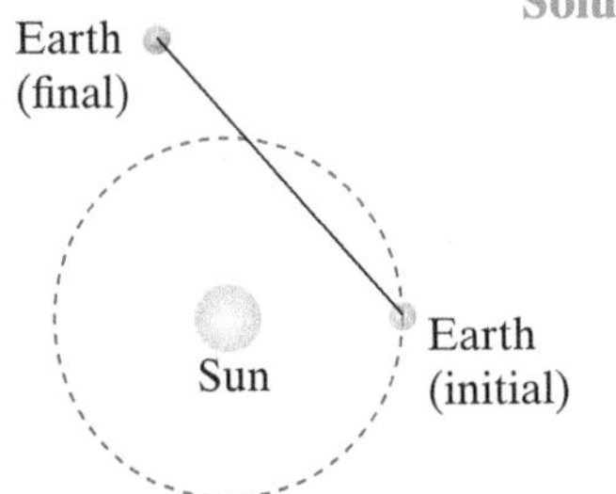

Figure 3.23 The displacement of the Earth predicted by a one-step calculation using the initial value of gravitational force, and the final velocity. The dashed line shows the actual path of the Earth around the Sun.

Solution System: Earth
Surroundings: Sun (ignore all other planets, stars)

$$\Delta t = (3\,\text{mo})\left(30\frac{\text{days}}{\text{mo}}\right)\left(24\frac{\text{hr}}{\text{day}}\right)\left(60\frac{\text{min}}{\text{hr}}\right)\left(60\frac{\text{sec}}{\text{min}}\right) = 7.8\times10^{6}\,\text{s}$$

$$\vec{p}_i = \langle 0, 1.8\times10^{29}, 0\rangle\,\text{kg}\cdot\text{m/s}$$

$$\vec{p}_f = \langle 0, 1.8\times10^{29}, 0\rangle\,\text{kg}\cdot\text{m/s} + \langle -3.6\times10^{22}, 0, 0\rangle\,\text{N}\cdot(7.8\times10^{6}\,\text{s})$$

$$= \langle -2.8\times10^{29}, 1.8\times10^{29}, 0\rangle\,\text{kg}\cdot\text{m/s}$$

$$\vec{v}_f \approx \frac{\vec{p}_f}{m} = \frac{\langle -2.8\times10^{29}, 1.8\times10^{29}, 0\rangle\,\text{kg}\cdot\text{m/s}}{6\times10^{24}\,\text{kg}}$$

$$= \langle -4.6\times10^{4}, 3\times10^{4}, 0\rangle\,\text{m/s}$$

$$\vec{r}_{E,f} = \langle 1.5\times10^{11}, 0, 0\rangle\,\text{m} + \langle -4.6\times10^{4}, 3\times10^{4}, 0\rangle\,\text{m/s}\cdot(7.8\times10^{6}\,\text{s})$$

$$= \langle -2e11, 2.3e11, 0\rangle\,\text{m}$$

Check: Units are ok (meters). Is the location reasonable? NO! The path we predicted, from initial location to final location, is shown in Figure 3.23. This disagrees very much with our more accurate calculation using a time step of 1 mo. Even our 1 mo calculation was imprecise; in a computer calculation we would choose a much shorter time step than 1 mo.

QUESTION What went wrong? In this calculation we used the same procedure we used before.

The major error was the tacit assumption that the force on the Earth by the Sun was constant in magnitude and direction over the 3 mo interval. The direction and magnitude of the gravitational force between two objects changes when the relative positions of the objects change. Also, because the time interval was so long, the final velocity was a poor approximation to the actual average velocity over the interval.

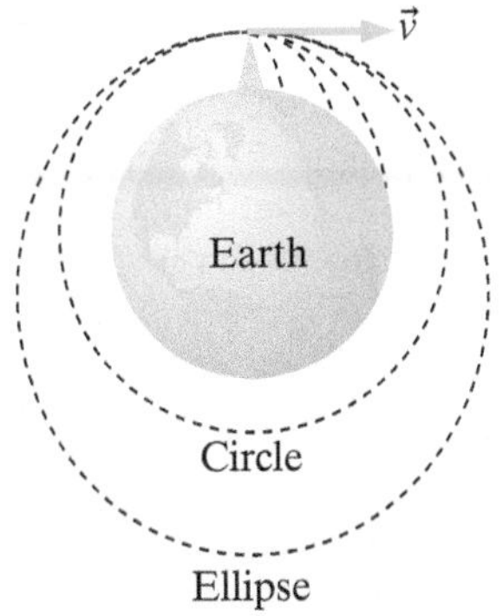

Figure 3.24 Newton imagined throwing a rock horizontally with different speeds from a high mountain resulting in different ellipses. Only one special speed makes a circular orbit.

Elliptical Orbits

The Earth's orbit around the Sun isn't really a circle; it's actually an ellipse that is very nearly circular. Newton himself drew an instructive diagram

(Figure 3.24) to show how the launch speed of an object affects the shape of the orbit, by imagining throwing a rock horizontally from the top of a mountain so tall as to stick up above the atmosphere (so there will be no air resistance). In this figure you can see that getting a circular orbit requires throwing the rock at just the right speed. Any other speed will give a noncircular orbit, which Newton was able to prove would be an ellipse (or portion of an ellipse if the rock hits the ground).

EXAMPLE

Analyzing an Elliptical Orbit

$\vec{p}_i$ A

Sun

B

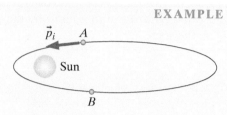

Figure 3.25 A comet approaches the Sun.

It is useful to "think iteratively" about a highly elliptical orbit like that of many comets that come near the Sun at one extreme of the orbit and go far beyond Pluto at the other extreme. At some instant the comet is at location A, headed toward the Sun (Figure 3.25). Graphically and qualitatively, carry out one step in the iterative updating of momentum and position, starting with the comet at position A with momentum $\vec{p}_i$ tangent to the orbit. Will the magnitude of the momentum (and therefore the speed) increase or decrease? (Think about the relative directions of $\vec{p}_{\text{now}}$ and $\vec{F}_{\text{net,now}}$.)

Solution Consider a short time interval Δt and apply the Momentum Principle to the system of the comet, which experiences a gravitational force exerted by the Sun. To update the momentum and position graphically, follow the procedure illustrated in Figure 3.26: (1) Draw the net force acting on the comet, due to the Sun. (2) Draw the impulse $\vec{F}_{\text{net}}\,\Delta t$ for a short time interval Δt. The change in momentum $\Delta\vec{p}$ is equal to the impulse. (3) Draw the new momentum $\vec{p}_f = \vec{p}_i + \Delta\vec{p}$. (4) Approximate the average velocity by the new velocity $\vec{v}_{\text{avg}} \approx \vec{p}_f/m$ and indicate the displacement $\Delta\vec{r} = \vec{r}_f - \vec{r}_i = (\vec{p}_f/m)\Delta t$.

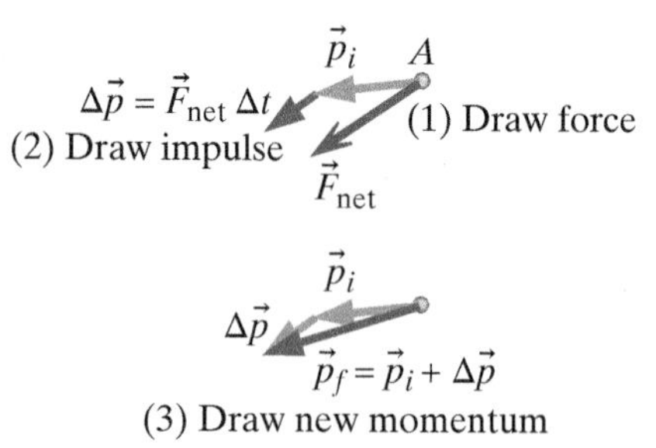

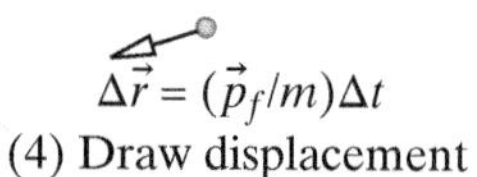

Figure 3.26 Steps in graphical updating of momentum and position. (1) Draw the net force, (2) draw the impulse, (3) draw the new momentum, (4) use the new momentum to draw the displacement.

Since part of the force acts in the same direction as the momentum of the comet, the magnitude of the comet's momentum increases.

QUESTION A student said, "The reason why the comet speeds up is because the force gets bigger as the comet moves toward the Sun. Bigger force, bigger speed." What's wrong with this analysis?

This statement ignores the fact that $\vec{p}_{\text{future}}$ depends on two things: $\vec{p}_{\text{now}}$ and $\vec{F}_{\text{net,now}}$, not just on $\vec{F}_{\text{net,now}}$. If, over the next Δt, $\vec{F}_{\text{net}}$ is at least partly in the same direction as the current momentum, then the new momentum will be larger. This would be true even if the magnitude of the net force on the comet were constant or decreasing.

Checkpoint 5 Now consider the situation when the comet is at position B in Figure 3.25, heading away from the Sun. Mentally take one step in the iterative update of momentum and position. What happens to the magnitude of the momentum, and the speed? What happens to the direction of the momentum? Explain briefly in terms of $\vec{p}_{\text{future}}$, $\vec{p}_{\text{now}}$, and $\vec{F}_{\text{net,now}}$.

3.6 GRAVITATIONAL FORCE IN COMPUTATIONAL MODELS

In Chapter 2 we saw how to structure an iterative prediction of motion on a computer, for a mass–spring system. To predict the motion of objects interacting gravitationally we will use the same basic structure, but we need to calculate the gravitational force instead of a spring force.

In writing a program to tell a computer how to calculate a gravitational force, we use exactly the same organization as we use when doing the calculation with a calculator. Let's assume that our program creates two objects named `rock1` and `rock2`, each represented by a sphere at an appropriate location in 3D space. The variables `m1` and `p1` represent the mass and momentum, respectively, of `rock1`; assume that they have been given appropriate values, and the relevant constants have been defined. Recall that the `pos` attribute of an object in VPython is a vector. A translation of the mathematical expressions (shown in red) into code (shown in blue) for calculating the gravitational force on `rock1` due to `rock2` into VPython code might look like this:

$\vec{r} = \vec{r}_1 - \vec{r}_2$	`r = rock1.pos - rock2.pos`
$\hat{r} = \vec{r}/\lvert\vec{r}\rvert$	`rhat = r / mag(r)`
$\lvert\vec{F}\rvert = Gm_1m_2/\lvert\vec{r}\rvert^2$	`Fmag = G * m1 * m2 / mag(r)**2`
$\vec{F} = \lvert\vec{F}\rvert \cdot (-\hat{r})$	`F_on1 = Fmag * (-rhat)`

QUESTION Is the quantity `r` calculated in the first line of code above a vector or a scalar?

It is a vector, since it is the difference of two position vectors.

QUESTION The code for calculating the force on `rock1` above was used in a program that generated the display shown in Figure 3.27. Suppose that you wanted to calculate and display the force on `rock2` due to `rock1`. What code would you write to do this? If the mass of `rock2` is less than the mass of `rock1`, how should the magnitudes and directions of the two forces compare?

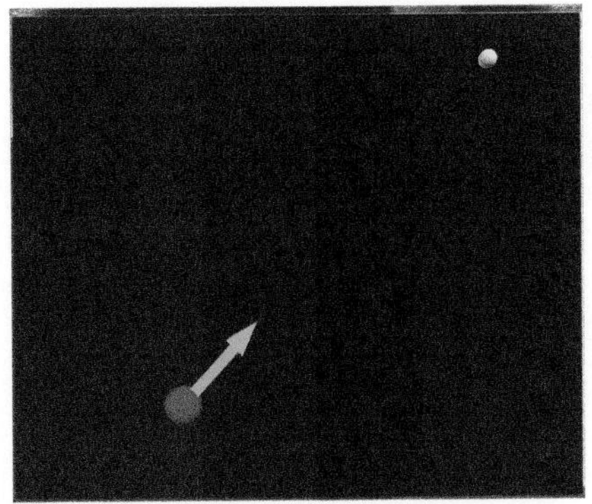

Figure 3.27 Display generated by a VPython program to calculate the gravitational force on one object due to another.

Because of the reciprocity of gravitational forces, the direction of the force on `rock2` will be opposite to the direction of the force on `rock1`, but their magnitudes will be the same, no matter what their masses may be. So we could say:

```
F_on2 = -F_on1
```

QUESTION If you write a program to model the motion of gravitationally interacting objects, should you put the calculation of the gravitational force before the iterative loop, or should it be done inside the loop?

The gravitational force calculation must be done inside the loop. Every time one of the objects moves, the relative positions of the objects change, and therefore the gravitational force changes. It is necessary to recalculate the gravitational force in every time step. The key lines of code inside the iterative loop might look like this:

```
while True:
    rate(100)
    r = rock1.pos - rock2.pos
    rhat = r/mag(r)
    Fmag = G * m1 * m2 / mag(r)**2
    F_on1 = Fmag * (-rhat)
    p1 = p1 + F_on1 * deltat
    rock1.pos = rock1.pos + (p1/m1) * deltat
    F_on2 = -F_on1
    p2 = p2 + F_on2 * deltat
    rock2.pos = rock2.pos + (p2/m2) * deltat
```

QUESTION In a program that incorporates the while loop shown above, will both objects move?

Yes, because the momentum and position of both rocks are updated.

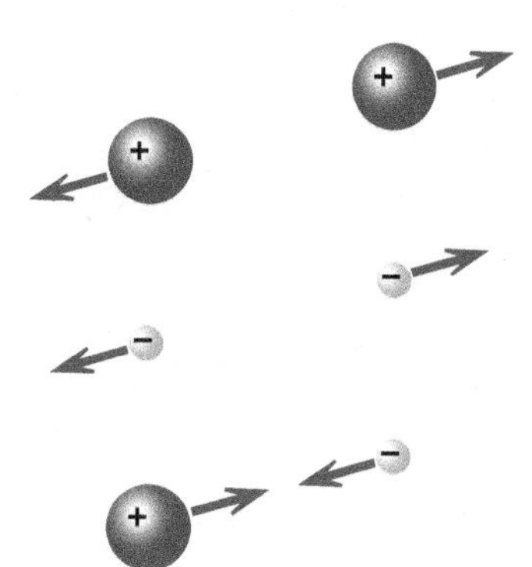

Figure 3.28 Protons repel each other; electrons repel each other; protons and electrons attract each other.

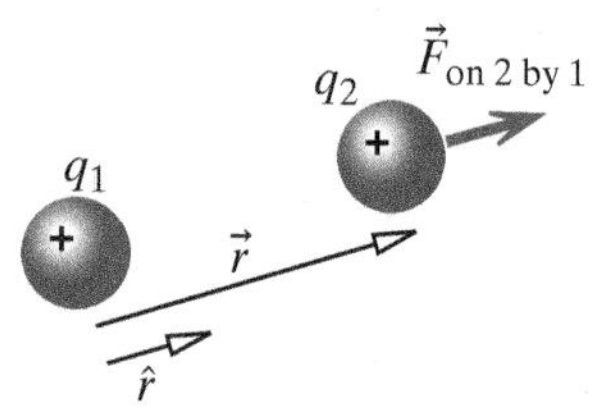

Figure 3.29 The electric force exerted on object 2 by object 1. (The force exerted on object 2 by object 1 has the same magnitude but opposite direction.)

3.7 THE ELECTRIC FORCE

Particles such as the protons and electrons found in atoms are electrically charged. It is observed that two protons repel each other, as do two electrons, while a proton and an electron attract each other (Figure 3.28). Protons are said to have "positive electric charge" and electrons have "negative electric charge."

The force corresponding to this electric interaction is known as "Coulomb's law" to honor the French scientist who established this behavior of electric forces in the late 1700s, during the same period when Cavendish measured the gravitational constant G. See Figure 3.29.

THE ELECTRIC FORCE (COULOMB'S LAW)

$$\vec{F}_{\text{elec on 2 by 1}} = \frac{1}{4\pi\varepsilon_0}\frac{q_1 q_2}{|\vec{r}|^2}\hat{r}$$

q_1 and q_2 are the electric charges of the objects.
$\vec{r} = \vec{r}_2 - \vec{r}_1$ is the position of 2 relative to 1.

The universal constant $\dfrac{1}{4\pi\varepsilon_0} = 9 \times 10^9 \dfrac{\text{N}\cdot\text{m}^2}{\text{C}^2}$.

The SI unit of electric charge is the coulomb, abbreviated as C. The proton has a charge of $+1.6 \times 10^{-19}$ C and the electron has a charge of -1.6×10^{-19} C. The symbol e is often used to represent the charge of a proton; $e = +1.6 \times 10^{-19}$ C. We can see from the equations describing the electric and gravitational forces that there are similarities between these forces. The most striking similarity is that the magnitude of both forces is inversely proportional to the square of the distance between the centers of the objects, and that the force acts along the line connecting the objects, as indicated by the presence of the unit vector $\hat{r}$:

$$\vec{F}_{\text{electric}} = \frac{1}{4\pi\varepsilon_0}\frac{q_1 q_2}{|\vec{r}|^2}\hat{r} \qquad \vec{F}_{\text{grav}} = -G\frac{m_1 m_2}{|\vec{r}|^2}\hat{r}$$

However, there are important differences between the electric and gravitational forces as well. Electric forces may be attractive or repulsive, whereas gravitational forces are always attractive. Electric charge may have positive or negative values, while mass has only positive values. As indicated in Figure 3.28, particles with the same sign of charge repel each other, and particles with opposite-signed charges attract each other.

QUESTION What quantities in the equation for electric force determine the direction of the force?

There are three quantities we need to consider: the unit vector $\hat{r}$, and the charges q_1 and q_2, which may be positive or negative. If the product of q_1 and q_2 is a positive number (that is, if q_1 and q_2 have the same signs), then the force will be in the direction of $\hat{r}$. If q_1 and q_2 have different signs, their product will be negative, and the force will be opposite to $\hat{r}$.

QUESTION Is the following quantity the magnitude of the electric force?

$$\frac{1}{4\pi\varepsilon_0}\frac{q_1 q_2}{|\vec{r}|^2}$$

No. It is a scalar quantity, but it may be negative, whereas a magnitude is always positive. When calculating an electric force, it is more useful to think about this scalar quantity than about the magnitude.

EXAMPLE

An Electron and an Alpha Particle

An electron is located at $\langle -3\times10^{-9}, 2\times10^{-9}, 0\rangle$ m and an alpha particle (two protons and two neutrons) is located at $\langle 6\times10^{-9}, -4\times10^{-9}, 0\rangle$ m (Figure 3.30). Find the force exerted on the electron by the alpha particle.

Solution

$$\vec{r} = \langle -9\times10^{-9}, 6\times10^{-9}, 0\rangle \text{ m}$$
$$|\vec{r}| = 1.08\times10^{-8}\text{ m}$$
$$\hat{r} = \langle -0.832, 0.555, 0\rangle$$
$$\vec{F} = \left(9\times10^{9}\frac{\text{N}\cdot\text{m}^2}{\text{C}^2}\right)\frac{(2\cdot1.6\times10^{-19}\text{ C})(-1.6\times10^{-19}\text{ C})}{(1.08\times10^{-8}\text{ m})^2}\langle -0.832, 0.555, 0\rangle$$
$$= \langle 3.28\times10^{-12}, -2.19\times10^{-12}, 0\rangle \text{ N}$$

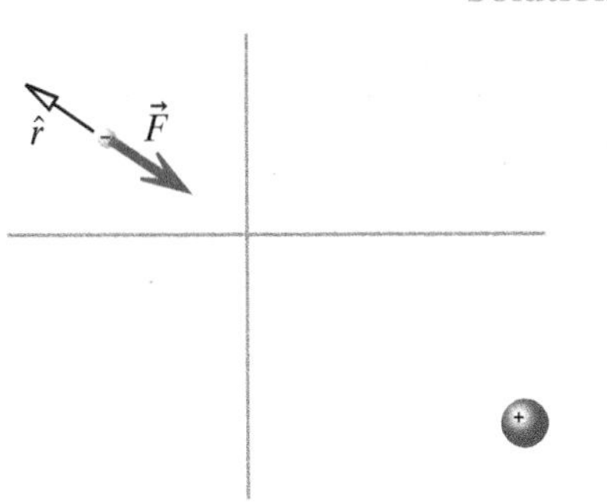

Figure 3.30 Force on an electron due to an alpha particle.

QUESTION Does the direction of the force make sense?

Yes, because the electron and the alpha particle have opposite signs, so the electron should be attracted to the alpha particle.

QUESTION Isn't this a very small force?

To be able to answer that question, we need to rephrase it in terms of a comparison—small relative to what? In the next example we explore this issue.

EXAMPLE

Comparison of Electric and Gravitational Forces

Two protons are a distance r from each other, center to center, as shown in Figure 3.31. Find the ratio of the magnitude of the electric force one proton exerts on the other to the magnitude of the gravitational force one proton exerts on the other.

Figure 3.31 Two protons a distance r apart.

Solution

$$F_{\text{elec}} = (9\times10^{9}\text{ N}\cdot\text{m}^2/\text{C}^2)\frac{(1.6\times10^{-19}\text{C})^2}{r^2}$$
$$F_{\text{grav}} = (6.7\times10^{-11}\text{ N}\cdot\text{m}^2/\text{ kg}^2)\frac{(1.7\times10^{-27}\text{ kg})^2}{r^2}$$
$$\frac{F_{\text{elec}}}{F_{\text{grav}}} = 1.2\times10^{36}$$

Notice that r canceled in this ratio, because both the electric and gravitational forces are proportional to the same factor $1/r^2$. The units of newtons also canceled.

This ratio is astoundingly huge. The electric repulsion of two protons is greater than their gravitational attraction by a factor of

1,200,000,000,000,000,000,000,000,000,000,000,000 !!

As a result, when considering the interactions of subatomic charges, it is reasonable to ignore their extremely tiny gravitational interactions with one another.

QUESTION If electric interactions are intrinsically so enormously much stronger than gravitational interactions, why is it that we are so much more aware of gravitational forces in our daily lives?

At the microscopic level, electric forces are extremely important. The electrons and protons that make up a material object are bound together by electric forces. The fact that you can stand up despite the gravitational pull of the entire Earth, instead of having all the electrons and protons in your body end up in a puddle on the Earth's surface, indicates how strong these electric interactions are. However, at the macroscopic level, it is rare to find large chunks of matter that have a nonzero electric charge, since under most circumstances matter is electrically "neutral"—that is, an object typically has equal numbers of electrons and protons.

It is occasionally possible to transfer some charge from one object to another by rubbing, leaving one object with a tiny excess of electrons and the other with a tiny excess of protons; when this occurs you may see electric effects such as "static cling." We'll explore such effects in depth in later chapters on electric interactions.

The forces that objects exert on each other when they touch are also due to electric interactions. Although matter is typically electrically neutral, atoms on the surfaces of two objects that come into contact deform each other in such a way that the electric forces do not entirely cancel out. In the next chapter we will examine interatomic forces in more detail.

Checkpoint 6 A moving electron passes near the nucleus of a gold atom, which contains 79 protons and 118 neutrons. At a particular moment the electron is a distance of 3×10^{-9} m from the gold nucleus. Recall that the electron charge is $-e = -1.6 \times 10^{-19}$ C, the proton charge is $+e = +1.6 \times 10^{-19}$ C, and the neutron charge is 0. **(a)** What is the magnitude of the force exerted by the gold nucleus on the electron? **(b)** What is the magnitude of the force exerted by the electron on the gold nucleus?

3.8 THE STRONG INTERACTION

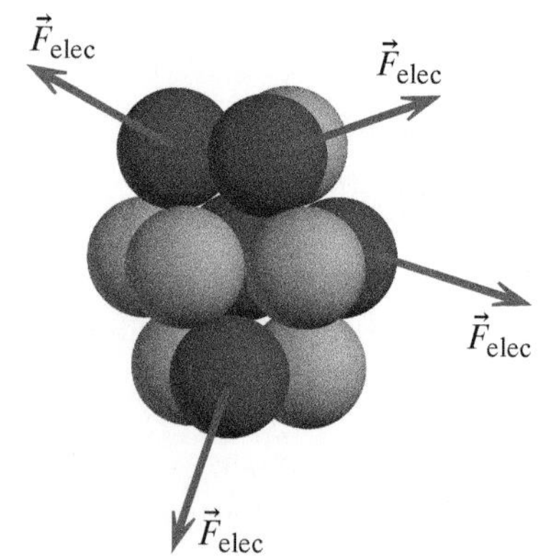

Figure 3.32 The strong force: the protons in the nucleus of an atom exert repulsive electric forces on each other as shown in the diagram. The strong interaction (which involves neutrons as well as protons) holds the nucleus together despite this electric repulsion.

The strong interaction involves particles made of quarks, so leptons such as electrons do not experience this force, because they are not made of quarks. The strong interaction holds the nucleus of an atom together. Inside an atomic nucleus the protons repel each other electrically, yet the protons and neutrons remain packed closely together due to the strong interaction (Figure 3.32). This is a very short-range interaction. The strong (or nuclear) interactions among protons and neutrons are essentially zero unless the particles touch each other, which means that the center-to-center distance must be about equal to the diameter of a proton or neutron (the radius of a proton or neutron is about 1×10^{-15} m, very much smaller than the radius of an atom, which is about 1×10^{-10} m). At these very short distances, the electric force is extremely large, since it is inversely proportional to the tiny distance squared, yet the strong force is even larger in stable nuclei.

There is a delicate balance between the electric repulsion and the strong attraction that determines whether a nucleus can exist at all. For example, a nucleus consisting of just two protons and no neutrons does not exist in nature because the attractive strong force between the two protons is smaller than the repulsive electric force. This unstable helium nucleus would be called He-2, meaning the total number of protons and neutrons is 2. He-3 (2 protons and 1 neutron) and He-4 (2 protons and 2 neutrons) are stable "isotopes" of helium (different isotopes of the same element have the same number of protons but different numbers of neutrons). In these stable nuclei the protons experience attractive strong interactions with the neutrons as well as with the other proton, and this is enough to overcome the electric repulsion between the two protons.

At the start of the periodic table you'll find about the same number of neutrons as protons in the nuclei, as in He-4. For example, the most common isotope of oxygen has 8 protons and 8 neutrons (O-16). As you go farther in the periodic table you'll see that more and more neutrons are needed to offset the proton repulsions. The most common isotope of uranium has

92 protons but many more neutrons, 146 in U-238. In uranium not only are there more neutrons for the protons to interact with but also the many neutrons dilute the concentration of protons, so that on average the protons are farther apart, thereby decreasing the electric repulsion.

While there are (relatively) simple equations for the gravitational force and the electric force, it is very difficult to write a "force law" equation for the strong interaction. Usually models of the strong interaction involve energy (which we will encounter in Chapter 6) rather than force, and require significant mathematical sophistication. We will not try to work quantitatively with the strong interaction in this text.

■ The large particle physics laboratory called CERN, near Geneva, Switzerland, is the home of many collaborative international physics projects that seek answers to questions about fundamental particles and interactions. You may have heard of the Large Hadron Collider (LHC) at CERN, which was featured in the movie *Particle Fever*, and is an important tool for these investigations. You can find out more at the CERN website.

QUESTION How strong is the "strong" interaction?

If we "rank" the fundamental interactions, we find that the gravitational force is intrinsically extremely weak. You certainly notice the gravitational force when you fall, but it is significant only because you're interacting with the entire Earth. The electric force, which we have seen to be very strong, and the strong force are comparable when two protons touch each other, but if the protons are not in contact the strong force is zero.

Antimatter

For every fundamental particle, there is an "antiparticle" whose mass is the same, but whose other properties (such as charge) are opposite. These particles are referred to as "antimatter." For example, an antiproton has the same mass as a proton, but it has a negative charge. When a particle meets its own antiparticle (for example, when a proton encounters an antiproton), the two particles annihilate each other, producing various other particles. A particle such as a proton is composed of quarks, whereas an antiproton is composed of antiquarks. However, there are interesting particles called "mesons" that are composed of a quark and a different kind of antiquark. An example of a meson is a pion, which is represented by the Greek letter π. Antiparticles and mesons, along with ordinary particles, may be produced in nuclear reactions.

One situation in which we can observe the strong interaction is the reaction in which a high-speed proton hits a stationary proton (which can be a proton in a nucleus) so hard that an additional particle, a pion, is produced. The pion has a mass about 15% that of a proton. Depending on which quark and antiquark are involved, a pion can have an electric charge of $-e$, 0, or e (where e is the electric charge of the proton: $e = +1.6 \times 10^{-19}$ C). Here are examples of such reactions, where p is a proton, n is a neutron, π is a pion, and a superscripted $+$ is shorthand for an amount of charge $+e$.

$$p^+ + p^+ \rightarrow p^+ + p^+ + \pi^0$$
$$p^+ + p^+ \rightarrow p^+ + n^0 + \pi^+$$

Conservation

Many fundamental ideas in physics can be expressed as "conservation" principles. In the nuclear reactions above, as in chemical reactions you may have studied in chemistry, charge is "conserved": the Universe has the same net (total) charge before and after the interaction. The total charge is calculated by adding up all the $+e$ and $-e$ charges. In a process involving any of the four fundamental interactions, not only is charge conserved, but the number of quarks is also conserved. A proton or a neutron consists of three quarks, and a pion consists of a quark and an antiquark. The Universe has the same "quark number" before and after an interaction, where this number is calculated by adding up all the quarks and subtracting all the antiquarks. In the reactions shown above, the newly created pion doesn't change the quark number, because it consists of a quark and an antiquark; one quark minus one antiquark is zero.

Later in this chapter we will see that momentum is also conserved in all processes involving any of the four fundamental interactions. The total momentum of the Universe does not change.

Checkpoint 7 Look at the periodic table on the inside front cover of this textbook. As is standard practice, each entry gives the "atomic number" (the number of protons in the nucleus; 8 for oxygen, 92 for uranium) and the "atomic mass" in grams per mole, which is equal to the number of protons plus neutrons in a nucleus, averaged over the various isotopes for that element, which contain different numbers of neutrons. **(a)** How many protons and neutrons are in a typical silicon nucleus (Si)? **(b)** In a tin nucleus (Sn)? **(c)** In a gold nucleus (Au)? **(d)** In a thorium nucleus (Th)? **(e)** Do you see a trend?

3.9 THE WEAK INTERACTION

The fourth fundamental interaction is called the "weak interaction" because it involves nuclear interactions that are much weaker than the strong interaction or the electromagnetic interaction. A distinctive feature of many weak interactions is the production of neutrinos, ghostly particles that interact so weakly with ordinary matter that almost all of the neutrinos emitted in fusion reactions in the Sun pass right through the Earth as though the Earth were completely transparent. Although about 100 trillion (1×10^{14}) solar neutrinos pass through your body every second, on average only one of the neutrinos will undergo a weak interaction with your body in your entire lifetime! A lowercase Greek letter nu (ν) is used to represent a neutrino.

The weak interaction is responsible for the instability of a "free neutron," a neutron that is isolated outside of a nucleus. With an average lifetime of about 15 minutes, a free neutron decays into a proton, an electron, and an antineutrino (the bar above the ν indicates an antiparticle):

$$n^0 \rightarrow p^+ + e^- + \bar{\nu}^0$$

This same weak interaction process occurs in some radioactive nuclei, in which a neutron turns into a proton with the emission of an electron and antineutrino, and the new proton remains bound into the nucleus. This means that the nucleus has changed into that of a different element, with one more proton (and one less neutron) than before. Such nuclear change is called "beta decay" because in early studies of this kind of radioactivity it wasn't at first clear that the negatively charged particles were electrons, and they were called "beta rays." Beta decay occurs in heavy elements that have excess neutrons.

In the Sun, fusion reactions lead to the emission of neutrinos. Solar neutrinos have been detected on Earth despite the extremely low probability of an interaction between a neutrino and an ordinary object. For this purpose, in the late 1960s the American physicist Raymond Davis built a huge vat containing the liquid dry-cleaning chemical perchlorethylene, which contains chlorine atoms, deep in the Homestake mine in Lead, South Dakota (to shield the detector from unwanted cosmic rays). Neutrinos from the Sun occasionally interacted with a chlorine nucleus to change a neutron into a proton, thereby changing the nucleus into an argon nucleus. Here is the reaction concerning the neutron, which you can see is related to the neutron decay reaction shown earlier:

$$\nu^0 + n^0 \rightarrow p^+ + e^-$$

The corresponding change in the nucleus from chlorine to argon was $\nu^0 + \text{Cl} \rightarrow \text{Ar} + e^-$. Davis collected the argon produced in the vat as an indication of how many neutrinos had interacted with chlorine nuclei in the vat.

■ Raymond Davis won the Nobel Prize in 2002 for his work on neutrinos. You can see a video of Davis's Nobel lecture here: http://www.nobelprize.org/

The Mystery of the Missing Neutrinos

Given what we know about the fusion reactions taking place in the Sun, it was possible to predict how many neutrino events Davis should have observed, but he detected only a third as many neutrinos as expected. This was a deep mystery for several decades until several theoretical physicists proposed that on their way from the Sun to the Earth the neutrinos repeatedly change identity, sometimes existing as "electron neutrinos," which Davis's apparatus could detect, and sometimes as "muon neutrinos," a different kind of neutrino that the apparatus could not detect.

At our particular distance from the Sun, these "neutrino oscillations" are such that an electron neutrino detector finds only one-third of the neutrinos that were headed our way. An electron neutrino detector closer to or farther from the Sun would find a different fraction of the total. You can read more about this scientific mystery and its resolution by searching the Internet for "Raymond Davis neutrinos." Davis shared the 2002 Nobel Prize for physics for his work.

Later experiments in many countries confirmed and extended Davis's work and provided solid support for the neutrino oscillation theory. For example, in the MINOS experiment begun in 2003, neutrinos produced by the Fermilab accelerator outside Chicago traveled through 735 km of rock to an underground detector in the Soudan mine in northern Minnesota, and the neutrino oscillation theory was again shown to be able to explain the observations.

As with the other three fundamental interactions, in the weak interaction charge is conserved, quark number is conserved, and lepton number is conserved. In the Universe the number of leptons (mainly electrons and neutrinos) minus the number of antileptons (mainly positrons and antineutrinos) does not change. However, a unique feature of the weak interaction is that it doesn't conserve "parity," a measure of right- or left-handedness. You can read about this by searching the Internet for "parity nonconservation."

3.10 CONSERVATION OF MOMENTUM

Now that we understand the reciprocity of gravitational and electric forces, we are in a position to understand the principle of Conservation of Momentum, which is a very general statement of the Momentum Principle.

The Momentum Principle, $\Delta\vec{p} = \vec{F}_{\text{net}}\Delta t$, predicts how the momentum of a *system* will change, due to forces exerted on the system by objects in the surroundings of that system. But what happens to the momentum of objects in the *surroundings*? For all four kinds of interactions, gravitational, electromagnetic, strong, and weak, it is observed that momentum gained by the system is lost by the surroundings, and we say that momentum is "conserved." That is, you can't create or destroy momentum, you can only move it from one object to another:

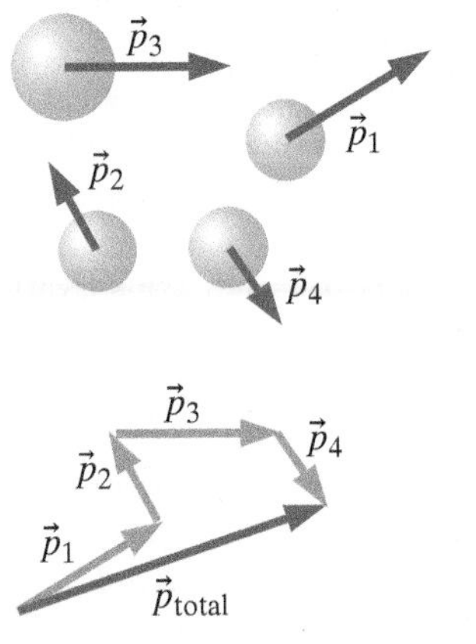

Figure 3.33 The total momentum of the system of four objects is the sum of the individual momenta of each object.

CONSERVATION OF MOMENTUM

$$\Delta\vec{p}_{\text{sys}} + \Delta\vec{p}_{\text{surr}} = \vec{0}$$

The momentum of the system is defined as the (vector) sum of the momenta of all objects in the system (Figure 3.33):

$$\vec{p}_{\text{sys}} = \vec{p}_1 + \vec{p}_2 + \vec{p}_3 + \cdots$$

Similarly, the momentum of the surroundings is the (vector) sum of the momenta of all the objects in the surroundings. This principle is true regardless of the speed of the objects involved; that is, whether or not $\gamma \approx 1$.

Momentum Can Flow Between System and Surroundings

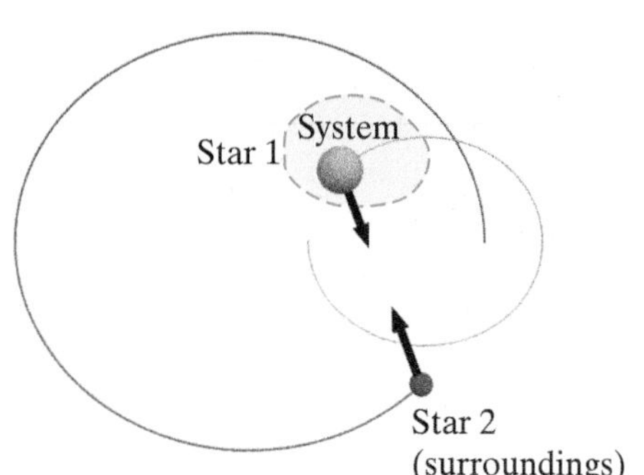

Figure 3.34 A binary star far from other objects. The gray lines show the trajectories of the individual stars. Choose star 1 as the system; star 2 is the surroundings.

Consider the case of two stars orbiting each other, a "binary star" far from all other objects (Figure 3.34). If you choose as the system of interest star 1, star 2 is the surroundings and exerts a force that changes the first star's momentum. Star 2, the surroundings, experiences a force of equal magnitude and opposite direction (reciprocity of gravitational forces). In an amount of time Δt, the momentum change of the system (star 1) is $\Delta\vec{p}_1 = \vec{F}_{\text{on 1 by 2}}\Delta t$, and the momentum change of the surroundings (star 2) is $\Delta\vec{p}_2 = -\vec{F}_{\text{on 1 by 2}}\Delta t = -\Delta\vec{p}_1$. Whatever momentum the system gains, the surroundings lose: $\Delta\vec{p}_{\text{sys}} + \Delta\vec{p}_{\text{surr}} = \vec{0}$. We say momentum is a "conserved" quantity.

The Momentum of an Isolated System Does Not Change

If on the other hand you choose both stars as your system (Figure 3.35), the surroundings consist of other stars, which may be so far away as to have negligible effects on the binary star. In that case the net force exerted by the surroundings on the system is nearly zero. The momentum of the combined system is $\vec{p}_{\text{sys}} = \vec{p}_1 + \vec{p}_2$. Let's see how the momentum of the system consisting of both stars changes:

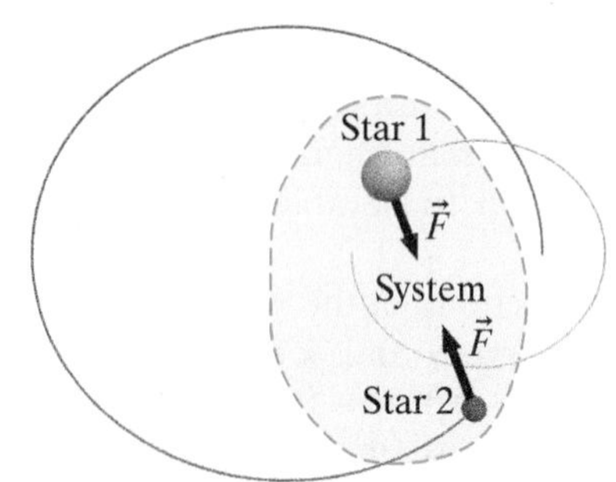

Figure 3.35 A binary star: choose both stars as the system. The momentum of the combined system doesn't change.

$$
\begin{aligned}
\Delta\vec{p}_{\text{sys}} &= \Delta\vec{p}_1 + \Delta\vec{p}_2 \\
&= \vec{F}_{\text{on 1 by 2}}\Delta t + \vec{F}_{\text{on 2 by 1}}\Delta t \\
&= \vec{F}_{\text{on 1 by 2}}\Delta t + (-\vec{F}_{\text{on 1 by 2}}\Delta t) \\
&= \vec{0}
\end{aligned}
$$

For the two-star system, the Momentum Principle $\vec{p}_f = \vec{p}_i + \vec{F}_{\text{net}}\Delta t$ reduces to $\vec{p}_{\text{sys},f} = \vec{p}_{\text{sys},i}$ and the momentum of the system $\vec{p}_{\text{sys}} = \vec{p}_1 + \vec{p}_2$ remains constant in magnitude and direction. This is an important special case: the momentum of an isolated system, a system with negligible interactions with the surroundings, doesn't change but stays constant. The momentum of each star changes, due to interactions with the other star, but momentum lost by one star is gained by the other star, and the total momentum of the system doesn't change. (In the case of the interactions of charged particles, some of the momentum may be associated with electric or magnetic "fields," in addition to the momentum of the particles.)

Stating the Momentum Principle in terms of the conservation of momentum highlights another aspect of the predictive power of this fundamental principle. It allows us to predict and explain aspects of the behavior of complex systems without knowing the details of the large number of interactions inside the systems. It can be particularly useful in analyzing the behavior of systems of many particles whose interactions with the surroundings are negligible during the time interval of interest.

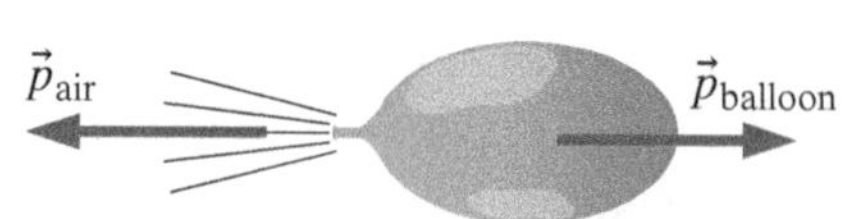

Figure 3.36 When the balloon is released, the total momentum of the system must still be zero.

EXAMPLE **A Balloon Rocket**

Suppose that you blow up a balloon, and hold it closed with your fingers. When you release your grip, air rushes out the opening in the balloon. If the balloon is oriented horizontally, as shown in Figure 3.36, so the outflowing air travels in the $-x$ direction, what must happen to the balloon?

Solution System: Balloon plus all the air initially inside it
Surroundings: Everything else
Initial time: Just before the balloon is released
Final time: Just after the balloon is released
Assumption: Negligible interactions with surroundings during this short Δt
Momentum Principle:

$$\Delta\vec{p}_{\text{sys}} + \Delta\vec{p}_{\text{surr}} = \vec{0}$$

Let $\vec{p}$ represent $\vec{p}_{\text{sys}}$, the total momentum of the system; that is, it is the sum of the momenta of the air inside the balloon plus the momentum of the rubber balloon itself.

$$\vec{p} = \vec{p}_{\text{air}} + \vec{p}_{\text{rubber}}$$

Initially the balloon and its contents are at rest, so

$$\vec{p}_i = \vec{0}$$

Because the interaction with the surroundings was negligible over this short interval, it follows that

$$\vec{p}_f - \vec{p}_i = \vec{0}$$
$$\vec{p}_f = \vec{0}$$

In the final state the total momentum of the system must still be zero. However, the momentum of the air coming out of the balloon is not zero; it is in the $-x$ direction. Therefore the momentum of the rubber balloon must be nonzero, in the $+x$ direction, so the balloon will move in the $+x$ direction. In this way the x components of the momenta of all of the objects in the system add up to zero.

EXAMPLE **Two Blocks Collide**

A system consists of a 1 kg block moving with velocity $\langle 5,2,0\rangle$ m/s and a 3 kg block moving with velocity $\langle -3,4,0\rangle$ m/s. The two-block system is nearly isolated from the surroundings. **(a)** What is the momentum of this two-block system? **(b)** The blocks collide and bounce off each other, changing both the magnitude and direction of the velocity of each block. What, if anything, can you conclude about the momentum of the two-block system after the collision?

Solution **(a)** The momentum of the system is simply the vector sum of the momenta of each piece of the system:

$$\begin{aligned}\vec{p}_{\text{system}} &= \vec{p}_1 + \vec{p}_2 = (1\text{ kg})(\langle 5,2,0\rangle\text{ m/s}) + (3\text{ kg})(\langle -3,4,0\rangle\text{ m/s})\\ &= \langle -4,14,0\rangle\text{ kg}\cdot\text{m/s}\end{aligned}$$

(b) The total momentum of the system is still $\langle -4,14,0\rangle$ kg · m/s.

EXAMPLE **The Weak Interaction: Neutron Decay**

A free neutron, one that is outside a nucleus, is unstable and decays into a positively charged proton, a negatively charged electron, and an uncharged antineutrino:

$$n \rightarrow p^+ + e^- + \bar{\nu}$$

This decay of a free neutron is caused by the weak interaction. The precise time that the neutron will decay is governed by quantum mechanical laws of probability and the nature of the weak interaction and is not predictable for any particular neutron, but the average lifetime of a neutron is about 15 minutes. Also probabilistic are the magnitudes and directions of the momenta of the decay products (proton, electron, and antineutrino), and Figure 3.37 shows some decays of an isolated free neutron that was at rest. Which of the decays shown in Figure 3.37 is actually impossible, if the neutron was at rest?

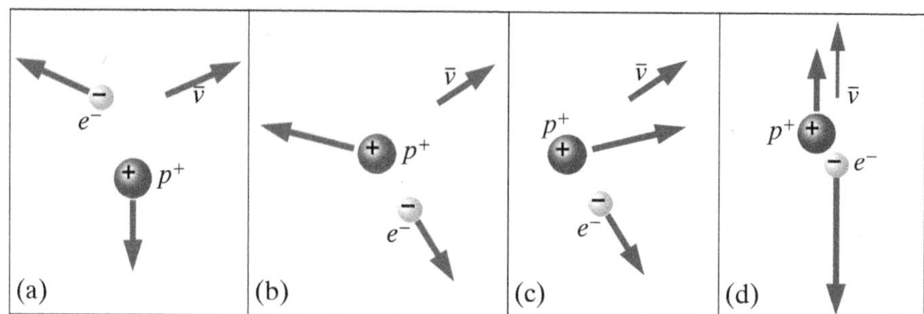

Figure 3.37 Which of these decays of a neutron at rest is impossible? The arrows represent momentum.

Solution The decay shown in Figure 3.37c is not actually possible. Conservation of momentum applies to all of the four fundamental interactions, including the weak interaction. The momentum of the isolated neutron was zero, so the momenta of the decay products must also add up to zero. In Figure 3.37c, however, the total momentum of the proton, electron, and antineutrino has a positive x component, which is not possible.

What's interesting about this is that we can rule out the possibility of the decay shown in Figure 3.37c without knowing anything about the details of the weak interaction. In fact, in quantum mechanical calculations at the atomic level concerned with electromagnetism, the strong interaction, or the weak interaction, force isn't even a particularly useful concept, yet conservation of momentum continues to play an important role even in the subatomic world.

Relativistic Momentum Conservation

Particle accelerators produce beams of particles such as electrons, protons, and pions at speeds very close to the speed of light. When these high-speed particles interact with other particles, experiments show that $\Delta\vec{p}_{\text{sys}} + \Delta\vec{p}_{\text{surr}} = \vec{0}$, but only if the momentum of each particle is defined in the way Einstein proposed, $\vec{p} \equiv \gamma m\vec{v}$. When the speed v approaches the speed of light c, the low-speed approximation $\gamma \approx 1$ ($\vec{p} \approx m\vec{v}$) is not valid. The quantity $m\vec{v}$ is not a conserved quantity when the speed v approaches the speed of light c, but the momentum $\vec{p} = \gamma m\vec{v}$ is conserved.

3.11 THE MULTIPARTICLE MOMENTUM PRINCIPLE

In our applications of the Momentum Principle we have often chosen a single object as the system of interest. However, we have seen that in some situations it can be very useful to choose a system that consists of two or more interacting objects. (Note that we can choose any object or set of objects as the system. Sometimes one choice is more useful than another, but any choice is valid.)

We will call forces "external" if they are exerted on objects in the system by objects in the surroundings. We call forces "internal" if they are forces exerted on objects in the system by other objects that are also in the system. What is remarkable is that the internal forces do not appear in the Momentum Principle for multiparticle systems:

MOMENTUM PRINCIPLE FOR MULTIPARTICLE SYSTEMS

$$\Delta\vec{p}_{\text{sys}} = (\vec{p}_{\text{sys},f} - \vec{p}_{\text{sys},i}) = \vec{F}_{\text{net}}\,\Delta t$$

$$\vec{p}_{\text{sys}} = \vec{p}_1 + \vec{p}_2 + \vec{p}_3 + \cdots$$
$$\vec{F}_{\text{net}} = \vec{F}_1 + \vec{F}_2 + \vec{F}_3 + \cdots \quad \text{(sum of all \textit{external} forces)}$$

This is an interesting statement of the Momentum Principle: it says that if we know the external forces acting on a multi-object system, forces exerted on objects in the chosen system by objects in the surroundings, we can draw conclusions about the change of momentum of the system over some time interval without worrying about any of the details of the interactions of the objects inside the system with each other. This can greatly simplify the analysis of the motion of some very complex systems. In a way, it should not be surprising that internal forces alone cannot change a system's total momentum. If they could, you could lift yourself off the ground by pulling up on your own feet!

We have implicitly been using this multiparticle version of the Momentum Principle when we have treated macroscopic objects like humans, spacecraft, and planets as if they were single point-like objects.

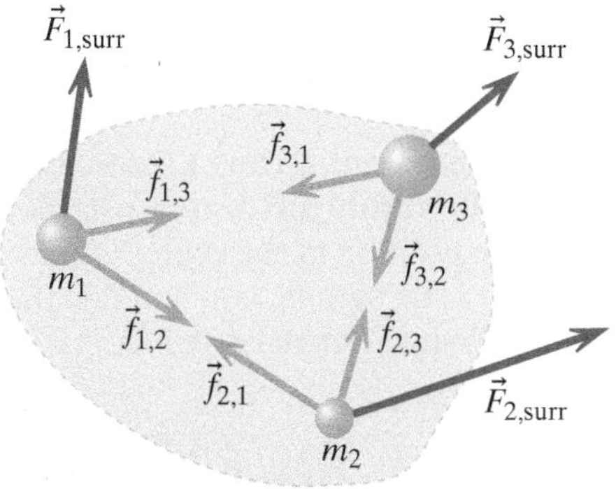

Figure 3.38 External and internal forces acting on a system of three particles.

Internal Forces Cancel

In a three-particle system (Figure 3.38) we show all of the forces acting on each particle, where the lowercase $\vec{f}$'s are forces the particles exert on each other (so-called "internal" forces), and the uppercase $\vec{F}$'s are forces exerted by objects in the surroundings that are not shown and are not part of our chosen system (these are so-called "external" forces, such as the gravitational attraction of the Earth, or a force that you exert by pulling on one of the particles).

We will use the following shorthand notation: $\vec{f}_{1,3}$ will denote the force exerted on particle 1 by particle 3; $\vec{F}_{2,\text{surr}}$ will denote the force on particle 2 exerted by objects in the surroundings, and so on. We start by applying the Momentum Principle to each of the three particles separately:

$$\Delta\vec{p}_1 = (\vec{F}_{1,\text{surr}} + \vec{f}_{1,2} + \vec{f}_{1,3})\Delta t$$
$$\Delta\vec{p}_2 = (\vec{F}_{2,\text{surr}} + \vec{f}_{2,1} + \vec{f}_{2,3})\Delta t$$
$$\Delta\vec{p}_3 = (\vec{F}_{3,\text{surr}} + \vec{f}_{3,1} + \vec{f}_{3,2})\Delta t$$

Nothing new so far—but now we add up these three equations. That is, we create a new equation by adding up all the terms on the left sides of the three equations, and adding up all the terms on the right sides, and setting them equal to each other:

$$\Delta\vec{p}_1 + \Delta\vec{p}_2 + \Delta\vec{p}_3 = (\vec{F}_{1,\text{surr}} + \vec{f}_{1,2} + \vec{f}_{1,3} + \vec{F}_{2,\text{surr}} + \vec{f}_{2,1} + \vec{f}_{2,3} + \vec{F}_{3,\text{surr}} + \vec{f}_{3,1} + \vec{f}_{3,2})\Delta t$$

Many of these terms cancel. By the principle of reciprocity (see Section 3.4), which is obeyed by gravitational and electric interactions, we have this:

$$\vec{f}_{1,2} = -\vec{f}_{2,1}$$
$$\vec{f}_{1,3} = -\vec{f}_{3,1}$$
$$\vec{f}_{2,3} = -\vec{f}_{3,2}$$

Thanks to reciprocity, all that remains after the cancellations is this:

$$\Delta\vec{p}_1 + \Delta\vec{p}_2 + \Delta\vec{p}_3 = (\vec{F}_{1,\text{surr}} + \vec{F}_{2,\text{surr}} + \vec{F}_{3,\text{surr}})\Delta t$$

The total momentum of the system is $\vec{p}_{\text{sys}} = \vec{p}_1 + \vec{p}_2 + \vec{p}_3$, so we have:

$$\Delta\vec{p}_{\text{sys}} = \vec{F}_{\text{net, surr}}\,\Delta t$$

The importance of this equation is that reciprocity has eliminated all of the internal forces (the forces that the particles in the system exert on each other); internal forces cannot affect the motion of the system as a whole. All that matters in determining the change of (total) momentum is the net external force. The equation has the same form as the Momentum Principle for a single particle.

The Center of Mass

The momentum of a multiparticle system such as a binary star can be expressed in terms of the velocity of the "center of mass" of the system. The position of the center of mass of a system of particles is defined as a weighted average of their positions:

$$\vec{r}_{\text{CM}} = \frac{m_1\vec{r}_1 + m_2\vec{r}_2 + m_3\vec{r}_3 + \cdots}{M_{\text{total}}}$$

$M_{\text{total}} = m_1 + m_2 + m_3 + \cdots$ is the total mass of the system. It can be shown that this definition is also valid if object 1 is a large object (not just a small particle) as long as m_1 is the mass of object 1 and $\vec{r}_1$ is the position of the center of mass of object 1. If you apply this definition to a binary star system, you find that the center of mass is between the two stars, closer to the more massive star.

EXAMPLE **Two Rocks**

A 10 kg rock is located at $\langle 3,0,0\rangle$ m, and a 2 kg rock is located at $\langle 8,2,0\rangle$ m (Figure 3.39). Find the center of mass of the two-rock system.

Solution

$$\vec{r}_{\text{CM}} = \frac{(10\text{ kg})\langle 3,0,0\rangle\text{ m} + (2\text{ kg})\langle 8,2,0\rangle\text{ m}}{(10\text{ kg} + 2\text{ kg})}$$
$$= \langle 3.83, 0.33, 0\rangle\text{ m}$$

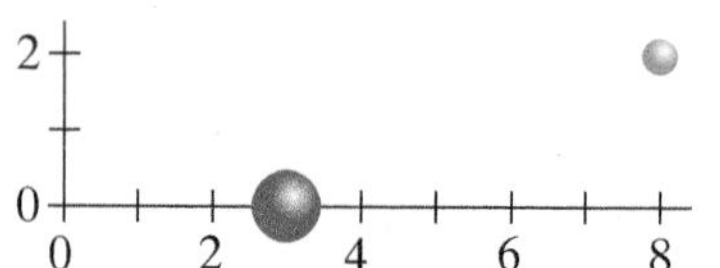

Figure 3.39 A 10 kg rock is located at $\langle 2,0,0\rangle$ m and a 2 kg rock is located at $\langle 8,2,0\rangle$ m. Where is the center of mass?

The center of mass of the system is closer to the more massive rock, as we would expect.

The Velocity of the Center of Mass

Taking the derivative with respect to time of the definition of center of mass, and assuming no changes in the masses, we find the following (since $\vec{v} = d\vec{r}/dt$):

$$\vec{v}_{\text{CM}} = \frac{m_1\vec{v}_1 + m_2\vec{v}_2 + m_3\vec{v}_3 + \cdots}{M_{\text{total}}}$$

Multiplying through by the total mass, we have this:

$$M_{\text{total}}\vec{v}_{\text{CM}} = m_1\vec{v}_1 + m_2\vec{v}_2 + m_3\vec{v}_3 + \cdots$$

However, if the speeds are small compared to the speed of light ($\gamma \approx 1$), the quantity $m_1\vec{v}_1 + m_2\vec{v}_2 + m_3\vec{v}_3 + \cdots$ is the total momentum of the system, so we can express the momentum of a system in terms of its center-of-mass velocity:

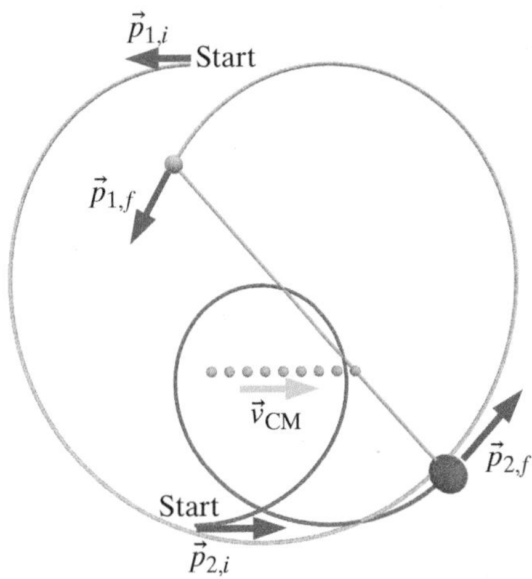

Figure 3.40 A binary star drifts through space. The center of mass moves with constant velocity. A line connecting the two stars passes through the center of mass, which is closer to the more massive star.

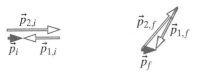

Figure 3.41 The total momentum of the two-star system does not change, as can be seen by the graphical sum of the vector momenta.

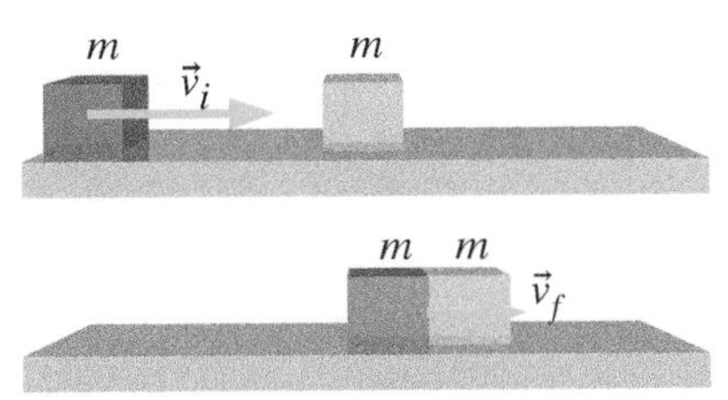

Figure 3.42 A sticking collision between blocks of equal mass. Top: Before the collision. Bottom: After the collision.

MOMENTUM AND CENTER OF MASS VELOCITY

$$\vec{p}_{sys} = M_{total}\vec{v}_{CM}$$

(If the speeds are small compared to the speed of light.)

If there are negligible interactions with the surroundings of a binary star, the velocity of the center of mass does not change but is constant as the binary star drifts through space (or is zero if the binary star's total momentum is zero). Figure 3.40 shows the motion of the center of mass for a binary star that is drifting through space. The two stars orbit around the (moving) center-of-mass point. A line connecting the two stars passes through the center of mass, which is closer to the more massive star.

The initial momentum of the two-star system is $\vec{p}_{1,i}+\vec{p}_{2,i}$, which is also equal to $(m_1+m_2)\vec{v}_{CM}$. Later, both momenta of the stars have changed, but the vector sum $\vec{p}_{1,f}+\vec{p}_{2,f}$ is still the same and still equal to $(m_1+m_2)\vec{v}_{CM}$. Figure 3.41 shows graphical addition of the momenta of the two stars to make the total system momentum, which does not change.

We can characterize the momentum of a multiparticle system simply as $\vec{p}_{sys} = M_{total}\vec{v}_{CM}$, as though the system were a point particle. Moreover, if an object is a sphere whose density is only a function of radius, the object exerts a gravitational force on other objects as though the sphere were a point particle. We can predict the motion of a system of star or a planet or an asteroid as though it were a single point particle of large mass.

Checkpoint 8 A system consists of a 2 kg block moving with velocity $\langle -3,4,0\rangle$ m/s and a 5 kg block moving with velocity $\langle 2,6,0\rangle$ m/s. Calculate the velocity of the center of mass of the two-block system, given that the momentum of the system is $M_{total}\vec{v}_{CM}$.

3.12 COLLISIONS: NEGLIGIBLE EXTERNAL FORCES

An event is called a "collision" if it involves an interaction that takes place in a relatively short time and has a large effect on the momenta of the objects compared to the effects of other interactions during that short time. A collision does not necessarily involve actual physical contact between objects, which may be interacting via long-range forces like the gravitational force or the electric force. A collision can involve attractive forces. For example, a spacecraft is deflected as it passes close to Jupiter on its way to Saturn, and Jupiter exerts a large gravitational force for a relatively short time. During that short time the effects of the other planets are negligible.

EXAMPLE **What Is *Not* Conserved in a Collision?**

In later chapters we will see that energy and angular momentum are also conserved quantities. However, most quantities, including velocity, are not conserved quantities. Consider a collision between two blocks on ice (so that there is low friction), each with mass m. One block is initially at rest and the other is moving with initial velocity $\langle v_i,0,0\rangle$. In the collision they stick together and move with final velocity $\langle v_f,0,0\rangle$ (Figure 3.42). What is the final velocity? Is velocity a conserved quantity?

Solution As the blocks slide on the ice there are no changes in momentum in the vertical direction (the y components of momentum are zero at all times). In the xz plane of the ice there are negligible interactions with the surroundings, so the momentum of the surroundings doesn't change, and $\Delta\vec{p}_{sys}=\vec{0}$. Therefore $\vec{p}_{sys} = \vec{p}_1+\vec{p}_2$ doesn't change. Before the collision the total momentum is

$$\vec{p}_{sys} = m\langle v_i,0,0\rangle + m\vec{0} = \langle mv_i,0,0\rangle$$

After the collision the total momentum is

$$\vec{p}_{\text{sys}} = m\langle v_f,0,0\rangle + m\langle v_f,0,0\rangle = \langle 2mv_f,0,0\rangle$$

Since $\vec{p}_{\text{sys}}$ doesn't change, we have

$$\langle 2mv_f,0,0\rangle = \langle mv_i,0,0\rangle$$

$$v_f = \frac{1}{2}v_i$$

The final speed (and velocity) is half the initial value. Nothing in the surroundings changed, so $\Delta\vec{v}_{\text{system}} + \Delta\vec{v}_{\text{surr}} \neq \vec{0}$. We say that velocity is not a conserved quantity, but momentum is. This is another reason why momentum is such an important physical quantity.

EXAMPLE **Different Choices of System**

Consider the head-on collision of two identical bowling balls (Figure 3.43). Each ball has a mass of 6 kg. Ball A with velocity $\langle 5,0,0\rangle$ m/s strikes ball B, which was at rest. Then ball A stops and ball B moves with the same velocity ($\langle 5,0,0\rangle$ m/s) that ball A had initially. For the following three different choices of system, find the momentum change of the system and the momentum change of the surroundings. **(a)** Choose a system consisting only of ball A. **(b)** Choose a system consisting only of ball B. **(c)** Choose a system consisting of both balls.

Solution **(a)** The initial momentum of ball A is $\vec{p}_{A,i} = \langle 30,0,0\rangle$ kg · m/s, so if we choose ball A as the system:

$$\Delta\vec{p}_{\text{sys}} = \langle 0,0,0\rangle \text{ kg}\cdot\text{m/s} - \langle 30,0,0\rangle \text{ kg}\cdot\text{m/s} = \langle -30,0,0\rangle \text{ kg}\cdot\text{m/s}$$

Since ball A is the system, ball B is the surroundings:

$$\Delta\vec{p}_{\text{surr}} = \langle 30,0,0\rangle \text{ kg}\cdot\text{m/s} - \langle 0,0,0\rangle \text{ kg}\cdot\text{m/s} = \langle +30,0,0\rangle \text{ kg}\cdot\text{m/s}$$

(b) If we choose ball B as the system:

$$\Delta\vec{p}_{\text{sys}} = \langle +30,0,0\rangle \text{ kg}\cdot\text{m/s}$$
$$\Delta\vec{p}_{\text{surr}} = \langle -30,0,0\rangle \text{ kg}\cdot\text{m/s}$$

(c) If we choose both balls as the system, then:

$$\vec{p}_{\text{sys},i} = \langle 30,0,0\rangle \text{ kg}\cdot\text{m/s} + \langle 0,0,0\rangle \text{ kg}\cdot\text{m/s} = \langle 30,0,0\rangle \text{ kg}\cdot\text{m/s}$$
$$\vec{p}_{\text{sys},f} = \langle 0,0,0\rangle \text{ kg}\cdot\text{m/s} + \langle 30,0,0\rangle \text{ kg}\cdot\text{m/s} = \langle 30,0,0\rangle \text{ kg}\cdot\text{m/s}$$
$$\Delta\vec{p}_{\text{sys}} = \langle 0,0,0\rangle \text{ kg}\cdot\text{m/s}$$

5 kg 5 kg
A B
< 6,0,0 > m/s Initially at rest
Initial state (before collision)

Final state (after collision)

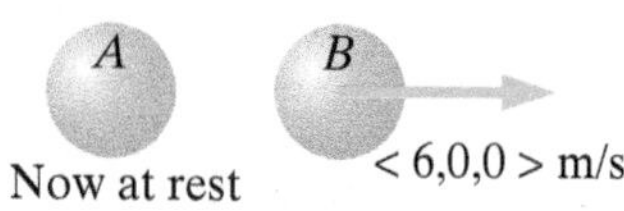

Figure 3.43 A head-on collision between two identical bowling balls.

Often it can be useful to analyze collisions by choosing a system that includes all of the colliding objects, as we will see in the following example.

EXAMPLE **Two Lumps of Clay Collide**

Two lumps of clay travel through the air toward each other, at speeds much less than the speed of light (Figure 3.44), rotating as they move. When the lumps collide they stick together. The mass of lump 1 is 0.2 kg and its initial velocity is $\langle 6,0,0\rangle$ m/s, and the mass of lump 2 is 0.5 kg and its initial velocity is $\langle -5,4,0\rangle$ m/s. What is the final velocity of the center of mass of the stuck-together lumps?

Solution The Momentum Principle applies to the motion of the center of mass, even though the objects are rotating. We can neglect air resistance and gravitational force (which are negligible compared to the large contact forces during the very short time of the collision).

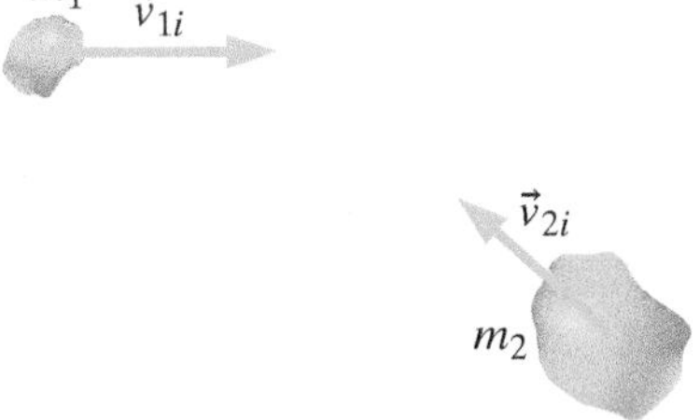

Figure 3.44 Two lumps of clay traveling through the air, just before colliding.

System: Both lumps
Surroundings: Air, Earth

$$\vec{p}_{\text{total},f} = \vec{p}_{\text{total},i} + \vec{F}_{\text{net}}\,\Delta t$$
$$\vec{p}_{\text{total},f} \approx m_1\vec{v}_{1i} + m_2\vec{v}_{2i} + \langle 0,0,0\rangle\Delta t$$
$$(m_1+m_2)\vec{v}_f = m_1\vec{v}_{1i} + m_2\vec{v}_{2i}$$
$$\vec{v}_f = \frac{m_1\vec{v}_{1i} + m_2\vec{v}_{2i}}{m_1+m_2}$$
$$\vec{v}_f = \frac{(0.2\text{ kg})\langle 6,0,0\rangle + (0.5\text{ kg})\langle -5,4,0\rangle}{(0.7\text{ kg})}\text{ m/s}$$
$$\vec{v}_f = \langle -1.86, 2.86, 0\rangle\text{ m/s}$$

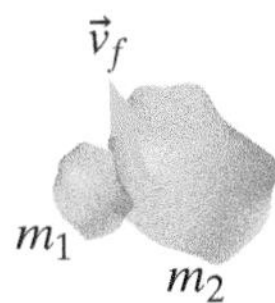

Figure 3.45 The momentum of the stuck-together lumps of clay just after the collision is equal to the sum of the initial momenta of the two lumps of clay. The object may rotate.

Check: The units are correct.

By choosing both lumps as the system, we were able to find the final momentum of the system without needing to know anything about the details of the complex forces that the lumps exerted on each other during the collision.

QUESTION Why doesn't it matter that the lumps may be rotating after the collision (Figure 3.45)?

The rotation of an object doesn't affect its momentum; the total momentum of the system is still the sum of the individual momenta of the objects within the system and equal to $M_{\text{total}}\vec{v}_{\text{CM}}$. However, the Momentum Principle does not tell us anything about how fast the stuck-together objects will be rotating; for this, we will have to apply the Angular Momentum Principle, which is discussed in a later chapter.

The initial velocities are the center-of-mass velocities of each lump, and the final velocity is the center-of-mass velocity of the combined system after the two lumps stick together. We can also speak of the velocity of the center of mass just before the collision, which is the velocity of a mathematical point between the two lumps, calculated as the weighted average of their positions.

QUESTION What was the center-of-mass velocity of the two lumps just before they collided?

It was $\langle -1.86, 2.86, 0\rangle$ m/s, the same as after the collision, since the total momentum can be calculated as $M_{\text{total}}\vec{v}_{\text{CM}}$ both before and after the collision.

Some Problems Require More than One Principle

Because the lumps in the previous example stuck together, we had enough information to solve for the final velocity of the stuck-together lumps—we had one equation with only one unknown (vector) quantity. In more complex situations, we will find that we sometimes do not have enough information to solve for all of the unknown quantities by applying only the Momentum Principle.

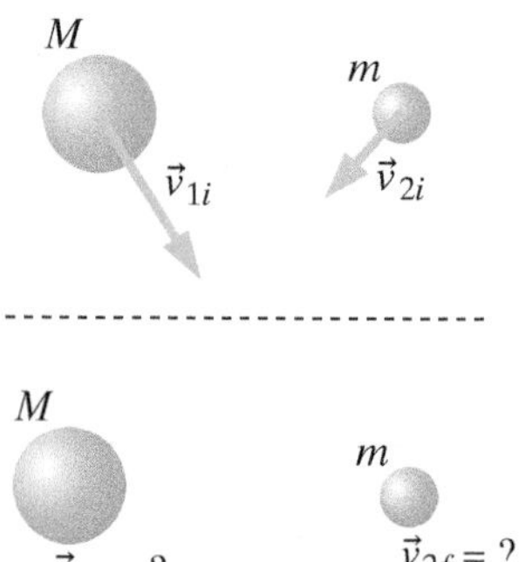

Figure 3.46 The initial momenta of the colliding objects are known, but the final momenta after the collision are unknown.

For example, consider a collision between two balls that bounce off each other, as shown in Figure 3.46. The Momentum Principle tells us that

$$\vec{p}_{\text{total},f} = \vec{p}_{\text{total},i}$$
$$\vec{p}_{1f} + \vec{p}_{2f} = \vec{p}_{1i} + \vec{p}_{2i}$$

but we are left with two unknown (vector) quantities, $\vec{p}_{1f}$ and $\vec{p}_{2f}$, and only one equation. We will not be able to analyze problems of this kind fully until we can invoke the Energy Principle (Chapter 6) and the Angular Momentum Principle (Chapter 11), along with the Momentum Principle.

Checkpoint 9 You and a friend each hold a lump of wet clay. Each lump has a mass of 20 g. You each toss your lump of clay into the air, where the lumps collide and stick together. Just before the impact, the velocity of one lump was $\langle 5, 2, -3 \rangle$ m/s, and the velocity of the other lump was $\langle -3, 0, -2 \rangle$ m/s. **(a)** What was the total momentum of the lumps just before impact? **(b)** What is the momentum of the stuck-together lump just after the impact? **(c)** What is its velocity?

3.13 NEWTON AND EINSTEIN

Newton devised a particular explanatory scheme in which the analysis of motion is divided into two distinct parts:

1. Quantify the interaction in terms of a concept called "force." Specific examples are Newton's gravitational force and Coulomb's electric force.
2. Quantify the change of motion in terms of the change in a quantity called "momentum." The change in the momentum is equal to the force times Δt (for sufficiently small values of Δt).

This scheme, called the "Newtonian synthesis," has turned out to be extraordinarily successful in explaining a huge variety of diverse physical phenomena, from the fall of an apple to the orbiting of the Moon. Yet we have no way of asking whether the Universe *really* works this way. It seems unlikely that the Universe actually uses the human concepts of force and momentum in the unfolding motion of an apple or the Moon.

We refer to the Newtonian synthesis both to identify and honor the particular, highly successful analysis scheme introduced by Newton, but also to remind ourselves that this is not the only possible way to view and analyze the Universe.

Einstein's Alternative View

Newton identified the gravitational force and represented it algebraically but could give no explanation for it. He was content with showing that it correctly predicted the motion of the planets, and this was a huge advance, the real beginning of modern science.

Einstein made a further huge advance by giving a deeper explanation for gravity, as a part of his Theory of General Relativity. He realized that it was possible to think of the massive Sun bending space and time (!) in such a way as to make the planets move the way they do. The equations in Einstein's General Theory of Relativity make it possible to calculate the curvature of space and time due to massive objects, and to predict how other objects will move in this altered space and time.

Moreover, Einstein's Theory of General Relativity accurately predicts some tiny effects that Newton's gravitational force does not, such as the slight bending of light as it passes near the Sun. General Relativity also explains some bizarre large-scale phenomena such as black holes and the observed expansion of the space between the galaxies.

Einstein's earlier Theory of Special Relativity established that nothing, not even information, can travel faster than light. Because Newton's gravitational force law depends only on the distance between objects, not on the time, something's wrong with Newton's law, since this implies that if an object were suddenly yanked away, its force on another object would vanish instantaneously, thus giving (in principle) a way to send information from one place to another instantaneously. Einstein's Theory of General Relativity doesn't have this problem.

Since the equations of General Relativity are very difficult to work with, and Newton's gravitational force works very well for most purposes, in this textbook we will use Newton's approach to gravity. However, you should be aware that for the most precise calculations one must use the Theory of General Relativity. For example, the highly accurate atomic clocks in the satellites that make up the Global Positioning System (GPS) have to be continually corrected using Einstein's Theory of General Relativity. Otherwise, GPS positions would be wrong by several kilometers after just one day of operation!

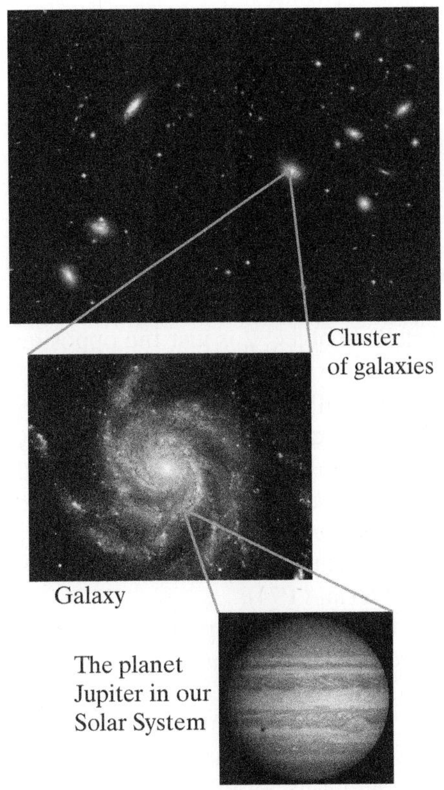

Figure 3.47 Our Solar System orbits around the center of our galaxy, which interacts with other galaxies. (Courtesy NASA/JPL-Caltech)

3.14 PREDICTING THE FUTURE OF COMPLEX SYSTEMS

We can use the Newtonian method and numerical integration to try to predict the future of a group of objects that interact with each other mainly gravitationally, such as the Solar System consisting of the Sun, planets, moons, asteroids, comets, and so on. There are other, nongravitational interactions present. Radiation pressure from sunlight makes a comet's tail sweep away from the Sun. Streams of charged particles (the "solar wind") from the Sun hit the Earth and contribute to the Northern and Southern Lights (auroras). However, the main interactions within the Solar System are gravitational.

The Solar System as a group is orbiting around the center of our Milky Way galaxy, and our galaxy is interacting gravitationally with other galaxies in a local cluster of galaxies (Figure 3.47), but changes in these speeds and directions take place very slowly compared to those within the Solar System. To a very good approximation we can neglect these interactions while studying the inner workings of the Solar System itself.

We have modeled the motion of planets and stars using the Momentum Principle and the gravitational force. We can use the same approach to model the gravitational interactions of more complex systems involving three, four, or any number of gravitationally interacting objects. All that is necessary is to include the forces associated with all pairs of objects.

In principle we could use these techniques to predict the future of our entire Solar System. Of course our prediction would be only approximate, because we take finite time steps that introduce numerical errors, and we are neglecting various small interactions. We would need to investigate whether the prediction of our model is a reasonable approximation to the real motion of our real Solar System.

Our Solar System contains a huge number of objects. Most of the planets have moons, and Jupiter and Saturn have many moons. There are thousands of asteroids and comets, not to mention an uncounted number of tiny specks of dust. This raises a practical limitation: the fastest computers would take a very long time to carry out even one time step for all the pieces of the Solar System.

The Three-Body Problem

As we have seen, in order to carry out a numerical integration we need to know the position and velocity of each object at some time t, from which we can

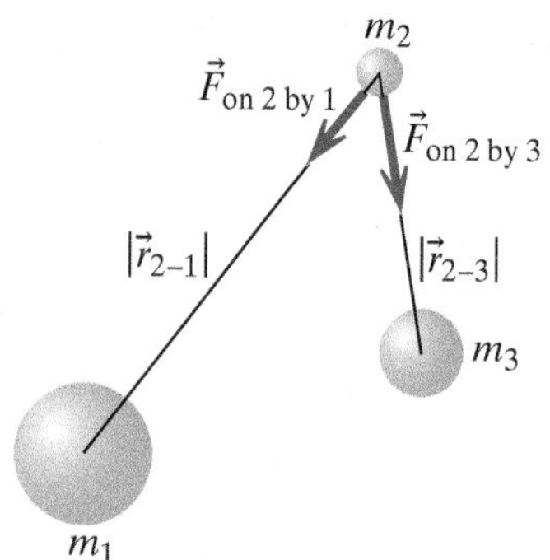

Figure 3.48 In the "three-body problem" each of the three objects interacts with two other objects.

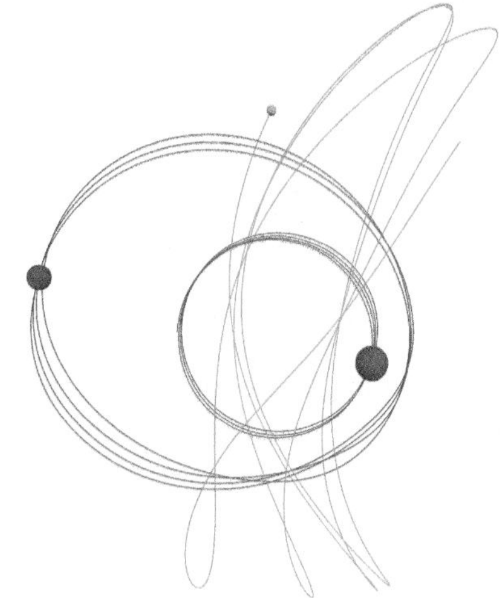

Figure 3.49 An example of three bodies interacting gravitationally.

calculate the forces on each object at that time. This enables us to calculate the position and momentum for each object at a slightly later time $t+\Delta t$. We need the net force on an object, which we get from vector addition of all the pairwise forces calculated from Newton's gravitational force.

Suppose that there are three gravitating objects, as shown in Figure 3.48. The force on object m_2, which is acted upon gravitationally by objects m_1 and m_3, is a natural extension of what we did earlier for the analysis of a star and a planet. We simply use the superposition principle to include the interaction with m_3 as well as the interaction with m_1:

$$\vec{F}_{\text{net on } m_2} = -G\frac{m_2 m_1}{|\vec{r}_{2-1}|^2}\hat{r}_{2-1} - G\frac{m_2 m_3}{|\vec{r}_{2-3}|^2}\hat{r}_{2-3}$$

We use this force to update the momentum of object m_2. Similar computations apply to the other masses, m_1 and m_3. These calculations are tedious, but let the computer do them! We can carry out a numerical integration for each mass: given the current locations of all the masses we can calculate the vector net force exerted on each mass at a time t, and then we can calculate the new values of momentum and position at a slightly later time $t+\Delta t$.

Reciprocity simplifies these calculations somewhat. For example, for the pair of interacting objects 2 and 3 you only need to calculate fully the force of object 2 on object 3. The force of object 3 on object 2 is just the opposite: $\vec{F}_{3\text{ on }2} = -\vec{F}_{2\text{ on }3}$.

While there exist analytical (nonnumerical) solutions for two-body gravitational motion, except for some very special cases the "three-body problem" has not been solved analytically. However, in a numerical integration, the computer doesn't mind that there are lots of quantities and a lot of repetitive calculations to be done. You yourself can solve the three-body problem numerically, by adding additional calculations to your two-body computations (see the homework problems P66 and P74).

For a single object orbiting a very massive stationary object, or for two objects orbiting each other (with zero total momentum), the possible trajectories are only a circle, ellipse, parabola, hyperbola, or straight line. The motion of a three-body system can be vastly more complex and diverse. Figure 3.49 shows the numerical integration of a complicated three-body trajectory for one particular set of initial conditions. Adding just one more object opens up a vast range of complex behaviors.

Imagine how complex the motion of a galaxy can be!

Checkpoint 10 Suppose you have four stars with given initial conditions (positions and momenta). The net force on each star is the vector sum of three forces exerted by the other three stars, for a total of 12 forces that must be evaluated in each iterative step. In each step, how many forces would you have to calculate fully, and how many forces could you calculate simply by reversing vectors already calculated?

Sensitivity to Initial Conditions

For a two-body system, slight changes in the initial conditions make only slight changes in the orbit, such as changing a circle into an ellipse, or an ellipse into a slightly different ellipse, but you never get anything other than a simple trajectory.

The situation is very different in systems with three or more interacting objects. Consider a low-mass object orbiting two massive objects, which we imagine somehow to be nailed down so that they cannot move. (This could also represent two positive charges that are fixed in position, with a low-mass negative charge orbiting around them.) In Figure 3.50 are two very different orbits (one shown in red, the other in blue), starting from rest with slightly

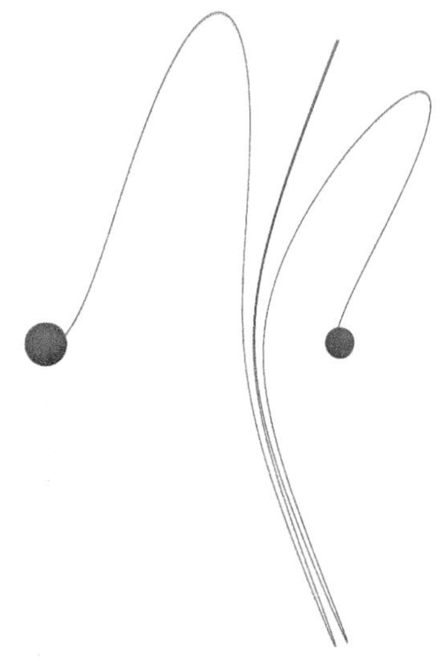

Figure 3.50 Two different orbits (one red, the other blue), starting from rest but from slightly different initial values of x for a low-mass object. The two orbits end with crashes on the two different massive objects. The mass of the red object is twice the mass of the blue object; both of these objects are somehow nailed down and cannot move.

different initial values of x. The two orbits end with crashes on the two different massive objects. The trajectories are wildly different!

This sensitivity to initial conditions generally becomes more extreme as you add more objects, and the Solar System contains lots of objects, big and small. Also, we can anticipate that if small errors in specifying the initial conditions can have large effects, so too it is likely that failing to take into account the tiny force exerted by a small asteroid might make a big difference after a long integration time.

The Solar System is actually fairly predictable, because there is one giant mass (the Sun), and the other, much smaller masses are very far apart. This is unlike the example shown here, which deliberately emphasizes the sensitivity observed when there are large masses near each other.

3.15 DETERMINISM

If you know the net force on an object as a function of its position and momentum (or velocity), you can predict the future motion of the object by simple step-by-step calculations based on the Momentum Principle. Newton's demonstration of the power of this approach induced many 17th and 18th century philosophers and scientists to adopt the view already proposed by Descartes, Boyle, and others, that the Universe is a giant clockwork device, whose future is completely determined by the present positions and momenta of all the macroscopic and microscopic objects in it. Just turn the crank, and predict the future! This point of view is called "determinism," and taken to its extreme it raises the question of whether humans actually have any free will, or whether all of our actions are predetermined. Scientific and technological advances in the 20th century, however, have led us to see that although the Newtonian approach can be used to predict the long-term future of a simple system or the short-term future of a complicated system, there are both practical and theoretical limitations on predictions in some systems.

Practical Limitations

One reason we may be unable to predict the long-term future of even a simple system is a practical limitation in our ability to measure its initial conditions with sufficient accuracy. Over time, even small inaccuracies in initial conditions can lead to large cumulative errors in calculations. This is a practical limitation rather than a theoretical one, because in principle if we could measure initial conditions more accurately, we could perform more accurate calculations.

Another practical limitation is our inability to account for all interactions in our model. Every object in the Universe interacts with every other object. In constructing simplified models, we ignore interactions whose magnitude is extremely small. However, over time, even very small interactions can lead to significant effects. Even the tiny effects of sunlight have been shown to affect noticeably the motion of an asteroid over a long time. With larger and faster computers, we can include more and more interactions in our models, but our models can never completely reflect the complexity of the real world.

A branch of current research that focuses on how the detailed interactions of a large number of atoms or molecules lead to the bulk properties of matter is called "molecular dynamics." For example, one way to test our understanding of the nature of the interactions of water molecules is to create a computational model in which the forces between molecules and the initial state of a large number of molecules—several million—are described, and then see whether running the model produces a virtual fluid that actually behaves like water. However, even with an accurate expression for the (electric) forces between atoms or molecules, it is not feasible to compute and track the motions of the 10^{25} molecules in a glass of water. Significant efforts are under way to build faster and faster computers, and to develop more efficient computer algorithms, to permit realistic numerical integration of more complex systems.

Chaos

A second kind of limitation occurs in systems that display an extreme sensitivity to initial conditions such as the situation seen in Figure 3.50, where small changes in initial conditions produce large changes in behavior. In recent years scientists have discovered systems in which *any* change in the initial conditions, no matter how small (infinitesimal), can lead to complete loss of predictability: the difference in the two possible future motions of the system diverges exponentially with time. Such systems are called "chaotic." It is thought that over a long time period the weather may be literally unpredictable in this sense.

An interesting popular science book about this new field of research is *Chaos: Making a New Science*, by James Gleick (Penguin, 1988). The issue of small changes in initial conditions is a perennial concern of time travelers in science fiction stories. Perhaps the most famous such story is Ray Bradbury's "The Sound of Thunder," in which a time traveler steps on a butterfly during the Jurassic era, and returns to an eerily changed present time.

Breakdown of Newton's Laws

There are other situations of high interest in which we cannot usefully apply Newton's laws of motion, because these classical laws do not adequately describe the behavior of physical systems. To model systems composed of very small particles such as protons and electrons, quarks (the constituents of protons and neutrons), and photons, it is necessary to use the laws of quantum mechanics. To model in detail the gravitational interactions between massive objects, it is necessary to apply the principles of General Relativity. Given the principles of quantum mechanics and General Relativity, one might assume that it would be possible to follow a procedure similar to the one we have followed in this chapter, and predict in detail the future of these systems. However, there appear to be more fundamental limitations on what we can know about the future.

Probability and Uncertainty

Our understanding of the atomic world of quantum mechanics suggests that there are fundamental limits to our ability to predict the future, because at the atomic and subatomic levels the Universe itself is nondeterministic. We cannot know exactly what will happen at a given time, but only the probability that certain events will occur within a given time frame.

A simple example may make this clear. A free neutron (one not bound into a nucleus) is unstable and eventually decays into a positively charged proton, a negatively charged electron, and an electrically neutral antineutrino:

$$n \rightarrow p^+ + e^- + \overline{\nu}$$

The average lifetime of the free neutron is about 15 minutes. Some neutrons survive longer than this, and some last a shorter amount of time. All of our experiments and all of our theory are consistent with the notion that *it is not possible to predict when a particular neutron will decay*. There is only a (known) probability that the neutron will decay in the next microsecond, or the next, or the next.

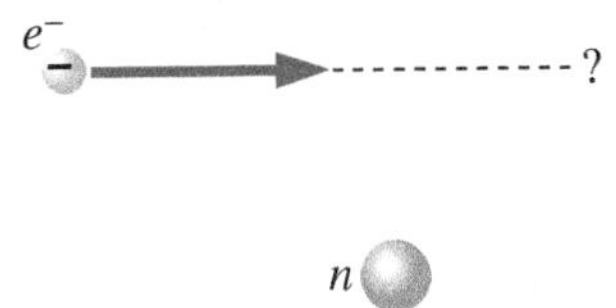

Figure 3.51 The electron may continue to travel in a straight line, but we cannot be certain that it will, because we do not know when the neutron will decay.

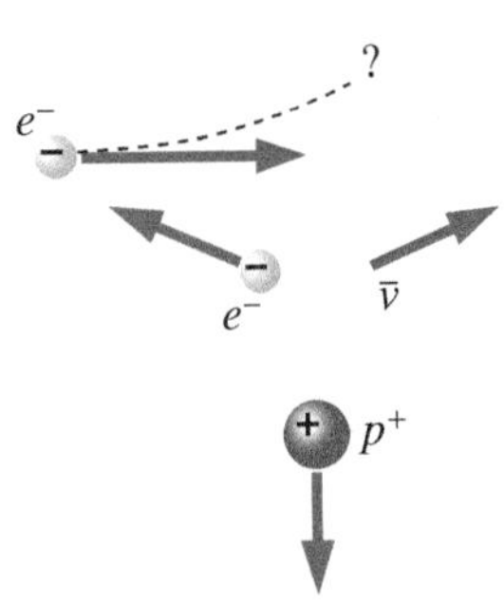

Figure 3.52 One possible path of the electron if the neutron does decay. The arrows represent momentum.

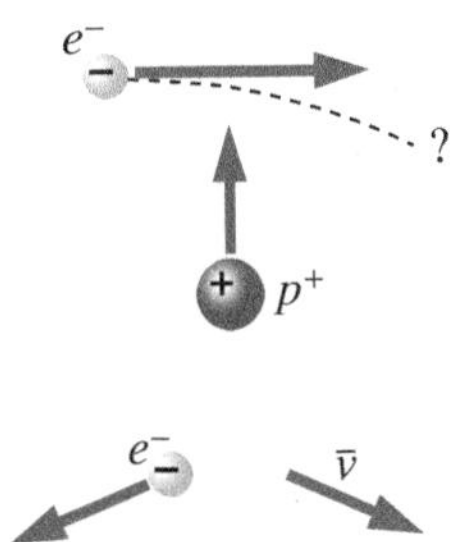

Figure 3.53 Another possible path of the electron if the neutron does decay. The trajectory of the electron depends on the directions of the momenta of the decay products. The arrows represent momentum.

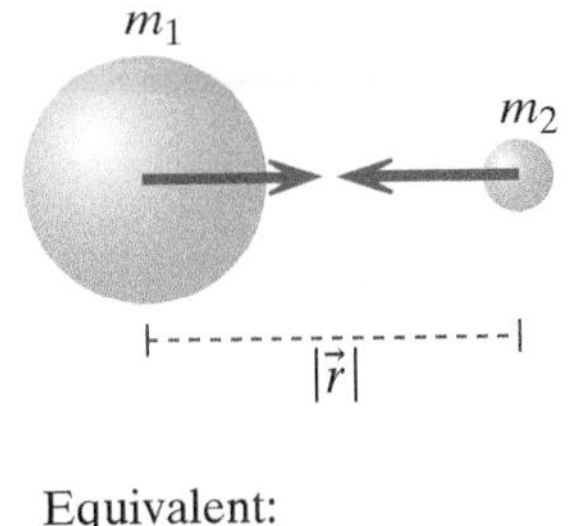

Figure 3.54 A sphere whose density depends only on distance from the center acts like a point particle with all the mass at the center.

If this is really how the Universe works, as a great deal of evidence demonstrates, then there is an irreducible lack of predictability and determinism of the Universe itself. As far as our own predictions using the Momentum Principle are concerned, consider the following simple scenario. An electron is traveling with constant velocity through nearly empty space, but there is a free neutron in the vicinity (Figure 3.51).

We might predict that the electron will move in a straight line, and some fraction of the time we will turn out to be right (the electrically uncharged neutron exerts no electric force on the electron, there is a magnetic interaction but it is quite small, and the gravitational interaction is tiny). However, if the neutron happens (probabilistically) to decay just as our electron passes nearby, suddenly our electron is subjected to large electric forces due to the decay proton and electron, and will deviate from a straight path. Moreover, the directions in which the decay products move are also only probabilistic, so we can't even predict whether the proton or the electron will come closest to our electron (Figures 3.52 and 3.53).

This is a simple but dramatic example of how quantum indeterminacy can lead to indeterminacy even in the context of the Momentum Principle.

The Heisenberg Uncertainty Principle

The Heisenberg uncertainty principle states that there are actual theoretical limits to our knowledge of the state of physical systems. One quantitative formulation states that the position and the momentum of a particle cannot both be simultaneously measured exactly (here Δ means variation, not difference):

$$\Delta x \Delta p_x \geq h$$

This relation says that the product of the uncertainty in position Δx and the uncertainty in momentum Δp_x is greater than or equal to a constant h, called Planck's constant. Planck's constant is tiny ($h = 6.6 \times 10^{-34}$ kg·m^2/s), so this limitation is not noticeable for macroscopic systems, but for objects small enough to require a quantum mechanical description, the uncertainty principle puts fundamental limits on how accurately we can know the initial conditions, and therefore how well we can predict the future.

3.16 POINTS AND SPHERES

The expression for the gravitational force applies to objects that are "point-like" (very small compared to the center-to-center distance between the objects). In later chapters we will be able to show that any hollow spherical shell with uniform density acts gravitationally on external objects as though all the mass of the shell were concentrated at its center. The density of the Earth is not uniform, because the central iron core has higher density than the outer layers. However, by considering the Earth as layers of hollow spherical shells, like an onion, with each shell of nearly uniform density, we get the result that the Earth can be modeled for most purposes as a point mass located at the center of the Earth (for very accurate calculations one must take into account small irregularities in the Earth's density from place to place). Similar statements can be made about other planets and stars. In Figure 3.54, the gravitational force is correctly calculated using the center-to-center distance $|\vec{r}|$.

This is not an obvious result. After all, in Figure 3.54 some of the atoms are closer and some farther apart than the center-to-center distance $|\vec{r}|$, but the net effect after adding up all the interactions of the individual atoms is as though the two objects had collapsed down to points at their centers. This is a very special property of $1/r^2$ forces, both gravitational and electric, and is not true for forces that have a different dependence on distance.

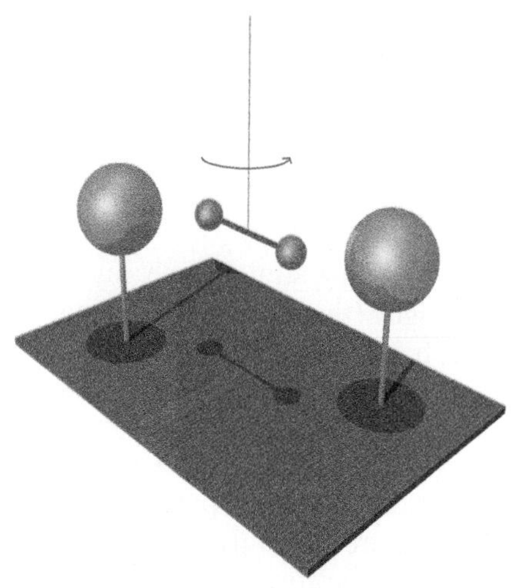

Figure 3.55 The Cavendish experiment.

3.17 MEASURING THE GRAVITATIONAL CONSTANT G

In order to make quantitative predictions and analyses of physical phenomena involving gravitational interactions, it is necessary to know the universal gravitational constant G. In 1797–1798 Henry Cavendish performed the first experiment to determine a precise value for G (Figure 3.55). In this kind of experiment, a bar with metal spheres at each end is suspended from a thin quartz fiber that constitutes a "torsional" spring. From other measurements, it is known how large a tangential force measured in newtons is required to twist the fiber through a given angle. Large balls are brought near the suspended spheres, and one measures how much the fiber twists due to the gravitational interactions between the large balls and the small spheres.

If the masses are measured in kilograms, the distance in meters, and the force in newtons, the gravitational constant G has been measured in such experiments to be

$$G = 6.7 \times 10^{-11} \frac{\text{N} \cdot \text{m}^2}{\text{kg}^2}$$

This extremely small number reflects the fact that gravitational interactions are inherently very weak compared with electromagnetic interactions. The only reason that gravitational interactions are significant in our daily lives is that objects interact with the entire Earth, which has a huge mass. It takes sensitive measurements such as the Cavendish experiment to observe gravitational interactions between two ordinary-sized objects.

SUMMARY

Four fundamental interactions:

- Gravitational interactions (all objects attract each other gravitationally)
- Electromagnetic interactions (electric and magnetic interactions, closely related to each other); interatomic forces are electric in nature
- "Strong" interactions (inside the nucleus of an atom)
- "Weak" interactions (neutron decay, for example)

The gravitational force (Figure 3.56)

$$\vec{F}_{\text{grav on 2 by 1}} = -G\frac{m_1 m_2}{|\vec{r}|^2}\hat{r}$$

$\vec{r} = \vec{r}_2 - \vec{r}_1$ is the position of 2 relative to 1.

G is a universal constant: $G = 6.7 \times 10^{-11} \frac{\text{N} \cdot \text{m}^2}{\text{kg}^2}$.

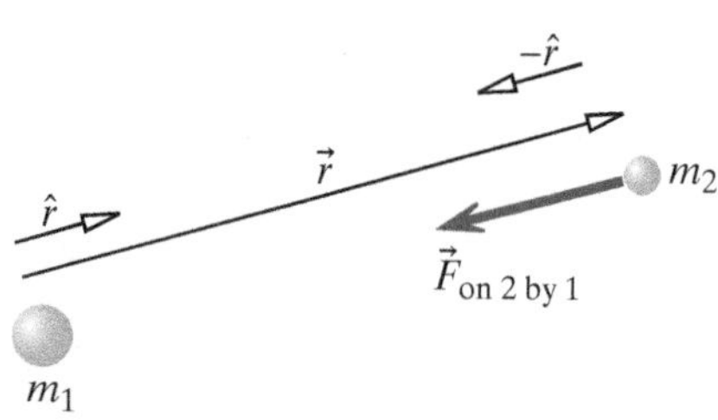

Figure 3.56 Near the Earth's surface, $|\vec{F}_{\text{grav}}| = mg$, where $g = +9.8$ N/kg.

The electric force (Figure 3.57)

$$\vec{F}_{\text{elec on 2 by 1}} = \frac{1}{4\pi\varepsilon_0}\frac{q_1 q_2}{|\vec{r}|^2}\hat{r}$$

$\vec{r} = \vec{r}_2 - \vec{r}_1$ is the position of 2 relative to 1.

$\frac{1}{4\pi\varepsilon_0}$ is a universal constant: $\frac{1}{4\pi\varepsilon_0} = 9 \times 10^9 \frac{\text{N} \cdot \text{m}^2}{\text{C}^2}$.

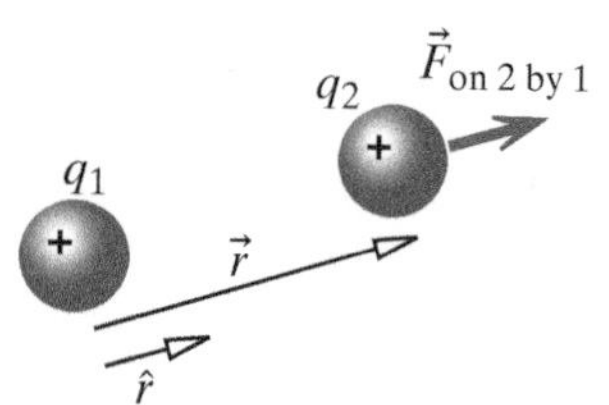

Figure 3.57

Reciprocity

$\vec{F}_{\text{on 1 by 2}} = -\vec{F}_{\text{on 2 by 1}}$ (gravitational and electric forces). This is also called "Newton's third law of motion."

Conservation of momentum

$$\Delta\vec{p}_{\text{sys}} + \Delta\vec{p}_{\text{surr}} = \vec{0}$$

The multiparticle Momentum Principle

$$\Delta\vec{p}_{\text{sys}} = \vec{F}_{\text{net,surr}}\Delta t$$

Additional results

Uniform-density spheres act as though all the mass were at the center.

The center of mass

$$\vec{r}_{CM} = \frac{m_1\vec{r}_1 + m_2\vec{r}_2 + m_3\vec{r}_3 + \cdots}{M_{total}}$$

$$\vec{v}_{CM} = \frac{m_1\vec{v}_1 + m_2\vec{v}_2 + m_3\vec{v}_3 + \cdots}{M_{total}}$$

$$\vec{p}_{sys} = M_{total}\vec{v}_{cm}$$

Complex systems; determinism; chaos

QUESTIONS

Q1 Which fundamental interaction (gravitational, electromagnetic, strong, or weak) is responsible for each of these processes? How do you know? **(a)** A neutron outside a nucleus decays into a proton, electron, and antineutrino. **(b)** Protons and neutrons attract each other in a nucleus. **(c)** The Earth pulls on the Moon. **(d)** Protons in a nucleus repel each other.

Q2 Why is the value of the constant g different on Earth and on the Moon? Explain in detail.

Q3 You hold a tennis ball above your head, then open your hand and release the ball, which begins to fall. At this instant (ball starting to fall, no longer in contact with your hand), what can you conclude about the relative magnitudes of the force on the ball by the Earth and the force on the Earth by the ball? Explain.

Q4 Suppose that you are going to program a computer to carry out an iterative calculation of motion involving electric forces. Assume that as usual we use the final velocity in each time interval as the approximate average velocity during that interval. Which of the following calculations should be done *before* starting the repetitive calculation loop? Which of the calculations listed above should go *inside* the repetitive loop, and in what order?

(a) Define constants such as $\frac{1}{4\pi\varepsilon_0}$.
(b) Update the (vector) position of each object.
(c) Calculate the (vector) forces acting on the objects.
(d) Specify the initial (vector) momentum of each object.
(e) Specify an appropriate value for the time step.
(f) Specify the mass of each object.
(g) Update the (vector) momentum of each object.
(h) Specify the initial (vector) position of each object.

Q5 A bullet traveling horizontally at a very high speed embeds itself in a wooden block that is sitting at rest on a very slippery sheet of ice. You want to find the speed of the block just after the bullet embeds itself in the block. **(a)** What should you choose as the system to analyze? **(b)** Which of the following statements is true? (1) After the collision, the speed of the block with the bullet stuck in it is the same as the speed of the bullet before the collision. (2) The momentum of the block with the bullet stuck in it is the same as the momentum of the bullet before the collision. (3) The initial momentum of the bullet is greater than the momentum of the block with the bullet stuck in it.

Q6 You hang from a tree branch, then let go and fall toward the Earth. As you fall, the y component of your momentum, which was originally zero, becomes large and negative. **(a)** Choose yourself as the system. There must be an object in the surroundings whose y momentum must become equally large, and positive. What object is this? **(b)** Choose yourself and the Earth as the system. The y component of your momentum is changing. Does the total momentum of the system change? Why or why not?

Q7 One kind of radioactivity is called "alpha decay." For example, the nucleus of a radium-220 atom can spontaneously split into a radon-216 nucleus plus an alpha particle (a helium nucleus containing two protons and two neutrons). Consider a radium-220 nucleus that is initially at rest. It spontaneously decays, and the alpha particle travels off in the $+z$ direction. What can you conclude about the motion of the new radon-216 nucleus? Be as precise as you can, and explain your reasoning.

Q8 A bowling ball is initially at rest. A Ping-Pong ball moving in the $+z$ direction hits the bowling ball and bounces off it, traveling back in the $-z$ direction. Consider a time interval Δt extending from slightly before to slightly after the collision. **(a)** In this time interval, what is the sign of Δp_z for the system consisting of both balls? **(b)** In this time interval, what is the sign of Δp_z for the system consisting of the bowling ball alone?

Q9 The windshield of a speeding car hits a hovering insect. Consider the time interval from just before the car hits the insect to just after the impact. **(a)** For which choice of system is the change of momentum zero? **(b)** Is the magnitude of the change of momentum of the bug bigger than, the same as, or smaller than that of the car? **(c)** Is the magnitude of the change of velocity of the bug bigger than, the same as, or smaller than that of the car?

PROBLEMS

Section 3.2

•P10 At a particular instant the magnitude of the gravitational force exerted by a planet on one of its moons is 3×10^{23} N. If the mass of the moon were three times as large, what would be the magnitude of the force? If instead the distance between the moon and the planet were three times as large (no change in mass), what would be the magnitude of the force?

•P11 Masses M and m attract each other with a gravitational force of magnitude F. Mass m is replaced with a mass $3m$, and it is moved four times farther away. Now what is the magnitude of the force?

•**P12** A 3 kg ball and a 5 kg ball are 2 m apart, center to center. What is the magnitude of the gravitational force that the 3 kg ball exerts on the 5 kg ball? What is the magnitude of the gravitational force that the 5 kg ball exerts on the 3 kg ball?

•**P13** The mass of the Earth is 6×10^{24} kg, and the mass of the Moon is 7×10^{22} kg. At a particular instant the Moon is at location $\langle 2.8 \times 10^8, 0, -2.8 \times 10^8 \rangle$ m, in a coordinate system whose origin is at the center of the Earth. **(a)** What is $\vec{r}_{M-E}$, the relative position vector from the Earth to the Moon? **(b)** What is $|\vec{r}_{M-E}|$? **(c)** What is the unit vector $\hat{r}_{M-E}$? **(d)** What is the gravitational force exerted by the Earth on the Moon? Your answer should be a vector.

•**P14** A star exerts a gravitational force of magnitude $|\vec{F}|$ on a planet. The distance between the star and the planet is r. If the planet were three times farther away (that is, if the distance between the bodies were $3r$), by what factor would the force on the planet due to the star change?

•**P15** A planet exerts a gravitational force of magnitude 7×10^{22} N on a star. If the planet were four times closer to the star (that is, if the distance between the star and the planet were 1/4 what it is now), what would be the magnitude of the force on the star due to the planet?

•**P16** A moon orbits a planet in the xy plane, as shown in Figure 3.58. You want to calculate the force on the moon by the planet at each location labeled by a letter (A, B, C, D). At each of these locations, what are: **1.** the unit vector $\hat{r}$, **2.** the unit vector $\hat{F}$ in the direction of the force?

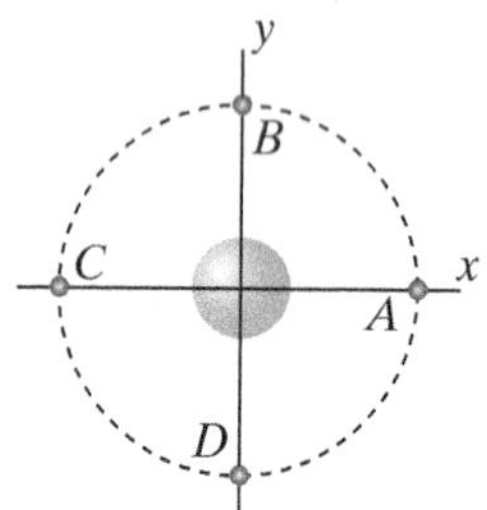

Figure 3.58

•**P17** The mass of the Sun is 2×10^{30} kg, and the mass of Mercury is 3.3×10^{23} kg. The distance from the Sun to Mercury is 4.8×10^{10} m. **(a)** Calculate the magnitude of the gravitational force exerted by the Sun on Mercury. **(b)** Calculate the magnitude of the gravitational force exerted by Mercury on the Sun.

•**P18** Measurements show that Jupiter's gravitational force on a mass of 1 kg near Jupiter's surface would be 24.9 N. The radius of Jupiter is 71500 km (71.5×10^6 m). From these data determine the mass of Jupiter.

•**P19** A planet of mass 4×10^{24} kg is at location $\langle 5e11, -2e11, 0 \rangle$ m. A star of mass 5×10^{30} kg is at location $\langle -2e11, 3e11, 0 \rangle$ m. It will be useful to draw a diagram of the situation, including the relevant vectors. **(a)** What is the relative position vector $\vec{r}$ pointing from the planet to the star? **(b)** What is the distance between the planet and the star? **(c)** What is the unit vector $\hat{r}$ in the direction of $\vec{r}$? **(d)** What is the magnitude of the force exerted on the planet by the star? **(e)** What is the magnitude of the force exerted on the star by the planet? **(f)** What is the (vector) force exerted on the planet by the star? **(g)** What is the (vector) force exerted on the star by the planet?

•**P20** A planet of mass 4×10^{24} kg is at location $\langle -6e11, 3e11, 0 \rangle$ m. A star of mass 6×10^{30} kg is at location $\langle 5e11, -3e11, 0 \rangle$ m. What is the force exerted on the planet by the star? (It will probably be helpful to draw a diagram, including the relevant vectors.)

••**P21** Two copies of this textbook are standing right next to each other on a bookshelf. Make a rough estimate of the magnitude of the gravitational force that the books exert on each other. Explicitly list all quantities that you had to estimate, and all simplifications and approximations you had to make to do this calculation. Compare your result to the gravitational force on the book by the Earth.

Section 3.3

•**P22** The mass of Mars is 6.4×10^{23} kg and its radius is 3.4×10^6 m. What is the value of the constant g on Mars?

•**P23** At what height above the surface of the Earth is there a 1% difference between the approximate magnitude of the gravitational field (9.8 N/kg) and the actual magnitude of the gravitational field at that location? That is, at what height y above the Earth's surface is $GM_E/(R_E + y)^2 = 0.99GM_E/R_E^2$?

•**P24** Calculate the approximate gravitational force exerted by the Earth on a human standing on the Earth's surface. Compare with the approximate gravitational force of a human on another human at a distance of 3 m. What approximations or simplifying assumptions must you make?

•••**P25** A steel ball of mass m falls from a height h onto a scale calibrated in newtons. The ball rebounds repeatedly to nearly the same height h. The scale is sluggish in its response to the intermittent hits and displays an *average* force F_{avg}, such that $F_{\text{avg}}T = F\Delta t$, where $F\Delta t$ is the brief impulse that the ball imparts to the scale on every hit, and T is the time between hits.

Calculate this average force in terms of m, h, and physical constants. Compare your result with the scale reading if the ball merely rests on the scale. Explain your analysis carefully (but briefly).

Section 3.5

•**P26** The mass of the Sun is 2×10^{30} kg, the mass of the Earth is 6×10^{24} kg, and their center-to-center distance is 1.5×10^{11} m. Suppose that at some instant the Sun's momentum is zero (it's at rest). Ignoring all effects but that of the Earth, what will the Sun's speed be after one day? (Very small changes in the velocity of a star can be detected using the "Doppler" effect, a change in the frequency of the starlight, which has made it possible to identify the presence of planets in orbit around a star.)

••**P27** At $t = 532.0$ s after midnight, a spacecraft of mass 1400 kg is located at position $\langle 3 \times 10^5, 7 \times 10^5, -4 \times 10^5 \rangle$ m, and at that time an asteroid whose mass is 7×10^{15} kg is located at position $\langle 9 \times 10^5, -3 \times 10^5, -12 \times 10^5 \rangle$ m. There are no other objects nearby. **(a)** Calculate the (vector) force acting on the spacecraft. **(b)** At $t = 532.0$ s the spacecraft's momentum was $\vec{p}_i$, and at the later time $t = 538.0$ s its momentum was $\vec{p}_f$. Calculate the (vector) change of momentum $\vec{p}_f - \vec{p}_i$.

••**P28** **(a)** In outer space, far from other objects, block 1 of mass 40 kg is at position $\langle 7, 11, 0 \rangle$ m, and block 2 of mass 1000 kg is at position $\langle 18, 11, 0 \rangle$ m. What is the (vector) gravitational force acting on block 2 due to block 1? It helps to make a sketch of the situation. **(b)** At 4.6 s after noon both blocks were at rest at the positions given above. At 4.7 s after noon, what is the (vector) momentum of block 2? **(c)** At 4.7 s after noon, what is the (vector) momentum of block 1? **(d)** At 4.7 s after noon, which one of the following statements is true? A. Block 1 and block 2 have the same speed. B. Block 2 is moving faster than block 1. C. Block 1 is moving faster than block 2.

••P29 A star of mass 7×10^{30} kg is located at $\langle 5 \times 10^{12}, 2 \times 10^{12}, 0 \rangle$ m. A planet of mass 3×10^{24} kg is located at $\langle 3 \times 10^{12}, 4 \times 10^{12}, 0 \rangle$ m and is moving with a velocity of $\langle 0.3 \times 10^4, 1.5 \times 10^4, 0 \rangle$ m/s. **(a)** At a time 1×10^6 s later, what is the new velocity of the planet? **(b)** Where is the planet at this later time? **(c)** Explain briefly why the procedures you followed in parts (a) and (b) were able to produce usable results but wouldn't work if the later time had been 1×10^9 s instead of 1×10^6 s after the initial time. Explain briefly how you could use a computer to get around this difficulty.

••P30 At $t = 0$ a star of mass 4×10^{30} kg has velocity $\langle 7 \times 10^4, 6 \times 10^4, -8 \times 10^4 \rangle$ m/s and is located at $\langle 2.00 \times 10^{12}, -5.00 \times 10^{12}, 4.00 \times 10^{12} \rangle$ m relative to the center of a cluster of stars. There is only one nearby star that exerts a significant force on the first star. The mass of the second star is 3×10^{30} kg, its velocity is $\langle 2 \times 10^4, -1 \times 10^4, 9 \times 10^4 \rangle$ m/s, and this second star is located at $\langle 2.03 \times 10^{12}, -4.94 \times 10^{12}, 3.95 \times 10^{12} \rangle$ m relative to the center of the cluster of stars. **(a)** At $t = 1 \times 10^5$ s, what is the approximate momentum of the first star? **(b)** Discuss briefly some ways in which your result for (a) is approximate, not exact. **(c)** At $t = 1 \times 10^5$ s, what is the approximate position of the first star? **(d)** Discuss briefly some ways in which your result for (b) is approximate, not exact.

•••P31 In June 1997 the NEAR spacecraft ("Near Earth Asteroid Rendezvous"; see http://near.jhuapl.edu/), on its way to photograph the asteroid Eros, passed within 1200 km of asteroid Mathilde at a speed of 10 km/s relative to the asteroid. (See Figure 3.59.) From photos transmitted by the 805 kg spacecraft, Mathilde's size was known to be about 70 km by 50 km by 50 km. The asteroid is presumably made of rock. Rocks on Earth have a density of about 3000 kg/m^3 (3 grams/cm^3). **(a)** Make a rough diagram to show qualitatively the effect on the spacecraft of this encounter with Mathilde. Explain your reasoning. **(b)** Make a very rough estimate of the change in momentum of the spacecraft that would result from encountering Mathilde. Explain how you made your estimate. **(c)** Using your result from part (b), make a rough estimate of how far off course the spacecraft would be, one day after the encounter. **(d)** From actual observations of the location of the spacecraft one day after encountering Mathilde, scientists concluded that Mathilde is a loose arrangement of rocks, with lots of empty space inside. What was it about the observations that must have led them to this conclusion?

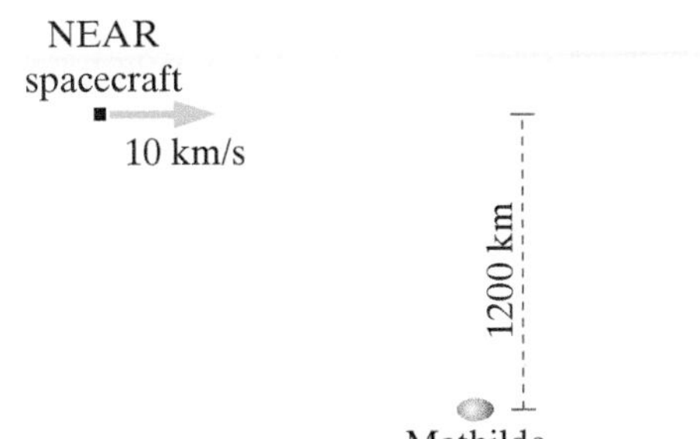

Figure 3.59

Experimental background: The position was tracked by very accurate measurements of the time that it takes for a radio signal to go from Earth to the spacecraft followed immediately by a radio response from the spacecraft being sent back to Earth. Radio signals, like light, travel at a speed of 3×10^8 m/s, so the time measurements had to be accurate to a few nanoseconds (1 ns $= 1 \times 10^{-9}$ s).

Section 3.7

•P32 Figure 3.60 shows two positively charged objects (with the same charge) and one negatively charged object. What is the direction of the net electric force on the negatively charged object? If the net force is zero, state this explicitly.

Figure 3.60

•P33 Figure 3.61 shows two negatively charged objects (with the same charge) and one positively charged object. What is the direction of the net electric force on the positively charged object? If the net force is zero, state this explicitly.

Figure 3.61

•P34 The left side of Figure 3.62 shows a proton and an electron. **(a)** What is the direction of the electric force on the electron by the proton? **(b)** What is the direction of the electric force on the proton by the electron? **(c)** How do the magnitudes of these forces compare? The right side of Figure 3.62 shows two electrons. **(d)** What is the direction of the electric force on electron A due to electron B? **(d)** What is the direction of the electric force on electron B due to electron A? **(d)** How do the magnitudes of these forces compare?

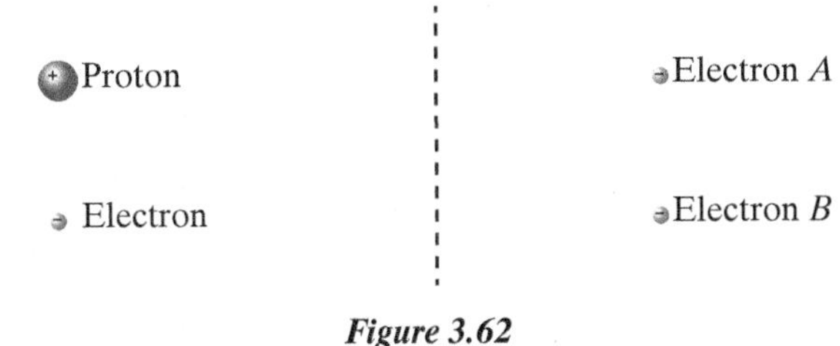

Figure 3.62

•P35 An alpha particle contains two protons and two neutrons, and has a net charge of $+2e$. The alpha particle is 1 mm away from a single proton, which has a charge of $+e$. Which statement about the magnitudes of the electric forces between the particles is correct? **(a)** The force on the proton by the alpha particle is larger than the force on the alpha particle by the proton. **(b)** The force on the alpha particle by the proton is larger than the force on the proton by the alpha particle. **(c)** The forces are equal in magnitude. **(d)** Not enough information is given.

••P36 A proton and an electron are separated by 1×10^{-10} m, the radius of a typical atom. Calculate the magnitude of the electric force that the proton exerts on the electron, and the magnitude of the electric force that the electron exerts on the proton.

••P37 Two thin hollow plastic spheres, about the size of a Ping-Pong ball with masses ($m_1 = m_2 = 2 \times 10^{-3}$ kg), have been rubbed with wool. Sphere 1 has a charge $q_1 = -2 \times 10^{-9}$ C and is at location $\langle 0.50, -0.20, 0\rangle$ m. Sphere 2 has a charge $q_2 = -4 \times 10^{-9}$ C and is at location $\langle -0.40, 0.40, 0\rangle$ m. It will be useful to draw a diagram of the situation, including the relevant vectors. **(a)** What is the relative position vector $\vec{r}$ pointing from q_1 to q_2? **(b)** What is the distance between q_1 and q_2? **(c)** What is the unit vector $\hat{r}$ in the direction of $\vec{r}$? **(d)** What is the magnitude of the gravitational force exerted on q_2 by q_1? **(e)** What is the (vector) gravitational force exerted on q_2 by q_1? **(f)** What is the magnitude of the electric force exerted on q_2 by q_1? **(g)** What is the (vector) electric force exerted on q_2 by q_1? **(h)** What is the ratio of the magnitude of the electric force to the magnitude of the gravitational force? **(i)** If the two masses were four times farther away (that is, if the distance between the masses were $4\vec{r}$), what would be the ratio of the magnitude of the electric force to the magnitude of the gravitational force now?

••P38 A proton is located at $\langle 0, 0, -2 \times 10^{-9}\rangle$ m, and an alpha particle (consisting of two protons and two neutrons) is located at $\langle 1.5 \times 10^{-9}, 0, 2 \times 10^{-9}\rangle$ m. **(a)** Calculate the force the proton exerts on the alpha particle. **(b)** Calculate the force the alpha particle exerts on the proton.

••P39 Use data from the inside back cover to calculate the gravitational and electric forces two electrons exert on each other when they are 1×10^{-10} m apart (about one atomic radius). Which interaction between two electrons is stronger, the gravitational attraction or the electric repulsion? If the two electrons are at rest, will they begin to move toward each other or away from each other? Note that since both the gravitational and electric forces depend on the inverse square distance, this comparison holds true at all distances, not just at a distance of 1×10^{-10} m.

••P40 At a particular instant a proton exerts an electric force of $\langle 0, 5.76 \times 10^{-13}, 0\rangle$ N on an electron. How far apart are the proton and the electron?

Section 3.10

•P41 Two balls of mass 0.3 kg and 0.5 kg are connected by a low-mass spring (Figure 3.63). This device is thrown through the air with low speed, so air resistance is negligible. The motion is complicated: the balls whirl around each other, and at the same time the system vibrates, with continually changing stretch of the spring. At a particular instant, the 0.3 kg ball has a velocity $\langle 4, -3, 2\rangle$ m/s and the 0.5 kg ball has a velocity $\langle 2, 1, 4\rangle$ m/s. **(a)** At this instant, what is the total momentum of the device? **(b)** What is the net gravitational (vector) force exerted by the Earth on the device? **(c)** At a time 0.1 s later, what is the total momentum of the device?

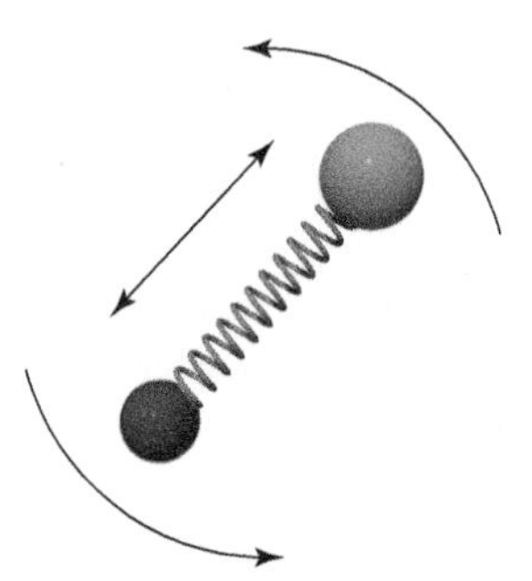

Figure 3.63

Section 3.11

•P42 At a certain instant object 1 is at location $\langle 10, -8, 6\rangle$ m, moving with velocity $\langle 4, 6, -2\rangle$ m/s. At the same instant object 2 is at location $\langle 3, 0, -2\rangle$ m, moving with velocity $\langle -8, 2, 7\rangle$ m/s. **(a)** What is the location of the center of mass of the two equal-mass objects? **(b)** What is the velocity of the center of mass?

•P43 The mass of the Earth is 6×10^{24} kg, the mass of the Moon is 7×10^{22} kg, and the center-to-center distance is 4×10^8 m. How far from the center of the Earth is the center of mass of the Earth–Moon system? Note that the Earth's radius is 6.4×10^6 m.

•P44 The mass of the Sun is 2×10^{30} kg, the mass of the Earth is 6×10^{24} kg, and the center-to-center distance is 1.5×10^{11} m. How far from the center of the Sun is the center of mass of the Sun–Earth system? Note that the Sun's radius is 7×10^8 m.

••P45 Two rocks are tied together with a string of negligible mass and thrown into the air. At a particular instant, rock 1, which has a mass of 0.1 kg, is headed straight up with a speed of 5 m/s, and rock 2, which has a mass of 0.25 kg, is moving parallel to the ground, in the $+x$ direction, with a speed of 7.5 m/s. **(a)** What is the total momentum of the system consisting of both rocks and the string? **(b)** What is the velocity of the center of mass of the system?

••P46 A tennis ball of mass 0.06 kg traveling at a velocity of $\langle 9, -2, 13\rangle$ m/s is about to collide with an identical tennis ball whose velocity is $\langle 4, 5, -10\rangle$ m/s. **(a)** What is the total momentum of the system of the two tennis balls? **(b)** What is the velocity of the center of mass of the two tennis balls?

Section 3.12

•P47 In outer space, far from other objects, two rocks collide and stick together. Before the collision their momenta were $\langle -10, 20, -5\rangle$ kg·m/s and $\langle 8, -6, 12\rangle$ kg·m/s. What was their total momentum before the collision? What must be the momentum of the combined object after the collision?

•P48 When they are far apart, the momentum of a proton is $\langle 3.4 \times 10^{-21}, 0, 0\rangle$ kg·m/s as it approaches another proton that is initially at rest. The two protons repel each other electrically, without coming close enough to touch. When they are once again far apart, one of the protons now has momentum $\langle 2.4 \times 10^{-21}, 1.55 \times 10^{-21}, 0\rangle$ kg·m/s. At that instant, what is the momentum of the other proton?

••P49 You and a friend each hold a lump of wet clay. Each lump has a mass of 30 g. You each toss your lump of clay into the air, where the lumps collide and stick together. Just before the impact, the velocity of one lump was $\langle 3, 3, -3\rangle$ m/s, and the velocity of the other lump was $\langle -3, 0, -3\rangle$ m/s. **(a)** What was the total momentum of the lumps just before the impact? **(b)** What is the momentum of the stuck-together lump just after the collision? **(c)** What is the velocity of the stuck-together lump just after the collision?

••P50 A car of mass 2800 kg collides with a truck of mass 4700 kg, and just after the collision the car and truck slide along, stuck together. The car's velocity just before the collision was $\langle 40, 0, 0\rangle$ m/s, and the truck's velocity just before the collision was $\langle -14, 0, 29\rangle$ m/s. **(a)** What is the velocity of the stuck-together car and truck just after the collision? **(b)** In your analysis in part (a), why can you neglect the effect of the force of the road on the car and truck?

••P51 A car of mass 2045 kg moving in the x direction at a speed of 29 m/s strikes a hovering mosquito of mass

2.5 mg, and the mosquito is smashed against the windshield. The interaction between the mosquito and the windshield is an electric interaction between the electrons and protons in the mosquito and those in the windshield. **(a)** What is the approximate momentum change of the mosquito? Give magnitude and direction. Explain any approximations you make. **(b)** At a particular instant during the impact, when the force exerted on the mosquito by the car is F, what is the magnitude of the force exerted on the car by the mosquito? **(c)** What is the approximate momentum change of the car? Give magnitude and direction. Explain any approximations you make. **(d)** Qualitatively, why is the collision so much more damaging to the mosquito than to the car?

••P52 A bullet of mass 0.105 kg traveling horizontally at a speed of 300 m/s embeds itself in a block of mass 2 kg that is sitting at rest on a nearly frictionless surface. What is the speed of the block after the bullet embeds itself in the block?

••P53 Object A has mass $m_A = 8$ kg and initial momentum $\vec{p}_{A,i} = \langle 20, -5, 0 \rangle$ kg·m/s, just before it strikes object B, which has mass $m_B = 11$ kg. Just before the collision object B has initial momentum $\vec{p}_{B,i} = \langle 5, 6, 0 \rangle$ kg·m/s. **(a)** Consider a system consisting of both objects A and B. What is the total initial momentum of this system just before the collision? **(b)** The forces that A and B exert on each other are very large but last for a very short time. If we choose a time interval from just before to just after the collision, what is the approximate value of the impulse applied to the two-object system due to forces exerted on the system by objects outside the system? **(c)** Therefore, what does the Momentum Principle predict that the total final momentum of the system will be just after the collision? **(d)** Just after the collision, object A is observed to have momentum $\vec{p}_{A,f} = \langle 18, 5, 0 \rangle$ kg·m/s. What is the momentum of object B just after the collision?

••P54 In outer space a small rock with mass 5 kg traveling with velocity $\langle 0, 1800, 0 \rangle$ m/s strikes a stationary large rock head-on and bounces straight back with velocity $\langle 0, -1500, 0 \rangle$ m/s. After the collision, what is the vector momentum of the large rock?

••P55 Two rocks collide in outer space. Before the collision, one rock had mass 9 kg and velocity $\langle 4100, -2600, 2800 \rangle$ m/s. The other rock had mass 6 kg and velocity $\langle -450, 1800, 3500 \rangle$ m/s. A 2 kg chunk of the first rock breaks off and sticks to the second rock. After the collision the 7 kg rock has velocity $\langle 1300, 200, 1800 \rangle$ m/s. After the collision, what is the velocity of the other rock, whose mass is 8 kg?

••P56 Two rocks collide with each other in outer space, far from all other objects. Rock 1 with mass 5 kg has velocity $\langle 30, 45, -20 \rangle$ m/s before the collision and $\langle -10, 50, -5 \rangle$ m/s after the collision. Rock 2 with mass 8 kg has velocity $\langle -9, 5, 4 \rangle$ m/s before the collision. Calculate the final velocity of rock 2.

••P57 In outer space two rocks collide and stick together. Here are the masses and initial velocities of the two rocks:

Rock 1: mass = 15 kg, initial velocity = $\langle 10, -30, 0 \rangle$ m/s
Rock 2: mass = 32 kg, initial velocity = $\langle 15, 12, 0 \rangle$ m/s

What is the velocity of the stuck-together rocks after colliding?

••P58 A bullet of mass m traveling horizontally at a very high speed v embeds itself in a block of mass M that is sitting at rest on a nearly frictionless surface. What is the speed of the block after the bullet embeds itself in the block?

••P59 A satellite that is spinning clockwise has four low-mass solar panels sticking out as shown. A tiny meteor traveling at high speed rips through one of the solar panels and continues in the same direction but at reduced speed. Afterward, calculate the v_x and v_y components of the center-of-mass velocity of the satellite. In Figure 3.64 $\vec{v}_1$, and $\vec{v}_2$ are the initial and final velocities of the meteor, and $\vec{v}$ is the initial velocity of the center of mass of the satellite, in the x direction.

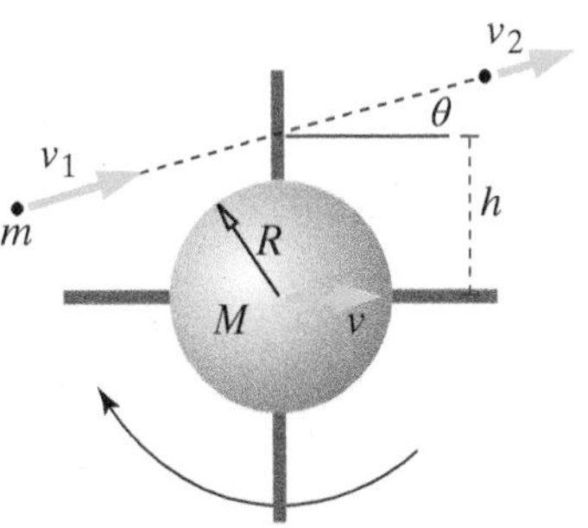

Figure 3.64

••P60 A tiny piece of space junk of mass m strikes a glancing blow to a spinning satellite. Before the collision the satellite was moving and rotating as shown in Figure 3.65. After the collision the space junk is traveling in a new direction and moving more slowly. The velocities of the space junk before and after the collision are shown in the diagram. The satellite has mass M and radius R. Just after the collision, what are the components of the center-of-mass velocity of the satellite (v_x and v_y)?

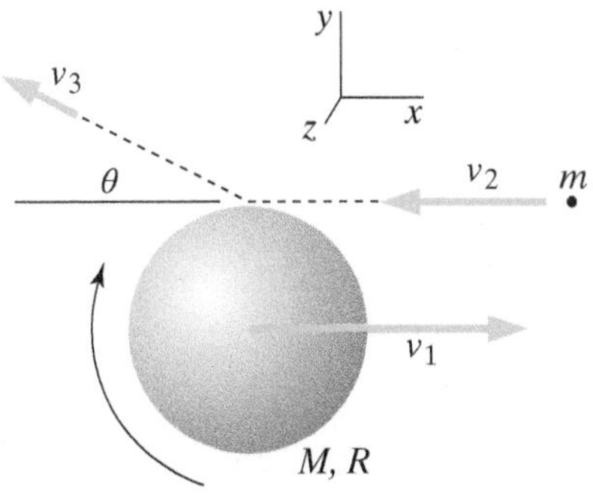

Figure 3.65

••P61 A space station has the form of a hoop of radius R, with mass M. Initially its center of mass is not moving, but it is spinning. Then a small package of mass m is thrown by a spring-loaded gun toward a nearby spacecraft as shown in Figure 3.66; the package has a speed v after launch. Calculate the center-of-mass velocity (a vector) of the space station after the launch.

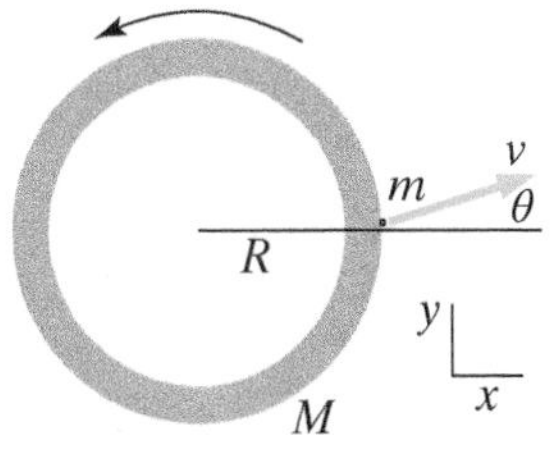

Figure 3.66

••**P62** A ball of mass 0.05 kg moves with a velocity $\langle 17,0,0\rangle$ m/s. It strikes a ball of mass 0.1 kg that is initially at rest. After the collision, the heavier ball moves with a velocity of $\langle 3,3,0\rangle$ m/s. **(a)** What is the velocity of the lighter ball after impact? **(b)** What is the impulse delivered to the 0.05 kg ball by the heavier ball? **(c)** If the time of contact between the balls is 0.03 s, what is the force exerted by the heavier ball on the lighter ball?

••**P63** Suppose that all the people of the Earth go to the North Pole and, on a signal, all jump straight up. Estimate the recoil speed of the Earth. The mass of the Earth is 6×10^{24} kg, and there are about 6 billion people (6×10^{9}).

COMPUTATIONAL PROBLEMS

More detailed and extended versions of some of these computational modeling problems may be found in the lab activities included in the *Matter & Interactions*, Fourth Edition, resources for instructors.

••**P64** To use an `arrow` object to visualize a force, it is usually necessary to scale the length of the arrow in order to make it fit on the screen with the objects exerting and experiencing the force. Watch VPython Instructional Video 5: Scalefactors, at vpython.org/video05.html to learn how to do this.

Now, write a VPython program to do the following: **(a)** Create a sphere representing a planet at the origin, with radius 6.4×10^{6} m. The mass of this planet is 6×10^{24} kg. **(b)** Create five spheres representing 5 spacecraft, at locations $\langle -13\times 10^{7}, 6.5\times 10^{7}, 0\rangle$ m, $\langle -6.5\times 10^{7}, 6.5\times 10^{7}, 0\rangle$ m, $\langle 0, 6.5\times 10^{7}, 0\rangle$ m, $\langle 6.5\times 10^{7}, 6.5\times 10^{7}, 0\rangle$ m, and $\langle 13\times 10^{7}, 6.5\times 10^{7}, 0\rangle$ m. You will have to exaggerate the radius of each spacecraft to make it visible; try 3×10^{6} m. The mass of each spacecraft is 15×10^{3} kg. **(c)** For each spacecraft, have your program calculate the gravitational force exerted on the spacecraft by the planet, and visualize it with an arrow whose tail is at the center of the spacecraft. Use the same scalefactor for all arrows. Check that the display your program produces makes physical sense. **(d)** For each spacecraft, have your program calculate the gravitational force exerted on the planet by the spacecraft, and visualize it with an arrow whose tail is at the center of the planet. Use the same scalefactor as you used in the previous part. Check that the display your program produces makes physical sense.

••**P65** Write a computer program to model the motion of a spacecraft of mass 15000 kg that is launched from a location 10 Earth radii from the center of the Earth (Figure 3.67). (Data for the Earth are given on the inside back cover of the textbook. Start with a Δt of 60 s and an initial speed of 2×10^{3} m/s in a direction perpendicular to the line between the spacecraft and the Earth.) **(a)** Vary the initial speed (but not the direction), and have the spacecraft leave a trail. What trajectories can you produce? **(b)** Find an initial speed that produces an elliptical orbit. **(c)** For the elliptical orbit, display arrows indicating the directions of the momentum of the spacecraft and the net force on the spacecraft as it moves. **(d)** Find an initial speed that produces a circular orbit. **(e)** Experiment by increasing and decreasing the time step Δt. What is the largest value of Δt that gives enough accuracy to produce a closed circular orbit?

$\vec{v}_i$

Earth

Spacecraft

Figure 3.67

••**P66** Extend the program you wrote in the previous problem by including the effect of the Moon, placing the Moon on the opposite side of the Earth from the spacecraft's initial location (Figure 3.68). (Relevant data are given on the inside back cover of the textbook.) To simplify the model, keep both the Earth and the Moon fixed. (This is called a "restricted three-body problem.") **(a)** Find an initial speed for the spacecraft that results in an orbit around both Earth and Moon. **(b)** By adjusting the initial speed of the spacecraft, can you produce a figure-eight trajectory? **(c)** What other interesting trajectories can you produce by varying the initial speed? (Small variations may produce large effects.)

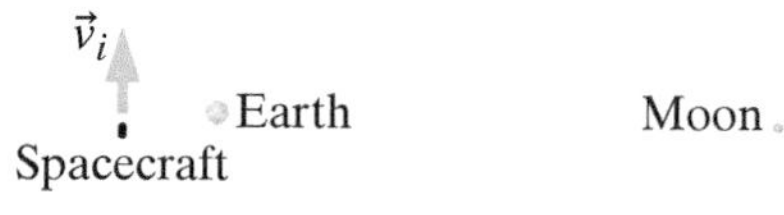

Figure 3.68

•••**P67** Once you have completed the previous problem, allow the Earth and Moon to move. Use what you know about the period of their orbits to determine appropriate initial velocities for the Earth and the Moon.

••**P68** Model the motion of Mars around the Sun. You can find data for masses and distances online. Use the known period of Mars's orbit to determine an approximate initial velocity. Display a trail, so you can see the shape of the orbit. Determine an appropriate value for Δt. Display arrows representing the momentum of Mars and the net force on Mars. Determine the period of the orbit of the planet in your model. How can you produce noncircular orbits?

••**P69** About a half of the visible "stars" are actually systems consisting of two stars orbiting each other, called "binary stars." In your computer model of a planet and Sun (Problem P68), replace the planet with a star whose mass is half the mass of our Sun, and take into account the gravitational effects that the second star has on the Sun. **(a)** Give the second star the speed of the actual Earth, and give the Sun zero initial momentum. What happens? Try a variety of other initial conditions. What kinds of orbits do you find? **(b)** Choose initial conditions so that the total momentum of the two-star system is zero, but the stars are not headed directly at each other. What is special about the motion you observe in this case?

•••**P70** Modify your orbit computation to use a different force, such as a force that is proportional to $1/r$ or $1/r^3$, or a constant force, or a force proportional to r (this represents the force of a spring whose relaxed length is nearly zero). How do orbits with these forces differ from the circles and ellipses that result from a $1/r^2$ force? If you want to keep the magnitude of the force roughly the same as before, you will need to adjust the force constant G.

••**P71** The first U.S. spacecraft to photograph the Moon close up was the unmanned *Ranger 7* photographic mission in 1964. The spacecraft, shown in Figure 3.69, contained television cameras that transmitted close-up pictures of the Moon back to Earth as the spacecraft approached the Moon. The spacecraft did not have retro-rockets to slow itself down, and it eventually simply crashed onto the Moon's surface, transmitting its last photos immediately before impact.

Figure 3.69 (Image courtesy of NASA)

Figure 3.70 is the first image of the Moon taken by a U.S. spacecraft, *Ranger 7*, on July 31, 1964, about 17 minutes before impact on the lunar surface. To find out more about the actual *Ranger* lunar missions, see http://nssdc.gsfc.nasa.gov/planetary/lunar/ranger.html.

Figure 3.70 (Image courtesy of NASA)

Create a computational model of the *Ranger 7* mission. Start your model when the spacecraft, whose mass is = 173 kg, has been brought to 50 km above the Earth's surface (5×10^4 m) by several stages of large rockets, and has a speed of around 1×10^4 m/s. All fuel has been used up, and the spacecraft now coasts toward the Moon. Data for the Earth and Moon may be found on the inside back cover of the textbook.

For this simple model, keep the Earth and Moon fixed in space during the mission, and ignore the effect of the Sun. **(a)** Compute and display the path of the spacecraft, having it leave a trail. **(b)** Determine experimentally the approximate *minimum* initial speed needed to reach the Moon, to three significant figures (this is the speed that the spacecraft obtained from the multistage rocket, at the time of release above the Earth's atmosphere). **(c)** Check your result by decreasing the time step size until your results do not change significantly. **(d)** Use a launch speed 10% larger than the approximate minimum value found in part (b). How long does it take to go to the Moon, in hours or days? **(e)** What is the "impact speed" of the spacecraft (its speed just before it hits the Moon's surface)? Make sure that your spacecraft crashes on the surface of the Moon, and not at its center!

•••P72 In the *Ranger 7* model, take into account the motion of the Moon around the Earth and the motion of the Earth around the Sun. In addition, the Sun and other planets exert gravitational forces on the spacecraft.

•••P73 In the *Ranger 7* analysis (the Moon voyage), you used a simplified model in which you neglected among other things the effect of Venus. An important aspect of physical modeling is making estimates of how large the neglected effects might be. Venus and the Earth have similar size and mass. At its closest approach to the Earth, Venus is about 40 million km away (4×10^{10} m). In the real world, Venus would attract the Earth and the Moon as well as the spacecraft, but to get an idea of the size of the effects, imagine that the Earth, the Moon, and Venus are all fixed in position. (See Figure 3.71.)

Figure 3.71

If we take Venus into account, make a rough estimate of whether the spacecraft will miss the Moon entirely. How large a sideways deflection of the crash site will there be? Explain your reasoning and approximations. If you expect a significant effect, modify your program to include the effects of Venus.

•••P74 Create a computational model of the motion of a three-body gravitational system, with all three objects free to move, and plot the trajectories, leaving trails behind the objects. Calculate all of the forces before using these forces to update the momenta and positions of the objects. Otherwise the calculations of gravitational forces would mix positions corresponding to different times.

Try different initial positions and initial momenta. Find at least one set of initial conditions that produces a long-lasting orbit, one set of initial conditions that results in a collision with a massive object, and one set of initial conditions that allows one of the objects to wander off without returning. Report the masses and initial conditions that you used.

ANSWERS TO CHECKPOINTS

1 (a) 4×10^{25} N, **(b)** 8×10^{25} N, **(c)** 4.4×10^{24} N
2 (a) (1) $\langle -1,0,0 \rangle$, (2) $\langle 1,0,0 \rangle$, (3) $\langle 1,0,0 \rangle$, **(b)** (1) $\langle 1,0,0 \rangle$, (2) $\langle -1,0,0 \rangle$, (3) $\langle -1,0,0 \rangle$
3 (a) 0.285 N/kg, **(b)** 19.9 N, **(c)** 0.029 as much (the astronaut's weight on the asteroid is about 3% of what it would be on Earth)
4 (a) 588 N, **(b)** 588 N
5 The magnitude of $\vec{p}_{\text{future}}$ will be less than the magnitude of $\vec{p}_{\text{now}}$ (so speed decreases) because part of $\vec{F}_{\text{net,now}}$ is opposite to $\vec{p}_{\text{now}}$. The direction of $\vec{p}_{\text{future}}$ will also be different (it will have a larger $+y$ component).
6 (a) 2.02×10^{-9} N, **(b)** 2.02×10^{-9} N
7 (a) Si: 14 p and 14 n, **(b)** Sn: 50 p and 69 n, **(c)** Au: 79 p and 118 n, **(d)** Th: 90 p and 142 n, **(e)** The farther you go in the periodic table, the more "excess" neutrons with their strong interactions are needed to offset the proton repulsions.
8 $\langle 0.57, 5.43, 0 \rangle$ kg·m/s
9 (a) $\langle 0.04, 0.04, -0.1 \rangle$ kg·m/s, **(b)** $\langle 0.04, 0.04, -0.1 \rangle$ kg·m/s, **(c)** $\langle 1, 1, -2.5 \rangle$ m/s
10 6, 6

CHAPTER

4

Contact Interactions

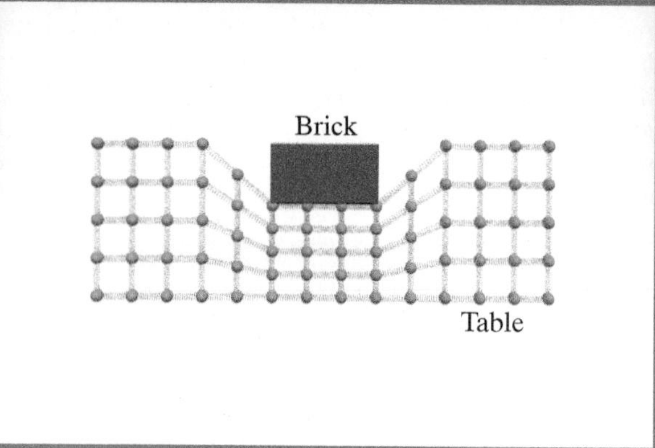

OBJECTIVES

After studying this chapter, you should be able to

- Using the ball–spring model, explain in words how an inanimate object can exert a force on an object that touches it.
- Mathematically relate the density of a solid and Young's modulus to the stiffness of an interatomic bond modeled as a spring, and to the speed of sound.
- Analyze systems subjected to tension, compression, friction, and buoyancy forces in terms of the derivative (instantaneous) form of the Momentum Principle.
- Mathematically describe the motion of an object that interacts with a spring, both analytically and computationally (iteratively).
- Determine the speed of sound in a solid in terms of the microscopic and macroscopic properties of the solid.

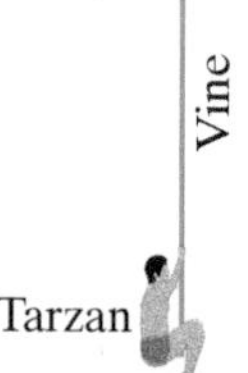

Figure 4.1 Tarzan hangs motionless from a vine, which does not break.

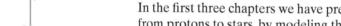

4.1 BEYOND POINT PARTICLES

In the first three chapters we have predicted the motion of a variety of systems, from protons to stars, by modeling them as point particles. However, when we consider the interactions of objects that are in physical contact with each other, such a model is no longer adequate. Objects that touch actually deform each other, stretching or compressing each other. We can't use a point particle model to understand contact interactions, because a point particle has no internal structure and can't be deformed. Therefore we'll need a more detailed model of the internal structure of solid objects. To frame our study of contact interactions and the structure of solids, we'll consider the following anecdote about Tarzan swinging from a vine.

Tarzan wants to use a vine to swing across a river. To make sure the vine is strong enough to support him, he tests it by hanging motionless on the vine for several minutes (Figure 4.1). The vine passes this test, so Tarzan grabs the vine and swings out over the river. He is annoyed and perplexed when the vine breaks midway through the swing (Figure 4.2), and he ends up drenched and shivering in the middle of the cold river, to the great amusement of the onlooking apes.

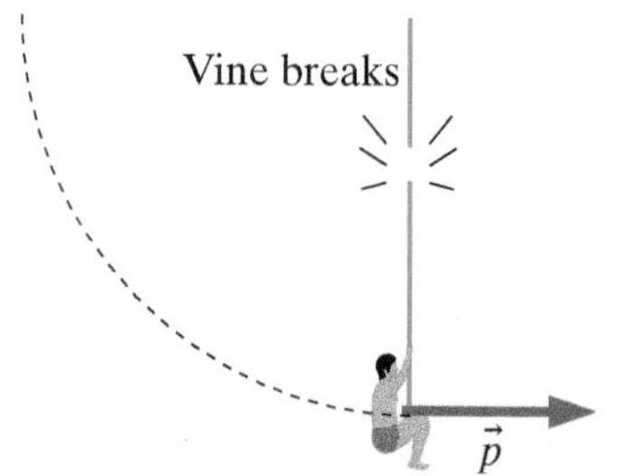

Figure 4.2 The vine breaks midway through Tarzan's swing.

Why did the vine break while Tarzan was swinging on it, but not while he hung motionless from it? Apparently what happens to the vine is different in these two situations. To understand this, we need to understand how objects that are in contact exert forces on each other; this is the topic of the current chapter. The issue of how much force the vine must exert when Tarzan swings on it, as opposed to hanging motionless, will be discussed in Chapter 5.

4.2 THE BALL–SPRING MODEL OF A SOLID

The great 20th-century American physicist Richard Feynman said:

> "If, in some cataclysm, all of scientific knowledge were to be destroyed, and only one sentence passed on to the next generations of creatures, what statement would contain the most information in the fewest words? I believe it is the atomic hypothesis (or the atomic fact, or whatever you wish to call it) that all things are made of atoms—little particles that move around in perpetual motion, attracting each other when they are a little distance apart, but repelling upon being squeezed into one another. In that one sentence, you will see, there is an enormous amount of information about the world, if just a little imagination and thinking are applied." (*The Feynman Lectures on Physics* by R. P. Feynman, R. B. Leighton, & M. Sands, 1965; Palo Alto, Cal.: Addison-Wesley.)

In the quote above, Feynman summarized the basic properties of atoms and of interatomic forces. The main properties of atoms that will be important to us in this chapter are these:

- All matter consists of atoms, whose typical radius is about 1×10^{-10} m.
- Atoms attract each other when they are close to each other but not too close.
- Atoms repel each other when they get too close to each other.
- Atoms in solids, liquids, and gases keep moving even at very low temperatures.

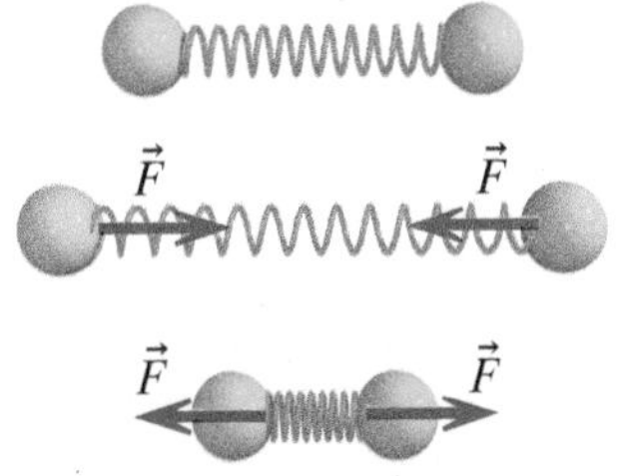

Figure 4.3 Two balls connected by a spring. Top: A relaxed spring exerts no forces on the balls. Middle: A stretched spring exerts forces to bring balls closer together. Bottom: A compressed spring exerts forces to move balls apart.

These properties were established during a century of intense study of atoms by physicists and chemists, using a wide variety of experimental techniques.

> QUESTION What properties of a block of aluminum, observable without special equipment, support the claims that "atoms attract each other when they are not too close to each other" and "atoms repel each other when they get too close to each other"?

Since the block of metal does not spontaneously fall apart or evaporate, the atoms in the block must be attracting each other. It is very difficult to compress the block, so the atoms must resist an attempt to push them closer together. Evidently the atoms in a block of aluminum are normally at just the right distance from each other, not too close and not too far away. This just-right distance is called the "equilibrium" distance between atoms.

A Chemical Bond Is Like a Spring

Two atoms linked by a chemical bond behave in a manner very similar to two macroscopic balls attached to the ends of a very low mass spring (Figure 4.3). The ball–spring system is a good model for the atomic system. In this model:

- Each ball in the model represents a massive atomic nucleus, surrounded by the inner electrons of the atom. Almost all the mass of an atom is concentrated in the tiny nucleus.
- The spring in the model represents the chemical bond, which is due to the shared outer electrons of both atoms.

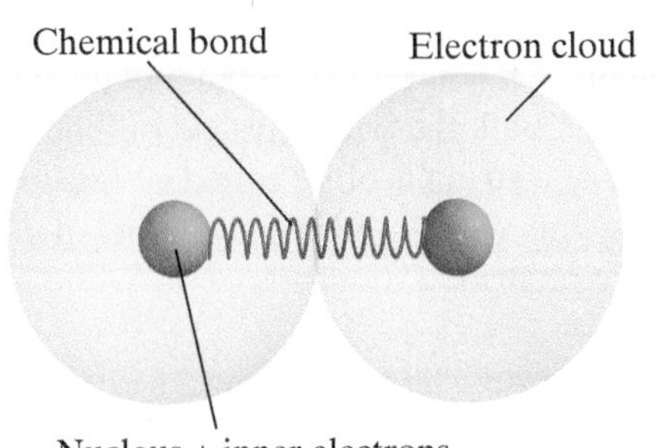

Figure 4.4 The ball–spring model for two atoms connected by a chemical bond (represented by a spring), superimposed on a space-filling model in which the entire electron clouds of the atoms are shown.

The microscopic atomic system behaves much like the macroscopic model system, as long as the stretch or compression is small, as it is in ordinary processes. If the atoms are moved farther apart, they experience forces resisting the separation. If the atoms are pushed closer together, they experience forces resisting the compression. In Figure 4.4 the ball–spring model of two bonded atoms is superimposed on a space-filling representation showing the full electron cloud of each atom.

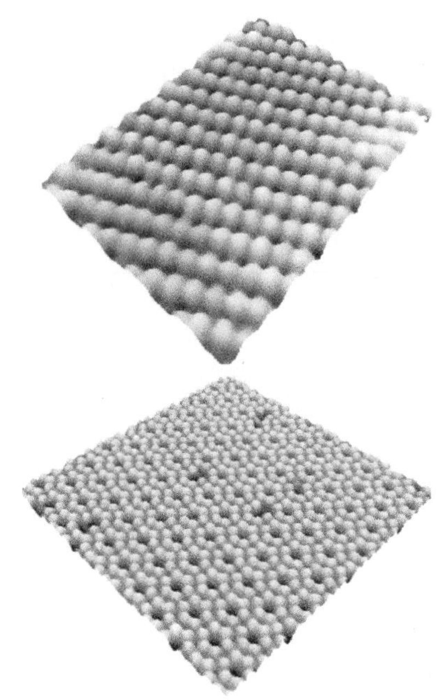

Figure 4.5 STM images of two surfaces of a silicon crystal. (Images courtesy of Randall Feenstra, IBM Corp.)

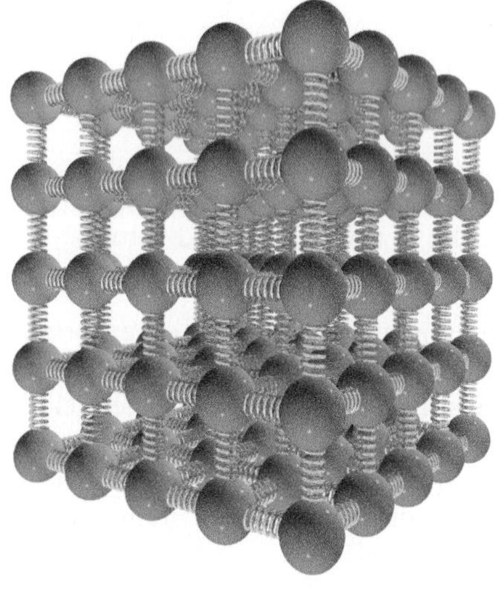

Figure 4.6 A simple model of a solid: tiny balls in constant motion, connected by springs. This figure shows only a small section of a solid object, which has many more atoms than are depicted here.

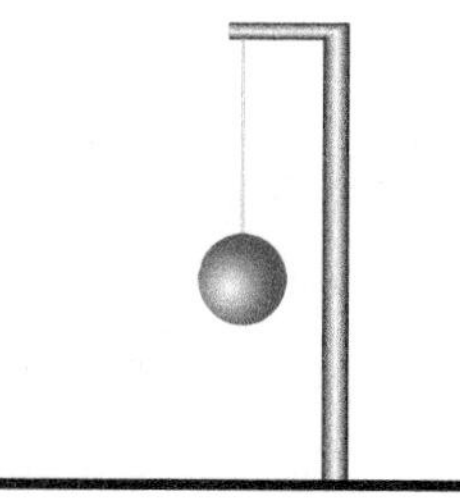

Figure 4.7 A heavy ball hangs motionless at the end of a thin wire.

A Ball–Spring Model for a Solid Object

A solid object contains many atoms, not just two. We know from x-ray studies and from images produced by scanning tunneling microscopes (STM) and atomic force microscopes (AFM) that many solid objects, such as metals, are crystals composed of regular arrays of atoms, as shown in the STM images of silicon surfaces in Figure 4.5. In these images, the spheres indicate the entire electron cloud associated with each atom, not just the nucleus and inner electrons represented by the balls in our model.

These STM images show only one surface of a solid, but inside the solid, atoms are arranged in 3D patterns, like the balls and springs in the model solid shown in Figure 4.6. Solids in which the atoms are arranged in regular "lattices" like those shown in Figure 4.6 are called crystals; these include metals, quartz, diamond, ice, and table salt (NaCl), but not most organic solids such as plastic or wood.

The lattice shown in Figure 4.6 is the simplest kind of crystal, and is called a "cubic" lattice because the atoms in the crystal are located at the corners of adjacent cubes. More complex lattice arrangements are possible; for example, in one common variant called "body-centered cubic" there is an additional atom at the center of each cube. In Figure 4.5 the lower image shows a hexagonal arrangement of atoms. Although most crystals are more complex than the simple cubic lattice, we will use this simple cubic ball-and-spring model because it incorporates all the important features we need.

In a solid at room temperature, the atoms are in motion, continually oscillating around their equilibrium positions. If we heat the solid, these atomic oscillations become more vigorous. One of the most important outcomes of research on the atomic nature of matter was the realization that for many materials, the temperature of an object, as measured by an ordinary thermometer, is just an indicator of the average energy of the atoms; the higher the temperature, the more vigorous the atomic motion. Figure 4.6 shows atoms in a model solid frozen in their equilibrium positions.

In later chapters we will see that the ball–spring model of a solid will also enable us to answer questions about energy (Chapter 6), energy dissipation and temperature changes (Chapter 7), and the irreversibility of certain processes (Chapter 11). By refining and extending the model in Chapter 14, we will also be able to model the electrical properties of metals and other conductors.

4.3 TENSION FORCES

By applying the Momentum Principle we can show that an object like a wire, string, or vine can exert a force on an object attached to it. Figure 4.7 shows a heavy iron ball hanging motionless on the end of a wire. The y momentum of the ball is not changing, so the y component of the net force on the ball must be zero. Since the Earth is pulling down on the ball, the wire must be pulling up. The force exerted by an object like a wire or a string is often called a "tension" force, or sometimes just "the tension in the wire," and usually labeled $\vec{F}_T$. A tension force always acts along the wire or string.

QUESTION If the mass of the ball is 1 kg, how large an upward force does the wire exert on the ball?

System: Ball
Surroundings: Earth, wire
Free-body diagram: Figure 4.8

Choose a value for Δt: 10 s (since the ball remains at rest indefinitely, we could choose any value of Δt other than zero)

As reflected in the free-body diagram, the only significant forces exerted on the ball by the surroundings are in the $\pm y$ direction, so we can focus on the y components of force and momentum.

$$\Delta p_y = (F_T - mg)(10\,\text{s})$$
$$0 = (F_T - mg)(10\,\text{s})$$
$$F_T = mg = (1\,\text{kg})(9.8\,\text{N/kg}) = 9.8\,\text{N}$$

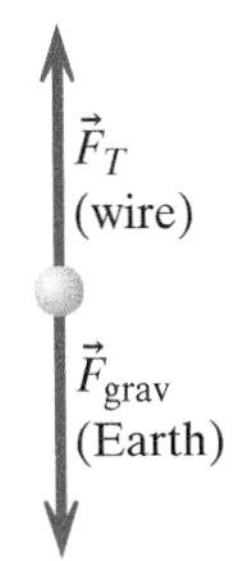

Figure 4.8 Free-body diagram for the ball.

QUESTION If the mass of the ball were 2 kg, how large an upward force would the wire exert on the ball?

$$F_T = (2\,\text{kg})(9.8\,\text{N/kg}) = 19.6\,\text{N}$$

This simple result has an interesting implication. Evidently the magnitude of the tension force F_T exerted by the wire depends on the mass of the ball.

QUESTION How does an inanimate object like a wire "know" how large a force to exert in a given situation?

We have already seen that the force exerted by a spring depends on how much the spring is stretched. A wire can be thought of as a very stiff spring. When a weight is hung on the end of a wire, the wire does stretch, although the stretch is usually very small. The ball–spring model of a solid helps us understand what happens when a wire stretches.

Microscopic View: Tension in a Wire

We can model a wire as a chain of balls and springs (atoms connected by spring-like chemical bonds). For simplicity, we'll consider a wire that is only one atom thick.

When the wire lies on a table, each of the springs (bonds) is relaxed. When the wire hangs vertically with nothing attached to it, the bonds stretch a tiny amount, just to support the weight of the atoms below them. However, when a massive object is hung from the wire, the spring-like bonds between atoms stretch significantly, because each bond must support the weight of everything below it (Figure 4.9). If we make the approximation that the total mass of all the atoms in the wire is negligible compared to the mass of the hanging object, then we can say that each bond in the wire is stretched by approximately the same amount, or equivalently, that the tension is the same throughout the wire.

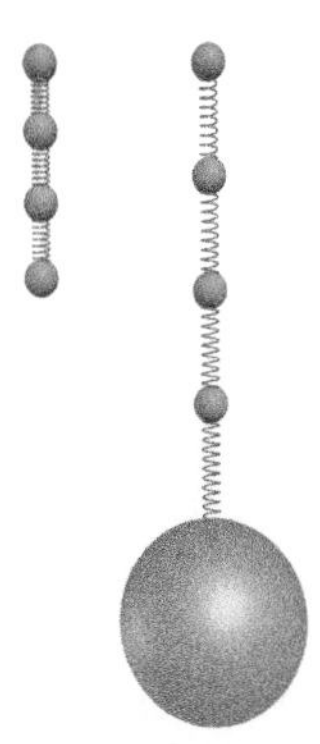

Figure 4.9 The interatomic bonds in a wire stretch when a heavy mass is hung from the wire. In this idealized diagram the wire is only one atom thick, and the stretch of each bond is greatly exaggerated.

QUESTION How does the wire manage to exert a larger upward force when a more massive object is hung from it?

The more massive object stretches the interatomic bonds more than a less massive object. Of course, there is a limit to how much the wire can stretch. A sufficiently heavy weight will break the wire.

4.4 LENGTH OF AN INTERATOMIC BOND

Our goal in the next sections is to determine the stiffness of an interatomic bond, considered as a spring. The first step is to determine the length of an interatomic bond in a particular material. Bond lengths will be slightly different in different materials (for example, aluminum vs. lead), depending on the size of the atoms. We will determine the interatomic bond length in solid copper.

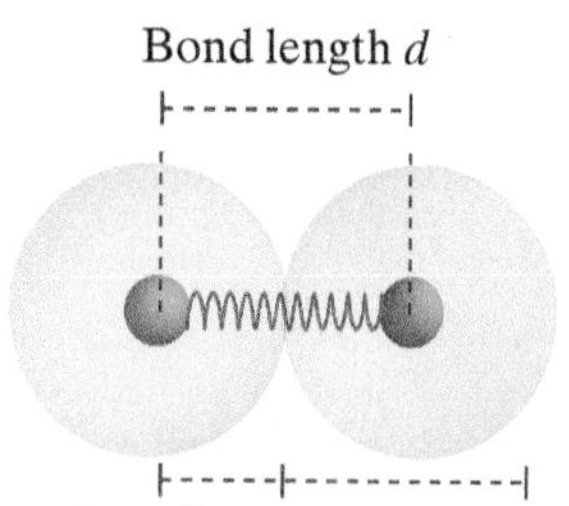

Figure 4.10 The length of an interatomic bond is defined as the center-to-center distance between adjacent atoms. This is the same as the diameter of an atom (including the full electron cloud).

We will define the length of one interatomic bond as the center-to-center distance between two adjacent atoms (Figure 4.10). If we consider the space-filling model of a solid, this distance is equal to twice the radius of an atom, or the diameter of an atom.

The diameter of an atom in a solid is one of the important properties of matter that plays a role in interactions. We can calculate atomic diameters for crystals of particular elements by using the measured density of the material in kilograms per cubic meter and Avogadro's number (the number of atoms in one mole of the material).

QUESTION Does the density of a block of aluminum depend on the dimensions of the block? Does it depend on the mass of the block?

Density does not depend on the size, shape, or mass of an object. Density is a property of the material itself, and the ratio of mass to volume will always be the same for objects made of the same solid material. The density of a wide range of materials has been measured and may easily be looked up in reference materials.

EXAMPLE **Diameter of a Copper Atom (Length of a Bond in Copper)**

One mole of copper (6.02×10^{23} atoms) has a mass of 64 g (see the periodic table on the inside front cover). The density of copper is 8.94 g/cm^3. What is the approximate diameter, in meters, of a copper atom in solid copper?

Solution Often densities are given in grams per cubic centimeter rather than in kilograms per cubic meter. We often need to convert these densities to SI units, like this:

$$\left(8.94\frac{\text{g}}{\text{cm}^3}\right)\frac{(1\,\text{kg})}{(1 \times 10^3\,\text{g})}\frac{(1 \times 10^2\,\text{cm})^3}{(1\,\text{m})^3} = 8.94 \times 10^3\frac{\text{kg}}{\text{m}^3} \quad \text{(SI units)}$$

Figure 4.11 shows a cube that contains $5 \times 5 \times 5 = 125$ atoms. The number of atoms on one side of the cube is $\sqrt[3]{125} = 5$.

In a cube that is 1 m on each side, how many copper atoms are there?

$$(8.94 \times 10^3\,\text{kg})\left(\frac{1\,\text{mol}}{0.064\,\text{kg}}\right)\frac{(6.02 \times 10^{23}\,\text{atoms})}{(1\,\text{mol})} = 8.41 \times 10^{28}\,\text{atoms}$$

Along one edge of the cube, which is 1 m long, there are

$$\sqrt[3]{8.41 \times 10^{28}} = 4.38 \times 10^9\,\text{atoms}$$

A row of 4.38×10^9 atoms is 1 m long, so the diameter of one atom is

$$d = \left(\frac{1\,\text{m}}{4.38 \times 10^9\,\text{atoms}}\right) = 2.28 \times 10^{-10}\,\text{m}$$

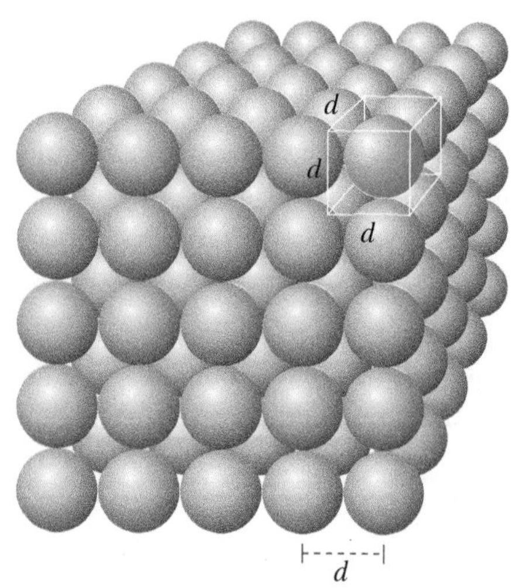

Figure 4.11 A simple cubic arrangement of atoms. The volume of space associated with each atom is a tiny cube d by d by d.

In a solid block of copper, the length of a bond between two adjacent copper atoms is 2.28×10^{-10} m.

An alternative approach to this calculation involves a "micro" view of density. Since density is independent of the amount of an object, the density of one copper atom ought to be the same as the density of a large block of copper. Figure 4.11 shows that the volume of space taken up by each atom is a tiny cube d on a side, where d is the diameter of an atom in the solid and is also the distance from the center of one atom to the center of a neighboring atom. The micro density is the mass of one atom divided by the volume of the tiny cube associated with each atom. This micro density is the same as the macro density, mass per volume:

$$\text{density} = \frac{\text{mass of } 8.41 \times 10^{28} \text{ atoms}}{\text{volume of } 8.41 \times 10^{28} \text{ atoms}} = \frac{\text{mass of 1 atom}}{\text{volume of 1 atom}} = \frac{m_a}{d^3}$$

The mass of one atom can be determined using the mass of one mole and the knowledge that one mole contains 6.02×10^{23} atoms (Avogadro's number):

$$m_a = \frac{\text{mass of one mole}}{6.02 \times 10^{23} \text{ atoms/mole}}$$

Knowing the macroscopic density and the mass of one atom, you can solve for the unknown atomic diameter d. It might be tempting to put these concepts together into an equation of the form $d = \ldots$, but try to avoid the temptation. It is much more memorable and much safer to think through these physical relationships each time you make the calculation, to avoid serious mistakes. Just set the macroscopic and microscopic densities equal to each other and solve for d.

Few solid elements actually have cubic lattices, because a cubic lattice is rather unstable. However, assuming that all crystalline solids have cubic lattices greatly simplifies our model, and gives adequately accurate results for our purposes.

The diameters of the smallest and largest atoms differ by only about a factor of 8. Most metallic elements have similar radii, on the order of 1.5×10^{-10} m. It is useful to remember that the radius of an "average" atom is on the order of 1×10^{-10} m.

Checkpoint 1 The density of aluminum is 2.7 g/cm^3 and the density of lead is 11.4 g/cm^3. What is the approximate diameter of an aluminum atom (length of a bond) in solid aluminum? What is the approximate diameter of a lead atom (length of a bond) in solid lead?

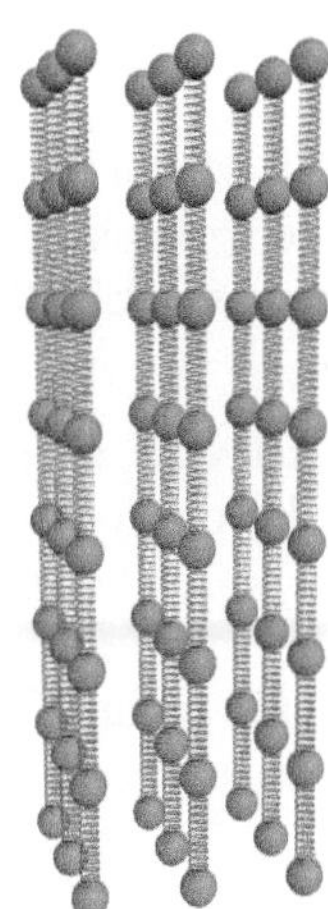

Figure 4.12 A solid wire considered as many parallel long chains of atoms connected by springs (interatomic bonds). For clarity, the horizontal bonds are not shown; they aren't significant when stretching the wire.

4.5 THE STIFFNESS OF AN INTERATOMIC BOND

Knowing the length of an interatomic bond in solid copper (and the diameter of a copper atom), we can now use experimental data to find the stiffness of the interatomic bond, considered as a spring. You have probably had some experience with ordinary macroscopic springs. A Slinky, for example, is a soft spring; its spring stiffness is small—around 1 N/m. The stiffness of the spring on a pogo stick is much larger—around 5000 N/m. It is difficult to measure the stiffness of an interatomic bond directly, but we can analyze data from macroscopic (large-scale) experiments to determine this quantity.

The basic idea is to hang heavy masses on a long wire and measure the stretch of the wire; then, by figuring out how many interatomic "springs" there are in the wire, we can determine the spring constant of a single interatomic bond. To do this, we need to be able to relate the stiffness of an object composed of many springs to the stiffness of each individual spring. We will consider the wire to be composed of many parallel long chains of atoms connected by springs (Figure 4.12). We will need to determine how the stiffness of one of the short springs (bonds) is related to the stiffness of an entire chain of springs, and how the stiffness of the entire wire is related to the stiffness of one chain of springs.

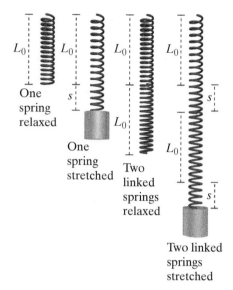

Figure 4.13 Two identical springs linked end to end stretch twice as much as one spring when the same force is applied. The combined spring therefore is only half as stiff as the individual springs.

Two Springs Linked End-to-End (in Series)

Suppose that we have a long, low-mass spring whose spring stiffness is 100 N/m. If we hang a block whose weight is 100 N (about a 10 kg mass) on the end of the spring, the spring will stretch 1 m. (This is a big stretch, so the relaxed spring must be several meters long; we are using simple numbers here to make our reasoning easier.)

> QUESTION We make a longer spring by linking two identical springs end-to-end (in "series"). Each shorter spring has a stiffness of 100 N/m. What will the stiffness of the longer spring be?

When we hang the 100 N block on the end of the longer spring, this longer spring will stretch 2 m, because each of the individual springs stretches 1 m (Figure 4.13). We can use the Momentum Principle to get the stiffness of this longer spring. When the block hangs motionless, its momentum is not changing. Choosing a Δt of 10 s, we find this:

System: Block
Surroundings: Spring, Earth
Free-body diagram: Figure 4.14

$$\begin{aligned}\Delta p_y &= F_{\text{net},y}\Delta t\\ 0 &= (k_s s - mg)(10\,\text{s})\\ k_s s &= mg\\ k_s(2\,\text{m}) &= 100\,\text{N}\\ k_s &= 50\,\text{N/m}\end{aligned}$$

A long spring made of two identical springs linked end-to-end is only half as stiff as each of the shorter springs.

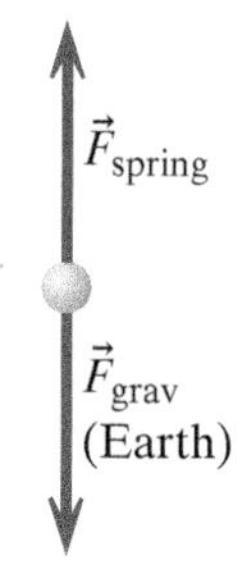

Figure 4.14 Free-body diagram for hanging block.

> **Checkpoint 2** If a chain of 20 identical short springs linked end-to-end has a stiffness of 40 N/m, what is the stiffness of one short spring?

Two Springs in Parallel (Side-by-Side)

> QUESTION Suppose that we support a 100 N weight with two identical springs side by side (Figure 4.15). If the spring stiffness of each spring is 100 N/m, how much will each spring stretch? What is the effective stiffness of the two-spring system?

Informally, we see that each spring supports only 50 N, so each spring will stretch only 0.5 m. You can work out the formal solution yourself using the Momentum Principle, remembering to include two separate upward spring forces.

We can think of the two springs as a single, wider spring. The effective spring stiffness of this "double-width" spring is:

$$k_{s,\text{effective}} = \frac{100\,\text{N}}{0.5\,\text{m}} = 200\,\text{N/m}$$

Two springs side by side are effectively twice as stiff as a single spring.

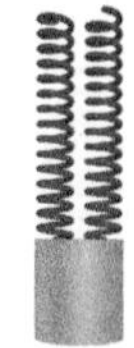

Figure 4.15 Two springs side by side supporting a block.

> **Checkpoint 3** Nine identical springs are placed side by side (in parallel, as in Figure 4.15), and connected to a large massive block. The stiffness of the nine-spring combination is 2700 N/m. What is the stiffness of one of the individual springs?

Cross-Sectional Area

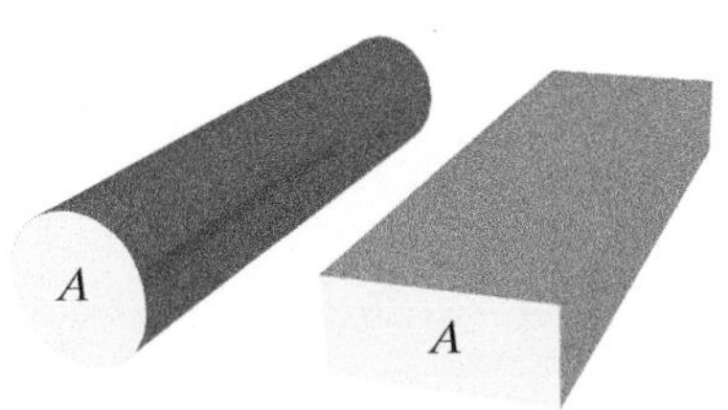

Figure 4.16 The cross-sectional area of a cylinder is the area of a circle. The cross-sectional area of a rectangular solid is the area of a rectangle.

The cross-sectional area of an object is the area of a flat surface made by slicing through the object (Figure 4.16). A cylindrical object like a round pencil has a circular cross-sectional area (imagine sawing crosswise through a pencil). A cylindrical pencil 10 cm long and 0.5 cm in diameter has a cross-sectional area of

$$A = \pi(0.0025\,\text{m})^2 = 1.96\times10^{-5}\,\text{m}^2$$

Note that the length of the object is irrelevant.

A long object with four flat sides, like a board, has a cross-sectional area that is rectangular, or perhaps square. A wooden board 7 m long, 5 cm wide, and 3 cm high, has a cross-sectional area of

$$A = (0.05\,\text{m})(0.03\,\text{m}) = 1.5\times10^{-3}\,\text{m}^2$$

EXAMPLE **Interatomic Bond Stiffness in Copper**

A copper wire is 2 m long. The wire has a square cross section (that is, it has four flat sides—it is not round). Each side of the wire is 1 mm in width. Making sure the wire is straight, you hang a 10 kg mass on the end of the wire. Careful measurement shows that the wire is now 1.67 mm longer. From these measurements, determine the stiffness of one interatomic bond in copper.

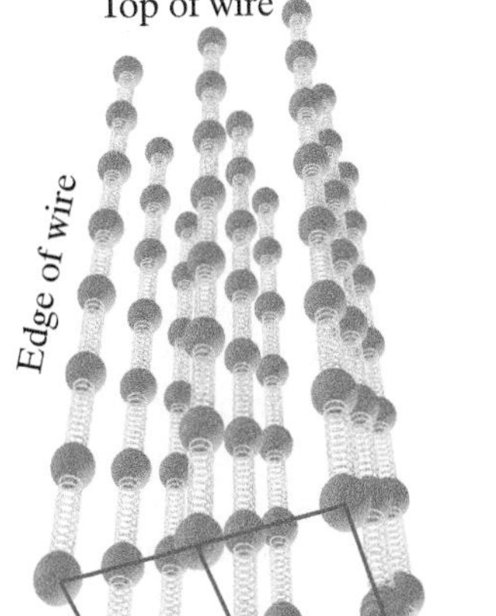

Figure 4.17 A wire modeled as an assembly of side-by-side chains of balls and springs, viewed from the bottom. Colored lines connect the bottom layer of atoms in the wire (only nine ball–spring chains are shown here).

Solution

1. What is the spring stiffness $k_{s,\text{wire}}$ of the entire wire, considered as a single macroscopic (large scale), very stiff spring?
   System: Mass
   Surroundings: Earth, wire
   Momentum Principle:

$$\Delta p_y = 0 = (k_{s,\text{wire}}s - mg)\Delta t$$

$$k_{s,\text{wire}} = \frac{mg}{s} = \frac{(10\,\text{kg})(9.8\,\text{N/kg})}{(1.67\times10^{-3}\,\text{m})} = 5.87\times10^4\,\text{N/m}$$

2. How many side-by-side atomic chains (long springs) are there in this wire? This is the same as the number of atoms on the bottom surface of the copper wire (see Figure 4.17; the horizontal bonds are not shown, as they aren't stretched significantly).
   Cross-sectional area of wire:

$$A_{\text{wire}} = (1\times10^{-3}\,\text{m})^2 = 1\times10^{-6}\,\text{m}^2$$

Cross-sectional area of one copper atom (even though an atom itself is spherical, each atom occupies a cubical space in the crystal):

$$A_{1\,\text{atom}} \approx (2.28\times10^{-10}\,\text{m})^2 = 5.20\times10^{-20}\,\text{m}^2$$

Number of side-by-side atomic chains in the wire:

$$N_{\text{chains}} = \frac{A_{\text{wire}}}{A_{1\,\text{atom}}} = 1.92\times10^{13}$$

3. How many interatomic bonds are there in one atomic chain running the length of the wire?

$$N_{\text{bonds in 1 chain}} = \frac{L_{\text{wire}}}{d} = \frac{2\,\text{m}}{2.28\times10^{-10}\,\text{m}} = 8.77\times10^9$$

4. What is the stiffness $k_{s,i}$ of a single interatomic spring?
Using our results about springs linked end to end and springs arrayed side by side:

$$k_{s,\text{wire}} = \frac{(k_{s,i})(N_{\text{chains}})}{N_{\text{bonds in 1 chain}}} \text{ (solve for } k_{s,i})$$

$$k_{s,i} = \frac{(5.87 \times 10^4\,\text{N/m})(8.77 \times 10^9)}{(1.92 \times 10^{13})} = 26.8\,\text{N/m}$$

An interatomic bond in copper is stiffer than a Slinky, but less stiff than a pogo stick. The stiffness of a single interatomic bond is very much smaller than the stiffness of the entire wire (which we found to be around 6×10^4 N/m).

Checkpoint 4 The 2 m copper wire with square cross section of 1 mm by 1 mm stretched 1.67 mm when it supported a 10 kg mass. Cut a length of this wire 0.2 m long and hang a 10 kg mass from it. How much will this short wire stretch?

4.6 STRESS, STRAIN, AND YOUNG'S MODULUS

QUESTION Suppose that we had used a different copper wire, which was 3 m long, and had a circular cross section, with a diameter of 0.9 mm. Would our result for the interatomic bond stiffness have been different?

The result should not be different, because the interatomic bond stiffness is a property of the material (solid copper). The macroscopic dimensions of the wire will not change the intrinsic properties of copper.

You will not find tables of interatomic spring stiffness for different solid materials in reference libraries. Instead, what is published is a macroscopic quantity called Young's modulus. Like density and interatomic spring stiffness, Young's modulus is a property of a particular material (for example, copper) and is independent of the shape or size of a particular object made of that material. Young's modulus is a macroscopic measure of the "stretchability" of a solid material. It relates the fractional change in length of an object to the force per square meter of cross-sectional area applied to the object.

Strain

The longer a wire is, the more atomic bonds it has along its length, and the more it will stretch when a force is applied to it. In order to define a quantity (Young's modulus) that is a property of the material and doesn't depend on the shape of the wire, we need to take into account not only the measured stretch, but also the original length of the wire. If the wire has a length L, we call the amount of stretch of the wire ΔL (a small increment in the length, what we have called s in a spring), and the fractional stretch $\Delta L/L$ is called the "strain."

DEFINITION OF STRAIN

$$\text{strain} = \frac{\Delta L}{L}$$

Stress

As we saw above, all the chains of atoms help hold up the weight attached to the wire. In order to define a quantity that doesn't depend on the thickness of the wire, we need to take into account not only the tension force F_T, but also the cross-sectional area of the wire A (Figure 4.18). The tension force per unit area is called the "stress."

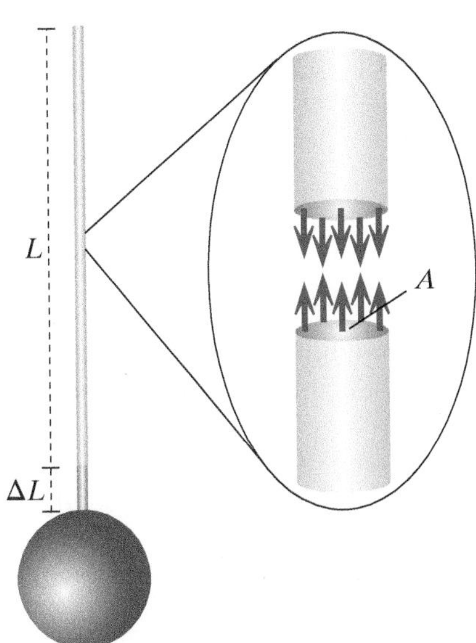

Figure 4.18 The heavy ball stretches the wire an amount ΔL. A is the cross-sectional area—the area of a slice through the wire, perpendicular to its length.

DEFINITION OF STRESS

$$\text{stress} = \frac{F_T}{A}$$

Young's Modulus

As long as the strain isn't too big, the strain is proportional to the stress. At the atomic level, the stress F_T/A can be related to the force that each long chain of atomic bonds must exert, and the strain $\Delta L/L$ can be related to the stretch of the interatomic bond. The ratio of stress to strain is a property of the material and differs for steel, aluminum, and so on, but doesn't depend on the length or thickness of the wire we use in our measurement of Young's modulus.

DEFINITION OF YOUNG'S MODULUS

$$Y = \frac{\text{stress}}{\text{strain}} = \frac{\left(\dfrac{F_T}{A}\right)}{\left(\dfrac{\Delta L}{L}\right)}$$

Young's modulus is the ratio of stress to strain for a particular material. Young's modulus is a property of the material, and does not depend on the size or shape of an object. The stiffer the material, the larger is Young's modulus.

Using Young's modulus, a quantitative form of the relation between stress and strain in a wire is written as follows:

$$\frac{F_T}{A} = Y\frac{\Delta L}{L}$$

Note the similarity to the spring force, $F_T = k_s s\ (=k_s \Delta L)$.

You may have the opportunity to measure Young's modulus in a laboratory experiment, and to use your measurements to determine the interatomic spring stiffness for a particular material.

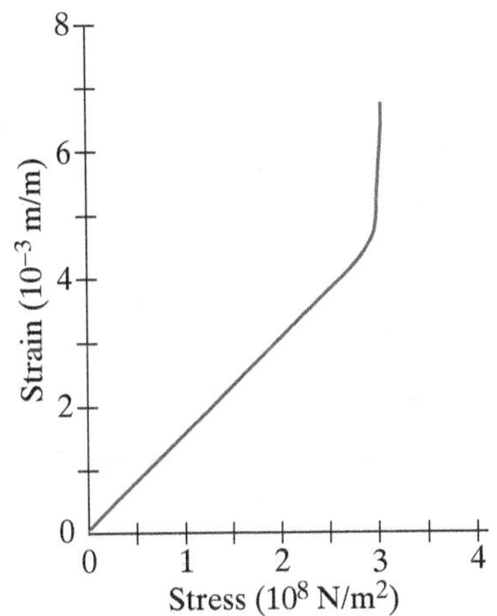

Figure 4.19 Strain vs. stress for a particular aluminum alloy.

Limit of Applicability of Young's Modulus

If you apply too large a stress, the wire "yields" (suddenly stretches a great deal) or breaks, and the proportionality of stress and strain is no longer even approximately true. Figure 4.19 shows a graph of the strain $\Delta L/L$ that results as a function of the applied stress F_T/A for a particular aluminum alloy.

You see that for moderate stress the resulting strain is proportional to the applied stress: double the stress, double the strain. However, once you reach the yield stress, a further slight increase in the stress leads to a very large increase in the length of the metal. This can be quite dramatic: as you add more weights to the end of a wire the wire gets slightly longer, then suddenly it starts to lengthen a lot, very rapidly, and then the wire breaks. In materials reference manuals, stress is usually plotted against strain, so that the slope of the line is equal to Young's modulus, stress divided by strain. We plot strain vs. stress here to emphasize that stress is the *cause* (the independent variable) and strain the *effect* (the dependent variable).

Checkpoint 5 You hang a 5 kg mass from the end of a rod that is 1.7 m by 2 mm by 2 mm. You observe that the rod's length increases to 1.74 m. **(a)** What is the strain? **(b)** What is the stress? **(c)** What is Young's modulus for this material?

Relating Young's Modulus to Interatomic Spring Stiffness

From macroscopic measurements of Young's modulus we can calculate an approximate spring stiffness $k_{s,i}$ for the interatomic bond by again making a "macro-micro" connection.

Consider a single stretched interatomic bond in a wire that is being stretched. Let d represent both the relaxed length of an interatomic bond and the diameter of one atom. The cross-sectional area occupied by one atom is d^2 (Figure 4.20) (recall that even though an atom is approximately spherical, it occupies a cube of space in the crystal lattice).

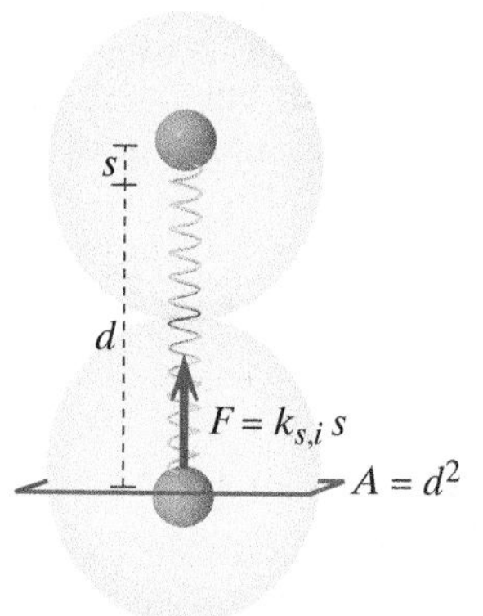

Figure 4.20 A single stretched bond in a solid wire. The effective cross-sectional area of one atom is d^2, and the force acting on that portion is $k_{s,i}s$.

QUESTION What is the magnitude of the stress on one atom in terms of the interatomic force $k_{s,i}s$?

If the stretch of the interatomic bond is s, the atomic stress (force per unit area) is $\dfrac{k_{s,i}s}{d^2}$.

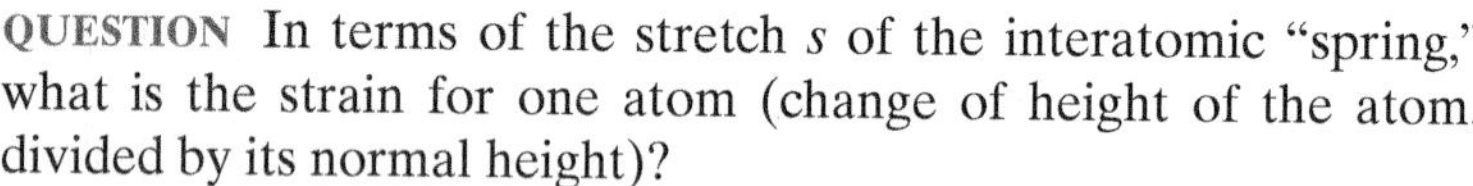

QUESTION In terms of the stretch s of the interatomic "spring," what is the strain for one atom (change of height of the atom, divided by its normal height)?

The strain for one atom is s/d, because its height was originally d, and the change in its height is s. Therefore we can express Young's modulus in terms of atomic quantities:

YOUNG'S MODULUS IN TERMS OF ATOMIC QUANTITIES

$$Y = \frac{(k_{s,i}s/d^2)}{(s/d)} = \frac{k_{s,i}}{d}$$

where $k_{s,i}$ is the stiffness of an interatomic bond in a solid, and d is the length of an interatomic bond (and the diameter of an atom).

If you know Young's modulus for some metal, you can calculate the effective stiffness of the interatomic bond, modeled as a spring: $k_{s,i} = Yd$, where d is the atomic diameter in the metal. This is another example of an important theme, that of relating macroscopic properties to microscopic (atomic-level) properties.

Tarzan and the Vine Revisited

We now know that interatomic bonds in the vine stretch when Tarzan hangs from the vine. It is not yet clear why the vine breaks when Tarzan swings on it but not when he hangs motionless. Apparently the vine must stretch more when he swings—perhaps it must apply a larger force? In Chapter 5 we will learn how to apply the Momentum Principle to Tarzan's situation.

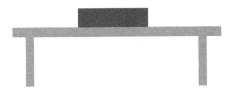

4.7 COMPRESSION (NORMAL) FORCES

Figure 4.21 A brick lies on a table.

A closely related situation is that of a brick lying at rest on a metal table (Figure 4.21). In this situation the table is compressed, while in the previous discussion the wire was stretched.

Take the brick as the system of interest (Figure 4.22). The Earth pulls down on the brick of mass M, and the magnitude of the gravitational force is Mg.

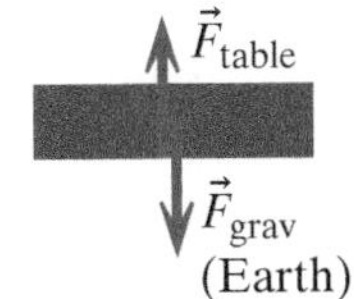

Figure 4.22 Forces acting on the brick.

Since the brick remains at rest, the vertical component of the net force must be zero:

$$0 = +F_{\text{Table}} - Mg$$

We conclude that the force F_{Table} exerted by the table is equal to the weight Mg of the brick, an unsurprising result.

How does the metal table exert a force on the brick? If the brick were extremely heavy, we might not be surprised to observe the table sagging under its weight. Even if the table does not sag visibly, the molecules in the top layers of the table are pushed down by molecules in the bottom layer of the brick. The table actually deforms somewhat—its interatomic "springs" are compressed. (Of course, atoms in the brick are compressed somewhat, too.)

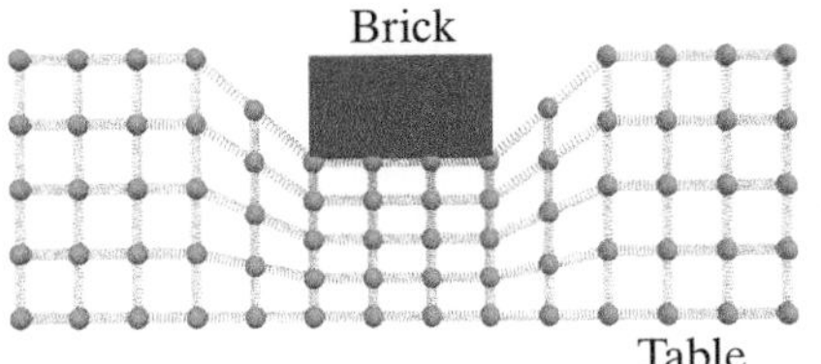

Figure 4.23 A schematic view of a portion of a brick resting on a portion of a table (scale greatly exaggerated). Interatomic "springs" in the contact region are compressed by the weight of atoms above.

Atoms in the table are squeezed closer than their normal equilibrium positions, compressing the spring-like interatomic bonds. Figure 4.23 schematically shows the contact region between the brick and the table.

It would be appropriate to call the upward force of the table F_C for "compression force," but it is common practice to label this kind of force as a "normal" force. The word is used in the mathematical sense, meaning "perpendicular to," because the direction of the force is perpendicular to the surface of the table. Usually "normal" forces are labeled as F_N.

Note: Neither normal force nor tension force is a different kind of fundamental force, like the gravitational force or electric force. These are merely descriptive names for particular kinds of interatomic electric interactions.

4.8 FRICTION

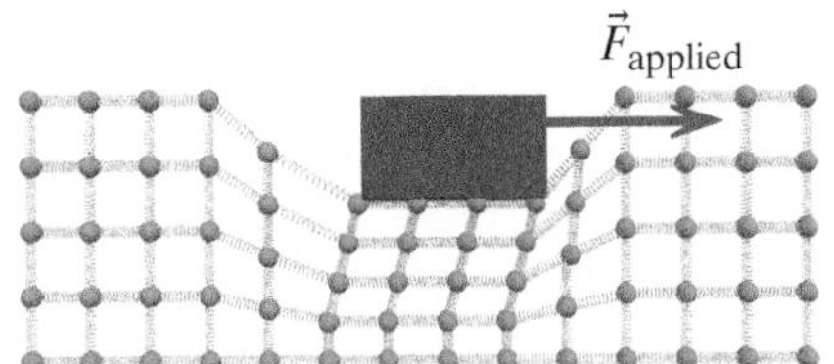

Figure 4.24 When you apply a force on the brick, it runs into atoms of the table, compressing interatomic bonds in front and stretching them behind (scale greatly exaggerated). Similar deformations occur in the block but for clarity are not shown.

Friction is another phenomenon that involves deformable objects; to understand it we again need to consider the internal structure of solid matter. As you've just seen, a brick sitting on a table sinks into the table, compressing the interatomic bonds, and the table pushes upward on the brick. If you then drag the brick across the table as in Figure 4.24, the brick continually runs into previously uncompressed regions of the table, compressing bonds ahead of the brick and stretching bonds behind, so we can consider the force of the table as the sum of two perpendicular components. The component that is parallel to the velocity (and opposite to the velocity in this case), is called a "frictional force," as shown in Figure 4.25. The full free-body diagram for the brick is shown in Figure 4.26. After the brick has passed by, the atoms tend to go back to their equilibrium positions.

However, the disturbance of the atoms away from their equilibrium positions leaves the atoms vibrating about their equilibrium positions, in both the table and in the block. As we will discuss in detail in later chapters, this increased vibrational motion or "internal energy" is associated with higher temperature. Dragging the brick across the table makes the temperature of the table and brick rise. The increase in internal energy means that you must continually apply a force and expend energy to do the dragging. If you stop pulling, the brick will slow down and stop as its motional ("kinetic") energy is dissipated away into internal energy in the form of agitation of atoms throughout the objects. We say that friction is a "dissipative" process.

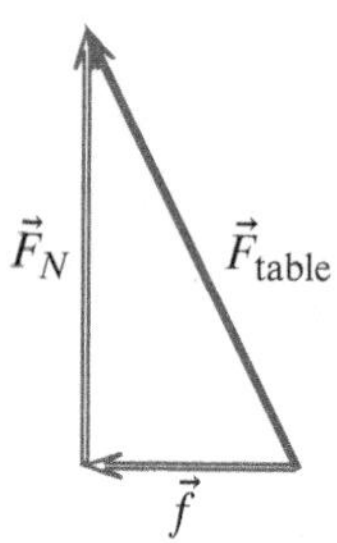

Figure 4.25 If the brick is pulled to the right, the table now exerts a force upward and to the left. Here the force $\vec{F}_{table}$ is expressed as the sum of two perpendicular components: the upward "normal" force $\vec{F}_N$ and the sideways "friction" force $\vec{f}$.

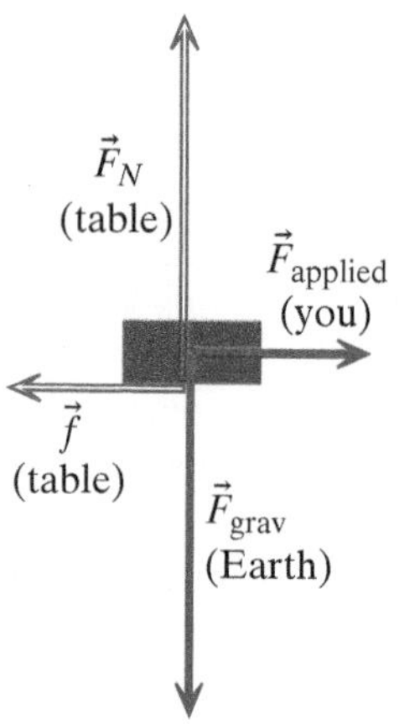

Figure 4.26 The free-body diagram for the brick now includes the components of the force by the table, as well as the force by you and the force by the Earth.

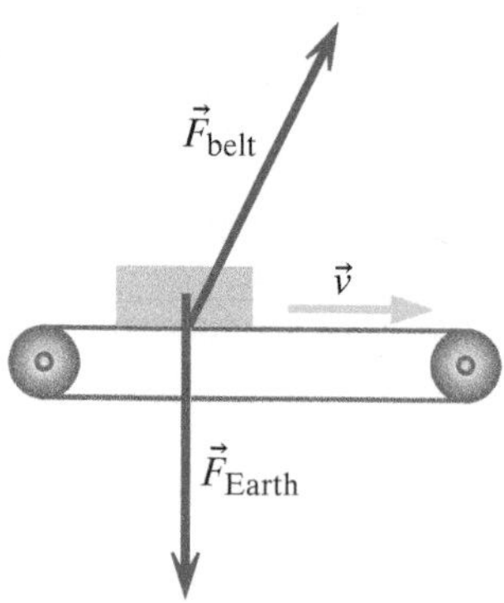

Figure 4.27 Place a box on a moving conveyor belt, and the horizontal component of $\vec{F}_{belt}$ accelerates the box.

Evidently for a full understanding of the block sliding on the table, we can't simply consider the block and table to be point particles. They are in fact multiparticle systems, and friction is very much a multiparticle phenomenon.

Friction is a complex process, and this picture of friction at the atomic level, based on the ball–spring model of solids, is necessarily simplified. In Chapter 9 we will discuss friction in more detail, after we have studied energy.

QUESTION To keep the brick moving at a constant speed, you have to keep pulling. It's tempting to conclude that a constant force is needed to keep an object moving at a constant speed, which sounds like a violation of Newton's first law of motion and of the Momentum Principle. What's going on here?

It is the net force that governs the motion of the brick, not just the force that you apply. Since the net force is zero, the brick moves with constant speed. You are pulling with a force whose magnitude is equal to that of the parallel component of the force of the table, so the net force $\vec{F}_{net}$ is zero.

It is often the case that the direction of the friction force is opposite to the velocity, tending to slow down the object. However, this is not always true. For example, if you drop a box onto a moving conveyor belt, bonds at the back of the box are compressed and the friction force pushes the box forward with increasing speed (Figure 4.27). Eventually the box reaches the speed of the moving belt, at which time the friction force drops to zero and the box moves with constant speed, subject to a zero net force.

When a sprinter speeds up, she pushes her feet backwards against the ground, and the horizontal component of the force of the ground on her feet points forward, accelerating her. On a very slick surface with almost no friction she can't accelerate.

Sliding Friction

When one object slides on another, the component of force exerted by one object on the other has a component parallel (or antiparallel) to the motion with the following properties:

THE SLIDING FRICTION FORCE

$$f_{friction} \approx \mu_k F_N$$

F_N is the normal force, the perpendicular component of the force that is squeezing the two objects into each other. μ_k (lowercase Greek mu) is the "coefficient of kinetic friction," which depends on the nature of the two materials in contact, with values typically in the range of 0 to 1. The sliding friction force does not depend on the speed of the object.

The more you squeeze the two objects together, the deeper their interpenetration and the more difficult it is to slide sideways, which is why the force is proportional to F_N.

It is interesting that the friction force does not depend on the speed. This is in contrast to air resistance, another dissipative force. Air resistance is larger at higher speeds, as you can easily verify by putting your hand out of the window of a moving car and noticing how much greater the air resistance force is on your hand at high speed than it is at low speed.

$f_{friction} \approx \mu_k F_N$ is only approximate and does not have the fundamental character of the gravitational and electric forces. Nevertheless, it is a good approximation to the effects of friction when one object slides on another.

QUESTION If you are pulling a brick at a constant speed, what happens if you increase your force to be bigger than $\mu_k F_N$?

If while the brick is sliding you apply a force bigger than $\mu_k F_N$, the net force becomes nonzero, so the momentum of the brick will increase.

Static Friction

QUESTION What will happen if you exert a horizontal force *less* than $\mu_k F_N$ on a brick lying on a table?

In this case, although the brick doesn't slide, there is a sideways compression of the atomic bonds, so there is a parallel component of the force of the table on the brick, equal to the force you're applying. This situation is called "static friction," where "static" means "not moving." The static friction force can have any value that is less than $\mu_s F_N$, where μ_s is called the "coefficient of static friction." As you apply a bigger and bigger force, but less than $\mu_s F_N$, the bonds are compressed more but the object doesn't slide.

For some pairs of materials it may be necessary to exert a force somewhat bigger than $\mu_k F_N$ in order to get the object to start to move, but once the object has started to slide, you only need a force of $\mu_k F_N$ to keep it moving. In such situations we say that the coefficient of static friction μ_s is larger than the one for sliding friction μ_k (*k* for kinetic, meaning with motion).

STATIC FRICTION

$$f_{\text{friction}} \leq \mu_s F_N$$

μ_s is the "coefficient of static friction." The parallel component f_{friction} of the force squeezing the two objects together can be less than $\mu_s F_N$. The maximum static friction force is $\mu_s F_N$.

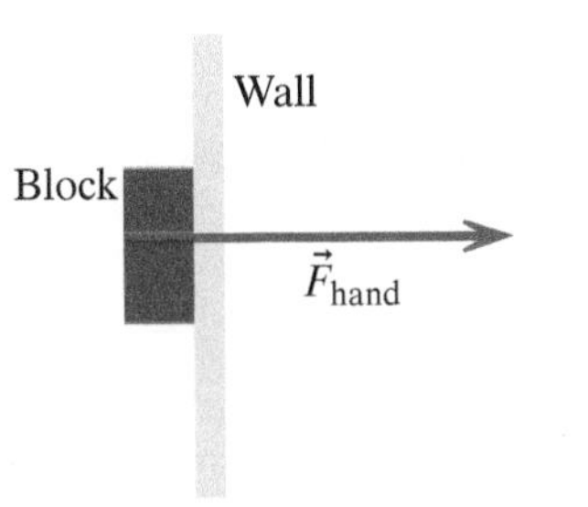

Figure 4.28 You press a block against a wall.

When μ_s is larger than μ_k, there can be "stick-slip" motion. You pull on a string attached to the brick and apply a large force $\mu_s F_N$ to start the motion, then the friction force decreases to $\mu_k F_N$, the net force becomes nonzero, and the brick accelerates, making the string go slack if you don't accelerate your hand. Now that the string has gone slack, the brick slows down and may stop, and you have to start over, hence the stick-slip motion.

EXAMPLE **Holding a Block Against a Wall**

You hold a 3 kg metal block against a wall by applying a horizontal force of 40 N, as shown in Figure 4.28. The coefficient of friction for the metal–wall pair of materials is 0.6 for both static friction and sliding friction. Does the block slip down the wall?

Solution

System: Block
Surroundings: Earth, wall, hand
Free-body diagram: Figure 4.29

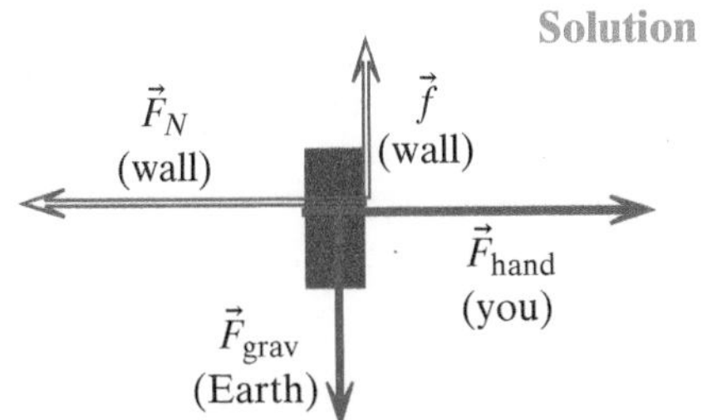

Figure 4.29 Free-body diagram for a block pressed against a wall.

$$x\text{: } \Delta p_x = (F_{\text{hand}} - F_N)\Delta t = 0$$

$$y\text{: } \Delta p_y = (\mu_k F_N - mg)\Delta t, \text{ assuming it slides}$$

Combining these two equations, we have

$$\Delta p_y = (\mu_k F_{\text{hand}} - mg)\Delta t = (0.6(40\,\text{N}) - (3\,\text{kg})(9.8\,\text{N/kg}))\Delta t$$

$$\Delta p_y = (-5.4\,\text{N})\Delta t$$

Since there is a nonzero impulse in the $-y$ direction, the block will slip downward with increasing speed.

To prevent the block from sliding downward, you would have to apply a force such that $F_{net,y} = 0$, in which case $\mu_s F_{hand} = mg$, so $F_{hand} = (3\,\text{kg})(9.8\,\text{N/kg})/0.6 = 49\,\text{N}$. Note that F_N isn't always equal to mg; here it is equal to the force of your hand.

If you apply a force greater than 49 N, how big is the static friction force f? If the block doesn't move, the y component of the net force $F_{net,y}$ is zero, so the static friction force will always be mg, no matter how hard you press the block against the wall, with $f_{friction} = mg < \mu_s F_N$.

EXAMPLE **Sliding to a Stop**

You take the same 3 kg metal block and slide it along the floor, where the coefficient of friction is only 0.4. You release the block with an initial velocity of $\langle 6,0,0 \rangle$ m/s. How long will it take for the block to come to a stop? How far does the block move?

Solution System: Block
Surroundings: Earth, floor
Free-body diagram: Figure 4.30
Initial state: $v_x = 6$ m/s
Final state: $v_x = 0$

$$x\text{: } \Delta p_x = -\mu_k F_N \Delta t$$
$$y\text{: } \Delta p_y = (F_N - mg)\Delta t = 0$$

Combining these two equations and writing $p_x = mv_x$, we have

$$\Delta(mv_x) = -\mu_k mg \Delta t$$
$$\Delta v_x = -\mu_k g \Delta t$$
$$\Delta t = \frac{0 - v_{xi}}{-\mu_k g} = \frac{v_{xi}}{\mu_k g}$$
$$\Delta t = \frac{6\,\text{m/s}}{0.4(9.8\,\text{N/kg})} = 1.53\,\text{s}$$

Since the net force was constant, $v_{x,avg} = (v_{xi} + v_{xf})/2$, so

$$\Delta x/\Delta t = ((6+0)/2)\,\text{m/s} = 3\,\text{m/s}$$
$$\Delta x = (3\,\text{m/s})(1.53\,\text{s}) = 4.5\,\text{m}$$

$\vec{F}_N$ (floor)

$\vec{v}$

$\vec{f}$ (floor)

$\vec{F}_{grav}$ (Earth)

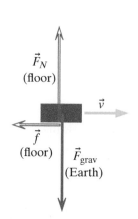

Figure 4.30 F_N is the normal component of the force of the floor and $f = \mu_k F_N$ is the parallel component.

Checkpoint 6 When you apply a horizontal force of 80 N to a block, the block moves across the floor at a constant speed, with $\mu_k = \mu_s$. **(a)** What happens if instead you apply a force of 60 N? **(b)** When you apply a force of 60 N, what is the magnitude of the horizontal component of the force that the floor exerts on the block? **(c)** What happens if instead you exert a force of 100 N? **(d)** When you apply a force of 100 N, what is the magnitude of the horizontal component of the force that the floor exerts on the block?

4.9 SPEED OF SOUND IN A SOLID AND INTERATOMIC BOND STIFFNESS

When we have considered objects that exert tension forces, such as vines, strings, and wires, we have made the (very good) approximation that the stretch

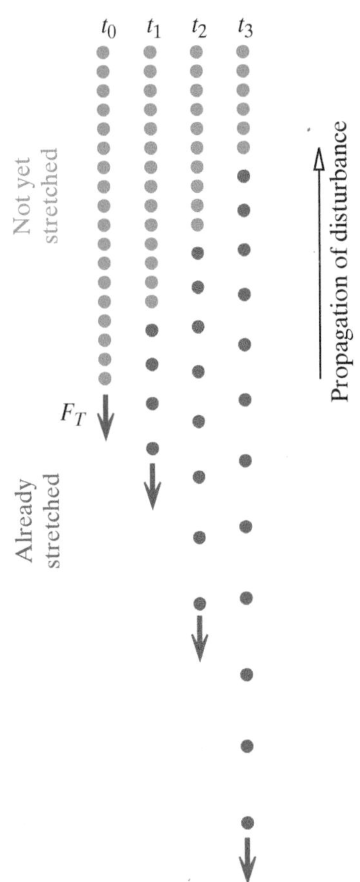

Figure 4.31 At time t_0 Tarzan starts pulling downward on the vine with a force F_T. The "disturbance" in the vine propagates upward. At successive times t_1, t_2, and t_3, more and more of the interatomic bonds in the vine are stretched. Atoms in the stretched region are indicated by colored dots. The amount of stretch is greatly exaggerated here.

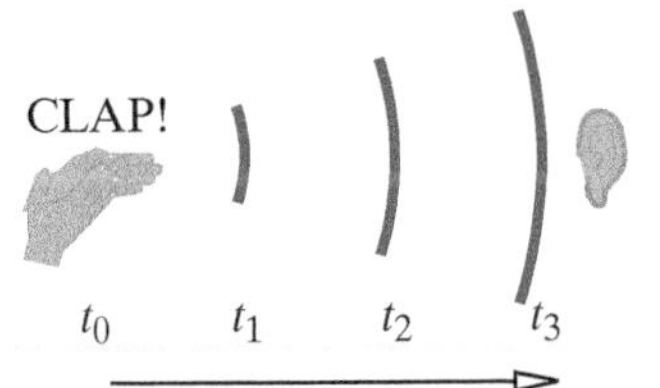

Figure 4.32 Clap your hands—sound travels to your ear as a pulse of air compression.

of the interatomic bonds is nearly the same throughout the length of the object. Likewise, we assume that inside objects exerting compression forces, the interatomic bonds are all compressed nearly the same amount.

The Propagation of Stretch or Compression

The equal stretch or compression of bonds does not, however, occur instantaneously. When Tarzan first starts pulling down on the end of a vine, he very slightly lengthens the interatomic bonds between neighboring atoms in the nearby section of the vine (Figure 4.31). As a result of their downward displacement, these atoms stretch the bonds of their neighbors, and very quickly this new interatomic bond-stretching propagates upward, all the way to the other end of the vine, and the whole vine is then in tension.

It is the boundary between the already stretched region of the vine and the newly stretched region that moves upward, along the length of the vine. Individual atoms move extremely short distances, though for clarity in the diagram we have enormously exaggerated the interatomic stretches and the distance any individual atom moves. The process is the same for compression.

This process is fast, but it is not instantaneous. The rate of propagation of the boundary between the stretched and unstretched regions is called the "speed of sound" in the material. There are various ways to measure this speed. One of the simplest methods is to place a microphone at one end of a metal bar, strike the other end of the bar with a hammer, and use a computer interface to measure how long it takes for the disturbance to travel from one end of the bar to the other.

QUESTION What does this have to do with sound?

What we call "sound" is a disturbance in a material. We are most familiar with sound traveling through air. When you clap your hands, a pulse of sound propagates through the air to your ear (Figure 4.32). This pulse starts as a momentary compression of the air (an increase in the air density) made by your hands. This compression is passed on to neighboring regions of the air, and the air near the hands relapses to its original density.

Eventually your eardrum is hit by the pulse and moves in response to the sudden increase in density (and pressure). The motion of the eardrum is detected by your inner ear and passed on to the brain, which interprets the signal as meaning that you clapped your hands. Note that no individual air molecule moves from near your hands to your ear. Rather it is the disturbance that moves from one place to another. The propagation of a disturbance is called a wave.

An Iterative Model of Sound Propagation in a Metal

We will model a copper rod as a single horizontal chain of one hundred copper atoms, each with mass $(64 \times 10^{-3}\ \text{kg})/(6.02 \times 10^{23})$. The initial distance between any two atoms is the length of the interatomic bond in solid copper (2.28×10^{-10} m, see Section 4.4), and the stiffness of each interatomic spring is 26.8 N/m (see Section 4.5). Initially each atom is at rest. The calculation is organized in a way similar to other iterative calculations we have done; the only difference is that we must update the position and momentum of 100 objects after each time step. An outline of the calculation is this:

Displace the leftmost atom by a small amount
Choose a very short time step ($\Delta t \approx 1 \times 10^{-14}$ s)
Set elapsed time = 0
Then, repeatedly:

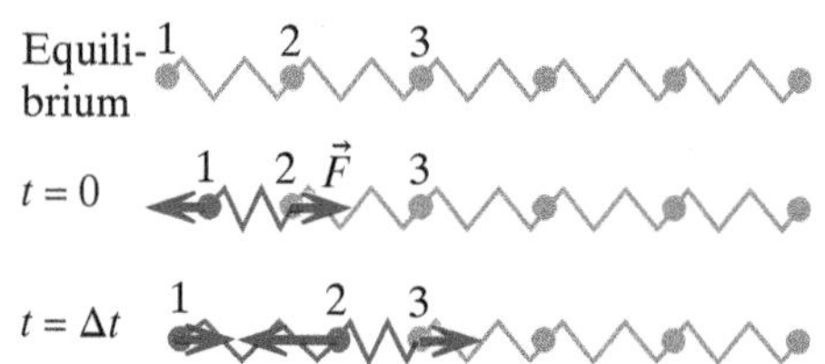

Figure 4.33 Top: A chain of atoms and bonds in their equilibrium position. At $t = 0$ one atom is displaced to the right, starting a wave that moves along the chain.

- From the current positions of the atoms, find the stretch of every spring.
- Find the net force on each atom, due to the two springs attached to it.
- Calculate the change in momentum of each atom.
- Calculate the new momentum of each atom.
- Update the position of each atom.
- Add Δt to the elapsed time.

Stop the calculation when the rightmost atom is displaced. Divide the distance traveled (100 atoms)(2.28×10^{-10} m/atom) by elapsed time to get the speed at which the disturbance traveled down the chain.

Figure 4.33 shows a visualization of the first step in this process. At the top is a chain of atoms and bonds in their equilibrium positions. At time $t = 0$ atom 1 is displaced to the right, compressing a bond, which then exerts a force on atom 2 (arrows denote the net forces on each atom). After applying the Momentum Principle to all atoms, and updating positions of all atoms over a time step Δt, both atoms 1 and 2 are displaced from equilibrium; the second bond is compressed, and the first is now stretched. Iterative updates continue, involving progressively more atoms after each time step.

Visualizing the propagation of the disturbance can be done in various ways. A particularly useful visualization is to plot the horizontal displacements (y_n) vertically, at the locations of the horizontal equilibrium locations (x_n). An example is shown in Figure 4.34, where you see that a displacement to the left is plotted below the axis, and a displacement to the right is plotted above the axis.

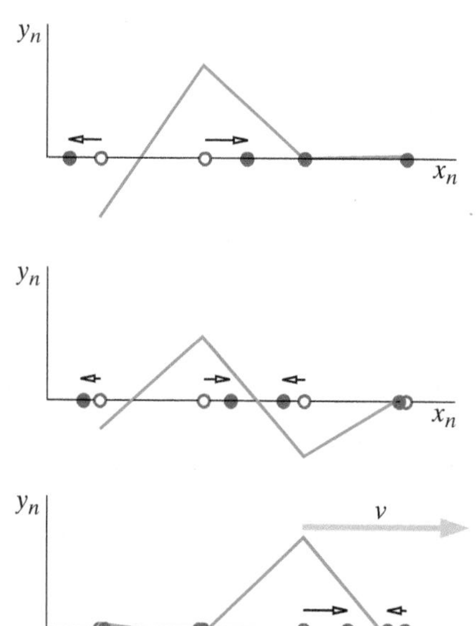

Figure 4.34 Visualizing the propagation by plotting the displacements y_n away from the equilibrium positions x_n.

It is clearly desirable to program a computer to do this calculation. When we carry out this calculation for a copper bar, using the values given above, we obtain a value for the speed of sound in copper of about 3750 m/s. Experimentally measured values given in engineering and science reference materials are around 3800 m/s, so our simple model gives a surprisingly good result.

Although predictions based on iterative calculations can assure us that our understanding of the underlying nature of a phenomenon is correct, it is also useful to have a simple algebraic expression for a central relationship such as that between interatomic bond stiffness and speed of sound. In order to be able to do this, we will need to derive an analytical solution for the oscillations of a single mass and spring. This will give us added insight into the propagation of a disturbance along a chain of masses and springs.

4.10 DERIVATIVE FORM OF THE MOMENTUM PRINCIPLE

The forms of the Momentum Principle we have used so far,

$$\Delta\vec{p} = \vec{F}_{\text{net}}\,\Delta t \quad \text{and} \quad \vec{p}_f = \vec{p}_i + \vec{F}_{\text{net}}\,\Delta t$$

are particularly useful when we know the momentum at a particular time and want to predict what the momentum will be at a later time. We use these forms in repetitive computer calculations to predict future motion.

In solving problems involving objects whose momentum was not changing at all, such as a ball hanging motionless on the end of a wire, we saw that although we had to choose a value for Δt, it didn't matter what value we chose, as long as it was greater than zero. In such problems, we could also have used a different form of the Momentum Principle, which we obtain by dividing by the time interval Δt:

$$\frac{\Delta\vec{p}}{\Delta t} = \vec{F}_{\text{net}}$$

If we let Δt get smaller and smaller, the ratio $\Delta\vec{p}/\Delta t$ becomes the time derivative of $\vec{p}$:

$$\lim_{\Delta t \to 0} \frac{\Delta\vec{p}}{\Delta t} = \frac{d\vec{p}}{dt}$$

The derivative of a vector is itself a vector:

$$\frac{d\vec{p}}{dt} = \left\langle \frac{dp_x}{dt}, \frac{dp_y}{dt}, \frac{dp_z}{dt} \right\rangle$$

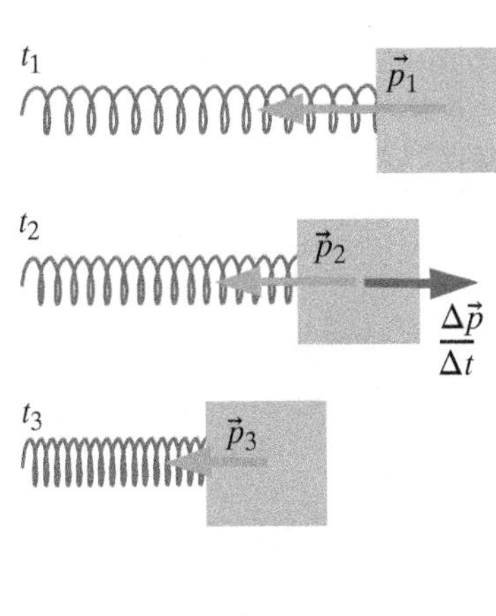

Figure 4.35 A block attached to a spring moves to the left, slowing down. To find $d\vec{p}/dt$ at time t_2, we take the difference of the two values of momentum at times t_1 and t_3 (at equal times before and after t_2), and divide by $\Delta t = t_3 - t_1$. As we take Δt to be smaller and smaller, $\Delta\vec{p}/\Delta t$ approaches $d\vec{p}/dt$. In the bottom diagram the collinear arrows $\vec{p}_1$ and $\vec{p}_3$ have been offset slightly to make the graphical subtraction clear.

We have previously encountered other vector derivatives: the instantaneous velocity $\vec{v} = d\vec{r}/dt$, and the acceleration $\vec{a} = d\vec{v}/dt$. Like any other vector, the derivative of a vector has both a magnitude and a direction.

QUESTION For a given physical system, such as the mass–spring system in Figure 4.35, how can we find $d\vec{p}/dt$?

Operationally, to find $d\vec{p}/dt$ at a particular instant we apply the definition of a derivative, as shown in Figure 4.35:

1. Find $\vec{p}$ at equal time intervals before and after the time of interest (t_2 in Figure 4.35).
2. Find $\Delta\vec{p} = \vec{p}_3 - \vec{p}_1$ (the bottom diagram in Figure 4.35).
3. Divide $\Delta\vec{p}$ by $\Delta t = t_3 - t_1$ to get $\Delta\vec{p}/\Delta t$ (whose direction is the same as the direction of $\Delta\vec{p}$).
4. Let Δt approach 0 by taking t_1 and t_3 closer to t_2. As $\Delta t \to 0$, $\Delta\vec{p}/\Delta t$ approaches $d\vec{p}/dt$.

We can now write the Momentum Principle in its derivative form:

THE MOMENTUM PRINCIPLE (DERIVATIVE FORM)

$$\frac{d\vec{p}}{dt} = \vec{F}_{\text{net}}$$

In words: The instantaneous time rate of change of the momentum of an object is equal to the net force acting on the object. In calculus terms, we can say that the derivative of the momentum with respect to time is equal to the net force acting on the object.

This form of the Momentum Principle is useful when we know something about the rate of change of the momentum at a particular instant. Knowing the rate of change of momentum, we can use this form of the Momentum Principle to deduce the net force acting on the object, which is numerically equal to the rate of change of momentum. Knowing the net force, we may be able to figure out particular contributions to the net force.

QUESTION What does the derivative form of the Momentum Principle tell us about the relationship between the direction of $d\vec{p}/dt$ and the direction of $\vec{F}_{\text{net}}$?

If two vectors are equal, both their magnitudes and their directions are the same. Consider the mass–spring system in Figure 4.35. The block is slowing down because the spring is being compressed, and consequently exerts a force to the right. The direction of this force is in fact the same as the direction of $d\vec{p}/dt$.

Checkpoint 7 At a certain instant the momentum of an object is changing at a rate of $\langle 0,0,4 \rangle$ kg·m/s/s. **(a)** At this instant, what is the net force on the object? **(b)** What, if anything, can you conclude about the direction of the object's momentum at this instant?

4.11 ANALYTICAL SOLUTION: SPRING–MASS SYSTEM

In Chapter 2 we applied the finite-time form of the Momentum Principle iteratively to predict the motion of a mass attached to a spring. Now, using the derivative form of the Momentum Principle, we can derive an analytical (algebraic) solution for the motion of an ideal mass–spring system. By "ideal" we mean that the spring can stretch or compress nearly indefinitely, that the mass of the spring is nearly zero, that the spring never gets hot or deforms, and that there is neither friction nor air resistance.

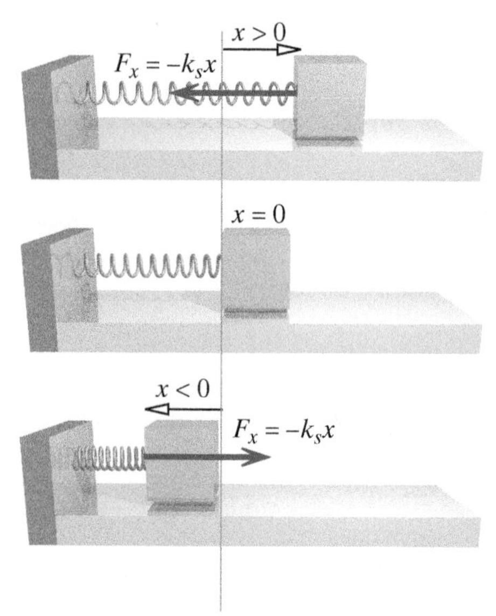

Figure 4.36 A block connected by a spring to a wall slides back and forth with little friction. The x component of the force that the spring exerts on the block is equal to $-k_s x$.

Applying the Momentum Principle to a Spring–Mass System

Consider a block of mass m connected to a spring whose stiffness is k_s (that is, the force exerted by the spring is $k_s s$ when the spring is stretched or compressed an amount s). The other end of the spring is attached to the wall (Figure 4.36). Our goal is to derive an equation for the position x as a function of time t,

$$x(t) = \ldots$$

which predicts the position of the block at any time in the future.

We will assume that:

- The block slides with almost no friction on an air table, where it is supported on a cushion of air.
- The mass of the spring is negligible compared to the mass of the block.
- Air resistance is negligible.

We can simplify our calculations by choosing to measure the position of the block with respect to an origin chosen to be at the equilibrium position, the location of the block when the spring is relaxed, so that the stretch s of the spring is equal to the position x of the block:

$$L = L_0 + x$$
$$s = L - L_0 = (L_0 + x) - L_0 = x$$

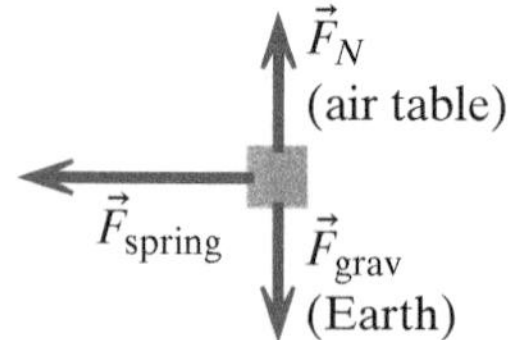

Figure 4.37 Free-body diagram.

Therefore the x component of the force on the block by the spring is $F_x = -k_s x$, as shown in Figure 4.36.

QUESTION What would be the equation for the spring force in terms of the position of the block if the origin were located at the wall, and the relaxed length of the spring is L_0?

In that case we would write $F_x = -k_s(x - L_0)$, because what matters in this expression is the stretch (change in length of the spring), which is $(x - L_0)$. A check on this is the fact that when the block is at location $x = L_0$, the stretch $(x - L_0)$ is zero.

System: Block
Surroundings: Spring, Earth, air table
Free-body diagram: Figure 4.37

Momentum Principle:

$$d\vec{p}/dt = \vec{F}_{\text{net}}$$

Since the block oscillates along the x axis, we know that the y component of the momentum of the block is zero and remains zero. Therefore we can deduce that the y component of the net force on the block must be zero. Given this, we can focus only on the x components of force and momentum.

Because $v_x = dx/dt$:

$$\frac{dp_x}{dt} = \frac{d(mv_x)}{dt} = m\frac{d^2x}{dt^2}$$

From the Momentum Principle:

$$\frac{dp_x}{dt} = -k_s x$$

$$m\frac{d^2x}{dt^2} = -k_s x$$

If we ignore the constants and signs for a moment, this equation (which is a "differential equation") says that if we know x as a function of time, taking the second derivative of that function will give us back the original function.

QUESTION What function appears in its own second derivative?

If we differentiate a cosine function twice, we get a cosine again. Further, since the block oscillates back and forth, cosine sounds like a reasonable guess. Guessing a function isn't as peculiar as it may seem. This is actually one of the standard approaches to trying to solve a differential equation.

However, $x(t) = \cos(t)$ doesn't look quite right. The cosine function has values between 1 and −1, but the block may not move 1 m. Let's multiply by a constant A to change the maximum displacement to a reasonable value.

Now we have $x(t) = A\cos(t)$, but that still isn't quite right, because the period of the oscillations might not be 2π s. We can add another constant, ω (lowercase Greek omega), to adjust the period of the oscillations, which gives us the equation

$$x(t) = A\cos(\omega t)$$

To see if this expression for $x(t)$ is in fact a solution, we will try substituting this expression for x back into the Momentum Principle differential equation:

$$m\frac{d^2}{dt^2}(A\cos(\omega t)) = -k_s A\cos(\omega t)$$

$$m\frac{d}{dt}(-A\omega\sin(\omega t)) = -k_s A\cos(\omega t)$$

$$-Am\omega^2\cos(\omega t) = -k_s A\cos(\omega t)$$

$$m\omega^2 = k_s$$

$$\omega = \sqrt{\frac{k_s}{m}}$$

We find that $x(t) = A\cos(\omega t)$ is an analytical solution for the position of the mass as a function of time, if we choose ω to have the value $\sqrt{k_s/m}$. This is an interesting result, because it says that the period of the oscillations depends on the physical properties of the mass and of the spring.

ANALYTICAL SOLUTION: IDEAL SPRING–MASS SYSTEM

$$x(t) = A\cos(\omega t), \text{ with } \omega = \sqrt{\frac{k_s}{m}}$$

A (the "amplitude") is a constant that depends on the initial conditions. A is equal to the maximum stretch of the spring during an oscillation. ω is called the "angular frequency" and has units of radians/s.

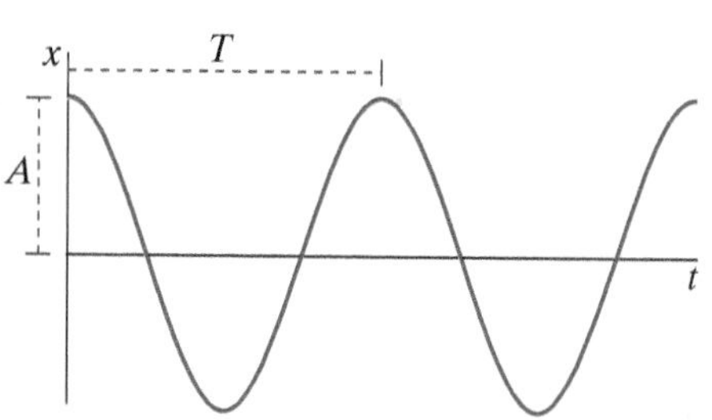

Figure 4.38 Amplitude A and period T, shown on a plot of x vs. t for the mass.

Often we write simply $x = A\cos(\omega t)$, since the time dependence of x is made clear by the right-hand side of the equation. The constant A is called the "amplitude" of the motion. It is equal to the maximum stretch of the spring. If the spring stretches or compresses a maximum of 5 cm, then A will have the value 0.05 m. The constant ω in $x = A\cos(\omega t)$ is called the "angular frequency" and is measured in radians per second (ω is lowercase Greek omega). When the argument ωt of the cosine increases by 2π rad, the motion repeats (one complete cycle), so if we call the "period" T (the time required for one complete cycle, Figure 4.38), we have $\omega T = 2\pi$. One other quantity often used to describe oscillating systems is the "frequency" f, which is the number of complete cycles per second: $f =$ (1 cycle) per T seconds. Cycles per second are also called "hertz."

ANGULAR FREQUENCY, PERIOD, AND FREQUENCY

$$\text{Angular frequency } \omega = \frac{2\pi}{T} \quad \text{(radians per second)}$$

$$\text{Period } T = \frac{2\pi}{\omega} \quad \text{(seconds)}$$

$$\text{Frequency } f = \frac{1}{T} = \frac{\omega}{2\pi} \quad \text{cycles per second or hertz}$$

QUESTION According to the analytical solution for the motion of an ideal mass–spring system, how does the amplitude of the motion change over time?

In the solution the amplitude A is a constant, and does not decrease over time. This makes sense because in an idealized system there is no friction. This idealized system will continue to oscillate forever with a constant amplitude.

QUESTION According to the analytical solution, if you decreased the amplitude of the oscillations (for example, by stretching the spring less initially), how would the period of the motion change?

From the analytical solution, we see that the angular frequency (and therefore the period) does not depend on the amplitude at all! It depends only on the spring stiffness and the mass of the block. Such an idealized spring–mass system is called a "harmonic oscillator," and its oscillatory motion is called "harmonic motion." The special features of a harmonic oscillator are:

HARMONIC OSCILLATOR

- The position of the mass as a function of time is given by a cosine function.
- The period (round-trip time) is independent of the amplitude of the oscillations.

The fact that the period of a harmonic oscillator does not depend on the amplitude is surprising. With a larger amplitude, the mass travels farther in one period. However, it also reaches higher speeds, and for an ideal mass–spring system, these two effects exactly cancel.

Not all oscillating systems are harmonic. If you drop a rubber ball and let it bounce, as the bounce amplitude decreases, the period decreases too: the time between bounces gets shorter and shorter as the height of the bounces gets lower and lower. The bouncing ball is an example of an "anharmonic oscillator." "Anharmonic" simply means "not harmonic."

QUESTION Does the analytical solution above for a horizontal oscillator apply to a mass–spring system hanging vertically?

Yes. Although we need to take the gravitational force into account for a vertical system, we can show that this analytical solution is also valid for a vertical mass–spring system, in which the gravitational force on the mass acts parallel or antiparallel to the force on the mass by the spring. The details of this analysis are given in Section 4.16.

EXAMPLE **Period Depends on Mass and Stiffness**

Suppose that the period of a particular ideal mass–spring system is 2 s. **(a)** What would be the period of the system if the mass were tripled? **(b)** What would be the period of the system with the original mass, but with a spring that was four times as stiff?

Solution

(a) Triple the mass:

$$T_{\text{original}} = \frac{2\pi}{\sqrt{\dfrac{k_s}{m}}} = 2\pi\sqrt{\frac{m}{k_s}}$$

$$T_{\text{new}} = 2\pi\sqrt{\frac{3m}{k_s}} = \sqrt{3}\,T_{\text{original}}$$

$$= (1.73)(2\text{ s}) = 3.46\text{ s}$$

(b) Quadruple the spring stiffness:

$$T_{\text{new}} = 2\pi\sqrt{\frac{m}{4k_s}} = \frac{1}{\sqrt{4}}T_{\text{original}}$$

$$= 0.5(2\text{ s}) = 1\text{ s}$$

EXAMPLE **Mass–Spring Oscillation (Analytical Solution)**

A 50 g mass is attached to a low-mass horizontal spring whose relaxed length is 25 cm and whose stiffness is 4 N/m. The block slides on a very slippery, low-friction surface. You stretch the spring so it is 30 cm long, and release the block. **(a)** What is the amplitude of the oscillations? **(b)** How long will it take the block to make one full oscillation? **(c)** What would be the amplitude of the oscillations if the initial stretched length of the spring were 35 cm instead of 30 cm? **(d)** How long would it take the block to make one full oscillation if the initial stretched length of the spring were 35 cm instead of 30 cm? **(e)** How many oscillations will the block make in 10 s? **(f)** The initial length of the stretched spring is 35 cm. What will be the location of the block 3.15 s after you release it?

Solution System: Mass
Surroundings: Spring, Earth, table

QUESTION Are the assumptions for an ideal mass–spring system satisfied?

Yes. The mass of the spring, friction, and air resistance are all negligible, so we can use the analytical solution $x(t) = A\cos(\omega t)$.

(a) $A = s_i = 0.3\,\text{m} - 0.25\,\text{m} = 0.05\,\text{m}$
(b) Time for one round-trip $= T$ (period)

$$\omega = \sqrt{\frac{k_s}{m}} = \sqrt{\frac{4\,\text{N/m}}{0.05\,\text{kg}}} = 8.94\,\text{s}^{-1}$$

$$T = \frac{2\pi}{\omega} = \frac{2\pi}{8.94\,\text{s}^{-1}} = 0.70\,\text{s}$$

(c) $A = s_i = 0.35\,\text{m} - 0.25\,\text{m} = 0.10\,\text{m}$
(d) $T = 0.70\,\text{s}$
(e)

$$f = \frac{1}{T} = \frac{1}{0.70\,\text{s}} = 1.43\ \text{cycles/s}$$

$$\text{number of cycles} = ft = (1.43\,\text{cycles/s})(10\,\text{s}) = 14.3$$

(f)

$$x(t) = A\cos(\omega t)$$

$$x(3.15\,\text{s}) = (0.10\,\text{m})\cos\left((8.94\,\text{s}^{-1})(3.15\,\text{s})\right) = -0.0994\,\text{m}$$

Checkpoint 8 You have a rubber band whose relaxed length is 8.5 cm. You hang a coffee cup, whose mass is 330 g, from the rubber band, which stretches to a length of 14 cm. Consider the rubber band to be a single spring. **(a)** What is the stiffness of the rubber band? **(b)** If you start this mass–spring system oscillating, how long does one round-trip take?

4.12 ANALYTICAL VS. ITERATIVE SOLUTIONS

In Chapter 2 we modeled the motion of a block on a vertical spring by iteratively applying the Momentum Principle and updating position. In the previous section of the current chapter, by solving a differential equation involving the derivative form of the Momentum Principle, we obtained an analytical solution for the position of an ideal 1D mass–spring system.

QUESTION What are the advantages of the analytical solution?

Our analytical solution for the position of an ideal harmonic oscillator as a function of time allows us to see important general features of the behavior of the ideal system:

- The period does not depend on the amplitude.
- Increasing the mass increases the period.
- Increasing the spring stiffness decreases the period.

Although we could have learned these things by experimenting with different values of mass, spring stiffness, or initial stretch in an iterative solution, it is particularly easy to see this relationship from examining the analytical solution, as shown in the example entitled "Period Depends on Mass and Stiffness."

QUESTION What are the advantages of the iterative (computational) solution?

Although we were able to obtain an analytical solution for an ideal 1D system, it is not straightforward to extend this solution to the general case of a 3D mass–spring system that may be moving in a variety of ways. Recall from Chapter 2 that if $\vec{r}_0$ is the location of the fixed end of the spring, then we can calculate the 3D spring force in VPython this way:

$\vec{L} = \vec{r} - \vec{r}_0$	`L = block.pos - spring.pos`
$\hat{L} = \vec{L}/\lvert\vec{L}\rvert$	`Lhat = L/mag(L)`
$s = \lvert\vec{L}\rvert - L0$	`s = mag(L) - L0`
$\vec{F} = -k_s \cdot s \cdot \hat{L}$	`F_spring = -ks * s * Lhat`

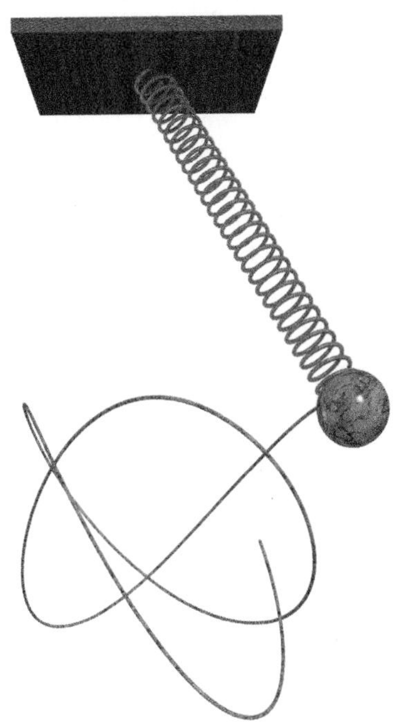

Figure 4.39 A snapshot from a computational model of a mass–spring system oscillating in 3D.

This force calculation forms the core of a 3D computational model of the system that is easy to refine and extend, in order to model more realistic systems. For example:

- Since we use the full 3D vector force in our computational model, by varying the initial conditions we can get 3D motion (Figure 4.39).
- We can easily add the effects of sliding friction.
- We can easily add the effects of air resistance.
- We can easily modify the spring force to model nonideal springs (for example, a spring whose force depended on $k_s s^3$ instead of $k_s s$).
- We can analyze the behavior of a spring whose mass is not negligible, such as a Slinky, by modeling it as a chain of tiny masses connected by ideal springs (Figure 4.40).

QUESTION Is it always possible to find an analytical solution?

Figure 4.40 A massive spring may be modeled as a chain of point masses connected by massless springs.

For many systems the differential equation corresponding to the derivative form of the Momentum Principle has a mathematical form that cannot be solved analytically and must be solved numerically. The power of iterative application of the Momentum Principle is that it can be used even in those situations where no analytical solution seems possible.

For example, $x = A\cos(\omega t)$ *cannot* solve the following "nonlinear" differential equation for any value of ω:

$$\frac{dp_x}{dt} = m\frac{d^2x}{dt^2} = -k_s x^3$$

There may or may not exist an analytical solution of this differential equation in terms of well-known mathematical functions, but *any* differential equation can be solved numerically, just as we did for the spring–mass equation. Given x and p at some instant in time, we can calculate the force $(-k_s x^3)$ and determine the new values of x and p at a time Δt later, then repeat as often as necessary to reach the time of interest.

Note that either an analytical or an iterative prediction of the future applies only to our *model* of the actual real-world situation. If our model is a good approximation to the real world, our prediction will be a good approximation to what will actually happen, but even an analytical prediction cannot be exact because of the practical impossibility of knowing exactly the initial conditions and the net force due to all the other objects in the Universe.

Of course, for both types of solution—iterative or analytical—we need to know the initial conditions (position and momentum of the mass) in order to predict the position of the system at some future time. In the previous example, the mass was initially at rest, and the spring was stretched. It is also possible to start an oscillator by placing it at its equilibrium position and giving it a kick—in this case its initial stretch is zero, but its initial momentum is nonzero. In an iterative solution, this is easy to take into account. A discussion at the end of this chapter explains how this is taken into account in the analytical solution.

4.13 ANALYTICAL EXPRESSION FOR SPEED OF SOUND

Under ordinary conditions, the speed of sound in air is about 340 m/s. If your friend stands 340 m away and yells, you will hear the sound a second later. The speed of sound in a solid object is much faster than it is in air. In aluminum, the speed of sound is about 4800 m/s. In lead, which is a much softer material than aluminum, the speed of sound is only about 1200 m/s. There are two significant differences between lead and aluminum at the atomic level. First, the mass of a lead atom is about eight times larger than the mass of an aluminum atom. Second, the interatomic bond stiffness in aluminum is greater than the interatomic bond stiffness in lead (see Problem P36). Since the propagation of a disturbance through a metal rod involves displacements of atoms, it seems reasonable that both of these factors should play a role in determining the speed with which the disturbance propagates. Our goal in the following discussion is to use the ball–spring model of a solid to predict this relationship.

Qualitative Guess

Think of a solid metal rod as a network of atoms (balls) connected by springs (interatomic bonds) as in Figure 4.41.

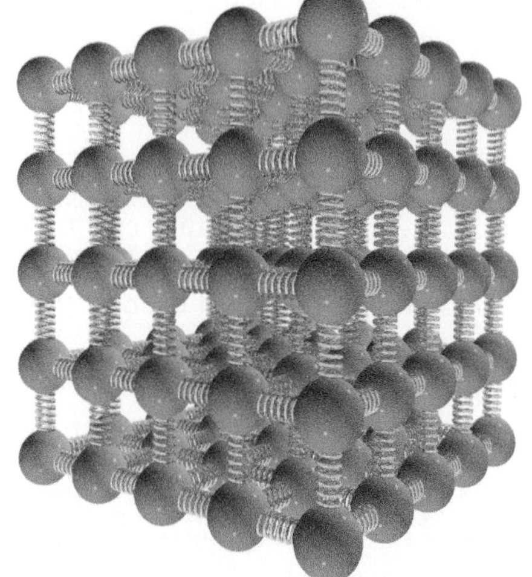

Figure 4.41 A model of a metal: atoms (balls) connected by springs (interatomic bonds).

QUESTION How should the speed of sound depend on m_a, the mass of an atom in a material? (Think of hitting one end of a metal rod with a hammer.)

A given applied force produces the same change of momentum for a massive atom like lead or a lighter atom like aluminum. However, the change in speed of the lead atom will be less than the change in speed of the aluminum atom. This suggests that the greater the mass of the atoms, the slower a sound pulse (disturbance) might propagate through a rod.

QUESTION How should the speed of sound depend on the stiffness of the interatomic bonds in a material?

The stiffer the bond, the greater the force the "spring" will exert on a neighboring atom in response to the same displacement. Stiffer interatomic bonds should make a sound pulse propagate faster through a rod.

These initial qualitative guesses don't tell us details of the mathematical relationships involved: for example, whether speed of sound is proportional to $1/m_a, 1/m_a^2, 1/\sqrt{m_a}$, or some other factor. We need to apply our understanding of mass–spring systems to refine these guesses.

Dimensional Analysis

We need to construct a mathematical expression that has the correct units, or "dimensions." Speed has units of m/s, so we need an equation involving interatomic bond stiffness $k_{s,i}$ and atomic mass m, ending up with meters in the numerator and seconds in the denominator:

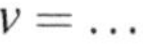

$$v = \ldots$$

The angular frequency $\omega = \sqrt{k_{s,i}/m_a}$ and has units of s^{-1}, so ω might appear in the numerator of the expression:

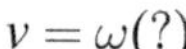

$$v = \omega(?)$$

QUESTION Is it reasonable that v should be directly proportional to ω?

Yes. The larger the angular frequency ω is, the faster an atom oscillates back and forth, compressing and stretching intermolecular bonds, which affect neighboring atoms.

QUESTION What other quantity should go in the expression?

Since we need to end up with m/s, we need to multiply by something that has units of meters. ω is related to the rate at which one atom hits its neighbor, propagating a change through a distance of one interatomic spacing. Multiplying by the interatomic bond length d might be appropriate.

$$v = \omega d$$

This informal reasoning is an example of "dimensional analysis," in which we identify meaningful quantities in the situation and combine them to guess an expression for some other quantity of interest. While this can be a fruitful procedure, we shouldn't be surprised if we find that the speed of sound is actually given by $v = 2\omega d$, or $v = \omega d/\pi$, for example. Moreover, note that atomic oscillations in the x direction involve two atomic "springs" on each side of an atom, not one, so we may have omitted a factor of 2.

Nevertheless, we can guess that $\left(\sqrt{k_{s,i}/m_a}\right) d$ is proportional to the actual speed of sound, with the correct dependence on $k_{s,i}$ and m_a. In a section on the "wave equation" in Supplement S3 there is a derivation showing that this is in fact the correct result.

SPEED OF SOUND IN A SOLID

$$v = \omega d = \sqrt{\frac{k_{s,i}}{m_a}} d$$

where v is the speed of sound (m/s), $k_{s,i}$ is the stiffness of the interatomic bond, m_a is the mass of one atom, and d is the length of the interatomic bond.

EXAMPLE **Speed of Sound in Copper**

We found (Section 4.5) that the stiffness of the interatomic bond in solid copper is about 26.8 N/m, and that the interatomic bond length in solid copper is about 2.28×10^{-10} m (Section 4.4). There are 64 g of copper in one mole. Use these data to calculate the speed of sound in solid copper.

Solution

$$m_a = \frac{0.064\,\text{kg}}{6.02 \times 10^{23}\,\text{atoms}} = 1.06 \times 10^{-25}\,\text{kg/atom}$$

$$v = \sqrt{\frac{26.8\,\text{N/m}}{1.06 \times 10^{-25}\,\text{kg}}}(2.28 \times 10^{-10}\,\text{m}) = 3620\,\text{m/s}$$

This agrees well with experimentally measured values of around 3800 m/s.

Speed of Sound in Different Materials

In aluminum, which is very stiff (large k_s) and has a small atomic mass m_a, the speed of sound is very high, about 5000 m/s. If a wire is half a meter long, it takes about

$$\frac{0.5\,\text{m}}{5000\,\text{m/s}} = 1 \times 10^{-4}\,\text{s} \quad \text{(a tenth of a millisecond)}$$

from the time you pull on one end for the other end to become tense. On the other hand, lead is quite soft (small k_s) and has a large atomic mass m_a, and the speed of sound in lead is only about 1200 m/s, much lower than in aluminum. (The speed of sound in solids is much higher than the speed of sound in air, which is about 340 m/s. Because air molecules are usually far from one another, the speed of sound in air is determined by the average speed of the air molecules, not by interatomic forces.)

More on Waves

Additional material on waves is provided in Supplement S3, from the point of view of the "wave equation," a partial differential equation that describes many kinds of wave motion.

Checkpoint 9 A certain metal with atomic mass 2×10^{-25} kg has an interatomic bond with length 2.1×10^{-10} m and stiffness 40 N/m. What is the speed of sound in a rod made of this metal?

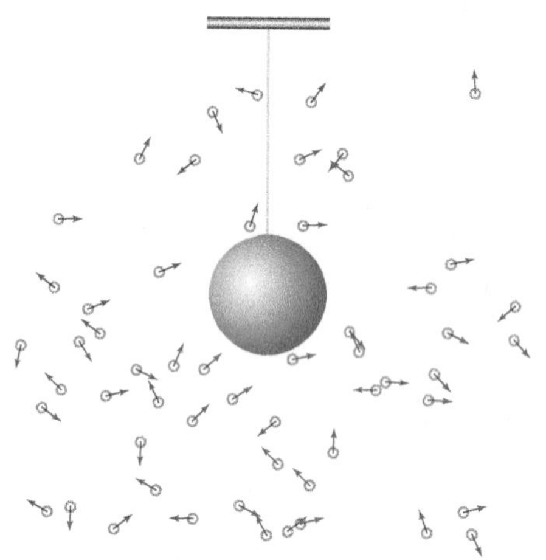

Figure 4.42 Air molecules in motion near a hanging ball. The density of the air beneath the ball is slightly greater than the air density above the ball, so there is a small upward force on the ball exerted by the air.

4.14 CONTACT FORCES DUE TO GASES

Buoyancy

Until now we have considered only contact forces due to solids. However, gases can also exert contact forces. In everyday circumstances, the most significant of these is the force that the air exerts on objects—even stationary objects. Consider the surrounding air that is in contact with a ball hanging from a wire, as shown in Figure 4.42. You might think that the air has no effect on the ball, especially since the ball is not moving, so we don't need to worry about air resistance. To be certain, let's consider this issue at the microscopic level.

Air is a gas, composed of about 80% nitrogen and 20% oxygen, with small amounts of other gases, especially argon (about 1%) and water vapor (which varies with the humidity). The molecules in the air are continuously in motion, moving in random directions. Molecules in the air near the ball will frequently bump into the ball, exerting a small force on it. If there is no wind, the motion of the gas molecules is random, so in a short time period there should be about as many molecules colliding with the right-hand side of the ball as with the left-hand side, and there should be on average no sideways force on the ball.

However, the situation is not quite the same for collisions with the top and bottom of the ball. The directions of molecular velocities are still random, but there are more air molecules per cubic meter below the ball than there are above it. The density of the Earth's atmosphere decreases as one moves upward away from the Earth's surface, until finally at a sufficient distance (approximately 50 km) there is essentially no air left. You may have observed this variation yourself if you have ever gone from sea level to a location several thousand meters higher; you may feel light-headed if you exercise at high altitude because the density of oxygen is significantly lower.

This variation in air density occurs over kilometers; can it possibly be important over a distance of a few centimeters? Interestingly enough, the surprising answer is yes! There is actually a "buoyant force" upward on the ball, because the number of air molecules hitting the bottom of the ball per second is slightly greater than the number of molecules hitting the top of the ball per second.

A Macroscopic View of Buoyancy

How can we determine the magnitude of the buoyant force on the ball due to the air? Let's consider a macroscopic viewpoint. The following discussion, based on the Momentum Principle, applies equally well to the buoyant force exerted on an object by any fluid (air, water, other liquids, other gases). The result we will obtain is called the Archimedes principle, for the Greek thinker who first understood it.

Imagine a box filled with air (or water). Consider a ball-shaped region that a ball will eventually occupy, at a moment when the ball is not yet there, so the ball-shaped region is filled with the fluid. Let's choose this ball-shaped mass of fluid as our system, and consider what forces must be acting on it (Figure 4.43). This may seem odd, but it is a perfectly legitimate choice of system.

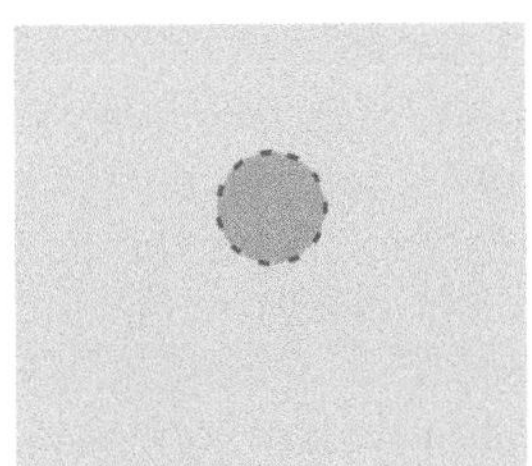

Figure 4.43 A system consisting of a ball-shaped spherical region of fluid, surrounded by more fluid.

QUESTION Is the momentum of the indicated sphere of fluid changing? What must be the net force on this sphere?

The momentum of the indicated sphere of fluid is not changing. The Earth exerts a downward gravitational force on it, but the sphere does not sink. Because its momentum isn't changing, the net force on the sphere must be zero, so there must be an upward force that balances the gravitational force.

QUESTION What exerts this upward force on the sphere?

The only object in contact with this sphere is the rest of the fluid, so it must be that the rest of the fluid exerts an upward force F_b ("buoyant force") whose magnitude is mg, where m is the mass of the sphere of fluid (Figure 4.44).

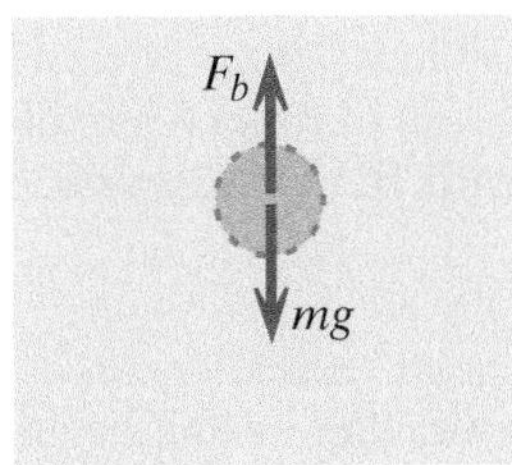

Figure 4.44 Forces on the sphere of fluid. Here m is the mass of the fluid (air or water) in the sphere.

Now remove the sphere of fluid and replace it with the ball. It must be that initially the rest of the fluid still exerts the same upward force F_b of magnitude mg (Figure 4.45). This force may not be large enough to counteract the downward gravitational force on the ball; in the case of a ball hanging in air it clearly is not, since if the ball were not supported by a wire it would begin accelerating downward.

The upward force is called a "buoyant" force but is simply an interatomic contact force due to fluid molecules striking atoms in the surface of the ball. The key point is that despite the complexity of the interatomic interactions, the net effect is simply an upward force whose magnitude is the mass of an equivalent volume of fluid, times g.

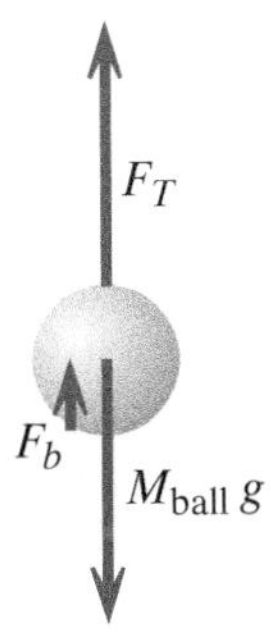

Figure 4.45 A ball hanging from a wire, surrounded by air. The tension force is actually less than the weight of the ball, because the buoyant force F_b due to the air is also in the upward direction.

When Can the Buoyant Force Be Neglected?

Often it is reasonable to neglect buoyant forces in air (though usually not in water). To see why, let's compare the buoyant force to the gravitational force F_g on a ball hanging in air, where we write V for the volume of the ball:

$$\frac{F_b}{F_g} = \frac{m_{\text{air}}\, g}{M_{\text{ball}}\, g} = \frac{m_{\text{air}}}{M_{\text{ball}}} = \frac{m_{\text{air}}/V}{M_{\text{ball}}/V} = \frac{\rho_{\text{air}}}{\rho_{\text{ball}}}$$

The ratio of the forces is equal to the ratio of ρ_{air} (the density of air) to ρ_{ball} (the density of the ball; ρ is the lowercase Greek letter rho). The density of air at STP (standard temperature and pressure, 0°C and 1 atmosphere) can be calculated if we remember from chemistry that one mole of a gas at STP occupies 22.4 liters. The molecular mass of N_2 is 28 and that of O_2 is 32, which gives an average molecular mass of about 29 for air (which is about 80% nitrogen and 20% oxygen):

$$\rho_{\text{air}} \approx \frac{29\text{ g/mol}}{22.4 \times 10^3\text{ cm}^3/\text{mol}} \approx 1.3 \times 10^{-3}\text{ g/cm}^3$$

What is the density of the ball? We can estimate it by noting that the density of water is 1 g/cm^3, and that most solids are more dense than water. Using 1 g/cm^3 as the density of the ball, we find that $F_b/F_g \approx 1.3 \times 10^{-3}/1$.

The buoyant force $m_{\text{air}}g$ of air on a solid object is therefore very small compared to the gravitational force mg. In many cases we can safely choose to neglect it. Since the upward force exerted by the air on the ball is small compared to the gravitational force or the tension force exerted by the wire, we can choose to neglect it in an analysis of the forces on a hanging ball. However, it is not always appropriate to neglect the buoyant force; in accurate analytical chemistry it is necessary to correct for buoyancy when weighing a sample of liquid or low-density solid on an analytical balance.

A final note: The buoyant force being equal to $m_{\text{air}}g$ is really a time average, because the buoyant force is due to random collisions with air molecules and therefore has an intermittent character. However, the collision rate is so high that the force seems nearly continuous and constant. If, however, the ball is of microscopic size, it may not be possible to ignore the intermittent nature of the collisions. In a microscope it is possible to observe small particles being jostled about by the random collisions with the water molecules. This effect is called "Brownian motion."

Checkpoint 10 Calculate the buoyant force in air on a kilogram of iron (whose density is about 8 g/cm^3). Compare with the weight mg of this much iron.

Pressure

QUESTION Why is it difficult to remove a suction cup from a surface?

There is an interesting case in which the effect of the air is very significant and cannot be neglected. Suppose that an object lies on a table and the surfaces of the table and object are so smooth that upon squeezing them together all air is expelled. Such a situation occurs when a suction cup is pressed down onto a smooth surface. Now there are no longer any air molecules striking the bottom surface of the suction cup, while air molecules continue to strike the top surface. The force of the air on the top surface is quite large, about 1×10^5 newtons on each square meter at sea level!

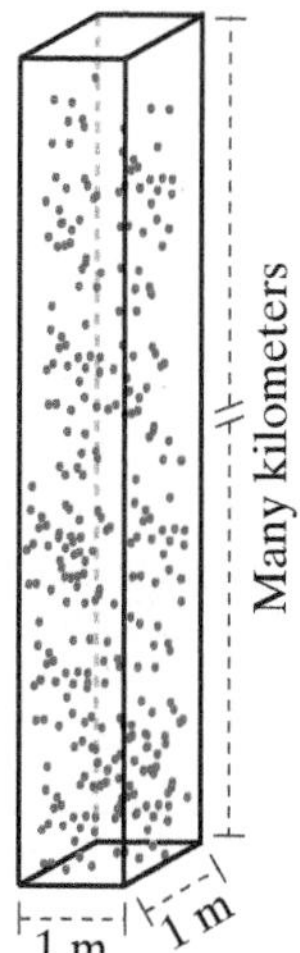

Figure 4.46 A one-square-meter column of air extending upward through the entire atmosphere.

This large force per unit area is called the "pressure" P of the air. The pressure can be thought of as due to the weight of a one-square-meter column of the entire atmosphere, extending upward many kilometers (although with diminishing density), as illustrated in Figure 4.46.

DEFINITION OF PRESSURE

Pressure $P = F/A$ (force per unit area)

Atmospheric pressure at sea level is about $1 \times 10^5\text{ N/m}^2$. In older units this is approximately 15 lb/in^2.

A Constant-Density Model of the Atmosphere

If the atmosphere had a constant density ρ like that near the surface (1.3 g/cm^3), we could calculate the height h of the atmosphere in the following way. Choose the column of air as the system. The volume of a column of air of height h and cross-sectional area A is Ah, and its mass is $M_{air} = \rho(Ah)$. Because the center of mass of the column is at rest, the Momentum Principle tells us that the upward force of the ground on the bottom of the air column must equal the downward gravitational force on the total mass of air in the column. By the reciprocity of interatomic electric forces (air molecules hitting the ground), the force of the air on the ground must also be equal to the weight of the air in the column. Therefore the atmospheric pressure on the ground is the weight $M_{air}g$ divided by the cross-sectional area: $P = M_{air}g/A = \rho gh$.

The height of a constant-density atmosphere, $h = P/(\rho g)$, would be about 8000 m (about 5 mi), about the height of Mount Everest. In reality air density is not constant but decreases as you go higher, so there is still some low-density air at much higher altitudes than this.

This macroscopic picture, taking the whole column of air as the system, is a useful one, but remember that the actual force exerted on a square meter at sea level is due to air molecules hitting an area, and they strike surfaces at all angles. Every square meter of the walls and ceiling of a room at sea level, as well as the floor, is subjected to the same rate of molecular bombardments. The momentum change per second $\Delta\vec{p}/\Delta t$ of all the molecules colliding with one square meter is equal in magnitude to the force of 1×10^5 N.

Checkpoint 11 Do the calculation and verify that the height of a constant-density atmosphere would be about 8000 m (8 km).

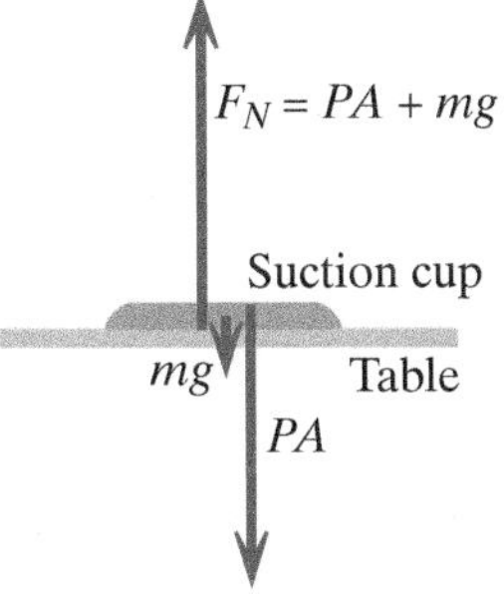

Figure 4.47 The weight of the suction cup is very small compared to the downward force by the air or the normal force by the table.

Suction

A free-body diagram for a suction cup looks like Figure 4.47, where PA represents the force of the air on the area A of the top surface of the suction cup. The normal force exerted upward by the table and the downward force of the air are huge compared to the weight of the suction cup mg.

The normal force $F_N = PA + mg \approx PA$, since the force of the air is so much larger than the weight mg of the suction cup. For example, a rubber suction cup with a diameter of 4 cm has an area of $\pi(0.02\,\text{m})^2 = 1.3 \times 10^{-3}\,\text{m}^2$, so the force of the air on the suction cup is

$$PA = (1 \times 10^5\,\text{N/m}^2)(1.3 \times 10^{-3}\,\text{m}^2) = 130\,\text{N}$$

whereas its own weight mg might be about 0.01 kg (10 g) times 9.8 N/kg, or only about 0.1 N. Note that a force of 130 N is the weight of an object whose mass is about 13 kg, or about 29 lb, which is why it isn't easy to pull even a small suction cup off a flat surface.

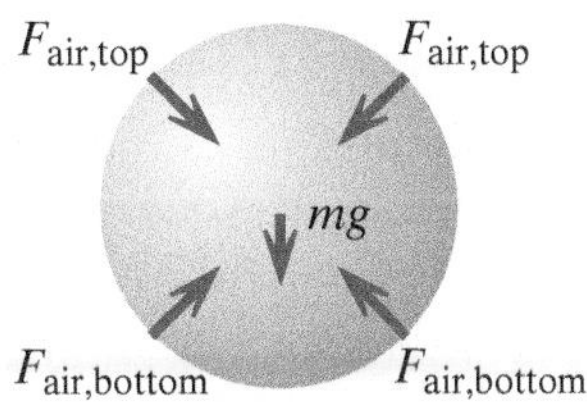

Figure 4.48 A ball hanging in air is subject to large compression forces due to the surrounding air.

Every solid object in air, such as a hanging ball, is subjected to large compression forces by the pressure of the surrounding air (Figure 4.48). The upward buoyant force is the tiny difference between these huge forces. A hollow object must be very strong to support this compression, if air is removed from the inside. There is no problem, however, if air can flow into and out of the container, because then it is just the solid walls that are compressed, since the air exerts comparable pressure on the inner and outer surfaces.

All of this is even more relevant under water, because the density of water is about 1000 times greater than the density of air, and the compression forces are about 1000 times as strong. A hollow submersible vehicle that dives very deep must have an extremely strong hull. When a scuba diver descends to greater depth where the water pressure is larger, the breathing apparatus automatically increases the pressure of the air supplied to the diver to equal the increased pressure of the outside water at the new depth, to prevent the lungs from being crushed.

4.15 *ACCELERATION

In previous studies you may have seen a simplified, approximate form of the Momentum Principle, written $F = ma$. This equation is valid under the following conditions:

- Only one force acts on the system.
- Motion occurs only in one dimension.
- The mass of the system is constant.
- The speed of the system is much less than the speed of light.

This equation can be derived from the derivative form of the Momentum Principle, using the low speed approximation $\vec{p} \approx m\vec{v}$. If we take the approximate derivative of $\vec{p}$ with respect to time, we get:

$$\frac{d\vec{p}}{dt} \approx \frac{d(m\vec{v})}{dt} \quad \text{(assuming } v \ll c\text{)}$$

$$\frac{d(m\vec{v})}{dt} = \left(\frac{dm}{dt}\right)\vec{v} + m\frac{d\vec{v}}{dt} \quad \text{(product rule for derivatives)}$$

$$\frac{d(m\vec{v})}{dt} = 0 + m\frac{d\vec{v}}{dt} \quad \text{(assuming system mass is constant)}$$

As noted in Chapter 1, the rate of change of velocity with respect to time is called acceleration, and is usually denoted by the symbol $\vec{a}$. Given this, we can write the approximate rate of change of momentum as

$$\frac{d\vec{p}}{dt} \approx m\vec{a}$$

and an approximate form of the Momentum Principle, usually referred to as Newton's second law, as

$$m\vec{a} \approx \vec{F}_{\text{net}} \quad \text{(nonrelativistic form; constant mass)}$$

A highly simplified scalar form of this equation, $F = ma$, involves a single force and one-dimensional motion. Although $F = ma$ is casually referred to as Newton's second law, Newton himself worked with the Momentum Principle in the forms emphasized in this textbook, $d\vec{p}/dt = \vec{F}_{\text{net}}$ and $\Delta\vec{p} = \vec{F}_{\text{net}}\Delta t$.

The definition of acceleration is useful, because it is occasionally important to know the acceleration of an object (the rate at which its velocity is changing). However, we will generally find that the full form of the Momentum Principle, $d\vec{p}/dt = \vec{F}_{\text{net}}$, is more appropriate for our analyses.

- $d\vec{p}/dt = \vec{F}_{\text{net}}$ is relativistically correct (correct at any speed), and we will sometimes deal with fast-moving particles.
- $d\vec{p}/dt = \vec{F}_{\text{net}}$ involves momentum, which is a conserved quantity; we need to work with momentum when we analyze situations such as collisions, as we did in Chapter 3.
- $d\vec{p}/dt = \vec{F}_{\text{net}}$ is correct even if the mass of an object is not constant, such as a rocket including its exhaust gases ejecting out the back.
- $d\vec{p}/dt = \vec{F}_{\text{net}}$ is a vector principle, and contains information about directions. The arrows over the symbols are extremely important; they remind us that there are really three separate component equations, for x, y, and z.
- $d\vec{p}/dt = \vec{F}_{\text{net}}$ reminds us that we have to add up all vector forces to give the net force, so we write $\vec{F}_{\text{net}}$, not just F.

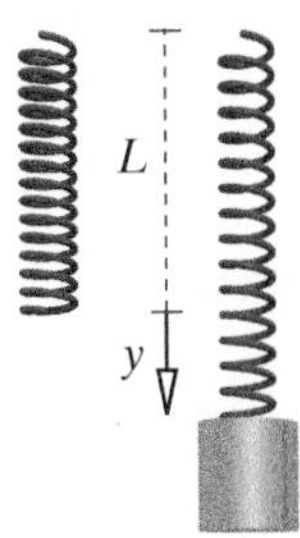

Figure 4.49 A vertical spring–mass system minimizes sliding friction.

4.16 *A VERTICAL SPRING–MASS SYSTEM

To reduce friction, it is easier to experiment with a mass oscillating vertically up and down on a spring (Figure 4.49) rather than sliding on an air table. We will show that we expect the same period whether the spring is vertical or horizontal.

We choose a coordinate system with the positive y axis pointing downward (Figure 4.49). We'll measure y downward from the location at which the end of the relaxed spring would be (the relaxed length of the spring is L).

Neglecting air resistance and gravitational interactions with objects other than the Earth, the free-body diagram of the forces acting on the hanging block is shown in Figure 4.50.

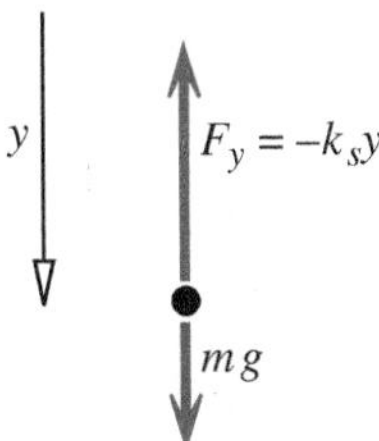

Figure 4.50 Free-body diagram for the hanging block.

We then have the following:

$$\frac{dp_y}{dt} = -k_s y + mg$$

We choose to analyze only simple vertical motion, with no bouncing side to side, so we know this:

$$p_x = 0 \text{ at all times, so } \frac{dp_x}{dt} = 0$$

We can reduce the p_y equation to a familiar form if we measure a new coordinate, w, from the equilibrium position, when the block is hanging motionless, rather than from the unstretched spring position (Figure 4.51). When the mass hangs motionless from the spring, the spring is stretched an amount $k_s s_0 = mg$, so $s_0 = mg/k_s$. We'll measure w downward from the end of the spring, so

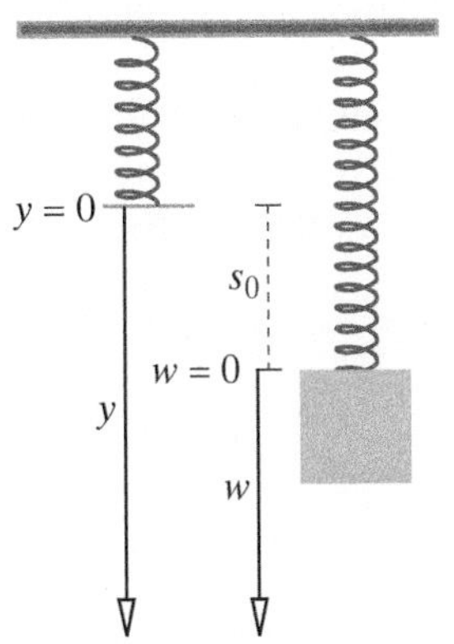

Figure 4.51 Measure w from the place where the block can hang motionless.

$$y = (s_0 + w) = \frac{mg}{k_s} + w$$

and therefore

$$\frac{dp_y}{dt} = -k_s\left(\frac{mg}{k_s} + w\right) + mg = -k_s w$$

This says that if we measure from the end of the spring at its equilibrium location, mg/k_s below the bottom of the unstretched spring, the equation for the motion of the mass is the same as in a horizontal spring–mass system, whose motion we already know. This means that the analytical solution we obtained for the horizontal case is also the solution for the vertical case (substituting y for x in the appropriate places).

The moving mass will oscillate up and down around this equilibrium position, as you can observe if you experiment with a vertical mass–spring system.

4.17 *GENERAL SOLUTION FOR THE MASS–SPRING SYSTEM

It is easy to show that the function $x = A\cos(\omega t + \phi)$ is a general solution for a spring–mass system with arbitrary k_s, m, and initial conditions, where $\omega = \sqrt{k_s/m}$, and where the amplitude A and the phase shift ϕ (measured in radians and denoted by the Greek letter phi) are constants determined by the initial conditions, as is explained below. The phase shift essentially shifts the starting point of the oscillation. For example, if $\phi = \pi/2$, the cosine function actually represents a negative sine function (Figure 4.52):

$$x = A\cos\left(\omega t + \frac{\pi}{2}\right) = -A\sin(\omega t)$$

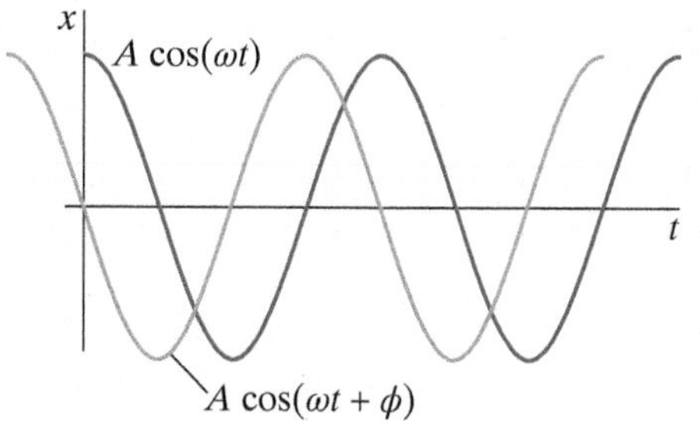

Figure 4.52 Phase shift ϕ.

Determining the Amplitude and Phase Shift

The amplitude A and phase shift ϕ can be determined from the initial values at time t_0 of position x_0 and velocity v_0:

$$x_0 = A\cos\left(\sqrt{\frac{k_s}{m}}t_0 + \phi\right)$$

$$v_0 = -\sqrt{\frac{k_s}{m}}A\sin\left(\sqrt{\frac{k_s}{m}}t_0 + \phi\right)$$

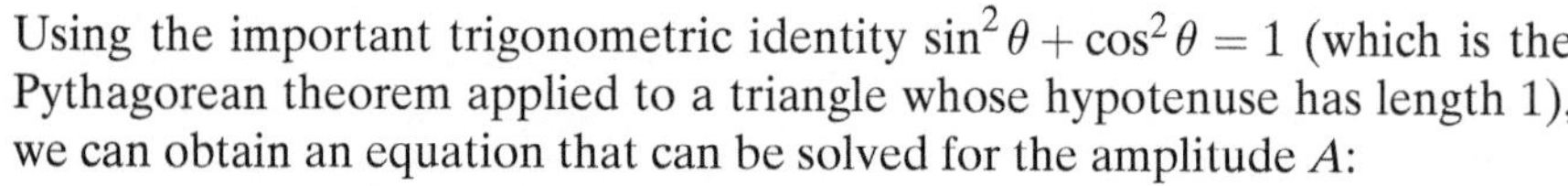

Using the important trigonometric identity $\sin^2\theta + \cos^2\theta = 1$ (which is the Pythagorean theorem applied to a triangle whose hypotenuse has length 1), we can obtain an equation that can be solved for the amplitude A:

$$x_0^2 + \frac{m}{k_s}v_0^2 = A^2, \text{ so we have } A = \sqrt{x_0^2 + \frac{m}{k_s}v_0^2}$$

For example, if the initial velocity is zero, the amplitude is simply equal to the initial stretch.

The value for A can be plugged back into either the x_0 equation or the v_0 equation to obtain the phase shift ϕ. Alternatively, by dividing the x_0 equation by the v_0 equation we obtain an equation that can be solved for the phase shift ϕ:

$$\frac{v_0}{x_0} = -\sqrt{\frac{k_s}{m}}\tan\left(\sqrt{\frac{k_s}{m}}t_0 + \phi\right)$$

$$\sqrt{\frac{k_s}{m}}t_0 + \phi = \arctan\left(-\sqrt{\frac{m}{k_s}}\frac{v_0}{x_0}\right)$$

$$\phi = \arctan\left(-\sqrt{\frac{m}{k_s}}\frac{v_0}{x_0}\right) - \sqrt{\frac{k_s}{m}}t_0$$

Checkpoint 12 Suppose that the system is not oscillating and you strike it with a hammer, applying a large force F for a *very* short time Δt. What is the initial speed v_0? Explain briefly. What will be the amplitude (the maximum displacement away from the equilibrium position)? What is the phase shift ϕ? (Does this make sense?)

Comment: Two Constants for a Second-Order Differential Equation

Note that whether we specify the starting conditions in terms of x_0 and v_0 or in terms of the amplitude A and the phase shift ϕ, two constants are required. You can see why two constants are needed by looking at the start of a numerical integration. You need an initial position x_0 in order to be able to calculate the initial force and impulse, so that you can use the Momentum Principle to update the initial momentum p_0, from which you get the new velocity that lets you update the position; then repeat.

We have been working with what is called a "second-order differential equation." That is, the highest derivative in the equation is a second derivative—the acceleration $dv/dt = d(dx/dt)/dt$. A second-order differential equation normally requires two constants to specify the starting conditions completely.

Linear Systems

We saw that doubling the amplitude of a spring–mass system simply multiplies the solution by a factor of two without changing the time dependence of the motion. This is an extremely important property of systems described by "linear" differential equations (equations in which the variables appear only to the first power, not squared, for example). More formally, in the Momentum Principle for a spring–mass system, let x be replaced by the quantity Rx, where R is a constant:

$$m\frac{d}{dt}\left(\frac{d(Rx)}{dt}\right) = -k_s(Rx)$$

$$Rm\frac{d}{dt}\left(\frac{d(x)}{dt}\right) = -Rk_s(x)$$

The R's cancel, leaving us with the original equation. Therefore starting with a different initial stretch will simply scale up the vertical axis of the graph of x vs. t. Also, since the motion is periodic, a change in the initial velocity corresponds to starting the motion at a different time. For example, instead of starting with an initial stretch and zero velocity, you could start with zero stretch and some initial velocity, and this would correspond to a point on the original graph where the function crosses the time axis.

In contrast, consider a differential equation in which there is an x^3, and try scaling by R:

$$m\frac{d}{dt}\left(\frac{d(Rx)}{dt}\right) = -b(Rx)^3$$

$$Rm\frac{d}{dt}\left(\frac{d(x)}{dt}\right) = -R^3bx^3$$

In this case the R's do *not* cancel. The solutions to this "nonlinear" equation do not scale up in the simple way that solutions to a linear equation do.

SUMMARY

Contact forces:

- A solid object may be modeled as a 3D lattice of balls (atoms) connected by springs (interatomic bonds).
- Based on this model, and given the density and atomic weight of an element, one can find the length of an interatomic bond in a solid metal.
- From macroscopic measurements of stress and strain, one can determine the stiffness of an interatomic bond in a solid material.
- Solid objects exert contact forces because the interatomic bonds in the solid are stretched (a tension force) or compressed (a normal force).
- Friction forces arise when one object slides on another.
- The speed of sound in a solid is the speed at which a disturbance propagates through the material.

Definitions:

For a long object like a rod, which is stretched or compressed:

$$\text{strain} = \frac{\Delta L}{L}$$
$$\text{stress} = \frac{F_T}{A}$$

L is the original length of the wire, ΔL is the change in length, F_T is the force applied to stretch or compress the object, and A is the cross-sectional area of the object.

$$\text{Young's modulus: } Y = \frac{\text{stress}}{\text{strain}} = \frac{\left(\frac{F_T}{A}\right)}{\left(\frac{\Delta L}{L}\right)}$$

Macro/micro connections:
Young's modulus in terms of microscopic quantities is

$$Y = \frac{k_{s,i}}{d}$$

Y is Young's modulus, $k_{s,i}$ is the stiffness of the interatomic bond, and d is the length of the interatomic bond (distance between adjacent nuclei).

$$\text{Speed of sound in a solid: } v = \sqrt{\frac{k_{s,i}}{m_a}}\, d$$

$k_{s,i}$ is the stiffness of the interatomic bond, d is the length of the interatomic bond (distance between adjacent nuclei), m_a is the mass of one atom.

Momentum Principle: derivative form

$$\frac{d\vec{p}}{dt} = \vec{F}_{\text{net}}$$

The instantaneous time rate of change of the momentum of an object is equal to the net force acting on the object.

A simplified version of the Momentum Principle is $\vec{F}_{\text{net}} = m\vec{a}$, valid for $v \ll c$ and constant mass.

Analytical solution for mass–spring oscillations

$$x = A\cos(\omega t), \text{ with } \omega = \sqrt{\frac{k_s}{m}}$$

assuming that the spring's mass is negligible, and there is no friction and no air resistance.

A (the "amplitude") is a constant equal to the maximum stretch of the spring during an oscillation. ω is the angular frequency, and its units are radians/s. k_s is the spring stiffness, and m is the mass of the object.

The period T is the time for one complete cycle, in seconds (Figure 4.53).

$$T = \frac{2\pi}{\omega}$$

The frequency f is the number of complete cycles per second.

$$f = \frac{1}{T} = \frac{\omega}{2\pi}$$

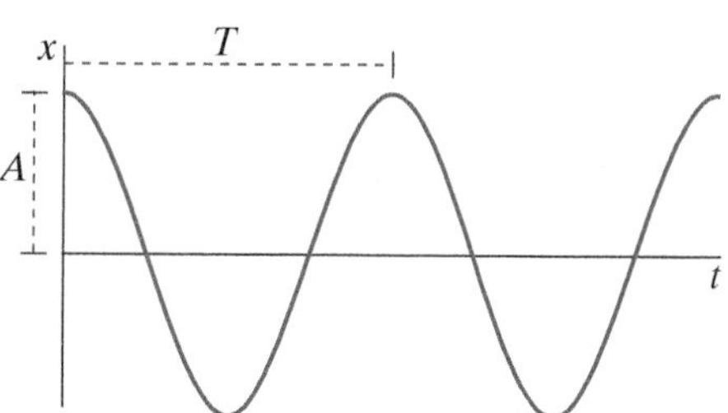

Figure 4.53

Contact forces due to gases
Upward buoyancy force = $m_{\text{fluid}}g$, where m_{fluid} is the mass of an equivalent volume of fluid.

Pressure is force per unit area; air pressure at sea level is about 1×10^5 N/m^2.

QUESTIONS

Q1 Why are two balls connected by a spring a good model for two atoms connected by a chemical bond? **(a)** If the two atoms get farther apart than the equilibrium interatomic distance, they attract each other. **(b)** Each spring represents a real microscopic coiled metal wire that connects two adjacent atoms. If the two atoms get closer together than the equilibrium interatomic distance, they repel each other. **(c)** The magnitude of the force one atom exerts on another is proportional to the stretch or compression of the bond between them.

Q2 Approximately what is the radius of a copper atom? Is it 1×10^{-15} m, 1×10^{-12} m, 1×10^{-10} m, 1×10^{-8} m, or 1×10^{-6} m?

Q3 **(a)** A climber whose mass is 55 kg hangs motionless from a rope. What is the tension in the rope? **(b)** Later, a different climber whose mass is 88 kg hangs from the same rope. Now what is the tension in the rope? **(c)** Compare the physical state of the rope when it supports the heavier climber to the state of the rope when it supports the lighter climber. Which statements about the physical state of the rope are true? Check all that apply. (1) Because the same rope is used, the tension in the rope must be the same in both cases. (2) The interatomic bonds in the rope are stretched more when the rope supports the heavier climber than when the rope supports the lighter climber. (3) The rope is slightly longer when it supports the heavier climber than when it supports the lighter climber.

Q4 You hang a 10 kg mass from a copper wire, and the wire stretches by 8 mm. **(a)** If you suspend the same mass from two copper wires, identical to the original wire, what happens? **(b)** If you suspend the same mass from a copper wire with half the cross-sectional area but the same length as the original wire, what happens? **(c)** If you suspend the same mass from a copper wire with the same cross-sectional area but twice the length of the original wire, what happens?

Q5 You hang a mass M from a spring, which stretches an amount s_1. Then you cut the spring in half, and hang a mass M from one half. How much does the half-spring stretch? **(a)** $2s_1$, **(b)** s_1, **(c)** $s_1/2$

Q6 A spring has stiffness k_s. You cut the spring in half. What is the stiffness of the half-spring? **(a)** $2k_s$, **(b)** k_s, **(c)** $k_s/2$

Q7 Lead is much softer than aluminum, and can be more easily deformed or pulled into a wire. What difference between the two materials best explains this? **(a)** Pb and Al atoms have different sizes. **(b)** Pb and Al atoms have different masses. **(c)** The stiffness of the interatomic bonds is different in Pb and Al.

Q8 Two wires are made of the same kind of metal. Wire A has a diameter of 2.4 mm and is initially 2.8 m long. You hang a 8 kg mass from wire A, measure the amount of stretch, and determine Young's modulus to be $Y_A = 1.37 \times 10^{12}\,\text{N/m}^2$.

Wire B, which is made of the same kind of metal as wire A, has the same length as wire A but twice the diameter. You hang the same 8 kg mass from wire B, measure the amount of stretch, and determine Young's modulus, Y_B.

Which one of the following is true? **1.** $Y_B = Y_A$, **2.** $Y_B > Y_A$, or **3.** $Y_B < Y_A$

Q9 Two wires with equal lengths are made of pure copper. The diameter of wire A is twice the diameter of wire B. When 6 kg masses are hung on the wires, wire B stretches more than wire A. You make careful measurements and compute Young's modulus for both wires. What do you find? **(a)** $Y_A > Y_B$, **(b)** $Y_A = Y_B$, **(c)** $Y_A < Y_B$

Q10 Suppose you attempt to pick up a very heavy object. Before you tried to pick it up, the object was sitting still—its momentum was not changing. You pull very hard, but do not succeed in moving the object. Is this a violation of the Momentum Principle? How can you be exerting a large force on the object without causing a change in its momentum? What does change when you apply this force?

Q11 **(a)** In outer space, a rod is pushed to the right by a constant force F (Figure 4.54). Describe the pattern of interatomic distances along the rod. Include a specific comparison of the situation at locations A, B, and C. Explain briefly in terms of fundamental principles.

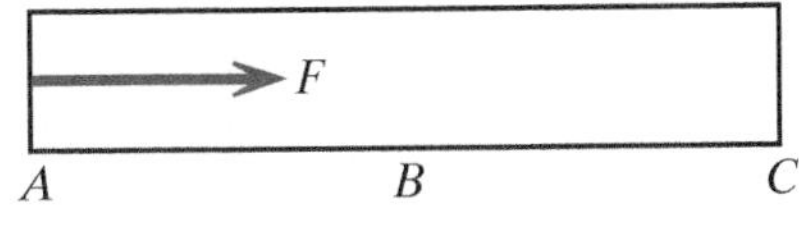

Figure 4.54

Hint: Consider the motion of an individual atom inside the rod, and various locations along the rod.

(b) After the rod in part (a) reaches a speed v, the object that had been exerting the force on the rod is removed. Describe the subsequent motion of the rod and the pattern of interatomic distances inside the rod. Include a specific comparison of the situation at locations A, B, and C. Explain briefly.

Q12 Bob is pushing a box across the floor at a constant speed of 1 m/s, applying a horizontal force whose magnitude is 20 N. Alice is pushing an identical box across the floor at a constant speed of 2 m/s, applying a horizontal force. **(a)** What is the magnitude of the force that Alice is applying to the box? **(b)** With the two boxes starting from rest, explain qualitatively what Alice and Bob did to get their boxes moving at different constant speeds.

Q13 In a spring–mass oscillator, when is the magnitude of momentum of the mass largest: when the magnitude of the net force acting on the mass is largest, or when the magnitude of the net force acting on the mass is smallest?

Q14 For a vertical spring–mass oscillator that is moving up and down, which of the following statements are true? (More than one statement may be true.) **(a)** At the lowest point in the oscillation, the momentum is zero. **(b)** At the lowest point in the oscillation, the rate of change of the momentum is zero. **(c)** At the lowest point in the oscillation, $mg = k_s s$. **(d)** At the lowest point in the oscillation, $mg > k_s s$. **(e)** At the lowest point in the oscillation, $mg < k_s s$.

Q15 The period of a particular spring–mass oscillator is 1 s when the amplitude is 5 cm. **(a)** What would be the period if we doubled the mass? **(b)** What would be the period if we replaced the original spring with a spring that is twice as stiff (keeping the original mass)? **(c)** What would be the period if we cut the original spring in half and use just one of the pieces (keeping the original mass)? **(d)** What would be the period if we increased the amplitude of the original system to 10 cm, so that the total distance traveled in one period is twice as large? **(e)** What would be the period if we took the original system to a massive planet where $g = 25$ N/kg?

Q16 How should you start the system going at $t = 0$ in order for the motion to be $A\cos(\omega t)$? How should you start the system going at $t = 0$ in order for the motion to be $A\sin(\omega t)$?

Q17 Describe two examples of oscillating systems that are *not* harmonic oscillators.

Q18 Two rods are both made of pure titanium. The diameter of rod A is twice the diameter of rod B, but the lengths of the rods are equal. You tap on one end of each rod with a hammer and measure how long it takes the disturbance to travel to the other end of the rod. In which rod did it take longer? **(a)** Rod A, **(b)** Rod B, **(c)** The times were equal.

Q19 A particular spring–mass oscillator oscillates with period T. Write out the general equation for the period of such an oscillator to use as a guide when answering the following questions. **(a)** If you double the mass but keep the stiffness the same, by what numerical factor does the period change? (That is, if the original period was T and the new period is bT, what is b?) **(b)** If, instead, you double the spring stiffness but keep the mass the same, what is the factor b? **(c)** If, instead, you double the mass and also double the spring stiffness, what is the factor b? **(d)** If, instead, you double the amplitude (keeping the original mass and spring stiffness), what is the factor b?

Q20 Uranium-238 (U^{238}) has three more neutrons than uranium-235 (U^{235}). Compared to the speed of sound in a bar of U^{235}, is the speed of sound in a bar of U^{238} higher, lower, or the same? Explain your choice, including justification for assumptions you make.

PROBLEMS

Section 4.4

•**P21** If in a certain material whose atoms are in a cubic array the interatomic distance is 1.7×10^{-10} m and the mass of one atom is 8.2×10^{-26} kg, what would be the density of this material?

•**P22** A block of one mole of a certain material whose atoms are in a cubic array has dimensions of 5 cm by 4 cm by 0.5 cm. What is the interatomic distance?

•**P23** The diameter of a copper atom is approximately 2.28×10^{-10} m. The mass of one mole of copper is 64 g. Assume that the atoms are arranged in a simple cubic array. Remember to convert to SI units. **(a)** What is the mass of one copper atom, in kg? **(b)** How many copper atoms are there in a cubical block of copper that is 4.6 cm on each side? **(c)** What is the mass of the cubical block of copper, in kg?

•**P24** One mole of tungsten (6.02×10^{23} atoms) has a mass of 184 g, as shown in the periodic table on the inside front cover of the textbook. The density of tungsten is $19.3\ \text{g/cm}^3$. What is the approximate diameter of a tungsten atom (length of a bond) in a solid block of the material? Make the simplifying assumption that the atoms are arranged in a cubic array.

Section 4.5

•**P25** If a chain of 50 identical short springs linked end to end has a stiffness of 270 N/m, what is the stiffness of one short spring?

•**P26** A certain spring has stiffness 190 N/m. The spring is then cut into two equal lengths. What is the stiffness of one of these half-length springs?

•**P27** Forty-five identical springs are placed side by side (in parallel) and connected to a large massive block. The stiffness of the 45-spring combination is 20,250 N/m. What is the stiffness of one of the individual springs?

•**P28** A certain spring has stiffness 140 N/m. The spring is then cut into two equal lengths. What is the stiffness of one of these half-length springs?

•**P29** Five identical springs, each with stiffness 390 N/m, are attached in parallel (that is, side by side) to hold up a heavy weight. If these springs were replaced by an equivalent single spring, what should be the stiffness of this single spring?

••**P30** A hanging titanium wire with diameter 2 mm (2×10^{-3} m) is initially 3 m long. When a 5 kg mass is hung from it, the wire stretches an amount 0.4035 mm, and when a 10 kg mass is hung from it, the wire stretches an amount 0.807 mm. A mole of titanium has a mass of 48 grams, and its density is $4.51\ \text{grams/cm}^3$. Find the approximate value of the effective spring stiffness of the interatomic force, and explain your analysis.

••**P31** A hanging copper wire with diameter 1.4 mm (1.4×10^{-3} m) is initially 0.95 m long. When a 36 kg mass is hung from it, the wire stretches an amount 1.83 mm, and when a 72 kg mass is hung from it, the wire stretches an amount 3.66 mm. A mole of copper has a mass of 63 g, and its density is $9\ \text{g/cm}^3$. Find the approximate value of the effective spring stiffness of the interatomic force.

••**P32** One mole of tungsten (6.02×10^{23} atoms) has a mass of 184 g, and its density is $19.3\ \text{g/cm}^3$, so the center-to-center distance between atoms is 2.51×10^{-10} m. You have a long thin bar of tungsten, 2.5 m long, with a square cross section, 0.15 cm on a side. You hang the rod vertically and attach a 415 kg mass to the bottom, and you observe that the bar becomes 1.26 cm longer. From these measurements, it is possible to determine the stiffness of one interatomic bond in tungsten. **(a)** What is the spring stiffness of the entire wire, considered as a single macroscopic (large scale), very stiff spring? **(b)** How many side-by-side atomic chains (long springs) are there in this wire (Figure 4.55)? This is the same as the number of atoms on the bottom surface of the tungsten wire. Note that the cross-sectional area of one tungsten atom is $(2.51 \times 10^{-10})^2\ \text{m}^2$.

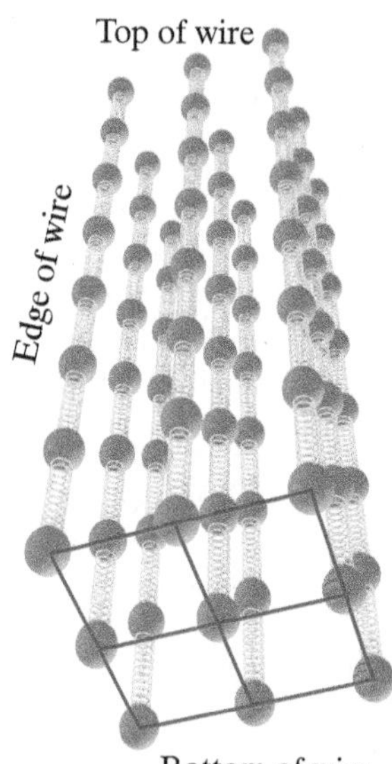

Figure 4.55

(c) How many interatomic bonds are there in one atomic chain running the length of the wire? **(d)** What is the stiffness of a single interatomic "spring"?

••**P33** A hanging iron wire with diameter 0.08 cm is initially 2.5 m long. When a 52 kg mass is hung from it, the wire stretches an amount 1.27 cm. A mole of iron has a mass of 56 g, and its density is $7.87\ \text{g/cm}^3$. **(a)** What is the length of an interatomic bond in iron (diameter of one atom)? **(b)** Find the approximate value of the effective spring stiffness of one interatomic bond in iron.

Section 4.6

•**P34** Steel is very stiff, and Young's modulus for steel is unusually large, $2 \times 10^{11}\ \text{N/m}^2$. A cube of steel 28 cm on a side supports a load of 85 kg that has the same horizontal cross section as the steel cube. **(a)** What is the magnitude of the normal force that the steel cube exerts on the load? **(b)** What is the compression of the steel cube? That is, what is the small change in height of the steel cube due to the load it supports? Give your answer as a positive number. The compression of a wide, stiff support can be extremely small.

•**P35** Suppose that you are going to measure Young's modulus for three rods by measuring their stretch when they are suspended vertically and weights are hung from them (Figure 4.56).

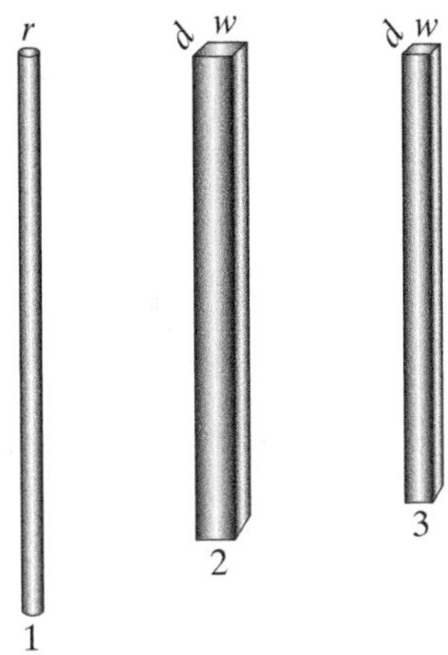

Figure 4.56

Rod 1 is 2.7 m long and cylindrical with radius 4 mm (1 mm is 0.001 m). Rod 2 is 3.2 m long by 12 mm wide by 6 mm deep. Rod 3 is 3 m long by 6 mm wide by 6 mm deep. The definition of Young's modulus, $Y = (F/A)/(\Delta L/L)$, includes the quantity A, the cross-sectional area. **(a)** What is the cross-sectional area of rod 1? **(b)** What is the cross-sectional area of rod 2? **(c)** What is the cross-sectional area of rod 3?

••**P36** Young's modulus for aluminum is 6.2×10^{10} N/m^2. The density of aluminum is 2.7 g/cm^3, and the mass of one mole (6.02×10^{23} atoms) is 27 g. If we model the interactions of neighboring aluminum atoms as though they were connected by springs, determine the approximate spring constant of such a spring. Repeat this analysis for lead: Young's modulus for lead is 1.6×10^{10} N/m^2, the density of lead is 11.4 g/cm^3, and the mass of one mole is 207 g. Make a note of these results, which we will use for various purposes later on. Note that aluminum is a rather stiff material, whereas lead is quite soft.

••**P37** Suppose that we hang a heavy ball with a mass of 10 kg (about 22 lb) from a steel wire 3 m long that is 3 mm in diameter. Steel is very stiff, and Young's modulus for steel is unusually large, 2×10^{11} N/m^2. Calculate the stretch ΔL of the steel wire. This calculation shows why in many cases it is a very good approximation to pretend that the wire doesn't stretch at all ("ideal nonextensible wire").

••**P38** You hang a heavy ball with a mass of 14 kg from a gold wire 2.5 m long that is 2 mm in diameter. You measure the stretch of the wire, and find that the wire stretched 0.00139 m. **(a)** Calculate Young's modulus for the wire. **(b)** The atomic mass of gold is 197 g/mole, and the density of gold is 19.3 g/cm^3. Calculate the interatomic spring stiffness for gold.

••**P39** A hanging wire made of an alloy of iron with diameter 0.09 cm is initially 2.2 m long. When a 66 kg mass is hung from it, the wire stretches an amount 1.12 cm. A mole of iron has a mass of 56 g, and its density is 7.87 g/cm^3. Based on these experimental measurements, what is Young's modulus for this alloy of iron?

•••**P40** A certain coiled wire with uneven windings has the property that to stretch it an amount s from its relaxed length requires a force that is given by $F = bs^3$, so its behavior is different from a normal spring. You suspend this device vertically, and its unstretched length is 25 cm. **(a)** You hang a mass of 18 g from the device, and you observe that the length is now 29 cm. What is b, including units? **(b)** Which of the following were needed in your analysis in part (a)? (1) The Momentum Principle, (2) The fact that the gravitational force acting on an object near the Earth's surface is approximately mg, (3) The force law for an ordinary spring ($F = k_s s$), (4) The rate of change of momentum being zero **(c)** Next you take hold of the hanging 18 g mass and throw it straight downward, releasing it when the length of the device is 33 cm and the speed of the mass is 5 m/s. After a very short time, 0.0001 s later, what is the stretch of the device, and what was the change in the speed of the mass (including the correct sign of the change) during this short time interval? It helps enormously to draw a diagram showing the forces that act on the mass after it leaves your hand.

Section 4.7

••**P41** Two blocks of mass m_1 and m_3, connected by a rod of mass m_2, are sitting on a low-friction surface, and you push to the left on the right block (mass m_1) with a constant force of magnitude F (Figure 4.57).

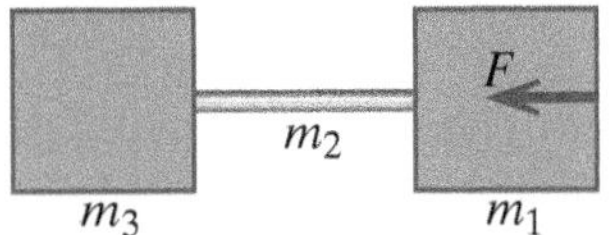

Figure 4.57

(a) What is the acceleration dv_x/dt of the blocks? **(b)** What is the vector force $\vec{F}_{\text{net}\,3}$ exerted by the rod on the block of mass m_3? ($|\vec{F}_{\text{net}\,3}|$ is approximately equal to the compression force in the rod near its left end.) What is the vector force $\vec{F}_{\text{net}\,1}$ exerted by the rod on the block of mass m_1? ($|\vec{F}_{\text{net}\,1}|$ is approximately equal to the compression force in the rod near its right end.) **(c)** Suppose that instead of pushing on the right block (mass m_1), you pull to the left on the left block (mass m_3) with a constant force of magnitude F. Draw a diagram illustrating this situation. Now what is the vector force $\vec{F}_{\text{net}\,3}$ exerted by the rod on the block of mass m_3?

Section 4.8

•**P42** A 5 kg box with initial speed 4 m/s slides across the floor and comes to a stop after 0.7 s. **(a)** What is the coefficient of kinetic friction? **(b)** How far does the box move? **(c)** You put a 3 kg block in the box, so the total mass is now 8 kg, and you launch this heavier box with an initial speed of 4 m/s. How long does it take to stop?

•**P43** A 3 kg block measures 5 cm by 10 cm by 20 cm. When it slides on a 10 cm by 20 cm face, it moves with constant speed when pulled horizontally by a force whose magnitude is 3 N. How big a horizontal force must be applied to pull it with constant speed if it slides on a 5 cm by 20 cm face?

•**P44** A 15 kg box sits on a table. The coefficient of static friction μ_s between table and box is 0.3, and the coefficient of kinetic friction μ_k is 0.2. **(a)** What is the force required to start the box moving? **(b)** What is the force required to keep it moving at constant speed? **(c)** What is the force required to maintain an acceleration of 2 m/s/s?

•**P45** A 20 kg box is being pushed across the floor by a constant force $\langle 90,0,0\rangle$ N. The coefficient of kinetic friction for the table and box is 0.25. At $t = 5$ s the box is at location $\langle 8,2,-1\rangle$ m, traveling with velocity $\langle 3,0,0\rangle$ m/s. What is its position and velocity at $t = 5.6$ s?

•**P46** You drag a block across a table while a friend presses down on the block. The coefficient of friction between the table and the block is 0.6. The vertical component of the force exerted by the table on the block is 190 N. How big is the horizontal component of the force exerted by the table on the block?

••P47 For this problem you will need measurements of the position vs. time of a block sliding on a table, starting with some initial velocity, slowing down, and coming to rest. If you do not have an appropriate laboratory setup for making these measurements, your instructor will provide you with such data. Analyze these data to determine the coefficient of friction and to see how well they support the assertion that the force of sliding friction is essentially independent of the speed of sliding.

•••P48 It is sometimes claimed that friction forces always slow an object down, but this is not true. If you place a box of mass 8 kg on a moving horizontal conveyor belt, the friction force of the belt acting on the bottom of the box speeds up the box. At first there is some slipping, until the speed of the box catches up to the speed of the belt, which is 5 m/s. The coefficient of kinetic friction between box and belt is 0.6. **(a)** How much time does it take for the box to reach this final speed? **(b)** What is the distance (relative to the floor) that the box moves before reaching the final speed of 5 m/s?

Section 4.10

•••P49 A chain of length L and mass M is suspended vertically by one end with the bottom end just above a table. The chain is released and falls, and the links do not rebound off the table, but they spread out so that the top link falls very nearly the full distance L. Just before the instant when the entire chain has fallen onto the table, how much force does the table exert on the chain? Assume that the chain links have negligible interaction with each other as the chain drops, and make the approximation that there is a very large number of links. *Hint*: Consider the instantaneous rate of change of momentum of the chain as the last link hits the table.

Section 4.11

•P50 A ball whose mass is 1.4 kg is suspended from a spring whose stiffness is 4 N/m. The ball oscillates up and down with an amplitude of 14 cm. **(a)** What is the angular frequency ω? **(b)** What is the frequency? **(c)** What is the period? **(d)** Suppose this apparatus were taken to the Moon, where the strength of the gravitational field is only 1/6 of that on Earth. What would be the period on the Moon? (Consider carefully how the period depends on properties of the system; look at the equation.)

•P51 A mass of 2.2 kg is connected to a horizontal spring whose stiffness is 8 N/m. When the spring is relaxed, $x = 0$. The spring is stretched so that the initial value of $x = +0.18$ m. The mass is released from rest at time $t = 0$. Remember that when the argument of a trigonometric function is in radians, on a calculator you have to switch the calculator to radians or convert the radians to degrees. Predict the position x when $t = 1.15$ s.

•P52 A bouncing ball is an example of an anharmonic oscillator. If you quadruple the maximum height, what happens to the period? (Assume that the ball keeps returning almost to the same height.)

••P53 Here on Earth you hang a mass from a vertical spring and start it oscillating with amplitude 1.7 cm. You observe that it takes 2.1 s to make one round-trip. You construct another vertical oscillator with a mass 6 times as heavy and a spring 10 times as stiff. You take it to a planet where $g_{\text{planet}} = 6.8$ N/kg. You start it oscillating with amplitude 3.3 cm. How long does it take for the mass to make one round-trip?

••P54 In the approximation that the Earth is a sphere of uniform density, it can be shown that the gravitational force it exerts on a mass m inside the Earth at a distance r from the center is $mg(r/R)$, where R is the radius of the Earth. (Note that at the surface, the force is indeed mg, and at the center it is zero). Suppose that there were a hole drilled along a diameter straight through the Earth, and the air were pumped out of the hole. If an object is released from one end of the hole, how long will it take to reach the other side of the Earth? Include a numerical result.

••P55 A spring suspended vertically is 18 cm long. When you suspend a 30 g weight from the spring, at rest, the spring is 22 cm long. Next you pull down on the weight so the spring is 23 cm long and you release the weight from rest. What is the period of oscillation?

••P56 It was found that a 20 g mass hanging from a particular spring had an oscillation period of 1.2 s. **(a)** When two 20 g masses are hung from this spring, what would you predict for the period in seconds? Explain briefly.

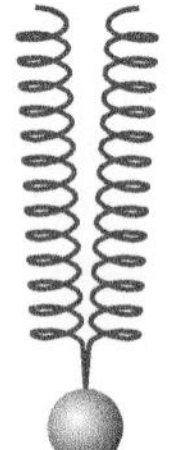

Figure 4.58

(b) When one 20 g mass is supported by two of these vertical, parallel springs (Figure 4.58), what would you predict for the period in seconds? Explain briefly. **(c)** Suppose that you cut one spring into two equal lengths, and you hang one 20 g mass from this half spring. What would you predict for the period in seconds? Explain briefly. **(d)** Suppose that you take a single (full-length) spring and a single 20 g mass to the Moon and watch the system oscillate vertically there. Will the period you observe on the Moon be longer, shorter, or the same as the period you measured on Earth? (The gravitational field strength on the Moon is about one-sixth that on the Earth.) Explain briefly.

••P57 A vertical mass–spring oscillator has an amplitude of 0.06 m and a period of 0.4 s. **(a)** What is the maximum speed of the mass? **(b)** What is the maximum acceleration of the mass?

••P58 In Problem P36 you can find the effective spring stiffness corresponding to the interatomic force for aluminum and lead. Let's assume for the moment that, *very* roughly, other atoms have similar values. **(a)** What is the (very) approximate frequency f for the vibration of H_2, a hydrogen molecule? **(b)** What is the (very) approximate frequency f for the vibration of O_2, an oxygen molecule? **(c)** What is the approximate vibration frequency f of the molecule D_2, both of whose atoms are deuterium atoms (that is, each nucleus has one proton and one neutron)? **(d)** Explain why the *ratio* of the deuterium frequency to the hydrogen frequency is quite accurate, even though you have estimated each of these quantities very approximately, and the effective spring stiffness is normally expected to be significantly different for different atoms. (*Hint:* What interaction is modeled by the effective "spring"?)

••**P59** Find a spring (or a rubber band) and one or more masses with which to study the motion of a mass hanging from a spring (Figure 4.59). **(a)** Measure the value of the spring stiffness k_s of the spring in N/m. The magnitude of a spring force is $k_s s$, where s is the stretch of the spring (change from the unstretched length). Explain briefly how you measured k_s. Include in your report the unstretched length of the spring. **(b)** Measure the period (the round-trip time) of a mass hanging from the vertical spring. Report the mass that you use, and the amplitude of the oscillation. Amplitude is the maximum displacement, plus or minus, from the equilibrium position (the position where the mass can hang motionless). **(c)** With twice the amplitude that you reported in part (b), measure the period again. Since the mass has to move twice as far, one might expect the period to lengthen. Does it?

Figure 4.59

There are unavoidable fluctuations in starting and stopping the timing, but you can minimize the error this contributes by timing many complete cycles so that the starting and stopping fluctuations are a small fraction of the total time measured.

It is good practice to count out loud *starting from zero, not one.* To count five cycles, say out loud "Zero, one, two, three, four, five." If on the other hand you say "One, two, three, four, five," you have actually counted only four cycles, not five.

Since the motion is periodic, you can start (say "zero") at any point in the motion. It is best to start and stop when the mass is moving fast past some marker near the equilibrium point, because it is difficult to estimate the exact time when the mass reaches the very top or the very bottom, because it is moving slowly at those turnaround points. Be sure to measure full round-trip cycles, not the half-cycles between returns to the equilibrium point (but going in the opposite direction).

•••**P60** An object of mass m is attached by two stretched springs (stiffness k_s and relaxed length L_0) to rigid walls, as shown in Figure 4.60. The springs are initially stretched by an amount $(L - L_0)$. When the object is displaced to the right and released, it oscillates horizontally. Starting from the Momentum Principle, find a function of the displacement x of the object and the time t that describes the oscillatory motion. **(a)** What is the period of the motion? **(b)** If L_0 were shorter (so that the springs are initially stretched more), would the period be larger, smaller, or the same?

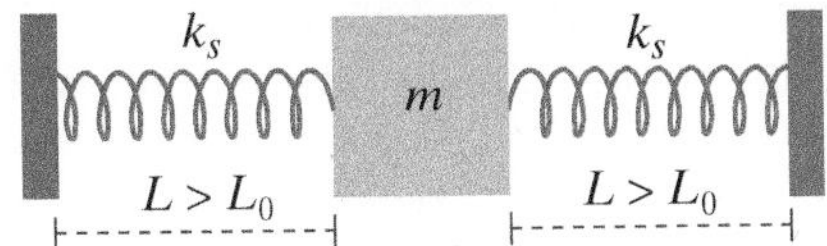

Figure 4.60

•••**P61** A simple pendulum (Figure 4.61) consists of a small mass of mass m swinging at the end of a low-mass string of length L (in contrast to a pendulum whose mass is distributed, such as a rod swinging from one end). **(a)** Show that the tangential momentum p of the small mass along the arc to the right (increasing θ) obeys the following equation, where s is the arc length $L\theta$:

$$\frac{dp}{dt} = -mg\sin\theta = -mg\sin\left(\frac{s}{L}\right)$$

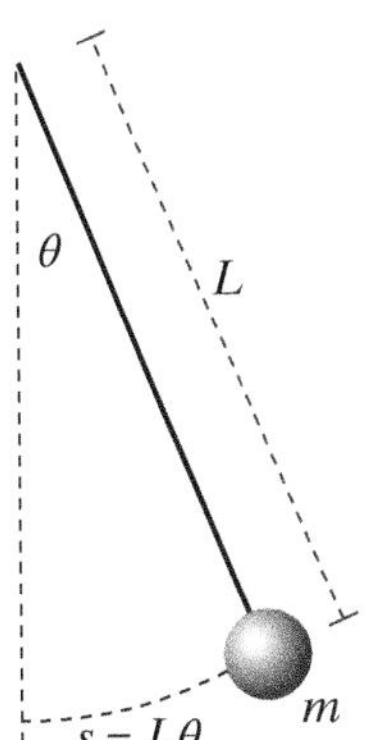

Figure 4.61

(b) What form does this equation take for small-amplitude swings? (Note that if θ is measured in radians, $\sin\theta \approx \theta$ for small angles. For example, even for an angle as large as $\theta = 30°$, $\theta = \pi/6 = 0.524$ radians, which is close to $\sin(30°) = 0.5$, and the approximation gets better for smaller angles.) **(c)** Compare the form of the approximate equation you obtained in part (b) with the form of the Momentum Principle for a spring–mass system. By comparing these equations for different systems, determine the period of a simple pendulum for small-amplitude swings. (Note that any time you can approximate the Momentum Principle for some system by an equation that looks like the equation for a mass on a spring, you know that the system will move approximately sinusoidally.) **(d)** Make a simple pendulum and predict its period, then measure the period. Do this for a long pendulum and for a short pendulum. Report your experimental data and results, with comparisons to your theoretical predictions.

Section 4.13

•**P62** Two metal rods are made of different elements. The interatomic spring stiffness of element A is three times larger than the interatomic spring stiffness for element B. The mass of an atom of element A is three times greater than the mass of an atom of element B. The atomic diameters are approximately the same for A and B. What is the ratio of the speed of sound in rod A to the speed of sound in rod B?

••**P63** You hang a heavy ball with a mass of 41 kg from a silver rod 2.6 m long by 1.5 mm by 3.1 mm. You measure the stretch of the rod, and find that the rod stretched 0.002898 m. Using these experimental data, what value of Young's modulus do you get? The atomic mass of silver is 108 g/mole, and the density of silver is 10.5 g/cm^3. Using this information along with the measured value of Young's modulus, calculate the speed of sound in silver.

••**P64** One mole of nickel (6.02×10^{23} atoms) has a mass of 59 g, and its density is 8.9 g/cm^3. You have a bar of nickel 2.5 m long, with a square cross section, 2 mm on a side. You hang the rod vertically and attach a 40 kg mass to the bottom, and you observe that the bar becomes 1.2 mm longer. Next you remove the 40 kg mass, place the rod horizontally, and strike one end with a hammer. How much time T will elapse before a microphone at the other end of the bar will detect a disturbance?

Section 4.14

•P65 It is hard to imagine that there can be enough air between a book and a table so that there is a net upward (buoyant) force on the book despite the large downward force on the top of the book. About how many air molecules are there between a textbook and a table, if there is an average distance of about 0.01 mm between the uneven surfaces of the book and table?

•P66 Calculate the buoyant force in air on a kilogram of lead (whose density is about 11 g/cm^3). The density of air is 1.3×10^{-3} g/cm^3. Remember that the buoyant force is equal to the weight of a volume of air that is equal to the volume of the object. (Compare with the weight mg of this much lead.)

•P67 Air pressure at the surface of a fresh water lake near sea level is about 1×10^5 N/m^2. At approximately what depth below the surface does a diver experience a pressure of 3×10^5 N/m^2? How would this be different in sea water, which has higher density than fresh water?

••P68 Here are two examples of floating objects: **(a)** A block of wood 20 cm long by 10 cm wide by 6 cm high has a density of 0.7 g/cm^3 and floats in water (whose density is 1.0 g/cm^3). How far below the surface of the water is the bottom of the block? Explain your reasoning. **(b)** An advertising blimp consists of a gas bag, the supporting structure for the gas bag, and the gondola hung from the bottom, with its cabin, engines, and propellers. The gas bag is about 30 m long with a diameter of about 10 m, and it is filled with helium (density about 4 g per 22.4 l under these conditions). Estimate the total mass of the blimp, including the helium.

COMPUTATIONAL PROBLEMS

More detailed and extended versions of some of these computational modeling problems may be found in the lab activities included in the *Matter & Interactions, 4th Edition*, resources for instructors.

••P69 Write an iterative computational model to predict and display the motion of a block of mass 60 g that sits on top of a vertical spring of stiffness 8 N/m and relaxed length 20 cm (Figure 4.62). **(a)** Use initial conditions that represent the block at rest on top of the spring, which has been pushed down until its total length is 10 cm. These values are the same as those used in an example in Chapter 2, Section 2.6, so you can use that example to check your work. **(b)** Add a graph of the y component of the block's position vs. time. What is the period of the block's motion? (It will be easiest to determine this if you stop the graph after two or three complete cycles. By holding down the mouse button while dragging over a VPython graph you will get crosshairs and a numerical readout of the coordinates.) **(c)** Double the stiffness of the spring. How does that affect the period of the motion? **(d)** Restore the stiffness of the spring to its original value, but double the mass. How does this affect the period of the motion?

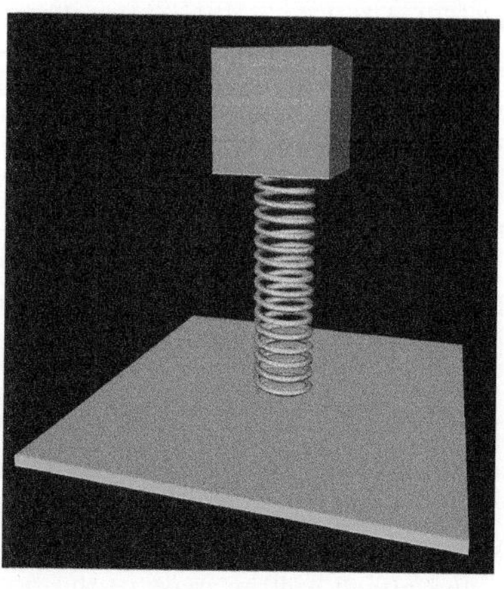

Figure 4.62

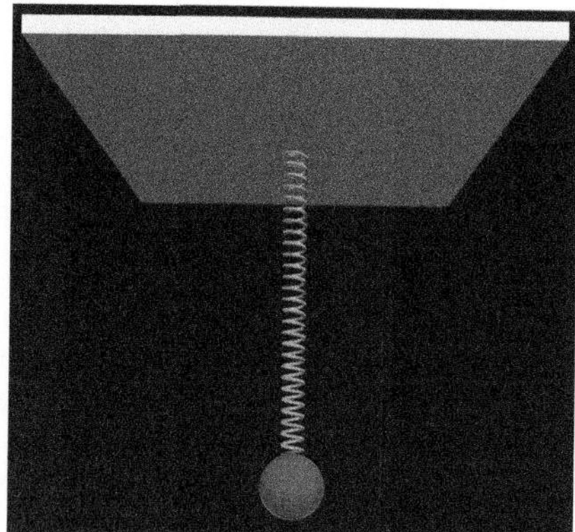

Figure 4.63

••P70 A ball of mass 20 g hangs from a spring whose relaxed length is 20 cm and whose stiffness is 0.9 N/m (Figure 4.63). **(a)** Write a VPython program to model the motion of this spring. **(b)** Add a graph of the y component of the ball's position vs. time. From the graph, determine the period of the ball's motion. (It will be easiest to determine this if you stop the graph after two or three complete cycles. By holding down the mouse button while dragging over a VPython graph you will get crosshairs and a numerical readout of the coordinates.) **(c)** How does doubling the mass affect the period of the oscillations? **(d)** How does keeping the mass unchanged but doubling the spring stiffness affect the period of the oscillations? **(e)** Make the ball leave a trail behind it as it moves. Find initial conditions that result in oscillations in all three dimensions. Rotate the display to make sure the motion is not planar.

••P71 Create a VPython program to model the motion of a 0.03 kg mass connected to two horizontal springs whose stiffness is 4 N/m, and whose relaxed length is 0.5 m, as shown in Figure 4.64. Neglect the effects of gravity (perhaps the system is in outer space). **(a)** Add a graph of x vs. t for the mass, and determine the period of the oscillator from this graph. It will be easiest to do this if you stop the graph after two or three complete cycles. If you hold down the mouse button while dragging over a VPython graph you will get crosshairs and a numerical readout of the coordinates. **(b)** Using the period you determined above, calculate the effective spring stiffness of this two-spring system. Is this the same as the spring stiffness you assigned to each spring in your model? Explain. **(c)** How does changing the amplitude of the oscillations affect the period?

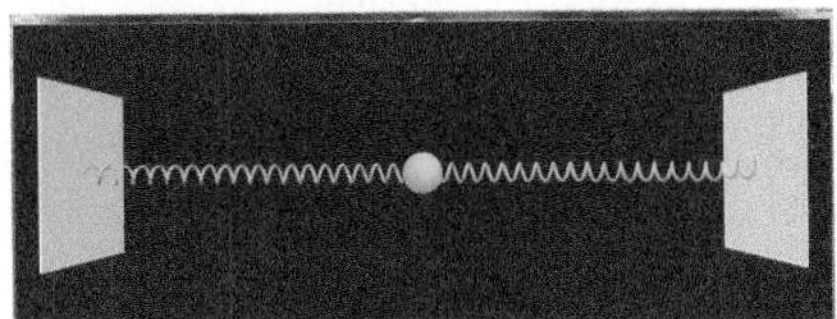

Figure 4.64

••**P72** On a space station in outer space a horizontal chain of two identical 3 kg masses and three identical springs of relaxed length 14 m and stiffness 50 N/m is suspended between two walls, as shown in Figure 4.65. Write a VPython program to model the motion of this system. **(a)** Start the motion by displacing the leftmost mass to the left, and observe the motion of the system. **(b)** Add a graph of x vs. t for each mass, making the graphs different colors so you can tell which is which. Is this system a simple harmonic oscillator? **(c)** Experiment with different initial conditions. Make a screen shot of a graph and report the initial conditions that led to this motion. **(d)** Find a set of initial conditions that results in simple sinusoidal motions of both masses. What are the periods of these motions? **(e)** (Optional) The pattern of motion you found in part (d) is called a "normal mode" of the system. This particular system has two such normal modes. Can you find initial conditions that produce the second? **(f)** (Optional) It is interesting to graph p.x for each mass as well as x. To do this you will need to make a separate gdisplay (graph window) for each graph. See the VPython help to learn how to do this. Explain how the plots of p.x and x are related.

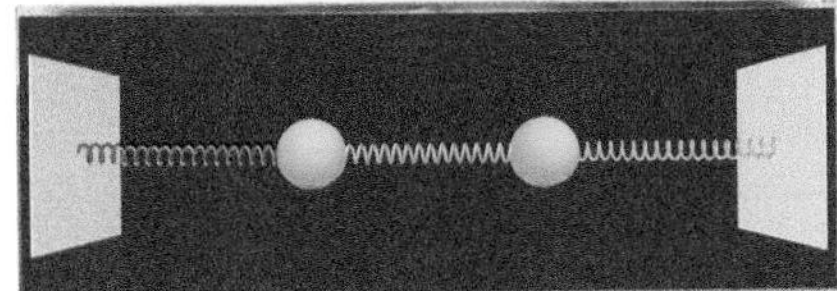

Figure 4.65

••**P73** Create a computational model of a system like the one shown in Figure 4.66, in which a ball of mass 30 g is connected by six identical springs of stiffness 30 N/m and relaxed length 50 cm to six fixed walls. **(a)** Find initial conditions that make the system oscillate only along the x axis. **(b)** Find initial conditions that make the system oscillate in such a way that the ball doesn't travel in a straight line. Have the ball leave a trail to check your results. **(c)** Is this system a simple harmonic oscillator? Explain how you determined this.

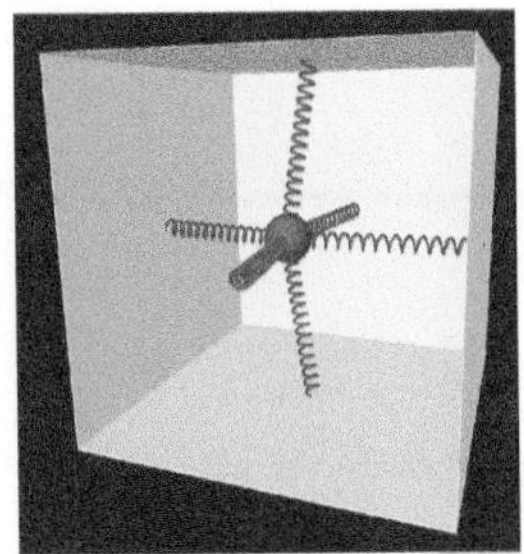

Figure 4.66

••**P74** An object of mass m slides with negligible friction on a table. The object is connected to a post by a spring whose stiffness is k_s and whose relaxed length is L_0. Make a computational model of the situation and investigate the various kinds of motions that result from choosing different initial conditions and different values of k_s and L_0. Leave a trail behind the moving object to help in visualizing the motion (Figure 4.67). The program that generated Figure 4.67 used $m = 0.2$ kg, $k_s = 2$ N/m, and $L_0 = 0.16$ m.

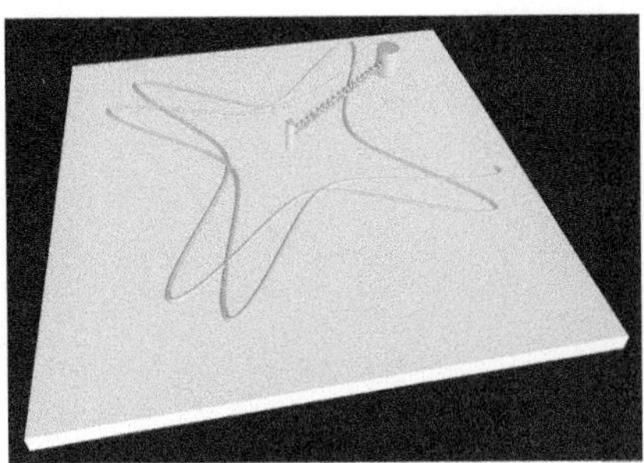

Figure 4.67

••**P75** A peculiar spring-like device exerts a force whose magnitude is $2 \times 10^9 \times s^7$ N, where s is the stretch in meters. The relaxed length is 0.1 m and the initial stretch is 0.02 m. The device is attached along the x axis to an object of mass 0.1 kg that is free to move with negligible friction on a horizontal surface. Write a program to model this device. A time step Δt of about 0.01 s is a reasonable choice. **(a)** Release the system from rest (the initial speed is zero) and display the motion as a function of time for the first 20 seconds. **(b)** Display on the same graph both the stretch s and the x component of velocity, using different colors. **(c)** How does the motion differ from that of an ordinary spring–mass system? **(d)** How does changing the initial stretch to 0.03 m affect the period of the system? Is this a harmonic oscillator?

•••**P76** As explained in Problem P61, a simple pendulum consisting of a small mass of mass m swinging at the end of a low-mass string of length L behaves like a harmonic oscillator when the amplitude of the swings is small. For large-amplitude swings, the small-angle approximation discussed in Problem P61 is not valid, but you can use an iterative model to calculate the motion. For a mass on a string, the largest possible amplitude is 90°, but if the string is replaced by a lightweight rod, the amplitude can be as large as 180° (standing upside-down, with the mass at the top). It turns out that such a pendulum can be modeled by the same equation as the equation derived in part (a) of Problem P61, but the concepts of torque and angular momentum are required to prove it. **(a)** Assume that the equation is valid, and write a computational model to predict the motion of the pendulum. Think of s and p as being like x and p_x so that your program is basically a one-dimensional calculation, with a force $-mg\sin(s/L)$ instead of $-k_s s$. **(b)** Plot the position s and momentum p as a function of time for a pendulum whose amplitude is nearly 180°. **(c)** How does the period of this system depend on the amplitude? Is it a harmonic oscillator?

ANSWERS TO CHECKPOINTS

1 2.6×10^{-10} m; 3.1×10^{-10} m
2 800 N/m
3 300 N/m
4 0.167 mm
5 (a) 0.0235, **(b)** 1.225×10^7 N/m^2, **(c)** 5.21×10^8 N/m^2
6 (a) The box would not move, **(b)** 60 N, **(c)** The box would accelerate, **(d)** 80 N
7 (a) $\langle 4, 0, 0 \rangle$ N, **(b)** Nothing. $d\vec{p}/dt$ gives no information about the instantaneous value of $\vec{p}$.
8 (a) 58.8 N/m, **(b)** 0.47 s
9 2970 m/s
10 $F_{\text{buoyant}} = 1.6 \times 10^{-3}$ N; $\dfrac{F_{\text{buoyant}}}{F_{\text{weight of iron}}} = 1.6 \times 10^{-4}$
12 $v_0 = \dfrac{F}{m}\Delta t$; $A = v_0\sqrt{\dfrac{m}{k}}$; 90° ($\pi/2$ radians), corresponding to a sine rather than a cosine (start at $x = 0$).

CHAPTER

5

Determining Forces from Motion

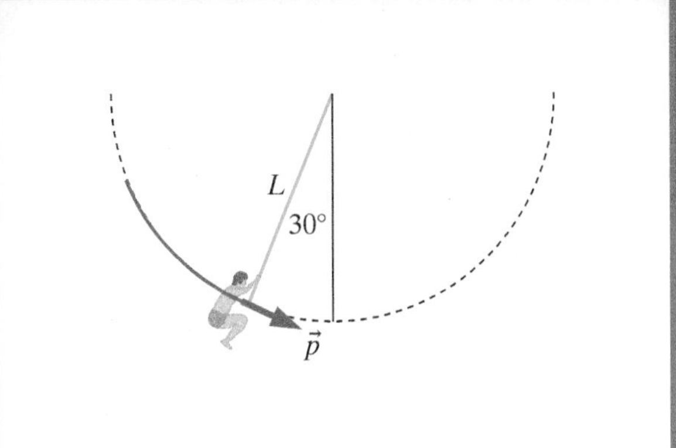

OBJECTIVES

After studying this chapter, you should be able to

- Identify systematically all the forces acting on a system.
- Determine the values of unknown 3D forces acting on a system whose motion is known.
- Analyze curving motion mathematically, and relate parallel and perpendicular components of $d\vec{p}/dt$ to the net force acting on the system.

5.1 UNKNOWN FORCES

In previous chapters we have applied the Momentum Principle to predict the motion of systems that are affected by a known net force. In this chapter we will do the opposite: knowing the motion of a system, we will deduce the net force acting on the system. In some cases, knowing the net force, and also knowing the values of some, but not all, of the individual forces contributing to the net force, we will be able to determine the forces whose values are unknown.

One important application of this procedure is to deduce the value of a contact force that we cannot calculate directly. In Chapter 4, what was the force exerted by the vine on Tarzan when he swung on it, just before it broke (Figure 5.1)? What is the force of the seat on an airplane passenger when the plane makes a sudden change of direction, and why is it different from the passenger's weight?

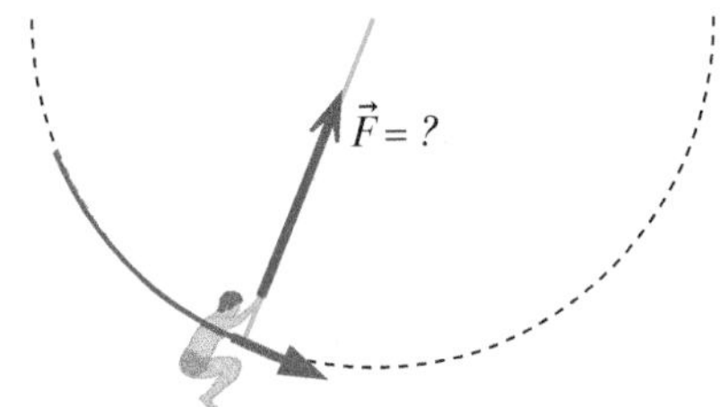

Figure 5.1 The force exerted by the vine on Tarzan is an unknown force.

5.2 IDENTIFYING ALL FORCES

It is surprisingly important to list all forces on a system explicitly when beginning a problem. Once we start doing mathematical operations, it is easy to forget a force, or to include it twice, so it is essential to have a written record for reference. Enumerating forces on a system is a little like following the rules of a game:

RULES FOR IDENTIFYING FORCES ON A SYSTEM

- Identify every force due to an *object* in the surroundings that interacts at a distance (gravitationally or electrically) with the system (*distance forces*).
- Identify every force due to an *object* in the surroundings that touches the system (*contact forces*).
- Draw a free-body diagram in which each force is labeled with the name of the object in the surroundings that is causing the force.

QUESTION What is the point of listing objects in the surroundings? Why should we say "the Earth" instead of "gravity"?

We know that an interaction involves two objects, so each force on the system must be due to an object in the surroundings. Naming the object is a way of avoiding double counting, because sometimes forces have more than one standard name. For example, if we included both "the gravitational force" and "the weight of the ball" we would be counting the same force twice. Identifying each force by the name of the interacting object helps avoid counting a force twice, or missing a force. If you can't name an *object* in the surroundings that is responsible for the force, there is no such force on the system as long as we use an inertial (nonaccelerating, nonrotating) reference frame, which we will always do in this textbook. If you find yourself identifying objects named "tension," "normal," "gravity," or "centripetal," you are in danger of double-counting forces, or including nonexistent forces.

Making Free-Body Diagrams

A free-body diagram can be very simple, as shown in Figure 5.2:

- If the system is modeled as a point particle, represent it by a dot.
- Represent each force by an arrow. If you don't know the magnitude of the force, that is okay—draw the arrow with an arbitrary length.
- Label each force. A label has two parts: a symbol that can be used in equations, and the name of the object in the surroundings that is responsible for the force.

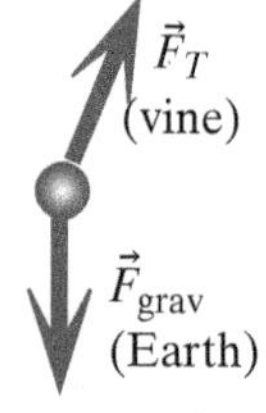

Figure 5.2 Free-body diagram for Tarzan at the instant shown in Figure 5.1. Tarzan is modeled as a point particle, and is represented by a dot. Each force is labeled with a symbol for use in equations and with the name of the interacting object in the surroundings. It is common to use the symbol $\vec{F}_T$ for a tension force. Since we do not yet know relative magnitudes, we have drawn the vine force with an arbitrary length.

If you would like to include more information in your free-body diagram, such as magnitudes of forces, direction of momentum, or values of angles or distances, that is fine. However, a simple free-body diagram, recording forces and their corresponding objects, is often quite adequate.

QUESTION In a free-body diagram should we show forces on objects in the surroundings?

No. We construct a free-body diagram showing forces on the chosen system to help us apply the Momentum Principle correctly to that system. Only forces on the system itself affect the momentum of the system.

QUESTION In Figure 5.2, why don't we include a force due to Tarzan's motion?

Force is a quantitative measure of interaction with an object in the surroundings. Motion isn't an object. The system's motion at this instant is certainly important, and it will be taken into account when we apply the Momentum Principle (remember that $\vec{p}_{\text{future}} = \vec{p}_{\text{now}} + \vec{F}_{\text{net}}\,\Delta t$).

Checkpoint 1 In each of the following cases identify all objects in the surroundings that exert forces on the system, and draw a free-body diagram for the system. Assume that air resistance is negligible. **(a)** You hit a baseball with a bat. Choose the baseball as the system, and consider the instant of contact with the bat. **(b)** You are playing with a yo-yo. Choose the yo-yo as the system. As the yo-yo moves downward, you pull up on the string.

5.3 DETERMINING UNKNOWN FORCES

Once we have identified all the objects that exert a force on a system, we may discover that we cannot calculate all of the forces directly. This is especially true for contact forces: we do not have an equation for calculating the force exerted by a table on a coffee cup or by a string on a yo-yo. We will need to use the Momentum Principle to deduce the values of unknown forces, by following this procedure:

FINDING UNKNOWN FORCES

- Explicitly choose a system, and stick to this choice.
- Systematically identify all the forces on the system due to objects in the surroundings, and draw and label a free-body diagram.
- Observe the rate at which the system's momentum changes ($d\vec{p}/dt$).
- Use the Momentum Principle ($d\vec{p}/dt = \vec{F}_{\text{net}}$) to deduce the net force on the system.
- Solve for the unknown contributions to the net force.

This procedure will work only if we know $d\vec{p}/dt$, and if we are careful to identify *all* of the contributions to the net force. It is important to be very clear and explicit about the choice of system, and to include all other objects in the surroundings. The free-body diagram is the basis for correctly accounting for all contributions to $\vec{F}_{\text{net}}$. The following example illustrates this process with a very simple example for which you already know the answer.

EXAMPLE **A Hanging Ball**

A 1 kg ball hangs motionless at the end of a wire (Figure 5.3). The top of the wire is attached to an iron support. How large an upward force does the wire exert on the ball?

Solution Here we use the convenient and widely used convention that F_T means the magnitude of the vector $\vec{F}_T$, for example,

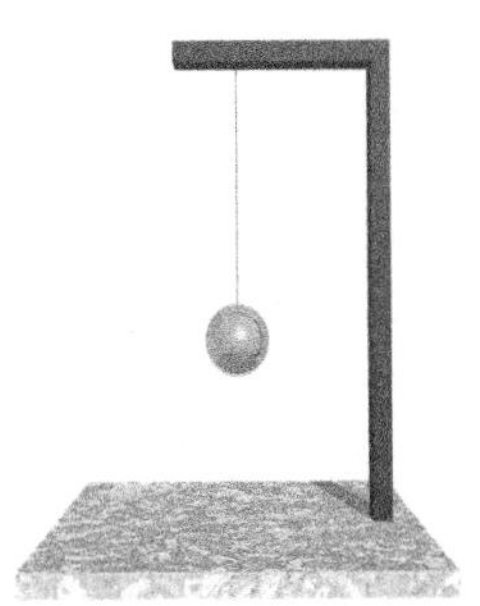

Figure 5.3 A heavy ball hangs motionless at the end of a thin wire.

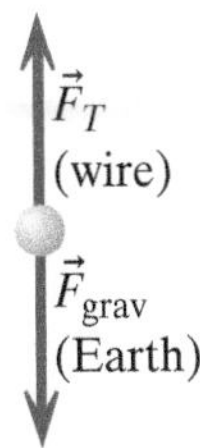

Figure 5.4 Free-body diagram for the hanging ball.

System: Ball
Surroundings: Earth, wire
Free-body diagram: Figure 5.4
Momentum Principle, y component:

$$\begin{aligned} dp_y/dt &= F_{\text{net},y} \\ 0 &= F_T - F_{\text{grav}} \\ 0 &= F_T - 1\,\text{kg}\cdot 9.8\,\text{N/kg} \\ F_T &= 9.8\,\text{N} \end{aligned}$$

You almost certainly knew before going through these careful steps that the tension in the wire would be equal to the weight of the ball. The reason for going through this analysis without omitting any steps is that this same procedure can be used in all situations, including situations much more complicated than this one.

QUESTION Why didn't we include the iron support in the surroundings? Why doesn't the free-body diagram show a force due to the support?

Because the support doesn't touch the chosen system (the ball), the interatomic bonds in the ball are not stretched or compressed by the support. Only two kinds of forces can act on the ball: forces involving interaction at a distance and contact forces. To exert a contact force, the atoms of one object must be in contact with the atoms of another object, which is not the case here. The support does exert a very tiny gravitational force on the ball (acting over a distance), but this force is too small to have a noticeable effect.

It is of course important for the iron support to be present, but the support does not interact directly with the ball. The support exerts a force on the wire, but the wire is not the system of interest; it does get taken into account indirectly, since the wire would not be under tension if it were not supported by something.

QUESTION What if we had chosen the ball plus the wire as the system?

Compare the following example to the previous one.

EXAMPLE **Ball and Wire as the System (A Hanging Ball)**

Again consider Figure 5.3, but choose the ball and the wire as the system. Find the tension force exerted by the support on the ball–wire system.

Solution System: Ball and wire, modeled as a point particle
Surroundings: Earth
Free-body diagram: Figure 5.5

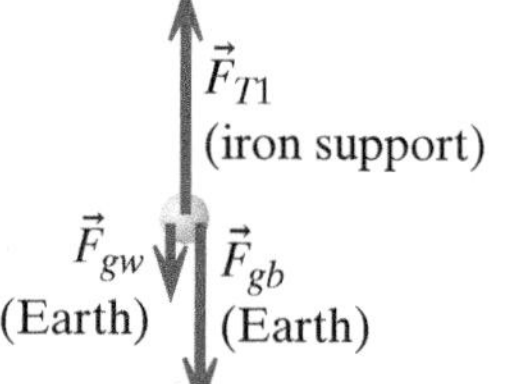

Figure 5.5 The free-body diagram for the system consisting of the ball plus the wire, modeled as a point particle. The symbol $\vec{F}_{gb}$ represents the gravitational force on the ball by the Earth, and the symbol $\vec{F}_{gw}$ represents the gravitational force on the wire by the Earth. The gravitational force of the Earth on the wire is exaggerated for clarity.

$$\frac{dp_y}{dt} = F_{T1} - F_{gb} - F_{gw}$$
$$0 = F_{T1} - m_b g - m_w g$$
$$F_{T1} = (m_b + m_w)g$$

where m_b and m_w are the masses of the ball and the wire, respectively. Note that the tension force by the support on the wire is slightly larger than the tension force by the wire on the ball, which we calculated in the previous example.

QUESTION Is the tension in the wire uniform throughout the wire?

Actually, the tension in the top of the wire is a little greater than the tension at the bottom of the wire, because the top of the wire supports not only the ball, but the rest of the wire as well. Of course if the mass of the wire is small compared to the mass of the ball, it is a good approximation to say that the interatomic bonds in the wire are stretched almost as much at the bottom as at the top.

5.4 UNIFORM MOTION

The simplest situation in which we can deduce the values of unknown forces is that of uniform motion. The momentum of a system in uniform motion is constant and does not change with time. Recall that a special case of uniform motion is a situation in which an object is at rest and remains at rest; this situation is called "static equilibrium." (Problems in which the system is at rest are often called "statics" problems. If you are studying engineering you may take an entire course on this topic.) In a typical problem involving uniform motion you will be asked to deduce the unknown magnitude of a force on the system by applying the Momentum Principle.

The underlying physics is the same for all systems in uniform motion (including those at rest):

QUESTION If the system's velocity is constant, what is $\frac{d\vec{p}}{dt}$?

The momentum of the system isn't changing, so the rate of change of the momentum is zero:

$$\frac{d\vec{p}}{dt} = \vec{0} = \langle 0,0,0 \rangle$$

QUESTION What does this imply about the net force acting on the system?

From the derivative form of the Momentum Principle, $d\vec{p}/dt = \vec{F}_{\text{net}}$, we deduce that the net force acting on a system that never moves must be zero:

$$\vec{F}_{\text{net}} = \vec{0} = \langle 0,0,0 \rangle$$

That's really all there is to it: knowing that the net force must be zero may allow us to deduce the values of some of the contributions to the net force. The previous example of a ball hanging from a wire is a simple 1D uniform motion problem.

Forces Exerted by Strings

Many uniform motion problems involve strings or wires. Because a string or wire is flexible, the tension force exerted by a string is always in the direction of the string. A string cannot exert a force perpendicular to the string. Also, of course, a string cannot push—it can only pull. It is common to refer to the tension force exerted by a string as "the tension in the string," and to use the symbol $\vec{F}_T$ to represent a tension force.

Component Equations

If a situation involves forces in two or three dimensions, it is frequently easiest to write one equation for each dimension:

$$\frac{dp_x}{dt} = F_{\text{net},x} \quad \text{and} \quad \frac{dp_y}{dt} = F_{\text{net},y} \quad \text{and} \quad \frac{dp_z}{dt} = F_{\text{net},z}$$

Typically we know something about the directions of unknown forces, but we don't know their magnitudes. In such cases, we will start by expressing the components of each unknown force in terms of their magnitudes. For example, suppose that we know that a string exerts a force $\vec{F}_1$ in the $-x$ direction on a system, but we don't know the magnitude of the force. Using the notation convention that $F_1 = |\vec{F}_1|$, we can write the force as

$$\vec{F}_1 = F_1 \langle -1,0,0 \rangle = \langle -F_1,0,0 \rangle$$

so the x component of the momentum principle becomes

$$\frac{dp_x}{dt} = -F_1 + \ldots$$

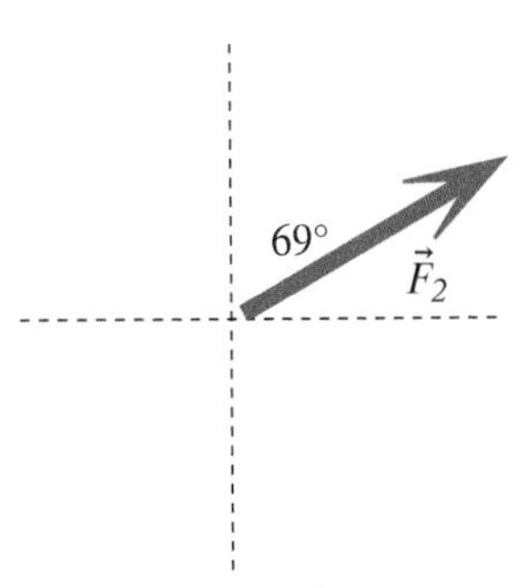

Figure 5.6 A force $\vec{F}_2$ is applied in the xy plane at an angle of 69° to the $+y$ axis.

If an unknown force is at some angle to the axes, the force will have more than one nonzero component. For example, suppose a spring pulls on a system at an angle of 69° to the $+y$ axis, as shown in Figure 5.6. If we call this unknown force $\vec{F}_2$, then by using direction cosines as we did in Chapter 1 we have this:

$$\vec{F}_2 = F_2 \langle \cos 21°, \cos 69°, \cos 90° \rangle$$

and the Momentum Principle can be written in component form as

$$\frac{dp_x}{dt} = F_2 \cos 21° + \ldots$$
$$\frac{dp_y}{dt} = F_2 \cos 69° + \ldots$$
$$\frac{dp_z}{dt} = 0 + \ldots$$

In the following examples, to simplify notation, we use the convention of writing F_1 instead of $|\vec{F}_1|$, etc. When doing this we need to be careful of signs, which must match the directions of forces.

EXAMPLE **Forces in Two Directions**

A metal block is suspended from a spring, which supports some, but not all, of the block's weight, so that the block rests lightly on a wooden board (Figure 5.7). The block is also pulled to the right by a spring and to the left by a copper wire. The vertical spring has a stiffness of 40 N/m and is stretched 3 cm beyond its relaxed length. The horizontal spring has a stiffness of 60 N/m and is stretched 1.5 cm. The mass of the block is 300 grams. **(a)** What is the tension force exerted on the block by the copper wire? **(b)** What is the normal force exerted on the block by the wooden board? **(c)** What force would the wooden board exert on the block if the vertical spring were removed?

Solution System: Block
Surroundings: Spring 1, spring 2, board, wire, Earth
Free-body diagram: Figure 5.8

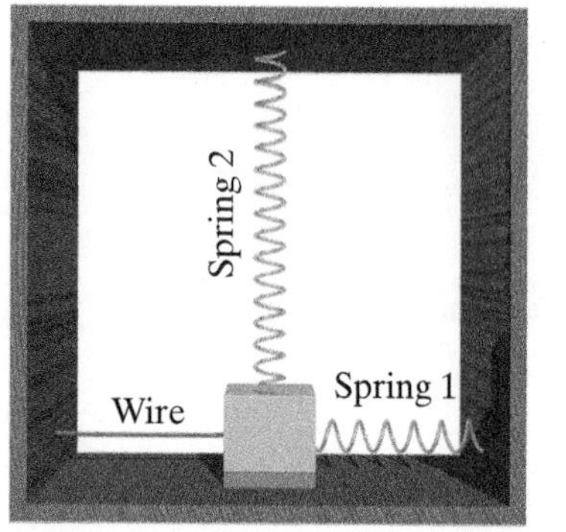

Figure 5.7 A block is partially supported by a vertical spring and rests lightly on a wooden surface. Another spring pulls the block from the right, and a wire pulls it from the left.

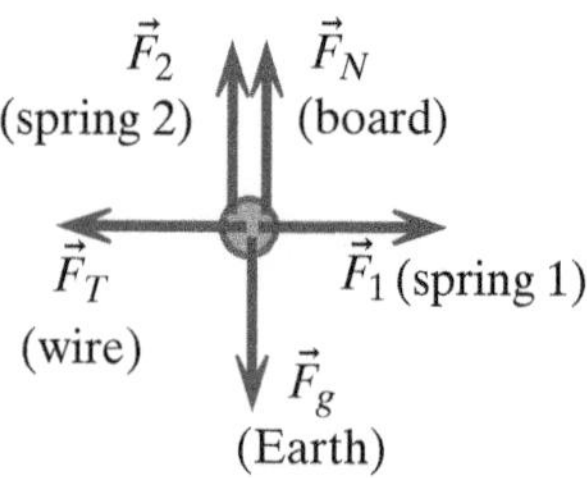

Figure 5.8 Free-body diagram for the block. Lengths of arrows representing unknown forces are arbitrary.

x components:

$$\frac{dp_x}{dt} = F_{\text{net,x}}$$
$$0 = F_1 + (-F_T)$$
$$0 = (60\text{ N/m})(0.015\text{ m}) - F_T$$
$$F_T = 0.9\text{ N}$$

(a) The rod exerts a force of $\langle -0.9, 0, 0\rangle$ N on the block.

y components:

$$\frac{dp_y}{dt} = F_{\text{net,y}}$$
$$0 = F_2 + F_N - F_g$$
$$0 = (40\text{ N/m})(0.03\text{ m}) + F_N - (0.3\text{ kg})(9.8\text{ N/kg})$$
$$F_N = 1.74\text{ N}$$

(b) The wood surface exerts a force of $\langle 0, 1.74, 0\rangle$ N on the block.
(c) If the vertical spring were removed the table would have to support the full weight of the block, which is 2.94 N, so the normal force would be $\langle 0, 2.94, 0\rangle$ N.

In the previous example every force was directed along the x, y, or z axis. In most problems, however, this isn't the case. Typically such problems involve relating angles to unit vectors in order to get the x, y, and z components of forces, as in the next example.

EXAMPLE **Pulling at an Angle**

A 3 kg block hangs from string 1 (Figure 5.9). Then you pull the block to the side by pulling horizontally with a second string (string 2). **(a)** What is the tension in string 2 in order for the block to hang motionless with string 1 at an angle of 33° to the vertical, as shown in Figure 5.9? **(b)** In this situation, what is the tension in string 1 (the force on the block by the string)? **(c)** In which situation are the interatomic bonds in string 1 stretched more: block hanging straight down, suspended from string 1, or block hanging at an angle, pulled on by string 1 and string 2?

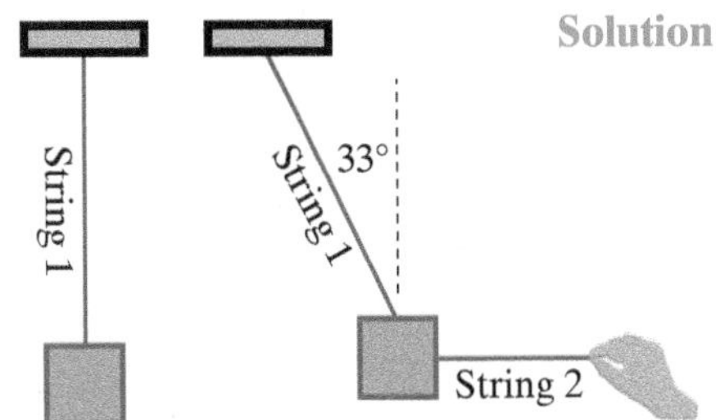

Figure 5.9 Left: A block hangs from a string. Right: Then you pull the block to the side with another string.

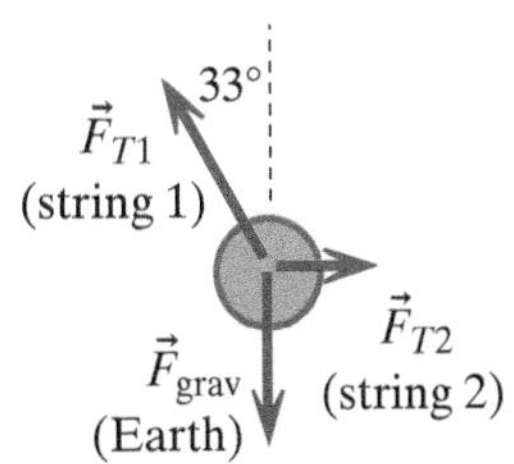

Figure 5.10 Free-body diagram.

Solution System: Block
Surroundings: String 1, string 2, and Earth
Free-body diagram: Figure 5.10

x components:

$$\frac{dp_x}{dt} = F_{\text{net},x}$$
$$0 = F_{T2} + F_{T1,x}$$

y components:

$$\frac{dp_y}{dt} = F_{\text{net},y}$$
$$0 = -mg + F_{T1,y}$$
$$0 = -(3\text{ kg})(9.8\text{N/kg}) + F_{T1,y}$$
$$0 = -29.4\text{ N} + F_{T1,y}$$

It may look as if we have two equations but three unknowns ($F_{T1,x}$, $F_{T1,y}$, and F_{T2}), which would make it impossible to solve the problem. However, we can write $F_{T1,x}$ and $F_{T1,y}$ in terms of F_{T1} by using direction cosines:

$$\hat{F}_{T1} = \langle \cos(90° + 33°), \cos 33°, 0\rangle$$
$$\vec{F}_{T1} = F_{T1}\langle \cos 123°, \cos 33°, 0\rangle$$

Now we have two equations in two unknowns:

$$0 = F_{T2} + F_{T1}\cos 123°$$
$$0 = -29.4\text{ N} + F_{T1}\cos 33°$$

and we can solve for the unknown force magnitudes (parts (a) and (b)):

$$F_{T1} = \frac{29.4\text{ N}}{\cos 33°} = 35.1\text{ N}$$
$$F_{T2} = -F_{T1}(\cos 123°) = -(35.1\text{ N})(-0.545) = 19.1\text{ N}$$

(c) When the block hangs straight down, the force exerted by string 1 must be equal in magnitude to the force by the Earth on the block, which is 29.4 N. Evidently the tension in string 1 is greater (35.1 N) when the block is pulled to the side than when the block hangs straight down. The interatomic bonds in string 1 must be stretched more when the block is pulled to the side.

It is interesting that in different situations the tension force exerted by an object such as a string or a wire may be different, associated with different amounts of stretch of interatomic bonds.

Check: The units are correct (N). Special cases: Consider other angles.

Call the angle between the string and the vertical θ (in the example above, $\theta = 33°$). We can rewrite the y component equation algebraically:

$$F_{T1} = \frac{mg}{\cos\theta}$$

If $\theta = 0°$, the block hangs vertically, and $F_{T1} = mg$, as we expect.
If $\theta = 90°$, string 2 is horizontal, and $F_{T1} = \infty$, which is impossible. This actually does make sense, since if both strings are horizontal, there is no y component to support the block—an impossible situation.

EXAMPLE **Tension and Young's Modulus (Pulling at an Angle)**

Suppose that we replace string 1 in the preceding example (Pulling at an Angle) with a thin steel wire whose length is 40 cm and thickness is 0.2 mm. How much will the wire stretch when the block is pulled to the side?

Solution Young's modulus for steel is approximately $2 \times 10^{11}\ \mathrm{N/m^2}$, though different kinds of steel differ in their stiffness.

$$Y = \frac{F/A}{\Delta L/L}$$

$$\Delta L = \frac{F/A}{Y/L} = \frac{(35.1\ \mathrm{N})/[\pi((0.2 \times 10^{-3}\ \mathrm{m})/2)^2]}{(2 \times 10^{11}\ \mathrm{N/m^2})/(0.4\,\mathrm{m})}$$

$$\Delta L = 2.2 \times 10^{-3}\ \mathrm{m} = 2.2\ \mathrm{mm}$$

EXAMPLE **Forces on a String in Tension (Pulling at an Angle)**

The horizontal string (string 2) in the previous example (Pulling at an Angle, Figure 5.9) applied a 19.1 N force to the right on the block. What was the force that your hand applied to the other end of that string?

Solution System: String 2 (Figure 5.9)
Surroundings: Block, hand (neglect Earth because the string mass is small)
Free-body diagram: Figure 5.11

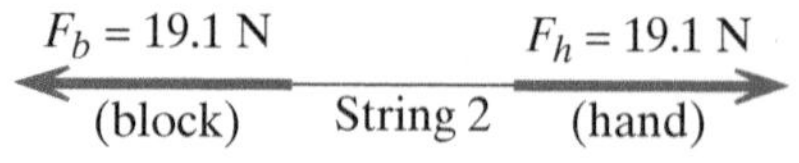

$F_b = 19.1$ N (block) String 2 $F_h = 19.1$ N (hand)

Figure 5.11 Free-body diagram for String 2 (Figure 5.9).

Choose the string as the system of interest, so the surroundings consist of the block, the Earth, and your hand. We can neglect the gravitational force on string 2, because in order for this force to be comparable to the tension, the string would have to have a mass of nearly 2 kg.

By reciprocity of (interatomic) electric forces, the force F_b that the block exerts to the left on the string is equal to F_T, the force that the string exerts on the block, so it has a magnitude of 19.1 N.

$$F_b = F_T$$

The momentum of string 2 is not changing, so we deduce that the force F_h exerted by your hand on the string must also have a magnitude of 19.1 N to the right on the string (Figure 5.11).

$$F_h = F_T$$

Evidently the forces at each end of a low-mass string in tension must be nearly equal in magnitude, and equal to the tension in the string.

EXAMPLE **Three Wires**

A 350 kg load is suspended from a vertical wire that is tied to two aluminum wires of radius 1.2 mm, as shown in Figure 5.12. Young's modulus for this aluminum alloy is $6.9 \times 10^{10}\ \mathrm{N/m^2}$. **(a)** Calculate the tension in each wire (that is, the magnitude of the tension force exerted by each wire). **(b)** What is the fractional stretch (the strain) of each wire?

Solution **(a)** This problem requires two steps, each with a different choice of system. In the first step we choose the load as the system, and find the tension force exerted on the load by the vertical wire. This is exactly the same problem as

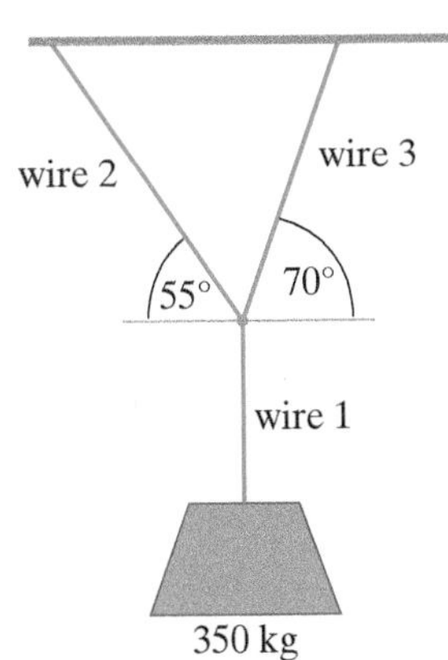

Figure 5.12 A load hangs from an aluminum wire that is tied to two other aluminum wires.

the one we solved in the example entitled A Hanging Ball. We find that the magnitude of the tension in wire 1 is equal to the weight of the ball:

$$F_{T1} = (350\,\text{kg})(9.8\,\text{N/kg}) = 3430\,\text{N}$$

The choice of system for the second step may seem counterintuitive: we choose the knot connecting the three wires as the system. This allows us to relate the known force $\vec{F}_{T1}$ to the two unknown forces by the other wires.

System: Knot
Surroundings: Wire 1, wire 2, wire 3
Free-body diagram: Figure 5.13

Write the forces as vectors, using direction cosines:

$$\vec{F}_{T2} = F_{T2}\langle \cos 125°, \cos 35°, 0\rangle$$
$$\vec{F}_{T3} = F_{T3}\langle \cos 70°, \cos 20°, 0\rangle$$
$$\vec{F}_{T1} = \langle 0, -3430, 0\rangle\ \text{N}$$

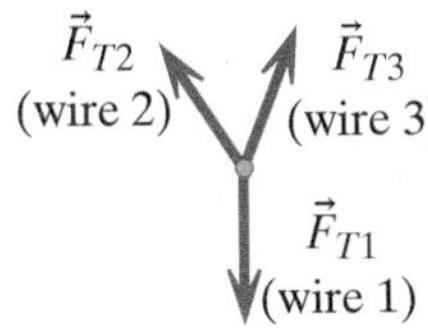

Figure 5.13 Free-body diagram for the knot.

Separate components:

$$\frac{dp_x}{dt} = F_{\text{net},x}$$
$$0 = F_{T2}\cos 125° + F_{T3}\cos 70° + 0$$
$$\frac{dp_y}{dt} = F_{\text{net},y}$$
$$0 = F_{T2}\cos 35° + F_{T3}\cos 20° - 3430\ \text{N}$$

We now have two equations in two unknowns (F_{T2} and F_{T3}). Solving for the unknown force magnitudes, we find that

$$F_{T2} = 1431\ \text{N}$$
$$F_{T3} = 2402\ \text{N}$$

Check: The units are correct (N). We can plug these results back into the momentum equations to verify that they satisfy those equations.

(b) Fractional stretch:

$$Y = \frac{F/A}{\Delta L/L}$$
$$A = \pi(1.2\times 10^{-3}\,\text{m})^2 = 4.52\times 10^{-6}\ \text{m}^2$$

$$\text{Wire 1:}\quad \Delta L/L = \frac{\left(\dfrac{1431\ \text{N}}{4.52\times 10^{-6}\ \text{m}^2}\right)}{6.9\times 10^{10}\,\text{N/m}^2} = 4.6\times 10^{-3}\quad \text{or about } 0.5\%$$

$$\text{Wire 2:}\quad \Delta L/L = \frac{\left(\dfrac{2402\ \text{N}}{4.52\times 10^{-6}\ \text{m}^2}\right)}{6.9\times 10^{10}\,\text{N/m}^2} = 7.7\times 10^{-3}\quad \text{or about } 0.8\%$$

Nonzero Velocity

Both systems at rest and systems moving with nonzero constant velocity exhibit uniform motion. The physical analysis process is the same in both cases.

EXAMPLE **Uniform Motion**

You place a 4 kg crate on the floor of an elevator that is temporarily stopped. Then the elevator starts moving, briefly speeds up, then moves upward at a constant speed of 2 m/s for many seconds (it is in a very tall building). While the elevator is moving upward at constant speed, what is the magnitude of the force exerted by the elevator floor on the crate?

Solution System: Crate
Surroundings: Earth, elevator floor
Free-body diagram: Figure 5.14
Momentum Principle:

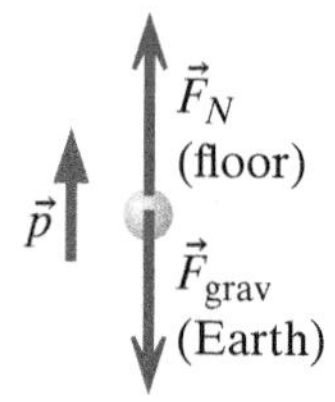

$\vec{F}_N$ (floor)
$\vec{p}$
$\vec{F}_{grav}$ (Earth)

Figure 5.14 Free-body diagram for the crate. The momentum of the crate is also shown on the diagram.

$$\frac{dp_y}{dt} = F_N - mg$$

$$0 = F_N - (4\,\text{kg})(9.8\,\text{N/kg})$$

$$F_N = 39.2\ \text{N}$$

QUESTION How can the crate keep moving upward if the net force on it is zero?

Let's apply the Momentum Principle at a specific instant, when the crate (and elevator) are moving at constant speed. The momentum of the crate a short time Δt in the future depends on two things:

$$\vec{p}_{\text{future}} = \vec{p}_{\text{now}} + \vec{F}_{\text{net,now}}\Delta t$$

The current momentum of the crate, $\vec{p}_{\text{now}}$, is in the upward direction. If the net force on the crate is zero, then we can conclude that $\vec{p}_{\text{future}}$ will be equal in magnitude and direction to $\vec{p}_{\text{now}}$; the speed of the crate will not change. A nonzero net force would change the momentum of the crate, causing it either to speed up or to slow down, depending on the direction of that net force. For example, during the brief time when the elevator was speeding up, the crate's momentum was increasing, and we can deduce that during that time the net force on the crate must have been nonzero.

It is interesting to think about this situation in terms of the principle of relativity. We have analyzed the system from the viewpoint of an observer standing on the ground, who sees the elevator and crate moving at constant velocity, and who concludes that the net force on the crate must be zero. Suppose, however, that there is a video camera inside the elevator. From the frame of reference of the camera, the elevator and crate appear motionless. Looking at the video, an observer would also conclude that the net force on the crate was zero, because the momentum of the crate was not changing. The Momentum Principle is valid in both frames of reference.

EXAMPLE **Pulling a Sled**

You pull a loaded sled whose mass is 40 kg at constant speed in the x direction with a rope at an angle of 35°. The coefficient of kinetic friction between the snow and the sled runners is 0.2. What is the tension in the rope?

Solution System: Sled
Surroundings: Earth, snow, rope
Free-body diagram: Figure 5.15
Write out the vector components of $\vec{F}_T$:

$$\vec{F}_T = F_T\langle \cos 35°, \cos(90° - 35°), 0\rangle$$

The momentum of the sled is constant, so:

$$\frac{dp_x}{dt} = F_{\text{net},x}$$
$$0 = F_T \cos 35^\circ - \mu_k F_N$$
$$\frac{dp_y}{dt} = F_{\text{net},y}$$
$$0 = F_N + F_T \cos(90^\circ - 35^\circ) - mg$$

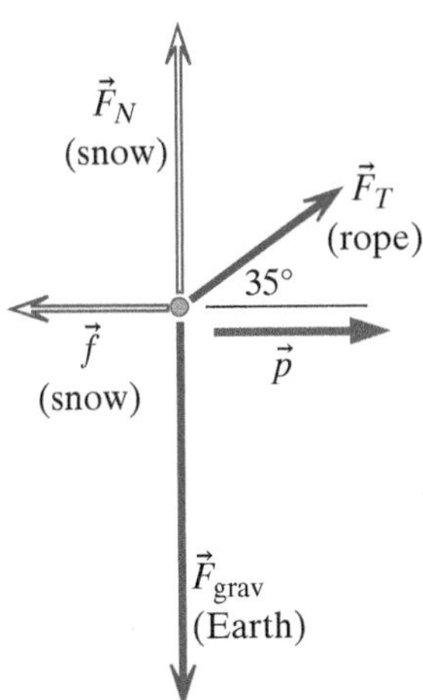

Figure 5.15 Free-body diagram for a sled pulled at constant speed. The sled's momentum is also shown.

Combine these two equations:

$$F_T \cos 35^\circ - \mu_k(mg - F_T \cos 55^\circ) = 0$$
$$F_T(\cos 35^\circ + \mu_k \cos 55^\circ) = \mu_k mg$$
$$F_T = \frac{\mu_k mg}{\cos 35^\circ + \mu_k \cos 55^\circ}$$
$$F_T = \frac{0.2(40\,\text{kg})(9.8\,\text{N/kg})}{0.819 + 0.2(0.574)} = 84.0\,\text{N}$$

Pulling at an angle rather than horizontally makes the force squeezing the sled and snow together less than mg, which it would be if you pulled horizontally. Essentially you've partly lifted the sled out of the snow, which reduces the vertical component of the force of the snow, F_N, and therefore reduces the horizontal component of the force of the snow, the friction force $\mu_k F_N$.

Checkpoint 2 An object is moving with constant momentum $\langle 10, -12, -8 \rangle$ kg·m/s. What is the rate of change of momentum $d\vec{p}/dt$? What is the net force acting on the object? The object is acted on by three objects in the surroundings. Two of the forces are $\langle 100, 50, -60 \rangle$ N and $\langle -130, 40, -70 \rangle$ N. What is the third force?

Torque and Angular Momentum

Modeling a system as a point particle, and applying the Momentum Principle, can allow us to predict motion (when forces are known) or to find unknown forces (when motion is known). However, there are situations in which such an analysis, though valid, does not tell us everything we need to know. For example, consider two children sitting motionless on a seesaw (Figure 5.16). Choose as the system the two children and the board. For this choice of system, the objects in the surroundings that exert significant (external) forces on the system are the Earth (pulling down) and the support pivot (pushing up). The momentum of the system isn't changing, so the Momentum Principle correctly tells us that the net force on the system must be zero, and the support pivot must be pushing up with a force equal to the weight of the children plus the board.

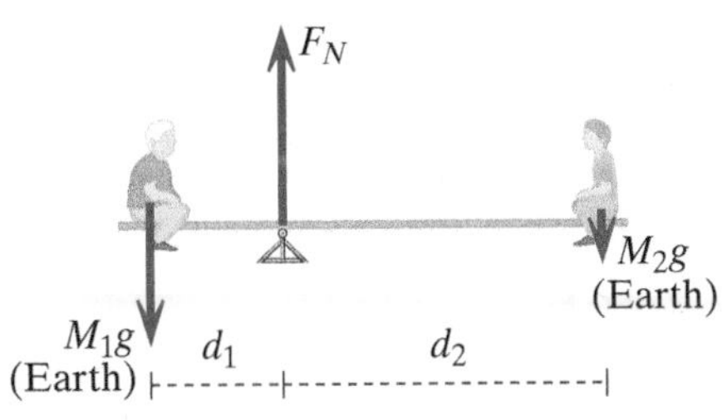

Figure 5.16 Two children sit on a seesaw in equilibrium.

However, the Momentum Principle alone doesn't tell us where the children have to sit in order to achieve balance and equilibrium. In order to learn more, we will need to model the system as an extended object (not a point particle) and to invoke the Angular Momentum Principle, which we will encounter in Chapter 11. By modeling the system as an extended object, we see that the force by the Earth on a child or on the unevenly balanced board exerts a twist (the technical term is "torque") about the pivot that tends to make the board turn. Torque has a magnitude and a direction, so it can be expressed as a vector. In order for the rate of rotation to be zero (or any other constant) the torques associated with the forces on the two children and the board must add up to zero. Torque is defined as force times lever arm, and a nonzero net *torque* causes changes in a quantity called angular momentum, just as a nonzero net *force* causes changes in the ordinary momentum we have been studying. The Angular Momentum Principle quantitatively relates the net torque on a system to the rate of change of the system's rotation.

In the present chapter we will continue to consider most objects (except things like wires, vines, and springs, which can stretch or compress) as point particles, which cannot deform or rotate.

5.5 CHANGING MOMENTUM

In the preceding sections we focused on systems whose momentum is constant. In such situations we know immediately that the value of $d\vec{p}/dt$ is zero, and that therefore the net force must also be zero. In this section we deal with systems consisting of several objects, in which the momentum may be changing. The motion of the center of mass is determined by the net force acting on the system, but we may want to know the forces that objects inside the system exert on each other.

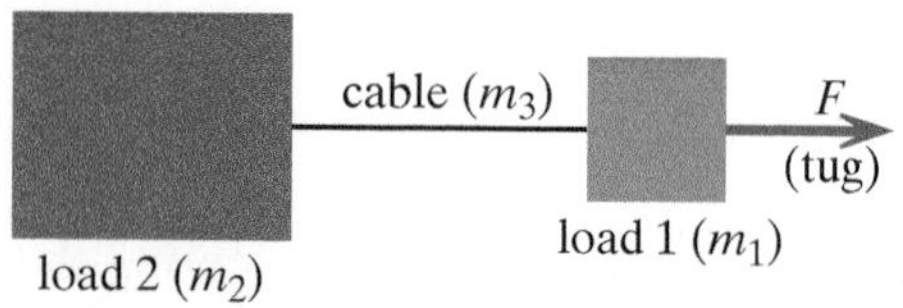

Figure 5.17 A space tug pulls two loads connected by a steel cable.

Consider the following example, which is typical of a class of problems called "dynamics" problems, in which it is often necessary to analyze different choices of system in order to determine some unknown forces, based on the motion. If you are studying engineering you may take an entire course on "dynamics."

EXAMPLE **Two Loads Connected by a Steel Cable**

You're an asteroid miner and your space tug is heading in the $+x$ direction toward a space station, pulling two loads connected by a steel cable. Your tug applies a force of $F = 10000$ N (Figure 5.17). The mass of load 1 is $m_1 = 500$ kg, the mass of load 2 is $m_2 = 2000$ kg, and the mass of the steel cable is $m_3 = 30$ kg. **(a)** What is the x component of the acceleration of the center of mass of the system? **(b)** What is the x component of the acceleration of load 1? Of load 2? Of the cable? **(c)** What is the tension in the cable? **(d)** Young's modulus for the steel used to make the cable is 2×10^{11} N/m^2. The cable is 5 m long and the diameter of the cable is 8 mm. How much is the cable stretched?

Solution **(a)** Choose the two loads and the cable as the system (Figure 5.18). The Momentum Principle:

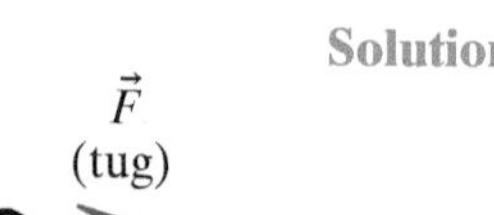

Figure 5.18 Free-body diagram for the two loads.

$$\frac{dp_{\mathrm{cm},x}}{dt} = F_{\mathrm{net,x}}$$

$$(m_1 + m_2 + m_3)\frac{v_{\mathrm{cm},x}}{dt} = 10000 \text{ N}$$

$$a_{\mathrm{cm},x} = \frac{10000 \text{ N}}{2530 \text{ kg}} = 3.95 \text{ m/s}^2$$

(b) As long as the cable remains taut, all of the parts of this multi-object system move together. Because the three objects have the same velocity but different masses, they have different momenta but the same acceleration, which is also the acceleration of the center of mass. This makes it convenient to analyze the motion in terms of acceleration.

(c) For the inclusive system of the two loads and the cable, the tension in the cable is internal to the system and doesn't contribute to the net force. To learn about the cable, we need to choose a different system to analyze, a system for which the cable is part of the surroundings. Let's choose load 2 as the system

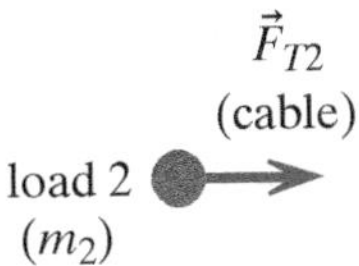

Figure 5.19 Free-body diagram for load 2.

(Figure 5.19) and let F_{T2} represent the magnitude of the unknown force that the cable exerts on load 2:

$$\frac{dp_{2,x}}{dt} = F_{\text{net,x}}$$

$$m_2\frac{dv_{2,x}}{dt} = F_{T2}$$

$$F_{T2} = (2000 \text{ kg})(3.95 \text{ m/s}^2) = 7905 \text{ N}$$

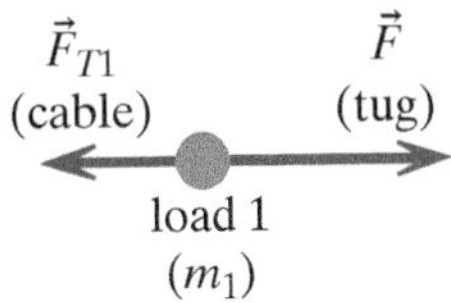

Figure 5.20 Free-body diagram for load 1.

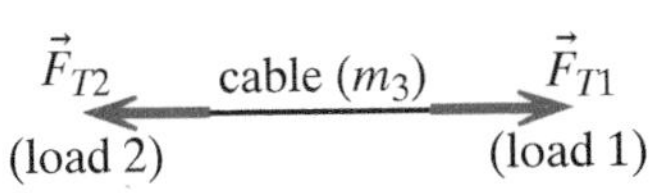

Do we get the same answer if we choose load 1 as the system? Let F_{T1} represent the magnitude of the unknown force that the cable exerts on load 1 (Figure 5.20):

$$\frac{dp_{1,x}}{dt} = F_{\text{net,x}}$$

$$m_1\frac{dv_{1,x}}{dt} = F - F_{T1}$$

$$F_{T1} = 10000 \text{ N} - (500 \text{ kg})(3.95 \text{ m/s}^2) = 8024 \text{ N}$$

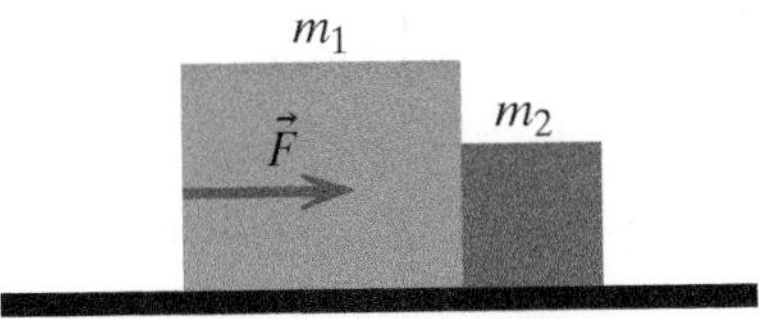

Figure 5.21 Free-body diagram for the cable.

We see that the tension at the two ends of the cable differs by $(8024 - 7905) =$ 119 N. The reason for this small difference is that we didn't neglect the mass of the cable. If you approximate the cable's mass as zero, you will find the same tension at both ends. A check on our calculations is to choose the cable as the system (Figure 5.21):

$$\frac{dp_{3,x}}{dt} = F_{\text{net,x}}$$

$$m_3\frac{dv_{3,x}}{dt} = F_{T1} - F_{T2}$$

$$F_{T1} - F_{T2} = (30 \text{ kg})(3.95 \text{ m/s}^2) = 119 \text{ N}$$

(d) Because the tension in the cable is somewhat greater on the right than on the left, the interatomic bonds are stretched more on the right than the left. To calculate the total stretch, an adequate approach is to use the average tension $(7905+8024)/2=7965$ N:

$$Y = \frac{F/A}{\Delta L/L}$$

$$\Delta L = \frac{FL}{YA} = \frac{(7965 \text{ N})(5 \text{ m})}{(2 \times 10^{11} \text{N/m}^2)(\pi(4 \times 10^{-3})^2 \text{ m}^2)}$$

$$\Delta L = 3.96 \times 10^{-3} \text{ m} = 3.96 \text{ mm}$$

The key idea seen in this example is that by choosing different systems to which we apply the Momentum Principle, we obtain information about otherwise unknown forces acting among multiple objects inside the full system.

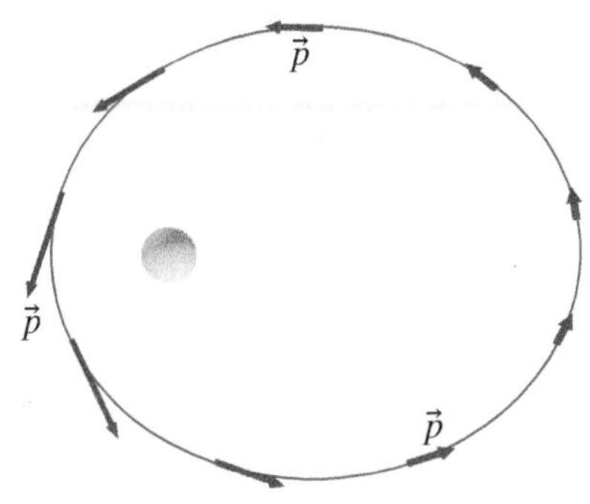

Figure 5.22 Push two blocks across a low-friction surface.

Checkpoint 3 You push two blocks across a low-friction surface, applying a known force $\vec{F}$ (Figure 5.22). **(a)** What system would you choose in order to determine the acceleration of the center of mass? **(b)** After determining the center-of-mass acceleration, what system would you choose in order to determine the magnitude of the force that block m_1 applies to block m_2?

Figure 5.23 The momentum of a comet (arrows) changes in magnitude and direction as it orbits a star.

5.6 FORCE AND CURVING MOTION

In some situations in which the momentum of a system is not constant, such as the case of an object moving along a curving path with varying speed (Figure 5.23), not only is $d\vec{p}/dt$ nonzero, but both $d\vec{p}/dt$ and $\vec{F}_{\text{net}}$ may be varying in magnitude and direction.

When analyzing the motion of systems moving in 2D or 3D with changing direction of momentum, it is often useful to divide both the $\vec{F}_{\text{net}}$ and $d\vec{p}/dt$ into two parts: a part parallel to the instantaneous momentum of the system and a part perpendicular to the momentum. Using this strategy, we will focus for the remainder of the chapter on understanding curving motion. We will find the values of unknown forces for such systems as a comet orbiting a star, Tarzan swinging on a vine, and a roller coaster looping the loop.

Parallel and Perpendicular Parts of Force

QUESTION What makes a moving object's path curve?

We know that a moving object whose path curves must be experiencing a nonzero net force due to objects in the surroundings, since

$$\vec{p}_{\text{future}} = \vec{p}_{\text{now}} + \vec{F}_{\text{net,now}}\Delta t$$

and if $\vec{F}_{\text{net,now}}$ is zero, $\vec{p}_{\text{future}}$ will be the same as $\vec{p}_{\text{now}}$; the system's momentum will not change. But not all forces result in curving paths.

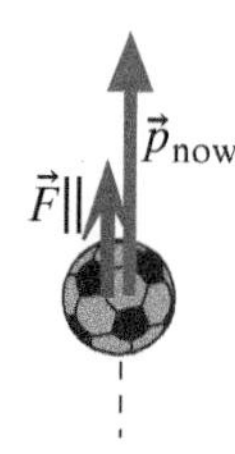

Figure 5.24 A force applied parallel to the momentum of a rolling ball.

To begin, let's consider what forces *do not* cause an object's path to curve. Suppose that a soccer ball is rolling in a straight line on a smooth surface. You kick the rolling ball gently from behind, so the force you exert is in the direction of the ball's momentum (Fig 5.24).

QUESTION Will a force in the direction of motion cause the ball's momentum to change direction?

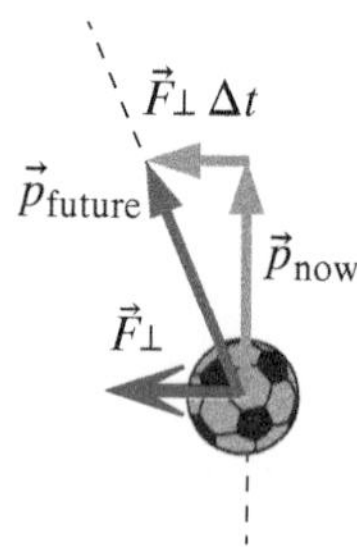

Figure 5.25 A kick perpendicular to the ball's momentum adds a sideways component to the ball's existing momentum.

Clearly not. A force exerted parallel to the ball's momentum will speed the ball up, or slow it down or even reverse its direction if the force is opposite to $\hat{p}$, but it will not change the line along which the ball is moving. To make the ball turn, part of the force applied to the ball must be in a direction that is not along the line of the ball's motion.

The most extreme case of a force "not along the line of the ball's motion" would be a force that is perpendicular to the ball's momentum (Fig 5.25). Such a force would certainly change $\hat{p}$, the direction of the ball's momentum.

QUESTION Suppose that you kick the rolling soccer ball in a direction exactly perpendicular to its momentum. What will be the new direction of the ball's motion?

This depends on the relative magnitudes of the impulse you apply ($\vec{F}_{\perp}\Delta t$) and the ball's current momentum ($\vec{p}_{\text{now}}$), as shown in Figure 5.25. If the impulse you apply is relatively small, the ball won't turn very much. (Δt is the time your foot is in contact with the ball.)

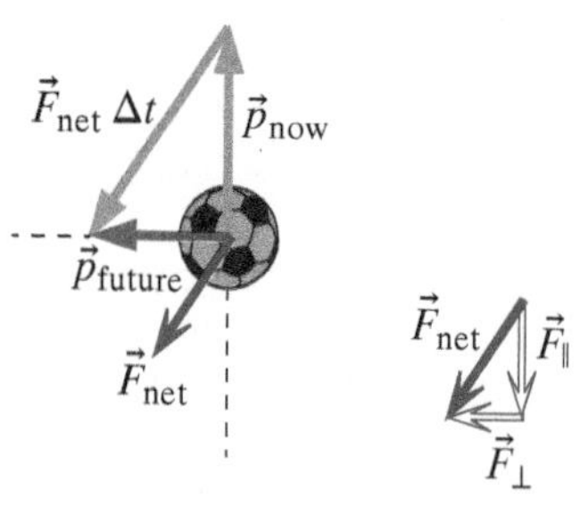

Figure 5.26 To make a 90° turn, you must kick backward at an angle. $\vec{F}_{\|}\Delta t$ reduces the forward component of momentum to zero, and $\vec{F}_{\perp}\Delta t$ imparts a perpendicular component of momentum to the ball.

QUESTION If you want the future momentum of the ball to be in a direction perpendicular to its momentum now, in what direction should you kick the ball?

To force the rolling ball to turn 90° to the left, your kick would actually need to be directed partly backwards as well as to the left: both $\vec{F}_{\|}$ and $\vec{F}_{\perp}$ would need to be nonzero, and $\vec{F}_{\|}\Delta t$ would need to be equal in magnitude and opposite in direction to $\vec{p}_{\text{now}}$, as shown in Figure 5.26.

QUESTION Why can't you just kick exactly in the direction you want the ball to go?

We need to think about both $\vec{p}_{\text{now}}$ and the impulse $\vec{F}_{\text{net,now}}\Delta t$.

$$\vec{p}_{\text{future}} = \vec{p}_{\text{now}} + \vec{F}_{\text{net,now}}\Delta t$$

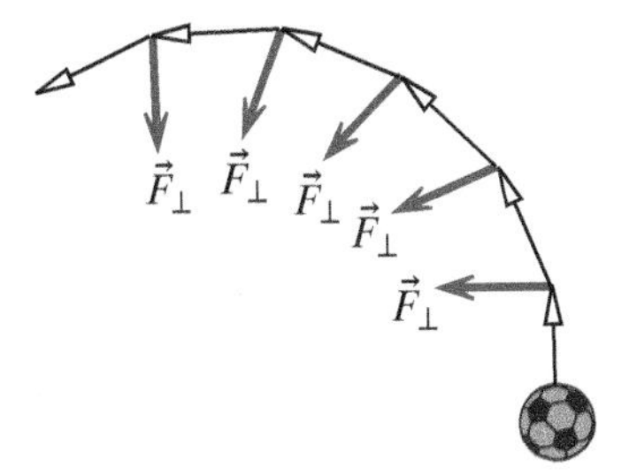

Figure 5.27 If you give a ball repeated big kicks perpendicular to its current heading, you can make the ball move in a roughly circular path.

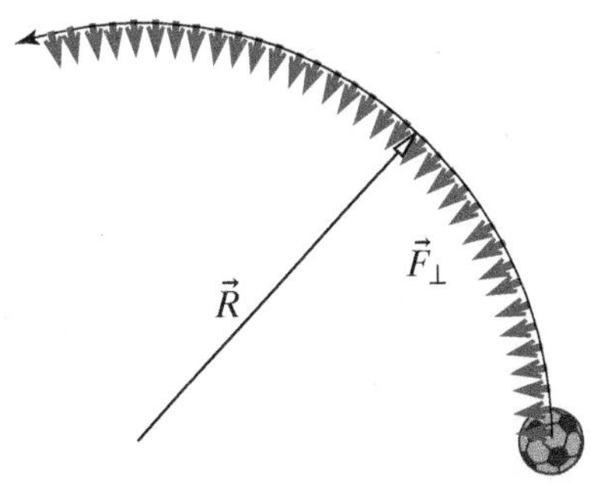

Figure 5.28 With a large number of small perpendicular kicks you can make a ball move along a path that is nearly indistinguishable from a circular path.

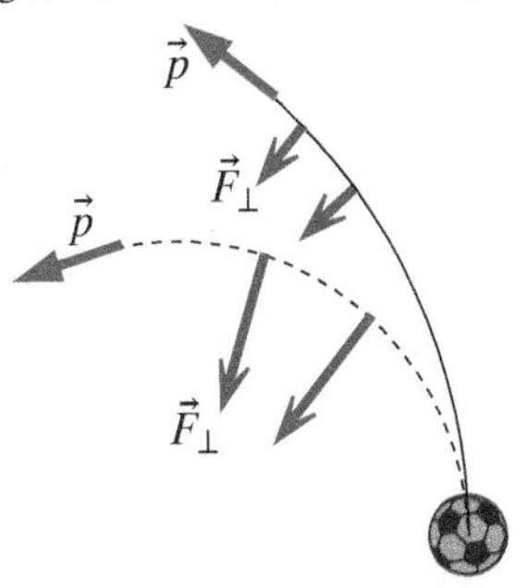

Figure 5.29 If you continually apply a bigger perpendicular force, you make the object turn more sharply, so the circle has a smaller radius.

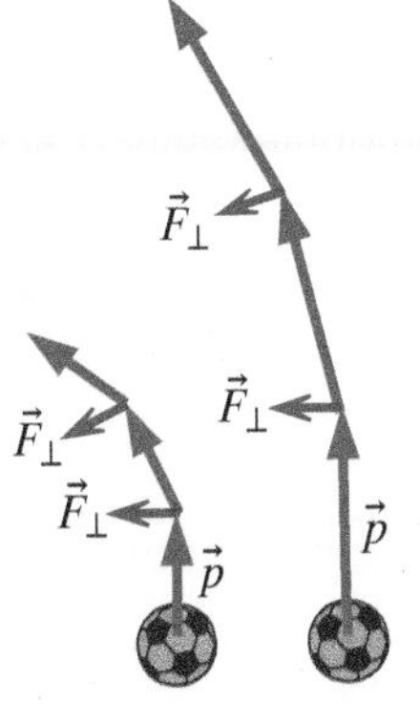

Figure 5.30 If the object's momentum is greater, the change due to the same perpendicular impulse is proportionally smaller, and the radius of curvature is larger.

If $\vec{p}_{\text{now}}$, the current momentum of the ball, is zero, then an impulse applied in the direction you want the ball to go would work. In the case discussed above, though, $\vec{p}_{\text{now}}$ is not zero, so we need to figure out what impulse would add the correct amount to $\vec{p}_{\text{now}}$ to get the desired $\vec{p}_{\text{future}}$ (Figure 5.26).

To express this idea generally, we note that any vector can be written as the sum of two other vectors that are perpendicular to each other. As shown in Figure 5.26, we can write the force $\vec{F}$ exerted on a moving system as the sum of two other forces: a force parallel to $\vec{p}$, called $\vec{F}_\parallel$, and a force perpendicular to $\vec{p}$, called $\vec{F}_\perp$:

$$\vec{F} = \vec{F}_\parallel + \vec{F}_\perp$$

We can write the Momentum Principle in a way that reflects this:

$$\Delta\vec{p} = (\vec{F}_\parallel + \vec{F}_\perp)\Delta t = \vec{F}_\parallel \Delta t + \vec{F}_\perp \Delta t$$

Circular Motion

Suppose that you deliver repeated kicks, at equal time intervals, to a rolling ball, each perpendicular to $\hat{p}$ at the current instant (Figure 5.27). If each perpendicular kick delivers the same impulse you can make the ball move along a roughly circular path, always turning to the left from its previous direction.

> QUESTION Suppose that instead of giving a series of infrequent big kicks you give a large number of small sideways kicks (with equal short times between these small kicks). What would the motion of the ball look like?

The ball would move along a path that is nearly indistinguishable from a circular path, as seen in Figure 5.28. We see that in order to make a ball move along a circular path, we have to continually push inward on the ball. If we always push perpendicular to the motion (and therefore toward the center of the circle), there is no parallel part of the force and so no change in the magnitude of the momentum, in which case the ball moves along a circular path with no change in speed, just a continuous change of direction.

> QUESTION Suppose that you kick just as often, but each of your kicks applies a bigger force than before. Would the radius of the circular path be larger or smaller?

Bigger kicks change the direction more, making sharper turns, so the radius of the circular path would be smaller (Figure 5.29). The larger the impulse, the larger $\vec{F}_\perp \Delta t$ is compared to $\vec{p}_{\text{now}}$, and the more the path will curve. Smaller impulses produce smaller changes; in fact, if you apply a very small sideways force, the motion will be nearly a straight line, which corresponds to an extremely large radius of curvature. In the limit of infinitesimal force, the radius will be infinite: the object moves in a straight line in the absence of a sideways force.

> QUESTION How does the speed of the ball affect the curvature of the path due to a perpendicular force of constant magnitude?

For a given impulse $\vec{F}_\perp \Delta t$, consider two different values of $\vec{p}_{\text{now}}$. The same impulse will produce a proportionally smaller change to a large initial momentum, so the higher the speed, the less a given force will make the path curve (Figure 5.30).

> QUESTION What would happen if you stopped kicking the ball? Would it continue to move in a circle?

The ball would move in a straight line, in the direction of its current momentum. Curving motion requires the continuous application of a force that is at least in part perpendicular to the instantaneous momentum of the object.

Calculating $\vec{F}_{\parallel}$ and $\vec{F}_{\perp}$

We have seen that the parallel part of a force, $\vec{F}_{\parallel}$, is associated with changes in the magnitude of the object's momentum (changes in $|\vec{p}|$) and the perpendicular part of the net force, $\vec{F}_{\text{net}\perp}$, is associated with changes in the direction of motion of the object (changes in $\hat{p}$). How do we calculate these vectors?

In Figure 5.31 the force $\vec{F}$ is expressed as the sum of two vectors: $\vec{F}_{\parallel}$ in the direction of $\hat{p}$ and $\vec{F}_{\perp}$ in the direction perpendicular to $\hat{p}$, in the plane defined by $\vec{F}$ and $\vec{p}$. Using direction cosines, we can see from the diagram that the part of $\vec{F}$ parallel to $\hat{p}$ is $|\vec{F}|\cos\theta$. This is the part of the force in the direction of $\hat{p}$, so:

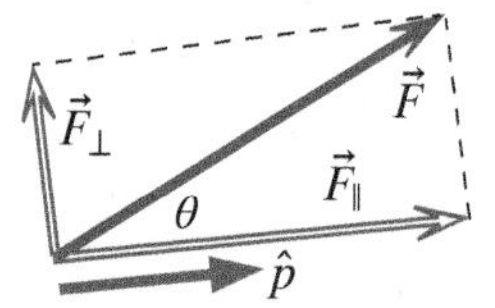

Figure 5.31 $\vec{F}$ can be expressed as the sum of $\vec{F}_{\parallel}$ and $\vec{F}_{\perp}$. We refer to the angle between $\vec{F}$ and $\hat{p}$ as θ.

$$\vec{F}_{\parallel} = |\vec{F}|\cos\theta\hat{p}$$

Since $\vec{F} = \vec{F}_{\parallel} + \vec{F}_{\perp}$ we can find $\vec{F}_{\perp}$ by subtraction:

$$\vec{F}_{\perp} = \vec{F} - \vec{F}_{\parallel}$$

EXAMPLE **Parallel Part of a Force**

A ball that is rolling in the $+z$ direction is kicked with a force of 50 N in a direction 33° from the $-z$ axis (Figure 5.32). Find $\vec{F}_{\parallel}$, the part of the force parallel to the ball's momentum. How will $\vec{F}_{\parallel}$ affect the magnitude of the ball's momentum?

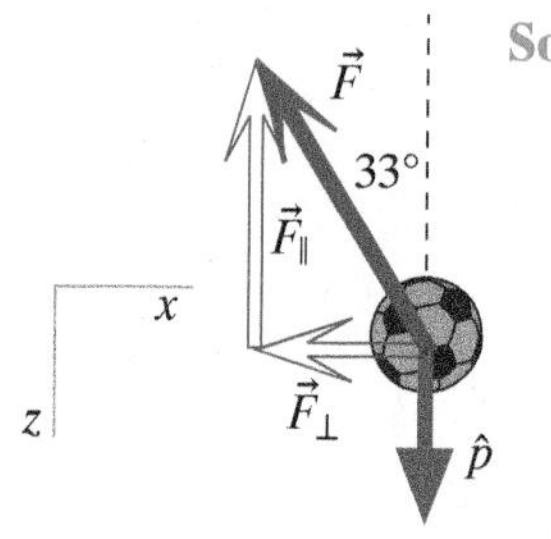

Figure 5.32 Top view, looking down on a ball rolling in the $+z$ direction. The ball is given a kick at an angle of 33° to the $-z$ axis.

Solution $\hat{p} = \langle 0,0,1\rangle$. The angle between $\vec{F}$ and $\hat{p}$ is $(180-33)° = 147°$.

$$\vec{F}_{\parallel} = (50\,\text{N})\cos(147°)\hat{p}$$
$$\vec{F}_{\parallel} = (-41.9\,\text{N})\hat{p} = \langle 0,0,-41.9\rangle\ \text{N}$$

The cosine of 147°, the angle between $\vec{F}$ and $\hat{p}$, is a negative number, so $\vec{F}_{\parallel}$ is opposite to $\hat{p}$. The magnitude of the ball's momentum will decrease.

QUESTION What would $\vec{F}_{\parallel}$ be if the angle between $\vec{F}$ and $\hat{p}$ were 90°?

The cosine of 90° is 0, so in this case $\vec{F}_{\parallel} = \vec{0}$, as it should.

The Vector Dot Product

The quantity $|\vec{F}|\cos\theta$ is an example of a more general mathematical operation called a vector dot product. For any two vectors $\vec{A}$ and $\vec{B}$, the dot product, which is written $\vec{A}\bullet\vec{B}$, is defined to be equal to $\vec{A}\bullet\vec{B} = |\vec{A}||\vec{B}|\cos\theta$, where θ is the angle between the two vectors (Figure 5.33):

$$\vec{A}\bullet\vec{B} = \vec{A}\bullet\vec{B} = |\vec{A}||\vec{B}|\cos\theta$$

The result of a vector dot product is a scalar, which may be positive or negative, depending on the angle θ.

We can write $\vec{F}_{\parallel}$ in terms of a vector dot product:

$$\vec{F}_{\parallel} = \left(\vec{F}\bullet\hat{p}\right)\hat{p} = |\vec{F}||\hat{p}|\cos\theta\hat{p}$$

Since $|\hat{p}| = 1$, the result is $|\vec{F}|\cos\theta\hat{p}$.

An alternative way of evaluating the vector dot product $\vec{A} \bullet \vec{B}$ uses the components of the vectors $\vec{A}$ and $\vec{B}$:

$$\vec{A} \bullet \vec{B} = A_x B_x + A_y B_y + A_z B_z$$

This result can be understood by expressing the two vectors in terms of the standard unit vectors $\hat{\imath} = \langle 1,0,0 \rangle$, $\hat{\jmath} = \langle 0,1,0 \rangle$, and $\hat{k} = \langle 0,0,1 \rangle$, and then multiplying out the product using the rules of ordinary algebra:

$$\langle A_x, A_y, A_z \rangle = (A_x\hat{\imath} + A_y\hat{\jmath} + A_z\hat{k})$$
$$\langle B_x, B_y, B_z \rangle = (B_x\hat{\imath} + B_y\hat{\jmath} + B_z\hat{k})$$
$$(A_x\hat{\imath} + A_y\hat{\jmath} + A_z\hat{k}) \bullet (B_x\hat{\imath} + B_y\hat{\jmath} + B_z\hat{k}) = A_x\hat{\imath} \bullet B_x\hat{\imath} + A_x\hat{\imath} \bullet B_y\hat{\jmath} + A_x\hat{\imath} \bullet B_z\hat{k} + \ldots$$

The dot product of any unit vector with itself is 1, because the angle between the vectors is 0°, and $\cos(0°) = 1$. Therefore:

$$\hat{\imath} \bullet \hat{\imath} = 1 \quad \text{and} \quad \hat{\jmath} \bullet \hat{\jmath} = 1 \quad \text{and} \quad \hat{k} \bullet \hat{k} = 1$$

In contrast, the dot product of any of these unit vectors with a different unit vector is zero, because the angles between them are 90°, and the cosine of 90° is zero. So

$$\hat{\imath} \bullet \hat{\jmath} = 0 \quad \text{and} \quad \hat{\jmath} \bullet \hat{k} = 0 \quad \text{and} \quad \hat{k} \bullet \hat{\imath} = 0$$

Therefore, any term such as $A_x\hat{\imath} \bullet B_y\hat{\jmath}$ is zero, and the only nonzero terms are those involving $\hat{\imath} \bullet \hat{\imath}$, $\hat{\jmath} \bullet \hat{\jmath}$, or $\hat{k} \bullet \hat{k}$:

$$(A_x\hat{\imath} + A_y\hat{\jmath} + A_z\hat{k}) \bullet (B_x\hat{\imath} + B_y\hat{\jmath} + B_z\hat{k}) = A_x B_x + A_y B_y + A_z B_z$$

Since in many circumstances we know the components of 3D vectors rather than knowing the angles between them (which are often not easy to calculate in 3D), the sum form of the dot product can be very useful.

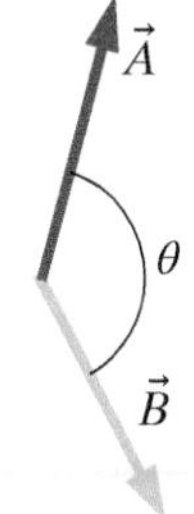

Figure 5.33 Two vectors $\vec{A}$ and $\vec{B}$, with an angle θ between them.

VECTOR DOT PRODUCT

Given two vectors $\vec{A} = \langle A_x, A_y, A_z \rangle$ and $\vec{B} = \langle B_x, B_y, B_z \rangle$, the vector dot product is defined as

$$\vec{A} \bullet \vec{B} = |\vec{A}||\vec{B}|\cos\theta = A_x B_x + A_y B_y + A_z B_z$$

The angle θ is the angle between the two vectors; $\theta \leq 180°$ (Figure 5.33). The dot product is sometimes also called the "scalar product," because its result is a scalar.

EXAMPLE **$\vec{F}_{\parallel}$ in a Planetary Orbit**

A planet orbits a star, following an elliptical path. At an instant when the momentum of the planet is $\langle 3 \times 10^{29}, -6.0 \times 10^{28}, 0 \rangle$ kg·m/s, the force on the planet by the star is $\langle -4 \times 10^{22}, 1 \times 10^{23}, 0 \rangle$ N (Figure 5.34). Find the parts of the gravitational force parallel to and perpendicular to the momentum of the planet. How will these forces affect the momentum of the planet in the next small time interval?

Solution

$$\hat{p} = \frac{\langle 3\times10^{29}, -6\times10^{28}, 0\rangle\ \text{kg}\cdot\text{m/s}}{\sqrt{(3\times10^{29})^2+(-6\times10^{28})^2}\ \text{kg}\cdot\text{m/s}}$$

$$= \langle 0.98, -0.20, 0\rangle$$

$$\vec{F}\bullet\hat{p} = (-4\times10^{22})(0.98)+(1\times10^{23})(-0.20)+(0)(0)\ \text{N}$$

$$= -5.9\times10^{22}\ \text{N}$$

$$\vec{F}_{\parallel} = (-5.9\times10^{22}\ \text{N})\langle 0.98, -0.20, 0\rangle$$

$$= \langle -5.8\times10^{22}, 1.2\times10^{22}, 0\rangle\ \text{N}$$

$$\vec{F}_{\perp} = \langle -4\times10^{22}, 1\times10^{23}, 0\rangle\ \text{N} - \langle -5.8\times10^{22}, 1.2\times10^{22}, 0\rangle\ \text{N}$$

$$= \langle 1.8\times10^{22}, 8.8\times10^{23}, 0\rangle\ \text{N}$$

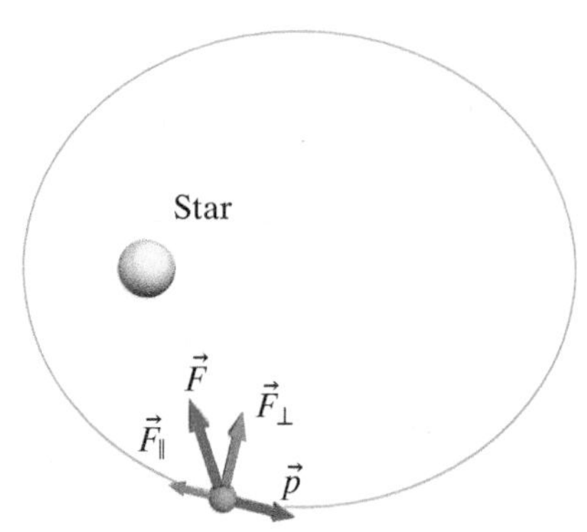

Figure 5.34 $\vec{F}$, $\vec{F}_{\parallel}$, and $\vec{F}_{\perp}$ at a particular instant for a planet orbiting a star.

Because the direction of $\vec{F}_{\parallel}$ is opposite to the direction of $\hat{p}$, the momentum of the planet will decrease during the next small time interval. Since $\vec{F}_{\perp}$ is not zero, the direction of the planet's momentum will also change.

In a computational model of curving motion, it can often be quite informative to display the parallel and perpendicular parts of the net force, since one determines changes in speed and the other changes in direction. The dot function in VPython provides a convenient way to calculate a dot product, and the norm function calculates a unit vector in the direction of a given vector. Assuming the current momentum and net force are the vectors `p` and `F`, one could use these VPython statements to calculate the force parts:

```
phat = norm(p)    # create unit vector from p
Fpara = dot(F, phat)*phat
Fperp = F - Fpara
```

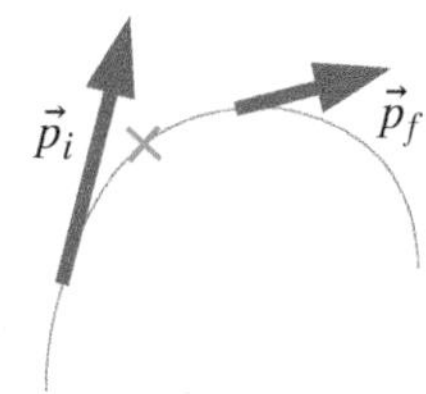

Figure 5.35 A system moves along the path shown. The arrows shown represent the momentum of the system at equal times before and after the time of interest, when the object is at the green x.

5.7 $d\vec{p}/dt$ FOR CURVING MOTION

It is not difficult to move from thinking about finite time steps to thinking about the instantaneous rate of change of momentum, $d\vec{p}/dt$, for curving motion. In Chapter 4 we used a three-step procedure to find $d\vec{p}/dt$ for a mass oscillating on a spring. We can use the same procedure, which is based on the definition of a derivative, to find $d\vec{p}/dt$ for an object whose path curves. Because we need to know the direction of $d\vec{p}/dt$ we often do this graphically, as in Figures 5.35–5.37.

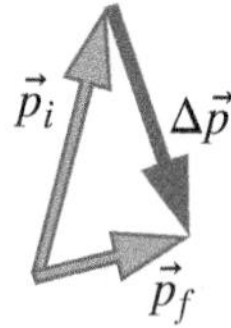

Figure 5.36 Graphically determine $\Delta\vec{p}$.

FINDING $d\vec{p}/dt$ GRAPHICALLY

1. Draw two arrows representing $\vec{p}$ at equal times before and after the time of interest (Figure 5.35).
2. Find $\Delta\vec{p} = \vec{p}_f - \vec{p}_i$ (Figure 5.36).
3. Divide $\Delta\vec{p}$ by Δt to get $\Delta\vec{p}/\Delta t$ (Figure 5.37). The smaller Δt is, the closer the result is to $d\vec{p}/dt$.

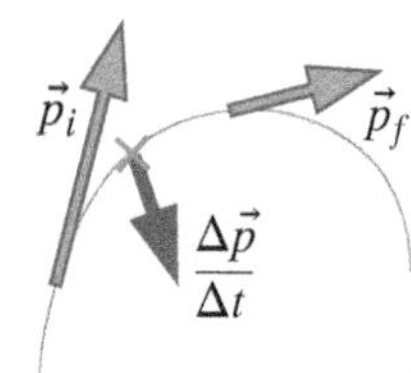

Figure 5.37 Divide $\Delta\vec{p}$ by Δt to get $\Delta\vec{p}/\Delta t$. Draw an arrow representing $\Delta\vec{p}/\Delta t$ at the location of interest. The smaller Δt is, the closer the result is to $d\vec{p}/dt$.

QUESTION For curving motion, does $d\vec{p}/dt$ always point toward the inside of the curve?

Yes. Refer to Figures 5.27–5.30, in which we saw how a force perpendicular to motion caused an object's path to curve in the direction of that force. $d\vec{p}/dt$ is in the same direction as $\vec{F}_{\text{net}}$, and must point at least partly toward the inside of the curve, although it may also have a part parallel to $\vec{p}$ if the speed is changing as well.

Parallel and Perpendicular Parts of $d\vec{p}/dt$

We have seen that at any instant the net force on an object can be expressed as the sum of two parts: one that is parallel to the object's momentum and one that is perpendicular to the system's momentum. The same is true for the vector $d\vec{p}/dt$:

$$\frac{d\vec{p}}{dt} = \frac{d\vec{p}}{dt}_{\parallel} + \frac{d\vec{p}}{dt}_{\perp}$$

We can see this mathematically by "factoring" momentum into the product of the magnitude of the momentum (a scalar) and a unit vector in the direction of the momentum, and then taking the derivative, by applying the product rule. The resulting expression has two pieces, which correspond to the two ways in which momentum can change, in magnitude and in direction. (We use the notation convention that $p = |\vec{p}|$ to make the following discussion easier to follow.)

$$\vec{p} = p\hat{p}$$

$$\frac{d\vec{p}}{dt} = \frac{dp}{dt}\hat{p} + p\frac{d\hat{p}}{dt}$$

This may look unfamiliar, but we can make sense of it by considering each term separately.

QUESTION Which term in the expression above involves a change in magnitude of momentum (change in speed)?

$\frac{dp}{dt}\hat{p}$ is nonzero if speed is changing, and zero otherwise.

QUESTION Which term involves a change in direction?

$p\frac{d\hat{p}}{dt}$ is nonzero if direction is changing, and zero otherwise.

We can relate the parallel and perpendicular parts of $d\vec{p}/dt$ to the parallel and perpendicular parts of the net force responsible for changing the momentum of a system:

PARALLEL AND PERPENDICULAR

$$\frac{d\vec{p}}{dt}_{\parallel} = \frac{dp}{dt}\hat{p} = \vec{F}_{\text{net}\,\parallel}$$

$$\frac{d\vec{p}}{dt}_{\perp} = p\frac{d\hat{p}}{dt} = \vec{F}_{\text{net}\,\perp}$$

Note that this division is useful only for smooth, continuous curving motion. At turning points or other times when $\hat{p}$ changes abruptly, $d\hat{p}/dt$ is undefined.

This way of splitting the Momentum Principle is useful in finding the values of unknown forces that cause curving motion. If we can find the value of either part of $d\vec{p}/dt$, this will allow us to deduce the value of the corresponding part of the net force. We can show that, if we know an object's momentum and the instantaneous radius of curvature of its curving path, we can calculate $d\vec{p}/dt_{\perp}$.

The Kissing Circle

Motion along a circular path is easily described by giving the radius of the circle. However, we need a way to describe the instantaneous curvature of an object's path when that path is not a circle.

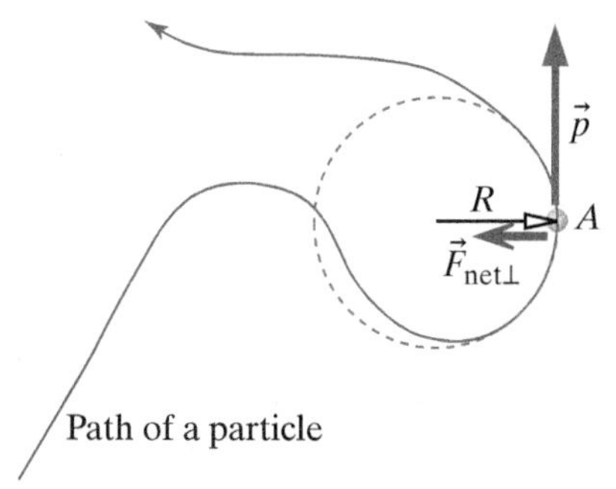

Figure 5.38 The "kissing circle" (dashed line) fits the trajectory at A as smoothly as possible.

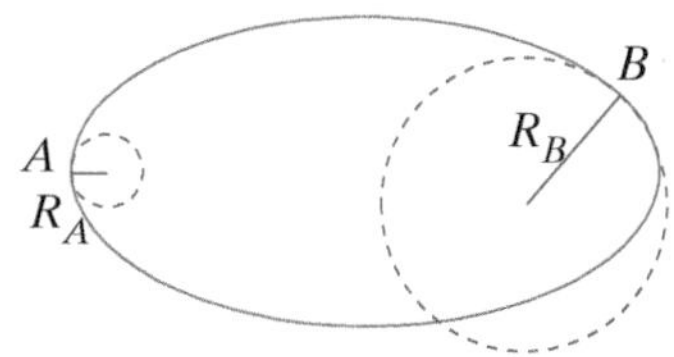

Figure 5.39 The radius of the kissing circle (dashed lines) can be different at different locations along a path, such as the elliptical path shown.

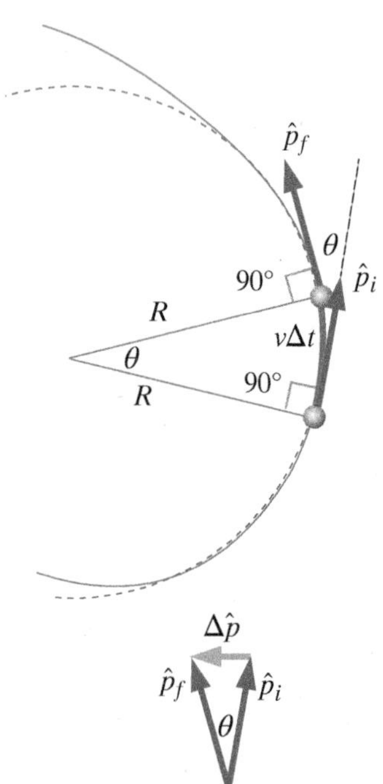

Figure 5.40 In a short time Δt, the particle moves a distance $v\Delta t$, through a small angle θ along a curve (red) whose radius of curvature is R. The angle θ, whose size is exaggerated in this diagram, is also the angle between $\hat{p}_i$ and $\hat{p}_f$. The kissing circle is indicated by a dashed line.

Consider motion along the curving path shown in Figure 5.38. A dashed circle has been drawn inside the path, tangent to the path at location A. This dashed circle of radius R is called the "osculatory" or "kissing" circle, because it just "kisses" the curving path at location A, fitting into the trajectory as smoothly as possible, with the circle and the trajectory sharing the same tangent ($\vec{p}$ is tangent to the path) and the same radius of curvature R at location A.

From the preceding discussion in which we considered applying a sequence of impulses, we see that both the perpendicular component of the net force, $\vec{F}_{net\perp}$, and the part of $d\vec{p}/dt$ perpendicular to the instantaneous momentum point toward the center of the kissing circle. If this force were bigger, the radius of the kissing circle would be smaller (a sharper turn). If on the other hand the momentum were bigger, the force would make a smaller change in direction in the same amount of time, and the kissing circle would be bigger. If the curvature of an object's path varies, then the radius of the kissing circle will be different at different locations along the path (Figure 5.39).

Calculating $d\vec{p}/dt_\perp$

We can start by using geometry to derive an expression for $d\hat{p}/dt$, the rate at which the direction of momentum changes. In Figure 5.40 we see a particle that in a short time Δt moves a distance $v\Delta t$, through a small angle θ along a curve whose radius of curvature is R. The small angle θ (measured in radians) is the arc length $v\Delta t$ divided by the radius R of the kissing circle: $\theta = v\Delta t/R$. (We've exaggerated the size of θ in order to be able to see it easily on the diagram.)

Because $\hat{p}$ is perpendicular to the radius of the kissing circle, as the particle moves through the small angle θ the vector $\hat{p}$ rotates through the same angle θ. At the bottom of Figure 5.40 we place the initial and final unit vectors tail to tail and show the change, $\Delta\hat{p}$. In this triangle, in the limit of a small angle θ, the angle θ is the length $|\Delta\hat{p}|$ divided by the length $|\hat{p}_i|$. Therefore

$$\theta = \frac{|\Delta\hat{p}|}{|\hat{p}_i|} = \frac{v\Delta t}{R}$$

Since $|\hat{p}_i| = 1$, we can rearrange to get this result:

$$\left|\frac{\Delta\hat{p}}{\Delta t}\right| = \frac{v}{R}$$

In the limit as Δt becomes infinitesimally small, we have the derivative:

$$\lim_{\Delta t\to 0}\left|\frac{\Delta\hat{p}}{\Delta t}\right| = \left|\frac{d\hat{p}}{dt}\right| = \frac{v}{R} \quad \text{on a smoothly curving path}$$

Check units: $\hat{p}$ is dimensionless, so $|d\hat{p}/dt|$ has units of $1/s$. Since v/R also has units of $1/s$, the units match.

RATE OF CHANGE OF DIRECTION

$$\left|\frac{d\hat{p}}{dt}\right| = \frac{v}{R}$$

R is the radius of the kissing circle. The direction of $d\hat{p}/dt$ is perpendicular to the particle's momentum, toward the center of the kissing circle. This result is valid *only* for a smoothly continuous curving motion, not for times when $\hat{p}$ changes abruptly, as in a collision or at a turning point.

We now have an expression for $|d\vec{p}/dt_{\perp}|$, the magnitude of the perpendicular part of $d\vec{p}/dt$:

MAGNITUDE OF $d\vec{p}/dt_{\perp}$

$$\left|\frac{d\vec{p}}{dt}_{\perp}\right| = p\left|\frac{d\hat{p}}{dt}\right| = p\left(\frac{v}{R}\right) = \frac{\gamma m v^2}{R}$$

At speeds much less than the speed of light:

$$\left|\frac{d\vec{p}}{dt}_{\perp}\right| \approx \frac{mv^2}{R}$$

The direction of $d\vec{p}/dt_{\perp}$ is toward the center of the kissing circle. R is the radius of the kissing circle. This result is valid *only* for a smoothly continuous curving motion, not for times when $\hat{p}$ changes abruptly, as in a collision or at a turning point.

If you examine the preceding derivation you'll see that it also applies to the rate of change of any unit vector that changes its direction at the same rate as the momentum unit vector, such as $\hat{v}$. Therefore we can write a similar expression for the perpendicular part of the acceleration $\vec{a} = d\vec{v}/dt$:

$$|\vec{a}_{\perp}| = \left|\left(\frac{d\vec{v}}{dt}\right)_{\perp}\right| = v\left|\frac{d\hat{v}}{dt}\right| = v\left(\frac{v}{R}\right) = \frac{v^2}{R}$$

Again, this result is valid *only* for a smoothly continuous curving motion, not for times when $\hat{v}$ changes abruptly, as in a collision or at a turning point.

EXAMPLE **Finding $\vec{F}_{\text{net}\perp}$**

Suppose that a 3 kg object is moving with a speed of 4 m/s and at a certain instant the radius of curvature of the kissing circle (the best fit to the curving motion at that point) is 2 m. What are the magnitude and direction of $\vec{F}_{\text{net}\perp}$?

Solution We don't know anything about the surroundings of the moving object, so we cannot calculate $\vec{F}_{\text{net}}$ directly. However, we can calculate $|d\vec{p}/dt_{\perp}|$, and use this to deduce the value of $|\vec{F}_{\text{net}\perp}|$:

$$\left|\frac{d\vec{p}}{dt}_{\perp}\right| = p\left(\frac{v}{R}\right) = (3\,\text{kg}\cdot 4\,\text{m/s})\frac{4\,\text{m/s}}{2\,\text{m}} = 24\,\text{kg}\cdot\text{m/s}^2$$

Therefore $|\vec{F}_{\text{net}\perp}| = 24\,\text{N}$

The direction of $(d\vec{p}/dt)_{\perp}$ is toward the center of the kissing circle, so the direction of $\vec{F}_{\text{net}\perp}$ is also toward the center of the kissing circle.

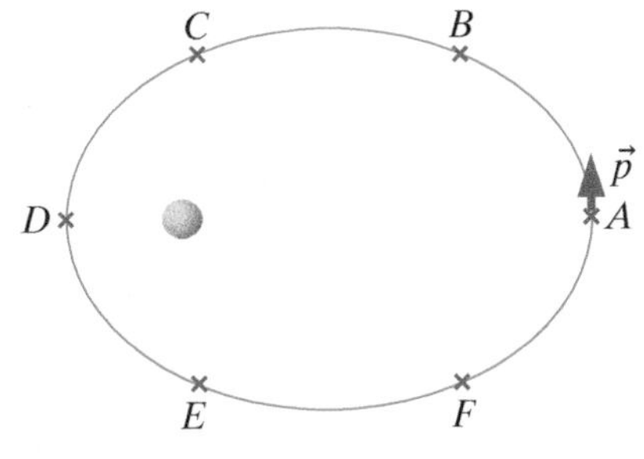

Figure 5.41 The elliptical orbit of a comet around a star. An arrow represents the momentum of the comet when it is at location A.

Checkpoint 4 A comet orbits a star in an elliptical orbit, as shown in Figure 5.41. The momentum of the comet at location A is shown in the diagram. At the instant the comet passes each location labeled A, B, C, D, E, and F, answer the following questions about the net force on the comet and the rate of change of the momentum of the comet: **(a)** Draw an arrow representing the direction and relative magnitude of the gravitational force on the comet by the star. **(b)** Is $\vec{F}_{\text{net}\perp}$ zero or nonzero? **(c)** Is $\vec{F}_{\text{net}\parallel}$ zero or nonzero? **(d)** Is $d|\vec{p}|/dt$ positive, negative, or zero? **(e)** Is $d\hat{p}/dt$ zero or nonzero?

"Centrifugal" and "Centripetal"

You may sometimes hear the terms "centrifugal" and "centripetal" in the context of discussions of forces and curving motion. These terms have simple meanings: "centrifugal" means "away from the center" and "centripetal" means "toward the center."

We have seen that both $d\vec{p}/dt_{\perp}$ and $\vec{F}_{\text{net}\,\perp}$ point toward the center of the kissing circle, so we could correctly refer to $\vec{F}_{\text{net}\,\perp}$ as a centripetal force, and $d\vec{p}/dt_{\perp}$ as a centripetal rate of change. However, there doesn't seem to be any advantage in using such terms.

> **QUESTION** Is there a "centrifugal" force involved in curving motion?

No. As long as we stay in our standard inertial reference frame, the force that causes an object's path to curve is directed toward the inside of the curve, not the outside. Probably it is the kinesthetic sensations that people have when riding in vehicles that turn that lead to confusion about this issue. For a discussion of how the kinesthetic sensations we experience are related to a physical analysis of motion, see Section 5.9. To avoid confusion, remember that each force included in a free-body diagram must be associated with an object in the surroundings.

Turning Points: System Momentarily at Rest

We have noted above that the quantity $d\hat{p}/dt$ is undefined at moments when an object's momentum changes abruptly, such as turning points where the direction of $\hat{p}$ is reversed. How do we analyze turning points, when an object is momentarily motionless?

EXAMPLE **Tarzan Swinging Back and Forth**

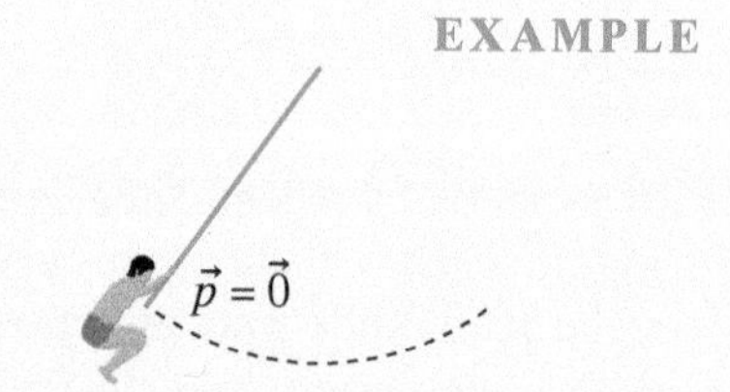

Figure 5.42 Tarzan is momentarily at rest.

Tarzan hangs from a vine, swinging back and forth. At the moment when he reaches his maximum displacement from the vertical, as indicated in Figure 5.42, Tarzan is momentarily at rest (a "turning point"). At this instant, is the rate of change of Tarzan's momentum zero or nonzero? If nonzero, what is the direction of $d\vec{p}/dt$? At this moment, is the net force on Tarzan zero or nonzero?

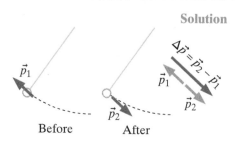

Solution We use the procedure for finding $d\vec{p}/dt$ shown in Figure 5.43. Both immediately before this moment and immediately after this moment, Tarzan's momentum is nonzero; this is a turning point. The directions of his momenta at these two instants are shown in Figure 5.43, as is $\Delta\vec{p} = \vec{p}_2 - \vec{p}_1$. Therefore, at the instant Tarzan is motionless, $d\vec{p}/dt$ is nonzero, and its direction is the direction of $\Delta\vec{p}$, as shown in Figure 5.44.

Figure 5.43 Tarzan's momentum an instant before and an instant just after he is momentarily at rest, and $\Delta\vec{p}$.

Note that it is important to consider a very short time interval in order to get the direction of $d\vec{p}/dt$. The direction of $\Delta\vec{p}$ would be different for a longer time interval (and not the same as the direction of $d\vec{p}/dt$).

To decide if the net force on Tarzan is zero or not, we can reason from the update form of the Momentum Principle:

$$\vec{p}_{\text{future}} = \vec{p}_{\text{now}} + \vec{F}_{\text{net,now}}\Delta t$$

Pick the moment when the system is at rest as "now," and a short time Δt later as "future." We know that $\vec{p}_{\text{now}}$ is zero but $\vec{p}_{\text{future}}$ will not be zero, so we deduce that $\vec{F}_{\text{net,now}}$ is nonzero.

We get the same answer reasoning from the derivative form of the Momentum Principle:

$$d\vec{p}/dt = \vec{F}_{\text{net}}$$

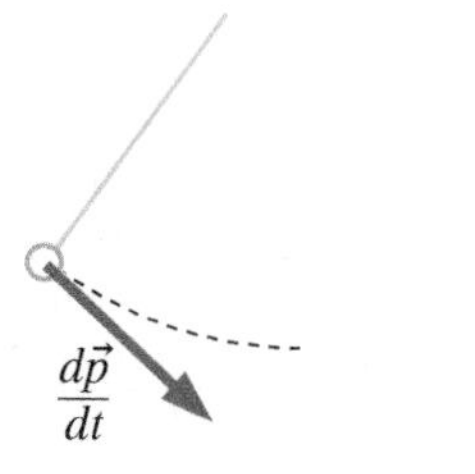

Figure 5.44 The direction of $d\vec{p}/dt$ at the instant Tarzan is motionless.

Since at a turning point $d\vec{p}/dt \neq \vec{0}$, then $\vec{F}_{\text{net}}$ must also be nonzero. The direction of $\vec{F}_{\text{net}}$ must be the same as the direction of $d\vec{p}/dt$.

QUESTION At a particular instant a block hanging from a spring is at rest. What is the net force on the block?

Without more information, we can't answer this question. If the block is at rest and remains at rest, this is a case of uniform motion, and the net force on the system is zero. However, if the block is only momentarily at rest, then the rate of change of its momentum is not zero, and the net force on the block is not zero.

- For a system that remains at rest, $d\vec{p}/dt = \vec{0}$.
- For a system that is only momentarily at rest, $d\vec{p}/dt \neq \vec{0}$.

5.8 UNKNOWN FORCES: CURVING MOTION

If we know the mass and speed of an object moving along a curving path whose instantaneous radius is known, we can calculate $|d\vec{p}/dt_\perp|$, and use this to find the magnitude of $\vec{F}_{\text{net}\,\perp}$. We know that at any instant the direction of $\vec{F}_{\text{net}\,\perp}$ is toward the center of the kissing circle.

In each problem we also need to think about whether or not $|d\vec{p}/dt_\parallel|$ is zero or nonzero at the particular instant of interest. If it is zero, then $|\vec{F}_{\text{net}\,\parallel}| = 0$, and $\vec{F}_{\text{net}} = \vec{F}_{\text{net}\,\perp}$; if not, we need more information in order to calculate $\vec{F}_{\text{net}\,\parallel}$.

EXAMPLE **A Child on a Merry-Go-Round**

A child whose mass is 30 kg sits on a merry-go-round at a distance of 3 m from the center (Figure 5.45). The merry-go-round makes one revolution every 8 s. What are the magnitude and direction of the net force acting on the child?

Solution System: Child
Surroundings: Earth, seat

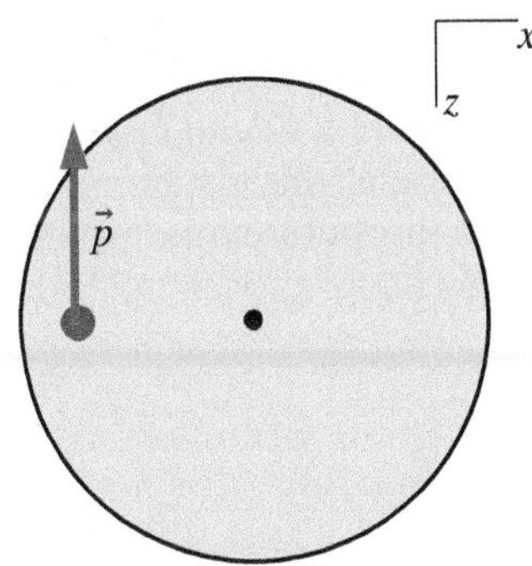

Figure 5.45 Top view of a child (red dot) on a merry-go-round that is rotating clockwise. At the instant shown the child's momentum is in the $-z$ direction.

Parallel: Because the merry-go-round revolves at a constant rate, the speed of the child, and the magnitude of the child's momentum, are constant. Therefore we know that:

$$\frac{d\vec{p}}{dt}_\parallel = \vec{0} \quad \text{and therefore} \quad \vec{F}_{\text{net}\,\parallel} = \vec{0}$$

Perpendicular: The direction of the child's momentum is changing continuously, so:

$$\left|\frac{d\vec{p}}{dt}_\perp\right| = \frac{mv^2}{R} = |\vec{F}_{\text{net}\,\perp}|$$

What is the speed v? In 8 s the child travels once around the circumference $2\pi R$, so the speed is $v = 2\pi(3\,\text{m})/(8\,\text{s}) = 2.36\,\text{m/s}$. So

$$|\vec{F}_{\text{net}\,\perp}| = \frac{(30\,\text{kg})(2.36\,\text{m/s})^2}{3\,\text{m}} = 55.5\,\text{N}$$

Since $\vec{F}_{\text{net}\,\perp}$ is the only nonzero part of the net force, we know that this is also the magnitude of the net force, which at any instant is directed toward the center of the circle. In the instant shown in Figure 5.45, this force would be in the $+x$ direction.

> **QUESTION** What object in the surroundings exerts this force on the child?

We listed only two objects in the surroundings that exert significant forces on the child. The Earth exerts a downward force, so that isn't it. The seat exerts a force on the child that is partly upward, balancing the Earth's force, and partly toward the center of the circle. This part of the force is a friction force—if the seat were extremely slippery, the child would slide off it instead of moving around in a circle.

Tension Forces and Curving Motion

We now have all the tools we need to answer the question about Tarzan and the vine posed at the beginning of Chapter 4: Why does the vine break when Tarzan swings on it, but not when he hangs motionless? In the following problem, we can calculate $d\vec{p}/dt$, but the magnitude of the tension force on Tarzan by the vine is unknown. We can find the unknown $|\vec{F}_{\text{vine}}|$ by applying the Momentum Principle.

EXAMPLE **Tarzan and the Broken Vine**

Tarzan, whose mass is 90 kg, wants to use a vine to swing across a river. To make sure the 8-m-long vine is strong enough to support him, he tests it by hanging motionless on the vine for several minutes. The vine passes this test, so Tarzan grabs the vine and swings out over the river. He is annoyed and perplexed when the vine breaks midway through the swing, when his speed is 12 m/s (Figure 5.46), and he ends up drenched and shivering in the middle of the cold river, to the great amusement of the onlooking apes. Why did the vine break while Tarzan was swinging on it, but not while he hung motionless from it?

Solution System: Tarzan
Surroundings: Vine, Earth (neglect air)
Free-body diagram: Figure 5.47

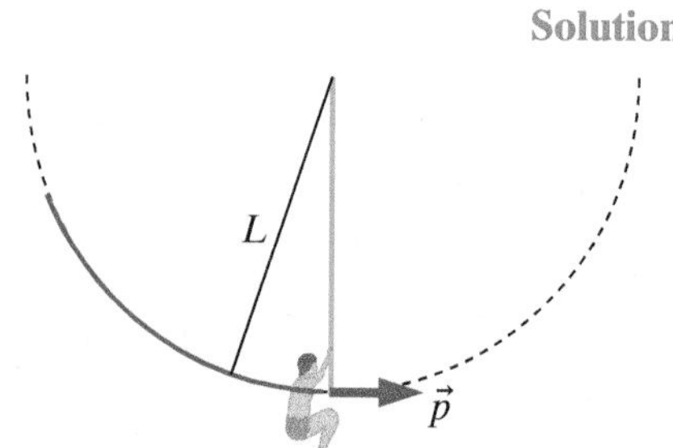

Figure 5.46 Tarzan at the bottom of his swing, the instant before the vine breaks. The colored line indicates his path up to this moment. The radius of the kissing circle (dashed line) is L, the length of the vine. At this moment Tarzan's speed is at a maximum.

$\vec{F}_T$ (vine)
$\vec{p}$
$\vec{F}_{\text{grav}}$ (Earth)

Figure 5.47 Free-body diagram.

Parallel: At the instant shown in Figure 5.46, we know that the only forces acting on Tarzan are in the $+y$ or $-y$ direction (Figure 5.47), so $F_{\text{net}\parallel} = 0$. This implies that $|d\vec{p}/dt_{\parallel}| = 0$. This makes sense because at this instant $|\vec{p}|$ is going through a maximum: Tarzan's speed is increasing just before this instant (as he descends), and would decrease just afterward (as he went back up, if the vine did not break). So

$$\frac{d\vec{p}}{dt}_{\parallel} = \vec{F}_{\text{net}\parallel} = \vec{0}$$

Perpendicular: The direction of $d\vec{p}/dt_{\perp}$ is upward, toward the center of the kissing circle.

$$\left|\frac{d\vec{p}}{dt}_{\perp}\right| = \left|\vec{F}_{\text{net}\perp}\right|$$

$$\frac{mv^2}{L} = F_T - mg$$

$$\frac{(90\,\text{kg})(12\,\text{m/s})^2}{(8\,\text{m})} = F_T - (90\,\text{kg})(9.8\,\text{N/kg})$$

$$F_T = 2502\,\text{N}$$

QUESTION When Tarzan hung motionless on the vine, what was the tension in the vine?

$$\frac{dp_y}{dt} = F_T - mg$$
$$0 = F_T - (90\,\text{kg})(9.8\,\text{N/kg})$$
$$F_T = 882\ \text{N}$$

The tension force exerted by the vine was nearly three times as great when Tarzan was midway through his swing as it was when he hung motionless. The interatomic bonds in the vine were stretched three times as much while he swung—evidently too much!

QUESTION Why must the vine exert a greater force when Tarzan's momentum is changing than when he hangs motionless?

It is not simply that Tarzan is moving—zero net force would be required if Tarzan's momentum were constant. However, to turn Tarzan's momentum from horizontal to upward requires an upward net force. The extra tension in the vine is necessary to change the direction of Tarzan's momentum. To swing upward Tarzan has to pull down harder on the vine, and therefore the vine stretches more, pulling up more on Tarzan.

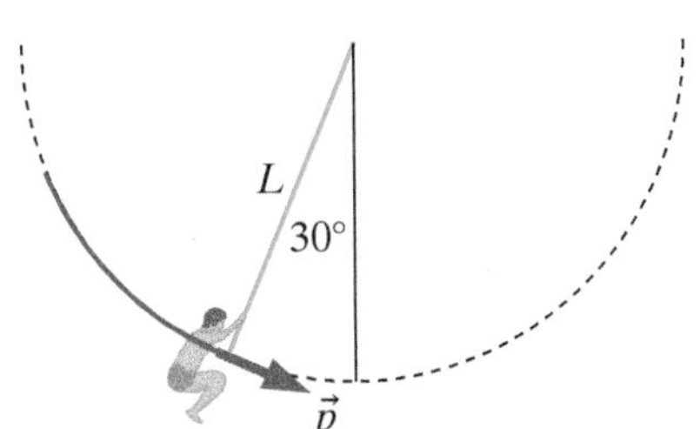

Figure 5.48 Tarzan swinging downward, with a speed of 11.1 m/s.

In the preceding example we picked a special instant when $|d\vec{p}/dt_{\parallel}|$ was zero. How would we go about finding the tension in the vine at some other instant?

EXAMPLE **Magnitude and Direction of Momentum Changing**

Consider the instant shown in Figure 5.48, before the vine breaks, when the angle between the vine and the vertical is 30°. At this instant Tarzan's speed is 11.1 m/s, and both the magnitude and direction of his momentum are changing. As before, Tarzan's mass is 90 kg, and the vine is 8 m long. **(a)** What is the tension in the vine at this moment? **(b)** What is the rate of change of the magnitude of Tarzan's momentum?

Solution System: Tarzan
Surroundings: Earth, vine (ignore the air)
Free-body diagram: Figure 5.49

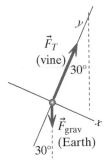

Figure 5.49 Free-body diagram. We have chosen to align the x axis with $\hat{p}$, and the y axis perpendicular to $\hat{p}$.

Here we have chosen to orient our coordinate axes so that the x axis is aligned with $\hat{p}$. Therefore the y axis is perpendicular to $\hat{p}$, and is aligned with the vine. This is done for convenience; we would get the same answers for the magnitudes of the quantities if we had chosen a different orientation for the axes, but the unit vectors $\hat{p}$ and $\hat{F}_T$ would be different.

(a) Perpendicular (y):

$$\left|\frac{d\vec{p}}{dt}_{\perp}\right| = \left|\vec{F}_{\text{net}\,\perp}\right|$$
$$\frac{mv^2}{L} = F_T + mg\cos(180° - 30°)$$
$$\frac{(90\,\text{kg})(11.1\,\text{m/s})^2}{(8\,\text{m})} = F_T + (90\,\text{kg})(9.8\,\text{N/kg})\cos 150°$$
$$F_T = 2150\ \text{N}$$
$$\vec{F}_T = \langle 0, 2150, 0\rangle\ \text{N}$$

(b) Parallel (x):

$$\left|\frac{d\vec{p}}{dt}_{\parallel}\right| = \left|\vec{F}_{\text{net}\,\parallel}\right|$$

$$\frac{d|\vec{p}|}{dt} = mg\cos(90-30)^\circ$$

$$= (90\,\text{kg})(9.8\,\text{N/kg})\cos 60^\circ$$

$$\frac{d|\vec{p}|}{dt} = 441\,\text{kg}\cdot\text{m/s}^2$$

$$\frac{d\vec{p}}{dt}_{\parallel} = \langle 441, 0, 0\rangle\ \text{kg}\cdot\text{m/s}^2$$

The tension in the vine is greater than when Tarzan was hanging motionless, but not as large as it will be at the bottom of the swing.

EXAMPLE **Swinging a Bucket in a Vertical Plane**

You swing a bucket of water in a vertical plane at the end of a rope, fast enough that the water stays in the bucket. The center of mass of the bucket and water moves approximately in a circle of radius 1.2 m. The speed of the bucket is faster at the bottom and slower at the top. The bucket with the water has a mass of 3 kg. At the instant when the rope is horizontal, the speed of the bucket is 9 m/s. What is the tension in the rope when the rope is horizontal? What is the rate of change of the speed at this moment?

Solution System: Bucket with water
Surroundings: Earth, rope
Free-body diagram: Figure 5.50

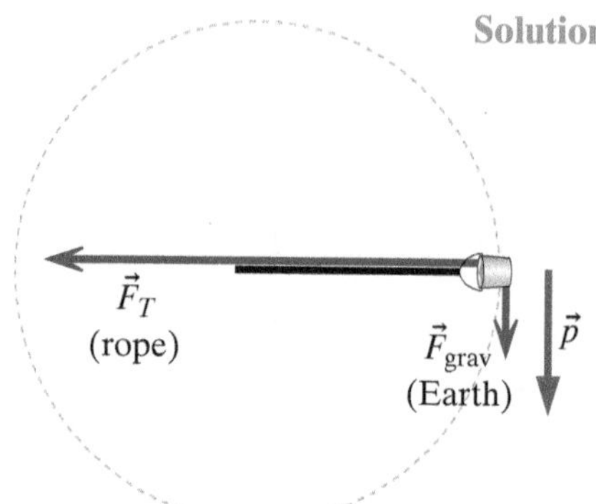

Figure 5.50 Swinging a bucket of water in a vertical plane.

Perpendicular:

$$\left|\frac{d\vec{p}}{dt}_{\perp}\right| = \left|\vec{F}_{\text{net}\,\perp}\right|$$

$$\frac{mv^2}{R} = F_T$$

$$\frac{(3\,\text{kg})(9\,\text{m/s})^2}{1.2\,\text{m}} = F_T$$

$$F_T = 202.5\,\text{N}$$

If the bucket simply hung motionless from the rope, the tension in the rope would be only $mg = (3\,\text{kg})(9.8\,\text{N/kg}) = 29.4\,\text{N}$, much less than 202.5 N. However, as we saw with Tarzan and the vine, during curving motion the tension in the vine might be greater than mg.

Parallel:

$$\left|\frac{d\vec{p}}{dt}_{\parallel}\right| = \left|\vec{F}_{\text{net}\,\parallel}\right|$$

$$m\frac{d|\vec{v}|}{dt} = mg$$

$$\frac{d|\vec{v}|}{dt} = |\vec{a}| = g = 9.8\,\text{m/s}^2$$

Figure 5.51 The bucket just barely makes it over the top with very little tension in the rope.

Note that this is only the case when the gravitational force is parallel to the momentum of the bucket. At other points in the swing, dv/dt will be different.

QUESTION How fast must the bucket be moving at the top of the swing?

There is a minimum speed necessary for the bucket to make it over the top without the rope going slack. In Figure 5.51 we see that as long as the rope is not slack there are two downward forces acting on the bucket, due to the rope and the Earth. If the bucket just barely makes it over the top of the circular loop with the rope just barely taut, the force of the rope is negligible compared to the force of the Earth, so we have $mv^2/R = mg$, or $v = \sqrt{gR} = \sqrt{(9.8\,\text{N/kg})(1.2\,\text{m})} = 3.4\,\text{m/s}$. Any speed at the top that is less than 3.4 m/s will mean that the rope has gone slack, and the motion won't be a circle any more.

QUESTION Why doesn't the water fall out of the bucket when the bucket is upside down?

As long as the speed at the top is at least 3.4 m/s the water will not fall out of the bucket, because its motion in a circle is perfectly compatible with the Momentum Principle. Of course the water actually is falling, moving downward and to the right, but you keep pulling inward on the bucket by pulling on the rope, and you're continually "catching" the water as it falls.

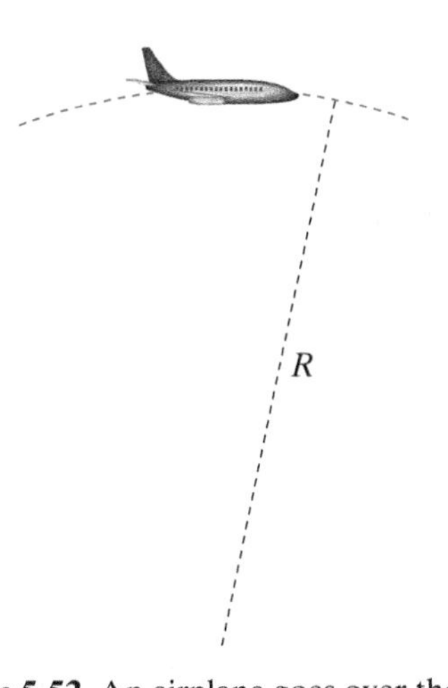

Figure 5.52 An airplane goes over the top of its path with speed v. The kissing circle has radius R at the top of the path.

Compression Forces and Curving Motion

Just as we have deduced the value of unknown tension forces, we can use the known motion of a system to deduce the value of an unknown compression force, such as the force that a seat exerts on a person sitting on it.

EXAMPLE **Compression Force**

A passenger of mass M rides in an airplane flying along a curved path at constant speed v as shown in Figure 5.52, where the radius of the kissing circle at the top of the path is R. At the instant shown, when the airplane is at the top of its arc, how large is the compression force that the seat bottom exerts on the passenger? Compare this force to the force the seat exerts on the passenger when the airplane moves in a straight line at constant speed.

Solution System: Passenger
Surroundings: Seat, Earth
Free-body diagram: Figure 5.53

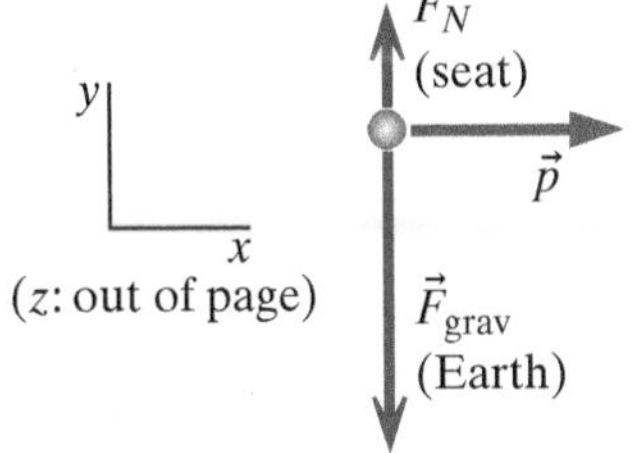

Figure 5.53 Free-body diagram.

Parallel: Because the passenger travels at constant speed, $|\vec{p}|$ is constant. Therefore we know that:

$$\frac{d\vec{p}}{dt}_{\parallel} = \vec{0} \quad \text{and therefore} \quad \vec{F}_{\text{net}\,\parallel} = \vec{0}$$

Perpendicular: We need to be careful of signs in this part. The direction of $d\hat{p}/dt$ is the $-y$ direction. Working with y components, we have:

$$\frac{dp_y}{dt} = F_{\text{net},y}$$

$$\frac{-Mv^2}{R} = F_N - Mg$$

$$F_N = Mg - \frac{Mv^2}{R}$$

QUESTION Why isn't $F_N = Mg$?

The minus sign in the result is important. The upward force of the seat F_N on the passenger's bottom is *less* than the passenger's weight Mg. The interatomic bonds in the seat are compressed less than they would be if the airplane's momentum were constant. The seat is being pulled down from under the passenger.

QUESTION Since $F_N = Mg - Mv^2/R$, if the airplane's speed v is big enough, the force exerted by the seat could be zero. Does that make any sense?

Yes, that's the case where the airplane changes direction so quickly that the seat is essentially jerked out from under the passenger, and there is no longer contact between the seat and the passenger. No contact, no force of the seat on the passenger. You may have experienced a similar situation on a roller coaster going over a hill.

5.9 KINESTHETIC SENSATIONS

EXAMPLE **A Turning Car**

Suppose that you're riding as the passenger in a convertible with left-hand drive (American or continental European), so you are sitting on the right (Figure 5.54). Assume that you have foolishly failed to fasten your seat belt. The driver makes a sudden right turn. Before studying physics you would probably have said that you are "thrown to the left," and in fact you do end up closer to the driver. But there's something wrong with this "thrown to the left" idea. What object in the surroundings exerted a force to the left to make you move to the left?

Solution Don't confuse your kinesthetic perceptions with the physical analysis. There is no object that exerts an outward (leftward) force. To understand what's happening, watch the turning car from a fixed vantage point above the convertible, and observe carefully what happens to the passenger. In the absence of forces to change the passenger's direction of motion, the passenger keeps moving ahead, in a straight line at the same speed as before (Figure 5.55). It is the car that is yanked to the right, out from under the passenger. The driver moves closer to the passenger; it isn't that the passenger moves toward the driver.

The driver may also feel "thrown to the left" against the door. However, it is the door that runs into the driver, forcing the driver to the *right*, not to the left.

Figure 5.54 You are a passenger in a car that makes a sudden right turn.

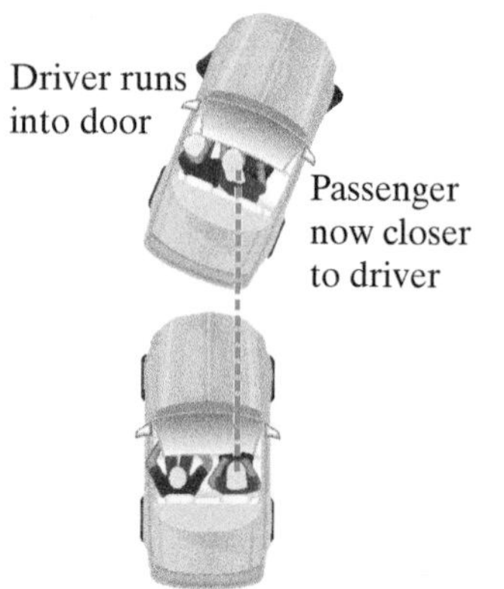

Figure 5.55 The passenger moves straight ahead and is now closer to the driver, who was thrown to the right by the door pushing the driver to the right.

There exists an advanced technique for using a noninertial rotating reference frame fixed to the car, in which case it is necessary to introduce additional "pseudoforces" not associated with objects in the surroundings, such as a centrifugal force. In this textbook we will always carry out our analyses in an inertial reference frame, a frame that is not accelerating, including not rotating. In an inertial reference frame there is no centrifugal force, and all forces are associated with objects in the surroundings.

In the example of the airplane traveling through an arc at constant speed (Section 5.8), what does the situation look and feel like to the passenger? From the passenger's point of view it is natural to think that the passenger has been thrown upward, away from the seat. Actually, though, the seat has been yanked out from under the passenger, in which case the passenger is falling toward the Earth (being no longer supported by the seat), but is changing velocity less rapidly than the airplane and so the ceiling comes closer to the passenger.

In the extreme case, if the airplane's maneuver is very fast, it may seem as though the passenger is "thrown up against the cabin ceiling." In actual fact, however, it is the cabin ceiling that is yanked down and hits the passenger!

You can see the value of wearing a seat belt, to prevent losing contact with the seat. When the airplane yanks the seat downward, the seat belt yanks the passenger downward. This may be uncomfortable, but it is a lot better than having the cabin ceiling hit you hard in the head.

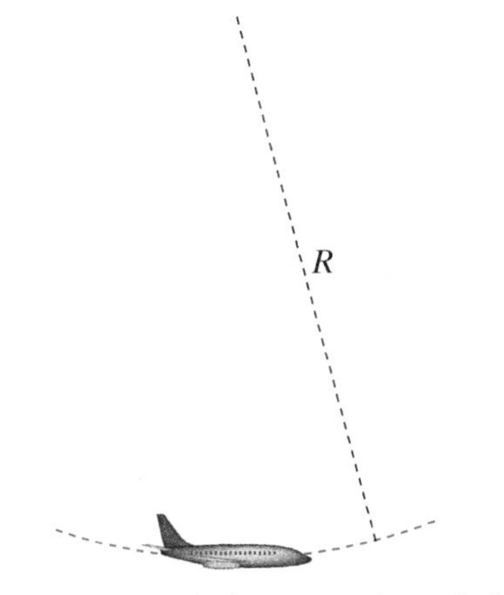

Figure 5.56 An airplane goes through the bottom of its path.

QUESTION How would this analysis change if the airplane goes through the bottom of a curve rather than the top (Figure 5.56)?

Now the center of the kissing circle is above the airplane, so $d\vec{p}/dt$ points upward. That means that the net force must also point upward.

QUESTION Which is larger, the force of the seat on the passenger or the gravitational force of the Earth on the passenger?

The force of the seat is greater than Mg, since the net force is upward. The passenger sinks deeper into the seat. Or to put it another way, the airplane yanks the seat up harder against the passenger's bottom, squeezing the seat against the passenger.

We Perceive Contact Forces

As you sit reading this text, you feel a sensation that you associate with weight. If someone suddenly yanks the chair out from under you, you feel "weightless," which feels funny and odd. It also feels scary, because you know from experience that whenever you lose the support of objects under you (chair, airplane seat, floor, mountain path), something bad is about to happen.

In reality, what we perceive as "weight" is actually not the gravitational force at all but rather the forces of atoms in the chair (or airplane seat or floor or mountain path) on atoms in your skin. You have nerve endings that sense the compression of interatomic bonds in your skin, and you interpret this as evidence for gravity acting on you. However, if you were placed in a spaceship that was accelerating, going faster and faster, its floor would squeeze against you in a way that would fool your nerve endings and brain into thinking you were subjected to gravity, even if there are no stars or planets nearby.

If you lose contact with the seat in an airplane that is going rapidly over the top of a curving path, your nerve endings no longer feel any contact forces, and you feel "weightless." Yet this is a moment when the only force acting on you is in fact the Earth's gravitational force. Weightlessness near the Earth paradoxically is associated with being subject only to your weight Mg.

Contact Forces Internal to the Body

It isn't just nerve endings in your skin that give you the illusion of "weight." As you sit here reading, your internal organs press upward on other organs above them, making the net force zero. You feel these internal contact forces. If you are suddenly weightless, a main reason for feeling funny is that the forces one organ exerts on another inside your body are suddenly gone. Your inner ear is also affected, which can produce dizziness.

To train astronauts, NASA has a cargo plane that is deliberately flown in a curving motion over the top, so that people in the padded interior lose contact with the floor and seem to float freely (in actual fact, they are accelerating toward the Earth due to the Mg forces acting on them, but the plane is also deliberately accelerating toward the Earth rather than flying level, so the people don't touch the walls and appear to be floating). The brain is crying out for the usual comforting signals from the nerve endings and not getting them. This airplane is sometimes called the "vomit comet" in honor of the effects it can have on trainees.

5.10 MORE COMPLEX PROBLEMS

Problems that ask more than "What is the magnitude of $\vec{F}_{\text{net}\perp}$?" are more complex to solve. However, typically the solution emerges if we simply begin the analysis from the Momentum Principle, exactly as we have done in all the preceding problems. Even if it is not obvious at first where the desired answer will come from, this kind of "forward reasoning," which is typical in physics problem solving, will almost always lead to a solution.

EXAMPLE **An Amusement Park Ride**

There is an amusement park ride that some people love and others hate in which a bunch of people stand against the wall of a cylindrical room of radius R, and the room starts to rotate at higher and higher speed (Figure 5.57). The surface of the wall is designed to maximize friction between the person and the wall, so it is fuzzy or sticky, not slick. When a certain critical speed is reached, the floor drops away, leaving the people stuck against the wall as they whirl around at constant speed. How fast must the ride go to keep a person from falling down?

Solution System: Person
Surroundings: Wall, Earth (neglect air resistance)
Free-body diagram: Figure 5.58

Figure 5.57 An amusement park ride.

At the instant shown in Figure 5.58 the person is in contact with the right wall, with momentum directed into the page ($-z$).

QUESTION Why doesn't the person slide down the wall?

We know the Earth pulls down on the person, so we deduce that there must be an upward force that is equal in magnitude to the downward gravitational force. The only object in contact with the person is the wall, so this force must be exerted by the wall. The free-body diagram reflects this deduction. In the diagram the force due to the wall has been separated into two parts: a normal part $\vec{F}_N$ that turns the direction of the person's momentum and a frictional part $\vec{f}$ that keeps the person from sliding downward.

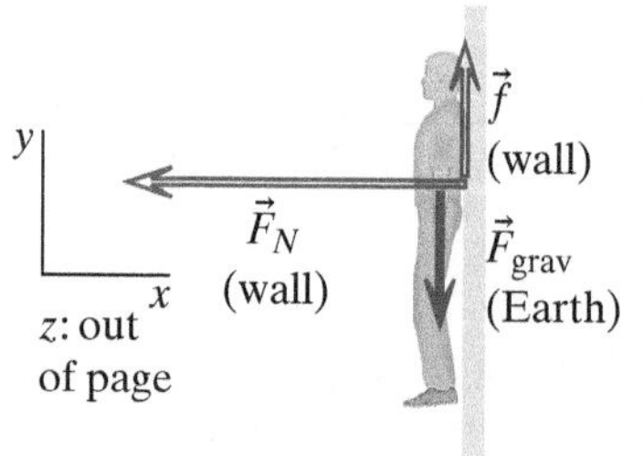

Figure 5.58 Free-body diagram.

Parallel (z): No change in speed, so $\dfrac{d\vec{p}}{dt}_{\parallel} = \vec{0}$.

There are two perpendicular directions, x and y. In the x direction the momentum of the person is changing:

$$\left|\frac{dp_x}{dt}\right| = |-F_N|$$

$$\frac{mv^2}{R} = F_N$$

This implies that the faster the ride goes, the larger the inward force (the normal force F_N) of the wall.

In the y direction there is no change in momentum, so

$$\frac{dp_y}{dt} = f - mg$$
$$0 = f - mg$$
$$f = mg$$

Recall from Chapter 4 that $\mu_s F_N \geq f$: a static friction force f (a friction force on an object that is not moving) has a maximum value of $\mu_s F_N$, where μ_s is the coefficient of static friction, and depends on the materials that are in contact with each other. If a force greater than $\mu_s F_N$ is applied to the object, it will begin to slide. In the case of the person on the ride, as long as $\mu_s F_N \geq mg$, the person will be supported. So the requirement is that

$$\mu_s F_N \geq mg$$
$$\mu_s \frac{mv^2}{R} \geq mg$$
$$\mu_s \geq \frac{gR}{v^2}$$

This predicts that the lower the speed v, the greater the coefficient of friction must be to hold the person up on the wall, which makes sense. Conversely, if the ride spins very fast, the wall doesn't have to have a very large coefficient of friction to hold the people against the wall.

EXAMPLE **Earth Satellite**

How long does it take a satellite in a circular orbit of radius R to make one complete revolution about the Earth?

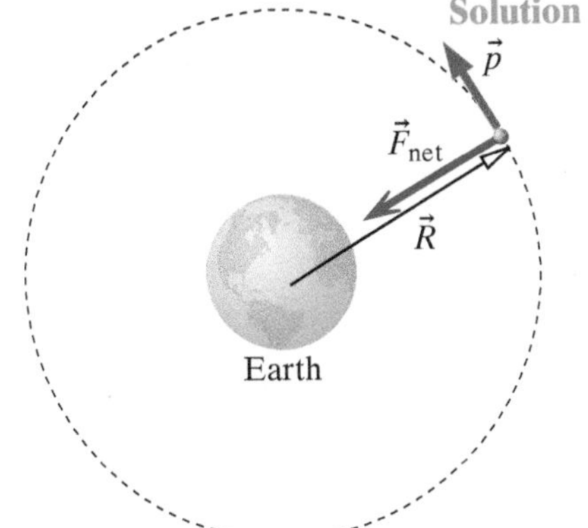

Figure 5.59 A satellite in a circular orbit around the Earth.

Solution

System: Satellite
Surroundings: Earth
Free-body diagram: Figure 5.59

Parallel:

Right-hand side of equation: $F_{net\parallel} = 0$

Therefore, left-hand side is $\frac{d|\vec{p}|}{dt} = 0$ (constant speed)

Perpendicular component (toward center):

Left-hand side of equation: $\left|\left(\frac{d\vec{p}}{dt}\right)_\perp\right| = \left(\frac{v}{R}\right)\gamma mv \approx \frac{mv^2}{R}$

Right-hand side of equation: $F_{net\perp} = \frac{GMm}{R^2}$

These two very different things, one the rate of change of the satellite's momentum and the other the gravitational attraction between the Earth and the satellite, are related to each other by the Momentum Principle to be numerically equal:

$$\frac{mv^2}{R} = \frac{GMm}{R^2}$$
$$v^2 = \frac{GM}{R}$$
$$v = \sqrt{\frac{GM}{R}}$$

The satellite goes a distance of the circumference $2\pi R$ in a time T, called the "period" of the orbit (and we say that the repetitive motion is "periodic"). So we have

$$\frac{2\pi R}{T} = \sqrt{\frac{GM}{R}}$$

$$T = \frac{2\pi R}{\sqrt{\frac{GM}{R}}} = \frac{2\pi R^{3/2}}{\sqrt{GM}}$$

Since the period is proportional to $R^{3/2}$, the smaller the radius of the orbit, the shorter the period T. The shortest period for an Earth satellite is for one in which the satellite orbits just high enough that air resistance isn't significant. This minimum height is about 50 km (5×10^4 m), so the radius of the orbit of a "near-Earth" satellite is $R_{\text{Earth}} = 6.4 \times 10^6$ m plus 5×10^4 m, which is not very different from the radius of the Earth. The period for a near-Earth satellite is therefore

$$T = \frac{2\pi(6.4 \times 10^6\,\text{m})^{3/2}}{\sqrt{(6.7 \times 10^{-11}\text{N}\cdot\text{m}^2/\text{kg}^2)(6 \times 10^{24}\,\text{kg})}} = 5074\,\text{s} = 85\,\text{min}$$

The speed of a near-Earth satellite is $v = 2\pi R/T = 2\pi(6.4 \times 10^6\,\text{m})/(5074\,\text{s}) = 7925\,\text{m/s} \approx 8$ km/s. A chemical rocket must accelerate from rest on the launch pad to 8 km/s in order to go into a minimal near-Earth orbit.

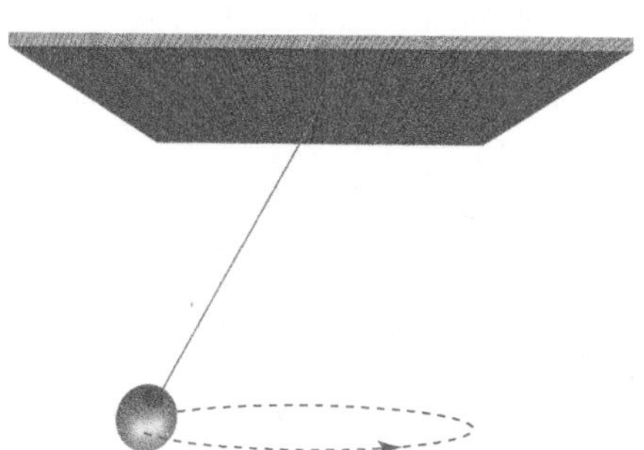

Figure 5.60 A circular pendulum, moving in the xz plane.

EXAMPLE **Circular Pendulum**

A ball is suspended from a string, and after being given a push, moves along a horizontal circular path, as shown in Figure 5.60. You measure the length of the string L and the angle θ, shown in Figure 5.61. You also time the motion, and find that it takes T s for the ball to make one complete circular trip. From these measurements, determine the gravitational constant g. (Newton used a large circular pendulum to measure g with considerable accuracy.)

Solution System: Ball

Surroundings: String, Earth (neglect air resistance)

Free-body diagram: Figure 5.61.

$$d\vec{p}/dt = \vec{F}_{\text{net}}$$

$$\hat{F}_T = \langle \cos(90° - \theta), \cos\theta, 0 \rangle$$

$$\vec{F}_T = F_T\langle \cos(90° - \theta), \cos\theta, 0 \rangle$$

$$\vec{F}_{\text{net}} = \langle F_T\cos(90° - \theta), (F_T\cos\theta - mg), 0 \rangle$$

Parallel component $(+z)$:

$$\frac{dp_z}{dt} = F_{\text{net},z} = 0$$

Perpendicular components:

$$y: \frac{dp_y}{dt} = F_T\cos\theta - mg$$

$$0 = F_T\cos\theta - mg$$

$$F_T = \frac{mg}{\cos\theta}$$

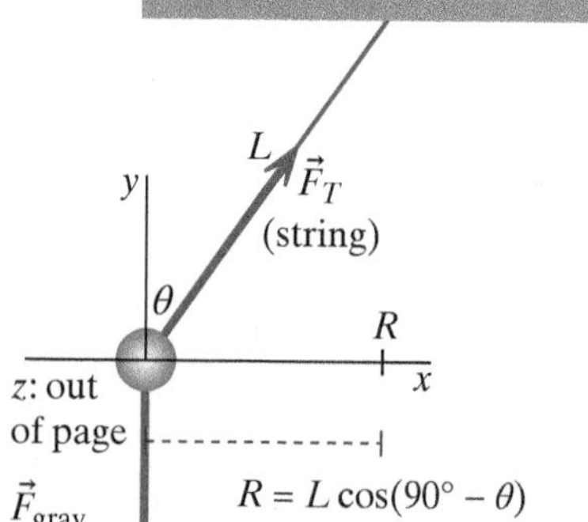

Figure 5.61 The circular pendulum at an instant when the string is in the xy plane, and the ball is moving in the $+z$ direction.

$$x: \frac{dp_x}{dt} = F_T \cos(90° - \theta)$$

$$\frac{mv^2}{R} = F_T \sin\theta$$

$$\frac{mv^2}{L\sin\theta} = \left(\frac{mg}{\cos\theta}\right)\sin\theta$$

$$g = \frac{v^2\cos\theta}{L(\sin\theta)^2}$$

Measuring the speed v, length L, and angle θ allows us to determine g. v can be found by measuring T, the time to go around once, since $v = (2\pi R)/T$, where $R = L\sin\theta$.

Special case: If $\theta = 0$ the string hangs vertically and there is no motion, so the tension in the string ought to be mg, which is consistent with the equation for F_T that we found above.

Note that $\vec{F}_{\text{net}}$ and $d\vec{p}/dt$ are not the same thing. In the example above, we knew the x and y components of $d\vec{p}/dt$ (the effect), and from this deduced the x and y components of $\vec{F}_{\text{net}}$ (the cause). In the z direction, it was the other way around: we knew that there were no forces in this direction, so we deduced that the z component of momentum was not changing.

There could in principle be many possible motions of this system. The ball could swing back and forth in a plane; this is the motion of a "simple pendulum." The ball could go around in a path something like an ellipse, in which case the ball's height would vary, just as with simple pendulum motion. The ball could move so violently that the string goes slack for part of the time, with abrupt changes of momentum every time the string suddenly goes taut. If the string can stretch noticeably, there can be a sizable oscillation superimposed on the swinging motion.

All of these motions are difficult to analyze. In principle, we could use a computer to predict the general motion of the ball hanging from the string if we had an equation for the tension force in the string, as a function of its length. Essentially, we would need the effective spring stiffness for this stiff "spring." However, the string may be so stiff that even a tiny stretch implies a huge tension (a nearly inextensible string). The observable length of the string is nearly constant, but the tension force that the string applies to the ball can vary a great deal. This makes it challenging to do a computer numerical integration, because a tiny error in position makes a huge change in the force. Special techniques have been developed to handle such situations.

SUMMARY

Determining unknown forces

Make a clear choice of the system; everything else is the surroundings.

Identify objects in the surroundings that can exert either a distant or contact force on the chosen system.

Draw the forces on a free-body diagram and include the name of the object responsible.

Apply the Momentum Principle to the chosen system.

Use what is known about $d\vec{p}/dt$, such as $d\vec{p}/dt = \vec{0}$ in the case of a system at rest or in uniform motion, or what can be obtained from the radius of the kissing circle.

Solve for the unknown forces. For a system containing several objects, it may be necessary to make a different choice of system in order to be able to determine forces between these objects.

Finding the rate of change of momentum graphically

1. Draw two arrows representing $\vec{p}$ at equal time intervals before and after the time of interest (Figure 5.35).
2. Find $\Delta\vec{p} = \vec{p}_f - \vec{p}_i$ (Figure 5.36).

3. Divide $\Delta\vec{p}$ by Δt to get $\Delta\vec{p}/\Delta t$ (Figure 5.37). The smaller Δt, the closer the result is to $d\vec{p}/dt$.

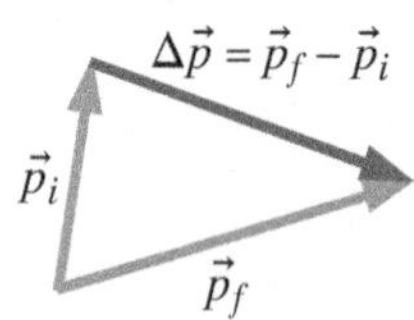

Figure 5.62

Momentarily at rest vs. uniform motion and equilibrium

For a system in equilibrium $d\vec{p}/dt = \vec{0}$.

For a system in uniform motion $d\vec{p}/dt = \vec{0}$.

For a system that is momentarily at rest, $d\vec{p}/dt \neq \vec{0}$.

The Momentum Principle: Parallel and perpendicular components

$$\frac{dp}{dt}\hat{p} = \vec{F}_{\|} = (\vec{F}_{\text{net}} \bullet \hat{p})\hat{p} \quad \text{and} \quad p\frac{d\hat{p}}{dt} = \vec{F}_{\perp} = \vec{F}_{\text{net}} - \vec{F}_{\|}$$

Valid only for a moving object, because $\hat{p}$ and $d\hat{p}/dt$ are undefined when $\vec{p} = \vec{0}$.

Component of net force parallel to momentum changes magnitude of momentum.

Component of net force perpendicular to momentum changes direction of momentum.

Effect of perpendicular component of net force on a particle moving along a curving path

$$p\left|\frac{d\hat{p}}{dt}\right| = p\frac{v}{R} = |\vec{F}_{\text{net}\perp}|$$

R is the radius of the kissing circle. The component of net force perpendicular to the particle's momentum changes its direction but not the magnitude of its momentum (speed). Must be a smoothly curving motion; not valid if there is an abrupt change of direction.

The dot product

$$\vec{A} \bullet \vec{B} = |\vec{A}||\vec{B}|\cos\theta = A_xB_x + A_yB_y + A_zB_z$$

QUESTIONS

Q1 You are riding in the passenger seat of an American car, sitting on the right side of the front seat. The car makes a sharp left turn. You feel yourself thrown to the right and your right side hits the right door. Is there a force that pushes you to the right? What object exerts that force? What really happens? Draw a diagram to illustrate and clarify your analysis.

Q2 A student said, "When the Moon goes around the Earth, there is an inward force due to the Moon and an outward force due to centrifugal force, so the net force on the Moon is zero." Give two or more physics reasons why this is wrong.

Q3 A space shuttle is in a circular orbit near the Earth. An astronaut floats in the middle of the shuttle, not touching the walls. On a diagram, draw and label **(a)** the momentum $\vec{p}_1$ of the astronaut at this instant, **(b)** all of the forces (if any) acting on the astronaut at this instant, **(c)** the momentum $\vec{p}_2$ of the astronaut a short time Δt later, and **(d)** the momentum change (if any) $\Delta\vec{p}$ in this time interval. **(e)** Why does the astronaut seem to "float" in the shuttle?

It is ironic that we say the astronaut is "weightless" despite the fact that the only force acting on the astronaut is the astronaut's weight (that is, the gravitational force of the Earth on the astronaut).

Q4 Tarzan swings back and forth on a vine. At the microscopic level, why is the tension force on Tarzan by the vine greater than it would be if he were hanging motionless?

Q5 Tarzan swings from a vine. When he is at the bottom of his swing, as shown in Figure 5.63, which is larger in magnitude: the force by the Earth on Tarzan, the force by the vine (a tension force) on Tarzan, or neither (same magnitude)? Explain how you know this.

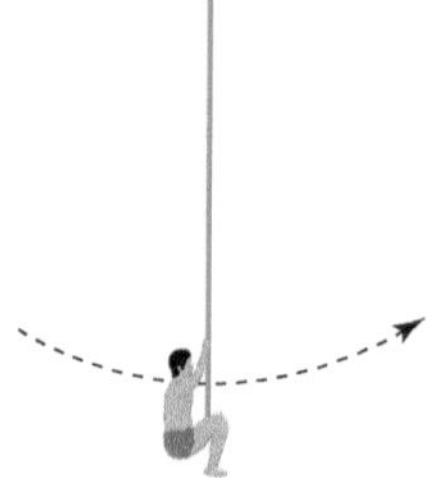

Figure 5.63

PROBLEMS

Section 5.4

•P6 A rope is attached to a block, as shown in Figure 5.64. The rope pulls on the block with a force of 210 N, at an angle of $\theta = 23°$ to the horizontal (this force is equal to the tension in the rope).

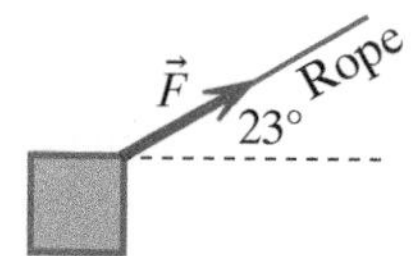

Figure 5.64

(a) What is the x component of the force on the block due to the rope? (b) What is the y component of the force on the block due to the rope?

••P7 A box of mass 40 kg hangs motionless from two ropes, as shown in Figure 5.65. The angle is 38 degrees. Choose the box as the system. The x axis runs to the right, the y axis runs up, and the z axis is out of the page.

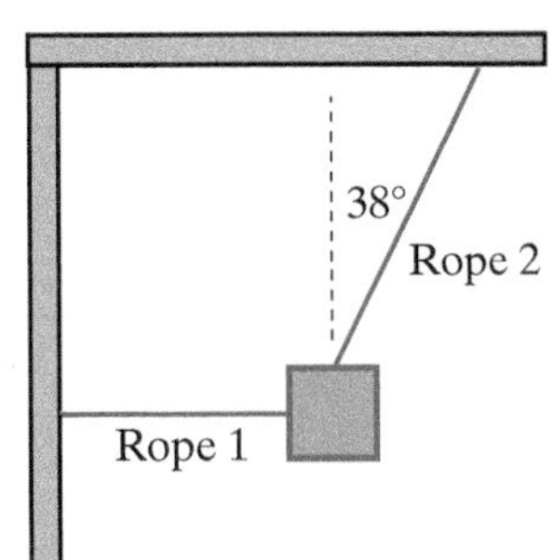

Figure 5.65

(a) Draw a free-body diagram for the box. (b) Is $d\vec{p}/dt$ of the box zero or nonzero? (c) What is the y component of the gravitational force acting on the block? (A component can be positive or negative). (d) What is the y component of the force on the block due to rope 2? (e) What is the magnitude of $\vec{F}_2$? (f) What is the x component of the force on the block due to rope 2? (g) What is the x component of the force on the block due to rope 1?

••P8 A helicopter flies to the right (in the $+x$ direction) at a constant speed of 12 m/s, parallel to the surface of the ocean. A 900 kg package of supplies is suspended below the helicopter by a cable as shown in Figure 5.66: the package is also traveling to the right in a straight line, at a constant speed of 12 m/s. The pilot is concerned about whether or not the cable, whose breaking strength is listed at 9300 N, is strong enough to support this package under these circumstances. (a) Choose the package as the system, and draw a free-body diagram. (b) What is the magnitude of the tension in the cable supporting the package? (c) Write the force exerted on the package by the cable as a vector. (d) What is the magnitude of the force exerted by the air on the package? (e) Write the force on the package by the air as a vector. (f) Is the cable in danger of breaking?

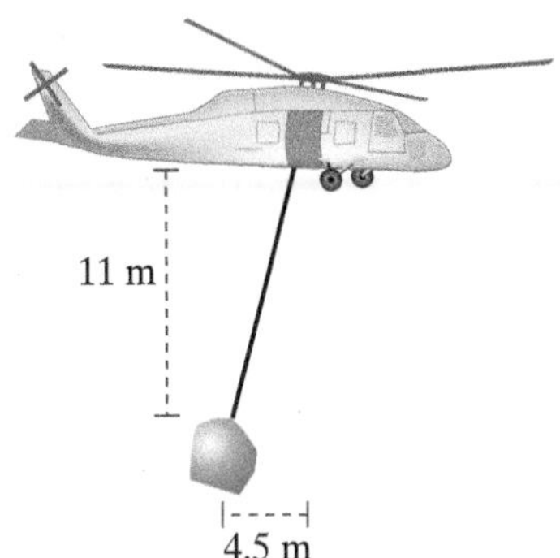

Figure 5.66

••P9 You pull with a force of 255 N on a rope that is attached to a block of mass 30 kg, and the block slides across the floor at a constant speed of 1.1 m/s. The rope makes an angle $\theta = 40°$ with the horizontal. Both the force and the velocity of the block are in the xy plane. See Figure 5.67.

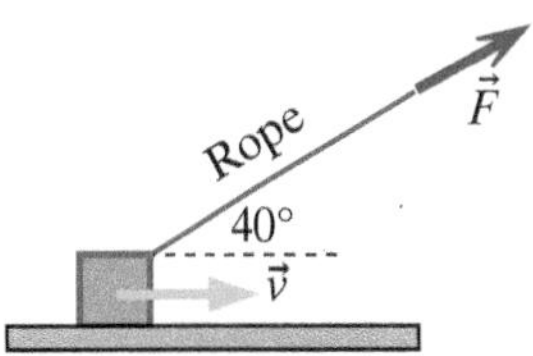

Figure 5.67

(a) Express the tension force exerted by the rope on the block as a vector. (b) Express the force exerted by the floor on the block as a vector.

••P10 A ball of mass 450 g hangs from a spring whose stiffness is 110 N/m. A string is attached to the ball and you are pulling the string to the right, so that the ball hangs motionless, as shown in Figure 5.68. In this situation the spring is stretched, and its length is 15 cm. What would be the relaxed length of the spring, if it were detached from the ball and laid on a table?

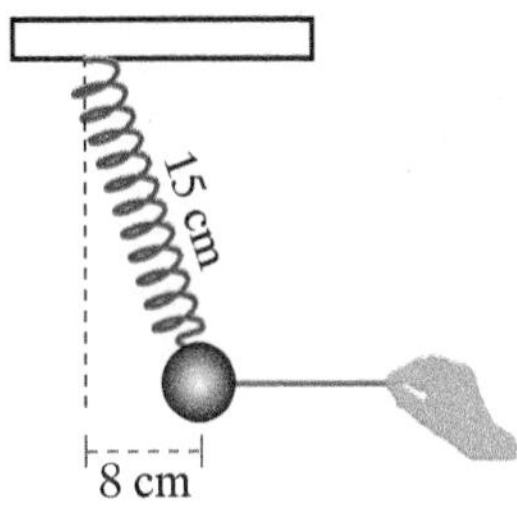

Figure 5.68

••P11 An 800 kg load is suspended as shown in Figure 5.69. (a) Calculate the tension in all three wires (that is, the magnitude of the tension force exerted by each of these wires). (b) These wires are made of a material whose value for Young's modulus is 1.3×10^{11} N/m^2. The diameter of the wires is 1.1 mm. What is the strain (fractional stretch) in each wire?

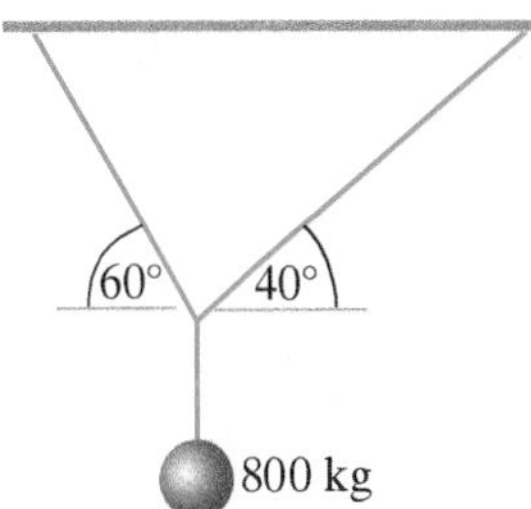

Figure 5.69

•••P12 A cylindrical steel rod with diameter 0.7 mm is 2 m long when lying horizontally (Figure 5.70). The density of this particular steel is 7.85 g/cm^3, and Young's modulus is 2×10^{11} N/m^2. (a) You now hang the rod vertically from a support and attach a 60 kg mass to the bottom end. How much does the rod stretch? (b) Next you remove the 60 kg mass so the rod hangs under its own weight. Describe qualitatively how the stretch of the interatomic bonds depends on position along the rod. (c) Calculate how much the rod stretches under its own weight. (This requires the use of integral calculus, because the tension and strain vary along the rod.) (d) If the rod were twice as thick, how much would it stretch?

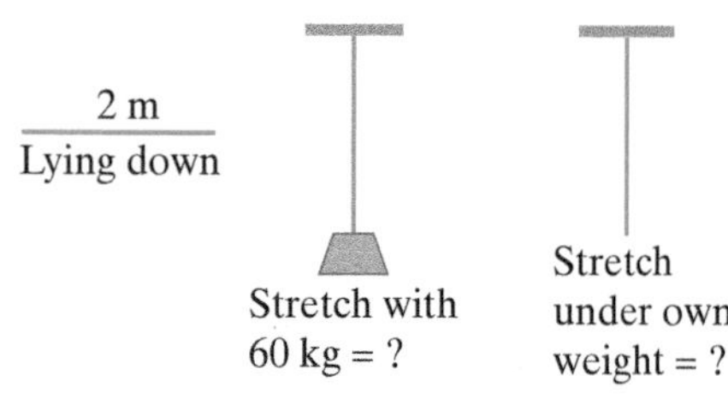

Figure 5.70

Section 5.5

••**P13** An elevator is accelerating upward at a rate of 3.2 m/s^2. A block of mass 30 kg hangs by a low-mass rope from the ceiling, and another block of mass 80 kg hangs by a low-mass rope from the upper block. **(a)** What are the tensions in the upper and lower ropes? **(b)** What are the tensions in the upper and lower ropes when the elevator accelerates downward at a rate of 3.2 m/s^2?

••**P14** A 10 kg block is placed on top of a 35 kg block (Figure 5.71). A force of 350 N is applied to the right on the lower block, and the upper block slips on the lower block (accelerating less than the lower block). The coefficient of kinetic friction between the upper block and the lower block is 0.2, and the coefficient of kinetic friction between the lower block and the floor is 0.7. **(a)** What is the acceleration of the upper block? **(b)** What is the acceleration of the lower block? **(c)** How big would the coefficient of static friction between the upper and lower block have to be so that the upper block would not slip on the lower block?

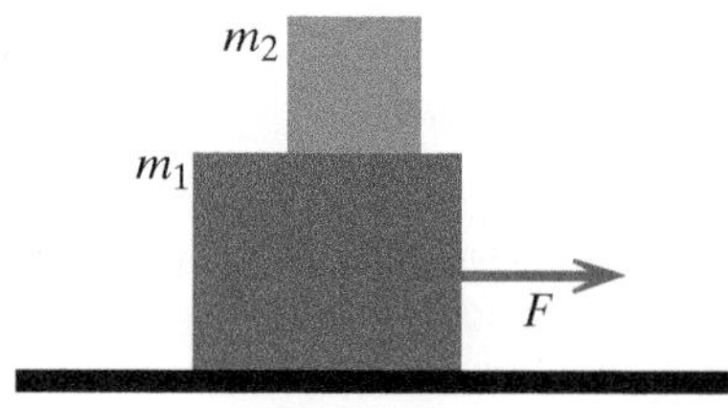

Figure 5.71

••**P15** In Figure 5.72 $m_1 = 12$ kg and $m_2 = 5$ kg. The kinetic coefficient of friction between m_1 and the floor is 0.3 and that between m_2 and the floor is 0.5. You push with a force of magnitude $F = 110$ N. **(a)** What is the acceleration of the center of mass? **(b)** What is the magnitude of the force that m_1 exerts on m_2?

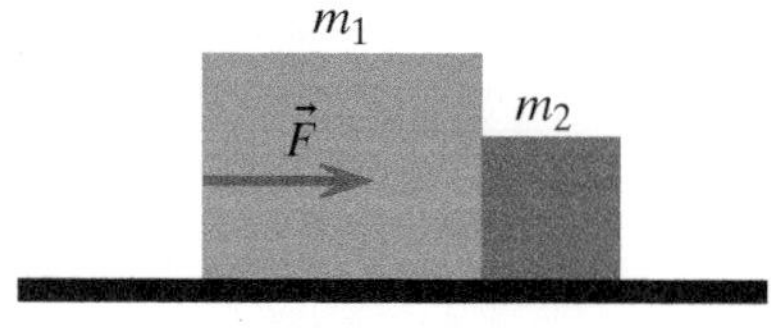

Figure 5.72

Section 5.6

•**P16** The radius of a merry-go-round is 11 m, and it takes 12 s to go around once. What is the speed of an atom in the outer rim?

•**P17** If the radius of a merry-go-round is 5 m, and it takes 14 s to go around once, what is the speed of an atom at the outer rim? What is the direction of the velocity of this atom: toward the center, away from the center, or tangential?

••**P18** At a particular instant the magnitude of the momentum of a planet is 2.3×10^{29} kg·m/s, and the force exerted on it by the star it is orbiting is 8.9×10^{22} N. The angle between the planet's momentum and the gravitational force exerted by the star is 123°. **(a)** What is the parallel component of the force on the planet by the star? **(b)** What will the magnitude of the planet's momentum be after 9 h?

••**P19** The angle between the gravitational force on a planet by a star and the momentum of the planet is 61° at a particular instant. At this instant the magnitude of the planet's momentum is 3.1×10^{29} kg·m/s, and the magnitude of the gravitational force on the planet is 1.8×10^{23} N. **(a)** What is the parallel component of the force on the planet by the star? **(b)** What will be the magnitude of the planet's momentum after 8 h?

••**P20** A planet orbits a star in an elliptical orbit. At a particular instant the momentum of the planet is $\langle -2.6 \times 10^{29}, -1.0 \times 10^{29}, 0\rangle$ kg·m/s, and the force on the planet by the star is $\langle -2.5 \times 10^{22}, -1.4 \times 10^{23}, 0\rangle$ N. Find $\vec{F}_{\parallel}$ and $\vec{F}_{\perp}$.

••**P21** A planet of mass 6×10^{24} kg orbits a star in a highly elliptical orbit. At a particular instant the velocity of the planet is $\langle 4.5 \times 10^4, -1.7 \times 10^4, 0\rangle$ m/s, and the force on the planet by the star is $\langle 1.5 \times 10^{22}, 1.9 \times 10^{23}, 0\rangle$ N. Find $\vec{F}_{\parallel}$ and $\vec{F}_{\perp}$.

Section 5.7

•**P22** An object moving at a constant speed of 23 m/s is making a turn with a radius of curvature of 4 m (this is the radius of the kissing circle). The object's momentum has a magnitude of 78 kg·m/s. What is the magnitude of the rate of change of the momentum? What is the magnitude of the net force?

•**P23** The radius of a merry-go-round is 7 m, and it takes 12 s to make a complete revolution. **(a)** What is the speed of an atom on the outer rim? **(b)** What is the direction of the momentum of this atom? **(c)** What is the direction of the rate of change of the momentum of this atom?

•**P24** A proton moving in a magnetic field follows the curving path shown in Figure 5.73. The dashed circle is the kissing circle tangent to the path when the proton is at location A. The proton is traveling at a constant speed of 7.0×10^5 m/s, and the radius of the kissing circle is 0.08 m. The mass of a proton is 1.7×10^{-27} kg. Refer to the directional arrows shown at the right in Figure 5.73 when answering the questions below.

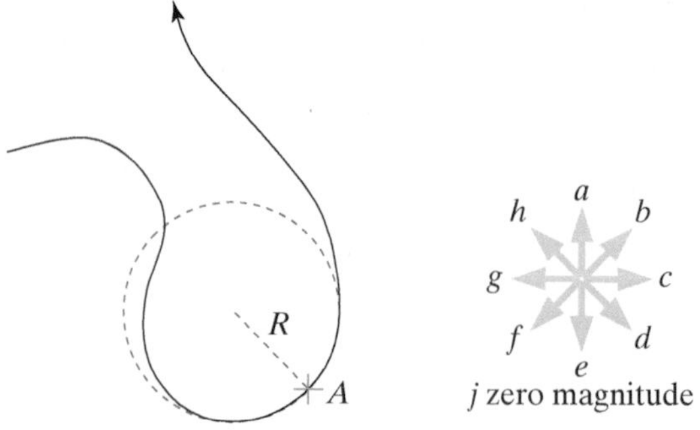

Figure 5.73

(a) When the proton is at location A, what are the magnitude and direction of $(d|\vec{p}|/dt)\hat{p}$, the parallel component of $d\vec{p}/dt$? **(b)** When the proton is at location A, what are the magnitude and direction of $|\vec{p}|d\hat{p}/dt$, the perpendicular component of $d\vec{p}/dt$?

•**P25** A proton moving in a magnetic field follows the curving path shown in Figure 5.74, traveling at constant speed in the direction shown. The dashed circle is the kissing circle tangent to the path when the proton is at location A. Refer to the directional arrows shown at the right in Figure 5.74 when answering the questions below.

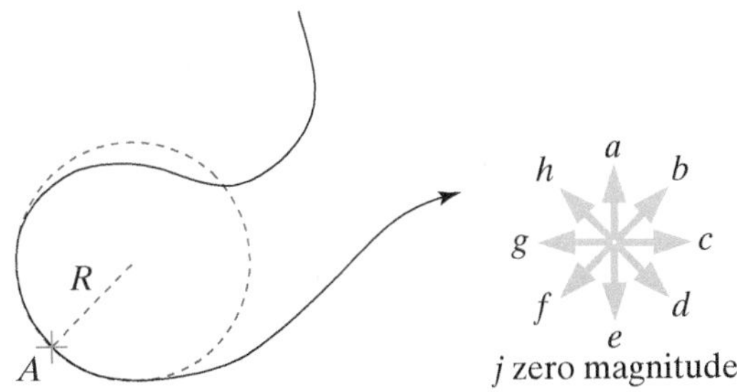

Figure 5.74

(a) When the proton is at location A, what is the direction of the proton's momentum? **(b)** When the proton is at location A, what is the direction of $d\vec{p}/dt$? **(c)** The mass of a proton is 1.7×10^{-27} kg. The proton is traveling at a constant speed of 6.0×10^5 m/s, and the radius of the kissing circle is 0.07 m. What is the magnitude of $d\vec{p}/dt$ of the proton?

••P26 A particle moving at nearly the speed of light ($v \approx c$) passes through a region where it is subjected to a magnetic force of constant magnitude that is always perpendicular to the momentum and has a magnitude of 2×10^{-10} N. As a result, the particle moves along a circular arc with a radius of 8 m. What is the magnitude of the momentum of this particle?

Section 5.8

•P27 A child of mass 40 kg sits on a wooden horse on a carousel. The wooden horse is 5 m from the center of the carousel, which completes one revolution every 90 s. What is $(d|\vec{p}|/dt)\hat{p}$ for the child, both magnitude and direction? What is $|\vec{p}|(d\hat{p}/dt)$ for the child? What is the net force acting on the child? What objects in the surroundings exert this force?

•P28 A child of mass 35 kg sits on a wooden horse on a carousel. The wooden horse is 3.3 m from the center of the carousel, which rotates at a constant rate and completes one revolution every 5.2 s. **(a)** What are the magnitude and direction (tangential in direction of velocity, tangential in the opposite direction of the velocity, radial outward, radial inward) of $(d|\vec{p}|/dt)\hat{p}$, the parallel component of $d\vec{p}/dt$ for the child? **(b)** What are the magnitude and direction of $|\vec{p}|d\hat{p}/dt$, the perpendicular component of $d\vec{p}/dt$ for the child? **(c)** What are the magnitude and direction of the net force acting on the child? **(d)** What objects in the surroundings contribute to this horizontal net force acting on the child? (There are also vertical forces, but these cancel each other if the horse doesn't move up and down.)

•P29 A 30 kg child rides on a playground merry-go-round, 1.4 m from the center. The merry-go-round makes one complete revolution every 5 s. How large is the net force on the child? In what direction does the net force act?

•P30 The orbit of the Earth around the Sun is approximately circular, and takes one year to complete. The Earth's mass is 6×10^{24} kg, and the distance from the Earth to the Sun is 1.5×10^{11} m. What is $(d|\vec{p}|/dt)\hat{p}$ of the Earth? What is $|\vec{p}|(d\hat{p}/dt)$ of the Earth? What is the magnitude of the gravitational force the Sun (mass 2×10^{30} kg) exerts on the Earth? What is the direction of this force?

••P31 You swing a bucket full of water in a vertical circle at the end of a rope. The mass of the bucket plus the water is 3.5 kg. The center of mass of the bucket plus the water moves in a circle of radius 1.3 m. At the instant that the bucket is at the top of the circle, the speed of the bucket is 4 m/s. What is the tension in the rope at this instant?

••P32 In outer space two identical spheres are connected by a taut steel cable, and the whole apparatus rotates about its center. The mass of each sphere is 60 kg. The distance between centers of the spheres is 3.2 m. At a particular instant the velocity of one of the spheres is $\langle 0,5,0 \rangle$ m/s and the velocity of the other sphere is $\langle 0,-5,0 \rangle$ m/s. What is the tension in the cable?

••P33 In the dark in outer space, you observe a glowing ball of known mass 2 kg moving in the xy plane at constant speed in a circle of radius 6.5 m, with the center of the circle at the origin ($\langle 0,0,0 \rangle$ m). You can't see what's making it move in a circle. At time $t = 0$ the ball is at location $\langle -6.5,0,0 \rangle$ m and has velocity $\langle 0,40,0 \rangle$ m/s.

On your own paper draw a diagram of the situation showing the circle and showing the position and velocity of the ball at time $t = 0$. The diagram will help you analyze the situation. Use letters a–j (Figure 5.75) to answer questions about directions ($+x$ to the right, $+y$ up).

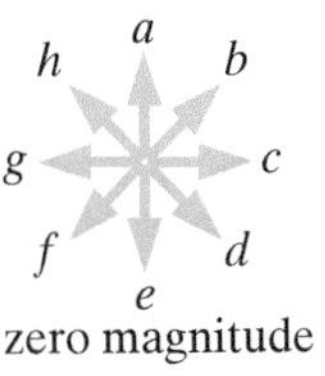

Figure 5.75

At time $t = 0$: **(a)** What is the direction of the vector $\vec{p}$? **(b)** What are the magnitude and direction of $(d|\vec{p}|/dt)\hat{p}$, the parallel component of $d\vec{p}/dt$? **(c)** What are the magnitude and direction of $|\vec{p}|d\hat{p}/dt$, the perpendicular component of $d\vec{p}/dt$? **(d)** Even though you can't see what's causing the motion, what can you conclude must be the direction of the vector $\vec{F}_{\text{net}}$? **(e)** Even though you can't see what's causing the motion, what can you conclude must be the vector $\vec{F}_{\text{net}}$? **(f)** You learn that at time $t = 0$, two forces act on the ball, and that at this instant one of these forces is $\vec{F}_1 = \langle 196,-369,0 \rangle$ N. What is the other force?

••P34 You're driving a vehicle of mass 1350 kg and you need to make a turn on a flat road. The radius of curvature of the turn is 76 m. The coefficient of static friction and the coefficient of kinetic friction are both 0.25. **(a)** What is the fastest speed you can drive and still make it around the turn? Invent symbols for the various quantities and solve algebraically before plugging in numbers. **(b)** Which of the following statements are true about this situation? (1) The net force is nonzero and points away from the center of the kissing circle. (2) The rate of change of the momentum is nonzero and points away from the center of the kissing circle. (3) The rate of change of the momentum is nonzero and points toward the center of the kissing circle. (4) The momentum points toward the center of the kissing circle. (5) The centrifugal force balances the force of the road, so the net force is zero. (6) The net force is nonzero and points toward the center of the kissing circle. **(c)** Look at your algebraic analysis and answer the following question. Suppose that your vehicle had a mass five times as big (6750 kg). Now what is the fastest speed you can drive and still make it around the turn? **(d)** Look at your algebraic analysis and answer the following question. Suppose that you have the original 1350 kg vehicle but the turn has a radius twice as large (152 m). What is the fastest speed you can drive and still make it around the turn? This problem shows why high-speed curves on freeways have very large radii of curvature, but low-speed entrance and exit ramps can have smaller radii of curvature.

••P35 What is the minimum speed v that a roller coaster car must have in order to make it around an inside loop and just barely lose contact with the track at the top of the loop (see Figure 5.76)? The center of the car moves along a circular arc of radius R. Include a carefully labeled force diagram. State briefly what approximations you make. Design a plausible roller coaster loop, including numerical values for v and R.

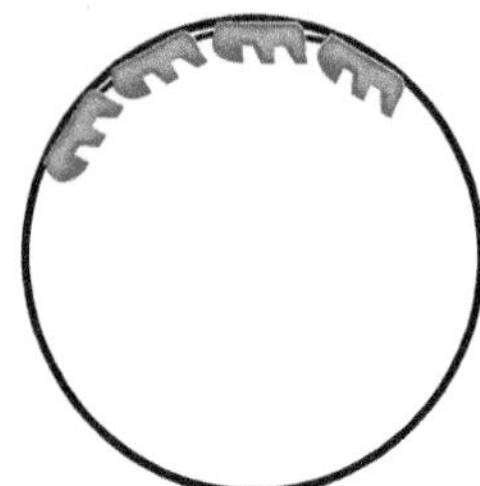

Figure 5.76

••P36 In outer space a rock of mass 4 kg is attached to a long spring and swung at constant speed in a circle of radius 9 m. The spring exerts a force of constant magnitude 760 N. **(a)** What is the speed of the rock? **(b)** What is the direction of the spring force? **(c)** The relaxed length of the spring is 8.7 m. What is the stiffness of this spring?

••P37 A child of mass 26 kg swings at the end of an elastic cord. At the bottom of the swing, the child's velocity is horizontal, and the speed is 12 m/s. At this instant the cord is 4.30 m long. **(a)** At this instant, what is the parallel component of the rate of change of the child's momentum? **(b)** At this instant, what is the perpendicular component of the rate of change of the child's momentum? **(c)** At this instant, what is the *net* force acting on the child? **(d)** What is the magnitude of the force that the elastic cord exerts on the child? (It helps to draw a diagram of the forces.) **(e)** The relaxed length of the elastic cord is 4.22 m. What is the stiffness of the cord?

••P38 An engineer whose mass is 70 kg holds onto the outer rim of a rotating space station whose radius is 14 m and which takes 30 s to make one complete rotation. What is the magnitude of the force the engineer has to exert in order to hold on? What is the magnitude of the net force acting on the engineer?

••P39 In June 1997 the NEAR spacecraft ("Near Earth Asteroid Rendezvous"; see http://near.jhuapl.edu/), on its way to photograph the asteroid Eros, passed near the asteroid Mathilde. After passing Mathilde, on several occasions rocket propellant was expelled to adjust the spacecraft's momentum in order to follow a path that would approach the asteroid Eros, the final destination for the mission. After getting close to Eros, further small adjustments made the momentum just right to give a circular orbit of radius 45 km (45×10^3 m) around the asteroid. So much propellant had been used that the final mass of the spacecraft while in circular orbit around Eros was only 500 kg. The spacecraft took 1.04 days to make one complete circular orbit around Eros. Calculate what the mass of Eros must be.

••P40 A Ferris wheel is a vertical, circular amusement ride. Riders sit on seats that swivel to remain horizontal as the wheel turns. The wheel has a radius R and rotates at a constant rate, going around once in a time T. At the bottom of the ride, what are the magnitude and direction of the force exerted by the seat on a rider of mass m? Include a diagram of the forces on the rider.

••P41 A block with mass 0.4 kg is connected by a spring of relaxed length 0.15 m to a post at the center of a low-friction table. You pull the block straight away from the post and release it, and you observe that the period of oscillation is 0.6 s. Next you stretch the spring to a length of 0.28 m and give the block an initial speed v perpendicular to the spring, choosing v so that the motion is a circle with the post at the center. What is this speed v?

••P42 When a particle with electric charge q moves with speed v in a plane perpendicular to a magnetic field B, there is a magnetic force at right angles to the motion with magnitude qvB, and the particle moves in a circle of radius r (see Figure 5.77). This equation for the magnetic force is correct even if the speed is comparable to the speed of light. Show that

$$p = \frac{mv}{\sqrt{1-(|\vec{v}|/c)^2}} = qBr$$

even if v is comparable to c.

This result is used to measure relativistic momentum: if the charge q is known, we can determine the momentum of a particle by observing the radius of a circular trajectory in a known magnetic field.

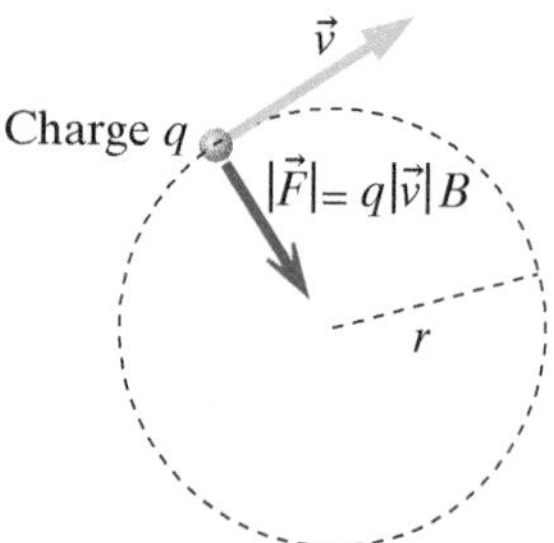

Figure 5.77

••P43 A ball of unknown mass m is attached to a spring. In outer space, far from other objects, you hold the other end of the spring and swing the ball around in a circle of radius 1.5 m at constant speed. **(a)** You time the motion and observe that going around 10 times takes 6.88 s. What is the speed of the ball? **(b)** Is the momentum of the ball changing or not? How can you tell? **(c)** If the momentum is changing, what interaction is causing it to change? If the momentum is not changing, why isn't it? **(d)** The relaxed length of the spring is 1.2 m, and its stiffness is 1000 N/m. While you are swinging the ball, since the radius of the circle is 1.5 m, the length of the spring is also 1.5 m. What is the magnitude of the force that the spring exerts on the ball? **(e)** What is the mass m of the ball?

••P44 A sports car (and its occupants) of mass M is moving over the rounded top of a hill of radius R. At the instant when the car is at the very top of the hill, the car has a speed v. You can safely neglect air resistance. **(a)** Taking the sports car as the system of interest, what object(s) exert nonnegligible forces on this system? **(b)** At the instant when the car is at the very top of the hill, draw a diagram showing the system as a dot, with force vectors whose tails are at the location of the dot. Label the force vectors (that is, give them algebraic names). Try to make the lengths of the force vectors be proportional to the magnitudes of the forces. **(c)** Starting from the Momentum Principle, calculate the force exerted by the road on the car. **(d)** Under what conditions will the force exerted by the road on the car be zero? Explain.

••P45 A small block of mass m is attached to a spring with stiffness k_s and relaxed length L. The other end of the spring is fastened to a fixed point on a low-friction table. The block slides on the table in a circular path of radius $R > L$. How long does it take for the block to go around once?

••P46 A person of mass 70 kg rides on a Ferris wheel whose radius is 4 m. The person's speed is constant at 0.3 m/s. The person's location is shown by a dot in Figure 5.78. **(a)** What is the magnitude of the rate of change of the momentum of the person at the instant shown?

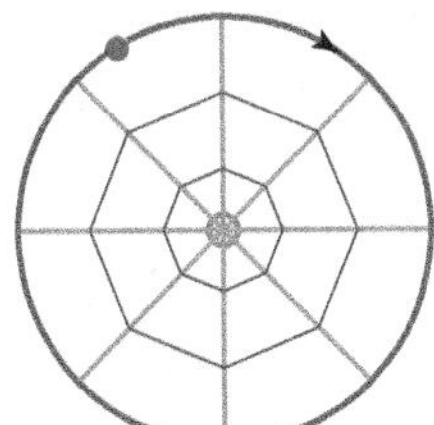

Figure 5.78

(b) What is the direction of the rate of change of momentum of the person at the instant shown? **(c)** What is the magnitude of the *net* force acting on the person at the instant shown? Draw the *net* force vector on the diagram at this instant, with the tail of the vector on the person.

••P47 The planets in our Solar System have orbits around the Sun that are nearly circular, and $v \ll c$. Calculate the period T (a year—the time required to go around the Sun once) for a planet whose orbit radius is r. This is the relationship discovered by Kepler and explained by Newton. (It can be shown by advanced techniques that this result also applies to elliptical orbits if you replace r by the semimajor axis, which is half the longer, major axis of the ellipse.) Use this analytical solution for circular motion to predict the Earth's orbital speed, using the data for Sun and Earth on the inside back cover of the textbook.

•••P48 In the 1970s the astronomer Vera Rubin made observations of distant galaxies that she interpreted as indicating that perhaps 90% of the mass in a galaxy is invisible to us ("dark matter"). She measured the speed with which stars orbit the center of a galaxy, as a function of the distance of the stars from the center. The orbital speed was determined by measuring the Doppler shift of the light from the stars, an effect that makes light shift toward the red end of the spectrum ("red shift") if the star has a velocity component away from us, and makes light shift toward the blue end of the spectrum if the star has a velocity component toward us.

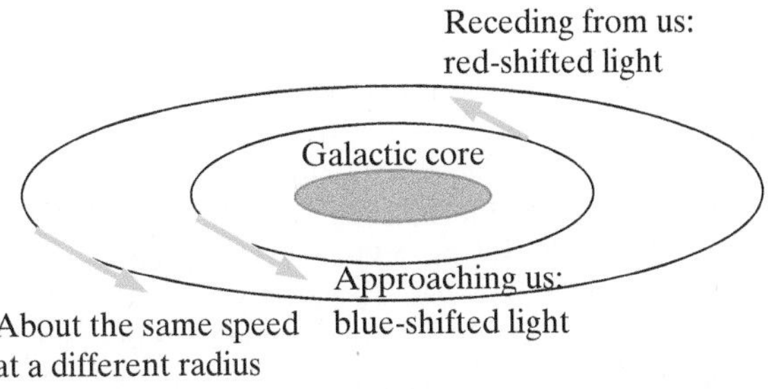

Figure 5.79

She found that for stars farther out from the center of the galaxy, the orbital speed of the star hardly changes with distance from the center of the galaxy, as is indicated in Figure 5.79. The visible components of the galaxy (stars, and illuminated clouds of dust) are most dense at the center of the galaxy and thin out rapidly as you move away from the center, so most of the visible mass is near the center. **(a)** Predict the speed v of a star going around the center of a galaxy in a circular orbit, as a function of the star's distance r from the center of the galaxy, assuming that almost all of the galaxy's mass M is concentrated at the center. **(b)** Construct a logical argument as to why Rubin concluded that much of the mass of a galaxy is not visible to us. Reason from principles discussed in this chapter, and your analysis of part (a). Explain your reasoning. You need to address the following issues: (i) Rubin's observations are not consistent with your prediction in (a). (ii) Most of the *visible* matter is in the center of the galaxy. (iii) Your prediction in (a) assumed that most of the mass is at the center.

This issue has not yet been resolved, and is still a current topic of astrophysics research. Here is a discussion by Rubin of her work: "Dark Matter in Spiral Galaxies" by Vera C. Rubin, *Scientific American*, June 1983 (96–108). You can find several graphs of the rotation curves for spiral galaxies on page 101 of this article.

Section 5.9

••P49 The Ferris wheel in Figure 5.80 is a vertical, circular amusement ride with radius 10 m. Riders sit on seats that swivel to remain horizontal. The Ferris wheel rotates at a constant rate, going around once in 10.5 s. Consider a rider whose mass is 56 kg.

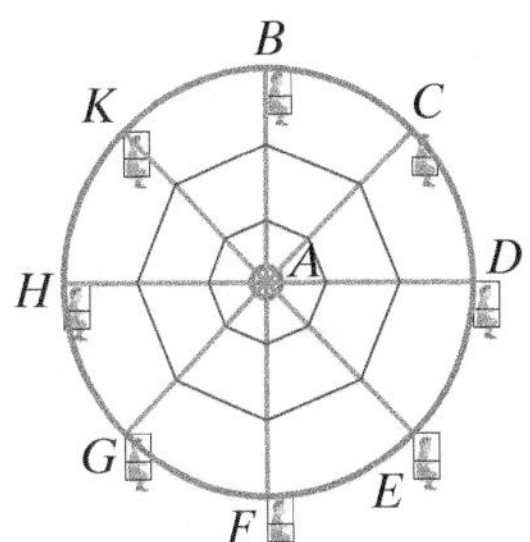

Figure 5.80

(a) At the bottom of the ride, what is the rate of change of the rider's momentum? **(b)** At the bottom of the ride, what is the vector gravitational force exerted by the Earth on the rider? **(c)** At the bottom of the ride, what is the vector force exerted by the seat on the rider? **(d)** Next consider the situation at the top of the ride. At the top of the ride, what is the rate of change of the rider's momentum? **(e)** At the top of the ride, what is the vector gravitational force exerted by the Earth on the rider? **(f)** At the top of the ride, what is the vector force exerted by the seat on the rider?

A rider *feels* heavier if the electric, interatomic contact force of the seat on the rider is larger than the rider's weight mg (and the rider sinks more deeply into the seat cushion). A rider *feels* lighter if the contact force of the seat is smaller than the rider's weight (and the rider does not sink as far into the seat cushion). **(g)** Does a rider *feel* heavier or lighter at the bottom of a Ferris wheel ride? **(h)** Does a rider *feel* heavier or lighter at the top of a Ferris wheel ride?

••P50 By weight we usually mean the gravitational force exerted on an object by the Earth. However, when you sit in a chair your own perception of your own weight is based on the contact force the chair exerts upward on your rear end rather than on the gravitational force. The smaller this contact force is, the less weight you perceive, and if the contact force is zero, you feel peculiar and "weightless" (an odd word to describe a situation when the only force acting on you is the gravitational force exerted by the Earth!). Also, in this condition pressure on your inner ear is released, which affects your sense of balance, and your internal organs no longer press on each other, all of which contributes to the odd sensation in your stomach. **(a)** How fast must a roller coaster car go over the top of a circular arc for you to feel "weightless"? The center of the car moves along a circular arc of radius R (see Figure 5.81). Include a carefully labeled force diagram.

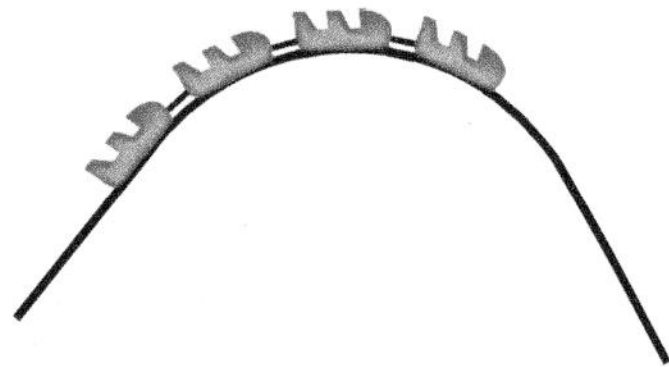

Figure 5.81

(b) How fast must a roller coaster car go through a circular dip for you to feel three times as "heavy" as usual, due to the upward force of the seat on your bottom being three times as large as usual? The center of the car moves along a circular arc of radius R (see Figure 5.82). Include a carefully labeled force diagram.

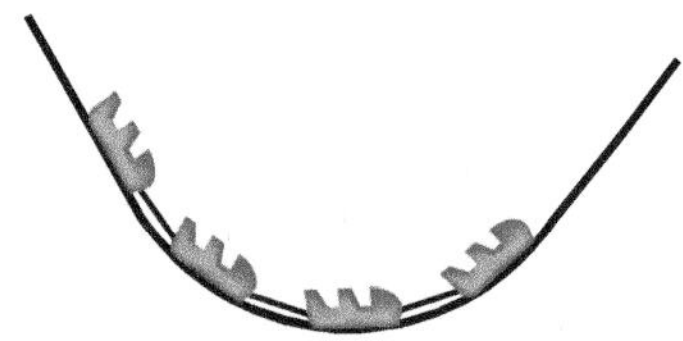

Figure 5.82

Section 5.10

•P51 A circular pendulum of length 1.1 m goes around at an angle of 28 degrees to the vertical. Predict the speed of the mass at the end of the string. Also predict the period, the time it takes to go around once. Remember that the radius of the circle is the length of the string times the sine of the angle that the string makes to the vertical.

••P52 Use a circular pendulum to determine g. You can increase the accuracy of the time it takes to go around once by timing N revolutions and then dividing by N. This minimizes errors contributed by inaccuracies in starting and stopping the clock. It is wise to start counting from zero (0, 1, 2, 3, 4, 5) rather than starting from 1 (1, 2, 3, 4, 5 represents only four revolutions, not five). It also improves accuracy if you start and stop timing at a well-defined event, such as when the mass crosses in front of an easily visible mark. This was the method used by Newton to get an accurate value of g. Newton was not only a brilliant theorist but also an excellent experimentalist. For a circular pendulum he built a large triangular wooden frame mounted on a vertical shaft, and he pushed this around and around while making sure that the string of the circular pendulum stayed parallel to the slanting side of the triangle.

••P53 **(a)** Many communication satellites are placed in a circular orbit around the Earth at a radius where the period (the time to go around the Earth once) is 24 h. If the satellite is above some point on the equator, it stays above that point as the Earth rotates, so that as viewed from the rotating Earth the satellite appears to be motionless. That is why you see dish antennas pointing at a fixed point in space. Calculate the radius of the orbit of such a "synchronous" satellite. Explain your calculation in detail. **(b)** Electromagnetic radiation including light and radio waves travels at a speed of 3×10^8 m/s. If a phone call is routed through a synchronous satellite to someone not very far from you on the ground, what is the minimum delay between saying something and getting a response? Explain. Include in your explanation a diagram of the situation. **(c)** Some human-made satellites are placed in "near-Earth" orbit, just high enough to be above almost all of the atmosphere. Calculate how long it takes for such a satellite to go around the Earth once, and explain any approximations you make. **(d)** Calculate the orbital speed for a near-Earth orbit, which must be provided by the launch rocket. (The advantages of near-Earth communications satellites include making the signal delay unnoticeable, but with the disadvantage of having to track the satellites actively and having to use many satellites to ensure that at least one is always visible over a particular region.) **(e)** When the first two astronauts landed on the Moon, a third astronaut remained in an orbiter in circular orbit near the Moon's surface. During half of every complete orbit, the orbiter was behind the Moon and out of radio contact with the Earth. On each orbit, how long was the time when radio contact was lost?

••P54 There is no general analytical solution for the motion of a three-body gravitational system. However, there do exist analytical solutions for very special initial conditions. Figure 5.83 shows three stars, each of mass m, which move in the plane of the page along a circle of radius r. Calculate how long this system takes to make one complete revolution. (In many cases three-body orbits are not stable: any slight perturbation leads to a breakup of the orbit.)

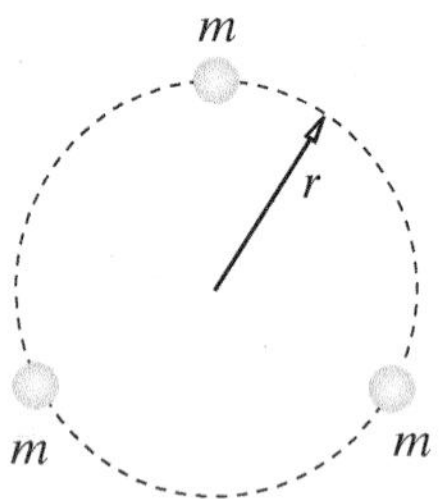

Figure 5.83

••P55 Remarkable data indicate the presence of a massive black hole at the center of our Milky Way galaxy. The W. M. Keck 10-m-diameter telescopes in Hawaii were used by Andrea Ghez and her colleagues to observe infrared light coming directly through the dust surrounding the central region of our galaxy (visible light is multiply-scattered by the dust, blocking a direct view). Stars were observed for several consecutive years to determine their orbits near a motionless center that is completely invisible ("black") in the infrared but whose precise location is known due to its strong output of radio waves, which are observed by radio telescopes. The data were used to show that the object at the center must have a mass that is huge compared to the mass of our own Sun, whose mass is a "mere" 2×10^{30} kg. Figure 5.84 shows positions from 1995 to 2012 of one of the stars, called S0-20, orbiting around the galactic center. The orbit appears nearly circular with the radius shown.

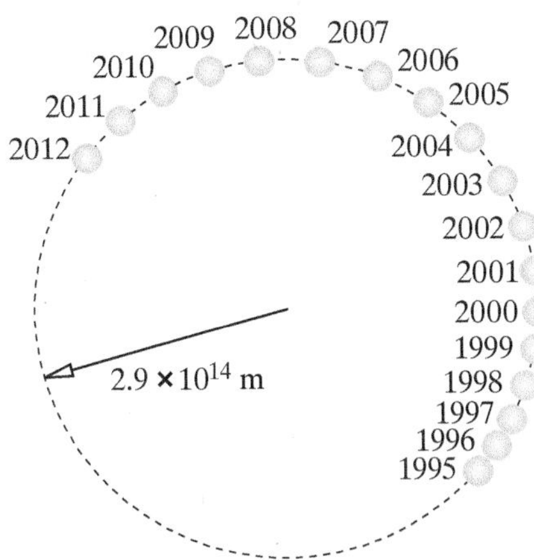

Figure 5.84 Positions on the sky of the star S0-20 near the center of our galaxy.

(a) Using the positions and times shown in Figure 5.84, what is the approximate speed of this star in m/s? Also express the speed as a fraction of the speed of light. **(b)** This is an extraordinarily high speed for a macroscopic object. Is it reasonable to approximate the star's momentum as mv? (Some other stars near the galactic center with highly elliptical orbits move even faster when they are closest to the center.) **(c)** Based on these data, estimate the mass of the massive black hole about which this star is orbiting. (You can see in Figure 5.84 that the speed appears greater in 2007–2008 than in 1995–1996. You're looking at a projection of an elliptical orbit, which only happens to look circular as viewed from Earth, and the black hole is not at the center of this circle. Therefore your result will be approximate and will differ from that obtained by Ghez and colleagues, who carefully made a consistent fit of the position data for seven different stars.) **(d)** How many of our Suns does this represent?

It is now thought that most galaxies have such a black hole at their centers, as a result of long periods of mass accumulation. When many bodies orbit each other, sometimes in an interaction an object happens to acquire enough speed to escape from the group, and the remaining objects are left closer together. Some simulations show that over time, as much as half the mass may be ejected, with agglomeration of the remaining mass. This could be part of the mechanism for forming massive black holes.

For more information, search the web for Andrea Ghez. You may see the term "arc seconds," which is an angular measure of how far apart objects appear in the sky, and "parsecs," which is a distance equal to 3.3 light-years (a light-year is the distance light goes in one year).

•••P56 You put a 10 kg object on a bathroom scale at the North Pole, and the scale reads exactly 10 kg (actually, it measures the force F_N that the scale exerts on the object, but displays a reading in kg). At the North Pole you are 6357 km from the center of the Earth. At the equator, the scale reads a different value due to two effects: (1) The Earth bulges out at the equator (due to its rotation), and you are 6378 km from the center of the Earth. (2) You are moving in a circular path due to the rotation of the Earth (one rotation every 24 hours). Taking into account ***both*** of these effects, what does the scale read at the equator?

COMPUTATIONAL PROBLEMS

More detailed and extended versions of some computational modeling problems may be found in the lab activities included in the *Matter & Interactions, 4th Edition,* resources for instructors.

••P57 Start with the program you wrote to model the motion of a spacecraft near the Earth (Chapter 3). Choose initial conditions that produce an elliptical orbit. **(a)** Create an arrow to represent the net force on the spacecraft, and place the tail of the arrow at the position of the spacecraft. Inside the calculation loop, update the arrow's position every time you update the craft's position. **(b)** Update the axis of the arrow inside the loop, so it always points in the direction of the net force. Find a scale factor that gives the arrow an appropriate length when the craft is near the Earth, and allows the arrow to be visible when the craft is far from the Earth. **(c)** Now create two additional arrows to represent $\vec{F}_{\parallel}$ and $\vec{F}_{\perp}$, the parallel and perpendicular parts of the net force on the spacecraft. In VPython you can take the dot product of two vectors, $\vec{A}$ and $\vec{B}$, like this: `C = dot(A,B)`. Make your program display $\vec{F}_{\text{net}}$, $\vec{F}_{\parallel}$, and $\vec{F}_{\perp}$ as the craft orbits the Earth. **(d)** In what part of the orbit does $\vec{F}_{\parallel}$ point in the same direction as $\vec{p}$? What effect does it have on the craft's momentum? **(e)** In what part of the orbit does $\vec{F}_{\parallel}$ point in the opposite direction to $\vec{p}$? What effect does it have on the craft's momentum? **(f)** Are there any locations in the orbit where $\vec{F}_{\parallel} = 0$? If so, what are they?

••P58 Start with the program you wrote to model the 3D motion of a mass hanging from a spring (Chapter 4). **(a)** Create an arrow to represent the net force on the mass, and place the tail of the arrow at the position of the mass. Update both the position and axis of the arrow inside the calculation loop, using an appropriate scale factor, so the arrow always shows the net force on the mass. **(b)** Now create two additional arrows to represent $\vec{F}_{\parallel}$ and $\vec{F}_{\perp}$, the parallel and perpendicular parts of the net force on the mass. Make your program display $\vec{F}_{\text{net}}$, $\vec{F}_{\parallel}$, and $\vec{F}_{\perp}$ as the mass oscillates. **(c)** Find initial conditions for which $\vec{F}_{\parallel}$ is zero and remains zero. Explain. **(d)** Find initial conditions for which $\vec{F}_{\perp}$ is zero and remains zero. Explain. **(e)** Find initial conditions for which both $\vec{F}_{\parallel}$ and $\vec{F}_{\perp}$ are nonzero most of the time. Explain.

•••P59 Start with the program you wrote to model the motion of a spacecraft near the Earth. Use initial conditions that produce an elliptical orbit. Modify the program so the kissing circle is continuously displayed as the craft orbits the Earth. You may want to use a `ring` object in VPython to display the circle.

ANSWERS TO CHECKPOINTS

1 **(a)** Baseball: left diagram in Figure 5.85, **(b)** Yo-yo: right diagram in Figure 5.85

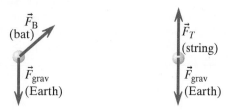

Figure 5.85 Left diagram: baseball. Right diagram: yo-yo.

2 $\vec{0}$; $\vec{0}$; $\langle 30, -90, 130 \rangle$ N

3 **(a)** Both blocks. **(b)** Either one of the blocks; note that the magnitude of the force that block m_2 exerts on m_1 is the same as the magnitude of the force that block m_1 exerts on block m_2.

4 **(a)** Figure 5.86 shows the forces:

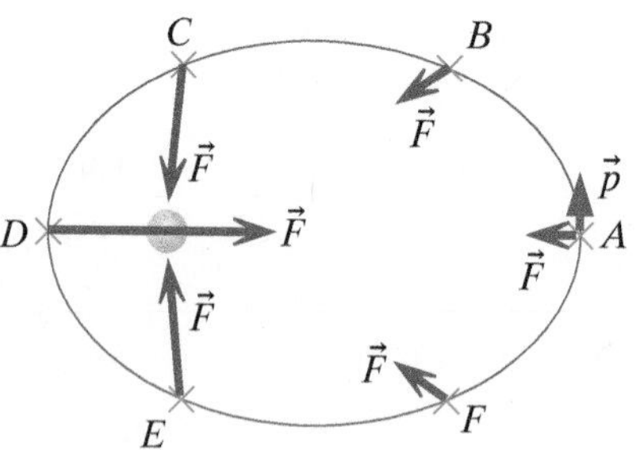

Figure 5.86

A: **(b)** zero, **(c)** nonzero, **(d)** zero, **(e)** changing; *B*: **(b)** nonzero, **(c)** nonzero, **(d)** positive, **(e)** changing; *C*: **(b)** nonzero, **(c)** nonzero, **(d)** positive, **(e)** changing; *D*: **(b)** zero, **(c)** nonzero, **(d)** zero, **(e)** changing; *E*: **(b)** nonzero, **(c)** nonzero, **(d)** negative, **(e)** changing; *F*: **(b)** nonzero, **(c)** nonzero, **(d)** negative, **(e)** changing

CHAPTER

6

The Energy Principle

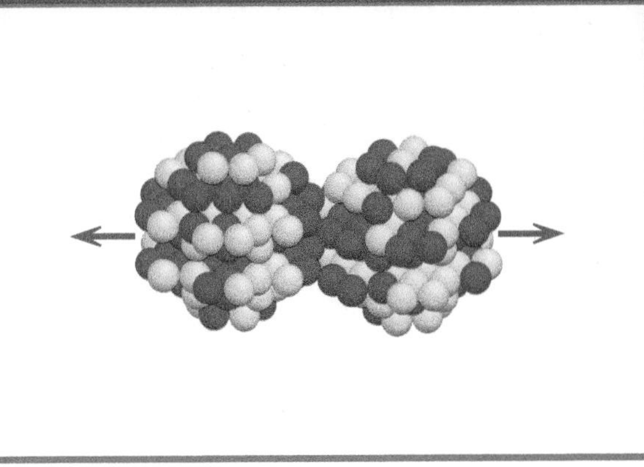

OBJECTIVES

After studying this chapter, you should be able to

- Calculate the total energy (rest energy + kinetic energy) of a single particle system.
- Calculate the total energy of a multiparticle system (rest energy, kinetic energy, and gravitational and electric potential energy).
- Mathematically relate changes in energy of a system to the work done by the surroundings.
- Analyze in detail processes involving changes in potential energy, kinetic energy, and rest energy.
- Construct and interpret graphs of multiparticle system energy as a function of separation.

6.1 THE ENERGY PRINCIPLE

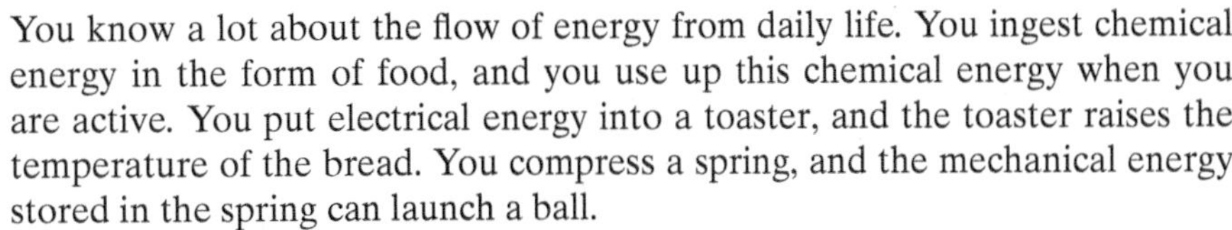

You know a lot about the flow of energy from daily life. You ingest chemical energy in the form of food, and you use up this chemical energy when you are active. You put electrical energy into a toaster, and the toaster raises the temperature of the bread. You compress a spring, and the mechanical energy stored in the spring can launch a ball.

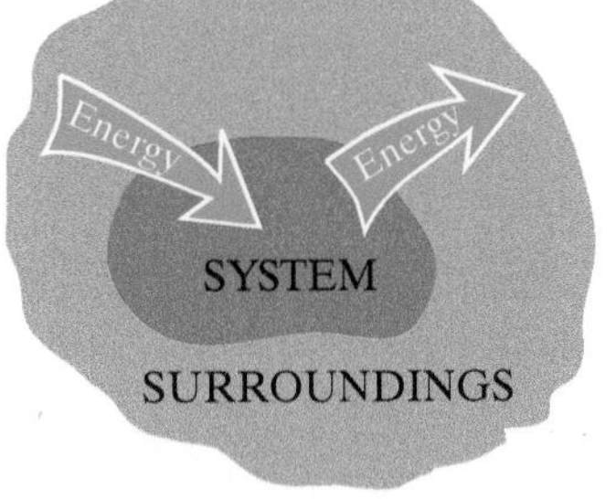

Figure 6.1 Energy may flow between system and surroundings.

Energy cannot be created or destroyed, but it can change form. The only way for a system to gain or lose energy is if the surroundings lose or gain the same amount of energy (Figure 6.1). Therefore the Energy Principle can be stated most generally in terms of conservation of energy:

CONSERVATION OF ENERGY

$$\Delta E_{\text{system}} + \Delta E_{\text{surroundings}} = 0$$

In a detailed energy analysis of a system, it is often useful to express the energy principle in terms of inputs of energy from the surroundings:

THE ENERGY PRINCIPLE

$$\Delta E_{\text{system}} = (\text{Energy inputs from surroundings})$$

The energy inputs from the surroundings may be positive or negative, corresponding to flows of energy from surroundings to system or from system to surroundings, respectively.

There are a number of mechanisms by which energy may be transferred between system and surroundings. In this chapter, we will focus on "work," the result of a force acting through a distance. Other mechanisms will be discussed in subsequent chapters.

The validity of the Energy Principle has been verified through a very wide variety of observations and experiments, involving large and small objects, moving slowly or at speeds near the speed of light, and even undergoing nuclear reactions that change the identity of the objects. It is a summary of the way energy flows in the real world.

The Energy Principle is a fundamental principle because:

- It applies to every possible system, no matter how large or small it is (from clusters of galaxies to subatomic particles) or how fast it is moving.
- It is true for any kind of interaction (gravitational, electromagnetic, strong, weak).
- It relates an effect (change in energy of a system) to a cause (an interaction with the surroundings).

6.2 ENERGY OF A SINGLE PARTICLE

The simplest possible system consists of a single point particle. "A point particle" might refer to an electron, but we can also model a baseball or even a planet as a point particle if during the process of interest there are no significant changes internal to the system such as changes of shape or rotation or vibration or temperature.

In the current chapter we will discuss energy and changes in energy for systems of point particles. We will defer the consideration of systems that include extended objects that undergo internal changes, and hence cannot be modeled as point particles, until Chapter 7.

A single particle can have only two kinds of energy: the energy associated with its rest mass, called "rest energy," and the energy associated with its motion, called "kinetic energy." Both of these are combined into a single compact expression, first expressed by Einstein in 1905:

ENERGY OF A SINGLE PARTICLE

$$E_{\text{particle}} = \gamma mc^2$$

This equation defines particle energy. m is the mass of the particle. As usual $\gamma = \dfrac{1}{\sqrt{1-\left(\frac{v}{c}\right)^2}}$, where c is the speed of light. The unit of energy is the joule, abbreviated J. One $\text{J} = \text{one kg}\cdot(\text{m/s})^2$.

Later in this chapter we'll see that this definition of particle energy is consistent with the definition of a particle's momentum, $\vec{p} = \gamma m\vec{v}$.

QUESTION What are the similarities and differences between the definition of particle energy and the definition of particle momentum?

Similarities:

- Both contain the same factor γ.
- Both are proportional to the mass m.

Differences:

- Particle energy is a scalar.
- Particle momentum is a vector.

Rest Energy

QUESTION If a particle is at rest, what is its energy?

If $v = 0$ then $\gamma = 1$, and we find the famous equation $E = mc^2$. This has the striking interpretation that a particle has energy even when it is sitting still, which was a revolutionary idea when Einstein proposed it. We call mc^2 the "rest energy," the amount of energy a particle has when it is sitting still.

REST ENERGY

$$E_{\text{rest}} = mc^2$$

The "rest energy" is the energy of a particle at rest.

Even more strikingly, Einstein realized that this means that the mass of an object at rest is its energy content divided by a constant: $m = E_{\text{rest}}/c^2$. For example, a hot object, with more internal energy, has very slightly more mass than a cold object. In a real sense, mass and energy are the same thing, although for historical reasons we use different units for them: mass in kilograms, energy in joules, with a constant factor of c^2 difference.

Note another difference between energy and momentum: A particle at rest has zero momentum but it has nonzero energy, its rest energy.

Kinetic Energy

QUESTION If a particle is moving with speed v, is its energy greater than mc^2?

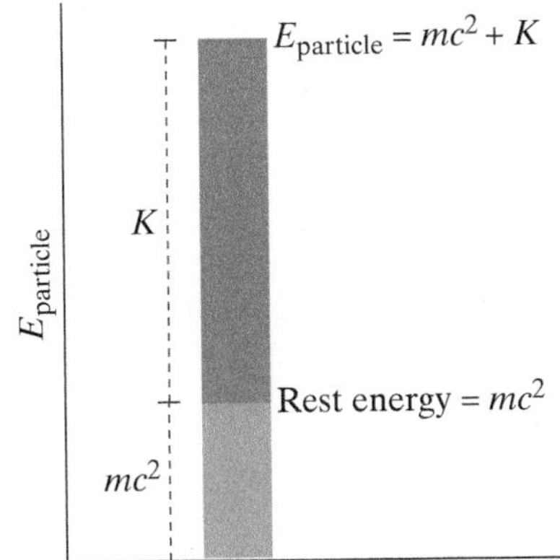

Figure 6.2 Particle energy is equal to rest energy plus kinetic energy K. This particle has high speed, and its kinetic energy is large compared to its rest energy.

If $v > 0$, $\sqrt{1-(v/c)^2}$ is less than 1, so $\gamma = 1/\sqrt{1-(v/c)^2}$ is greater than 1, and the particle energy γmc^2 increases with increasing speed. As a consequence of its motion, the particle has additional, "kinetic" energy K:

KINETIC ENERGY K OF A PARTICLE

$$K = \gamma mc^2 - mc^2$$

The kinetic energy K of a particle is the energy a moving particle has in addition to its rest energy. It is often useful to turn this equation around:

$$E_{\text{particle}} = mc^2 + K$$

If a particle is at rest, its particle energy is just its rest energy mc^2. If the particle moves, it has not only its rest energy mc^2 but also an additional amount of energy, which we call the kinetic (motional) energy K (Figure 6.2).

EXAMPLE **Energy of a Fast-Moving Proton**

A proton in a particle accelerator has a speed of 2.91×10^8 m/s. **(a)** What is the energy of the proton? **(b)** What is the rest energy of the proton? **(c)** What is the kinetic energy of the moving proton?

Solution **(a)** Moving proton:

$$\gamma = \frac{1}{\sqrt{1-\left(\dfrac{2.91\times10^8}{3\times10^8}\right)^2}} = 4.11$$

$$E = (4.11)(1.7\times10^{-27}\text{ kg})(3\times10^8\text{ m/s})^2$$

$$E = 6.288\times10^{-10}\text{ J}$$

(b) Proton at rest:

$$E_{\text{rest}} = mc^2$$
$$E_{\text{rest}} = (1.7 \times 10^{-27}\text{ kg})(3 \times 10^8\text{ m/s})^2$$
$$E_{\text{rest}} = 1.53 \times 10^{-10}\text{ J}$$

(c) Kinetic energy of proton:

$$K = E - E_{\text{rest}}$$
$$K = 6.288 \times 10^{-10}\text{ J} - 1.53 \times 10^{-10}\text{ J} = 4.76 \times 10^{-10}\text{ J}$$

In this example, the kinetic energy of the fast-moving proton turns out to be about three times as large as the rest energy of the proton—about the ratio shown in Figure 6.2.

QUESTION Is there a limit to the amount of energy a particle can have?

Delivering more and more energy to a particle increases its energy more and more, and there is no theoretical limit to how much energy a particle can acquire. As the speed approaches the speed of light $c = 3 \times 10^8$ m/s, the denominator $\sqrt{1-(v/c)^2}$ becomes very small, and the energy becomes very large.

However, even though you can increase the energy, it becomes very difficult to increase the speed v, because as you near the speed of light a tiny speed increase would require a huge energy increase. There is effectively a universal speed limit: a particle can't reach the speed of light c because an infinite amount of energy input would be required to achieve this.

Checkpoint 1 An electron has mass 9×10^{-31} kg. If the electron's speed is $0.988c$ (that is, $v/c = 0.988$), what is its particle energy? What is its rest energy? What is its kinetic energy?

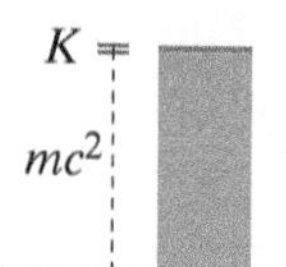

Figure 6.3 At low speeds the kinetic energy K is small compared to the rest energy.

Approximate Kinetic Energy at Low Speeds

If the speed v of a particle is small compared to the speed of light ($v \ll c$), we can find a simple approximate expression for the kinetic energy of the particle. In such a case, as shown in Figure 6.3, the kinetic energy is very small compared to the rest energy, and we can derive an approximate expression for the kinetic energy at low speeds by using the binomial expansion. (See Section 6.19 at the end of this chapter for a more detailed explanation.)

$$\frac{mc^2}{\sqrt{1-(v/c)^2}} = mc^2\left[1 + \frac{1}{2}\left(\frac{v}{c}\right)^2 + \frac{3}{8}\left(\frac{v}{c}\right)^4 + \frac{5}{16}\left(\frac{v}{c}\right)^6 + \cdots\right]$$
$$\frac{mc^2}{\sqrt{1-(v/c)^2}} \approx mc^2 + \frac{1}{2}mv^2 \quad \text{if } v/c \ll 1$$

Since we define kinetic energy through the relation $E_{\text{particle}} = mc^2 + K$, at low speeds the kinetic energy is approximately the following:

APPROXIMATE KINETIC ENERGY OF A PARTICLE

$$K \approx \frac{1}{2}mv^2 \quad (\text{if } v \ll c)$$

Approximate kinetic energy can also be written in terms of approximate magnitude of momentum.

$$K \approx \frac{1}{2}mv^2 \approx \frac{1}{2}m\left(\frac{p}{m}\right)^2 = \frac{p^2}{2m} \quad (\text{if } v \ll c)$$

EXAMPLE **Approximate Kinetic Energy**

The solar wind consists of charged particles streaming away from the Sun. The speed of a proton in the solar wind can be as high as 1×10^6 m/s. What is the kinetic energy of such a proton?

Solution

QUESTION Is it appropriate to use the approximate expression for kinetic energy in this case?

$$\left(\frac{v}{c}\right)^2 = \left(\frac{1 \times 10^6 \text{ m/s}}{3 \times 10^8 \text{ m/s}}\right)^2 = (3.33 \times 10^{-3})^2 = 1.11 \times 10^{-5}$$

so $\gamma \approx 1$.
It is appropriate to use the approximate expression for K.

$$K \approx \frac{1}{2}mv^2 = \frac{1}{2}(1.7 \times 10^{-27} \text{ kg})(1 \times 10^6 \text{ m/s})^2 = 8.5 \times 10^{-16} \text{ J}$$

The kinetic energy of this relatively slow-moving proton is much less (nearly a million times less) than its rest energy, which was calculated in the previous example.

EXAMPLE **Pull a Block**

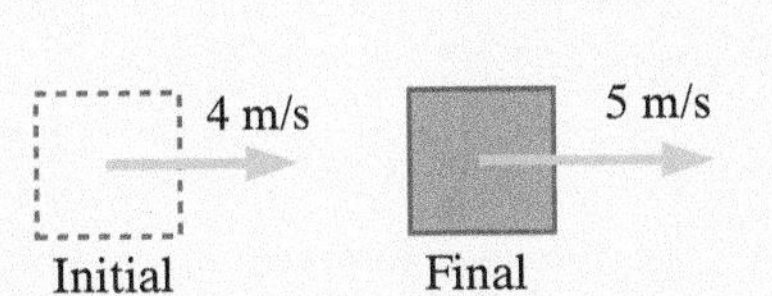

Figure 6.4 You pull a block, changing its speed from 4 m/s to 5 m/s.

If you pull on a 3 kg block and change its speed from 4 m/s to 5 m/s (Figure 6.4), what is the minimum change of chemical (food) energy in you? Give magnitude, sign, and units. (Your temperature rises when you do this, involving an increase in thermal energy, which we'll study in the next chapter, so there is more actual chemical cost than the minimum amount you calculate.)

Solution The speed is tiny compared to c, so the change in kinetic energy of the block is

$$\Delta K = \frac{1}{2}mv_f^2 - \frac{1}{2}mv_i^2 = \frac{1}{2}(3 \text{ kg})[(5 \text{ m/s})^2 - (4 \text{ m/s})^2]$$

$$\Delta K = 13.5 \text{ J}$$

Because you increased the kinetic energy of the block by 13.5 J, and your temperature also rose a bit, the change of your store of chemical (food) energy is at least -13.5 J.

Checkpoint 2 An automobile traveling on a highway has an average kinetic energy of 1.1×10^5 J. Its mass is 1.5×10^3 kg; what is its average speed? Convert your answer to miles per hour to see whether it makes sense. If you could use all of the mc^2 rest energy of some amount of fuel to provide the car with its kinetic energy of 1.1×10^5 J, what mass of fuel would you need?

Relativistic Energy and Momentum

Because both relativistic energy and momentum involve the factor

$$\gamma = \frac{1}{\sqrt{1-\left(\frac{v}{c}\right)^2}}$$

it seems likely that they can be related in some way. As you can show in Problem P8, the relationship turns out to be the following, where p represents $|\vec{p}|$:

RELATIVISTIC ENERGY AND MOMENTUM OF A PARTICLE

$$E^2 - (pc)^2 = (mc^2)^2$$

where $E = \gamma mc^2$; $m =$ particle mass.

This relation is true in all reference frames, for all particle speeds. Both energy and momentum will change if a phenomenon is viewed in a different reference frame, but this relation will still be true. The quantity $E^2 - (pc)^2$ is therefore called an "invariant" quantity—a quantity that does not vary when the reference frame is changed.

This relation turns out to be $E^2 - (pc)^2 = 0$ for those particles such as photons that always travel at the speed of light. Such particles have zero mass, and can never be at rest. The relation for so-called "massless" particles reduces to $E^2 = (pc)^2$, or $E = pc$. This relation is also approximately true for any particle that is traveling at very high speed, because in that case mc^2 is small compared to E. Neutrinos have extremely small mass and travel at nearly the speed of light.

The relation $E^2 - (pc)^2 = (mc^2)^2$ can be used to find another expression for kinetic energy, valid for both high and low speeds:

$$E^2 - (mc^2)^2 = (pc)^2$$
$$(E + mc^2)(E - mc^2) = p^2c^2$$
$$(\gamma mc^2 + mc^2)(K) = p^2c^2$$
$$K = \frac{p^2}{(\gamma + 1)m}$$

At low speeds, $\gamma \approx 1$, and we find as before that $K \approx p^2/(2m)$.

EXAMPLE **Proton Momentum and Energy**

A proton has a mass of 1.7×10^{-27} kg. If its total energy is three times its rest energy, what is the magnitude of its momentum?

Solution

$$mc^2 = (1.7 \times 10^{-27}\text{ kg})(3 \times 10^8\text{ m/s})^2 = 1.5 \times 10^{-10}\text{ J}$$
$$E_{\text{particle}} = mc^2 + K = 3mc^2 \text{ and } (pc)^2 = (E_{\text{particle}})^2 - (mc^2)^2$$
$$pc = \sqrt{(E_{\text{particle}})^2 - (mc^2)^2} = \sqrt{(3mc^2)^2 - (mc^2)^2}$$
$$= \sqrt{8(mc^2)^2} = \sqrt{8}(mc^2)$$
$$p = \sqrt{8}(mc) = \sqrt{8}(1.7 \times 10^{-27}\text{ kg})(3 \times 10^8\text{ m/s}) = 1.4 \times 10^{-18}\text{ kg} \cdot \text{m/s}$$

6.3 WORK: MECHANICAL ENERGY TRANSFER

We are interested in changes in the energy of a system that are due to the interaction of the system with objects in its surroundings. Our goal is to figure out how to describe such changes quantitatively.

Displacement $\Delta\vec{r}$ Instead of Time Interval Δt

The Momentum Principle relates a change in momentum to the net impulse applied to a system. Impulse is the product of force and time.

$$\text{(change of momentum)} = \text{(force)(time)}$$

We might ask what property of a system changes due to forces acting not for a time but through a distance, something like this:

$$\text{(change of ?)} = \text{(force)(distance)}$$

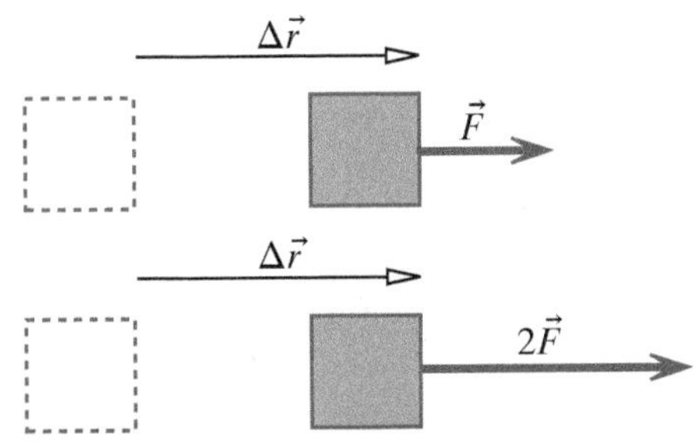

Figure 6.5 Pulling a block over a nearly frictionless surface with a force twice as big (through the same distance) means you expend twice as much chemical energy.

The quantity that changes turns out to be the energy of the system, γmc^2 in the case of a single particle. The quantity involving force times distance is called "work."

In Section 6.6 we will show that this is correct. For now we'll assume this to be true, and after explaining how to calculate work we'll use the Energy Principle for a particle to analyze several kinds of processes.

QUESTION What is the effect of increasing the applied force?

Suppose that you apply a force with constant magnitude F in a constant direction to a block that was initially at rest on a nearly frictionless surface, pulling it through a displacement Δr (Figure 6.5). The block moves faster and faster, acquiring a certain amount of kinetic energy. If you were to pull twice as hard ($2F$) through the same displacement Δr, the block would presumably acquire twice as much kinetic energy (and you would expend twice as much chemical energy). Note that the rest energy mc^2 of the block doesn't change, just the kinetic energy.

QUESTION What is the effect of increasing the displacement?

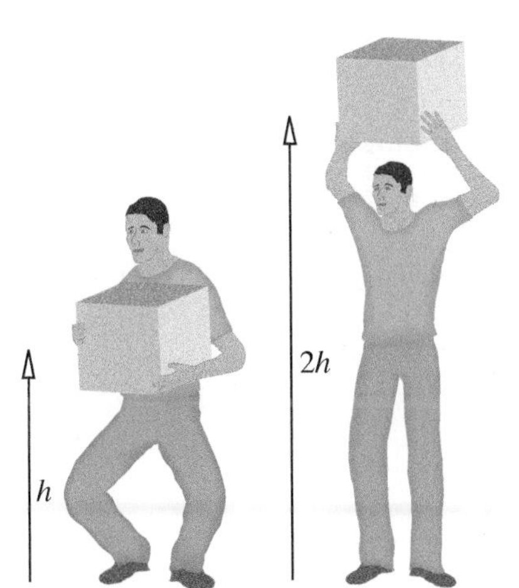

Figure 6.6 Lifting a box twice as high (with the same force) means you expend twice as much chemical energy.

If you lift a heavy box to a height h above the ground, you will have expended a certain amount of chemical energy (Figure 6.6). If you lift the box twice as high (a height $2h$ above the ground), you will presumably have had to expend twice as much energy.

Work

Because the amount of energy expended appears to be proportional both to the magnitude of the force exerted and to the distance through which an object is moved, we'll tentatively define the amount of energy transfer between you and the object as the product of force times displacement, $F\Delta r$. This type of mechanical energy transfer from the surroundings into the system is called "work." Here we use the word "work" as a technical term in the context of physics. This technical meaning is only loosely connected with everyday uses of the word.

The unit of work is newton · meter. From the Momentum Principle we know that a newton is the same as a $\text{kg}\cdot(\text{m/s})/\text{s}$, so a newton · meter is the same as a $\text{kg}\cdot\text{m}^2/\text{s}^2$, which is a joule (same units as mc^2). Work has the same units as energy, as it should if we expect the change in energy of a system to be equal to the work.

QUESTION Is doubling the displacement the same as doubling the time?

No, it isn't. If you pull a block through twice the displacement, you do twice as much work, and the block acquires twice as much kinetic energy. However, because the block is moving faster in the second half of the doubled displacement, the time interval is less than doubled (and the impulse is less than doubled), so the momentum change is not doubled.

Note carefully: work and kinetic energy change are proportional to displacement, whereas impulse and momentum change are proportional to time interval. In symbols: work $F\Delta r$ is not the same as impulse $F\Delta t$.

Work in 3D

When calculating the work done on a system by an applied force, we are interested in the effect of the component of the force that is parallel to the displacement of the system. This sounds familiar—we saw in Chapter 5 that it was useful to divide an applied force into two parts: $\vec{F}_{\parallel}$, the component of a force parallel to a system's momentum, which can change the magnitude of the momentum (and hence the speed), and $\vec{F}_{\perp}$, the perpendicular component of the force, which can change the direction of momentum but not the magnitude. For example, in the circular orbit of a planet around a star, the planet's speed is constant, since the force by the star is always perpendicular to the planet's momentum.

In calculating work we see a similar pattern: the part of an applied force that is parallel to $\Delta\vec{r}$, the displacement, does nonzero work on the system, but the perpendicular part of the force does not contribute to changing the system's energy. In order to calculate the work done on a system by a force $\vec{F}$, we want to use the parallel component of $\vec{F}$. This suggests using a dot product:

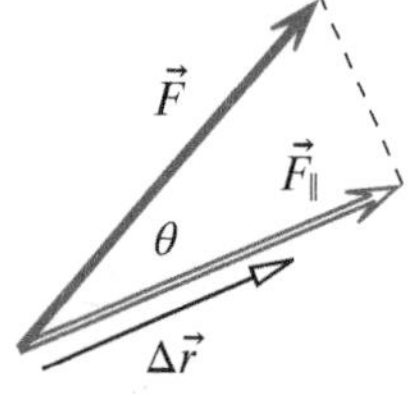

Figure 6.7 The part of a force that is parallel to the displacement does work on a system.

WORK DONE BY A CONSTANT FORCE

$$W = \vec{F} \bullet \Delta\vec{r}$$
$$= \left|\vec{F}\right| \left|\Delta\vec{r}\right| \cos\theta$$
$$= F_x\Delta x + F_y\Delta y + F_z\Delta z$$

The displacement $\Delta\vec{r} = \langle \Delta x, \Delta y, \Delta z \rangle$. The angle θ is the angle between $\vec{F}$ and $\Delta\vec{r}$, as shown in Figure 6.7.

As we'll discuss in more detail in a moment, work can be positive or negative. If you push or pull in the direction of displacement, you do positive work and you increase the energy of the system. If you push or pull in the direction opposite to the displacement, you do negative work and you decrease the energy of the system.

EXAMPLE **Pull a Block**

You pull a block 1.5 m across a table in the $-x$ direction while exerting a force of $\langle -0.3, 0.25, 0 \rangle$ N (Figure 6.8). How much work do you do on the block?

Solution

$$W = \vec{F} \bullet \Delta\vec{r} = \langle -0.3, 0.25, 0 \rangle \text{ N} \bullet \langle -1.5, 0, 0 \rangle \text{ m}$$
$$W = (-0.3\text{ N})(-1.5\text{ m}) + (0.25\text{ N})(0) + 0 \cdot 0 = 0.45\text{ J}$$

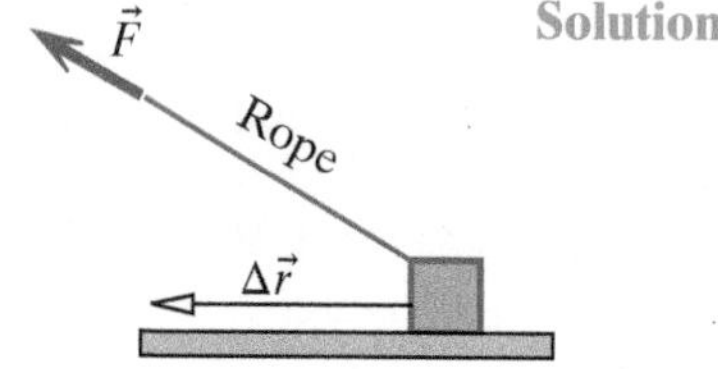

Figure 6.8 You pull a block across a table.

Checkpoint 3 A paper airplane flies from position $\langle 6, 10, -3 \rangle$ m to $\langle -12, 2, -9 \rangle$ m. The net force acting on it during this flight, due to the Earth and the air, is nearly constant at $\langle -0.03, -0.04, -0.09 \rangle$ N. What is the total work done on the paper airplane by the Earth and the air?

Positive and Negative Work

The Energy Principle states that the change in energy of a system (in this case, a single particle) is equal to the energy input from work done by the surroundings.

$$\Delta E_{\text{sys}} = W_{\text{surr}}$$

Since the kinetic energy of a particle might increase or decrease, it must be the case that the work done by the surroundings can be positive or negative.

QUESTION What is the meaning of negative work?

Suppose that you want to slow down a moving object. You push in a direction opposite to the object's motion, and although the object keeps moving in the original direction, it gradually slows down. The object's kinetic energy decreased, so you must have done negative work on the object. Evidently a force acting in a direction opposite to the displacement of a system does negative work.

It is important to determine the correct sign of the work done on a system, because increasing energy is associated with positive work, and decreasing energy is associated with negative work.

EXAMPLE **Positive and Negative Work**

Consider four simple situations that bring out the main issues. In Figure 6.9 a box is shown moving 2 m to the right or left, acted on by a force of magnitude 3 N to the right or left. In each case, calculate the work done on the box by the applied force, and state whether the box speeded up or slowed down.

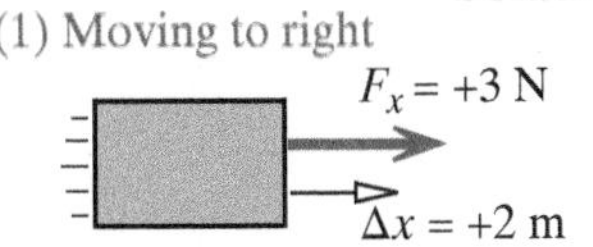

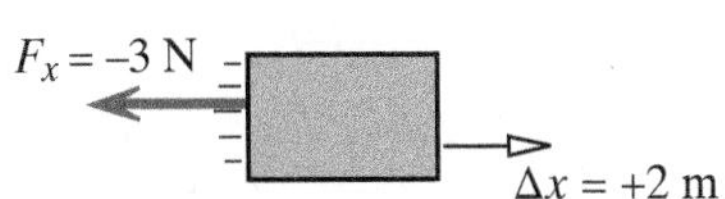

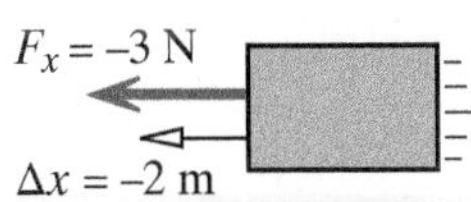

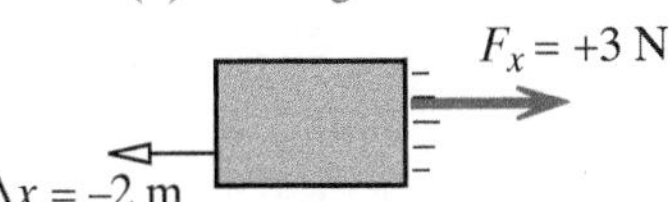

Figure 6.9 Force and displacement in various directions.

Solution Since the forces and displacements are all in the $+x$ or $-x$ direction, in each case $W = F_x\Delta x + F_y\Delta y + F_z\Delta z$ reduces to $W = F_x\Delta x$.

Case 1: Force and displacement both in $+x$ direction:

$$W = F_x\Delta x = (+3\,\text{N})(+2\,\text{m}) = +6\,\text{J}$$

The work done is positive, so the kinetic energy of the box increases. The box speeds up.

Case 2: Force in $-x$ direction and displacement in $+x$ direction:

$$W = F_x\Delta x = (-3\,\text{N})(+2\,\text{m}) = -6\,\text{J}$$

The work done is negative, so the kinetic energy of the box decreases. The box slows down.

Case 3: Force and displacement both in $-x$ direction:

$$W = F_x\Delta x = (-3\,\text{N})(-2\,\text{m}) = +6\,\text{J}$$

The work done is positive, so the kinetic energy of the box increases. The box speeds up.

Case 4: Force in $+x$ direction and displacement in $-x$ direction:

$$W = F_x\Delta x = (+3\,\text{N})(-2\,\text{m}) = -6\,\text{J}$$

The work done is negative, so the kinetic energy of the box decreases. The box slows down.

From these four cases we can infer that work is positive if the force and the displacement are in the same direction, and work is negative if force is opposite to displacement.

QUESTION Is it possible to determine whether work is positive or negative simply by looking at the direction of the force alone?

No, we need to know both the direction of the force and of the displacement. A force in the $-x$ direction can do positive work if the displacement of the system is also in the $-x$ direction.

EXAMPLE **Throw a Ball Straight Up**

You throw a 250 g ball straight up, and it rises 8 m into the air (Figure 6.10). Neglecting air resistance, is the work done on the ball positive or negative? How much work is done by the gravitational force?

Solution System: Ball
Surroundings: Earth

Physical reasoning: The ball slows down on its way up, so its kinetic energy is decreasing. Therefore the work must be negative. Calculate work:

$$F_y = -mg = -(0.25\,\text{kg})(+9.8\,\text{N/kg}) = -2.45\,\text{N}$$
$$W = F_y \Delta y = (-2.45\,\text{N})(8\,\text{m}) = -19.6\,\text{J}$$

Negative work is done on the ball, and the ball slows down, losing some kinetic energy.

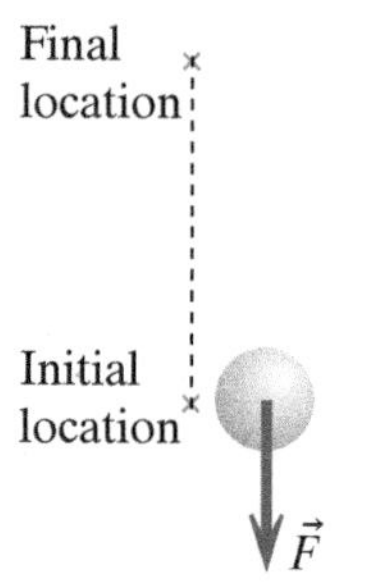

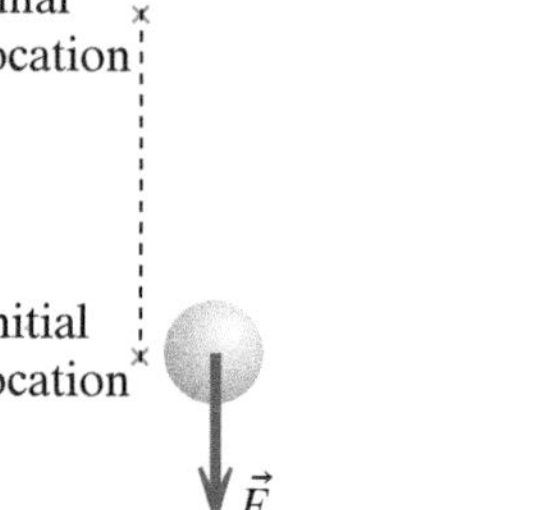

Figure 6.10 A ball travels straight up.

Zero Work

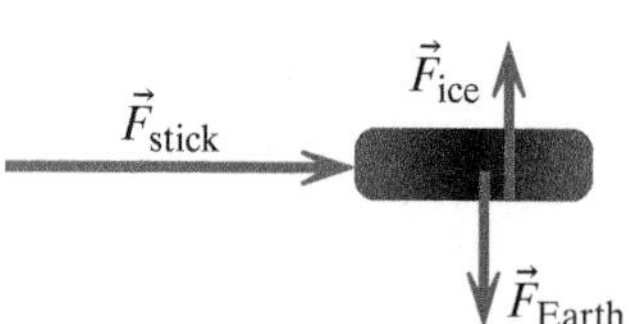

Figure 6.11 A hockey puck slides to the right with low friction on ice, sped up by a hockey stick pushing on the puck.

In Figure 6.11 a puck is sliding with low friction to the right on ice, being sped up by a hockey stick pushing on it. There are three different objects in the surroundings that are interacting with the puck. The hockey stick pushes horizontally, doing positive work to speed up the puck; the Earth pulls down, and the ice pushes up, supporting the puck.

QUESTION How much work is done by the force exerted downward by the Earth? Does this force speed up or slow down the puck? Is the work done by the Earth positive or negative?

The work done by the Earth's gravitational force is zero. It doesn't make the puck speed up or slow down. We can check this:

$$W = F_x \Delta x + F_y \Delta y + F_z \Delta z = (0)\Delta x + (-mg)(0) + (0)(0) = 0$$

A force perpendicular to the motion does zero work.

QUESTION How much work is done by the force exerted upward by the ice? Does this force speed up or slow down the puck? Is the work done by the ice positive or negative?

Again, the work done by the force exerted by the ice is zero. It doesn't make the puck speed up or slow down (remember that we're neglecting any horizontal, frictional component of the force that the ice exerts on the puck).

A particularly striking case of zero work is circular motion at constant speed, such as the Moon going around the Earth in a (nearly) circular orbit under the influence of the Earth's gravitational force, which always acts perpendicular to the motion if the orbit is a circle (Figure 6.12).

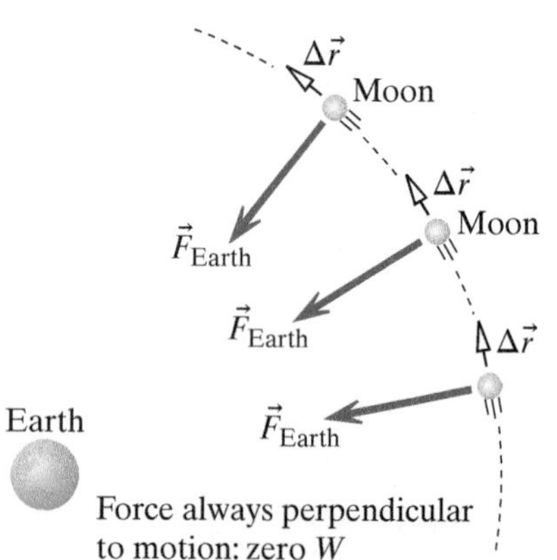

Figure 6.12 The gravitational force of the Earth on the Moon in a (nearly) circular orbit is always (nearly) perpendicular to the motion, so the work is zero.

QUESTION Is the Moon speeding up or slowing down? Therefore what can you say about the work done on the Moon by the Earth's gravitational attraction? Is this consistent with the definition of work?

Figure 6.13 The floor exerts an upward force through zero distance and does zero work.

The Moon's speed and kinetic energy aren't changing (and its rest energy isn't changing), so its energy isn't changing. Therefore the work done should be zero. This makes sense, because the force is always perpendicular to the motion.

> QUESTION Then what does the Earth's gravitational force do, if it doesn't do any work?

It changes the momentum of the Moon. The momentum is a vector quantity, and change of direction represents a change of the vector momentum. A force is required to change the momentum of the Moon, otherwise it would move in a straight line, not a circle.

Again we see important differences between energy and momentum, and between work and impulse. There is no change in the Moon's energy, since that depends on the *magnitude* of the velocity, but there is a change of the Moon's momentum, since that depends on the *direction* of the velocity.

EXAMPLE **Work Done on a Jumper by the Floor**

Suppose a person crouches down and jumps straight up (Figure 6.13). How much work is done by the force of the floor on the feet of the jumper?

Solution Consider the jumper as the system. The surroundings are the Earth, the floor, and the air (but we'll neglect the small air resistance force). As the jumper rises, the Earth pulls down, doing negative work.

> QUESTION What is the displacement through which the force exerted by the floor acts?

The distance that the atoms in the jumper's shoes move while they are in contact with the atoms of the floor is nearly zero (especially if the floor is quite rigid). Therefore the displacement through which the force acts is zero. This means that the floor does zero work!

> QUESTION If the contact force by the floor does no work, what *does* this force do?

The force of the floor contributes to the net force that determines the change in the jumper's momentum, but the floor force does no work. The only significant work done by the surroundings is the negative work done by the force of the Earth (correspondingly, the chemical energy of the jumper decreases).

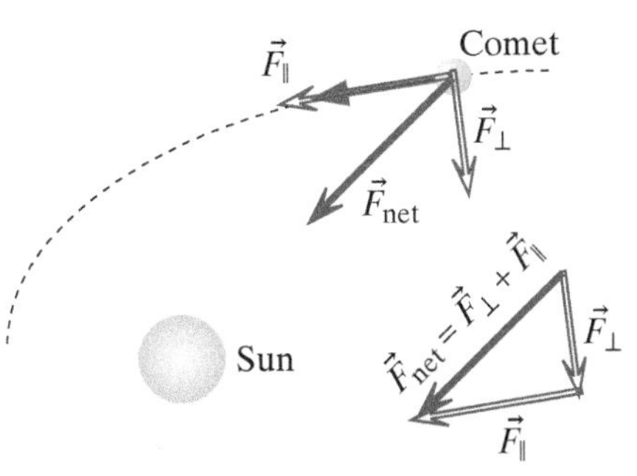

Figure 6.14 The parallel part of the force by the Sun does work, changing the comet's kinetic energy. The perpendicular part does zero work but changes the direction of the comet's momentum.

Of course, in many situations the forces applied to a system have parts both parallel and perpendicular to $\Delta\vec{r}$, as shown in Figure 6.14. $\vec{F}_\parallel$ does either positive or negative work, and $\vec{F}_\perp$ does zero work, but changes the direction of the system's momentum. Figure 6.15 summarizes the relationship between the relative directions of $\vec{F}$ and $\Delta\vec{r}$ and the sign of the work done.

Force in direction of motion: $W > 0$

Force opposite the motion: $W < 0$

$\vec{F}$ $\Delta\vec{r}$ $\Delta\vec{r}$ $\Delta\vec{r}$ $\vec{F}$ $\vec{F}$

Force perpendicular to motion: $W = 0$

Figure 6.15 Force in direction of motion, opposite the motion, and perpendicular to motion.

SIGN OF WORK

- A force in the direction of motion does positive work.
- A force opposite to the direction of motion does negative work.
- A force perpendicular to the direction of motion does zero work.
- A force that acts through zero displacement does zero work.

Checkpoint 4 In each of the following cases state whether the work done by the specified force is positive, negative, or zero. Also state whether the kinetic energy of the object in question increases, decreases, or remains the same. **(a)** A ball is moving upward, acted on by a downward gravitational force. **(b)** A ball is falling downward, acted on by a downward gravitational force. **(c)** A car is moving rapidly to the left, and Superman exerts a force on it to the right to slow it down, backing up to the left as he pushes to the right. **(d)** You throw a ball downward. Consider the force exerted by your hand on the ball while they are in contact. **(e)** In a 6 month period the Earth travels halfway around its nearly circular orbit of the Sun. Consider the gravitational force exerted on the Earth by the Sun during this period.

Work Done by a Nonconstant Force

If a force changes in magnitude or direction during a process, we can't calculate the work simply by multiplying a constant force times the net displacement. The nonconstant force acts on the object along a path, and we split the path into small increments $\Delta\vec{r}$ (Figure 6.16). This is similar to the way in which we took small time steps Δt when computing the trajectories of objects.

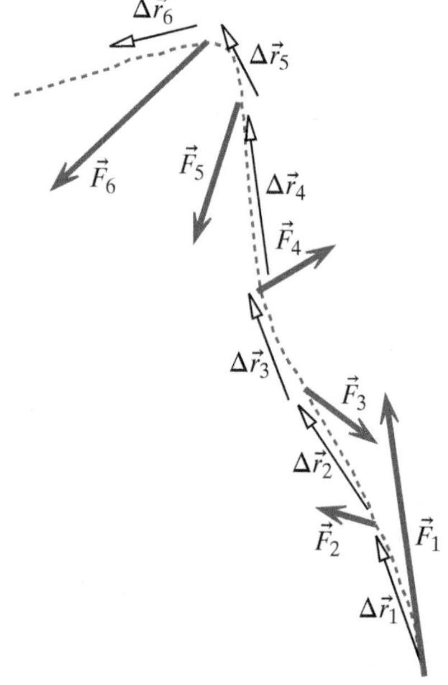

Figure 6.16 A varying force is exerted on an object moving along a curved path.

If our increments along the path are sufficiently small, the force is approximately constant in magnitude and direction within one increment. We can write the total work done as a sum along the path:

$$W = \vec{F}_1 \bullet \Delta\vec{r}_1 + \vec{F}_2 \bullet \Delta\vec{r}_2 + \vec{F}_3 \bullet \Delta\vec{r}_3 + \cdots \quad \text{(small steps)}$$

More compactly, we can use the symbol Σ (Greek capital sigma) to mean "sum":

$$W = \Sigma(\vec{F} \bullet \Delta\vec{r})$$

Checkpoint 5 You push a heavy crate out of a carpeted room and down a hallway with a waxed linoleum floor. While pushing the crate 2.3 m out of the room you exert a force of 30 N; while pushing it 8 m down the hallway you exert a force of 15 N. How much work do you do in all?

We have refined our definition of work to allow us to calculate the work done by varying forces. This more complete definition of work allows us to apply the Energy Principle in situations where forces vary.

Work as an Integral

If we go to the limit where the increments are infinitesimal ($\Delta\vec{r} \to d\vec{r}$), the sum used for calculating work turns into a sum of an infinite number of infinitesimal contributions. Such a summation is called a "definite integral." It is written as a distorted "S" for "sum," with an indication of the initial position "i" and the final position "f":

$$W = \int_i^f \vec{F} \bullet d\vec{r}$$

If you have already studied integration in a calculus course, you already know some techniques for evaluating definite integrals such as this; if not, you will learn them soon in calculus. However, in many real-world cases it is not possible to find an analytical form for evaluating an integral, in which case one uses numerical integration, approximating the infinite sum by a finite sum, with finite increments $\Delta\vec{r}$.

GENERAL DEFINITION OF WORK

$$W \equiv \int_i^f \vec{F} \bullet d\vec{r}$$

This general definition applies to any force, whether constant or not. The integral must be evaluated over the path along which the force is applied. The integral may be approximated computationally by the finite sum $\Sigma(\vec{F} \bullet \Delta\vec{r})$.

Later in this chapter we'll see that in some cases the work integral has the same value no matter what path is followed between the initial and final positions. This is called "path independence."

EXAMPLE **Work Done by a Spring**

A ball traveling horizontally runs into a horizontal spring, whose stiffness is 100 N/m (Figure 6.17). As the spring is compressed, the ball slows down. If the spring is compressed until its length is 20 cm shorter than its relaxed length, how much work does the spring do on the ball?

Solution Initial state: $s = 0$

Final state: $s = -0.2\,\text{m}$

Choose the origin at the end of the relaxed spring, so $F_x = -k_s x$.

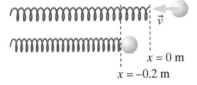

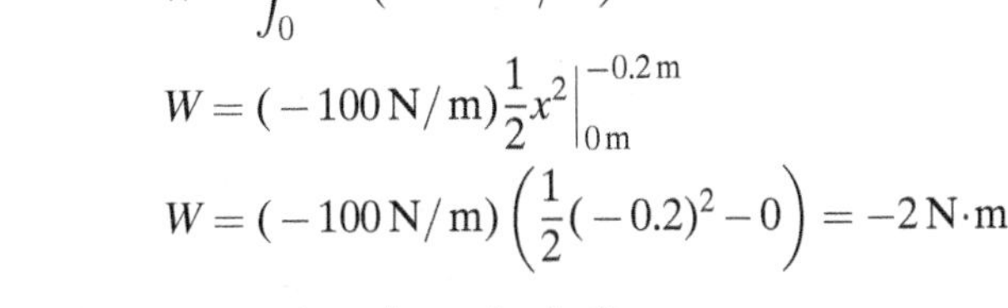

$$W = \int_0^{-0.2} (-100\,\text{N/m})x\,dx$$

$$W = (-100\,\text{N/m})\frac{1}{2}x^2\Big|_{0\,\text{m}}^{-0.2\,\text{m}}$$

$$W = (-100\,\text{N/m})\left(\frac{1}{2}(-0.2)^2 - 0\right) = -2\,\text{N}{\cdot}\text{m}$$

$x = 0$ m

$x = -0.2$ m

Figure 6.17 A ball runs into a horizontal spring.

The spring does -2 J of work on the ball.

Check: Sign ok—negative work decreases kinetic energy of ball. Units ok ($\text{N}{\cdot}\text{m} = \text{J}$).

6.4 WORK AND ENERGY

The Energy Principle involves a difference in the energy of a system in its final state and the energy in its initial state. We will refer to this as the "difference form" of the Energy Principle.

THE ENERGY PRINCIPLE: DIFFERENCE FORM

$$\Delta E_{\text{sys}} = W_{\text{surr}} + \text{other inputs}$$

$\Delta E_{\text{sys}} = E_{\text{sys},f} - E_{\text{sys},i}$, so we can also write the Energy Principle this way:

THE ENERGY PRINCIPLE: UPDATE FORM

$$E_{\text{sys},f} = E_{\text{sys},i} + W_{\text{surr}} + \text{other inputs}$$

Sometimes it's easier to think about the situation using the difference form of the Energy Principle, and sometimes the update form is more appropriate. Use whichever form you're comfortable with, but make sure you understand both forms.

A simple and effective procedure for solving problems involving energy is this:

SOLVING PROBLEMS INVOLVING ENERGY

- Write out all terms in the Energy Principle.
- Identify terms that don't change and therefore cancel.
- Substitute in all numerical values you know.
- Solve for the unknown quantity.

EXAMPLE **Pushing a Box in Space**

You're outside a spacecraft, pushing on a heavy box whose mass is $m =$ 2000 kg. You exert a force $\vec{F} = \langle 300, 500, 0 \rangle$ N while the box moves through a displacement $\Delta\vec{r} = \langle 0.1, -0.3, 0.2 \rangle$ m. Initially the box had a speed $|\vec{v}_i| =$ 0.7 m/s. **(a)** How much work do you do? **(b)** What is the final kinetic energy of the box? **(c)** What is the final speed of the box? **(d)** What is the direction of the final velocity of the box?

Solution System: Box (considered to be a particle)
Surroundings: You
Initial state: When you start pushing the box
Final state: When the box has been displaced the amount $\Delta\vec{r}$

$$E_f = E_i + W$$
$$(\cancel{mc^2} + K_f) = (\cancel{mc^2} + K_i) + W$$
$$K_f = K_i + W$$
$$K_f = K_i + (F_x\Delta x + F_y\Delta y + F_z\Delta z)$$
$$K_f = K_i + (300\,\text{N})(0.1\,\text{m}) + (500\,\text{N})(-0.3\,\text{m}) + (0\,\text{N})(0.2\,\text{m})$$
$$K_f = K_i + (-120\,\text{J})$$
$$K_f = \left(\frac{1}{2}mv_i^2\right) + (-120\,\text{J})$$
$$K_f = \frac{1}{2}(2000\,\text{kg})(0.7\,\text{m/s})^2 + (-120\,\text{J})$$
$$K_f = (490\,\text{J}) + (-120\,\text{J}) = 370\,\text{J}$$

Final speed:

$$\frac{1}{2}mv_f^2 = K_f$$
$$v_f = \sqrt{\frac{2K_f}{m}}$$
$$v_f = \sqrt{\frac{2(370\,\text{J})}{2000\,\text{kg}}} = 0.608\,\text{m/s}$$

(a) $W = -120$ J
(b) $K_f = 370$ J. This is less than K_i, as it should be since you did negative work on the system.
(c) $v_f = 0.608$ m/s
(d) We don't have enough information to answer this.

We answered all the questions except for (d), which asked for the direction of the final velocity of the box. We can't answer that question! The Energy

Principle is a scalar principle, and only the magnitude of the velocity appears in the particle energy, not the direction. In order to figure out the direction of the final velocity we would have to use the Momentum Principle, which is a vector principle. However, note that it was not stated how long the process took, so we can't apply the Momentum Principle directly. We would have to determine the approximate Δt from the known displacement and initial speed. (The actual trajectory is a curve; we've approximated the small displacement by a straight line.)

Note that we knew that the rest energy of the box did not change (it did not, for example, turn into solid gold), so we were able to eliminate mc^2 from both sides of the energy equation.

EXAMPLE

Throwing a Ball

You hold a ball of mass 0.5 kg at rest in your hand, then throw it forward, underhand, so that as it leaves your hand its speed is 20 m/s. How much work did you do on the ball? Note that you do not know how big a force you applied, nor do you know the distance through which you applied this force.

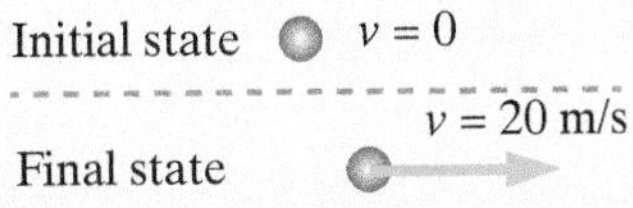

Initial state $v = 0$

Final state $v = 20$ m/s

Figure 6.18 Throwing a ball.

Solution System: Ball

Surroundings: You (we can neglect the gravitational effect of the Earth in this short process)

Initial state: At the start of the throw, when the ball's speed is zero

Final state: When the ball just leaves your hand at 20 m/s (Figure 6.18)

$$
\begin{aligned}
E_f &= E_i + W \\
(mc^2 + K_f) &= (mc^2 + K_i) + W \\
K_f &= K_i + W \\
\frac{1}{2}mv_f^2 &= \frac{1}{2}mv_i^2 + W \\
\frac{1}{2}(0.5\,\text{kg})(20\,\text{m/s})^2 &= \frac{1}{2}m(0)^2 + W \\
100\,\text{J} &= W
\end{aligned}
$$

There is no change of identity, so no change in rest energy; the mc^2 terms cancel. The energy of the ball increased; it sped up from 0 to 20 m/s. You did positive work, pushing the ball to speed it up.

Check: Work is positive; this makes sense because the final speed is larger than the initial speed. The units are correct.

We were able to calculate the work you did on the ball even though we didn't know how large a force you applied or how large was the displacement of your hand in the throwing motion. This is one aspect of the power of the Energy Principle—that simply by determining the change in energy of a system you know how much energy input there was to that system.

QUESTION In the previous example, what was the change in the energy of the surroundings? Can you say what kind of energy it was in the surroundings that changed?

Energy is a conserved quantity: $\Delta E_{\text{system}} + \Delta E_{\text{surroundings}} = 0$. The energy of the system (the ball) increased by 100 J. Evidently the change in energy of the surroundings was -100 J. This decrease of energy in the surroundings corresponds to a decrease in the chemical energy stored in your body, coming from food you consumed. In throwing the ball you converted some stored chemical energy into kinetic energy of the ball.

Actually, you need to expend more than 100 J of chemical energy (an energy decrease), because in addition to increasing the kinetic energy of the ball by 100 J, you also raised your body temperature a bit (thermal energy increase; more on this in Chapter 7). Your muscles aren't 100% efficient; it takes more than 100 J of chemical energy to give the ball 100 J of kinetic energy.

The net energy change of the surroundings is −100 J, which could consist (say) of −180 J of chemical energy change and +80 J of thermal energy rise in your body.

Checkpoint 6 In the example above, suppose that you wanted the final speed of the ball to be twice as big (40 m/s). How much work would you have to do?

EXAMPLE **The Energy Principle at High Speed**

An electron (mass 9×10^{-31} kg) in a particle accelerator is acted on by a constant electric force $\langle 5 \times 10^{-13}, 0, 0 \rangle$ N. (This is much greater than the force of gravity, which is only $mg \approx 1 \times 10^{-29}$ N, so we can neglect the effect of the Earth on the electron.) The initial velocity of the electron is $\langle 0.995c, 0, 0 \rangle$, where c as usual is the speed of light. The electron moves through a displacement of $\langle 5, 0, 0 \rangle$ m. What is its final speed, expressed as a fraction of the speed of light?

Solution System: Electron

Surroundings: Accelerator (which applies the electric force); neglect the gravitational effect of the Earth in this short process.

Initial state: Beginning of given displacement

Final state: End of given displacement

The unknown quantity, final speed, is part of γ_f. It's easiest to find γ_f first, then solve for v_f. Because the electron's speed is near the speed of light, we can't use an approximate expression for K. We need to use the exact, relativistic equation for particle energy.

$$E_f = E_i + W$$

$$\gamma_f mc^2 = \gamma_i mc^2 + W$$

$$\gamma_f mc^2 = \gamma_i mc^2 + (F_x \Delta x + F_y \Delta y + F_z \Delta z)$$

$$\gamma_f mc^2 = \gamma_i mc^2 + (5 \times 10^{-13}\,\text{N})(5\,\text{m}) + (0\,\text{N})(0\,\text{m}) + (0\,\text{N})(0\,\text{m})$$

$$\gamma_f mc^2 = \gamma_i mc^2 + (2.5 \times 10^{-12}\,\text{J})$$

$$\gamma_f mc^2 = \frac{1}{\sqrt{1-(0.995c/c)^2}} mc^2 + (2.5 \times 10^{-12}\,\text{J})$$

$$\gamma_f mc^2 = 10.0 mc^2 + (2.5 \times 10^{-12}\,\text{J})$$

$$\gamma_f = \frac{10.0 mc^2 + (2.5 \times 10^{-12}\,\text{J})}{mc^2}$$

$$\gamma_f = 10.0 + \frac{2.5 \times 10^{-12}\,\text{N}}{(9 \times 10^{-31}\,\text{kg})(3 \times 10^8\,\text{m/s})^2} = 40.9$$

$$\sqrt{1-(v_f/c)^2} = \frac{1}{40.9} = 2.44 \times 10^{-2}$$
$$1-(v_f/c)^2 = (2.44 \times 10^{-2})^2 = 5.98 \times 10^{-4}$$
$$(v_f/c)^2 = 1 - 5.98 \times 10^{-4} = 0.9994$$
$$v_f/c = \sqrt{0.9994} = 0.9997$$
$$v_f = 0.9997c$$

Check: The final speed is greater than the initial speed, consistent with positive work. The units look right; in particular, the values for γ were dimensionless, as they should be.

The speed increased only slightly, from $0.995c$ to $0.9997c$, even though the energy changed a lot, from $10.0mc^2$ to $40.9mc^2$. The electron's energy increased by about a factor of 4, but the electron's speed increased very little. When a particle is traveling at nearly the speed of light, adding a lot of energy to the particle doesn't change its speed very much.

Checkpoint 7 In the preceding example, at the final speed, $0.9997c$, what was the particle energy as a multiple of the rest energy mc^2? (That is, if it was twice mc^2, write $2mc^2$.) What was the kinetic energy as a multiple of mc^2? Was the kinetic energy large or small compared to the rest energy? At low speeds, is the kinetic energy large or small compared to the rest energy?

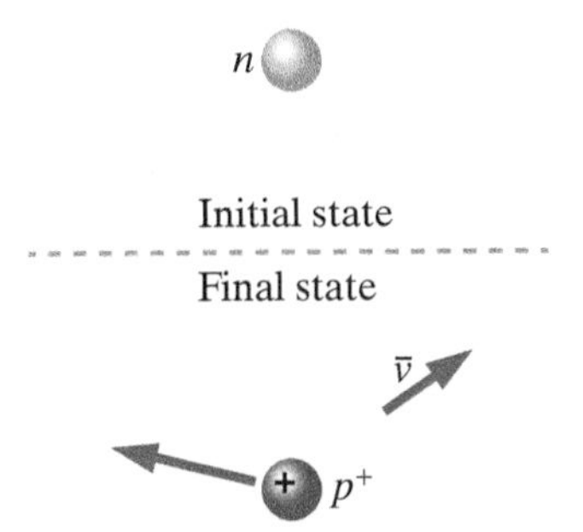

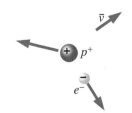

Figure 6.19 A stationary neutron (n) decays into a proton (p^+), an electron (e^-), and a nearly massless antineutrino ($\overline{\nu}$) in a reaction that can be written $n \to p^+ + e^- + \overline{\nu}$. This is an example of a process in which there is a change of identity and therefore a change in rest energy.

6.5 CHANGE OF REST ENERGY

As we saw in the preceding examples, if a particle doesn't change its identity, both the initial and the final energies of the system include mc^2 rest-energy terms and these cancel out in the energy equation:

$$\cancel{mc^2} + K_f = \cancel{mc^2} + K_i + W$$

This is an important and common special case, but it is not the only case. In some processes the mass of a particle can change because the identity of the particle changes.

Neutron decay is a good example to illustrate the issues. A free neutron (one not bound into a nucleus) is unstable, with an average lifetime of about 15 min. As shown in Figure 6.19, it decays into a proton, an electron, and a nearly massless antineutrino, which travels at nearly the speed of light. All three of these particles have kinetic energy. That is, they have energy above and beyond their rest energy; the energy of the antineutrino is nearly all kinetic, because it has almost no rest energy. This is a situation with a change of particle identity, and a change in particle rest energy.

An Electron Volt (eV) Is a Unit of Energy

The mass of a neutron is 1.6749×10^{-27} kg, so its rest energy mc^2 is about 1.51×10^{-10} J. Because the rest energy of a single particle is such a small number of joules, it is usually expressed in different units of a million electron volts, abbreviated MeV. One electron volt is the amount of energy an electron acquires when moving through an electric potential difference of one volt. (Electric potential difference is introduced in a later chapter; an ordinary flashlight battery has an electric potential difference across it of 1.5 V.)

ELECTRON VOLT (eV) AND MeV

$$1 \text{ eV} = 1.6 \times 10^{-19} \text{ J}$$

eV is the abbreviation for electron volt. 1 MeV $= 1 \times 10^6$ eV

The rest energy of a neutron is 939.6 MeV, and the rest energy of a proton is 938.3 MeV—slightly less than that of a neutron. An electron is much less massive; its rest energy is 0.511 MeV.

Scientists studying elementary particles often use the unit MeV/c^2 as a unit of mass, because dividing energy by c^2 gives mass. In these units the mass of a proton is 938.3 MeV/c^2. The mass of a neutron is 939.6 MeV/c^2. The mass of the electron is 0.511 MeV/c^2. A neutrino or antineutrino is nearly massless—almost all of its energy is kinetic energy.

EXAMPLE **Neutron Decay**

In the decay of a free neutron at rest (Figure 6.19), how much kinetic energy (in MeV) do the decay products (proton, electron, and antineutrino) share? Why can't a proton decay into a neutron?

Solution System: All the particles, both before and after the decay (as long as we include all particles in the system, it is okay if some of them change identity during the process)
Surroundings: Nothing (the Earth can be neglected, since the process is so fast that work done by gravitational forces is negligible)

Initial state: Neutron at rest
Final state: Proton (p), electron (e), and antineutrino ($\overline{\nu}$), far apart

Energy Principle:

$$E_f = E_i + W$$
$$(m_P c^2 + K_P) + (m_e c^2 + K_e) + (K_{\overline{\nu}}) = (m_N c^2 + K_N) + W$$
$$(938.3 \text{ MeV} + K_P) + (0.511 \text{ MeV} + K_e) + (K_{\overline{\nu}}) = (939.6 \text{ MeV} + 0) + 0$$
$$(K_P + K_e + K_{\overline{\nu}}) + 938.8 \text{ MeV} = 939.6 \text{ MeV}$$
$$(K_P + K_e + K_{\overline{\nu}}) = (939.6 - 938.8) \text{ MeV}$$
$$(K_P + K_e + K_{\overline{\nu}}) = 0.8 \text{ MeV}$$

Here the neutron undergoes a change of identity, so there is a change in rest energy, and the mc^2 terms do not cancel. Because the sum of the rest energies of the three decay particles is less than the rest energy of the parent neutron, there is energy left over that shows up as kinetic energy shared among the three decay particles. How this kinetic energy is divided up among the three particles varies from one neutron decay to the next, governed by probabilities that can be calculated using the science of quantum mechanics and the properties of the "weak interaction" that is responsible for neutron decay. We found the total kinetic energy of all the particles in the final state, but we don't know the separate kinetic energies of the individual particles.

QUESTION Why can't a proton decay into a neutron?

A proton can't decay into a neutron, because the neutron has more rest energy than a proton. In such a decay the kinetic energy of the decay products would turn out to be a negative number, which is impossible. Kinetic energy is always positive.

In a chemical reaction such as carbon combining with oxygen in the fire of a coal-burning power plant, the resulting kinetic energy (which is responsible for heating water to make steam) is on the order of 1 eV per molecule. Nuclear reactions produce vastly more kinetic energy, in the case of neutron decay almost a million times more per nucleus (1 MeV/1 eV is one million). This is why some power plants use nuclear rather than chemical reactions.

Checkpoint 8 In the preceding example, what fraction of the original neutron's rest energy was converted into kinetic energy?

Intelligent "Plug and Chug"

Look back at the last four worked-out examples (pushing a box, throwing a ball, energy at high speed, and neutron decay) and you'll see that the same procedure was used to solve each one. Applying the Energy Principle is basically an intelligent sort of "plug and chug" procedure.

- Specify system and surroundings.
- Specify the initial and final states.
- Write out the Energy Principle in detail for this system.
- Use the given information to evaluate all the terms you can.
- Solve for the unknown quantity, which may be inside one of the remaining terms.
- Check that signs and values make physical sense.

The four preceding examples were quite different, but the same procedure worked in each case. The central idea is to let the terms in the Energy Principle tell you what to calculate, and then go ahead and calculate them.

Often students have the impression that physics consists of "knowing what formula to use" for a particular type of problem, with a different formula for every problem. But look again at the last four examples. There weren't four "formulas" giving the four answers. Rather there was one logical procedure that worked in every case, based on the fundamental Energy Principle that applies to all situations.

In following this procedure, you still need to think, but let your thinking be organized by the principle itself. You need to decide whether to try using the Momentum Principle or the Energy Principle, you need to choose a system and initial and final states intelligently, and so on. However, the basic framework never changes.

QUESTION How do you decide whether to use the Momentum Principle or the Energy Principle?

Think about what quantities are given, and what you must find: Work involves a distance and affects the energy, whereas impulse involves a time and affects the momentum. Momentum is a vector, and involves information about directions; energy is a scalar, and has no directional information.

You should never find yourself saying, "I don't have any idea even where to start on this problem!"

If you do find yourself saying this, take a deep breath, choose a system, identify the objects in the surroundings that exert forces on the system, and write down the Momentum Principle or the Energy Principle for this situation. If you try using the Energy Principle, choose initial and final states, and flesh out the various terms in the Energy Principle for the particular situation at hand, then solve for the unknown quantities.

Physics can be easier than most people think, if you go with the flow. If on the other hand you try to invent a new and different approach for every new problem you encounter, or search futilely for some "formula," you'll waste huge amounts of time and get very frustrated.

6.6 PROOF OF THE ENERGY PRINCIPLE FOR A PARTICLE

We have asserted that the following is true for a particle:

$$\Delta E = W_{\text{surr}}$$

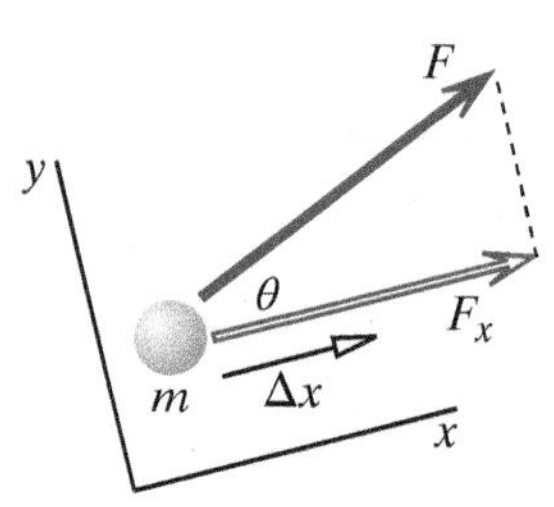

Figure 6.20 A particle of mass m. We choose the x axis to be in the direction of the motion. The net force acting on the particle has a component F_x in the direction of motion.

To prove this, we need to show that $\Delta E = W_{\text{surr}}$ is true given our definitions of particle energy and work. Consider a single pointlike particle of mass m. For convenience in the reasoning we are about to do, choose coordinate axes so that the x axis is in the direction of the motion at this instant, so the displacement in the next short time interval is simply Δx (Figure 6.20). There is a net force $\vec{F}$ (magnitude F) acting at an angle θ to the displacement. The physical results we will obtain do not depend on our choice of coordinate axes; we are free to choose axes that simplify our calculations.

The component of the net force that is parallel to the displacement, F_x, affects the speed and hence the kinetic energy of the particle. The perpendicular component F_y merely changes the direction of the momentum, and does not change speed or energy.

In the very short displacement Δx, the energy of the particle E_{particle} changes by a small amount, and this is supposed to be equal to the work done on the particle:

$$\Delta E = F_x \Delta x$$

According to the Momentum Principle $F_x = \dfrac{\Delta p_x}{\Delta t}$, so

$$\Delta E = \left(\frac{\Delta p_x}{\Delta t}\right)\Delta x$$

Dividing by Δx, we have this:

$$\frac{\Delta E}{\Delta x} = \frac{\Delta p_x}{\Delta t}$$

In the limit as $\Delta t \to 0$, this equation relates the spatial rate of change of energy to the time rate of change of momentum:

$$\frac{dE}{dx} = \frac{dp_x}{dt}$$

In words, this says that the change in energy of the particle per unit distance (in the direction of the motion) is equal to the change in the particle's momentum in that direction per unit time. This equation reflects the fact that a change in position (Δx or dx) is associated with a change in energy, whereas a change in time (Δt or dt) is associated with a change in momentum. If we can show that this equation is true if $E = \gamma mc^2$, we will have proven that $\Delta E = W$ is correct.

Proof for Low Speeds

In the special case that the speed of a particle is small compared to the speed of light, the proof is simple. We need to show that if particle energy is given by the approximate equation:

$$E \approx mc^2 + \frac{1}{2}mv^2 \quad (v \ll c)$$

then the equation derived above:

$$\frac{dE}{dx} = \frac{dp_x}{dt}$$

is satisfied.

We chose the x axis to be in the direction of the motion, so the particle energy is:

$$E \approx mc^2 + \frac{1}{2}mv_x^2 \quad (v \ll c)$$

Left-hand side of equation:

$$\begin{aligned}\frac{dE}{dx} &= \frac{d}{dx}\left(\frac{1}{2}mv_x^2\right)\\ &= \frac{1}{2}m\frac{d}{dx}(v_x^2) = mv_x\frac{dv_x}{dx} = m\frac{dx}{dt}\frac{dv_x}{dx} = m\frac{dv_x}{dt}\end{aligned}$$

Right-hand side of equation (since mc^2 is constant, its derivative is zero):

$$\frac{dp_x}{dt} = \frac{d}{dt}(mv_x) = m\frac{dv_x}{dt} \quad (v \ll c)$$

So since the left-hand and right-hand sides of the equation are equal,

$$\frac{dE}{dx} = \frac{dp_x}{dt}$$

is indeed true with our definitions of energy and work. We have proved that at least at low speed the Energy Principle for a particle $\Delta E = W$ is correct.

General Proof (Relativistically Correct)

The proof that the relativistically correct expression $E = \gamma mc^2$ satisfies the relation

$$\frac{dE}{dx} = \frac{dp_x}{dt}$$

is given in Section 6.20, at the end of the chapter. It follows the same reasoning as was used to prove the low-speed case but is algebraically more complex.

6.7 POTENTIAL ENERGY IN MULTIPARTICLE SYSTEMS

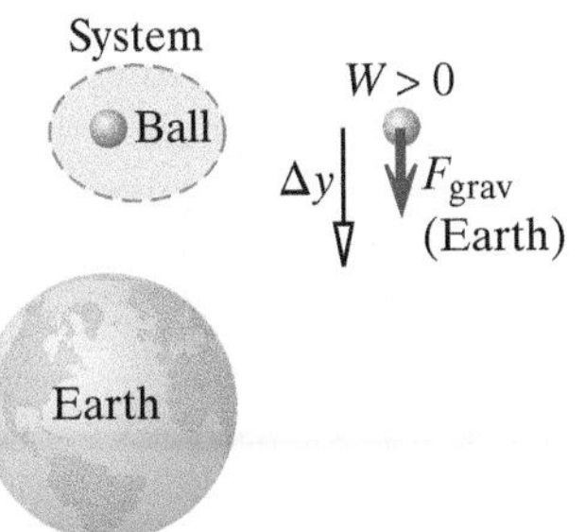

Figure 6.21 Ball alone: Earth is part of the surroundings and does positive work on the ball, increasing the kinetic energy of the ball.

Up to this point, our discussions of energy have been confined to systems consisting of a single particle (or systems that we could consider to be equivalent to a point particle because there were no internal changes). However, we saw in the case of momentum that it was often productive to choose a system containing two or more objects. We'll look at the effect of choice of system on the energy analysis of a simple situation: a ball falls from rest.

If the ball alone is the system, the Earth is part of the surroundings (Figure 6.21). The kinetic energy of the system (ball) increases, due to positive work done by the Earth on the system. The gravitational force acts in the same direction as the displacement of the ball, so the work done by the Earth on the ball is positive.

EXAMPLE **Ball and Earth: Ball Alone as the System**

In Figure 6.22 a ball of mass 100 g is released from rest 7 m above the ground. When the ball has fallen to 4 m above the ground, what is its kinetic energy K_f?

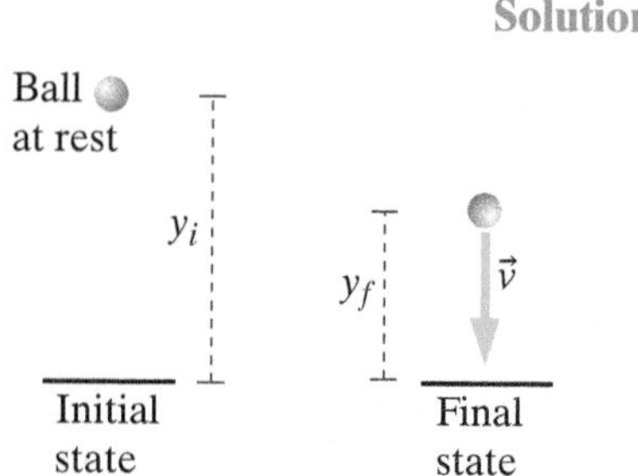

Figure 6.22 A ball falls from rest, and at a lower location its kinetic energy has increased.

Solution System: Ball
Surroundings: Earth
Initial State: Ball at rest 7 m above ground
Final State: Ball falling, 4 m above ground

$$\Delta(\cancel{m_{\text{ball}}c^2}) + \Delta K_{\text{ball}} = W_{\text{by Earth}}$$
$$\Delta K_{\text{ball}} = F_y \Delta y = -mg\Delta y$$
$$K_f - 0 = -(0.1\,\text{kg})(9.8\,\text{N/kg})(-3\,\text{m}) = 2.94\,\text{J}$$
$$K_f = 2.94\,\text{J}$$

The Earth did an amount of work +2.94 J on the ball, and the kinetic energy of the ball increased from zero to 2.94 J.

QUESTION What if we choose the ball plus the Earth as the system (Figure 6.23)?

In this case there is nothing significant in the surroundings, so:

$$\Delta K_{\text{sys}} = W_{\text{surr}}$$
$$\Delta K_{\text{ball}} + \Delta K_{\text{Earth}} = 0 \quad \text{But this can't be right!}$$

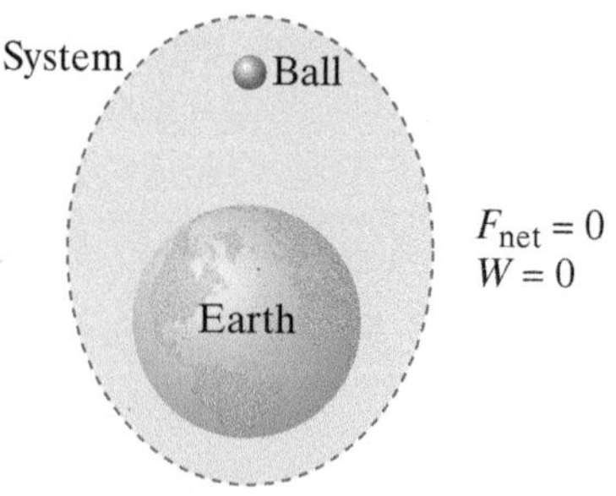

Figure 6.23 As the ball falls no work is done on the system of (Ball + Earth) because there are no significant interactions with the surroundings.

We know that the kinetic energy of the ball increased, and that the kinetic energy of the Earth also increased a tiny amount as it was drawn toward the ball. But the surroundings did no work! The change in kinetic energy of the system is not equal to the work done by the surroundings.

Before deciding that the Energy Principle has been violated, we should consider the possibility that we have overlooked a kind of energy that is present in systems containing more than one interacting object. In fact, this is the case.

Potential Energy Belongs to Pairs of Interacting Objects

In any system containing two or more interacting particles, such as compressed or stretched springs, galaxies of stars interacting gravitationally, or atoms in which the protons and electrons interact electrically, there is energy associated with the interactions between pairs of particles inside the system. This interaction energy is not the same as the rest energies or kinetic energies of the individual particles. We call this pairwise interaction energy "potential energy." Traditionally potential energy is represented by the symbol U.

Since the system of (Ball + Earth) contains two interacting objects, its energy is really this:

$$E_{\text{sys}} = (m_{\text{ball}}c^2 + K_{\text{ball}}) + (m_{\text{Earth}}c^2 + K_{\text{Earth}}) + U_{\text{ball-Earth}}$$

A change in potential energy is associated with a change in the separation of interacting objects. We can think of separation changes as change in the shape of the multiparticle system, such as the ball and Earth moving closer together, a spring stretching or compressing, or an electron moving away from a proton.

In the case above, as the ball and the Earth get closer together, the kinetic energy of the system increases, but we will see that the potential energy (interaction energy) decreases. The net change in the energy of the system is in fact zero, which is consistent with the fact that no work was done on the system by the surroundings:

$$\Delta K_{\text{ball}} + \Delta K_{\text{Earth}} + \Delta U_{\text{ball-Earth}} = 0$$

QUESTION How do we calculate changes in potential energy?

When we take the ball + Earth as the system, the force acting on the ball is exerted by another object inside the system. We call such a force an "internal force," which does "internal work." A simple rearrangement of the Energy Principle for the system of the ball alone leads to a useful way to account for energy in the ball + Earth system:

$$\Delta K_{\text{ball}} = W_{\text{by Earth}}$$

Rewrite this by moving the work to the left side of the equation:

$$\Delta K_{\text{ball}} + (-W_{\text{by Earth}}) = 0$$

For a system of more than one object, we define "the change of potential energy ΔU" to be the negative of the internal work: $\Delta U = -W_{\text{by Earth}}$. For the system of the ball + Earth, $\Delta U = -(F_y \Delta y) \approx -(-mg)\Delta y = \Delta(mgy)$. Actually, because we approximated the gravitational force by a constant mg, this is an approximate result valid only for height changes near the Earth's surface that are small compared to the radius of the Earth.

APPROXIMATE ΔU NEAR EARTH'S SURFACE

$$\Delta U_{\text{grav}} \approx \Delta(mgy)$$

for a system consisting of (object + Earth) near the surface of the Earth

Writing the Energy Principle for the ball + Earth system as

$$\Delta K_{\text{ball}} + \Delta K_{\text{Earth}} + \Delta U_{\text{ball-Earth}} = 0$$

has the advantage of putting on the left side of the equation quantities associated with the system (kinetic energy of the ball, interaction energy of the ball + Earth), and leaving on the right side of the equation only those quantities associated with the surroundings (in this case, nothing).

EXAMPLE **Ball and Earth: Ball + Earth as the System**

In Figure 6.22 a ball of mass 100 g is 7 m above the ground, initially at rest ($K_i = 0$). When the ball is 4 m above the ground, what is its kinetic energy K_f? Choose the ball + Earth as the system.

Solution System: Ball + Earth
Surroundings: Nothing significant

$$\begin{aligned}
\Delta(\cancel{m_{\text{Earth}}c^2}) + \Delta K_{\text{Earth}} + \Delta(\cancel{m_{\text{ball}}c^2}) + \Delta K_{\text{ball}} + \Delta U &= 0 \\
0 + (K_{\text{ball,f}} - 0) + \Delta(mgy) &= 0 \\
K_{\text{ball,f}} + (0.1\text{ kg})(9.8\text{ N/kg})(-3\text{ m}) &= 0 \\
K_{\text{ball,f}} + (-2.94\text{ J}) &= 0 \\
K_{\text{ball,f}} &= 2.94\text{ J}
\end{aligned}$$

In the second step we assume that $\Delta K_{\text{Earth}} \approx 0$. This makes sense because although the Earth is attracted to the ball, the displacement of the Earth is negligibly small, so essentially zero work is done on the Earth.

When we chose the ball alone as the system, the Earth was in the surroundings and did +2.94 J of work on the ball. When we choose the ball + Earth as the system, no work is done on the system, but the potential energy of the ball–earth pair of interacting objects decreases; $\Delta U = -2.94$ J. We of course get the same result for the final kinetic energy of the ball no matter which system we choose, but the viewpoint is rather different in the two cases.

QUESTION Why didn't we have to include potential energy when we chose the ball alone as the system?

A system consisting of a single object has no potential energy. Potential energy is associated with the interactions of pairs of objects inside the system, and there are no such pairs of objects in a one-particle system.

Interaction Energy in a Multiparticle System

Having illustrated the basic idea of potential energy in a situation where the internal force was constant (mg), we now offer a more general analysis. Consider a system of three particles that exert forces on each other and that are also acted upon by objects in the surroundings, outside this system, as shown in Figure 6.24. For example, these could be three charged particles that exert electric forces on each other but also have springs connecting them to objects in the surroundings. The forces that are exerted by other particles in the system are called "internal" forces, and are denoted here by lowercase fs. The forces exerted by objects in the surroundings are called "external" forces, and are denoted here by uppercase Fs.

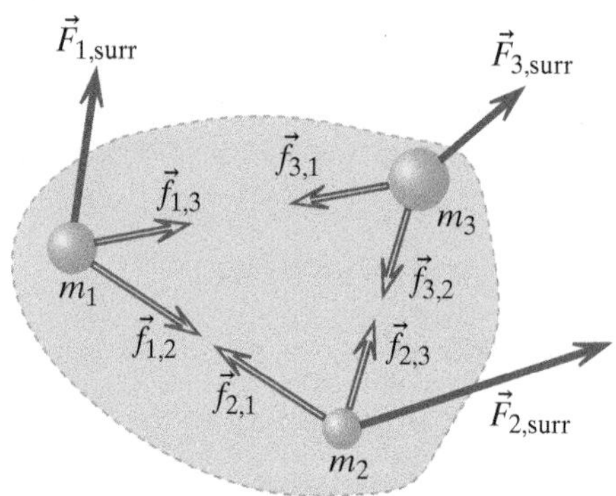

Figure 6.24 A system consisting of three particles. They exert "internal" forces on each other (lowercase fs), and additional "external" forces (black arrows, uppercase Fs) are exerted by objects in the surroundings, which are not shown.

The work done by these internal and external forces on each particle changes the particle energy E_{particle} of that particle. Using the symbol W_{int} for work done by internal forces, and W_{surr} for work done by external forces, we can write the Energy Principle for each particle:

$$\Delta E_1 = (\vec{f}_{1,2} + \vec{f}_{1,3} + \vec{F}_{1,\text{surr}}) \bullet \Delta\vec{r}_1 = W_{1,\text{int}} + W_{1,\text{surr}}$$
$$\Delta E_2 = (\vec{f}_{2,1} + \vec{f}_{2,3} + \vec{F}_{2,\text{surr}}) \bullet \Delta\vec{r}_2 = W_{2,\text{int}} + W_{2,\text{surr}}$$
$$\Delta E_3 = (\vec{f}_{3,1} + \vec{f}_{3,2} + \vec{F}_{3,\text{surr}}) \bullet \Delta\vec{r}_3 = W_{3,\text{int}} + W_{3,\text{surr}}$$

Adding up the changes in the particle energy for each of the three particles (the left sides of the equations) and the work done on each particle (the right sides of the equations), we can write this:

$$\Delta(E_1 + E_2 + E_3) = W_{\text{int}} + W_{\text{surr}}$$

where W_{int} is the sum of the work done by internal forces on each of the particles, and W_{surr} is the sum of the work done by external forces on each of the particles. Added together, these two terms equal the sum of the changes in the individual particle energies. Note that the work done by each individual external force involves the displacement of that force (equivalently, the displacement of the particle on which that particular force acts).

Separating Internal and External Energy Terms

As we saw with the falling ball, it can be useful to have all terms relating to energy changes internal to the system on one side of the equation and terms relating to objects in the surroundings on the other side. As we did with the falling ball, we can do this by making a simple algebraic rearrangement, moving the work done on the particles by the internal forces, W_{int}, to the other side of the equation:

$$\Delta(E_1 + E_2 + E_3) + (-W_{\text{int}}) = W_{\text{surr}}$$

We relabel the term $(-W_{\text{int}})$ as a change of potential energy U, energy associated with the interactions of the particles with each other.

CHANGE IN POTENTIAL ENERGY

$$\Delta U = -W_{int}$$

The change in potential energy (ΔU) of a system is defined to be the negative of the work done by forces internal to the system.

The left side of the Energy Principle equation for a multiparticle system now represents all changes in the energy of the system, and the right side represents energy inputs from the surroundings (work). It is customary to drop the subscript and write W_{surr} simply as W, with the understanding that the work W is mechanical energy transfer from the surroundings to the system, with a resulting change in the energy of the system.

Because the "internal work" term involves the forces exerted on each particle by each other particle in the system, the potential energy U is really the sum of the interaction energies of all pairs of particles in the system:

$$U = U_{12} + U_{23} + U_{31}$$

In general, we can now extend our definition of energy to a system of interacting particles:

ENERGY OF A SYSTEM OF INTERACTING PARTICLES

$$E_{sys} = (m_1c^2 + K_1) + (m_2c^2 + K_2) + (m_3c^2 + K_3) + \cdots$$
$$+ U_{12} + U_{23} + U_{13} + \cdots$$

And we can write the Energy Principle in a way that explicitly includes this interaction energy:

ENERGY PRINCIPLE FOR A MULTIPARTICLE SYSTEM

$$\Delta(E_1 + E_2 + E_3 + \cdots) + \Delta(U_{12} + U_{13} + U_{23} + \cdots) = W$$

$E_1 = m_1c^2 + K_1$ is the particle energy of object 1, etc.

Checkpoint 9 If a system contains four particles, how many potential energy pairs U_{12}, etc., are there? List them.

Multiparticle Systems vs. Single-Particle Systems

The Energy Principle is fundamentally the same for a multiparticle system as it is for a single-particle system. To the left of the equals sign are quantities that are internal to the system, while to the right of the equals sign are the external effects on the system due to objects in the surroundings. The interpretation, as indicated in (Figure 6.25), is:

(Change in energy of a system) = (Energy inputs from the surroundings)

The difference is in the kinds of energy found inside each kind of system:

- A single-particle system has only particle energy (rest energy and kinetic energy).
- A multiparticle system has particle energy and potential energy (interaction energy).

Work done by the surroundings, which may be positive or negative, is one form of energy input to a system. Work can affect both particle energy and potential energy inside a multiparticle system.

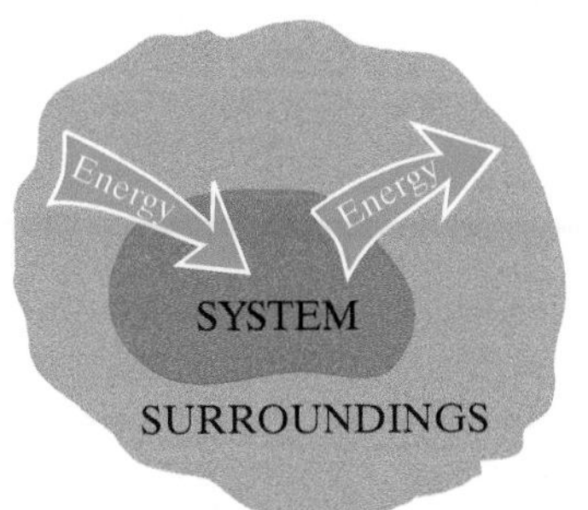

Figure 6.25 Change of system energy is equal to energy inputs from the surroundings, which can be either positive or negative. Work done by the surroundings is one form of energy input.

You get to choose how to define the system you want to analyze, but once you have specified the portion of the Universe you're considering to be the system of interest, you must be consistent about what energy flows into or out of the system (work done by the surroundings), versus what energy changes occur inside your system (possible changes in particle energies or potential energy).

The Experimental Basis for the Energy Principle for Multiparticle Systems

We were able to make a mathematical proof of the Energy Principle for a single particle, starting from the Momentum Principle. For a complex, multiparticle system, the more general Energy Principle ultimately is based on over a hundred years of experiments starting in the early 1800s. Slowly scientists learned how to identify flows of energy in many different kinds of systems, and once the energy was accounted for, they found that conservation of energy, $\Delta E_{sys} + \Delta E_{surr} = 0$, was always true, not only in physics but also in chemistry and biology.

The Discovery of Neutrinos

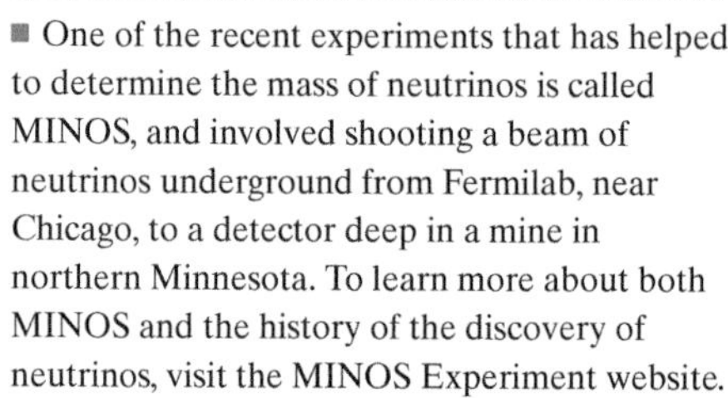

■ One of the recent experiments that has helped to determine the mass of neutrinos is called MINOS, and involved shooting a beam of neutrinos underground from Fermilab, near Chicago, to a detector deep in a mine in northern Minnesota. To learn more about both MINOS and the history of the discovery of neutrinos, visit the MINOS Experiment website.

Neutrinos were discovered because of the Energy Principle. In the early part of the 20th century, scientists found that apparently energy was not conserved in nuclear reactions called "beta decays," in which certain radioactive nuclei spontaneously emit an electron, and a neutron in the nucleus changes into a proton. (Early in the history of atomic physics, energetic electrons were called "beta rays.") It was expected that the emitted electrons would all have exactly the same energy, corresponding to the change in mc^2 of the nucleus, but instead it was observed that the emitted electrons had a wide range of energies. In many cases the energies didn't add up. Was the Energy Principle in fact not true? The Swiss physicist Wolfgang Pauli made the bold proposal that an unseen particle, later named the "neutrino," was also emitted, with energy shared between it and the electron.

Pauli was so convinced of the universal correctness of energy conservation that he was willing to postulate an unseen particle rather than discard energy conservation. It took many years before experiments were sensitive enough to detect the elusive neutrinos, but eventually Pauli's analysis was confirmed. For many years it was thought that neutrinos had only kinetic energy, with zero rest energy (and zero mass). However, recent careful experiments have verified theoretical predictions that neutrinos should have a very tiny mass.

6.8 GRAVITATIONAL POTENTIAL ENERGY

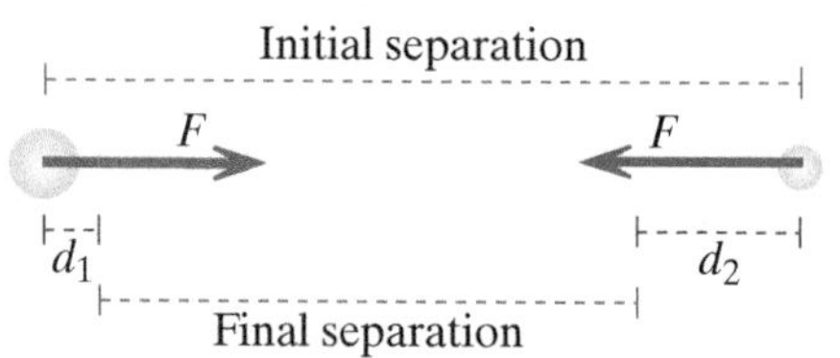

Figure 6.26 Two stars move toward each other during a short time interval. The internal work for the two-star system is F times the change in the separation distance of this pair of objects. The displacements are shown overly large for clarity.

We found that for a system of a ball falling near the surface of the Earth, where the force of the Earth was a constant mg, the change of gravitational potential energy for the ball + Earth system was $\Delta U \approx \Delta(mgy)$. Next we will find an expression for gravitational potential energy in the more general case where the force depends on distance, $\vec{F}_{grav} = -(GMm/r^2)\hat{r}$.

A big star and a small star are initially at rest (Figure 6.26). They start to move toward each other along the x axis due to the gravitational forces they exert on each other, with the same magnitude F (reciprocity). In a short time the big star moves a distance d_1 and the small star moves a distance d_2 (the displacements are shown overly large for clarity). The increase in the magnitude of the momentum of each star is the same, because the magnitude of the impulse $F\Delta t$ is the same, so the small star reaches a higher speed ($v \approx p/m$) and moves farther: $d_2 > d_1$, as indicated in Figure 6.26. Therefore more work is done on the small star.

The impulse (and momentum change) have the same magnitude for both stars because both forces act for the same time interval Δt. However, the work done is different because the two forces act through different distances. This is an important difference between impulse and work, and therefore between the Momentum Principle and the Energy Principle.

Consider the system consisting of both stars. The internal work W_{int} is the work done by forces internal to the system (not exerted by objects in the surroundings):

$$W_{\text{int}} = Fd_1 + Fd_2 = F(d_1 + d_2)$$

Change of potential energy is defined as the negative of the internal work:

$$\Delta U = -W_{\text{int}} = -F(d_1 + d_2)$$

Since ΔU for the two stars is negative, the potential energy decreased, which would imply that the kinetic energy increased, which is indeed what happened.

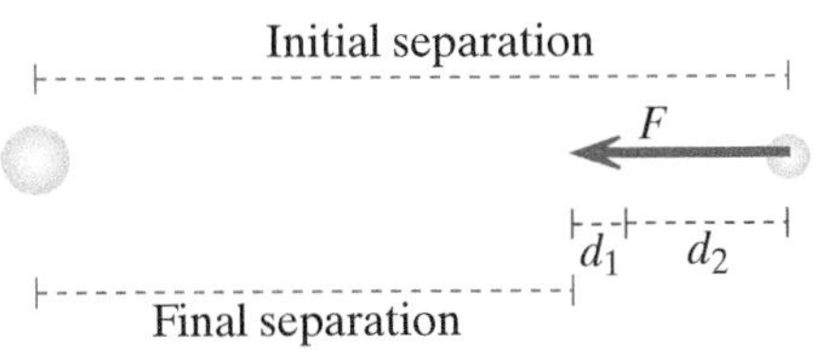

Figure 6.27 We can calculate internal work as though one of the stars were stationary and the other moved.

Change in Separation of Two Objects

The stars got closer to each other; their separation decreased by an amount $(d_1 + d_2)$. We can calculate internal work by multiplying the force one object exerts on another by the change in the separation between the two objects. The individual positions of the objects don't matter. The way the algebra works out, the internal work is calculated as though a single force F acted through a distance equal to the change in the separation of the pair, as in Figure 6.27.

Next we'll prove this more generally. Again consider a system of two interacting stars. We define the relative position vector $\vec{r}$ to be the difference between the positions of two stars, as shown in Figure 6.28. When the stars move a little closer together, star 1 does work on star 2, and star 2 does work on star 1. Since both stars are inside the system, the change in gravitational potential energy of the system is the negative of the sum of these work terms. Assume that the changes in position are small enough that we can consider the forces to be constant:

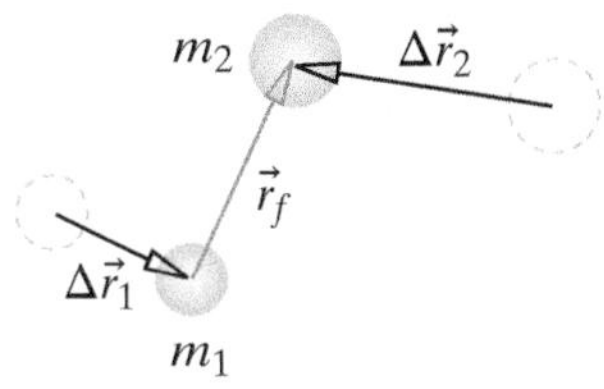

Figure 6.28 Two stars whose relative positions are changing (in this case, the stars move closer together).

$$\Delta U = -W_{\text{int}} = -(\vec{f}_{1,2} \bullet \Delta\vec{r}_1 + \vec{f}_{2,1} \bullet \Delta\vec{r}_2)$$
$$\vec{f}_{1,2} = -\vec{f}_{2,1}, \text{ so } \Delta U = -\vec{f}_{2,1} \bullet (\Delta\vec{r}_2 - \Delta\vec{r}_1)$$

The term $(\Delta\vec{r}_2 - \Delta\vec{r}_1)$ involves the change of position of each star (Figure 6.28). We can rewrite this in terms of $\Delta\vec{r}$, the change in relative position of the two stars:

$$\begin{aligned}\Delta\vec{r}_2 - \Delta\vec{r}_1 &= (\vec{r}_{2f} - \vec{r}_{2i}) - (\vec{r}_{1f} - \vec{r}_{1i})\\ &= (\vec{r}_{2f} - \vec{r}_{1f}) - (\vec{r}_{2i} - \vec{r}_{1i})\\ &= \vec{r}_f - \vec{r}_i\\ &= \Delta\vec{r}\end{aligned}$$

where

$$\vec{r}_i = \vec{r}_{2i} - \vec{r}_{1i}$$
$$\vec{r}_f = \vec{r}_{2f} - \vec{r}_{1f}$$

Using this result we can rewrite the change in potential energy:

$$\Delta U = -\vec{f}_{2,1} \bullet \Delta\vec{r}$$

$\vec{r}$ is the position of star 2 relative to star 1. Therefore we can calculate the amount of change of potential energy, ΔU, in terms of the *relative displacements* or separation between *pairs of particles*, and any change of potential energy is associated with a change of "shape," with pairs of particles getting closer together or farther apart. The algebra works out in such a way that the force we use in calculating ΔU is just one of the pairs of forces, not both. The result is the same if the interaction is an electric interaction instead of a gravitational interaction, since the reasoning above depended only on reciprocity of forces, and not on any other details of the force.

We write U_{12} for the potential energy of the pair consisting of particles 1 and 2, U_{23} for pair 2 and 3, and so on. If a system consists of two particles, there are three energy terms: two particle energies E_1 and E_2, and one potential energy term, U_{12}, since there is only one pair of particles. (Of course the particle energies can be further subdivided into rest energy plus kinetic energy.)

If a system consists of three particles, there are six energy terms: three particle energies E_1, E_2, and E_3, and three potential energy terms, U_{12}, U_{23}, and U_{31}, since there are three pairs of particles. Change of potential energy is associated with change of distance between pairs of particles.

Force Is the Negative Gradient of U

Since it is the relative displacement that matters, even if both objects move we can pretend that object 1 remains stationary, and just calculate the work done on object 2. The force is in a line with $\vec{r}$, so we can evaluate the dot product and write the following, where F_r is the component of the gravitational force on object 2 in the direction of $\vec{r}$, and r is the distance from object 1 to object 2 (Figure 6.29):

$$dU = -F_r dr$$

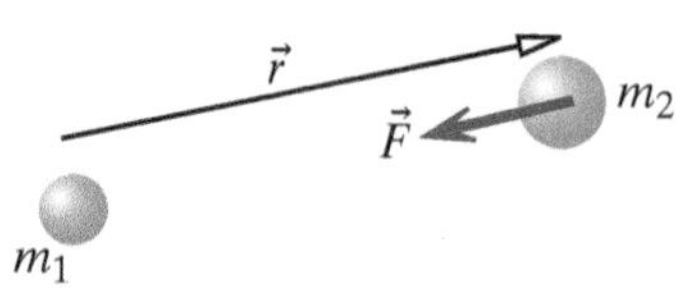

Figure 6.29 The r component of the gravitational force, F_r.

Therefore if we already knew the expression for potential energy, we could calculate the associated force like this: $F_r = -dU/dr$. A rate of change of a quantity with respect to position such as this is called a "gradient." For example, if a hill is said to have "a 7% grade," this means that it rises 7 m vertically for every 100 m horizontally—the tangent of the angle of the hill (the slope) is 0.07 (Figure 6.30).

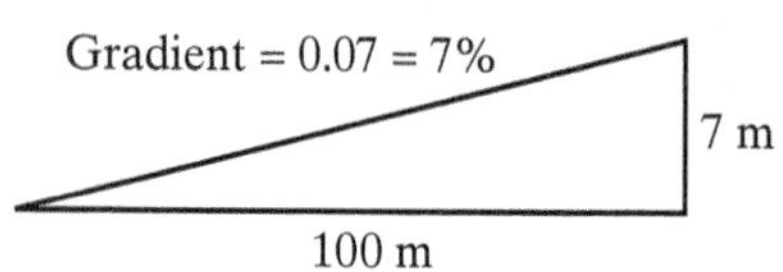

Figure 6.30 The gradient is the rate of change of y with respect to x.

FORCE IS THE NEGATIVE GRADIENT OF POTENTIAL ENERGY

$$F_r = -\frac{dU}{dr}$$

All that remains to be done to find an expression for gravitational energy is to think of an expression whose (negative) derivative is the component of the gravitational force acting on m_2 in the direction of $\vec{r}$. From Figure 6.29 we see that this component of the gravitational force is negative: $F_r = -G(m_1 m_2/r^2)$, where r is the center-to-center distance. The minus sign reflects the fact that the attractive gravitational force on object 2 points back toward object 1.

QUESTION Can you think of a function of r whose (negative) derivative with respect to r has this value of F_r?

The derivative of r^n with respect to r is nr^{n-1}. We have a r^{-2} factor in the gravitational force, so $n = -1$, and the "antiderivative" of r^{-2} (the function whose derivative gives r^{-2}) must be $-r^{-1}$.

QUESTION Therefore, what is the equation for gravitational energy?

Being careful about the signs, we find that $U = -Gm_1m_2/r$. To check, we can differentiate this expression with respect to r to get the gravitational force.

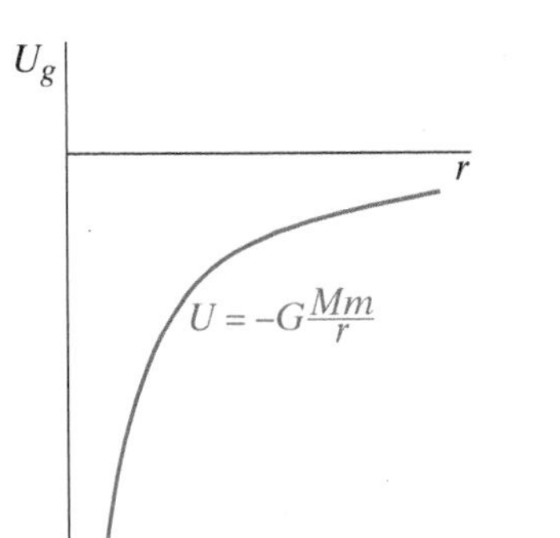

Figure 6.31 Gravitational potential energy is negative. It increases with increasing separation r.

QUESTION What about the minus sign on U? Is it correct?

Yes, because the potential energy decreases as the particles get closer together. When stars fall toward each other, their kinetic energies increase and the pair-wise potential energies must decrease to more and more negative values. Figure 6.31 shows that although U is negative, it does increase with increasing separation r, which means that you have to do work to move two gravitationally attracting objects farther apart. It may seem odd that potential energy is negative, but it makes physical sense. We'll examine how this negative quantity contributes to the total energy of a multiparticle system in Section 6.12.

In summary,

GRAVITATIONAL POTENTIAL ENERGY

$$U = -G\frac{m_1m_2}{r}$$

r is the center-to-center separation of m_1 and m_2.

QUESTION Could there be an additional constant in this equation?

It is true that the gradient of $U = (-Gm_1m_2/r + \text{constant})$ would also give us the correct gravitational force, because the derivative of a constant is zero. However, as we will soon discuss in more detail, when the objects are very far apart the total energy of the system must be equal just to the sum of the particle energies, and the potential energy must be zero.

QUESTION When the particle separation r is very large, what is the value of $-Gm_1m_2/r$?

When r is very large, $-Gm_1m_2/r$ is nearly zero. Therefore, for the potential energy to be zero when the separation is very large, the value of the constant in $U = (-Gm_1m_2/r + \text{constant})$ must be zero, in order for U to be zero at large separations. The pair-wise gravitational potential energy must be simply $U = -Gm_1m_2/r$, with no added constant.

EXAMPLE **A Meteor Falls Toward the Earth**

In February 2013 a large meteor, whose mass has been estimated to be 1.2×10^7 kg, fell to Earth near Chelyabinsk, Russia. (This meteor exploded spectacularly at a height of about 30 km, doing significant damage to objects on the ground.) Consider a meteor of the same mass falling toward the Earth. Choose the Earth plus the meteor as the system. As the meteor falls from a distance of 1×10^8 m from the center of the Earth to 1×10^7 m, what is the change in the kinetic energy of the meteor? Explain the signs of the changes in kinetic and potential energy of the system,

Solution System: Earth and meteor
Surroundings: Nothing significant (because the distance of the objects from the Sun doesn't change much)

Initial state: Center to center distance of 1×10^8 m
Final state: Center to center distance of 1×10^7 m

To get a sense of the magnitudes and signs of the gravitational potential energy, we will calculate the value of U_{grav} explicitly for each state instead of making algebraic simplifications.

$$U_i = -G\frac{M_E m_m}{r_i} = -6.7\times 10^{-11}\,\text{Nm}^2/\text{kg}^2\frac{(6\times 10^{24}\,\text{kg})(1.2\times 10^7\,\text{kg})}{1\times 10^8\,\text{m}}$$
$$= -4.82\times 10^{13}\,\text{J}$$
$$U_f = -G\frac{M_E m_m}{r_i} = -6.7\times 10^{-11}\,\text{Nm}^2/\text{kg}^2\frac{(6\times 10^{24}\,\text{kg})(1.2\times 10^7\,\text{kg})}{1\times 10^7\,\text{m}}$$
$$= -4.82\times 10^{14}\,\text{J}$$

Both the initial and final values of the gravitational potential energy are negative. U_f, the value when the meteor is closer to Earth, is a larger negative number, so $U_f - U_i < 0$. This is consistent with Figure 6.31, which shows that as the separation between two objects decreases, the gravitational potential energy of the two-object system becomes more negative.

Applying the Energy Principle, we find that:

$$\Delta K_{\text{meteor}} + (U_f - U_i) = 0$$
$$\Delta K_{\text{meteor}} + (-4.82\times 10^{14}\,\text{J} - -4.82\times 10^{13}\,\text{J}) = 0$$
$$\Delta K_{\text{meteor}} + -4.34\times 10^{14}\,\text{J} = 0$$
$$\Delta K_{\text{meteor}} = 4.34\times 10^{14}\,\text{J}$$

EXAMPLE

A Robot Spacecraft Leaves an Asteroid

A robot spacecraft lands on an asteroid, picks up a sample, and blasts off to return to Earth; its total mass is 1500 kg. When it is 200 km (2×10^5 m) from the center of the asteroid, its speed is 5.0 m/s, and the rockets are turned off. At the moment when it has coasted to a distance 500 km (5×10^5 m) from the center of the asteroid, its speed has decreased to 4.1 m/s. **(a)** Draw and label a diagram showing your choice of initial and final states. **(b)** Calculate the mass of the asteroid.

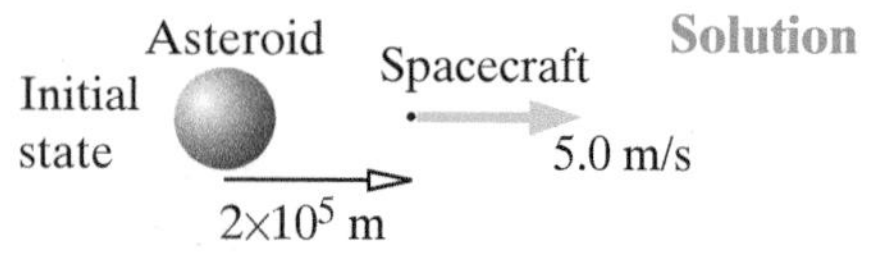

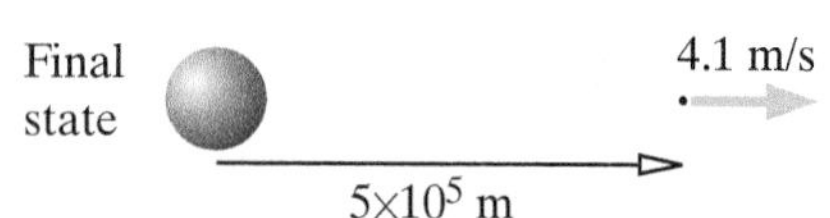

Figure 6.32 Initial and final states.

Solution

(a) Figure 6.32 shows the situation. It is important to keep in mind that r is measured center to center in $U = -Gm_1m_2/r$.

(b) We'll use M for the mass of the asteroid and m for the mass of the spacecraft.

System: Asteroid and spacecraft

Surroundings: Nothing significant

Initial state: Rockets off, 5.0 m/s, 200 km from center of asteroid, asteroid at rest

Final state: Speed 4.1 m/s, 500 km from center of asteroid

$$E_f = E_i + W$$
$$\cancel{Mc^2} + \cancel{K_{M,f}} + \cancel{mc^2} + K_{m,f} + U_f = \cancel{Mc^2} + \cancel{K_{M,f}} + \cancel{mc^2} + K_{m,i} + U_i + W$$
$$K_{m,f} + U_f = K_{m,i} + U_i + 0$$
$$\frac{1}{2}\cancel{m}v_f^2 + \left(-G\frac{M\cancel{m}}{r_f}\right) = \frac{1}{2}\cancel{m}v_i^2 + \left(-G\frac{M\cancel{m}}{r_i}\right)$$
$$\frac{1}{2}v_f^2 + \left(-G\frac{M}{r_f}\right) = \frac{1}{2}v_i^2 + \left(-G\frac{M}{r_i}\right)$$
$$GM\left(\frac{1}{r_i} - \frac{1}{r_f}\right) = \frac{1}{2}(v_i^2 - v_f^2)$$

$$M = \frac{\frac{1}{2}(v_i^2 - v_f^2)}{G(1/r_i - 1/r_f)}$$

$$M = \frac{\frac{1}{2}(5.0^2 - 4.1^2)\ \text{m/s}^2}{(6.7 \times 10^{-11}\ \text{N}\cdot\text{m}^2/\text{kg}^2)\left(\frac{1}{2 \times 10^5\ \text{m}} - \frac{1}{5 \times 10^5\ \text{m}}\right)} = 2 \times 10^{16}\ \text{kg}$$

By choosing both the asteroid and the spacecraft as the system, we didn't have to calculate work, since there are no significant external forces. We solved the problem algebraically before putting in numbers, and found that m canceled—the answer doesn't depend on the mass of the spacecraft.

QUESTION Could we have used the Momentum Principle instead of the Energy Principle to solve this problem?

We could not have used the Momentum Principle to carry out an accurate analysis without writing an iterative computer program because we didn't know how much time the process took. The velocity was changing at a nonconstant rate, so knowing the distance wouldn't give us the time, since we didn't know the average velocity. We were able to use the Energy Principle because we knew something about distances.

QUESTION We made the approximation that K_M, the kinetic energy of the asteroid, did not change. Was this reasonable? Don't the spacecraft and asteroid exert equal magnitude forces on each other?

It is true that the spacecraft pulls just as hard on the asteroid as the asteroid pulls on the spacecraft (reciprocity of gravitational forces). However, although the forces act during equal time intervals, they do not act through equal displacements, because the speeds of the objects are different.

We can use Conservation of Momentum for the system of asteroid + spacecraft to get the actual final speed of the asteroid (V_A). We'll assume all velocities have only x components, so the following equations are for the x component of the system's momentum:

$$(MV_A - 0) + (mv_f - mv_i) = 0$$

$$V_A = \frac{m}{M}(v_i - v_f)$$

$$= \frac{1.5 \times 10^3\ \text{kg}}{2 \times 10^{16}\ \text{kg}}(5.0 - 4.1)\ \text{m/s}$$

$$= 6.75 \times 10^{-14}\ \text{m/s}$$

Since the asteroid's speed is so low, its displacement during the interval of interest is tiny, and therefore the work done on it by the spacecraft is negligible. The change in kinetic energy of the asteroid is:

$$\Delta K_A = \frac{1}{2}(2 \times 10^{16}\ \text{kg})(6.75 \times 10^{-14}\ \text{m/s})^2 - 0$$

$$= 4.6 \times 10^{-11}\ \text{J}$$

In contrast, the kinetic energy change of the spacecraft was

$$\Delta K_s = \frac{1}{2}(1.5 \times 10^3\ \text{kg})(4.1^2 - 5.0^2)(\text{m/s})^2 = -6143\ \text{J}$$

More Than Two Interacting Objects

We have been applying energy considerations to two-particle systems. A homework problem in Chapter 3 involves predicting the motion of the *Ranger 7* spacecraft as it travels from the Earth to the Moon (where it crash lands). In our simple model for this voyage there are three "particles": the spacecraft with mass m, the Earth with mass M_{Earth}, and the Moon with mass M_{Moon}.

QUESTION How many interaction pairs are there in this three-particle system?

In a three-particle system there are three interaction pairs: U_{12}, U_{13}, and U_{23}. We can write the gravitational potential energy like this:

$$U = \left[-G\frac{M_{\text{Earth}}m}{r_{\text{to Earth}}}\right] + \left[-G\frac{M_{\text{Moon}}m}{r_{\text{to Moon}}}\right] + \left[-G\frac{M_{\text{Earth}}M_{\text{Moon}}}{r_{\text{Earth to Moon}}}\right]$$

In our simple model, the Earth and Moon are fixed in space, so the Earth–Moon term of this expression doesn't change. Also, there is no change of identity of the three particles, so their masses do not change. Therefore the energy equation for the system can be written like this, where v is the speed of the spacecraft, the only moving particle in the model system (and $v \ll c$):

$$\Delta\left(\frac{1}{2}mv^2\right) + \Delta\left[-G\frac{M_{\text{Earth}}m}{r_{\text{to Earth}}}\right] + \Delta\left[-G\frac{M_{\text{Moon}}m}{r_{\text{to Moon}}}\right] = 0$$

The change in the energy of the system is zero because no external work is done on the system (we're ignoring the effects of other objects, including the Sun). We can rearrange the terms and write the energy equation in terms of initial and final energies, which must be equal:

$$\frac{1}{2}mv_f^2 + \left[-G\frac{M_{\text{Earth}}m}{r_{\text{to Earth},f}}\right] + \left[-G\frac{M_{\text{Moon}}m}{r_{\text{to Moon},f}}\right] =$$
$$\frac{1}{2}mv_i^2 + \left[-G\frac{M_{\text{Earth}}m}{r_{\text{to Earth},i}}\right] + \left[-G\frac{M_{\text{Moon}}m}{r_{\text{to Moon},i}}\right]$$

That is, the final (kinetic plus potential) energy of the system is equal to the initial (kinetic plus potential) energy of the system, because no external work is done on the system (and the masses don't change; no change of identity).

Gravitational Potential Energy Near Earth's Surface

In Section 6.7 we found that for motion near the surface of the Earth, $\Delta U \approx \Delta(mgy)$. However, in Section 6.8 we found that the general expression $\Delta U_g = \Delta(-Gm_1m_2/r)$ describes a change in gravitational potential energy.

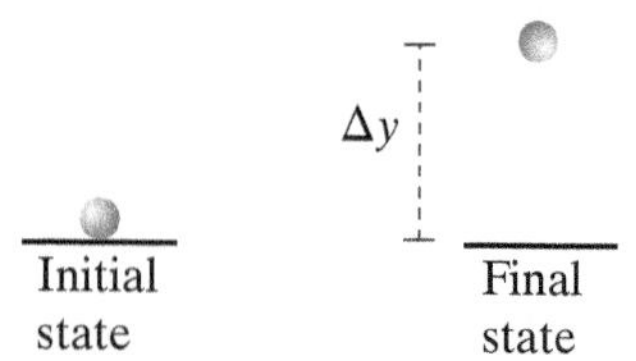

Figure 6.33 An object is lifted from the surface of the Earth to a distance Δy above the surface.

QUESTION How are $\Delta(mgy)$ and $\Delta(-Gm_1m_2/r)$ related?

The expression $\Delta(mgy)$ is an approximate expression valid only near the Earth's surface. We can derive it from the full expression for gravitational potential energy. Consider a system of the Earth plus an object of mass m located on the surface of the Earth. Suppose you move the object from the surface of the Earth to a location a small distance Δy above the Earth's surface (Figure 6.33). If R_E is the radius of the Earth, then the change in potential energy of the system is:

$$\Delta U_g = \left(-G\frac{M_E m}{R_E + \Delta y}\right) - \left(-G\frac{M_E m}{R_E}\right)$$

$$\Delta U_g = (-GM_E m)\frac{R_E - (R_E + \Delta y)}{R_E(R_E + \Delta y)}$$

$$\Delta U_g = (-GM_E m)\frac{-\Delta y}{R_E^2 + R_E \Delta y}$$

$$(R_E^2 + R_E\Delta y) \approx R_E^2 \text{ since } \Delta y \ll R_E\text{, so}$$

$$\Delta U_g \approx G\frac{M_E}{R_E^2} m\Delta y$$

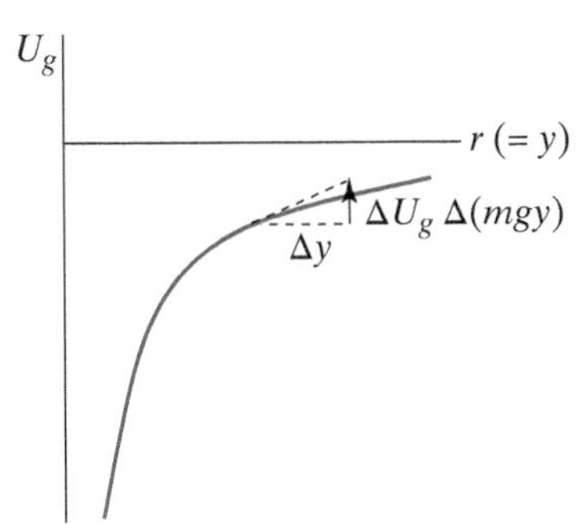

Figure 6.34 A small increase Δy in height y corresponds to a small increase in the (negative) gravitational energy.

In Chapter 3 we found that $G\frac{M_E}{R_E^2} = g$, so

$$\Delta U_g \approx gm\Delta y \quad \text{near the Earth's surface}$$

As expected, the sign of ΔU_g is correct; as you go up (farther from the center of the Earth), the change $\Delta(mgy)$ is positive; if you go down, it is negative (Figure 6.34).

QUESTION Is mgy the actual potential energy of the (Earth + object) system?

No. The actual potential energy of the system in a given state is $(-GM_E m/R_E + mgy)$, which is a negative quantity, as we will see in the following example.

EXAMPLE

ΔU Near the Earth

A 500 kg horse stands on the ground. **(a)** What is the potential energy of a system consisting of the Earth plus the horse? **(b)** Suppose the horse jumps over a fence. What is the potential energy of the system when the horse is 1.5 m above the ground?

Solution **(a)** The actual potential energy of the system in the initial state is

$$U_g = -G\frac{M_E m_{\text{horse}}}{R_E}$$

$$U_g = -6.7 \times 10^{-11}\,\text{Nm}^2/\text{kg}^2 \frac{(6 \times 10^{24}\,\text{kg})(5 \times 10^2\,\text{kg})}{6.4 \times 10^6\,\text{m}}$$

$$U_g = -3.15 \times 10^{10}\,\text{J}$$

(b) The *change* of the system's potential energy is

$$mg\Delta y = (500\,\text{kg})(9.8\,\text{N/kg})(1.5\,\text{m}) = 7350\,\text{J}$$

so the new potential energy of the system is

$$U_{g,f} = -3.15 \times 10^{10}\,\text{J} + 7350\,\text{J}$$

To indicate that the actual potential energy of a system contains a large negative constant term we could write the energy equation for systems near Earth's surface like this:

$$K_f + (mgy_f + -GmM_E/R_E) = K_i + (mgy_i + -GmM_E/R_E) + W$$

However, it is more convenient to omit the constant, since it will not contribute to a change, and write

$$K_f + mgy_f = K_i + mgy_i + W$$

QUESTION In what situations is the approximation $\Delta U_g \approx \Delta mgy$ a good approximation?

As long as Δy is small compared to the radius of the Earth (6400 km), this is a good approximation. The gravitational force is inversely proportional to the square of the distance from the center of the Earth, and decreases as distance from Earth increases. However, if y doesn't change much, neither does gravity. If you go up 0.1 km (100 m, the length of a football field), the gravitational force decreases by a tiny factor:

$$\left(\frac{6400\,\text{km}}{6400.1\,\text{km}}\right)^2 \approx 0.99997$$

EXAMPLE **A Falling Ball**

At a certain instant a ball is falling with a speed of 6 m/s, and its position is $y = 35$ m above the Earth's surface. How fast is the ball falling when it has fallen to a position $y = 20$ m above the Earth's surface, assuming that we can neglect air resistance?

Solution System: Ball + Earth
Surroundings: Nothing significant (neglect air resistance)

Initial state: Speed 6 m/s, 35 m above ground
Final state: 20 m above ground

Energy Principle:

$$K_f + U_f = K_i + U_i + W$$

$$\frac{1}{2}mv_f^2 + mgy_f = \frac{1}{2}mv_i^2 + mgy_i + 0$$

$$\frac{1}{2}mv_f^2 = \frac{1}{2}mv_i^2 - (mgy_f - mgy_i)$$

$$v_f^2 = v_i^2 - 2g(y_f - y_i)$$

$$v_f = \sqrt{v_i^2 - 2g(y_f - y_i)}$$

$$v_f = \sqrt{(6\,\text{m/s})^2 - 2(9.8\,\text{N/kg})[(20\,\text{m}) - (35\,\text{m})]} = 18.2\,\text{m/s}$$

There is nothing in our equations to specify whether the ball is initially heading upward or downward. In fact, it doesn't matter. In either case, if we neglect air resistance, the speed will be 18.2 m/s when the ball reaches a height of 20 m above the ground. (If it was initially headed upward, it will again have a speed of 6 m/s when it returns to a height of 35 m.)

Checkpoint 10 Suppose that in this situation we measure y from 25 m above the surface, so that $y_{\text{initial}} = +10$ m, and $y_{\text{final}} = -5$ m. Why do we get the same value for the final speed?

Including Horizontal Motion

Moving vertically up or down involves a change in gravitational energy approximately equal to $\Delta(mgy)$. What about sideways horizontal movements? There is negligible change in the gravitational energy of the Universe when an object is moved a short distance horizontally (that is, tangent to the Earth's surface). Consider moving a block along a surface with it rolling very easily on low-friction wheels, or a hockey puck sliding easily along the ice. You hardly have to use any energy at all to move the block or puck horizontally, as long as there is little friction.

What if a block is moved horizontally, but at a constant height of 3 m above the surface? You could support the block with a tall cart, and again there would be little effort involved in moving the block horizontally. Or you could pile up dirt 3 m deep and place a slick surface on top, and then you could easily move the block horizontally. We conclude that horizontal position doesn't affect the gravitational energy. (Again, we're considering movements over distances small compared with the radius of the Earth, so we can ignore the Earth's curvature.)

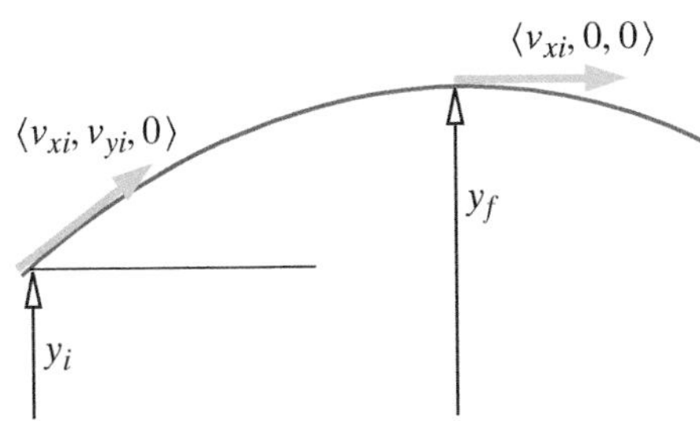

Figure 6.35 Path of a ball thrown at an angle to the horizontal.

Checkpoint 11 Suppose that you throw a ball at an angle to the horizontal, and just after it leaves your hand at a height y_i its velocity is $\langle v_{xi}, v_{yi}, 0\rangle$ (Figure 6.35). Assuming that we can neglect air resistance, at the top of its trajectory, when it is momentarily traveling horizontally, its velocity is $\langle v_{xi}, 0, 0\rangle$. What is the height y_f at the top of the trajectory, in terms of the other known quantities? Use the Energy Principle.

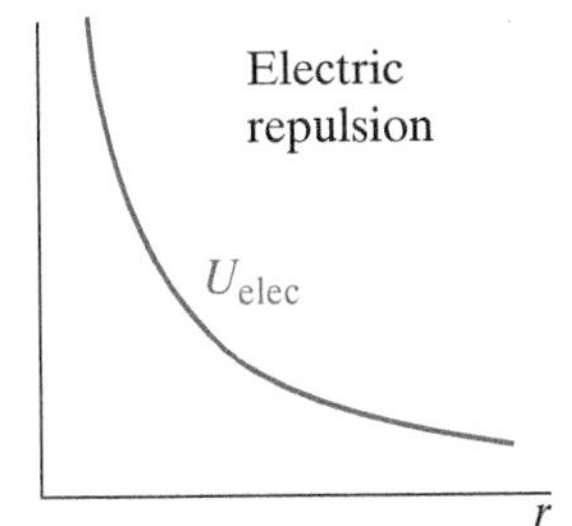

Figure 6.36 Electric potential energy is positive when like charged objects repel each other.

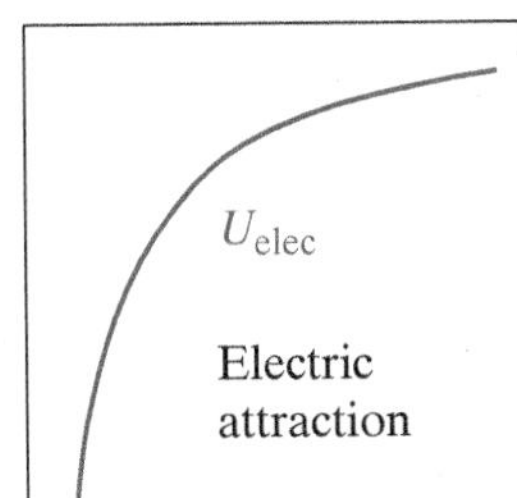

Figure 6.37 Electric potential energy is negative (like gravitational potential energy) when objects attract.

6.9 ELECTRIC POTENTIAL ENERGY

In Chapter 3 we saw that the electric force law for two charged particles is similar to the gravitational force law:

$$F_{\text{elec}} = \frac{1}{4\pi\varepsilon_0}\frac{|q_1 q_2|}{r^2} \quad \text{where } \frac{1}{4\pi\varepsilon_0} = 9\times 10^9 \frac{\text{N}\cdot\text{m}^2}{\text{C}^2}$$

The constant $1/(4\pi\varepsilon_0)$ is read as "one over four pi epsilon-zero." The quantities q_1 and q_2 represent the amount of charge on the two particles (measured in coulombs), and r is the distance between the two particles. Because this is an inverse square force, like gravitation, the electric potential energy has an equation very similar to that for gravitational potential energy:

ELECTRIC POTENTIAL ENERGY U_{elec}

$$U_{\text{elec}} = \frac{1}{4\pi\varepsilon_0}\frac{q_1 q_2}{r}$$

There is no minus sign in the expression for electric potential energy, because when the electric interaction is attractive, like gravitation, q_1 and q_2 have different signs, and the minus sign shows up automatically. However, if the two charges have the same sign, the interaction is repulsive, and there is a plus sign in the expression for the electric energy (Figure 6.36). If the two charges have opposite signs, the interaction is attractive, and there is a minus sign in the expression for electric energy (Figure 6.37).

EXAMPLE **Ionization Energy**

A certain amount of energy is required to remove an outer electron from a neutral atom, leaving behind a positive "ion" that is missing one electron. The charge of the ion is $+e$ (where $e = 1.6\times 10^{-19}$ C); the electron's charge is of course $-e$. The minimum energy to remove one outer electron is called the "ionization" energy. A typical atom has a radius of about 1×10^{-10} m. Estimate a typical ionization energy in joules, then convert your result to electron volts (eV).

Solution System: Two particles, the outer electron with charge $-e$, and the remaining part of the atom, the ion with charge $+e$.

Surroundings: An object that exerts an external force to remove the electron from the atom.

Initial state: The electron is bound to the atom, at a distance r_{atom} from the center.
Final state: The two particles are very far from each other, at rest.

We don't know the initial K of the electron, so we will assume it is zero, to get an approximate result. We will assume that the energy input to the system is in the form of work.

$$E_f = E_i + W$$
$$K_{\text{ion},f} + K_{e,f} + \cancel{m_i c^2} + \cancel{m_e c^2} + U_f = K_{\text{ion},i} + K_{e,i} + \cancel{m_i c^2} + \cancel{m_e c^2} + U_i + W$$
$$0 + 0 + U_f = 0 + 0 + U_i + W$$
$$W = U_f - U_i$$

$$W = \frac{1}{4\pi\varepsilon_0}\frac{q_1 q_2}{r_\infty} - \frac{1}{4\pi\varepsilon_0}\frac{q_1 q_2}{r_{\text{atom}}}$$
$$W = -\frac{1}{4\pi\varepsilon_0}\frac{(+e)(-e)}{r_{\text{atom}}}$$
$$W = \left(9 \times 10^9 \frac{\text{N}\cdot\text{m}^2}{\text{C}^2}\right)\frac{(1.6 \times 10^{-19}\,\text{C})^2}{(1 \times 10^{-10}\,\text{m})}$$
$$W = 2.3 \times 10^{-18}\,\text{J}$$
$$W = (2.3 \times 10^{-18}\,\text{J})\left(\frac{1\,\text{eV}}{1.6 \times 10^{-19}\,\text{J}}\right) = 14\,\text{eV}$$

This is an estimate because we don't know the electron's initial kinetic energy K_i, but typical ionization energies are indeed a few electron volts.

Checkpoint 12 Two protons are hurled straight at each other, each with a kinetic energy of 0.1 MeV. You are asked to calculate the separation between the protons when they finally come to a stop. **(a)** Write out the Energy Principle for this system, using the update form and including all relevant terms. **(b)** Which term of the equation contains the unknown quantity?

6.10 PLOTTING ENERGY vs. SEPARATION

A graphical representation of energy can be exceptionally helpful in reasoning about a process. In Figure 6.38 we show an energy graph for the spacecraft leaving the asteroid, the process analyzed in Section 6.8 (the graph continues beyond the separation r_f = 500 km). We plot the pair-wise gravitational potential energy U, the spacecraft's kinetic energy K, and the sum $K + U$, as a function of the separation distance r between the center of the asteroid and the spacecraft. We omit the rest energies, which are not changing during the motion.

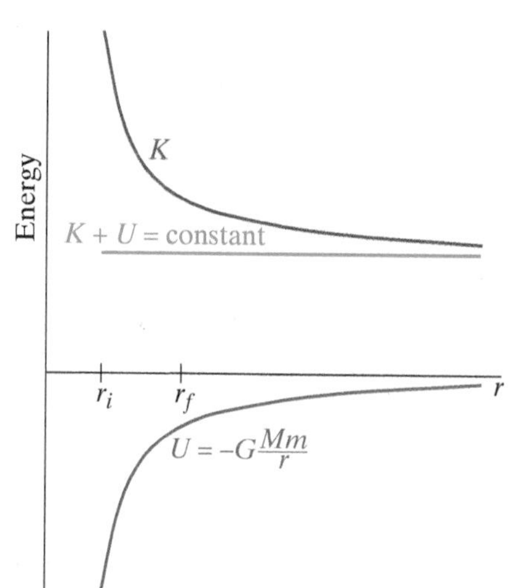

Figure 6.38 Energy vs. the separation distance between the asteroid and the spacecraft. (We omit the rest energies, which aren't changing.)

Important features of the graph are these: The gravitational potential energy U is negative because the interaction is attractive, not repulsive, and the potential energy increases with increasing separation r. As the potential energy increases, the kinetic energy K of the spacecraft decreases (the gravitational force exerted on the spacecraft by the asteroid is slowing it down). Because there is negligible work done on the combined system of asteroid plus spacecraft, $K_f + U_f = K_i + U_i$, so $K + U$ has a constant value at all times, and the graph of $K + U$ is a straight horizontal line on the graph, representing a constant positive value.

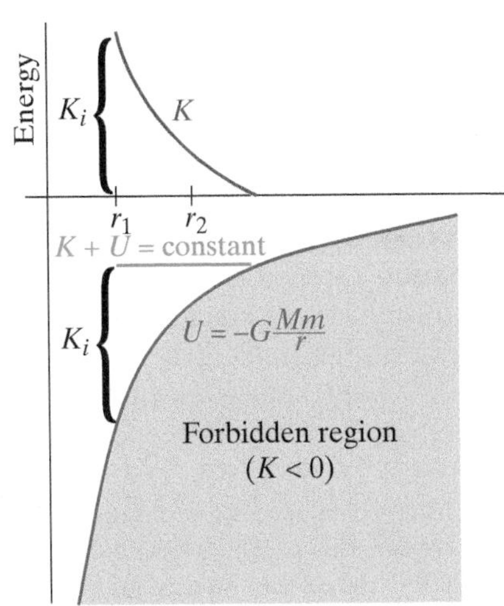

Figure 6.39 Energy vs. the separation distance between the asteroid and the spacecraft. (We omit the rest energies, which aren't changing.) At r_1 the potential energy is U_1 and the kinetic energy is K_1.

As the spacecraft gets farther and farther from the asteroid, you can see from the graph that its kinetic energy K will continue to fall, and the graph of K will approach the $K+U$ line at large separation. When the two objects are very far apart, U is nearly zero, and $K+U$ is just K.

This is an example of an "unbound" system, because the separation between spacecraft and asteroid will increase without bound. A gravitationally unbound system has $K+U$ greater than zero.

As an example of a "bound" system, again consider the spacecraft and the asteroid but this time give the spacecraft a smaller initial velocity, headed away from the asteroid. As a function of the center-to-center asteroid-to-spacecraft separation r, we again plot a graph of the potential energy $-GMm/r$ (Figure 6.39) and a graph of kinetic energy K. As the spacecraft moves away from the asteroid, its kinetic energy K falls as the pair-wise potential energy U rises. You can see on the graph that K falls to zero, and it cannot go negative, because kinetic energy is always a positive quantity. What happens next? The spacecraft momentarily comes to a stop (a turning point), then turns around and speeds up toward the asteroid; eventually it will crash onto the asteroid's surface.

We add to the graph a horizontal line whose height above or below the axis represents the total kinetic plus potential energy of the two-object system. The line on the graph for $K+U$ is horizontal because no external work is done on this isolated system, and there is no change of the masses (no change of identity), so $K+U$ is constant during the motion.

QUESTION Is $K+U$ positive or negative for the system shown in Figure 6.39?

The horizontal line representing the constant value of $K+U$ is below the axis representing $U=0$, so $K+U$ is negative. This corresponds to a "bound" state of the spacecraft and asteroid; the spacecraft and asteroid cannot get completely away from each other.

The vertical distance from the U curve up to the $K+U$ line is equal to K, the total kinetic energy of the spacecraft and asteroid. At all separations between the spacecraft and the asteroid, the height of the $K+U$ line above the U curve tells you the total kinetic energy. Notice that when K falls to zero, $K+U=U$ and the right end of the $K+U$ line touches the U graph.

The shaded area in Figure 6.39 represents a forbidden region, where the kinetic energy K would be negative, no matter what the value of $K+U$ (more about this in the following discussion).

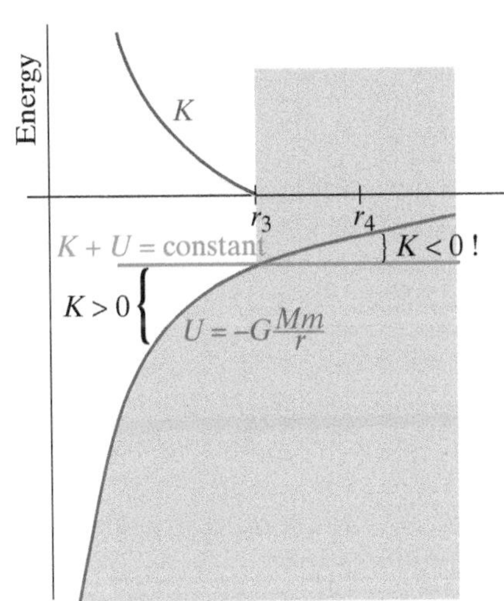

Figure 6.40 The asteroid and spacecraft cannot get farther away from each other than r_3; at r_4 the kinetic energy would be negative. (We omit the rest energies, which aren't changing.)

Limits on Possible Motion

The energy graph shows at a glance the limits on the motion for a given energy. In Figure 6.40, the asteroid and spacecraft with $K+U<0$ cannot get farther away from each other than r_3, a bound that is set by the potential-energy curve. At a larger separation r_4, the kinetic energy would have to be negative, which is "classically forbidden" (that is, impossible according to the laws of classical, prequantum mechanics). Kinetic energy cannot be negative. At low speeds, $K \approx \frac{1}{2}mv^2$, which is always positive. At high speeds, $K=\gamma mc^2 - mc^2$, which is always positive, because γ is always greater than 1.

If the system has the total $K+U$ shown ($K+U<0$), the separation of the asteroid and spacecraft can never be larger than r_3, and we say that this is a "bound" system. An energy graph shows at a glance the range of possible positions achievable during the motion, for a given total $K+U$. (We should mention, however, that in the world of atoms where a full analysis requires quantum mechanics, in some cases a system can "tunnel through" a region that is classically forbidden!)

An energy graph can show at a glance whether a system is bound or unbound. An example of a bound system is a planet in circular or elliptical orbit around a star: the planet cannot escape. Another example is an electron bound to an atom.

An example of an unbound system is the Earth plus a spacecraft whose initial speed is great enough that the spacecraft will get away from the Earth and never come back. For another example, if an amount of energy greater than or equal to the ionization energy is supplied to an atom, an electron can become unbound and escape from the atom. A gravitationally or electrically unbound system has total $K+U \geq 0$, which is easy to see on an energy graph, where it is clear that arbitrarily large separations are possible. Only when $K+U \geq 0$ can the system become separated by large distances, because K is never negative, and U goes to zero at large separations. (The following Checkpoint shows a different kind of potential energy situation, where there can be a bound state even when $K+U \geq 0$.)

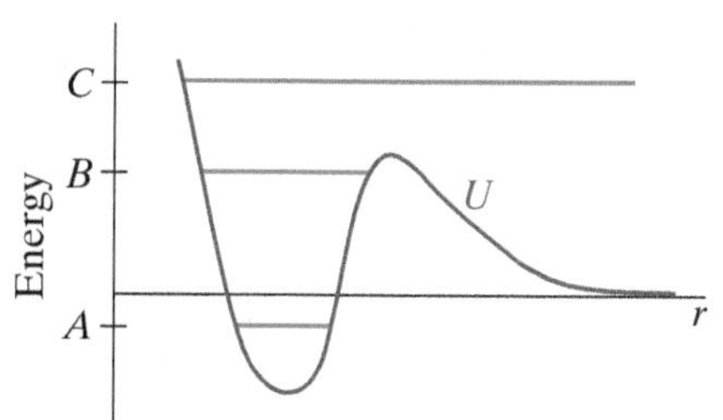

Figure 6.41 An energy graph for a two-object system. Which states are bound?

Checkpoint 13 (a) In Figure 6.39, as the asteroid–spacecraft separation increases from r_1 to r_2 does the kinetic energy of the asteroid–spacecraft system increase, decrease, or remain constant? **(b)** In the energy graph for a two-object system shown in Figure 6.41, consider the various energy states indicated. Which of these values for the energy $K+U$ (A, B, or C) represents a bound state? Which represents an unbound state (the particle can escape)? Which represents a bound state with enough energy to be unbound but with a barrier that (classically) prevents escape?

Drawing Energy Graphs

It is important to be able to draw (create) energy graphs as well as to be able to read them. Here is a practical scheme:

- Draw U vs. r for the particular interaction (gravitational, electric).
- At some r where you happen to know K, plot the point (r, K).
- Add that value of K to the value U has at the same separation r.
- Plot $K+U$ at that r, then draw a horizontal line through the point.
- At some other r, find a K that when added to U at that r gives $K+U$.
- Given these two points on the K graph, sketch the behavior of K vs. r.

EXAMPLE **Making an Energy Graph**

Let's follow through these steps for the energy graph for the spacecraft leaving the asteroid, which is shown in Figure 6.42.

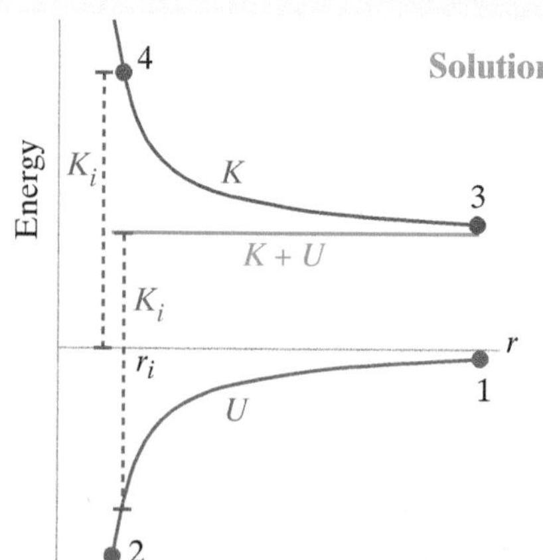

Figure 6.42 Energy vs. the separation distance between the asteroid and the spacecraft.

Solution

- We know the shape of U for gravity; it is $-GMm/r$. If r is small, U is a large negative number, so we plot point 1 on the graph. If r is large, U is nearly zero, so we plot point 2 on the graph. Now we can sketch the behavior of U vs. r.
- We know that K is nonzero at large r, because the spacecraft escapes, so at large r, we plot a positive K (point 3 on the graph).
- At that distant location, U is nearly zero, so the K you just plotted is nearly the same as $K+U$. Draw a horizontal line representing $K+U$ for all separations.
- At the initial separation r_i, determine graphically the value K_i that when added to the value of U_i at r_i yields the known value of $K+U$ (this is the height of the $K+U$ line about the U value at r_i). This is point 4 on the graph.
- Sketch the behavior of K vs. r. Note that its shape is the mirror image of U vs. r.

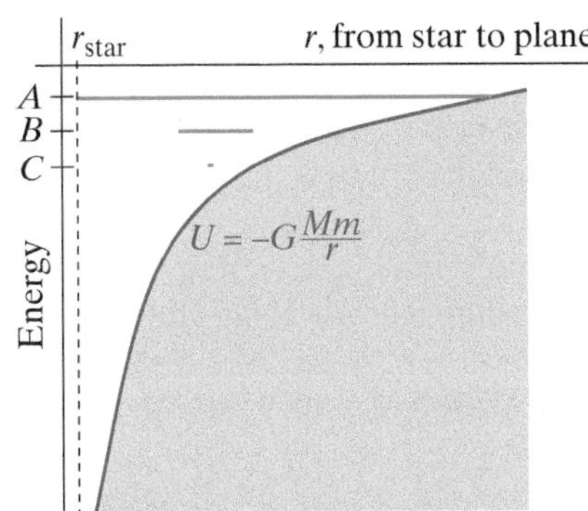

Figure 6.43 What kind of motion is represented by each of these situations?

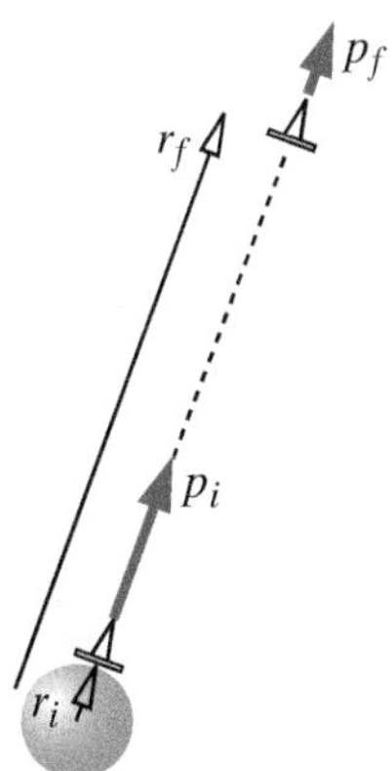

Figure 6.44 A spacecraft is launched with momentum p_i from near an airless planet's surface; it has momentum p_f when it has traveled far away.

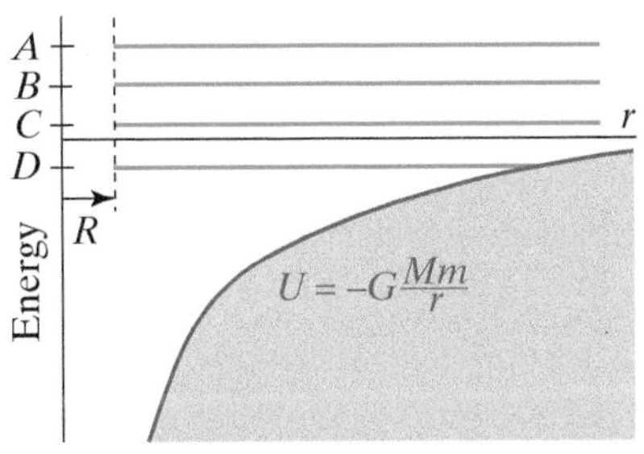

Figure 6.45 What is the minimum K required to escape?

Checkpoint 14 In Figure 6.43, what kind of motion is represented by the situation with $K + U = A$? B? C? Think about the range of r in each situation. For example, C represents a circular orbit (constant r).

Escape Speed

An analysis based on the Energy Principle will allow us to answer the following question: With what minimum speed v must a spacecraft leave the surface of an airless planet (no air resistance) of mass M and radius R in order that it can coast away without ever coming back (Figure 6.44)? Assume that there are no other objects nearby.

If you did the Moon voyage problem in Chapter 3 you found that there was a minimum initial speed required to reach the Moon. With an initial speed smaller than this, the spacecraft fell back to Earth without reaching the Moon. Your determination of the minimum initial speed was done by trial and error. However, we can now use energy relationships to calculate such minimum speeds directly.

As is often the case, an energy diagram can be very helpful. Consider Figure 6.45. Motions with $K + U = A$, B, C, or D all start from the surface of an airless planet, with a separation R between the center of the planet and the spacecraft. (Remember that the gravitational force outside a uniform sphere is exactly the same as it would be if all the mass collapsed to the center of the sphere.)

QUESTION Which of these motions starts with the largest kinetic energy K? Which starts with the least K?

Each of the horizontal lines represents motion with $K + U$ constant. Evidently A is the highest value of $K + U$, and therefore involves the largest initial kinetic energy. D is the lowest value of $K + U$, and starts with the least kinetic energy.

QUESTION For which of these motions does the spacecraft "escape" and never come back?

For motions with $K + U = A$, B, or C, no matter how far from the planet the spacecraft gets, there is still some kinetic energy, so the spacecraft will never return. This is another example of an unbound state being associated with a positive value of $K + U$.

In contrast, with motion $K + U = D$ the spacecraft will reach a maximum separation from the planet, at which point its kinetic energy has fallen to zero. The spacecraft will fall back to the planet.

QUESTION How would you describe the motion that requires the least possible initial kinetic energy to achieve escape?

Evidently the least costly escape is to have $K = 0$ when the spacecraft has reached a distance very far from the planet. At a large separation, $U = 0$, so $K + U = 0$. This corresponds on the diagram to a horizontal line lying on the axis. Because $K + U$ doesn't change (no external work on the planet–spacecraft system), it must also be true that $K + U = 0$ at the start of the motion:

MINIMAL CONDITION FOR ESCAPE

$$K + U = 0$$

The minimal initial kinetic plus potential energy of the system composed of planet plus spacecraft is this, assuming that the kinetic energy of the planet is negligible:

$$K_i + U_i = \frac{1}{2}mv_{esc}^2 + \left(-G\frac{Mm}{R}\right) = 0 \quad (\text{for } v \ll c)$$

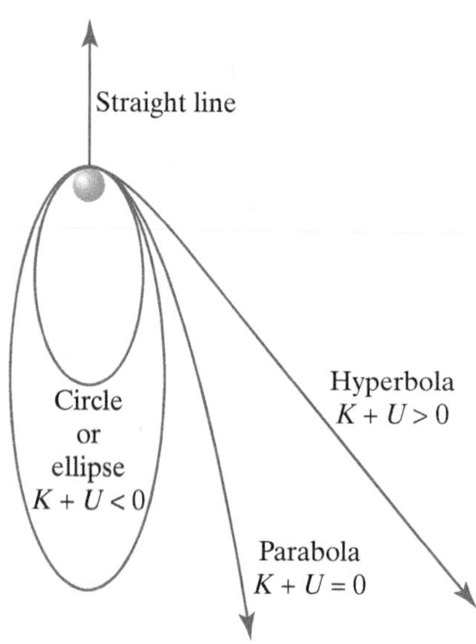

Figure 6.46 Possible orbital trajectories.

This lets us calculate the minimum speed, which is called the "escape speed" v_{esc}, for a planet with mass M and radius R. Though we speak of escape speed, it is wrong to say that the spacecraft has "escaped from gravity." As long as the spacecraft is a finite distance from the planet, it feels a finite gravitational attraction to the planet.

If you give the spacecraft more than the minimum kinetic energy, it not only escapes but also has nonzero kinetic energy when it is far away (for example, see motions A and B in Figure 6.45).

It is interesting to note that the initial direction of motion doesn't matter. An energy analysis is insensitive to direction.

Bound vs. Unbound States

If the initial velocity points directly away from the planet, the path is a straight line. If the initial velocity points in some other direction, the path can be shown to be a parabola if the initial speed is exactly equal to escape speed and a hyperbola for higher speeds.

If the speed is below escape speed, there is no escape. The various possible orbits (Figure 6.46) can be classified on the basis of their total kinetic plus potential energy, with initial speed v_i:

$$\frac{1}{2}mv_i^2 + \left(-G\frac{Mm}{R}\right) > 0: \quad \text{escape with } v_\infty > 0\text{; straight line or hyperbola}$$

$$\frac{1}{2}mv_i^2 + \left(-G\frac{Mm}{R}\right) = 0: \quad \text{escape with } v_\infty = 0\text{; straight line or parabola}$$

$$\frac{1}{2}mv_i^2 + \left(-G\frac{Mm}{R}\right) < 0: \quad \text{no escape; straight line, circular, or elliptical}$$

This is a specific example of an important general principle. If the kinetic plus potential energy is negative, the system is bound: the objects cannot become widely separated, because that would require that the kinetic energy become negative, which is impossible (classically). If the kinetic plus potential energy is positive, the system is unbound or free (unless the particle is in a "trap" as in Figure 6.41): the objects can become widely separated, with net kinetic energy (and zero potential energy).

BOUND AND UNBOUND STATES

If $K + U < 0$, the system is in a bound state.

If $K + U \geq 0$, the system is unbound (if not in a "trap").

EXAMPLE **Escape Speed**

What is escape speed from Earth?

Solution System: Earth + spacecraft
Surroundings: Nothing significant (ignore Sun and Moon)

Initial state: At surface of Earth, with speed
Final state: Spacecraft extremely far away from Earth, nearly at rest
Energy Principle (negligible change in kinetic energy of Earth; no work by surroundings):

$$K_{s,f} + \cancel{K_{E,f}} + U_f = K_{s,i} + \cancel{K_{E,i}} + U_i + W$$

$$0 + 0 = \frac{1}{2}mv_i^2 + \left(-G\frac{Mm}{r_i}\right) + 0$$

$$v_i = \sqrt{\frac{2GM}{r_i}}$$

$$v_{esc} = \sqrt{\frac{2(6.7 \times 10^{-11} \text{N} \cdot \text{m}^2/\text{kg}^2)(6 \times 10^{24}\,\text{kg})}{(6.4 \times 10^6\,\text{m})}} = 1.12 \times 10^4\,\text{m/s}$$

Checkpoint 15 Turn the argument around. If an object falls to Earth starting from rest a great distance away, what is the speed with which it will hit the upper atmosphere? (Actually, a comet or asteroid coming from a long distance away might well have an even larger speed, due to its interaction with the Sun.) Small objects vaporize as they plunge through the atmosphere, but a very large object can penetrate and hit the ground at very high speed. Such a massive impact is thought to have killed off the dinosaurs (see *T. Rex and the Crater of Doom*, Walter Alvarez, Princeton University Press, 1997).

6.11 GENERAL PROPERTIES OF POTENTIAL ENERGY

There are some general properties of potential energy that are true not only for gravitational interactions but also for other kinds of interactions as well, including electric interactions. Here is a list of important properties:

- Potential energy depends on the separation between pairs of particles, not on their individual positions. (As we saw with gravitational interactions, this depends on the reciprocity of gravitational and electric forces.)
- Potential energy must approach zero as the separation between particles becomes very large, as we will soon show.
- If an interaction is attractive, potential energy becomes negative as the distance between particles decreases. (We saw this in the case of gravitational interactions.)
- If an interaction is repulsive, potential energy becomes positive as the distance between particles decreases.

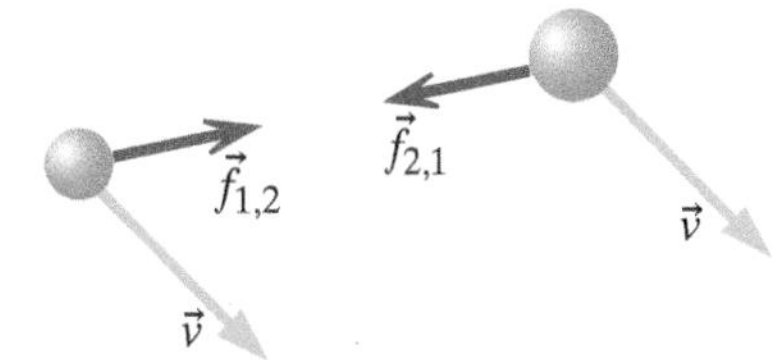

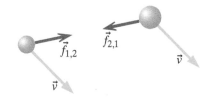

Figure 6.47 The two particles move together, with the same velocity.

Let's look in a bit more detail at the consequences of the fact that potential energy depends on the separation between the interacting objects.

QUESTION Suppose that two particles are part of a rigid system and move together, with the same velocity (Figure 6.47). In that case, what is $\Delta\vec{r}_{21}$ for this two-particle system? What is ΔU_{12}?

In this case the relative positions don't change, $\Delta\vec{r}_{21}$ is zero, and there is no change in the pair-wise potential energy.

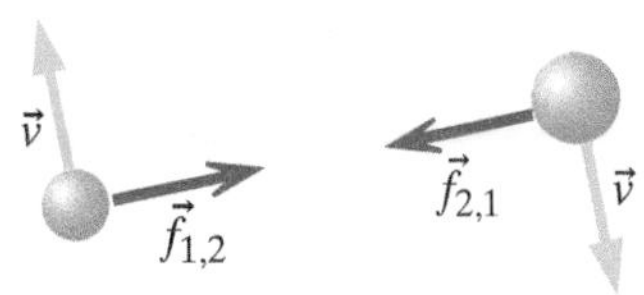

Figure 6.48 The two particles rotate around each other.

QUESTION Suppose instead that the two particles rotate around each other (Figure 6.48). What can you say about ΔU_{12}?

This is a bit trickier, because if there is rotation, $\Delta\vec{r}_{21}$ isn't zero, because $\vec{r}_{21}$ is now a rotating vector. In rotation, however, $\Delta\vec{r}_{21}$ is perpendicular to $\vec{r}_{21}$ and to the force $\vec{f}_{2,1}$, so the dot product $\vec{f}_{2,1} \bullet \Delta\vec{r}_{21}$ is zero. So again there is no change in the pair-wise potential energy.

QUESTION What can you conclude about potential energy for a rigid system?

Evidently U is constant for a rigid system, one whose shape doesn't change. For the potential energy to change, there must be a change of shape or size. Conversely, a change of shape or size is an indicator that U may have changed.

$U \to 0$ When the Particles Are Very Far Apart

Consider a system consisting of two identical stars, which attract each other gravitationally. If the stars are at rest, very, very far apart—almost infinitely far apart, so that the force one exerts on the other is nearly zero—then the total energy of the two-star system must be simply the sum of their rest energies:

$$E_{sys} = mc^2 + mc^2$$

This implies that the potential energy associated with the interaction of these two stars must be zero if the stars are "infinitely" far apart.

More generally, the total energy of a multiparticle system is given by the expression $E_{system} = (E_1 + E_2 + \cdots) + U$, where $E_1 = \gamma m_1 c^2$ is the particle energy of particle 1, and so on. When the particles are very far apart, we must define U to be zero, so that the total energy of these far-apart particles is simply the sum of the particle energies:

POTENTIAL ENERGY OF A PAIR OF PARTICLES THAT ARE VERY FAR APART

$$U \to 0 \quad \text{as} \quad r \to \infty$$

Attractive Interaction: U Becomes Negative as r Decreases

If the particles start to come closer together, we find that U starts to become negative if the interactions are attractive, while U starts to become positive if the interactions are repulsive. To see this, first consider an isolated system consisting of two stars that are very far apart, initially at rest, so the initial energy of the system is just the sum of the rest energies of the two stars. At first slowly, then more and more rapidly, their kinetic energies increase due to their mutual gravitational attractions (Figure 6.49).

Figure 6.49 Initially two stars are very far apart, at rest, but they attract each other.

QUESTION U was initially zero, so what must be the sign of U when the stars have acquired significant kinetic energies?

If the system is isolated, there are no external forces and hence no external work, so E_{sys} does not change. If the individual particle energies increase (so the total kinetic energy increases), the pair-wise potential energy must decrease. Because U was defined to be zero initially, U must become negative, as we found for gravitational potential energy. A graph of the energies is shown in Figure 6.50, omitting the very large rest energies of the stars. Since the rest energies don't change, $K + U$ is a constant.

Figure 6.50 The energies of two stars (with rest energies omitted) as a function of separation r. U must be zero at large r. As r decreases, K increases and U decreases. The total energy of the two-star system is constant, and $K + U$ is constant.

Does this negative sign make sense in terms of the definition of potential energy change, $\Delta U = -W_{int}$? The attractive gravitational forces that the stars exert on each other certainly do a positive amount of work on each star, so the potential energy change must indeed be negative.

Repulsive Interaction: U Becomes Positive as r Decreases

Next consider an isolated system consisting of two protons that are very far apart, and suppose that initially they have high speeds, heading inward (Figure 6.51). Their mutual electric repulsions make the protons slow down (their electric repulsion is enormously larger than their gravitational attraction).

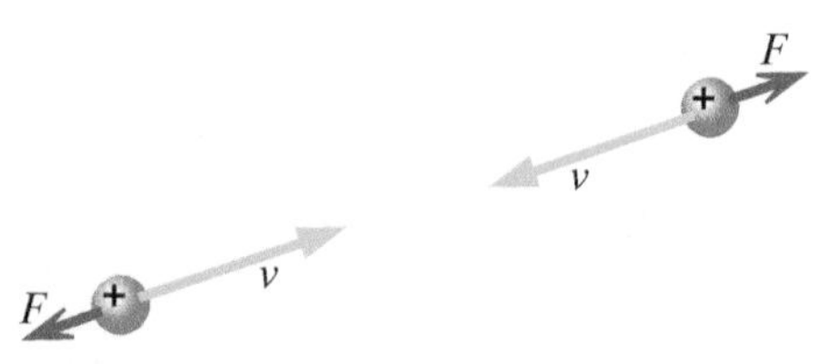

Figure 6.51 Initially two protons are very far apart, heading inward. The mutual repulsions slow the protons down.

QUESTION U was initially zero, so what must be the sign of U when the protons have slowed down somewhat?

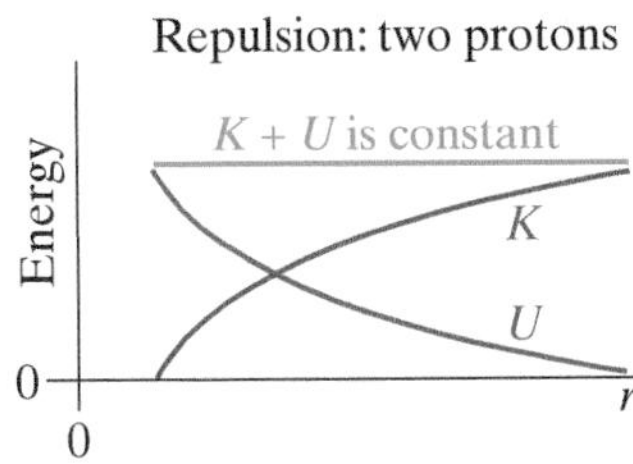

Figure 6.52 The energies of two protons as a function of separation (with rest energies omitted). U must be zero at large r. As r decreases, K decreases and U increases. The total energy of the two-proton system is constant, and $K+U$ is constant.

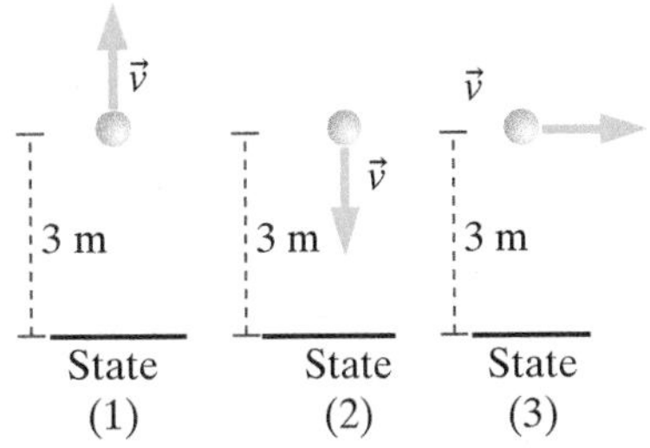

Figure 6.53 The potential energy of the system of (ball + Earth) is the same in each of these states, since it depends only on the separation of the ball and the Earth.

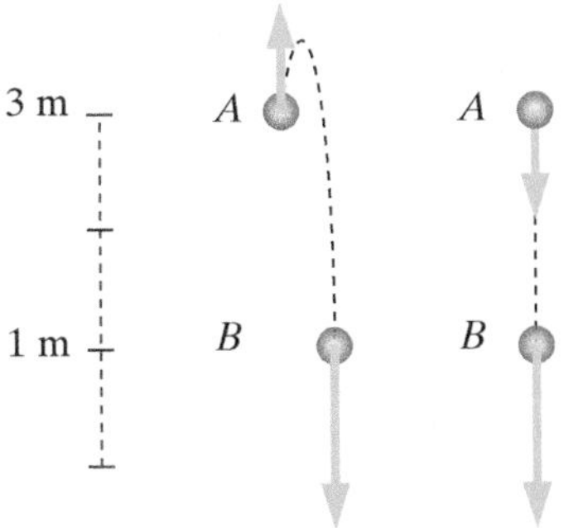

Figure 6.54 ΔU for the system of (ball + Earth) does not depend on the path taken to get from initial to final state.

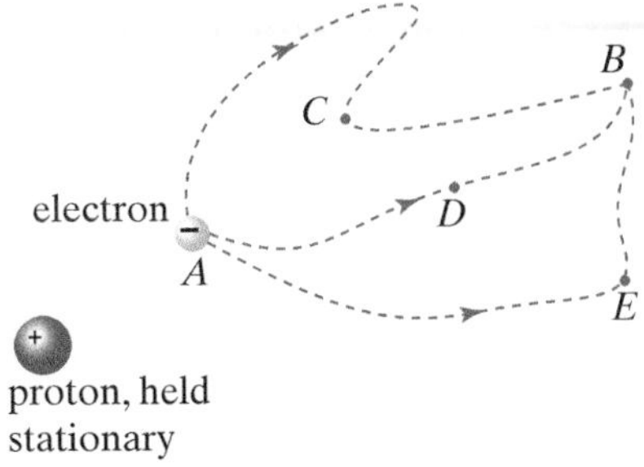

Figure 6.55 A proton is held stationary while an electron is moved from location A to location B along different paths.

Again, there is no external work done on this isolated multiparticle system, so E_{sys} does not change. The particle energies decrease, so the potential energy must increase; U becomes positive (Figure 6.52). Also, the repulsive electric forces do negative work on each other, so $\Delta U = -W_{int}$ is positive. The graph in Figure 6.52 omits the rest energies of the protons. Since the rest energies don't change, $K+U$ is a constant.

Note that the change in potential energy for the two stars or the two protons was associated with a change in shape (change in configuration) of the two-particle system. If there is no change in shape or size, there is no change in potential energy.

Path Independence

One of the most interesting things about the expressions for different kinds of potential energy is what is *not* in these equations: history. Since the potential energy of two interacting point particles depends only on the relative positions of the particles, it does not depend on how the particles got to those positions. If a tennis ball is 3 m above the surface of the Earth, it doesn't matter how the tennis ball got there (Figure 6.53). It could be moving up, down, or horizontally—only the relative positions of the ball and the Earth matter. This may seem obvious and unremarkable, but it has some important implications.

One consequence is that since U depends only on separation, any "trip" between two locations, or separations, for a two-particle system, results in the same ΔU. For the system of (ball + Earth) in Figure 6.54, the difference between the potential energy of the system in the initial state shown (3 m above the ground) and the final state in which the ball is only 1 m above the ground does not depend on whether the ball goes straight down or goes up, stops, and descends. The fact that the difference in potential energy of a multiparticle system depends only on the states chosen, and not on the path taken by the particles to get to those states, is called "path independence."

$$U_B - U_A = \text{constant}$$

independent of path taken between A and B

A special case of a trip between two states is a round trip, in which the initial state is the same as the final state. The potential energy change associated with a round trip must be zero, since

$$\Delta U_{\text{round trip}} = U_A - U_A = 0$$

A second interesting consequence is that the dependence of potential energy on configuration, not history, allows us to make deductions about the amount of work required to change the state of a system no matter how that change happens. This is easiest to see if we consider processes in which the kinetic energy of the system does not change. Suppose that a proton remains fixed at the origin while we move an electron from location A to location B at constant speed (Figure 6.55). The electron is attracted to the proton, so we need to do positive work to move it farther from the proton.

To move the electron along the path that goes through location D we will have to do positive work to separate the particles. In contrast, on part of the path that goes through location C the electron is moving back toward the proton, and on that path segment we will have to do negative work to keep it from speeding up. In the end, though, all the positive and negative work we do to move the electron along any path from A to B will add up to the same amount. We can see this from the Energy Principle:

$$\begin{aligned} K_{e,f}+K_{p,f}+U_B &= K_{e,i}+K_{p,i}+U_A+W \\ \cancel{K_{e,f}}+0+U_B &= \cancel{K_{e,i}}+0+U_A+W \\ U_B-U_A &= W \end{aligned}$$

It also follows from this that the total amount of work required to move the electron from A to B along any path, and then back to A along any other path, without changing its kinetic energy, must add up to zero:

$$W_{\text{round trip}} = 0$$

What About Friction?

We've defined potential energy for systems of interacting point particles. Point particles are a remarkably good model for many physical situations, but not for situations in which there is friction.

> QUESTION Why isn't the point particle model appropriate when there is friction?

As you know from everyday experience, when there is significant friction between two moving objects, the objects can get hot. A temperature change indicates that the amount of energy inside an object has changed. We'll see in Chapter 9 that friction results from the deformation of extended objects that are in contact with each other. However, we can't model such objects as point particles; since a point particle, by definition, has no internal structure, it can't deform or get hot.

Further, the work done by friction is clearly not independent of the path taken; the longer the path, the hotter the objects get, regardless of direction. We call forces like friction and air resistance "dissipative forces," and there isn't a way to define anything like potential energy for such interactions. We'll consider dissipative interactions further in Chapter 7.

You may have heard friction described as a "nonconservative" force, an older term used to denote interactions in which a potential energy cannot be defined. This isn't a very informative term, because friction itself is simply a consequence of the electric interactions of electrons and protons in two extended objects in contact with each other. The real issue is the fact that these electric forces lead to deformations of the objects, and that energy can flow into the vibrations of the atoms inside the objects.

6.12 THE MASS OF A MULTIPARTICLE SYSTEM

A surprising prediction of Einstein's Special Theory of Relativity (1905) was that mass and energy are closely related. This prediction has been confirmed in many different kinds of experiments. The most dramatic example of mass–energy equivalence is the annihilation reaction that occurs between matter and antimatter, as in the case of the annihilation of an electron and a positron (a positive electron) into two photons (high-energy quanta of light denoted by lowercase Greek gamma, γ):

$$e^- + e^+ \rightarrow \gamma + \gamma$$

Before the reaction, an electron and a positron can be almost at rest far from each other, a situation we might be inclined to describe as one in which there is mass but seemingly no energy. After the reaction there is light (in the form

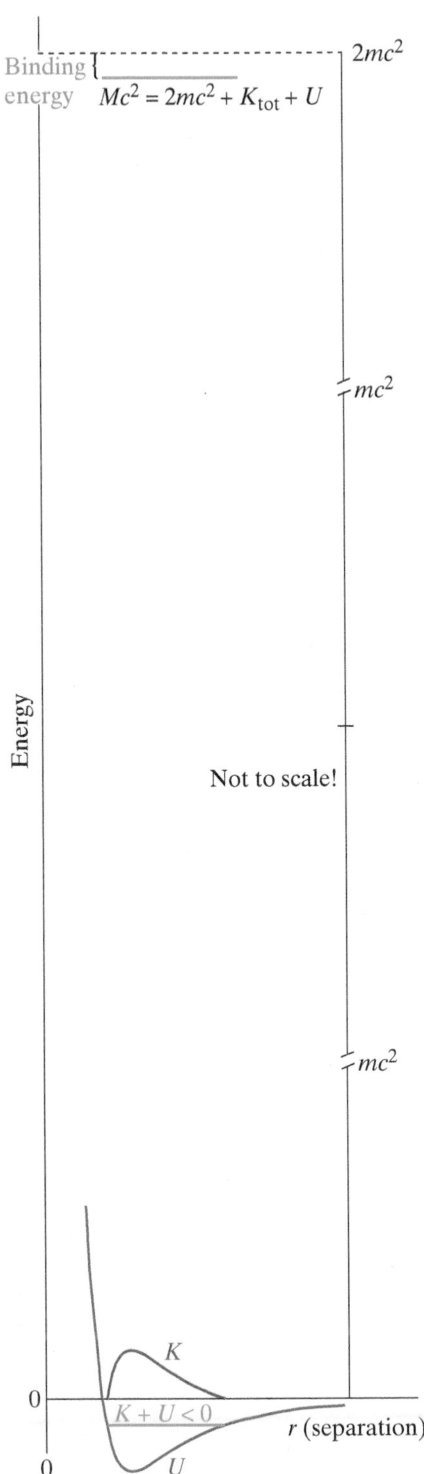

Figure 6.56 A plot of rest energy, kinetic energy, potential energy, and total energy (Mc^2) for O_2. Note that the total energy is less than the sum of the rest energies of the two individual atoms.

of two photons), a situation that we would naturally describe as pure energy but seemingly no mass. However, from the point of view of Special Relativity, the mass of the electron–positron system *is* energy, and the energy of the two-photon system *is* mass.

The equivalence of mass and energy is of course visible in Einstein's equation for the energy of a particle:

$$E = \gamma mc^2 = mc^2 + K \quad (\textit{definition} \text{ of kinetic energy } K)$$

In this equation m is the mass of the particle and is a constant. The particle has rest energy $E_{rest} = mc^2$ even if it is not moving. If it is moving, it has additional, kinetic energy K.

What is the mass of a multiparticle system consisting of interacting particles? Consider the simple case where the multiparticle system is overall at rest. The individual particles are moving around, but the system as a whole is not moving (more technically, the total momentum of the system is zero, and its center of mass is at rest). An example would be a box containing air molecules that are moving around rapidly in random directions, but the box isn't going anywhere. Another example is a diatomic molecule such as O_2 that is vibrating but has no overall translational motion.

It is natural to suppose that the total mass of such a system is just the sum of the masses of the particles: $M = m_1 + m_2 + \cdots$. However, this is not quite right. Rather, as predicted by the theory of relativity, the rest energy of the multiparticle system is Mc^2, and therefore its mass is this:

$$M = \frac{E_{system}}{c^2} = \frac{m_1c^2 + K_1 + m_2c^2 + K_2 + \cdots + U}{c^2}$$

$$M = (m_1 + m_2 + \cdots) + \left(\frac{K_1 + K_2 + \cdots + U}{c^2}\right)$$

The mass M of the system may be greater or less than the sum of the individual particle masses, because the final term in this equation may be positive or negative.

In most everyday situations the kinetic energies of the individual particles are very small compared to their rest energies (that is, the speeds of the particles are small compared to the speed of light), and the interaction energy U (the sum of all the pair-wise interactions) is also small compared to the rest energies. In that case, the mass of the multiparticle system is very nearly equal to the sum of the individual masses. However, this is really only an approximation, and it is often not a good approximation in nuclear and particle physics.

The Total Energy of a Multiparticle System

Consider an oxygen molecule (O_2) at rest. Instead of plotting only the kinetic and potential energy for this two-atom system, let's include the rest energy of the atoms, too, as we attempt to show in Figure 6.56. We can't draw the full vertical scale, because the rest energies are enormous compared to the kinetic and potential energies. In order to show the rest energy to the same scale as K and U, we would need a page stretching from the Earth to the Moon!

The total energy of the system (indicated by the green horizontal line near the top of the graph) is the sum of the rest energies of the two atoms, the potential energy U (which is negative), and K_{tot}, the total kinetic energy of the two atoms. For the bound state drawn on the graph you can see that the rest energy of the two-atom O_2 system is very slightly less than the sum of the individual atomic rest energies, because $K + U < 0$.

Binding Energy

When the diatomic molecule O_2 forms, about 5 eV of energy are given off (for example, by emitting photons, which carry energy). Correspondingly, it takes an input of about 5 eV of energy to break O_2 apart into two oxygen atoms. This is called the "binding energy" of the molecule and is shown at the top of Figure 6.56.

Two Ways of Thinking About the Energy of the System

Given these aspects of particle energy, there are two rather different ways to think about the energy of a multiparticle system:

- The energy of a multiparticle system consists of the individual particle energies of the particles that make up the system, plus their pair-wise interaction energies.
- A multiparticle system itself has energy, rather like a particle, and if the system is at rest (its center of mass is not moving; its net momentum is zero) its energy E is simply Mc^2, where M is the mass of the system.

The Mass of the System

The mass M may be greater or smaller than the sum of the masses of the individual particles, depending on whether the total $K+U$ of the particles is positive or negative. The graph in Figure 6.56 indicates that in the case of the O_2 molecule the quantity Mc^2 (where M is the mass of the whole molecule) is less than $2mc^2$, and the difference is the binding energy of the molecule. From the graph we can see that:

$$Mc^2 = 2mc^2 + K + U < 2mc^2$$

We analyzed the O_2 molecule in terms of the masses of the individual oxygen atoms and the molecular binding energy. We could make a more detailed analysis by considering the energy terms that contribute to the mass of each oxygen atom. The oxygen atom is itself a multiparticle system consisting of protons and neutrons bound together in the nucleus, plus eight electrons. These particles have rest energy and kinetic energy, and there is potential energy associated with pairs of particles. One could even go further, recognizing that the protons and neutrons are made of quarks and gluons. However, if we're just interested in the binding energy of the O_2 molecule, we can start with the rest mass of each oxygen atom, because all of the internal complexity of each atom is summarized by the value of the atom's mass.

Mass of a Hot Object

This way of thinking about mass leads us to conclude that a hot block has a greater mass than a cold block. As we'll discuss in more detail in the next chapter, a hot metal block has more thermal energy than a cold metal block, with more kinetic energy of the atoms and more potential energy in the interactions with neighboring atoms (represented as springs in our simple model of a solid). Although the effect is much too small to measure directly, the equivalence of mass and energy embodied in the special theory of relativity, $M = E/c^2$, leads to the conclusion that a hot block is actually more massive than a cold block! The same impulse $\vec{F}\Delta t$ applied to the two blocks should change the velocity of the cold block very slightly more than the velocity of the hot block.

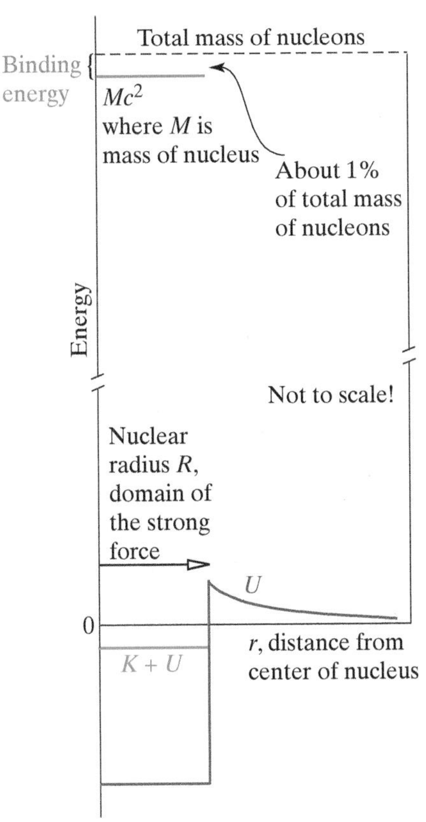

Figure 6.57 An energy diagram for a proton in an iron nucleus, not to scale. The mass of the nucleus is about 1% less than the sum of the masses of the individual nucleons (protons and neutrons).

Checkpoint 16 Calculate the approximate ratio of the binding energy of O_2 (about 5 eV) to the rest energy of O_2. Most oxygen nuclei contain 8 protons and 8 neutrons, and the rest energy of a proton or neutron is about 940 MeV. Do you think you could use a laboratory scale to detect the difference in mass between a mole of molecular oxygen (O_2) and two moles of atomic oxygen?

Nuclear Binding

In chemical reactions it is nearly impossible to measure the change of mass directly, but in nuclear interactions the changes are easily measurable. Consider the nuclear binding energy of an iron nucleus, which contains 26 protons and 30 neutrons, a total of 56 nucleons. When the protons and neutrons bind to each other through the "strong" (nuclear) force, the resulting iron nucleus has a mass that is about 1% smaller than the total mass of its individual constituents when separated. There is a sizable negative potential energy U associated with the strong force, which makes the multiparticle system's rest energy measurably smaller than the sum of the individual rest energies.

The average binding energy per nucleon is largest in iron. Lighter nuclei (hydrogen through manganese) are less tightly bound and can in principle fuse together to form the more tightly bound iron, with the release of energy. In fact, fusion reactions in stars build up from hydrogen to heavier and heavier nuclei, stopping at iron (heavier elements are made in supernova explosions).

Heavier nuclei (cobalt through uranium and beyond) are also less tightly bound than iron and can in principle fission to form the more tightly bound iron. However, only a few nuclei such as uranium and plutonium do actually spontaneously fission.

Figure 6.57 shows an energy diagram (not to scale) for one of the protons in an iron nucleus, subjected to the forces of all the other protons and neutrons. There is a deep potential well corresponding to the strong or nuclear force, which is essentially the same for protons and neutrons. This potential well has a very steep side corresponding to the very strong nuclear force (force is negative gradient of potential energy). The well is so steep it can be described approximately as a "square well." This deep well extends only out to the radius R of the nucleus, because the nuclear force has a very short range. A proton (but not a neutron) in addition experiences the repulsive electric force of the other protons (the electric U curve outside the nucleus). The total mass of the nucleus is about 1% less than the sum of the masses of the individual nucleons (protons and neutrons).

EXAMPLE **Nuclear Binding Energy**

The rest energy of an iron nucleus (26 protons and 30 neutrons) is 52,107 MeV. Show that the average binding energy per nucleon is about 1% of the rest energy of a nucleon, which is easily measurable. The rest energy of a proton is 938.3 MeV; of a neutron, 939.6 MeV.

Solution As shown in Figure 6.57, the binding energy is essentially the difference between the rest energy of the multiparticle system (the nucleus) and the sum of the rest energies of its constituent particles.

$$\begin{aligned}\text{binding energy} &= ((26)(938.3)\,\text{MeV} + (30)(939.6)\,\text{MeV}) - 52107\,\text{MeV}\\ &= 494.8\,\text{MeV}\end{aligned}$$

$$\text{energy per nucleon} = \frac{494.8\,\text{MeV}}{26+30} = 8.8\,\text{MeV}$$

This is about 1% of the rest energy of one nucleon.

Checkpoint 17 The deuteron, the nucleus of the deuterium atom ("heavy" hydrogen), consists of a proton and a neutron. It is observed experimentally that a high-energy photon ("gamma ray") with a minimum energy of 2.2 MeV can break up the deuteron into a free proton and a free neutron; this process is called "photo-dissociation." About what fraction of the deuteron rest energy corresponds to its binding energy? The result shows that the deuteron is very lightly bound compared to the iron nucleus.

Nuclear Fission

The nuclear fission of uranium leads to an easily measured change in masses. The average binding energy per nucleon in uranium is about 7.8 MeV, so the uranium nucleus is less tightly bound than an iron nucleus, whose binding is 8.8 MeV per nucleon. This suggests the possibility of obtaining energy from the fission of a uranium nucleus, corresponding to roughly 0.1% (0.7 MeV per nucleon) of its rest energy (about 940 MeV per nucleon).

In other words, the "mass deficiency" of a medium-mass nucleus is about 1%, with a variation of about a tenth of that. Even though there is only a small difference in average binding energy between iron and uranium, the energy involved is vast compared to the energy of a chemical reaction, which only involves the outer electrons of an atom, not the nucleus.

A uranium-235 nucleus contains 92 protons and 143 neutrons. The 92 protons exert strong electric repulsions on each other, but the large number of neutrons spreads the protons throughout a larger volume and reduces the electric repulsions. The neutrons and the protons also attract neighboring protons with strong nuclear forces. (Low-mass nuclei don't have an overabundance of neutrons; for example, ordinary carbon has 6 protons and 6 neutrons, and iron has only 30 neutrons with its 26 protons.)

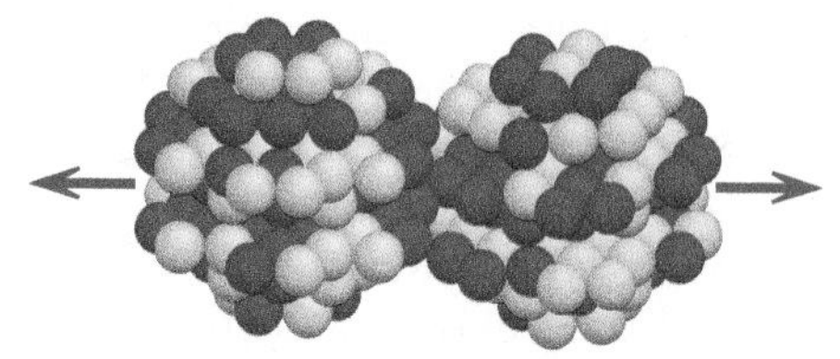

Figure 6.58 A U-235 nucleus in the process of fissioning.

Uranium-235 has an unusual property that few other nuclei have. If you hit a U-235 nucleus with a slowly moving neutron, the nucleus may deform into a dumbbell-like form and then split apart because the electric repulsion of the protons can now overcome the short-range attraction of the nucleons for each other (Figure 6.58). (In addition to the two main fission fragments there are typically one or more free neutrons in the final state.)

The fission fragments quickly acquire very high speed, and they leave the original atom's 92 electrons behind. These highly charged nuclei storm through the block of uranium metal, tearing up the material and violently heating the metal. In a nuclear power reactor the reaction is controlled in such a way that the metal merely gets hot enough to boil water. With many uranium-235 nuclei fissioning, the metal can boil water in a nuclear power plant to produce steam and power a turbine to drive an electric generator.

The masses of the two fission-fragment nuclei total less than the mass of the uranium-235 nucleus. The mass difference Δm is only about one-tenth of one percent of the original mass, but this mass change is easily measurable, and c^2 times this mass change is indeed found to be equal to the observed kinetic energy of the fission fragments. This 0.1% change is about ten million times as big an effect as the mass changes that occur in chemical reactions, which only involve rearrangements of the outer electrons in the atoms, and do not affect the nucleus.

Nuclear Waste

The problem of radioactive waste from nuclear power plants comes from the fact that the fission products are nuclei with a large overabundance of neutrons, and they emit energetic radiations. For example, uranium can fission into two palladium nuclei, each with 46 protons and 71 or 72 neutrons, whereas ordinary

$$n \rightarrow p^+ + e^- + \overline{\nu}$$

Figure 6.59 A neutron in a neutron-rich nucleus can change into a proton with emission of an electron and an antineutrino.

$$N^* \rightarrow N + \gamma$$

Figure 6.60 A nucleus in an excited state (called N^*) can drop to a lower energy state and emit a gamma ray (a high-energy photon).

stable palladium has only 60 neutrons (see the inside front cover of this book). A neutron in such a neutron-rich nucleus can change into a proton (Figure 6.59) with the emission of an energetic electron and an antineutrino (the antineutrinos almost never interact with matter, so they aren't dangerous). This process is called "beta decay" (energetic electrons were initially called "beta rays"). After a beta decay the new nucleus, in which a neutron has changed into a proton, may be in an "excited state" and emit a gamma ray (a high-energy photon, see Figure 6.60). Heavy elements such as plutonium also emit alpha particles (the nuclei of helium atoms) and there may be heavy elements mixed in with the rest of the nuclear waste.

When energetic electrons or gamma rays or alpha particles from radioactive nuclei go through matter, energy is absorbed. If the radiation goes through living tissue there can be damage to cells, and to the genetic information in the DNA in cells, through ionization of the biological molecules. The range of the radiation is relatively short and it can be stopped by a layer of concrete or other shielding. The big concern is that radioactive wastes might leak out of storage facilities, spread through the ground water, and get into the water supply. Drinking contaminated water would bring radioactive nuclei into the body, close to cells, where damage could occur.

A great deal of engineering and scientific study is under way to try to find a secure storage scheme that would prevent leakage for hundreds or even thousands of years, because some of the radioactive nuclei have very long "half-lives" (the time it takes for half of the nuclei to decay).

6.13 REFLECTION: WHY ENERGY?

It is appropriate at this point to reflect on why we introduced the topic of energy. Energy methods often make it possible to analyze some aspect of a phenomenon with much less effort than is required by direct application of the Momentum Principle. A striking example is the ability to calculate escape speed, or the impact speed of the *Ranger* spacecraft on the Moon, without carrying out a lengthy numerical integration.

Limits on the Possible

Another powerful capability of energy analyses is that they can easily predict whether some process can occur or not. For example, if there is insufficient energy, the rocket can't escape from the planet, or the atom cannot be ionized, or the biochemical reaction cannot proceed, or a proton can't decay into a neutron.

Less Detail

However, energy analyses give less detailed information than is given by a step-by-step numerical integration of the Momentum Principle. In particular, energy analyses tell us nothing about how fast a process will proceed; time does not appear in the statement of conservation of energy. Also, since the Energy Principle is a scalar principle, it doesn't tell us about change in the direction of motion.

Here is an example that illustrates the strengths and weaknesses of energy analyses. Consider snowboarding down the three hills shown in Figure 6.61, starting from rest. If we can neglect friction, the final speed at the bottom should be the same in all three cases, because the final kinetic energy should be equal to *mgh*, the amount by which the gravitational energy decreases.

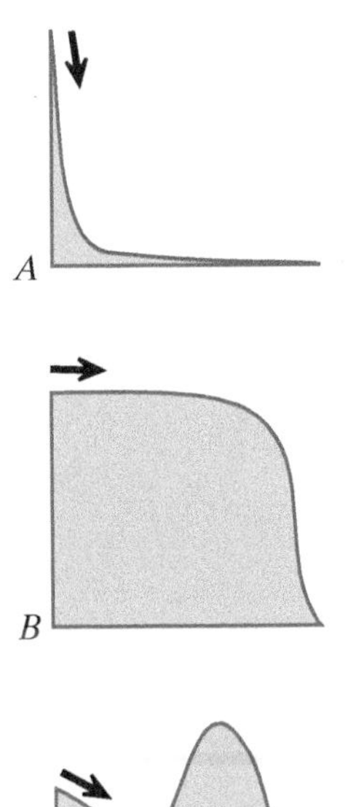

Figure 6.61 The final speed of the snowboarder will be the same at the bottom of each hill.

QUESTION But how will the amount of *time* compare for the ride down the hill?

On the first hill you get going very fast very quickly, and the trip is short. On the second hill you move very slowly for a long time before dropping to the bottom, and the total trip takes a long time. On the third hill you never get to the bottom, despite the fact that the change in potential energy would be negative, because you would have to pass through a region that is energetically forbidden. So energy methods are powerful, but at the cost of losing some information, including the amount of time required for a process.

6.14 IDENTIFYING INITIAL AND FINAL STATES

Some processes involve several identifiable states, not just two. Any two states can be chosen as initial and final states; the rationale for making the choice depends on what unknown quantity you are trying to determine. In this section we offer two examples of situations involving more than two states, one simple case and one more complex process.

EXAMPLE **Throwing a Ball: Three States**

Throw a metal ball with mass 0.3 kg straight up with initial speed 10 m/s; air resistance is negligible. Consider these three states (Figure 6.62): (1) $y_1 = 7\,\text{m}$, $v_{y1} = 10\,\text{m/s}$, (2) $y_2 = ?$, $v_{y2} = 0$ (maximum height; a turning point in the motion), (3) $y_3 = 4$ m, $v_{y3} = ?$ (after coming back down, to a lower level). Calculate the maximum height y_2 and the final speed, $|v_{y3}|$.

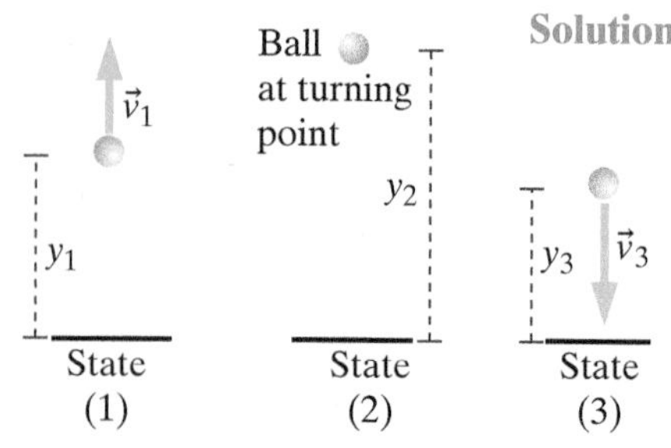

Figure 6.62 Throw a ball straight up. Find the maximum height and the final y component of the velocity.

Solution System: Ball + Earth

Initial state: 1
Final state: 2 (because the unknown is y_2)

$$K_2 + U_2 = K_1 + U_1 + W$$
$$0 + mgy_2 = K_1 + mgy_1 + 0$$
$$y_2 = y_1 + \frac{\frac{1}{2}mv_{y1}^2}{mg}$$
$$y_2 = 7\,\text{m} + \frac{\frac{1}{2}(10\,\text{m/s})^2}{9.8\,\text{N/kg}}$$
$$y_2 = 12.1\,\text{m}$$

Initial state: 2
Final state: 3 (because the unknown is v_3)

$$K_3 + U_3 = K_2 + U_2 + W$$
$$K_3 + mgy_3 = 0 + mgy_2 + 0$$
$$\frac{1}{2}mv_{y3}^2 = mg(y_2 - y_3)$$

$$v_{y3} = \pm\sqrt{2g(y_2 - y_3)}$$
$$|v_{y3}| = \sqrt{2(9.8\,\text{N/kg})(12.1 - 4)\,\text{m}}$$
$$|v_{y3}| = 12.6\,\text{m/s}$$

(See below for a discussion of the $\pm$ issue.)

QUESTION Could we have used states 1 and 3 instead of 2 and 3 to find the final speed?

Yes. We should get the same result for the final speed:

Initial state: 1
Final state: 3

$$K_3 + U_3 = K_1 + U_1 + W$$
$$K_3 + mgy_3 = K_1 + mgy_1 + 0$$
$$\frac{1}{2}mv_{y3}^2 = \frac{1}{2}mv_{y1}^2 + mg(y_1 - y_3)$$
$$v_{y3} = \pm\sqrt{v_{y1}^2 + 2g(y_1 - y_3)}$$
$$|v_{y3}| = \sqrt{(10\,\text{m/s})^2 + 2(9.8\,\text{N/kg})(7-4)\,\text{m}}$$
$$|v_{y3}| = 12.6\,\text{m/s}$$

QUESTION Aren't there two possible solutions to the quadratic that gives v_{y3}?

Yes. Unlike the Momentum Principle, the Energy Principle doesn't give us information about the direction of the velocity. In taking the square root above, v_{y3} could be either +12.6 m/s or −12.6 m/s. Which is correct? Both!

On the way down, $v_{y3} = -12.6$ m, which corresponds to the situation we analyzed. However, the mathematics claims that another possible value is $v_{y3} = +12.6$ m. This solution corresponds to the past, before the time of state 1. If the ball was thrown upward from the ground, v_y was +12.6 m/s as it rose past the height we've called $y_3 = 4$ m. It kept on rising, and v_y was +10 m/s as it rose past the height we've called $y_1 = 7$ m. Whenever you take a square root in a physics equation, you need to use physical sense to decide which root is the one you want.

QUESTION How do we decide which two states to use?

When using the Energy Principle, you get to choose which two states to use, just as you get to choose the system. The energy of one of the states must involve the unknown quantity whose value you want to find—for example, a distance or a speed. The choice of the other state is often a matter of convenience. Selecting different pairs of initial and final states often allows us to answer different questions about the process.

EXAMPLE **Reaction Between an Alpha Particle and a Carbon Nucleus**

If an energetic alpha particle (a helium nucleus: two protons and two neutrons) collides with a carbon nucleus (containing six protons and six neutrons), a nuclear reaction can occur, forming an oxygen nucleus whose rest energy is higher than an ordinary oxygen nucleus (an excited state, a concept discussed in Chapter 8). After a short time, the excited oxygen nucleus can emit its excess energy as a photon (a particle of light) called a "gamma ray" whose energy is 10.352 MeV. A photon (γ) has no rest energy, only kinetic energy. The emission of the photon reduces the rest energy of the nucleus to that of an ordinary oxygen nucleus. The oxygen nucleus recoils with momentum equal and opposite to the photon's momentum, but negligible kinetic energy.

This reaction won't occur unless the original particles have enough kinetic energy. Let's find out how much kinetic energy they must have. To simplify the analysis we will consider a situation in which the alpha particle and carbon nucleus start with equal magnitudes of momenta p_1, so the total momentum of the system is zero (Figure 6.63). The speeds involved are nonrelativistic.

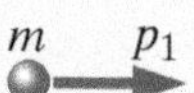

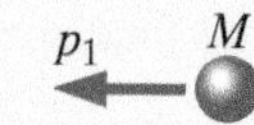

Figure 6.63 State 1: Alpha and C far apart, moving.

Masses of the ground (not excited) states of selected nuclei, where $1\ \text{u} = 1.6605 \times 10^{-27}$ kg, are:

$$\text{He-4 } (2\text{p} + 2\text{n}) = 4.00040868\ \text{u}$$
$$\text{C-12 } (6\text{p} + 6\text{n}) = 11.99670852\ \text{u}$$
$$\text{O-16 } (8\text{p} + 8\text{n}) = 15.99052636\ \text{u}$$

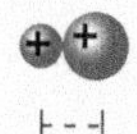

Figure 6.64 State 2: Alpha and C touch.

A proton or neutron has a radius of roughly 1×10^{-15} m, and a nucleus is a tightly packed collection of nucleons. Experiments show that the radius of a nucleus containing N nucleons is approximately $(1.3 \times 10^{-15}\ \text{m}) \times N^{1/3}$.

Figure 6.65 State 3: Excited O* nucleus.

(a) What is the minimum magnitude of momentum p_1 that the alpha particle must have in order for this sequence of phenomena to occur?
(b) Since both particles are positively charged, they repel each other. If the particles do not have sufficient kinetic energy to start with, they will not get close enough to touch, and the nuclear reaction will not occur. Find the minimum initial momentum p_2 each particle must have to get close enough to touch.
(c) Compare your answers to parts (a) and (b) to determine what momentum the particles must have initially to make this reaction occur.

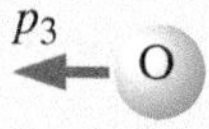

Figure 6.66 State 4: O nucleus and photon.

Solution

In this process there are four identifiable states:

- State 1: Alpha particle and carbon nucleus moving toward each other (Figure 6.63)
- State 2: Alpha particle and carbon nucleus just touch (Figure 6.64)
- State 3: Excited oxygen nucleus at rest (extra rest energy) (Figure 6.65)
- State 4: Oxygen nucleus and photon, heading in opposite directions (Figure 6.66)

(a) Find p_1.

System: All particles in existence (alpha + C, or excited O, or O + photon)
Surroundings: No external objects

Initial state: State 1, Figure 6.63, because it involves the unknown momentum
Final state: State 4, Figure 6.66, because particle energies are known

$$E_f = E_i + W$$
$$(m_O c^2 + K_O) + K_\gamma + U_f = (m_\alpha c^2 + K_\alpha) + (m_C c^2 + K_C) + U_i$$
$$m_O c^2 + 0 + K_\gamma + 0 = m_\alpha c^2 + m_C c^2 + \frac{p_1^2}{2m_\alpha} + \frac{p_1^2}{2m_C} + 0$$
$$p_1 = \sqrt{\frac{(m_O - m_\alpha - m_C)c^2 + K_\gamma}{\frac{1}{2m_\alpha} + \frac{1}{2m_C}}}$$

$U_f = 0$ because the photon has no charge. $K_O \approx 0$ since there is negligible recoil kinetic energy for this comparatively massive particle. $U_i = 0$ because particles are initially far apart.

We need to use many significant figures when solving for p_1, because the mass change is very small:

$$(m_O - m_\alpha - m_C) = (15.99052636 - 4.00040868 - 11.99670852)\,\text{u}$$
$$= -0.00659084\,\text{u}$$

$$(-0.00659084\,\text{u})(1.6605 \times 10^{-27}\frac{\text{kg}}{\text{u}})(3 \times 10^{8}\frac{\text{m}}{\text{s}})^2 = -9.8497 \times 10^{-13}\,\text{J}$$

$$K_\gamma = (10.352 \times 10^6 \text{eV})(1.60 \times 10^{-19} \text{eV/J}) = 16.5632 \times 10^{-13}\,\text{J}$$

$$p_1 = \sqrt{\frac{(16.5632 \times 10^{-13} - 9.8497 \times 10^{-13})\,\text{J}}{\frac{1}{2}\left(\frac{1}{4.000} + \frac{1}{11.997}\right)\left(\frac{1}{1.6605 \times 10^{-27}\,\text{kg}}\right)}}$$

$$p_1 = 8.18 \times 10^{-20}\,\text{kg} \cdot \text{m/s}$$

(b) Find p_2.
System: alpha particle and carbon nucleus.
Surroundings: No external objects.

Initial state: State 1, Figure 6.63, because it involves the unknown momentum
Final state: State 2, Figure 6.64, because this is the final state of interest

$$E_f = E_i + W$$

$$\cancel{m_a c^2} + K_{a,f} + \cancel{m_C c^2} + K_{C,f} + U_f = \cancel{m_a c^2} + K_{a,i} + \cancel{m_C c^2} + K_{C,i} + U_i + W$$

$$0 + 0 + \frac{1}{4\pi\varepsilon_0}\frac{(2e)(6e)}{r_f} = \frac{p_2^2}{2m_a} + \frac{p_2^2}{2m_C} + 0 + 0$$

$$p_2 = \sqrt{\frac{\frac{1}{4\pi\varepsilon_0}\frac{12e^2}{r_f}}{\frac{1}{2}\left(\frac{1}{m_a} + \frac{1}{m_C}\right)}}$$

There is no change of rest energies yet, so the rest energies cancel. In the final state the particles are at rest, just touching (remember that we want to know the minimum kinetic energy required to do this).

Final center-to-center distance:

$$r_f = (1.3 \times 10^{-15}\,\text{m})(4^{1/3} + 12^{1/3}) = 5.04 \times 10^{-15}\,\text{m}$$

$$p_2 = \sqrt{\frac{\left(9 \times 10^9 \frac{\text{N} \cdot \text{m}^2}{\text{C}^2}\right)\frac{12(1.6 \times 10^{-19}\text{C})^2}{(5.04 \times 10^{-15}\,\text{m})}}{\frac{1}{2}\left(\frac{1}{4} + \frac{1}{12}\right)\left(\frac{1}{1.66 \times 10^{-27}\,\text{kg}}\right)}}$$

$$p_2 = 7.39 \times 10^{-20}\,\text{kg} \cdot \text{m/s}$$

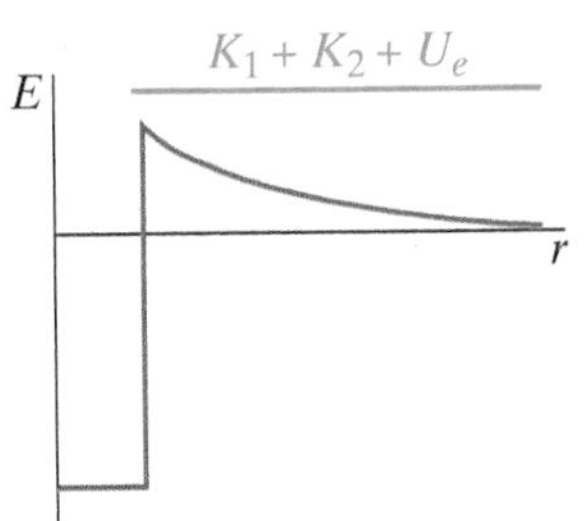

Figure 6.67 The initial kinetic energies are a bit greater than what is needed to overcome the electric potential energy barrier.

(c) The momentum required to touch is less than the momentum required to initiate the reaction.

The potential energy graph for this process is shown in Figure 6.67. The electric potential energy rises (due to a repulsive interaction) as the particles get closer together, up to the point at which they touch. After this, the attractive strong (nuclear) force dominates the interaction, and the electric repulsion between the particles is insignificant.

This kind of potential energy curve is sometimes described as having a "Coulomb barrier." This means that $K + U$ total must be larger than the maximum value of U_{elec}.

We found that the incoming momentum p_i required to create the excited oxygen nucleus is greater than the momentum p_2 required just to overcome the electric potential energy barrier and bring the alpha particle and carbon nucleus in contact. It is necessary to bring nuclei into contact for a reaction to proceed, since the nuclear force has a very short range. However, in this reaction just making contact is insufficient. A bit more energy is needed in order to produce the larger-mass oxygen nucleus.

We never did use state 3 in our analysis. However, if we needed to find the mass of the excited O* nucleus, we would use this state as either the final or initial state.

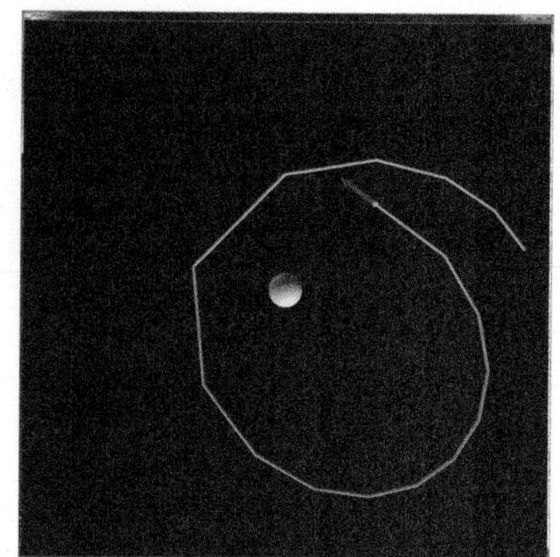

Figure 6.68 An elliptical orbit computed with a time step of two weeks.

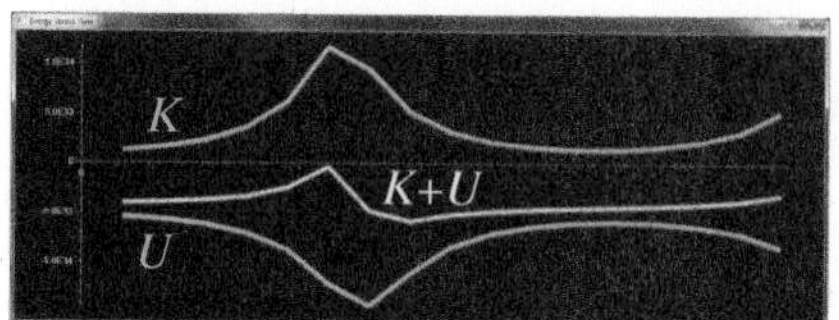

Figure 6.69 Energy versus time. Kinetic energy (cyan line), gravitational potential energy (green line), and the sum of $K+U$ (yellow line), as a function of time for the elliptical orbit shown in Figure 6.68 (Δt of two weeks).

6.15 ENERGY IN COMPUTATIONAL MODELS

In this chapter we have analyzed a wide variety of systems, involving objects as diverse as alpha particles, asteroids, and horses. We have been able to draw conclusions such as "When the objects are this far apart they will have that much kinetic energy," or "These objects do not have enough kinetic energy to get that far apart." However, we have not predicted the detailed trajectory of any moving object. We have put limits on what can happen, but we have not made detailed predictions of motion.

Because the Energy Principle involves neither direction nor time, it does not tell us which way objects are moving, or how long it will take them to get to a particular separation. However, energy can still play an important role in iterative computational models.

Monitoring Step Size

We have seen that if we use a time step size that is too large in an iterative model, we predict trajectories that don't look right. In such computational models, tracking the kinetic and potential energy of such a system can provide important quantitative information about the performance of the model.

Figure 6.68 shows the display from a computational model of a planet orbiting a star. This model iteratively invoked the Momentum Principle, using a Δt of two weeks. The trajectory is jagged and the orbit does not close. Figure 6.69 shows plots of K_{planet}, U_g, and $(K_{\text{planet}} + U_g)$ as a function of time, computed and plotted by the same program.

> QUESTION What is wrong with the energy plots shown in Figure 6.69?

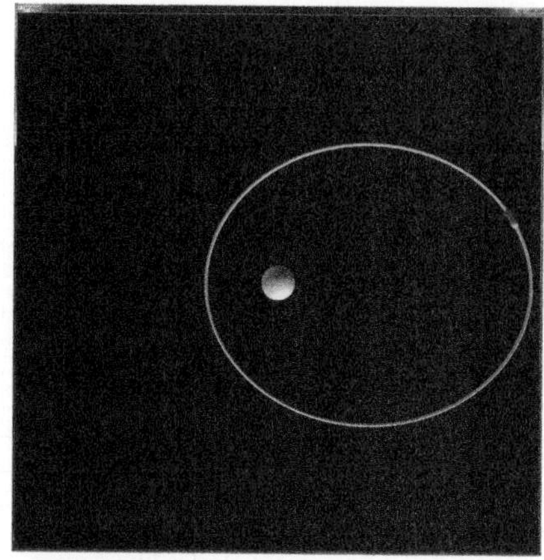

Figure 6.70 An elliptical orbit computed with a time step of one day.

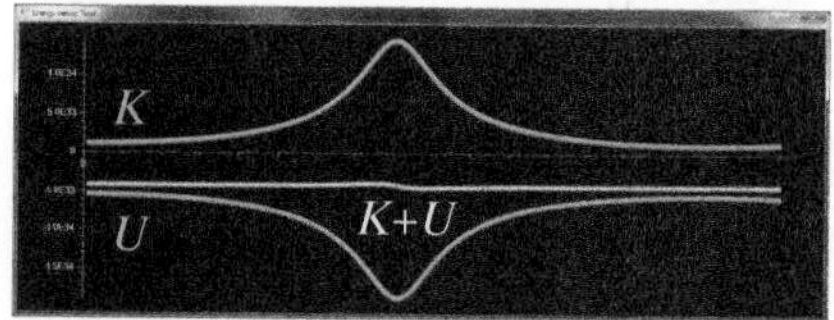

Figure 6.71 Kinetic energy (cyan line), gravitational potential energy (green line), and the sum of $K+U$ (yellow line), for an elliptical orbit computed with a time step of one day.

The graph of $K+U$ in Figure 6.69 shows an apparent violation of the Energy Principle. In the model there are no objects in the surroundings, and therefore the sum of $K+U$ for the system should be constant. The fact that $K+U$ is not constant tells us that our calculations are insufficiently accurate. This is important information that we can get even without knowing what the motion of the system should really look like.

When we watch the running program, we can see that the calculated sum of $K+U$ varies most when the planet is close to the star. Since the force is largest when the planet is closest, this is the region in which the assumption of constant force over the time Δt is most problematic for a long time step.

To confirm the hypothesis that too large a value for Δt is producing large errors in our calculations, we can try reducing the time step and re-running the program. Figure 6.70 shows the orbit predicted by the same model, using a Δt of one day instead of two weeks, and Figure 6.71 shows the associated plots of energy versus time. Although the trajectory now appears smooth, and the orbit closes, there is still a small wiggle in the graph of the sum of $K+U$. It is necessary to reduce the time step even further to get $K+U$ to remain constant, as shown in Figure 6.72, in which a time step of one hour was used.

Evidently monitoring the energy of a system in a computational model can provide a sensitive check on the accuracy of the computation.

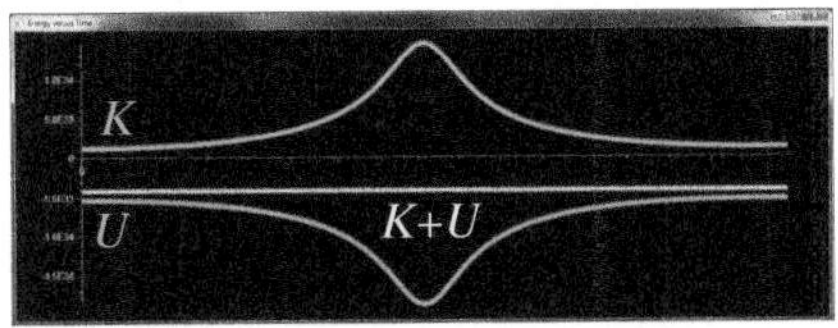

Figure 6.72 Kinetic energy (cyan line), gravitational potential energy (green line), and the sum of $K+U$ (yellow line), for an elliptical orbit computed with a time step of one hour.

Energy Graphs in VPython

Adding graphs of energy is straightforward in VPython. In the partial program below, line numbers have been added to make it easy to refer to particular lines. In addition to the usual calculations done in a program modeling motion, which are indicated by comments, these things are done:

- Line 02 imports the graph module.
- Lines 04–06 create and assign names to three gcurve objects, one for each plot desired.
- Each time through the loop, K and U are calculated (lines 13–14).
- Each time through the loop, the statements in lines 15–17 are executed to add a point to the graph of each quantity of interest.

```
01 from visual import *
02 from visual.graph import *
03 ## create some objects ...
04 kineticenergy = gcurve(color=color.cyan)
05 potentialenergy = gcurve(color=color.green)
06 kplusu = gcurve(color=color.yellow)
07 ## set initial values...
08 while t < 60*60*24*365:
09     rate(100)
10     ## calculate forces...
11     ## update momentum...
12     ## update position...
13     K = ## calculate kinetic energy
14     U = ## calculate potential energy
15     kineticenergy.plot( pos=(t,K) )
16     potentialenergy.plot( pos=(t,U) )
17     kplusu.plot( pos=(t,K+U) )
18     t = t + delta_t
```

6.16 *A PUZZLE

The principle of conservation of energy states that when a system's energy increases, the energy of the surroundings decreases by the same amount. For example, take a block as the system of interest. If you push the block away from you and raise its kinetic energy, your store of chemical energy decreases by the same amount. The flow of energy from you (the surroundings) to the block takes the form of mechanical energy transfer we call "work." We can demonstrate that there is a puzzle concerning the energy in the surroundings of a star whose kinetic energy increases due to an interaction with another star.

Consider two stars of equal mass initially at rest, far apart (Figure 6.73). The two stars do work on each other:

Figure 6.73 Two stars with equal mass attract each other, starting from rest.

$$\Delta E_1 = W_{\text{on }1} \quad \text{star 1 as system}$$
$$\Delta E_2 = W_{\text{on }2} \quad \text{star 2 as system}$$

If we add these two energy equations for the two stars, we have this:

$$\Delta E_1 + \Delta E_2 = W_{\text{on }1} + W_{\text{on }2} \quad \text{(sum of two energy equations)}$$

Moving $(W_{\text{on}\,1} + W_{\text{on}\,2})$ to the left side of the equation and renaming the negative of this quantity ΔU, we have this familiar result:

$$\Delta E_1 + \Delta E_2 + \Delta U = 0 \quad \text{(energy equation for two stars)}$$

Our interpretation of this equation is that the gravitational potential energy U of the two-star system (calculated from the work on the individual particles) decreases as the particle energies increase. This point of view is valid, yields physically correct results, and is the view that we will take in this introductory textbook.

However, it is not difficult to show that there is a puzzle concerning energy flow between one of the stars and its surroundings. Consider star 1 as the system. Since it is a single object, there is no potential energy U, no pair-wise interaction energy. The energy of star 1 increases due to work done on it: $\Delta E_1 = E_{\text{on}\,1}$. There are very strong reasons to believe in energy conservation, so we expect that when star 1 gains an amount of energy ΔE_1 the surroundings of star 1 should lose that much energy.

> **QUESTION** Part of the surroundings is star 2. Does star 2 lose energy?

No! Star 2 doesn't lose energy, it *gains* energy, and of the same amount $\Delta E_2 = \Delta E_1$, due to the symmetry of the situation, with the two stars having equal mass.

What else is there in the surroundings that could experience a change in energy? The only other matter is star 2, but its energy change is inconsistent with the principle of energy conservation. Evidently our model of the world is incomplete. To be able to analyze this situation fully, we need to introduce the abstract idea of a "field." Fields, which are the subject of later chapters of this textbook, are associated with objects having mass (gravitational fields) or charge (electric and magnetic fields), and fields extend throughout all space. It is energy stored in the gravitational field that accounts for the energy in the surroundings of star 1. Fields can also have momentum, as we will see in the chapters that deal with electric and magnetic interactions.

Despite the fact that we haven't yet defined what a field is, we can determine indirectly how much the field energy must change. Since the energy in the surroundings of star 1 changes by an amount $-\Delta E_1$, and the energy of star 2 increases by an amount $+\Delta E_1$, it must be that the energy stored in the gravitational field changes by an amount $-2\Delta E_1$. It is an indication of the power of the principle of conservation of energy that we can calculate the energy change of an entity that we know nothing about! This energy change is precisely the amount of change in the potential energy of the two-star system: $\Delta U = -(\Delta E_1 + \Delta E_2) = -2\Delta E_1$.

Because the potential energy U correctly accounts for the energy changes, we don't have to consider field energy in our calculations. Our analyses in terms of U that focus on particles and their interactions, not their surroundings, are correct. However, the difficulty in accounting for the energy in the surroundings of a star suggests that eventually we will need to include fields in our models of the world.

6.17 *GRADIENT OF POTENTIAL ENERGY

Because the (negative) gradient of U is equal to the force, the force is the (negative) slope of a graph of potential energy. This means that you can mentally read the force right off a potential energy graph (Figure 6.74). Where the curve is steep, the force is large (large gradient), and where the curve is shallow, the force is small. Where the slope of the curve is horizontal, the force is zero, and this represents a possible equilibrium position for the system. The component of the force is given by the negative gradient, so a negative slope corresponds to a force to the right, while a positive slope corresponds to a force to the left.

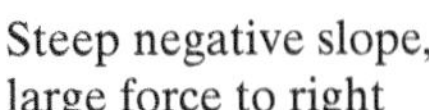
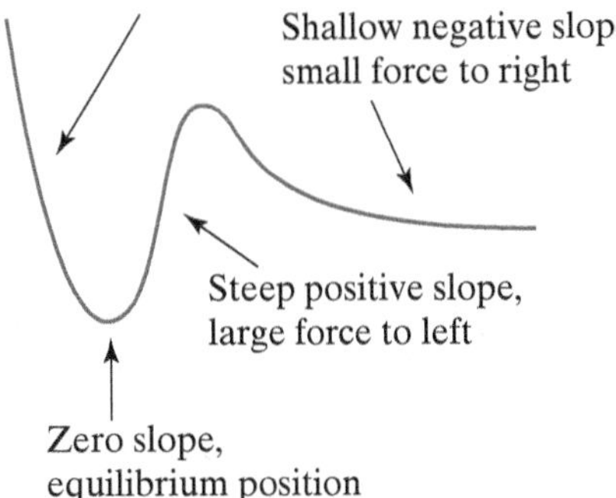

Figure 6.74 The negative of the slope (gradient) of a graph of potential energy vs. separation gives the force.

Actually, in determining the x component of the force, we differentiate with respect to x while holding y and z constant. This kind of derivative is called a "partial derivative" and has a special symbol to emphasize that everything else is held constant:

$$F_x = -\frac{\partial U}{\partial x}$$

To get all three components of the force we take partial derivatives of the potential energy with respect to x, y, and z:

FORCE IS NEGATIVE GRADIENT OF POTENTIAL ENERGY

$$F_x = -\frac{\partial U}{\partial x}, F_y = -\frac{\partial U}{\partial y}, F_z = -\frac{\partial U}{\partial z}$$

$$\vec{F} = \left\langle -\frac{\partial U}{\partial x}, -\frac{\partial U}{\partial y}, -\frac{\partial U}{\partial z} \right\rangle$$

A standard mathematical notation for this relation is $\vec{F} = -\vec{\nabla} U$.

Checkpoint 18 With y pointing upward from the center of the Earth, find the y component of the gravitational force from $-\partial U/\partial y$, where $U = -GMm/y$. Find the x component of the force from $-\partial U/\partial x$.

6.18 *INTEGRALS AND ANTIDERIVATIVES

In determining the expression for gravitational potential energy, we started from these complementary expressions:

$$\Delta U_g = -F_r \Delta r \quad \text{and} \quad F_r = -\frac{dU_g}{dr}$$

We calculated the change in potential energy from the negative of the internal work:

$$U_{s,f} - U_{s,i} = \Delta\left(-G\frac{m_1 m_2}{r}\right) = -\Sigma F_r \Delta r$$

However, if we had let Δr approach zero, the finite sum would have turned into an integral:

$$\Delta\left(-G\frac{m_1 m_2}{r}\right) = -\int_i^f F_r dr = \int_i^f G\frac{m_1 m_2}{r^2} dr$$

You can see that a definite integral (the infinitesimal version of a finite sum) can be evaluated in terms of the antiderivative. The antiderivative of Gm_1m_2/r^2 is $-Gm_1m_2/r$, and the change in the value of the antiderivative turns out to be equal to the definite integral (the infinite sum of infinitesimal elements).

The reason why this works this way can be traced back to the fact that the force is the derivative of the energy, so the energy must be the antiderivative of the force, yet the energy is also the sum of lots of $F_r\Delta r$'s. Hence the sum is given by the change in the antiderivative.

When the quantity to be integrated (here, the force) is a simple function, it may be easy to find the antiderivative. However, in many practical cases there is no known simple function that is the antiderivative, in which case an integral must be evaluated by numerical integration, as a finite sum.

6.19 *APPROXIMATION FOR KINETIC ENERGY

An approximation for the kinetic energy of a particle at low speeds can be obtained through use of the mathematics of the "binomial expansion":

$$(1+\varepsilon)^n = 1 + n\varepsilon + \frac{n(n-1)}{2}\varepsilon^2 + \frac{n(n-1)(n-2)}{3\cdot 2}\varepsilon^3 + \cdots$$

This expression comes from working through what happens when you multiply the quantity $(1+\varepsilon)$ times itself n times. The first term is 1 multiplied by itself n times (1^n); the next term is made up of all those products, n of them, in which ε appears only once [$n\varepsilon$]; and so on.

Now for the important point that makes the binomial expansion useful in obtaining approximate results. If ε is a number that is very small compared to 1, each term in the expansion is much smaller than the preceding term:

$$n\varepsilon \text{ is much smaller than } 1, \frac{n(n-1)}{2}\varepsilon^2 \text{ is much smaller than } n\varepsilon, \text{ etc.}$$

Therefore we can get a good approximation if we keep just the first few terms and ignore the rest.

Although we have developed the binomial expansion for the case where n is an integer, it can be shown that the expression is valid even for noninteger values of n. In particular,

$$\frac{1}{\sqrt{1-\varepsilon}} = (1-\varepsilon)^{-1/2}$$
$$= 1 + \left(-\frac{1}{2}\right)(-\varepsilon) + \frac{(-\frac{1}{2})(-\frac{3}{2})}{2}(-\varepsilon)^2 + \frac{(-\frac{1}{2})(-\frac{3}{2})(-\frac{5}{2})}{6}(-\varepsilon)^3 + \cdots$$

If v/c is small compared to 1, then $(v/c)^2$ is even smaller compared to 1. For example, 1/10 is small compared to one, but $(1/10)^2 = 1/100$ is very small indeed. Therefore we can write this for low speeds:

$$\frac{mc^2}{\sqrt{1-(v/c)^2}} = mc^2\left[1 + \frac{1}{2}\left(\frac{v}{c}\right)^2 + \frac{3}{8}\left(\frac{v}{c}\right)^4 + \frac{5}{16}\left(\frac{v}{c}\right)^6 + \cdots\right]$$
$$\frac{mc^2}{\sqrt{1-(v/c)^2}} \approx mc^2 + \frac{1}{2}mv^2$$

The last term can be rewritten in terms of momentum by inserting $v \approx p/m$ into $K \approx \frac{1}{2}mv^2$:

$$K \approx \frac{1}{2}mv^2 \approx \frac{1}{2}m\left(\frac{p}{m}\right)^2 = \frac{p^2}{2m}$$

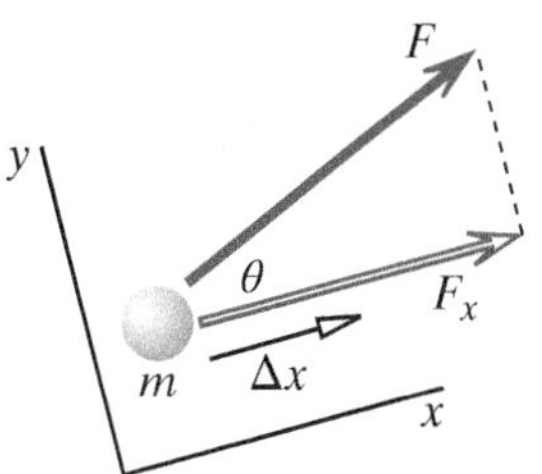

Figure 6.75 Choose the x axis to be in the direction of the motion. The net force acting on the particle has a component F_x in the direction of the motion.

6.20 *FINDING THE EXPRESSION FOR PARTICLE ENERGY

If we choose our axes so that x is in the direction of the motion at this instant (Figure 6.75) we have

$$\Delta E = F_x \Delta x = \left(\frac{\Delta p_x}{\Delta t}\right)\Delta x$$

$$\text{so } \frac{\Delta E}{\Delta x} = \frac{\Delta p_x}{\Delta t}$$

for the change in energy of a point particle acted on by a net force. In the limit of small displacement and short time interval, the energy E should satisfy the following relationship between two derivatives:

$$\frac{dE}{dx} = \frac{dp_x}{dt}$$

We will show that if we choose the following function for E, the relationship between the two derivatives will be satisfied:

$$E = \frac{mc^2}{\sqrt{1-(v/c)^2}}$$

The proof proceeds by calculating dE/dx, calculating dp_x/dt, and showing that the two derivatives are equal to each other. First, here is dE/dx:

$$\frac{dE}{dx} = \frac{d}{dx}\left(\frac{mc^2}{\sqrt{1-(v/c)^2}}\right)$$

$$= \frac{mc^2}{(1-(v/c)^2)^{3/2}}\left(-\frac{1}{2}\right)\left(-\frac{2v}{c^2}\right)\frac{dv}{dx}$$

$$\frac{dE}{dx} = \frac{mv}{(1-(v/c)^2)^{3/2}}\frac{dv}{dx}$$

Next we calculate dp_x/dt, keeping in mind that $v_x = v$, because we chose our axes so that x is in the direction of the motion at this instant:

$$\frac{dp_x}{dt} = \frac{d}{dt}\left(\frac{mv}{\sqrt{1-(v/c)^2}}\right)$$

$$= \frac{m}{\sqrt{1-(v/c)^2}}\frac{dv}{dt} + \frac{mv}{(1-(v/c)^2)^{3/2}}\left(-\frac{1}{2}\right)\left(-\frac{2v}{c^2}\right)\frac{dv}{dt}$$

$$\frac{dp_x}{dt} = \frac{m}{(1-(v/c)^2)^{3/2}}\left[\left(1-\frac{v^2}{c^2}\right)+\frac{v^2}{c^2}\right]\frac{dv}{dt}$$

$$= \frac{m}{(1-(v/c)^2)^{3/2}}\frac{dv}{dt}$$

This is almost in the same form as the derivative dE/dx. Now note this:

$$\frac{dv}{dt} = \left(\frac{dv}{dt}\right)\left(\frac{dx}{dx}\right)$$

$$= \left(\frac{dv}{dx}\right)\left(\frac{dx}{dt}\right) = \frac{dv}{dx}v \quad \text{because } v = \frac{dx}{dt}$$

Substituting this result for dv/dt into the derivative dp_x/dt, we find that this derivative is indeed equal to the derivative dE/dx if E is defined as

$$E = \frac{mc^2}{\sqrt{1-(v/c)^2}}$$

Hence the relativistic expression for energy does follow from the relativistic expression for momentum.

It might seem that we could define the particle energy with an additive constant, which would drop out when we took the derivative dE/dx. However, when one considers a variety of particle reactions, it is found that to account for energy in a consistent way requires that the additive constant be zero.

This approach to solving a "differential equation" (an equation containing derivatives), by proposing a solution and showing that it satisfies the equation, is a legitimate technique for solving differential equations, and in fact is used quite frequently.

6.21 *FINDING AN ANGLE FROM THE DOT PRODUCT

The alternative forms of the dot product offer a way to determine the angle between any two given vectors $\vec{a} = \langle a,b,c\rangle$ and $\vec{q} = \langle q,r,s\rangle$.

$$\vec{a}\bullet\vec{q} = (aq+br+cs) \quad \text{and also} \quad \vec{a}\bullet\vec{q} = |\vec{a}||\vec{q}|\cos\theta$$

So $\cos\theta = \dfrac{(aq+br+cs)}{|\vec{a}||\vec{q}|}$, from which θ can be determined.

SUMMARY

The unit of energy is the joule, abbreviated J.

Energy of a single particle
The energy of a particle with mass m is:

$$E_{\text{particle}} \equiv \gamma mc^2 \quad \text{where } \gamma = \frac{1}{\sqrt{1-(v/c)^2}}$$

Particle energy has two parts: rest energy and kinetic energy:

$$E_{\text{particle}} = mc^2 + K$$

The rest energy is the energy of a particle at rest.

$$E_{\text{rest}} = mc^2$$

The kinetic energy K of a particle is the energy a moving particle has in addition to its rest energy.

$$K = \gamma mc^2 - E_{\text{rest}}$$

For a particle whose speed is low compared to the speed of light, kinetic energy is approximately:

$$K \approx \frac{1}{2}mv^2 = \frac{p^2}{2m} \quad (\text{if } v \ll c)$$

For a particle with any speed:

$$E^2 - (pc)^2 = (mc^2)^2$$

For a massless particle this reduces to

$$E = pc$$

The Energy Principle

$$\Delta E_{\text{sys}} = W_{\text{surr}}$$

The change in energy of a system is equal to the work done by the surroundings.

Work
Work W_F done by a single constant force $\vec{F}$:

$$W_F = \vec{F}\bullet\Delta\vec{r} = F_x\Delta r_x + F_y\Delta r_y + F_z\Delta r_z$$
$$= F\Delta r\cos\theta$$

Work done by a varying force:

$$W_F = \Sigma\vec{F}\bullet\Delta\vec{r} \quad \text{or} \quad W = \int_i^f \vec{F}\bullet d\vec{r}$$

Work can be positive, negative, or zero. The sign of work depends on the relative directions of the force and the displacement of the point of application of the force.

Conservation of energy

$$\Delta E_{sys} + \Delta E_{surr} = 0$$

Potential energy
Multiparticle system energy consists of particle energy and potential energy:

$$E_{sys} \equiv (K_1 + m_1c^2) + (K_2 + m_2c^2) + \cdots + (U_{12} + \cdots)$$

Potential energy U is the energy of interaction of a pair of objects, and change in U is the negative of work done by internal forces:

$$\Delta U = -W_{int}$$

Force is the negative gradient of potential energy:

$$F_x = -\frac{dU}{dx}$$

Potential energy U depends on the separation r between two objects.
Potential energy U goes to zero as objects get very far apart:

$$U \to 0 \quad \text{as } r \to \infty$$

For an attractive interaction, U becomes negative as separation decreases.

For a repulsive interaction, U becomes positive as separation decreases.
The potential energy change between two states is independent of the path taken. The change in potential energy in a round trip is zero.

Gravitational potential energy

$$U = -G\frac{m_1 m_2}{r}$$

$$G = 6.7 \times 10^{-11} \text{N} \cdot \text{m}^2/\text{kg}^2$$

m_1 and m_2 are the masses of the two interacting objects, and r is the center-to-center distance between them.

$$\Delta U \approx \Delta(mgy) \quad \text{near Earth's surface}$$

Electric potential energy

$$U_{elec} = \frac{1}{4\pi\varepsilon_0}\frac{q_1 q_2}{r}$$

$$\frac{1}{4\pi\varepsilon_0} = 9 \times 10^9 \text{ N} \cdot \text{m}^2/\text{C}^2$$

q_1 and q_2 are the charges of the two interacting particles, in coulombs, and r is the center-to-center distance between them.

Energy graphs are highly useful tools.

Nuclear binding
Heavy nuclei can fission, light nuclei can fuse, with energy released in both cases.

$$\text{Electron volt: } 1\,\text{eV} = 1.6 \times 10^{-19}\,\text{J}$$

QUESTIONS

Q1 Show that the units of $p^2/(2m)$ and mc^2 are indeed joules. (Note that 1 N is $1\,\text{kg} \cdot (\text{m/s})^2$.)

Q2 Give brief explanations for your answers to each of the following questions: **(a)** You hold a 1 kg book in your hand for 1 min. How much work do you do on the book? **(b)** In a circular pendulum, how much work is done by the string on the mass in one revolution? **(c)** For a mass oscillating horizontally on a spring, how much work is done by the spring on the mass in one complete cycle? In a half cycle?

Q3 You pull a block of mass m across a frictionless table with a constant force. You also pull with an equal constant force a block of larger mass M. The blocks are initially at rest. If you pull the blocks through the same distance, which block has the greater kinetic energy, and which block has the greater momentum? If instead you pull the blocks for the same amount of time, which block has the greater kinetic energy, and which block has the greater momentum?

Q4 One often hears the statement, "Nuclear energy production is fundamentally different from chemical energy production (such as burning of coal) because the nuclear case involves a change of mass." Critique this statement. Discuss the similarities and differences of the two kinds of energy production.

Q5 Figure 6.76 shows the path of a comet orbiting a star.

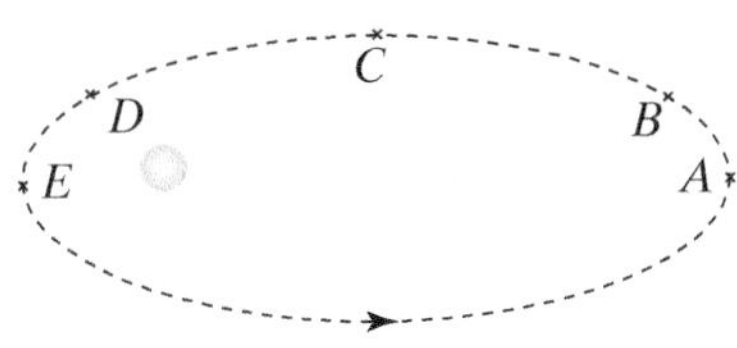

Figure 6.76

(a) Rank-order the locations on the path in terms of the magnitude of the comet's momentum at each location, starting with the location at which the magnitude of the momentum is the largest. **(b)** Rank-order the locations on the path in terms of the comet's kinetic energy at each location, starting with the location at which the kinetic energy is the largest. **(c)** Rank-order the locations on the path in terms of the potential energy of the system at each location, largest first. **(d)** Rank-order the locations on the path in terms of the sum of the kinetic energy and the potential energy of the system at each location, largest first.

Q6 Figure 6.77 is a graph of the energy of a system of a planet interacting with a star. The gravitational potential energy U_g is shown as the thick curve, and plotted along the vertical axis are various values of $K + U_g$.

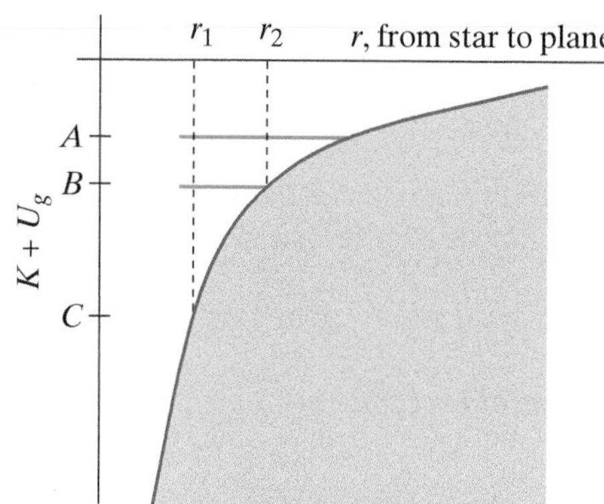

Figure 6.77

Suppose that $K + U_g$ of the system is A. Which of the following statements are true? **(a)** The potential energy of the system decreases as the planet moves from r_1 to r_2. **(b)** When the separation between the two bodies is r_2, the kinetic energy of the system is $(A - B)$. **(c)** The system is a bound system; the planet can never escape. **(d)** The planet will escape. **(e)** When the separation between the two bodies is r_2, the kinetic energy of the system is $(B - C)$. **(f)** The kinetic energy of the system is greater when the distance between the star and planet is r_1 than when the distance between the two bodies is r_2.

Suppose instead that $K + U_g$ of the system is B. Which of the following statements are true? **(a)** When the separation between the planet and star is r_2, the kinetic energy of the system is zero. **(b)** The planet and star cannot get farther apart than r_2. **(c)** This is not a bound system; the planet can escape. **(d)** When the separation between the planet and star is r_2, the potential energy of the system is zero.

Q7 A particle moves inside a circular glass tube under the influence of a tangential force of constant magnitude F (Figure 6.78). Explain why we cannot associate a potential energy with this force. How is this situation different from the case of a block on the end of a string, which is swung in a circle?

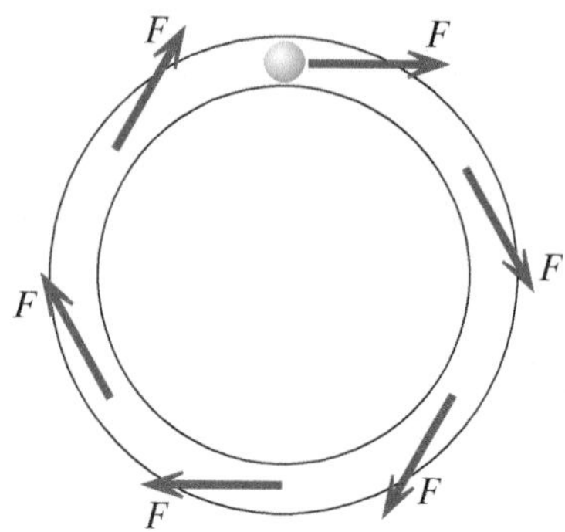

Figure 6.78

Q8 Show the validity of the relation $E^2_{\text{particle}} - (pc)^2 = (mc^2)^2$ when $m \neq 0$, by making these substitutions:

$$E_{\text{particle}} = \frac{mc^2}{\sqrt{1-(v/c)^2}} \quad \text{and} \quad p = \frac{mv}{\sqrt{1-(v/c)^2}}$$

PROBLEMS

Section 6.2

•P9 It is not very difficult to accelerate an electron to a speed that is 99% of the speed of light, because the electron has such a very small mass. What is the ratio of the kinetic energy K to the rest energy mc^2 in this case? In the definition of what we mean by kinetic energy K, $E = mc^2 + K$, you must use the full relativistic expression for E, because v/c is not small compared to 1.

•P10 A pitcher can throw a baseball at about 100 mi/h (about 44 m/s). What is the ratio of the kinetic energy to the rest energy mc^2? (Can you use $K \approx \frac{1}{2}mv^2$?)

•P11 What is the speed of an electron whose total energy is equal to the total energy of a proton that is at rest? What is the kinetic energy of this electron?

•P12 The point of this question is to compare rest energy and kinetic energy at low speeds. A baseball is moving at a speed of 17 m/s. Its mass is 145 g (0.145 kg). **(a)** What is its rest energy? **(b)** Is it okay to calculate its kinetic energy using the expression $\frac{1}{2}mv^2$? **(c)** What is its kinetic energy? **(d)** Which is true? A. The kinetic energy is approximately equal to the rest energy. B. The kinetic energy is much bigger than the rest energy. C. The kinetic energy is much smaller than the rest energy.

•P13 The point of this question is to compare rest energy and kinetic energy at high speeds. An alpha particle (a helium nucleus) is moving at a speed of 0.9993 times the speed of light. Its mass is 6.40×10^{-27} kg. **(a)** What is its rest energy? **(b)** Is it okay to calculate its kinetic energy using the expression $\frac{1}{2}mv^2$? **(c)** What is its kinetic energy? **(d)** Which is true? A. The kinetic energy is approximately equal to the rest energy. B. The kinetic energy is much bigger than the rest energy. C. The kinetic energy is much smaller than the rest energy.

•P14 A runner whose mass is 60 kg runs in the $+x$ direction at a speed of 7 m/s. **(a)** What is the kinetic energy of the runner? **(b)** The runner turns around and runs in the $-x$ direction at the same speed. Now what is the kinetic energy of the runner?

•P15 A baseball of mass 144 g has a velocity of $\langle 22, 23, -11 \rangle$ m/s. What is the kinetic energy of the baseball?

•P16 One mole of helium atoms has a mass of 4 grams. If a helium atom in a balloon has a kinetic energy of 1.437×10^{-21} J, what is the speed of the helium atom? (The speed is much lower than the speed of light.)

•P17 You throw a ball of mass 160 g upward (Figure 6.79). When the ball is 2 m above the ground, headed upward (the initial state), its speed is 19 m/s. Later, when the ball is again 2 m above the ground, this time headed downward (the final state), its speed is 19 m/s. What is the change in the kinetic energy of the ball from initial to final state?

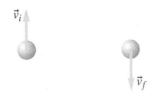

Figure 6.79

•**P18** A fan cart of mass 0.8 kg initially has a velocity of $\langle 0.9,0,0\rangle$ m/s. Then the fan is turned on, and the air exerts a constant force of $\langle -0.3,0,0\rangle$ N on the cart for 1.5 s. **(a)** What is the change in momentum of the fan cart over this 1.5 s interval? **(b)** What is the change in kinetic energy of the fan cart over this 1.5 s interval?

Section 6.3

•**P19** You push a crate 3 m across the floor with a 40 N force whose direction is 30° below the horizontal. How much work do you do?

•**P20** You pull your little sister across a flat snowy field on a sled. Your sister plus the sled have a mass of 20 kg. The rope is at an angle of 35 degrees to the ground. You pull a distance of 50 m with a force of 30 N. How much work do you do?

•**P21** A 2-kg ball rolls off a 30-m-high cliff, and lands 25 m from the base of the cliff. Express the displacement and the gravitational force in terms of vectors and calculate the work done by the gravitational force. Note that the gravitational force is $\langle 0,-mg,0\rangle$, where g is a *positive* number (+9.8 N/kg).

•**P22** A boat is coasting toward a dock you're standing on, and as it comes toward you, you push back on it with a force of 300 N. As you do this, you back up a distance of 2 m. How much work do you do on the boat?

•**P23** A jar of honey with a mass of 0.5 kg is nudged off the kitchen counter and falls 1 m to the floor. What force acts on the jar during its fall? How much work is done by this force?

•**P24** An object that is originally at location $\langle -17,0,0\rangle$ m moves to location $\langle -25,0,0\rangle$ m as shown in Figure 6.80. While it is moving it is acted on by a constant force of $\langle 22,0,0\rangle$ N.

Figure 6.80

(a) How much work is done on the object by this force? **(b)** Does the kinetic energy of the object increase or decrease?

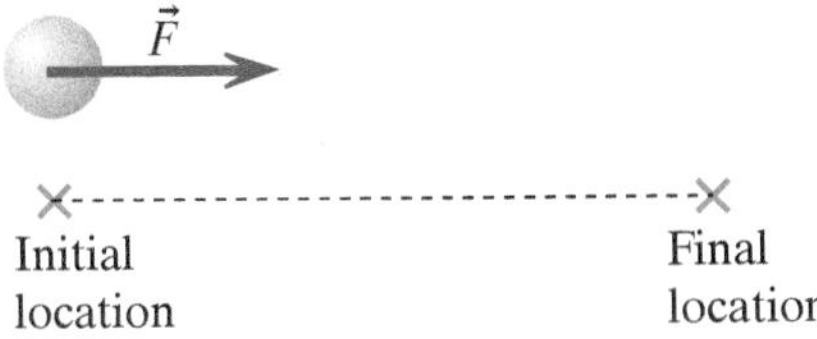

Figure 6.81

(c) A different object moves from location $\langle -25,0,0\rangle$ m to location $\langle -17,0,0\rangle$ m, as shown in Figure 6.81. While it is moving it is acted on by a constant force of $\langle 22,0,0\rangle$ N. How much work is done on the second object by this force? **(d)** Does the kinetic energy of the object increase or decrease?

•**P25** A constant force $\langle 23,-12,32\rangle$ N acts through a displacement $\langle 0.12,0.31,-0.24\rangle$ m. How much work does this force do?

•**P26** One end of a spring whose spring constant is 20 N/m is attached to the wall, and you pull on the other end, stretching it from its equilibrium length of 0.2 m to a length of 0.3 m. Estimate the work done by dividing the stretching process into two stages and using the average force you exert to calculate work done during each stage.

•**P27** An electron traveling through a curving wire in an electric circuit experiences a constant force of 5×10^{-19} N, always in the direction of its motion through the wire. How much work is done on the electron by this force as it travels through 0.5 m of the wire?

•**P28** You bring a boat toward the dock by pulling on a rope with a force of 130 N through a distance of 6 m. **(a)** How much work do you do? (Include the appropriate sign.) **(b)** Then you slow the boat down by pushing against it with a force of 40 N, opposite to the boat's movement of 5 m. How much work do you do? (Include the appropriate sign.) **(c)** What is the total amount of work that you do?

••**P29** You push a box out of a carpeted room and along a hallway with a waxed linoleum floor. While pushing the crate 2 m out of the room you exert a force of 34 N; while pushing it 6 m along the hallway you exert a force of 13 N. To slow it down you exert a force of 40 N through a distance of 2 m, opposite to the motion. How much work do you do in all?

Section 6.4

•**P30** A ball of mass 0.7 kg falls downward, as shown in Figure 6.82. Initially you observe it to be 4.5 m above the ground. After a short time it is just about to hit the ground.

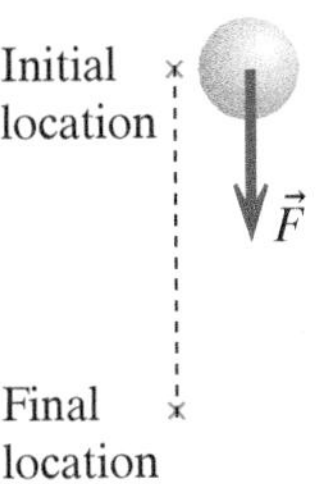

Figure 6.82

(a) During this interval how much work was done on the ball by the gravitational force? **(b)** Does the kinetic energy of the ball increase or decrease?

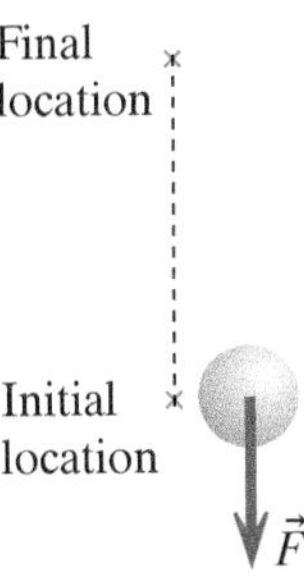

Figure 6.83

(c) The ball hits the ground and bounces back upward, as shown in Figure 6.83. After a short time it is 4.5 m above the ground again. During this second interval (between leaving the ground and reaching a height of 4.5 m) how much work was done on the ball by the gravitational force? **(d)** Does the kinetic energy of the ball increase or decrease?

•P31 A lithium nucleus has mass 5.1×10^{-27} kg. If its speed is 0.984c (that is, $|\vec{v}|/c = 0.984$), what are the values of the particle energy, the rest energy, and the kinetic energy?

Next an electric force acts on the lithium nucleus and does 4.7×10^{-9} J of work on the particle. Now what are the values of the particle energy, the rest energy, and the kinetic energy?

•P32 A space probe in outer space has a mass of 111 kg, and it is traveling at a speed of 29 m/s. When it is at location $\langle 445, 535, -350 \rangle$ m it begins firing two booster rockets. The rockets exert constant forces of $\langle 90, 150, 195 \rangle$ N and $\langle 90, -90, -585 \rangle$ N, respectively. The rockets fire until the spacecraft reaches location $\langle 449, 541, -354 \rangle$ m. Now what is its speed? There is negligible mass loss due to the rocket exhaust.

••P33 An object with mass 100 kg moved in outer space. When it was at location $\langle 9, -24, -4 \rangle$ its speed was 3.5 m/s. A single constant force $\langle 250, 400, -170 \rangle$ N acted on the object while the object moved from location $\langle 9, -24, -4 \rangle$ m to location $\langle 15, -17, -8 \rangle$ m. Then a different single constant force $\langle 140, 250, 150 \rangle$ N acted on the object while the object moved from location $\langle 15, -17, -8 \rangle$ m to location $\langle 19, -24, -3 \rangle$ m. What is the speed of the object at this final location?

••P34 Outside the space shuttle, you and a friend pull on two ropes to dock a satellite whose mass is 700 kg. The satellite is initially at position $\langle 3.5, -1, 2.4 \rangle$ m and has a speed of 4 m/s. You exert a force $\langle -400, 310, -250 \rangle$ N. When the satellite reaches the position $\langle 7.1, 3.2, 1.2 \rangle$ m, its speed is 4.01 m/s. How much work did your friend do?

••P35 A crate with a mass of 100 kg glides through a space station with a speed of 3.5 m/s. An astronaut speeds it up by pushing on it from behind with a force of 180 N, continually pushing with this force through a distance of 6 m. The astronaut moves around to the front of the crate and slows the crate down by pushing backward with a force of 170 N, backing up through a distance of 4 m. After these two maneuvers, what is the speed of the crate?

••P36 An object with mass 120 kg moved in outer space from location $\langle 11, -24, -5 \rangle$ to location $\langle 17, -28, -10 \rangle$ m. A single constant force $\langle 250, 440, -220 \rangle$ N acted on the object while the object moved. Its final speed was 7.9 m/s. What was the speed of the object at the initial location?

••P37 An object with mass 120 kg moved in outer space. When it was at location $\langle 7, -20, -8 \rangle$ its speed was 12 m/s. A single constant force $\langle 250, 490, -160 \rangle$ N acted on the object while the object moved to location $\langle 10, -29, -13 \rangle$ m. What is the speed of the object at this final location?

••P38 Jack and Jill are maneuvering a 3000 kg boat near a dock. Initially the boat's position is $\langle 2, 0, 3 \rangle$m and its speed is 1.3 m/s. As the boat moves to position $\langle 4, 0, 2 \rangle$m, Jack exerts a force $\langle -400, 0, 200 \rangle$ N and Jill exerts a force $\langle 150, 0, 300 \rangle$ N. **(a)** How much work does Jack do? **(b)** How much work does Jill do? **(c)** Without doing any calculations, say what is the angle between the (vector) force that Jill exerts and the (vector) velocity of the boat. Explain briefly how you know this. **(d)** Assuming that we can neglect the work done by the water on the boat, what is the final speed of the boat?

••P39 An electron traveling at a speed 0.99c encounters a region where there is a constant electric force directed opposite to its momentum. After traveling 3 m in this region, the electron's speed was observed to decrease to 0.93c. What was the magnitude of the electric force acting on the electron?

••P40 A mass of 0.12 kg hangs from a vertical spring in the lab room. You grab hold of the mass and throw it vertically downward. The speed of the mass just after leaving your hand is 3.40 m/s. **(a)** While the mass moved downward a distance of 0.07 m, how much work was done on the mass by the Earth? **(b)** The speed of the mass has decreased to 2.85 m/s. How much work was done on the mass by the spring?

••P41 An electron (mass 9×10^{-31} kg) is traveling at a speed of 0.91c in an electron accelerator. An electric force of 1.6×10^{-13} N is applied in the direction of motion while the electron travels a distance of 2 m. You need to find the new speed of the electron. Which of the following steps must be included in your solution to this problem? **(a)** Calculate the initial particle energy $\gamma_i mc^2$ of the electron. **(b)** Calculate the final particle energy $\gamma_f mc^2$ of the electron. **(c)** Determine how much time it takes to move this distance. **(d)** Use the expression $\frac{1}{2}m|\vec{v}|^2$ to find the kinetic energy of the electron. **(e)** Calculate the net work done on the electron. **(f)** Use the final energy of the electron to find its final speed.

What is the new speed of the electron as a fraction of c?

••P42 SLAC, the Stanford Linear Accelerator Center, located at Stanford University in Palo Alto, California, accelerates electrons through a vacuum tube 2 mi long (it can be seen from an overpass of the Junipero Serra freeway that goes right over the accelerator). Electrons that are initially at rest are subjected to a continuous force of 2×10^{-12} N along the entire length of two mi (1 mi is 1.6 km) and reach speeds very near the speed of light. A similar analysis in a previous chapter required numerical integration, but with the new techniques of this chapter you can analyze the motion analytically. **(a)** Calculate the final energy, momentum, and speed of the electron. **(b)** Calculate the approximate time required to go the 2 mi distance.

Section 6.5

•P43 In positron-emission tomography (PET) used in medical research and diagnosis, compounds containing unstable nuclei that emit positrons are introduced into the brain, destined for a site of interest in the brain. When a positron is emitted, it goes only a short distance before coming nearly to rest. It forms a bound state with an electron, called "positronium," which is rather similar to a hydrogen atom. The binding energy of positronium is very small compared to the rest energy of an electron. After a short time the positron and electron annihilate. In the annihilation, the positron and the electron disappear, and all of their rest energy goes into two photons (particles of light) that have zero mass; all their energy is kinetic energy. These high-energy photons, called "gamma rays," are emitted at nearly 180° to each other. What energy of gamma ray (in MeV, million electron volts) should each of the detectors be made sensitive to? (The mass of an electron or positron is 9×10^{-31} kg.)

••P44 A nucleus whose mass is 3.499612×10^{-25} kg undergoes spontaneous alpha decay. The original nucleus disappears and there appear two new particles: a He-4 nucleus of mass 6.640678×10^{-27} kg (an "alpha particle" consisting of two protons and two neutrons) and a new nucleus of mass

3.433132×10^{-25} kg (note that the new nucleus has less mass than the original nucleus, and it has two fewer protons and two fewer neutrons). **(a)** When the alpha particle has moved far away from the new nucleus (so the electric interactions are negligible), what is the combined kinetic energy of the alpha particle and new nucleus? **(b)** How many electron volts is this? In contrast to this nuclear reaction, chemical reactions typically involve only a few eV.

••**P45** In a location in outer space far from all other objects, a nucleus whose mass is 3.894028×10^{-25} kg and that is initially at rest undergoes spontaneous alpha decay. The original nucleus disappears, and two new particles appear: a He-4 nucleus of mass 6.640678×10^{-27} kg (an alpha particle consisting of two protons and two neutrons) and a new nucleus of mass 3.827555×10^{-25} kg. These new particles move far away from each other, because they repel each other electrically (both are positively charged).

Because the calculations involve the small difference of (comparatively) large numbers, you need to keep seven significant figures in your calculations, and you need to use the more accurate value for the speed of light, 2.99792e8 m/s.

Choose all particles as the system. Initial state: Original nucleus, at rest. Final state: Alpha particle + new nucleus, far from each other. **(a)** What is the rest energy of the original nucleus? *Give seven significant figures.* **(b)** What is the sum of the rest energies of the alpha particle and the new nucleus? *Give seven significant figures.* **(c)** Did the portion of the total energy of the system contributed by rest energy increase or decrease? **(d)** What is the sum of the kinetic energies of the alpha particle and the new nucleus?

••**P46** A proton (1.6726×10^{-27} kg) and a neutron (1.6749×10^{-27} kg) at rest combine to form a deuteron, the nucleus of deuterium or "heavy hydrogen." In this process, a gamma ray (high-energy photon) is emitted, and its energy is measured to be 2.2 MeV (2.2×10^6 eV). **(a)** Keeping all five significant figures, what is the mass of the deuteron? Assume that you can neglect the small kinetic energy of the recoiling deuteron. **(b)** Momentum must be conserved, so the deuteron must recoil with momentum equal and opposite to the momentum of the gamma ray. Calculate approximately the kinetic energy of the recoiling deuteron and show that it is indeed small compared to the energy of the gamma ray.

••**P47** Many heavy nuclei undergo spontaneous "alpha decay," in which the original nucleus emits an alpha particle (a helium nucleus containing two protons and two neutrons), leaving behind a "daughter" nucleus that has two fewer protons and two fewer neutrons than the original nucleus. Consider a radium-220 nucleus that is at rest before it decays to radon-216 by alpha decay.

The mass of the radium-220 nucleus is 219.96274 u (unified atomic mass units) where 1 u = 1.6603×10^{-27} kg (approximately the mass of one nucleon).

The mass of a radon-216 nucleus is 215.95308 u, and the mass of an alpha particle is 4.00151 u. Radium has 88 protons, radon has 86 protons, and an alpha particle has 2 protons. **(a)** Make a diagram of the final state of the radon-216 nucleus and the alpha particle when they are far apart, showing the momenta of each particle to the same relative scale. Explain why you drew the lengths of the momentum vectors the way you did. **(b)** Calculate the final kinetic energy of the alpha particle. For the moment, assume that its speed is small compared to the speed of light. **(c)** Calculate the final kinetic energy of the radon-216 nucleus. **(d)** Show that the nonrelativistic approximation was reasonable.

•••**P48** A nucleus whose mass is 3.917268×10^{-25} kg undergoes spontaneous alpha decay. The original nucleus disappears and there appear two new particles: a He-4 nucleus of mass 6.640678×10^{-27} kg (an alpha particle consisting of two protons and two neutrons) and a new nucleus of mass 3.850768×10^{-25} kg. (Note that the new nucleus has less mass than the original nucleus, and it has two fewer protons and two fewer neutrons.) **(a)** What is the total kinetic energy of the alpha particle and the new nucleus? **(b)** Use the Conservation of Momentum in order to determine the kinetic energy of the alpha particle and the kinetic energy of the new nucleus.

Section 6.8

•**P49** You throw a ball straight up, and it reaches a height of 20 m above your hand before falling back down. What was the speed of the ball just after it left your hand?

•**P50** A 1 kg block rests on the Earth's surface. How much energy is required to move the block very far from the Earth, ending up at rest again?

•**P51** An object with mass 7 kg moves from a location $\langle 22,43,-41\rangle$ m near the Earth's surface to location $\langle -27,11,46\rangle$ m. What is the change in the potential energy of the system consisting of the object plus the Earth?

•**P52** The radius of the Moon is 1750 km, and its mass is 7×10^{22} kg. What would be the escape speed from an isolated Moon? Why was a small rocket adequate to lift the lunar astronauts back up from the surface of the Moon?

•**P53** Use energy conservation to find the approximate final speed of a basketball dropped from a height of 2 m (roughly the height of a professional basketball player). Why don't you need to know the mass of the basketball?

•**P54** Under certain conditions the interaction between a "polar" molecule such as HCl located at the origin and an ion located along the x axis can be described by a potential energy $U = -b/x^2$, where b is a constant. What is F_x, the x component of the force on the ion? What is F_y, the y component of the force on the ion?

•**P55 (a)** A 0.5 kg teddy bear is nudged off a window sill and falls 2 m to the ground. What is its kinetic energy at the instant it hits the ground? What is its speed? What assumptions or approximations did you make in this calculation? **(b)** A 1.0 kg flowerpot is nudged off a window sill and falls 2 m to the ground. What is its kinetic energy at the instant it hits the ground? What is its speed? How do the speed and kinetic energy compare to that of the teddy bear in part (a)?

•**P56** You throw a ball of mass 1.2 kg straight up. You observe that it takes 3.1 s to go up and down, returning to your hand. Assuming we can neglect air resistance, the time it takes to go up to the top is half the total time, 1.55 s. Note that at the top the momentum is momentarily zero, as it changes from heading upward to heading downward. **(a)** Use the Momentum Principle to determine the speed that the ball had just after it left your hand. **(b)** Use the Energy Principle to determine the maximum height above your hand reached by the ball.

•**P57** Suppose that a pitcher can throw a ball straight up at 100 mi/h (about 45 m/s). Use energy conservation to calculate how high the baseball goes. Explain your work. Actually, a pitcher can't attain this high a speed when throwing straight up, so your result will be an overestimate of what a human can do; air resistance also reduces the achievable height.

••P58 The radius of Mars (from the center to just above the atmosphere) is 3400 km (3400×10^3 m), and its mass is 0.6×10^{24} kg. An object is launched straight up from just above the atmosphere of Mars. **(a)** What initial speed is needed so that when the object is far from Mars its final speed is 1000 m/s? **(b)** What initial speed is needed so that when the object is far from Mars its final speed is 0 m/s? (This is called the escape speed.)

••P59 The radius of an airless planet is 2000 km (2×10^6 m), and its mass is 1.2×10^{23} kg. An object is launched straight up from the surface. **(a)** What initial speed is needed so that when the object is far from the planet its final speed is 900 m/s? **(b)** What initial speed is needed so that when the object is far from the planet its final speed is 0 m/s? (This is called the escape speed.)

••P60 The escape speed from an asteroid whose radius is 10 km is only 10 m/s. If you throw a rock away from the asteroid at a speed of 20 m/s, what will be its final speed?

••P61 The escape speed from a very small asteroid is only 24 m/s. If you throw a rock away from the asteroid at a speed of 35 m/s, what will be its final speed?

••P62 Calculate the speed of a satellite in a circular orbit near the Earth (just above the atmosphere). If the mass of the satellite is 200 kg, what is the minimum energy required to move the satellite from this near-Earth orbit to very far away from the Earth?

••P63 A spacecraft is coasting toward Mars. The mass of Mars is 6.4×10^{23} kg and its radius is 3400 km (3.4×10^6 m). When the spacecraft is 7000 km (7×10^6 m) from the center of Mars, the spacecraft's speed is 3000 m/s. Later, when the spacecraft is 4000 km (4×10^6 m) from the center of Mars, what is its speed? Assume that the effects of Mars's two tiny moons, the other planets, and the Sun are negligible. Precision is required to land on Mars, so make an accurate calculation, not a rough, approximate calculation.

••P64 A comet is in an elliptical orbit around the Sun. Its closest approach to the Sun is a distance of 4×10^{10} m (inside the orbit of Mercury), at which point its speed is 8.17×10^4 m/s. Its farthest distance from the Sun is far beyond the orbit of Pluto. What is its speed when it is 6×10^{12} m from the Sun? (This is the approximate distance of Pluto from the Sun.)

••P65 An electron is traveling at a speed of $0.95c$ in an electron accelerator. An electric force of 1.6×10^{-13} N is applied in the direction of motion while the electron travels a distance of 2 m. What is the new speed of the electron?

••P66 You stand on a spherical asteroid of uniform density whose mass is 2×10^{16} kg and whose radius is 10 km (10^4 m). These are typical values for small asteroids, although some asteroids have been found to have much lower average density and are thought to be loose agglomerations of shattered rocks. **(a)** How fast do you have to throw the rock so that it never comes back to the asteroid and ends up traveling at a speed of 3 m/s when it is very far away? **(b)** Sketch graphs of the kinetic energy of the rock, the gravitational potential energy of the rock plus asteroid, and their sum, as a function of separation (distance from center of asteroid to rock). Label the graphs clearly.

•••P67 In the rough approximation that the density of the Earth is uniform throughout its interior, the gravitational field strength (force per unit mass) inside the Earth at a distance r from the center is gr/R, where R is the radius of the Earth. (In actual fact, the outer layers of rock have lower density than the inner core of molten iron.) Using the uniform-density approximation, calculate the amount of energy required to move a mass m from the center of the Earth to the surface. Compare with the amount of energy required to move the mass from the surface of the Earth to a great distance away.

••P68 This problem is closely related to the spectacular impact of the comet Shoemaker-Levy with Jupiter in July 1994:

http://www.jpl.nasa.gov/sl9/sl9.html

A rock far outside our Solar System is initially moving very slowly relative to the Sun, in the plane of Jupiter's orbit around the Sun. The rock falls toward the Sun, but on its way to the Sun it collides with Jupiter. Calculate the rock's speed just before colliding with Jupiter. Explain your calculation and any approximations that you make.

$M_{\text{Sun}} = 2 \times 10^{30}$ kg, $M_{\text{Jupiter}} = 2 \times 10^{27}$ kg

Distance, Sun to Jupiter $= 8 \times 10^{11}$ m

Radius of Jupiter $= 1.4 \times 10^8$ m

•••P69 A pendulum (see Figure 6.84) consists of a very light but stiff rod of length L hanging from a nearly frictionless axle, with a mass m at the end of the rod. **(a)** Calculate the gravitational potential energy as a function of the angle θ, measured from the vertical. **(b)** Sketch the potential energy as a function of the angle θ, for angles from $-210°$ to $+210°$. **(c)** Let $s = L\theta$ = the arc length away from the bottom of the arc. Calculate the tangential component of the force on the mass by taking the (negative) gradient of the energy with respect to s. Does your result make sense? **(d)** Suppose that you hit the stationary hanging mass so it has an initial speed v_i. What is the minimum initial speed needed for the pendulum to go over the top ($\theta = 180°$)? On your sketch of the potential energy (part b), draw and label energy levels for the case in which the initial speed is less than, equal to, or greater than this critical initial speed.

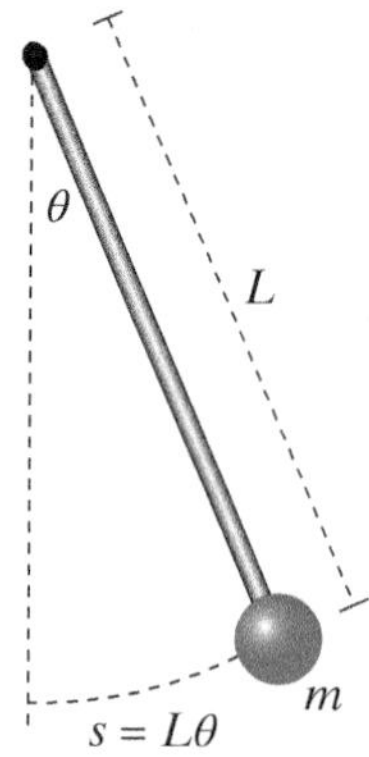

Figure 6.84

Section 6.9

•P70 **(a)** Two protons are a distance 4×10^{-9} m apart. What is the electric potential energy of the system consisting of the two protons? If the two protons move closer together, will the electric potential energy of the system increase, decrease, or remain the same? **(b)** A proton and an electron are a distance 4×10^{-9} m apart. What is the electric potential energy of the system consisting of the proton and the electron? If the proton and the electron move closer together, will the electric potential energy of the system increase, decrease, or remain the same? **(c)** Which of the following statements are true? A. In some situations charged particles released from rest would move in a direction that increases electric potential energy, but not in other situations. B. If released from rest two protons would move closer together, increasing the potential energy of the system. C. If any two charged particles are released from rest, they will spontaneously move in the direction in which the potential energy of the system will be decreased.

••P71 **(a)** A particle with mass M and charge $+e$ and its antiparticle (same mass M, charge $-e$) are initially at rest, far from each other. They attract each other and move toward each other. Make a graph of energy terms vs. separation distance r between the two particles. Label the various energies involved in this process. Include the rest energy of the particles, assuming that the other energy terms are comparable to the rest energy. **(b)** When the particle and antiparticle collide, they annihilate and produce two new particles, one with mass m (much smaller than M) and charge $+e$, and its antiparticle (same mass m, charge $-e$). When these two particles have moved far away from each other, how fast are they going? Is this speed large or small compared to c? **(c)** Now take the specific case of a proton and antiproton colliding to form a positive and negative pion. Each pion has a mass of 2.5×10^{-28} kg. When the pions have moved far away, how fast are they going? **(d)** How far apart must the two pions be (in meters) for their electric potential energy to be negligible compared to their kinetic energy? Be explicit and quantitative about your criterion and your result.

••P72 Four protons, each with mass M and charge $+e$, are initially held at the corners of a square that is d on a side. They are then released from rest. What is the speed of each proton when the protons are very far apart?

Section 6.11

•P73 There are three different ways to get from location A to location E in Figure 6.85. Along path 1, you take an elevator directly from location A straight up to location E. Along path 2, you walk from A to B, climb a rope from B to C, then walk from C to E. Along path 3, you walk from location A to D, then climb a ramp up to location E. The following questions focus on the work done on a 80 kg person by the Earth while following each of these paths.

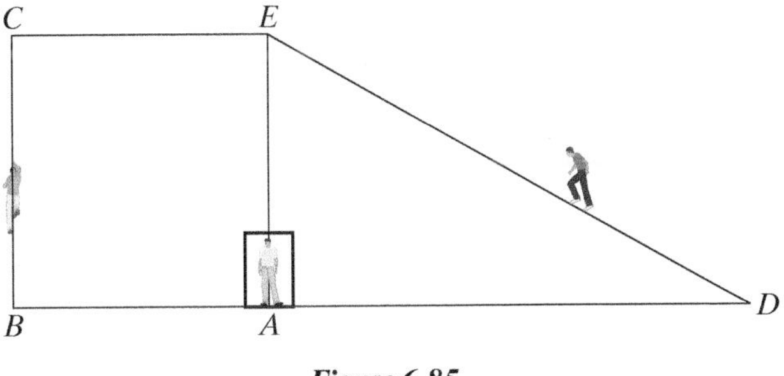

Figure 6.85

Taking the origin at location A, the coordinates of the other locations are: $B\ \langle 0,-3,0\rangle$; $C\ \langle -3,13,0\rangle$; $D\ \langle 33,0,0\rangle$; $E\ \langle 0,13,0\rangle$. **(a)** Path 1 (A to E): What is the displacement vector $\Delta\vec{r}$ for the path from A to E on the elevator? **(b)** What is the gravitational force acting on the person along path 1, as a vector? **(c)** What is the work done by the gravitational force of the Earth on the 80 kg person during the elevator ride from A to E? **(d)** Path 2 (A to B to C to E): What is the work done by the gravitational force of the Earth on the person as he walks from A to B? **(e)** What is the work done by the gravitational force of the Earth on the person as he climbs from B to C? **(f)** What is the work done by the gravitational force of the Earth on the person as he walks from C to E? **(g)** What is the total work done by the gravitational force of the Earth on the person as the person goes from A to E along path 2? **(h)** Path 3 (A to D to E): What is the work done by the gravitational force of the Earth on the person as he walks from A to D? **(i)** What is the work done by the gravitational force of the Earth on the person as he climbs from D to E? **(j)** What is the total work done by the gravitational force of the Earth on the person as the person goes from A to E along path 3? **(k)** You calculated the work done by the gravitational force of the Earth on the person along three different paths. How do these quantities compare with each other?

•P74 Refer to Figure 6.86. Calculate the change in electric energy along two different paths in moving charge q away from charge Q from A to B along a radial path, then to C along a circle centered on Q, then to D along a radial path. Also calculate the change in energy in going directly from A to D along a circle centered on Q. Specifically, what are $U_B - U_A$, $U_C - U_B$, $U_D - U_C$, and their sum? What is $U_D - U_A$? Also, calculate the round-trip difference in the electric energy when moving charge q along the path from A to B to C to D to A.

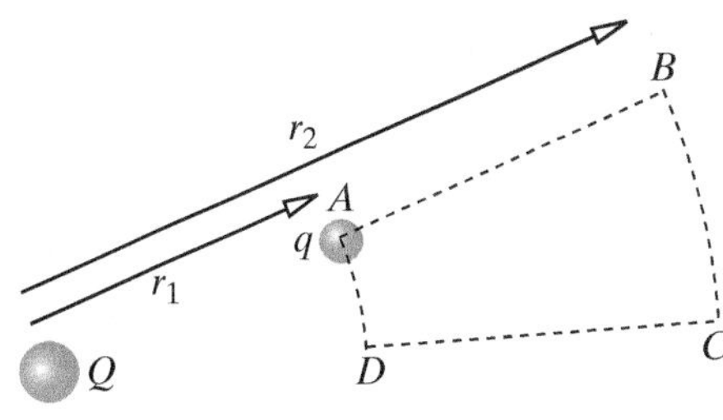

Figure 6.86

Section 6.12

••P75 One of the possible fission modes of Uranium-235 involves nearly equal fragments, palladium nuclei with $Q_1 = Q_2 = 46e$. The masses of the two palladium nuclei add up to less than the mass of the original nucleus. (In addition to the two main fission fragments there are typically one or more free neutrons in the final state; in your analysis make the simplifying assumption that there are no free neutrons, just two palladium nuclei.)

The mass of the U-236 nucleus is 235.996 u (unified atomic mass units), and the mass of each Pd-118 nucleus is 117.894 u, where 1 u $= 1.7 \times 10^{-27}$ kg (approximately the mass of one nucleon). Although in most problems you solve in this course it is adequate to use values of constants rounded to two or three significant figures, in this problem you must keep at least six significant figures throughout your calculation. Problems involving mass changes require many significant figures because the changes in mass are small compared to the total mass. **(a)** Calculate the final speed v, when the palladium nuclei have moved far apart (due to their mutual electric repulsion). Is this speed small enough that $p^2/(2m)$ is an adequate approximation for the kinetic energy of one of the palladium nuclei? (It is all right to go ahead and make the nonrelativistic assumption first, but you then must check that the calculated v is indeed small compared to c.) **(b)** Using energy considerations, calculate the distance between centers of the palladium nuclei just after fission, when they are starting from rest. **(c)** A proton or neutron has a radius of roughly 1×10^{-15} m, and a nucleus is a tightly packed collection of nucleons. Experiments show that the radius of a nucleus containing N nucleons is approximately $(1.3 \times 10^{-15}\text{ m}) \times N^{1/3}$. What is the approximate radius of a palladium nucleus? Draw a sketch of the two palladium nuclei in part (b), and label the distances you calculated in parts (b) and (c). If the two palladium nuclei are nearly touching, this would be consistent with our model of fission, in which the uranium nucleus fissions into two pieces that are nearly at rest. **(d)** The kinetic energy of the fast-moving daughter nuclei is eventually absorbed in the surrounding material and raises the temperature of that material. In a fission power plant, this thermal energy is used to boil water to drive a steam turbine and generate electricity. If a mole of uranium undergoes this fission reaction, how much kinetic energy is generated? For comparison, only around 1×10^6 J are obtained from burning a mole of gasoline, which is why energy from fission is of great interest, if some way can be found to deal with the waste products.

••P76 A proton (^{1}H) and a deuteron (^{2}H, "heavy" hydrogen) start out far apart. An experimental apparatus shoots them toward each other (with equal and opposite momenta). If they get close enough to make actual contact with each other, they can react to form a helium-3 nucleus and a gamma ray (a high-energy photon, which has kinetic energy but zero rest energy):

$$^1\text{H} + {}^2\text{H} \rightarrow {}^3\text{He} + \gamma$$

This is one of the thermonuclear or fusion reactions that takes place inside a star such as our Sun.

The mass of the proton is 1.0073 u (unified atomic mass unit, 1.7×10^{-27} kg), the mass of the deuteron is 2.0136 u, the mass of the helium-3 nucleus is 3.0155 u, and the gamma ray is massless. Although in most problems you solve in this course it is adequate to use values of constants rounded to two or three significant figures, in this problem you must keep at least six significant figures throughout your calculation. Problems involving mass changes require many significant figures because the changes in mass are small compared to the total mass. **(a)** The strong interaction has a very short range and is essentially a contact interaction. For this fusion reaction to take place, the proton and deuteron have to come close enough together to touch. The approximate radius of a proton or neutron is about 1×10^{-15} m. What is the approximate initial total kinetic energy of the proton and deuteron required for the fusion reaction to proceed, in joules and electron volts ($1\text{ eV} = 1.6 \times 10^{-19}$ J)? **(b)** Given the initial conditions found in part (a), what is the kinetic energy of the ^{3}He plus the energy of the gamma ray, in joules and in electron volts? **(c)** The net energy released is the kinetic energy of the ^{3}He plus the energy of the gamma ray found in part (b), minus the energy input that you calculated in part (a). What is the net energy release, in joules and in electron volts? Note that you do get back the energy investment made in part (a). **(d)** Kinetic energy can be used to drive motors and do other useful things. If a mole of hydrogen and a mole of deuterium underwent this fusion reaction, how much kinetic energy would be generated? (For comparison, around 1×10^6 J are obtained from burning a mole of gasoline.) **(e)** Which of the following potential energy curves (1–4) in Figure 6.87 is a reasonable representation of the interaction in this fusion reaction? Why?

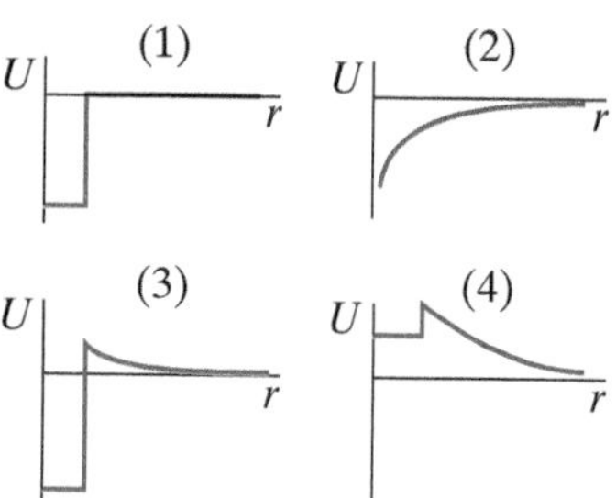

Figure 6.87

As we will study later, the average kinetic energy of a gas molecule is $\frac{3}{2}k_BT$, where k is the "Boltzmann constant," 1.4×10^{-23} J/K, and T is the absolute or Kelvin temperature, measured from absolute zero (so that the freezing point of water is 273 K). The approximate temperature required for the fusion reaction to proceed is very high. This high temperature, required because of the electric repulsion barrier to the reaction, is the main reason why it has been so difficult to make progress toward thermonuclear power generation. Sufficiently high temperatures are found in the interior of the Sun, where fusion reactions take place.

COMPUTATIONAL PROBLEMS

More detailed and extended versions of some computational modeling problems may be found in the lab activities included in the *Matter & Interactions, 4th Edition*, resources for instructors.

••P77 Start with the program you wrote to model the motion of a spacecraft near the Earth, in Chapter 3 (Problem P65). Add the statements necessary to plot graphs of K, U, and $(K+U)$ versus time, like the graphs in Figure 6.71. Use initial conditions that produce an elliptical orbit. **(a)** Refer to the shapes of the plots to explain the flow of energy within the system. **(b)** Use the plot of $K+U$ versus time produced by the program to determine the maximum reasonable value of Δt for this model. **(c)** Predict the shape each plot should have for a perfectly circular orbit. **(d)** Use your energy plots to find (by trial and error) the initial speed of the spacecraft that produces a circular orbit, and explain what information from the plots you used.

••P78 As in Problem P77, start with the program you wrote to model the motion of a spacecraft near the Earth in Chapter 3 (Problem P64). Add the statements necessary to plot graphs of graphs of K, U, and $(K+U)$ versus r, the separation between the spacecraft and the Earth. Use initial conditions that produce an elliptical orbit. In VPython, when you create a `gcurve` object, it can be helpful to add a moving dot to clarify what is happening in the plot, like this:

```
ugraph = gcurve(color=color.green, dot=True)
```

(a) Refer to the shapes of the plots to explain the flow of energy within the system. **(b)** What effect does changing the initial speed of the spacecraft have on these three plots? **(c)** Predict the shape each plot should have for a perfectly circular orbit. **(d)** Use your energy plots to find (by trial and error) the initial speed of the spacecraft that produces a circular orbit, and explain what information from the plots you used. **(e)** Use your energy plots to help you find the escape speed for this system. **(f)** Describe your plot of $K+U$ when the spacecraft starts with this escape speed.

••P79 Start with the program you wrote to model the motion of a spacecraft near the Earth and a stationary Moon, in Chapter 3 (Problem P66). Add the statements necessary to plot graphs of K, U, and $(K+U)$ versus time, like the graphs in Figure 6.71. Give the spacecraft an initial velocity of $\langle 0, 3.27\times 10^3, 0\rangle$ m/s. **(a)** Refer to the shapes of the plots to explain the flow of energy within the system. **(b)** Use the plot of $K+U$ versus time produced by the program to determine the maximum reasonable value of Δt for this model. **(c)** What changes in the energy graphs when you vary the initial speed?

••P80 As in Problem P79, start with the program you wrote to model the motion of a spacecraft near the Earth and a stationary Moon, in Chapter 3 (Problem P66). Add the statements necessary to plot graphs of graphs of K, U, and $(K+U)$ versus r, the separation between the spacecraft and the Earth. Give the spacecraft an initial velocity of $\langle 0, 3.27\times 10^3, 0\rangle$ m/s. In VPython, when you create a `gcurve` object, it can be helpful to add a moving dot to clarify what is happening in the plot, like this:

```
ugraph = gcurve(color=color.green, dot=True)
```

(a) Why are the shapes of the graphs of U and K versus separation different for this three-body system than they were for the two-body (spacecraft + Earth) system? **(b)** Refer to the shapes of the plots to explain the flow of energy within the system. **(c)** What effect does changing the initial speed of the spacecraft have on these three plots?

••P81 Start with the program you wrote in Problem P71 in Chapter 3 to model the Ranger 7 mission to the Moon. (If you have not already written this program, do it now.) **(a)** Add a calculation of the work done by the gravitational forces of the Earth and the Moon to your analysis of sending a spacecraft to the Moon. You need to approximate the work by adding up the amount of work done by gravitational forces along each step of the path:

$$W = \Sigma\vec{F}\bullet\Delta\vec{r} = \vec{F}_1\bullet\Delta\vec{r}_1+\vec{F}_2\bullet\Delta\vec{r}_2+\vec{F}_3\bullet\Delta\vec{r}_3+\cdots$$

(b) Compare the numerical value of the work with the change in the kinetic energy (final kinetic energy just before crashing on the Moon, minus initial kinetic energy when released above the Earth's atmosphere). **(c)** Modify your program to make graphs of two quantities: the kinetic energy of the spacecraft, and the work done by the Earth and the Moon on the spacecraft, as a function of time. **(d)** Find a way to determine the maximum value of Δt you can use without introducing significant errors into your computation.

••P82 Energy conservation is a powerful check on the accuracy of a numerical integration. Modify the program for the Chapter 3 problem on the Ranger 7 mission to the Moon to plot graphs of kinetic energy, of gravitational potential energy, and of the sum of the kinetic energy and the gravitational potential energy, vs. position. Does the kinetic plus potential energy remain constant? What if you vary the step size (which varies the accuracy of the numerical integration)? Vary the launch speed, and explain the effect that this has on your graphs.

••P83 Use energy conservation to calculate analytically (that is, without doing a numerical integration) the final speed of the spacecraft just before it hits the Moon. Include the gravitational effect of the Moon. Use a launch speed of 1.3×10^4 m/s. Modify your program for the Chapter 3 problem on the Ranger 7 mission to the Moon to print out the speed of the spacecraft when it hits the surface of the Moon, and compare this value to your analytical result. What questions that could be addressed in the numerical integration are you *not* able to answer by doing this energy calculation?

ANSWERS TO CHECKPOINTS

1 Particle energy = 5.2×10^{-13} J, rest energy = 8.1×10^{-14} J; kinetic energy = 4.4×10^{-13} J
2 12 m/s; 27 miles per hour; 1.2×10^{-12} kg (an extremely small amount of mass!)
3 1.4 J
4 (a) decreases, **(b)** increases, **(c)** decreases, **(d)** increases, **(e)** stays the same
5 189 J
6 400 J
7 40.8; 39.8; very large; very small
8 8.5×10^{-4}, approximately 0.1%
9 Six pairs: $U_{12}+U_{13}+U_{14}+U_{23}+U_{24}+U_{34}$
10 Same ΔU
11 $\frac{1}{2}m(v_{xi}^2)-\frac{1}{2}m(v_{xi}^2+v_{yi}^2)+mg(y_f-y_i)=0$, so $y_f=y_i+v_{yi}^2/(2g)$
12 (a) $\dfrac{1}{4\pi\varepsilon_0}\dfrac{e^2}{r} = (0.2\times 10^6\text{ eV})(1.6\times 10^{-19}\text{ J/eV})$, **(b)** the electric potential energy term
13 (a) decrease, **(b)** A is bound; C is unbound; B is trapped. In a quantum system in state B there is some probability of "tunneling" through the barrier and getting out of the trap!
14 A: go out in a straight line and stop, then fall back; B: elliptical (motion with varying separation); C: nearly circular (nearly constant separation)
15 1.12×10^4 m/s
16 About 1.7×10^{-10}; accuracy of laboratory scale is far below what would be needed to detect this
17 About 0.1%
18 $-GMm/y^2$; 0

CHAPTER

7

Internal Energy

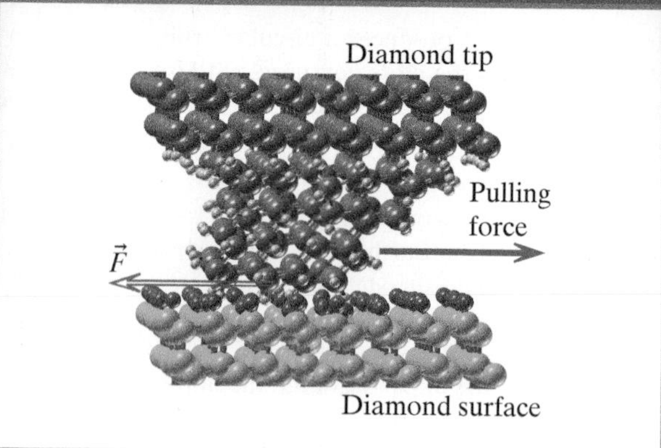

OBJECTIVES

After studying this chapter, you should be able to

- Explain changes in the internal energy of a solid object at a microscopic level in terms of the ball–spring model.
- Calculate internal energy changes in extended objects.
- Mathematically relate temperature change, energy transfer due to a temperature difference, and specific heat.
- Calculate and use the potential energy of a system containing a spring.
- Create a computational model of a system in which there is dissipation (for example, air resistance or sliding friction).

7.1 EXTENDED OBJECTS

In Chapter 4 we considered contact forces between extended objects—objects that can be stretched, compressed, and deformed, and that therefore cannot be modeled as point particles. We found that at the microscopic level a ball–spring model of an extended solid object allowed us to understand the origin of tension and compression forces, and to predict quantities such as the speed of sound in an object composed of a particular metal.

When considering energy changes in systems consisting of extended objects, we again find that it is not possible to model such systems as point particles that have no internal structure. We know that, as a result of interactions with their surroundings, macroscopic objects can get hotter or colder. We will see that a change in the temperature of an object is correlated with the amount of energy inside the object. In order to understand the nature of internal energy, we will return to the ball–spring model of the microscopic structure of a solid object (Figure 7.1).

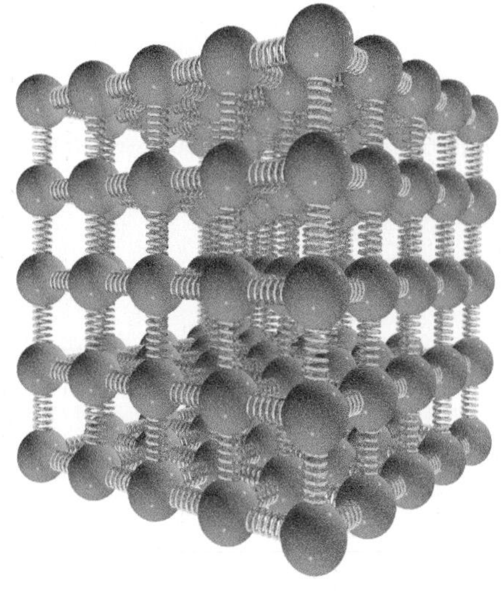

Figure 7.1 The ball–spring model of a solid object.

Recall that in this model, the balls represent atoms, and the springs represent the spring-like covalent chemical bonds between the atoms. Because the atoms inside a solid object can move, they can have kinetic energy. Since work is required to compress or stretch a spring, it is plausible that potential energy can be associated with springs. In order to think about internal energy in the context of the ball–spring model of a solid, we need to find an expression for the potential energy associated with a spring, which is discussed in the following section.

7.2 POTENTIAL ENERGY OF MACROSCOPIC SPRINGS

A real spring has a nonzero mass and can be damaged in various ways: it can be deformed or broken, the metal can fatigue, and so on. There is a limit to how much a real spring can be stretched (it will eventually break) or compressed (the coils will eventually touch each other, turning the spring into a stiff hollow rod). Initially, to reduce the complexity of our model, we will consider an idealized version of a spring, which we will call an ideal spring.

An Ideal Spring

An idealized spring has none of the problems of a real spring. There are no limits on the stretch s in the ideal spring force equation (here we're calling the axis of the spring the s axis):

$$F_s = -k_s s$$

Despite its limitations, an ideal spring proves to be a useful model for both real macroscopic springs and for spring-like interatomic bonds.

To determine the potential energy associated with the spring force, recall that the (negative) gradient of potential energy is equal to the associated force. So to find the spring potential energy U_s we need a solution to this equation:

$$-\frac{dU_s}{ds} = -k_s s$$

which simplifies to

$$\frac{dU_s}{ds} = k_s s$$

QUESTION Can you think of a function of s whose derivative with respect to s is $k_s s$?

The derivative with respect to s of $\frac{1}{2}k_s s^2$ is $k_s s$, and the negative gradient of $\frac{1}{2}k_s s^2$ is the force $-k_s s$.

IDEAL SPRING POTENTIAL ENERGY

$$U_s = \frac{1}{2}k_s s^2$$

s is stretch, measured from the equilibrium point.

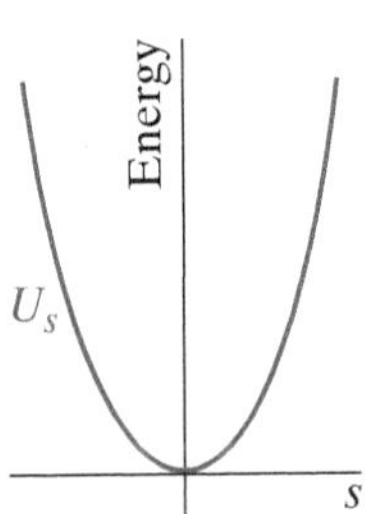

Figure 7.2 Potential energy as a function of stretch s for an ideal spring.

The function $\frac{1}{2}k_s s^2$ is an upturned parabola (Figure 7.2). Note that the stretch s appears squared in the expression for spring energy. This reflects the fact that either a lengthening (positive stretch) or a compression (negative stretch) of the spring involves increased potential energy.

Peculiar Features of the Ideal Spring Potential Energy

This ideal spring potential energy curve has some peculiar features:

- As the absolute value of stretch becomes large, U becomes infinite, instead of approaching zero. (This curve is sometimes called an "infinite potential energy well.")
- For a system of a mass and ideal spring, all states are bound states; there are no unbound states.

As a result of these peculiarities, this ideal potential energy function doesn't fully represent the nature of a macroscopic spring you can hold in your hand, or of an interatomic bond.

One possible modification to the equation would be to subtract a constant E_S. Such a function would still satisfy the gradient equation above, but it would allow for U to be negative in places, as in Figure 7.3.

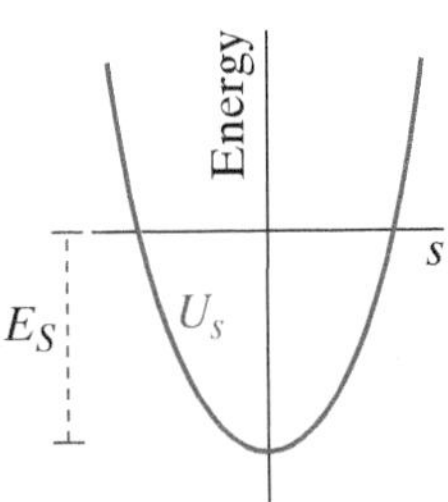

Figure 7.3 Subtracting a constant E_S shifts the ideal spring potential energy curve downward.

$$U_s = \frac{1}{2}k_s s^2 - E_S$$

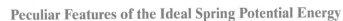

For the ideal spring, the constant is conventionally set to zero, but the negative offset may be needed when analyzing real systems.

Real Springs

We are primarily interested in helical (spiral) springs. A modest force applied to a helical spring gives a substantial change in the length of the spring, even though the total length of the coiled wire hardly changes. A large stretch of the spring with a small force implies a much smaller spring stiffness than the wire would have when straightened out.

If you stretch a helical spring far enough, it straightens out into a long straight wire, and it suddenly becomes very difficult to lengthen the wire any further. If you continue to stretch the straight wire, eventually it yields and then breaks. If you compress a helical spring so far that its coils run into each other, it suddenly gets extremely hard to compress the spring any further. Therefore, the linear relation between stretch and force, $F_s = -k_s s$, is valid only over a limited region of stretch, although this linear range is much wider than for the microscopic "spring."

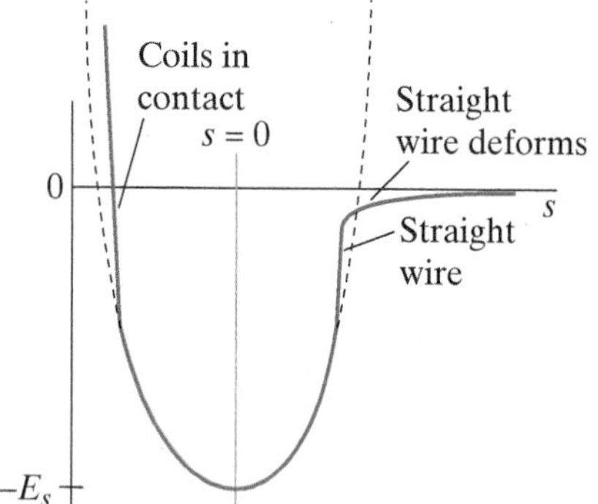

Figure 7.4 The dashed curve represents the spring potential energy

$$U_s = \frac{1}{2}k_s s^2 - E_S$$

within the linear range where $F_s = -k_s s$. The colored solid curve represents a real macroscopic spring.

In many situations the entire phenomenon of interest takes place within the valid range of the linear force approximation. One can speak of an "infinite" potential energy well as shown as the dashed curve in Figure 7.4, which is an idealization of the real situation. Within the linear range where $F_s = -k_s s$ is the negative gradient of $\frac{1}{2}k_s s^2$, we approximate the real curve by a parabola. In our calculations with macroscopic springs we're almost always interested only in *changes* in the potential energy, so we shift the origin to the bottom of the well by an amount $-E_s$ and measure $U_s = \frac{1}{2}k_s s^2$ from there.

In the blue curve shown in Figure 7.4 (which is still an idealization of the real situation) there are very steep sides of the potential energy well, corresponding to the large forces required to compress a spring whose coils are in contact and the large forces required to stretch the spring after it has straightened out into a straight wire. Since force is the negative gradient of the potential energy, a steep slope corresponds to a large force. The region marked "Straight wire deforms" in Figure 7.4 refers to the fact that a wire may "yield" (lengthen with little additional force required) or break.

Energy in an Oscillating Spring–Mass System

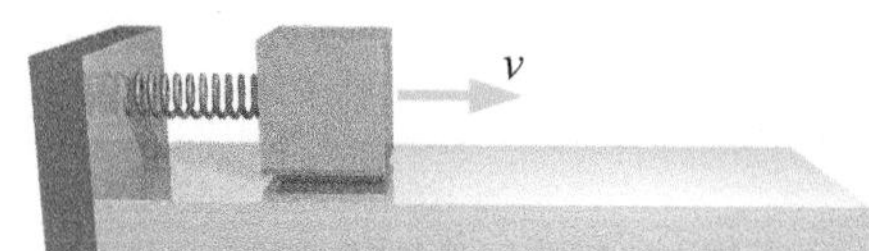

Figure 7.5 As the compressed spring expands the kinetic energy of the block increases, and spring potential energy decreases.

In an oscillating spring–mass system such as the one discussed in Chapter 4, the energy of the system is continually changing from kinetic energy to potential energy, and back again.

Consider a block on a low-friction surface, connected by a helical spring to a wall. If you compress the spring and then let go, the block acquires kinetic energy (Figure 7.5), with a loss of spring potential energy. If you stretch the spring and then let go, the block also acquires kinetic energy, again with a loss of spring potential energy (Figure 7.6).

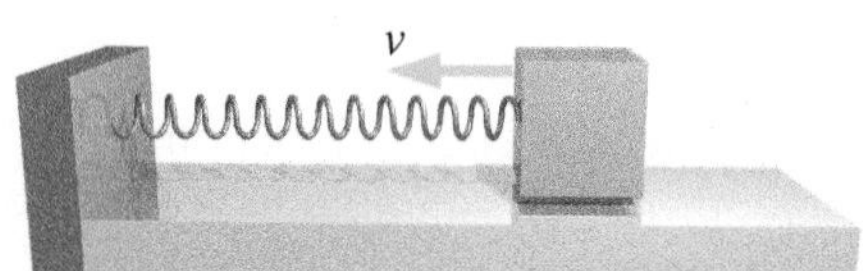

Figure 7.6 As the stretched spring contracts, the kinetic energy of the block increases and spring potential energy decreases.

> QUESTION At the moment that the spring is released, what is the kinetic energy of the spring–mass system? What is the total energy of the system?

Since the block is released from rest, its kinetic energy is zero. The energy of the system is equal to the energy stored in the spring, $\frac{1}{2}k_s s^2$ (omitting the constant rest energy and constant $-E_S$).

> QUESTION At what point in the oscillation is the kinetic energy of the system highest? At that moment, what is the potential energy of the system?

When the contracting spring reaches its equilibrium length, the energy stored in the spring is a minimum, $\frac{1}{2}k_s s^2 = 0$. The maximum possible amount of spring potential energy has now been converted to kinetic energy, so this is the instant when the kinetic energy of the system is highest.

QUESTION At what point in the oscillation will the kinetic energy of the system be lowest?

At either turning point of the oscillation—spring fully compressed or spring fully extended—the instantaneous speed of the mass is zero. The kinetic energy of the system is lowest here (zero in fact), and all the energy has been momentarily converted back into spring potential energy. Since the absolute value of the stretch is highest at these locations, the potential energy $\frac{1}{2}k_s s^2$ is also highest.

Does the Wall Do Work on the System?

For the system of spring + block, the surroundings include the Earth, the low-friction table, and the wall. Both the table and the wall touch the system and therefore exert contact forces on it.

The table and Earth exert forces perpendicular to the block's motion, so they do zero work on the system (assuming negligible friction).

QUESTION The force by the wall on the spring is parallel to the block's motion. Does it do nonzero work on the system?

This force does zero work, because the point at which the force acts (the left end of the spring) undergoes zero displacement. So

$$F_{\text{wall},x}\Delta x = 0 \quad \text{because } \Delta x = 0 \text{ for the left end of the spring}$$

This force does have an effect on the momentum of the system, however, because the time interval over which it acts is not zero. The momentum of the mass–spring system is changed by the force on the spring by the wall. (Think about what would happen if the wall were not there.) This is another example of the difference between the Momentum Principle (all forces act for the same time interval) and the Energy Principle (different forces may act over different distances).

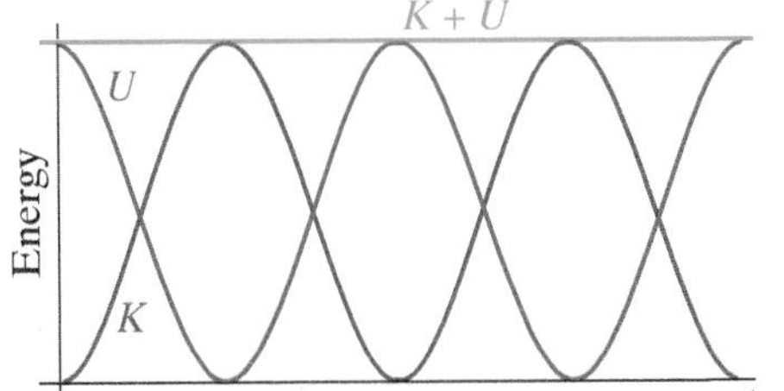

Figure 7.7 Kinetic energy (red), potential energy (blue), and the sum of kinetic + potential energy (green) for an oscillating spring–mass system.

Flow of Energy in the Mass–Spring System

Since the surroundings do no work on the system, the total energy of the system must be constant (in the absence of friction), and energy flows from potential to kinetic and back again, as shown in Figure 7.7.

Here we can see a nice example of "path independence," which was discussed in Chapter 6. Consider the round trip starting with the block moving at maximum speed as it passes the equilibrium point, where the potential energy is zero, moving in the $+x$ direction. The block slows down due to the spring force that acts in the $-x$ direction, momentarily comes to a stop, then moves with increasing speed in the $-x$ direction due to the spring force, which is still in the $-x$ direction.

The block passes the equilibrium point with the same kinetic energy it had when it was previously at this location, but now moving in the $-x$ direction. The round trip of the spring–mass system, from equilibrium configuration and return, with no change in the kinetic energy, required zero work on our part (path independence). The key to this is that the spring force and associated potential energy are purely functions of position. At any location you choose on the x axis, the spring force is the same in magnitude and direction when the block is moving to the right at that location and when it is returning to the left at that location.

If there were significant friction, the frictional force would be in the $-x$ direction outward bound but in the $+x$ direction during the return, unlike the spring force which is always in the $-x$ direction during the entire round trip. The frictional force depends on the direction of motion, the velocity, whereas the spring force depends solely on the position.

Checkpoint 1 During one complete oscillation of a mass on a spring (one period), what is the change in potential energy of the mass + spring system, in the absence of friction?

EXAMPLE **Energy in a Spring–Mass System**

A mass of 0.2 kg is attached to a horizontal spring whose stiffness is 12 N/m. Friction is negligible. At $t = 0$ the spring has a stretch of 3 cm and the mass has a speed of 0.5 m/s. **(a)** What is the amplitude (maximum stretch) of the oscillation? **(b)** What is the maximum speed of the block?

Solution System: Mass and spring
Surroundings: Earth, table, wall (neglect friction, air)

(a) The stretch is maximum when the speed of the mass is zero, so:
Initial state: $s_i = 3$ cm, $v_i = 0.5$ m/s
Final state: $v_f = 0$

$$K_f + U_f = K_i + U_i + W$$
$$0 + \frac{1}{2}k_s s_f^2 = \frac{1}{2}mv_i^2 + \frac{1}{2}k_s s_i^2 + 0$$
$$\frac{1}{2}k_s s_f^2 = \frac{1}{2}(0.2\text{ kg})(0.5\text{ m/s})^2 + \frac{1}{2}(12\text{ N/m})(0.03\text{ m})^2$$
$$\frac{1}{2}k_s s_f^2 = 0.03\text{ J}$$
$$s_f = \sqrt{2(0.03\text{ J})/(12\text{ N/m})} = 0.07\text{ m}$$
$$s_f = 0.07\text{ m}$$

(b) The speed is maximum when the stretch is zero, so:
Initial state: $s_i = 3$ cm, $v_i = 0.5$ m/s
Final state: $s_f = 0$

$$K_f + U_f = K_i + U_i + W$$
$$K_f + 0 = 0.03\text{ J} + 0$$
$$\frac{1}{2}mv_f^2 = 0.03\text{ J}$$
$$v_{max} = \sqrt{2(0.03\text{ J})/(0.2\text{ kg})} = 0.55\text{ m/s}$$

Assuming negligible energy dissipation by friction or air resistance, the energy of the system never changes. If we can find $K + U$ for one state, we can always use this as the initial state in any calculation.

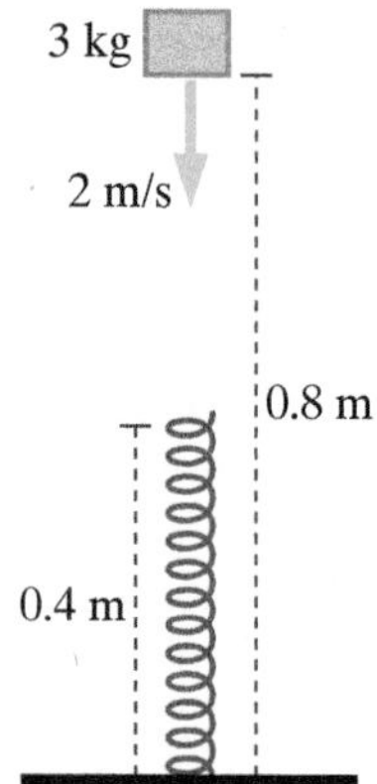

Figure 7.8 A block falls onto a vertical spring, compresses the spring, then rebounds upward.

EXAMPLE **A Rebounding Block**

A metal block of mass 3 kg is moving downward with speed 2 m/s when the bottom of the block is 0.8 m above the floor (Figure 7.8). When the bottom of the block is 0.4 m above the floor, it strikes the top of a relaxed vertical spring 0.4 m in length. The stiffness of the spring is 2000 N/m. **(a)** The block continues downward, compressing the spring. When the bottom of the block is 0.3 m above the floor, what is its speed? **(b)** The block eventually heads back upward, loses contact with the spring, and continues upward. What is the maximum height reached by the bottom of the block above the floor? **(c)** What approximations did you make?

Solution **(a)** Find the speed of the block when it is 0.3 m above the floor.
System: Earth, block, spring
Surroundings: Nothing significant
Initial state: Block 0.8 m above floor, moving downward, $v_i = 2$ m/s, spring relaxed
Final state: Block 0.3 m above floor, spring compressed

$$K_f + U_{s,f} + U_{g,f} = K_i + U_{s,i} + U_{g,i} + W$$
$$\frac{1}{2}mv_f^2 + \frac{1}{2}k_s s_f^2 + mgy_f = \frac{1}{2}mv_i^2 + 0 + mgy_i + 0$$
$$v_f^2 = v_i^2 + 2g(y_i - y_f) - \frac{k_s}{m}s_f^2$$
$$v_f = \sqrt{(2\text{ m/s})^2 + 2(9.8\text{ N/kg})(0.5\text{ m}) - \frac{2000\text{ N/m}}{3\text{ kg}}(0.1\text{ m})^2}$$
$$v_f = 2.7\text{ m/s}$$

(b) Find the maximum height.
Initial state: Same as in part (a)
Final state: Block at highest point, $v_f = 0$, spring relaxed

$$K_f + U_{s,f} + U_{g,f} = K_i + U_{s,i} + U_{g,i} + W$$
$$0 + 0 + mgy_f = \frac{1}{2}mv_i^2 + 0 + mgy_i + 0$$
$$y_f = y_i + \frac{v_i^2}{2g} = (0.8\text{ m}) + \frac{(2\text{ m/s})^2}{2(9.8\text{ N/kg})} = 1.0\text{ m}$$

(c) Approximations: Air resistance and dissipation in the spring are negligible. $U_g \approx mgy$ near Earth's surface. ΔK_{Earth} is negligible.

Both Earth and spring are part of the system, so there are two potential energy terms, one for (block + spring) and one for (block + Earth). Since the mass of the spring is negligible, there is not a (spring + Earth) potential energy term. In part (b) we could equally well have used the final state from part (a) as the initial state.

Checkpoint 2 How many joules of energy can you store in a spring whose stiffness is 0.6 N/m, by starting from a relaxed spring and stretching it 20 cm?

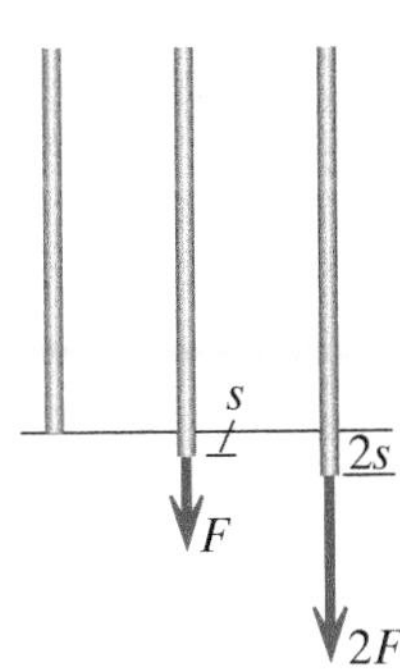

Figure 7.9 A bar of metal can be treated as a spring. For a limited range, double the force gives double the stretch.

Nonhelical Macroscopic Springs

We saw in Chapter 4 that a bar of metal can be treated as a spring, since stretching or compressing the bar stretches or compresses the interatomic bonds (Figure 7.9). Double the force produces double the change in length, as long as we're in the range where the spring approximation to the interatomic force is adequate. This is usually described for a bar in terms of Young's modulus Y, which we studied in Chapter 4, where the tension force F_T (per unit area) is proportional to the stretch ΔL (per unit length):

$$\frac{F_T}{A} = Y\frac{\Delta L}{L}$$

Since ΔL, like s, represents a change in length, and F_T is a spring-like force (tension force), this equation can be rewritten to parallel the spring force equation:

$$F_T = \left(\frac{YA}{L}\right)|\Delta L| \quad \text{is like} \quad F = (k_s)|s|$$

The effective "spring stiffness" of the metal bar is YA/L.

If you stretch the bar too much, two things can make the bar stop behaving in a spring-like manner. You might exceed the interatomic stretch for which $k_s s$ is a good approximation to the magnitude of the interatomic force. Also, the regular array of atoms may be disrupted by dislocations of the crystal structure, leading to large-scale slippage of crystal planes. Suddenly the bar "yields" and grows very much longer with little applied force.

A straight object is not usually used as a longitudinal spring but may be used in a bending mode. For example, when a diving board bends under a load, atomic bonds in the upper part of the board are stretched, and atomic bonds in the lower part are compressed (Figure 7.10). The net effect is a spring-like behavior of the diving board, with the amount of bend proportional to the applied force.

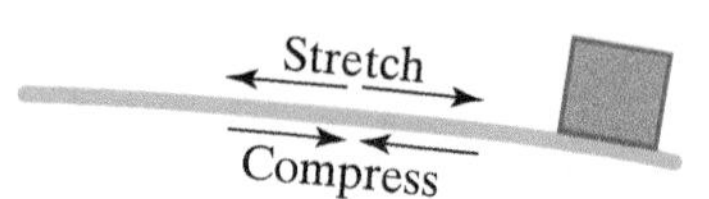

Figure 7.10 The bending of a diving board stretches and compresses interatomic bonds.

7.3 POTENTIAL ENERGY OF A PAIR OF NEUTRAL ATOMS

In Chapter 6 we found an equation for electric potential energy for pairs of charged particles. In Chapter 4 we studied the spring-like forces between electrically *neutral* atoms (a neutral object has equal amounts of positive and negative charge, so it has zero net charge). These interatomic forces are the superposition of electric attractions and repulsions among the positively charged protons and negatively charged electrons of which the atoms are made.

Figure 7.11 Two neutral atoms interact very little when far apart, attract each other at intermediate distances, and repel each other at very short distances.

When two electrically neutral atoms are far from each other (Figure 7.11), they exert almost no force on each other because the attractions between unlike charges (electrons in atom one with protons in atom two, and vice versa) are nearly equal to the repulsions between like charges. However, when the atoms come quite near each other, the electron clouds distort in such a way that the two atoms attract each other. Molecules and solid objects made of two or more atoms are bound together by these electric forces.

However, if you try to push two atoms even closer together, you eventually encounter a rapidly increasing repulsion. The attraction at a small distance and the repulsion when very close are due ultimately to the superposition of the electric forces of the various protons and electrons in the atoms, with the probable locations of these particles governed by the laws of quantum mechanics.

Figure 7.12 The interatomic potential energy is approximately spring-like if the bond is not stretched or compressed too much.

Potential Energy for Spring-like Interatomic Bonds

Since an interatomic bond behaves like a spring as long as it is not stretched or compressed excessively, we can guess that the equation for interatomic potential energy has the same form as the potential energy for an ideal spring (Figure 7.12):

APPROXIMATE INTERATOMIC POTENTIAL ENERGY

$$U = \frac{1}{2}k_s s^2 - E_M$$

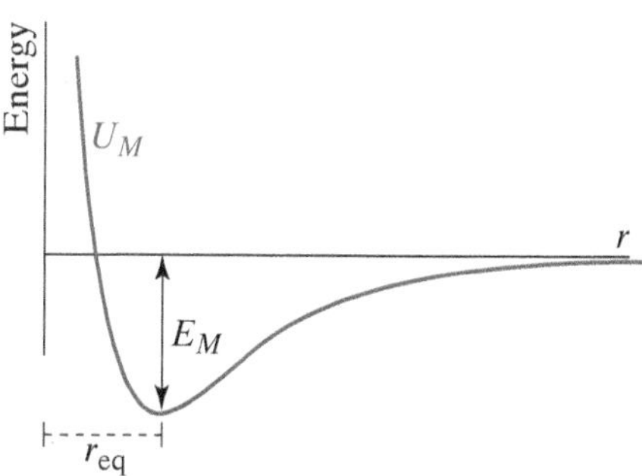

Figure 7.13 The Morse potential energy function U_M, representing the potential energy of two interacting neutral atoms as a function of separation.

What About Large Separations?

Of course, the potential energy of a real interatomic bond can't become infinitely large at large separation—it needs to approach zero. A more realistic potential energy function for the interatomic bond is called the Morse potential energy and is shown in Figure 7.13. The Morse potential energy is described by this equation:

$$U_M = E_M[1 - e^{-\alpha(r-r_{eq})}]^2 - E_M$$

The Morse approximation is designed to model the basic behavior of the interatomic force and is in approximate agreement with experimental measurements of the interatomic force.

The parameter E_M gives the depth of the potential energy at its minimum, and the parameter r_{eq} (on the order of 1×10^{-10} m) is the distance at which the potential energy curve is a minimum. The parameter α is adjusted to give the curve the right width. When the interatomic separation r is very large, U_M goes to zero, as it should.

The depth of the potential energy well, E_M, is typically a few electron volts (recall that 1 eV is equal to 1.6×10^{-19} J). For example, the potential energy well for oxygen, O_2, is about 5 eV deep. (As you saw in Chapter 6, a typical ionization energy is also in the range of a few electron volts.) However, for the noble gases such as helium and argon the well is so shallow that these atoms don't form stable diatomic molecules at room temperature, because collisions easily provide enough energy to break the molecules apart.

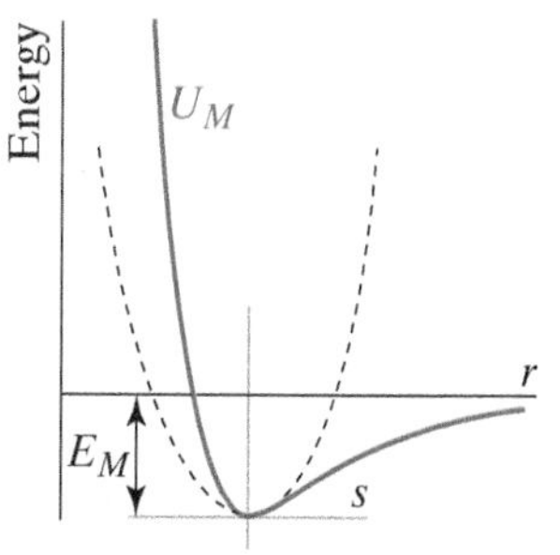

Figure 7.14 The bottom of the interatomic potential energy curve may be approximated by a parabola.

QUESTION What is the physical significance of the minimum of the curve?

At the minimum of the potential energy curve, the slope is zero, so the force between the two atoms is zero ($F = -dU/dr$). This is the position of stable equilibrium.

QUESTION What is the direction of the force on an atom at a distance smaller than the equilibrium distance?

The left side of the curve has a large negative slope, so the force $F = -dU/dr$ will be large and to the right (in the $+r$ direction), away from the other atom. This makes sense; when the atoms are too close together they push each other apart.

Likewise, if the atom moves to the right (farther away) there should be an attractive restoring force to the left. The slope of this region of the curve is positive, so the sign of the force $F = -dU/dr$ is negative, meaning its direction is to the left, as expected.

As indicated by the dashed curve in Figure 7.14, the bottom of the interatomic potential energy curve can be approximated by the spring-like potential energy function:

$$U = \frac{1}{2}k_s s^2 - E_M$$

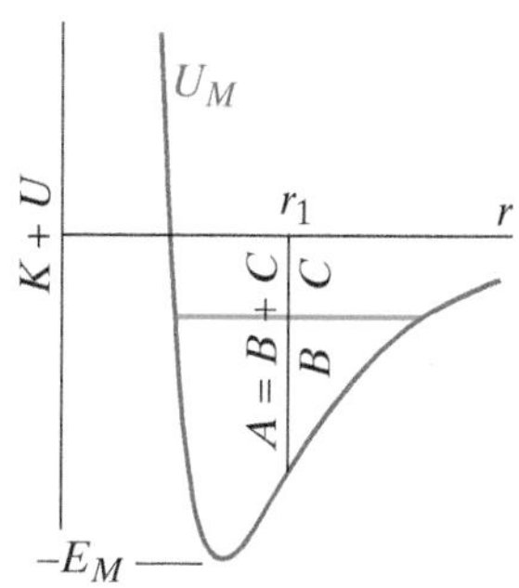

Figure 7.15 A horizontal line represents constant kinetic plus potential energy.

Because we can make this approximate fit, we know that for ordinary small oscillations around the equilibrium point the two atoms will act as though they are connected by a spring with a $-k_s s$ force. This is the rationale for modeling a solid as a collection of balls connected by springs.

Bound and Unbound States with Interatomic Potential Energy

In Figure 7.15, a horizontal line represents motion in which the sum of the kinetic energies and the potential energy of a two-atom system does not change. A horizontal line corresponds to a possible state of the system when it is isolated from external forces.

Checkpoint 3 At separation r_1 in Figure 7.15, what is the physical significance of the quantity A? Of the quantity B? Of the quantity C? In Figure 7.16, which of the states are bound states of a two-atom system? Which are unbound states?

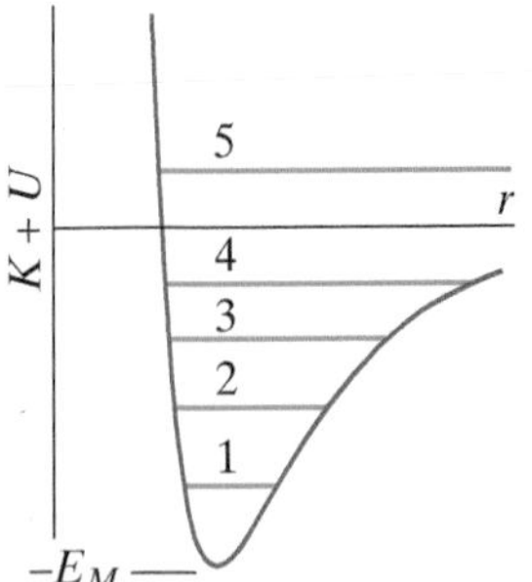

Figure 7.16 Which are bound states? Which are unbound states?

7.4 INTERNAL ENERGY

Consider two massive balls connected by a spring. If we model such an object as a point particle, in addition to its rest energy it can have only kinetic energy, which is associated with the motion of the center of mass of the object. However, by modeling the object as a point particle, we ignore the fact that in addition to the motion of the center of mass of the system, the balls could be oscillating back and forth, compressing and stretching the spring, and that the entire object could be rotating. We refer to the energy associated with the vibration and rotation of the system as "internal energy," and it is possible for the system to have such internal energy because it is not a point particle, and has some internal structure.

For example, Figure 7.17 shows two identical objects each consisting of two balls connected by identical springs, one relaxed, and the other compressed (and kept compressed by a string with negligible mass tied around it). If these two objects move at the same speed v, they have the same kinetic energy $K = \frac{1}{2}Mv^2$, where M is the total mass, but the object with the compressed spring has more internal energy, associated with the increase in spring potential energy.

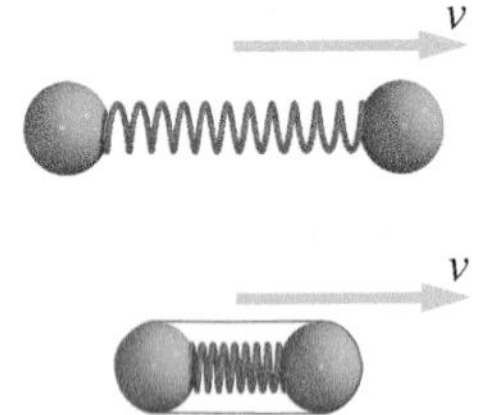

Figure 7.17 The object with the compressed spring has more "internal energy." It is kept compressed by a string of negligible mass.

It is often useful to split the total internal energy into various categories of interest to us in a particular situation. For example, if an object rotates about its center of mass, this internal energy is called "rotational energy" (Figure 7.18). If two balls connected by a spring oscillate relative to the center of mass of the system, this internal energy is called "vibrational energy" (Figure 7.19). If an object is stretched or compressed, there is an associated internal energy change, as in Figure 7.17. When you eat, you increase your internal energy in the form of more "chemical energy." If the temperature of an object increases, there is an increase in the internal energy that is called "thermal energy."

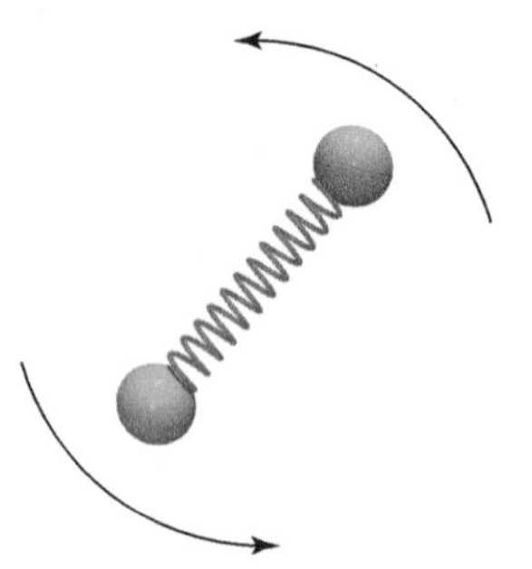

Figure 7.18 More rotation means more internal energy in the form of "rotational energy."

INTERNAL ENERGY

$$\text{Internal energy} = E_{\text{thermal}} + E_{\text{rotational}} + E_{\text{vibrational}} + E_{\text{chemical}} + \cdots$$

It is important to note that internal energy, thermal energy, rotational energy, vibrational energy, chemical energy, nuclear energy, and so on are not new kinds of energy. The energy of a multiparticle system has only three fundamental components:

- rest energy
- kinetic energy
- potential energy

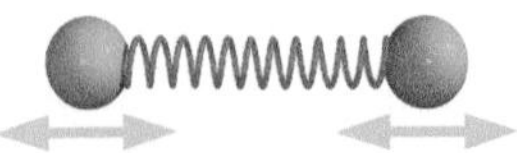

Figure 7.19 More vibration means more internal energy in the form of "vibrational energy."

We classify certain kinds of energy as internal energy mostly as a matter of convenience in analyzing energy changes in a complicated system. In fact, from a fundamental perspective, all of these internal energy increases can be thought of as contributing to an increase in the mass of the system: $M_{\text{sys}} = E_{\text{sys}}/c^2$. However, for many practical macroscopic calculations it makes sense to focus just on the changes in the thermal energy, or vibrational energy, or rotational energy, or deformational energy, or chemical energy, because these changes are extremely small compared to the huge total rest energy Mc^2.

Internal Energy at the Microscopic Level

At the microscopic level, a solid object is composed of atoms that are held together by spring-like chemical bonds. By considering the ball–spring model of a solid, we can see that an extended object can have internal kinetic energy associated with the motion of its component atoms and internal potential energy associated with the compression and stretch of interatomic bonds. Since all of the atoms in the solid are interconnected, this internal energy can flow throughout the system, and it ends up being distributed randomly among the many atoms and spring-like bonds. We call this energy "thermal energy." As the name suggests, this randomized energy at the microscopic level is associated with the temperature of an object—the greater the amount of thermal energy, the more internal motion there is, and the higher the temperature of the object.

Changes in the temperature of a macroscopic object are associated only with changes in its randomized thermal energy, and not with macroscopic rotations or vibrations. For example, if we increase the spin rate of a wheel that is spinning with negligible friction and air resistance, the rotational energy of the wheel increases, but the temperature of the wheel doesn't change. However, in some processes macroscopic kinetic energy may be converted to thermal energy. If two objects collide and stick to each other, some of their initial kinetic energy may be converted into random motions at the atomic level, with a corresponding increase in the temperature of the objects.

We can measure macroscopic spring potential energy and kinetic energy with simple measurements that don't require a microscope. However, when the temperature of a solid object increases, the increased energy inside the solid is not visible to the naked eye, and we face the problem of how to evaluate the increase in the thermal energy.

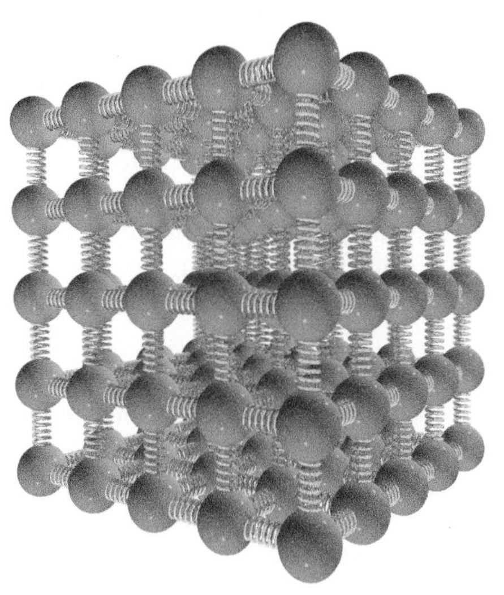

Figure 7.20 In our model solid we must account for the kinetic energy of every ball and the spring energy of every spring.

Microscopic Kinetic and Spring Potential Energy

The increased microscopic energy in a solid is in two forms. First, the atoms on average are moving around faster, with increased kinetic energy $\frac{1}{2}mv^2$. Second, there is increased spring potential energy $\frac{1}{2}k_s s^2$ in the interatomic bonds ("springs" in our model of solids), where k_s is the stiffness of an interatomic bond modeled as a spring (Figure 7.20).

In principle the internal energy in a solid could be evaluated by simultaneously measuring at some instant the momenta of *all* of the atoms, and the stretches or compressions of *all* of the springs (that is, changes in atomic positions away from their equilibrium positions, together with knowledge of the spring stiffness). Since there are about 1×10^{23} atoms (and even more "springs") in a macroscopic object, it is in practice impossible to evaluate the energy this way.

We might try to make microscopic measurements of just one atom (and its attached "springs"), assume that this is the average energy of each atom, and then multiply by the number of atoms there are in the object to get the total internal energy. That doesn't work either, because at a given instant we may choose an atom that is momentarily sitting motionless at its equilibrium position (zero $K + U$), but an instant later will be displaced away from its equilibrium position and moving rapidly. Energy keeps getting passed back and forth among the many atoms and springs, and measuring the energy of just one atom and its attached springs at one instant doesn't give us the correct average energy we would need to determine the total energy of the object.

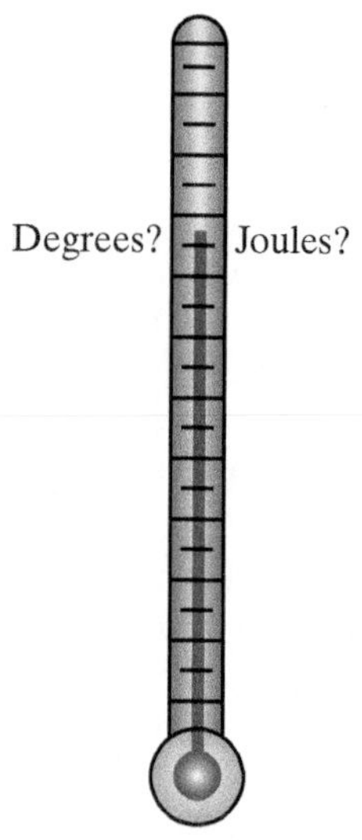

Figure 7.21 A thermometer might be calibrated in joules instead of in degrees.

Temperature

In the 1800s it was realized that for many systems the temperature is essentially a measure of the average random energy of the atoms in the system. Therefore we can use a thermometer to measure thermal energy. A familiar example is the mercury thermometer (silver colored) or the alcohol thermometer (tinted red for visibility). In these thermometers, a very thin column of liquid expands more than the glass when heated. By tradition, marks on the thermometer are placed to represent "degrees of temperature," but we could just as well place marks to indicate the average energy of the molecules in the thermometer, in joules (Figure 7.21). One kelvin (or one Celsius degree of temperature) is equivalent to an average molecular energy of about 1×10^{-23} J.

This is not the whole story. In Chapter 12 we will make more precise the relationship between temperature and thermal energy, and we will see that temperature is more directly related to a quantity called "entropy" than it is to energy. Nevertheless, for most ordinary systems at room temperature, the temperature of a system is a good measure of the average thermal energy of its atoms.

How does a thermometer work? When two objects are in good contact with each other, the average kinetic energies of the molecules in the two objects slowly come to be equal. At a location where two objects touch each other, molecules collide with each other, and the faster molecules on average lose kinetic energy to the slower molecules in a collision. Once the average kinetic energies of molecules in both objects have come to be equal, additional collisions on average make no further change.

To measure the thermal energy of an object, we place a thermometer in contact with the object. We wait for the temperatures (average molecular kinetic energies) of the object and thermometer to equilibrate, then we read the thermometer. A useful thermometer must have a relatively small mass compared to the object of interest, so that attaching the thermometer doesn't add or subtract much energy to or from the object. Another advantage of a small thermometer is that it will reach thermal equilibrium quicker than a large thermometer will.

How was the calibration between one kelvin (or one Celsius degree) and energy established? In one of a series of classic experiments performed by Joule in the 1840s, a paddle wheel in water was turned by a falling weight. The work done by the Earth turned the paddle wheel, which stirred the water. When the paddle wheel stopped, the increase in the internal energy of the water was equal to the work done by the Earth, assuming little rise in internal energy of the paddle wheel and container, and little energy transfer to the surroundings.

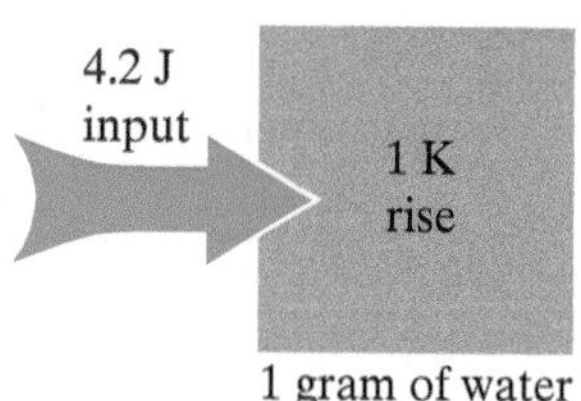

Figure 7.22 Energy input of 4.2 J into a gram of water raises the temperature by 1 K. We say that the specific heat of water is 4.2 J/K/g.

It was found for water that the energy required to raise the temperature of one gram of water by one kelvin (1 K) is 4.2 J. The "heat capacity" of an object is the amount of energy required to raise its temperature one kelvin. The "specific heat" is the heat capacity on a per-gram or per-mole or per-atom basis, and is a property of the material. The specific heat of water is 4.2 J/K/g (Figure 7.22). Other materials have different specific heats. We will study an atomic theory of specific heat in a later chapter.

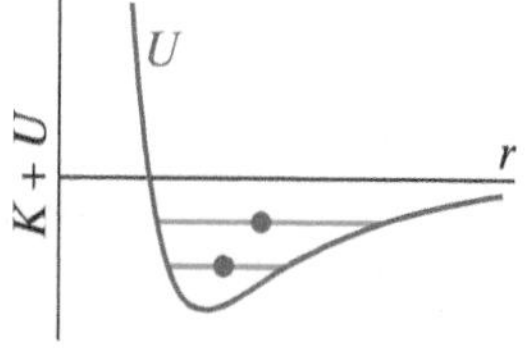

Figure 7.23 As $K + U$ (horizontal lines) increases, the average separation of the two bound atoms also increases.

Volume and Temperature

A liquid thermometer works because an ordinary liquid has a larger volume at higher temperature. Why does this happen?

For large oscillations of atoms whose interaction is described by a potential energy such as that shown in Figure 7.23, the average stretch increases. Because the actual potential energy curve is not really symmetric around the equilibrium point, with increasing energy the average interatomic bond length gets slightly longer. This is why an object typically expands at higher temperature. Higher temperature implies larger amplitude of oscillations and higher energy, and there is a slight shift in the center of the oscillations at higher energy.

In the interior of a solid the potential energy curve for an atom is symmetric, not asymmetric, because the potential energy is associated with forces exerted by atoms to the left and to the right. However, since the bond lengthening can start at the surface and propagate into the interior, a solid ultimately expands at higher temperatures.

Other Kinds of Thermometers

There are many other kinds of thermometers. All materials change in some way when they get hotter or colder, and some of these effects are the basis for useful temperature indicators. Liquid crystals are increasingly used in thermometers. In a liquid crystal there is some molecular order over long distances in the liquid, unlike a true liquid in which there is hardly any long-range order. In one form of liquid crystal thermometer, helical molecules remain roughly parallel to each other, and the distance between coils in the helix changes with temperature. Because the distance between coils is comparable to the wavelength of visible light, there is an effect on the way light is reflected by or transmitted through the liquid crystal. The result is that a change in temperature affects the appearance of the material.

Calculations Using Specific Heat

Joule found that it takes 4.2 J of energy input to a gram of water to raise its temperature 1 K. If such experiments are done on other materials, the temperature rise is found to be different. For example, the specific heat of ethanol is found to be 2.4 J/K/g, and the specific heat of copper is only 0.4 J/K/g. One of the many unusual and useful properties of water is that it has a very large specific heat on a per-gram basis, which means that it is difficult to change its temperature. Here is the meaning of specific heat on a per-gram basis, where $\Delta E_{\text{thermal}}$ is the rise in the internal energy of the system, in the form of increased atomic kinetic and potential energy:

SPECIFIC HEAT *C* ON A PER-GRAM BASIS

$$C = \frac{\Delta E_{\text{thermal}}}{m\Delta T} \quad (m \text{ in grams})$$

In principle, it should be possible to predict the specific heat of a material if something is known about its atomic structure, since a temperature rise is a measure of increased energy at the atomic level. Comparisons of calculations with experimental values of the specific heat data are a good test of our understanding of atomic models of solids, liquids, and gases. In later chapters you will learn how to predict the specific heat of a solid (based on the ball-and-spring model of a solid) and of a gas.

If the specific heat of a material is known, the amount of energy transfer into an object can be determined by measuring the temperature rise.

EXAMPLE **Energy Input Raises the Temperature**

You stir 12 kg of water vigorously, doing 36,000 J of work. If the container is well insulated (so that all of your energy input goes into increasing the energy of the water), what temperature rise would you expect?

Solution

$$\Delta E_{sys} = W$$
$$mC\Delta T = W$$
$$\Delta T = \frac{W}{mC} = \left(\frac{36{,}000\text{ J}}{12{,}000\text{ g}}\right)\left(\frac{1}{4.2\text{ J/K/g}}\right) = 0.7\text{ K}$$

Including units helps avoid making calculational mistakes. Note that a lot of work is required to achieve a small temperature change.

EXAMPLE **Thermal Equilibrium**

A 300 g block of aluminum at temperature 500 K is placed on a 650 g block of iron at temperature 350 K in an insulated enclosure. At these temperatures the specific heat of aluminum is approximately 1.0 J/K/g, and the specific heat of iron is approximately 0.42 J/K/g. Within a few minutes the two metal blocks reach the same common temperature T_f. Calculate T_f.

Solution System: The two blocks
Surroundings: Due to the insulation, no objects exchange energy with the chosen system
Initial state: Different temperatures, not in contact
Final state: Blocks have come to thermal equilibrium, same T_f
Energy Principle: $\Delta E_{Al} + \Delta E_{Fe} = 0$ (The total energy of the two blocks does not change, because there is no energy transfer from or to the surroundings.)

$$m_1C_1(T_f - T_{1i}) + m_2C_2(T_f - T_{2i}) = 0$$

$$(300\text{g})(1.0\text{ J/K/g})(T_f - 500) + (650\text{ g})(0.42\text{ J/K/g})(T_f - 350) = 0$$

Solving for the final temperature, we find $T_f = 429$ K.

In words, what happens in this process is that the aluminum temperature falls from 500 K to 429 K, and the iron temperature rises from 350 K to 429 K. The thermal energy decrease in the aluminum is equal to the thermal energy increase in the iron.

What Part of Internal Energy is "Thermal"?

What we mean by a change of "thermal" energy is that part of the internal energy that is associated with a temperature change. When two cars hit head-on and stop dead, their lost macroscopic kinetic energies show up as increased internal energy. A thermometer will register a temperature increase in the metal parts, and we say there is increased thermal energy $\Delta E_{thermal} = mC\Delta T$. This "random" energy does not account for all of the increased internal energy. In such collisions some of the internal energy rise is in the form of locked-in potential energy in compressed interatomic bonds, a "nonrandom" form of internal energy.

Suppose you compress a spring, then hold it compressed with a rubber band. Simply sliding the rubber band off the spring will release the energy stored in the spring. Similarly, one can get back some of the stored energy in deformed metal by heating the metal to free up the locked-in "springs."

In many situations it isn't possible to say how much of the internal energy is thermal, but if the heat capacity is known, we can use a thermometer to measure a *change* in the thermal energy.

Checkpoint 4 300 g of water whose temperature is 25°C are added to a thin glass containing 800 g of water at 20°C (about room temperature). What is the final temperature of the water? What simplifying assumptions did you have to make in order to determine your approximate result?

7.5 ENERGY TRANSFER DUE TO A TEMPERATURE DIFFERENCE

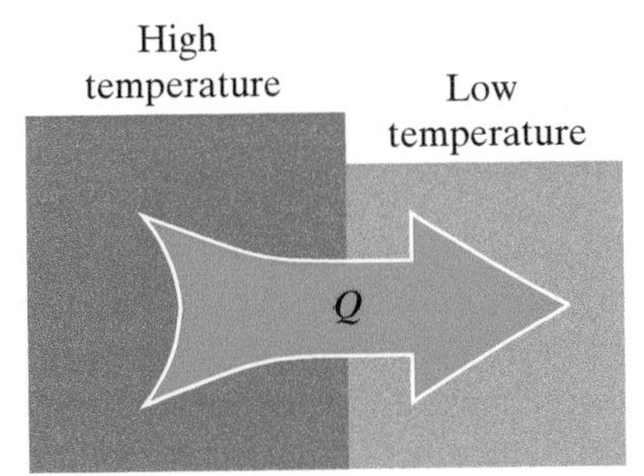

Figure 7.24 A hot block in contact with a cold block.

Work is mechanical energy transfer into or out of a system involving forces acting through macroscopic displacements. When a hot object is placed in contact with a cold object, energy is transferred from the hot object to the cold one, but there are no macroscopic forces or displacements, so we don't refer to this kind of energy transfer as work.

Instead, we speak of "energy transfer Q due to a temperature difference." At the microscopic level there is actual work; when a hot block is placed in contact with a cold block (Figure 7.24), at the interface the atoms in the two blocks collide with each other, and do work on each other. The atoms in the hot block have greater average kinetic energy than the atoms in the cold block, so in an individual collision it is likely that a fast-moving atom in the hot block loses energy to a slow-moving atom in the cold block.

It can happen that the atom in the hot block happens to be moving slowly and gains energy when it is hit by an atom in the cold object that happens to be moving fast, but this is less likely. On average there is energy flow ("microscopic work") from the hot block to the cold block. These energy changes will diffuse throughout the blocks. If left to themselves, the two blocks will eventually come to the same temperature, intermediate between their two initial temperatures.

Q = ENERGY TRANSFER DUE TO A TEMPERATURE DIFFERENCE

The symbol Q represents an amount of energy that flows from the surroundings into a system, due to a temperature difference between the system and the surroundings. One can also call this process "microscopic work." Q can be positive or negative.

If we choose the cold block as our system of interest, this system gains energy (at the expense of the surroundings, the hot block). At the atomic level, this energy increase is the result of work done on the atoms at the surface of the cold block. Usually we are unable to observe these atomic-level interactions directly. We can infer the amount of energy transfer into the system by observing the temperature rise of that system, if we know the specific heat of the material.

It is common practice to denote energy transfer due to a temperature difference (microscopic work) by the letter Q and work (macroscopic, mechanical energy transfer) by the letter W. The energy equation for an "open" (not isolated) system is then

$$\Delta E_{\text{sys}} = Q + W$$

That is, energy transfer due to a temperature difference and mechanical energy transfers into or out of the system produce a change in the energy of the system. The energy of the system includes particle energy and potential energy, at the microscopic or macroscopic level.

Sign of Q

Like W, Q can be negative. If there is energy transfer out of a system, we give Q a negative sign, because the change in the energy of the system is a decrease. See Figures 7.25 and 7.26. Typically such transfer would be due to the system having a higher temperature than the surroundings. Similarly, if the system does work on its surroundings rather than the other way around, the sign of W is negative. (Some older textbooks on thermal physics reverse the sign of W, counting energy outputs from the system as positive.)

It is thought that for the Universe as a whole Q and W are zero, and the energy of the Universe does not change. However, even a small system may be effectively closed, with its energy unchanging, if it is thermally well insulated (to prevent the flow of energy due to a temperature difference) and if there are no other energy transfers from the surroundings.

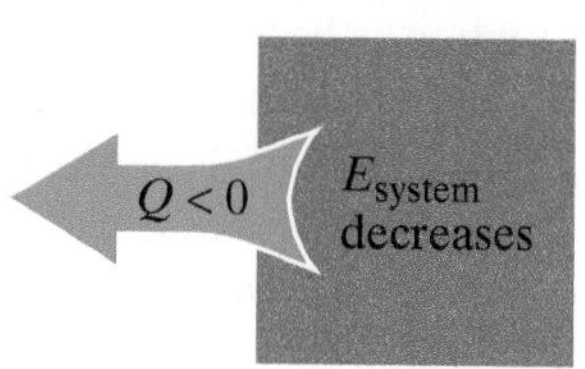

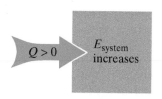

Figure 7.25 If energy flows from the surroundings into the system, Q is positive.

$Q < 0$
E_{system} decreases

Figure 7.26 If energy flows from the system into the surroundings, Q is negative.

Terminology: The Word *Heat*

The technical meaning of the word "heat" in science is not the same as its everyday meaning, just as the scientific meaning of the word "work" is different from its everyday meaning. In science the word "heat" is reserved to refer to energy transfer across a system boundary due to a temperature difference (microscopic work); this quantity is represented by the symbol Q. One cannot say "there is heat in the object"; instead, one says there is thermal energy E_{thermal} in the object. The amount of energy in the object might increase due to heat Q (microscopic work). However, even professional scientists sometimes slip in their usage of the word heat!

To avoid confusion, we will avoid the use of the word heat as a noun. Instead, we will speak of "energy transfer Q due to a temperature difference" between system and surroundings, and change of "thermal energy" inside the system $\Delta E_{\text{thermal}}$ (in the form of increased atomic kinetic energy and potential energy).

> **Checkpoint 5** Suppose that you warm up 500 g of water (half a liter, or about a pint) on a stove while doing 5×10^4 J of work on the water with an electric beater. The temperature of the water is observed to rise from 20°C to 80°C. What was the change in the thermal energy of the water? Taking the water as the system, how much transfer of energy Q due to a temperature difference was there across the system boundary? What was the energy change of the surroundings?

Other Kinds of Energy Transfers

Work W and energy transfer Q due to a temperature difference are very common kinds of energy transfers into or out of an open system, but there are several other kinds of energy transfer. Here is a list, together with examples of each type.

- W, mechanical work $\vec{F} \bullet \Delta\vec{r}$, in which a force acts through a macroscopic distance: Throw a ball, increasing the ball's kinetic energy; compress a macroscopic spring.
- Q, energy transfer due to a temperature difference between system and surroundings: Boil water over a hot flame; a warm house loses energy to the cold outside air.
- Matter transfer: Put gasoline into a car; natural gas is piped into the home.
- Mechanical waves: Sound enters a microphone and moves a plate to generate electric signals; ocean waves drive an electric generator.
- Electricity: Electrical current enters the house and is used to run appliances; electrical current stores energy in the batteries of an electric car.
- Electromagnetic radiation, including visible light, infrared light, gamma rays, x-rays, and ultraviolet light: Sunlight absorbed by the Earth raises the temperature (and the thermal energy) of the daylight side of the Earth. We will usually account for this kind of energy transfer in terms of photons

(packets of electromagnetic energy) crossing the system boundary, to be discussed in a later chapter.

A general form of the Energy Principle is this:

THE ENERGY PRINCIPLE

$$\Delta E_{sys} = W + Q + \text{other energy transfers}$$

Other energy transfers into a system may include matter transfer, mechanical waves, electric current, and electromagnetic radiation.

For historical reasons, many textbooks dealing with thermal physics define work W as the work done *by* a system rather than work done *on* a system, so the sign of W is changed. We define W as the work done on the system to be consistent with other, non-thermal uses of the Energy Principle.

Checkpoint 6 An electric hot plate raises its own internal energy and the internal energy of a cup of water by 8000 J, and there is at the same time 1000 J transferred to the cooler air (that is, $Q = -1000$ J). How much energy was transferred to the hot plate in the form of electricity?

The Steady State

A system is in a *steady state* if its energy does not change despite the fact that there are energy transfers between the system and the surroundings. In the steady state, the rates of energy flow into and out of the system must be equal.

EXAMPLE **An Electric Heater**

An electric heater receives an energy input of 5000 J of electric energy. During this time the heating element is maintained at a constant high temperature. What is $\Delta E_{thermal}$ for the heater, and what is Q, the energy transfer between the heater's hot heating element and the cooler air?

Solution System: Heater
Surroundings: Electric wires, air
Energy Principle: $\Delta E_{sys} = W + Q + \text{other energy transfers}$

Since there is no internal energy change, $\Delta E_{thermal} = \Delta E_{sys} = 0$. This is a steady-state situation.

$$\Delta E_{sys} = W + Q + \text{electric energy input}$$
$$0 = 0 + Q + 5000\,\text{J}$$
$$Q = -5000\,\text{J}$$

In this situation the sign of Q is negative, reflecting the fact that energy is transferred from the heating element to the air. We describe this situation as a steady state, not equilibrium, because there are energy transfers into and out of the system. However, these energy transfers, plus and minus, add up to zero, so there is no change internally.

If the heater were in an insulating enclosure and no energy could flow into the air, as electric energy continually flowed into the heater the temperature of the heating element would keep increasing until the element melted.

QUESTION Is Q the same as $\Delta E_{thermal}$?

No. This is an important distinction. Q represents an energy transfer between system and surroundings due to a temperature difference. $\Delta E_{thermal}$, the change of the thermal energy of a system, can differ from Q, because there may be other positive or negative energy inputs to the system. In the previous example of the electric heater, the thermal or internal energy of the heating element did not change ($\Delta E_{sys} = 0$, steady state), but $Q = -5000$ J.

Checkpoint 7 In a certain time interval, natural gas with energy content of 10,000 J was piped into a house during a winter day. In the same time interval sunshine coming through the windows delivered 1000 J of energy into the house. The temperature of the house didn't change. What was $\Delta E_{thermal}$ of the house, and what was Q, the energy transfer between the house and the outside air?

7.6 POWER: ENERGY PER UNIT TIME

In technical usage, the word "power" is defined to mean "energy per unit time." From the point of view of energy usage, it makes no difference whether you take a minute or a month to climb a flight of stairs, yet from a practical point of view you certainly do notice the *rate* of energy usage, called power. The units of power are joules per second, or watts (honoring James Watt, the developer of the first efficient steam engine).

POWER

Energy per unit time (J/s or W)

We can construct an equation for the instantaneous power associated with the work done by a force, through the following reasoning. If a force $\vec{F}$ does an amount of work $\vec{F} \bullet \Delta\vec{r}$ in a time Δt, then:

$$\text{power} = \frac{\vec{F} \bullet \Delta\vec{r}}{\Delta t} \to \vec{F} \bullet \frac{d\vec{r}}{dt}$$

INSTANTANEOUS POWER

$$\text{power} = \vec{F} \bullet \vec{v}$$

EXAMPLE **Light Bulb Power**

How much energy is required to run a 100 W light bulb for an hour?

Solution

$$(100\ \text{J/s})(1\ \text{h})\left(60\frac{\text{min}}{\text{h}}\right)\left(60\frac{\text{s}}{\text{min}}\right) = 3.6 \times 10^5\ \text{J}$$

Checkpoint 8 A vehicle with a mass of 1000 kg has an engine whose maximum power output is 50 kW (about 67 hp; 1 hp is 746 W). At a speed of 20 m/s (about 45 mi/h) and maximum power, determine the maximum acceleration by calculating the force that is acting.

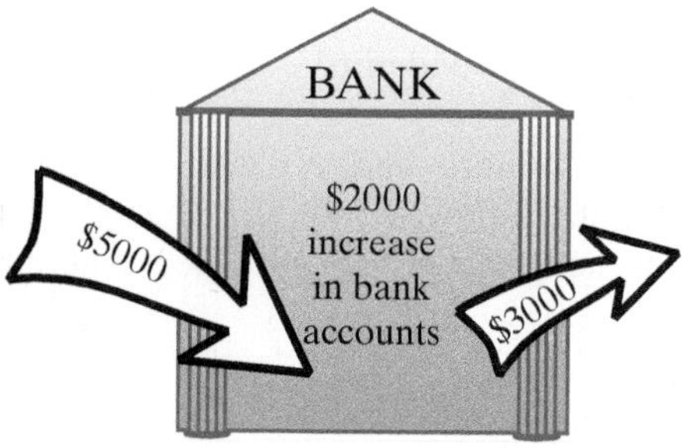

Figure 7.27 A bank is an open system.

7.7 OPEN AND CLOSED SYSTEMS

We have already had some experience with the difference between a closed system and an open system. Because of the importance of this distinction, in this section we will go more deeply into the issues. Figure 7.27 shows transfers of money into and out of a bank during a particular time period, and the corresponding change in accounts inside the bank.

QUESTION We have shown a situation in which more money is transferred into the bank than is transferred out. During these transactions, does the total amount of money inside the bank remain unchanged?

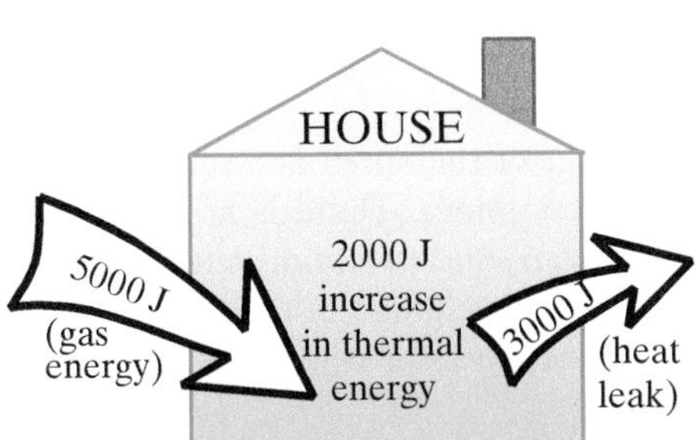

Figure 7.28 A house is an open system.

In Figure 7.27 \$5000 came in and \$3000 went out, for a net gain of \$2000. We say that the bank is an "open system"—a portion of the world open to transfers in and out, and therefore subject to changes in its internal amount of money. During times when the bank does not permit transfers, the bank is temporarily a "closed system" and its total internal amount of money is unchanged, although there may be changes in the form of money inside the bank such as transfers between checking accounts and savings accounts. Now consider Figure 7.28, which shows energy transfers during a particular time period in the winter, into and out of a house that is heated by gas, and the corresponding change in the energy inside the house. In this situation more energy comes into the house than goes out, and as a result the temperature inside the house rises.

QUESTION During these energy transactions, the total amount of energy in the whole Universe is unchanged, but is the amount of energy inside the house unchanged?

The amount of energy inside the house increased by 2000 J, and the energy of the surroundings decreased by 2000 J. We say that the house is an "open system"—a portion of the Universe open to energy transfers, and therefore subject to changes in its internal amount of energy. If the house could be perfectly insulated against heat leakage (and solar radiation) and the gas (and electricity) turned off, the house would be a "closed system" with respect to energy and its total internal amount of energy would be unchanged, although there may be changes in the form of energy inside the house, such as the family dog converting chemical energy into kinetic energy when chasing its tail.

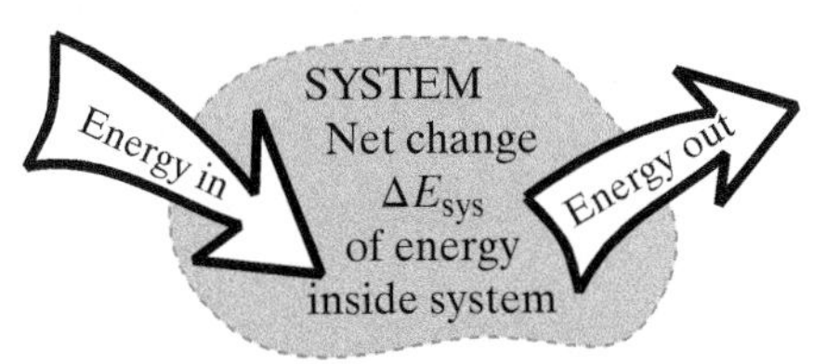

Figure 7.29 The change of energy of the system is equal to energy transferred into the system across the system boundary minus energy transferred out of the system across the system boundary.

These considerations lead to the following scheme for keeping track of energy. Choose some portion of the Universe and mentally surround it by a dashed line marking the boundary (Figure 7.29). Then the energy transfer into the system across the system boundary minus the energy transfer out of the system across the system boundary is equal to the change of the energy inside the system.

The change in energy *inside* the system, ΔE_{sys}, can be either a positive or a negative amount of energy. For example, take an electric car as the system of interest:

Charge the battery: ΔE_{sys} is positive (more energy stored in the battery).

Do work on the car by pushing it: ΔE_{sys} is positive (more kinetic energy).

Run the headlights (radiate electromagnetic energy): ΔE_{sys} is negative (less energy stored in the battery).

The car is an open system whose total amount of energy changes due to energy transfers into or out of the system: the energy of an *open* system is *not* constant.

In contrast, the total energy of the Universe remains constant, because there are compensating energy changes in the surroundings of the car: the electric company lost a store of chemical energy by running its generators to charge the car, you used up some chemical energy to push the car, and the light from the headlights is absorbed by the surroundings and leads to a rise in temperature of the surroundings. The measurements that we are able to make confirm the premise that the Universe is a closed system whose total amount of energy never changes, although the form of the energy may change.

For any closed system, *inflow* = *outflow* = 0, so $\Delta E_{sys} = 0$.
The energy of a closed system does not change.

The Universe as a whole is the most important example of a closed system, but it is often the case that a portion of the Universe can be considered to be a closed system, at least approximately. For example, put hot water and ice cubes into a very well insulated container. During the short time that the ice melts and the water gets somewhat cooler, we can neglect the small amount of energy leakage through the walls of the insulated container. There is an increase in the energy of the ice cubes (which changed from solid to liquid), and a decrease in the energy of the hot water, but negligible net change in the energy inside the container, which is approximately a closed system whose total energy is (approximately) unchanged.

7.8 THE CHOICE OF SYSTEM AFFECTS ENERGY ACCOUNTING

In Section 6.7 we analyzed a falling ball in two ways. When we chose the ball alone as the system, there was no potential energy, but the surroundings did work on the ball. Alternatively, when we chose the ball plus the Earth as the system, no external work was done, but the potential energy of the system changed as the ball fell. The form of the energy equation depended on our choice of system, but the physical results were consistent.

Let's explore this idea further in the case of a complicated system in which internal energy is important. We will analyze the same problem three times, using three different choices of system. As before, we will find that the form of the Energy Principle will be different in each case, but the results of each analysis are consistent.

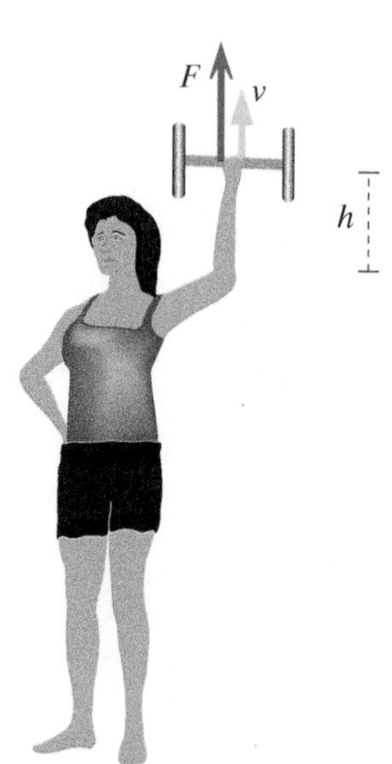

Figure 7.30 A woman lifts a barbell a distance h, exerting a force F. In the final state the barbell is moving upward with speed v.

The Woman, the Barbell, and the Earth

Consider a somewhat complicated situation: a woman applying a constant force F to lift a barbell of mass m from rest through a distance h, at which point the barbell is not only higher above the Earth but also has acquired some speed v (Figure 7.30). We're considering a time when the barbell is still headed upward, before the woman has brought the barbell to a stop above her.

For convenience in the following discussion, let E_w represent the energy of the woman, with the following energy terms:

E_w = internal energy of woman +
kinetic energy of woman (moving arm) +
gravitational energy of woman and Earth

System: Woman + Barbell + Earth

System: Woman + barbell + Earth (Figure 7.31)
Surroundings: Nothing significant
Initial state: Barbell at rest
Final state: Barbell has moved upward a distance h, and has speed v

$$E_f = E_i + W$$
$$K_f + U_f + E_{w,f} = K_i + U_i + E_{w,i} + W$$
$$\frac{1}{2}mv^2 + mgh + E_{w,f} = 0 + 0 + E_{w,i} + 0$$
$$\Delta E_w = -\left(\frac{1}{2}mv^2 + mgh\right)$$

Figure 7.31 System: woman + barbell + Earth.

Because the woman was part of the chosen system, we were able to calculate the energy term ΔE_w, which represents changes in energy associated with the woman. Her internal energy change involves both a decrease in the chemical energy stored in her body and an increase in her thermal energy. The woman must expend more energy than the rest of the system gets, since she not only has to supply energy to the rest of the system but also has to supply her own energy increases: increased kinetic energy and gravitational potential energy associated with her moving, raised arm, and increased thermal energy, because her temperature rises a bit when she exerts herself.

System: Barbell

System: Barbell (Figure 7.32)
Surroundings: Woman and Earth
Initial state: Barbell at rest
Final state: Barbell has moved upward a distance h, and has speed v

Figure 7.32 System: barbell only.

$$E_f = E_i + W$$
$$K_f = K_i + W_{\text{woman}} + W_{\text{Earth}}$$
$$\frac{1}{2}mv^2 = 0 + Fh - mgh$$

QUESTION Why didn't we include potential energy in the energy equation?

Since the chosen system consists of a single object (the barbell), there are no interactions between objects within the system, and hence no potential energy.

QUESTION Which is bigger, F or mg? Why?

Since the barbell's upward momentum increases, the force F exerted by the woman must be larger than mg, since the net force in the upward direction is $F - mg$.

By choosing the barbell alone as the system, we were able to calculate the change in kinetic energy of the barbell—a quantity we could not find when we chose all three objects as the system.

QUESTION Can we relate ΔE_w to the force exerted by the woman?

We can compare our results for system 1 (woman + barbell + Earth) and system 2 (barbell only).

$$\Delta E_w = -\left(\frac{1}{2}mv^2 + mgh\right)$$
$$Fh = \frac{1}{2}mv^2 + mgh$$
$$\text{So } \Delta E_w = -Fh$$

The sign makes sense: we had concluded earlier that her energy change must be negative.

System: Barbell + Earth

System: Barbell + Earth
Surroundings: Woman
Initial state: Barbell at rest

Figure 7.33 System: barbell + Earth.

Final state: Barbell has moved upward a distance h, and has speed v

$$E_f = E_i + W$$
$$K_f + U_f = K_i + U_i + W_{\text{woman}}$$
$$\frac{1}{2}mv^2 + mgh = 0 + 0 + Fh$$

If we choose the system to consist of the barbell plus the Earth (Figure 7.33), this system changes configuration as the woman pushes the two pieces of the system apart. The result we get with this choice of system is consistent with our previous results, even though in the previous case (barbell alone) the term $-mgh$ corresponded to work done by the surroundings, whereas when we chose the Earth plus the barbell we found a potential energy change of $+mgh$.

QUESTION Does the woman do any work on the Earth? Why or why not?

The woman's feet push down on the Earth but do negligible work, because there is essentially no displacement there (the Earth hardly moves). Work done by a force is calculated by taking into account the displacement of the point where the force is applied. In this case, the point of application hardly moves, so the work done by the force is negligible.

Looking back over the equations obtained for different choices of system, you can see how energy terms that represent transfers across the system boundary for one choice of system become changes of energy *inside* the system for a different choice of system. Comparing equations for different choices of system can be useful in determining an unknown quantity (such as the energy change in the woman, $-Fh$, in the example you just worked through). Also, analyzing a process for more than one choice of system is a good check on your calculations and your understanding.

Checkpoint 9 Consider a harmonic oscillator (mass on a spring without friction). Taking the mass alone to be the system, how much work is done on the system as the spring of stiffness k_s contracts from its maximum stretch A to its relaxed length? What is the change in kinetic energy of the system during this motion? For what choice of system does energy remain constant during this motion?

7.9 THE CHOICE OF REFERENCE FRAME AFFECTS ENERGY ACCOUNTING

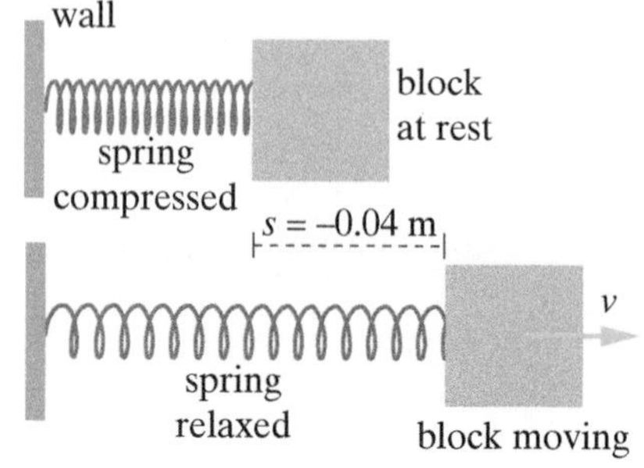

Figure 7.34 A block at rest is launched by a compressed spring.

We have just seen how the choice of system affects energy accounting. The Energy Principle is valid independent of our choice of system, but the terms that appear in the energy equation are different for different choices of system. In a similar way, the Energy Principle is independent of our choice of (inertial) reference frame, but the terms that appear in the energy equation are different for different choices of reference frame. It is instructive to look at the specific example of a compressed spring launching a block, as seen from two different reference frames.

Consider a spring with stiffness 400 N/m attached on the left to a wall, with a block of mass 1 kg held at rest against the spring (Figure 7.34). The spring is compressed, with $s = -0.04$ m. You release the block from rest, and the spring expands, accelerating the block on a low-friction surface. Choose the spring plus the block as the system.

A Stationary Reference Frame

First, let's view the system from a stationary reference frame, as usual. Assume that you, the observer, are standing motionless beside the wall.

QUESTION How much work is done on the spring–block system?

The upward force of the surface and the downward force of the Earth do no work, because there is no displacement in the y direction. Similarly, the wall does no work, because the force that the wall exerts on the left end of the spring acts through zero distance and does no work ($F_{\text{wall}}\Delta x = 0$). The wall force does increase the momentum of the system by providing an impulse $F_{\text{wall}}\Delta t$, but it does not affect the energy of the system. Here is the Energy Principle, with the initial state being just after release and the final state being when the spring reaches its relaxed length:

$$K_f + U_f = K_i + U_i$$
$$\frac{1}{2}mv^2 + 0 = 0 + \frac{1}{2}k_s s^2$$
$$\frac{1}{2}(1\text{ kg})v^2 = \frac{1}{2}(400\text{ N/m})(0.04^2\text{ m}^2) = 0.32\text{ J}$$
$$v = 0.8\text{ m/s}$$

A Moving Reference Frame

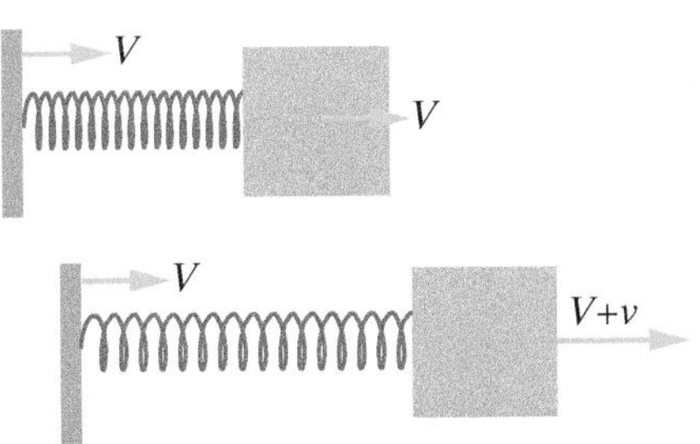

Figure 7.35 In a reference frame moving to the left with speed V, the wall moves to the right with speed V, and the initial speed of the block is also V. At a later moment the block is moving faster than the wall, with some speed $V+v$.

Now let's view the same system from a moving reference frame. Suppose you (the observer) move to the left with constant speed $V = 10$ m/s. In this altered perspective, the wall has a constant velocity of 10 m/s to the right, and the block's initial velocity is also 10 m/s to the right (Figure 7.35). You would see the system the same way if you stood still and the apparatus moved to the right. We again choose the spring plus the block as the system.

QUESTION In this moving reference frame, is the work done by the wall on the left end of the spring negative, zero, or positive?

In the original stationary reference frame, with the wall always at rest, the wall force did no work because the point of application of the wall force did not move; there was zero displacement. Now, however, the wall is moving at a speed of 10 m/s, and during the expansion of the spring the wall force acts through a significant distance to the right, doing positive work W on the spring.

In this moving reference frame the initial speed of the block is not zero but $V = 10$ m/s, and, using the results of the previous analysis, we know that its final speed is $V + v = 10.8$ m/s, because it is moving 0.8 m/s faster to the right than the wall, which is moving at 10 m/s to the right. We can calculate the work done by the wall in this reference frame:

$$K_f + U_f = K_i + U_i + W$$
$$\frac{1}{2}m(V+v)^2 + 0 = \frac{1}{2}mV^2 + \frac{1}{2}k_s s^2 + W$$
$$\frac{1}{2}(1\text{ kg})(10.8\text{ m/s})^2 = \frac{1}{2}(1\text{ kg})(10\text{ m/s})^2 + \frac{1}{2}(400\text{ N/m})(0.04^2\text{ m}^2) + W$$
$$58.32\text{ J} = 50\text{ J} + 0.32\text{ J} + W$$
$$W = 8\text{ J}$$

Notice that the initial and final potential energies of the spring were the same in both reference frames because spring potential energy only depends on the stretch of the spring, which is the same in both reference frames.

Checking Our Assumptions

In our calculation of the work done by the wall in the moving reference frame, we assumed that the Energy Principle functions correctly in both reference frames. To check this assumption, we can integrate $F_x dx$ to calculate the work done by the wall directly, in the moving reference frame, without using results obtained from the stationary reference frame. If the work is indeed 8 J, our assumption will be justified.

The force exerted by the wall on the spring is $-k_s s$. During the process of interest the stretch of the spring depends on time, and can be expressed as

$$s = (-0.04 \text{ m})\cos\left(\sqrt{\frac{k_s}{m}}t\right)$$

Initially the spring is compressed: $s = -0.04$ m. At the end of the process $s = 0$, because the spring has reached its relaxed length. The duration of this process is one-quarter of the total oscillation period T of the spring–mass system, which is

$$T = 2\pi\sqrt{\frac{m}{k_s}} = 2\pi\sqrt{\frac{1 \text{ kg}}{400 \text{ N/m}}} = 0.314 \text{ s}$$

During each time interval dt the wall moves a distance $dx = Vdt$. So the work done by the wall on the spring–mass system is:

$$\begin{aligned} W &= \int_{x_i}^{x_f} F_x dx \\ &= \int_0^{T/4} -k_s s V dt \\ &= \int_0^{0.314/4} (-400 \text{ N/m})(-0.04 \text{ m})\cos\left(\sqrt{\frac{400 \text{ N/m}}{1 \text{ kg}}}t\right)(10 \text{ m/s})dt \\ &= 8 \text{ J} \end{aligned}$$

This direct calculation of the work done by the wall agrees with our previous result.

In changing from the original reference frame to the moving one, the work done by the wall changed from 0 to 8 J, the initial kinetic energy of the block changed from 0 to 50 J, and the final kinetic energy changed from 0.32 J to 58.32 J. All of these individual energy values changed a lot, but the Energy Principle is still valid. This is an example of the relativity principle, that physics principles are valid in all uniformly moving reference frames.

7.10 ENERGY DISSIPATION

The total energy of the Universe does not change, but useful energy is often "dissipated" into forms less useful to us. Push a chair across the floor, and some of the work that you do goes into raising the temperature of the floor and the chair, rather than into increasing the macroscopic kinetic energy of the chair. Throw a ball up into the air, and some of the initial energy is dissipated into increased microscopic energy of the air. Sliding friction, air resistance, viscous friction—all of these phenomena are examples of energy dissipation discussed in the rest of this chapter.

Air Resistance

Air resistance isn't a major factor when you drop a metal ball a short distance. A video sequence of a falling metal ball shows that the ball moves faster and faster as it falls (an effect that is hard to observe by eye alone). The time between adjacent frames in Figure 7.36 is 1/15 s, and the increasing distances between heights of the ball in adjacent frames show that the speed is getting faster and faster. (Also note the increasing blur due to faster motion of the ball.) The gravitational attraction of the Earth acting on the ball makes the momentum of the ball continually increase. A curve is drawn along the tops of the ball images. The major visible marks on the vertical meter sticks are 10 cm apart. In 7/15 s the ball falls about 1 m.

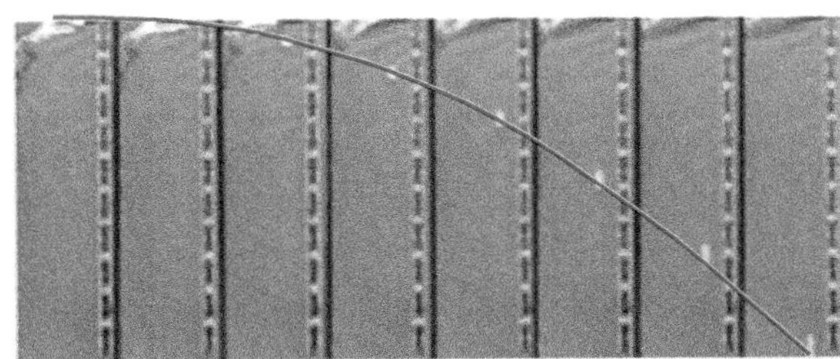

Figure 7.36 A sequence of video frames of a falling metal ball, which travels farther in each frame, reflecting the increase in its speed. (Image courtesy of Stacy Benson)

In contrast, if you drop a flat-bottomed paper coffee filter, a video sequence of the falling filter (Figure 7.37) shows that the filter's speed does increase at first, but instead of continuing to gain speed it quickly reaches a constant speed despite the gravitational force acting on it.

In the video sequence shown in Figure 7.37, the time between adjacent frames is again 1/15 s. In 16/15 s the filter falls about 1 m (the dense ball took only 7/15 s to fall that far). A curve is drawn through the centers of the coffee filter images. The nearly straight line in the later part of the motion indicates motion at constant speed.

Figure 7.37 A sequence of video frames of a falling coffee filter, which speeds up briefly, then falls at constant speed. (Image courtesy of Stacy Benson)

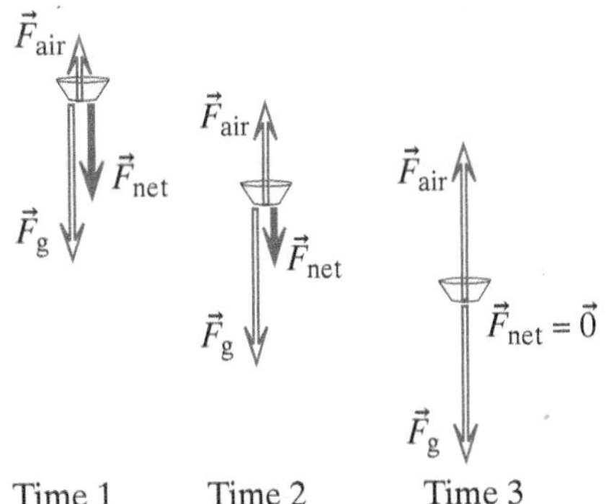

Figure 7.38 Forces acting on a falling coffee filter at three times during its fall.

Dependence of Air Resistance on Speed

The restraining effect of the air is called "air resistance" or "drag." You have probably observed that it is harder to move something quickly through a fluid such as water or air than to move it slowly. This suggests that air resistance acting on a falling object might depend on the speed of the object.

> QUESTION If the air resistance force on a falling object increases as the speed of the object increases, how will this affect the net force on the object?

Since the downward gravitational force on the falling object does not change, while the upward air resistance force increases as the object speeds up, the net downward force on the object should get smaller and smaller, and eventually reach zero (Figure 7.38).

> QUESTION If the net force on the falling coffee filter becomes zero, why doesn't the filter just float motionless in the air?

Consider the Momentum Principle:

$$\vec{p}_{\text{future}} = \vec{p}_{\text{now}} + \vec{F}_{\text{net}}\,\Delta t$$

At the instant when $\vec{F}_{\text{net}}$ becomes zero, the object is falling, so $\vec{p}_{\text{now}}$ is nonzero. Therefore a short time Δt in the future, the momentum $\vec{p}_{\text{future}}$ of the coffee filter will be equal to $\vec{p}_{\text{now}} + \vec{0}$, and the filter will continue to fall at constant speed.

Experiments show that the magnitude of the air resistance force is approximately proportional to the square of an object's speed. You can do a simple experiment to confirm this approximate dependence (see Problem P40).

Terminal Speed

A sky diver initially speeds up due to the gravitational force acting downward, but the sky diver's speed doesn't keep getting bigger and bigger, because the upward air resistance force increases with speed. Eventually the sky diver reaches a "terminal speed" and falls thereafter at constant speed despite the gravitational force. This terminal speed for falling humans is so high (about 60 m/s, or about 135 mi/h) that hitting the ground without a parachute is normally fatal, although there have been cases of people hitting deep snow at terminal speed and surviving.

Checkpoint 10 A sky diver whose mass is 90 kg is falling at a terminal speed of 60 m/s. What is the magnitude of the force of the air on the sky diver?

Dependence of Air Resistance on Cross-Sectional Area

It is easy to observe that the effect of air resistance increases as the cross-sectional area of an object increases—a sky diver falls much more slowly with an open parachute than with a closed parachute. A low-density paper coffee filter has a large cross-sectional area, so air resistance has a large effect on a falling coffee filter. In contrast, a high-density metal ball has a small cross-sectional area, but a large gravitational force acts on it, and air resistance may be negligible in comparison with gravity.

Dependence of Air Resistance on Shape

QUESTION From your own experience, would you expect larger air resistance (at a particular speed) for a pointed object or a blunt object?

As you might expect, the air resistance force depends not only on density, cross-sectional area, and speed, but also on the shape of the object. There is a larger air resistance force on a blunt object such as a coffee filter than on a ball with the same cross-sectional area. Streamlined objects such as arrows have reduced air resistance.

Dependence of Air Resistance on Air Density

QUESTION If you could double the density of the air through which a coffee filter falls, what change in the air resistance force would you expect?

You would probably expect a larger air resistance force in denser air. In fact, measurements of a variety of ordinary-sized objects moving through air or other fluids show that air resistance is proportional to the density ρ of the air.

For example, there is less air resistance at higher altitudes, where the air is less dense. Balls travel farther in Denver, Colorado, a mile above sea level, than they do in New York City.

An Empirical Equation for Air Resistance

An equation describing the air resistance force on a moving object must incorporate all the effects we have listed above: dependence on speed v, cross-sectional area A, shape of the object, and density of the air ρ. The shape effect is captured by a parameter denoted by the symbol C that reflects the sharpness or bluntness of the object. The parameter C is called the "drag coefficient," and is typically determined empirically. Typically $0.3 \leq C \leq 1.0$; blunter objects have higher values of C. The direction of the force is $(-\hat{v})$, opposite to the velocity.

APPROXIMATE AIR RESISTANCE FORCE (EMPIRICAL)

$$\vec{F}_{\text{air}} \approx -\frac{1}{2}C\rho A v^2 \hat{v}$$

for blunt objects at ordinary speeds. A is the cross-sectional area of the object; ρ is the density of the air; C reflects the bluntness of the object.

How important is the effect of air resistance in the everyday world? In Chapter 2 we analyzed the motion of a ball thrown through the air. However, we neglected air resistance, which can have a sizable effect. In Problem P47 you are asked to include the air resistance force given above. You should find that a baseball thrown at high speed by a professional baseball pitcher goes only about half as far in air as it would go in a vacuum.

However, even when we include air resistance in our predictions of the motion of a baseball, we still have ignored a force that can have a major effect on the trajectory. If a ball has spin, there is an effect of fluid flow around the ball that raises the air pressure on the side where the rotational motion is in the same direction as the ball's velocity, and lowers the air pressure on the other side, where the rotational motion is in the opposite direction to the velocity. If the force points to the side, the ball curves; this is an important effect in baseball pitching. If the force points upward (due to "backspin," as in Figure 7.39), it lifts the ball and extends the range. If the force points downward (due to "topspin") it decreases the range. The calculations involved in modeling this effect are quite complex, and involve "fluid dynamics" (the study of fluid flow).

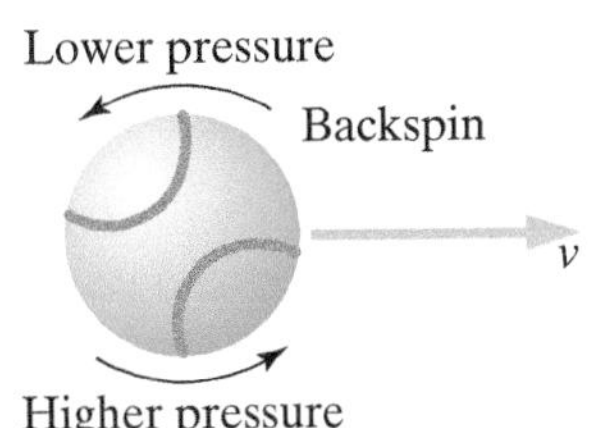

Figure 7.39 A spinning ball experiences additional forces.

QUESTION Is air resistance a fundamental force like the gravitational and electric forces?

No, air resistance is a result of the electric interactions between the molecules on the surface of a moving object and gas molecules in the air. Air resistance is the average result of a huge number of momentary contact electric interactions of air molecules hitting atoms on the surfaces of the falling object, and the average net air resistance is very difficult to calculate from fundamental principles at the molecular level.

Checkpoint 11 A coffee filter of mass 1.4 g dropped from a height of 2 m reaches the ground with a speed of 0.8 m/s. How much kinetic energy K_{air} did the air molecules gain from the falling coffee filter?

Mechanism of Air Resistance

We would like to understand the details of the interaction of the air with a falling object. In particular, we'd like to find a reason that the air resistance would depend on the square of the speed of the object, the cross-sectional area of the object, and the density of the air.

Let's consider the interactions of the air and object from a microscopic viewpoint. When you hold a coffee filter stationary, air molecules hit it nearly equally from above and below, as seen in Figure 7.40. (Actually there are slightly more collisions per second on the lower surface than on the upper surface, leading to a small upward buoyant force that is very small compared to air resistance.) However, when the filter is moving downward due to the gravitational attraction of the Earth, the bottom side of the filter is running into the air molecules, while the top side is moving away from the air molecules.

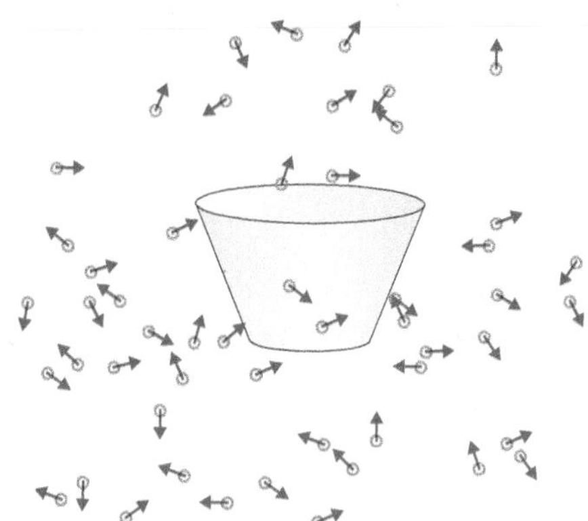

Figure 7.40 Air molecules collide with a falling coffee filter.

On average, the bottom side will have an increased number of collisions per second with air molecules, and with greater impact. On the top side there will be a reduced number of collisions per second, and with less impact. There is a net upward push on the filter. This air resistance force increases with increasing downward speed of the filter, because higher speed increases the rate and impact of collisions on the bottom and decreases the rate and impact of collisions on the top.

Eventually, when the downward speed of the coffee filter has become big enough, the net upward push by the air molecules becomes as large as the downward gravitational pull of the Earth. From then on the filter falls at a constant speed (terminal speed).

While this model captures the qualitative aspects of the situation, a quantitative analysis that predicts the v^2 dependence of the force is beyond the scope of this introductory course. We can accept the approximate force law as expressing a summary of experimental data.

> **Checkpoint 12** If you let a mass at the end of a string start swinging, at first the maximum swing decreases rather quickly, but once the swing has become small it takes a long time for further significant decrease to occur. Try it! Explain this simple observation.

Viscous Friction

Very small particles such as dust or droplets of fog falling slowly through air, or small objects moving through a thick liquid such as honey, experience a friction force that is proportional to v rather than v^2. This is called "viscous" friction.

The dependence of the friction force on v (for viscous friction) or v^2 (for air resistance) is approximately valid only in specific situations. The expressions for friction forces do not have the wide applicability of those for the gravitational and electric forces. However, even approximate expressions for friction forces allow us to make much better predictions than we could make if we ignored friction completely.

Viscous Friction in a Mass–Spring System

One of the most obvious things about the motion of a real macroscopic spring–mass system is that the amplitude gets smaller and smaller with time. Our model of a spring–mass system that we developed in Chapter 4 is too simple: we should include the effects of friction. See Problem P45 on numerical integration of a spring–mass system with friction.

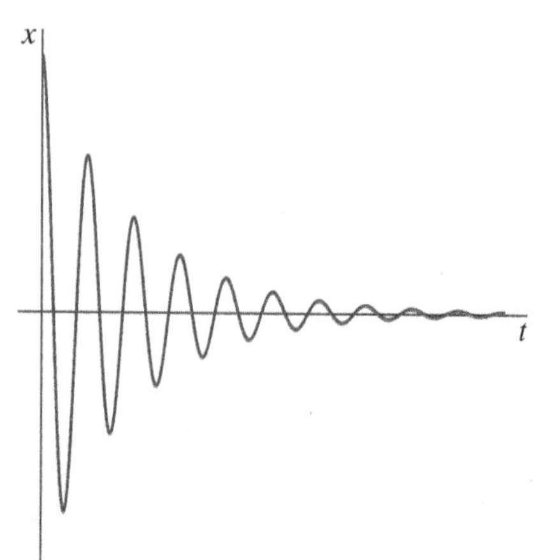

Figure 7.41 Position vs. time for an oscillator with viscous friction.

In the case of viscous friction, where the friction force is $\vec{f} = -c\vec{v}$ with c constant, it is possible with elaborate math (or by guessing the solution and plugging it into the Momentum Principle for the system) to find an analytical solution for the position x as a function of time.

$$x(t) = A\,e^{-(c/2m)t}\cos(\omega t)$$

The amplitude dies away exponentially with time (Figure 7.41). The angular frequency ω of this damped oscillator is

$$\omega = \sqrt{\frac{k_s}{m} - \left(\frac{c}{2m}\right)^2}$$

which is approximately $\sqrt{k_s/m}$ for low friction.

There is no such simple analytical solution if the friction force is independent of the speed (sliding friction) or proportional to v^2 (air resistance for large blunt objects), but such frictional forces are easily handled in a computational model.

Dissipation with Sliding Friction

In Chapter 4 we saw that when a block slides on a table, the table exerts a force on the block associated with compression of the interatomic bonds in the table and block. This force has two components: $\vec{F}_N$ normal to the surface, and $\vec{f}$ parallel to the surface, called the frictional force. Atoms of the block run into atoms in the table, and the resulting interactions stretch and compress interatomic bonds in the objects.

The frictional force is nearly independent of the speed of the block, and $f \approx \mu_k F_N$, where μ_k is the kinetic (motional) coefficient of friction.

If the block does not slide across the table, the frictional force is just big enough to make the net force be zero. In this case, $f \leq \mu_s F_N$, where μ_s is called the coefficient of static friction.

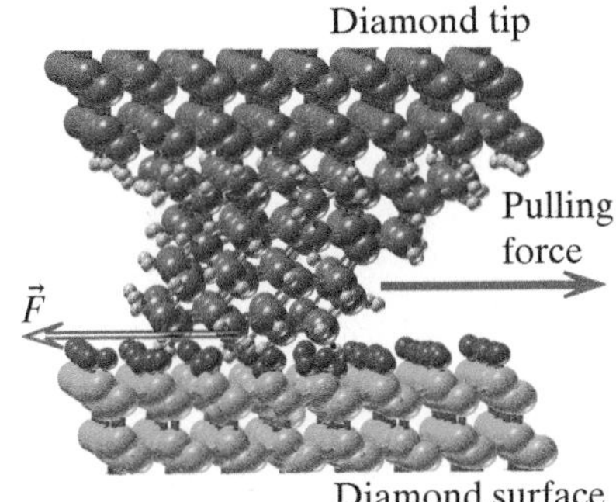

Figure 7.42 A computational model of sliding friction between a diamond tip and diamond surface. The vertical forces are not shown. (© U.S. Government. Courtesy of Professor Judith A. Harrison, U.S. Naval Academy.)

Figure 7.42 provides an atomic-level picture of friction. This image is a single frame from a molecular dynamics computer simulation of sliding friction between two objects, created by Judith A. Harrison and coworkers at the United States Naval Academy in Annapolis, Maryland. The computations are similar to the numerical integrations you have done, but they involve a large number of objects (atoms). The forces between the atoms are modeled by spring-like forces.

In this image, a diamond tip is dragged to the right across a diamond surface. The large spheres represent carbon atoms, and the small ones are hydrogen atoms on the carbon surfaces. Note that the projecting tip has caught on the lower surface, causing interatomic bonds in the two objects to deform. The stretching and compression of interatomic bonds is responsible for the horizontal component of the force exerted on the moving object by the lower object. The collisions between neighboring atoms lead to an increase in the motion of the atoms, with an increase in the internal energy. Some of the work is dissipated throughout the objects.

Dissipation and Path Independence

In Chapter 6 we saw that the work required to change the configuration of gravitationally or electrically interacting objects that could be considered to be point particles was independent of the path along which the work was done. We cannot consider a block sliding with friction to be a point particle because its internal energy changes.

Consider sliding a heavy block in a round trip on an uphill and downhill surface where there is considerable friction. The Energy Principle for the round-trip movement of the block contains a term corresponding to a rise in the internal energy of the block:

$$\Delta K + \Delta U_{\text{grav}} + \Delta E_{\text{internal}} = W$$
$$0 + 0 + \Delta E_{\text{internal}} = W$$

The longer the round-trip path taken, the more increase in internal energy. The total work W is nonzero and is greater for longer paths. Path independence does not apply. Some energy is dissipated/shared among the many atoms of the block.

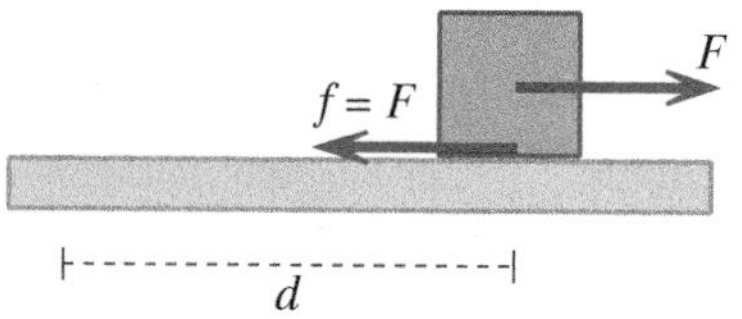

Figure 7.43 Forces in the x direction on a block pulled with constant speed across a table. (Forces in the y direction are not shown.)

A Paradox Involving Friction

If we are careful in our energy bookkeeping for a system involving motion with friction, we encounter a paradox that cannot easily be resolved. Consider the situation shown in Figure 7.43.

You exert a constant force F, and pull a block a distance d across a table. Because there is friction between the block and the table, the block travels at constant speed. Since the momentum of the block does not change, we conclude correctly that the net force on the block is zero, and that therefore the magnitude of the friction force must also be F.

In terms of energy inputs, you have done an amount of work Fd on the block. The friction force, acting in the opposite direction, has apparently done an amount of work $-Fd$. The net work done on the block therefore would seem to be zero. This is consistent with the fact that the kinetic energy of the block did not seem to change.

It is not, however, consistent with the fact that both the block and the table get hotter as they rub together! If the net work done on the block was zero, where did that increased internal energy come from?

This is not a "trick question," but a genuine puzzle. We will return to this issue in a later chapter, when we will deal with it in enough detail to resolve the paradox.

> **Checkpoint 13** You move a block slowly to the right a distance of 0.2 m on a table. You apply a constant force of 20 N. How much work do you do? You then move the block slowly back to the starting point; how much work do you do in this second move? What change has there been in the internal energy of the block and table?

7.11 ENERGY DISSIPATION IN COMPUTATIONAL MODELS

We saw in the previous chapter that although we cannot predict the detailed motion of a system directly by iteratively applying the Energy Principle (because neither direction nor time appears in the Energy Principle), it can be extremely useful to monitor and display the kinetic and potential energy of a system whose motion we predict by the iterative application of the Momentum Principle. This is especially true when we include dissipation in our models in order to produce a more realistic picture of the behavior of everyday systems. Such a model enables us to answer questions such as how rapidly energy dissipation occurs, whether energy is dissipated more rapidly at the beginning or end of a process, and so on. Incorporating dissipation into a computational model of motion requires that we translate our algebraic expressions for forces such as air resistance, sliding friction, or viscous friction into computational program statements.

In the previous section we saw that the magnitude of the air resistance force depended on the shape of an object, its speed, and the density of the air. If we wanted to model the motion of a baseball, including the effects of air resistance, we would need values not only for the mass of the baseball, but also its cross-sectional area A (which you can calculate given data for a standard baseball), and also a value for the drag coefficient C, which you can look up (0.3 is a reasonable value). You would also need the density of the air ρ at the altitude of interest, in SI units. Given all this, the calculation of the net force inside your loop might look like this:

```
while ball.pos.y > ground.pos.y:
    rate(100)
    F_grav = m_ball * vector(0, -g, 0)
    v = p_ball / m_ball
```

```
vhat = v / mag(v)
F_air = 0.5*C*rho_air*A_ball*mag(v)**2*(-vhat)
F_net = F_grav + F_air

## update momentum and update position...
```

If you wanted to model the damping of the motion of an object moving through a viscous liquid, the net force would have to include a viscous friction force proportional to the speed, and proportional to a damping coefficient, which is usually symbolized by lowercase c (not to be confused with the speed of light).

```
F_viscous = c * (-p_object/m_object)
```

Modeling dissipation due to sliding friction can be a little tricky. As we saw in Chapter 4, the sliding friction force is in a direction opposite to an object's velocity, but is independent of speed, depending only on how much two objects compress each other (represented in the equation by F_N, the magnitude of the normal force):

$$f_{\text{friction}} \approx \mu_k F_N$$

Calculating the sliding friction force in VPython is not difficult:

```
F_sliding = mu_k * F_N * (-p/mag(p))
```

The tricky part is sensing when the moving object has come nearly to a stop, and exiting the computational loop so the friction force doesn't start the object moving again!

QUESTION Why is it difficult to detect when a moving object has come exactly to a stop in a computational model?

There are two reasons. The first is that the time step Δt is finite; it is unlikely that the object's speed will be zero exactly at the end of a time step, when we can test to see if v is zero. The second reason, which was discussed in Chapter 2, is that the internal representations of floating-point numbers in computers and calculators are necessarily imprecise, and there is always a small but finite round-off error. So instead of testing to see if v is exactly zero, we need to test to see if v is very close to zero. Such a test in a computational loop might look like this:

```
while True:
    rate(100)
    F_sliding = mu_k * F_N * (-p/mag(p))
    ## calculate net force
    ## update momentum and update position...
    if mag(p_object / m_object) < 0.001:
        break
```

In VPython (as in many computational languages), a logical test comparing the value of a variable to a target value can be done using an `if` statement, as in the program fragment above. If the result of the comparison is true, then the statements indented below the `if` statement will be executed; if the result is false, then they will be skipped. The test is at the end of the loop to make it possible to start the system with $v = 0$.

The `break` command terminates the computational loop. Because it is indented below the `if` statement in the program fragment above, the `break` command is executed only if the test is true—that is, only if $v < 0.001$ m/s.

In the Computational Problems section at the end of this chapter, you will find several problems involving adding dissipative forces to models of motion.

7.12 *RESONANCE

Imagine pushing somebody in a swing. If you time your pushes to match the natural frequency of the swing, you can build up an increasingly large amplitude of the swing. However, larger amplitude means higher speed and a higher rate of energy dissipation.

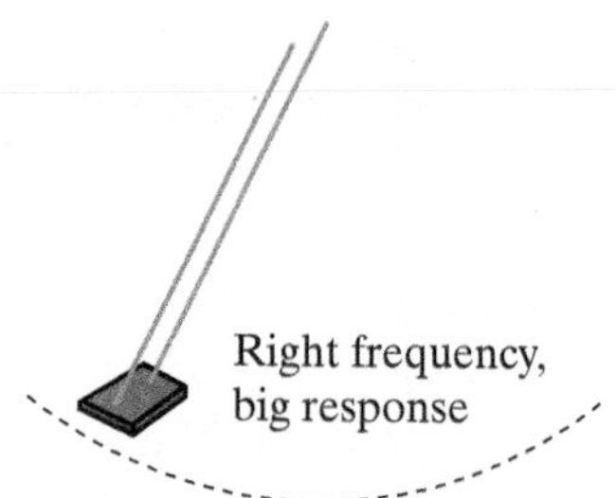

Figure 7.44 Small pushes at the right times lead to a big response.

> QUESTION You keep pushing with constant (small) amplitude, and the swing eventually reaches a "steady state" with constant (large) amplitude. What determines the size of this steady-state amplitude?

The steady state is established when the energy dissipated through air resistance on each cycle has grown to be equal to the energy input you make on each cycle. The large steady-state oscillation needs only little pushes at the right times from you to make up for energy dissipation, and the amplitude of the oscillation can be much larger than the amplitude of the pushes (Figure 7.44).

On the other hand, if you deliberately push at the wrong times, not matching the natural frequency of the swing, the amplitude of the swing won't continually build up (Figure 7.45). To get a large response from an oscillating system, inputs should be made at the natural, free-oscillation frequency of that system. This important feature of driven oscillating systems is called "resonance."

Figure 7.45 Pushes at the wrong times lead to a small response.

*Analytical Treatment of Resonance

Problem P50 is a numerical integration of a sinusoidally driven spring–mass oscillator, as shown in Figure 7.46.

This problem lets you study the buildup of the steady state as well as the steady state itself. For the steady-state portion of the motion, it is possible to obtain an analytical solution, if the energy dissipation is due to viscous friction. Here is the Momentum Principle for the driven oscillator in the diagram, with viscous friction force $-cv$:

$$\frac{dp}{dt} = -k_s[x - D\sin(\omega_D t)] - cv$$

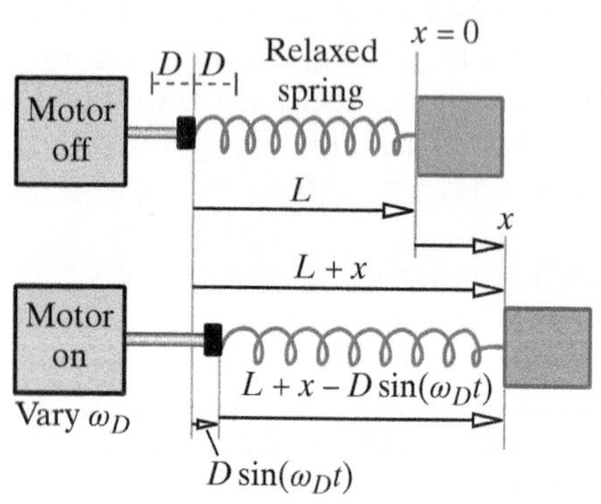

Figure 7.46 A sinusoidally driven oscillator.

where x is the position of the mass, D is the amplitude of the sinusoidal motion, and ω_D is the variable angular frequency of the driven end of the spring. For more details on how this equation was obtained, see the discussion in Problem P50.

In the steady state x must be a sinusoidal function, with amplitude A and phase shift ϕ depending strongly on the driving angular frequency ω_D:

$$x = A\sin(\omega_D t + \phi)$$

By taking derivatives of this expression to obtain $v = dx/dt$ and dv/dt as a function of time t and plugging these derivatives into the Momentum Principle, it can be shown that this expression for x as a function of time is a solution in the steady state if the amplitude A and phase shift ϕ have values determined by the following expressions, where $\omega_F^2 = k_s/m$:

$$A = \frac{\omega_F^2}{\sqrt{(\omega_F^2 - \omega_D^2)^2 + \left(\frac{c}{m}\omega_D\right)^2}} D$$

$$\cos\phi = \frac{(\omega_F^2 - \omega_D^2)}{\sqrt{(\omega_F^2 - \omega_D^2)^2 + \left(\frac{c}{m}\omega_D\right)^2}}$$

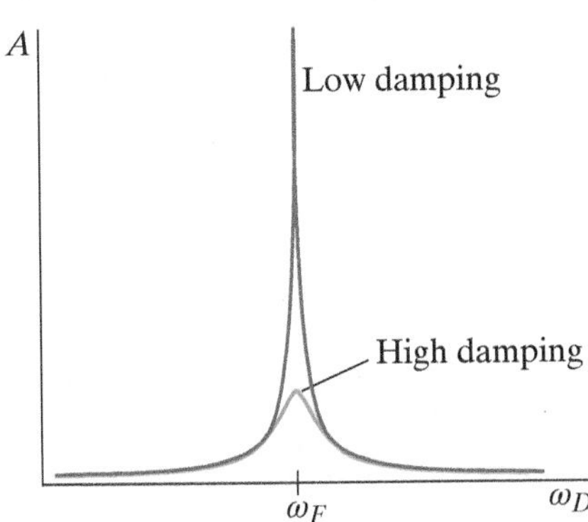

Figure 7.47 Amplitude as a function of driving frequency. The damping in the lower curve is five times greater than in the upper curve.

The detailed proof is rather complicated, so we omit it. Instead, in the exercises that follow we give you the opportunity to see that these expressions for amplitude and phase shift are in agreement with your experimental observations and the computer modeling of Problem P50. Figure 7.47 is a graph of the amplitude A as a function of the driving angular frequency ω_D for two different values of the viscous friction parameter c.

Checkpoint 14 (a) Using the equation for the amplitude A, show that if the viscous friction is small, the amplitude is large when ω_D is approximately equal to ω_F. Using the equation involving the phase shift ϕ, show that the phase shift is approximately 0° for very low driving frequency ω_D, approximately 180° for very high driving frequency ω_D, and 90° at resonance, consistent with your experiment. (b) Show that with small viscous friction, the amplitude A drops to $1/\sqrt{2}$ of the peak amplitude when the driving angular frequency differs from resonance by this amount:

$$|\omega_F - \omega_D| \approx \frac{c}{2m}\omega_F$$

(*Hint:* Note that near resonance $\omega_D \approx \omega_F$, so $\omega_F + \omega_D \approx 2\omega_F$.)

Given these results, how does the width of the resonance peak depend on the amount of friction? What would the resonance curve look like if there were very little friction?

Resonance in Other Systems

We have discussed resonance in the context of a particular system—a mass on a spring driven by a motor. We study resonance because it is an important phenomenon in a great variety of situations.

When you choose a radio or television station, you adjust the parameters of an electrically oscillating circuit so that the circuit has a narrow resonance at the chosen station's main frequency, and other stations with significantly different frequencies have little effect on the circuit.

Magnetic resonance imaging (MRI) is based on the phenomenon of nuclear magnetic resonance (NMR) in which nuclei acting like toy tops precess in a strong magnetic field. The precessing nuclei are significantly affected by radio waves only at the precession frequency. Because the precession frequency depends slightly on what kind of atoms are nearby, resonance occurs for different radio frequencies, which makes it possible to identify different kinds of tissue and produce remarkably detailed images.

SUMMARY

Potential energy of an ideal spring–mass system:

$$U_s = \frac{1}{2}k_s s^2$$

Approximate interatomic potential energy:

$$U = \frac{1}{2}k_s s^2 - E_M$$

A better approximation to interatomic potential energy:

$$U_M = E_M[1 - e^{-\alpha(r - r_{\text{eq}})}]^2 - E_M$$

Thermal energy: related to the temperature of the object
Specific heat: energy per gram to raise temperature 1 K

$$\Delta E_{\text{thermal}} = mC\Delta T$$

$$C = \frac{\Delta E_{\text{thermal}}}{m\Delta T} \quad (C \text{ in J/K/g}, m \text{ in grams})$$

Energy transfer due to a temperature difference, Q
Other energy transfers: work W, matter transfer, mechanical waves, electricity, electromagnetic radiation

Power $= \frac{\text{energy}}{\text{time}}$ (J/s = watts); also calculable as $\vec{F} \bullet \vec{v}$
Open system: energy flows into and/or out of system.
Closed system: no energy flows into or out of system.
The choice of system (open or closed) affects energy accounting, as does the choice of reference frame.
Dissipation of energy
Terminal speed
Air resistance: collisions between air molecules and moving object
Air resistance is approximately proportional to speed squared, and in a direction opposite to the velocity:

$$\vec{F}_{\text{air}} \approx -\frac{1}{2}C\rho A v^2 \hat{v}$$

For small particles ("viscous" friction), fluid friction is approximately proportional to v.

Sliding friction force due to forming and breaking of bonds between two solid objects
Impossibility of describing friction in terms of point particles and potential energy
*Analytical solution for driven spring–mass system: $x = A\sin(\omega_D t + \phi)$, where $\omega_F = \sqrt{k_s/m}$; viscous friction cv, driven at ω_D:

$$A = \frac{\omega_F^2}{\sqrt{(\omega_F^2 - \omega_D^2)^2 + \left(\frac{c}{m}\omega_D\right)^2}} D$$

$$\cos\phi = \frac{(\omega_F^2 - \omega_D^2)}{\sqrt{(\omega_F^2 - \omega_D^2)^2 + \left(\frac{c}{m}\omega_D\right)^2}}$$

Resonance: sensitivity of driven oscillator to driving frequency

QUESTIONS

Q1 Write an equation for the total energy of a system consisting of a mass suspended vertically from a spring, and include the Earth in the system. Place the origin for gravitational energy at the equilibrium position of the mass, and show that the changes in energy of a vertical spring–mass system are the same as the changes in energy of a horizontal spring–mass system.

Q2 Substance A has a large specific heat (on a per gram basis), while substance B has a smaller specific heat. If the same amount of energy is put into a 100 g block of each substance, and if both blocks were initially at the same temperature, which one will now have the higher temperature?

Q3 An oil company included in its advertising the following phrase: "Energy—not just a force, it's power!" In technical usage, what are the differences among the terms energy, force, and power?

Q4 Electricity is billed in kilowatt-hours. Is this energy or power? How much is one kilowatt-hour in standard physics units? (The typical cost of one kilowatt-hour is 5 to 10 cents.)

Q5 State which of the following are open systems with respect to energy, and which are closed: a car, a person, an insulated picnic chest, the Universe, the Earth. Explain why.

Q6 When a falling object reaches terminal speed, its kinetic energy reaches a constant value. However, the gravitational energy of the system consisting of object plus Earth continues to decrease. Does this violate the principle of conservation of energy? Explain why or why not.

Q7 You lift a heavy box. We'll consider this process for different choices of system and surroundings. **(a)** Choose the box as the system of interest. What objects in the surroundings exert significant forces on this system? **(b)** Choose you and the box as the system of interest. What objects in the surroundings exert significant forces on this system? **(c)** Choose you, the box, and the Earth as the system of interest. What objects in the surroundings exert significant forces on this system?

Q8 Consider the process of a woman lifting a barbell discussed in Section 7.8. Analyze the energy changes in this process, choosing the woman alone as the system. What quantities can be calculated with this choice of system?

Q9 A block of mass m is projected straight upward by a strong spring whose stiffness is k_s. When the block is a height y_1 above the floor, it is traveling upward at speed v_1, and the spring is compressed an amount s_1. A short time later the block is at height y_2, traveling upward at speed v_2, and the spring is compressed an amount s_2. Assume that thermal transfer of energy (microscopic work) Q between the block and the air is negligible. For each of the following choices of system, write the Energy Principle in the update form $E_f = E_i + W$. **(a)** The block, the spring, and the Earth; **(b)** the block plus spring; **(c)** the block alone.

Q10 A horse whose mass is M gallops at constant speed v up a long hill whose vertical height is h, taking an amount of time t to reach the top. The horse's hooves do not slip on the rocky ground, so the work done by the force of the ground on the hooves is zero (no displacement of the force). When the horse started running, its temperature rose quickly to a point at which from then on, heat transferred from the horse to the air keeps the horse's temperature constant.

(a) First consider the horse as the system of interest. In the initial state the horse is already moving at speed v. In the final state the horse is at the top of the hill, still moving at speed v. Write out the Energy Principle $\Delta E_{\text{sys}} = W + Q$ for the system of the horse alone. The terms on the left-hand side should include only energy changes for the system, while the terms on the right-hand side should relate to the surroundings (everything else). Which of the terms are equal to 0? Which of the terms should go on the system (left) side (ΔE_{sys})? Which of the terms should go on the surroundings (right) side?

(b) Next consider the Universe as the system of interest. Write out the energy principle for this system. Remember that terms on the left-hand side should include all energy changes for the system, while the terms on the right-hand side should relate to the surroundings. Which of the terms are equal to 0? Which of the terms should go on the system (left) side (ΔE_{sys})? Which of the terms should go on the surroundings (right) side?

Q11 Describe a situation in which it would be appropriate to neglect the effects of air resistance.

Q12 Describe a situation in which neglecting the effects of air resistance would lead to significantly wrong predictions.

Q13 Throw a ball straight up and catch it on the way down, at the same height. Taking into account air resistance, does the ball take longer to go up or to come down? Why?

Q14 You drag a block with constant speed v across a table with friction. Explain in detail what you have to do in order to change to a constant speed of $2v$ on the same surface. (That is, the puzzle is to explain how it is possible to drag a block with sliding friction at different constant speeds.)

Q15 Figure 7.48 is a portion of a graph of energy terms vs. time for a mass on a spring, subject to air resistance. Identify and label the three curves as to what kind of energy each represents. Explain briefly how you determined which curve represented which kind of energy.

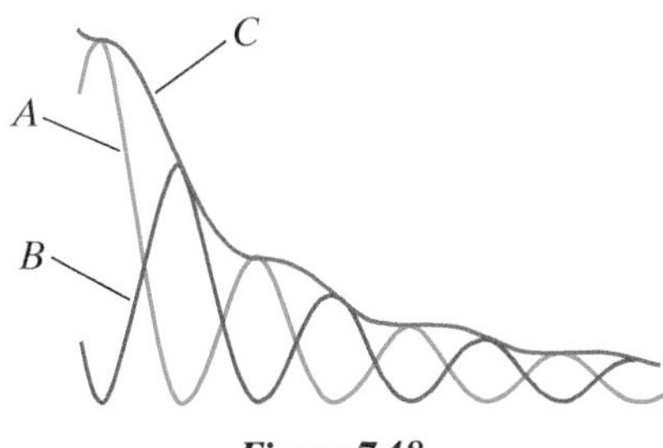

Figure 7.48

PROBLEMS

Section 7.2

•P16 A spring has a relaxed length of 6 cm and a stiffness of 100 N/m. How much work must you do to change its length from 5 cm to 9 cm?

•P17 A horizontal spring with stiffness 0.5 N/m has a relaxed length of 15 cm. A mass of 20 g is attached and you stretch the spring to a total length of 25 cm. The mass is then released from rest and moves with little friction. What is the speed of the mass at the moment when the spring returns to its relaxed length of 15 cm?

•P18 A spring whose stiffness is 800 N/m has a relaxed length of 0.66 m. If the length of the spring changes from 0.55 m to 0.96 m, what is the change in the potential energy of the spring?

••P19 A spring with stiffness k_s and relaxed length L stands vertically on a table. You hold a mass M just barely touching the top of the spring. **(a)** You *very slowly* let the mass down onto the spring a certain distance, and when you let go, the mass doesn't move. How much did the spring compress? How much work did *you* do? **(b)** Now you again hold the mass just barely touching the top of the spring, and then let go. What is the maximum compression of the spring? State what approximations and simplifying assumptions you made. **(c)** Next you push the mass down on the spring so that the spring is compressed an amount s, then let go, and the mass starts moving upward and goes quite high. When the mass is a height of $2L$ above the table, what is its speed?

••P20 A horizontal spring–mass system has low friction, spring stiffness 200 N/m, and mass 0.4 kg. The system is released with an initial compression of the spring of 10 cm and an initial speed of the mass of 3 m/s. **(a)** What is the maximum stretch during the motion? **(b)** What is the maximum speed during the motion? **(c)** Now suppose that there is energy dissipation of 0.01 J per cycle of the spring–mass system. What is the average power input in watts required to maintain a steady oscillation?

••P21 A package of mass 9 kg sits on an airless asteroid of mass 8.0×10^{20} kg and radius 8.7×10^5 m. We want to launch the package in such a way that it will never come back, and when it is very far from the asteroid it will be traveling with speed 226 m/s. We have a large and powerful spring whose stiffness is 2.8×10^5 N/m. How much must we compress the spring?

••P22 A mass of 0.3 kg hangs motionless from a vertical spring whose length is 0.8 m and whose unstretched length is 0.65 m. Next the mass is pulled down so the spring has a length of 0.9 m and is given an initial speed upward of 1.2 m/s. What is the maximum length of the spring during the following motion? What approximations or simplifying assumptions did you make?

•••P23 A relaxed spring of length 0.15 m stands vertically on the floor; its stiffness is 1000 N/m. You release a block of mass 0.4 kg from rest, with the bottom of the block 0.8 m above the floor and straight above the spring. How long is the spring when the block comes momentarily to rest on the compressed spring?

•••P24 Design a "bungee jump" apparatus for adults. A bungee jumper falls from a high platform with two elastic cords tied to the ankles. The jumper falls freely for a while, with the cords slack. Then the jumper falls an additional distance with the cords increasingly tense. You have cords that are 10 m long, and these cords stretch in the jump an additional 24 m for a jumper whose mass is 80 kg, the heaviest adult you will allow to use your bungee jump (heavier customers would hit the ground). You can neglect air resistance. **(a)** Make a series of five simple diagrams, like a comic strip, showing the platform, the jumper, and the two cords at various times in the fall and the rebound. On each diagram, draw and label vectors representing the forces acting on the jumper, and the jumper's velocity. Make the relative lengths of the vectors reflect their relative magnitudes. **(b)** At what instant is there the greatest tension in the cords? How do you know? **(c)** What is the jumper's speed at this instant? **(d)** Is the jumper's momentum changing at this instant or not? (That is, is $d\vec{p}/dt$ nonzero or zero?) Explain briefly. **(e)** Focus on this instant, and use the principles of this chapter to determine the spring stiffness k_s for each cord. Explain your analysis. **(f)** What is the maximum tension that each cord must support without breaking? **(g)** What is the maximum acceleration (in g's) that the jumper experiences? What is the direction of this maximum acceleration? **(h)** State clearly what approximations and estimates you have made in your design.

Section 7.3

••P25 Figure 7.49 is a potential energy curve for the interaction of two neutral atoms. The two-atom system is in a vibrational state indicated by the green horizontal line.

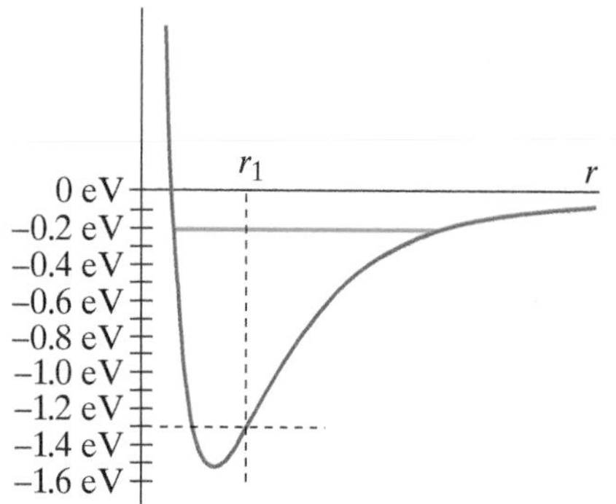

Figure 7.49

(a) At $r = r_1$, what are the approximate values of the kinetic energy K, the potential energy U, and the quantity $K + U$? **(b)** What minimum energy must be supplied to cause these two atoms to separate? **(c)** In some cases, when r is large, the interatomic potential energy can be expressed approximately as $U = -a/r^6$. For large r, what is the algebraic form of the magnitude of the force the two atoms exert on each other in this case?

Section 7.4

•P26 Niagara Falls is about 50 m high. What is the temperature rise in kelvins of the water from just before to just after it hits the rocks at the bottom of the falls, assuming negligible air resistance during the fall and that the water doesn't rebound but just splats onto the rock? It is helpful (but not essential) to consider a 1 g drop of water.

••P27 400 g of boiling water (temperature 100°C, specific heat 4.2 J/K/g) are poured into an aluminum pan whose mass is 600 g and initial temperature 20°C (the specific heat of aluminum is 0.9 J/K/g). After a short time, what is the temperature of the water? Explain. What simplifying assumptions did you have to make?

••P28 Here are questions about human diet. **(a)** A typical candy bar provides 280 calories (one "food" or "large" calorie is equal to 4.2×10^3 J). How many candy bars would you have to eat to replace the chemical energy you expend doing 100 sit-ups? Explain your work, including any approximations or assumptions you make. (In a sit-up, you go from lying on your back to sitting up.) **(b)** How many days of a diet of 2000 large calories are equivalent to the gravitational energy difference for you between sea level and the top of Mount Everest, 8848 m above sea level? (However, the body is not anywhere near 100% efficient in converting chemical energy into change in altitude. Also note that this is in addition to your basal metabolism.)

••P29 You observe someone pulling a block of mass 43 kg across a low-friction surface. While they pull a distance of 3 m in the direction of motion, the speed of the block changes from 5 m/s to 7 m/s. Calculate the magnitude of the force exerted by the person on the block. What was the change in internal energy (chemical energy plus thermal energy) of the person pulling the block?

Section 7.5

••P30 You place into an insulated container a 1.5 kg block of aluminum at a temperature of 45°C in contact with a 2.1 kg block of copper at a temperature of 18°C. The specific heat of aluminum is 0.91 J/g and the specific heat of copper is 0.39 J/g. What is the final temperature of the two blocks?

••P31 180 g of boiling water (temperature 100°C, heat capacity 4.2 J/K/g) are poured into an aluminum pan whose mass is 1050 g and initial temperature 26°C (the heat capacity of aluminum is 0.9 J/K/g). **(a)** After a short time, what is the temperature of the water? **(b)** What simplifying assumptions did you have to make? (1) The thermal energy of the water doesn't change. (2) The thermal energy of the aluminum doesn't change. (3) Energy transfer between the system (water plus pan) and the surroundings was negligible during this time. (4) The heat capacities for both water and aluminum hardly change with temperature in this temperature range. **(c)** Next you place the pan on a hot electric stove. While the stove is heating the pan, you use a beater to stir the water, doing 29,541 J of work, and the temperature of the water and pan increases to 86.9°C. How much energy transfer due to a temperature difference was there from the stove into the system consisting of the water plus the pan?

Section 7.6

•P32 A certain motor is capable of doing 3000 J of work in 11 s. What is the power output of this motor?

•P33 In the Niagara Falls hydroelectric generating plant, the energy of falling water is converted into electricity. The height of the falls is about 50 m. Assuming that the energy conversion is highly efficient, approximately how much energy is obtained from one kilogram of falling water? Therefore, approximately how many kilograms of water must go through the generators every second to produce a megawatt of power (1×10^6 W)?

••P34 Here are questions dealing with human power. **(a)** If you follow a diet of 2000 food calories per day (2000 kC), what is your average rate of energy consumption in watts (power input)? (A food or "large" calorie is a unit of energy equal to 4.2×10^3 J; a regular or "small" calorie is equal to 4.2 J.) Compare with the power input to a table lamp. **(b)** You can produce much higher power for short periods. Make appropriate measurements as you run up some stairs, and report your measurements. Use these measurements to estimate your power output (this is in addition to your basal metabolism—the power needed when resting). Compare with a horsepower (which is about 750 W) or a toaster (which is about 1000 W).

•••P35 Humans have about 60 ml (60 cm^3) of blood per kilogram of body mass, and blood makes a complete circuit in about 20 s, to keep tissues supplied with oxygen. Make a crude estimate of the *additional* power output of your heart (in watts) when you are standing compared with when you are lying down. Note that we're not asking you to estimate the power output of your heart when you are lying down, just the change in power when you are standing up. You will have to estimate the values of some of the relevant parameters. Because we're only looking for a crude estimate, try to construct a model that is as simple as possible. Describe the approximations and estimates you made.

Section 7.7

••P36 During three hours one winter afternoon, when the outside temperature was 0°C (32°F), a house heated by electricity was kept at 20°C (68°F) with the expenditure of

45 kWh (kilowatt·hours) of electric energy. What was the average energy leakage in joules per second through the walls of the house to the environment (the outside air and ground)?

The rate at which energy is transferred between two systems is often proportional to their temperature difference. Assuming this to hold in this case, if the house temperature had been kept at 25°C (77°F), how many kWh of electricity would have been consumed?

Section 7.9

•••P37 A man sits with his back against the back of a chair, and he pushes a block of mass $m = 2$ kg straight forward on a table in front of him, with a constant force $F = 30$ N, moving the block a distance $d = 0.3$ m. The block starts from rest and slides on a low-friction surface. **(a)** How much work does the man do on the block? **(b)** What is the final kinetic energy K of the block? **(c)** What is the final speed v of the block? **(d)** How much time Δt does this process take? **(e)** Consider the system of the man plus the block: how much work does the chair do on the man? **(f)** What is the internal energy change of the man?

Now suppose that the man is sitting on a train that is moving in a straight line with speed $V = 15$ m/s, and you are standing on the ground as the train goes by, moving to your right. From your perspective (that is, in your reference frame), answer the following questions: **(g)** What is the initial speed v_i of the block? **(h)** What is the final speed v_f of the block? **(i)** What is the initial kinetic energy K_i of the block? **(j)** What is the final kinetic energy K_f of the block? **(k)** What is the change in kinetic energy $\Delta K = K_f - K_i$, and how does this compare with the change in kinetic energy in the man's reference frame? **(l)** How far does the block move (Δx)? **(m)** How much work does the man do on the block, and how does this compare with the work done by the man in his reference frame, and with ΔK in your reference frame? **(n)** How far does the chair move? **(o)** Consider the system of the man plus the block: how much work does the chair do on the man, and how does this compare with the work done by the chair in the man's reference frame? **(p)** What is the internal energy change of the man, and how does this compare with the internal energy change in his reference frame?

Section 7.10

•P38 A coffee filter of mass 1.8 g dropped from a height of 4 m reaches the ground with a speed of 0.8 m/s. How much kinetic energy K_{air} did the air molecules gain from the falling coffee filter? Start from the Energy Principle, and choose as the system the coffee filter, the Earth, and the air.

•P39 You are standing at the top of a 50 m cliff. You throw a rock in the horizontal direction with speed 10 m/s. If you neglect air resistance, where would you predict it would hit on the flat plain below? Is your prediction too large or too small as a result of neglecting air resistance?

••P40 A simple experiment can allow you to determine approximately the dependence of the air resistance force on the speed of a falling object. The basic logic of the experiment is this: if we can determine the air resistance force on two objects of the same shape as they fall at different speeds, we can figure out a mathematical relationship between the air resistance force and speed. For example, if the air resistance force is eight times as large when the speed is only twice as large, then we may conclude that $F \propto v^3$. With only a timer and a ruler, it is possible to measure the terminal speed of a falling object. A simple way to increase the terminal speed of an object is to retain its shape but make it more massive. This will increase the gravitational force on the object, and therefore in order to get to a speed at which $\vec{F}_{net} = \vec{0}$ and the object falls at a constant terminal speed, the air resistance force will need to be larger. If you use a paper object (Figure 7.50) as the falling object, you can make it more massive without changing its shape simply by stacking two or more of these objects together.

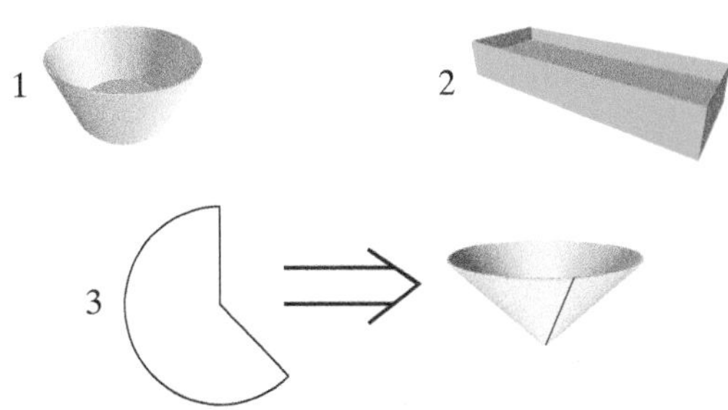

Figure 7.50 A flat-bottomed coffee filter, a folded dollar bill, or a paper cone can be used in Problem P40. The mass of a dollar bill is 1 g.

(a) Measure the terminal speed of a single falling paper object. Drop the object from as high a starting position as possible. Start timing when you think the object has reached terminal speed (think of a way to test this). Do this several times and average your results. **(b)** At terminal speed, what is the value of the air resistance force on a single falling object? If you don't have the equipment to measure the mass of a filter, leave your answer in terms of m. **(c)** Repeat this measurement for stacks of different numbers of objects of the same shape; one of the measurements should be with a stack of four objects. Qualitatively, as the mass of the falling object increases, what happens to the terminal speed? Explain why. **(d)** Now determine the mathematical form of the dependence of the air resistance force on speed, using your measured terminal speeds. (There are many ways to do this.)

•P41 A box with its contents has a total mass of 36 kg. It is dropped from a very high building. **(a)** After reaching terminal speed, what is the magnitude of the air resistance force acting upward on the falling box? **(b)** The box survived the fall and is returned to the top of the building. More objects are put into the box, and the box with its contents now has a total mass of 71 kg. The box is dropped, and it reaches a higher terminal speed than before. After reaching terminal speed, what is the magnitude of the air resistance force acting upward on the falling box? (The fact that the heavier object reaches a higher terminal speed shows that the air resistance force increases with increasing speed.)

••P42 You drop a single coffee filter of mass 1.7 g from a very tall building, and it takes 52 s to reach the ground. In a small fraction of that time the coffee filter reached terminal speed. **(a)** What was the upward force of the air resistance while the coffee filter was falling at terminal speed? **(b)** Next you drop a stack of five of these coffee filters. What was the upward force of the air resistance while this stack of coffee filters was falling at terminal speed? **(c)** Again assuming that the stack reaches terminal speed very quickly, about how long will the stack of coffee filters take to hit the ground? (*Hint:* Consider the relation between speed and the force of air resistance.)

Section 7.12

••P43 You can observe the main effects of resonance with very simple experiments. Hold a spring vertically with a mass suspended at the other end, and observe the frequency of "free" oscillations with your hand kept still. Then stop the oscillations, and move your hand *extremely* slowly up and down in a kind of slow sinusoidal motion. You will see that the mass moves up and down with the same very low frequency. **(a)** How does the amplitude (plus or minus displacement from the center location) of the mass compare with the amplitude of your hand? (Notice that the phase shift of the oscillation is 0°; the mass moves up when your hand moves up.) **(b)** Next move your hand up and down at a significantly higher frequency than the free-oscillation frequency. How does the amplitude of the mass compare to the amplitude of your hand? (Notice that the phase shift of the oscillation is 180°; the mass moves down when your hand moves up.) **(c)** Finally, move your hand up and down at the free-oscillation frequency. How does the amplitude of the mass compare with the amplitude of your hand? (It is hard to observe, but the phase shift of the oscillation is 90°; the mass is at the midpoint of its travel when your hand is at its maximum height.) **(d)** Change the system in some way so as to increase the air resistance significantly. For example, attach a piece of paper to increase drag. At the free-oscillation frequency, how does this affect the size of the response? A strong dependence of the amplitude and phase shift of the system to the driving frequency is called resonance.

COMPUTATIONAL PROBLEMS

More detailed and extended versions of some computational modeling problems may be found in the lab activities included in the *Matter & Interactions* resources for instructors.

••P44 Energy graphs with springs: Start with the program you wrote to model the motion of a mass hanging from a spring (Chapter 4, Problem P70). Add the statements necessary to plot graphs of K, U_{spring}, and $K+U_{\text{spring}}$ vs. time. **(a)** Use initial conditions that produce 1D vertical oscillations. Is the sum $K+U_{\text{spring}}$ constant over time? Should it be? Explain. **(b)** Use your graphs to explain the flow of energy within this system. If you need to plot an additional quantity in order to support your explanation, do so. **(c)** Now experiment with different initial conditions that produce 2D or 3D oscillations. What do your graphs show about energy flow within this system?

••P45 Air resistance: Start with the program you wrote to model the motion of a mass hanging from a spring (Chapter 4, Problem P70). Add the statements necessary to plot graphs of K, U, and $K+U$ vs. time. Use initial conditions that produce 1D vertical oscillations. **(a)** Make sure that the plots generated by your program make sense. Have you included all potential energy terms? **(b)** Add an air resistance force to your model. You may wish to modify the hanging mass to be a flat disk. Choose an approximate value for the drag coefficient C in your model, and adjust this value until the model behaves appropriately for a system experiencing the effects of air resistance. **(c)** Consider the energy graphs produced by the model. Is the rate of energy dissipation constant? If not, at what point(s) in the oscillation cycle is this rate largest? Explain this on physical grounds.

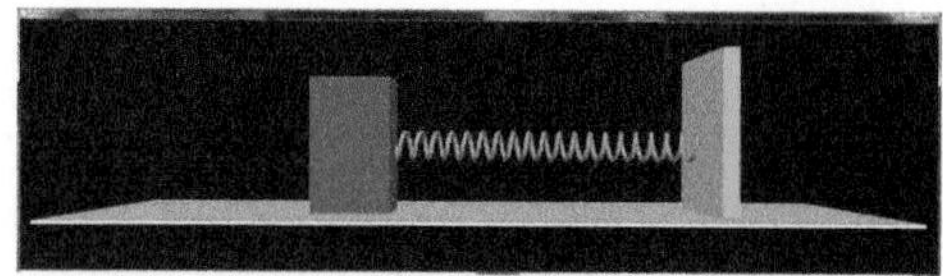

Figure 7.51

••P46 Sliding and viscous friction: Write a program to model a horizontal mass–spring system, like the one shown in Figure 7.51. Give the spring a relaxed length of 0.9 m and a stiffness of 1.5 N/m. Give the block a mass of 0.02 kg. **(a)** Compare the period of the oscillations of your model system (without any friction) to the period you would expect, to make sure your model behaves correctly. **(b)** Add graphs of K, U, and $K+U$. Again, check to make sure these make physical sense, in the absence of friction. **(c)** Now add a viscous friction force to the model. Adjust the viscous friction coefficient to make the behavior of the system clear. (A reasonable starting value is about 0.03). **(d)** Is the rate of energy loss constant over time? Use the energy graphs produced by your program to explain physically why the rate of energy loss varies as observed. **(e)** Replace the viscous friction force with a sliding friction force. A reasonable value for the coefficient in this model is about 0.15. **(f)** Compare the shape of the graph of $K+U$ for sliding friction to the shape of the $K+U$ graph for viscous friction. Explain the difference in the shapes in physical terms (consider the mathematical expressions for these forces).

••P47 Effect of air resistance on a baseball: Write a computational model to predict the motion of a baseball that is hit at a speed of 44 m/s (100 mi/h) at an angle of 45° to the horizontal. A baseball has a mass of 155 g and a diameter of 7 cm. The drag coefficient C for a baseball is about 0.35, and the density of air at sea level is about 1.3×10^{-3} g/cm^3, or 1.3 kg/m^3. **(a)** In your initial model, neglect air resistance. How far does the ball go? Is this distance reasonable? (A baseball field is about 400 ft from home plate to the fence in center field. An outfielder cannot throw a baseball in the air from the fence to home plate.) **(b)** Plot a graph of $K+U_g$ vs. time for the system of baseball plus Earth. **(c)** Now add air resistance to your model. How far does the ball go? **(d)** Compare the graphs of $K+U_g$ with and without air resistance. **(e)** In Denver, a mile above sea level, the air is about 83% as dense as the air at sea level. Including the effect of air resistance, use your computer model to predict the trajectory and range of a baseball thrown in Denver. How does the predicted range compare with the predicted range at sea level?

••P48 Determining a drag coefficient: Write a VPython program to model the motion of a paper object falling straight down, including the effects of air resistance. If you did the experiments in Problem P40, use your own data. Otherwise, you can use these data from an actual experiment that used coffee filters: filter mass = 8.52×10^{-3} kg, outer radius = 5.8×10^{-2} m, initial height = 2.03 m, drop time for one filter = 1.44 s, drop time for a stack of four coffee filters = 0.88 s. This experiment was performed in Santa Fe, New Mexico (altitude 2100 m), where the air density is only 1.045 kg/m^3; the air density at an altitude of H meters is $e^{-H/8500}$ times the air density at sea level (see Chapter 12).

(a) Add a plot of v vs. t to your program. Set the air resistance force to zero and use this plot to check that your program is working correctly. What should this plot look like without air resistance? **(b)** Now add the air resistance force to your program. What is the initial slope of your graph of speed vs. time? (By dragging the mouse across a VPython graph you can accurately read coordinates off the graph.) What should be the initial slope? Why? **(c)** Adjust the drag coefficient C until the drop time predicted by your model matches your experimental data for the case of the smallest mass. (Run your loop until the y coordinate of the moving object reaches zero, then print the computed time.) **(d)** What terminal speed does your model predict for this case? **(e)** Using the value of C that you determined with the smallest mass, what drop time does your model predict for an object of four times the mass but the same cross section? Does this agree with your measurement? **(f)** Using the value of C that you determined with the smallest mass, what terminal speed does your model predict for an object of four times the smallest mass but the same cross section? Is this what you expected? Why or why not?

••P49 Terminal speed for a falling sky diver has been measured to be about 60 m/s. Write a computational model to determine how far (in meters) and how long (in seconds) a sky diver falls before reaching terminal speed. (You will need to think about the meaning of "terminal speed" and how to test for it.) Plot graphs of speed vs. time and position vs. time.

•••P50 You can study resonance in a driven oscillator in detail by modifying the program you wrote for Problem P46. Let one end of the spring be moved back and forth sinusoidally by a motor, with a motion given by $D\sin(\omega_D t)$ (see Figure 7.52). Here D is the amplitude of the motor motion and ω_D is the angular frequency of the motor, which can be varied. (The free-oscillation angular frequency $\omega_F = \sqrt{k_s/m}$ of the spring–mass system has a fixed value, determined by the spring stiffness k_s and the mass m.)

We need an expression for the stretch s of the spring in order to be able to calculate the force $-k_s s$ of the spring on the mass. We have to do a bit of geometry to figure out the length of the spring when the mass is displaced a distance x from the equilibrium position, and the motor has moved the other end of the spring by an amount $D\sin(\omega_D t)$. In the figure, the spring gets longer when the mass moves to the right ($+x$), but the spring gets shorter when the motor moves to the right ($-D\sin(\omega_D t)$). The new length of the spring is $L + x - D\sin(\omega_D t)$, and the net stretch of the spring is this quantity minus the unstretched length L, yielding ($s = x - D\sin(\omega_D t)$). A check that this is the correct expression for the net stretch of the spring is that if the motor moves to the right the same distance as the mass moves to the right, the spring will have zero stretch. Replace x in your computer computation with the quantity $[x - D\sin(\omega_D t)]$.

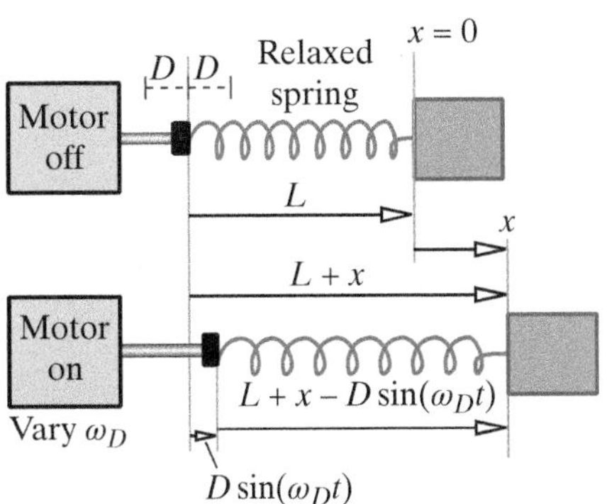

Figure 7.52

Use viscous damping (friction proportional to v). The damping should be small. That is, without the motor driving the system ($D = 0$), the mass should oscillate for many cycles. Set ω_D to $0.9\omega_F$ (that is, 0.9 times the free-oscillation angular frequency, $\omega_F = \sqrt{k_s/m}$), and let x_0 be 0. Graph position x vs. time for enough cycles to show that there is a transient buildup to a "steady state." In the steady state the energy dissipation per cycle has grown to exactly equal the energy input per cycle.

Vary ω_D in the range from 0 to $2.0\omega_F$, with closely spaced values in the neighborhood of $1.0\omega_F$. Have the computer plot the position of the motor end as well as the position of the mass as a function of time. For each of these driving frequencies, record for later use the steady-state amplitude and the phase shift (the difference in angle between the sinusoids for the motor and for the mass). If a harmonic oscillator is lightly damped, it has a large response only for driving frequencies near its own free-oscillation frequency.

Is the steady-state angular frequency equal to ω_F or ω_D for these various values of ω_D? (Note that during the transient buildup to the steady state the frequency is not well-defined, because the motion of the mass isn't a simple sinusoid.)

Sketch graphs of the steady-state amplitude and phase shift vs. ω_D. Note that when $\omega_D = \omega_F$ the amplitude of the mass can be much larger than D, just as you observed with your hand-driven spring–mass system. Also note the interesting variation of the phase shift as you go from low to high driving frequencies—does this agree with the phase shifts you observed with your hand-driven spring–mass system?

Repeat the analysis with viscous friction twice as large. What happens to the resonance curve (the graph of steady-state amplitude vs. angular frequency ω_D)?

ANSWERS TO CHECKPOINTS

1 No change; same stretch
2 0.012 J
3 A is $|U|$; B is K; C is $|K + U|$; 1–4 are bound; 5 is unbound.
4 21.36°C; neglect the material of the thin glass.
5 1.26×10^5 J; 7.6×10^4 J; -1.26×10^5 J
6 9000 J
7 0, since no temperature change; −11,000 J
8 2.5 m/s^2
9 $\frac{1}{2}k_s A^2$; $\frac{1}{2}k_s A^2$; spring plus mass
10 (90 kg)(9.8 N/kg) = 882 N; the net force must be zero.
11 2.7×10^{-2} J
12 Air resistance is proportional to v^2, so the energy dissipation rate is higher at higher speeds.
13 +4 J; another +4 J; +8 J
14 Denominator gets very small when $\omega_D = \omega_F$.

Low frequency:

$$\cos\phi \approx \frac{(\omega_F^2)}{\sqrt{(\omega_F^2)^2 + \left(\frac{c}{m}\omega_D\right)^2}} \approx 1 \text{ so } \phi \approx 0°$$

High frequency:

$$\cos\phi \approx \frac{(-\omega_D^2)}{\sqrt{(-\omega_D^2)^2 + \left(\frac{c}{m}\omega_D\right)^2}} \approx -1,\ \phi \approx 180°$$

Resonance frequency, $\cos\phi = 0$, so $\phi \approx 90°$

14 When $\omega_D = \omega_F$, denominator $= \frac{c}{m}\omega_D$.

Denominator $= \sqrt{2}\frac{c}{m}\omega_D$ when $|(\omega_F^2 - \omega_D^2)| = \frac{c}{m}\omega_D$.
However, since near resonance $\omega_D \approx \omega_F$, we have

$$(\omega_F^2 - \omega_D^2) = (\omega_F + \omega_D)(\omega_F - \omega_D)$$
$$\approx 2\omega_D(\omega_F - \omega_D)$$

Hence when $|\omega_F - \omega_D| \approx \frac{c}{2m}\omega_D \approx \frac{c}{2m}\omega_F$, the amplitude is down by a factor of $1/\sqrt{2}$. Larger friction, wider resonance peak. If friction is very small, the resonance peak is a very narrow spike.

CHAPTER

8

Energy Quantization

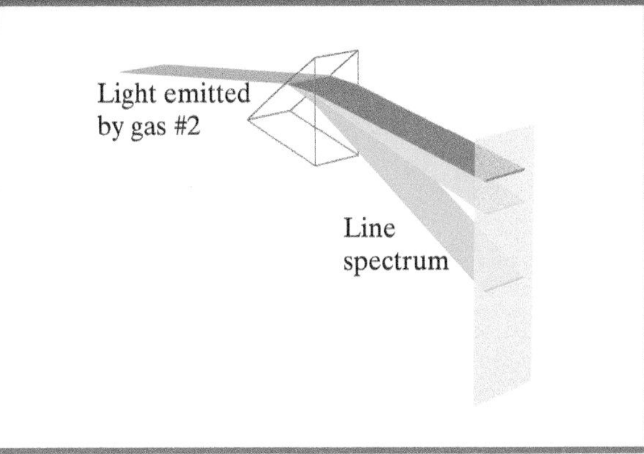

OBJECTIVES

After studying this chapter, you should be able to

- Calculate the energy of a photon emitted or absorbed in a transition between given energy levels.
- Given an emission spectrum, construct a possible energy-level diagram for a system.
- Calculate the energy levels of a quantized harmonic oscillator from its microscopic properties (stiffness, mass).

The Energy Principle applies to microscopic as well as macroscopic systems. In this chapter we will apply the Energy Principle to systems such as atoms and molecules, in both gas and solid form. In particular, an understanding of the energy of vibrations of atoms in a solid lattice and in diatomic molecules will provide an important foundation for our study of entropy in Chapter 12.

Increasing energy
10^6 eV — Gamma ray
10^4 eV — X-ray
10^2 eV — Ultra violet
1 eV — Visible
Infrared
10^{-2} eV
Microwave
10^{-6} eV — Radio
10^{-10} eV

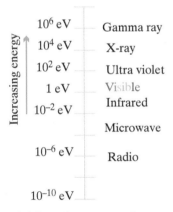

Figure 8.1 The electromagnetic spectrum. See Figure 8.2 for an expanded view of the portion of this spectrum that is visible to the human eye.

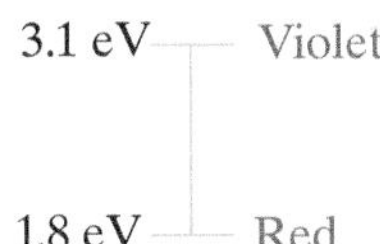

Figure 8.2 An expanded view of the portion of the electromagnetic spectrum that is visible to the human eye.

8.1 PHOTONS

One of the things that we will consider in this chapter is the interaction of light and matter. For the purposes of this chapter, we are interested only in the fact that light is a form of energy—we will defer considerations of the optical properties of light until later. In a contemporary model of light, a beam of light is composed of many particles of light, called "photons." A photon is a particle that has only kinetic energy—its rest energy is zero. A photon can never be at rest; it may be thought of as a packet of energy, traveling at the speed of light. An atom can gain energy by absorbing a photon; if this happens, all of the photon's energy is transferred to the object, and the photon disappears. Conversely, an object can lose energy by emitting a photon with a particular amount of energy. The general term applied to photons is "electromagnetic radiation" (or simply "light"), and in later chapters on electricity and magnetism we will go more deeply into the electric and magnetic aspects of electromagnetic radiation, and we will also discuss its wave-like aspects.

Photons also have momentum, and the Momentum Principle applies to absorption and emission of photons. In Chapter 6 we saw that for any particle $E^2 - (pc)^2 = (mc^2)^2$. Since the photon's mass is zero, we have $E^2 = (pc)^2$, and the magnitude of the momentum of a photon whose energy is E is $p = E/c$ (this result is approximately true for any particle traveling so fast that mc^2 is negligible compared to E).

Historically, photons in different energy ranges have been given different names, usually because the instruments needed to detect and measure electromagnetic radiation depend on the photon energies involved. The entire range of possible photon energies is called the electromagnetic spectrum

(Figure 8.1). Only photons whose energies fall between about 1.8 eV (red light) and 3.1 eV (violet light) can be detected by the human eye (Figure 8.2), so photons in this energy range are referred to as visible light. Colloquially we often refer to all photons as "light," whether they are high-energy gamma rays or low-energy radio waves.

Although earlier scientists, including Isaac Newton, had speculated that light might be made up of discrete particles, it was not until early in the 20th century that clear evidence for the particle nature of light was obtained. As in the case of atoms, the noncontinuous, lumpy nature of light was not noticed for a long time because the amount of energy carried by one photon in an ordinary beam of light is extremely small compared to the total energy of the beam. It has now become routine to detect individual photons in a very weak beam of light and to measure the energy of an individual photon. Observations of the interaction of photons and ordinary matter have provided key insights into the internal structure of molecules, atoms, and nuclei.

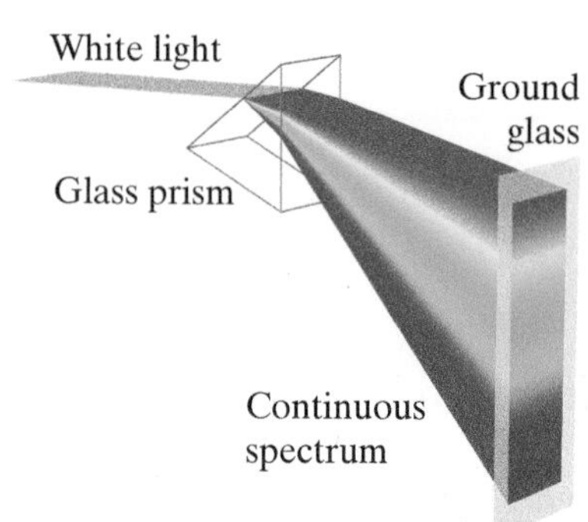

Figure 8.3 A prism separates a narrow beam of white light into a continuous rainbow of colors. The color of visible light is correlated with the energy of the photons; a red photon has less energy than a blue photon.

One can separate a beam of visible light into its component energies by passing a narrow beam of light through a prism. You have probably seen the continuous rainbow spectrum produced when ordinary white light goes through a prism, as in Figure 8.3. Your eye reacts differently to photons of different energies, which is why you see a colorful spectrum.

Checkpoint 1 The photon energy for green light lies between the values for red and violet light. What is the approximate energy of the photons in green light? The intensity of sunlight above the Earth's atmosphere is about 1400 W (J/s) per square meter. That is, when sunlight hits perpendicular to a square meter of area, about 1400 W of energy can be absorbed. Using the photon energy of green light, about how many photons per second strike an area of one square meter? (This is why the lumpiness of light was not noticed for so long.)

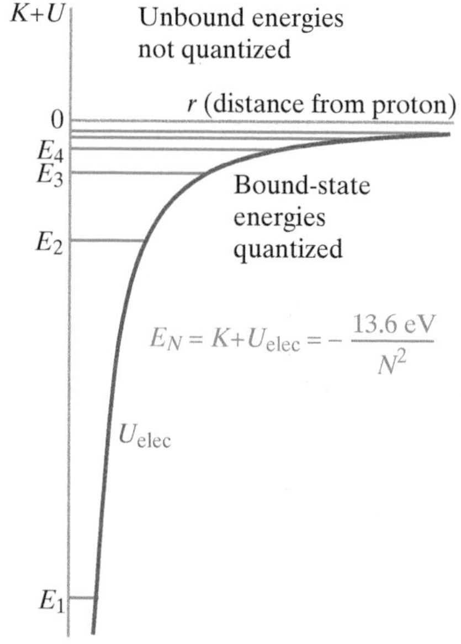

Figure 8.4 Electronic energy levels for a hydrogen atom. The potential energy curve describes the potential energy of the electron–proton system as a function of separation. Each colored horizontal line corresponds to a stable value of the internal energy of the atom.

8.2 ELECTRONIC ENERGY LEVELS

One of the most exciting discoveries of the early 20th century was that for the internal energy of an atom to be stable, there can be only certain discrete values of the internal energy, called energy "levels." An example of discrete energy levels is found in the "electronic" energy associated with the motion of electrons around the nucleus of an atom. You probably know from your study of chemistry that electronic energy levels in an atom are discrete—the only stable bound states of a system of nucleus plus electrons are those with certain specific values of $K+U$. The electronic energy levels in atomic hydrogen are shown in Figure 8.4 (atomic hydrogen is a single H atom, composed of one proton and one electron, not the usual diatomic molecule H_2). The term "quantized" is often used colloquially as a synonym for "discrete." We will use it this way, although the technical meanings of the two words are not quite identical.

Figure 8.4 shows the familiar electric potential energy curve for the proton–electron system (an atom of hydrogen). The horizontal lines represent the possible, stable bound energy states for the hydrogen atom; stable bound states with other values of $(K+U)$ are not observed. Unbound states $(K+U \geq 0)$, however, are not quantized.

The electric potential energy of an electron and a proton bound together in a hydrogen atom is negative for all separations:

$$U_{elec} = \frac{1}{4\pi\varepsilon_0}\frac{q_1 q_2}{r} = \frac{1}{4\pi\varepsilon_0}\frac{(+e)(-e)}{r} = -\frac{1}{4\pi\varepsilon_0}\frac{e^2}{r}$$

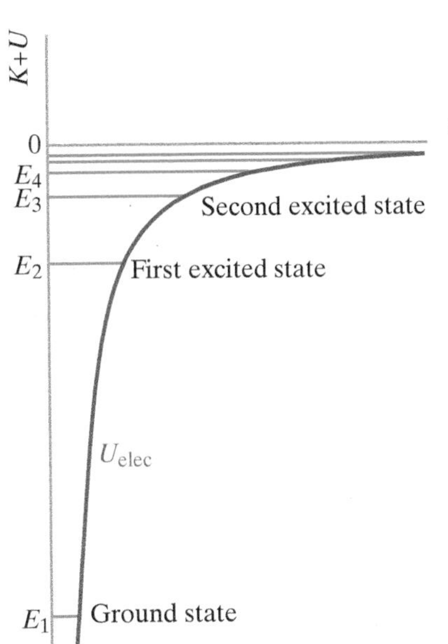

Figure 8.5 The lowest stable energy state is called the ground state. States with higher energy are called excited states.

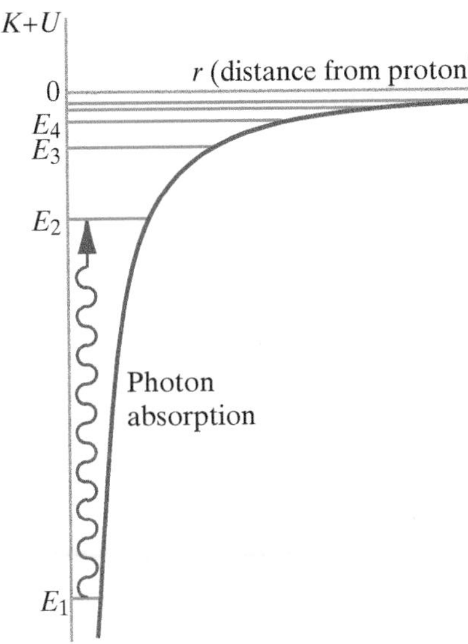

Figure 8.6 An atom can gain energy by absorbing a photon whose energy corresponds to the difference between two energy levels. By convention a squiggly line with an arrowhead indicates a photon. The upward arrow indicates a transition to a higher energy level.

A quantum mechanical calculation of the energy levels of a system involves solving a differential equation called the Schrödinger equation, a form of the Energy Principle that can be applied to atomic phenomena. This equation includes both a potential energy term and a kinetic energy term. Solving this equation for the hydrogen atom predicts that the discrete energy levels for bound states of hydrogen (ignoring the rest energies of the proton and the electron) will be these:

ELECTRONIC ENERGY LEVELS OF A HYDROGEN ATOM

$$E_N = K + U_{\text{elec}} = -\frac{13.6\text{ eV}}{N^2}, \quad N = 1,2,3,\text{etc.}$$

This arrangement of hydrogen energy levels has been verified by observations of the interaction of light and hydrogen, as we will see.

These energy levels denote different bound states of the electron + proton system. Recall that a bound state has a negative value of $K + U$, since a potential energy of zero is associated with a very large separation between the electron and the proton. We have seen that bound gravitational orbits are also associated with negative values of $K + U$. As you can see from the energy diagram (Figure 8.4), in higher-energy states the average location of the electron is farther from the proton.

The Ground State and Excited States

In the energy level diagram shown in Figure 8.5 there is a lowest-energy state, called the "ground state." This is the lowest possible internal energy that a single hydrogen atom can have: $K + U$ cannot be less than $-13.6\,\text{eV}$. Stable energy states whose energy is higher than the energy of the ground state are called "excited states."

A note on potentially confusing terminology: the ground state of the hydrogen atom is called E_1, corresponding to a value of $N = 1$ in the equation above. This means that for this system, E_2 (corresponding to $N = 2$) is called the "first excited state," because it is the first state above the ground state. Similarly, E_3 ($N = 3$) is the second excited state, and so on.

Raising the Internal Energy of an Atom

If an atom is in its ground state, how can it gain enough energy to get to one of its excited states? We will consider two important mechanisms for this process:

- An atom (or molecule) can gain energy by absorbing a photon.
- An atom (or molecule) can gain energy through a collision with a particle such as an energetic electron, or another fast-moving atom.

Photon Absorption

In order for an atom to gain energy by absorbing a photon, the energy of the photon must match the difference in energy between two energy states of the atom (Figure 8.6). When a photon is absorbed by an atom, nothing is left over—the photon no longer exists, and the atom has more energy. We can describe the absorption of a photon by a hydrogen atom in this compact way (the asterisk is used to denote an atom with higher internal energy):

$$\text{H} + \text{photon} \rightarrow \text{H}^*$$

Applying the Energy Principle to this process, and choosing all particles as the system, we can write this equation:

$$E_{H^*,f} = E_{H,i} + K_{photon}$$

EXAMPLE **A Hydrogen Atom Absorbs a Photon**

A hydrogen atom is originally in its ground (lowest) electronic energy state ($N = 1$). It absorbs a photon, which raises the hydrogen atom's energy to the next energy level ($N = 2$). What was the energy of the photon, in eV?

Solution System: All particles
Surroundings: Nothing significant

$$E_f = E_i + K_{photon}$$

$$\left(-\frac{13.6\text{ eV}}{2^2}\right) = \left(-\frac{13.6\text{ eV}}{1^2}\right) + K_{photon}$$

$$K_{photon} = 10.2\text{ eV}$$

It is not possible to add less than 10.2 eV to a hydrogen atom in its lowest energy level. Any smaller amount fails to be absorbed because there is no stable energy level between the ground state and the first excited state.

Collisional Excitation

If a fast-moving particle collides with an atom, and if the kinetic energy of the particle is equal to or greater than the difference between two energy levels of the atom, some of the particle's kinetic energy can be converted to internal energy of the atom, raising it to an excited state. Any excess kinetic energy is retained by the particle. For example, a beam of fast-moving electrons may be used to excite atoms in a gas to a higher energy state. If we consider the system of a moving electron plus the atom:

$$E_{H^*,f} + K_{e,f} = E_{H,i} + K_{e,i}$$

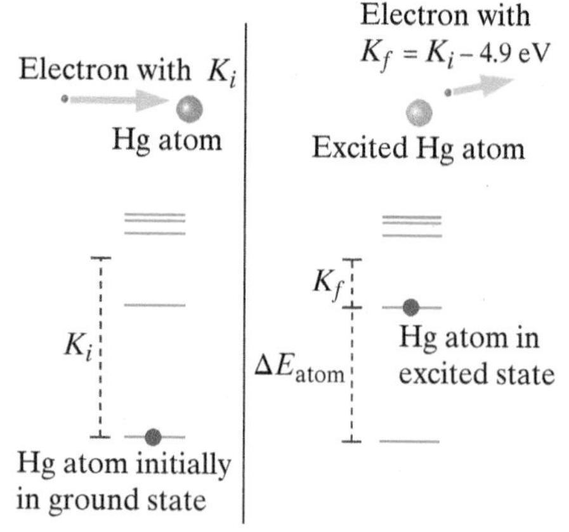

Figure 8.7 An electron with initial kinetic energy of more than 4.9 eV can excite a mercury atom (Hg) from the ground state to the first excited state. The electron loses 4.9 eV of kinetic energy.

Because electrons are (negatively) charged electrically, they interact strongly with atoms, and an electron that passes near an atom may excite the atom from the ground state to an excited state whose internal energy is ΔE_{atom} higher, with a corresponding loss of kinetic energy of the electron. Unlike photon absorption, which almost always occurs only if the photon energy exactly matches the difference between energy levels, an electron can give up some kinetic energy and have some kinetic energy left over. The process is illustrated in Figure 8.7 for an electron colliding with a mercury atom, whose first excited state is 4.9 eV above the ground state.

QUESTION For example, suppose an electron with kinetic energy of 6 eV collides with a mercury atom that is in the ground state. What can happen?

The first excited state of mercury is 4.9 eV above the ground state. This 6 eV electron has enough kinetic energy to excite the mercury atom to the first excited state, giving up 4.9 eV of energy. The electron moves away with 1.1 eV of kinetic energy remaining. The mercury atom will quickly drop back down to the ground state, emitting a 4.9 eV photon. If a beam containing a large number of energetic electrons passes through a gas of mercury vapor, there will be continual collisional excitation and continual photon emission. However, as soon as you turn off the electron beam, the mercury will stop emitting photons.

> QUESTION What happens if you send a beam of electrons with kinetic energy of 4 eV through the mercury vapor?

4 eV isn't enough energy to go from the ground state to the first excited state in mercury, so the electrons can pass right through the gas without changing the electronic energy states of the atoms. (An electron may collide with an atom, but because the mass of an electron is several thousand times smaller than the mass of a mercury atom, the electron will bounce off the atom like a Ping-Pong ball bouncing off a bowling ball—the kinetic energy of the atom will not change significantly.)

The German physicists James Franck and Gustav Hertz performed a collisional excitation experiment in 1914 that provided early dramatic evidence for discrete energy levels in atoms. Electrons with varying amounts of initial kinetic energy were sent through a gas of mercury vapor, and the number of electrons per second making it through the gas was measured. When the initial kinetic energy of the electrons was less than 4.9 eV, many electrons reached the detector. However, when the initial kinetic energy reached 4.9 eV, few electrons reached the detector. The Franck–Hertz experiment was correctly interpreted to mean that the first excited state of a mercury atom is 4.9 eV above the ground state (Figure 8.7), and an electron with kinetic energy of 4.9 eV can excite the atom, leaving the electron with almost no kinetic energy, so the number of electrons that made it through the gas became very small.

> QUESTION What happens if the initial kinetic energy of the electron is 9.8 eV?

In this case the electron could excite one mercury atom and still have 4.9 eV of kinetic energy available to excite a second mercury atom. In the Franck–Hertz experiment it was observed that the number of electrons reaching the detector again dropped sharply when the initial kinetic energy of the electrons was increased past 9.8 eV.

Every element has a different set of discrete energy levels. The first excited state in mercury is 4.9 eV above the ground state, whereas the first excited state in hydrogen is 10.2 eV above the ground state. These differences make it possible to identify elements from emission or absorption spectra, or measurements of the Franck–Hertz type.

> QUESTION What happens if you send a beam of electrons with kinetic energy of 25 eV through a gas of atomic hydrogen?

Because the ground state of hydrogen has $K + U = -13.6$ eV, only 13.6 eV is needed to raise a hydrogen atom to any of its excited states, or even to break the atom apart into a proton and an electron. Therefore a beam containing lots of 25 eV electrons passing through the atomic hydrogen gas will keep many excited states populated, and you'll observe many different photon energies being emitted as atoms drop from excited states to lower excited states or to the ground state.

EXAMPLE

An Electron Collides with a Hydrogen Atom

An electron with kinetic energy 12.5 eV collides with a hydrogen atom that is initially in its ground state. After the collision the hydrogen atom is its second excited state ($N = 3$). **(a)** What is the kinetic energy of the electron after the collision? **(b)** Would it have been possible for this electron to excite the hydrogen atom to its third excited state instead?

Solution **(a)** System: Electron and H atom
Surroundings: Nothing

$$E_{\mathrm{H}^*,f}+K_{\mathrm{e},f}=E_{\mathrm{H},i}+K_{\mathrm{e},i}$$
$$\left(-\frac{13.6\text{ eV}}{3^2}\right)+K_{\mathrm{e},f}=\left(-\frac{13.6\text{ eV}}{1^2}\right)+12.5\text{ eV}$$
$$K_{\mathrm{e},f}=0.411\text{ eV}$$

(b) The energy difference between the ground state and the third excited state ($N=4$) is

$$\Delta E=\left(-\frac{13.6\text{ eV}}{4^2}\right)-\left(-\frac{13.6\text{ eV}}{1^2}\right)$$
$$=12.75\text{ eV}$$

The initial kinetic energy of the electron is less than 12.75 eV, so it is not possible for the collision to excite the atom to the third excited state.

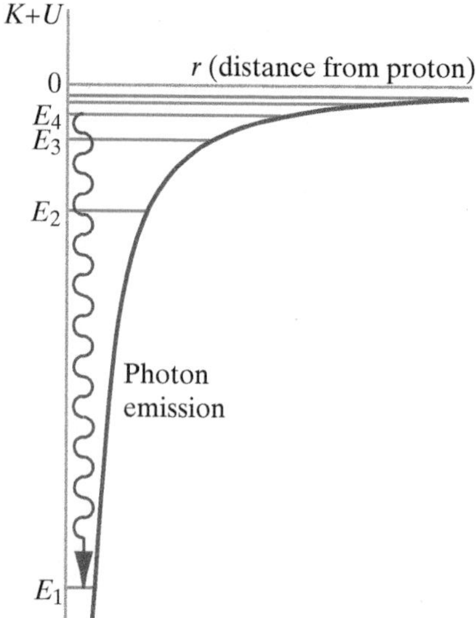
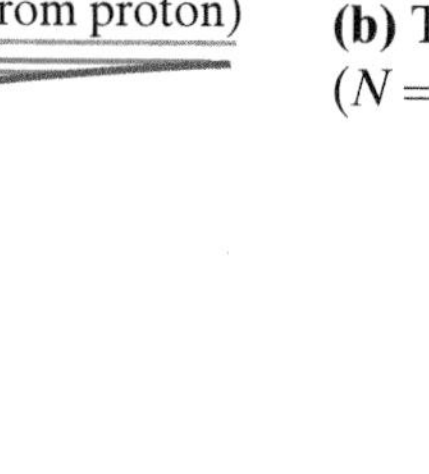

Figure 8.8 An atom can drop to a lower energy level by emitting a photon whose energy corresponds to the difference between the energies of the levels.

Checkpoint 2 The first excited state of a mercury atom is 4.9 eV above the ground state. A moving electron collides with a mercury atom, and excites the mercury atom to its first excited state. Immediately after the collision the kinetic energy of the electron is 0.3 eV. What was the kinetic energy of the electron just before the collision?

Decreasing the Internal Energy of an Atom

Just as an atom or molecule can absorb a photon, ending up in an excited state (with more internal energy), it can lose energy by emitting a photon (Figure 8.8), which comes into existence specifically to carry away a particular amount of energy. For a hydrogen atom, emission of a photon can be described this way:

$$\mathrm{H}^*\rightarrow \mathrm{H}+\text{photon}$$
$$E_{\mathrm{H},f}+K_{\text{photon}}=E_{\mathrm{H}^*,i}$$

E_0+8.5 eV — 2nd excited state
E_0+6.0 eV — 1st excited state
E_0 — Ground state

Figure 8.9 Photon emission from an atom.

EXAMPLE **Photon Emission from an Atom**

Different atoms have different quantized energy levels. Consider a particular atom (not hydrogen) whose first excited state is 6.0 eV above the ground state, and the second excited state is 8.5 eV above the ground state (Figure 8.9). What is the energy of a photon emitted when such an atom drops from the second excited state to the first excited state?

Solution System: All particles
Surroundings: Nothing

$$E_f=E_i$$
$$(E_0+6.0\text{ eV})+K_{\text{photon}}=(E_0+8.5\text{ eV})$$
$$K_{\text{photon}}=2.5\text{ eV}$$

The energy of the emitted photon (2.5 eV) is the difference between the energies of the second and third energy levels.

Figure 8.10 shows a portion of a neon sign, in which neon atoms in a clear glass tube have gained energy through collisions with energetic electrons, and are emitting orange photons whose energy is characteristic of neon.

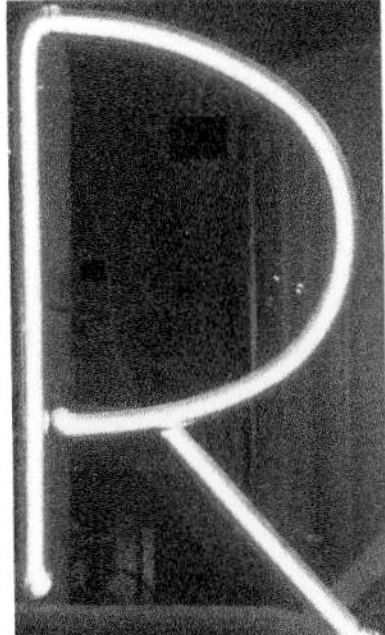

Figure 8.10 Collisional excitation of neon gas in a neon sign leads to emission of photons with the characteristic orange color. (Ruth Chabay/Bruce Sherwood)

Excited States Have Limited Lifetimes

If an isolated hydrogen atom is initially in a high energy level, over time it will spontaneously emit photons and drop to lower and lower states, eventually ending up in the lowest energy level, which is called the "ground state." Once an atom has dropped to the ground state, it cannot lose any more energy, or emit photons, because there is no lower energy level than the ground state.

Excited states are not in fact completely stable and eventually transition to a lower energy state. When a hydrogen atom drops from a higher energy level E_4 to a lower energy level E_1 it emits a photon with the associated energy $E_4 - E_1$ (Figure 8.8). In the energy diagrams we worked with for gravitational orbits, the orbit was stable and the energy was constant, but in an atom in a (nearly) stable state a photon can be emitted and the atom's energy drop to a lower level. When a change from one energy level to another happens, the change is instantaneous and takes no time, which is why these transitions are often called "quantum jumps" or "quantum leaps."

Another distinctive feature of photon emission from an excited state is that there is no way to predict exactly when it will happen. The probability that a hydrogen atom in the first excited state will remain in that state a time t after its formation is an exponential function of t, $e^{-t/\tau}$, where τ is approximately 2 ns (2 nanoseconds = 2×10^{-9} s). The quantity τ is called the "mean lifetime," the average time the excited state will persist. At the time $t = \tau$, the probability of the excited state still existing is $e^{-1} = 1/e = 0.368$. That is, when $t = \tau$, there is only a 36.8% probability that the atom is still in the first excited state (Figure 8.11). Most other excited states of hydrogen also have mean lifetimes of a few nanoseconds. Quantum mechanics can be used to predict the value of the mean lifetime τ, but when the emission from an excited state will actually occur is probabilistic and cannot be predicted. This is very different from the situation in our everyday macroscopic world.

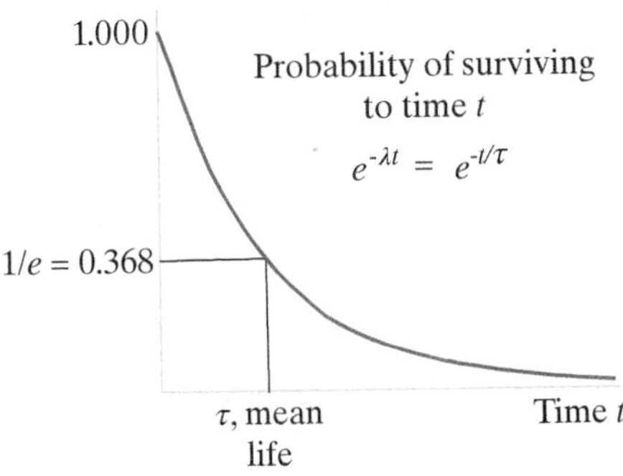

Figure 8.11 The probability of an excited atom remaining excited without emitting a photon, as a function of time.

Observations Typically Involve Many Atoms

It is much easier to observe the behavior of many atoms interacting with many photons than to observe a single atom absorbing or emitting a single photon. For example, a tube full of a gas (such as a fluorescent light tube) contains many atoms, which can be collisionally excited by a beam of electrons, and which emit many easily observable photons as they transition back down to the ground state. Instead of considering a single atom, therefore, another way of thinking about the probabilistic nature of photon emission is to consider a collection of 100 hydrogen atoms, all of them initially in the first excited state.

Different atoms in this collection will emit photons at different times. After a time $t = \tau$, there will on average be only $100e^{-1} \approx 37$ atoms still in the first excited state; the other 63 atoms will have dropped to the ground state with the emission of 63 photons of energy 10.2 eV. Because emission comes at random times, at $t = \tau$ there might still be 40 atoms in the excited state, or 35, but the average number in repeated experiments will be $100/e$. (For more details, see Section 8.8.)

When one of the original 100 atoms emits a 10.2 eV photon, it is possible that an atom that is already in the ground state might absorb that photon and

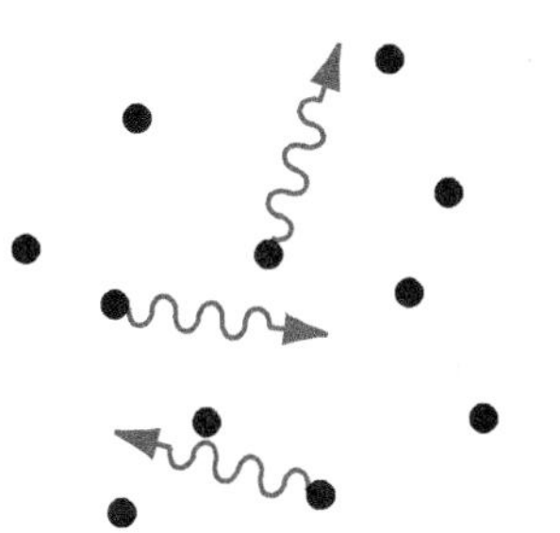

Figure 8.12 In a collection of atoms originally in the first excited state, many emitted photons will escape.

jump back up to the first excited state. However, in hydrogen gas at ordinary densities the atoms are rather far apart, so most photons will escape from a glass container of excited hydrogen atoms (Figure 8.12). Moreover, just because a photon is headed toward an atom in the ground state doesn't necessarily mean that it will interact with that atom, as this kind of interaction is also probabilistic.

EXAMPLE

Emission from Hydrogen Atoms

Consider a collection of many hydrogen atoms. Initially each individual atom is in one of the lowest three energy states, $N=1$ (the ground state), $N=2$ (the first excited state), or $N=3$ (the second excited state). What will be the energies of photons emitted by this collection of hydrogen atoms?

Solution There are three possibilities for atoms in the $N=3$ state: transitions from $N=3$ to $N=2$, from $N=2$ to $N=1$, or from $N=3$ to $N=1$ (Figure 8.13). Any atoms that are in the ground state ($N=1$) cannot emit a photon. Any atoms in the $N=2$ state can only emit a 10.2 eV photon. The photon energies for the three transitions are these:

$$N=3 \text{ to } N=2: \quad \left(-\frac{13.6\text{ eV}}{3^2}\right)-\left(-\frac{13.6\text{ eV}}{2^2}\right)=1.9\text{ eV}$$

$$N=2 \text{ to } N=1: \quad \left(-\frac{13.6\text{ eV}}{2^2}\right)-\left(-\frac{13.6\text{ eV}}{1^2}\right)=10.2\text{ eV}$$

$$N=3 \text{ to } N=1: \quad \left(-\frac{13.6\text{ eV}}{3^2}\right)-\left(-\frac{13.6\text{ eV}}{1^2}\right)=12.1\text{ eV}$$

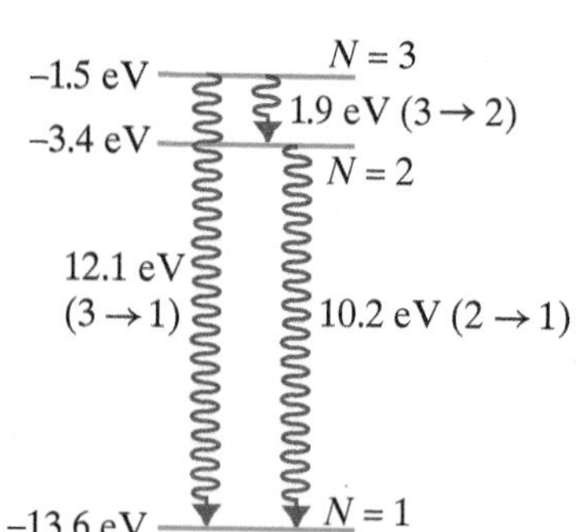

Figure 8.13 Photon emissions involving the ground state and first two excited states of a hydrogen atom. Only the $K+U$ energy levels are shown; for simplicity one often omits the potential energy curve.

Here again probability plays a distinctive quantum mechanical role. If a hydrogen atom is in the $N=3$ state, it can spontaneously (at an unpredictable time) change instantaneously to the ground state ($N=1$) with the emission of a 12.1 eV photon, or it can change instantaneously to the $N=2$ state with the emission of a 1.9 eV photon, followed shortly thereafter by the emission of a photon of 10.2 eV as the state changes instantaneously to the ground state.

Which of these events will happen? Will the atom emit one photon of 12.1 eV or two photons of 1.9 eV and 10.2 eV? With quantum mechanics it is only possible to calculate the probabilities of which event will happen; it is not possible to predict which event will in fact happen. This is very different from the situation in pre-1900 "classical mechanics."

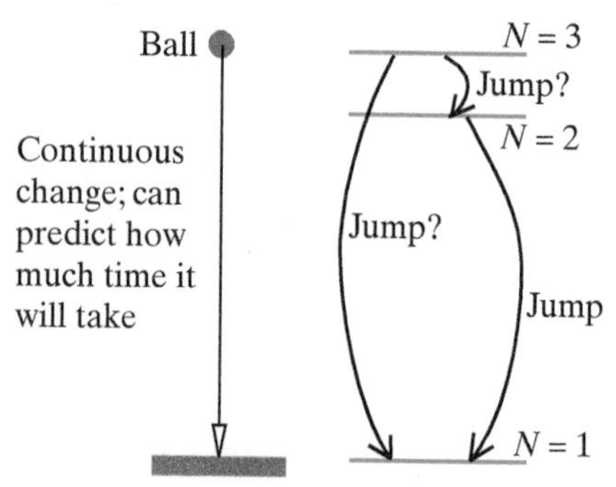

Figure 8.14 Classical mechanics vs. quantum mechanics: a ball falls in a continuous way; an atom changes state instantaneously, at unpredictable times.

Figure 8.14 illustrates this striking difference between classical mechanics and quantum mechanics. A ball falls to Earth in a continuous way, starting to move downward at a known time, and taking a predictable amount of time to reach the ground; the change in the Earth + ball system is continuous. In contrast, consider a hydrogen atom that starts out in the $N=3$ state at a known time. It is impossible to predict when the atom will drop to a lower state, and when it does drop, it does so instantaneously, a "quantum jump." There is no continuous change from the $N=3$ state to a lower state (with photon emission).

Moreover, it is impossible to predict whether this particular atom will make an instantaneous quantum jump from the $N=3$ state directly to the ground state (with the emission of a 12.1 eV photon) or whether it will get to the ground state in two jumps, stopping briefly in the $N=2$ state. Quantum mechanics can determine the probability of jumping directly to the ground state but cannot predict whether this particular atom will do so.

Checkpoint 3 How many different photon energies would emerge from a collection of hydrogen atoms that occupy the lowest four energy states ($N = 1, 2, 3, 4$)? (You need not calculate the energies of these transitions.)

Emission Spectra

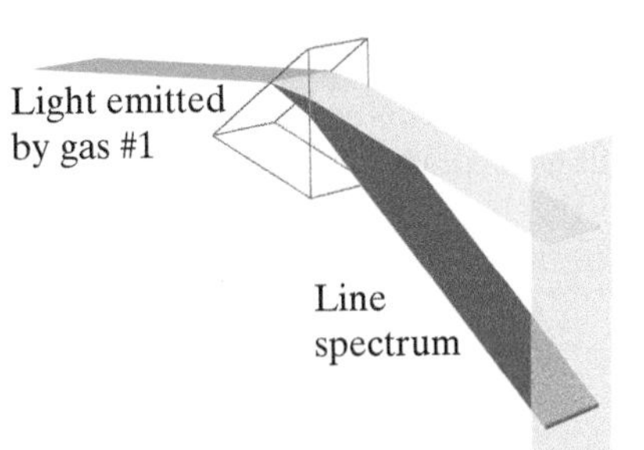

Figure 8.15 The light emitted by atoms of an excited gas.

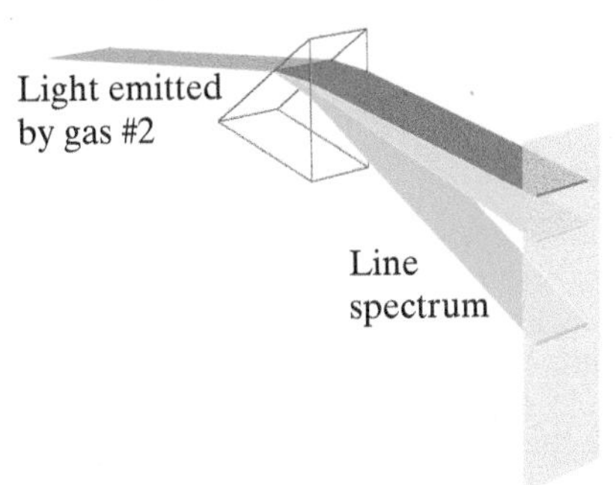

Figure 8.16 The light emitted by atoms of a second kind of excited gas.

Suppose we have a glass tube containing a gas, through which it is possible to run a beam of electrons that can transfer some of their kinetic energy to the atoms through collisions. These atoms remain in excited states for a few nanoseconds, then emit photons and return to the ground state, where they can repeat the process by gaining energy from collisions, and so on. Some of the photons emitted in this process may have energies in the visible range of the spectrum, so if we direct the emitted light through a slit and onto a prism, we will see a "line spectrum" like the one in Figure 8.15.

The light emitted by an excited gas is called an "emission spectrum," and the bright lines of light of particular energies in the spectrum are called "spectral lines" due to the line-like shape of the light coming through a slit in the apparatus. The energies of the photons emitted will depend on the particular gas, because the energy levels of every atom are different. Figure 8.16 shows an emission spectrum from a gas that is different from the one in Figure 8.15. The energy of the photons in any spectral line correspond to the differences in energy between two electronic energy levels of the particular atom emitting the photons.

The observed energies of light emitted by excited atomic hydrogen are consistent with the energy level differences predicted by quantum mechanics. Other atoms have different quantized energy levels and emit different photon energies, so a spectrum can be used to identify the emitting atoms by the pattern of emitted photon energies.

Absorption Spectra

Suppose that we have a container of hydrogen atoms, all in their ground state.

QUESTION Could we observe an emission spectrum from this gas?

No. If all the atoms are in the ground state, there is no lower energy state for them to go to, so they cannot lose energy by emitting photons. However, such a collection of atoms could gain energy by absorbing photons. If a beam of light composed of photons with a continuous distribution of energies (white light) is directed at a transparent container of gas, some of the photons in the beam will be absorbed by gas atoms.

QUESTION What should we expect to see if we pass a beam of white light through a transparent container of gas atoms in their ground state?

Only photons whose energy is the difference of atomic energy levels will be strongly absorbed, while other photons can pass right through the material. Energy absorption is by far most likely to occur if the photon energy corresponds exactly to the difference between two energy levels of the atom. What we will observe in this case is essentially the opposite of what we see in an emission spectrum. If we inspect the beam of light after it has passed through the container of gas, we will observe photons that were not absorbed, but photons whose energies match differences between energy levels of the gas will be missing because they were absorbed by the gas. In this case, the spectral lines are dark lines (missing energies) on a light background, instead of the bright lines on a dark background observed in an emission spectrum. This is called an "absorption spectrum" (Figure 8.17).

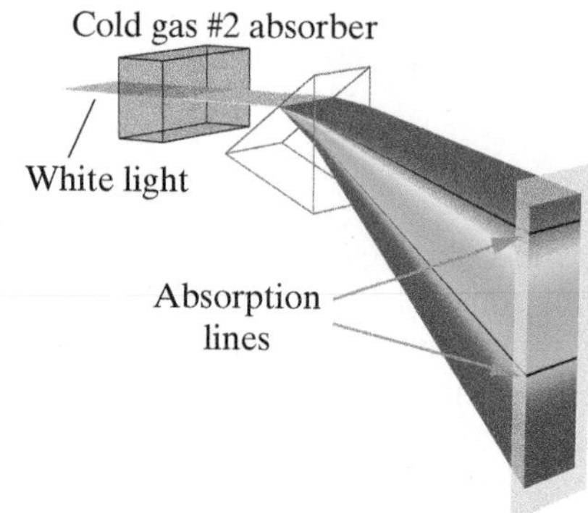

Figure 8.17 In an absorption spectrum, a beam of photons of many energies is directed through a gas. Although most photons pass through, photons matching the differences between energy levels are absorbed by gas atoms, leaving dark lines in the spectrum.

Photon absorption may be followed almost immediately by emission from the excited state to a lower state, so at any instant there are few atoms in excited states that could absorb photons. The result is that in dark-line absorption spectra only absorption from the ground state is observed, not from higher-level states.

The atoms that have jumped to higher energy levels through photon absorption will eventually drop to lower energy levels with the emission of photons. However, these photons are emitted in all directions, not just in the direction of the original beam of light, so the light that has passed through the material will have little intensity at photon energies corresponding to the strongly absorbed energies (Figure 8.18).

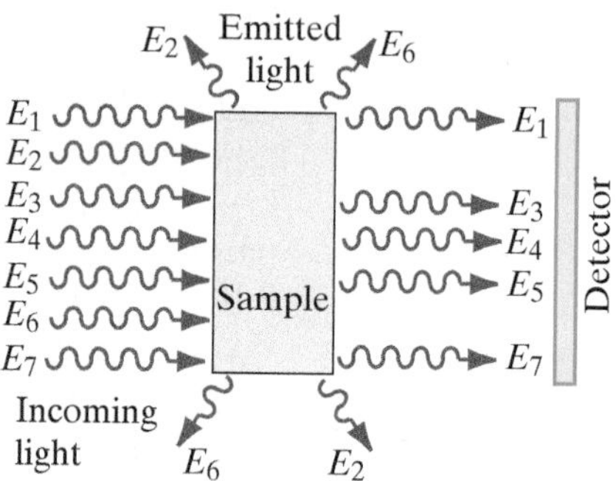

Figure 8.18 Measuring an absorption spectrum. A sample of a material is placed in a transparent container. Some energies of incoming light are absorbed. The outgoing light is depleted of the absorption energies (in this case E_2 and E_6).

QUESTION The gas that produced the three line emission spectrum in Figure 8.16 is the same gas that produced the two line absorption spectrum in Figure 8.17. Why do we observe three emission lines but only two absorption lines?

In an absorption spectrum the gas starts out in the ground state, so only transitions from the ground state to an excited state are possible. Evidently the middle line in the emission spectrum corresponds to a transition between two excited states, and won't appear in an absorption spectrum.

EXAMPLE **Why Is Hydrogen Gas Transparent?**

At room temperature in a gas of atomic hydrogen almost all of the atoms are in the ground state. Why might you expect the gas to be completely transparent to visible light?

Solution As we saw in an earlier example, the energy required to go from the ground state of atomic hydrogen to the first excited state is 10.2 eV, but visible light consists of photons whose energies lie between 1.8 eV and 3.1 eV. So visible light cannot be absorbed by ground-state hydrogen atoms. (In a collision between a photon of visible light and a hydrogen atom, the atom can acquire some kinetic energy, but no internal energy change is possible.)

Checkpoint 4 Suppose you had a collection of hypothetical quantum objects whose individual energy levels were −4.0 eV, −2.3 eV, and −1.6 eV. If nearly all of the individual objects were in the ground state, what would be the energies of dark spectral lines in an absorption spectrum if visible white light (1.8 to 3.1 eV) passes through the material?

EXAMPLE

Determining the Energy Levels from Photon Energies

You observe photon emissions from a collection of quantum objects, each of which is known to have just four quantized energy levels. The collection is continually bombarded by an electron beam, and you detect emitted photons using a detector sensitive to photons in the energy range from 2.5 eV to 30 eV. With this detector, you observe photons emitted with energies of 3 eV, 6 eV, 8 eV, and 9 eV, but no other energies.

(a) It is known that the ground-state energy of each of these objects is −10 eV. Propose two possible arrangements of energy levels that are consistent with the experimental observations. Explain in detail, using diagrams.

(b) You obtain a second detector that is sensitive to photon energies in the energy range from 0.1 eV to 2.5 eV. What additional photon energies do you observe to be emitted? Explain briefly.

(c) You turn off the electron beam so that essentially all the objects are in the ground state. Then you send a beam of photons with a wide range of energies through the material. Using both detectors to determine the absorption spectrum, what are the photon energies of the dark lines? How can this information be used to choose between your two proposed energy level schemes? Explain briefly.

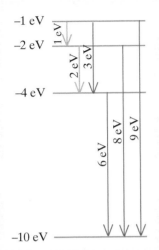

Figure 8.19 A possible energy-level scheme that fits the emission data. Lighter arrows indicate photons not detectable by this detector.

Solution

System: Everything—some unknown object and a photon
Surroundings: Nothing significant
Initial state: Object in excited state (no photon)
Final state: Object in lower energy state, plus photon
$E_{\text{obj},f} + K_{\text{photon}} = E_{\text{obj},i}$

The energy of any emitted photon must be equal to the difference in energy between an excited state and a lower-energy state.

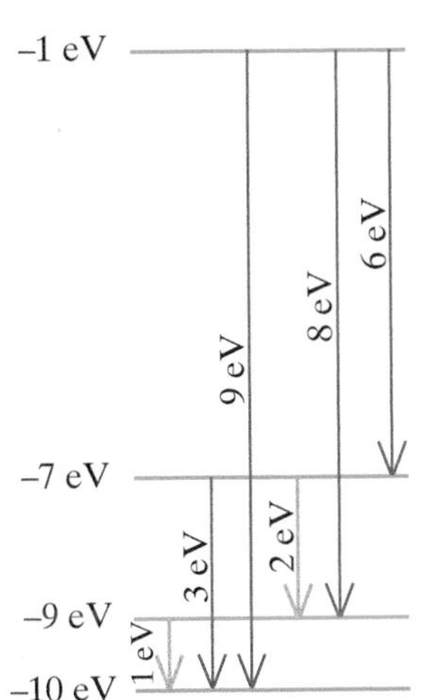

Figure 8.20 Another possible energy level scheme that fits the emission data.

(a) Figure 8.19 shows a possible energy-level scheme that fits the emission data. The highest photon energy (9 eV) tells us the difference between the highest and lowest states. The rest is trial and error, checking that the correct photon energies would be produced. The arrows are labeled with the energies of photons emitted in transitions between the two states indicated. The 1 eV and 2 eV emissions would not be detected by a detector that is sensitive to the photon energy range 2.5 eV to 30 eV.

Figure 8.20 shows another possible scheme that fits the emission data. This is somewhat less likely, because it is more common to find the energy-level spacing decreasing at higher energies, as shown in Figure 8.19.

(b) With this detector we observe the 1 eV and 2 eV photon emissions.

(c) With the electron beam turned off, almost all of the objects are in the ground state. If the energy levels are as shown in Figure 8.19, the dark lines will be at 6 eV, 8 eV, and 9 eV, corresponding to absorption in the ground state. If the energy levels are as shown in Figure 8.20, the dark lines will be at 1 eV, 3 eV, and 9 eV. Therefore the absorption spectrum allows us to distinguish between the two possible energy level schemes.

Note that since this was an energy analysis, we didn't need to know what kind of object this was, or whether the energy levels were electronic energy levels or some other kind.

Figure 8.20 is an example of an energy-level diagram, with photon emissions indicated by arrows that start on the upper level and point downward, with arrowheads on the lower level. Photon absorptions are indicated by arrows pointing up from the ground state to an upper level.

Checkpoint 5 The energy levels of a particular quantum object are −8 eV, −3 eV, and −2 eV. If a collection of these objects is bombarded by an electron beam so that there are some objects in each excited state, what are the energies of the photons that will be emitted?

Selection Rules

For simplicity, we've been assuming that photon emissions can occur from any discrete energy level to a lower level. Real atomic and molecular systems may be more complex, and some photon emissions are "forbidden." In a later chapter we will study angular momentum, a quantity that is related to rotational motion and is a conserved quantity, like momentum and energy. Photon emission and absorption, like all other physical processes, must conserve angular momentum, and as a result transitions between two particular energy levels may be forbidden. Such constraints are called "selection rules" and govern photon emission and absorption (collisional excitation is not constrained by selection rules). Selection rules are beyond the scope of this introductory course, but you will learn about them if you take a later course in quantum mechanics. For the purposes of this introduction to quantized energy, we will ignore selection rules and assume that all energetically possible transitions are allowed.

We also ignore questions of how probable an allowed transition is, which requires full quantum mechanics to calculate. Suppose that an object is in the second excited state, from which it could drop either to the first excited state or directly to the ground state. One of these two transitions could be much more probable than the other one, even if the low-probability transition isn't forbidden by a selection rule.

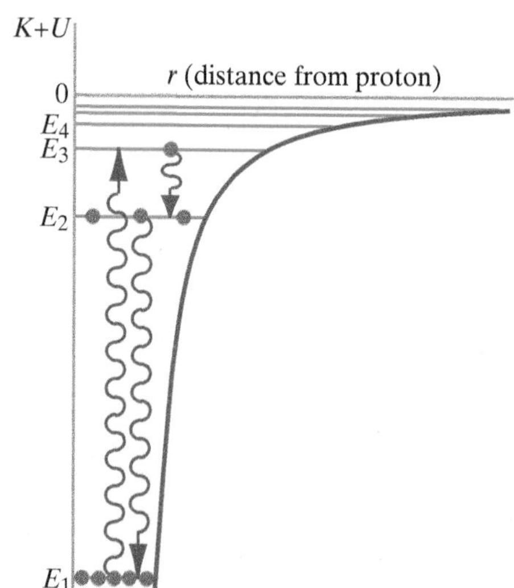

Figure 8.21 A dot represents the energy state of one hydrogen atom in a group of atoms. Energy is exchanged with the surroundings through photon emission and absorption.

8.3 THE EFFECT OF TEMPERATURE

When we observe atoms we are usually observing a collection of atoms, not an isolated single atom. Each atom in the collection (the "sample") interacts with other atoms and perhaps with a container. If an atom absorbs radiation coming from elsewhere, or absorbs energy from being jostled by a neighboring atom (collisional excitation), it may jump to a higher energy level, and eventually emit photons and drop again to a lower level (Figure 8.21 shows emissions as downward transitions and absorptions as upward transitions for a collection of hydrogen atoms). In thermal equilibrium with its surroundings, an atom's energy continually goes up and down, but with an average energy determined by the temperature of the collection of atoms.

In Figures 8.21 and 8.22 each dot represents one hydrogen atom in a container of atomic hydrogen. If three dots are on a line, this means three atoms are in this particular state. Each hydrogen atom has the same discrete energy levels, and at any particular instant some of the atoms will be in the ground state, some in the first excited state, and so on.

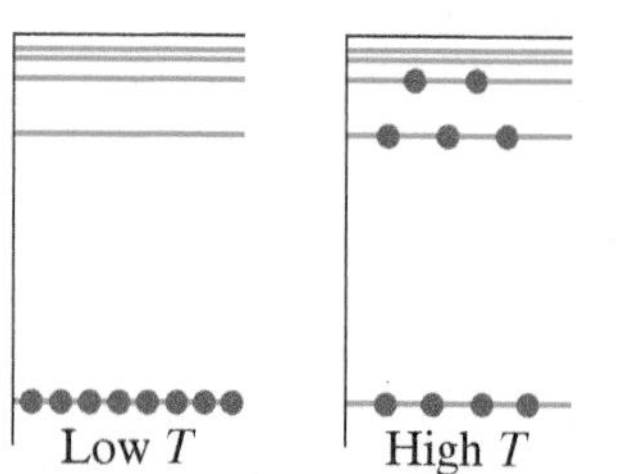

Figure 8.22 At a low temperature (system on the left) all atoms are in low-energy states. At a high temperature (system on the right) some atoms are in higher-energy states, though the largest fraction will still be in the ground state.

For a particular temperature, at any instant there is a certain probability that one particular atom will be in state 1, 2, 3, and so on. At low temperature almost all atoms are in the ground state (Figure 8.22). At high temperature the various atoms are in various states, from the ground state up to a high energy level, though the largest fraction will still be in the ground state. As we will see in a later chapter on statistical mechanics, in a collection of atoms the fraction of the atoms in a state whose energy is E above the ground state is proportional to the "Boltzmann factor" e^{-E/k_BT}, where k_B is the Boltzmann constant (1.38×10^{-23} J/K) and T is the absolute temperature in kelvins (K), where 0 °C is 273.15 K.

QUESTION Can you observe an emission spectrum if the material is very cold? Can you observe an absorption spectrum if the material is very cold?

In very cold matter the atoms will normally be in the ground state and will not emit photons. However, an absorption spectrum will show gaps (dark lines) corresponding to absorption of photons of energies appropriate to lift an atom from its ground state to an excited state. In order to observe an absorption spectrum you would of course need to use a beam of photons whose energies are sufficiently high to raise the system energy at least to the first excited state above the ground state.

Cold Samples

In a sample at room temperature (about 300 K), almost all atoms are in their electronic ground state. For example, if the first excited state is 1 eV above the ground state, the Boltzmann factor is

$$e^{-E/k_BT} = e^{-\frac{1.6\times10^{-19}\,\text{J}}{(1.4\times10^{-23}\,\text{J/K})(300\,\text{K})}} = e^{-38} = 3\times10^{-17}$$

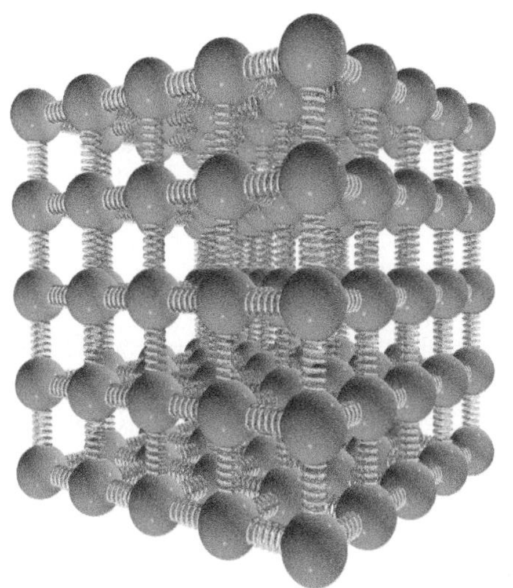

Figure 8.23 Model of a solid as a network of balls and springs.

This extremely tiny number represents the fact that at room temperature an insignificant fraction of the atoms will be excited thermally to 1 eV above the ground state. Therefore, it is reasonable to assume that essentially all the atoms in an ordinary room-temperature absorption measurement start out in the electronic ground state.

On the other hand, the surface temperature of the Sun is about 5000 K, and with $E = 1$ eV the Boltzmann factor is $e^{-E/k_BT} = e^{-2.3} = 0.1$, so a sizable fraction of atoms can be in an excited state 1 eV above the ground state, due to thermal excitation.

8.4 VIBRATIONAL ENERGY LEVELS

For concreteness we have described energy quantization for the electronic states of atomic hydrogen. Many other kinds of systems have quantized energy. We will next discuss vibrational energy quantization in atomic oscillators.

Vibrational Energy in a Classical Oscillator

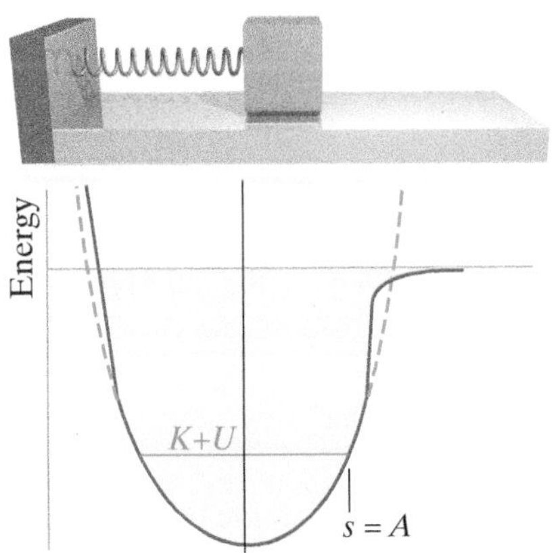

Figure 8.24 A classical harmonic oscillator (spring–mass system) can vibrate with any amplitude, and hence can have any energy. $K+U$ of the system could have any value in the shaded region.

We have modeled a solid as a network of classical harmonic oscillators, balls and springs (Figure 8.23). Although this classical model explains many aspects of the behavior of solids, the predictions it makes regarding the thermal properties of solids do not correspond to experimental measurements at low temperatures. The classical model also fails to explain some aspects of the interaction of light with solid matter. To explain these phenomena we need to make our model of a solid more sophisticated, by incorporating the quantum nature of these atomic oscillators into our model.

The vibrational energy of a classical (nonquantum) spring–mass system can have *any* value within the colored region in Figure 8.24. Equivalently, the amplitude A (maximum stretch) can have *any* value within the allowed range. We can express the energy as $E = \frac{1}{2}k_sA^2$, because at the maximum stretch A the speed is zero, so the kinetic energy is zero, and the energy $K+U$ is just the potential energy at that instant. Classically, the amplitude A can have continuous rather than quantized values.

Vibrational Energy in a Quantum Mechanical Oscillator

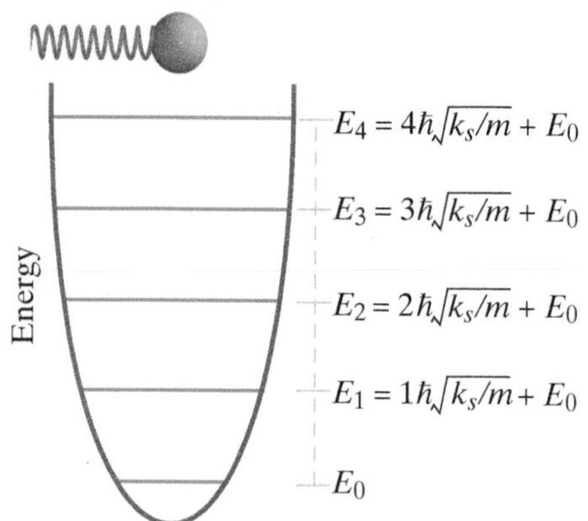

Figure 8.25 A quantum harmonic oscillator can have only certain energies, indicated by the horizontal lines on the graph. This idealized potential energy well represents the situation near the bottom of a real well (which is at a negative energy).

Historically, the first hint that energy might be quantized came from an attempt by the German physicist Max Planck in 1900 to explain some puzzling features of the spectrum of light emitted through a small hole in a hot furnace (so-called "blackbody radiation"). Planck found that he could correctly predict the properties of blackbody radiation if he assumed that transfers of energy in the form of electromagnetic radiation were quantized—that is, an energy transfer could take place only in fixed amounts, rather than having a continuous range of possible values.

Later, with the full development of quantum mechanics, it was recognized that one could model an atom in the solid wall of the furnace as a tiny spring–mass system whose energy levels are discrete (Figure 8.25). Classically, the energy of a mass on a spring is related to the amplitude of its oscillation ($E = \frac{1}{2}k_sA^2$), so saying that the stable or nearly stable energy states of a spring–mass oscillator are discrete is equivalent to saying that only certain amplitudes correspond to these states. To see this, note that we have

$$K + U = \frac{1}{2}\frac{p^2}{m} + \frac{1}{2}k_s s^2 = \frac{1}{2}k_s A^2$$

since when the stretch s is equal to the amplitude A (the maximum stretch), the momentum is zero.

By solving the quantum mechanical Schrödinger equation for a system involving an atom and a chemical bond, one can show that the spacing ΔE between the quantized energies has the following value:

$$\Delta E = \hbar\sqrt{\frac{k_s}{m}} \quad \text{or} \quad \Delta E = \hbar\omega_0, \quad \text{where } \omega_0 = \sqrt{\frac{k_s}{m}}$$

In this case, the quantity $k_s = k_{s,i}$ is the interatomic spring stiffness, which we estimated in Chapter 4, $m = m_a$ is the mass of one atom, and $\omega_0 = \sqrt{k_s/m}$ is the angular frequency of the oscillator (in radians per second), as defined in Chapter 4. The quantity $\hbar$, pronounced "h-bar," is equal to h, "Planck's constant," divided by 2π. Planck's constant h has an astoundingly small value:

$$h = 6.6 \times 10^{-34}\ \text{J}\cdot\text{s}$$

The energy levels for an atomic harmonic oscillator are these:

VIBRATIONAL ENERGY LEVELS

$$E_N = N\hbar\omega_0 + E_0,\ N = 0,1,2,\ldots, \quad \text{where } \omega_0 = \sqrt{k_s/m}$$

In other words, the quantized energies have the following values:

$$E = E_0,\ 1\hbar\omega_0 + E_0,\ 2\hbar\omega_0 + E_0,\ 3\hbar\omega_0 + E_0,\ldots.$$

For an atom and a spring-like interatomic bond: $k_s = k_{s,i}$ (interatomic spring stiffness) and $m = m_a$ (mass of an atom)

$$\hbar = \frac{h}{2\pi} = 1.05 \times 10^{-34}\ \text{J}\cdot\text{s ("h-bar")}$$

It is an unusual and special feature of the quantized harmonic oscillator that the spacing between adjacent energy levels is the same for all levels ("uniform spacing"). This is quite different from the uneven spacing of energy levels in the hydrogen atom. These energy levels are predicted by quantum mechanics, and experiments confirm this prediction.

It is important to understand that in both the quantum and the classical picture, the angular frequency $\omega_0 = \sqrt{k_s/m}$ of a particular oscillator is fixed and is the same whether the energy of the oscillator is large or small. A useful way to think about the situation is to say that the frequency is determined by the parameters k_s and m, while the energy is determined by the initial conditions (or equivalently by the amplitude A). In the quantum world, it is as though the amplitude can have only certain specific values.

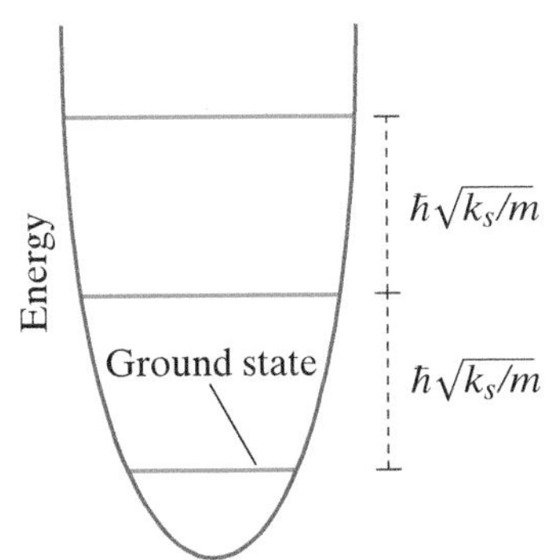

Figure 8.26 A quantum oscillator with a large k_s has widely spaced energy levels.

Checkpoint 6 Suppose that a collection of quantum harmonic oscillators occupies the lowest four energy levels, and the spacing between levels is 0.4 eV. What is the complete emission spectrum for this system? That is, what photon energies will appear in the emissions? Include all energies, whether or not they fall in the visible region of the electromagnetic spectrum.

Spring Stiffness k_s

The energy quantum depends on the stiffness k_s of the "spring." A stiffer spring (larger value of k_s, Figure 8.26) means a steeper, narrower potential curve and larger spacing between the allowed energies. A smaller value of k_s (Figure 8.27) means a shallower, wider potential curve and smaller spacing between the allowed energies.

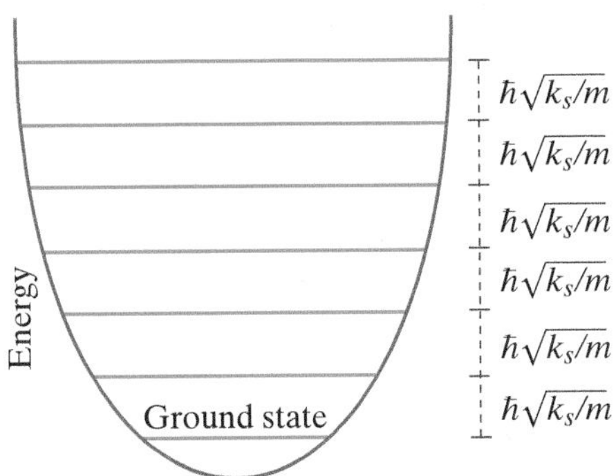

Figure 8.27 A quantum oscillator with a small k_s has closely spaced energy levels.

Checkpoint 7 You may have measured the properties of a simple spring–mass system in the lab. Suppose that you found $k_s = 0.7$ N/m and $m = 0.02$ kg, and you observed an oscillation with an amplitude of 0.2 m. What is the approximate value of N, the "quantum number" for this oscillator? (That is, how many levels above the ground state is this oscillator?)

The Ground State of a Quantum Oscillator

The lowest possible energy level E_0 is not zero relative to the bottom of the potential curve, but is $\frac{1}{2}(\hbar\sqrt{k_s/m})$ above the bottom of the well (Figures 8.26 and 8.27). This is closely related to the "Heisenberg uncertainty principle." When applied to the harmonic oscillator, this principle implies that if you knew that the mass on the spring were exactly at the center of the potential, its speed would be completely undetermined, whereas if you knew that the speed of the mass were exactly zero, the position of the mass could be literally anywhere! The minimum energy $\frac{1}{2}(\hbar\sqrt{k_s/m})$ above the bottom of the well can be viewed as a kind of a compromise, with neither the speed nor the position fixed at zero. The minimum energy is called the ground state of the system.

In our work with the quantized oscillator we will measure energy from the ground state rather than from the bottom of the potential well, so the offset from the bottom of the well won't come into our calculations.

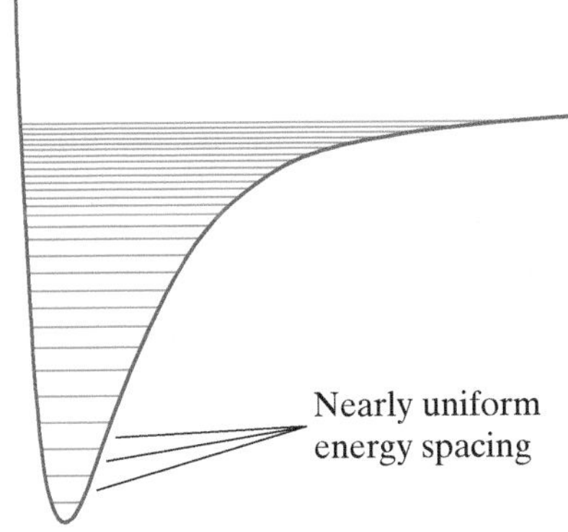

Figure 8.28 Energy levels for the interatomic potential energy, which describes the interaction of two neighboring atoms. This diagram is not to scale; the number of vibrational energy levels in the uniform region may be very large.

Energy Levels for the Interatomic Potential Energy

If we model the atoms in a solid as simple harmonic oscillators, whose potential energy curve is a parabola, then the energy levels of these systems are uniformly spaced. However, if we use a more realistic interatomic potential energy such as is shown in Figure 8.28, the energy levels of the system follow a more complex pattern. The lowest energy levels are nearly evenly spaced, corresponding to the fact that for small amplitude the bottom of the potential energy curve can be fit rather well by a curve of the form $\frac{1}{2}k_s s^2$. However, for large energies the energy spacing is quite small. Of course there are no bound states for energies greater than zero in Figure 8.28 (where zero is the value of the system energy at very large separations). The spacing of the vibrational energy levels

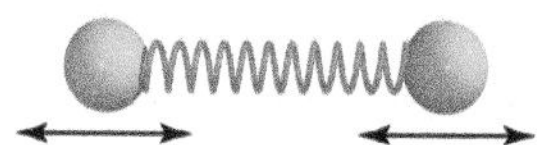

Figure 8.29 The vibrational energy levels of a diatomic molecule are discrete.

of a diatomic molecule (Figure 8.29) typically corresponds to photons in the infrared region of the spectrum (see Problem P26).

Energy Spacing in Various Systems

Most quantum objects have energy-level spacings that decrease as you go to higher levels. The harmonic oscillator potential energy is unusual in leading to uniform energy-level spacing. Even less common is the situation where the energy-level spacing actually increases with higher energy. One such exception is the so-called "square well" with very steep sides and a flat bottom. The energy levels in a square well actually get farther and farther apart for higher energies.

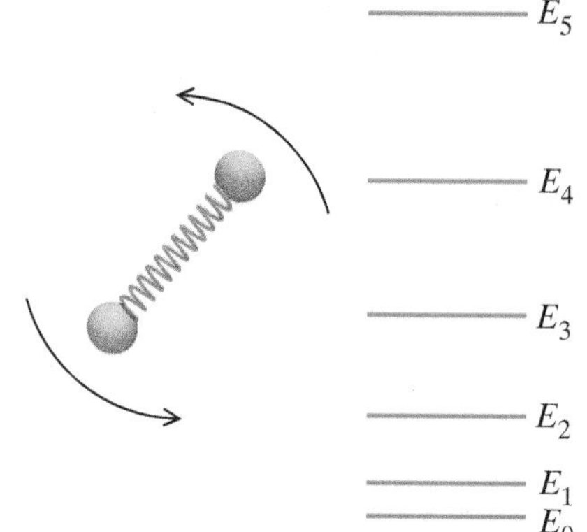

Figure 8.30 Rotational kinetic energy levels of a diatomic molecule.

8.5 ROTATIONAL ENERGY LEVELS

A molecule in the gas phase has not only electronic energy and vibrational energy but rotational energy as well. Like electronic and vibrational energy levels, rotational energy levels are also discrete—only certain rotational energies are permitted. As we will see in a later chapter, the angular momentum of a molecule is quantized, and this quantization gives rise to discrete rotational energy levels. In this case all the energy of rotation is kinetic energy—there is no potential energy involved. Rotational energy levels for a diatomic molecule such as N_2 are shown in Figure 8.30. Note that the spacing between levels actually increases with increasing energy, which is the opposite of the trend seen in many atomic systems. Rotational energy levels are typically close together; transitions between these levels usually involve photons in the microwave region of the electromagnetic spectrum. Microwave ovens excite rotational energies of water molecules in food. These excited molecules then give up their energy to the food, making the food hotter.

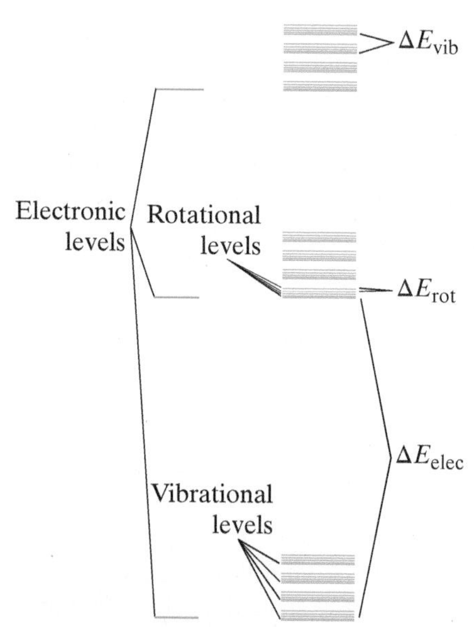

Figure 8.31 The energy bands of a diatomic molecule (not to scale). This energy-level structure gives rise to a band spectrum of emitted photons.

Case Study: A Diatomic Molecule

Analyzing a diatomic molecule such as N_2 provides a good review of three important types of discrete energy levels: electronic, vibrational, and rotational.

There are different configurations of the electron clouds in a diatomic molecule that correspond to discrete electronic energy levels. The spacings of the low-lying molecular electronic energy levels are typically in the range of 1 eV or more, like the spacings of the low-lying energy levels in the hydrogen atom.

For each major configuration of the electron clouds, many different vibrational energy states are possible, with energy spacings of the order of 1×10^{-2} eV (see Problem P26).

For each major configuration of the electron clouds and each vibrational energy state, many different rotational kinetic energy states are possible, with energy spacings of the order of 1×10^{-4} eV, as we will see in a later chapter.

These three different families of energy levels, with their very different energy-level spacings, give rise to energy "bands" and a distinctive "band" spectrum of emitted photons. See Figure 8.31, which is not to scale due to the large differences in level spacing for the three kinds of quantized energy levels. An emitted photon can have the energy of an electronic transition plus a smaller-energy vibrational transition plus an even smaller-energy rotational transition. The various possible energy differences give rise to "bands" in the observed emission spectrum.

8.6 OTHER ENERGY LEVELS

It is not only electronic, vibrational, and rotational energies in atoms and molecules that are quantized. Small objects such as nuclei and hadrons (particles made of quarks) also have discrete energy levels.

Nuclear Energy Levels

Just as atoms have discrete, stable or nearly stable states corresponding to allowed configurations of the electron cloud, so nuclei have discrete states corresponding to allowed configurations of the nucleons (protons and neutrons). An excited nucleus can drop to a lower state with the emission of a photon. The energy spacing of nuclear levels is very large, of the order of 1×10^6 eV (1 MeV), so the emitted photons have energies in the range of millions of electron volts. Such energetic photons are called "gamma rays."

Checkpoint 8 A gamma ray with energy greater than or equal to 2.2 MeV can dissociate a deuteron (the nucleus of deuterium, heavy hydrogen) into its constituents, a proton and a neutron. What is the nuclear binding energy of the deuteron?

Hadronic Energy Levels

Atoms have energy levels associated with the configurations of electrons in the atom, and nuclei have energy levels associated with the configurations of nucleons in the nucleus. Hadrons are particles made of quarks. In a simple model we represent baryons (particles like protons) as a combination of three quarks, and mesons (particles like pions) as a combination of a quark and an antiquark. Different configurations of quarks correspond to different energy levels of the multiquark system. These different configurations are the different hadrons we observe, and the corresponding energy levels are the observed rest energies of these particles. The energy spacing is of the order of hundreds of MeV, or about 1×10^8 eV. (Leptons, which include electrons, muons, and neutrinos, are not made of quarks.)

An example of a hadron is the Δ^+ particle, which has a rest energy of 1232 MeV. It can be considered to be an excited state of the three-quark system whose ground state is the proton, whose rest energy is 938 MeV. The Δ^+ particle can decay into a proton with the emission of a very high energy photon. However, this "electromagnetic" decay occurs in only about 0.5% of the decays. In most cases the Δ^+ decays into a proton plus a pion, a strong-interaction decay that has a much higher rate, corresponding to the much stronger interaction.

An excited electronic state of a hydrogen atom is still called hydrogen, but for historical reasons we give excited quark states distinctive names, such as Δ^+.

Checkpoint 9 What is the approximate energy of the photon emitted when the Δ^+ decays electromagnetically?

Type of state	Typical energy level spacing
hadronic	10^8 eV
nuclear	10^6 eV
electronic (atoms, molecules)	1 eV
vibrational (molecules)	10^{-2} eV
rotational (molecules)	10^{-4} eV

Figure 8.32 Typical energy level spacings in various quantum objects.

8.7 COMPARISON OF ENERGY-LEVEL SPACINGS

Figure 8.32 is a summary of typical energy scales for various kinds of objects. It is important to distinguish among hadronic, nuclear, electronic, vibrational, rotational, and other kinds of quantized energy situations. We can generalize the main issues to any kind of generic quantum object. The key features of all these different kinds of objects are these:

- discrete energy levels
- discrete emissions whose energies equal differences in energy levels

Some of the homework problems at the end of the chapter ask you to analyze generic objects, independent of the details of what gave rise to the particular discrete energy levels.

8.8 *RANDOM EMISSION TIME

Suppose we initially have an atom in its first excited state. In a short time interval dt there is a probability λdt that this atom will emit a photon and fall to the ground state, where λ is the average emission rate for this kind of atom. If we initially have $N = N_0$ atoms each in their first excited state, in a short time interval dt there is a probability of $N\lambda dt$ atoms emitting photons and falling to the ground state, so that the change in the number of excited atoms is $dN = -N\lambda dt$ or $dN/dt = -N\lambda$, where the negative sign indicates that N decreases. The equation shows that emissions are proportional to the remaining number of excited atoms, which makes sense. This differential equation can be solved:

$$\frac{dN}{dt} = -N\lambda$$

$$\frac{dN}{N} = -\lambda dt$$

$$\int \frac{dN}{N} = \int -\lambda dt$$

$$\ln N - \ln N_0 = -\lambda t$$

$$\ln N = \ln N_0 - \lambda t$$

$$N = N_0 e^{-\lambda t}$$

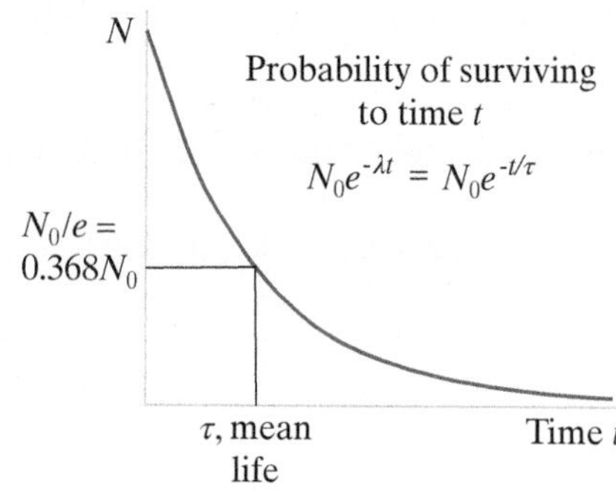

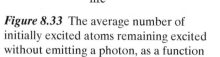

Figure 8.33 The average number of initially excited atoms remaining excited without emitting a photon, as a function of time.

The final equation is the result of equating e raised to the expression on the left to e raised to the expression on the right, where $e^{\ln x} = x$. The result shows that, on average, we expect the number of atoms remaining in the excited state to decrease exponentially with time t. The number remaining after a time $\tau = 1/\lambda$ is $1/e$ times the original number, and we often write $N = e^{-t/\tau}$, where τ is called the "mean lifetime" of the excited state (Figure 8.33). These same equations are used to describe radioactive decay, in which the number of radioactive nuclei decreases exponentially with time.

Problems P29 and P30 illustrate the full probabilistic character of emission, and the fact that the smooth exponential behavior shown in Figure 8.33 is seen only if the number of atoms is very large.

8.9 *CASE STUDY: HOW A LASER WORKS

A laser (Light Amplification by Stimulated Emission of Radiation) offers an interesting case study of the interaction of light and atomic matter. We will give a basic introduction to lasers, but we can only scratch the surface of a vast topic that continues to be an active field of research and development. Laser action depends on a process called "stimulated emission" and on creating an "inverted population" of quantum states, which can lead to a chain reaction and produce a special kind of light, called "coherent" light.

Stimulated Emission

If an atom can be placed into an excited state—that is, a state of higher energy than the ground state—normally it drops quickly to a lower-energy state and emits a photon in the process. For example, in a neon sign high-energy electrons collide with ground-state neon atoms and raise them to an excited state. An excited neon atom quickly drops back to the ground state, emitting a photon whose energy is the difference in energy between the two quantized energy levels, and you see the characteristic orange color corresponding to these photons.

We call this emission process "spontaneous emission." Spontaneous emission happens at an unpredictable time. All that quantum mechanics can predict is the *probability* that the excited atom will drop to the ground state in the next small time interval. It is truly not possible to predict the exact time of the emission, just as it is impossible to predict exactly when a free neutron will decay into a proton, electron, and antineutrino.

A collection of excited atoms such as those in a neon sign emit spontaneously, at random times. In the full quantum-mechanical description of light, light has both the properties of a particle (the photon picture) and the properties of a wave; see Section 8.10. Emissions at random times have random phases in a wave description: sine, cosine, and all phases between a sine and a cosine. We say that such light is "incoherent."

"Stimulated emission" of light is a different process that produces coherent light. Consider an atom in an energy level E above the ground state. A photon of this same energy E can interact with the atom in a remarkable way, causing the atom to drop immediately to the ground state with the emission of a second, "clone" photon. Where there had been one photon of energy E, there are now two photons of exactly the same energy E, and in a wave description they have exactly the same phase (Figure 8.34). That is, the two waves are both sines, or both cosines, or both with exactly the same phase somewhere between a sine and a cosine. Such light is called coherent light, and it has valuable properties for such applications as making holograms.

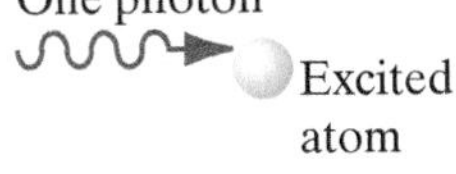

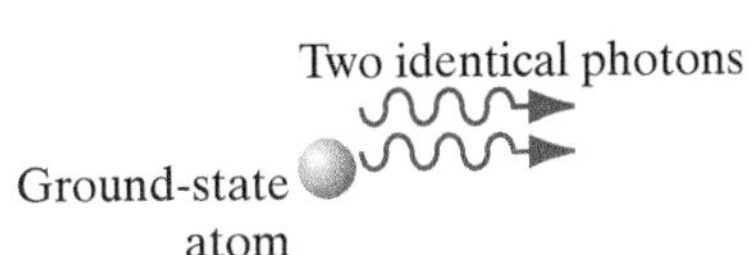

Figure 8.34 Stimulated emission produces two identical photons from one.

Chain Reaction

The phenomenon of stimulated emission can be exploited to create a chain reaction. Start with one photon of the right energy E to match the energy difference between the ground state and an excited state of the atoms in the system. If that photon causes stimulated emission from an excited atom, there are now two identical photons, both capable of triggering stimulated emission of other excited atoms, which yields four identical photons, then 8, 16, 32, 64, 128, 256, 512, 1024, and so on.

However, the chain reaction won't proceed if all of the photons escape from the system, so some kind of containment mechanism is required. In a common type of gas laser the gas is in a long tube with mirrors at both ends (Figure 8.35). Photons that happen to be going in the direction of the tube are continually reflected back and forth through the tube, which increases the probability that they will interact with excited atoms and cause stimulated emission. Since the stimulated-emission photons are clones, they too are headed in the direction of the tube and they too are likely to interact. One of the mirrors is deliberately made to be an imperfect reflector, so a fraction of the photons leaks out and provides a coherent beam of light. Unfortunately, the stimulated-emission photons also have just the right energy E to be absorbed by an atom in the ground state, in

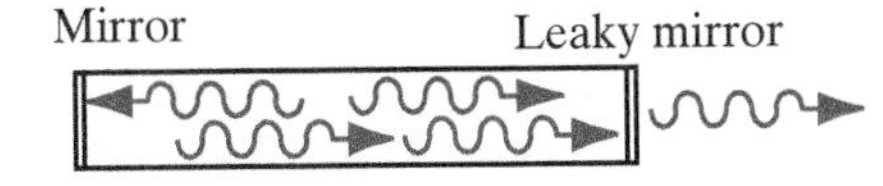

Figure 8.35 A gas laser with mirrors at the ends. Photons leak through one of the mirrors.

which case they can no longer contribute to the chain reaction. For this reason we need an "inverted population," with few atoms in the ground state to absorb the photons and stop the chain reaction. We discuss this issue next.

Inverted Population

As we have discussed earlier in this chapter, in a normal collection of atoms more atoms are in the ground state than in any other state, and the number of atoms decreases in each successively higher state. In fact, for a collection of atoms in thermal equilibrium the fraction of atoms in a particular energy state decreases exponentially with the energy above the ground state, as we will see in a later chapter.

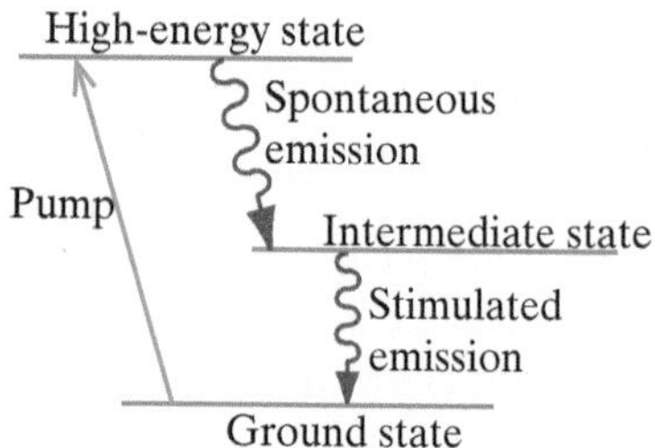

Figure 8.36 A three-state scheme for sustaining an inverted population.

By clever means it is possible to invert this population scheme, so that there are more atoms in an excited state than in the ground state (this is not a state of thermal equilibrium). An inverted population is crucial to laser action, so that stimulated emission with cloning of photons will dominate over simple absorption of photons by atoms in the ground state.

There are many different schemes for creating an inverted population. For example, the scheme used in the common helium–neon laser is to "pump" atoms to a high energy level (Figure 8.36). There are diverse energy input schemes, including electric discharges and powerful flashes of ordinary light. Atoms in this high energy level rapidly drop to an intermediate energy level, higher than the ground state. This intermediate state happens to have a rather long lifetime for spontaneous emission, so that it is possible to sustain an inverted population, with more atoms in the intermediate state than in the ground state. Stimulated emission drives these excited atoms down to the ground state, from which they are again pumped up to the high-energy state. Spontaneous emission from the high-energy state to the intermediate state maintains the inverted population.

8.10 *WAVELENGTH OF LIGHT

In this chapter we have emphasized the particle model of light, in terms of photons and their energy. However, as we have hinted earlier, some aspects of the behavior of light (of any energy) can best be understood by considering light to be a wave. For completeness we will mention a connection between the particle model and the wave model of light. Light consisting of photons whose energy is E can also be treated as a wave, and the energy E and wavelength λ are related as follows:

ENERGY AND WAVELENGTH OF LIGHT

$$E_{\text{photon}} = \frac{hc}{\lambda_{\text{light}}}$$

where h is Planck's constant.

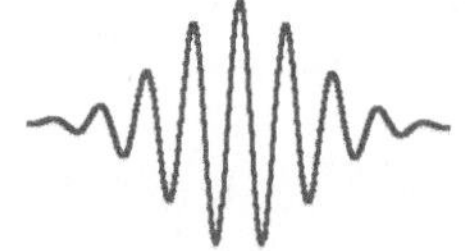

Figure 8.37 A wave packet is a traveling wave of finite spatial extent.

A model for light that incorporates both the features of a wave and a localized particle is a "wave packet," a wave whose magnitude is nonzero only in a small region of space (Figure 8.37). We will return to this topic in a later chapter on waves and particles.

SUMMARY

Light consists of particles called photons with energy and momentum but no rest mass, so $E = K$. Their speed is always 3×10^8 m/s.

Stable or nearly stable quantum systems can have only certain bound-state energies (unbound states have a continuous range of energies).

Three important processes involving energy-level change are collisional excitation, photon emission, and photon absorption.

The temperature is a measure of the average energy of a collection of objects, so high temperature is associated with significant probability of finding particles in any of many states, from the ground state up to a high energy level. Very low temperature is associated with little probability of the particles being in any state other than the ground state.

Examples were given of electronic, vibrational, and rotational energy levels, and of nuclear and hadronic energy levels.

Discrete hydrogen atom energy levels

$E_N = -(13.6 \text{ eV})/N^2$, where N is a nonzero positive integer (1,2,3...)

Discrete harmonic oscillator energy levels

$$E_N = N\hbar\omega_0 + E_0, \quad N = 0,1,2,\ldots, \quad \text{where}$$
$$\omega_0 = \sqrt{k_s/m}$$

Planck's constant $h = 6.6 \times 10^{-34}$ J · s

$$\hbar = \frac{h}{2\pi} = 1.05 \times 10^{-34} \text{ J} \cdot \text{s}$$

$1 \text{ eV} = 1.6 \times 10^{-19}$ J

Photon energy and wavelength: $E_{\text{photon}} = \dfrac{hc}{\lambda_{\text{light}}}$

QUESTIONS

Q1 Match the description of a process with the corresponding arrow in Figure 8.38: **(a)** Absorption of a photon whose energy is $E_1 - E_0$. **(b)** Absorption from an excited state (a rare event at ordinary temperatures). **(c)** Emission of a photon whose energy is $E_3 - E_1$. **(d)** Emission of a photon whose energy is $E_2 - E_0$. **(e)** In drawing arrows to represent energy transitions, which of the following statements are correct? (1) It doesn't matter in which direction you draw the arrow as long as it connects the initial and final states. (2) For emission, the arrow points down. (3) For absorption, the arrow points up. (4) The tail of the arrow is drawn on the initial state. (5) The head of the arrow is drawn on the final state. (6) It is not necessary to draw an arrowhead.

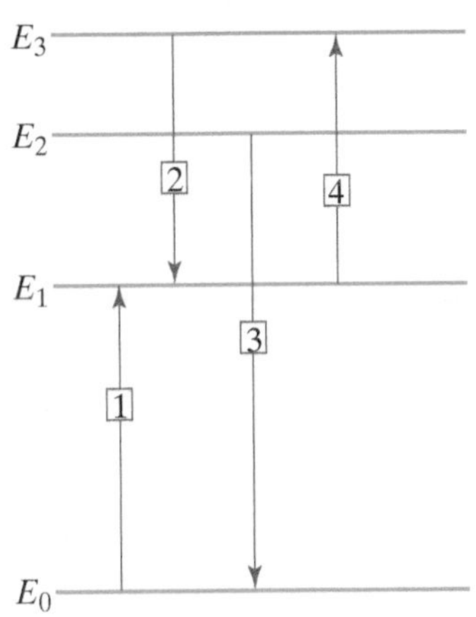

Figure 8.38

Q2 When starlight passes through a cold cloud of hydrogen gas, some hydrogen atoms absorb energy, then reradiate it in all directions. As a result, a spectrum of the star shows dark absorption lines at the energies for which less energy from the star reaches us. How does the spectrum of dark absorption lines for very cold hydrogen differ from the spectrum of bright emission lines from very hot hydrogen?

Q3 Which energy diagram in Figure 8.39 is appropriate for each of the following situations? **(a)** Hadronic (such as Δ^+). **(b)** Vibrational states of a diatomic molecule such as O_2. **(c)** Idealized quantized spring–mass oscillator. **(d)** Nuclear (such as the nucleus of a carbon atom). **(e)** Electronic, vibrational, and rotational states of a diatomic molecule such as O_2. **(f)** Rotational states of a diatomic molecule such as O_2. **(g)** Electronic states of a single atom such as hydrogen.

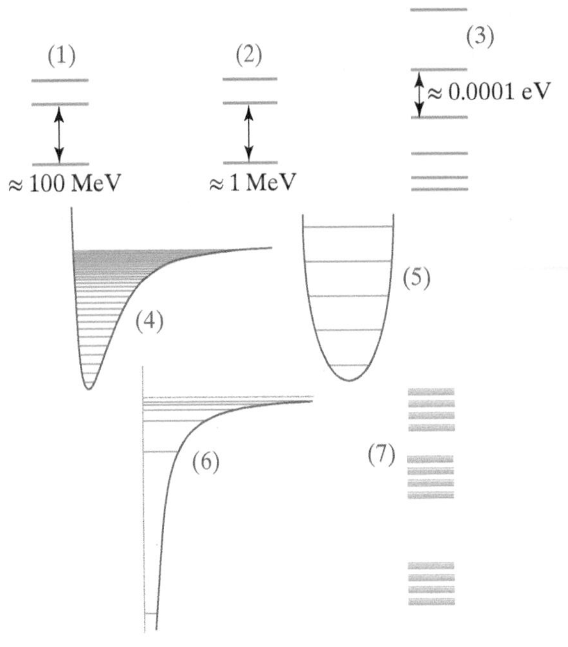

Figure 8.39

Q4 If you try to increase the energy of a quantum harmonic oscillator by adding an amount of energy $\frac{1}{2}\hbar\sqrt{k_s/m}$, the energy doesn't increase. Why not?

Q5 If you double the amplitude, what happens to the frequency in a classical (nonquantum) harmonic oscillator? In a quantum harmonic oscillator?

Q6 What is the energy of the photon emitted by a harmonic oscillator with stiffness k_s and mass m when it drops from energy level 5 to energy level 2?

Q7 Summarize the differences and similarities between different energy levels in a quantum oscillator. Specifically, for the first two levels in Figure 8.26, compare the angular frequency $\sqrt{k_s/m}$, the amplitude A, and the kinetic energy K at the same value of s. (In a full quantum-mechanical analysis the concepts of angular frequency and amplitude require reinterpretation. Nevertheless there remain elements of the classical picture. For example, larger amplitude corresponds to a higher probability of observing a large stretch.)

PROBLEMS

Section 8.1

•**P8** A certain laser outputs pure red light (photon energy 1.8 eV) with power 700 mW (0.7 W). How many photons per second does this laser emit?

Section 8.2

•**P9** How much energy in electron volts is required to ionize a hydrogen atom (that is, remove the electron from the proton), if initially the atom is in the state $N = 2$? (Remember that $N = 1$ if the atom is in the lowest energy level.)

•**P10** The mean lifetime of a certain excited atomic state is 5 ns. What is the probability of the atom staying in this state for 10 ns or more?

•**P11** At $t = 0$ all of the atoms in a collection of 10000 atoms are in an excited state whose lifetime is 25 ns. Approximately how many atoms will still be in this excited state at $t = 12$ ns?

•**P12** $N = 1$ is the lowest electronic energy state for a hydrogen atom. **(a)** If a hydrogen atom is in state $N = 4$, what is $K + U$ for this atom (in eV)? **(b)** The hydrogen atom makes a transition to state $N = 2$. Now what is $K + U$ in electron volts for this atom? **(c)** What is the energy (in eV) of the photon emitted in the transition from level $N = 4$ to $N = 2$? **(d)** Which of the arrows in Figure 8.40 represents this transition?

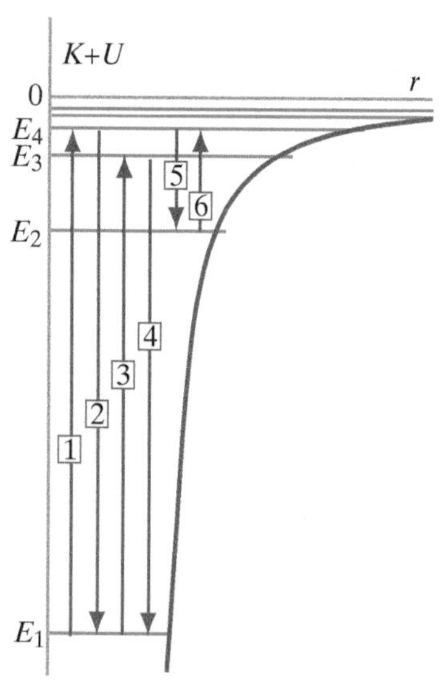

Figure 8.40

•**P13** The Franck–Hertz experiment involved shooting electrons into a low-density gas of mercury atoms and observing discrete amounts of kinetic energy loss by the electrons. Suppose that instead a similar experiment is done with a very cold gas of atomic hydrogen, so that all of the hydrogen atoms are initially in the ground state. If the kinetic energy of an electron is 11.6 eV just before it collides with a hydrogen atom, how much kinetic energy will the electron have just after it collides with and excites the hydrogen atom?

••**P14** Hydrogen atoms: **(a)** What is the minimum kinetic energy in electron volts that an electron must have to be able to ionize a hydrogen atom that is in its ground state (that is, remove the electron from being bound to the proton)? **(b)** If electrons of energy 12.8 eV are incident on a gas of hydrogen atoms in their ground state, what are the energies of the photons that can be emitted by the excited gas? **(c)** If instead of electrons, photons of all energies between 0 and 12.8 eV are incident on a gas of hydrogen atoms in the ground state, what are the energies at which the photons are absorbed?

••**P15** Suppose we have reason to suspect that a certain quantum object has only three quantum states. When we excite such an object we observe that it emits electromagnetic radiation of three different energies: 2.48 eV (green), 1.91 eV (orange), and 0.57 eV (infrared). **(a)** Propose two possible energy-level schemes for this system. **(b)** Explain how to use an absorption measurement to distinguish between the two proposed schemes.

••**P16** Predict how many emission lines will be seen by a human in the visible spectrum of atomic hydrogen. Give the energies of the emitted photons, and specify the energy levels involved in the transitions that are responsible for these lines.

••**P17** Assume that a hypothetical object has just four quantum states, with the following energies:

−1.0 eV (third excited state)
−1.8 eV (second excited state)
−2.9 eV (first excited state)
−4.8 eV (ground state)

(a) Suppose that material containing many such objects is hit with a beam of energetic electrons, which ensures that there are always some objects in all of these states. What are the six energies of photons that could be strongly emitted by the material? (In actual quantum objects there are often "selection rules" that forbid certain emissions even though there is enough energy; assume that there are no such restrictions here.) List the photon emission energies. **(b)** Next, suppose that the beam of electrons is shut off so that all of the objects are in the ground state almost all the time. If electromagnetic radiation with a wide range of energies is passed through the material, what will be the three energies of photons corresponding to missing ("dark") lines in the spectrum? Remember that there is hardly any absorption from excited states, because emission from an excited state happens very quickly, so there is never a significant number of objects in an excited state. Assume that the detector is sensitive to a wide range of photon energies, not just energies in the visible region. List the dark-line energies.

••**P18** Energy graphs: **(a)** Figure 8.41 shows a graph of potential energy vs. interatomic distance for a particular molecule. What is the direction of the associated force at location A? At location B? At location C? Rank the magnitude of the force at locations A, B, and C. (That is, which is greatest, which is smallest, and are any of these equal to each other?) For the energy level shown on the graph, draw a line whose height is the kinetic energy when the system is at location D.

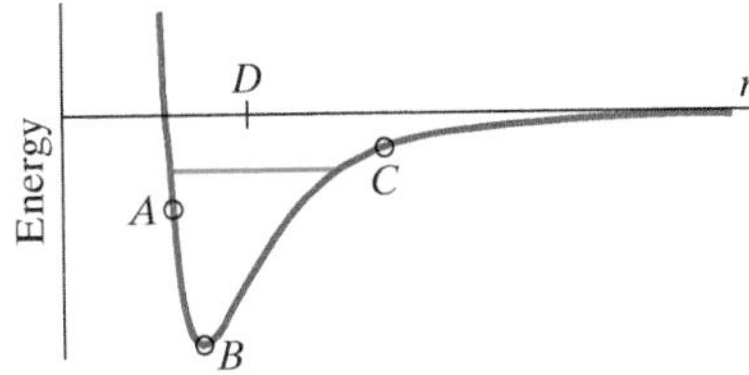

Figure 8.41

(b) Figure 8.42 shows all of the quantized energies (bound states) for one of these molecules. The energy for each state is given on the graph, in electron volts ($1 \text{ eV} = 1.6 \times 10^{-19}$ J). How much energy is required to break a molecule apart, if it is initially in the ground state? (Note that the final state must be an unbound state; the unbound states are not quantized.)

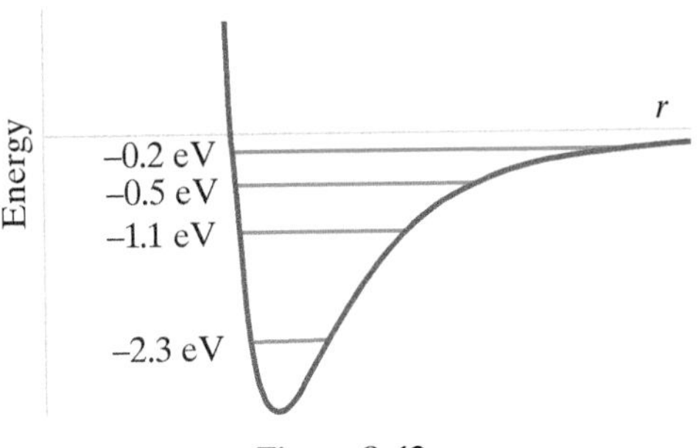

Figure 8.42

(c) At high enough temperatures, in a collection of these molecules there will be at all times some molecules in each of these states, and light will be emitted. What are the energies in electron volts of the emitted light? **(d)** The "inertial" mass of the molecule is the mass that appears in Newton's second law, and it determines how much acceleration will result from applying a given force. Compare the inertial mass of a molecule in the ground state and the inertial mass of a molecule in an excited state 10 eV above the ground state. If there is a difference, briefly explain why and calculate the difference. If there isn't a difference, briefly explain why not.

••**P19** A bottle contains a gas with atoms whose lowest four energy levels are −12 eV, −6 eV, −3 eV, and −2 eV. Electrons run through the bottle and excite the atoms so that at all times there are large numbers of atoms in each of these four energy levels, but there are no atoms in higher energy levels. List the energies of the photons that will be emitted by the gas.

Next, the electron beam is turned off, and all the atoms are in the ground state. Light containing a continuous spectrum of photon energies from 0.5 eV to 15 eV shines through the bottle. A photon detector on the other side of the bottle shows that some photon energies are depleted in the spectrum ("dark lines"). What are the energies of the missing photons?

••**P20** Suppose we have reason to suspect that a certain quantum object has only three quantum states. When we excite a collection of such objects we observe that they emit electromagnetic radiation of three different energies: 0.3 eV (infrared), 2.0 eV (visible), and 2.3 eV (visible). **(a)** Draw a possible energy-level diagram for one of the quantum objects, which has three bound states. On the diagram, indicate the transitions corresponding to the emitted photons, and check that the possible transitions produce the observed photons and no others. The energy $K + U$ of the ground state is −4 eV. Label the energies of each level ($K + U$, which is *negative*). **(b)** The material is now cooled down to a very low temperature, and the photon detector stops detecting photon emissions. Next, a beam of light with a continuous range of energies from infrared through ultraviolet shines on the material, and the photon detector observes the beam of light after it passes through the material. What photon energies in this beam of light are observed to be significantly reduced in intensity ("dark absorption lines")? Remember that there is hardly any absorption from excited states, because emission from an excited state happens very quickly, so there is never a significant number of objects in an excited state. **(c)** There exists another possible set of energy levels for these objects that produces the same photon emission spectrum. On an alternative energy-level diagram, *different from the one you drew in part (a),* indicate the transitions corresponding to the emitted photons, and check that the possible transitions produce the observed photons and no others. Label the energies of each level ($K + U$, which is *negative*). **(d)** For your second proposed energy-level scheme, what photon energies would be observed to be significantly reduced in intensity in an absorption experiment ("dark absorption lines")? (Given the differences from part (b), you can see that an absorption measurement can be used to tell which of your two energy-level schemes is correct.)

••**P21** Assume that a hypothetical object has just four quantum states, with the energies shown in Figure 8.43.

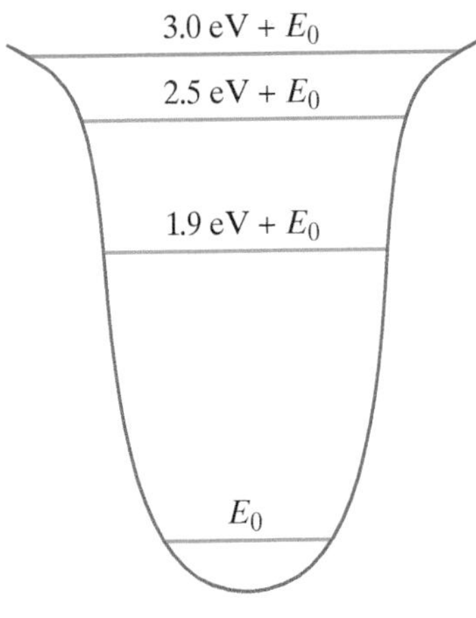

Figure 8.43

(a) Suppose that the temperature is high enough that in a material containing many such objects, at any instant some objects are found in all of these states. What are all the energies of photons that could be strongly emitted by the material?

(In actual quantum objects there are often "selection rules" that forbid certain emissions even though there is enough energy; assume that there are no such restrictions here.) **(b)** If the temperature is very low and electromagnetic radiation with a wide range of energies is passed through the material, what will be the energies of photons corresponding to missing ("dark") lines in the spectrum? (Assume that the detector is sensitive to a wide range of photon energies, not just energies in the visible region.)

Section 8.3

••P22 A certain material is kept at very low temperature. It is observed that when photons with energies between 0.2 and 0.9 eV strike the material, only photons of 0.4 eV and 0.7 eV are absorbed. Next, the material is warmed up so that it starts to emit photons. When it has been warmed up enough that 0.7 eV photons begin to be emitted, what other photon energies are also observed to be emitted by the material? Explain briefly.

••P23 Some material consisting of a collection of microscopic objects is kept at a high temperature. A photon detector capable of detecting photon energies from infrared through ultraviolet observes photons emitted with energies of 0.3 eV, 0.5 eV, 0.8 eV, 2.0 eV, 2.5 eV, and 2.8 eV. These are the only photon energies observed. **(a)** Draw and label a possible energy-level diagram for one of the microscopic objects, which has four bound states. On the diagram, indicate the transitions corresponding to the emitted photons. Explain briefly. **(b)** Would a spring–mass model be a good model for these microscopic objects? Why or why not? **(c)** The material is now cooled down to a very low temperature, and the photon detector stops detecting photon emissions. Next, a beam of light with a continuous range of energies from infrared through ultraviolet shines on the material, and the photon detector observes the beam of light after it passes through the material. What photon energies in this beam of light are observed to be significantly reduced in intensity ("dark absorption lines")? Explain briefly.

Section 8.4

•P24 For a certain diatomic molecule, the lowest-energy photon observed in the vibrational spectrum is 0.17 eV. What is the energy of a photon emitted in a transition from the 5th excited vibrational energy level to the 2nd excited vibrational energy level, assuming no change in the rotational energy?

••P25 Consider a microscopic spring–mass system whose spring stiffness is 50 N/m, and the mass is 4×10^{-26} kg. **(a)** What is the smallest amount of vibrational energy that can be added to this system? **(b)** What is the difference in mass (if any) of the microscopic oscillator between being in the ground state and being in the first excited state? **(c)** In a collection of these microscopic oscillators, the temperature is high enough that the ground state and the first three excited states are occupied. What are possible energies of photons emitted by these oscillators?

••P26 Molecular vibrational energy levels: **(a)** A HCl molecule can be considered to be a quantized harmonic oscillator, with quantized vibrational energy levels that are evenly spaced. Make a rough estimate of this uniform energy spacing in electron volts (where $1 \text{ eV} = 1.6 \times 10^{-19}$ J). You will need to make some rough estimates of atomic properties based on prior work. For comparison with the spacing of these vibrational energy states, note that the spacing between quantized energy levels for "electronic" states such as in atomic hydrogen is of the order of several electron volts. **(b)** List several photon energies that would be emitted if a number of these vibrational energy levels were occupied due to collisional excitation. To what region of the spectrum (x-ray, visible, microwave, etc.) do these photons belong? (See Figure 8.1 at the beginning of the chapter.)

••P27 A hot bar of iron glows a dull red. Using our simple ball-spring model of a solid (Figure 8.23), answer the following questions, explaining in detail the processes involved. You will need to make some rough estimates of atomic properties based on prior work. **(a)** What is the approximate energy of the lowest-energy spectral emission line? Give a numerical value. **(b)** What is the approximate energy of the highest-energy spectral emission line? Give a numerical value. **(c)** What is the quantum number of the highest-energy occupied state? **(d)** Predict the energies of two other lines in the emission spectrum of the glowing iron bar. (*Note*: Our simple model is too simple—the actual spectrum is more complicated. However, this simple analysis gets at some important aspects of the phenomenon.)

COMPUTATIONAL PROBLEMS

The VPython programs used in these computational problems are available on the Wiley student website for this textbook.

•P28 There is a "random" module for Python that contains a function (also named "random") that generates a random number, which we can use to model photon emission at random times. Here is a little program that graphs and prints successive values generated by this function (the graph is shown in Figure 8.44):

```
from visual import *
from visual.graph import *
from random import random
gg = gcurve(color=color.yellow)
n = 0
while n < 20:
    rate(10)
    a = random()
    print(a)
    gg.plot(pos=(n,a))
    n= n + 1
```

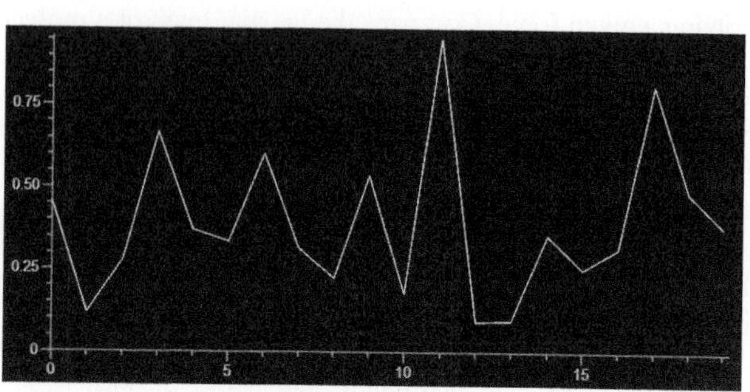

Figure 8.44

(a) Run this program several times, looking at the graph produced each time. What is being plotted in this graph? Is the graph the same each time you run the program? **(b)** What is the maximum random number generated by the `random()` function? **(c)** What is the minimum number generated by the `random()` function?

••P29 Refer to Problem P28 for an explanation of the `random()` function. The VPython program below models random photon emission in a collection of excited atoms. Read through the program, then answer the questions below. **(a)** In which line of code is it decided whether a particular atom will emit or not? How is a random number used in this decision? **(b)** What is the probability that a given atom will emit a photon in one nanosecond? **(c)** In which line of code is the count of excited atoms decreased after an atom emits a photon? **(d)** Start with 5 excited atoms, and run the program 10 times. In your observations, what is the longest time it takes for every atom to emit a photon? What is the shortest time? **(e)** Increase the number of atoms until the results of every run look the same. Approximately how many atoms are required? (You may wish to use a larger value for `rate()`). **(f)** With 10,000 atoms, drag the mouse across the graph and find a vertical bar whose height is $10000/e = 10000 * 0.368$. What is the value of t at this location? This value is called the "mean lifetime" and can be shown to be equal to the reciprocal of the emission rate (emissions per second, which is `P/dt`). How does your mean lifetime compare with `dt/P`)?

```
from visual import *
from visual.graph import *
from random import random

Natoms = 5
# P is the probability for an atom to emit
# during a time interval dt
P = 0.1
dt = 0.2   # ns
t = 0
tmax = 5*dt/P # 5 mean lifetimes
# Create a bar graph
gdisplay(xtitle='t, ns',
         ytitle='Atoms in excited state')
excited     =     gvbars(color=color.yellow,
delta=dt/2)

while t < tmax:
    rate(10)
    # Show number of excited atoms remaining
    excited.plot(pos=(t,Natoms))
    emissions = 0
    atom = 0
    while atom < Natoms:
        if random() < P: # emits?
            # count emissions in this dt
            emissions = emissions + 1
        atom = atom + 1
    Natoms = Natoms - emissions
    t = t + dt
```

Figure 8.45 shows sample output of this program.

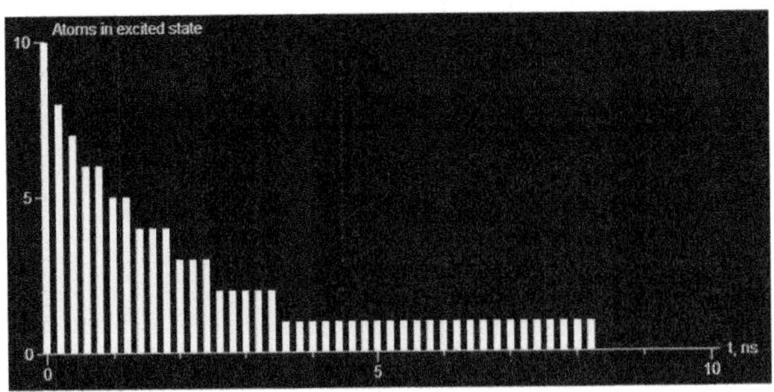

Figure 8.45

••P30 Refer to Problem P28 for an explanation of the `random()` function. The VPython program below models the random emission of red and green photons from a collection of atoms in their second excited state. These atoms can either emit a green photon and drop to the ground state, or they can emit a red photon and drop to the first excited state. Read the program carefully before running it, then answer the questions below. **(a)** What quantity is plotted in each of the graphs generated by the program? **(b)** Which is higher: the probability of emission of a red photon or a green photon? **(c)** Run the program repeatedly. Approximately, how much does the number of green emissions vary in repeated trials? **(d)** It can be shown that statistically one expects that the number of green emissions in repeated trials lies in the range $N \pm \sqrt{N}$, where N is the average number (which we expect to be `Pgreen*Natoms`). Do 30 or more trials and determine the experimental average number N and the fraction of trials in which the number of green emissions is within the range $N \pm \sqrt{N}$.

```
from visual import *
from visual.graph import *
from random import random

# Start with 100 atoms in an excited state
# (try larger or smaller numbers)

Natoms = 100
# P is the probability for an atom to emit
# during a time interval dt
P = 0.1
# Pgreen is the probability that when an
# atom emits, it emits a green photon
Pgreen = 0.3
dt = 0.2   # ns
t = 0
tmax = 5*dt/P # 5 mean lifetimes
# Create bar graphs
gdisplay(xtitle='t, ns', xmax=tmax,
         ytitle='Emissions of green photons')
greeng = gvbars(color=color.green, delta=dt/2)
gdisplay(y=400, xtitle='t, ns', xmax=tmax,
        ytitle='Emissions of red photons')
redg = gvbars(color=color.red, delta=dt/2)
greens = reds = 0
```

```
while t < tmax:
    rate(100)
    atom = 0
    g = r = 0
    while atom < Natoms:
        if random() < P: # emits?
            if random() < Pgreen: # green?
                g = g + 1
            else:                  # emits red
                r = r + 1
        atom = atom + 1
    greeng.plot(pos=(t,g))
    redg.plot(pos=(t,r))
    greens = greens + g
    reds = reds + r
    Natoms = Natoms - (g + r)
    t = t + dt

print(greens,'green emissions,',
      reds,'red emissions')
```

Figure 8.46 shows sample output of this program.

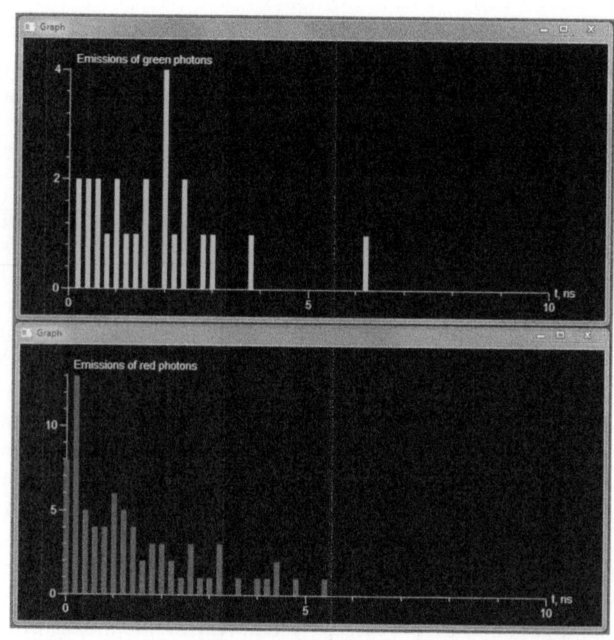

Figure 8.46 The vertical scales in the two graphs are different, due to the different probabilities of green and red emissions.

ANSWERS TO CHECKPOINTS

1 About 2.5 eV; about 3.5×10^{21} per second per square meter
2 5.2 eV
3 Six different photon energies, corresponding to transitions from 4 to 3, 4 to 2, 4 to 1, 3 to 2, 3 to 1, and 2 to 1
4 Just one dark line at 2.4 eV
5 1 eV, 5 eV, and 6 eV
6 0.4 eV (three transitions), 0.8 eV (two transitions), and 1.2 eV (one transition)
7 $N \approx 2 \times 10^{31}$. A system in a very high quantum state behaves like a classical system.
8 2.2 MeV
9 About 294 MeV

CHAPTER

9

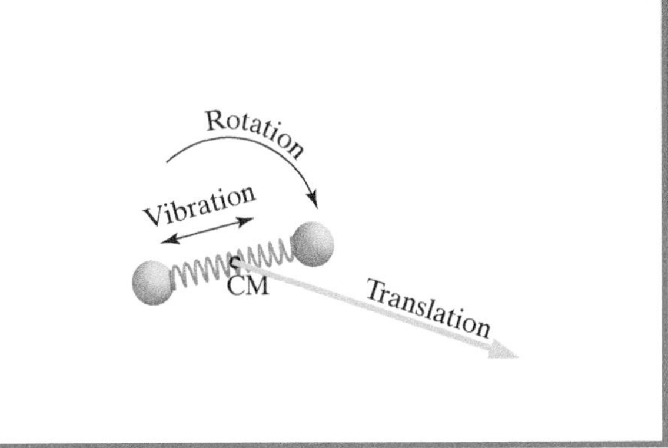

Translational, Rotational, and Vibrational Energy

OBJECTIVES

After studying this chapter, you should be able to

- Analyze the kinetic energy of a multiparticle system in terms of translational kinetic energy and kinetic energy relative to the center of mass.
- Combine the results of modeling a multiparticle system as a point particle and as an extended system to obtain information about particular forms of internal energy.

Suppose you throw a flying disc (such as a Frisbee®) of mass 140 g in such a way that it leaves your hand with a speed of 15 m/s. What is the kinetic energy of the disc?

Although this sounds like a straightforward question, we actually don't have enough information to answer it, because the motion of the disc is complicated—it is traveling through the air, but it is also spinning around its center. The kinetic energy associated with the motion of the center of mass of the object is called "translational kinetic energy," because in mathematics a "translation" is a function that moves every point in an object a constant distance in a specified direction. In addition to its translational kinetic energy, the spinning disc also has rotational kinetic energy, associated with the rotation of the atoms of the disc around the disc's center. A different kind of object, such as two balls connected by a spring, might be vibrating as it travels through the air—it would have vibrational kinetic energy as well.

This chapter will focus on determining how a system's kinetic energy is divided between translational kinetic energy and energy associated with motion relative to the center of mass of a system, such as rotation. Along the way we will find that by modeling a system two different ways—as a point particle, and as an extended system—we can predict how much of the work done on a system will go into different kinds of energy. This analysis method will allow us to solve some very complex problems in a very simple way.

9.1 SEPARATION OF MULTIPARTICLE SYSTEM ENERGY

In many situations we find that the energy of a multiparticle system can be analyzed simply, by dividing the kinetic energy of the system into two parts: "translational kinetic energy" K_{trans}, kinetic energy associated with the motion of the center of mass, and K_{rel}, the kinetic energy relative to the center of mass.

K_{rel} includes the kinetic energy associated with rotation of the object and the kinetic energy associated with vibration of the object.

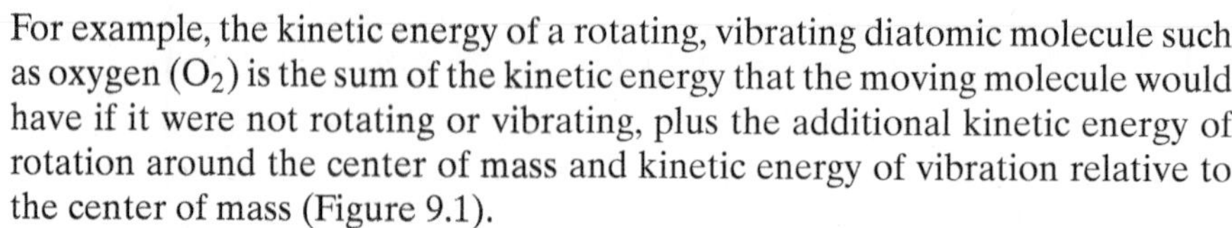

MULTIPARTICLE KINETIC ENERGY

$$K_{\text{tot}} = K_{\text{trans}} + K_{\text{rel}}$$

where $K_{\text{rel}} = K_{\text{vib}} + K_{\text{rot}}$.

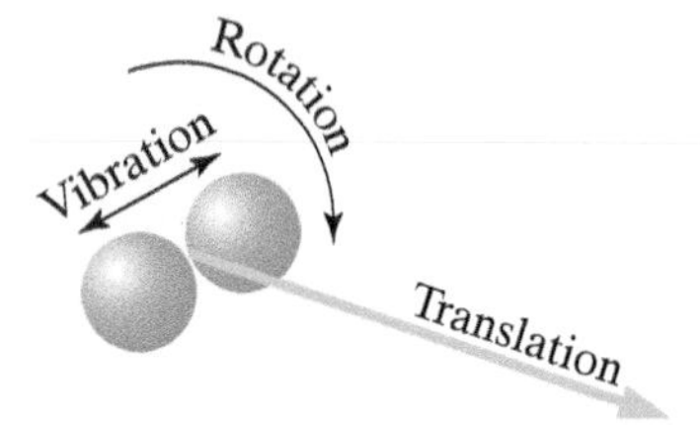

Figure 9.1 A diatomic molecule such as O_2 can vibrate and rotate while also translating.

For example, the kinetic energy of a rotating, vibrating diatomic molecule such as oxygen (O_2) is the sum of the kinetic energy that the moving molecule would have if it were not rotating or vibrating, plus the additional kinetic energy of rotation around the center of mass and kinetic energy of vibration relative to the center of mass (Figure 9.1).

This split of the total kinetic energy into different parts sounds plausible. A formal derivation of this result is given in Section 9.5 at the end of this chapter.

Translational Kinetic Energy

In physics and mathematics, the word "translate" means "move from one location to another." Translational motion is the ordinary kind of motion we have studied most so far. The translational motion of a macroscopic object is described by $\vec{v}_{\text{CM}}$, the velocity of the object's center of mass. We use the term "translate" here in order to distinguish between translational motion, in which the center of mass of a system moves from one location to another, and other kinds of motion such as rotation and vibration, in which parts of a system move relative to the center of mass.

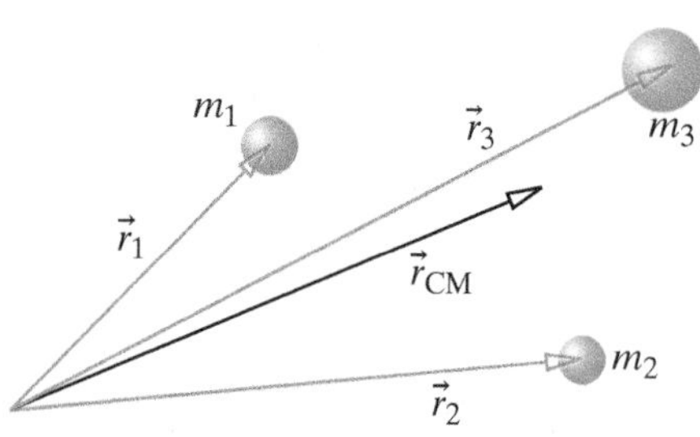

Figure 9.2 Location of the center of mass of a multiparticle system.

In Section 3.11 we defined the center of mass of a multiparticle system to be the weighted average of the positions of all atoms making up an object (Figure 9.2):

$$\vec{r}_{\text{CM}} = \frac{m_1\vec{r}_1 + m_2\vec{r}_2 + m_3\vec{r}_3 + \cdots}{M_{\text{total}}}$$

If each of the objects is itself a large extended object (Figure 9.3), we can calculate the center of mass by pretending that we shrink each object to a point at its own center of mass:

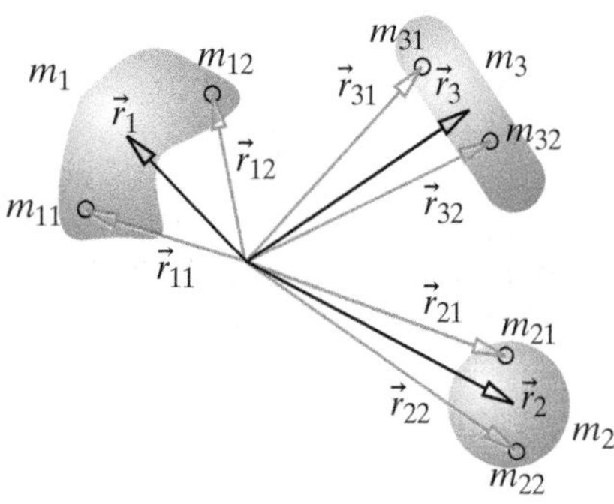

Figure 9.3 Finding the center of mass of a system consisting of several large objects.

$$\vec{r}_{\text{CM}} = \frac{(m_{11}\vec{r}_{11} + m_{12}\vec{r}_{12} + \cdots) + (m_{21}\vec{r}_{21} + m_{22}\vec{r}_{22} + \cdots) + \cdots}{(m_{11} + m_{12} + \cdots) + (m_{21} + m_{22} + \cdots) + \cdots}$$

$$\vec{r}_{\text{CM}} = \frac{m_1\vec{r}_1 + m_2\vec{r}_2 + m_3\vec{r}_3 + \cdots}{M_{\text{total}}}$$

This looks just like the equation for finding the center of mass of a bunch of atoms, except that the $\vec{r}$ vectors refer to the centers of mass of large objects.

By differentiating the expression for $\vec{r}_{\text{CM}}$ with respect to time, we found that the velocity of the center of mass is

$$\vec{v}_{\text{CM}} = \frac{m_1\vec{v}_1 + m_2\vec{v}_2 + m_3\vec{v}_3 + \cdots}{M_{\text{total}}}$$

and that therefore if $v \ll c$,

$$\vec{p}_{\text{sys}} \approx M_{\text{total}}\vec{v}_{\text{CM}}$$

Using the center-of-mass concept, we define the "translational kinetic energy" of a system as kinetic energy associated with translational motion, which does not include kinetic energy associated with rotation or vibration:

TRANSLATIONAL KINETIC ENERGY

$$K_{\text{trans}} = \frac{1}{2} M_{\text{total}} v_{\text{CM}}^2 = \frac{p_{\text{sys}}^2}{2M_{\text{total}}}$$

for an object with speed that is small compared to the speed of light ($v \ll c$).

EXAMPLE **Translational Kinetic Energy of a Frisbee®**

What is the translational kinetic energy of a spinning 140 g Frisbee that is traveling at a speed of 15 m/s? Is this the total kinetic energy of the Frisbee?

Solution

$$K_{\text{trans}} = \frac{1}{2}(0.14\,\text{kg})(15\ \text{m/s})^2 = 15.8\ \text{J}$$

This does not include the kinetic energy associated with the rotation of the disk.

Vibrational Energy

One form of energy that is internal to a system is vibrational energy, both elastic and kinetic. We already know how to calculate vibrational energy. Consider an oxygen molecule (O_2) that has no translational motion (that is, the center of mass is stationary), but is vibrating (Figure 9.4), with elastic energy and kinetic energy continually interchanging, but with the sum of the two energies remaining constant, like the mass and spring you modeled in earlier chapters.

Figure 9.4 A vibrating oxygen molecule whose center of mass is at rest.

VIBRATIONAL ENERGY

$$E_{\text{vib}} = K_{\text{vib}} + U_{\text{spring}}$$

If in addition the oxygen molecule is translating (Figure 9.5), the total energy of the molecule is the sum of the translational kinetic energy, the vibrational kinetic energy (in terms of velocities of the two atoms relative to the center of mass), and the elastic energy of the "spring" holding them together, and the rest energies of the constituent atoms:

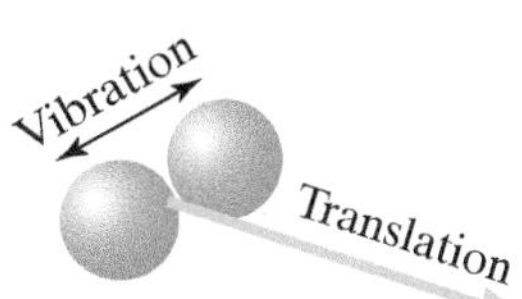

Figure 9.5 A translating, vibrating oxygen molecule.

$$E_{\text{tot}} = \frac{1}{2}(2m)v_{\text{CM}}^2 + K_{\text{vib}} + \frac{1}{2}k_s s^2 + 2mc^2$$

Note that the elastic energy $\frac{1}{2}k_s s^2$ is clearly unaffected by the translational motion of the center of mass, because it depends only on the stretch of the distance between the two atoms.

The same reasoning applies to the motion of a hot block of metal. In addition to its translational kinetic energy $\frac{1}{2}Mv_{\text{CM}}^2$ there is vibrational kinetic and elastic energy of the atoms around their equilibrium positions, which we call thermal energy, with more thermal energy if the block's temperature is higher.

Rotational Kinetic Energy

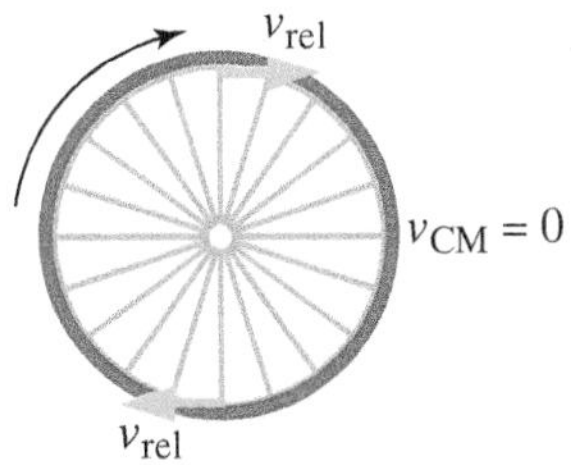

Figure 9.6 A bicycle wheel spinning on its axle, which is at rest. Atoms in the rim have a speed v_{rel} relative to the center of mass.

Just as a vibrating object has kinetic energy associated with vibration, even if its center of mass is at rest, so a rotating object has kinetic energy associated with rotation, even if its center of mass is at rest.

As an example, suppose that you spin a bicycle wheel on its axis, and hold the axle stationary, as in Figure 9.6. The spinning wheel has some kinetic energy K_{rel}, which in this case we will call K_{rot}, because the atoms are moving in rotational motion relative to the center of mass. Almost all of the mass M of the wheel is in the rim, so if the rim is traveling at a (tangential) speed v_{rel}, the kinetic energy of the wheel is approximately $K_{\text{rot}} \approx \frac{1}{2}Mv_{\text{rel}}^2$. (This isn't exact, because there is some mass in the spokes, and the atoms in the spokes are moving at speeds smaller than v_{rel}.)

In Section 9.2 we will see how to calculate the rotational kinetic energy of other kinds of rotating objects.

Rotation, Vibration, and Translation

The most general motion of an oxygen molecule involves rotation, vibration, and translation. It is easy to show that the internal (nontranslational) kinetic energy separates cleanly into rotational and vibrational contributions. Consider one atom with momentum $\vec{p}$ relative to the center of mass, and resolve its momentum into "tangential" and "radial" components (Figure 9.7). By "radial" component we mean a component in the direction of the line from the center of mass to the particle. By "tangential" component we mean a component perpendicular to that line. If the particle is in circular motion around the center of mass, the radial component is zero, so tangential momentum is associated with rotation. Radial momentum is associated with vibration.

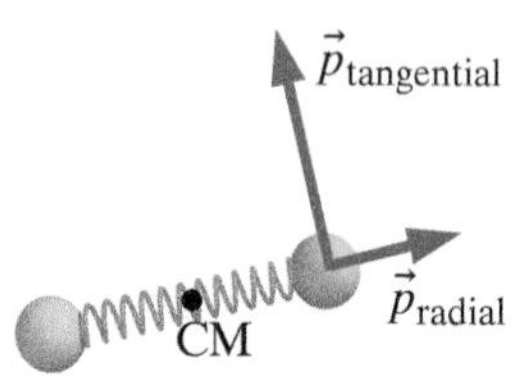

Figure 9.7 Radial and tangential components of momentum of a single atom, relative to the center of mass of an object.

Find the kinetic energy in terms of these two momentum components:

$$\frac{p^2}{2m} = \frac{(p^2_{tangential} + p^2_{radial})}{2m} = K_{rot} + K_{vib}$$

Therefore the kinetic energy relative to the center of mass splits into two terms, rotational energy and vibrational energy:

$$K_{rel} = K_{rot} + K_{vib}$$

The total energy of a translating, rotating, vibrating oxygen molecule can be written like this:

$$E_{tot} = K_{trans} + K_{rot} + K_{vib} + \frac{1}{2}k_s s^2 + 2mc^2$$

If the object can fly apart into separate pieces, we might call the kinetic energy relative to the center of mass $K_{explosion}$! For example, a fireworks rocket has translational kinetic energy that doesn't change from just before to just after its explosion, and the pieces move at high speeds relative to the center of mass of the rocket. These large kinetic energies relative to the center of mass come from the chemical energy used up in the explosion.

Vibrational and rotational kinetic energy are common types of K_{rel} but they aren't the only ones. For example, a person walking around on a moving train contributes relative kinetic energy to the total kinetic energy of the train and its passengers.

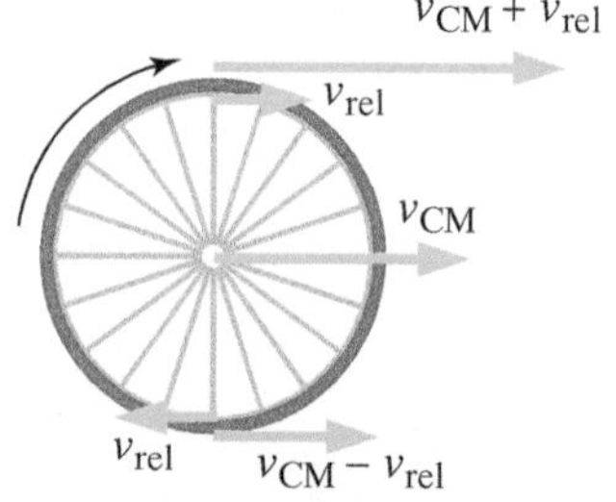

Figure 9.8 The velocity of an atom in the rim is the vector sum of the center-of-mass velocity and the velocity relative to the center of mass.

Why Separating Kinetic Energy Is Useful

If we do not separate kinetic energy into $K_{trans} + K_{rel}$, adding up the kinetic energies of all the particles in a system may be a very complex task. Consider the bicycle wheel shown in Figure 9.8. The fact that this wheel is both translating to the right, and also spinning on its axis, means that particles at different locations in the wheel have quite different speeds. For example, an atom at the bottom of the wheel is actually moving at a speed less than v_{CM}, while an atom at the top of the wheel is moving at a speed greater than v_{CM}. In order to calculate the kinetic energy of the whole system, we would have to figure out the instantaneous speed of each constituent particle!

Gravitational Energy of a Multiparticle System

In previous chapters we have calculated the gravitational potential energy of a system consisting of the Earth and a multiparticle object simply by treating the object as if all its mass were concentrated at a single point (its center of mass). Why is this correct?

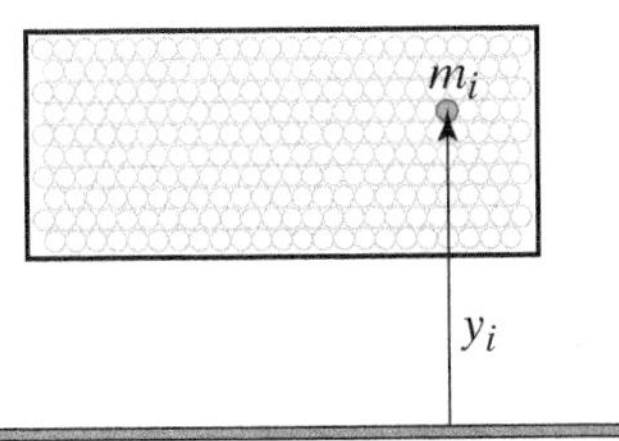

Figure 9.9 The i-th atom in the block is a distance y_i from the Earth's surface.

Consider the gravitational energy of a block of mass M and the Earth, near the Earth's surface. The gravitational energy associated with the interaction of the Earth with the i-th atom in the block is $m_i g y_i$ (Figure 9.9).

We can calculate the total gravitational energy by adding up the gravitational energy associated with the interaction of each of the N atoms with the Earth:

$$U_g = m_1 g y_1 + m_2 g y_2 + m_3 g y_3 + \cdots = g(m_1 y_1 + m_2 y_2 + m_3 y_3 + \cdots)$$

However, the last expression in parentheses appears in the calculation of the y component of the location of the center of mass:

$$M y_{\text{CM}} = m_1 y_1 + m_2 y_2 + m_3 y_3 + \cdots$$

Therefore for a block made up of many atoms, near the Earth's surface, we have this:

GRAVITATIONAL ENERGY U_g

$$U_g = M g y_{\text{CM}} \quad \text{near the Earth's surface}$$

In other words, the gravitational energy of a block of material and the Earth, near the Earth's surface, can be evaluated as though the block were a tiny particle of mass M located at the center of mass of the block.

This simple equation is not valid if the object is so large that the strength of the gravitational field is significantly different at different locations in the object. In that case mg would not be a good approximation for the gravitational force, nor mgy a good approximation for the gravitational energy associated with each atom.

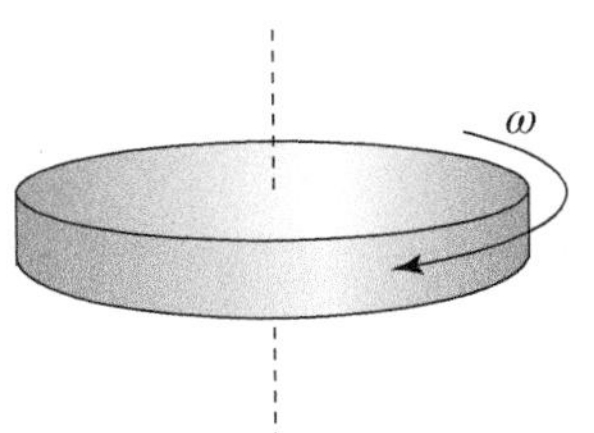

Figure 9.10 A rigid disk rotating about the y axis with angular speed ω.

> QUESTION In a calculation of gravitational energy in the voyage of a spacecraft to the Moon, is it a valid approximation to treat the spacecraft as if all its mass were concentrated at one point?

This is a good approximation, because the spacecraft is very small compared to the distances to the Earth and Moon, so the gravitational force on each atom is nearly the same. Also, if the spacecraft can be well approximated by a sphere of constant density, or nested spherical shells of constant density, it turns out that the gravitational force acts on it as though all the mass were concentrated at the center of mass.

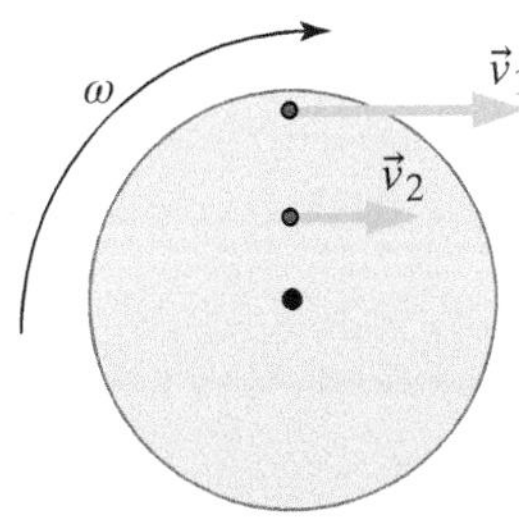

Figure 9.11 A top view of a disk rotating about an axis. A particle near the edge has a higher speed than a particle nearer to the axis. In each case, $v_i = \omega r_i$.

9.2 ROTATIONAL KINETIC ENERGY

A common instance of rotational motion is that in which a rigid system is rotating on an axis, as shown in Figure 9.10. In this situation all the atoms in the system share the same "angular speed" in radians per second. However, they have different linear speeds in meters per second, depending on their distances from the axis, because an atom near the edge must travel farther in one revolution (Figure 9.11).

Angular speed, normally denoted by ω (lowercase Greek omega), is a measure of how fast something is rotating. If an object makes one complete turn of 360 degrees (2π radians) in a time T, we say that its angular speed is $\omega = 2\pi/T$ radians per second. The time T is called the period.

PERIOD AND ANGULAR SPEED

$$\omega = \frac{2\pi}{T}$$

The period T is the time it takes for a rotating object to make one complete revolution. The angular speed ω has units of radians per second.

An atom rotating at a constant rate around an axis at a distance r from the axis goes a distance of the circumference $2\pi r$ in the period T, so the speed v of the atom can be expressed in terms of the angular speed:

$$v = \frac{2\pi r}{T} = \left(\frac{2\pi}{T}\right) r = \omega r$$

We measure angular speed ω in the "natural" units of radians per second rather than degrees per second, because the fundamental geometrical relationship between angle and arc length (arc length $= r\theta$) is valid only if the angle θ is measured in radians.

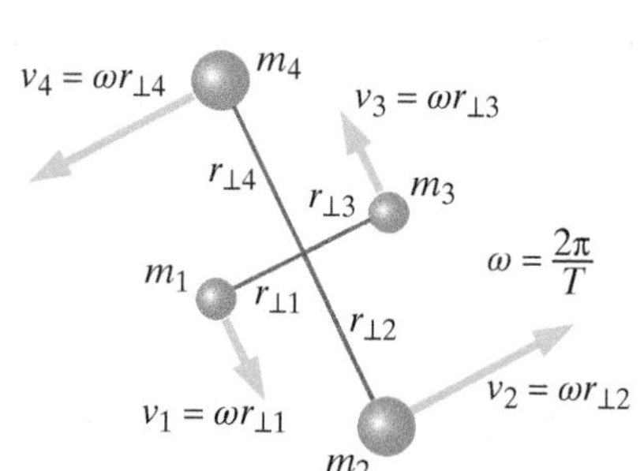

Figure 9.12 A case of rigid rotation about an axis with angular speed ω. Note the different speeds, with $v_i = \omega r_i$.

Moment of Inertia

The kinetic energy of the system shown in Figure 9.12, which is a rotating rigid object, is $\frac{1}{2}m_1v_1^2 + \frac{1}{2}m_2v_2^2 + \frac{1}{2}m_3v_3^2 + \cdots$. Since the speed of each mass is $v = \omega r_\perp$ (where $r_\perp$ is the perpendicular distance of the mass from the axis of rotation), we can write the rotational kinetic energy like this:

$$K_{\text{rot}} = \frac{1}{2}\left[m_1(\omega r_{\perp 1})^2 + m_2(\omega r_{\perp 2})^2 + m_3(\omega r_{\perp 3})^2 + \cdots\right]$$
$$= \frac{1}{2}\left[m_1 r_{\perp 1}^2 + m_2 r_{\perp 2}^2 + m_3 r_{\perp 3}^2 + \cdots\right]\omega^2$$

The quantity in brackets is called the "moment of inertia" and is usually denoted by the letter I:

MOMENT OF INERTIA

$$I = m_1 r_{\perp 1}^2 + m_2 r_{\perp 2}^2 + m_3 r_{\perp 3}^2 + m_4 r_{\perp 4}^2 + \cdots$$

The units of moment of inertia are $\text{kg}\cdot\text{m}^2$.

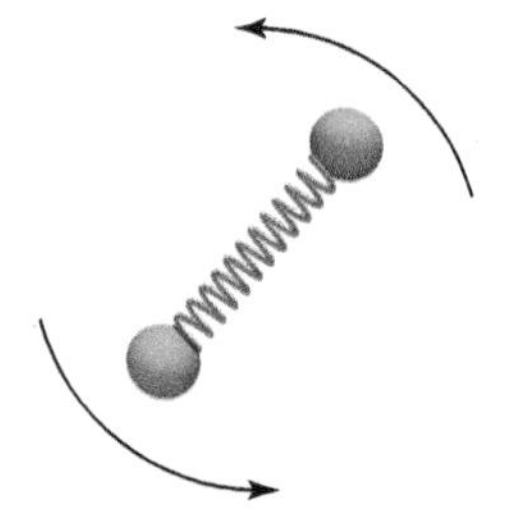

Figure 9.13 A rotating diatomic molecule.

Using this definition, we have a compact expression for rotational kinetic energy:

ROTATIONAL KINETIC ENERGY

$$K_{\text{rot}} = \frac{1}{2}I\omega^2$$

EXAMPLE **The Moment of Inertia of a Diatomic Molecule**

What is the moment of inertia of a diatomic nitrogen molecule N_2 around its center of mass (Figure 9.13)? The mass of a nitrogen atom is 2.3×10^{-26} kg and the average distance between nuclei is 1.5×10^{-10} m. Use the definition of moment of inertia carefully.

Solution For two masses, $I = m_1 r_{\perp 1}^2 + m_2 r_{\perp 2}^2$. The distance between masses is d, so the distance of each object from the center of mass is $r_{\perp 1} = r_{\perp 2} = d/2$. Therefore

$$I = M(d/2)^2 + M(d/2)^2 = 2M(d/2)^2$$

$$I = 2 \cdot (2.3 \times 10^{-26}\,\text{kg})(0.75 \times 10^{-10}\text{m})^2$$

$$I = 2.6 \times 10^{-46}\,\text{kg} \cdot \text{m}^2$$

EXAMPLE **The Moment of Inertia of a Bicycle Wheel**

A bicycle wheel has almost all its mass M located in the outer rim at radius R (Figure 9.14). What is the moment of inertia of the bicycle wheel about its center of mass? (*Hint:* It's helpful to think of dividing the wheel into the atoms it is made of and think about how much each atom contributes to the moment of inertia.)

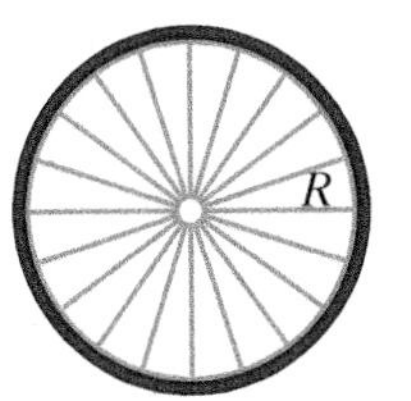

Figure 9.14 Most of the mass of a bicycle wheel is in the rim.

Solution Let m represent the mass of one atom in the rim. The moment of inertia is

$$I = m_1 r_{\perp 1}^2 + m_2 r_{\perp 2}^2 + m_3 r_{\perp 3}^2 + m_4 r_{\perp 4}^2 + \cdots$$

$$I = m_1 R^2 + m_2 R^2 + m_3 R^2 + m_4 R^2 + \cdots$$

$$I = [m_1 + m_2 + m_3 + m_4 + \cdots]R^2$$

$$I = MR^2$$

We've assumed that the mass of the spokes is negligible compared to the mass of the rim, so that the total mass is just the mass of the atoms in the rim.

EXAMPLE **Rotational Kinetic Energy and Work**

In Figure 9.15 a wheel is mounted on a stationary axle, which is nearly frictionless so that the wheel turns freely. The wheel has an inner ring with mass 5 kg and radius 10 cm and an outer ring with mass 2 kg and radius 25 cm; the spokes have negligible mass. A string with negligible mass is wrapped around the outer ring and you pull on it, increasing the rotational speed of the wheel. During the time that the wheel's rotation changes from 4 revolutions per second to 7 revolutions per second, how much work do you do?

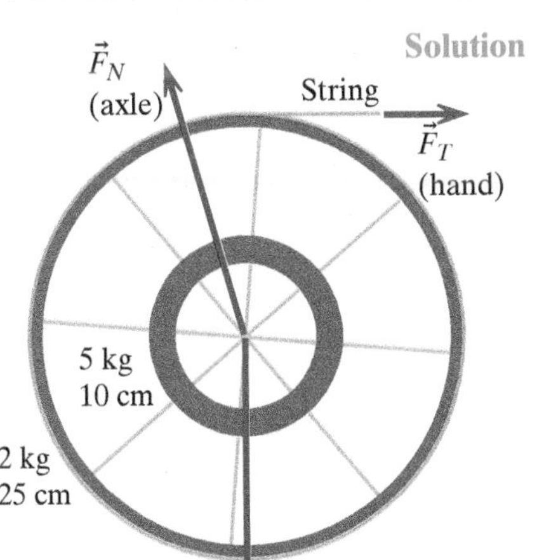

Figure 9.15 A rotating wheel with two rings.

Solution System: Wheel and string
Surroundings: Your hand, axle, Earth
Free-body diagram: Figure 9.15
Principle: Energy Principle

$$E_f = E_i + W$$

$$\frac{1}{2}I\omega_f^2 = \frac{1}{2}I\omega_i^2 + W$$

$$W = \frac{1}{2}I\left(\omega_f^2 - \omega_i^2\right)$$

What is the moment of inertia $I = m_1 r_{\perp 1}^2 + m_2 r_{\perp 2}^2 + m_3 r_{\perp 3}^2 + m_4 r_{\perp 4}^2 + \cdots$? Group this sum into a part that includes just the atoms of the inner ring and another part that includes just the atoms of the outer ring:

$$I = (m_1 r_{\perp 1}^2 + m_2 r_{\perp 2}^2 + \cdots)_{\text{inner}} + (m_1 r_{\perp 1}^2 + m_2 r_{\perp 2}^2 + \cdots)_{\text{outer}}$$

$$I = I_{\text{inner}} + I_{\text{outer}}$$

This is a general result: The moment of inertia of a composite object is the sum of the moments of inertia of the individual pieces, because we have to add up all the contributions of all the atoms. We already determined that the moment of inertia of a ring is MR^2 (all the atoms are at the same perpendicular distance R from the center), so the moment of inertia of this wheel is

$$I = M_{\text{inner}}R^2_{\text{inner}} + M_{\text{outer}}R^2_{\text{outer}}$$
$$I = (5\,\text{kg})(.1\text{m})^2 + (2\,\text{kg})(.25\text{m})^2 = (0.050 + 0.125)\,\text{kg}\cdot\text{m}^2 = 0.175\,\text{kg}\cdot\text{m}^2$$

We need to convert revolutions per second into radians per second:

$$\omega_i = \left(4\frac{\text{rev}}{\text{s}}\right)\left(\frac{2\pi\ \text{rad}}{\text{rev}}\right) = 25.1\ \text{rad/s}$$
$$\omega_f = \left(7\frac{\text{rev}}{\text{s}}\right)\left(\frac{2\pi\ \text{rad}}{\text{rev}}\right) = 44.0\ \text{rad/s}$$

You, the Earth, and the axle all exert forces on the system. How much work does the Earth do? Zero, because the center of mass of the wheel doesn't move. How much work does the axle do? If there is negligible friction between the axle and the wheel, the axle does no work, because there is no displacement of the axle's force. Therefore only you do work, and the work that you do is

$$W = \frac{1}{2}(0.175\,\text{kg}\cdot\text{m}^2)(44.0^2 - 25.1^2)\frac{\text{rad}^2}{\text{s}^2} = 114\ \text{J}$$

Although the outer ring has a smaller mass than the inner ring, the outer ring has a larger moment of inertia because its radius is 2.5 times as big, so the factor R^2 is 6.25 times as big. Applying the Momentum Principle to the wheel, we see that the three forces must add up to zero, because the momentum $M\vec{v}_{\text{CM}}$ of the wheel isn't changing.

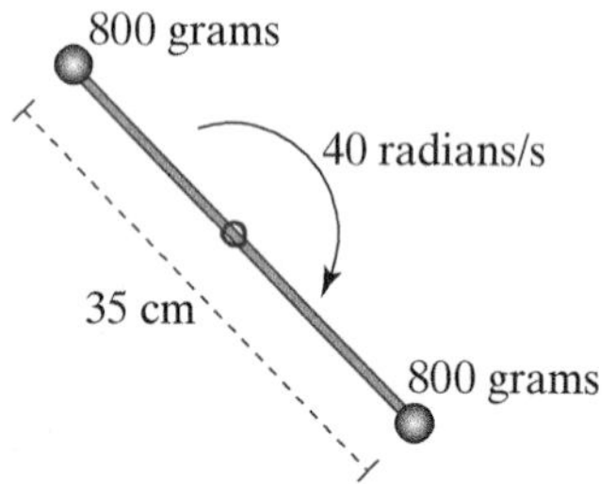

Figure 9.16 A rotating barbell.

Checkpoint 1 A barbell spins around a pivot at its center (Figure 9.16). The barbell consists of two small balls, each with mass 800 g, at the ends of a very low mass rod whose length is 35 cm. The barbell spins with angular speed 40 rad/s. Calculate K_{rot}.

Calculating Moment of Inertia

We've seen how to calculate the moment of inertia in a few simple special cases. We will calculate the moment of inertia of a long thin rod about its center to illustrate a more general technique. Consider a rod of mass M and length L whose thickness is small compared to its length, and that rotates around an axis perpendicular to the page (Figure 9.17). The density of the rod is the same everywhere ("uniform density"); every centimeter along the rod has the same mass. We need to evaluate the sum

$$I = m_1 r^2_{\perp 1} + m_2 r^2_{\perp 2} + m_3 r^2_{\perp 3} + m_4 r^2_{\perp 4} + \cdots$$

for all the atoms in the rod.

Divide into Small Slices As shown in Figure 9.17, we've chosen the origin of our axes to be at the center of the rod, and we've oriented the x axis to lie along the rod. We mentally divide the rod into N small slices of equal length $\Delta x = L/N$. Each slice has a mass $\Delta M = M/N$, because the rod's density is the same everywhere. We need to add up all the contributions to I made by all the slices.

The Mass of One Slice Concentrate on one representative slice, whose center in Figure 9.17 is located at $\langle x_n, 0, 0\rangle$ and whose mass is $\Delta M = M/N$. Since $\Delta x = L/N$,

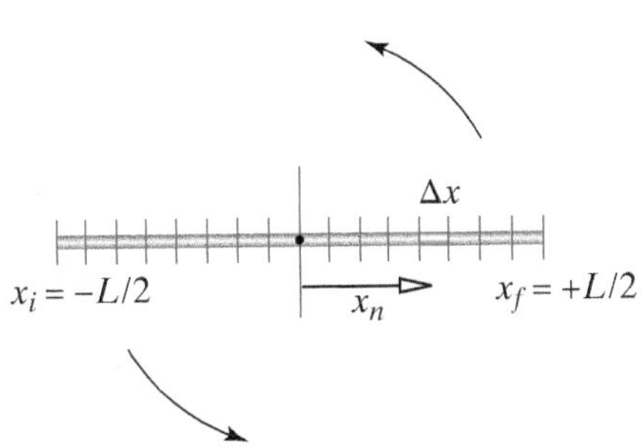

Figure 9.17 A uniform-density rod of length L rotates around an axis perpendicular to the page. We divide it into short slices and choose the origin to be at the center of the rod, with the x axis lying along the rod.

$$N = \frac{L}{\Delta x}$$

$$\Delta M = \frac{M}{N} = \frac{M}{L/\Delta x} = M\frac{\Delta x}{L}$$

Another way to see this is that $\Delta x/L$ is the fraction of the rod contained in this slice of length Δx, so the mass of this slice should be $M(\Delta x/L)$.

The Contribution of One Slice Now we're able to write down an expression for the contribution ΔI to the moment of inertia I of the rod that is made by this one representative slice of the rod. We make the approximation that all the atoms of this slice are approximately the same perpendicular distance $r_\perp \approx x_n$ from the center of the rod:

$$\Delta I = (\Delta M)\, x_n^2 = \left(M\frac{\Delta x}{L}\right) x_n^2 = \frac{M}{L} x_n^2 \Delta x$$

Adding Up the Contributions We need to add up all such contributions by all such slices. One way to do this is numerically, by dividing the rod into, say, $N = 50$ slices, evaluating $(M/L)x_n^2\Delta x$ for each slice, and adding up all the contributions ($\sum$ means "summation"):

$$I = \sum_{n=1}^{N} \Delta I = \frac{M}{L}\sum_{n=1}^{N} x_n^2 \Delta x$$

The Finite Sum Becomes a Definite Integral In some cases, including this case of a long thin rod, it is possible to use integral calculus to evaluate the summation. In the limit of large N (which means small Δx), the sum becomes an integral (the $\int$ symbol is a distorted S for Summation). The finite slice length Δx turns into the infinitesimal quantity dx, and the sum from 1 to N turns into a definite integral with lower and upper limits on x (x_i and x_f):

$$I = \frac{M}{L} \lim_{N\to\infty} \sum_{n=1}^{N} x_n^2 \Delta x = \frac{M}{L}\int_{x_i}^{x_f} x^2 dx$$

The Limits of Integration Since we put the origin at the center of the rod in Figure 9.17, the lower limit x_i is $-L/2$ and the upper limit x_f is $+L/2$:

$$I = \frac{M}{L}\int_{-L/2}^{+L/2} x^2 dx = \left(\frac{1}{12}\right) ML^2$$

Because it is easy to make a mistake in this kind of calculation, it is important to check whether the results are reasonable. First, the units are correct, since $(\frac{1}{12})ML^2$ has units of $\text{kg}\cdot\text{m}^2$, which are units of moment of inertia. Second, if we double the mass M, I doubles, which makes sense, and if we double the length L, I gets four times bigger, which also makes sense given the factors of $r_\perp^2$ that appear in the definition of moment of inertia.

QUESTION Why is this expression valid only for a *thin* rod?

We made the approximation that all the atoms in one slice had the same perpendicular distance x_n from the center of the rod. If the rod is thick, this isn't a valid approximation, even in the limit of infinitesimally short slices, because now those short slices are thick.

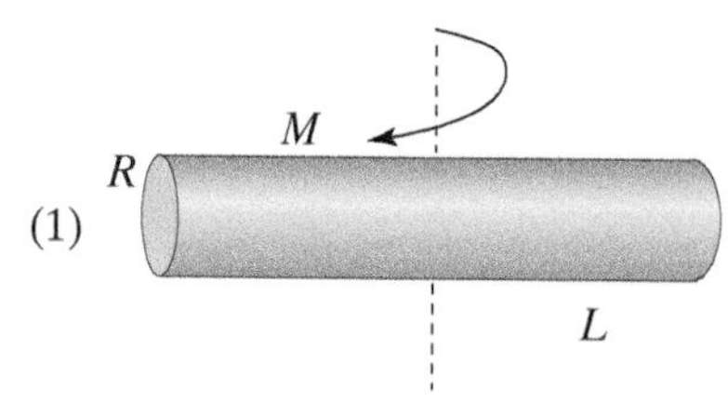

Checkpoint 2 A thin uniform-density rod whose mass is 1.2 kg and whose length is 0.7 m rotates around an axis perpendicular to the rod, with angular speed 50 radians/s. Its center moves with a speed of 8 m/s. **(a)** What is its rotational kinetic energy? **(b)** What is its total kinetic energy?

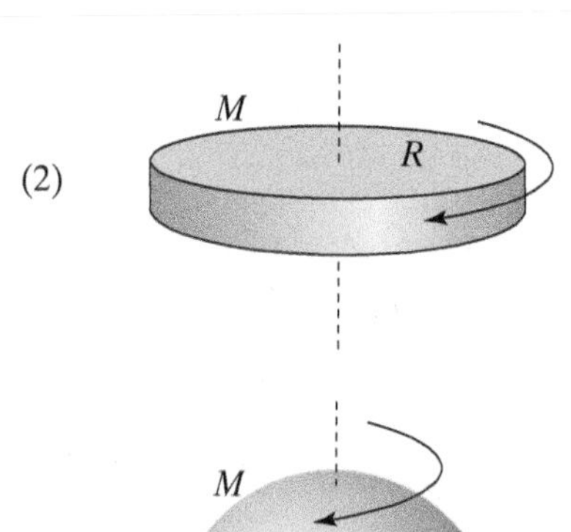

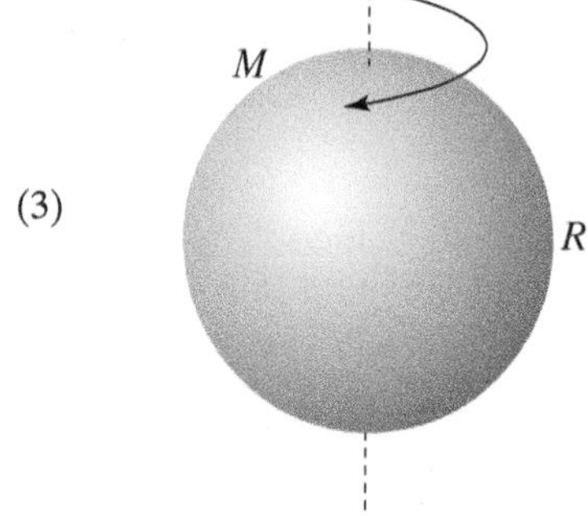

Figure 9.18 Some common uniform-density solids whose moments of inertia are known.

Moments of Inertia of Simple Shapes

By applying integral calculus as we did above in the case of a long thin uniform-density rod, it is possible to calculate the moments of inertia of some other simple shapes. Some useful results for objects of uniform density (see Figure 9.18) are as follows:

(1) For a cylinder of length L, radius R, rotating about an axis perpendicular to the cylinder, through center of cylinder:

$$I_{\text{cylinder}} = \frac{1}{12}ML^2 + \frac{1}{4}MR^2$$

(2) For a disk or cylinder rotating about its axis:

$$I_{\text{disk}} = I_{\text{cylinder}} = \frac{1}{2}MR^2$$

(3) For a sphere rotating around an axis passing through the center of the sphere:

$$I_{\text{sphere}} = \frac{2}{5}MR^2$$

A thin rod is like a cylinder with radius $R \approx 0$, in which case the moment of inertia for rotation about an axis perpendicular to the rod reduces to $\frac{1}{12}ML^2$.

Because only the perpendicular distances of atoms from the axis matter ($r_\perp$), the moment of inertia for rotation about the axis of a long cylinder has exactly the same form as that of a disk, $I_{\text{disk}} = \frac{1}{2}MR^2$. It doesn't matter how long the cylinder is (or to put it another way, how thick the disk is). See Figure 9.19.

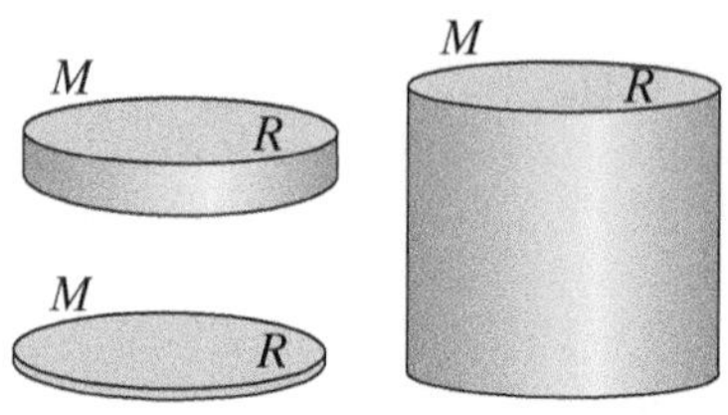

Figure 9.19 These disks and cylinders all have moment of inertia $\frac{1}{2}MR^2$ about their axes (but if they have the same density, they have different masses).

Checkpoint 3 A uniform-density sphere whose mass is 10 kg and radius is 0.4 m makes one complete rotation every 0.2 s. What is the rotational kinetic energy of the sphere?

Rigid Rotation about a Point Not the Center of Mass

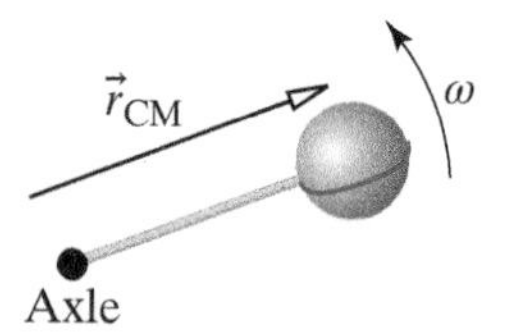

Figure 9.20 An object with known I_{CM} is connected to a low-mass rod and rotates around an axle. The red stripe on the object rotates at the same rate ω as the rod.

In Figure 9.20 a rigid object with known moment of inertia I_{CM} about its center of mass is connected to a low-mass rod and rotates about an axle that is a distance r_{CM} from the center of mass. What is the kinetic energy of this rotating object?

We know that in general $K_{\text{tot}} = K_{\text{trans}} + K_{\text{rel}}$, and in this case K_{rel} is just K_{rot}, since there is no vibration or explosion. The red line painted on the object in Figure 9.20 rotates with the same angular speed ω as the rod, which rotates around the axle; both the line and the object make one revolution in the same amount of time, so the kinetic energy that is relative to the (moving) center of mass is, as usual, $K_{\text{rot}} = \frac{1}{2}I_{\text{CM}}\omega^2$.

QUESTION What is K_{trans}, the kinetic energy associated with the motion of the center of mass?

The center of mass of the object rotates about the axle with angular speed ω at a distance r_{CM} from the axle, so the speed of the center of mass (in meters per second) is $v_{CM} = \omega r_{CM}$. Therefore we have

$$K_{trans} = \frac{1}{2}Mv_{CM}^2 = \frac{1}{2}M(\omega r_{CM})^2$$
$$K_{trans} = \frac{1}{2}(Mr_{CM}^2)\omega^2$$
$$K_{tot} = \frac{1}{2}(Mr_{CM}^2 + I_{CM})\omega^2$$

We see that when the center of mass rotates around a fixed point, K_{trans} brings into the total kinetic energy an additional term $\frac{1}{2}(Mr_{CM}^2)\omega^2$. We can summarize this result by saying that the total kinetic energy is $\frac{1}{2}I\omega^2$, where $I = Mr_{CM}^2 + I_{CM}$. This result is called the parallel axis theorem.

RIGID ROTATION ABOUT A POINT NOT THE CENTER OF MASS

$$K_{tot} = K_{trans} + K_{rot} = \tfrac{1}{2}(Mr_{CM}^2 + I_{CM})\omega^2$$

This is valid only for a rigid body, one whose rotation rate is the same as the rotation rate about the center of mass.

EXAMPLE **A Rod Rotating Not Around Its Center**

A thin rod with mass 140 g and 60 cm long rotates at an angular speed of 25 rad/s about an axle that is 20 cm from one end of the rod (Figure 9.21). What is its kinetic energy?

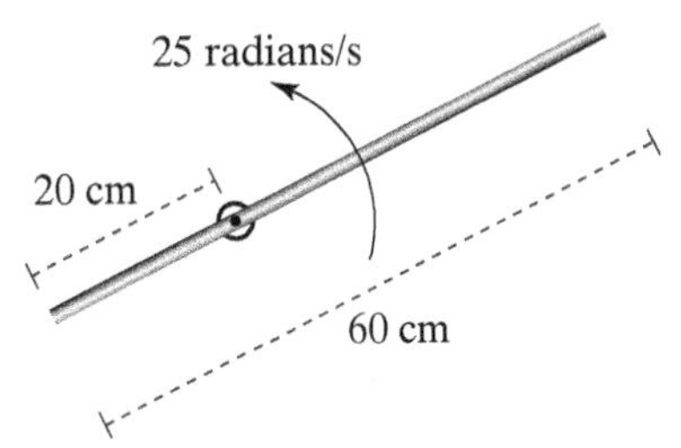

Figure 9.21 A rod rotates around an axle that is not located at the center of the rod.

Solution There is translational kinetic energy because the center of mass is moving, and there is rotational kinetic energy because there is rotation about the center of mass. From what we just showed, we have this ($r_{CM} = 0.1$ m):

$$K_{tot} = K_{trans} + K_{rot} = \frac{1}{2}(Mr_{CM}^2 + I_{CM})\omega^2$$
$$K_{tot} = \frac{1}{2}\left(Mr_{CM}^2 + \frac{1}{12}ML^2\right)\omega^2 = \frac{1}{2}M\left(r_{CM}^2 + \frac{1}{12}L^2\right)\omega^2$$
$$K_{tot} = \frac{1}{2}(.140\text{ kg})\left((0.1\text{ m})^2 + \frac{1}{12}(0.6\text{ m})^2\right)(25\text{ rad/s})^2$$
$$K_{tot} = 1.75\text{ J}$$

Checkpoint 4 A solid uniform-density sphere is tied to a rope and moves in a circle with speed v. The distance from the center of the circle to the center of the sphere is d, the mass of the sphere is M, and the radius of the sphere is R. **(a)** What is the angular speed ω? **(b)** What is the rotational kinetic energy of the sphere? **(c)** What is the total kinetic energy of the sphere?

9.3 COMPARING TWO MODELS OF A SYSTEM

In analyzing the motion and energy of multiparticle systems, we encounter situations in which it would be extremely convenient to be able to calculate only the change in the translational kinetic energy K_{trans} of the system. Perhaps this might be all we are interested in, or perhaps we might wish to subtract this energy change from the change of the total energy of the system to find out how much energy has gone into vibration or rotation. Fortunately, it is quite simple to calculate ΔK_{trans} by itself.

Modeling a System as a Point Particle

Because $d\vec{p}_{\text{sys}}/dt = \vec{F}_{\text{net}}$, the center of mass of a multiparticle system moves exactly like a simple point particle whose mass is the mass of the entire system, under the influence of the net external force applied to the entire system. Imagine that you could crush the system that is modeled as an extended object (the "extended system") down to a very tiny ball. Next, apply to this system, modeled as a point particle (the "point particle system"), forces with the same magnitudes and directions as those that acted on the extended system, but applied directly to the point particle rather than at their original locations. What would the motion of this system modeled as a point particle look like? It would look exactly like the motion of the center of mass of the extended system (Figure 9.22).

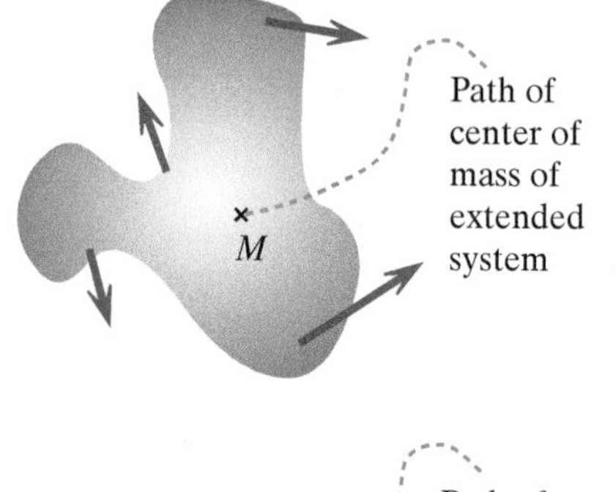

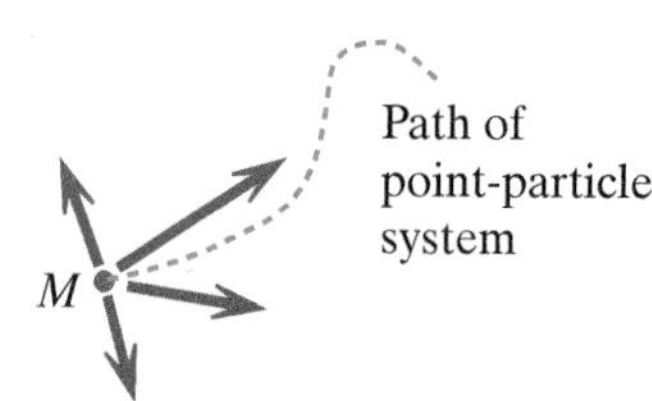

Figure 9.22 The path of the system modeled as a point particle is exactly the same as the path of the center of mass of the system modeled as an extended object.

These two different systems have the same total mass M, and both are acted on by the same net force, so the two paths are exactly the same. The difference is that the system when modeled as an extended object may rotate and stretch and vibrate due to the effects of the forces acting at different locations on the extended object. In contrast, the system when modeled as a point particle has no rotational motion, no vibrational motion, no internal energy of any kind. All of the forces act at the location of the point particle, and these forces don't stretch or rotate it. The only energy the point particle system can have is translational kinetic energy, and this is exactly the same as the translational kinetic energy of the extended system:

$$K_{\text{trans}} = \frac{p_{\text{sys}}^2}{2M} = \frac{1}{2}Mv_{\text{CM}}^2 \quad \text{(extended system or point particle system)}$$

An analysis of the point particle system gives the translational kinetic energy of the extended system. This is the reason it is useful to model an extended system as a point particle.

WORK AND ENERGY FOR A POINT PARTICLE

$$\Delta K_{\text{trans}} = \vec{F}_{\text{net}} \bullet \Delta\vec{r}_{\text{CM}}$$

This assumes that $\vec{F}_{\text{net}}$ is constant during the displacement $\Delta\vec{r}_{\text{CM}}$. If this is not the case, one must calculate the work as an integral of $\vec{F}_{\text{net}} \bullet d\vec{r}_{\text{CM}}$, analytically or numerically.

In Chapter 6 we showed that starting from the Momentum Principle for a single point particle we could derive the result that the change in the kinetic energy of the particle is equal to the work done on the particle by the net force. By modeling the system as a point particle, we assume that the net force is applied at the center-of-mass point, doing work to increase the kinetic energy of the point particle. Although this equation involves energy, it is actually derived from the Momentum Principle. A formal derivation may be found at the end of this chapter.

Modeling a System as an Extended Object

Because an extended object can rotate, vibrate, and change shape, not every part of the system necessarily moves the same distance as the center of mass moves. We need to consider this in calculating the work done on a system modeled as an extended object, because it matters where each force is applied.

To find the total work done on the system we need to consider separately the displacement of every point $\vec{r}_i$ where a force $\vec{F}_i$ is applied:

WORK AND ENERGY FOR AN EXTENDED SYSTEM

$$\Delta E_{\text{sys}} = \sum \vec{F}_i \bullet \Delta \vec{r}_i$$

Again, this assumes that each force $\vec{F}_i$ is constant during the displacement $\Delta\vec{r}_i$. If this is not the case, one must calculate the work of each force as an integral of $\vec{F}_i \bullet d\vec{r}_i$, analytically or numerically.

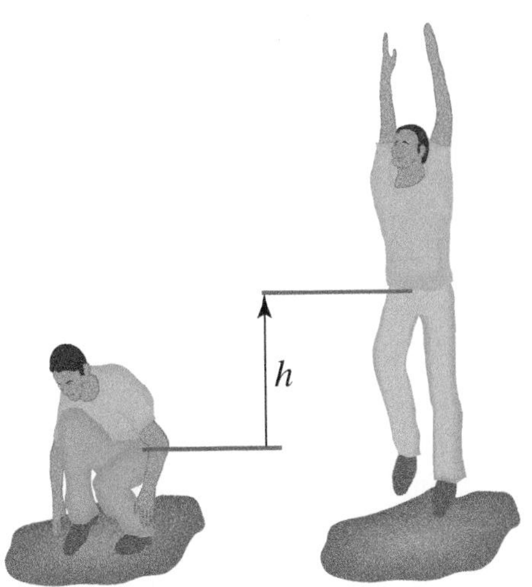

Figure 9.23 As the jumper leaves the floor, the center of mass has risen h. The system changes shape.

A Jumper Modeled as a Point Particle

As an example of how to exploit the point particle model, we'll analyze jumping (Figure 9.23). A jumper crouches down and then jumps straight upward. We want to calculate how fast the jumper's center of mass is moving at the instant that the feet leave the floor, v_{CM}. Let h be the distance through which the center of mass rises in this process. Note that there is a change of shape of the system.

Let F_N be the electric contact force that the atoms in the floor exert on the bottom of your foot, and let M be the jumper's total mass. We don't know exactly how the floor force F_N varies with time, but to get an idea of what happens let's make the crude approximation that it is constant as long as the feet are in contact with the floor (of course it falls suddenly to zero as the foot leaves the floor). Consider the system modeled as a point particle in Figure 9.24, which does not change shape.

QUESTION What is the net force on the jumper?

As indicated in Figure 9.24 the net force in the upward direction is $(F_N - Mg)$. Therefore the energy equation for the point particle system is

Figure 9.24 Free-body diagram for a jumper's body, modeled as a point particle. This system does not change shape.

$$\Delta K_{\text{trans}} = \vec{F}_{\text{net}} \bullet \Delta\vec{r}_{\text{CM}}$$

$$\frac{1}{2}Mv_{\text{CM}}^2 - 0 = (F_N - Mg)h$$

The work done on the point particle system involves the net force multiplied by the displacement of the center-of-mass point. As we will see, this is not the same as the energy equation for the system when modeled as an extended object, consisting of the jumper. The reason F_N shows up in the point particle energy equation is that it contributes to the net force and impulse, which affects the momentum (which is related to the motion of the center of mass).

A Jumper Modeled as an Extended Object

By modeling a jumper as a point particle, we were able to find out the change in the person's translational kinetic energy.

QUESTION What are some of the energy changes that occur during the jump that are not represented in the energy equation for the system when modeled as a point particle?

There is change of kinetic energy of the moving arms and legs relative to the center of mass, decrease of chemical energy, and increase of thermal energy (the jumper's temperature rises somewhat).

QUESTION Is the work done on an extended system the same as the work done on a point particle?

No. When calculating the work done on an extended system we must take into account the fact that the individual forces acting on the system are applied at different locations, as shown in the free-body diagram in Figure 9.25. Typically some of these forces act through very different distances than the displacement of the center of mass of the system, because different parts of the system move different distances.

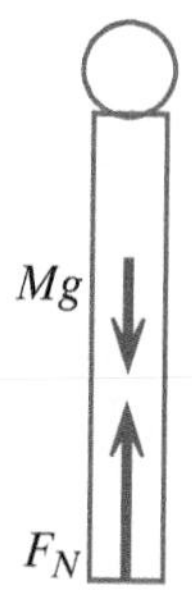

Figure 9.25 Free-body diagram for a jumper's body, modeled as an extended system. This system changes shape.

The energy equation for the extended system (the jumper) looks something like the following, ignoring Q, transfer of energy due to a temperature difference between the jumper and the surrounding air:

$$\Delta E_{\text{sys}} = \sum \vec{F}_i \bullet \Delta \vec{r}_i$$
$$\Delta K_{\text{trans}} + \Delta K_{\text{rel}} + \Delta E_{\text{internal}} = -Mg \cdot h + F_N \cdot 0$$

QUESTION Why does the force F_N exerted by the floor on the jumper's foot do zero work?

The floor force F_N is applied to the bottom of the foot, and the contact point does not move during the entire process leading to lift-off, so there is no work done by the floor force. The definition of the work done by a force involves (the parallel component of) the displacement of the point of contact, at the place where the force is applied to the object. No displacement, no work done on the system. Of course, F_N does affect the momentum of the system, by contributing to the net impulse.

ΔK_{rel} includes the rotation of the upper and lower legs and the swinging of the arms. $\Delta E_{\text{internal}}$ includes both the increase of thermal energy in the jumper's body due to the exertion, and the decrease of chemical energy stored in the jumper's body.

None of these terms appear in the energy equation for the point particle system, because the point particle equation deals just with the motion of the mathematical center of mass point, which doesn't rotate or stretch or get hot. A crucial difference is that the extended system changes shape (legs unbend, etc.), but the point particle system doesn't.

Combining the Two Models

Replace ΔK_{trans} in the energy equation for the extended system with the value of ΔK_{trans} we found for the point particle system:

$$\Delta K_{\text{trans}} + \Delta K_{\text{rel}} + \Delta E_{\text{internal}} = -Mgh$$
$$(F_N - Mg)h + \Delta K_{\text{rel}} + \Delta E_{\text{internal}} = -Mgh$$
$$\Delta K_{\text{rel}} + \Delta E_{\text{internal}} = -F_N h$$

QUESTION What are the differences between the energy equations for the system modeled as a point particle and as an extended system?

Most energy terms do not appear at all in the point particle equation, and the work that appears in the energy equation for the extended system is not the same as the work that appears in the energy equation for the point particle system.

Stretching a Spring

A simple situation—stretching a spring—illustrates why the work done on a system modeled as a point particle can be different from the work done on an extended system. Suppose that you pull on one end of a spring with a force $\vec{F}_L$ to the left, making a short displacement $\Delta \vec{r}_L$ to the left, and you pull on the other end with an equal but opposite force $\vec{F}_R$ and equal but opposite

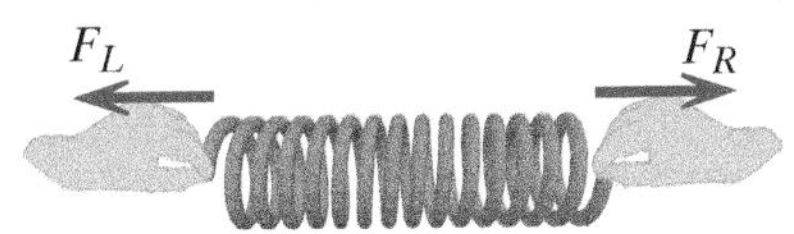

Figure 9.26 You pull on both ends of a spring with equal and opposite forces. The system (the spring) changes shape.

short displacement $\Delta\vec{r}_R$ (Figure 9.26). The system (the spring) changes shape. Our analysis should yield the expected result: the spring should not gain translational kinetic energy!

A Spring Modeled as a Point Particle

To begin, we will model the system as a point particle, located at the center of mass of the spring (Figure 9.27). This point particle system cannot change shape.

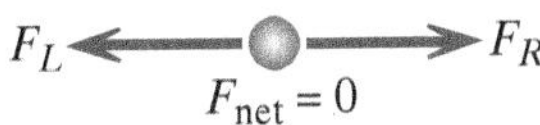

Figure 9.27 Free-body diagram for a spring pulled from each end, modeled as a point particle.

> QUESTION What is the net force $\vec{F}_{net}$? How far does the center of mass move? How much work is done on the point particle system? What is the change in K_{trans}?

$$\vec{F}_{net} = \vec{F}_L + \vec{F}_R = \vec{0}$$
$$\Delta\vec{r}_{CM} = \vec{0}$$
$$\Delta K_{trans} = \vec{0}\bullet\vec{0} = 0$$

This result is consistent with what we would expect: zero work is done, and the translational kinetic energy of the system does not change.

A Spring Modeled as an Extended Object

Now we will model the spring as a deformable extended system (Figure 9.28).

> QUESTION Does your right hand do any work on the spring? Positive or negative? Does your left hand do any work on the spring? Positive or negative? Is there any change in the energy of the spring?

Figure 9.28 Free-body diagram for a spring modeled as an extended object.

$$\Delta E_{sys} = \sum \vec{F}_i \bullet \Delta\vec{r}_i$$
$$\Delta K_{trans} + \Delta U_{spring} = \vec{F}_L \bullet \Delta\vec{r}_L + \vec{F}_R \bullet \Delta\vec{r}_R$$
$$\Delta K_{trans} + \Delta U_{spring} = F_L \Delta r_L + F_R \Delta r_R$$

Both hands do positive work, so the energy of the system increases. Combining the models, we find that

$$0 + \Delta U_{spring} = F_L \Delta r_L + F_R \Delta r_R$$
$$\Delta\left(\frac{1}{2}k_s s^2\right) = F_L \Delta r_L + F_R \Delta r_R$$

Checkpoint 5 A runner whose mass is 50 kg accelerates from a stop to a speed of 10 m/s in 3 s. (A good sprinter can run 100 m in about 10 s, with an average speed of 10 m/s.) **(a)** What is the average horizontal component of the force that the ground exerts on the runner's shoes? **(b)** How much displacement is there of the force that acts on the sole of the runner's shoes, assuming that there is no slipping? Therefore, how much work is done on the extended system (the runner) by the force you calculated in the previous exercise? How much work is done on the point particle system by this force? **(c)** The kinetic energy of the runner increases—what kind of energy decreases? By how much?

Analyzing Point Particle and Extended Systems

We have seen that we can often extract significant information about the energy distribution within a system by comparing the results of two models: a model of the system as a point particle, and a model of the system as an extended object. Modeling the system as a point particle gives us information about the translational kinetic energy K_{trans} of the system—the only kind of energy a point particle can have other than rest energy. Knowing K_{trans} can allow us to resolve other kinds of energy when we model the system as an extended object. The basic approach is to:

(a) Model the system as a point particle.

- Assume that the entire mass of the system is concentrated into a point at the location of the system's center of mass.
- Assume that all forces applied to the system act at this point.
- Find the displacement of the center of mass.
- Calculate the net work done on the point particle.
- Find ΔK_{trans}, the change in the system's translational energy.

(b) Model the system as an extended object.

- Draw a diagram showing the point of application of each force acting on the system. Typically some of these forces will not act at the center of mass of the system.
- Identify the distance through which each of these forces acts on the system. Typically these forces will act through different distances, especially if the object changes shape.
- Separately calculate the work done on the system by each force, acting through the appropriate distance.
- Apply the Energy Principle to find ΔE, the change in the extended system's energy.

(c) Find $\Delta E - \Delta K_{trans}$. This is the change in kinds of energy other than translational energy for the extended system.

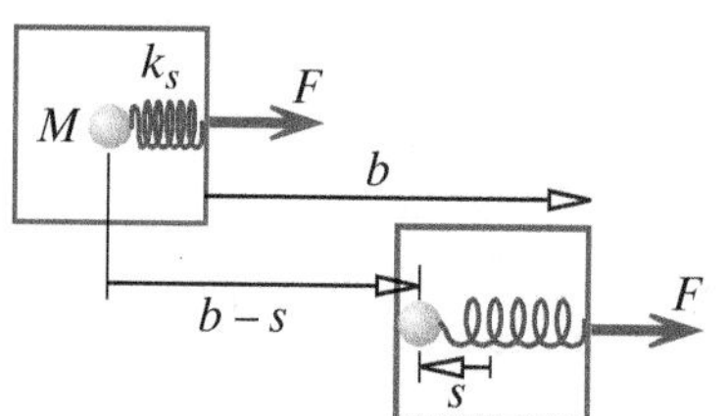

Figure 9.29 A constant force acts on a thin box containing a spring and a ball of clay. The box moves a distance b, but the ball (center of mass of the system) moves only a distance $b-s$. Tails of arrows representing forces are drawn at the point at which they act.

The following examples apply this method to answer questions about changes in internal energy and relative kinetic energy.

EXAMPLE **A Box Containing a Spring**

A thin box in outer space contains a large ball of clay of mass M, connected to an initially relaxed spring of stiffness k_s (Figure 9.29). The mass of the box and spring are negligible compared to M. The apparatus is initially at rest. Then a force of constant magnitude F is applied to the box. When the box has moved a distance b, the clay makes contact with the left side of the box and sticks there, with the spring stretched an amount s. See the diagram for distances. **(a)** Immediately after the clay sticks to the box, how fast is the box moving? **(b)** What is the increase in internal energy of the clay? Assume that the process takes place so quickly that there isn't time for any significant energy transfer Q due to a temperature difference between the (extended) system and surroundings.

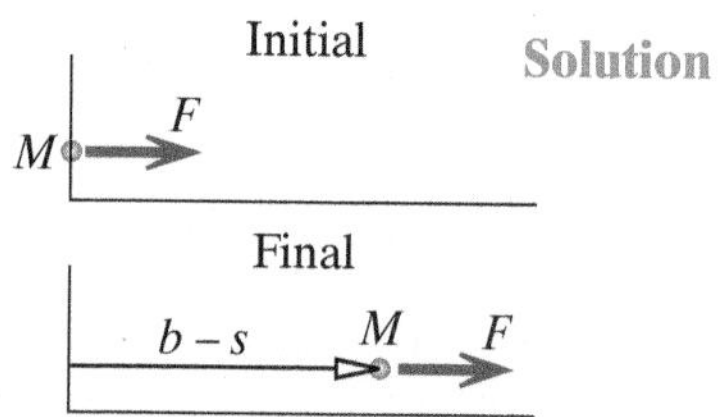

Figure 9.30 The point particle system for the apparatus. Tails of arrows representing forces are attached to the point particle, which moves the same distance as the center of mass of the extended system.

Solution **(a)** Point particle system
System: Point particle of mass M
Surroundings: External object exerting force F
Initial state: Point particle at rest
Final state: Point particle moving with (unknown) speed v
Free-body diagram: Figure 9.30
Energy Principle (point particle only has K_{trans}):

$$\Delta K_{trans} = F\Delta x_{CM}$$

$$\Delta x_{CM} = (b-s) \quad \text{(from the diagram)}$$

$$\frac{1}{2}Mv^2 - 0 = F(b-s)$$

$$v = \sqrt{\frac{2F(b-s)}{M}}$$

(b) Extended system
System: Mass, box, and spring
Surroundings: External object exerting force F
Initial state: System at rest
Final state: Box moving with (known) speed v (same as clay)
Free-body diagram: Figure 9.29
Energy Principle:

$$\Delta E_{\text{sys}} = W$$

$$\Delta(K_{\text{trans}} + U_{\text{spring}} + E_{\text{internal}}) = Fb$$

$$\Delta K_{\text{trans}} = F(b-s) \quad \text{(from part (a))}$$

$$F(b-s) + \frac{1}{2}k_s s^2 + \Delta E_{\text{internal}} = Fb$$

$$\Delta E_{\text{internal}} = Fs - \frac{1}{2}k_s s^2$$

In the energy equation for the point particle system there is nothing about the spring or internal energy, because the point particle system doesn't stretch or get hot. In Figure 9.30 the point at which the force vector is applied moves a distance $(b-s)$, moving with the point particle, so the work done on the point particle system is $F(b-s)$.

In the extended system, the point at which the force vector is applied moves a longer distance b, so the work done on the extended system is Fb. More work is done, and the extra work goes into spring potential energy and internal energy of the clay.

Good labeled diagrams of the two choices of system are extremely important, in order to determine the distances through which forces act in the two cases. Review the procedure at the beginning of the section and see how these steps were implemented in the example given above.

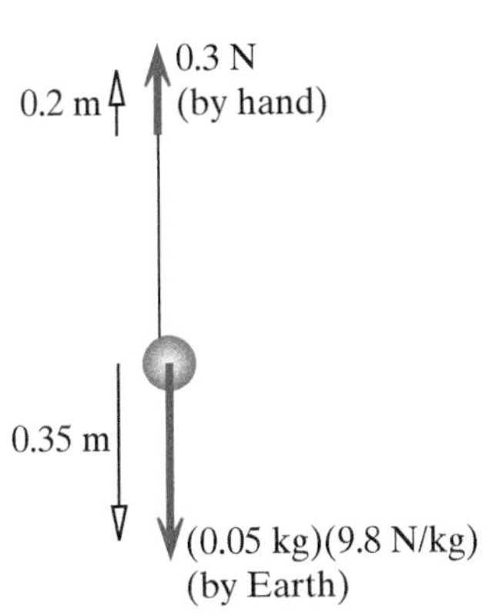

Figure 9.31 You pull up with a force of magnitude 0.3 N through a distance of 0.2 m on the string of a yo-yo while the yo-yo moves downward a distance 0.35 m.

EXAMPLE

A Yo-yo

You're playing with a yo-yo whose mass is 0.05 kg attached to a low-mass string (Figure 9.31). You pull up on the string with a force of magnitude 0.3 N, and your hand moves up a distance of 0.2 m. During this time the mass falls a distance 0.35 m and some of the string reels off the yo-yo's axle. **(a)** What is the change in the translational kinetic energy of the yo-yo? **(b)** What is the change in the rotational kinetic energy of the yo-yo, which spins faster?

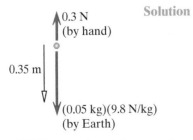

Solution

(a) Point particle system
System: Point particle
Surroundings: Earth and hand
Initial state: Point particle with initial translational kinetic energy
Final state: Point particle with final translational kinetic energy
Free-body diagram: Figure 9.32
Energy Principle (point particle only has K_{trans}):

$$\Delta K_{\text{trans}} = F_{\text{net},y}\Delta y_{\text{CM}}$$

$$\Delta K_{\text{trans}} = [0.3\text{ N} - (0.05\text{ kg})(9.8\text{ N/kg})](-0.35\text{ m}) = 0.0665\text{ J}$$

0.3 N
(by hand)
0.35 m
(0.05 kg)(9.8 N/kg)
(by Earth)

Figure 9.32 The point particle system for the yo-yo. Tails of arrows representing forces are attached to the point particle, which moves the same distance as the center of mass of the extended system.

(b) Extended system
System: Mass and string
Surroundings: Earth and hand

Initial state: Initial rotational and translational kinetic energy
Final state: Final rotational and translational kinetic energy
Free-body diagram: Figure 9.31
Energy Principle:

$$\begin{aligned}\Delta E_{\text{sys}} &= W_{\text{hand}} + W_{\text{Earth}} \\ \Delta K_{\text{trans}} + \Delta K_{\text{rot}} &= (0.3\text{ N})(0.2\text{ m}) + [-(0.05\text{ kg})(9.8\text{ N/kg})](-0.35\text{ m}) \\ &= 0.2315\text{ J} \\ \Delta K_{\text{trans}} &= 0.0665\text{ J} \quad \text{(from part (a))} \\ 0.0665\text{ J} + \Delta K_{\text{rot}} &= 0.2315\text{ J} \\ \Delta K_{\text{rot}} &= 0.165\text{ J}\end{aligned}$$

The key issue is that the distance through which the force of your hand moves is different in the extended system and the point particle system. In the extended system, your hand applies an upward force of magnitude 0.3 N through an upward displacement of 0.2 m, doing a positive amount of work, 0.06 J. However, in the "crushed" system, where all the mass is located at the center of mass of the extended system and all the force vectors are moved to that point, the upward-pointing force of your hand is applied to the point particle as the point particle drops a distance of 0.35 m. Therefore the work done by this force on the point particle is negative: (0.3 N)(− 0.35 m) = −0.105 J. The work done by the Earth's force is (0.05 kg)(9.8 N/kg)](0.35 m) = +0.1715 J for both systems, because this force is applied to the center of mass of the extended system, which is also the center of mass of the crushed system. The diagrams labeled with distances are critical to calculating the work correctly for the two different systems.

It is important to understand that the point particle or "crushed" system is a different model of the situation than is the extended-system model. The forces acting on the point particle model are equal in magnitude and direction to the forces acting on the extended-system model, but applied directly to the point particle, so the displacements through which these forces act are different from the displacements of the forces acting on the extended system.

In this yo-yo example the difference between the two models is quite striking, because the work done by the force of your hand on the extended system is positive, but the work done by this force on the point particle system is negative. Recall that the point particle equation was derived from the Momentum Principle; if we think about momentum, we note that the force of the hand acts to oppose the downward motion of the center of mass of the yo-yo.

It is interesting to note that the change of the rotational kinetic energy, 0.165 J, is the same as the work that you would do to increase the purely rotational kinetic energy of a yo-yo whose axle was fixed in space. The distance (0.2+0.35) m = 0.55 m is the length of the string that you would reel off the axle, and the force of your hand would do an amount of work (0.3 N)(0.55 m) = 0.165 J, which is ΔK_{rot}.

We chose to analyze this situation with specific numerical values in order to make the sign issues particularly clear. It is instructive to analyze the problem again symbolically, calling the upward force of your hand F, the distance you pull up d, the mass m, and the distance the yo-yo falls h.

EXAMPLE

An Ice Skater Pushes Off from a Wall

Consider a woman on ice skates who pushes on a wall with a nearly constant contact force to the left of magnitude F_N (normal to the wall in Figure 9.33). By the reciprocity of electric interatomic forces, the wall exerts a force of magnitude F_N to the right on the skater. As a result, she moves backward with increasing speed. Her center of mass is marked on each frame of the sequence. Look closely at Figure 9.33 and you'll see that as the skater straightens her arms, the location in her body of the center of mass shifts a little toward her outstretched hands and arms; it is not a point fixed at some place in her body. What can we learn from analyzing the **(a)** point particle and **(b)** extended versions of the system of the skater?

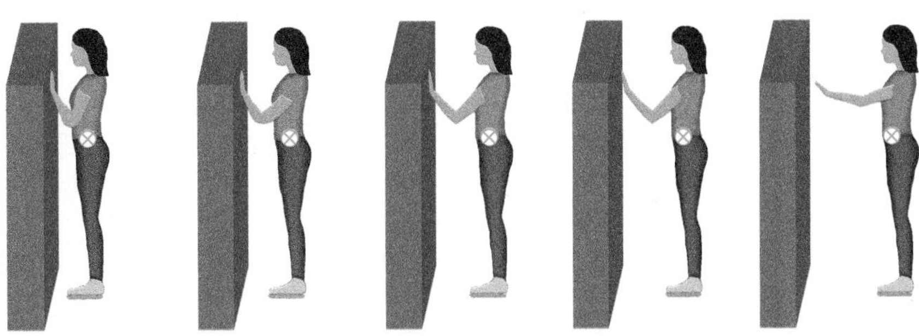

Figure 9.33

Solution Let's make the approximation that the change in the position of the center of mass in the skater's body is negligible. This is equivalent to neglecting the mass of the skater's arms and hands. Her center of mass moves to the right a distance d, which is approximately the length of her arms.

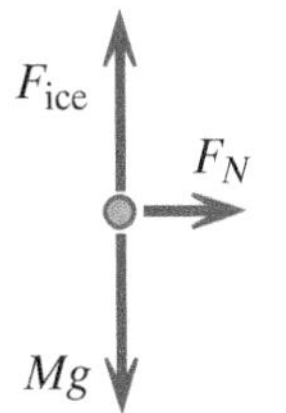

Figure 9.34 Free-body diagram for the point particle system.

(a) Point particle system
System: Point particle of mass M
Surroundings: Wall, ice, Earth
Initial state: Point particle at rest
Final state: Point particle moving with (unknown) speed v
Free-body diagram: Figure 9.34
Energy Principle (point particle only has K_{trans}):

$$\Delta K_{trans} = Fd$$
$$\frac{1}{2}Mv^2 - 0 = Fd$$
$$v = \sqrt{\frac{2Fd}{M}}$$

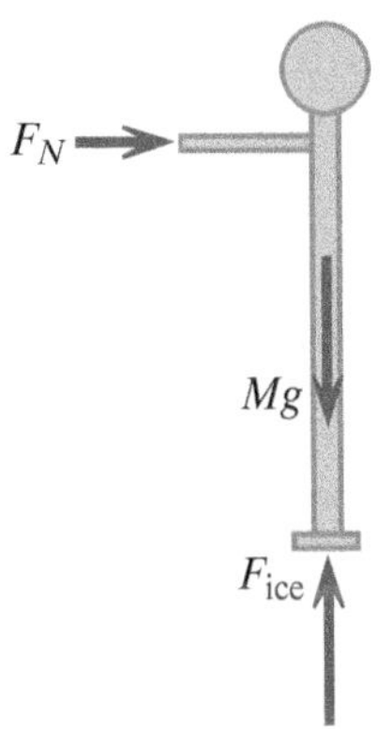

Figure 9.35 Free-body diagram for the extended system.

(b) Extended system
System: The skater
Surroundings: Wall, ice, Earth
Initial state: System at rest
Final state: Skater moving with speed v, with internal energy change
Free-body diagram: Figure 9.35
Energy Principle:

$$\Delta E_{sys} = W$$
$$\Delta(K_{trans} + E_{internal}) = F(0)$$
$$\Delta K_{trans} = Fd \quad \text{(from part (a))}$$
$$Fd + \Delta E_{internal} = 0$$
$$\Delta E_{internal} = -Fd$$

From the analysis of the point particle system we are able to determine the final speed of the skater. From the analysis of the extended system we are able to determine the internal energy change of the skater, which is negative, corresponding to a decrease in stored chemical energy (and a smaller increase in her thermal energy associated with a higher body temperature).

EXAMPLE **Pull on Two Hockey Pucks**

Tie a string to the center of a hockey puck, and wrap another string around the outside of a second hockey puck, as shown in Figure 9.36. Then two people pull on the two strings so that both strings have the same tension F_T. It may be surprising that the center of mass of both pucks moves the same distance, in the same direction, but remember that the x component of the Momentum Principle for speeds small compared to the speed of light can be written as $M\Delta v_{\text{CM},x} = F_{\text{net},x}\Delta t$. Here the net force is the same for both pucks, so the motion of the center of mass is the same for both pucks.

As shown in the figure, in a time Δt hand 1 has pulled a distance Δx_{CM} while hand 2 has pulled the bottom puck an additional distance d, because some string has unrolled from the puck. What can we learn from analyzing the point particle and extended versions of the system for each puck?

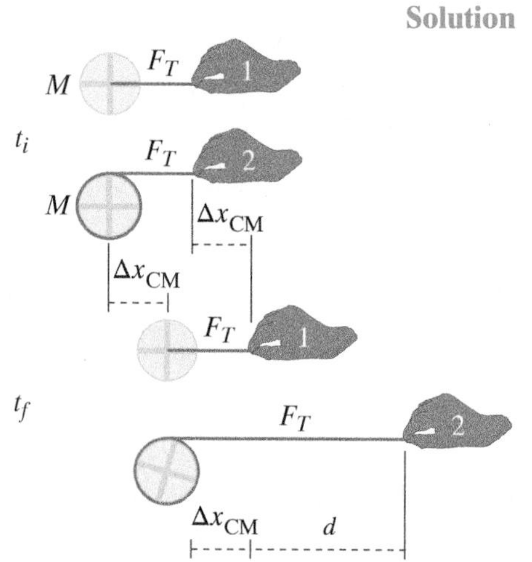

Figure 9.36 One puck is pulled by a string attached to its center. The other puck is pulled by a string wrapped around its edge, which unrolls a distance d as the puck is pulled.

Solution For either puck the point particle version of the system of puck plus string yields the result $\Delta K_{\text{trans}} = F\Delta x_{\text{CM}}$, and the same result applies to the case of the extended system of the puck and string pulled by hand 1.

What remains to be done is to analyze the extended version of the system consisting of the puck plus string that is pulled by hand 2:

$$\begin{aligned}
\Delta E_{\text{sys}} &= W \\
\Delta(K_{\text{trans}} + K_{\text{rot}}) &= F(\Delta x_{\text{CM}} + d) \\
\Delta K_{\text{trans}} &= F\Delta x_{\text{CM}} \quad \text{(from the point particle analysis)} \\
F\Delta x_{\text{CM}} + \Delta K_{\text{rot}} &= F(\Delta x_{\text{CM}} + d) \\
\Delta K_{\text{rot}} &= Fd
\end{aligned}$$

This makes sense: hand 2 pulls through a distance d farther than does hand 1, and this extra work Fd is equal to the increase in rotational kinetic energy.

A Reflection

If you look back over the various examples in this section, you can see that the Energy Principle for the point particle system and the Energy Principle for the extended system differ if any of the forces acting on the extended system acts through a displacement that is different from the displacement of the center of mass point. Or to put it the other way, the Energy Principle will be the same in both analyses only if all the atoms in the system have the same displacements (which will necessarily be the same as the displacement of the center of mass).

A change of shape or configuration, including deformation, or rotation with forces exerted at some location that isn't the center of mass, are examples of situations where you can expect to find that the Energy Principle for the two analyses will give different information, and where you can expect to find changes in the internal energy.

9.4 MODELING FRICTION IN DETAIL

In Chapter 7 we described a seeming paradox involving friction—that when you drag a block at constant speed across a table it seems that no work is done on the block, since you do an amount of work Fd, and the table apparently does an amount of work $-Fd$. Yet the block's temperature increases, indicating an increase in its internal energy.

In this section we'll see that the paradox arose from mistakenly treating the sliding block only as a point particle, which is inadequate for fully analyzing the energy aspects of the situation. Evidently there must be some kind of deformation of the block—some change of shape—and the friction force that the table exerts on the block must act through a displacement that is different from the displacement of the center of mass.

Sliding Block: Point Particle System

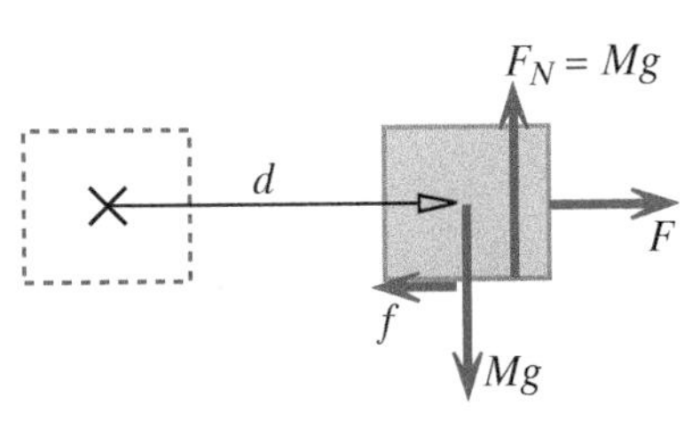

Figure 9.37 The center of mass of the block moves a distance d under the influence of a net force $F-f$.

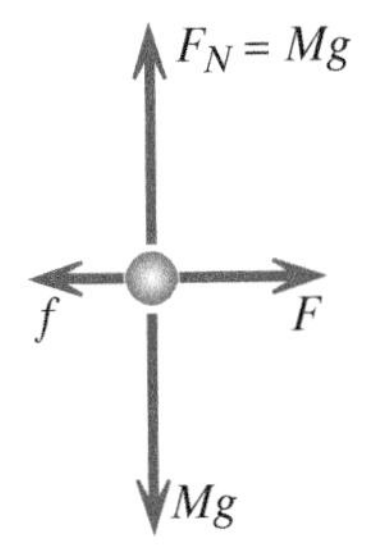

Figure 9.38 The block considered as a point particle.

A key issue that we will address is the question of how much work is done on the block by the friction force f exerted by the table. One of the new tools is the energy equation for the point particle system (Figure 9.37 and Figure 9.38). You apply a force of magnitude F to the right and the table exerts a force of magnitude f to the left. (Pretend for the moment that you don't know the magnitude of the friction force f.) The change in the kinetic energy of the fictitious center-of-mass point particle is given by the product of the net force $(F-f)$ times the displacement d of the center-of-mass point:

$$\Delta K_{\text{trans}} = (F-f)d$$

Since the speed of the block is constant, $\Delta K_{\text{trans}} = 0$, so $F-f=0$, which shows that the friction force is equal and opposite to the force you apply $f=F$. The internal energy does not appear in this energy equation for the point particle system, because this equation deals merely with the motion of the (mathematical) center of mass, not with the many kinds of energy in the extended system.

We also could have obtained the result $f=F$ directly from the Momentum Principle: $dp_{\text{sys}}/dt = F-f=0$. The Momentum Principle is closely related to the energy equation for the point particle system, because both involve the net force acting on a system and the motion of the center of mass.

Sliding Block: Extended System

The energy equation for the extended system of the block can be written as follows, where Fd is the work done by you, and W_{fric} is the work done by the table:

$$\Delta K_{\text{trans}} + \Delta E_{\text{internal}} = Fd + W_{\text{fric}}$$

where $\Delta E_{\text{internal}}$ is the rise in the internal energy of the block. We're assuming that the process takes a short enough time that there is negligible transfer of energy Q due to a temperature difference between the system of the block and the surroundings (the table), because energy transfer due to a temperature difference between table and block is a relatively slow process. (To remove such energy transfer as a possible complication, we could consider a block sliding not on a table but on an identical block. Then the symmetry of the situation is such that there cannot be any net transfer of energy due to a temperature difference into or out of the upper block. This tactic—reducing the complexity of a model—is a useful approach to complex problems.)

Since the speed of the center of mass v_{CM} does not change, $\Delta K_{\text{trans}} = 0$, and the energy equation reduces to

$$\Delta E_{\text{internal}} = Fd + W_{\text{fric}}$$

If we conclude that the friction force does an amount of work $-Fd$, then we would have to conclude that the internal energy of the block doesn't change, which is absurd. The block definitely gets hotter, indicating an increase in its internal energy. We need to find a way around the entirely plausible but apparently incorrect conclusion that the friction force does an amount of work $-Fd$.

QUESTION Considering the sign of the change of internal energy, must the magnitude of the work done by the friction force be greater or less than Fd?

Since the internal energy change $\Delta E_{\text{internal}}$ is surely positive (the block gets hotter, not colder), the magnitude of the work done by the friction force must be less than Fd. Since the friction force is definitely equal to F, the friction force must act through some effective distance d_{eff} *that must be less than* d!

Evidently the energy equation for the extended system has the following form, with d_{eff} less than d:

$$\Delta E_{\text{internal}} = Fd - Fd_{\text{eff}}$$

A Physical Model for Dry Friction

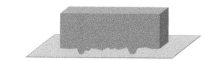

Figure 9.39 A block slides on a small number of protruding teeth (vertical scale exaggerated).

How can the effective distance through which the friction force does work be less than the distance through which the block moves? We can understand this by looking at a microscopic picture of what happens on the surfaces in contact.

When a metal block slides on a metal surface, the block is supported by as few as three protruding "teeth," called "asperities" in the scientific literature on friction (Figure 9.39). The very high load per unit area on these teeth makes the material partially melt and flow, and high local temperatures produced during sliding lead to adhesion (welding) in the contact regions. The frictional force divided by the tiny contact area corresponds to the large "shear" (sideways) stress required to break these welds. This shearing of contact welds is the dominant friction mechanism for a dry metal sliding on the same metal.

Because the tooth tips can become stronger than the bulk metal due to a process called "work-hardening" (which introduces dislocations into the otherwise regular geometrical arrangements of atoms), shearing often occurs in the weaker regions of the teeth, away from the tip. This is a major effect when the two objects are made of the same material, and chunks of metal can break off and embed in the other surface. Nevertheless, this wear will be ignored in the further discussion. It is in any case a symmetrical effect for identical blocks. If the metal surfaces have oxide coatings, this can reduce the shear stress required to break the temporary weld (which reduces the friction force) and can prevent the breaking off of chunks of metal, if the oxide contact area is the weakest section.

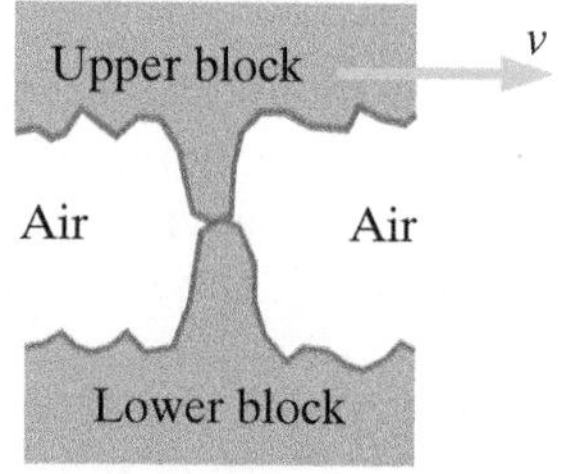

Figure 9.40 Two projecting teeth of identical blocks, which have temporarily adhered to each other. (Exaggerated vertical scale.)

This is the model of dry friction developed in a classic treatise on friction, *The Friction and Lubrication of Solids*, Part 1 and Part II, by F. P. Bowden and D. Tabor (Oxford University Press, 1950 and 1964). The physics and chemistry of friction continues to be an active field of research, because the effects of friction can be quite complex, and there is high practical interest in controlling friction. A more recent textbook is *Friction, Wear, Lubrication*, by K. C. Ludema (CRC Press, 1996).

Having briefly reviewed a basic model of dry friction, we proceed to use this model to calculate the work done by frictional forces exerted at the contact points. The key issue is that the surface is deformable, which leads to differences in the energy equation for the extended system compared with the energy equation for the point particle system.

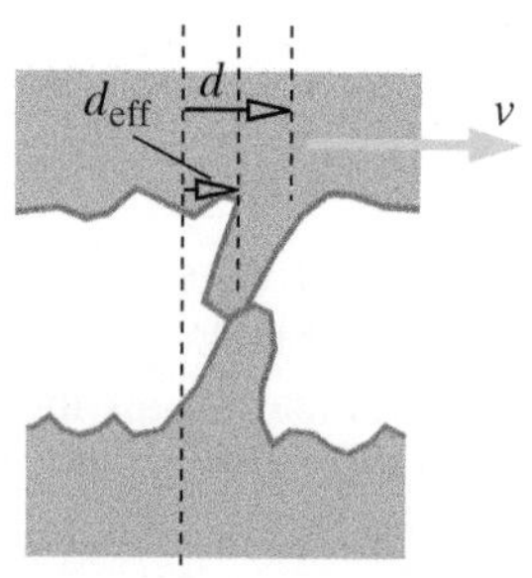

Figure 9.41 The top block has moved a distance d to the right, but the point of contact has moved only d_{eff}.

Actual Work Done by Friction Forces

We'll consider a microscopic picture of the contact region between two identical blocks. Figure 9.40 shows in a schematic way two teeth that have temporarily adhered to each other. The vertical scale has been greatly exaggerated for clarity—machined surfaces have much gentler slopes.

As the top block is dragged to the right, the teeth continue to stick together for a while. Both teeth must deform as a result; the top tooth is stretched backward, and the bottom tooth is stretched forward. In Figure 9.41 we see that when the top block has moved a distance d to the right, the point where the teeth touch each other has moved a distance of only d_{eff}, because the bottom block hasn't moved. (For two identical blocks made of the same material, on average the two teeth will bend the same amounts, and d_{eff} will be equal to $d/2$.)

> *This is the key issue:* The point of contact where the friction force is applied moves a shorter distance (d_{eff}) than the block itself moves (d). This means that the work done by the friction force is less than one might expect.

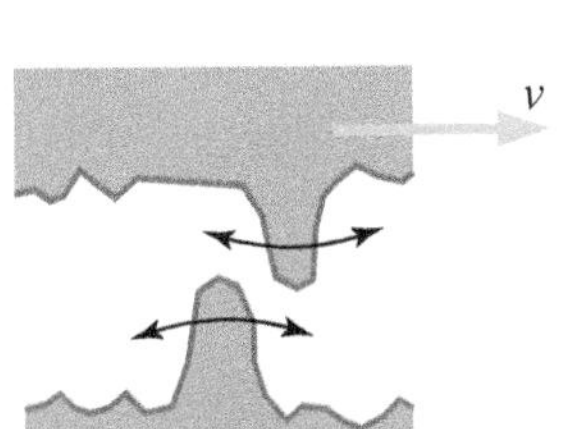

Figure 9.42 After the temporary weld breaks, the top tooth catches up to the rest of the block. Now no force acts on the teeth. Both teeth vibrate.

Eventually the weld breaks, and the top tooth snaps forward, so all the atoms in the tooth now catch up with the rest of the top block, but during this part of their journey, there is no external force acting on them (Figure 9.42).

The time average of the contact forces is indeed F (as determined by the fact that the net force must be zero, $F - F$), but the effective displacement d_{eff} at the point of contact of the frictional force is less than the displacement of the center of mass of the upper block. For two identical blocks we expect to find that $d_{eff} = d/2$. For other combinations of two materials in contact, all we know is that d_{eff} must be less than d but not necessarily equal to $d/2$.

Internal Energy of the Blocks

Once the weld has broken, the teeth can vibrate, as we saw in Figure 9.42. The vibration of the tooth belonging to the top block and thermal conduction upward from the hot tip into the main body of the block contribute to the increase in the internal energy of the upper block. Similarly, vibration and thermal conduction in the tooth belonging to the bottom block end up as increase in the internal energy of the bottom block.

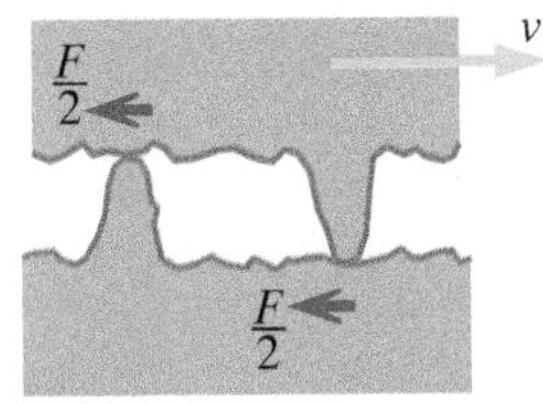

Figure 9.43 A more representative picture of contact between two surfaces, with projecting teeth in contact with flat areas.

A More Physical Picture

Having two long teeth in contact is rather unphysical. A better picture is the one shown in Figure 9.43 (again with a greatly exaggerated vertical scale), where we show one of the top block's longest teeth in contact with the average surface level of the bottom block, and one of the bottom block's longest teeth similarly in contact with the average surface level of the top block. The two teeth shown here are to be taken as representative of time and space averages of the sliding friction. Take the special case of identical blocks. By symmetry, the frictional force F exerted on the top block is divided on the average into two forces each of magnitude $F/2$ at the ends of the two sets of long teeth.

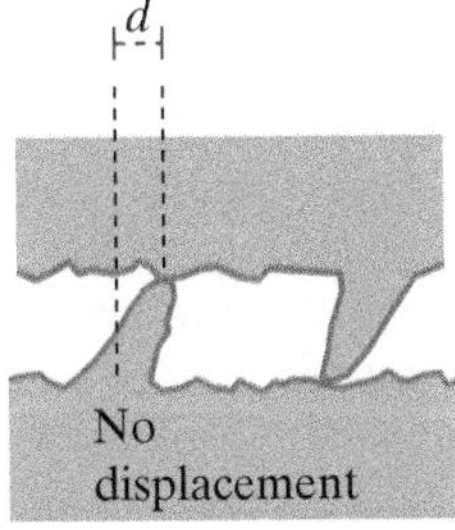

Figure 9.44 During this time interval the upper contact moves a distance d, while the lower contact does not move at all.

In Figure 9.44 we see that when the top block moves a distance d to the right the upper contact also moves a distance d, whereas the lower contact does not move at all. The frictional work is therefore $(-F/2)(0)+(-F/2)(d)$, which is $-Fd/2 = -Fd_{eff}$, so d_{eff} is equal to $d/2$ in the special case of two identical blocks.

Summary of Dry Friction

The fundamental reason why the friction force can act through a distance less than d is that the block is deformable. All atoms in the top block eventually move the same distance d, but because of the stick/slip contact between the blocks the friction force F acts only for a portion of the displacement ($d/2$ in the special case of identical blocks). The point of contact for the friction force does not move in the same way as the center of mass. Note again that the energy equation for the point particle system is not the same as the energy equation for the extended system if the system changes shape.

A Model-Independent Calculation of the Effective Distance

In the special case of a block sliding on an identical block we can calculate d_{eff}, independent of the particular model of the surfaces. Consider a system consisting of both blocks together (Figure 9.45). The bottom block is held stationary by applying a force to the left (to prevent it from being dragged to the right). This constraining force acts through no distance (the bottom block stands still), so this force does no work. The only work done on the two-block system is done by you, of magnitude Fd. So the energy equation for the extended two-block system is this:

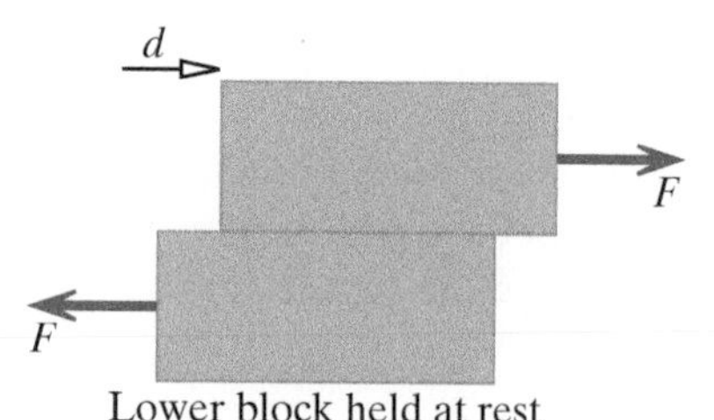

Figure 9.45 The lower block is held at rest, while the upper block slides a distance d. Forces in the vertical direction are not shown.

$$\Delta E_{\text{internal},1} + \Delta E_{\text{internal},2} = Fd$$

Since the two blocks are identical, half of this increased internal energy shows up in the upper block ($Fd/2$), and half in the lower block ($Fd/2$). We can plug this into our earlier energy equation for the upper block:

$$\Delta E_{\text{internal},1} = \frac{Fd}{2} = Fd - Fd_{\text{eff}}$$

QUESTION Calculate d_{eff}.

We find that the effective distance through which the friction force acts is half the distance d that the block moves: $d_{eff} = d/2$. This is in agreement with the model of dry friction with bending teeth that we examined earlier, but our result is a general one that applies to any kind of model of friction surfaces for two symmetrical blocks. Bear in mind that if the blocks are not made of the same material, d_{eff} need not be equal to $d/2$.

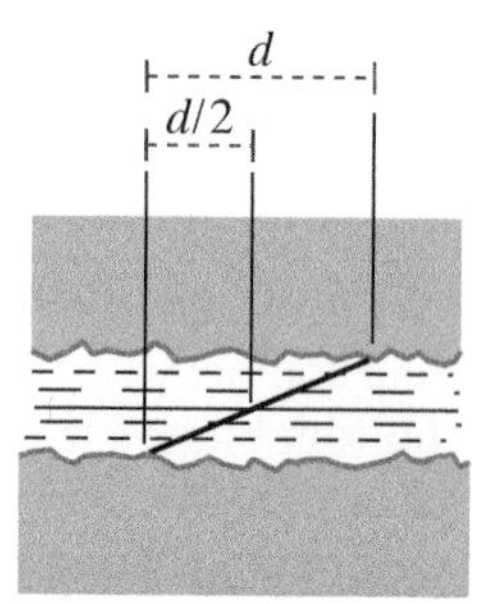

Figure 9.46 Two objects separated by a film of oil, which is modeled as a stack of fluid layers that can move relative to each other.

Lubricated Friction

It is interesting to see that the model-independent result for two identical blocks, $d_{eff} = d/2$, is also consistent with the case of lubricated friction. We separate the identical two blocks with a film of viscous lubricating oil, so that the two blocks do not make direct contact, as shown in Figure 9.46.

It is a property of simple fluid flow that fluid layers immediately adjacent to the blocks are constrained to share the motion of the blocks. Also, for a common type of flow called "laminar" flow, the displacement profile in the oil is linear. In particular, at the midplane the fluid moves half as far as the top block moves. As a result, as the top block is pulled a distance d to the right, the top layer of the oil is dragged along and moves a distance d to the right, the bottom layer of oil doesn't move, and the layer in the midplane (halfway between the blocks) moves a distance $d/2$.

If we take as the symmetrical systems of interest the top block with the upper half of the oil and the bottom block with the lower half of the oil, we see that the shear force between the two systems (at the midplane in the oil) acts through a distance that is again half the displacement of the top block: $d_{eff} = d/2$. (Of course the magnitude of the friction force is much reduced by the lubrication, and the applied force F must be much smaller if the velocity is to be constant.)

This discussion is based on an article "Work and heat transfer in the presence of sliding friction," by B. Sherwood and W. Bernard, *American Journal of Physics*, volume 52, number 11, Nov. 1984, pages 1001–1007, which in turn draws on an earlier article "Real work and pseudowork," by B. Sherwood, *American Journal of Physics,* volume 51, number 7, July 1983, pages 597–602.

9.5 *DERIVATION: KINETIC ENERGY OF A MULTIPARTICLE SYSTEM

In this section we derive the important result that $K_{\text{tot}} = K_{\text{trans}} + K_{\text{rel}}$, where $K_{\text{trans}} = \frac{1}{2}Mv_{\text{CM}}^2$ (for $v_{\text{CM}} \ll c$) and K_{rel} is the kinetic energy relative to the center of mass. This result sounds entirely plausible, but the formal proof is rather difficult.

As in the case of calculating the gravitational energy of a multiparticle object, the derivation of the kinetic energy of a multiparticle system hinges on the definition of the center of mass point of a collection of atoms. The (vector) location of the i-th atom of the object can be expressed as the sum of two vectors, one from the origin to the center of mass ($\vec{r}_{\text{CM}}$) plus another from the center of mass to the i-th atom ($\vec{r}_{i,\text{CM}}$): $\vec{r}_i = \vec{r}_{\text{CM}} + \vec{r}_{i,\text{CM}}$ (Figure 9.47).

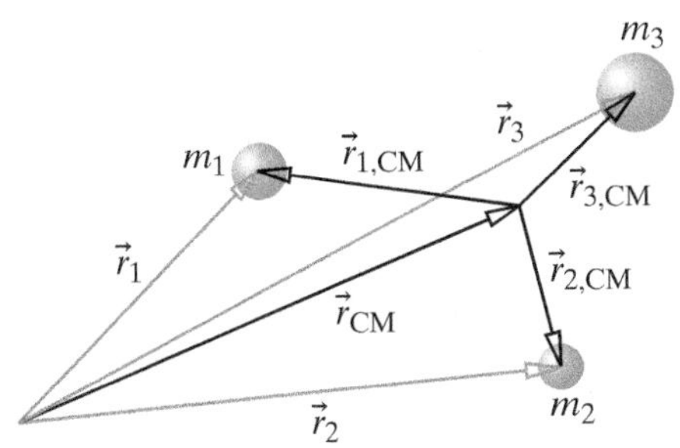

Figure 9.47 Center of mass of a multiparticle system.

The kinetic energy $K_i = \frac{1}{2}mv_i^2$ of the i-th atom is this, where we write $\vec{v}_{i,\text{CM}} = d\vec{r}_{i,\text{CM}}/dt$, the velocity of the i-th particle relative to the center of mass:

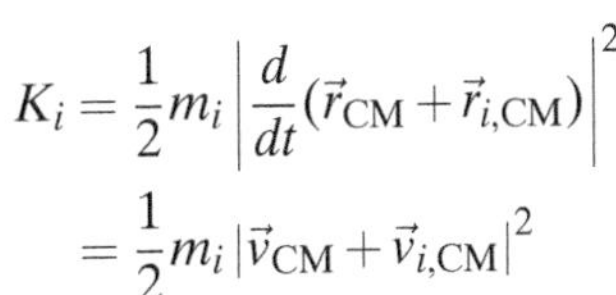

$$K_i = \frac{1}{2}m_i\left|\frac{d}{dt}(\vec{r}_{\text{CM}} + \vec{r}_{i,\text{CM}})\right|^2$$
$$= \frac{1}{2}m_i\left|\vec{v}_{\text{CM}} + \vec{v}_{i,\text{CM}}\right|^2$$

This can be expanded, using the fact that the magnitude squared of a vector can be written as a vector dot product, $|\vec{C}|^2 = \vec{C} \bullet \vec{C} = C^2\cos(0) = C^2$:

$$K_i = \frac{1}{2}m_i[\vec{v}_{\text{CM}} + \vec{v}_{i,\text{CM}}] \bullet [\vec{v}_{\text{CM}} + \vec{v}_{i,\text{CM}}]$$
$$= \frac{1}{2}m_i v_{\text{CM}}^2 + 2\frac{1}{2}m_i\vec{v}_{\text{CM}} \bullet \vec{v}_{i,\text{CM}} + \frac{1}{2}m_i v_{i,\text{CM}}^2$$

Now that we have the kinetic energy of the i-th atom in terms of its velocity $\vec{v}_{i,\text{CM}}$ relative to the center of mass, we need to add up the total kinetic energy of all the atoms. The total kinetic energy K_{tot} can be written in the following compact way, where Σ (Greek capital sigma) means "sum," and the sum goes from $i = 1$ through $i = N$:

$$K_{\text{tot}} = \underbrace{\sum_{i=1}^{N}\frac{1}{2}m_i v_{\text{CM}}^2}_{\textbf{First term}} + \underbrace{\sum_{i=1}^{N}2\left(\frac{1}{2}m_i\right)\vec{v}_{\text{CM}} \bullet \vec{v}_{i,\text{CM}}}_{\textbf{Second term}} + \underbrace{\sum_{i=1}^{N}\frac{1}{2}m_i v_{i,\text{CM}}^2}_{\textbf{Third term}}$$

First Term

The first term in this summation turns out to be the kinetic energy of a single particle of mass M, moving at the speed of the center of mass:

$$\sum_{i=1}^{N}\frac{1}{2}m_i v_{\text{CM}}^2 = \frac{1}{2}v_{\text{CM}}^2\left(\sum_{i=1}^{N}m_i\right) = \frac{1}{2}v_{\text{CM}}^2(m_1 + m_2 + m_3 + \cdots)$$
$$= \frac{1}{2}Mv_{\text{CM}}^2$$

Second Term

The second term can be shown to be zero:

$$\sum_{i=1}^{N}2\left(\frac{1}{2}m_i\right)\vec{v}_{\text{CM}} \bullet \vec{v}_{i,\text{CM}} = \vec{v}_{\text{CM}} \bullet \sum_{i=1}^{N}m_i\vec{v}_{i,\text{CM}} = \vec{v}_{\text{CM}} \bullet \frac{d}{dt}\left(\sum_{i=1}^{N}m_i\vec{r}_{i,\text{CM}}\right) = 0$$

This result follows from the way that we calculate the location of the center of mass of a system:

$$\vec{r}_{CM} = \frac{m_1\vec{r}_1 + m_2\vec{r}_2 + m_3\vec{r}_3 + \cdots}{m_1 + m_2 + m_3 + \cdots} = \frac{\sum_{i=1}^{N} m_i\vec{r}_i}{M}$$

We are measuring the location $\vec{r}_{i,CM}$ of the i-th atom relative to the center of mass, and since the distance from the center of mass to the center of mass is of course zero, we have

$$\vec{0} = \frac{\sum_{i=1}^{N} m_i\vec{r}_{i,CM}}{M} \quad \text{and therefore} \quad \sum_{i=1}^{N} m_i\vec{r}_{i,CM} = \vec{0}$$

Third Term

The third term by definition is the kinetic energy of the atoms relative to the (possibly moving) center of mass:

$$\sum_{i=1}^{N} \frac{1}{2}m_i v_{i,CM}^2 = \frac{1}{2}m_1 v_{1,CM}^2 + \frac{1}{2}m_2 v_{2,CM}^2 + \cdots = K_{rel}$$

Putting the three pieces together, we find that the total kinetic energy splits into two parts: a term associated with the overall motion of the center of mass plus the kinetic energy relative to the center of mass,

$$K_{tot} = K_{trans} + K_{rel}$$

where $K_{trans} = \frac{1}{2}Mv_{CM}^2 = \frac{p_{sys}^2}{2M}$.

9.6 *DERIVATION: THE POINT PARTICLE ENERGY EQUATION

In this chapter we showed that the energy equation for the point particle version of the system follows from the fact that the motion of the center of mass is just like that of a point particle with the total mass of the extended system and subjected to the *net* force acting on the extended system. Here we give a more formal derivation of this important result.

Start from the x component of the Momentum Principle for a multiparticle system whose center of mass is moving at nonrelativistic speed:

$$\frac{dp_{sys,x}}{dt} = M_{tot}\frac{dv_{CM,x}}{dt} = F_{net,x}$$

Integrate through the x displacement of the center of mass:

$$M_{tot}\int_i^f \frac{dv_{CM,x}}{dt}\,dx_{CM} = \int_i^f F_{net,x}\,dx_{CM}$$

Switch dv and dx:

$$M_{\text{tot}} \int_i^f \frac{dx_{\text{CM}}}{dt} dv_{\text{CM},x} = \int_i^f F_{\text{net},x}\, dx_{\text{CM}}$$

However, dx_{CM}/dt is the x component of the center of mass velocity:

$$M_{\text{tot}} \int_i^f v_{\text{CM},x}\, dv_{\text{CM},x} = \int_i^f F_{\text{net},x}\, dx_{\text{CM}}$$

The integral on the left can be carried out:

$$M_{\text{tot}} \left[\frac{1}{2} v_{\text{CM},x}^2\right]_i^f = \int_i^f F_{\text{net},x}\, dx_{\text{CM}}$$

$$\Delta\left[\frac{1}{2} M_{\text{tot}} v_{\text{CM},x}^2\right] = \int_i^f F_{\text{net},x}\, dx_{\text{CM}}$$

We could repeat exactly the same argument for the y and z motions:

$$\Delta\left[\frac{1}{2} M_{\text{tot}} v_{\text{CM},y}^2\right] = \int_i^f F_{\text{net},y}\, dy_{\text{CM}}$$

$$\Delta\left[\frac{1}{2} M_{\text{tot}} v_{\text{CM},z}^2\right] = \int_i^f F_{\text{net},z}\, dz_{\text{CM}}$$

Note that

$$\frac{1}{2} M_{\text{tot}} v_{\text{CM},x}^2 + \frac{1}{2} M_{\text{tot}} v_{\text{CM},y}^2 + \frac{1}{2} M_{\text{tot}} v_{\text{CM},z}^2 = \frac{1}{2} M_{\text{tot}} v_{\text{CM}}^2$$

Moreover,

$$F_{\text{net},x}\, dx_{\text{CM}} + F_{\text{net},y}\, dy_{\text{CM}} + F_{\text{net},z}\, dz_{\text{CM}} = \vec{F}_{\text{net}} \bullet d\vec{r}_{\text{CM}}$$

Adding the three equations together (the sum of the left sides is equal to the sum of the right sides), we have the following equation for the translational kinetic energy of the system:

$$\Delta\left[\frac{1}{2} M_{\text{tot}} v_{\text{CM}}^2\right] = \int_i^f \vec{F}_{\text{net}} \bullet d\vec{r}_{\text{CM}}$$

In words: the change in the translational kinetic energy of a system is equal to the integral of the *net* force acting through the displacement of the center of mass point.

The derivation that we have just carried out shows that although this equation looks like an energy equation, it is actually closely related to the Momentum Principle from which it was derived. The common element is the *net* force. Note that there are three separately valid equations, for x, y, and z, which indicates that we're really dealing with momentum (a vector) rather than energy (a scalar).

In contrast, the actual energy equation for the extended system involves the work done by each individual force through the displacement of the point of application of that force. If the system deforms or rotates, these displacements of the individual forces need not be the same as the displacement of the center of mass.

SUMMARY

Center of mass

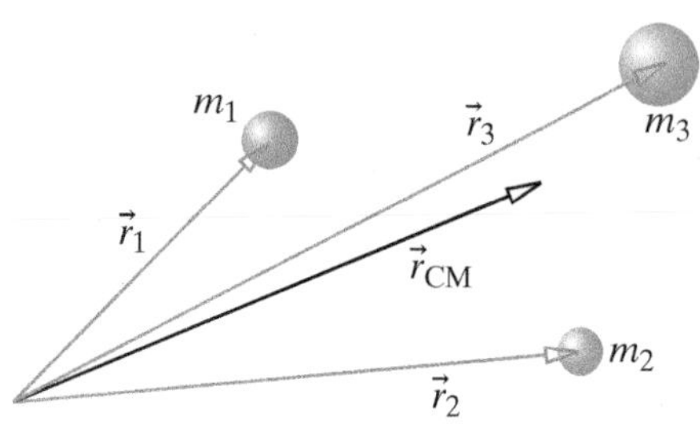

Figure 9.48

$$\vec{r}_{CM} \equiv \frac{m_1\vec{r}_1 + m_2\vec{r}_2 + m_3\vec{r}_3 + \cdots}{m_1 + m_2 + m_3 + \cdots}$$

Problem-solving techniques

Combined use of the energy equations for two different models, the extended system and the point particle system, in analyzing complex phenomena involving multiparticle systems.

Point particle system: Information on change of translational kinetic energy; $\Delta K_{trans} = \vec{F}_{net} \bullet \Delta\vec{r}_{CM}$

Extended system: Information on change of total energy; $\Delta E_{sys} = \sum \vec{F}_i \bullet \Delta\vec{r}_i$

The Momentum Principle for multiparticle systems, derived from Newton's second and third laws of motion, was extended by defining the center of mass:

$$\frac{d\vec{p}_{sys}}{dt} = \vec{F}_{net}$$

$\vec{p}_{sys} \approx M\vec{v}_{CM}$ (if $v \ll c$, and constant mass).

Gravitational energy of multiparticle systems plus Earth, near the Earth's surface:

$$U_g = Mgy_{CM}$$

Kinetic energy of multiparticle systems:

$$K_{tot} = K_{trans} + K_{rel}$$

where $K_{trans} = \frac{p_{sys}^2}{2M} = \frac{1}{2}Mv_{CM}^2$.

Kinetic energy relative to center of mass can be split into two terms:

$$K_{rel} = K_{rot} + K_{vib}$$

Moment of inertia about an axis of rotation:

$$I = m_1 r_{\perp 1}^2 + m_2 r_{\perp 2}^2 + m_3 r_{\perp 3}^2 + \cdots$$

Kinetic energy relative to center of mass:

$$K_{rot} = \frac{1}{2}I\omega^2$$

Moments of inertia:

$$I_{disk} = I_{cylinder} = \frac{1}{2}MR^2$$

about center of disk, rotating around axis of cylinder

$$I_{sphere} = \frac{2}{5}MR^2$$

for axis passing through center of sphere.

Uniform solid cylinder of length L, radius R, about axis perpendicular to cylinder, through center of cylinder:

$$I = \frac{1}{12}ML^2 + \frac{1}{4}MR^2$$

For a rigid object, $K_{tot} = K_{trans} + K_{rot} = \frac{1}{2}(Mr_{CM}^2 + I_{CM})\omega^2$

Point particle system:

$$K_{trans} = \Delta\left(\frac{1}{2}Mv_{CM}^2\right) = \int_i^f \vec{F}_{net,ext} \bullet d\vec{r}_{CM}$$

Sliding friction can deform the contact surfaces, with the result that the frictional force may act through a distance that is different from the distance through which the center-of-mass point moves. As with any deformable system, the energy equations for the extended system and for the point particle system differ (though both are correct).

QUESTIONS

Q1 Discuss qualitatively the motion of the atoms in a block of steel that falls onto another steel block. Why and how do large-scale vibrations damp out?

Q2 Can you give an example of a system that has no atoms located at its center of mass?

Q3 The Momentum Principle for multiparticle systems would seem to say that the center of mass of a system moves just as though the system were a point particle. This led Chris to ask, "Wouldn't that mean that a piece of paper ought to fall in the same way as a small metal ball, if they have the same mass?" Explain carefully to Chris the resolution of this puzzle.

Q4 Consider the voyage to the Moon that you studied in Chapter 3. Would it make any difference, even a very tiny difference, whether the spacecraft is long or short, if the mass is the same? Explain briefly.

Q5 Under what conditions does the energy equation for the point particle system differ from the energy equation for the extended system? Give two examples of such a situation. Give one example of a situation where the two equations look exactly alike.

Q6 Consider the acceleration of a car on dry pavement, if there is no slipping. The axle moves at speed v, and the outside of the tire moves at speed v relative to the axle. The instantaneous velocity of the bottom of the tire is zero. How much work is done by the force exerted on the tire by the road? What is the source of the energy that increases the car's translational kinetic energy?

Q7 Two people with different masses but equal speeds slide toward each other with little friction on ice with their arms extended straight out to the side (so each has the shape of a "T"). Her right hand meets his right hand, they hold hands and spin 90°, then release their holds and slide away. Make a rough sketch of the path of the center of mass of the system consisting of the two people, and explain briefly. (It helps to mark equal time intervals along the paths of the two people and of their center of mass.)

PROBLEMS

Section 9.1

•P8 Three uniform-density spheres are positioned as follows:

- A 3 kg sphere is centered at $\langle 10,20,-5\rangle$ m.
- A 5 kg sphere is centered at $\langle 4,-15,8\rangle$ m.
- A 6 kg sphere is centered at $\langle -7,10,9\rangle$ m.

What is the location of the center of mass of this three-sphere system?

•P9 Relative to an origin at the center of the Earth, where is the center of mass of the Earth–Moon system? The mass of the Earth is 6×10^{24} kg, the mass of the Moon is 7×10^{22} kg, and the distance from the center of the Earth to the center of the Moon is 4×10^{8} m. The radius of the Earth is 6400 km. One can show that the Earth and Moon orbit each other around this center of mass.

•P10 A meter stick whose mass is 300 g lies on ice (Figure 9.49). You pull at one end of the meter stick, at right angles to the stick, with a force of 6 N. The ensuing motion of the meter stick is quite complicated, but what are the initial magnitude and direction of the rate of change of the momentum of the stick, dp_{sys}/dt, when you first apply the force? What is the magnitude of the initial acceleration of the center of the stick?

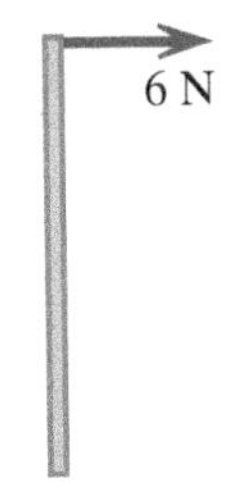

Figure 9.49

•P11 Determine the location of the center of mass of an L-shaped object whose thin vertical and horizontal members have the same length L and the same mass M. Use the formal definition to find the x and y coordinates, and check your result by doing the calculation with respect to two different origins, one in the lower left corner at the intersection of the horizontal and vertical members and one at the right end of the horizontal member.

••P12 A man whose mass is 80 kg and a woman whose mass is 50 kg sit at opposite ends of a canoe 5 m long, whose mass is 30 kg. **(a)** Relative to the man, where is the center of mass of the system consisting of man, woman, and canoe? (*Hint:* Choose a specific coordinate system with a specific origin.) **(b)** Suppose that the man moves quickly to the center of the canoe and sits down there. How far does the canoe move in the water? Explain your work and your assumptions.

•P13 If an object has a moment of inertia 19 kg · m² and rotates with an angular speed of 70 rad/s, what is its rotational kinetic energy?

•P14 A group of particles of total mass 35 kg has a total kinetic energy of 340 J. The kinetic energy relative to the center of mass is 85 J. What is the speed of the center of mass?

••P15 By calculating numerical quantities for a multiparticle system, one can get a concrete sense of the meaning of the relationships $\vec{p}_{sys}=M_{tot}\vec{v}_{CM}$ and $K_{tot}=K_{trans}+K_{rel}$. Consider an object consisting of two balls connected by a spring, whose stiffness is 400 N/m. The object has been thrown through the air and is rotating and vibrating as it moves. At a particular instant the spring is stretched 0.3 m, and the two balls at the ends of the spring have the following masses and velocities:

1: 5 kg, $\langle 8,14,0\rangle$ m/s
2: 3 kg, $\langle -5,9,0\rangle$ m/s

(a) For this system, calculate $\vec{p}_{sys}$. **(b)** Calculate $\vec{v}_{CM}$. **(c)** Calculate K_{tot}. **(d)** Calculate K_{trans}. **(e)** Calculate K_{rel}. **(f)** Here is a way to check your result for K_{rel}. The velocity of a particle relative to the center of mass is calculated by subtracting $\vec{v}_{CM}$ from the particle's velocity. To take a simple example, if you're riding in a car that's moving with $v_{CM,x}=20$ m/s, and you throw a ball with $v_{rel,x}=35$ m/s, relative to the car, a bystander on the ground sees the ball moving with $v_x=55$ m/s. So $\vec{v}=\vec{v}_{CM}+\vec{v}_{rel}$, and therefore we have $\vec{v}_{rel}=\vec{v}-\vec{v}_{CM}$. Calculate $\vec{v}_{rel}=\vec{v}-\vec{v}_{CM}$ for each mass and calculate the corresponding K_{rel}. Compare with the result you obtained in part (e).

••P16 Consider a system consisting of three particles:

$$m_1=2\text{ kg},\ \vec{v}_1=\langle 8,-6,15\rangle\text{ m/s}$$
$$m_2=6\text{ kg},\ \vec{v}_2=\langle -12,9,-6\rangle\text{ m/s}$$
$$m_3=4\text{ kg},\ \vec{v}_3=\langle -24,34,23\rangle\text{ m/s}$$

What is K_{rel}, the kinetic energy of this system relative to the center of mass?

••P17 Binary stars: **(a)** About half of the visible "stars" are actually binary star systems, two stars that orbit each other with no other objects nearby (the small, dim stars called "red dwarfs" however tend to be single). Describe the motion of the center of mass of a binary star system. Briefly explain your reasoning. **(b)** For a particular binary star system, telescopic observations repeated over many years show that one of the stars (whose unknown mass we'll call M_1) has a circular orbit with radius $R_1=6\times10^{11}$ m, while the other star (whose unknown mass we'll call M_2) has a circular orbit of radius $R_2=9\times10^{11}$ m about the same point. Make a sketch of the orbits, and show the positions of the two stars on these orbits at some instant. Label the two stars as to which is which, and label their orbital radii. Indicate on your sketch the location of the center of mass of the system, and explain how you know its location, using the concepts and results of this chapter. **(c)** This double-star system is observed to complete one revolution in 40 years. What are the masses of the two stars? (For comparison, the distance from Sun to Earth is about 1.5×10^{11} m, and the mass of the Sun is about 2×10^{30} kg.) This method is often used to determine the masses of stars. The mass of a star largely determines many of the other properties of a star, which is why astrophysicists need a method for measuring the mass.

Section 9.2

•P18 If an object's rotational kinetic energy is 50 J and it rotates with an angular speed of 12 rad/s, what is its moment of inertia?

•P19 A uniform-density disk of mass 13 kg, thickness 0.5 m, and radius 0.2 m makes one complete rotation every 0.6 s. What is the rotational kinetic energy of the disk?

•P20 A sphere of uniform density with mass 22 kg and radius 0.7 m is spinning, making one complete revolution every 0.5 s. The center of mass of the sphere has a speed of 4 m/s. **(a)** What is the rotational kinetic energy of the sphere? **(b)** What is the total kinetic energy of the sphere?

•P21 A uniform-density disk whose mass is 10 kg and radius is 0.4 m makes one complete rotation every 0.2 s. What is the rotational kinetic energy of the disk?

•P22 A cylindrical rod of uniform density is located with its center at the origin, and its axis along the x axis. It rotates about its center in the xy plane, making one revolution every 0.03 s. The rod has a radius of 0.08 m, length of 0.7 m, and mass of 5 kg. It makes one revolution every 0.03 s. What is the rotational kinetic energy of the rod?

•P23 A uniform-density 6 kg disk of radius 0.3 m is mounted on a nearly frictionless axle. Initially it is not spinning. A string is wrapped tightly around the disk, and you pull on the string with a constant force of 25 N through a distance of 0.6 m. Now what is the angular speed?

••P24 A high diver tucks himself into a ball and spins rapidly (Figure 9.50). Make estimates of the relevant parameters and calculate K_{rot}, the rotational kinetic energy of the diver.

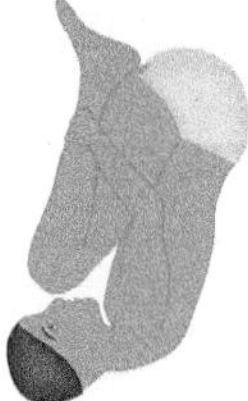

Figure 9.50

••P25 The Earth is 1.5×10^{11} m from the Sun and takes a year to make one complete orbit. It rotates on its own axis once per day. It can be treated approximately as a uniform-density sphere of mass 6×10^{24} kg and radius 6.4×10^{6} m (actually, its center has higher density than the rest of the planet, and the Earth bulges out a bit at the equator). Using this crude approximation, calculate the following: **(a)** What is v_{CM}? **(b)** What is K_{trans}? **(c)** What is ω, the angular speed of rotation around its own axis? **(d)** What is K_{rot}? **(e)** What is K_{tot}?

••P26 Show that the moment of inertia of a disk of mass M and radius R is $\frac{1}{2}MR^2$. Divide the disk into narrow rings, each of radius r and width dr. The contribution to I by one of these rings is simply r^2dm, where dm is the amount of mass contained in that particular ring. The mass of any ring is the total mass times the fraction of the total area occupied by the area of the ring. The area of this ring is approximately $2\pi r dr$. Use integral calculus to add up all the contributions.

Section 9.3

•P27 You pull straight up on the string of a yo-yo with a force 0.235 N, and while your hand is moving up a distance 0.18 m, the yo-yo moves down a distance 0.70 m. The mass of the yo-yo is 0.025 kg, and it was initially moving downward with speed 0.5 m/s and angular speed 124 rad/s. **(a)** What is the increase in the translational kinetic energy of the yo-yo? **(b)** What is the new speed of the yo-yo? **(c)** What is the increase in the rotational kinetic energy of the yo-yo? **(d)** The yo-yo is approximately a uniform-density disk of radius 0.02 m. What is the new angular speed of the yo-yo?

••P28 A string is wrapped around a disk of mass 2.1 kg (its density is not necessarily uniform). Starting from rest, you pull the string with a constant force of 9 N along a nearly frictionless surface. At the instant when the center of the disk has moved a distance 0.11 m, your hand has moved a distance of 0.28 m (Figure 9.51).

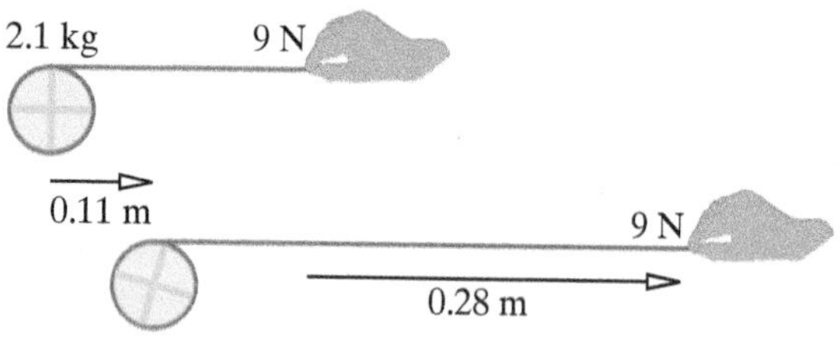

Figure 9.51

(a) At this instant, what is the speed of the center of mass of the disk? **(b)** At this instant, how much rotational kinetic energy does the disk have relative to its center of mass? **(c)** At this instant, the angular speed of the disk is 7.5 rad/s. What is the moment of inertia of the disk?

••P29 A chain of metal links with total mass $M = 7$ kg is coiled up in a tight ball on a low-friction table (Figure 9.52). You pull on a link at one end of the chain with a constant force $F = 50$ N. Eventually the chain straightens out to its full length $L = 2.6$ m, and you keep pulling until you have pulled your end of the chain a total distance $d = 4.5$ m.

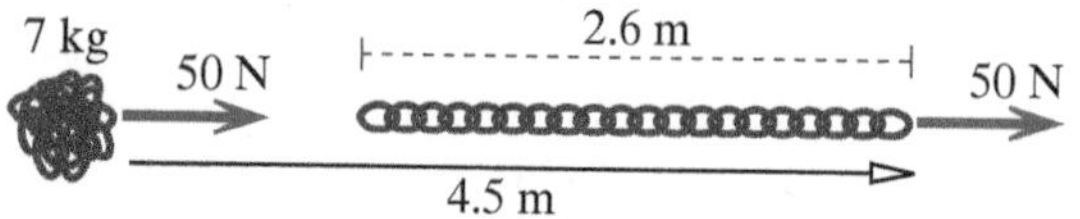

Figure 9.52

(a) Consider the point particle system. What is the speed of the chain at this instant? **(b)** Consider the extended system. What is the change in energy of the chain? **(c)** In straightening out, the links of the chain bang against each other, and their temperature rises. Assume that the process is so fast that there is insufficient time for significant transfer of energy from the chain to the table due to the temperature difference, and ignore the small amount of energy radiated away as sound produced in the collisions among the links. Calculate the increase in internal energy of the chain.

••**P30** Tarzan, whose mass is 100 kg, is hanging at rest from a tree limb. Then he lets go and falls to the ground. Just before he lets go, his center of mass is at a height 2.9 m above the ground and the bottom of his dangling feet are at a height 2.1 above the ground. When he first hits the ground he has dropped a distance 2.1, so his center of mass is (2.9 − 2.1) above the ground. Then his knees bend and he ends up at rest in a crouched position with his center of mass a height 0.5 above the ground. **(a)** Consider the point particle system. What is the speed v at the instant just before Tarzan's feet touch the ground? **(b)** Consider the extended system. What is the net change in internal energy for Tarzan from just before his feet touch the ground to when he is in the crouched position?

••**P31** Here is an experiment on jumping up you can do. **(a)** Crouch down and jump straight up, as high as you can. Estimate the location of your center of mass, and measure its height at three stages in this process: in the crouch, at lift-off, and at the top of the jump. Report your measurements. You may need to have a friend help you make the measurements. **(b)** Analyze this process as fully as possible, using all the theoretical tools now available to you, especially the concepts in this chapter. Include a calculation of the average force of the floor on your feet, the change in your internal energy, and the approximate time of contact from the beginning of the jump to lift-off. Be sure to explain clearly what approximations and simplifying assumptions you made in modeling the process.

••**P32** Consider an ice skater who pushes away from a wall. **(a)** Estimate the speed she can achieve just after pushing away from the wall. Then estimate the average acceleration during this process. How many g's is this? (That is, what fraction or multiple of 9.8 m/s^2 is your estimate?) Be sure to explain clearly what approximations and simplifying assumptions you made in modeling the process. **(b)** For this process, choose the woman as the system of interest and discuss the energy transfers, and the changes in the various forms of energy. Estimate the amount of each of these, including the correct signs.

••**P33** A hoop of mass M and radius R rolls without slipping down a hill, as shown in Figure 9.53. The lack of slipping means that when the center of mass of the hoop has speed v, the tangential speed of the hoop relative to the center of mass is also equal to v_{CM}, since in that case the instantaneous speed is zero for the part of the hoop that is in contact with the ground ($v - v = 0$). Therefore, the angular speed of the rotating hoop is $\omega = v_{CM}/R$.

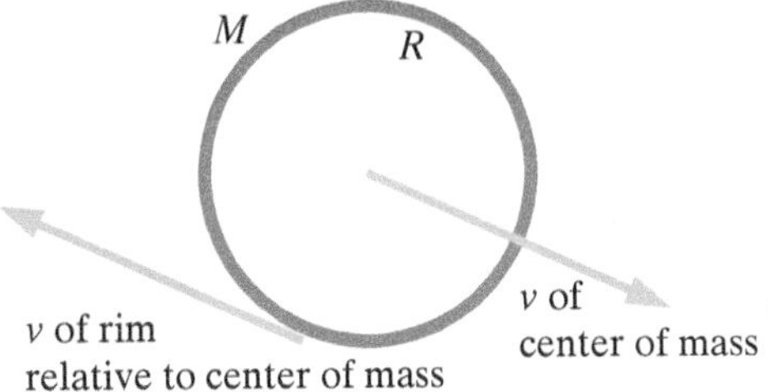

Figure 9.53

(a) The initial speed of the hoop is v_i, and the hill has a height h. What is the speed v_f at the bottom of the hill? **(b)** Replace the hoop with a bicycle wheel whose rim has mass M and whose hub has mass m, as shown in Figure 9.54. The spokes have negligible mass. What would be the bicycle wheel's speed at the bottom of the hill?

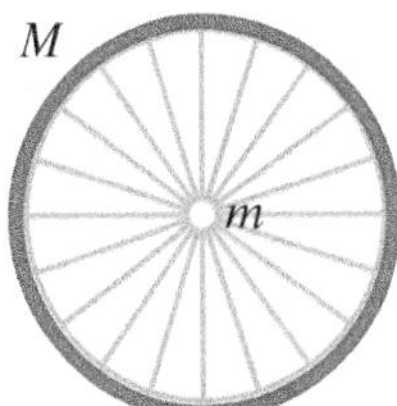

Figure 9.54

••**P34** A sphere or cylinder of mass M, radius R, and moment of inertia I rolls without slipping down a hill of height h, starting from rest. As explained in Problem P33, if there is no slipping $\omega = v_{CM}/R$. **(a)** In terms of the given variables (M, R, I, and h), what is v_{CM} at the bottom of the hill? **(b)** If the object is a thin hollow cylinder, what is v_{CM} at the bottom of the hill? **(c)** If the object is a uniform-density solid cylinder, what is v_{CM} at the bottom of the hill? **(d)** If the object is a uniform-density sphere, what is v_{CM} at the bottom of the hill? An interesting experiment that you can perform is to roll various objects down an inclined board and see how much time each one takes to reach the bottom.

••**P35** Two disks are initially at rest, each of mass M, connected by a string between their centers, as shown in Figure 9.55. The disks slide on low-friction ice as the center of the string is pulled by a string with a constant force F through a distance d. The disks collide and stick together, having moved a distance b horizontally.

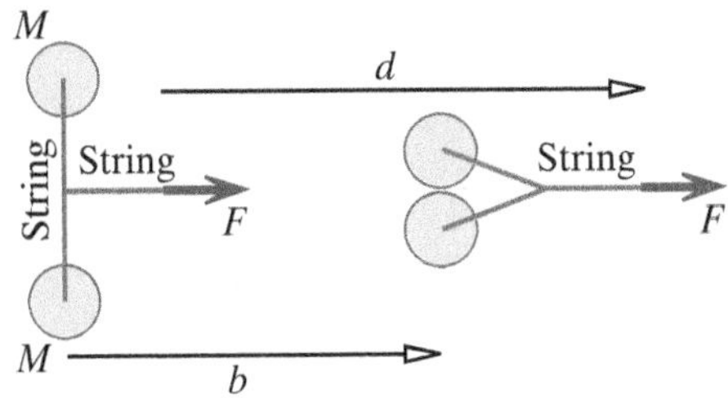

Figure 9.55

(a) What is the final speed of the stuck-together disks? **(b)** When the disks collide and stick together, their temperature rises. Calculate the increase in internal energy of the disks, assuming that the process is so fast that there is insufficient time for there to be much transfer of energy to the ice due to a temperature difference. (Also ignore the small amount of energy radiated away as sound produced in the collisions between the disks.)

••**P36** You hang by your hands from a tree limb that is a height L above the ground, with your center of mass a height h above the ground and your feet a height d above the ground, as shown in Figure 9.56. You then let yourself fall. You absorb the shock by bending your knees, ending up momentarily at rest in a crouched position with your center of mass a height b above the ground. Your mass is M. You will need to draw labeled physics diagrams for the various stages in the process.

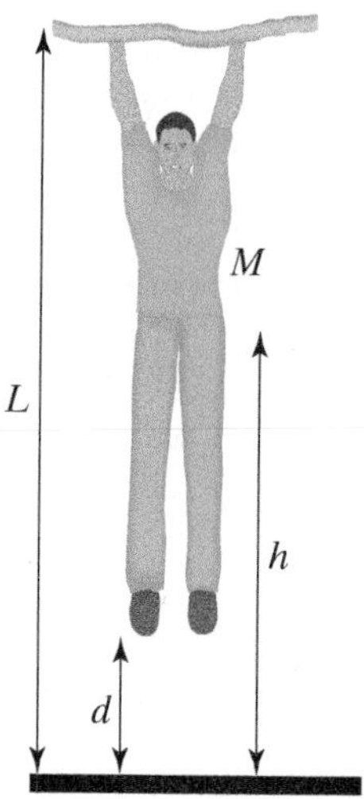

Figure 9.56

(a) What is the net internal energy change ΔE_{int} in your body (chemical plus thermal)? **(b)** What is your speed v at the instant your feet first touch the ground? **(c)** What is the approximate average force F exerted by the ground on your feet during the time when your knees are bending? **(d)** How much work is done by this force F?

••P37 A box and its contents have a total mass M. A string passes through a hole in the box (Figure 9.57), and you pull on the string with a constant force F (this is in outer space—there are no other forces acting).

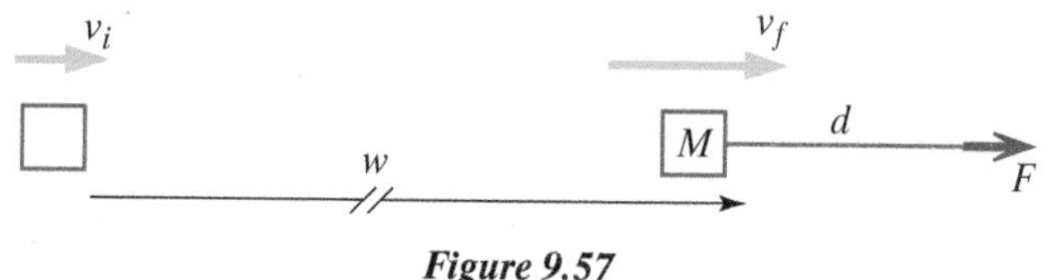

Figure 9.57

(a) Initially the speed of the box was v_i. After the box had moved a long distance w, your hand had moved an additional distance d (a total distance of $w+d$), because additional string of length d came out of the box. What is now the speed v_f of the box? **(b)** If we could have looked inside the box, we would have seen that the string was wound around a hub that turns on an axle with negligible friction, as shown in Figure 9.58. Three masses, each of mass m, are attached to the hub at a distance r from the axle. Initially the angular speed relative to the axle was ω_i. In terms of the given quantities, what is the final angular speed relative to the axis, ω_f?

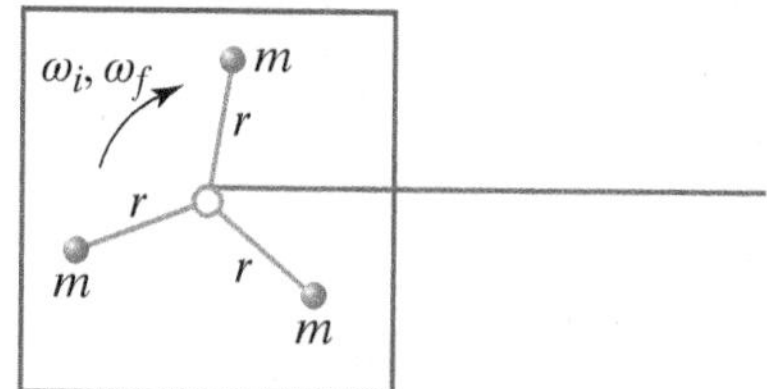

Figure 9.58

••P38 Two identical 0.4 kg blocks (labeled 1 and 2) are initially at rest on a nearly frictionless surface, connected by an unstretched spring, as shown in the upper portion of Figure 9.59.

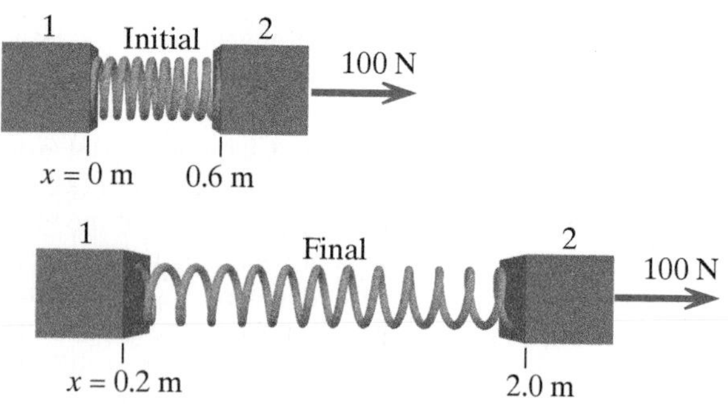

Figure 9.59

Then a constant force of 100 N to the right is applied to block 2, and at a later time the blocks are in the new positions shown in the lower portion of Figure 9.59. At this final time, the system is moving to the right and also vibrating, and the spring is stretched. **(a)** The following questions apply to the system modeled as a point particle. (i) What is the initial location of the point particle? (ii) How far does the point particle move? (iii) How much work was done on the particle? (iv) What is the change in translational kinetic energy of this system? **(b)** The following questions apply to the system modeled as an extended object. (1) How much work is done on the right-hand block? (2) How much work is done on the left-hand block? (3) What is the change of the total energy of this system? **(c)** Combine the results of both models to answer the following questions. (1) Assuming that the object does not get hot, what is the final value of $K_{\text{vib}} + U_{\text{spring}}$ for the extended system? (2) If the spring stiffness is 50 N/m, what is the final value of the vibrational kinetic energy?

••P39 You hold up an object that consists of two blocks at rest, each of mass $M = 5$ kg, connected by a low-mass spring. Then you suddenly start applying a larger upward force of constant magnitude $F = 167$ N (which is greater than $2Mg$). Figure 9.60 shows the situation some time later, when the blocks have moved upward, and the spring stretch has increased.

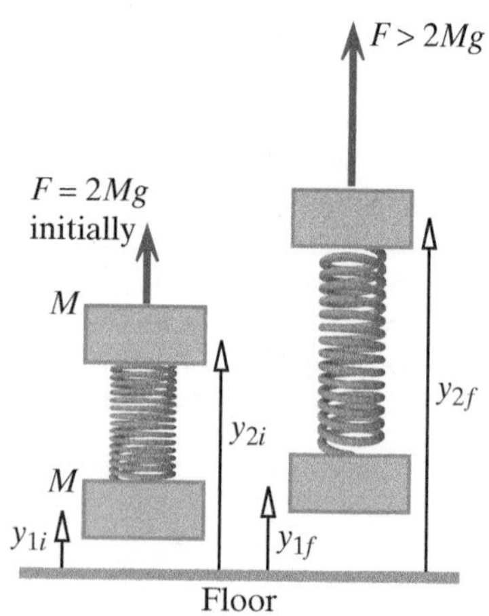

Figure 9.60

The heights of the centers of the two blocks are as follows:

Initial and final positions of block 1: $y_{1i} = 0.3$ m, $y_{1f} = 0.5$ m
Initial and final positions of block 2: $y_{2i} = 0.7$ m, $y_{2f} = 1.2$ m

It helps to show these heights on a diagram. Note that the initial center of mass of the two blocks is $(y_{1i} + y_{2i})/2$, and the final center of mass of the two blocks is $(y_{1f} + y_{2f})/2$. **(a)** Consider the point particle system corresponding to the two blocks and the spring. Calculate the increase in the total translational kinetic energy of the two blocks. It is important to draw a diagram showing all of the forces that are acting, and through what distance each force acts. **(b)** Consider the extended system corresponding to the two blocks and the spring. Calculate the increase of $(K_{vib} + U_s)$, the vibrational kinetic energy of the two blocks (their kinetic energy relative to the center of mass) plus the potential energy of the spring. It is important to draw a diagram showing all of the forces that are acting, and through what distance each force acts.

••P40 A box contains machinery that can rotate. The total mass of the box plus the machinery is 7 kg. A string wound around the machinery comes out through a small hole in the top of the box. Initially the box sits on the ground, and the machinery inside is not rotating (left side of Figure 9.61). Then you pull upward on the string with a force of constant magnitude 130 N. At an instant when you have pulled 0.6 m of string out of the box (indicated on the right side of Figure 9.61), the box has risen a distance of 0.2 m and the machinery inside is rotating.

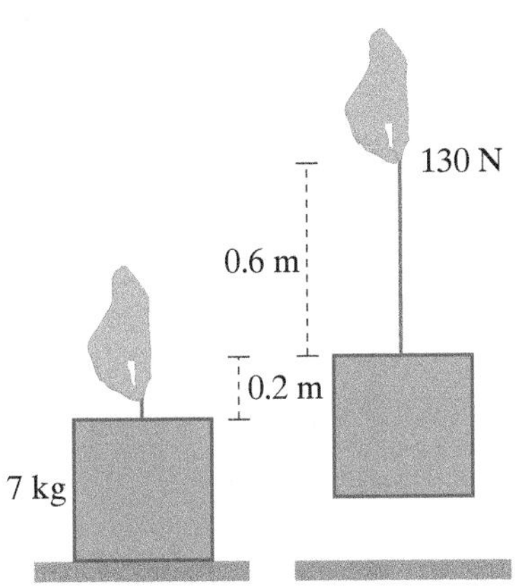

Figure 9.61

POINT PARTICLE SYSTEM **(a)** List all the forms of energy that change for the point particle system during this process. **(b)** What is the y component of the displacement of the point particle system during this process? **(c)** What is the y component of the net force acting on the point particle system during this process? **(d)** What is the distance through which the net force acts on the point particle system? **(e)** How much work is done on the point particle system during this process? **(f)** What is the speed of the box at the instant shown in the right side of Figure 9.61? **(g)** Why is it not possible to find the rotational kinetic energy of the machinery inside the box by considering only the point particle system?

EXTENDED SYSTEM **(h)** The extended system consists of the box, the machinery inside the box, and the string. List all the forms of energy that change for the extended system during this process. **(i)** What is the translational kinetic energy of the extended system, at the instant shown in the right side of Figure 9.61? **(j)** What is the distance through which the gravitational force acts on the extended system? **(k)** How much work is done on the system by the gravitational force? **(l)** What is the distance through which your hand moves? **(m)** How much work do you do on the extended system? **(n)** At the instant shown in the right side of Figure 9.61, what is the total kinetic energy of the extended system? **(o)** What is the rotational kinetic energy of the machinery inside the box?

••P41 String is wrapped around an object of mass M and moment of inertia I (the density of the object is not uniform). With your hand you pull the string straight up with some constant force F such that the center of the object does not move up or down, but the object spins faster and faster (Figure 9.62). This is like a yo-yo; nothing but the vertical string touches the object.

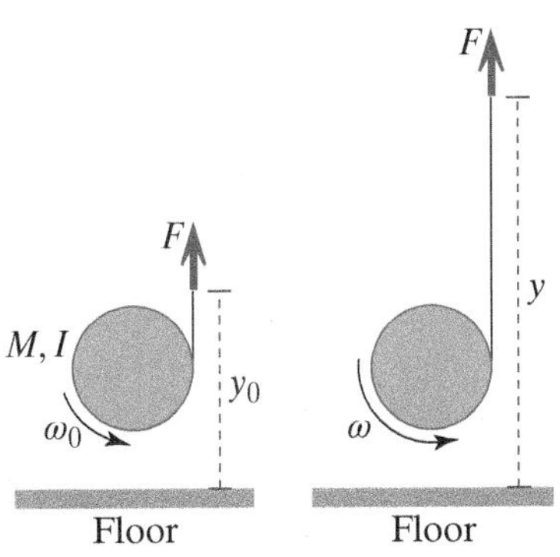

Figure 9.62

When your hand is a height y_0 above the floor, the object has an angular speed ω_0. When your hand has risen to a height y above the floor, what is the angular speed ω of the object?

Your result should not contain F or the (unknown) radius of the object. Explain the physics principles you are using.

••P42 String is wrapped around an object of mass 1.2 kg and moment of inertia 0.0015 $kg \cdot m^2$ (the density of the object is not uniform). With your hand you pull the string straight up with some constant force F such that the center of the object does not move up or down, but the object spins faster and faster (Figure 9.62). This is like a yo-yo; nothing but the vertical string touches the object. When your hand is a height $y_0 = 0.25$ m above the floor, the object has an angular speed $\omega_0 = 12$ rad/s. When your hand has risen to a height $y = 0.35$ m above the floor, what is the angular speed ω of the object? Your answer must be numeric and not contain the symbol F.

••P43 A string is wrapped around a uniform disk of mass M and radius R. Attached to the disk are four low-mass rods of radius b, each with a small mass m at the end (Figure 9.63).

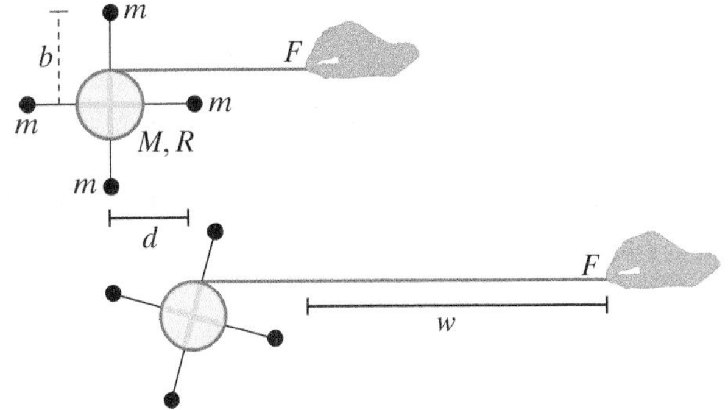

Figure 9.63

The apparatus is initially at rest on a nearly frictionless surface. Then you pull the string with a constant force F. At the instant when the center of the disk has moved a distance d, an additional length w of string has unwound off the disk. **(a)** At this instant, what is the speed of the center of the apparatus? Explain your approach. **(b)** At this instant, what is the angular speed of the apparatus? Explain your approach.

••P44 A string is wrapped around a uniform disk of mass $M = 1.2$ kg and radius $R = 0.11$ m (Figure 9.63). Attached to the disk are four low-mass rods of radius $b = 0.14$ m, each with a small mass $m = 0.4$ kg at the end. The device is initially at rest on a nearly frictionless surface. Then you pull the string with a constant force $F = 21$ N. At the instant that the center of the disk has moved a distance $d = 0.026$ m, an additional length $w = 0.092$ m of string has unwound off the disk. **(a)** At this instant, what is the speed of the center of the apparatus? Explain your approach. **(b)** At this instant, what is the angular speed of the apparatus? Explain your approach.

••P45 A rod of length L and negligible mass is attached to a uniform disk of mass M and radius R (Figure 9.64). A string is wrapped around the disk, and you pull on the string with a constant force F. Two small balls each of mass m slide along the rod with negligible friction. The apparatus starts from rest, and when the center of the disk has moved a distance d, a length of string s has come off the disk, and the balls have collided with the ends of the rod and stuck there. The apparatus slides on a nearly frictionless table. Here is a view from above:

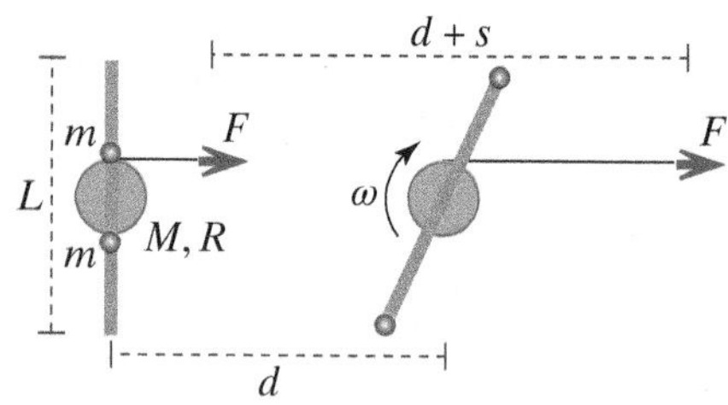

Figure 9.64

(a) At this instant, what is the speed v of the center of the disk? **(b)** At this instant the angular speed of the disk is ω. How much internal energy change has there been?

Section 9.4

•••P46 It is sometimes claimed that friction forces always slow an object down, but this is not true. If you place a box of mass M on a moving horizontal conveyor belt, the friction force of the belt acting on the bottom of the box speeds up the box. At first there is some slipping, until the speed of the box catches up to the speed v of the belt. The coefficient of friction between box and belt is μ. **(a)** What is the distance d (relative to the floor) that the box moves before reaching the final speed v? Use energy arguments, and explain your reasoning carefully. **(b)** How much time does it take for the box to reach its final speed? **(c)** The belt and box of course get hot. Is the effective distance through which the friction force acts on the box greater than or less than d? Give as quantitative an argument as possible. You can assume that the process is quick enough that you can neglect transfer of energy Q due to a temperature difference between the belt and the box. Do not attempt to use the *results* of the friction analysis in this chapter; rather, apply the *methods* of that analysis to this different situation. **(d)** Explain the result of part (c) qualitatively from a microscopic point of view, including physics diagrams.

ANSWERS TO CHECKPOINTS

1 39.2 J
2 (a) 61.3 J; **(b)** 99.7 J
3 316 J
4 (a) v/d; **(b)** $\frac{1}{2}(\frac{2}{5}MR^2)(v/d)^2$; **(c)** $\frac{1}{2}(Md^2 + \frac{2}{5}MR^2)(v/d)^2$

5 (a) 167 N; **(b)** zero displacement, so no work done on the extended system, but on the point particle system there is 2500 J of work done; **(c)** internal energy (especially chemical energy) decreases by 2500 J

CHAPTER 10

Collisions

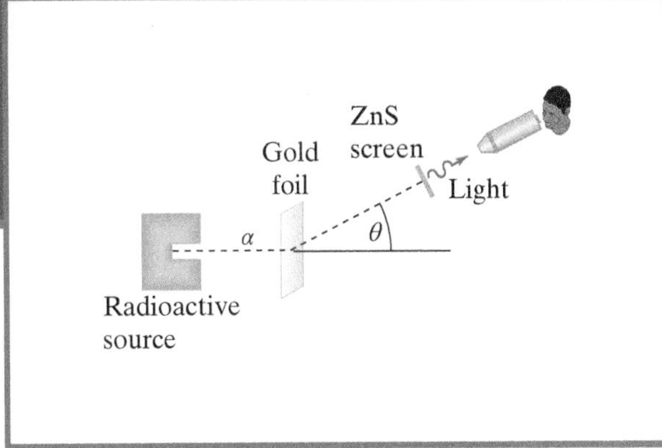

OBJECTIVES

After studying this chapter, you should be able to

- Compare the information provided by an analytical treatment of a collision to the information obtained from an iterative computational model of the process.
- Distinguish between elastic and inelastic collisions.
- Analyze both elastic and inelastic collisions using a combination of the Energy Principle and the Momentum Principle.
- Analyze collisions in multiple reference frames.

10.1 COLLISIONS

Collisions are common events. Neutral molecules in a gas frequently collide with each other—the interatomic electric force is effectively short range, so there is little interaction until the molecules come nearly into contact. Similarly, collisions between macroscopic objects are quite common: before and after a ball hits a wall it has little interaction with the wall, and the contact between ball and wall lasts a very short time.

Colliding objects need not come into actual contact during the collision, if they interact via long-range forces. Two magnets can repel each other without touching, yet we can treat the interaction between fast-moving magnets as a "collision." A collision is any process in which there is little interaction before and after a short time interval, and large interactions during that short time interval. Since the electric force acts at a distance, a moving electron can collide with a moving proton without physical contact, as shown in Figure 10.1.

Electron

Proton

Figure 10.1 A collision between moving charged particles, such as a proton and an electron, need not involve actual contact.

In Chapters 2 and 3 we applied the Momentum Principle to collisions, interactions involving large forces acting for a short time. Many kinds of collisions cannot be fully analyzed without also using the Energy Principle, and this chapter deals with such situations. We will also see that depending on the information we have and on what we want to know about a collision, an analytical solution may not be possible at all, and we will need to model the process computationally.

Choice of System: Both Objects

In analyzing a collision, we typically choose both colliding objects as the system. With this choice of system, the large forces between the colliding objects are internal forces. Because these internal forces are very large—much larger than forces due to objects in the surroundings—the effect of external forces is negligible during the collision. Since a collision occurs during a brief time, energy flow due to any temperature differences between system and surroundings is negligible also. Because we can neglect interactions with the surroundings, both the total momentum and the total energy of the system must remain constant during the collision, although both momentum and energy can flow between the objects in the system.

$$\begin{aligned} \Delta\vec{p}_{\text{sys}} + \cancel{\Delta\vec{p}_{\text{surr}}} &= 0 \\ \Delta\vec{p}_{\text{sys}} &= 0 \\ \vec{p}_f &= \vec{p}_i \end{aligned}$$

$$\begin{aligned} \Delta E_{\text{sys}} + \cancel{\Delta E_{\text{surr}}} &= 0 \\ \Delta E_{\text{sys}} &= 0 \\ E_f &= E_i \end{aligned}$$

External Forces Are Negligible

In Chapter 2 we examined the forces and interaction times involved in two collisions. When a hockey stick struck a hockey puck hard enough that the stick broke, the force exerted by the stick on the puck was about 1000 N, and the interaction lasted only about 0.005 s. (For comparison, the gravitational force by the Earth on a 160 g hockey puck is only about 1.6 N.) We also analyzed a head-on collision between two running students; we found the interaction time to be about 0.02 s, and the force exerted by one student on the other to be about 21000 N—35 times the weight of one student. Such brief, intense interactions are characteristic of collisions.

These characteristics justify an approach to the analysis of collisions that proves fruitful. Up until the collision, and after the collision, we do need to take into account external forces, which affect the total momentum of the system. For example, consider the collision of two baseballs in midair, shown in Figure 10.2. We can tell from the curving trajectories of the balls that the gravitational force by the Earth had a significant effect on the balls before and after the collision. However, during the very short time of the collision, the Earth's gravitational force had a negligible effect, and there was little time for energy transfer from the balls to the air due to a temperature difference, so both the total momentum and total energy of the system were nearly constant during the impact.

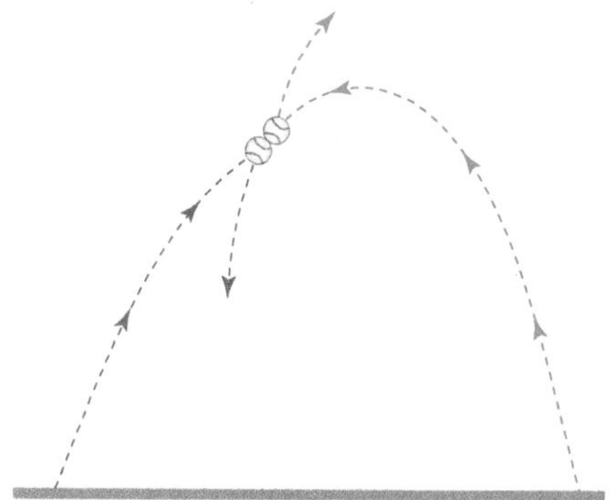

Figure 10.2 Midair collision between two baseballs.

Limitations of This Approach

In the following sections we'll see that applying the Momentum Principle and the Energy Principle will allow us to analyze several interesting types of collisions using an analytical (algebraic) approach. However, not all problems can be solved this way. We'll explore the power and the limitations of this approach in a later section.

10.2 ELASTIC AND INELASTIC COLLISIONS

Often we want to focus on the changes in translational kinetic energy that occur during a collision between two objects. We will use the term "internal energy" to refer to the other kinds of energy objects may have—for example, electronic energy, vibrational energy, rotational energy, or the randomized thermal energy of a solid object whose atoms and chemical bonds are vibrating.

Elastic Collisions: Internal Energy Doesn't Change

We call a collision "elastic" if there is no change in internal energy of the interacting objects: no thermal energy rise (associated with a temperature increase), no springs newly compressed, no lasting deformations, no new rotations or vibrations, and so on.

ELASTIC COLLISION

In an elastic collision, the internal energy of the objects in the system does not change: $\Delta E_{\text{int}} = 0$. Since no kinetic energy is converted to E_{int}, in an elastic collision $K_f = K_i$. In any collision $\vec{p}_f = \vec{p}_i$, because external forces are negligible.

When atomic systems with quantized energies collide, the collision will be perfectly elastic if there is insufficient energy available to raise the systems to excited quantum states (Figure 10.3). For example, as discussed in Chapter 8, the energy of the first excited electronic state of a mercury atom is 4.9 eV above the ground state. If an electron whose kinetic energy is only 3 eV collides with a ground-state mercury atom, the collision will be elastic, because there is not enough energy available to change the internal energy of the mercury atom.

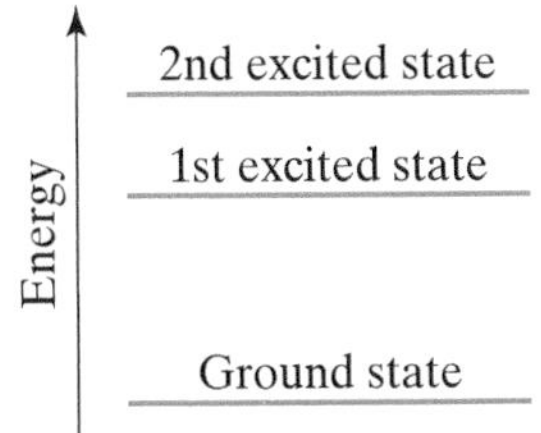

Figure 10.3 Energy levels in an atomic system. If the energy input is too small to reach an excited state, the energy cannot be absorbed.

With macroscopic systems there are no perfectly elastic collisions, because there is always some dissipation, but many macroscopic collisions are very nearly elastic. For example, collisions between billiard balls or steel balls can be nearly elastic.

Carts used to demonstrate collisions in physics lectures often contain magnets to repel each other without actually making contact, and such collisions are nearly elastic. Collisions between carts that interact through soft springs may also be nearly elastic.

Inelastic Collisions: Internal Energy Does Change

We call a collision "inelastic" if it isn't elastic; that is, there is some change in the internal energy of the colliding objects. They get hot, or deform, or rotate, or vibrate, and so on.

INELASTIC COLLISION

In an inelastic collision, the internal energy of the objects in the system changes: $\Delta E_{\text{int}} \neq 0$. Since some kinetic energy is converted to internal energy, in an inelastic collision $K_f \neq K_i$. In any collision $\vec{p}_f = \vec{p}_i$, because external forces are negligible.

As stated above, collisions of macroscopic objects are always at least a bit inelastic, but in many common situations the collision may be nearly elastic. When atomic systems collide, one or more of the colliding objects may undergo a change of quantized energy level, and we call that an inelastic collision. For example, if an electron with kinetic energy 5 eV collides with a mercury atom in its ground state, the mercury atom may be excited to a higher electronic state 4.9 eV above the ground state, leaving only 0.1 eV of kinetic energy in the system.

Maximally Inelastic Collisions ("Sticking Collisions")

The most extreme inelastic collision is the "maximally inelastic" collision, in which there is maximum dissipation. That doesn't necessarily mean that the objects stop dead, because momentum must be conserved. It can be shown that a collision in which the objects stick together is a maximally inelastic collision: the only remaining kinetic energy is present only because the total momentum can't change.

The simplest example of a maximally inelastic collision is that of two objects that have equal and opposite momenta, so that before the collision the total momentum of the combined system is zero. The momentum afterward must also be zero, and if they are stuck together that means that they must be at rest.

Identifying Inelastic Collisions

When deciding whether or not it is reasonable to treat a collision as approximately elastic, one can look for indications of inelasticity, such as:

- Objects are stuck together after the collision.
- An object is deformed after the collision.
- Objects are hotter after the collision.
- There is more vibration or rotation after the collision.
- An object is in an excited state after the collision.

10.3 A HEAD-ON COLLISION OF EQUAL MASSES

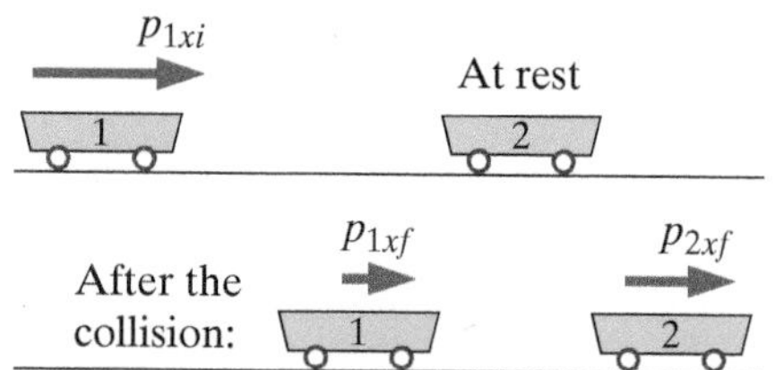

Figure 10.4 Cart 1 runs into stationary cart 2. The carts have equal masses. Friction and air resistance are negligible.

We will start by considering simple 1-D (head-on) collisions in order to see clearly how to apply both the Energy Principle and the Momentum Principle. Consider a head-on collision between two carts rolling or sliding on a track with low friction (Figure 10.4). Similar situations are billiard balls or hockey pucks or vehicles hitting each other head-on. Cart 1 with mass m moves to the right with x momentum p_{1xi}, and it runs into cart 2, which has the same mass m and is initially sitting still.

Taking the two carts as the system, and neglecting the small frictional force exerted by the track, all that the Momentum Principle can tell us is that after the collision the total final x momentum $p_{1xf}+p_{2xf}$ must equal the initial total x momentum p_{1xi}. By itself, the Momentum Principle doesn't tell us how this momentum will be divided between the two carts. We need to consider conservation of energy as well as conservation of momentum in order to predict what will happen.

Before the collision, nonzero energy terms included the kinetic energy of cart 1, K_{1i}, and the internal energies of both carts. After the collision, there is internal energy of both carts and kinetic energy of both carts, $K_{1f}+K_{2f}$.

> QUESTION What types of energy may have changed as a result of the collision?

K_1+K_2 might change, and internal energy may have changed. The temperature of the carts might be a bit higher now, with some of the initial kinetic energy having been dissipated into thermal energy. How much energy dissipation occurs depends on the details of the contact between the two interacting carts.

EXAMPLE **Elastic Collision of Two Identical Carts**

Cart 1 collides with stationary cart 2, which is identical (Figure 10.4). Suppose that the collision is (nearly) elastic, as it will be if the carts repel each other magnetically or interact through soft springs. In this case there is no change of internal energy. What are the final momenta of the two carts?

Solution Since the y and z components of momentum don't change, we can work with only x components.

System: Both carts
Surroundings: Earth, track, air (neglect friction and air resistance)

Initial situation: Just before collision
Final situation: Just after collision

Momentum Principle (negligible effect of surroundings; only x components change):

$$p_{xf} = p_{xi} + F_{\text{net},x}\Delta t$$
$$p_{1xf} + p_{2xf} = p_{1xi} + 0$$

Energy Principle (negligible effect of surroundings; negligible change in internal energy of the carts):

$$E_f = E_i + W + Q$$
$$(K_{1f} + \cancel{E_{\text{int1}f}}) + (K_{2f} + \cancel{E_{\text{int2}f}}) = (K_{1i} + \cancel{E_{\text{int1}i}}) + (K_{2i} + \cancel{E_{\text{int2}i}}) + 0 + 0$$
$$K_{1f} + K_{2f} = K_{1i} + 0$$

Since $K = p^2/2m$, we can combine the momentum and energy equations:

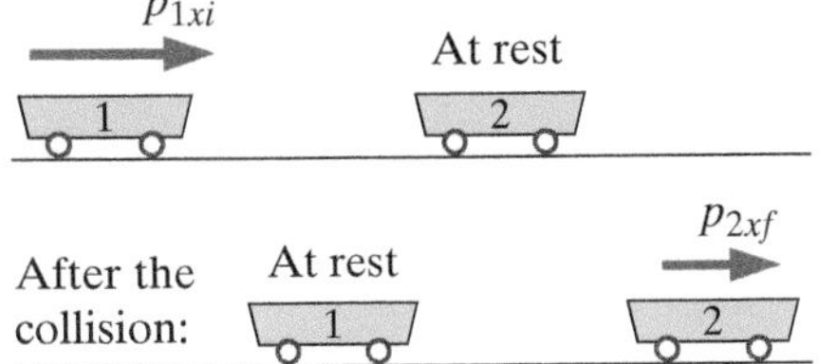

Figure 10.5 If the collision is elastic, cart 1 stops, and cart 2 has all the momentum cart 1 used to have.

$$\frac{p_{1xf}^2}{2m} + \frac{p_{2xf}^2}{2m} = \frac{(p_{1xf} + p_{2xf})^2}{2m}$$
$$p_{1xf}^2 + p_{2xf}^2 = p_{1xf}^2 + 2p_{1xf}p_{2xf} + p_{2xf}^2$$
$$2p_{1xf}p_{2xf} = 0$$

There are two possible solutions to this equation. The term $p_{1xf}p_{2xf}$ can be zero if $p_{1xf} = 0$ or if $p_{2xf} = 0$.

If $p_{1xf} = 0$, the physical situation is that cart 1 came to a complete stop. In that case we see from the momentum equation $p_{1xf} + p_{2xf} = p_{2xf} = p_{1xi}$ that cart 2 now has the same momentum that cart 1 used to have. There has been a complete transfer of momentum from cart 1 to cart 2. There has also been a complete transfer of kinetic energy from cart 1 to cart 2. See Figure 10.5.

If $p_{2xf} = 0$, what is the physical situation? In that case $p_{1xf} + p_{2xf} = p_{1xf} = p_{1xi}$, and cart 1 just keeps going, missing cart 2! Of course that won't happen if the carts are on the same narrow track, but the algebra doesn't know that.

It is not possible for both final momenta to be zero, since the total final momentum of the system must equal the nonzero total initial momentum of the system.

EXAMPLE **Maximally Inelastic Collision of Two Identical Carts**

Consider the opposite extreme—a maximally inelastic collision of the two identical carts, one initially at rest. That means the carts stick together (perhaps they have sticky material on their ends), and each has the same final momentum $p_{1xf} = p_{2xf}$ (Figure 10.6). **(a)** Find the final momentum, final speed, and final kinetic energy of the carts in terms of their initial values. **(b)** What is the change in internal energy of the two carts?

Solution Since the y and z components of momentum don't change, we can work with only x components.

System: Both carts
Surroundings: Earth, track, air (neglect friction and air resistance)

(a) Momentum Principle (x components):

$$p_{1xf} + p_{2xf} = p_{1xi}$$
$$2p_{1xf} = p_{1xi}$$
$$p_{1xf} = \frac{1}{2}p_{1xi}$$

The final speed of the stuck-together carts is half the initial speed:

$$v_f = \frac{1}{2}v_i.$$

Final translational kinetic energy:

$$(K_{1f} + K_{2f}) = 2\left(\frac{1}{2}mv_f^2\right)$$
$$(K_{1f} + K_{2f}) = 2\left(\frac{1}{2}m\left(\frac{1}{2}v_i\right)^2\right) = \frac{1}{4}mv_i^2$$
$$(K_{1f} + K_{2f}) = \frac{K_{1i}}{2}$$

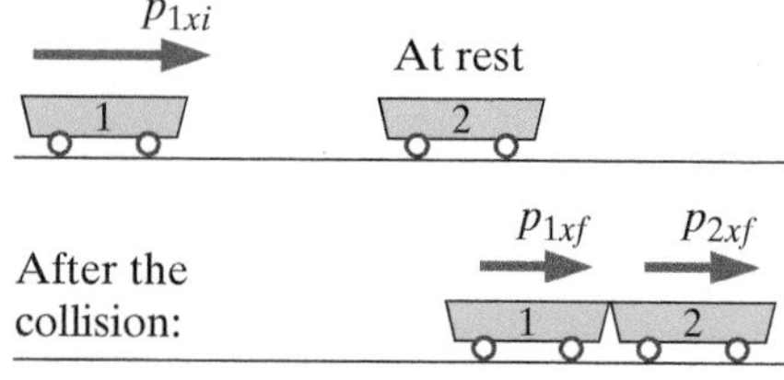

Figure 10.6 In a maximally inelastic collision, cart 1 sticks to cart 2, so the carts have the same final momentum.

(b) Energy Principle:

$$K_{1f} + K_{2f} + E_{\text{int},f} = K_{1i} + E_{\text{int},i}$$
$$E_{\text{int},f} - E_{\text{int},i} = K_{1i} - (K_{1f} + K_{2f})$$
$$\Delta E_{\text{int}} = K_{1i} - \frac{K_{1i}}{2}$$
$$\Delta E_{\text{int}} = \frac{K_{1i}}{2}$$

The Momentum Principle is still valid even though the collision is inelastic. Fundamental principles apply in all situations. The final kinetic energy of the system is only half of the original kinetic energy, which means that the other half of the original kinetic energy has been dissipated into increased internal energy ΔE_{int} of the two carts.

In both cases (elastic collision and inelastic collision) we needed to apply both the Momentum Principle and the Energy Principle to find all the unknown quantities. One equation alone was not enough.

We have shown the two extremes of energy conservation in the head-on collision of the moving cart with the stationary cart:

- If the collision is elastic, cart 1 stops and cart 2 moves with the speed cart 1 used to have.
- If the collision is maximally inelastic, the carts stick together and move with half the original speed. Half of the original kinetic energy is dissipated into increased internal energy.

Between these two extremes there can be an inelastic but not maximally inelastic collision, in which some amount of the original kinetic energy (less than half) is dissipated into increased internal energy.

Checkpoint 1 A 6 kg mass traveling at speed 10 m/s strikes a stationary 6 kg mass head-on, and the two masses stick together. **(a)** What was the initial total kinetic energy? **(b)** What is the final speed? **(c)** What is the final total kinetic energy? **(d)** What was the increase in internal energy of the two masses?

10.4 HEAD-ON COLLISIONS BETWEEN UNEQUAL MASSES

Suppose that a Ping-Pong ball hits a stationary bowling ball head-on, nearly elastically. Imagine that this takes place in outer space or floating in an orbiting spacecraft, so there is no friction—no significant external forces act on the combined system during the collision.

> QUESTION What do you expect? How will the Ping-Pong ball move after hitting the bowling ball? How will the bowling ball move after the collision?

You'll see the Ping-Pong ball bounce straight back with very little change of speed (Figure 10.7), and we'll show this is consistent with conservation of momentum and conservation of energy. Less obvious is that the bowling ball does move, although very slowly.

EXAMPLE **A Ping-Pong Ball Hits a Stationary Bowling Ball Head-On**

In an orbiting spacecraft a Ping-Pong ball of mass m (object 1) traveling in the $+x$ direction with initial momentum $\vec{p}_{1i}$ hits a stationary bowling ball of mass M (object 2) head on, as shown in Figure 10.7. What are the **(a)** momentum, **(b)** speed, and **(c)** kinetic energy of each object after the collision? Assume little change in the speed of the Ping-Pong ball, and assume that the collision is elastic.

Solution System: Ping-Pong ball and bowling ball
Surroundings: Nothing that exerts significant forces
Momentum Principle:

$$\vec{p}_{1f}+\vec{p}_{2f}=\vec{p}_{1i}+\vec{p}_{2i}$$

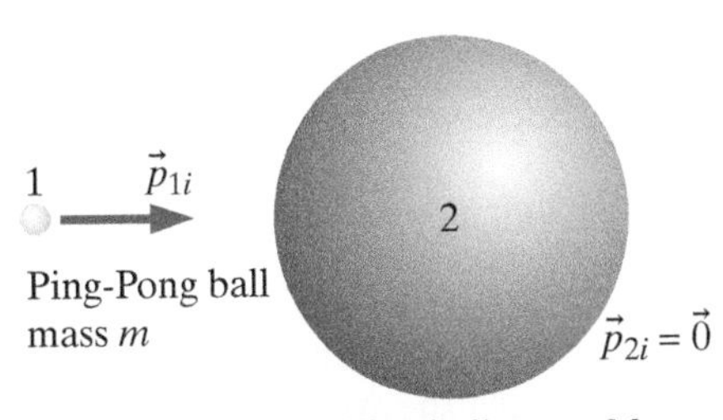

Figure 10.7 A Ping-Pong ball hits a bowling ball head-on and bounces off elastically.

Assume that the speed of the Ping-Pong ball does not change significantly in the collision, so $\vec{p}_{1f}\approx-\vec{p}_{1i}$.

$$-\vec{p}_{1i}+\vec{p}_{2f}=\vec{p}_{1i}$$
$$\vec{p}_{2f}=2\vec{p}_{1i}$$

(a) The final momentum of the bowling ball is twice the initial momentum of the Ping-Pong ball. It may be surprising that the bowling ball ends up with about twice the momentum of the Ping-Pong ball. One way to understand this is that the final momentum of the Ping-Pong ball is approximately $-\vec{p}_{1i}$, so the change in the Ping-Pong ball's momentum is approximately

$$-\vec{p}_{1i}-\vec{p}_{1i}=-2\vec{p}_{1i}$$

The Ping-Pong ball's speed hardly changed, but its momentum changed a great deal. Because momentum is a vector, a change of direction is just as much a change as a change of magnitude. This big change is of course due to the interatomic electric contact forces exerted on the Ping-Pong ball by the bowling ball. By reciprocity, the same magnitude of interatomic contact forces are exerted by the Ping-Pong ball on the bowling ball, which undergoes a momentum change of $+2\vec{p}_{1i}$.

(b) Final speed of bowling ball:

$$v_{2f}\approx\frac{p_{2f}}{M}=\frac{2p_{1i}}{M}=\frac{2mv_{1i}}{M}=2\left(\frac{m}{M}\right)v_{1i}$$

This is a very small speed since $m \ll M$. For example, if the mass of the bowling ball is about 5 kg, and the 2 g Ping-Pong ball is initially traveling at 10 m/s, the final speed of the bowling ball will be 0.008 m/s.

(c) Kinetic energies:

$$K_{2f} = \frac{(2p_{1i})^2}{2M} \quad \text{and} \quad K_{1f} = \frac{p_{1i}^2}{2m}$$

Because the mass of the bowling ball is much larger than the mass of the Ping-Pong ball, the kinetic energy of the bowling ball is much smaller than the kinetic energy of the Ping-Pong ball. The kinetic energy of the 2 g Ping-Pong ball, traveling at 10 m/s, is about 0.1 J, while the 5 kg bowling ball has acquired a kinetic energy of 1.6×10^{-4} J—nearly 1000 times less.

Although the kinetic energy of the slowly moving bowling ball is very small, it is not zero, which means that the Ping-Pong ball does lose a little speed in the collision. That's why our assumption that the Ping-Pong ball's speed hardly changes is an approximation.

We made the plausible assumption that the Ping-Pong ball bounced back with nearly its original speed, and showed that this is consistent with momentum conservation if the bowling ball gets twice the momentum of the Ping-Pong ball, and also consistent with energy conservation, because the bowling ball acquires very little kinetic energy. With a lot of algebra, it is also possible to solve the momentum and energy equations exactly for the unknown final momentum of the Ping-Pong ball, and when one makes the approximation that the mass of the bowling ball is very much larger than the mass of the Ping-Pong ball, the results are those given above (see Problem P15).

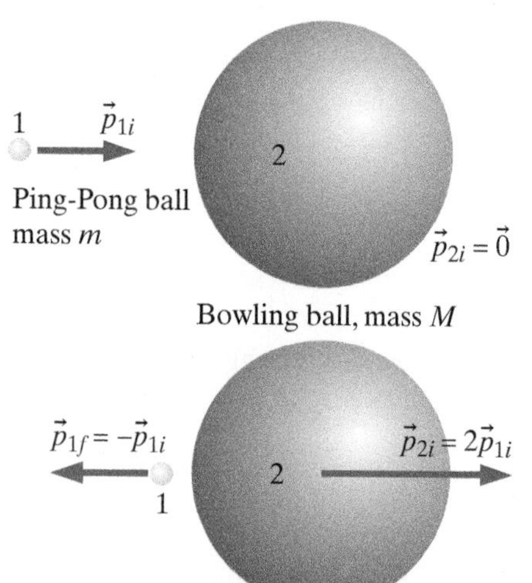

Figure 10.8 The bowling ball gets twice the momentum of the Ping-Pong ball (but its speed and kinetic energy are very small).

When a Ping-Pong ball hits a stationary bowling ball head-on, elastically:

- The Ping-Pong ball bounces back with almost the same speed.
- The Ping-Pong ball has a large change of momentum $-2\vec{p}_{1i}$.
- The bowling ball has a large change of momentum $+2\vec{p}_{1i}$ (Figure 10.8).
- The bowling ball's final speed is very small, because it has large mass but comparable momentum.
- The bowling ball's final kinetic energy is very small, because it has large mass but comparable momentum.

Similar situations include a ball bouncing off a wall, or falling and bouncing off the ground, in which case the entire Earth plays the role of the bowling ball. The Earth gets twice the momentum of the ball but an extremely small speed and extremely small kinetic energy. If a bowling ball sits on a table and is hit by a Ping-Pong ball, and the table exerts a large enough frictional force to prevent the bowling bowl from moving, the bowling ball is essentially part of the Earth.

Although we've looked at the very special case of a head-on collision, it is generally true that when low-mass projectiles collide elastically with stationary large masses, the low-mass projectile bounces off with little change of speed and kinetic energy but may undergo a large change of momentum. The large-mass target can pick up lots of momentum, but little speed or kinetic energy.

Checkpoint 2 We can use our results for head-on elastic collisions to analyze the recoil of the Earth when a ball bounces off a wall embedded in the Earth. Suppose that a professional baseball pitcher hurls a baseball ($m = 155$ g) with a speed of 100 mi/h ($v_1 = 44$ m/s) at a wall that is securely anchored to the Earth, and the ball bounces back with little loss of kinetic energy. **(a)** What is the approximate recoil speed of the Earth ($M = 6 \times 10^{24}$ kg)? **(b)** calculate the approximate recoil kinetic energy of the Earth and compare it to the kinetic energy of the baseball. The Earth gets lots of momentum (twice the momentum of the baseball) but very little kinetic energy.

10.5 FRAME OF REFERENCE

A frame of reference is a perspective from which a system is observed. A frame of reference may be stationary or it may move at constant velocity. (If the velocity of a reference frame is not constant, the frame is "noninertial," and the laws of physics as we have stated them do not apply. We'll stick to inertial frames in this textbook.) In Chapter 7 we saw that our choice of reference frame affected energy accounting. It is sometimes significantly simpler to analyze collisions from a moving reference frame.

As a simple example, imagine that you are standing on a street corner and a bus passes you. In your stationary frame of reference, a passenger sitting on the bus is moving with the same velocity as the bus. However, the bus driver may observe the passenger from a moving reference frame "anchored" to the bus. In the moving bus reference frame, the driver sees the passenger sitting still.

QUESTION How does the velocity of an object depend on the frame of reference we choose?

Suppose that the bus mentioned in the previous paragraph is heading in the $+x$ direction at a speed of 14 m/s. A passenger on the bus walks toward the front of the bus at a speed of 1.2 m/s. To the bus driver, the velocity of the passenger is $\langle 1.2, 0, 0\rangle$ m/s. However, as you stand on the street corner, you see the passenger moving past you slightly faster than the sitting passengers on the bus. You observe the walking passenger's velocity to be $\langle 15.2, 0, 0\rangle$ m/s, the sum of the velocity of the bus ($\langle 14, 0, 0\rangle$ m/s) and his velocity relative to the bus ($\langle 1.2, 0, 0\rangle$ m/s).

Evidently, the velocity $\vec{v}$ of an object viewed from a stationary reference frame is the sum of the velocity $\vec{v}_{\text{frame}}$ of a moving reference frame plus the velocity $\vec{v}'$ ("v-prime") of the object as viewed from the moving frame.

VELOCITY IN A MOVING REFERENCE FRAME

$$\vec{v} = \vec{v}' + \vec{v}_{\text{frame}}$$

where $\vec{v}$ is the velocity in a stationary frame, $\vec{v}_{\text{frame}}$ is the velocity of the moving reference frame, and $\vec{v}'$ is the velocity observed in the moving frame.

Changing frames of reference is a very powerful technique that is discussed in more depth later in this chapter. A change of reference frame often makes a problem much simpler to analyze, or reduces the problem to one you've already analyzed.

Bowling Ball Hits Ping-Pong Ball: Change of Reference Frame

In the previous example we analyzed a collision in which a moving Ping-Pong ball hit a stationary bowling ball head-on. What happens when a bowling ball of large mass M moving to the right with speed V hits elastically head-on a stationary Ping-Pong ball of small mass m? What is the final speed of the Ping-Pong ball?

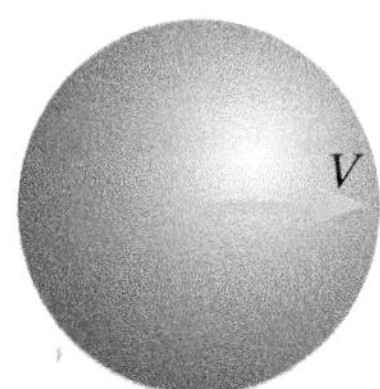

Figure 10.9 In a stationary reference frame, Jack sees a bowling ball with speed V move to the right toward a stationary Ping-Pong ball.

We can perform a change of reference frame that reduces this problem to the previous one. Suppose that this experiment is being done in a spacecraft. Jack and Jill are two crew members in the spacecraft. Jack hangs onto a wall, remaining stationary, and watches the bowling ball go by (Figure 10.9), while Jill launches herself away from the wall, moving to the right with speed V, and glides along beside the bowling ball.

According to the principle of relativity, in any reference frame in uniform motion one can apply physics principles and make correct predictions, even though the details of the calculations look different from the details in some other reference frame in uniform motion. If the walls are dark, and the Ping-Pong ball and bowling ball glow in ultraviolet light, Jill truly cannot tell whether she is moving and the Ping-Pong ball is stationary relative to the walls, or whether she is stationary and the Ping-Pong ball is moving relative to the walls. This is at the heart of the principle of relativity.

QUESTION What is the speed of the bowling ball as measured by Jill?

For Jill, the bowling ball isn't moving. It is always the same distance from her. In Jill's reference frame, the bowling ball is at rest, and the Ping-Pong ball is moving toward her with speed V (Figure 10.10).

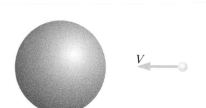

Figure 10.10 Moving with the bowling ball, at speed V, Jill sees the bowling ball at rest and the Ping-Pong ball moving toward her with speed V.

QUESTION What do Jack and Jill see during and after the collision?

Jill sees the moving Ping-Pong ball bounce off the stationary bowling ball. She observes the Ping-Pong ball traveling to the right with speed $\approx V$ (Figure 10.11). To Jill the process looks exactly the same as in the example in the previous section (A Ping-Pong Ball Hits a Stationary Bowling Ball Head-On).

Jack, however, sees the moving bowling ball hit the stationary Ping-Pong ball. After the collision he sees the bowling ball still moving to the right with speed $\approx V$, but he sees the Ping-Pong ball moving to the right with speed $\approx 2V$ (Figure 10.12).

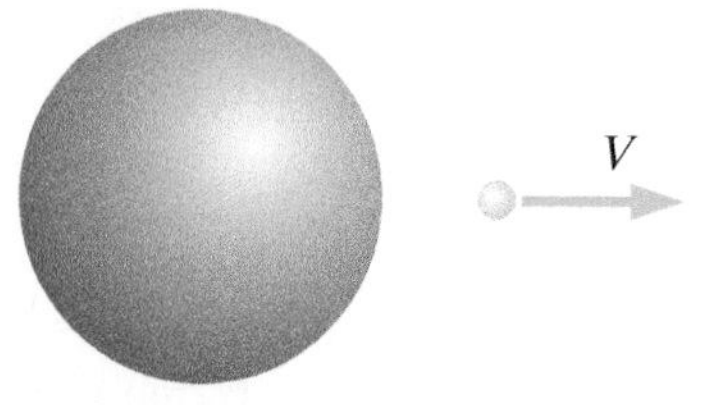

Figure 10.11 Jill (in the moving frame of reference) sees the moving Ping-Pong ball hit the bowling ball and bounce off. After the collision, she sees the Ping-Pong ball moving to the right with nearly its original speed V.

QUESTION How does this emerge from changing reference frames?

We need to apply the equation $\vec{v} = \vec{v}' + \vec{v}_{\text{frame}}$ to the velocities of both balls: $\vec{v}_B$ (bowling ball) and $\vec{v}_P$ (Ping-Pong ball). After the collision:

$$\begin{aligned}
\vec{v}_B &= \vec{v}_B{}' + \vec{v}_{\text{frame}} \\
&= \langle 0,0,0\rangle + \langle V,0,0\rangle \\
&= \langle V,0,0\rangle \\
\vec{v}_P &= \vec{v}_P{}' + \vec{v}_{\text{frame}} \\
&= \langle V,0,0\rangle + \langle V,0,0\rangle \\
&= \langle 2V,0,0\rangle
\end{aligned}$$

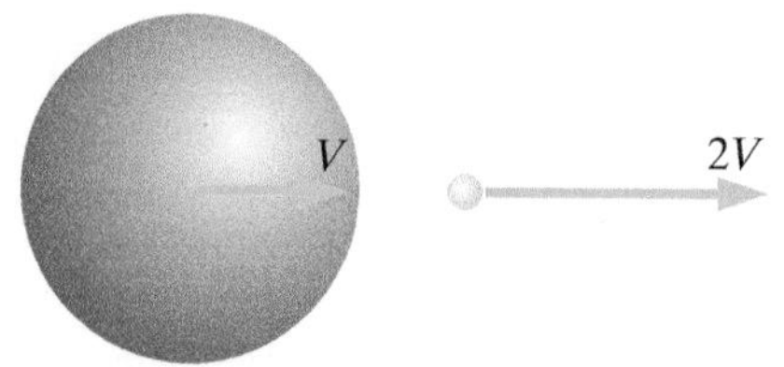

Figure 10.12 In Jack's stationary reference frame, the Ping-Pong ball's speed after the collision is nearly $2V$.

By transforming into a moving reference frame and back again, we were able to show that when a moving bowling ball hits a stationary Ping-Pong ball head-on, elastically, the Ping-Pong ball ends up moving with nearly twice the speed of the bowling ball. The change in speed of the bowling ball is insignificant. It just plows straight ahead, having lost very little kinetic energy to the Ping-Pong ball. The small amount of momentum lost by the bowling ball was gained by the Ping-Pong ball ($2mV$).

Checkpoint 3 A uranium atom traveling at speed 4×10^4 m/s collides elastically with a stationary hydrogen molecule, head-on. What is the approximate final speed of the hydrogen molecule?

10.6 SCATTERING: COLLISIONS IN 2D AND 3D

Scattering experiments are used to study the structure and behavior of atoms, nuclei, and other small particles. In a scattering experiment, a beam of particles collides with other particles. In atomic or nuclear collisions, we can't observe in detail the curving trajectories inside the tiny interaction region. We only observe the trajectories before and after the collision, when the particles are far apart and their mutual interaction is very weak, so they are traveling in nearly straight lines. An example of scattering is the collision of an alpha particle (helium nucleus) with the nucleus of a gold atom.

Momentum and Energy Equations

An alpha particle collides with a gold nucleus, and is deflected through some final angle θ. The gold nucleus recoils at some other final angle ϕ. The final angles are measured relative to the initial direction of the alpha particle (Figures 10.13 and 10.14). We say that the alpha particle "scattered" through an angle θ.

We take the "before" time early enough and the "after" time late enough that the two particles are sufficiently far away from each other that we can neglect their electric potential energy. As for conservation of momentum, we must write two equations, for the x and y components of momentum conservation.

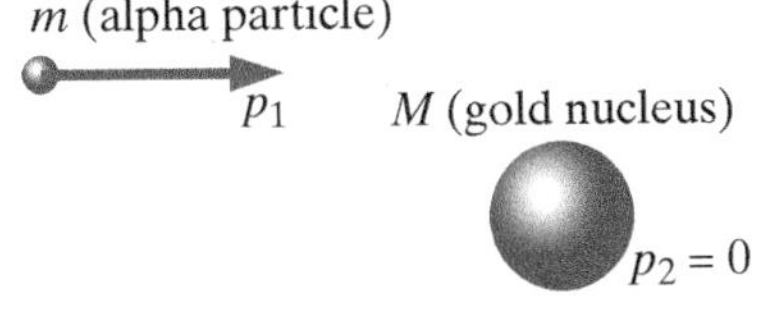

Figure 10.13 Before the collision, the alpha particle (mass m) approaches the gold nucleus (mass M).

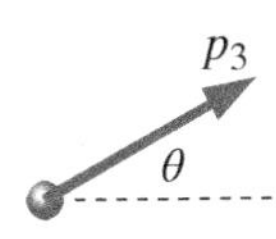

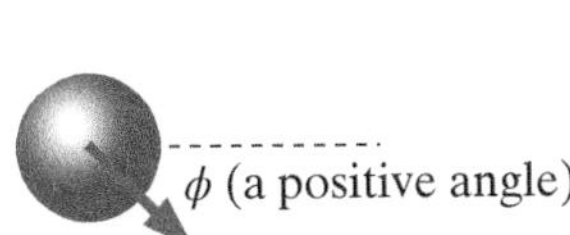

Figure 10.14 After the collision, when the particles are far apart and moving with nearly uniform velocities. Final angles are measured relative to the initial direction of the alpha particle.

QUESTION What are the momentum and the energy conservation equations for this system?

All the speeds are small compared to the speed of light. Choose x in the direction of the incoming alpha particle, and y at right angles to it. Consider ϕ to be a positive angle. Write component equations in terms of the magnitudes of the momenta, p_1, p_3, and p_4; the gold nucleus is initially at rest.

$$p_{xi} = p_{xf}$$
$$p_1 = p_3 \cos\theta + p_4 \cos\phi$$

$$p_{yi} = p_{yf}$$
$$0 = p_3 \cos(90^\circ - \theta) + p_4 \cos(90^\circ + \phi)$$

$$K_f = K_i$$
$$\frac{p_1^2}{2m} = \frac{p_3^2}{2m} + \frac{p_4^2}{2M}$$

Note that in the equations above, direction cosines are used to express vector components in terms of the magnitudes of momenta and the angles to the x axis. For example, the final momentum of the gold nucleus is $\vec{p}_4 = |\vec{p}_4|\langle \cos\phi, \cos(90^\circ + \phi), 0\rangle$.

QUESTION Which quantities in these three equations are unknown? Can we solve for the unknown quantities?

We know the masses of the alpha particle and the gold nucleus, and we know the initial kinetic energy of the alpha particle (so we know its speed and magnitude of momentum). You should have identified the two final momenta and the two final angles as unknown quantities (p_3, p_4, θ, and ϕ, all of which represent positive numbers). But if there are four unknowns and only three equations, how can we solve for the unknowns? The simple answer is, we can't!

Nevertheless, there are many situations in which these three equations can be exploited. For example, if we measure the final direction of motion of the alpha particle, the angle θ, we can solve for the other three unknown quantities and thereby predict the final momenta p_3 and p_4 of the two particles and the direction of motion ϕ of the gold nucleus, even without measuring all of these quantities.

It is worth mentioning that in experimental particle physics the momentum is often measured in a very direct way. Charged particles curve in a magnetic field due to magnetic forces that act on moving charges, and the radius of curvature is a direct measure of the momentum of the particle, if you know how much charge it has, which you do know if you know what kind of particle it is. Usually the momentum of a particle can be measured more accurately than the velocity, though once you have measured the momentum you can deduce the velocity if you know what kind of particle it is, since in that case you know its mass.

EXAMPLE

Elastic Collision: Identical Particles, One Initially at Rest

An interesting special case of an elastic collision that can be analyzed this way is one involving particles of equal mass, such as an elastic collision between two alpha particles, or two gold nuclei, or two billiard balls (whose collision is nearly elastic), in which one of the two objects (the "target") is initially at rest (Figure 10.15 and Figure 10.16). In this special case, show that the angle A must be 90° if the speeds are small compared to the speed of light.

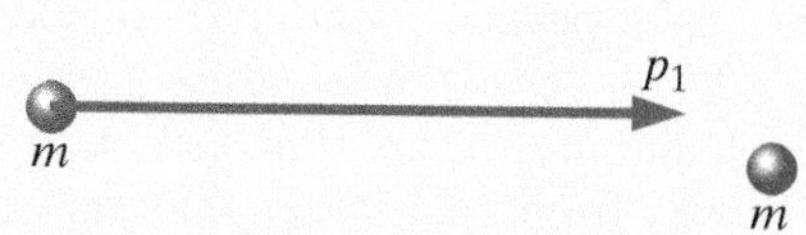

Figure 10.15 Before a collision of two identical particles, one initially at rest.

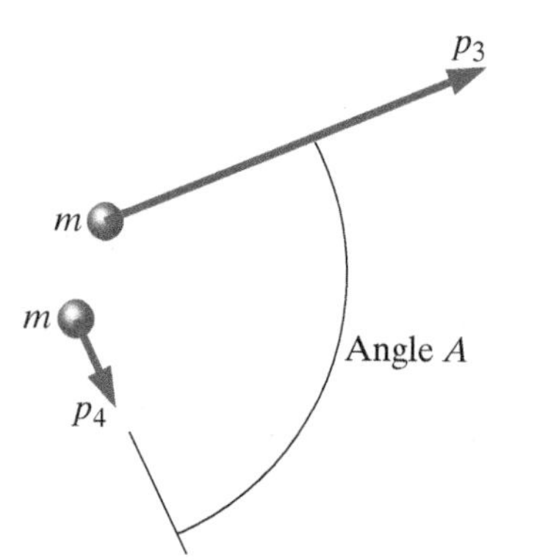

Figure 10.16 After the collision of two identical particles, one initially at rest.

Solution Because the vector momentum of the system is constant, we can write $\vec{p}_1 = \vec{p}_3 + \vec{p}_4$, then calculate p_1^2 by expanding the dot product:

$$p_1^2 = \vec{p}_1 \bullet \vec{p}_1 = (\vec{p}_3 + \vec{p}_4) \bullet (\vec{p}_3 + \vec{p}_4) = p_3^2 + p_4^2 + 2p_3p_4 \cos A$$

Divide by $2m$:

$$\frac{p_1^2}{2m} = \frac{p_3^2}{2m} + \frac{p_4^2}{2m} + \frac{2p_3p_4 \cos A}{2m}$$

Since the kinetic energy K at low speed can be written as $p^2/2m$,

$$K_1 = K_3 + K_4 + \frac{2p_3p_4 \cos A}{2m}$$

However, energy conservation for this elastic collision says that $K_1 = K_3 + K_4$, so we have

$$\cos A = 0 \text{ or } A = 90°$$

In this special situation—an elastic collision at low speeds between equal particles, one of which is initially at rest—the angle between the final velocities must be 90° as long as neither p_3 nor p_4 is zero, in which case the equation reduces to $0 \cos A = 0$, and $\cos A$ is indeterminate. If one of the final momenta is indeed zero, the other particle must have all the momentum of the initial particle. This can happen either because the incoming particle missed the target entirely and kept going in the original direction, or because the incoming particle hit the target dead center, in which case the only consistent solution of the equations is for the incoming particle to come to rest and the target particle to move forward with the initial momentum, as we saw in the head-on collision of two carts earlier in the chapter.

Impact Parameter

The distance between centers perpendicular to the incoming velocity is called the "impact parameter" and is often denoted by b. A head-on collision has an impact parameter of zero. Here in Figures 10.17–10.20 are possible elastic collisions between two billiard balls, for various impact parameters:

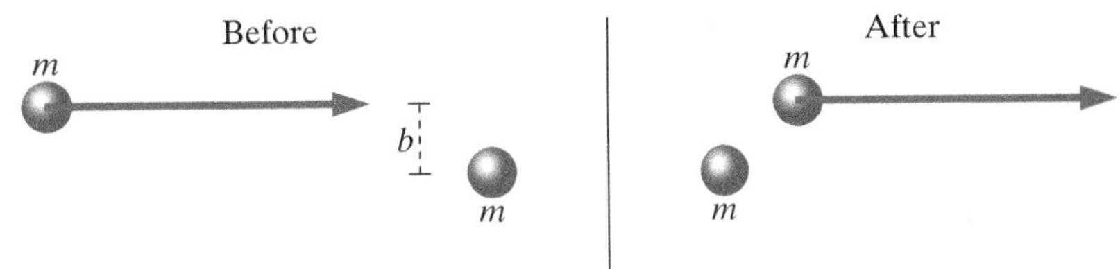

Figure 10.17 If the impact parameter, b, is too large, there is no interaction at all between two billiard balls.

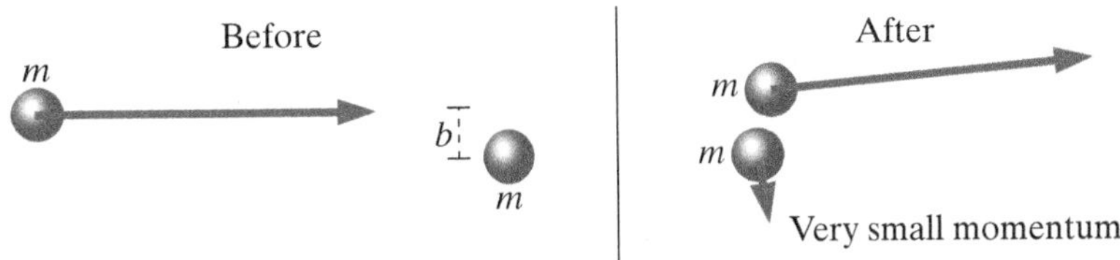

Figure 10.18 A grazing impact results in small-angle scattering.

Smaller-impact parameters lead to larger scattering unless $b = 0$:

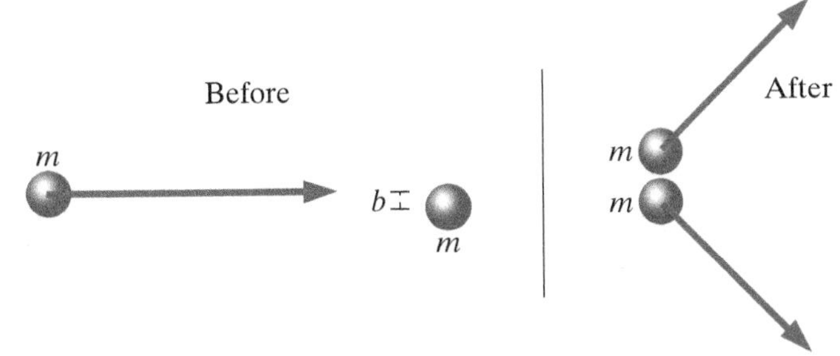

Figure 10.19 A medium-impact parameter results in symmetric scattering.

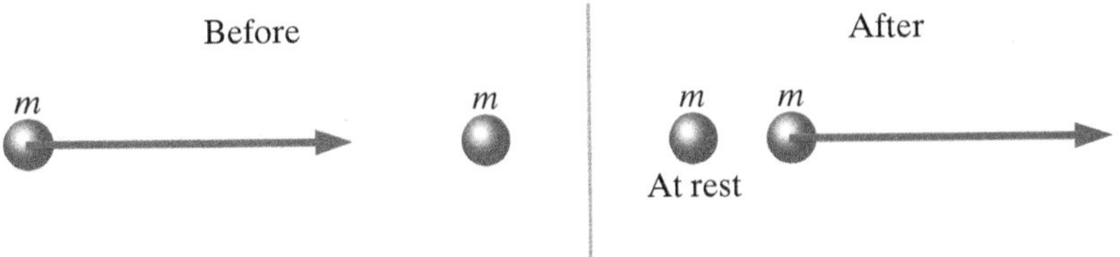

Figure 10.20 In a head-on collision, the impact parameter $b = 0$. In this case all the momentum is transferred to the target.

The smaller the impact parameter, the more severe is the collision, and the larger the deflection angle of the incoming particle (larger "scattering"), except for a head-on collision, where if the masses are equal the incoming ball stops dead and the target ball gets the entire momentum.

10.7 DISCOVERING THE NUCLEUS INSIDE ATOMS

A famous experiment involving collisions led to the discovery of the nucleus inside the atom. It was carried out in 1911 by a group in England led by the New Zealander Ernest Rutherford. In the experiment, high-speed alpha particles (now known to be helium nuclei, consisting of two protons and two neutrons) were shot at a thin gold foil (whose nuclei consist of 79 protons and 118 neutrons).

When Rutherford and his coworkers decided to perform a scattering experiment in order to probe the internal structure of matter, they were familiar with the relationships of momenta in collisions that we have just studied, and they reasoned that by shooting microscopic particles at a thin layer

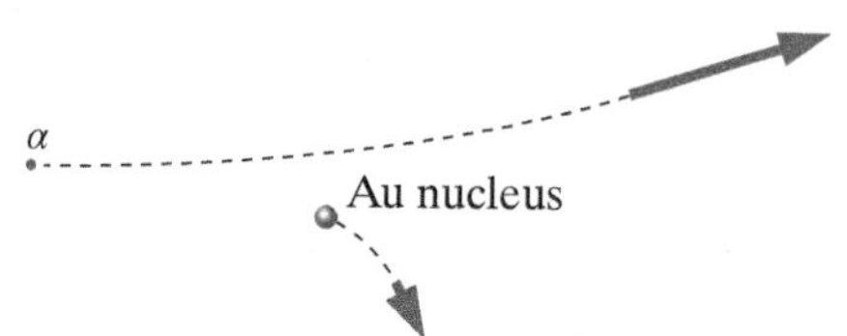

Figure 10.21 A collision that is not head-on.

of metal, they might be able to learn something about the microscopic structure of the metal by looking at patterns of scattering.

In the Rutherford experiment, the velocity of the alpha particles was not high enough to allow them to make actual contact with the nucleus. It is nonetheless appropriate to think of the process as a collision, because there are sizable electric forces acting between the alpha particle and the gold nucleus for a very short time. We call any process of this kind a collision, even if there is no direct contact but only gravitational, electric, or magnetic interactions at a distance. Figure 10.21 shows what the trajectories look like. We say that the alpha particle projectile is "scattered" (deflected) by the interaction with the initially stationary gold nucleus target, and such experiments are called "scattering" experiments.

Elastic Collisions in the Rutherford Experiment

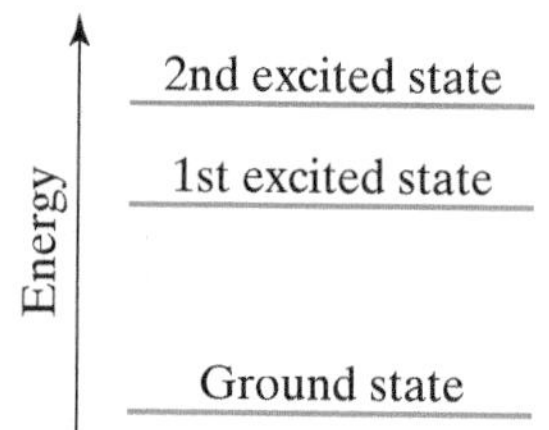

Figure 10.22 Energy levels in a nucleus. The potential energy curve is not shown.

The conditions of this experiment were such that there was no change in the internal energy of either the alpha particle or the gold nucleus (elastic collision). There do exist "excited" states of nuclei, in which there is a change in the configuration of the nucleons corresponding to a higher energy than the ground state, the normal configuration of the nucleus. The internal energy of the nucleus is quantized, and there are no stable or semi-stable states with energy intermediate between these quantized energies (Figure 10.22). In the Rutherford experiment, the interaction did not raise the gold nucleus or the alpha particle to an excited state; both nuclei remained in their ground states. This happy accident made it much easier for Rutherford to analyze the experiment.

Electric Potential Energy in the Rutherford Experiment

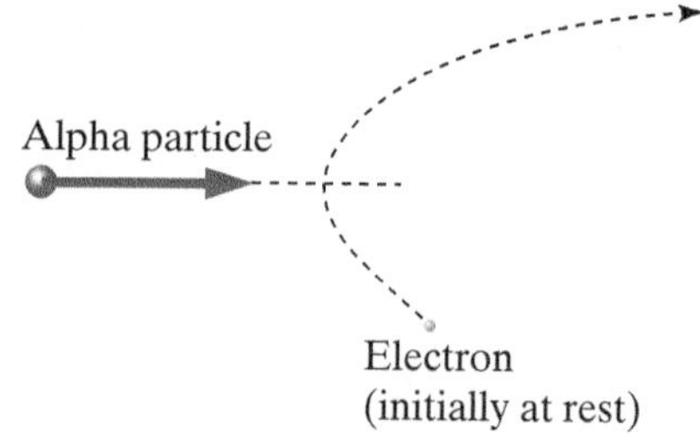

Figure 10.23 The alpha particle sweeps away the low-mass electrons and plows straight on ("bowling ball hits Ping-Pong ball").

What about the electric potential energy associated with the electric forces? Long before the collision the two nuclei were far from each other, and long after the collision they are again far from each other. Like gravitational energy, electric energy is proportional to $1/r$, where r is the distance between the two charged objects. Therefore, from long before the collision to long after the collision the change in electric energy would be very nearly zero ($\Delta U_{el} \approx 0$) if the alpha particle and gold nucleus are completely isolated from other particles. However, what about all the electrons surrounding a gold nucleus?

As the alpha particle plows through the gold foil it knocks low-mass electrons away (Figure 10.23), much as a moving bowling ball knocks Ping-Pong balls away (the alpha particle does of course lose a little energy in each such interaction). Therefore we can start our analysis when the alpha particle is deep inside the atom, with the electrons of the gold atom mostly far outside the region of interest. An atomic diameter is about 1×10^{-10} m, whereas the diameter of a gold nucleus is much smaller, less than 1×10^{-14} m, so we might start our analysis of the collision when the particles are about 1×10^{-12} m apart, a distance that is very small compared to an atom and very large compared to a nucleus. At this distance the electric potential energy is

$$U_{\text{elec}} = \frac{1}{4\pi\varepsilon_0}\frac{(2e)(79e)}{r} = \left(9\times10^9\frac{\text{N}\cdot\text{m}^2}{\text{C}^2}\right)\frac{(2)(79)(1.6\times10^{-19}\,\text{C})^2}{1\times10^{-12}\,\text{m}}$$

$$U_{\text{elec}} = (3.6\times10^{-14}\,\text{J})\frac{1\,\text{eV}}{1.6\times10^{-19}\,\text{J}} = 0.2\times10^6\,\text{eV} = 0.2\,\text{MeV}$$

The radioactive source used by the Rutherford group provided alpha particles whose kinetic energy was 10 MeV, so if we take as the initial state a separation of 1×10^{-12} m the electric energy is only 2% of the kinetic energy. If we also take as the final state a separation of 1×10^{-12} m we have $\Delta U_{elec} = 0$. Evidently the only significant energy changes in this experiment are changes in the kinetic energies of the alpha particle and the gold nucleus.

The gold nuclei in the gold foil are about 2×10^{-10} apart, the approximate diameter of a typical atom in a metal, so when an alpha particle is 1×10^{-12} m from a gold nucleus, it is comparatively very far from any other gold nucleus. This explains why Rutherford could analyze the experiment as though isolated alpha particles scattered off isolated gold nuclei; electrons and other gold nuclei were far away.

The Plum Pudding Model of an Atom

For about a decade, the prevailing model of an atom had been what was called a "plum pudding" model proposed by J. J. Thomson. Electrons were discovered in 1897 by Thomson, using a cathode ray tube, a device similar to an old-style television picture tube. Thomson was able to show that the "cathode rays" that he and others had observed in electric gas discharges were in fact particles that were negatively charged, by the already existing convention for labeling electrical charges. This was the first conclusive demonstration of the existence of what we now call "electrons," one of the major constituents of all atoms.

Since these electrons were extracted from ordinary matter, which is normally not electrically charged, Thomson's discovery implied that matter also contained something having a positive charge. Because the electrons apparently had little mass, it seemed likely that most of the mass in ordinary matter was charged positively. Thomson proposed what was called a plum pudding model of atoms: a positively charged substance of uniform, rather low density, with tiny electrons distributed like raisins throughout it. Rutherford's group expected to gather evidence supporting this model from their collision experiments. Rutherford was extremely surprised by the results of these experiments, which led to a radical revision of the model.

The Rutherford Experiment

A new source of fast-moving particles was provided through the discoveries of natural radioactivity by the French scientists Henri Becquerel and Marie and Pierre Curie. These newly discovered particles included high-speed, massive, positively charged "alpha particles," which we now know to be the nuclei of helium atoms, consisting of two protons and two neutrons bound together. Alpha particles are emitted in the radioactive decay of heavy elements. Alpha particles were convenient microscopic projectiles, and Rutherford's group decided to shoot alpha particles at a very thin piece of gold foil and observe where they reappeared.

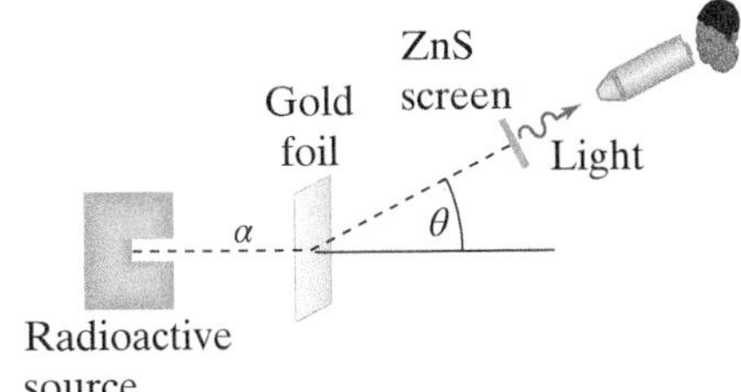

Figure 10.24 Schematic diagram of the Rutherford experiment. A radioactive source emits alpha particles, which are directed at a thin piece of gold foil. Particles scattered by gold nuclei in the gold foil hit a screen coated with ZnS, which fluoresces, giving off light when hit by an energetic particle. The light was observed by the experimenter.

> QUESTION Based on the plum pudding model, what did Rutherford and his colleagues expect to see in their collision experiments?

The researchers expected the alpha particles to be deflected only very slightly by interactions with the low-mass electrons and with the low-density "pudding" of positively charged matter in the gold atoms. They could detect a deflected ("scattered") alpha particle by a pulse of light that was emitted when the particle struck a fluorescent material (zinc sulfide). They moved their detector around at various angles to the incoming beam of alpha particles and measured what fraction of the alpha particles were scattered through various angles (Figure 10.24).

Rutherford and his coworkers were astounded to observe that an alpha particle, which is much more massive than an electron, sometimes bounced straight backward, as though it had encountered an even more massive yet very tiny positive particle inside the gold foil, rather than being slightly deflected in its passage through the "pudding" of positive charge. In other words, the experimenters thought they were throwing bowling balls at Ping-Pong balls, and they were amazed to see what they thought were bowling balls come bouncing back.

"It was quite the most incredible event that has ever happened to me in my life," Rutherford said. "It was almost as incredible as if you fired a 15-inch [artillery] shell at a piece of tissue paper and it came back and hit you."

He continued, "On consideration, I realized that this scattering backward must be the result of a single collision, and when I made the calculations I saw that it was impossible to get anything of that order of magnitude unless you took a system in which the greater part of the mass of the atom was concentrated in a minute nucleus. It was then that I had the idea of an atom with a minute massive center, carrying a charge." (See Figure 10.25.)

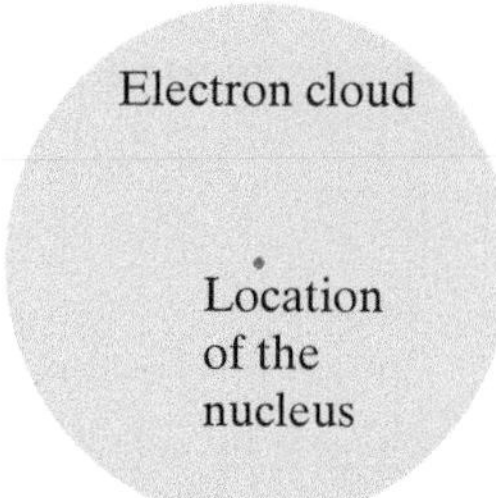

Figure 10.25 This diagram of an atom is not to scale; the radius of the nucleus is around 1×10^{-15} m, which is only 1/100,000 the radius of the electron cloud (about 1×10^{-10} m), so the nucleus would not be visible at all on this scale.

Computer Modeling of the Rutherford Experiment

We can make a computational model of the Rutherford experiment; see Problems P34 and P35. Remember that Coulomb's law for electric forces is very similar to Newton's law of gravitation. The electric force $\vec{F}_{elec}$ between two small charged particles carrying amounts of charge q_1 and q_2, and located a distance r apart, is

$$\vec{F}_{elec} = \frac{1}{4\pi\varepsilon_0}\frac{q_1 q_2}{r^2}\hat{r}$$

The charges q_1 and q_2 are measured in units of coulombs (C), and the proportionality factor is measured to be $1/(4\pi\varepsilon_0) = 9 \times 10^9\ \mathrm{N \cdot m^2/C^2}$. The electric charge of the proton is $e = +1.6 \times 10^{-19}$ C, and the electric charge of the electron is $-e = -1.6 \times 10^{-19}$ C. In the Rutherford experiment the charge on the alpha particle is $2e$ (2 protons and 2 neutrons), and the charge on the gold nucleus is $79e$ (79 protons and 118 neutrons).

10.8 DISTRIBUTION OF SCATTERING ANGLES

In general, we have three equations describing an elastic collision—two momentum equations (in the x and y directions) and one energy equation, but there are four unknown quantities (the outgoing speeds and directions for each particle). The concept of impact parameter provides an additional quantity, and you saw earlier how the impact parameter determines the scattering angle θ of the incoming particle, from which you can predict the other final values (p_3, p_4, and ϕ).

In a macroscopic situation, such as shooting billiard balls, you can choose an impact parameter simply by carefully aiming at a point a perpendicular distance b from the center of the target. Unfortunately, in experiments where you fire alpha particles or other subatomic projectiles at nuclei, you can't aim at a specific impact parameter because the target is so incredibly small. All you can do is fire lots of projectiles at the subatomic target and expect a probabilistic distribution of impact parameters.

When we fire a projectile at this microscopic target, the probability of hitting one particular ring (whose radius corresponds to the impact parameter) is proportional to the area of that ring (Figure 10.26). If you calculate the range of scattering angles that corresponds to hitting somewhere within a particular ring, you can predict the angular distribution of scattering angles that will be observed.

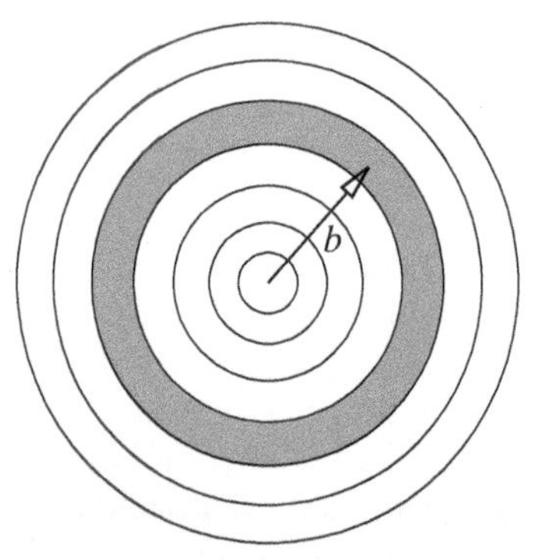

Figure 10.26 The probability of hitting a particular ring is proportional to the area of the ring.

Cross Section

If a target particle such as a gold nucleus has a radius R, we say that its "geometrical cross section" is πR^2. A high-speed alpha particle that hits within this cross-sectional area will interact strongly with the gold nucleus. Quantum-mechanical effects can make the effective target size bigger or smaller than the geometrical cross section, and the effective target size is called the "cross section" for interactions of this kind. Cross sections are measured experimentally by observing how often projectiles are deflected by an interaction, instead of passing straight through the material without interacting. Cross sections have units of square meters.

A "differential" cross section for scattering within a particular range of angles is the area in square meters of a ring bounded by impact parameters corresponding to the given angular range.

Conservation Laws vs. Details of the Interaction

We used general principles to determine the final momenta of two colliding particles (conservation of momentum, conservation of energy). We didn't have to say what kind of force was involved. However, doesn't it make a difference whether the interaction is electric or nuclear? Yes, the type of interaction determines the *distribution* of scattering angles. For example, in what fraction of the collisions is the scattering angle between 0 and 10 degrees? Between 10 and 20 degrees? Between 20 and 30 degrees? These fractions are different for different forces.

To put it another way, if the scattering angle is 20 degrees, the momenta are determined. The question is, how often is the scattering angle in fact near 20 degrees? Measurement of the distribution of scattering angles in collisions at modern particle accelerators is one of the ways used to study the nature of the strong and weak interactions.

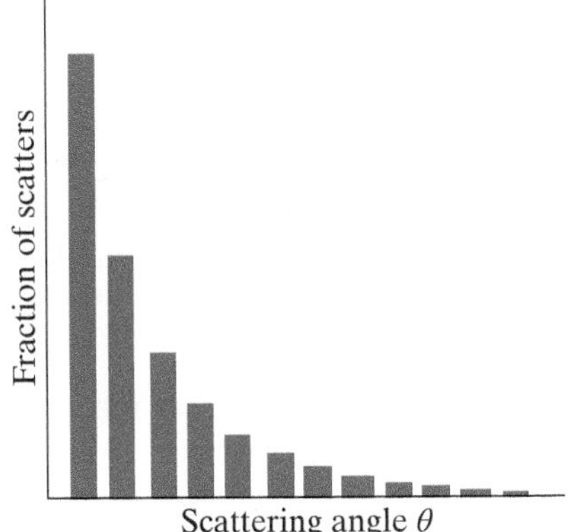

Figure 10.27 For an electric interaction, the fraction of scatters at different scattering angles falls rapidly with angle.

Rutherford used Coulomb's law of electric force to make a detailed prediction of the distribution of scattering angles, assuming that the positive nucleus was pointlike and very massive. That is, he predicted what fraction of the alpha particles would be scattered by less than 10 degrees, by an angle between 10 and 20 degrees, by an angle between 20 and 30 degrees, and so on. The observed distribution of scattering angles agreed with the predicted distribution (Figure 10.27 shows a portion of this distribution).

The calculation that Rutherford did involves a lot of complicated geometry, algebra, and probability (with a random distribution of impact parameters), so we won't go through it. The main features of Figure 10.27 are easily understood, however. Because the gold nucleus is very small, most of the time the alpha particle will miss the nucleus by a large distance. With a large impact parameter the electric force is very small, and the deflection (scattering angle) is very small. For that reason small scattering angles are common. Large scattering angles occur only for nearly head-on collisions (small impact parameter), and these events are rare.

Since Rutherford's prediction assumed a pointlike nucleus, the observations were consistent with the nucleus being very small. Also, the good fit between prediction and observation strengthened the belief that the interaction was purely electric in nature (the strong interaction, which was unknown at the time, played no role because it is a very short-range interaction, and the alpha particle was not going fast enough to overcome the electric forces and contact the gold nucleus).

If the alpha particle has high enough energy, it can overcome the electric repulsion and make contact with the gold nucleus. In that case the strong interaction comes into play, and the angular distribution changes. The radius of the gold nucleus can be determined by finding out how much energy an alpha particle has to have in order to obtain an angular distribution that deviates from what is expected due just to the electric interaction.

> QUESTION For a particular scattering angle, can we use momentum and energy conservation to determine the unknown angle and momenta, even if the alpha particle makes contact with the gold nucleus?

Yes. These conservation principles are completely general and apply to all types of interactions. However, they cannot by themselves predict the angular distribution—that is, how often a particular range of scattering angles will occur. To do that you have to have a mathematical description for the specific interaction, such as the $1/r^2$ force law for electric interactions.

10.9 COMPUTATIONAL AND ANALYTICAL APPROACHES

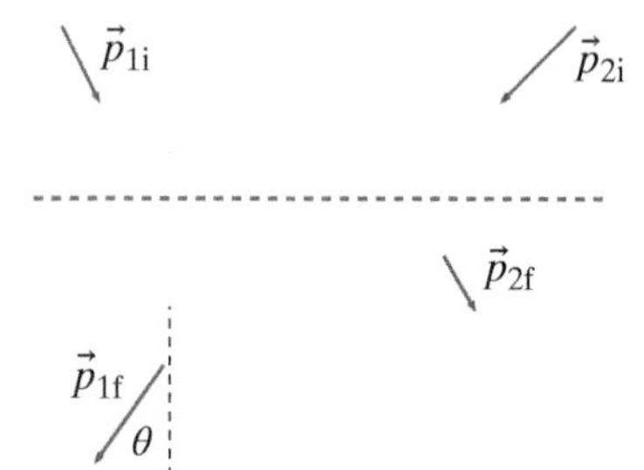

Figure 10.28 Using momentum and energy conservation let's predict the final momenta of the objects, but only if we have partial information about the outcome (for example, an experimental measurement of the direction of one of the final momenta, indicated here by the angle θ).

In the previous sections we have analyzed collisions analytically (algebraically) by choosing both objects as the system, neglecting external forces, and applying Conservation of Momentum and Conservation of Energy to determine the outcome of the process. It is also possible to model collisions computationally, by doing exactly what we have done before in modeling interacting objects such as a spacecraft and a planet or a mass and a spring—using a known force law, and updating position and momentum iteratively. These approaches are complementary; depending on what information we know, and what questions we are trying to answer, one approach may be preferable to the other.

Using Momentum and Energy Conservation

This approach is useful in analyzing the results of elastic collisions in which there are experimental observations of the outcome of the collision–at least the unit vector $\hat{p}$ in the direction of one of the final momenta (Figure 10.28). Except in special cases (such as collisions in 1D) this approach isn't useful in predicting full outcomes.

- This approach does not require knowledge of the particular force law involved in the interaction.
- This approach does not predict the full trajectories of the objects.
- As we observed in Section 10.6, in general (2D or 3D) it will not be possible to predict the final momenta of both objects without additional information, such as experimental observations of some of the final quantities.

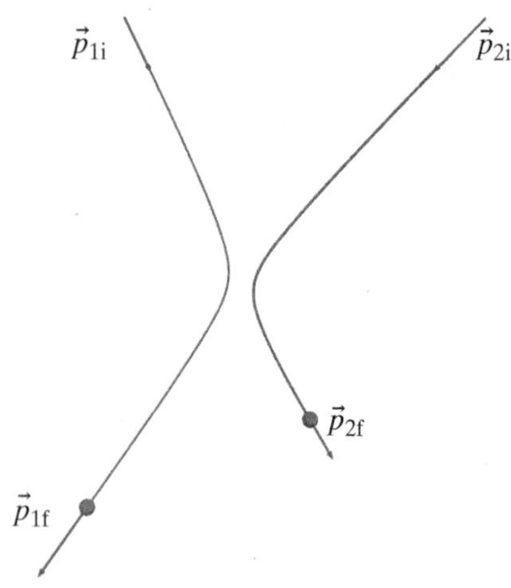

Figure 10.29 A computational approach predicts the full trajectories of the objects, if we know their initial positions and the appropriate force law.

Calculating Full Trajectories

This is a predictive approach for elastic collisions involving a known force law—it allows the prediction of full trajectories (Figure 10.29).

- This approach requires knowledge of initial positions as well as initial momenta.
- This approach requires knowledge of the force law involved in the interaction.
- This approach allows us to predict the full trajectories of the objects.
- This approach is always possible if we know forces, initial positions, and momenta.

Analyzing Inelastic Collisions

It is difficult to predict ahead of time how much initial kinetic energy will end up as internal energy after an inelastic collision. Most analyses of inelastic collisions depend on having detailed information about final momenta. For example, if both the magnitude and direction of $\vec{p}_{1f}$ are measured experimentally, it is then possible to use momentum conservation to determine $\vec{p}_{2f}$, and hence to determine the final kinetic energies. Given initial and final kinetic energies, we can determine the change in total internal energy of the system.

10.10 RELATIVISTIC MOMENTUM AND ENERGY

Rutherford's original experiments were carried out with alpha particles whose kinetic energy was about 10 MeV (10 million electron-volts), with a corresponding speed that was small compared to the speed of light. In order to probe the nucleus more deeply, physicists later built special machines, called "particle accelerators," which use electric forces to accelerate charged particles to very much higher energies, with speeds that may be as much as 99.9999% of the speed of light. For such high speeds we cannot use the low-speed approximate expressions for momentum and energy.

Physicists analyzing collisions produced by high-energy accelerators routinely do calculations based on the relativistic energy and momentum expressions. Consider a collision in which a particle of mass m_1 and magnitude of momentum p_1 strikes a stationary particle of mass m_2 (Figure 10.30). There is a change of identity, with the production of two new particles of mass m_3 and m_4, with momenta p_3 and p_4, at angles θ and ϕ to the horizontal (Figure 10.31).

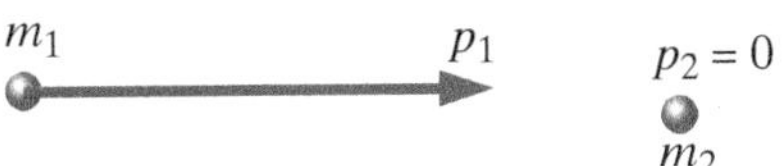

Figure 10.30 Before a high-energy collision.

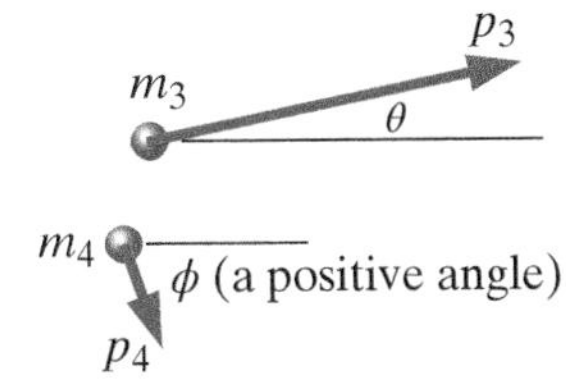

Figure 10.31 After the collision.

We will write down relativistically correct equations for momentum conservation and energy conservation, in terms of momentum p and energy E, knowing that we can determine E from p, since $E^2 - (pc)^2 = (mc^2)^2$:

$$p_1 = p_3 \cos\theta + p_4 \cos\phi, \text{ } x \text{ momentum conservation}$$

$$0 = p_3 \cos(90^\circ - \theta) - p_4 \cos(90^\circ - \phi), \text{ } y \text{ momentum conservation}$$

$$E_1 + m_2c^2 = E_3 + E_4, \text{ energy conservation (scalar)}$$

The energy equation is equivalent to

$$\sqrt{(p_1c)^2 + (m_1c^2)^2} + m_2c^2 = \sqrt{(p_3c)^2 + (m_3c^2)^2} + \sqrt{(p_4c)^2 + (m_4c^2)^2}$$

> QUESTION Suppose that the initial momentum p_1 of the projectile is known, and all four masses are known (because we were able to identify the particles). How many unknown quantities are there, considering that we can write the energy equation in terms of p?

There are four unknown quantities, the magnitudes of the momenta and the directions of the two outgoing particles (p_3, p_4, θ, and ϕ), but only three equations. A typical situation in a high-energy particle experiment is that one of these unknown quantities is measured, and momentum and energy conservation are used to calculate the other three quantities.

As in the Rutherford experiment, measurements of the outgoing particle momenta alone do not tell us anything about the detailed nature of the interaction involved in the collision, because they are determined by general momentum and energy conservation laws. The identity changes of the particles tell us a lot, and the distribution of scattering angles also tells us a lot.

Photons Have Momentum

The relation $E^2 - (pc)^2 = (mc^2)^2$ is valid even for particles with zero mass, so that a photon with energy E has momentum of magnitude $p = E/c$, which is verified by experiment. One cannot write $E = \gamma mc^2$ or $p = \gamma mv$ for a photon, because when $v = c$, γ is infinite, and γm (infinity times zero) is undefined. Nevertheless, both the energy and the momentum of a photon are well-defined.

EXAMPLE **Photodissociation of the Deuteron**

A photon of energy E is absorbed by a stationary deuteron. As a result, the deuteron breaks up into a proton and a neutron. The photon energy is sufficiently low that the proton and neutron have speeds small compared to the speed of light, but there is a change of rest mass. The proton is observed to move in a direction at an angle θ to the direction of the incoming photon. Write equations that could be solved to determine the unknown magnitudes of the proton momentum p_p and neutron momentum p_n and the positive angle α of the neutron to the direction of the incoming photon. Let the precise masses of the particles be M_d, M_p, and M_n for the deuteron, proton, and neutron. Explain clearly on what principles your equations are based. Provide a physics diagram of the reaction and label the angles. Do not attempt to solve the equations.

Solution System: All of the particles (Figure 10.32)
Surroundings: Nothing significant

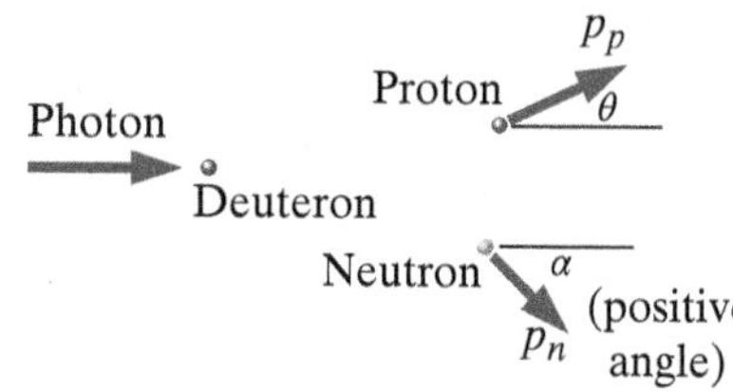

Figure 10.32 Physics diagram for photodissociation of deuteron. The angle θ was measured.

Momentum conservation (Figure 10.32): $\vec{p}_{\text{photon}} = \vec{p}_p + \vec{p}_n$

$$\langle E/c,0,0\rangle = \langle p_p \cos\theta + p_n \cos\alpha, p_p \cos(90^\circ - \theta) - p_n \cos(90^\circ - \alpha), 0\rangle$$

This is equivalent to two component equations:

$$E/c = p_p \cos\theta + p_n \cos\alpha$$

$$0 = p_p \cos(90^\circ - \theta) - p_n \cos(90^\circ - \alpha)$$

Energy conservation (small proton and neutron speeds):

$$E + M_d c^2 = \left(M_p c^2 + \frac{p_p^2}{2M_p}\right) + \left(M_n c^2 + \frac{p_n^2}{2M_n}\right)$$

We have three equations (two momentum component equations and one energy equation) with three unknowns: p_p, p_n, and α (θ was measured). In principle we can solve for the unknown quantities, though the algebra is messy.

We could also have written the energy equation relativistically, using the relationship $E^2 - (pc)^2 = (mc^2)^2$, which gives $E = \sqrt{(pc)^2 + (mc^2)^2}$:

$$E + M_d c^2 = \sqrt{(p_p c)^2 + (M_p c^2)^2} + \sqrt{(p_n c)^2 + (M_n c^2)^2}$$

Identifying a Particle

A variation on the relativistic collision analysis can be used to identify one of the outgoing particles. Suppose that we measure p_3 and θ, and we know m_3 but not m_4. We can use our three equations with the three unknowns (m_4, p_4, and ϕ) to solve for the unknown mass m_4. This is a very powerful technique when we are unable to observe the particle directly. For example, neutrinos don't have electromagnetic or strong interactions and therefore typically fail to register in ordinary detectors. However, if we measure the other outgoing particle we can determine the properties of the unseen particle.

The Discovery of the Neutrino

A specific example of such analyses was the original motivation for proposing the existence of the neutrino. The neutron (either free or when bound into some nuclei) was observed to decay with the creation of a proton and an electron, both charged particles that are easy to detect. However, the amount of energy and momentum carried by the proton and electron varied from one decay to the next, which made no sense:

$$n \to p^+ + e^-, \text{ violating momentum and energy conservation?}$$

It was proposed that there was an unseen third particle, called the neutrino (or antineutrino in neutron decay), which carried energy and momentum that would salvage the conservation laws:

$$n \to p^+ + e^- + \bar{\nu}, \text{ conserving momentum and energy}$$

It was possible to solve for the properties of the unseen particle by the techniques outlined above, and it was found that the neutrino must have nearly zero mass and must travel at nearly the speed of light. (Recent experiments show that the neutrino has a very small nonzero mass and must travel at slightly less than the speed of light.)

For many years the neutrino was just a theoretical construct that balanced the accounts, but eventually the neutrino was detected directly. The neutrino is very hard to detect because it does not have electromagnetic or strong interactions, only weak interactions, so the most probable thing for it to do is to pass right through any detector you place in its way. In fact, neutrinos emitted in nuclear reactions in the Sun mostly pass right through the entire Earth! However, with extremely intense neutrino beams produced in nuclear reactors or accelerators, or with gigantic underground detectors used to identify neutrinos coming from the Sun, it is possible to have a measurable rate of neutrino reactions.

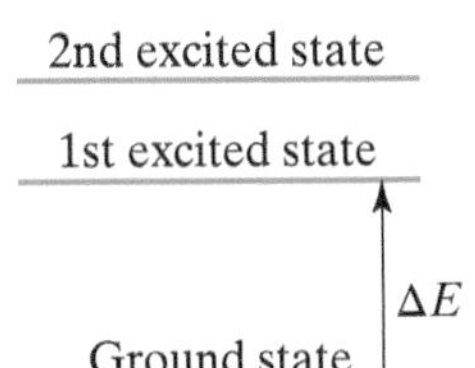

Figure 10.33 Energy levels for a gold nucleus. A sufficiently energetic alpha particle may have enough energy to raise the nucleus above its ground state.

10.11 INELASTIC COLLISIONS AND QUANTIZED ENERGY

The Rutherford experiment involved elastic collisions, because the internal energies of the gold nucleus and the alpha particle didn't change during the interaction. When there is an internal energy change, or the production of additional particles or the emission of light, we call the collision "inelastic."

If we shoot alpha particles at a gold nucleus with higher energy than the energy of alpha particles in the Rutherford experiment, we may be able to excite the gold nucleus to a higher quantum energy level, an amount ΔE above the ground state of the gold nucleus (Figure 10.33). Such a state corresponds to a different configuration of the nucleons inside the gold nucleus.

Suppose that you know the incoming momentum p_1 of the alpha particle, and you are able to measure its outgoing momentum p_3 and scattering angle θ (Figure 10.34).

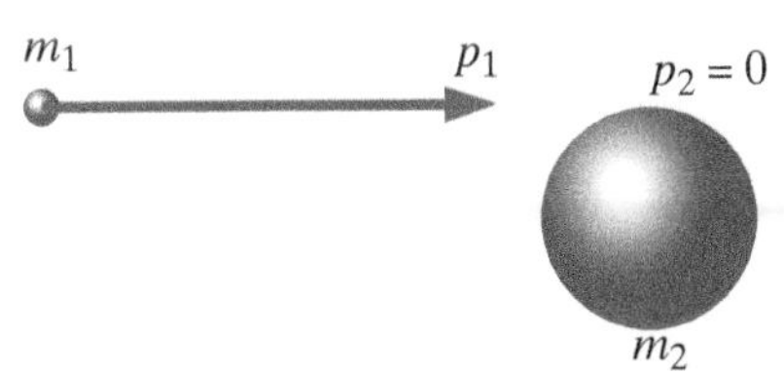

Figure 10.34 You measure the incoming and outgoing momenta of the alpha particle and deduce something about the energy levels in the gold nucleus.

QUESTION What is the form of the energy conservation and momentum conservation equations in this case? Don't assume that the speeds are small compared to the speed of light. Note that the rest energy of the recoiling gold nucleus is $M_4c^2 + \Delta E$ because it is in an excited state.

Here are the relevant equations:

$$p_1 = p_3 \cos\theta + p_4 \cos\phi,$$

where p_4 and ϕ refer to the gold nucleus, which is not observed in the experiment

$$0 = p_3 \cos(90° - \theta) - p_4 \cos(90° - \phi)$$

$$E_1 + Mc^2 = E_3 + E_4, \text{ where } E_4 = \sqrt{(p_4c)^2 + (M_4c^2 + \Delta E)^2}$$

> **QUESTION** How many unknown quantities are there in the energy and momentum equations? Is there enough information to be able to determine the internal energy difference ΔE? (This is one of the methods used to study the energy levels of nuclei. This energy level determination can be verified by measuring the energy of a gamma-ray photon emitted when the nucleus drops back to the ground state.)

There are three unknowns (p_4, ϕ, and ΔE), and there are three equations (energy, x and y momentum), so we could determine ΔE.

Change of Identity

Instead of merely exciting the target particle to a higher internal energy state, suppose that there is a change of identity in which the target particle turns into some other particle. An example is the following particle reaction, which was used in the 1950s to discover the Δ^+ particle of mass m_Δ, which is a particle more massive than a proton:

$$\pi^- + p^+ \to \pi^- + \Delta^+$$

A beam of negative pions (π^-) of known mass m_π and known kinetic energy, produced by a particle accelerator, is directed at a container of hydrogen gas, and the targets for the collisions are the protons of mass m_p that are the nuclei of the hydrogen atoms. Sometimes an incoming pion interacts through the strong interaction with a proton to produce a Δ^+ particle. The momentum of the scattered pion can be measured by measuring the radius of curvature of its curving trajectory in a magnetic field.

> **QUESTION** What is the form of the energy conservation and momentum conservation equations in this case? (Don't assume that the speeds are small compared to the speed of light.)

Here are the relativistically correct equations:

$$p_1 = p_3 \cos\theta + p_4 \cos\phi$$
$$0 = p_3 \cos(90° - \theta) - p_4 \cos(90° - \phi)$$
$$E_1 + m_p c^2 = E_\pi + E_\Delta, \text{ where } E^2 - (pc)^2 = (mc^2)^2$$

> **QUESTION** How many unknown quantities are there in the energy and momentum equations? Is there enough information to be able to determine the mass of the Δ^+ particle?

There are three unknowns (p_4, ϕ, and E_Δ) and three equations, so it is possible to determine the mass of the Δ^+ particle.

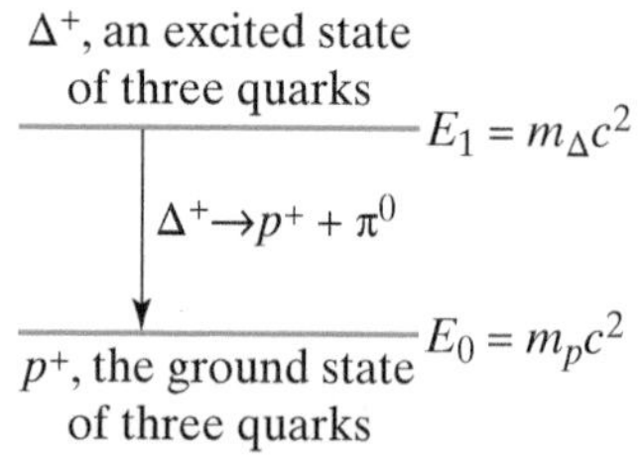

Figure 10.35 The Δ^+ particle can be considered to be an excited state of three quarks, with the proton representing the ground state. A decay to the ground state can occur with the emission of a neutral pion (π^0).

What We Learn from Inelastic Collisions

The analysis for finding the mass change in the preceding exercises is essentially the same as the analysis for determining the internal energy change of an excited gold nucleus. In fact, we can say that the Δ^+ particle is an excited state of three quarks, with the proton representing the ground state of three quarks (Figure 10.35). The excited state (the Δ^+ particle) can decay to the ground state (the proton) with the emission of a neutral pion (π^0), which is a quark–antiquark pair. The extra kinetic energy is equal to $(m_\Delta c^2 - m_p c^2)$. Scattering experiments have been an important tool for discovering new particles (or new excited states of quark systems).

10.12 COLLISIONS IN OTHER REFERENCE FRAMES

In the case of a bowling ball colliding head-on with a Ping-Pong ball, we saw that it can be useful to carry out the analysis in a moving reference frame. In this section we will explore this further by analyzing a two-dimensional sticking collision in two different reference frames.

A Sticking Collision

We'll consider an inelastic collision in which a truck and a car collide in an icy intersection and stick together (Figure 10.36). In the figure we have shown velocity vectors rather than momentum vectors; a typical truck would have much more momentum than the car, given the truck's much greater mass.

Let M be the mass of the truck, m the mass of the car, and θ the angle of their mutual final velocity, as shown in the diagram. Also let v_1 be the initial speed of the truck, v_2 the initial speed of the car, and v_3 the final speed when they're stuck together and sliding on the ice.

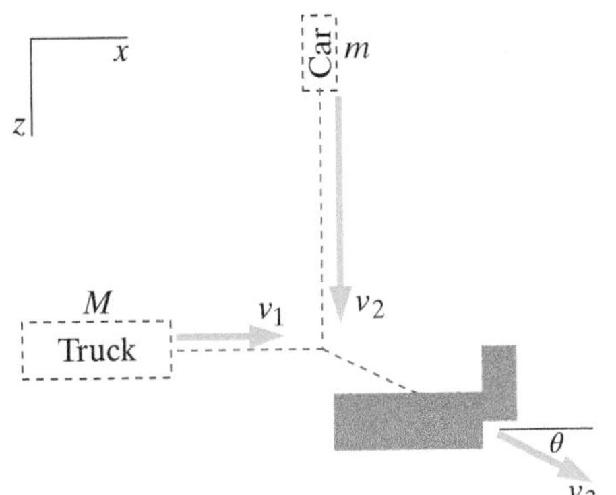

Figure 10.36 A sticking collision between a car and a truck on an icy road. This is a top view, looking down on the xz plane from a location on the $+y$ axis.

QUESTION How does the fact that the intersection is icy affect our ability to apply conservation of momentum to the truck–car collision?

Friction with the road is small if the intersection is icy, so we can presumably ignore road friction and air resistance during the short time of the actual collision. Therefore the total momentum of the truck + car system should be (approximately) unchanged. Momentum conservation is really three equations, one each for the x, y, and z directions. In the y direction (vertically upward) the ice simply supports the weights of the vehicles, so the net force is zero, and the zero y component of momentum remains zero throughout the collision.

QUESTION In terms of the named quantities, write two equations for the x and z components of the momentum conservation equation, equating momentum components before the collision to momentum components after the vehicles stick together.

The unit vector $\hat{v}_3$ giving the final direction of motion is:

$$\hat{v}_3 = \langle \cos\theta, 0, \cos(90^\circ - \theta) \rangle = \langle \cos\theta, 0, \sin\theta \rangle$$

so

$$\vec{v}_3 = v_3 \langle \cos\theta, 0, \sin\theta \rangle = \langle v_3 \cos\theta, 0, v_3 \sin\theta \rangle$$

The momentum conservation equation is:

$$\langle Mv_1, 0, 0 \rangle + \langle 0, 0, mv_2 \rangle = \langle (M+m)v_3 \cos\theta, 0, (M+m)v_3 \sin\theta \rangle$$

Separating this equation into x and z components:

$$x: \quad Mv_1 = (M+m)v_3 \cos\theta$$
$$z: \quad mv_2 = (M+m)v_3 \sin\theta$$

Assume that we know the initial speeds of the truck and car (v_1 and v_2). Then we have two equations with two unknowns (v_3 and θ), and we can solve for the final speed and angle. To find θ, divide the z equation by the x equation:

$$\frac{(M+m)v_3 \sin\theta}{(M+m)v_3 \cos\theta} = \frac{mv_2}{Mv_1} = \tan\theta$$

so

$$\theta = \arctan\left(\frac{mv_2}{Mv_1}\right)$$

To find v_3, square the equations and add them:

$$(Mv_1)^2 + (mv_2)^2 = [(M+m)v_3]^2[(\sin\theta)^2 + (\cos\theta)^2]$$

$$\sqrt{(Mv_1)^2 + (mv_2)^2} = (M+m)v_3$$

$$v_3 = \frac{\sqrt{M^2v_1^2 + m^2v_2^2}}{M+m}$$

We used the trig identity $\sin^2\theta + \cos^2\theta = 1$, which corresponds to the Pythagorean theorem applied to a triangle whose hypotenuse has length 1.

Energy in a Sticking Collision

A sticking collision is unusual in that momentum conservation alone gives us enough equations (two) to be able to solve for the unknown final speed and unknown direction. If the car and truck bounce off each other instead of sticking, there are four unknowns instead of two, because there are two final velocities, each with a speed and a direction to be calculated. In that case momentum conservation alone doesn't provide enough equations to be able to analyze the situation fully.

There is an important issue associated with energy in a sticking collision such as this one. Using our result for the final speed, we can write the kinetic energy after the collision:

$$\frac{1}{2}(M+m)v_3^2 = \frac{1}{2}(M+m)\left[\frac{(Mv_1)^2 + (mv_2)^2}{(M+m)^2}\right]$$

After minor rearrangements we get the following:

$$\frac{1}{2}(M+m)v_3^2 = \left(\frac{M}{M+m}\right)\left(\frac{1}{2}Mv_1^2\right) + \left(\frac{m}{M+m}\right)\left(\frac{1}{2}mv_2^2\right)$$

Since $M/(M+m)$ is less than 1, and $m/(M+m)$ is also less than 1, the final kinetic energy is less than the initial kinetic energy.

QUESTION Where is the "missing" kinetic energy?

Part of the energy was radiated away as sound waves in the air, produced when the vehicles collided with a bang. Some of the energy went into raising the temperature of the vehicles, so that their thermal energy is now higher than before (larger random motions of the atoms).

Some of the energy went into deforming the metal bodies of the truck and car. A bent fender has a higher internal energy, associated with a change in the configuration of the atoms. Sometimes you can get such energy back (as when you compress a spring, changing the configuration of the atoms in the spring, and later let it push back, returning to its original configuration), and we speak of "elastic" energy. Sometimes, however, you can't get the energy back, because the configuration change gets "locked in."

A sticking collision is an example of an "inelastic" collision, referring to the effective "loss" of kinetic energy into forms that are inaccessible.

Checkpoint 4 Car 1 headed north and car 2 headed west collide. They stick together and leave skid marks on the pavement, which show that car 1 was deflected 30° (so car 2 was deflected 60°). What can you conclude about the cars before the collision?

The Collision in a Different Reference Frame

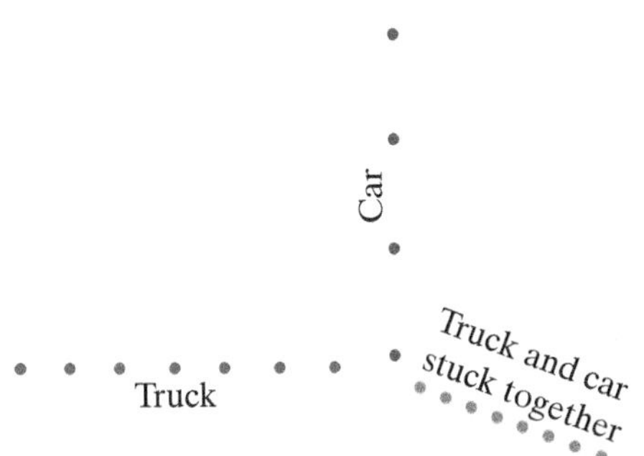

Figure 10.37 View of collision from hovering (stationary) helicopter.

We can show that a truck–car collision process looks a lot simpler from the vantage point of a helicopter flying over the scene with the final velocity of the truck + car combined system. This will lead to new insights. Suppose the accident takes place at night, and from the helicopter all you can see are lights mounted on top of the two vehicles. Figure 10.37 shows a multiple exposure of the scene that you would see from a helicopter that is hovering over the intersection.

Suppose instead that your helicopter flies over the intersection with the same velocity that the stuck-together truck and car share after their collision, as shown in Figure 10.38. The view thus obtained is in an important sense "simpler," as shown in Figure 10.39. From the moving observation platform you see the truck and car run into each other and then stand still (since you and they now have the same velocity). In particular, in this (moving) reference frame it is obvious that there is a loss of kinetic energy into other forms of energy, because after the collision the kinetic energy is clearly zero. In contrast, we had to go through a fair amount of algebra in the nonmoving reference frame in order to establish the loss of kinetic energy.

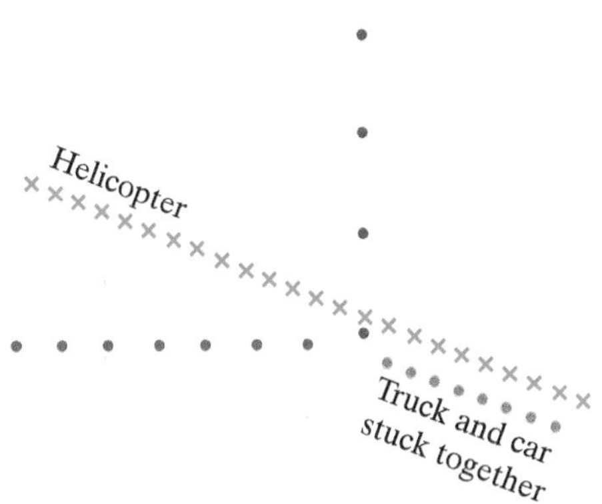

Figure 10.38 Path of moving helicopter that produces the simple view seen in Figure 10.39.

The reason that we imagine making these observations at night is so that you won't be distracted by the moving background. If you actually view this collision from a moving helicopter in daylight, there is a strong tendency to compensate mentally for your motion and to put yourself (mentally) into the reference frame of the road, in which case you could miss the important fact that *relative to you*, the car and truck are stationary after the collision.

You may have had an experience related to this situation. You're sitting in a bus or train, waiting to leave. You think you've started forward, but then you realize that you were fooled by a neighboring bus or train moving backward. For a moment you were mentally in the reference frame of the moving vehicle.

Total Momentum Is Zero in the Center-of-Mass Frame

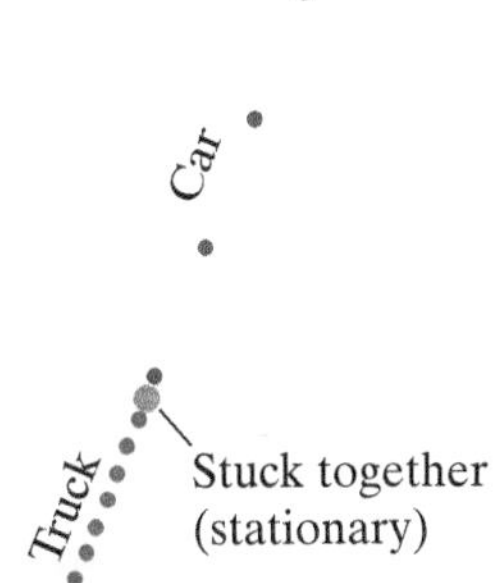

Figure 10.39 View of collision from helicopter moving with velocity of stuck-together wreck.

Viewing a collision from a moving reference frame can provide insights into all kinds of collisions, not just sticking collisions. The key point that leads to a useful generalization is that the total momentum of the truck + car system after the collision is zero when the vehicles are motionless relative to your helicopter.

> QUESTION In this moving reference frame, if the total momentum is zero after the collision, what was the total momentum before the collision? Why?

The helicopter is moving at constant velocity (constant speed and direction), and we know that the Momentum Principle is valid in this uniformly moving reference frame. Thus, the principle of momentum conservation still holds, and the initial total momentum of the car + truck system must be zero if the final total momentum is zero. That's what makes the process look so simple: the total momentum is zero before and after the collision in the moving reference frame.

In viewing the scene from the moving helicopter, we have subtracted from each of the velocities an unchanging, constant vector velocity, the velocity $\vec{v}_h$ of the helicopter, and this subtraction makes no change in the Momentum Principle for the car (or truck):

$$\frac{d[m(\vec{v}-\vec{v}_h)]}{dt} = \frac{d[m\vec{v}]}{dt} = \vec{F}_{\text{net}}$$

The net force is the same in both reference frames because the contact electric interaction forces depend on interatomic distances, which are the same in both reference frames.

Evidently we gain insight into a collision by viewing the collision in a moving reference frame where the total momentum is zero. How do we determine the appropriate velocity for such a reference frame? Let this velocity be $\vec{v}_{\text{fr}}$, which is to be subtracted from all the original velocities to obtain the velocities as viewed in the moving reference frame. We want the total momentum to add up to zero in the moving frame:

$$\vec{p}_{\text{sys,new frame}} = m_1(\vec{v}_1 - \vec{v}_{\text{fr}}) + m_2(\vec{v}_2 - \vec{v}_{\text{fr}}) = \vec{0}$$
$$(m_1 + m_2)\vec{v}_{\text{fr}} = m_1\vec{v}_1 + m_2\vec{v}_2$$
$$\vec{v}_{\text{fr}} = \frac{m_1\vec{v}_1 + m_2\vec{v}_2}{m_1 + m_2}$$

We have solved for a particular velocity for the moving reference frame that makes the total momentum in that frame be zero—but look at the form of this equation.

QUESTION Does this equation look familiar?

Evidently $\vec{v}_{\text{fr}} = \vec{v}_{\text{CM}}$, the velocity of the center of mass of the system, since

$$\vec{r}_{\text{CM}} = \frac{m_1\vec{r}_1 + m_2\vec{r}_2}{m_1 + m_2}$$

We have obtained an important result. In a reference frame moving with the center of mass of a system, the total momentum of the system is zero. If no external forces act, the total momentum is unchanged, and the total momentum is zero before and after a collision.

We can also express the frame velocity directly in terms of the total momentum:

$$\vec{v}_{\text{CM}} = \frac{m_1\vec{v}_1 + m_2\vec{v}_2}{m_1 + m_2} = \frac{\vec{p}_{\text{sys}}}{M_{\text{tot}}}$$

This provides another way to calculate the frame velocity and shows why this is often called the "center-of-momentum" velocity.

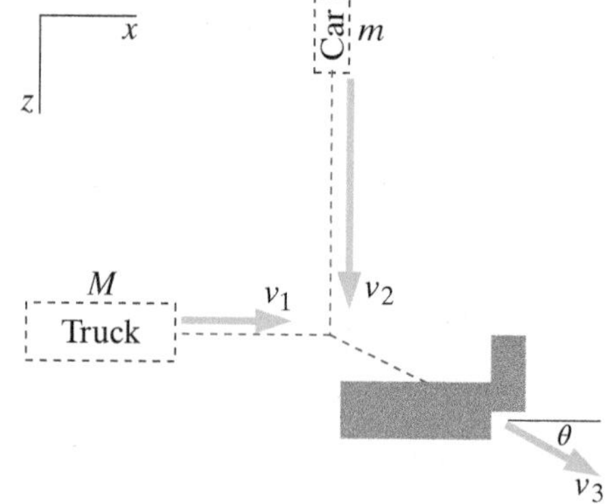

Figure 10.40 Collision of car and truck.

Let's reanalyze the car–truck collision from this new point of view. Again let M be the mass of the truck, m the mass of the car, and θ the angle of their mutual final velocity, as shown in Figure 10.40. Also let v_1 be the initial speed of the truck, v_2 the initial speed of the car, and v_3 the final speed when they're stuck together and sliding on the ice.

QUESTION What are the x and z components of $\vec{v}_{\text{CM}}$ in terms of v_1 and v_2?

$$\vec{v}_{\text{CM}} = \left\langle \frac{Mv_1}{M+m}, 0, \frac{mv_2}{M+m} \right\rangle$$

QUESTION In the moving center-of-momentum reference frame, what are the x and z components of the car's velocity before the collision? Remember that you find the car's velocity (or its components) in the center-of-momentum frame by subtracting the center-of-momentum velocity (or its components) from that of the car. (That is, the view from the helicopter is obtained by subtracting the helicopter's velocity from the velocities of the car and truck.)

$$\left\langle -\frac{Mv_1}{M+m}, 0, v_2 - \frac{mv_2}{M+m} \right\rangle = \left\langle -\frac{Mv_1}{M+m}, 0, \frac{Mv_2}{M+m} \right\rangle$$

QUESTION In the moving center-of-momentum reference frame, what are the x and z components of the truck's velocity before the collision? Show that this velocity is in a direction opposite to the velocity of the car, and that the total momentum is zero.

$$\left\langle v_1 - \frac{Mv_1}{M+m}, 0, -\frac{mv_2}{M+m} \right\rangle = \left\langle \frac{mv_1}{M+m}, 0, -\frac{mv_2}{M+m} \right\rangle$$

The x and z components of velocity are in the opposite direction to those of the car (multiply the car's velocity by $-m/M$ and you have the truck's velocity). Also, the total momentum is zero in the center-of-momentum frame:

$$\left\langle \left[m\left(-\frac{Mv_1}{M+m}\right) + M\left(\frac{mv_1}{M+m}\right)\right], 0, \left[m\left(\frac{Mv_2}{M+m}\right) + M\left(-\frac{mv_2}{M+m}\right)\right] \right\rangle = \vec{0}$$

QUESTION In the moving center-of-momentum reference frame, what is the speed of the car after the collision? Of the truck? Why?

Both are zero; in the center-of-mass frame the total momentum is zero, and with the car and truck stuck together the only way for this to be true is for them to be at rest. Because of this, the kinetic energy loss to other kinds of energy is equal to the total kinetic energy in the center-of-momentum frame before the collision.

QUESTION Now, after the collision, transform back to the (nonmoving) reference frame of the intersection, by adding back in the $\vec{v}_{CM}$ that had been subtracted. What are the speed and direction of the car after the collision? Of the truck? Compare with the results obtained in our earlier analysis of this collision.

$$\vec{v}_3 = \left\langle \frac{Mv_1}{M+m}, 0, \frac{mv_2}{M+m} \right\rangle$$

which yields
$$v_3 = \frac{\sqrt{M^2v_1^2 + m^2v_2^2}}{M+m}; \quad \theta = \arctan\left(\frac{mv_2}{Mv_1}\right)$$

These are the same results we obtained in our earlier analysis.

Energy in the Two Reference Frames

As far as energy is concerned, note that the kinetic energy of the car + truck system is, as usual,

$$K_{tot} = K_{trans} + K_{rel}$$

where $K_{trans} = \dfrac{p_{sys}^2}{2M_{tot}}$. In the original reference frame, $K_{trans} > 0$ but is the same before and after the collision, since p_{sys} doesn't change. In the center-of-momentum reference frame, $K_{trans} = 0$ both before and after the collision, because $p_{sys} = 0$ in the center-of-momentum reference frame.

In either reference frame, the kinetic energy relative to the center of mass, K_{rel}, decreases to zero in the collision, because the car and truck after the collision are traveling with the center of mass. It is ΔK_{rel} that represents the loss of kinetic energy in this inelastic collision, and this quantity is the same in both reference frames. The increase in internal energy is relatively easy to calculate in the center-of-momentum reference frame, as it is simply

$$\Delta E_{int} = -\Delta K_{rel} = 0 - (K_{truck,i} + K_{car,i})$$

Also note that distances between atoms are the same in both reference frames, so potential energy associated with bending of metal parts is the same in both frames.

Checkpoint 5 A 1000 kg car moving east at 30 m/s runs head-on into a 3000 kg truck moving west at 20 m/s. The vehicles stick together. Use the concept of the center-of-momentum frame to determine how much kinetic energy is lost.

SUMMARY

New concepts
Elastic and inelastic collisions
Scattering, and scattering distributions
Impact parameter (0 if head-on)
Reference frame transformations:

$$\vec{v} = \vec{v}' + \vec{v}_{\text{frame}}$$

The center-of-mass reference frame

$$\vec{v}_{\text{cm}} = \frac{m_1\vec{v}_1 + m_2\vec{v}_2}{m_1 + m_2} = \frac{\vec{p}_{\text{sys}}}{M_{\text{tot}}}$$

Simple results for head-on elastic collisions between Ping-Pong balls and bowling balls

Evidence for the nuclear model of an atom

Problem-solving techniques
Conservation of momentum and conservation of energy applied to collisions
Simplification of calculations in the center-of-mass reference frame
Computational models can predict full trajectories of colliding objects. Applying conservation of momentum and energy can predict only final momenta, and in most cases additional information is required for such an analysis.

QUESTIONS

Q1 Under what conditions is the momentum of a system constant? Can the x component of momentum be constant even if the y component is changing? In what circumstances? Give an example of such behavior.

Q2 In a collision between an electron and a hydrogen atom, why is it useful to select both objects as the system? Pick all that apply: (1) The total momentum of the system does not change during the collision. (2) The sum of the final kinetic energies must equal the sum of the initial kinetic energies for a two-object system. (3) The kinetic energy of a two-object system is nearly zero. (4) The forces the objects exert on each other are internal to the system and don't change the total momentum of the system. (5) During the time interval from just before to just after the collision, external forces are negligible.

Q3 Two asteroids in outer space collide and stick together. The mass of each asteroid, and the velocity of each asteroid before the impact, are known. To find the momentum of the stuck-together asteroids after the impact, what approach would be useful? (1) Use the Energy Principle. (2) Use the Momentum Principle. (3) It depends on whether or not the speed of the asteroids was near the speed of light. (4) Use the relationship among velocity, displacement, and time. (5) It depends on whether the collision was elastic or inelastic.

Q4 You know that a collision must be "elastic" if: (1) The colliding objects stick together. (2) The colliding objects are stretchy or squishy. (3) The sum of the final kinetic energies equals the sum of the initial kinetic energies. (4) There is no change in the internal energies of the objects (thermal energy, vibrational energy, etc.). (5) The momentum of the two-object system doesn't change.

Q5 What happens to the velocities of the two objects when a high-mass object hits a low-mass object head-on? When a low-mass object hits a high-mass object head-on?

Q6 It has been proposed to propel spacecraft through the Solar System with a large sail that is struck by photons from the Sun. **(a)** Which would be more effective, a black sail that absorbs photons or a shiny sail that reflects photons back toward the Sun? Explain briefly. **(b)** Suppose that N photons hit a shiny sail per second, perpendicular to the sail. Each photon has energy E. What is the force on the sail? Explain briefly.

Q7 In an elastic collision involving known masses and initial momenta, how many unknown quantities are there after the collision? How many equations are there? In a sticking collision involving known masses and initial momenta, how many unknown quantities are there after the collision? Explain how you can determine the amount of kinetic energy change.

Q8 What properties of the alpha particle and the gold nucleus in the original Rutherford experiment were responsible for the collisions being elastic collisions?

Q9 Give an example of what we can learn about matter through the use of momentum and energy conservation applied to scattering experiments. Explain what it is that we cannot learn this way, for which we need to measure the distribution of scattering angles.

Q10 What is it about analyzing collisions in the center-of-mass frame that simplifies the calculations?

Q11 In order to close a door, you throw an object at the door. Which would be more effective in closing the door, a 50 g tennis ball or a 50 g lump of sticky clay? Explain clearly what physics principles you used to draw your conclusion.

Q12 Consider a head-on collision between two objects. Object 1, which has mass m_1, is initially in motion, and collides head-on with object 2, which has mass m_2 and is initially at rest. Which of the following statements about the collision are true? (1) $\vec{p}_{1,\text{initial}} = \vec{p}_{1,\text{final}} + \vec{p}_{2,\text{final}}$. (2) $|\vec{p}_{1,\text{final}}| < |\vec{p}_{1,\text{initial}}|$. (3) If $m_2 \gg m_1$, then $|\Delta\vec{p}_1| > |\Delta\vec{p}_2|$. (4) If $m_1 \gg m_2$, then the final speed of object 2 is less than the initial speed of object 1. (5) If $m_2 \gg m_1$, then the final speed of object 1 is greater than the final speed of object 2.

Q13 In a nuclear fission reactor, each fission of a uranium nucleus is accompanied by the emission of one or more high-speed neutrons, which travel through the surrounding material. If one of these neutrons is captured in another uranium nucleus, it can trigger fission, which produces more fast neutrons, which could make possible a chain reaction (Figure 10.41).

However, fast neutrons have low probability of capture and usually scatter off uranium nuclei without triggering fission. In order to sustain a chain reaction, the fast neutrons must be slowed down in some material, called a "moderator." For reasons having to do with the details of nuclear physics, slow neutrons have a high probability of being captured by uranium nuclei.

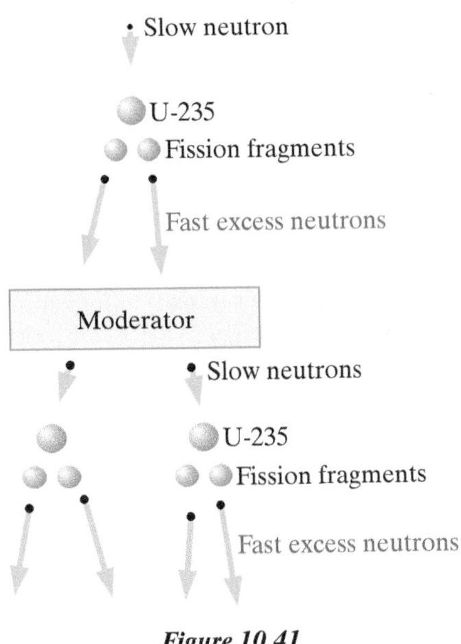

Figure 10.41

A slow neutron induces fission of U-235, with the emission of additional (fast) neutrons. The moderator is some material that slows down the fast neutrons, enabling a chain reaction.

In the following analyses, remember that neutrons have almost no interaction with electrons. Neutrons do, however, interact strongly with nuclei, either by scattering or by being captured and made part of the nucleus. Therefore you should think about neutrons interacting with nuclei (through the strong force), not with entire atoms. **(a)** Based on what you now know about collisions, explain why fast neutrons moving through a block of uranium experience little change in speed. **(b)** Explain why carbon should be a much better moderator of fast neutrons than uranium. **(c)** Should water be a better or worse moderator of fast neutrons than carbon? Explain briefly.

Background: The first fission reactor was constructed in 1941 in a squash court under the stands of Stagg Field at the University of Chicago by a team led by the physicist Enrico Fermi. The moderator consisted of blocks of graphite, a form of carbon. The graphite had to be exceptionally pure because certain kinds of impurities have nuclei that capture neutrons with high probability, removing them from contributing to the chain reaction. Many reactors use ordinary "light" water as a moderator, though sometimes a proton captures a neutron and forms a stable deuterium nucleus, in which case the neutron is lost to contributing to the chain reaction. Heavy water, D_2O, in which the hydrogen atoms are replaced by deuterium atoms, actually works better as a moderator than light water, because the probability of a deuterium nucleus capturing a neutron to form tritium is quite small.

PROBLEMS

Section 10.1

•P14 A ball whose mass is 0.2 kg hits the floor with a speed of 8 m/s and rebounds upward with a speed of 7 m/s. The ball was in contact with the floor for 0.5 ms (0.5×10^{-3} s). **(a)** What was the average magnitude of the force exerted on the ball by the floor? **(b)** Calculate the magnitude of the gravitational force that the Earth exerts on the ball. **(c)** In a collision, for a brief time there are forces between the colliding objects that are much greater than external forces. Compare the magnitudes of the forces found in parts (a) and (b).

Section 10.3

••P15 A projectile of mass m_1 moving with speed v_1 in the $+x$ direction strikes a stationary target of mass m_2 head-on. The collision is elastic. Use the Momentum Principle and the Energy Principle to determine the final velocities of the projectile and target, making no approximations concerning the masses. After obtaining your results, see what your equations would predict if $m_1 \gg m_2$, or if $m_2 \gg m_1$. Verify that these predictions are in agreement with the analysis in this chapter of the Ping-Pong ball hitting the bowling ball, and of the bowling ball hitting the Ping-Pong ball.

Section 10.6

•P16 Object A has mass $m_A = 7$ kg and initial momentum $\vec{p}_{A,i} = \langle 17, -5, 0 \rangle$ kg·m/s, just before it strikes object B, which has mass $m_B = 11$ kg. Object B has initial momentum $\vec{p}_{B,i} = \langle 4, 6, 0 \rangle$ kg·m/s. After the collision, object A is observed to have final momentum $\vec{p}_{A,f} = \langle 13, 3, 0 \rangle$ kg·m/s. In the following questions, "initial" refers to values before the collisions, and "final" refers to values after the collision. Consider a system consisting of both objects A and B. Calculate the following quantities: **(a)** The total initial momentum of this system. **(b)** The final momentum of object B. **(c)** The initial kinetic energy of object A. **(d)** The initial kinetic energy of object B. **(e)** The final kinetic energy of object A. **(f)** The final kinetic energy of object B. **(g)** The total initial kinetic energy of the system. **(h)** The total final kinetic energy of the system. **(i)** The increase of internal energy of the two objects. **(j)** What assumption did you make about Q (energy flow from surroundings into the system due to a temperature difference)?

•P17 In outer space a rock whose mass is 3 kg and whose velocity was $\langle 3900, -2900, 3000 \rangle$ m/s struck a rock with mass 13 kg and velocity $\langle 220, -260, 300 \rangle$ m/s. After the collision, the 3 kg rock's velocity is $\langle 3500, -2300, 3500 \rangle$ m/s. **(a)** What is the final velocity of the 13 kg rock? **(b)** What is the change in the internal energy of the rocks? **(c)** Which of the following statements about Q (transfer of energy into the system because of a temperature difference between system and surroundings) are correct? (1) $Q \approx 0$ because the duration of the collision was very short. (2) $Q = \Delta E_{\text{thermal}}$ of the rocks. (3) $Q \approx 0$ because there are no significant objects in the surroundings. (4) $Q = \Delta K$ of the rocks.

••P18 A spring has an unstretched length of 0.32 m. A block with mass 0.2 kg is hung at rest from the spring, and the spring becomes 0.4 m long (Figure 10.42). Next the spring is stretched to a length of 0.43 m and the block is released from rest. Air resistance is negligible.

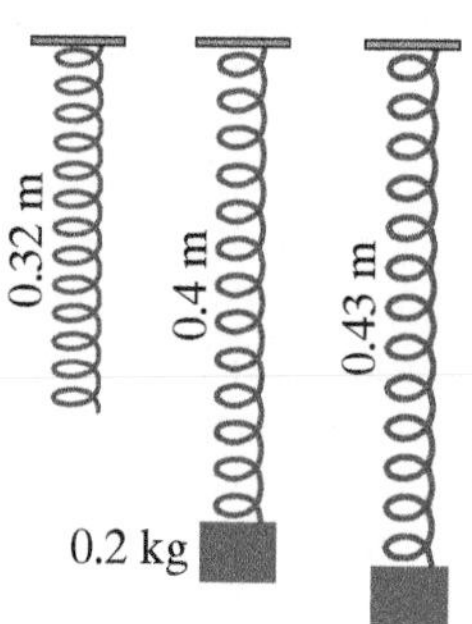

Figure 10.42

(a) How long does it take for the block to return to where it was released? **(b)** Next the block is again positioned at rest, hanging from the spring (0.4 m long) as shown in Figure 10.43. A bullet of mass 0.003 kg traveling at a speed of 200 m/s straight upward buries itself in the block, which then reaches a maximum height h above its original position. What is the speed of the block immediately after the bullet hits? **(c)** Now write an equation that could be used to determine how high the block goes after being hit by the bullet (a height h), but you need not actually solve for h.

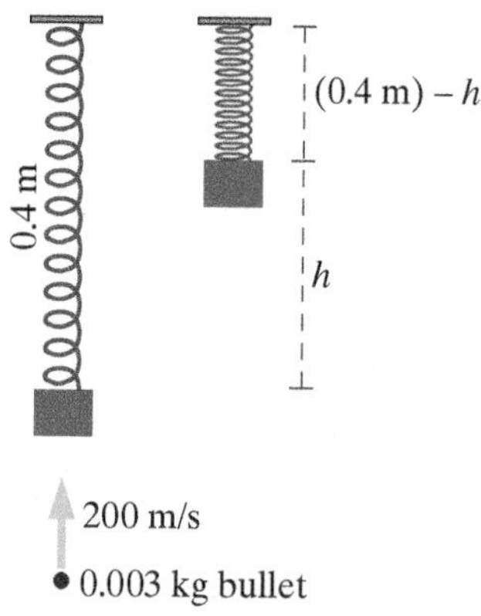

Figure 10.43

•P19 In outer space, rock 1 whose mass is 5 kg and whose velocity was $\langle 3300, -3100, 3400\rangle$ m/s struck rock 2, which was at rest. After the collision, rock 1's velocity is $\langle 2800, -2400, 3700\rangle$ m/s. **(a)** What is the final momentum of rock 2? **(b)** Before the collision, what was the kinetic energy of rock 1? **(c)** Before the collision, what was the kinetic energy of rock 2? **(d)** After the collision, what is the kinetic energy of rock 1? **(e)** Suppose that the collision was elastic (that is, there was no change in kinetic energy and therefore no change in thermal or other internal energy of the rocks). In that case, after the collision, what is the kinetic energy of rock 2? **(f)** On the other hand, suppose that in the collision some of the kinetic energy is converted into thermal energy of the two rocks, where $\Delta E_{\text{thermal},1} + \Delta E_{\text{thermal},2} = 7.16 \times 10^6$ J. What is the final kinetic energy of rock 2? **(g)** In this case (some of the kinetic energy being converted to thermal energy), what was the transfer of energy Q (microscopic work) from the surroundings into the two-rock system during the collision? (Remember that Q represents energy transfer due to a temperature difference between a system and its surroundings.)

•P20 A bullet of mass 0.102 kg traveling horizontally at a speed of 300 m/s embeds itself in a block of mass 3.5 kg that is sitting at rest on a nearly frictionless surface. **(a)** What is the speed of the block after the bullet embeds itself in the block? **(b)** Calculate the kinetic energy of the bullet plus the block before the collision. **(c)** Calculate the kinetic energy of the bullet plus the block after the collision. **(d)** Was this collision elastic or inelastic? **(e)** Calculate the rise in internal energy of the bullet plus block as a result of the collision.

••P21 A car of mass 2300 kg collides with a truck of mass 4300 kg, and just after the collision the car and truck slide along, stuck together, with no rotation. The car's velocity just before the collision was $\langle 38, 0, 0\rangle$ m/s, and the truck's velocity just before the collision was $\langle -16, 0, 27\rangle$ m/s. **(a)** Your first task is to determine the velocity of the stuck-together car and truck just after the collision. What system and principle should you use? (1) Energy Principle (2) Car plus truck (3) Momentum Principle (4) Car alone (5) Truck alone **(b)** What is the velocity of the stuck-together car and truck just after the collision? **(c)** In your analysis in part (b), why can you neglect the effect of the force of the road on the car and truck? **(d)** What is the increase in internal energy of the car and truck (thermal energy and deformation)? **(e)** Is this collision elastic or inelastic?

Section 10.7

••P22 A gold nucleus contains 197 nucleons (79 protons and 118 neutrons) packed tightly against each other. A single nucleon (proton or neutron) has a radius of about 1×10^{-15} m. Remember that the volume of a sphere is $\frac{4}{3}\pi r^3$. **(a)** Calculate the approximate radius of the gold nucleus. **(b)** Calculate the approximate radius of the alpha particle, which consists of 4 nucleons, 2 protons and 2 neutrons. **(c)** What kinetic energy must alpha particles have in order to make contact with a gold nucleus?

Rutherford correctly predicted the angular distribution for 10-MeV (kinetic energy) alpha particles colliding with gold nuclei. He was lucky: if the alpha particle had been able to touch the gold nucleus, the strong interaction would have been involved and the angular distribution would have deviated from that predicted by Rutherford, which was based solely on electric interactions.

•••P23 An alpha particle (a helium nucleus, containing 2 protons and 2 neutrons) starts out with kinetic energy of 10 MeV (10×10^6 eV), and heads in the $+x$ direction straight toward a gold nucleus (containing 79 protons and 118 neutrons). The particles are initially far apart, and the gold nucleus is initially at rest. Assuming that all speeds are small compared to the speed of light, answer the following questions about the collision. **(a)** What is the final momentum of the alpha particle, long after it interacts with the gold nucleus? **(b)** What is the final momentum of the gold nucleus, long after it interacts with the alpha particle? **(c)** What is the final kinetic energy of the alpha particle? **(d)** What is the final kinetic energy of the gold nucleus? **(e)** Assuming that the movement of the gold nucleus is negligible, calculate how close the alpha particle will get to the gold nucleus in this head-on collision.

Section 10.10

••P24 A Fe-57 nucleus is at rest and in its first excited state, 14.4 keV above the ground state (14.4×10^3 eV, where 1 eV $= 1.6 \times 10^{-19}$ J). The nucleus then decays to the ground state with the emission of a gamma ray (a high-energy photon). **(a)** What is the recoil speed of the nucleus? **(b)** Calculate the slight difference in eV between the gamma-ray energy and the 14.4 keV difference between the initial and final nuclear states. **(c)** The "Mössbauer effect" is the name given to a related phenomenon discovered by Rudolf Mössbauer in 1957, for which he received the 1961 Nobel Prize for physics. If the Fe-57 nucleus is in a solid block of iron, occasionally when the nucleus emits a gamma ray the entire solid recoils as one object. This can happen due to the fact that neighboring atoms and nuclei are connected by the electric interatomic force. In this case, repeat the calculation of part (b) and compare with your previous result. Explain briefly.

••P25 There is an unstable particle called the "sigma-minus" (Σ^-), which can decay into a neutron and a negative pion (π^-): $\Sigma^- \rightarrow n + \pi^-$. The mass of the Σ^- is 1196 MeV/c^2, the mass of the neutron is 939 MeV/c^2, and the mass of the π^- is 140 MeV/c^2. Write equations that could be used to calculate the momentum and energy of the neutron and the pion. You do not need to solve the equations, which would involve some messy algebra. However, be clear in showing that you have enough equations that you could in principle solve for the unknown quantities in your equations.

It is advantageous to write the equations not in terms of v but rather in terms of E and p; remember that $E^2 - (pc)^2 = (mc^2)^2$.

••P26 A beam of high-energy π^- (negative pions) is shot at a flask of liquid hydrogen, and sometimes a pion interacts through the strong interaction with a proton in the hydrogen, in the reaction $\pi^- + p^+ \rightarrow \pi^- + X^+$, where X^+ is a positively charged particle of unknown mass.

The incoming pion momentum is 3 GeV/c (1 GeV = 1000 MeV = 1×10^9 electron-volts). The pion is scattered through 40°, and its momentum is measured to be 1510 MeV/c (this is done by observing the radius of curvature of its circular trajectory in a magnetic field). A pion has a rest energy of 140 MeV, and a proton has a rest energy of 938 MeV.

What is the rest mass of the unknown X^+ particle, in MeV/c^2? Explain your work carefully.

It is advantageous to write the equations not in terms of v but rather in terms of E and p; remember that $E^2 - (pc)^2 = (mc^2)^2$.

••P27 A particle of mass m, moving at speed $v = (4/5)c$, collides with an identical particle that is at rest. The two particles react to produce a new particle of mass M and nothing else. **(a)** What is the speed V of the composite particle? **(b)** What is its mass M?

••P28 A charged pion ($m_\pi c^2 = 139.6$ MeV) at rest decays into a muon ($m_\mu = 105.7$ MeV) and a neutrino (whose mass is very nearly zero). Find the kinetic energy of the muon and the kinetic energy of the neutrino, in MeV (1 MeV = 1×10^6 eV).

••P29 A hydrogen atom is at rest, in the first excited state, when it emits a photon of energy 10.2 eV. **(a)** What is the speed of the ground-state hydrogen atom when it recoils due to the photon emission? Remember that the magnitude of the momentum of a photon of energy E is $p = E/c$. Make the initial assumption that the kinetic energy of the recoiling atom is negligible compared to the photon energy. **(b)** Calculate the kinetic energy of the recoiling atom. Is this kinetic energy indeed negligible compared to the photon energy?

•••P30 At the PEP II facility at the Stanford Linear Accelerator Center (SLAC) in California and at the KEKB facility in Japan, electrons with momentum 9.03 GeV/c were made to collide head-on with positrons whose momentum is 3.10 GeV/c (1 GeV = 10^9 eV); see Figure 10.44. That is, pc for the electron is 9.03 GeV and pc for the positron is 3.10 GeV. The values of pc and the corresponding energies are so large with respect to the electron or positron rest energy (0.5 MeV = 0.0005 GeV) that for the purposes of this analysis you may, if you wish, safely consider the electron and positron to be massless.

e^- $p_1 = 9.027$ GeV/c e^+ $p_2 = 3.10$ GeV/c Before
$\Upsilon(4S)$ $v = ?$ After

Figure 10.44

(a) The electron–positron collision produces in an intermediate state a particle called the $\Upsilon(4S)$ ("Upsilon 4S"), in the reaction $e^- + e^+ \rightarrow \Upsilon(4S)$. Show that the rest energy of the $\Upsilon(4S)$ is 10.58 GeV. **(b)** What is the speed of the $\Upsilon(4S)$ produced in the collision? **(c)** The $\Upsilon(4S)$ decays almost immediately into two "B" mesons: $\Upsilon(4S) \rightarrow B^0 + \overline{B^0}$. The $\overline{B^0}$ is the antiparticle of the B^0 and has the same rest energy $Mc^2 = 5.28$ GeV as the B^0. Consider the case in which both "B" mesons are emitted at the same angle θ to the direction of the moving $\Upsilon(4S)$, as shown in Figure 10.45. Calculate this angle θ.

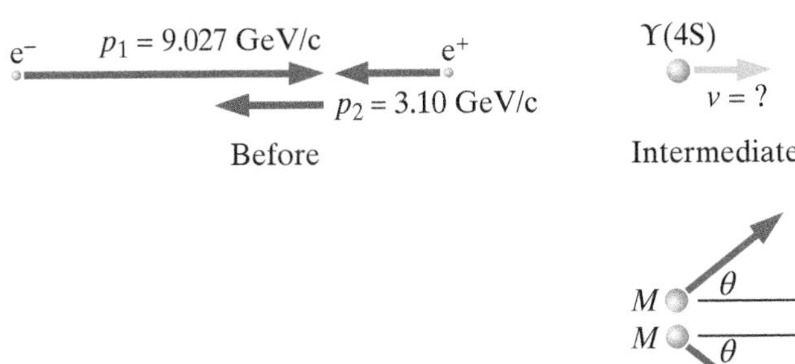

Figure 10.45

•••P31 In a reference frame where a Δ^+ particle is at rest, it decays into a proton and a high-energy photon (a "gamma ray"): $\Delta^+ \rightarrow p^+ + \gamma$. The mass of the Δ^+ particle is 1232 MeV/c^2 and the mass of the proton is 938 MeV/c^2 (1 MeV = 1×10^6 eV). Calculate the energy of the gamma ray and the speed of the proton.

Section 10.12

••P32 Redo Problem P21, this time using the concept of the center-of-momentum reference frame.

••P33 Redo the analysis of the Rutherford experiment, this time using the concept of the center-of-momentum reference frame. Let m = the mass of the alpha particle and M = the mass of the gold nucleus. Consider the specific case of the alpha particle rebounding straight back. The incoming alpha particle has a momentum p_1, the outgoing alpha particle has a momentum p_3, and the gold nucleus picks up a momentum p_4. **(a)** Determine the velocity of the center of momentum of the system. **(b)** Transform the initial momenta to that frame (by subtracting the center-of-momentum velocity from the original velocities). **(c)** Show that if the momenta in the center-of-momentum frame simply turn around (180°), with no change in their magnitudes, both momentum and energy conservation are satisfied, whereas no other possibility satisfies both conservation principles. (Try drawing some other momentum diagrams.) **(d)** After the collision, transform back to the original reference frame (by adding the center-of-momentum velocity to the velocities of the particles in the center-of-mass frame). Although using the center-of-momentum frame may be conceptually more difficult, the algebra for solving for the final speeds is much simpler.

COMPUTATIONAL PROBLEMS

More detailed and extended versions of some of these computational modeling problems may be found in the lab activities included in the *Matter & Interactions*, 4th Edition, resources for instructors.

••P34 The following code is the skeleton of a computational model of a collision between a single moving alpha particle and a single gold nucleus which is initially stationary. Some pieces of the physical model are missing. An alpha particle consists of 2 protons and 2 neutrons. A gold nucleus contains 79 protons and 118 neutrons.

```
from visual import *
from visual.graph import *
scene.width = 1024
scene.height = 600
q_e = 1.6e-19
m_p = 1.7e-27
oofpez = 9e9
m_Au = (79+118) * m_p
m_Alpha = (2+2) * m_p
qAu = 2 * q_e
qAlpha = 79 * q_e
deltat = 1e-23
Au = sphere(pos=vector(0,0,0), radius=4e-15,
              color=color.yellow, make_trail=True)
Alpha = sphere(pos=vector(-1e-13,5e-15,0),
                 radius=1e-15, color=color.magenta,
                 make_trail=True)
p_Au = m_Au*vector(0,0,0)
p_Alpha = vector(1.043e-19,0,0)
t = 0
while t<1.3e-20:
    rate(100)
    Alpha.pos = Alpha.pos+(p_Alpha/m_Alpha) *
deltat
    t = t + deltat
```

(a) Read the code, and predict what will happen when the unmodified skeleton program is run. Then run the program to check your prediction. **(b)** Modify the program so that it represents a reasonable physical model of this interaction. **(c)** Does the gold nucleus move during these collisions? Should it? You can see the trail left by the gold nucleus better if you make this object slightly transparent by adding this line of code before the loop: `Au.opacity = 0.7` **(d)** What is the value of the impact parameter b in this program? Experiment with different impact parameters, and report what you observe. **(e)** After the loop, add to your program a calculation of the scattering angle θ (the angle between the final and initial momenta of the alpha particle). The VPython function for the dot product is `dot(A,B)`. The function `acos(D)` returns the angle (in radians) whose cosine is D. Check to make sure the angle you calculate makes sense in terms of what you observe on the screen. **(f)** Find values of the impact parameter that lead to the following scattering angles: (1) 90°, (2) 168°, (3) 38°, (4) 13°.

•P35 Start with the program you wrote for Problem P34. Add graphs to display the values of the x and y components of the momentum of each particle, and the sums of all x components and of all y components, as a function of time. Is momentum conserved during the collision? What should you change in your program if you find that momentum is not conserved? By adding the following code before the loop you can create two different graphing windows, and plot all curves simultaneously:

```
gdx = gdisplay(x=0,y=600, width=500,
                title='p_x')
p_Au_x_graph = gcurve(color=color.yellow)
p_Alpha_x_graph = gcurve(color=color.magenta)
## other gcurves if needed
gdy = gdisplay(x=500,y=600,width=500,title='p_y')
p_Au_y_graph = gcurve(color=color.yellow)
## other gcurves if needed
```

•••P36 The *Voyager 1* spacecraft was launched in 1977 and has recently left the Solar System and entered interstellar space. No existing rocket is powerful enough to launch a spacecraft with enough speed to coast that far against the Sun's gravitational pull, so a "gravity boost" was obtained from Jupiter (and another from Saturn). The basic idea is that it was possible to arrange *Voyager's* trajectory in such a way that in a brief noncontact collision with Jupiter the spacecraft gained momentum, with a corresponding loss of the giant planet's momentum, which due to Jupiter's huge mass corresponded to a tiny change in Jupiter's speed.

Construct a computational model of this maneuver. The collision takes only a few days, so it is okay to approximate Jupiter's motion around the Sun with a straight line (Jupiter takes 12 Earth years to orbit the Sun). Jupiter's mass is 1.9×10^{27} kg. Give your spacecraft the same mass as the *Voyager* spacecraft (722 kg). Jupiter's radius is 7×10^7 m and it is 7.8×10^{11} m from the Sun, with an orbital speed of 1.3×10^4 m/s. **(a)** Place Jupiter at the location $\langle 0,0,0 \rangle$ m, with a velocity of $\langle -1.3 \times 10^4,0,0 \rangle$ m/s. Choose a starting location for the spacecraft at location $\langle x_i,-25R,0 \rangle$, where R is the radius of Jupiter. You will need to find a value for x_i that leads to the spacecraft passing close to the right side of Jupiter. Let the initial velocity of the spacecraft be $\langle 0,1 \times 10^4,0 \rangle$ m/s. (From this perspective you are looking down on the North pole of Jupiter, with the outward bound spacecraft heading toward Jupiter.) **(b)** Exit the computational loop when the spacecraft is as far from Jupiter as it was at the start. (Put this test at the end of your loop, so the loop doesn't stop immediately.) Print the change in the speed, $v_f - v_i$, and the time in Earth days for the process. Modify your starting position x_i until you find that the speed change is approximately 1.6×10^4 m/s. **(c)** Include a graph of the speed of the spacecraft as a function of time. **(d)** Find a different value of x_i for the spacecraft that results in the spacecraft losing a significant amount of speed instead of gaining speed.

Suggestions: If you precede a `gcurve` statement with `gdisplay(y=600)` the graph will be placed conveniently beneath your graphics window. If you specify `scene.autocenter = True` the camera will continually shift to point at the center of the scene, which is useful in this situation.

•••**P37** Start with the program you wrote to model a spacecraft's interaction with Jupiter (Problem P36). Use initial conditions that result in the spacecraft gaining speed. **(a)** Change to a center-of-mass reference frame by subtracting Jupiter's velocity from the initial velocities of the spacecraft and Jupiter. (Because Jupiter is so massive, its velocity is almost exactly the velocity of the center of mass of the combined system.) What do you observe when you re-run your program? **(b)** What is the speed change of the spacecraft in the center-of-mass reference frame?

ANSWERS TO CHECKPOINTS

1 (a) 300 J; **(b)** 5 m/s; **(c)** 150 J; **(d)** 150 J

2 (a) 2.3×10^{-24} m/s (!), so it is an excellent approximation to say that the Earth doesn't recoil. **(b)** 1.6×10^{-23} J; $K_{\text{baseball}} = 150$ J

3 8×10^4 m/s

4 Car 1's momentum was $\sqrt{3}$ greater than the momentum of car 2.

5 We outline the entire solution.

First find $v_{\text{cm},x}$:

$$v_{\text{cm},x} = \frac{(1000\,\text{kg})(30\,\text{m/s}) + (3000\,\text{kg})(-20\,\text{m/s})}{(4000\,\text{kg})} = -7.5\,\text{m/s}$$

Transform to center-of-momentum frame:

$$v_{\text{car},x} = (30\,\text{m/s}) - (-7.5\,\text{m/s}) = 37.5\,\text{m/s}$$
$$v_{\text{truck},x} = (-20\,\text{m/s}) - (-7.5\,\text{m/s}) = -12.5\,\text{m/s}$$

Check to make sure that the momenta are equal and opposite in this frame:

$$p_{\text{car},x} = (1000\,\text{kg})(37.5\,\text{m/s}) = 37500\,\text{kg}\cdot\text{m/s}$$
$$p_{\text{truck},x} = (3000\,\text{kg})(-12.5\,\text{m/s}) = -37500\,\text{kg}\cdot\text{m/s}$$

In this frame the kinetic energy before the collision is this:

$$K = \frac{(37500\,\text{kg}\cdot\text{m/s})^2}{2(1000\,\text{kg})} + \frac{(37500\,\text{kg}\cdot\text{m/s})^2}{2(3000\,\text{kg})} = 9.4 \times 10^5\,\text{J}$$

After the collision, the velocities are zero in this frame, so kinetic energy in this frame goes to zero, and there is an increase in the internal energy of the mangled car and truck of amount 9.4×10^5 J. There must be the same internal energy increase in the original reference frame, so the kinetic energy lost in the original reference frame was 9.4×10^5 J.

CHAPTER

11

Angular Momentum

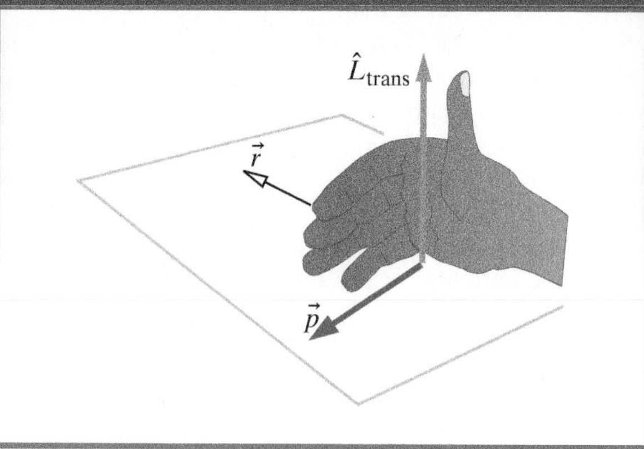

OBJECTIVES

After studying this chapter, you should be able to

- Calculate both translational and rotational angular momentum in 3D.
- Apply the Angular Momentum Principle to predict the new angular momentum of a system subject to a torque.
- Apply Conservation of Angular Momentum to a system consisting of more than one particle.
- Predict the position of a rotating object.

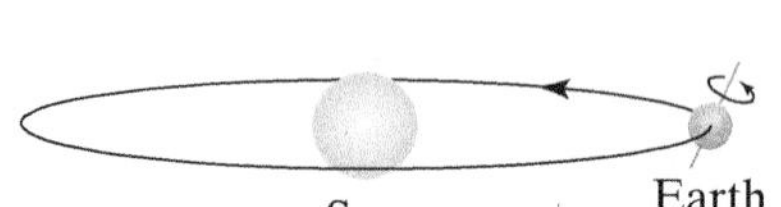

Figure 11.1 The Earth orbits the Sun, and it also spins about its own axis (which is tilted with respect to the orbital plane). The sizes of the Earth and Sun are exaggerated in this diagram.

Angular momentum is a measure of rotational motion. Translational (or "orbital") angular momentum describes motion such as the orbit of the Earth around the Sun. Rotational (or "spin") angular momentum describes motion such as the revolution of the Earth around its own axis (Figure 11.1). Just as the Momentum Principle relates a change in momentum to the net force on a system, the Angular Momentum Principle relates a change in angular momentum to the net torque, or twist, applied to a system.

11.1 TRANSLATIONAL ANGULAR MOMENTUM

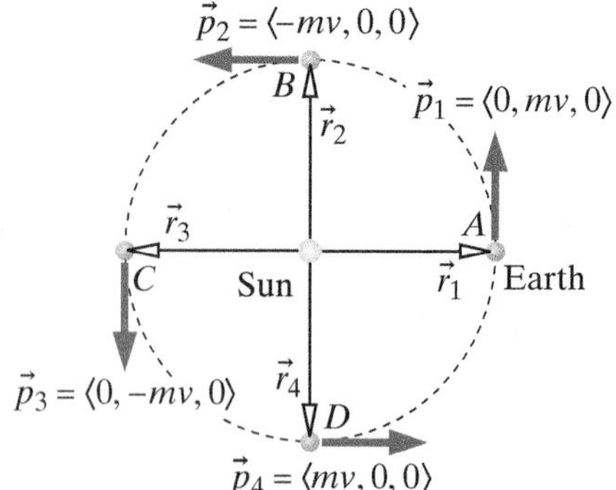

Figure 11.2 The Earth goes around the Sun in a nearly circular orbit. The position vector and the momentum vector change continuously.

As the Earth goes around the Sun in a nearly circular orbit, the position vector (from Sun to Earth) is constantly changing in direction, and the momentum of the Earth is also constantly changing in direction, as shown in Figure 11.2. Both the position vector and the momentum vector continually change, though their magnitudes are (nearly) constant. In Figure 11.2, as the Earth goes from A to B in three months (one quarter of a year), the Earth's momentum changes from $\langle 0,mv,0\rangle$ to $\langle -mv,0,0\rangle$, and the change in the momentum is $\Delta\vec{p} = \langle -mv,0,0\rangle - \langle 0,mv,0\rangle = \langle -mv,-mv,0\rangle$. This change in the Earth's momentum is of course due to the gravitational force exerted on the Earth by the Sun.

Nevertheless, it feels like something is constant here. For example, the angle between the position vector and the momentum vector is always 90° for this circular orbit. Also, the direction of rotation of the circular motion stays constant (counterclockwise in Figure 11.2). The quantity that is constant is $\vec{L}_{\text{trans,Sun}}$, the translational angular momentum of the Earth relative to the Sun.

Magnitude of Translational Angular Momentum

The magnitude of the Earth's orbital or translational angular momentum, relative to the center of the Sun, involves $|\vec{r}\,|$, $|\vec{p}|$, and the angle θ between the two. It is defined like this:

$$\left|\vec{L}_{\text{trans,Sun}}\right| = |\vec{r}\,|\,|\vec{p}|\sin\theta$$

EXAMPLE **Earth's Translational Angular Momentum**

Calculate the magnitude of the Earth's translational (orbital) angular momentum relative to the Sun when the Earth is at location A and when the Earth is at location B as shown in Figure 11.2. The mass of the Earth is 6×10^{24} kg and its distance from the Sun is 1.5×10^{11} m (see the table of values at the end of this book).

Solution The Earth makes one complete orbit of the Sun in 1 y, so its average speed is:

$$v = \frac{2\pi(1.5 \times 10^{11}\ \text{m})}{(365)(24)(60)(60)\ \text{s}} = 3.0 \times 10^{4}\ \text{m/s}$$

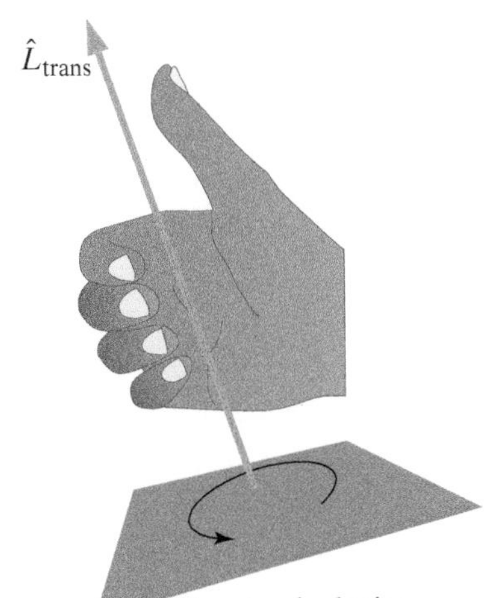

Figure 11.3 The right-hand rule.

At location A

$$\vec{p} = \langle 0, 6 \times 10^{24}\ \text{kg} \cdot 3.0 \times 10^{4}\ \text{m/s}, 0\rangle = \langle 0, 1.8 \times 10^{29}, 0\rangle\ \text{kg} \cdot \text{m/s}$$

$$|\vec{p}| = 1.8 \times 10^{29}\ \text{kg} \cdot \text{m/s}$$

$$\left|\vec{L}_{\text{trans,Sun}}\right| = (1.5 \times 10^{11}\ \text{m})(1.8 \times 10^{29}\ \text{kg} \cdot \text{m/s}) \sin 90^\circ$$

$$= 2.7 \times 10^{40}\ \text{kg} \cdot \text{m}^2/\text{s}$$

At location B, $|\vec{r}|$, $|\vec{p}|$, and θ are the same as they were at location A, so $\left|\vec{L}_{\text{trans,Sun}}\right|$ also has the same value it had at location A.

Direction of Translational Angular Momentum

The magnitude of the translational angular momentum doesn't tell us everything we need to know about the orbital motion of the Earth. To describe the motion indicated in Figure 11.2 we could say something like this:

> If you stand at a location on the $+z$ axis and look toward the origin, the Earth appears to be moving counterclockwise in the xy plane.

However, this is annoyingly verbose and not useful mathematically; for example, one can't add "counterclockwise in the xy plane" to "clockwise in the yz plane." Instead, we specify the direction this way:

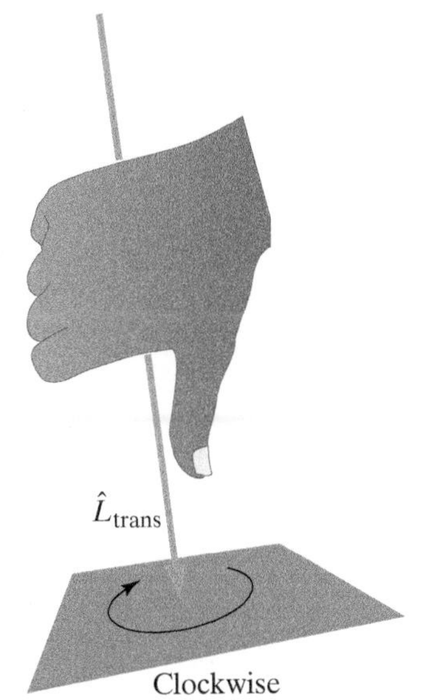

Figure 11.4 The right-hand rule.

- The plane in which both $\vec{r}$ and $\vec{p}$ lie can be indicated by a unit vector perpendicular to that plane.
- The direction of motion within the plane (clockwise or counterclockwise) can be indicated by establishing a "right-hand" rule for this unit vector.
- Right-hand rule: Curl the fingers of your right hand in the direction of the rotational motion in the plane and extend your thumb, as shown in Figure 11.3. The unit vector representing the direction of the angular momentum is defined to point in the direction of your thumb.
- If the rotational motion is counterclockwise, the unit vector (and your right thumb) point out of the plane, as in Figure 11.3. If the rotational motion is clockwise, the unit vector (and your right thumb) points into the plane, as in Figure 11.4.

EXAMPLE **Direction of Translational Angular Momentum**

Assuming the usual coordinate system, what is the direction of the Earth's translational angular momentum, as indicated in Figure 11.2?

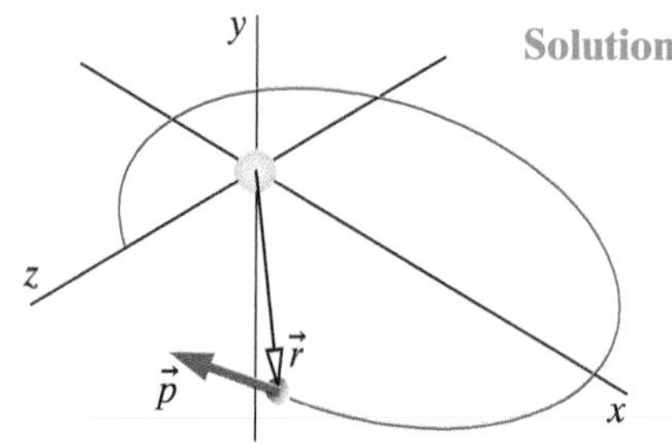

Figure 11.5 A comet orbits the Sun. The momentum vector and relative position vector lie in the xz plane.

Solution Applying the right-hand rule we find that the Earth's translational angular momentum is in the $+z$ direction: out of the page toward you.

Angular Momentum as a Vector

Putting together the results from the previous two examples, we can express the translational angular momentum of the Earth with respect to the Sun as a vector:

$$\vec{L}_{\text{trans,Sun}} = (2.7 \times 10^{40}\ \text{kg} \cdot \text{m}^2/\text{s}) \cdot \langle 0,0,1 \rangle = \langle 0,0,2.7 \times 10^{40} \rangle\ \text{kg} \cdot \text{m}^2/\text{s}$$

Checkpoint 1 What is the direction of the orbital (translational) angular momentum of the comet shown in Figure 11.5, relative to the Sun?

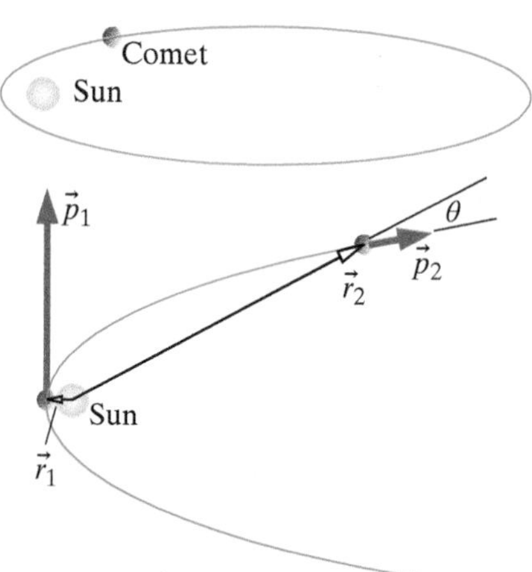

Figure 11.6 The orbit of Halley's comet, shown here in the xy plane. The top figure shows the full orbit. The bottom figure shows a magnified view of the portion of the orbit nearest the Sun.

An Elliptical Orbit: Halley's Comet

The previous exercise raises an interesting question: What if $|\vec{r}|$, $|\vec{p}|$, and θ are not constant (as they were in a circular orbit)? Is the angular momentum, relative to the Sun, of a comet in an elliptical orbit around the Sun constant, or does it change? During this orbit, all three parameters ($|\vec{r}|$, $|\vec{p}|$, and θ) change continuously.

Halley's comet orbits the Sun with a period of 75.3 y, and will next be visible on Earth in the year 2061. Figure 11.6 displays two views of the orbit, here shown in the xy plane. From the upper diagram it is clear that the orbit of the comet is highly elliptical. The lower diagram shows a magnified view of the portion of the orbit nearest to the Sun, and displays the momentum vectors and position vectors for the comet at two different points in its orbit. From available data we can calculate the angular momentum of Halley's comet relative to the Sun at two different locations.

EXAMPLE **Angular Momentum of Halley's Comet**

The highly elliptical orbit of Halley's comet is shown in Figure 11.6. When the comet is closest to the Sun, at the location specified by the position vector $\vec{r}_1$ ("perihelion"), it is 8.77×10^{10} m from the Sun, and its speed is 5.46×10^4 m/s. When the comet is at the location specified by the position vector $\vec{r}_2$, its speed is 1.32×10^4 m/s. At that location the distance between the comet and the Sun is 1.19×10^{12} m, and the angle θ is 17.81°. The mass of the comet is estimated to be 2.2×10^{14} kg. Calculate the translational (orbital) angular momentum of the comet, relative to the Sun, at both locations.

Solution Direction: At both locations, the direction of the translational angular momentum of the comet is in the $-z$ direction (into the page); determined by using the right-hand rule.

At location 1:

$$\left|\vec{L}_{\text{trans,Sun}}\right| = (8.77 \times 10^{10}\ \text{m})(2.2 \times 10^{14}\ \text{kg})(5.46 \times 10^4\ \text{m/s}) \sin 90°$$
$$= 1.1 \times 10^{30}\ \text{kg} \cdot \text{m}^2/\text{s}$$
$$\vec{L}_{\text{trans,Sun}} = \langle 0,0,-1.1 \times 10^{30} \rangle\ \text{kg} \cdot \text{m}^2/\text{s}$$

At location 2:

$$\left|\vec{L}_{\text{trans,Sun}}\right| = (1.19\times10^{12}\text{ m})(2.2\times10^{14}\text{ kg})(1.32\times10^{4}\text{ m/s})\sin 17.81°$$
$$= 1.1\times10^{30}\text{ kg}\cdot\text{m}^2/\text{s}$$
$$\vec{L}_{\text{trans,Sun}} = \langle 0,0,-1.1\times10^{30}\rangle\text{ kg}\cdot\text{m}^2/\text{s}$$

Even in this highly elliptical orbit, the comet's translational angular momentum is constant throughout the orbit, despite the fact that its position, its momentum, and the angle between them change continuously. It was observations like this that first suggested that angular momentum might be a conserved quantity, and should not change if there is no twist exerted on the system by the surroundings.

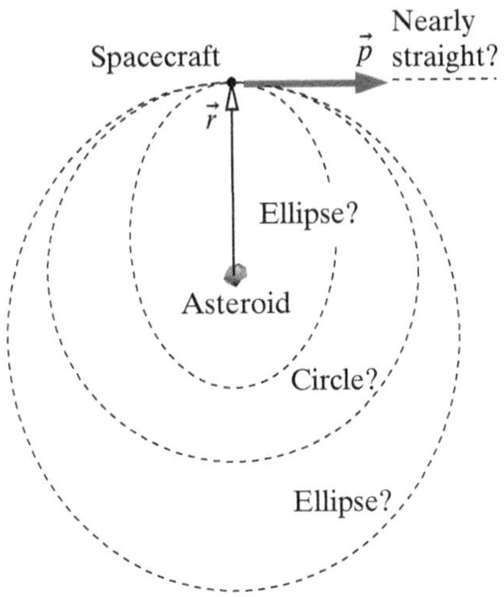

Figure 11.7 A snapshot of a spacecraft passing an asteroid. Its future motion might be a circle or an ellipse, or even a nearly straight line if the asteroid's mass is small.

A Straight Path

In calculating translational (orbital) angular momentum, we use the position and momentum of an object at a particular instant. We do not necessarily have any information about how the object's path will curve in the future or whether it has curved in the past. Figure 11.7 illustrates this point by showing many possible paths of a spacecraft passing an asteroid. It is possible that the spacecraft will be nearly unaffected by the asteroid, and will travel nearly in a straight line, but we can still calculate its orbital angular momentum at the instant shown.

Later in this chapter we will see why it may be physically meaningful to attribute angular momentum to an object that moves in a straight line.

Choosing a Location

Translational or orbital angular momentum is defined relative to a particular location. If we choose a location A, the magnitude of the angular momentum includes the distance r_A to the moving object. The translational angular momentum relative to some other location B would have a different magnitude because r_B would be a different distance. We will see later that it is usually advantageous to choose a location around which there is little or no twist applied by objects in the surroundings.

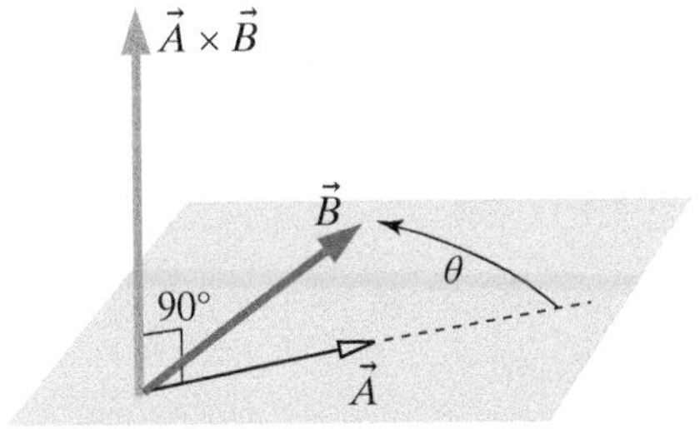

Figure 11.8 The cross product produces a vector that is perpendicular to the two original vectors.

The Vector Cross Product

The formal definition of translational angular momentum involves a vector cross product. In the mathematics of vectors, the cross product is an operation that combines two vectors to produce a third vector that is perpendicular to the plane defined by the original vectors (Figure 11.8). The cross product of two vectors $\vec{A}$ and $\vec{B}$ is written as $\vec{A}\times\vec{B}$.

We already know one way of evaluating a vector cross product, by finding the magnitude and direction separately. The quantity

$$\left|\vec{L}_{\text{trans,Sun}}\right| = |\vec{r}|\,|\vec{p}|\sin\theta$$

is the magnitude of the cross product $\vec{r}\times\vec{p}$:

$$|\vec{r}\times\vec{p}| = \left|\vec{L}_{\text{trans,Sun}}\right| = |\vec{r}|\,|\vec{p}|\sin\theta$$

where θ is the angle between the two vectors when they are placed tail to tail. θ is always less than or equal to 180°. The direction of $\vec{r}\times\vec{p}$ is given by a right-hand rule:

- Draw the vectors $\vec{r}$ and $\vec{p}$ so their tails are in the same location.

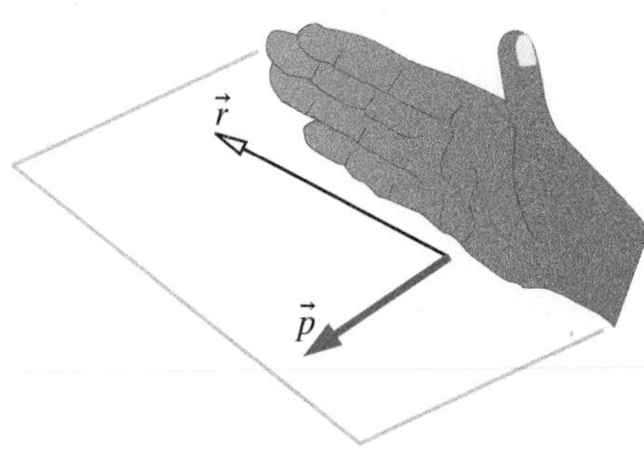

Figure 11.9 With the tails of $\vec{r}$ and $\vec{p}$ together, point your fingers along $\vec{r}$.

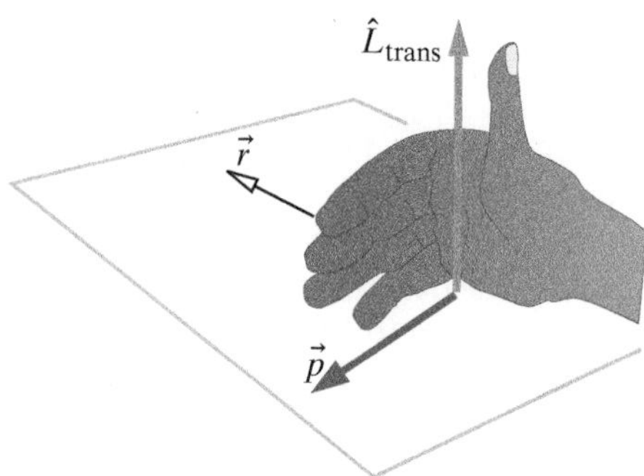

Figure 11.10 Fold your fingers toward $\vec{p}$. Your extended thumb points in the direction of $\vec{r} \times \vec{p}$.

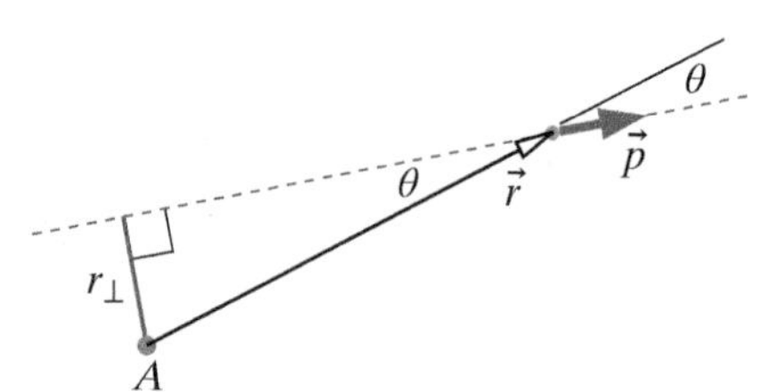

Figure 11.11 $r_\perp = r\sin\theta$ is the perpendicular distance from the tail of $\vec{r}$ to a line collinear with $\vec{p}$.

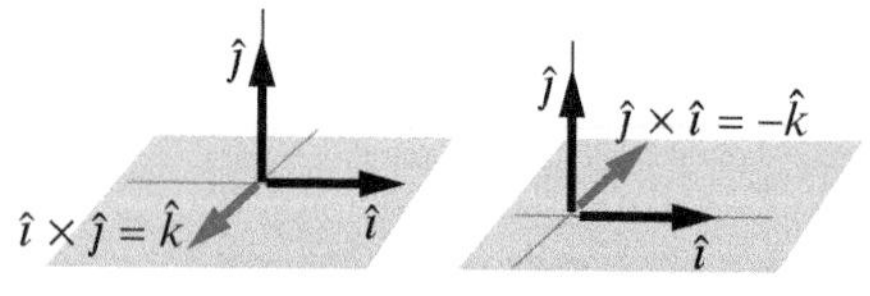

Figure 11.12 Cross products of unit vectors.

- Point the fingers of your right hand in the direction of $\vec{r}$ (Figure 11.9), then fold them through the angle θ toward $\vec{p}$ (Figure 11.10). You may need to turn your hand over to be able to do this.
- Stick your thumb out. Your thumb points in the direction of the angular momentum vector.

The Perpendicular Distance $r_\perp$

The quantity $r\sin\theta$ has a geometrical meaning. As shown in Figure 11.11, it is the component of $\vec{r}$ perpendicular to $\vec{p}$. If you draw a line perpendicular to $\vec{p}$, through the point about which $\vec{L}$ is to be calculated, $r_\perp = r\sin\theta$ is the distance along this line from the tail of $\vec{r}$ to the line along which the momentum or velocity points. $r_\perp$ is closely related to the impact parameter introduced in the previous chapter on collisions.

PERPENDICULAR DISTANCE $r_\perp$

$$r_\perp = r\sin\theta$$

Vector Components in the Cross Product

Given the components of two vectors, it is possible to evaluate the cross product in terms of unit vectors along the three axes (Figure 11.12). Recall that $\hat{\imath} = \langle 1,0,0\rangle$, and so on.

First, note that $\hat{\imath} \times \hat{\imath} = 0$, $\hat{\jmath} \times \hat{\jmath} = 0$, and $\hat{k} \times \hat{k} = 0$, since when we cross a vector with itself the angle between the two vectors is zero, and $\sin 0° = 0$.

Second, $\hat{\imath} \times \hat{\jmath} = \hat{k}$, since the angle is 90° and the right-hand rule gives a result in the $+z$ direction (out of the page; Figure 11.12). On the other hand, $\hat{\jmath} \times \hat{\imath} = -\hat{k}$, because the right-hand rule gives a result in the $-z$ direction (into the page). Similarly, $\hat{\jmath} \times \hat{k} = \hat{\imath}$, $\hat{k} \times \hat{\jmath} = -\hat{\imath}$, $\hat{k} \times \hat{\imath} = \hat{\jmath}$, and $\hat{\imath} \times \hat{k} = -\hat{\jmath}$.

If $\vec{A}$ and $\vec{B}$ lie in the xy plane, we can use these results for the unit vectors to calculate the cross product, which will be in the $+z$ or $-z$ direction:

$$\begin{aligned}
\vec{A} \times \vec{B} &= (A_x\hat{\imath} + A_y\hat{\jmath}) \times (B_x\hat{\imath} + B_y\hat{\jmath}) \\
&= (A_xB_x)\hat{\imath} \times \hat{\imath} + (A_yB_y)\hat{\jmath} \times \hat{\jmath} + (A_xB_y)\hat{\imath} \times \hat{\jmath} + (A_yB_x)\hat{\jmath} \times \hat{\imath} \\
&= 0 + 0 + A_xB_y\hat{k} - A_yB_x\hat{k} \\
&= (A_xB_y - A_yB_x)\hat{k}
\end{aligned}$$

Similar results hold for components of the cross product in the x and y directions, so that in general we have the following result:

CROSS PRODUCT $\vec{A} \times \vec{B}$

$$\vec{A} \times \vec{B} = \langle A_yB_z - A_zB_y, A_zB_x - A_xB_z, A_xB_y - A_yB_x\rangle$$
$$|\vec{A} \times \vec{B}| = |\vec{A}||\vec{B}|\sin\theta$$

The direction of the vector is given by the right-hand rule.

For example,

$$\begin{aligned}
\langle 2,3,4\rangle \times \langle 5,6,7\rangle &= \langle 3\cdot 7 - 4\cdot 6, 4\cdot 5 - 2\cdot 7, 2\cdot 6 - 3\cdot 5\rangle \\
&= \langle -3, 6, -3\rangle
\end{aligned}$$

This approach to calculating a cross product is particularly useful in computer calculations. Note the cyclic nature of the subscripts: $x{:}yz, y{:}zx, z{:}xy$.

If you have studied determinants, you may recognize that the components of $\vec{A} \times \vec{B}$ can be thought of as the "minors" of a 3 by 3 determinant in which $\vec{A}$ and $\vec{B}$ are the second and third rows, and the first row is $\hat{\imath}$, $\hat{\jmath}$, $\hat{k}$.

Applying this general formalism to translational angular momentum, we can define the angular momentum of a particle, relative to location A, as:

TRANSLATIONAL ANGULAR MOMENTUM RELATIVE TO LOCATION A

$$\vec{L}_{\text{trans,A}} = \vec{r}_A \times \vec{p}$$

$\vec{r}_A$ extends from location A to the object.

The vector dot product $\vec{r}_A \times \vec{p}$ may be evaluated directly:

$$\vec{L}_{\text{trans,A}} = \langle r_y p_z - r_z p_y, r_z p_x - r_x p_z, r_x p_y - r_y p_x \rangle$$

Alternatively, magnitude and direction may be found separately:

$$\left|\vec{L}_{\text{trans,A}}\right| = |\vec{r}\,||\vec{p}|\sin\theta$$

The direction of $\vec{L}_{\text{trans,A}}$ may be found using the right-hand rule.

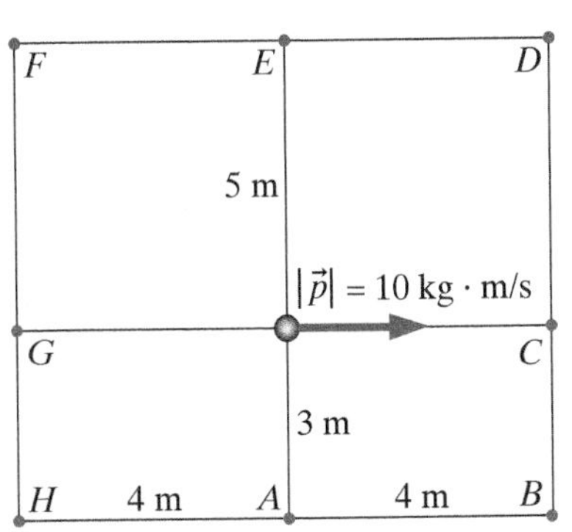

Figure 11.13 Determine the magnitude and direction of the angular momentum of the particle relative to various locations. All points lie in the xy plane.

Note that this formal definition allows us to calculate the translational (orbital) angular momentum of an object at any instant, even if we do not know the direction of the path it will follow in the future.

Two-Dimensional Projections

Because it is more difficult to sketch a situation in three dimensions, whenever possible we will work with two-dimensional projections onto a plane. If $\vec{A}$ and $\vec{B}$ lie in the xy plane, the cross product vector $\vec{A} \times \vec{B}$ points in the $+z$ direction (out of the page, ⊙) or in the $-z$ direction (into the page, ⊗). The symbol ⊙ is supposed to suggest the tip of an arrow pointing toward you (angular momentum pointing toward you). The symbol ⊗ is supposed to suggest the feathers on the end of an arrow that is heading away from you (angular momentum pointing away from you).

EXAMPLE **Calculating Angular Momentum**

Use three different methods to calculate the angular momentum of the particle in Figure 11.13 relative to location B. How does this value compare to the angular momentum of the same particle relative to location A and to location C?

Solution Method 1: Vector cross product

$$\begin{aligned}
\vec{r}_B &= \langle -4, 3, 0 \rangle \text{ m} \\
\vec{p} &= \langle 10, 0, 0 \rangle \text{ kg} \cdot \text{m/s} \\
\vec{L}_B &= \vec{r}_B \times \vec{p} \\
&= \langle r_y p_z - r_z p_y, r_z p_x - r_x p_z, r_x p_y - r_y p_x \rangle \\
&= \langle (3 \cdot 0 - 0 \cdot 0), (0 \cdot 10 - -4 \cdot 0), (-4 \cdot 0 - 3 \cdot 10) \rangle \\
&= \langle 0, 0, -30 \rangle \text{ kg} \cdot \text{m}^2/\text{s}
\end{aligned}$$

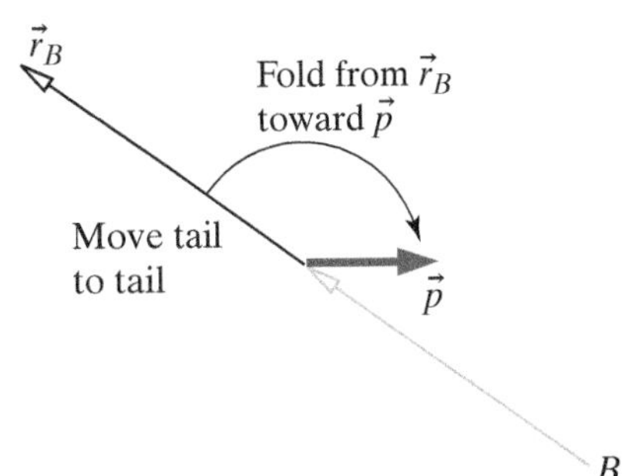

Figure 11.14 Place $\vec{r}_B$ and $\vec{p}$ tail to tail, then fold your right fingers from $\vec{r}_B$ toward $\vec{p}$.

Method 2: Direction and magnitude using $|\vec{r}_B||\vec{p}|\sin\theta$

In Figure 11.14 we have placed $\vec{r}_B$ and $\vec{p}$ tail to tail. The vectors $\vec{r}_B$ and $\vec{p}$ define a plane, which is the xy plane (the plane of the paper). Now fold the fingers of your right hand from $\vec{r}_B$ toward $\vec{p}$ (Figure 11.14). To be able to do this, your thumb must point into the page (⊗), so the angular momentum is in the $-z$ direction (into the plane of the paper). Note too that you fold your fingers clockwise, which is associated with angular momentum into the page.

$$|\vec{L}_B| = |\vec{r}_B||\vec{p}|\sin\theta$$
$$|\vec{r}_B| = \sqrt{3^2+4^2} = 5\text{ m}$$
$$|\vec{p}| = 10\,\text{kg}\cdot\text{m/s}$$
$$\sin\theta = 3/5 \quad \text{(See Figure 11.15)}$$
$$|\vec{L}_B| = (5)(10)(3/5) = 30\,\text{kg}\cdot\text{m}^2\text{/s}$$
$$\vec{L}_B = \langle 0,0,-30\rangle\,\text{kg}\cdot\text{m}^2\text{/s}$$

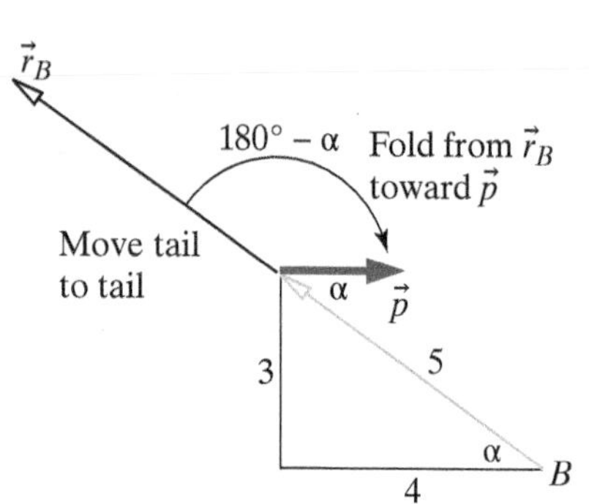

Figure 11.15 Place $\vec{r}_B$ and $\vec{p}$ tail to tail, then rotate $\vec{r}_B$ toward $\vec{p}$. The sine of this angle is $\sin(180° - \alpha) = \sin\alpha = 3/5$.

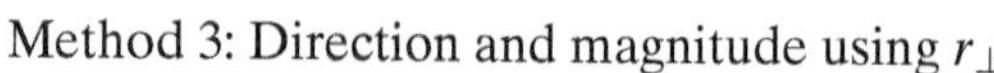

Method 3: Direction and magnitude using $r_\perp$

Determine the direction using the right-hand rule, as in Method 2 above.

Figure 11.15 shows that $r_\perp = 3$ m. Therefore

$$|\vec{L}_B| = (3)(10) = 30\,\text{kg}\cdot\text{m}^2\text{/s}$$

By examining Figure 11.13 we can see that the translational angular momentum relative to location A is exactly the same as the angular momentum relative to location B, because the perpendicular distance $r_\perp$ is the same (3 m).

About location C, $r_\perp$ is zero because the momentum points straight toward C (the impact parameter is zero). Therefore $|\vec{L}_C| = 0$, and the direction of $\vec{L}_C$ is undefined.

Complete the following checkpoint before going on, because you need to be able to use the concept of cross product quickly and easily in later discussions. One way to check your work is to try more than one method to find the angular momentum.

> **Checkpoint 2** Determine both the direction and magnitude of the angular momentum of the particle in Figure 11.13, relative to the locations D, E, F, G, and H. We've already analyzed the angular momentum relative to A, B, and C in the example given above. Notice how the magnitude and direction of the angular momentum relative to the different locations differ in magnitude and direction.

11.2 ROTATIONAL ANGULAR MOMENTUM

We've just seen how to calculate the (translational) angular momentum of a point particle. However, an extended object spinning about its own axis (like the Earth) can't be modeled as a point particle. By modeling such an object as a collection of point particles, and adding up their translational angular momentum, we will see that there is an easy way to calculate the angular momentum of a system such as a spinning bicycle wheel (Figure 11.16).

In Chapter 9 we found that it is useful to express the kinetic energy of a multiparticle system as $K = K_{\text{trans}} + K_{\text{rel}}$. We'll see that it is similarly useful to express the angular momentum of a multiparticle system as a translational part $\vec{L}_{\text{trans},A}$ associated with the motion of the center of mass relative to a location A, plus a rotational part $\vec{L}_{\text{rot}}$ that represents the angular momentum relative to the center of mass, which we'll focus on in this section.

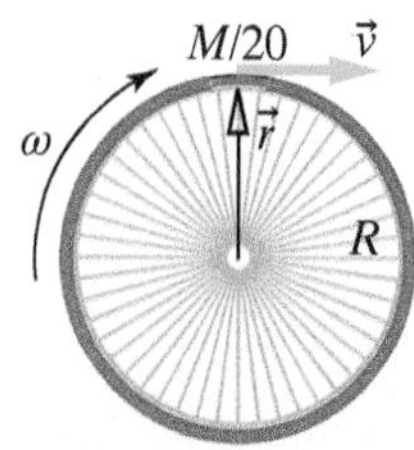

Figure 11.16 A bicycle wheel spinning with angular speed ω about a stationary axle. Almost all the mass of the wheel is in the rim.

Consider a bicycle wheel spinning about its center of mass with angular speed ω (Figure 11.16) about a stationary axle. Most of the mass is in the rim of the wheel, so we will neglect the low-mass spokes and the hub. Imagine dividing the rim into 20 equal-sized pieces, each with mass $M/20$, and consider each piece to be an object orbiting the axis. The speed of one piece is

$$v = \omega R$$

so the magnitude of the translational angular momentum of one piece, relative to the center of mass, is:

$$\begin{aligned}\left|\vec{L}_{\text{CM, one piece}}\right| &= R(M/20)v\sin 90^\circ \\ &= R(M/20)\omega R = (M/20)R^2\omega\end{aligned}$$

By the right-hand rule, the direction of the translational angular momentum of each piece is into the page (the $-z$ direction), so to get the magnitude of the angular momentum of the entire wheel, we can add the contribution of each piece:

$$\begin{aligned}\left|\vec{L}_{\text{CM}}\right| &= 20\cdot(M/20)R^2\omega \\ &= MR^2\omega\end{aligned}$$

Because every piece of the wheel is the same distance from the center of mass, the moment of inertia of the hoop-like wheel is $I = MR^2$. We can therefore rewrite the angular momentum of the wheel as

$$\left|\vec{L}_{\text{CM}}\right| = I\omega$$

The direction of the angular momentum of the entire wheel is $-z$ (into the page). Note that the result would have been the same if we had divided the wheel into 1000 pieces, or into its component atoms.

To convert the result to a vector, we define the angular velocity $\vec{\omega}$ to be a vector with magnitude ω and direction given by a variant of the right-hand rule (Figure 11.17):

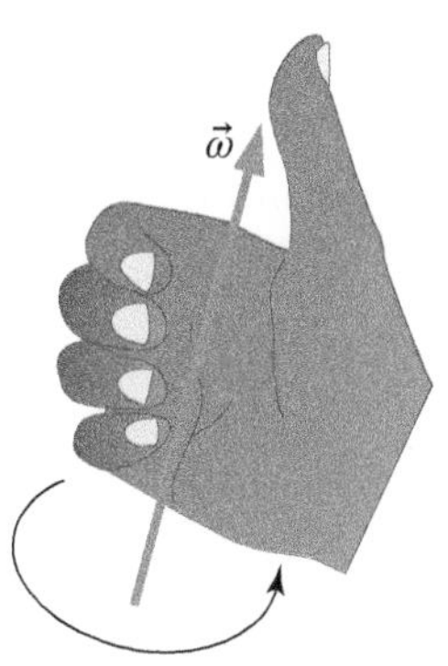
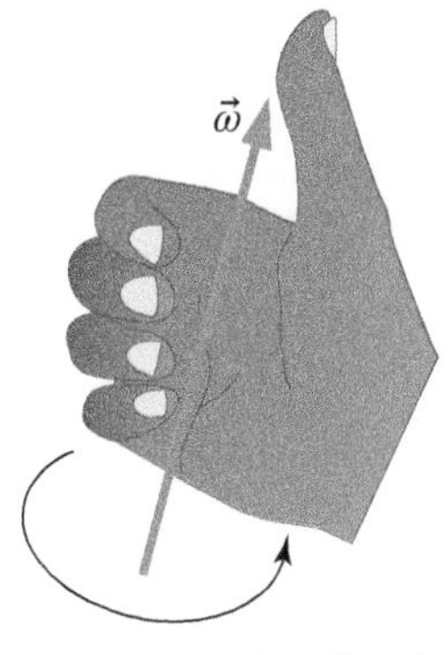

Figure 11.17 Finding the direction of the angular velocity vector. Curl your fingers in the direction of rotation; your thumb sticks out in the direction of $\vec{\omega}$.

ANGULAR VELOCITY VECTOR

$$\vec{\omega}$$

Magnitude: radians/second.
Direction: see Figure 11.17.

We call the angular momentum of a system relative to its center of mass the "rotational" angular momentum $\vec{L}_{\text{rot}}$. For the bicycle wheel we found that $\vec{L}_{\text{rot}} = I\vec{\omega}$, where $I = MR^2$.

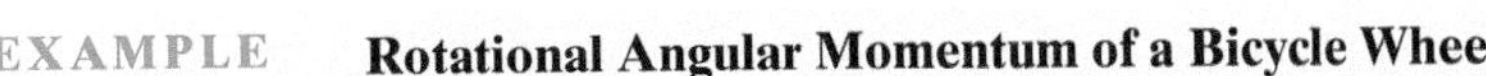

EXAMPLE **Rotational Angular Momentum of a Bicycle Wheel**

A bicycle wheel has a mass of 0.8 kg and a radius of 32 cm. If the wheel rotates in the xz plane, spinning clockwise when viewed from the $+y$ axis, and making one full revolution in 0.75 s, what is the rotational angular momentum of the wheel?

Solution The direction of $\vec{\omega}$ is $-y$.

$$\begin{aligned}I &= MR^2 = (0.8\,\text{kg})(0.32\text{ m})^2 = 0.082\text{ kg}\cdot\text{m}^2 \\ \omega &= \frac{2\pi}{0.75\text{ s}} = 8.38\text{ s}^{-1} \\ \left|\vec{L}_{\text{rot}}\right| &= (0.082\text{ kg}\cdot\text{m}^2)(8.38\text{ s}^{-1}) = 0.69\text{ kg}\cdot\text{m}^2/\text{s} \\ \vec{L}_{\text{rot}} &= \langle 0, -0.69, 0\rangle\text{ kg}\cdot\text{m}^2/\text{s}\end{aligned}$$

$\vec{L}_{\text{rot}}$ in Terms of $\vec{\omega}$

The result that $\vec{L}_{\text{rot}} = I\vec{\omega}$ is a general one, and is true for any object whose atoms all share the same angular velocity $\vec{\omega}$ in radians per second but with

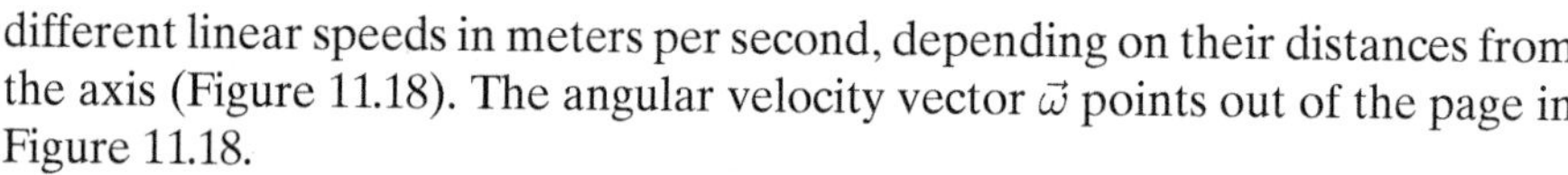

different linear speeds in meters per second, depending on their distances from the axis (Figure 11.18). The angular velocity vector $\vec{\omega}$ points out of the page in Figure 11.18.

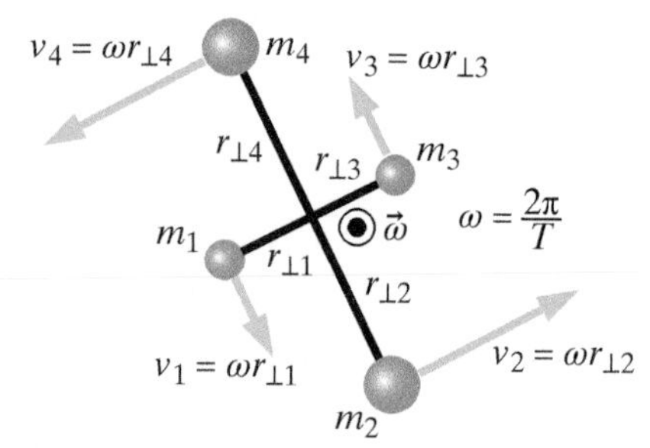

Figure 11.18 A case of rotation about an axis with all particles sharing the same angular speed ω. Note the different linear speeds, with $v_i = \omega r_{\perp i}$. $\vec{\omega}$ points out of the page.

The magnitude of the angular momentum of each piece is this, where $r_\perp$ is the perpendicular distance of a piece from the axis:

$$|\vec{r} \times \vec{p}| = r_\perp m v = r_\perp m(\omega r_\perp) = m r_\perp^2 \omega$$

So the magnitude of the rotational angular momentum is this:

$$|\vec{L}_{\text{rot}}| = [m_1 r_{\perp 1}^2 + m_2 r_{\perp 2}^2 + m_3 r_{\perp 3}^2 + m_4 r_{\perp 4}^2]\omega$$

The quantity in brackets is the moment of inertia:

$$I = m_1 r_{\perp 1}^2 + m_2 r_{\perp 2}^2 + m_3 r_{\perp 3}^2 + m_4 r_{\perp 4}^2 + \cdots$$

Since the rotational angular momentum points in the direction of $\vec{\omega}$, we have the following for the angular momentum relative to the center of mass, even if the center of mass is moving in some complicated way:

ROTATIONAL ANGULAR MOMENTUM

$$\vec{L}_{\text{rot}} = I\vec{\omega}$$

for a system, such as a rigid object, whose components all rotate around its center of mass with the same angular velocity $\vec{\omega}$.

This way of evaluating rotational angular momentum is often easier to use than the more basic expression in terms of cross products $\vec{r} \times \vec{p}$, especially because the moments of inertia for common objects can be looked up in reference books. For example, we saw in Chapter 9 that the moment of inertia about an axis passing through the center of a uniform-density solid disk of radius R is $\frac{1}{2}MR^2$, and the moment of inertia about an axis passing through the center of a uniform-density solid sphere of radius R is $\frac{2}{5}MR^2$.

Since $L_{\text{rot}} = I\omega$, we can express K_{rot} in terms of the rotational angular momentum:

K_{rot} and L_{rot}

$$K_{\text{rot}} = \frac{1}{2}I\omega^2 = \frac{1}{2}\frac{(I\omega)^2}{I} = \frac{L_{\text{rot}}^2}{2I}$$

In this discussion we have considered only simple, symmetric rotating systems. The optional Section 11.13 deals with more complicated situations.

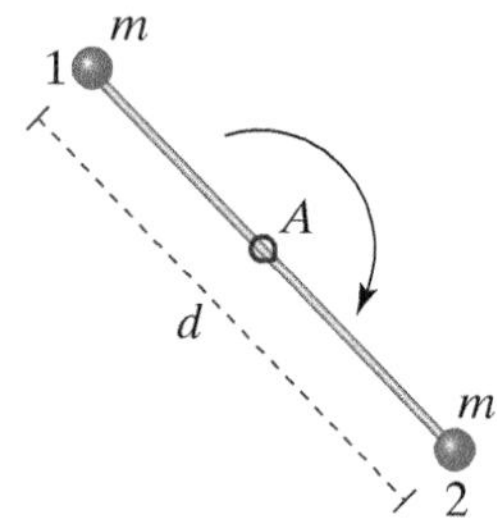

Figure 11.19 A rotating barbell.

Checkpoint 3 A barbell spins around a pivot at its center at A (Figure 11.19). The barbell consists of two small balls, each with mass $m = 0.4$ kg, at the ends of a very low mass rod of length $d = 0.6$ m. The barbell spins clockwise with angular speed $\omega_0 = 20$ radians/s. **(a)** Consider the two balls separately, and calculate $\vec{L}_{\text{trans},1,A}$ and $\vec{L}_{\text{trans},2,A}$ (both direction and magnitude in each case). **(b)** Calculate $\vec{L}_{\text{tot},A} = \vec{L}_{\text{trans},1,A} + \vec{L}_{\text{trans},2,A}$ (both direction and magnitude). **(c)** Next, consider the two balls together and calculate I for the barbell. **(d)** What is the direction of the angular velocity $\vec{\omega}_0$? **(e)** Calculate $\vec{L}_{\text{rot}} = I\vec{\omega}_0$ (both direction and magnitude). **(f)** How does $\vec{L}_{\text{rot}}$ compare to $\vec{L}_{\text{tot},A}$? The point is that the form $I\omega$ is just a convenient way of calculating the (rotational) angular momentum of a multiparticle system. In principle one can always calculate the angular momentum simply by adding up the individual angular momenta of all the particles. **(g)** Calculate K_{rot}.

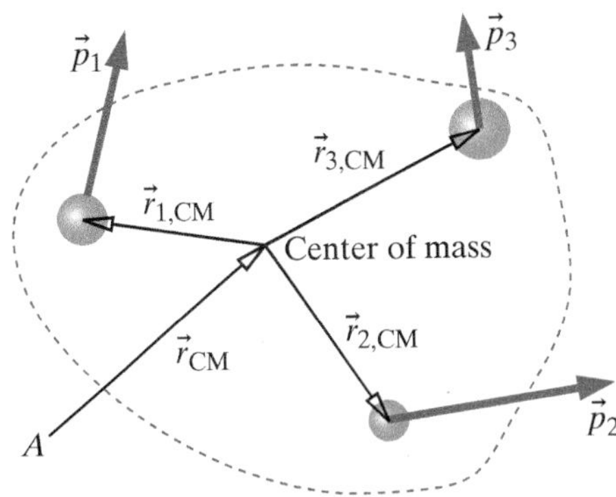

Figure 11.20 Position vectors relative to the center of mass.

11.3 TOTAL ANGULAR MOMENTUM

Kinetic energy can be usefully expressed as $K_{\text{tot}} = K_{\text{trans}} + K_{\text{rel}}$, where "rel" means "relative to the center of mass." Similarly, angular momentum can be usefully expressed as $\vec{L}_{\text{tot},A} = \vec{L}_{\text{trans},A} + \vec{L}_{\text{rel}}$, where we refer to the portion that is relative to the center of mass as rotational angular momentum, $\vec{L}_{\text{rot}}$.

We will calculate the total angular momentum relative to location A of a three-mass system (Figure 11.20) and prove that $\vec{L}_{\text{tot},A} = \vec{L}_{\text{trans},A} + \vec{L}_{\text{rot}}$, where the system need not be rigid. We specify the positions of each mass relative to location A by going from A to the center of mass (vector $\vec{r}_{\text{CM}}$) and then to the individual masses. The location A need not be in the same plane as the masses, nor are the momentum vectors necessarily in the plane of the masses.

The position of mass m_1 relative to location A is $\vec{r}_1 = \vec{r}_{\text{CM}} + \vec{r}_{1,\text{CM}}$, the position of mass m_2 is $\vec{r}_2 = \vec{r}_{\text{CM}} + \vec{r}_{2,\text{CM}}$, and the position of mass m_3 is $\vec{r}_3 = \vec{r}_{\text{CM}} + \vec{r}_{3,\text{CM}}$ (Figure 11.21). Using these vectors we calculate the total angular momentum $\vec{L}_A$ relative to location A:

$$\begin{aligned}\vec{L}_A &= (\vec{r}_{\text{CM}} + \vec{r}_{1,\text{CM}}) \times \vec{p}_1 + (\vec{r}_{\text{CM}} + \vec{r}_{2,\text{CM}}) \times \vec{p}_2 + (\vec{r}_{\text{CM}} + \vec{r}_{3,\text{CM}}) \times \vec{p}_3 \\ &= [\vec{r}_{\text{CM}} \times (\vec{p}_1 + \vec{p}_2 + \vec{p}_3)] + [\vec{r}_{1,\text{CM}} \times \vec{p}_1 + \vec{r}_{2,\text{CM}} \times \vec{p}_2 + \vec{r}_{3,\text{CM}} \times \vec{p}_3]\end{aligned}$$

Since $(\vec{p}_1 + \vec{p}_2 + \vec{p}_3)$ is the total momentum $\vec{P}_{\text{tot}}$, we have the following:

$$\vec{L}_A = \underbrace{[\vec{r}_{\text{CM}} \times \vec{P}_{\text{tot}}]}_{\text{translational}} + \underbrace{[\vec{r}_{1,\text{CM}} \times \vec{p}_1 + \vec{r}_{2,\text{CM}} \times \vec{p}_2 + \vec{r}_{3,\text{CM}} \times \vec{p}_3]}_{\text{rotational}}$$

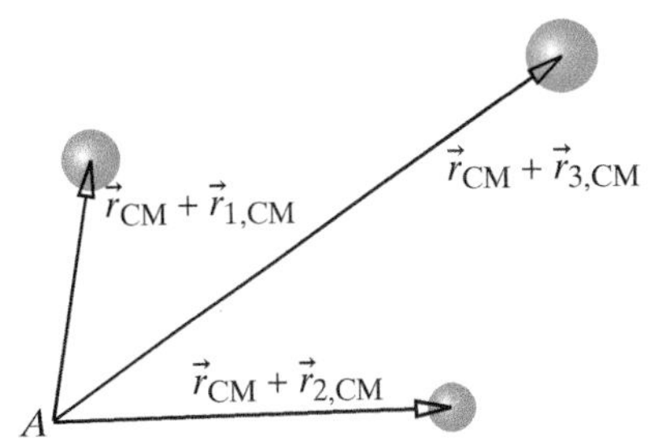

Figure 11.21 Position vectors relative to location A.

The first term in the total angular momentum, $\vec{r}_{\text{CM}} \times \vec{P}_{\text{tot}}$, is the angular momentum the system would have if all the mass were concentrated at the center of mass. It is the angular momentum that a "collapsed" point-particle version of the system would have. This is the translational (or orbital) angular momentum $\vec{L}_{\text{trans},A}$, which is already familiar.

The second term in the total angular momentum,

$$\vec{r}_{1,\text{CM}} \times \vec{p}_1 + \vec{r}_{2,\text{CM}} \times \vec{p}_2 + \vec{r}_{3,\text{CM}} \times \vec{p}_3$$

is the angular momentum of the system relative to the center of mass of the system, which we call the rotational angular momentum $\vec{L}_{\text{rot}}$. An example of rotational angular momentum is the angular momentum due to the rotation of the Earth on its axis. This term is also called the "spin" angular momentum. In the previous section we saw that $\vec{L}_{\text{rot}}$ can be written as $I\vec{\omega}$ if the object is rigid, with all particles sharing the same ω.

In Figure 11.22 we show both the translational (orbital) angular momentum of the Earth relative to the location of the Sun, and the rotational (spin) angular momentum of the Earth relative to the Earth's center of mass. The total angular momentum of the Earth relative to the Sun is the vector sum of the Earth's translational and rotational angular momenta.

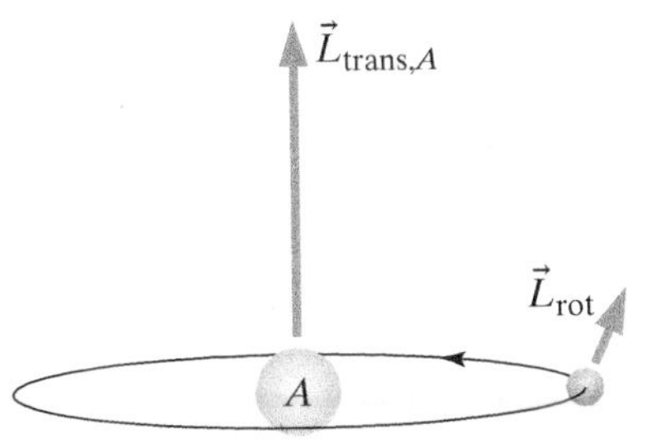

Figure 11.22 Translational angular momentum of the Earth (relative to the location of the Sun), and rotational angular momentum of the Earth (relative to the Earth's center of mass). At this instant the *linear* momentum $\vec{p}$ of the Earth is into the page. The Earth's axis is tilted 23° away from a perpendicular to the plane of the orbit, so the two angular momenta are not parallel.

TOTAL ANGULAR MOMENTUM

$$\vec{L}_A = \vec{L}_{\text{trans},A} + \vec{L}_{\text{rot}}$$

Translational: $\vec{L}_{\text{trans},A} = \vec{r}_{\text{CM},A} \times \vec{P}_{\text{tot}}$

Rotational: $\vec{L}_{\text{rot}} = \vec{r}_{1,\text{CM}} \times \vec{p}_1 + \vec{r}_{2,\text{CM}} \times \vec{p}_2 + \cdots$

The translational angular momentum differs for different choices of the location A, so we need the "A" subscript on the translational angular momentum and the total angular momentum to remind us of this dependence. The rotational angular momentum doesn't need a subscript because it is calculated relative to the center of mass, and this calculation is unaffected by our choice of the point A and by the motion of the center of mass, which may even be accelerating.

For an ideal point particle, or the "collapsed" point-particle version of a multiparticle system, the translational angular momentum is the only kind of angular momentum there is, just as translational kinetic energy is the only kind of kinetic energy a point particle can have.

In Chapter 9 we saw that a rigid object that rotates about an axle that is a distance r_{CM} from the center of mass of the object has translational plus rotational kinetic energy equal to $\frac{1}{2}I\omega^2$, where $I = Mr_{CM}^2 + I_{CM}$ (this is also referred to as the "parallel axis theorem" for moment of inertia). It is easy to show that the angular momentum of a rotating rigid object is similar: the translational plus rotational angular momentum is $(Mr_{CM}^2 + I_{CM})\vec{\omega}$.

Checkpoint 4 Pinocchio rides a horse on a merry-go-round turning counterclockwise as viewed from above, with his long nose always pointing forward, in the direction of his velocity. Is Pinocchio's translational angular momentum relative to the center of the merry-go-round zero or nonzero? If nonzero, what is its direction? Is his rotational angular momentum zero or nonzero? If nonzero, what is its direction?

11.4 TORQUE

The (linear) momentum of a system can be changed by a force exerted by the surroundings.

QUESTION What can change angular momentum?

The angular momentum of a rotating or orbiting object can be changed by a twist—a force applied off-center. Such a twist is called a "torque," a word of Latin origin that simply means "twist." Torque is usually symbolized by τ, the lowercase Greek letter tau. The effectiveness of the torque depends on where the force is applied, so the definition of torque includes not only the magnitude and direction of the force, but also the position at which the force acts, relative to some location which we must specify.

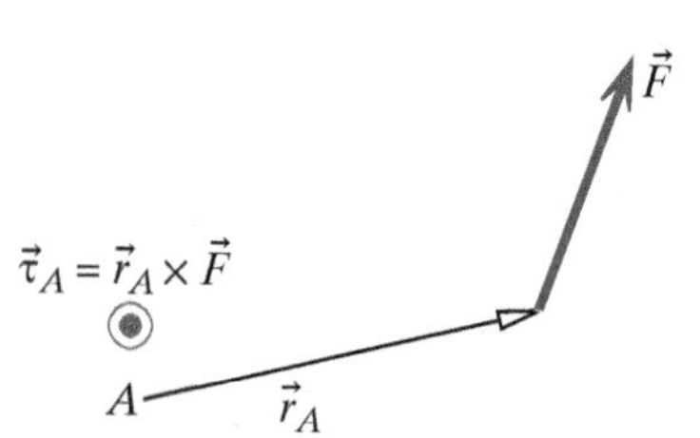

Figure 11.23 Torque relative to location A is defined as $\vec{\tau}_A = \vec{r}_A \times \vec{F}$.

DEFINITION OF TORQUE

$$\vec{\tau}_A = \vec{r}_A \times \vec{F}$$

Torque due to a force $\vec{F}$ is defined about a particular location, A. $\vec{r}_A$ is a vector from location A to the point of application of the force (Figure 11.23). The units of torque are N · m.

Applying a torque, or twist, to a system changes the angular momentum of the system. Just as we call the vector sum of all the forces acting on a system $\vec{F}_{net}$, we call the vector sum of all the torques acting on a system $\vec{\tau}_{net,A}$. Torque is a vector because it has both magnitude and direction. It is calculated with a cross product because it depends on both the direction of the applied force and the point of application.

Direction of the Force

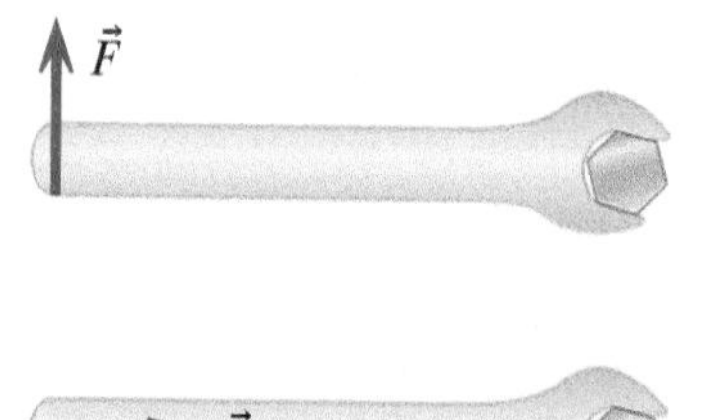

Figure 11.24 Twist a nut, or push straight at it?

QUESTION In Figure 11.24, which way of pushing on a wrench is more effective in twisting the nut?

You can see that a force parallel to the handle is completely ineffective in twisting the nut. Evidently it is only the perpendicular component of a force that is effective in twisting an object.

Point of Application of the Force

> QUESTION In Figure 11.25, where should you push on the wrench—at the end of the wrench or near the nut?

The farther away from the nut we push, the more effective we are in twisting the nut. The advantage of a long wrench is to provide more leverage in twisting.

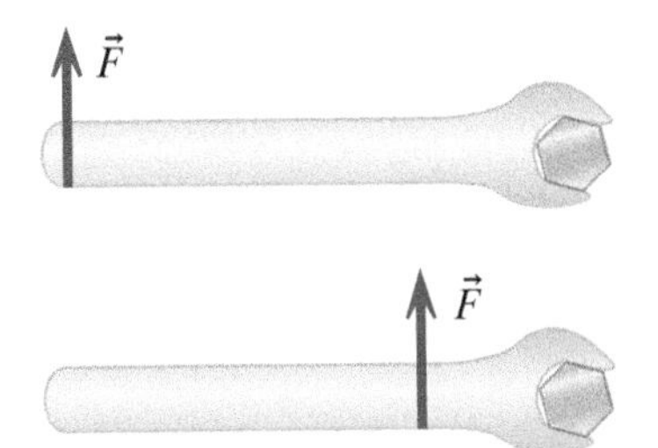

Figure 11.25 Push at the end, or near the nut?

Magnitude of Torque

Suppose that we apply a force at an angle θ to the radius (Figure 11.26). Evidently the component of the force that is effective in twisting is $F\sin\theta$, the perpendicular component. Moreover, the bigger the lever arm (r_A, the distance from the nut at location A), the more effective we'll be. To capture both effects, we define the magnitude of torque (which means twist) exerted by a force $\vec{F}$ relative to a location A as follows:

MAGNITUDE OF TORQUE RELATIVE TO LOCATION A

$$\tau = r_A F \sin\theta$$

For a purely perpendicular force, $\theta = 90°$, $\sin\theta = 1$, and the torque is $r_A F$. For a force that is parallel to the lever arm, $\theta = 0°$, $\sin\theta = 0$, and the torque is zero (Figure 11.24).

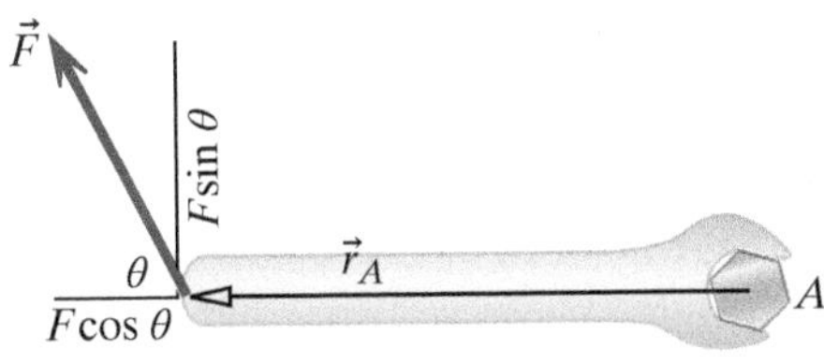

Figure 11.26 Push on the wrench at an angle.

Direction of Torque

The direction of the torque is given by the cross product and associated right-hand rule.

> QUESTION In Figure 11.26, is the torque vector into the page or out of the page? Use the right-hand rule.

Move the $\vec{r}$ and $\vec{F}$ vectors tail to tail (Figure 11.27), and align your hand so you can fold the fingers of your right hand from $\vec{r}$ toward $\vec{F}$. The only way you can do this results in your extended right thumb pointing into the page. If you try to keep your extended right thumb pointing out of the page, you will find that you can't fold your fingers toward the direction of the force. To put it another way, folding your fingers from the direction of $\vec{r}$ toward the direction of $\vec{F}$ is a clockwise rotation, which means the angular momentum points into the page.

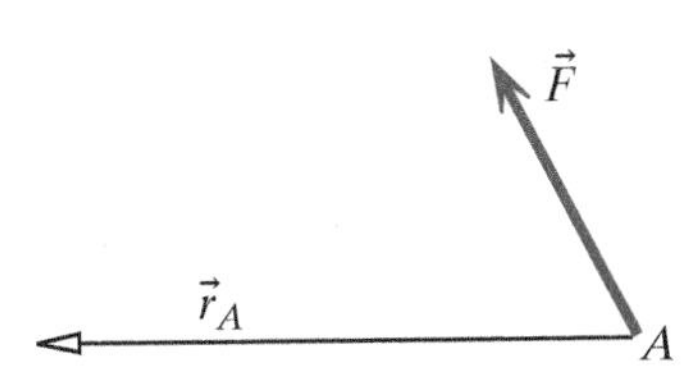

Figure 11.27 To find the direction of the cross product, move the vectors tail to tail.

Most nuts and bolts have what are called "right-handed threads." In Figure 11.26 a right-handed nut will advance in the direction of the torque vector, into the page. If you reverse the direction of the force, so the torque vector points out of the page, a right-handed nut will advance out of the page. (Exceptions include left-handed threads on gas cylinders containing explosive gases; the unexpected behavior alerts users to danger.)

The fact that a physical right-handed nut advances in the direction of the torque vector is another justification for drawing torque cross-product vectors along an axis of rotation, perpendicular to the plane of the motion.

Figure 11.28 Two children on a seesaw.

EXAMPLE **Torque on a Seesaw**

Two children sit on a seesaw, each at a distance of 1.2 m from the pivot point (Figure 11.28). The mass of child 1 is 30 kg, and the mass of child 2 is 20 kg. Assume that the mass of the seesaw itself is small. What is the net torque on the system of children plus seesaw, about the pivot point?

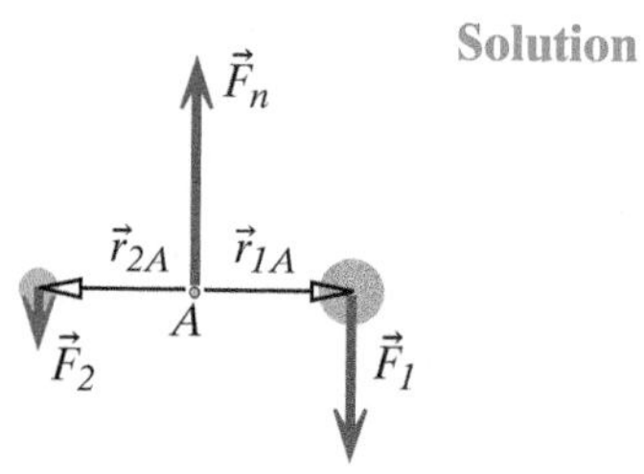

Figure 11.29 Forces on the see-saw and children.

Solution The diagram in Figure 11.29 shows all the forces acting on the system. We'll use the standard coordinate system, and we'll choose the location of the pivot as location A. To get the net torque we'll need to calculate all the torques about this location and add them up.

The torque due to child 1 is:

$$\begin{aligned}\vec{\tau}_{1,A} &= \vec{r}_{1,A} \times \vec{F}_1 \\ &= \langle 1.2,0,0\rangle \times \langle 0,-(30\,\text{kg})(9.8\text{N/kg}),0\rangle \\ &= \langle 0,0,-352.8\rangle\ \text{N}\cdot\text{m}\end{aligned}$$

Alternatively, if we put the vectors $\vec{r}_{1,A}$ and $\vec{F}_1$ tail to tail and apply the right-hand rule, we find that $\vec{\tau}_{1,A}$ is into the page $(-z)$. Its magnitude is:

$$\begin{aligned}|\vec{\tau}_{1,A}| &= (1.2\ \text{m})(30\,\text{kg})(9.8\text{N/kg})\sin 90^\circ \\ &= 352.8\,\text{N}\cdot\text{m}\end{aligned}$$

Combining these results for magnitude and direction, we again find that

$$\vec{\tau}_{1,A} = \langle 0,0,-352.8\rangle\ \text{N}\cdot\text{m}$$

The torque due to child 2 is:

$$\begin{aligned}\vec{\tau}_{2,A} &= \vec{r}_{2,A} \times \vec{F}_2 \\ &= \langle -1.2,0,0\rangle \times \langle 0,-(20\,\text{kg})(9.8\text{N/kg}),0\rangle \\ &= \langle 0,0,235.2\rangle\ \text{N}\cdot\text{m}\end{aligned}$$

The normal force $\vec{F}_n$ is important in holding up the seesaw. Its contribution to the net torque about A is:

$$\begin{aligned}\vec{\tau}_{n,A} &= \vec{r}_{n,A} \times \vec{F}_n \\ &= \langle 0,0,0\rangle\ \text{m} \times \langle 0,(50\,\text{kg})(9.8\text{N/kg}),0\rangle \\ &= \langle 0,0,0\rangle\ \text{N}\cdot\text{m}\end{aligned}$$

The torque due to the normal force is zero because the force acts at location A, so it can't twist the seesaw. The net torque is:

$$\begin{aligned}\vec{\tau}_{\text{net},A} &= \vec{\tau}_{n,A} + \vec{\tau}_{1,A} + \vec{\tau}_{2,A} \\ &= \langle 0,0,0\rangle\ \text{N}\cdot\text{m} + \langle 0,0,-352.8\rangle\ \text{N}\cdot\text{m} + \langle 0,0,235.2\rangle\ \text{N}\cdot\text{m} \\ &= \langle 0,0,-117.6\rangle\ \text{N}\cdot\text{m}\end{aligned}$$

Checkpoint 5 In Figure 11.26, if $r_A = 3$ m, $F = 4$ N, and $\theta = 30^\circ$, what is the magnitude of the torque about location A, including units? If the force in Figure 11.26 were perpendicular to $\vec{r}_A$ but gave the same torque as before, what would be its magnitude?

11.5 THE ANGULAR MOMENTUM PRINCIPLE

The Angular Momentum Principle relates a change in the angular momentum of a system to the net torque on the system due to the surroundings. Like the Momentum Principle, it can be written in several different forms.

THE ANGULAR MOMENTUM PRINCIPLE FOR A PARTICLE

$$\vec{L}_{f,A} = \vec{L}_{i,A} + \vec{\tau}_{\text{net},A}\Delta t \quad \text{(update form)}$$

$$\frac{d\vec{L}_A}{dt} = \vec{\tau}_{\text{net},A} \quad \text{(derivative form)}$$

$$\Delta\vec{L}_{A,\text{system}} + \Delta\vec{L}_{A,\text{surroundings}} = 0 \quad \text{(conservation form)}$$

In each case, angular momentum and torque are defined relative to a location A.

These statements of the Angular Momentum Principle apply to a system modeled as a single particle. We'll see a little later how to extend these equations to apply to an extended system.

By considering the derivative form of the Angular Momentum Principle, we can show that it is consistent with the Momentum Principle. For a single particle, the Momentum Principle says that the time derivative of the momentum is the net force: $d\vec{p}/dt = \vec{F}_{\text{net}}$. In the following derivation we will use these facts:

$$\frac{d\vec{r}_A}{dt} = \vec{v}, \quad \vec{p} = \gamma m\vec{v}, \quad \vec{v}\times\vec{v} = \vec{0}, \quad \frac{d\vec{p}}{dt} = \vec{F}_{\text{net}}, \quad \vec{r}_A \times \vec{F}_{\text{net}} = \vec{\tau}_{\text{net},A}$$

We'll start with the definition of angular momentum of a particle, and take its time derivative. Then we'll apply the product rule for derivatives:

$$\begin{aligned}
\vec{L}_A &= \vec{r}_A \times \vec{p} \\
\frac{d\vec{L}_A}{dt} &= \frac{d(\vec{r}_A \times \vec{p})}{dt} \\
&= \left(\frac{d\vec{r}_A}{dt}\times\vec{p}\right) + \left(\vec{r}_A \times \frac{d\vec{p}}{dt}\right) \\
&= (\vec{v}\times\gamma m\vec{v}) + \left(\vec{r}_A \times \frac{d\vec{p}}{dt}\right) \\
&= \vec{0} + \left(\vec{r}_A \times \frac{d\vec{p}}{dt}\right) \\
&= \vec{0} + \vec{r}_A \times \vec{F}_{\text{net}} \\
\frac{d\vec{L}_A}{dt} &= \vec{\tau}_{\text{net},A}
\end{aligned}$$

EXAMPLE **Momentum Change for a Seesaw**

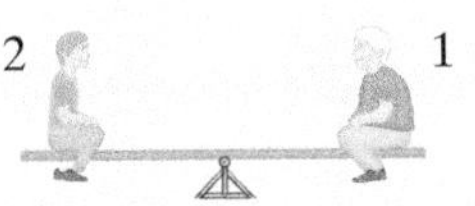

Figure 11.30 Two children on a seesaw.

In the previous example we calculated the net torque on the system of a seesaw plus two children (Figure 11.30). Apply the Angular Momentum Principle to calculate the angular momentum of the system at a time 0.4 s after it is released from rest. Will the seesaw be rotating clockwise or counterclockwise?

Solution We found the net torque on the seesaw to be $\langle 0,0,-117.6\rangle$ N·m.

$$\begin{aligned}
\vec{L}_{f,A} &= \vec{L}_{i,A} + \vec{\tau}_{\text{net},A}\Delta t \\
&= \vec{0} + \langle 0,0,-117.6\rangle \text{ N}\cdot\text{m}(0.4\text{ s}) \\
&= \langle 0,0,-47.0\rangle \text{ kg m}^2/\text{s}
\end{aligned}$$

Since $\vec{L}_A$ is in the $-z$ direction, the seesaw will rotate clockwise (that is, child 1 will go down and child 2 will go up). If you point your right thumb in the direction of $\vec{L}_A$, your fingers curl in the direction of rotation.

Three Fundamental Principles

There are parallels between the three fundamental principles we have encountered so far: the Momentum Principle, the Energy Principle, and the Angular Momentum Principle. Each principle relates a change in a property of the system (momentum, energy, or angular momentum) to interactions with objects in the surroundings:

$$\Delta\vec{p} = \vec{F}_{\text{net}}\,\Delta t$$
$$\Delta E = W + Q$$
$$\Delta\vec{L}_A = \vec{\tau}_{\text{net},A}\,\Delta t$$

Momentum, energy, and angular momentum are all conserved quantities:

$$\Delta\vec{p}_{\text{sys}} + \Delta\vec{p}_{\text{surr}} = \vec{0}$$
$$\Delta E_{\text{sys}} + \Delta E_{\text{surr}} = 0$$
$$\Delta\vec{L}_{\text{sys},A} + \Delta\vec{L}_{\text{surr},A} = \vec{0}$$

All three of these fundamental principles apply to any choice of system and surroundings. In complicated situations it may be necessary to use all three principles in order to predict or explain the behavior of a system.

11.6 MULTIPARTICLE SYSTEMS

The Angular Momentum Principle has already given us new insight into the motion of a single particle acted on by a torque due to a single force. The equation becomes a really powerful tool when we extend it to apply to multiparticle systems acted on by multiple torques. For example, we will be able to understand the counterintuitive behavior of spinning tops and gyroscopes.

The derivation of the multiparticle version of the Angular Momentum Principle follows closely the derivation in Chapter 3 of a multiparticle version of the Momentum Principle. The basic idea is that, due to reciprocity of forces, torques internal to a multiparticle system cancel, and only torques due to objects in the surroundings can change the system's angular momentum.

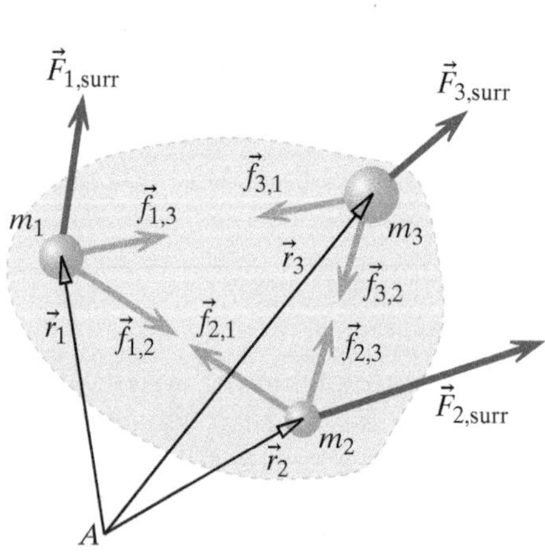

Figure 11.31 The Angular Momentum Principle in a multiparticle system.

To be as concrete as possible, we'll consider a system consisting of just three particles. It is easy to see how this generalizes to larger systems, including a block consisting of an astronomically huge number of atoms.

In Figure 11.31 we show all of the forces acting on each particle, where the lowercase $\vec{f}$'s are internal forces, and the uppercase $\vec{F}$'s are external forces applied by objects in the surroundings, such as the Earth that applies a gravitational attraction. We write the Angular Momentum Principle applied to each of the three particles. We measure angular momenta and torques relative to location A, but to avoid clutter we don't write the subscript A in any of these equations until the end. ($\vec{L}_1$ is the angular momentum of m_1 relative to A.)

$$\frac{d\vec{L}_1}{dt} = \vec{r}_1 \times (\vec{F}_{1,\text{surr}} + \vec{f}_{1,2} + \vec{f}_{1,3})$$
$$\frac{d\vec{L}_2}{dt} = \vec{r}_2 \times (\vec{F}_{2,\text{surr}} + \vec{f}_{2,1} + \vec{f}_{2,3})$$
$$\frac{d\vec{L}_3}{dt} = \vec{r}_3 \times (\vec{F}_{3,\text{surr}} + \vec{f}_{3,1} + \vec{f}_{3,2})$$

Nothing new so far—but now we add up these three equations. That is, we create a new equation by adding up all the terms on the left sides of the three equations, adding up all the terms on the right sides, and setting them equal to each other. In doing so, we take into account the fact that many of these terms cancel. By the reciprocity of electric and gravitational forces (Newton's third law), we have

$$\vec{f}_{1,2} = -\vec{f}_{2,1}$$
$$\vec{f}_{1,3} = -\vec{f}_{3,1}$$
$$\vec{f}_{2,3} = -\vec{f}_{3,2}$$

Therefore the torques of the internal forces cancel. For example, consider this piece of the sum:

$$\vec{r}_1 \times \vec{f}_{1,2} + \vec{r}_2 \times \vec{f}_{2,1} = \vec{r}_1 \times \vec{f}_{1,2} - \vec{r}_2 \times \vec{f}_{1,2} = (\vec{r}_1 - \vec{r}_2) \times \vec{f}_{1,2}$$

The vector $\vec{r}_1 - \vec{r}_2$ points from m_2 toward m_1. Because electric and gravitational forces act along the line connecting the particles, the angle between $\vec{r}_1 - \vec{r}_2$ and $\vec{f}_{1,2}$ is either 0° or 180°, and $\sin\theta$ is zero. Consequently all of the torques associated with the internal forces cancel, and all that remains is this:

$$\frac{d(\vec{L}_1 + \vec{L}_2 + \vec{L}_3)}{dt} = \vec{r}_1 \times \vec{F}_{1,\text{surr}} + \vec{r}_2 \times \vec{F}_{2,\text{surr}} + \vec{r}_3 \times \vec{F}_{3,\text{surr}}$$

The right side of this equation represents the net torque due to external forces, $\vec{\tau}_{\text{net},A}$. The great importance of this equation is that the reciprocity of gravitational and electric forces causes all the torques due to internal forces to cancel. We rewrite our result like this:

THE ANGULAR MOMENTUM PRINCIPLE FOR A MULTIPARTICLE SYSTEM

$$\frac{d\vec{L}_{\text{tot},A}}{dt} = \vec{\tau}_{\text{net},A}$$

or

$$\Delta\vec{L}_{\text{tot},A} = \vec{\tau}_{\text{net},A}\Delta t$$

In words, the rate of change of the "total angular momentum" of a system relative to a location A, $\vec{L}_{\text{tot},A} = \vec{L}_{1,A} + \vec{L}_{2,A} + \vec{L}_{3,A} + \cdots$, is equal to the net torque due to forces exerted on that system by objects in the surroundings, relative to location A. (Or, the change in the angular momentum is equal to the angular impulse.)

Often it pays to choose location A to be at the place where the center of mass happens to be at that instant. In that case, the Angular Momentum Principle simplifies:

$$\frac{d\vec{L}_{\text{CM}}}{dt} = \frac{d}{dt}[(\vec{r}_{\text{CM,CM}} \times \vec{P}_{\text{tot}}) + \vec{L}_{\text{rot}}] = \frac{d\vec{L}_{\text{rot}}}{dt}$$

since $\vec{r}_{\text{CM,CM}} = \vec{0}$. That is, the position of the center of mass, relative to the center of mass itself, is of course a zero vector. Therefore we have the following important and useful special case for the angular momentum relative to the center of mass:

THE ANGULAR MOMENTUM PRINCIPLE RELATIVE TO THE CENTER OF MASS

$$\frac{d\vec{L}_{\text{rot}}}{dt} = \vec{\tau}_{\text{net,CM}}$$

or

$$\Delta\vec{L}_{\text{rot}} = \vec{\tau}_{\text{net,CM}}\Delta t$$

11.7 SYSTEMS WITH ZERO TORQUE

In a situation in which there is zero external torque, the angular momentum of a system does not change, even if the moment of inertia of the system does change, due to a change in the shape or configuration of the system. Athletes like figure skaters, divers, and dancers can use this fact to change the rotation of their bodies.

Figure 11.32 The skater spins slowly with leg and arms stretched out.

Figure 11.33 The skater pulls in her leg and arms and spins much faster.

A Spinning Skater

You may have seen an ice skater spin vertically on the tip of one skate, with her arms and one leg outstretched, then pull her leg in and bring her arms to a vertical position above her head. She then spins much faster (Figures 11.32 and 11.33).

What's going on? There is some frictional force of the ice on the tip of the skate, but this force is applied so close to the axis of rotation that the torque is small.

> QUESTION During the short time when the skater quickly changes her configuration, what can you say about the rotational angular momentum? Why?

Small torque implies small change in angular momentum per unit time, so in a short time the angular momentum of the skater will hardly change.

> QUESTION If the rotational angular momentum hardly changes, how can the skater spin faster?

When she changes her configuration, the moment of inertia decreases, because some parts of her body are now closer to the axis of rotation. For the rotational angular momentum not to change, the angular velocity must increase to compensate for the decreased moment of inertia: $I_1\omega_1 = I_2\omega_2$, and therefore $\omega_2 = I_1\omega_1/I_2$.

Now you may quite legitimately be puzzled about how this actually works! *Why* does moving her arms and leg closer to the spin axis increase her angular speed? At one level of discussion, you can close your eyes (and maybe hold your nose) and say, "Well, we did all that general analysis of the effect of torques on the angular momentum of multiparticle systems, so if that's what we get when we apply these general principles, I guess that's that." At another level, though, it would be very nice to get a better sense of the detailed mechanisms involved in this odd phenomenon.

One approach is to analyze the changes in energy involved in this skating maneuver. The angular momentum L_{rot} of the skater doesn't change, so as her moment of inertia decreases, the kinetic energy of the skater $\frac{1}{2}I\omega^2 = L_{\text{rot}}^2/(2I)$ actually increases.

> QUESTION Where does this increased energy come from?

Evidently the skater has to expend chemical energy in order to increase her kinetic energy. In fact, at high spin rates it takes a noticeable effort to pull her arms and leg toward the spin axis. This is even more dramatic when holding heavy weights.

A popular physics demonstration is to sit on a rotating stool holding a dumbbell in each hand. Start spinning slowly with the arms held out, then pull your hands in toward your chest. It requires considerable effort to pull the dumbbells in. After all, the dumbbells would tend to move in straight lines, and you must exert a radial force just to keep turning them in a circle (though in that case you do no mechanical work, because the motion of the dumbbells has no radial component in the direction of the force you apply). To move the dumbbells into a smaller radius requires applying an even larger force that does work on the dumbbells, because there is now a radial component of the motion, in the direction of the radial force you exert.

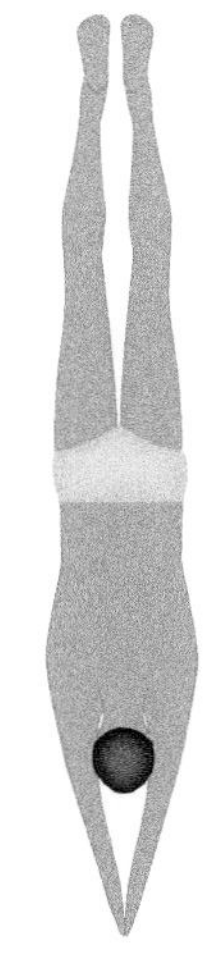

Figure 11.34 A diver's moment of inertia is large when his body is extended.

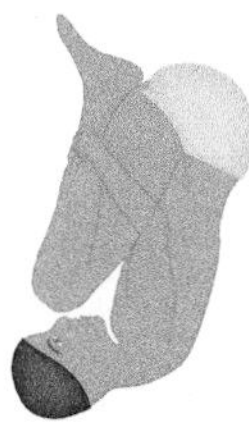

Figure 11.35 A diver's moment of inertia is smaller when he is curled into a ball.

A High Dive

You may have seen a skilled diver leap off a high board, tuck himself into a tight ball (holding onto his ankles), and rotate quite fast for a few turns. Then he straightens out and enters the water like a knife, hardly ruffling the surface (Figure 11.34).

> QUESTION What can you say about the diver's rotational angular momentum while he is in the air? Why?

We can probably neglect air resistance, because the diver is very far from reaching terminal speed. Although air resistance is distributed over the entire body, and there is a gravitational force acting on every atom in the body, torques to the left and right of the center of mass tend to cancel each other out (this cancellation is exact for the gravitational forces). Therefore there is negligible torque about the center of mass, and the rotational angular momentum $\vec{L}_{rot}$ does not change.

> QUESTION Early in the dive the diver is spinning rapidly (Figure 11.35). If rotational angular momentum is constant during the dive, where did that rapid spin come from?

When the diver jumps off the board, he must thrust with his feet in such a way that the force of the board on his feet has a sizable lever arm about his center of mass, so that by the time his feet lose contact with the board, the diver already has acquired a sizable rotational angular momentum.

> QUESTION When the diver pulls out of the tucked position, why does he stop spinning rapidly?

There is a large increase in the moment of inertia, because many atoms are now much farther from the center of mass than they were when the diver was in the tucked position. Larger moment of inertia implies smaller angular speed, since the magnitude of the rotational angular momentum $I\omega$ does not change.

> QUESTION Can his body go straight from then on?

His body can't really go completely straight, because he still has rotational angular momentum. However, his angular speed may be so small that you hardly notice it, especially in comparison with the very rapid spin in the preceding tucked position. Moreover, his body could approximately follow the curving path of his center of mass, with the body rotating just enough to stay tangent to the trajectory. This enhances the illusion of straight motion. The most important aspect for good form is to arrange for the body to rotate into the vertical position at the time of entering the water, so as to make little splash.

See Problem P52, which asks you to analyze a diver's motion.

EXAMPLE **A Satellite with Solar Panels**

A satellite has four low-mass solar panels sticking out (Figure 11.36). The satellite can be considered to be approximately a uniform solid sphere. Originally it is traveling to the right with speed v and rotating clockwise with angular speed ω_i.

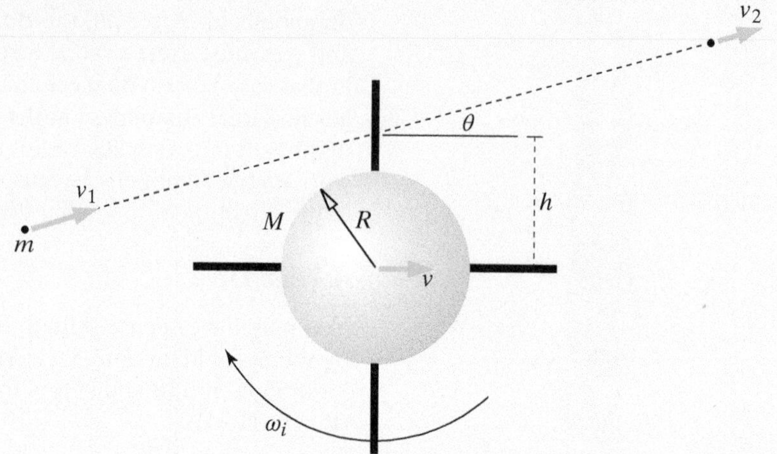

Figure 11.36 A satellite is struck by a tiny meteor.

A tiny meteor traveling at high speed v_1 rips through one of the solar panels and continues in the same direction but at reduced speed v_2. Afterward calculate the v_x and v_y components of the center-of-mass velocity of the satellite and its angular velocity ω_f (magnitude and direction). Additional data are provided on the diagram.

Solution What fundamental principles are useful starting points? The Momentum Principle and the Angular Momentum Principle. The Energy Principle would not be a useful starting point, because we don't know how much internal energy increase is associated with the meteor ripping through the panel. Take the satellite and meteor together as the system of interest, in which case the momentum and the angular momentum will remain constant, there being no external forces or torques.

System: Both the satellite and the meteor
Surroundings: Nothing during the collision

Initial State: Just before collision
Final state: Just after collision

Momentum Principle:
x components:

$$p_{xf} = p_{xi}$$
$$Mv_x + mv_2 \cos\theta = Mv + mv_1 \cos\theta$$
$$v_x = v + \frac{m}{M}(v_1 - v_2)\cos\theta$$

y components:

$$p_{yf} = p_{yi}$$
$$Mv_y + mv_2 \cos(90^\circ - \theta) = m_1 v_1 \cos(90^\circ - \theta)$$
$$v_y = \frac{m}{M}(v_1 - v_2)\sin\theta$$

Angular Momentum Principle, about the location in space where the center of the satellite was initially:

Component into the page ($-z$):

$$I\omega_i + hmv_1 \sin(90° - \theta) = I\omega_f + hmv_2 \sin(90° - \theta)$$

$$I = \frac{2}{5}MR^2$$

$$\omega_f = \omega_i + \frac{hm}{\left(\frac{2}{5}MR^2\right)}(v_1 - v_2)\cos\theta, \quad \text{clockwise, into the page}$$

Choosing times just before and just after the collision makes it particularly easy to see how to calculate the translational angular momentum of the meteor, since it is directly above the center of the satellite at those instants.

A Child Jumps onto a Playground Ride

A playground ride consists of a disk mounted on a low-friction axle (Figure 11.37). A child runs on a line tangential to the disk and jumps onto the outer edge of the disk. The disk spins, but at what rate? How many revolutions per minute, or radians per second?

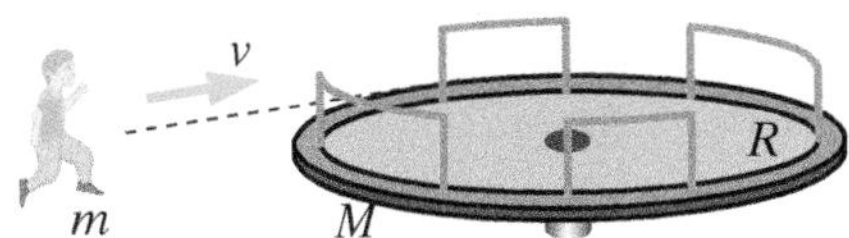

Figure 11.37 A child jumps onto a playground ride, and the disk spins. What is the rate of spin?

Neither the Momentum Principle nor the Energy Principle is sufficient to analyze this collision. If we know the speed of the child just after jumping onto the disk, we can use the Momentum Principle to determine the impulse exerted by the axle on the disk, which prevents the disk from moving sideways, but we need to know the spin rate in order to be able to calculate the new speed of the child. The Energy Principle can be used to determine how much internal energy is produced in the inelastic sticking collision of the child and the disk, but only if we know the spin rate. We need another principle that will yield the spin rate: the Angular Momentum Principle.

It is important to see that even though the child is running in a straight line in Figure 11.37, the child does have angular momentum about the center of the disk, because the disk is caused to rotate when hit off-center by the child.

EXAMPLE **A Child Jumps onto a Playground Ride**

A playground ride consists of a uniform-density disk of mass 300 kg and radius 2 m mounted on a low-friction axle (Figure 11.37). Starting from a distance of 5 m from the disk, a child of mass 40 kg runs at 3 m/s on a line tangential to the disk and jumps onto the outer edge of the disk. If the disk was initially at rest, how fast does it rotate just after the collision?

Solution System: Disk plus child
Surroundings: Earth, ground, axle
Initial state: Child 5 m away, running at 3 m/s; disk not rotating
Final state: Just after collision; child now stuck to the disk

Free-body diagram: Figure 11.38. The axle exerts a large impulsive force at the time of impact in response to the child ramming the disk against the axle; this force prevents the disk from moving to the right, and it changes the total momentum of the combined system.

Figure 11.38 The free-body diagram for the system of disk plus child. The axle is in the surroundings and exerts a large impulsive force to the left when the child rams the disk against the axle.

Principle: The Angular Momentum Principle $\vec{L}_{A,f} = \vec{L}_{A,i} + \vec{\tau}_{net}\Delta t$

We can choose any point A about which to calculate torque and angular momentum. To simplify the analysis, we choose A to be at the location of the axle, so that the large impulsive force of the axle points toward A and therefore exerts no torque. The force of the Earth is downward and the force of the ground is upward, with equal magnitude, so the net torque on our chosen system is zero, and $\vec{L}_{A,f} = \vec{L}_{A,i}$. We consider just the component of angular momentum that is perpendicular to the xz plane of the motion, $L_{A,y}$.

The initial angular momentum of the system consists solely of the child's angular momentum, since the disk isn't rotating. Here is the magnitude:

$$\begin{aligned}|L_{A,y,i}| &= |\vec{r}_A||\vec{p}|\sin\theta = (|\vec{r}_A|\sin\theta)|\vec{p}| \\ &= (2\text{ m})(40\text{ kg})(3\text{ m/s}) \\ &= 240\text{ kg}\cdot\text{m}^2/\text{s}\end{aligned}$$

Here we used the fact that in Figure 11.39 ($|\vec{r}_A|\sin\phi$) is the side opposite the angle ϕ, and that $\sin\phi = \sin(180° - \theta) = \sin\theta$.

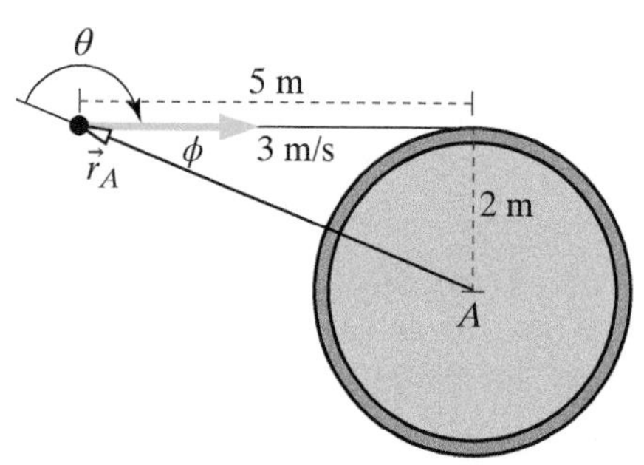

Figure 11.39 Applying the definition $L_A = r_A p \sin\theta$.

Another way to calculate the magnitude of the initial angular momentum for this motion in the xz plane is to use the general expression for components of angular momentum, which gives the same result and is a check on our work:

$$|L_{A,y,i}| = |zp_x - xp_z| = |zp_x| = (2\text{ m})(40\text{ kg})(3\text{ m/s}) = 240\text{ kg}\cdot\text{m}^2/\text{s}$$

The final angular momentum of the system consists of spin (rotational) angular momentum of the disk plus orbital (translational) angular momentum of the child on the disk. If the disk rotates with angular speed ω, the child's speed is $\omega r = \omega(2\text{ m})$.

$$\begin{aligned}|L_{A,y,f}| &= |L_{A,y,i}| = 240\text{ kg}\cdot\text{m}^2/\text{s} \\ &= I\omega + Rmv = \frac{1}{2}MR^2\omega + Rm(\omega R) \\ &= \frac{1}{2}(300\text{ kg})(2\text{ m})^2\omega + (40\text{ kg})(2\text{ m})^2\omega \\ \omega &= \frac{240\text{ kg}\cdot\text{m}^2/\text{s}}{\frac{1}{2}(300\text{ kg})(2\text{ m})^2 + (40\text{ kg})(2\text{ m})^2} \\ \omega &= 0.316\text{ rad/s}\end{aligned}$$

Before the collision there is only straight-line motion of the child, who nevertheless has (translational) angular momentum relative to the axle. There is obviously angular momentum after the collision, so it makes sense that there had to have been angular momentum before the collision, even though the child was running in a straight line.

Because the axle exerts a large impulsive force to the left, the Momentum Principle predicts that the child's speed should be smaller just after the collision (the rotating disk has zero linear momentum, because its center of mass doesn't move). In fact, the child's speed just after the collision is ωR, which is only 0.63 m/s.

Concerning the Energy Principle, this is an inelastic collision. The kinetic energy decreases and the internal energy increases; the child's feet and the disk get warmer.

Note that the perpendicular distance from A to the line of the child's momentum is $r_\perp = 2$ m, and $|L_{A,y}| = r_\perp|\vec{p}| = 240\text{ kg}\cdot\text{m}^2/\text{s}$.

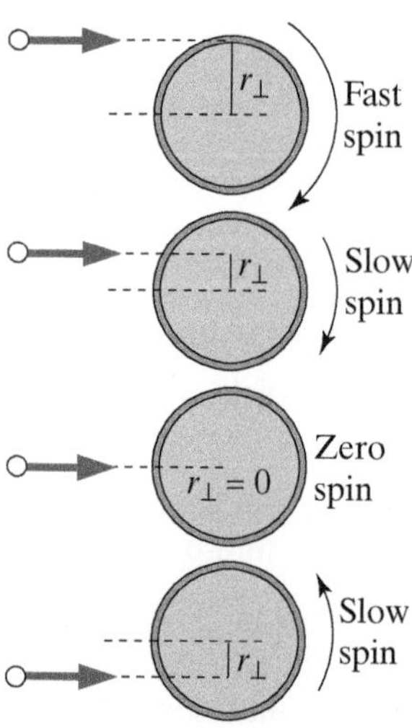

Figure 11.40 How much the disk spins, and in what direction, depends on the impact parameter $r_\perp$.

QUESTION What was the magnitude of the (translational) angular momentum of the child just before hitting the disk?

It was $240\text{ kg}\cdot\text{m}^2/\text{s}$, because $r_\perp = 2$ m all along the path, and the child's momentum $\vec{p}$ didn't change until hitting the disk.

In Figure 11.40 we see that how much the disk spins, and in what direction, depends on the impact parameter—how far off center the child hits the disk.

QUESTION Wait—doesn't the child have rotational angular momentum as well as translational angular momentum?

Yes. We've skipped over a subtle point in calculating the final angular momentum of the system consisting of the disk plus the child. After the collision the child is stuck to the disk and has not only orbital (translational) angular momentum $Rmv = Rm(R\omega) = mR^2\omega$ but also spin (rotational) angular momentum. As the disk rotates about the axle at a rate ω, the child also rotates about the child's own center of mass at a rate ω. If the child's moment of inertia about the child's center of mass is I_{CM}, there is an additional contribution to the final angular momentum of amount $I_{CM}\omega$.

We can make an approximate model of the child as a cylinder whose radius is about $r = 15$ cm, in which case the magnitude of the child's translational plus rotational angular momentum is

$$mR^2\omega + \frac{1}{2}mr^2\omega = m(2^2 + 0.5(0.15)^2)\omega$$

The quantity $(2^2 + 0.5(0.15)^2) = 4 + .011 \approx 4$, which is why in this case it makes only a very slight difference whether we include the child's rotational angular momentum around the child's own center of mass.

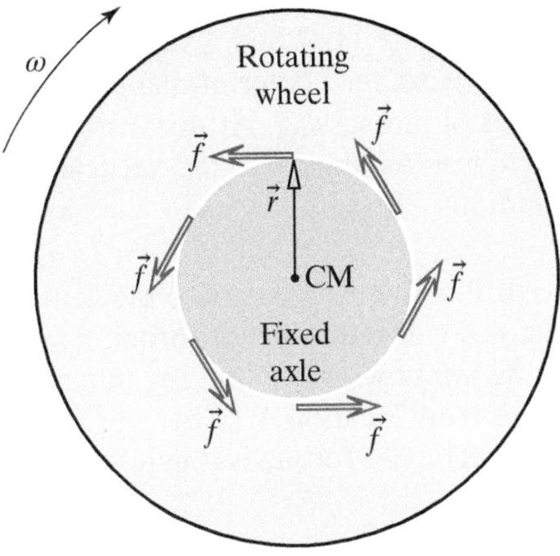

Figure 11.41 A wheel rotates on a fixed axle. There are frictional forces exerted by the axle on the inner surface of the wheel (other forces are omitted).

Frictional Torque

In the real world, friction is important. After the child in the preceding example has jumped onto the disk and made it rotate, the rotation rate will slowly decrease due to friction between the disk and the axle. Figure 11.41 shows a wheel that rotates on a fixed axle. There are frictional forces exerted by the axle on the inner surface of the wheel (other forces are omitted from the diagram). The frictional forces exert torques $(\vec{r} \times \vec{f})$ about the center of mass of the wheel, which make the angular speed ω of the wheel steadily decrease.

Figure 11.42 shows the opposite case: an axle rotates inside a bearing, which is fixed in place. For example, the axle of a wagon may pass through bearings attached to the bottom of the wagon, and there are wheels rigidly attached to each end of the axle. There are frictional forces exerted by the bearing on the outer surface of the axle (other forces are omitted from the diagram). We again see that frictional forces exert torques $(\vec{r} \times \vec{f})$ about the center of the axle, which make the angular speed ω of the axle steadily decrease.

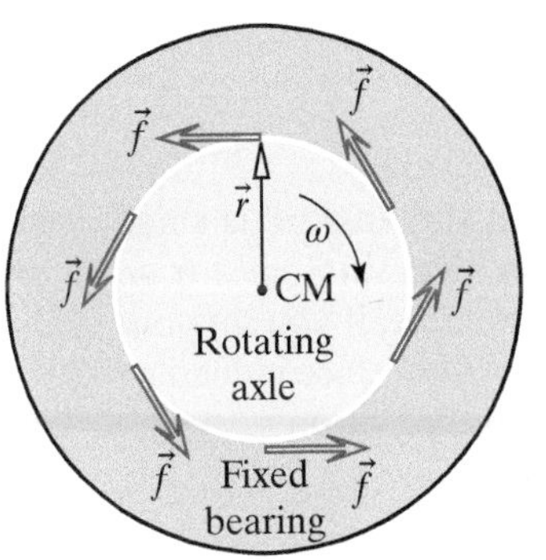

Figure 11.42 An axle rotates inside a fixed bearing. There are frictional forces exerted by the bearing on the outer surface of the axle (other forces are omitted).

There are three situations in which it may be a good approximation to neglect frictional torques as we will often do in this chapter:

- If the coefficient of friction μ between the axle and the wheel is small, the frictional force may be negligible, so the frictional torque may be negligible. In rotating machinery μ is often reduced with lubricating oil.
- In Figure 11.41, if the axle and the inside surface of the wheel have small radii, the distance $|\vec{r}|$ from the center of mass to the point of contact between axle and wheel is small, and the frictional torque $\vec{r} \times \vec{f}$ is small. For example, in a magnetic compass the needle sits on a sharp point, which has a very small radius, so any frictional force has a very small lever arm and torque. When an ice skater spins on the tip of the skate, the frictional torque may be very small because the lever arm $|\vec{r}|$ is very small. Although a thin axle has the advantage of reducing frictional torque on a wheel, there may be practical limits imposed by the requirement that the axle be able to support the wheel or the weight of a vehicle.
- In a rotational collision like that of the child jumping on the playground ride, during the brief time interval of the collision any frictional angular impulse $\vec{\tau}\Delta t$ of the axle is small because Δt is small.

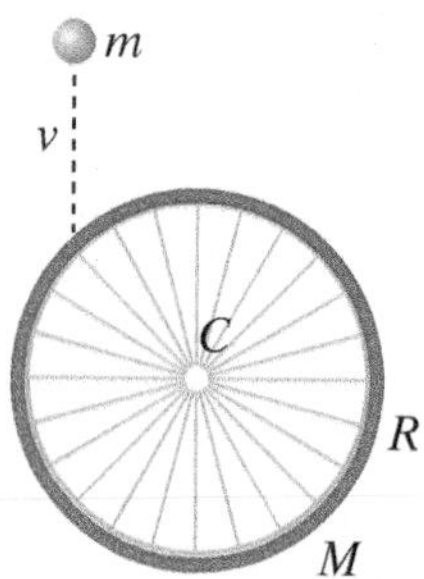

Figure 11.43 A lump of clay falls onto a wheel mounted on a frictionless axle.

Checkpoint 6 A stationary bicycle wheel of radius R is mounted in the vertical plane on a horizontal low-friction axle (Figure 11.43). The wheel has mass M, all concentrated in the rim (the spokes have negligible mass). A lump of clay with mass m falls and sticks to the outer edge of the wheel at the location shown. Just before the impact the clay has a speed v. **(a)** Just before the impact, what is the angular momentum of the combined system of wheel plus clay about the center C? **(b)** Just after the impact, what is the angular momentum of the combined system of wheel plus clay about the center C, in terms of the angular speed of the wheel? **(c)** Just after the impact, what are the magnitude and direction of the angular velocity of the wheel? **(d)** Qualitatively, what happens to the linear momentum of the combined system? Why?

EXAMPLE **A Physical Pendulum**

The name "physical pendulum" is given to hanging objects that swing back and forth and cannot be modeled as a point particle at the end of a string. An example is the object of mass M shown in Figure 11.44 that is supported by a low-friction axle located a distance r_{CM} from the center of mass. The moment of inertia about the object's center of mass is I_{cm}. At this instant the line from the axle to the center of mass is at an angle θ to the vertical. Calculate the period of small-angle oscillations.

unknown force by axle
A axle
θ
r_{CM}
$x = d_{CM}\sin\theta$
x
CM
Mg

Figure 11.44 A physical pendulum consisting of an object hanging from a low-friction axle.

Solution Apply the Angular Momentum Principle to the object. We choose the center of the axle for location A, about which we will calculate the net torque $\vec{\tau}_{net,A}$. The advantage of this choice for A is that the unknown force that the thin axle exerts on the object has a nearly zero distance from location A, so the axle force contributes almost nothing to the net torque. The net torque is due just to the gravitational force acting at the center of mass.

System: the hanging object

Surroundings: axle, Earth

Angular Momentum Principle: We'll use the parallel axis theorem, that the moment of inertia about a location A a distance r_{CM} from the center of mass is equal to the moment of inertia about the center of mass plus Mr_{CM}^2. We write the z component of the Angular Momentum Principle:

$$I_A\frac{d\omega_z}{dt} = -Mgr_{CM}\sin\theta$$

where $I_A = I_{CM} + Mr_{CM}^2$.

It is clear that the object will oscillate. If the angle θ is small, it can be shown that it is a good approximation to say that $\sin\theta \approx \theta$ for small angles, where the angle is measured in radians. For example, if you set your calculator to radians you'll find that $\sin(0.01)$ is 0.0099998, very close indeed to 0.01. Since $\omega_z = d\theta/dt$, $d\omega_z/dt = d^2\theta/dt^2$, so to a good approximation

$$I_A\frac{d^2\theta}{dt^2} = -Mgr_{CM}\theta$$

QUESTION Does this differential equation remind you of a similar differential equation we encountered in an earlier chapter?

This differential equation has the same form as the differential equation for the spring–mass system:

$$m\frac{d^2x}{dt^2} = -k_s x$$

We know that the solution for the spring–mass equation is of the form

$$x = \cos\left(\sqrt{\frac{k_s}{m}}t\right)$$

Comparing the differential equation for the physical pendulum and that for the spring–mass system, we see that the following is a solution for the physical pendulum, for small-angle oscillations:

$$\theta = \cos\left(\sqrt{\frac{Mgr_{CM}}{I_A}}\,t\right)$$

For the spring–mass system we could calculate the period T from the fact that $2\pi/T = \sqrt{k_s/m}$. For the physical pendulum we have $2\pi/T = \sqrt{Mgr_{CM}/I_A}$. Hence the period is $T = 2\pi\sqrt{I_A/Mgr_{CM}}$.

If we needed to know the force that the axle exerts on the hanging object we would need to use the Momentum Principle, which involves the net force.

QUESTION An important special case is the "simple pendulum" where there is a small mass M at the end of a string of length L. What is the period of small-angle oscillations of a simple pendulum?

The moment of inertia of the small point-like object is $I_{CM} + ML^2$, or simply ML^2 since the moment of inertia about the center of mass of the point-like object is nearly zero. Hence $T = 2\pi\sqrt{ML^2/(MgL)} = 2\pi\sqrt{L/g}$.

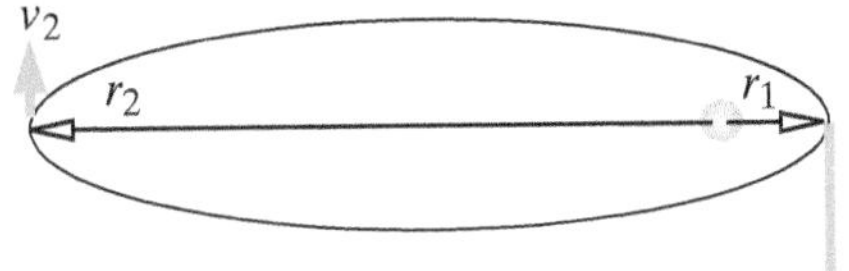

Figure 11.45 A long elliptical orbit of a comet around the Sun.

Comets and the Angular Momentum Principle

If the net torque around a location A is zero, the angular momentum about that location doesn't change. There may be forces acting, causing changes in the linear momentum, but if these forces don't exert any torques, the rate of change of angular momentum is zero, in which case angular momentum is constant.

Comet orbits provide a nice example. Most comets have very long elliptical orbits around the Sun (Figure 11.45). In our earlier discussion of Halley's comet (Section 11.1) we saw that the translational angular momentum of the comet relative to the Sun had the same value at different times along the orbit. We can now understand this observation in terms of the Angular Momentum Principle.

EXAMPLE **Torque on a Comet**

Relative to the center of the Sun, explain why the torque exerted by the Sun's gravitational force on the comet is zero at every point along the orbit (Figure 11.45). What does that say about the angular momentum of the comet relative to the center of the Sun? What if we choose location B instead of location A as the reference point?

Solution The gravitational force on the comet always points directly at the Sun, at location A (Figure 11.46). In $|\vec{\tau}_A| = |r_\perp \times \vec{F}_{grav}|$, the lever arm $r_\perp$ is always zero, so there is no torque relative to location A that can change the angular momentum of the comet. Therefore the angular momentum of the comet, relative to the location of the Sun, cannot change.

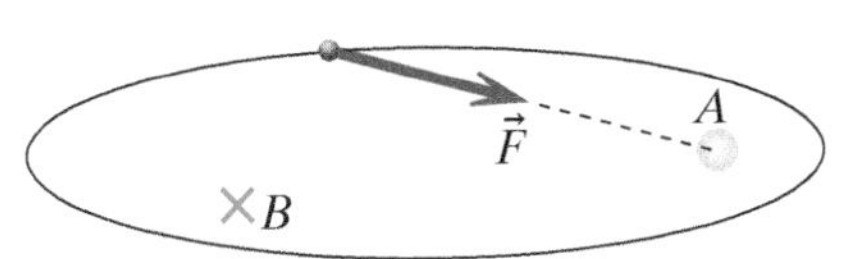

Figure 11.46 The gravitational force acting on the comet points directly at the center of the Sun (location A).

About location B in Figure 11.46, the torque acting on the comet is *not* zero, and the angular momentum of the comet does *not* stay constant. If on the other hand you consider the system consisting of both Sun and comet, the gravitational forces that the two objects exert on each other make torques about location B, or any location, that vectorially add up to zero, and the total angular momentum of the combined system will stay constant. The Sun actually moves a little due to the force that the comet exerts on it, so the Sun's angular momentum relative to location B changes, but its change is the opposite of the change of the angular momentum relative to B of the comet.

QUESTION When a comet is closest to the Sun, its speed is v_1 and its distance from the Sun is r_1. What is the comet's speed when it is farthest from the Sun (a distance r_2)?

The comet's angular momentum about the center of the Sun is constant, because the torque on the comet is zero. Therefore:

$$L_{2,A} = L_{1,A}$$

$$r_2 v_2 = r_1 v_1$$

$$v_2 = v_1 \frac{r_1}{r_2}$$

This equation tells us that the speed of a comet at its farthest point can be very small, since r_1/r_2 is a small quantity. Some comets go far beyond Pluto and spend most of their time there, because they're traveling very slowly. We see them when they come near the Sun, but only for a few months, because they're now traveling fast. The result $r_2 v_2 = r_1 v_1$ at the closest and farthest points in the orbit is correct no matter what kind of "central" force is involved. For any force that acts along a line connecting two objects, there is no torque about a point at the center of one of the objects, so angular momentum is constant about that location.

Kepler and Elliptical Orbits

Historically, the behavior of comets was described before either the Momentum Principle or the Angular Momentum Principle was discovered. Based on careful, accurate naked-eye measurements made by Tycho Brahe before the invention of the telescope, Johannes Kepler in 1609 announced his discovery that the planets follow elliptical orbits around the Sun. Kepler also stated that he had found that "a radius vector joining any planet to the Sun sweeps out equal areas in equal lengths of time." This is equivalent to conservation of angular momentum, as can be seen with the aid of Figure 11.47. The area swept out in a time Δt is the area of a triangle whose base is $v\Delta t$ and whose altitude is $r\sin\theta$. This area is $\frac{1}{2}(rv\sin\theta)\Delta t$, which is proportional to what we now call the angular momentum $rmv\sin\theta$.

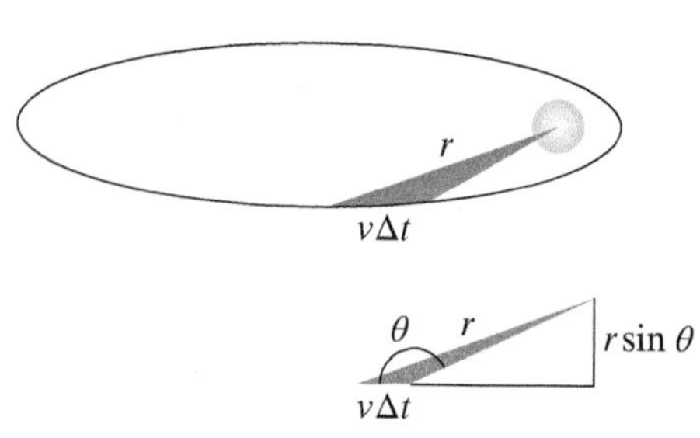

Figure 11.47 "Equal area in equal time."

In 1618 Kepler announced his discovery that the square of the time it takes a planet to go around the Sun is proportional to the cube of the mean distance from the Sun (a result you could derive in Chapter 5 for the simpler case of circular orbits). Later in the 1600s Newton explained all three of these discoveries as derivable from his second law of motion (the Momentum Principle) plus his universal law of gravitation. Kepler's insights provided important tests for Newton's theories.

Checkpoint 7 Because the Earth is nearly perfectly spherical, gravitational forces act on it effectively through its center. Explain why the Earth's axis points at the North Star all year long. Also explain why the Earth's rotation speed stays the same throughout the year (one rotation per 24 h). In your analysis, does it matter that the Earth is going around the Sun?

In actual fact, the Earth is not perfectly spherical. It bulges out a bit at the equator, and tides tend to pile up water at one side of the ocean. As a result, there are small torques exerted on the Earth by other bodies, mainly the Sun and the Moon. Over many thousands of years there are changes in what portion of sky the Earth's axis points toward (change of direction of rotational angular momentum), and changes in the length of a day (change of magnitude of rotational angular momentum). See Section 11.12.

11.8 SYSTEMS WITH NONZERO TORQUES

Of course, the angular momentum of a system is not always constant. External torques (due to objects in the surroundings) can change the angular momentum of a system.

EXAMPLE

A Meter Stick on the Ice

Consider a meter stick whose mass is 300 g and that lies on ice (in Figure 11.48 we're looking down on the meter stick). You pull at one end of the meter stick, at right angles to the stick, with a force of 6 N. Assume that friction with the ice is negligible. What is the rate of change of the center-of-mass speed v_{CM}? What is the rate of change of the angular speed ω? (In Chapter 9 we showed that the moment of inertia I around the center of mass of a uniform rod of mass M and length L is $ML^2/12$; $L = 1\,\text{m}$ here.)

Solution

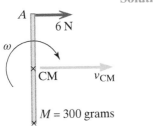

Figure 11.48 Pull one end of a meter stick that is lying on ice (negligible friction).

System: Stick
Surroundings: Your hand (pulling); ice (negligible effect)

Momentum Principle:
$d\vec{P}/dt = d(m\vec{v}_{CM})/dt = \vec{F}_{net}$
$dv_{CM}/dt = (6\text{N})/(0.3\,\text{kg}) = 20\,\text{m/s}^2$

Angular Momentum Principle about center of mass:
$d\vec{L}_{rot}/dt = \vec{\tau}_{net,CM}$

Component into page ($-z$ direction):
$I d\omega/dt = (0.5\,\text{m})(6\,\text{N})\sin 90° = 3\,\text{N}\cdot\text{m}$
$d\omega/dt = (3\,\text{N}\cdot\text{m})/[(0.3\,\text{kg}\cdot\text{m}^2)/12] = 120\ \text{rad/s}^2$

In vector terms, $d\vec{\omega}/dt$ points into the page, corresponding to the fact that the angular velocity points into the page and is increasing.

Alternative Analysis—Taking Torques Around the End of the Stick

It is interesting to re-analyze the motion of the meter stick by calculating torque and angular momentum about the end of the stick where the force is applied (location A in Figure 11.48), rather than about the center of mass. To be cautious and correct, we should say that we are taking torques about a location fixed in the ice next to the place where the end of the stick is momentarily located. This is a fixed location, not tied to the moving stick.

System: Stick
Surroundings: Your hand (pulling); ice (negligible effect)

Momentum Principle unchanged, which gives this:
$dv_{CM}/dt = (6\text{N})/(0.3\,\text{kg}) = 20\,\text{m/s}^2$

Angular Momentum Principle about location A (there is zero torque about location A, because there is zero distance from A to where the force is applied.):
$d\vec{L}_A/dt = \vec{\tau}_{net,A}$

Component into page ($-z$ direction):

$$\frac{d}{dt}\left[\frac{ML^2}{12}\omega - \left(\frac{L}{2}\right)Mv_{CM}\right] = 0$$

$$\frac{d\omega}{dt} = \frac{6}{L}\frac{dv_{CM}}{dt} = \frac{6}{(1\,\text{m})}(20\,\text{m/s}^2) = 120\,\text{rad/s}^2$$

This agrees with our analysis in which we took torques about the center of mass of the stick, but the details of the calculation are rather different. It is a good check on an analysis involving the Angular Momentum Principle to do the problem for two different choices of the location about which to calculate the net torque.

A Puck with String Wound Around It

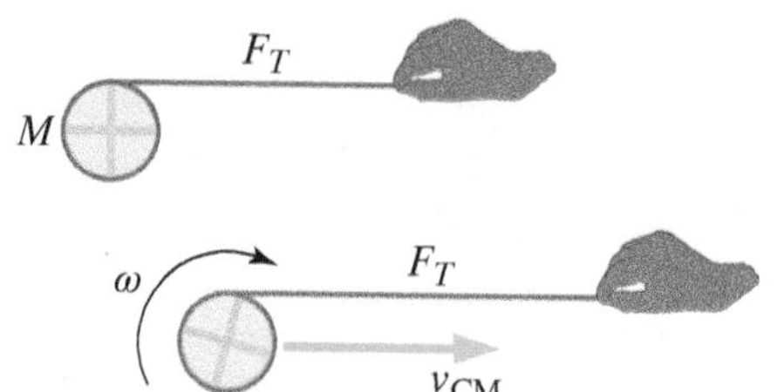

Figure 11.49 Wrap a string around a hockey puck, then pull it along the ice with negligible friction, with a constant tension F_T.

Wrap a string around the outside of a hockey puck. Then pull on the string with a constant tension F_T (Figure 11.49). The puck has mass M and radius R. Assume that friction with the ice rink is negligible. Evidently $dv_{CM}/dt = F_T/M$.

QUESTION The moment of inertia of a uniform solid puck of mass M and radius R is $MR^2/2$. What is the initial rate of change $d\omega/dt$ of the angular speed ω?

The torque about the center of mass is RF_T, into the page (down into the ice). The rotational angular momentum is $(MR^2/2)\omega$, into the page (down into the ice). Therefore $(MR^2/2)d\omega/dt = RF_T$, and we have $d\omega/dt = (2F_T)/(MR)$.

Checkpoint 8 Redo the analysis, calculating torque and angular momentum relative to a fixed location in the ice anywhere underneath the string (similar to the analysis of the meter stick around one end). Show that the two analyses of the puck are consistent with each other.

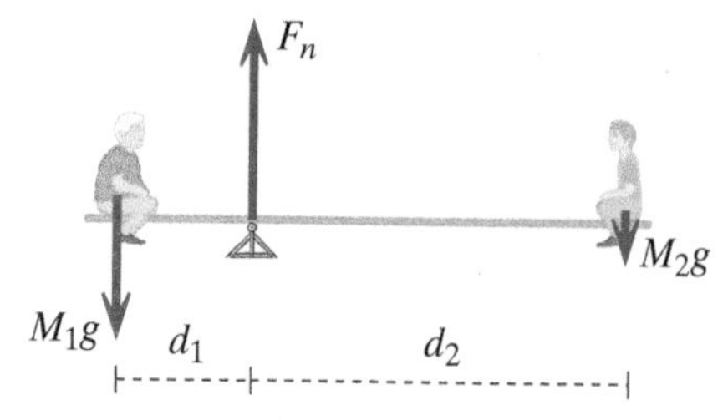

Figure 11.50 A seesaw in equilibrium.

Equilibrium

If the net torque on a system is zero, its angular momentum is constant. Conversely, if we know that the angular momentum is not changing, we can conclude that the net torque must be zero. For example, if a system is in equilibrium, not only must the net force on the system be zero; the net torque must also be zero. This allows us to make conclusions about the individual torques whose vector sum is zero.

EXAMPLE **Two People on a Seesaw**

Figure 11.50 shows two persons on a seesaw. The person on the left has mass $M_1 = 90$ kg and sits at a distance $d_1 = 1.2$ m from the nearly frictionless axle. The person on the right has mass $M_2 = 40$ kg. What is the upward force that the axle must exert? Where must the person on the right sit in order that the seesaw not rotate? The seesaw itself has negligible mass.

Solution System: The two persons and the seesaw
Surroundings: Earth and axle

Principle: The Momentum Principle, $\frac{d\vec{p}_{sys}}{dt} = \vec{F}_{net}$

Since the system is not moving, the momentum of the system isn't changing, so $d\vec{p}_{sys}/dt = \vec{0}$, and therefore the net force $\vec{F}_{net} = \vec{0}$. Consider the y forces:

$$F_N - M_1g - M_2g = 0$$

$$F_N = (M_1 + M_2)g = ((90+40)\,\text{kg})(9.8\,\text{N/kg}) = 1274\,\text{N}$$

It isn't surprising to find that the axle must exert an upward force equal to the combined weights of the two persons. Next we use a different principle to determine where the person on the right should sit.

Principle: The Angular Momentum Principle, $\dfrac{d\vec{L}_A}{dt} = \vec{\tau}_{\text{net}}$

Since the angular momentum isn't changing, $d\vec{L}_A/dt = \vec{0}$ about any point whatsoever, and therefore $\vec{\tau}_{\text{net}} = \vec{0}$ about any point whatsoever. It is convenient to choose point A to be at the location of the axle, because then the upward force of the axle exerts no torque about point A. Consider the z component of the net torque:

$$M_1 g d_1 - M_2 g d_2 = 0$$

$$d_2 = \frac{M_1}{M_2} d_1 = \frac{90\,\text{kg}}{40\,\text{kg}}(1.2\text{ m}) = 2.7\text{ m}$$

This analysis would also be valid if the seesaw were rotating at a constant angular velocity, because it would still be the case that the angular momentum would not be changing, and the net torque would be zero.

This is an example of a category of "statics" problems in which it is necessary to consider both the Momentum Principle and the Angular Momentum Principle in order to carry out a complete analysis.

Checkpoint 9 Write a different torque equation around a location fixed in space where the person on the left is sitting, and show that it is in fact equivalent to the torque equation around the axle, when you take the force equation into consideration. It is often convenient to choose your fixed location so that some of the forces create no torque around that location and therefore don't appear in the Angular Momentum Principle.

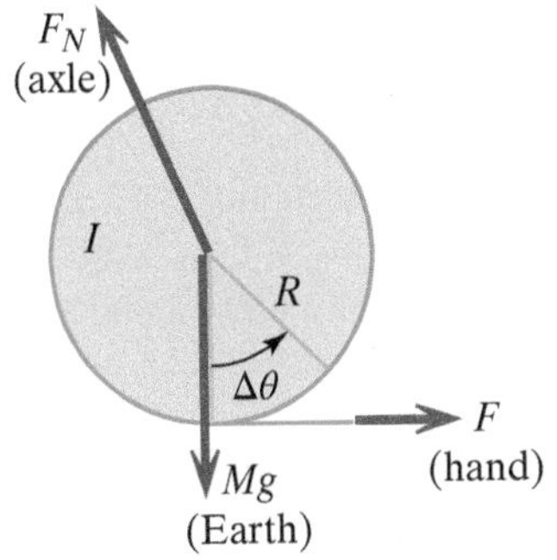

Figure 11.51 A wheel rotates on a low-friction axle, pulled by a string wrapped around the edge.

11.9 PREDICTING POSITIONS WHEN THERE IS ROTATION

In order to model the motion of a rotating system, we need to be able to predict its position. This usually means predicting the angle through which the object will rotate. If we know this angle, we can also figure out how far a given point has moved.

EXAMPLE **Predicting Position with Rotation**

A wheel of radius R and moment of inertia I is mounted on a low-friction axle (Figure 11.51). A string is wrapped around the edge, and you pull on it with a force F. At a certain time the angular speed is ω_i. **(a)** After a time interval Δt, what is the angular speed ω_f? **(b)** How far did your hand move during this time interval?

Solution System: Wheel plus string
Surroundings: Earth, axle, hand
Free-body diagram: Figure 11.51

Principle: The Momentum Principle tells us that the net force must be zero, because the center of mass does not move. To obtain information about rotation, we need to use the Angular Momentum Principle. We'll use the update form, about the center of mass:

$$\vec{L}_{\text{rot},f} = \vec{L}_{\text{rot},i} + \vec{\tau}_{\text{net}}\Delta t$$

(a) Calculate the final angular speed (only the force of the hand exerts a nonzero torque about the axle):

$$I\omega_f = I\omega_i + RF\Delta t$$
$$\omega_f = \omega_i + \frac{RF\Delta t}{I}$$

(b) To calculate how far your hand moved, first we will determine through how large an angle the wheel turned during the time interval Δt. Suppose that there is a blue line drawn radially from the center of the wheel to the edge, as shown in Figure 11.51. Let θ be the angle that this blue line makes with the y axis. This angle changes at an instantaneous rate of $\omega = d\theta/dt$, in radians per second. The change in the angle θ is

$$\Delta\theta = \omega_{avg}\Delta t$$

Because the angular speed ω changes at a constant rate RF/I, the average angular speed is

$$\omega_{avg} = \frac{\omega_i + \omega_f}{2}$$

Therefore the change in the angle of the blue line is

$$\Delta\theta = \frac{\omega_i + \omega_f}{2}\Delta t$$

Since we calculated ω_f from the Angular Momentum Principle, now we know through how big an angle the wheel turned. What remains to be calculated is how far your hand moved. If the angle through which the wheel rotates is 2π radians (360°), an amount of string $2\pi R$ will come off the edge of the wheel, since the circumference of the wheel is $2\pi R$. More generally, if the angle through which the wheel rotates is $\Delta\theta$ (in radians), the amount of string that comes off is $R\Delta\theta$.

The length of string that comes off the wheel is equal to the distance your hand moves:

$$\Delta x_{hand} = R\frac{\omega_i + \omega_f}{2}\Delta t$$

The rate of change of the angular velocity $d\vec{\omega}/dt$ is often called the "angular acceleration" $\vec{\alpha}$, and the Angular Momentum Principle applied to pure rotation of an object whose moment of inertia is not changing is written as $I\vec{\alpha} = \vec{\tau}_{net}$. In this form, the equation looks very similar to $M\vec{a} = \vec{F}_{net}$. Just as a constant acceleration $\vec{a}$ means that the average velocity is $(\vec{v}_i + \vec{v}_f)/2$, so also if the angular acceleration $\vec{\alpha}$ is constant the average angular velocity is $(\vec{\omega}_i + \vec{\omega}_f)/2$.

We could also determine the distance the hand moves by using the Energy Principle in combination with our result for ω_f, which we obtained using the Angular Momentum Principle:

$$\frac{1}{2}I\omega_f^2 = \frac{1}{2}I\omega_i^2 + F\Delta x_{hand}$$

Checkpoint 10 A uniform-density wheel of mass 6 kg and radius 0.3 m rotates on a low-friction axle. Starting from rest, a string wrapped around the edge exerts a constant force of 15 N for 0.6 s. **(a)** What is the final angular speed? **(b)** What is the average angular speed? **(c)** Through how big an angle did the wheel turn? **(d)** How much string came off the wheel?

11.10 COMPUTATION AND ANGULAR MOMENTUM

Computationally, modeling the motion of a rotating object subject to a nonzero net torque is very similar to modeling the motion of a moving object subject to a nonzero net force. After choosing a location A and specifying initial conditions, we iteratively update torque, angular momentum, and position (angle).

Repeat:

- Calculate the net torque $\vec{\tau}_{\text{net},A}$ acting on the system.
- Update the angular momentum of the system: $\vec{L}_f = \vec{L}_i + \vec{\tau}_{\text{net},A}\Delta t$.
- Update the position: $\theta_f = \theta_i + \omega\Delta t$.

In VPython, an object can be rotated by specifying the angle through which to rotate, the axis of rotation, and the origin about which to rotate:

```
wheel.rotate(angle=dtheta, axis=vector(0,1,0),
             origin=wheel.pos)
```

The angle `dtheta` is a signed scalar. Assuming the angular momentum and the axis of rotation are along the same line, we can use a vector dot product to find the magnitude and sign of $\vec{\omega}$, and use this to find `dtheta`:

```
omega_scalar = dot(omega, norm(axis_of_rotation))
dtheta = omega_scalar * deltat
```

In VPython, the `cross` function calculates vector cross products:

```
torque = cross(r_A, F)
```

The calculational loop in a program modeling the motion of a wheel subject to a torque might look something like this:

```
axis_of_rotation = vector(0,1,0)
while True:
    rate(100)
    torque = cross(r_a,F)
    L = L + torque * deltat
    omega = L/I
    omega_scalar = dot(omega, norm(axis_of_rotation))
    dtheta = omega_scalar * deltat
    wheel.rotate(angle=dtheta, axis=axis_of_rotation,
                 origin=wheel.pos)
```

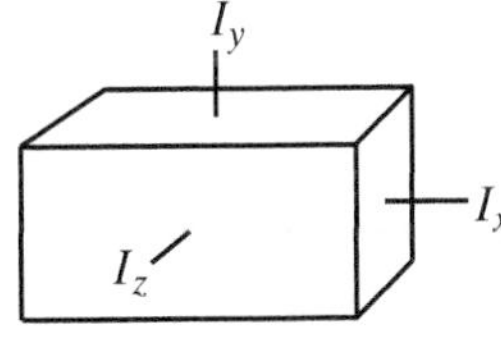

Figure 11.52

The approach outlined above is valid for situations in which the angular momentum and the axis of rotation lie along the same line. However, as discussed in the optional Section 11.13, the angular momentum vector $\vec{L}$ need not be in the same direction as the angular velocity vector $\vec{\omega}$. For example, the box shown in Figure 11.52 has three different moments of inertia about the three axes, and for rotations about some arbitrary axis $\vec{L}$ may not be in the direction of $\vec{\omega}$.

This complexity is beyond the scope of this textbook. In the computational modeling problems at the end of this chapter, we limit the complexity to situations where the net torque $\vec{\tau}_{\text{net}}$, the angular momentum $\vec{L}$, and the angular velocity $\vec{\omega}$ all lie along the same axis (the torque and angular momentum may be in opposite directions along that axis, for example if the torque is slowing down the rotation).

11.11 ANGULAR MOMENTUM QUANTIZATION

Many elementary particles have rotational angular momentum. Electrons bound in an atom may have translational angular momentum relative to the nucleus as well as their own rotational angular momentum. Atoms as a whole may have angular momentum, as do many nuclei. The surprising thing about angular momentum at the atomic and subatomic level is that in stable energy states the angular momentum is quantized:

THE ANGULAR MOMENTUM QUANTUM

$$\hbar = \frac{h}{2\pi} = 1.05 \times 10^{-34}\ \text{J} \cdot \text{s}$$

The constant h is "Planck's constant," 6.63×10^{-34} J · s. The constant $\hbar$ is called "h-bar." Whenever you measure a vector component of the angular momentum (that is, along the x or y or z axis), you get either a half-integer or integer multiple of this quantum of angular momentum.

Checkpoint 11 Show that $\hbar$ and angular momentum have the same units.

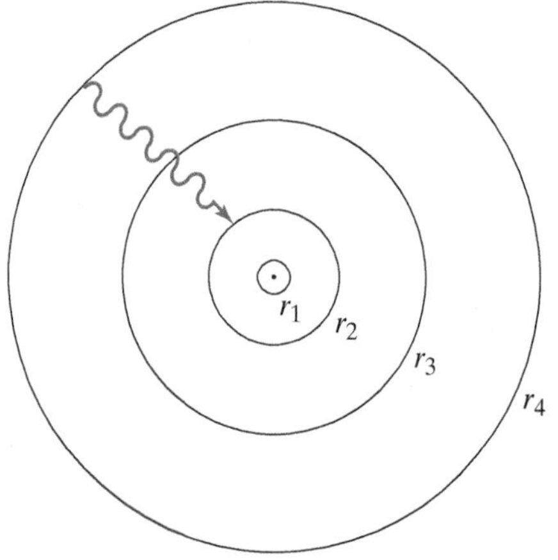

Figure 11.53 In the Bohr model of the hydrogen atom, electrons can be in only those circular orbits for which angular momentum is a multiple of $\hbar$. In a transition between allowed orbits, a photon is emitted.

The Bohr Model of the Hydrogen Atom

As discussed in the previous chapter, in 1911 Rutherford and his group discovered the nucleus in atoms. Stimulated by this discovery, in 1913 the Danish physicist Niels Bohr made a bold conjecture that a hydrogen atom could be modeled as an electron going around a proton in circular orbits, but only in those orbits whose translational angular momentum is an integer multiple of $\hbar$. Consequently, these orbits would have only certain radii, and only certain values of energy. The differences in energies between these quantized orbits match the observed energies of photons emitted by atomic hydrogen (Figure 11.53).

Bohr's basic hypothesis—that the angular momentum of electrons in an atom is quantized—has proven to be an important insight, and has been retained as a fundamental tenet of the quantum mechanical model of an atom. As this model has been refined, a probabilistic view of the motion of electrons has been adopted. In more sophisticated models electrons do not have precise trajectories, but only a "probability density"—a probability of being found at a particular location. These more complex quantum mechanical models explain a much wider range of atomic and molecular phenomena than does Bohr's original model. Additionally, the actual quantization rules in a hydrogen atom are more complex than those that are assumed in the Bohr model. For example, the translational angular momentum in the ground state ($N = 1$) is actually zero, not $\hbar$ as the Bohr model predicts, and for the next higher state ($N = 2$), the z component of translational angular momentum can be either 0 or $\hbar$. The photon itself has angular momentum $\hbar$, so this must also be taken into account in photon emissions or absorptions.

The simple Bohr model does, however, predict correctly the allowed electronic energy levels for atomic hydrogen (as determined from the emission and absorption spectra of hydrogen atoms). We will work with this model to see how quantization of angular momentum leads to the prediction of electronic energy levels in a hydrogen atom.

Allowed Radii of Electron Orbits

Because the proton has much more mass than an electron (about 2000 times as much), we'll make the approximation that the proton is at rest, with the electron in a circular orbit around the proton.

QUESTION If the electron momentum is p, and the radius of the circular orbit is r, what is the translational angular momentum?

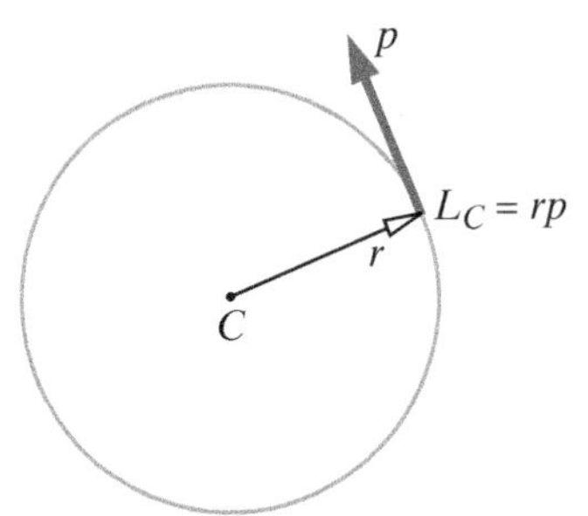

Figure 11.54 The angular momentum of the electron in a circular Bohr orbit relative to the proton. L_C is out of the page.

The translational angular momentum of the electron relative to the location of the proton is $L_{\text{trans},C} = rp\sin(90°) = rp$ (Figure 11.54). Bohr proposed that the only possible states of the hydrogen atom are those where the electron is in a circular orbit whose translational angular momentum is an integer multiple of $\hbar$:

BOHR: ANGULAR MOMENTUM IS QUANTIZED

$$|\vec{L}_{\text{trans},C}| = rp = N\hbar,$$

where N is an integer $(1, 2, 3, \ldots)$

The Electric Force and a Circular Orbit

QUESTION What is the magnitude of the electric force that the proton exerts on the electron?

$$F_{\text{el}} = \frac{1}{4\pi\varepsilon_0}\frac{e^2}{r^2}$$

where $+e$ is the electric charge on the proton ($-e$ for the electron).

QUESTION Use the Momentum Principle to relate this force to the circular motion of the electron.

In a circular orbit at constant speed we know that the rate of change of the momentum is $(v/r)p$, so we have in the approximation that $v \ll c$, where m is the mass of the electron,

$$\left(\frac{v}{r}\right)p = \frac{mv^2}{r} = F = \frac{1}{4\pi\varepsilon_0}\frac{e^2}{r^2}$$

Solving for r

We are looking for the allowed values of the orbit radius r, so we look for ways to express v in terms of r. Bohr's angular momentum condition gives us v in terms of r:

$$|\vec{L}_{\text{trans},C}| = rp = N\hbar \quad \text{leads to} \quad v = \frac{p}{m} = \frac{N\hbar}{mr}$$

Putting this result for r into the result obtained from the Momentum Principle, $mv^2/r = F$, we have

$$\frac{m}{r}\left(\frac{N\hbar}{mr}\right)^2 = \frac{1}{4\pi\varepsilon_0}\frac{e^2}{r^2}$$

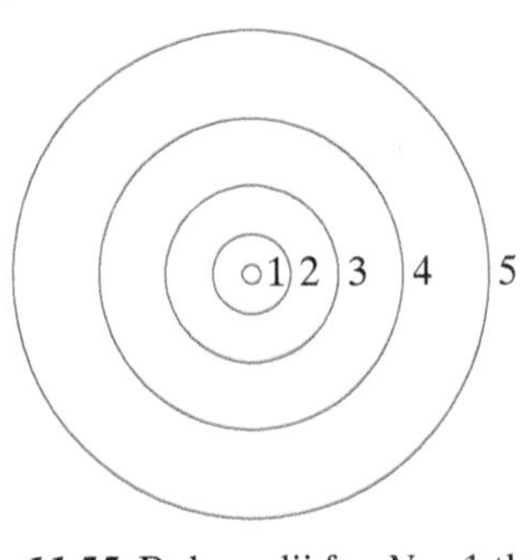

Figure 11.55 Bohr radii for $N = 1$ through 5. The nucleus is not shown.

Solving the latter relation for r, we obtain the following result (Figure 11.55):

ALLOWED BOHR RADII FOR ELECTRON ORBITS

$$r = N^2\frac{\hbar^2}{\left(\frac{1}{4\pi\varepsilon_0}\right)e^2 m}$$

Let's evaluate this result numerically, so that later we can compare with measurements of the spectrum of light emitted by excited atomic hydrogen. Using

$\hbar = 1.05 \times 10^{-34}$ J · s,
$1/4\pi\varepsilon_0 = 8.99 \times 10^9$ N · m^2/C^2,
$e = 1.60 \times 10^{-19}$ C,
and the mass of the electron $m = 9.11 \times 10^{-31}$,

we find this:

$$r = N^2(0.53 \times 10^{-10}\,\text{m})$$

This is a striking result. The simple Bohr model predicts that the smallest permissible electron radius ($N = 1$) is 0.53×10^{-10} m, and atoms are in fact observed to have radii of approximately this size.

Energy for a Circular Orbit

We cannot observe directly the radius of the orbit of an electron in an atom. What we can observe are photons emitted by excited atoms, which tell us the differences in energies between various energy levels of the atom. Using the Bohr model we can predict the energies of these energy levels, and see whether the differences between these levels match the energies of photons observed in an atomic spectrum.

Given a set of possible values for r, we can calculate the possible values for energy of the electron, starting from this observation:

$$\text{The result } mv^2/r = F \quad \text{leads to} \quad mv^2 = \frac{1}{4\pi\varepsilon_0}\frac{e^2}{r}.$$

QUESTION Now write an expression for the kinetic energy plus electric potential energy of the hydrogen atom, using the Bohr model.

Kinetic energy:

$$K = \frac{1}{2}mv^2 = \frac{1}{2}\left(\frac{1}{4\pi\varepsilon_0}\frac{e^2}{r}\right) \quad \text{(from the previous equation)}$$

Electric potential energy:

$$U_{\text{el}} = -\frac{1}{4\pi\varepsilon_0}\frac{e^2}{r}$$

Therefore the energy is this (omitting the rest energies):

$$E = K + U_{\text{el}} = -\frac{1}{2}\left(\frac{1}{4\pi\varepsilon_0}\frac{e^2}{r}\right)$$

According to the Bohr model only certain radii will actually occur. Inserting our previous expression for r, we get:

BOHR MODEL ENERGY LEVELS

$$E = -\frac{\left(\dfrac{1}{4\pi\varepsilon_0}\right)^2 e^4 m}{2N^2\hbar^2}$$

Evaluating this expression we get

$$E = -\frac{2.17 \times 10^{-18}\,\text{J}}{N^2}$$

Converting to electron-volts ($1\,\text{eV} = 1.6 \times 10^{-19}$ J) we have (Figure 11.56)

$$E = -\frac{13.6\,\text{eV}}{N^2}, \quad N = 1, 2, 3, \ldots$$

This prediction for the quantized energies of atomic hydrogen agrees well with the observed electronic spectrum of hydrogen. We quoted this result for the quantized energy levels in Chapter 8. The energies are negative, corresponding to bound states.

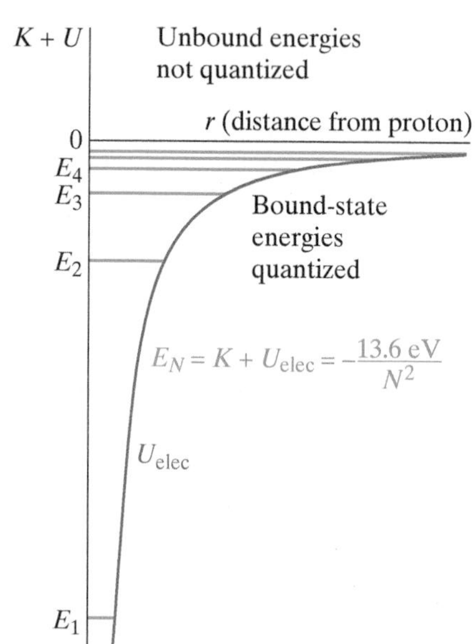

Figure 11.56 Bohr model prediction of electronic energy levels for a hydrogen atom. The potential energy curve for the electron–proton system is also shown.

Photons Emitted by Atomic Hydrogen

Bohr proposed that electromagnetic radiation would be emitted when there was a sudden change from a higher energy level to a lower one. We saw in Chapter 8 that the Bohr energy expression does correctly predict the energies of photons emitted by excited atomic hydrogen.

> QUESTION In the Bohr model, does photon emission correspond to the radius of the circular orbit getting larger or smaller? Does the quantum number N increase or decrease?

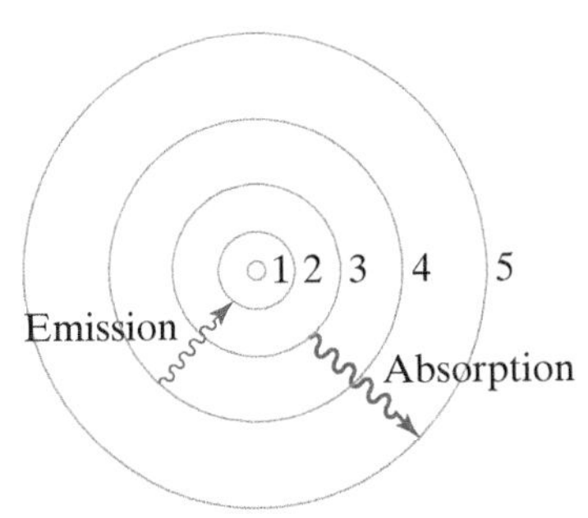

Figure 11.57 A photon is emitted in a transition from a higher to a lower energy level. A photon is absorbed in a transition from a lower to a higher energy level.

The kinetic plus potential energy is negative, corresponding to a bound state, so a higher energy is one that is less negative and therefore corresponds to a larger radius (since the energy is proportional to $-1/r$). Also, higher energy corresponds to a larger value of N, which is also associated with a larger radius. Therefore photon emission is associated with a decrease in r and a decrease in N (Figure 11.57). Photon absorption involves an increase in r and N.

When Bohr explained his model to Rutherford, Rutherford immediately realized that atoms display probabilistic behavior. For example, in the Bohr model an atom in the second excited state can either drop to the first excited state (with emission of a photon of energy 1.9 eV) or directly to the ground state (with emission of a photon of energy 12.4 eV). Both of these photon energies are observed in the spectrum of atomic hydrogen, so evidently some fraction of the time one process occurs, and the rest of the time the other process occurs. This is radically different from the deterministic behavior familiar in classical mechanics.

Particle Spin

We know that many elementary particles have rotational angular momentum because of observations of the interaction of these particles with other particles having nonzero angular momentum; in these interactions total angular momentum is constant.

For example, the electron, muon, and neutrino are said to have spin 1/2, because measurement of a component of their rotational angular momentum always yields $\pm(1/2)\hbar$. Every quark has spin 1/2, and particles built out of three quarks, such as protons and neutrons, necessarily have half-integral spin. The spin of the proton and neutron is 1/2, as though two of the quarks have their spins opposed (⇑⇓⇑), with no translational angular momentum of the quarks. Some short-lived three-quark particles have spin 3/2. One way to get this is for all the quarks to have their spins aligned (⇑⇑⇑), or there can be some translational angular momentum of the quarks. While rotational angular momentum can be half-integral, translational angular momentum is always integral (0, 1, 2, etc.).

Particles made out of one quark and one antiquark, called mesons, have integral spin. For example, the pion spin is zero (quark and antiquark spins opposed, ⇑⇓), and the spin of the "rho" meson is 1 (quark and antiquark spins aligned, ⇑⇑).

Presumably the angular momentum of a macroscopic object such as a baseball is also quantized, but the angular momentum of a baseball is huge compared to Planck's constant, and you don't notice the quantization, which is on an exceedingly fine scale (1×10^{-34} J · s!). Only at the atomic level are the effects of quantization really evident.

Particles, Nuclei, Atoms, and Molecules

Quantum mechanics predicts that the x, y, or z component of the angular momentum (L_x, L_y, or L_z) can only be an integer or half-integer multiple of $\hbar$, whereas the square magnitude of the angular momentum has the quantized values $l(l+1)\hbar^2$, where l has integer or half-integer values, depending on whether a component has integer or half-integer values:

QUANTIZED VALUES OF L^2

$$L^2 = l(l+1)\hbar^2$$

where l is integer or half-integer.

There is a very deep connection between angular momentum and the statistical behavior of particles. Particles such as the electron and the proton that have half-integral spin are called "fermions" and exhibit the peculiar behavior that two fermions cannot be in the exact same quantum state. This is why only two electrons can be in the lowest energy state of an atom, with their spins opposed. Their spin directions are forbidden to be the same, because then their energy and angular momentum would be exactly the same, which is forbidden for fermions. This prohibition is called the "Pauli exclusion principle."

On the other hand, particles with integral spin, called "bosons," are not subject to the Pauli exclusion principle, and there is no limit to the number of bosons that can be in the same energy state. In fact, there is a special state of matter called a "Bose–Einstein condensate" in which very large numbers of particles end up in exactly the same quantum state. This state of matter was predicted long before it was actually created and observed in 1995.

All nuclei that have an even number of protons and an even number of neutrons ("even-even" nuclei) have a total angular momentum of zero in their nuclear ground state. Examples include carbon-12 (6 protons and 6 neutrons) and oxygen-16 (8 protons and 8 neutrons). The spins of protons and neutrons in even-even nuclei are paired up in the lowest-energy state of a nucleus with each other in such a way as to produce zero net angular momentum. This would be an exceedingly unlikely outcome if angular momentum weren't quantized.

Angular momentum quantization plays a major role in determining the structure of atoms and the nature of the chemical periodic table. The two lowest-energy-state electrons in an atom always have zero translational angular momentum and zero rotational angular momentum (because the two rotational angular momenta are always oppositely aligned and add up exactly to zero).

In a diatomic gas molecule such as oxygen (O_2), the kinetic energy associated with rotation of the molecule is of course $L_{\text{rot}}^2/(2I)$. However, the rotational angular momentum L_{rot} is quantized, so the rotational energies of oxygen are quantized. The phenomenon of angular momentum quantization affects the specific heat of diatomic gases at low temperatures, as will be discussed in Chapter 12.

11.12 *GYROSCOPES

A gyroscope is a fascinating device, and its unusual properties have been exploited to stabilize ships and spacecraft. Its behavior provides a good model and analogy for some important aspects of the quantum mechanical behavior of atoms and nuclei. Magnetic resonance imaging (MRI) is based on the gyroscope-like behavior of spinning nuclei.

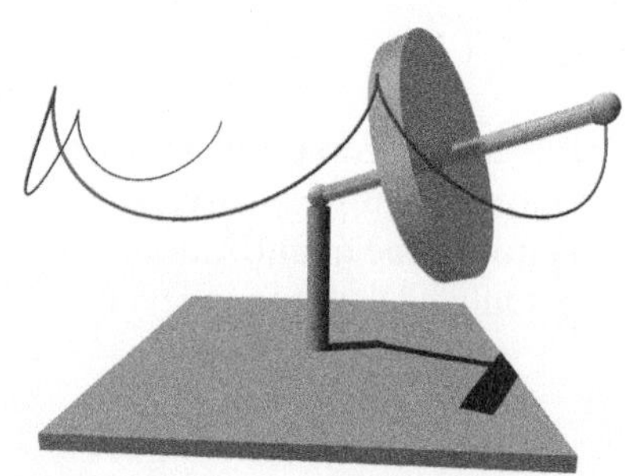

Figure 11.58 A gyroscope can exhibit both "nutation" and "precession."

A gyroscope has a spinning disk mounted on an axle. If you place one end of the axle on a vertical support, you may observe a very complex motion (Figure 11.58). The gyroscope rises and falls ("nutation") as it revolves around the support ("precession").

One fruitful approach in modeling a complex physical system is to try to analyze the simplest motion that the system is capable of. For example, we can observe a pure precession of a gyroscope, with no nutation. It is possible with care to start the gyroscope with special initial conditions so that it merely precesses, without bobbing up and down. We'll try to analyze this special kind of gyroscope motion. In fact, we'll start with a particularly simple form of precession, with the rotational axis horizontal (Figure 11.59).

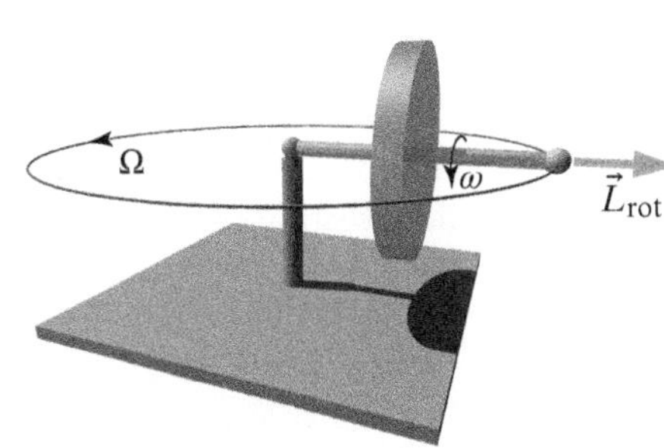

Figure 11.59 A gyroscope that is precessing horizontally, with no nutation.

We define some variables to make it easier to describe and discuss the situation. The gyroscope disk rotates around its own axis with some spin angular speed that we'll call ω, and the disk has a moment of inertia I. The rotational angular momentum $\vec{L}_{rot}$ of the disk always points horizontally but is continually changing direction as the gyroscope precesses (if we neglect friction, the angular speed ω is constant). The gyroscope revolves around the support with angular speed Ω (uppercase Greek omega), constant in magnitude and direction (if we neglect friction). Note that $v_{CM} = \Omega r$, so $\Omega = (v_{CM}/r)$, where r is the distance from the support to the center of mass.

In Chapter 5 we derived the expression for the time rate of change of the unit momentum vector $\hat{p}$, which had the magnitude (v/r), which is the same as angular speed ω. In general, the magnitude of the time rate of change of any rotating unit vector $\hat{X}$ is $|d\hat{X}/dt| = \omega$, and the magnitude of the rate of change of a rotating vector $\vec{X} = |\vec{X}|\hat{X}$ whose magnitude isn't changing is $\omega|\vec{X}|$. Since the angular speed of the gyroscope's precession is Ω, the magnitude of the rate of change of the (linear) momentum P of the gyroscope is ΩP, and the magnitude of the rate of change of the angular momentum is ΩL.

> QUESTION What forces act on the gyroscope? What relationships are there among these forces?

The support pushes up with a force we'll call F_N, and the Earth pulls down with a force Mg through the center of mass (Figure 11.60). Since the center of mass stays at the same height all the time in the case of simple precession, the vertical component of the net force must be zero. Therefore $F_N = Mg$.

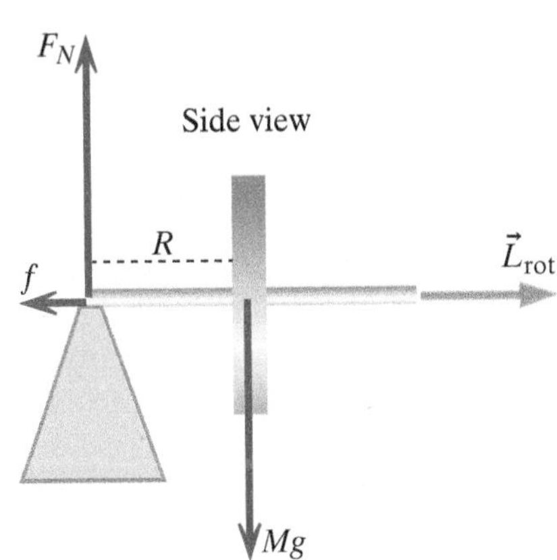

Figure 11.60 Side view showing external forces acting on the gyroscope and the rotational angular momentum vector.

This isn't quite the whole story, though. Note that the center of mass of the gyroscope is moving in a circle, which requires that there be a radially inward force to make the momentum vector rotate (with angular speed Ω). The only object that can exert this force is the support. There must be a small horizontal frictional force f such that $|d\vec{P}/dt| = \Omega P = f$, where $P = Mv_{CM} = M(\Omega R)$; R is the radius of the circle traveled by the center of mass. This force, $f = MR\Omega^2$, is small if the precession rate Ω is small.

It is easy to observe with a toy gyroscope that as the spin of the disk slows down due to friction (smaller ω), the precession actually speeds up (larger Ω). If we could predict the precession speed for a given spin, we would largely understand this simple example of gyroscope motion. What can we use to attempt a prediction?

> QUESTION What is the magnitude of the rotational angular momentum of the gyroscope? What is the magnitude of the translational angular momentum of the gyroscope around the support?

We said that the rotating disk has moment of inertia I and angular speed ω, so the rotational angular momentum is simply $L_{rot} = I\omega$. The magnitude of the translational angular momentum is this:

$$L_{support} = |\vec{R} \times \vec{P}| = RP$$

Evidently the magnitudes of both the rotational and translational angular momenta are constant, not changing with time. However, what about the directions of these angular momenta? Do they change with time?

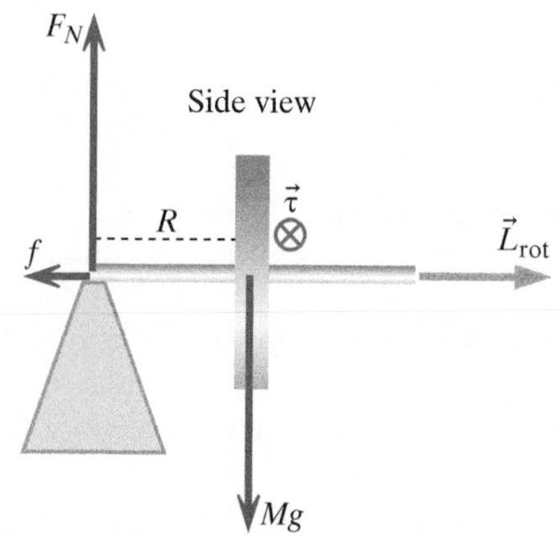

Figure 11.61 Side view showing external forces acting on the gyroscope and the rotational angular momentum vector. There is a torque around the center of mass, into the page, due to the F_N force.

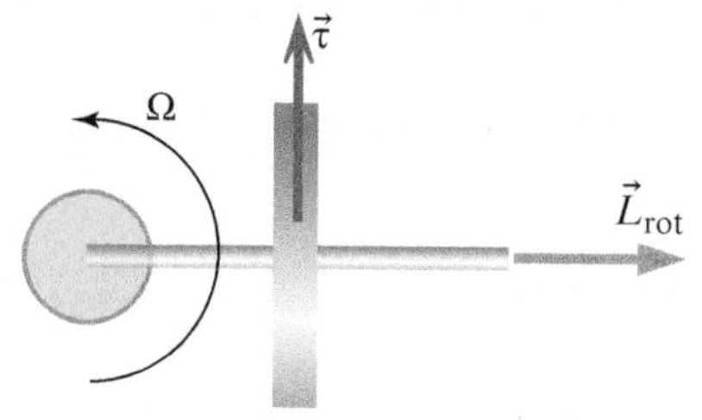

View from above

Figure 11.62 View from above, showing the torque vector and the rotational angular momentum vector.

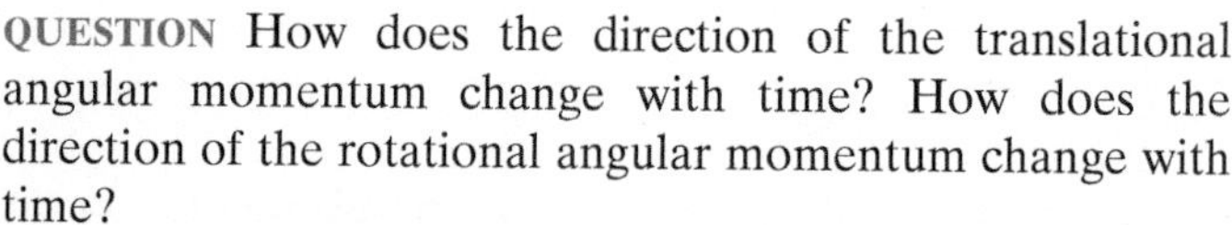

QUESTION How does the direction of the translational angular momentum change with time? How does the direction of the rotational angular momentum change with time?

Neither the magnitude nor the direction of the translational angular momentum changes (neglecting friction): it has constant magnitude and points vertically upward at all times. However, the direction of the rotational angular momentum is constantly changing as the gyroscope precesses. We need to calculate the rate at which the rotational angular momentum vector is changing, $d\vec{L}_{rot}/dt$. Since the rotational angular momentum vector is a rotating vector, which rotates with the precession angular speed Ω, we have this:

$$\left|\frac{d\vec{L}_{rot}}{dt}\right| = \Omega L_{rot}$$

QUESTION Therefore, what is the Angular Momentum Principle for the gyroscope?

The remaining element we need for the Angular Momentum Principle is the torque that acts around the center of mass, at the center of the spinning disk. Look again at the force diagram (Figure 11.61), and you see that the magnitude of the torque about the center of mass is $RF_N = RMg$, and the direction of the torque is into the page. Seen from above, the torque points at a right angle to the rotational angular momentum (Figure 11.62). Therefore the Angular Momentum Principle yields

$$\left|\frac{d\vec{L}_{rot}}{dt}\right| = \Omega L_{rot} = \tau_{CM}$$

where $L_{rot} = I\omega$, and $\tau_{CM} = RMg$. Solving for the precession angular speed Ω, we have

$$\Omega = \frac{\tau_{CM}}{L_{rot}} = \frac{RMg}{I\omega}$$

This is a surprisingly simple result for such a complicated system, and it agrees at least qualitatively with observations. A smaller spin ω is associated with a larger precession rate Ω. Conversely, a gyroscope that spins very fast precesses very slowly. If possible, your instructor will provide an opportunity to test the theory quantitatively by observing an actual gyroscope whose spin and precession rates can be measured.

Reflection

Reflect on how we arrived at this result. Look over the steps involved. The key point is that the rotational angular momentum varies in direction without varying in magnitude, and the magnitude of the rate of change of the rotational angular momentum is simply Ω times L_{rot}. This rate of change of the rotational angular momentum is equal to the net torque acting on the system (around the center of mass).

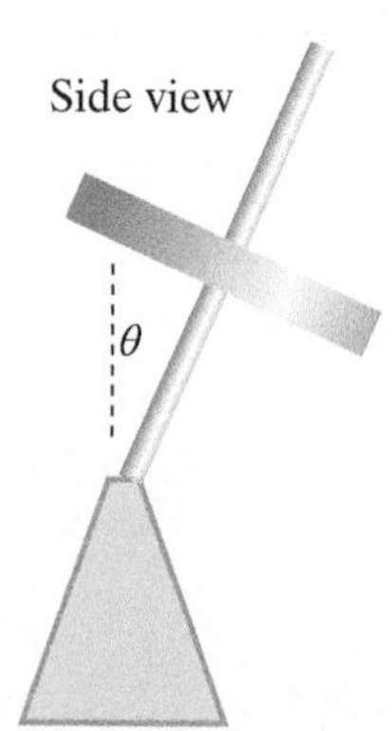

Figure 11.63 What is the precession rate when the spin axis is at an angle θ to the vertical?

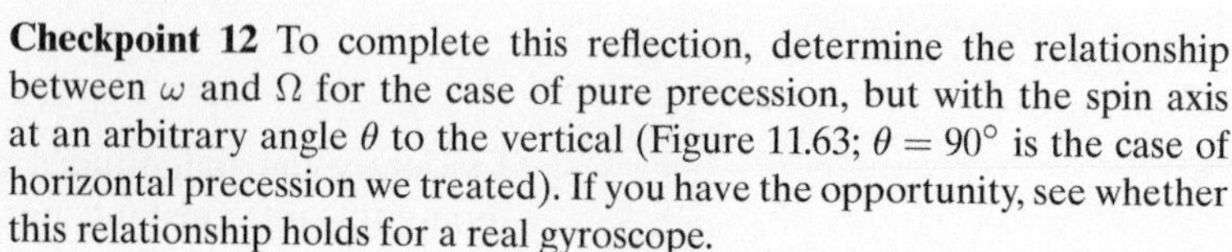

Checkpoint 12 To complete this reflection, determine the relationship between ω and Ω for the case of pure precession, but with the spin axis at an arbitrary angle θ to the vertical (Figure 11.63; $\theta = 90°$ is the case of horizontal precession we treated). If you have the opportunity, see whether this relationship holds for a real gyroscope.

Uses of Gyroscopes

The precession we calculated is the result of a nonzero torque acting around the center of mass. Through clever mounting of the gyroscope it is possible to make this torque vanishingly small, in which case the axis does not precess but always maintains the same direction, which has been useful in navigation. The satellite-based Global Positioning System (GPS) can tell you where your airplane is; gyroscopically based instruments can tell you the orientation of your airplane at this known location.

Large gyroscopes have been used to stabilize ships against rolling in the sea. A gyroscope with very high rotational angular momentum (large moment of inertia and large angular speed) is hard to turn quickly, and this can provide mechanical stability. Gyroscopes are also used to stabilize the orientation of spacecraft and satellites.

Friction

We deliberately neglected friction in the calculation. Friction has two main effects on a gyroscope. Friction in the spin axis makes ω decrease with time, which leads to an increased precession rate Ω. Friction on the top of the support slows down the precession, which means that Ω is no longer equal to RMg/L_{rot} as it should be for pure precession. As a result of these effects, what you observe with a gyroscope that starts out in pure precession is that the spin axis eventually starts to tip lower, accompanied by faster precession. The gyroscope starts out with a stately, dignified, slow precession, but toward the end gives the impression of motion that is more and more frantic.

Magnetic Resonance Imaging (MRI)

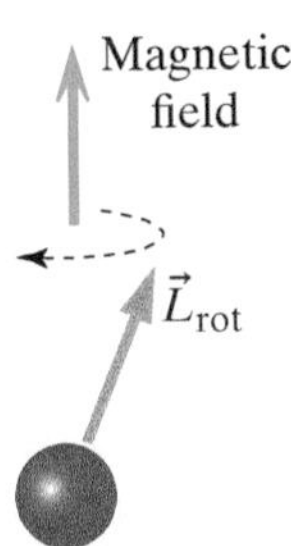

Figure 11.64 Precession of proton spin in an external magnetic field.

Electrically charged atomic and subatomic objects that have angular momentum act like little bar magnets with north and south poles along the angular momentum axis (we say that they have "magnetic moments"). It is as though the spinning charged particle constituted little loops of current, and current loops produce a magnetic field. Even some electrically neutral particles such as the neutron have magnetic moments, because they are built out of electrically charged quarks that have magnetic moments.

In the presence of an applied magnetic field (typically produced by other current-carrying coils of wire), a bar magnet tends to twist to line up with the applied magnetic field. That is, a magnetic field exerts a torque on a bar magnet. A well-known example is that a compass needle aligns with the Earth's magnetic field, which is useful for navigation. (The compass needle is magnetized and is itself a little bar magnet.)

Since a magnetic field applies a torque to a bar magnet, or to atomic or subatomic particles that have angular momentum, such particles precess in the presence of an applied magnetic field (Figure 11.64). This phenomenon is exploited in a technique important in physics, chemistry, and biology, called NMR—nuclear magnetic resonance. A particularly useful application of NMR is in magnetic resonance imaging (MRI).

In MRI, a large, steady, uniform magnetic field created by large direct-current-carrying coils of wire surrounding the patient makes protons in the person's body precess. Many of the most common nuclei in the body are even-even nuclei (in particular, the most common isotopes of carbon and oxygen) and have zero angular momentum and therefore no magnetic moment, and these do not precess. However, hydrogen nuclei (protons) are very common in the body, have rotational angular momentum, and do precess. The stronger the magnetic field, the larger the torque acting on the proton nuclear magnets, and the faster their precession. (Note that for a gyroscope, $\Omega = \tau/L_{rot}$; the precession rate is proportional to the applied torque τ.)

With the protons precessing in the presence of the large steady magnetic field, a small high-frequency (time-varying) magnetic field tuned to match exactly the precession frequency of the protons will flip the spins upside down. This is called a "resonance" phenomenon. The high-frequency signal is turned off, and the protons revert back to being aligned with the steady magnetic field. The act of flipping back emits radiation (at the precession frequency) that can be detected by coils connected to a receiver, and the strength of the signal indicates how many protons were affected.

That's the basic physical mechanism, but this by itself would not yield spatial detail about the interior of the body. The trick is to superimpose on the large steady magnetic field a small nonuniform magnetic field, which has the effect of establishing slightly different torques on protons at different locations in the body. As a result, when the protons flip back into alignment they radiate signals whose frequencies indicate their locations. A computer algorithm calculates how much of each frequency is present in the signal, and this indicates how many protons are at each location. In this way a very detailed image is built up of the interior of the body.

Precession of Spin Axes in Astronomy

As we mentioned earlier, the Earth is subject to small torques due to gravitational forces of the Sun and Moon acting on its nonspherical, "oblate" shape. There is a torque on the Earth's equatorial bulge, perpendicular to the spin rotation axis, and this causes the Earth's axis to precess very slowly, once around about every 26,000 years, so that the "North Star" hasn't always been the star we call Polaris. This effect is called the "precession of the equinoxes" because it leads to a change in what month the spring and fall equinoxes occur. To see why there is such an effect, consider Figure 11.65.

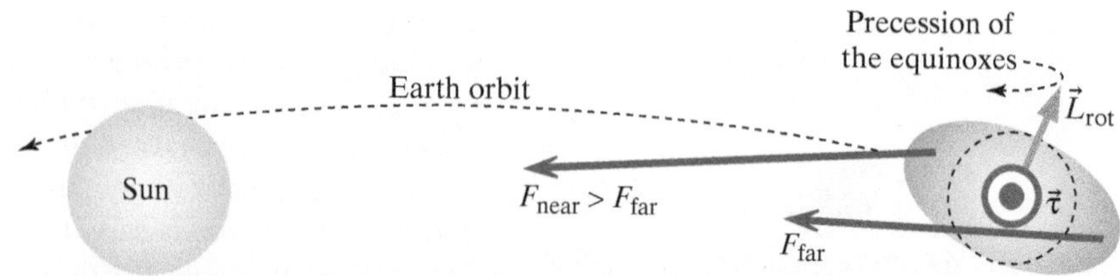

Figure 11.65 Forces on the Earth due to the Sun. The diagram is not to scale; the bulges of the Earth are greatly exaggerated. The net torque about the center of mass of the Earth is out of the page.

The Sun (or Moon) exerts a slightly larger force on the closer bulge due to the $1/r^2$ dependence of the force, so there is a nonzero net torque around the center of mass of the Earth, pointing toward you, out of the page. This makes the rotational angular momentum $\vec{L}_{rot}$ of the Earth precess in a "retrograde" way—that is, in the opposite direction to most rotations in our Solar System. The direction of the axis of the Earth slowly changes due to the torque. (The size of the equatorial bulge, and the differences in F_{near} and F_{far}, have been greatly exaggerated in Figure 11.65.)

Since the Earth's axis is tipped 23° away from perpendicular to the Earth–Sun orbit, the precession of the axis makes a big change in the location of the North Star. Thirteen thousand years from now the North Star will be a star that is 46° away from Polaris in the night sky.

The Moon going around the Earth is a kind of gyroscope. The Earth–Moon orbit is inclined a few degrees to the plane of the Earth–Sun orbit, and the Sun exerts a nonzero torque perpendicular to the angular momentum of the Earth–Moon system. To see why, consider a simpler case in which the Earth and Moon have equal mass, and you see the same effect as with the equatorial bulge of the Earth (Figure 11.66).

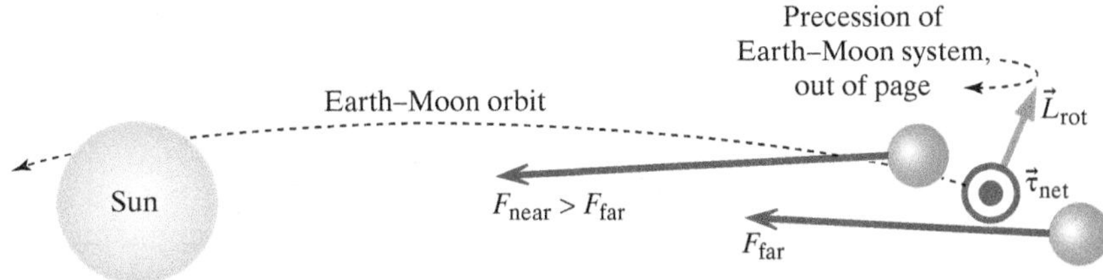

Figure 11.66 Forces on the Earth–Moon system due to the Sun. The diagram is not to scale; the Earth and Moon are shown as if their masses were equal. The net torque on the Earth–Moon system about its center of mass is out of the page.

This torque varies in magnitude during a month, but the averaged effect is that the Earth–Moon system precesses once around in about 18 y (this precession is also "retrograde"). This has an effect on the timing of eclipses, which can occur only when the Moon in its orbit is passing through the plane of the Earth–Sun orbit.

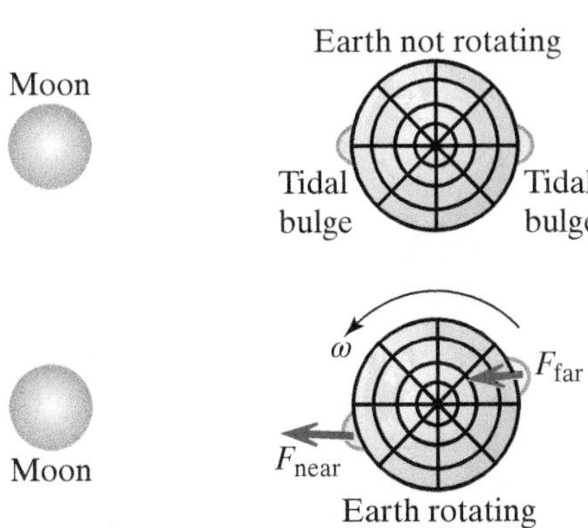

Figure 11.67 Tidal torques due to the Moon, viewed from a point above the North Pole. In the top image the Earth is not rotating; in the bottom image it is rotating. (Not to scale; the tidal effects are greatly exaggerated.)

Tidal Torques

If the Earth did not rotate, the Moon would create tides in the oceans that would pile up in line with the Moon. However, the rotating Earth drags these tidal bulges so that they are no longer in line with the Moon, and the Moon exerts a small torque (Figure 11.67). This torque is directed along the axis and acts to slow down the spin rate, so that the day is getting longer than 24 hours. This effect is called "tidal friction." This interaction with the Moon has the effect that as the rotational angular momentum of the Earth decreases, the translational angular momentum of the Earth–Moon system increases (conservation of total angular momentum), with the result that the Earth and Moon are getting farther apart. (These tidal forces have a small net tangential component that acts on both Earth and Moon to increase their translational angular momenta.)

Something like this presumably happened in the past to the Moon, when it was molten. The rotational angular momentum of the Moon decreased due to tidal torques exerted by the Earth, which are much larger than the tidal torques that the low-mass Moon exerts on the Earth. When the Moon's spin angular speed had decreased to be the same as its translational angular speed (currently about 2π rad or 360° per month), the process terminated. As a result, now the Moon always displays nearly the same face to the Earth.

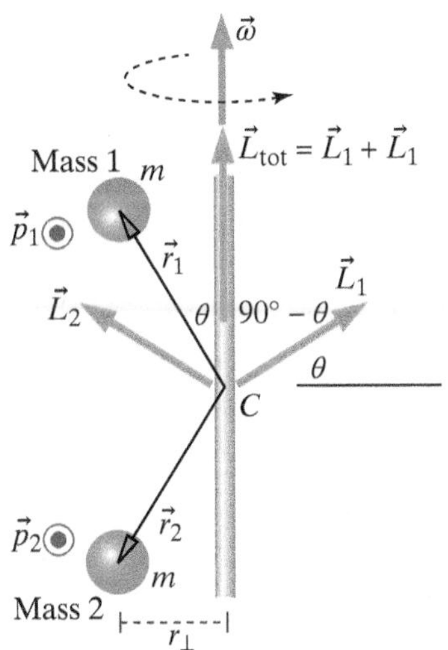

Figure 11.68 The off-axis components of the angular momenta of the two masses cancel. Their total angular momentum is along the axis of rotation.(The two masses at this instant are in the xy plane.)

11.13 *MORE ON MOMENT OF INERTIA

In Figure 11.18 all the masses lie in a plane that was perpendicular to the axis of rotation. What if they don't all lie in this plane? We'll show that for simple rigid systems the moment of inertia is still calculated by adding up terms like $mr_\perp^2\omega$, where the perpendicular distances are measured from the axis of rotation.

In Figure 11.68 two equal masses are mounted to a rotating shaft, and we want to determine the angular momentum relative to the center of the device, marked C. We show the directions of the angular momentum $\vec{L}_C = \vec{r}_C \times \vec{p}$ of each mass. You can see that the individual contributions have components that are not along the shaft. However, these components cancel for pairs of masses above and below each other. The effect is that the total angular momentum vector lies along the shaft, and can be calculated by adding up $mr_\perp^2\omega$ for each mass, where $r_\perp$ is the perpendicular distance from the axis.

Here is the proof: the magnitude of the angular momentum of mass 1 about location C is $|\vec{L}_1| = r_1 m v_1 = r_1 m \omega r_\perp$. The y component of this angular momentum is $|\vec{L}_1|\cos(90° - \theta) = r_1 m\omega r_\perp \sin\theta = m\omega r_\perp^2$ since $r_1 \sin\theta = r_\perp$. Similarly, the y component of the angular momentum of mass 2 about location C is also $m\omega r_\perp^2$. Therefore the total angular momentum of the two masses is in the y direction and is given by

$$L_{C,y} = mr_\perp^2\omega + mr_\perp^2\omega = (mr_\perp^2 + mr_\perp^2)\omega = I\omega$$

since
$$I = mr_\perp^2 + mr_\perp^2$$

For example, the moment of inertia of a uniform-density thin disk rotating on its axle is $I_{\text{disk}} = \frac{1}{2}MR^2$, where M is the mass of the disk and R is its radius. Because only the distances of atoms from the axle matter, the moment of inertia of a long cylinder has exactly the same form, $I_{\text{disk}} = \frac{1}{2}MR^2$, and it doesn't matter how long the cylinder is (or to put it another way, how thick the disk is).

In this introductory treatment of angular momentum we will deal with simple, symmetrical rotating systems for which the moment of inertia can be calculated simply as $I = m_1 r_{\perp 1}^2 + m_2 r_{\perp 2}^2 + m_3 r_{\perp 3}^2 + m_4 r_{\perp 4}^2 + \cdots$. For a more complicated situation, see the following discussion.

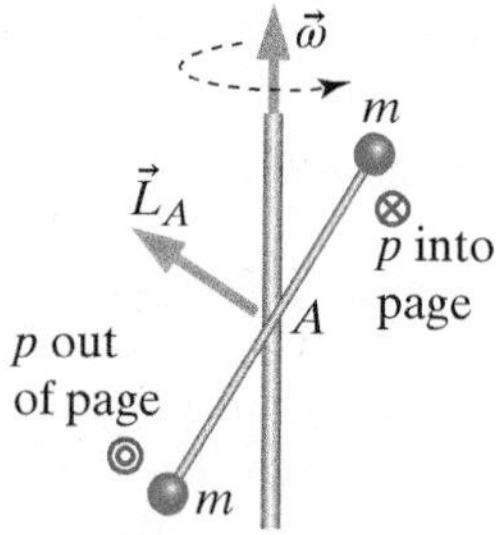

Figure 11.69 If the object lacks axial symmetry, the angular momentum and angular velocity may have different directions!

Lack of Symmetry

In more advanced courses you may study nonsymmetric rotational situations like the one in Figure 11.69. The angular velocity points along the axis of rotation, but the (translational) angular momentum of each mass is $\vec{r}_A \times \vec{p}$, so the total angular momentum $\vec{L}_A$ does *not* point along the axis! Moreover, the angular momentum vector continually changes direction. This change of angular momentum requires a nonzero torque, which is applied to the axle by the bearings in which the axle rotates. The effect of this "dynamic imbalance" is to cause severe wear on the axle and bearings. Car tires must be carefully balanced to prevent this.

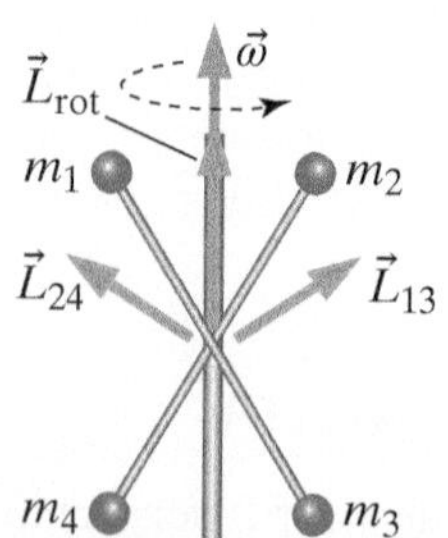

Figure 11.70 If the object is symmetric, the angular momentum is in the direction of the angular velocity.

A symmetric object like the one shown in Figure 11.70 has an angular momentum that does point along the axis, because the perpendicular components of the angular momenta cancel each other. This is the simpler kind of situation we have dealt with in this chapter. In the symmetric case shown in Figure 11.70 it is true that $I = m_1 r_{\perp 1}^2 + m_2 r_{\perp 2}^2 + m_3 r_{\perp 3}^2 + m_4 r_{\perp 4}^2 + \cdots$ about the axis of rotation, and the simple expressions $\vec{L}_{\text{rot}} = I\vec{\omega}$ and $K_{\text{rot}} = \frac{1}{2}I\omega^2$ are valid.

We calculated moment of inertia as a sum (integral) over $r_\perp^2$, the perpendicular distance to an axis. To deal with rotational motion in general requires expressing the moment of inertia as a "tensor," a 3 by 3 array of numbers representing integrals over $x^2, xy, xz, y^2, yx, yz, z^2, zx$, and zy. Matrix multiplication of this tensor times the angular velocity vector $\vec{\omega}$ yields an angular momentum vector that need not point in the same direction as $\vec{\omega}$. This concept is beyond the scope of this textbook.

SUMMARY

The Angular Momentum Principle for a system:

$$\frac{d\vec{L}_{\text{tot},A}}{dt} = \vec{\tau}_{\text{net},A} \quad \text{or} \quad \Delta\vec{L}_{\text{tot},A} = \vec{\tau}_{\text{net},A}\Delta$$

Angular momentum of a particle about location A (see Figure 11.71):

$$\vec{L}_A = \vec{r}_A \times \vec{p}$$
$$L_A = r_\perp p = r_A p \sin\theta$$

The direction is given by the right-hand rule.

$$\vec{L}_A = \langle (yp_z - zp_y), (zp_x - xp_z), (xp_y - yp_x) \rangle$$

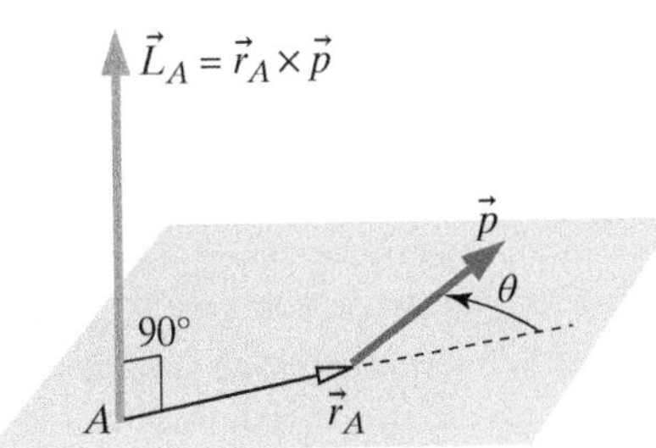

Figure 11.71

Torque is a measure of the twist imparted by a force about location A:

$$\vec{\tau}_A = \vec{r}_A \times \vec{F}$$

Angular momentum for a multiparticle system about some location A can be divided into translational and rotational angular momentum:

$$\vec{L}_A = \vec{L}_{\text{trans},A} + \vec{L}_{\text{rot}}$$
$$\vec{L}_{\text{trans},A} = \vec{r}_{\text{CM},A} \times \vec{P}_{\text{tot}}$$
$$\vec{L}_{\text{rot}} = \vec{r}_{1,\text{CM}} \times \vec{p}_1 + \vec{r}_{2,\text{CM}} \times \vec{p}_2 + \cdots$$

(See Figure 11.72.)

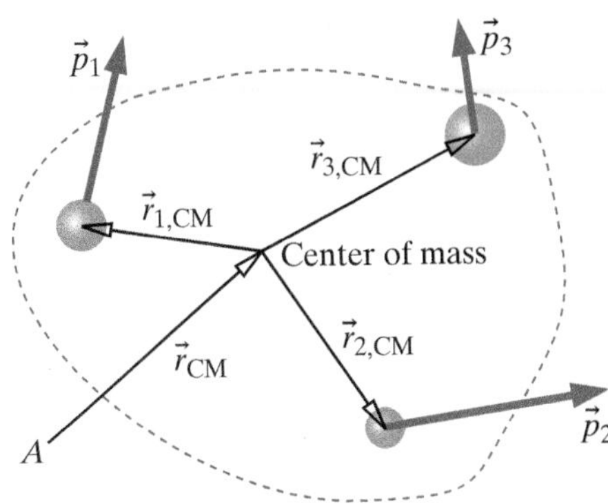

Figure 11.72

Rotational angular momentum in terms of moment of inertia:

$$\vec{L}_{\text{rot}} = I\vec{\omega}$$

Rate of change of rotational angular momentum:

$$\frac{d\vec{L}_{\text{rot}}}{dt} = \vec{\tau}_{\text{net,CM}}$$

Kinetic energy relative to center of mass in terms of moment of inertia and rotational angular momentum:

$$K_{\text{rot}} = \frac{1}{2}I\omega^2 = \frac{L_{\text{rot}}^2}{2I}$$

Moments of inertia:

$$I_{\text{disk}} = I_{\text{cylinder}} = \frac{1}{2}MR^2$$

about center of disk, rotating around axis of disk.

$$I_{\text{sphere}} = \frac{2}{5}MR^2$$

for axis passing through center of sphere.

$$I_{\text{cylinder}} = \frac{1}{12}ML^2 + \frac{1}{4}MR^2$$

for a uniform solid cylinder of length L, radius R, about axis perpendicular to cylinder, through center of cylinder.

$$I_{\text{cylinder}} \approx \frac{1}{12}ML^2$$

if the cylinder is very thin (R is very small).

Kinetic energy and angular momentum for a rigid object rotating about an axle that does not go through the center of mass:

$$K_{\text{tot}} = K_{\text{trans}} + K_{\text{rot}} = \frac{1}{2}(Mr_{\text{CM}}^2 + I_{\text{CM}})\omega^2$$
$$\vec{L}_A = \vec{L}_{\text{trans},A} + \vec{L}_{\text{rot}} = (Mr_{\text{CM}}^2 + I_{\text{CM}})\omega$$

Predicting positions when there is rotation:

$$\Delta\theta = \omega_{\text{avg}}\Delta t$$

Quantized angular momentum in microscopic systems:

A component of angular momentum is an integer or half-integer multiple of $\hbar$.

The square magnitude of angular momentum has quantized values

$$L^2 = l(l+1)\hbar^2$$

where l has integer or half-integer values.

Bohr model of hydrogen:

$$E = -\frac{13.6\,\text{eV}}{N^2}, \quad N = 1,2,3,\ldots$$

Angular speed of precession for a gyroscope:

$$\Omega = \frac{\tau}{L_{\text{rot}}} = \frac{RMg}{I\omega}$$

QUESTIONS

Q1 Give an example of a situation in which an object is traveling in a straight line, yet has nonzero angular momentum.

Q2 What are the units of moment of inertia? Of angular speed ω? Of angular momentum? Of linear momentum?

Q3 Under what circumstances is angular momentum constant? Give an example of a situation in which the x component of angular momentum is constant, but the y component isn't.

Q4 Give examples of translational angular momentum and rotational angular momentum in our Solar System.

Q5 Give an example of a physical situation in which the angular momentum is zero yet the translational and rotational angular momenta are both nonzero.

Q6 Under what conditions is the torque about some location equal to zero?

Q7 Make a sketch showing a situation in which the torque due to a single force about some location is 20 N · m in the positive z direction, whereas about another location the torque is 10 N · m in the negative z direction.

Q8 What is required for the angular momentum of a system to be constant? **(a)** zero net torque, **(b)** zero impulse, **(c)** no energy transfers, **(d)** zero net force

Q9 Consider a rotating star far from other objects. Its rate of spin stays constant, and its axis of rotation keeps pointing in the same direction. Why?

Q10 A device consists of eight balls, each of mass M, attached to the ends of low-mass spokes of length L, so the radius of rotation of the balls is $L/2$. The device is mounted in the vertical plane, as shown in Figure 11.73. The axle is held up by supports that are not shown, and the wheel is free to rotate on the nearly frictionless axle. A lump of clay with mass m falls and sticks to one of the balls at the location shown, when the spoke attached to that ball is at 45° to the horizontal. Just before the impact the clay has a speed v, and the wheel is rotating counterclockwise with angular speed ω.

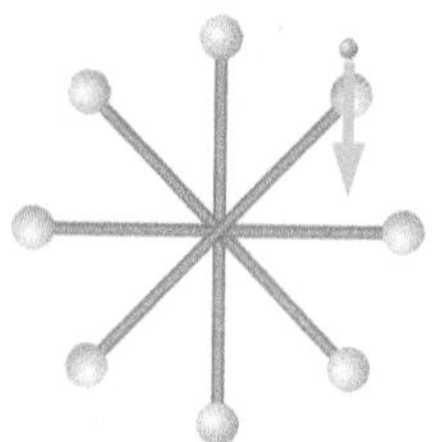

Figure 11.73

(a) Which of the following statements are true about the device and the clay, for angular momentum relative to the axle of the device? (1) The angular momentum of the device + clay just after the collision is equal to the angular momentum of the device + clay just before the collision. (2) The angular momentum of the falling clay is zero because the clay is moving in a straight line. (3) Just before the collision, the angular momentum of the wheel is 0. (4) The angular momentum of the device is the sum of the angular momenta of all eight balls. (5) The angular momentum of the device is the same before and after the collision. **(b)** Just before the impact, what is the (vector) angular momentum of the combined system of device plus clay about the center C? (As usual, x is to the right, y is up, and z is out of the screen, toward you.) **(c)** Just after the impact, what is the angular momentum of the combined system of device plus clay about the center C? **(d)** Just after the impact, what is the (vector) angular velocity of the device? **(e)** Qualitatively, what happens to the total linear momentum of the combined system? Why? (1) Some of the linear momentum is changed into energy. (2) Some of the linear momentum is changed into angular momentum. (3) There is no change because linear momentum is always conserved. (4) The downward linear momentum decreases because the axle exerts an upward force. **(f)** Qualitatively, what happens to the total kinetic energy of the combined system? Why? (1) Some of the kinetic energy is changed into linear momentum. (2) Some of the kinetic energy is changed into angular momentum. (3) The total kinetic energy decreases because there is an increase of internal energy in this inelastic collision. (4) There is no change because kinetic energy is always conserved.

Q11 A rod rotates in the vertical plane around a horizontal axle. A wheel is free to rotate on the rod, as shown in Figure 11.74. A vertical stripe is painted on the wheel. As the rod rotates clockwise, the vertical stripe on the wheel remains vertical. Is the translational angular momentum of the wheel relative to location A zero or nonzero? If nonzero, what is its direction? Is the rotational angular momentum of the wheel zero or nonzero? If nonzero, what is its direction? Consider a similar system, but with the wheel welded to the rod (not free to turn). As the rod rotates clockwise, does the stripe on the wheel remain vertical? Is the translational angular momentum of the wheel relative to location A zero or nonzero? If nonzero, what is its direction? Is the rotational angular momentum of the wheel zero or nonzero? If nonzero, what is its direction?

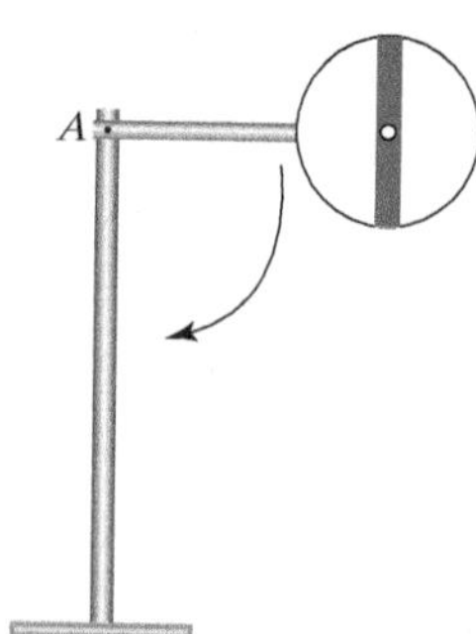

Figure 11.74 A rod rotating in the vertical plane with a wheel attached.

Q12 What features of the Bohr model of hydrogen are consistent with the later, full quantum mechanical analysis? What features of the Bohr model had to be abandoned?

PROBLEMS

Section 11.1

•**P13** Evaluate the cross product $(5\hat{\imath} + 3\hat{\jmath}) \times (-4\hat{\imath} + 2\hat{\jmath})$, which expands to $-20\hat{\imath}\times\hat{\imath} + 10\hat{\imath}\times\hat{\jmath} - 12\hat{\jmath}\times\hat{\imath} + 6\hat{\jmath}\times\hat{\jmath}$.

•**P14** What is the angular momentum $\vec{L}_A$ if $\vec{r}_A = \langle 9, -9, 0\rangle$ m and $\vec{p} = \langle 12, 10, 0\rangle$ kg·m/s?

•**P15** At a particular instant the location of an object relative to location A is given by the vector $\vec{r}_A = \langle 6, 6, 0\rangle$ m. At this instant the momentum of the object is $\vec{p} = \langle -11, 13, 0\rangle$ kg·m/s. What is the angular momentum of the object about location A?

•**P16** Figure 11.75 shows seven particles, each with the same magnitude of momentum $|\vec{p}| = 25$ kg·m/s but with different directions of momentum and different positions relative to location A. The distances shown in the diagram have these values: $w = 18$ m, $h = 28$ m, and $d = 27$ m.

Calculate the z component of angular momentum L_{Az} for each particle (x to the right, y up, z out of the page). Make sure you give the correct sign.

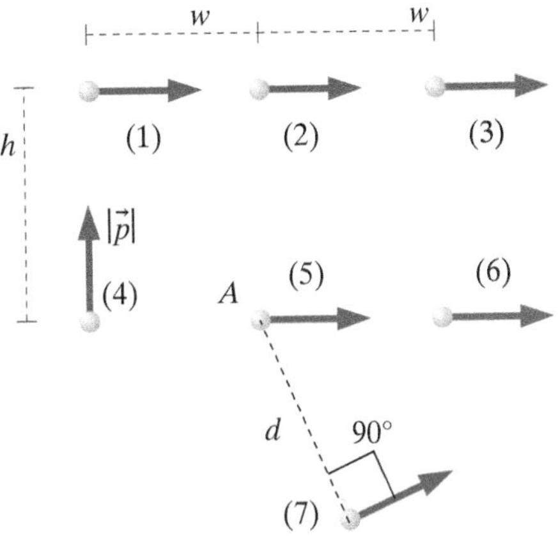

Figure 11.75

•**P17** A comet orbits the Sun (Figure 11.76). When it is at location 1 it is a distance d_1 from the Sun, and has magnitude of momentum p_1. Location A is at the center of the Sun. When the comet is at location 2, it is a distance d_2 from the Sun, and has magnitude of momentum p_2. **(a)** When the comet is at location 1, what is the direction of $\vec{L}_A$? **(b)** When the comet is at location 1, what is the magnitude of $\vec{L}_A$? **(c)** When the comet is at location 2, what is the direction of $\vec{L}_A$? **(d)** When the comet is at location 2, what is the magnitude of $\vec{L}_A$? Later we'll see that the Angular Momentum Principle tells us that the angular momentum at location 1 must be equal to the angular momentum at location 2.

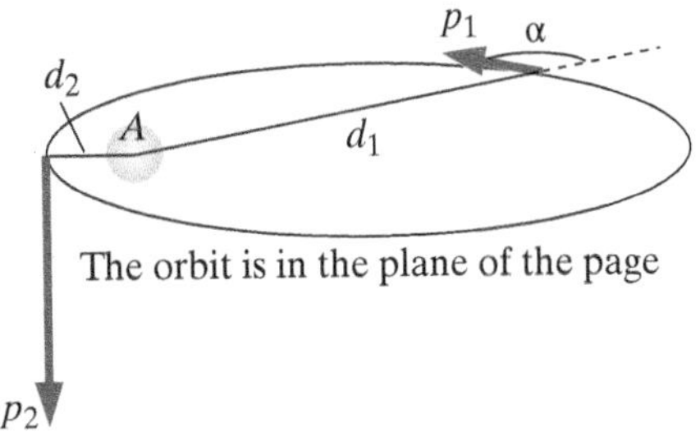

Figure 11.76

•**P18** A common amusement park ride is a Ferris wheel (see Figure 11.77, which is not drawn to scale). Riders sit in chairs that are on pivots so they remain level as the wheel turns at a constant rate. A particular Ferris wheel has a radius of 24 meters, and it makes one complete revolution around its axle (at location A) in 20 s. In all of the following questions, consider location A (at the center of the axle) as the location around which we will calculate the angular momentum. At the instant shown in the diagram, a child of mass 40 kg, sitting at location F, is traveling with velocity $\langle 7.5, 0, 0\rangle$ m/s.

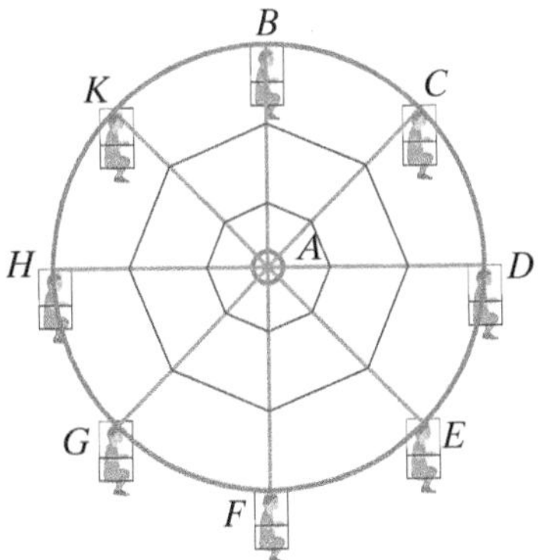

Figure 11.77

(a) What is the linear momentum of the child? **(b)** In the definition $\vec{L} = \vec{r} \times \vec{p}$, what is the vector $\vec{r}$? **(c)** What is $\vec{r}_\perp$? **(d)** What is the magnitude of the angular momentum of the child about location A? **(e)** What is the plane defined by $\vec{r}$ and $\vec{p}$ (that is, the plane containing both of these vectors)? **(f)** Use the right-hand rule to determine the z component of the angular momentum of the child about location A. **(g)** You used the right-hand rule to determine the z component of the angular momentum, but as a check, calculate in terms of position and momentum: What is xp_y? What is yp_x? Therefore, what is the z component of the angular momentum of the child about location A? **(h)** The Ferris wheel keeps turning, and at a later time, the same child is at location E, with coordinates $\langle 16.971, -16.971, 0\rangle$ m relative to location A, moving with velocity $\langle 5.303, 5.303, 0\rangle$ m/s. Now what is the magnitude of the angular momentum of the child about location A?

Section 11.2

•**P19** The moment of inertia of a sphere of uniform density rotating on its axis is $\frac{2}{5}MR^2$. Use data given at the end of this book to calculate the magnitude of the rotational angular momentum of the Earth.

•**P20** Calculate the angular momentum for a rotating disk, sphere, and rod: **(a)** A uniform disk of mass 13 kg, thickness 0.5 m, and radius 0.2 m is located at the origin, oriented with its axis along the y axis. It rotates clockwise around its axis when viewed from above (that is, you stand at a point on the $+y$ axis and look toward the origin at the disk). The disk makes one complete rotation every 0.6 s. What is the rotational angular momentum of the disk? What is the rotational kinetic energy of the disk? **(b)** A sphere of uniform density, with mass 22 kg and radius 0.7 m, is located at the origin and rotates around an axis parallel with the x axis. If you stand somewhere on the $+x$ axis and look toward the origin at the sphere, the sphere spins counterclockwise. One complete revolution takes 0.5 s. What is the rotational angular momentum of the sphere? What is the rotational kinetic energy of the sphere? **(c)** A cylindrical rod of uniform density is located with its center at the origin, and its axis along the z axis. Its radius is 0.06 m, its length is 0.7 m, and its mass is 5 kg. It makes one revolution every 0.03 s. If you stand on the $+x$ axis and look toward the origin at the rod, the rod spins clockwise. What is the rotational angular momentum of the rod? What is the rotational kinetic energy of the rod?

•P21 If an object has a moment of inertia 19 $\text{kg}\cdot\text{m}^2$ and the magnitude of its rotational angular momentum is 36 $\text{kg}\cdot\text{m}^2/\text{s}$, what is its rotational kinetic energy?

•P22 Mounted on a low-mass rod of length 0.32 m are four balls (Figure 11.78). Two balls (shown in red on the diagram), each of mass 0.82 kg, are mounted at opposite ends of the rod. Two other balls, each of mass 0.29 kg (shown in blue on the diagram), are each mounted a distance 0.08 m from the center of the rod. The rod rotates on an axle through the center of the rod (indicated by the "x" in the diagram), perpendicular to the rod, and it takes 0.9 s to make one full rotation.

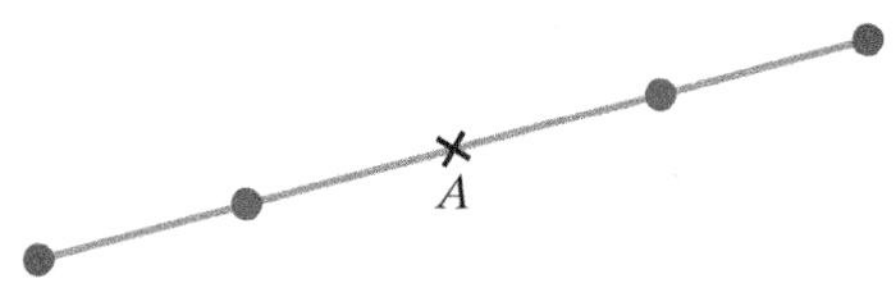

Figure 11.78

(a) What is the moment of inertia of the device about its center? **(b)** What is the angular speed of the rotating device? **(c)** What is the magnitude of the angular momentum of the rotating device?

•P23 The moment of inertia of a uniform-density disk rotating about an axle through its center can be shown to be $\frac{1}{2}MR^2$. This result is obtained by using integral calculus to add up the contributions of all the atoms in the disk. The factor of 1/2 reflects the fact that some of the atoms are near the center and some are far from the center; the factor of 1/2 is an average of the square distances. A uniform-density disk whose mass is 16 kg and radius is 0.15 m makes one complete rotation every 0.5 s. **(a)** What is the moment of inertia of this disk? **(b)** What is its rotational kinetic energy? **(c)** What is the magnitude of its rotational angular momentum?

•P24 In Figure 11.79 a barbell spins around a pivot at its center at A. The barbell consists of two small balls, each with mass 500 g (0.5 kg), at the ends of a very low mass rod of length $d = 20$ cm (0.2 m; the radius of rotation is 0.1 m). The barbell spins clockwise with angular speed 80 rad/s.

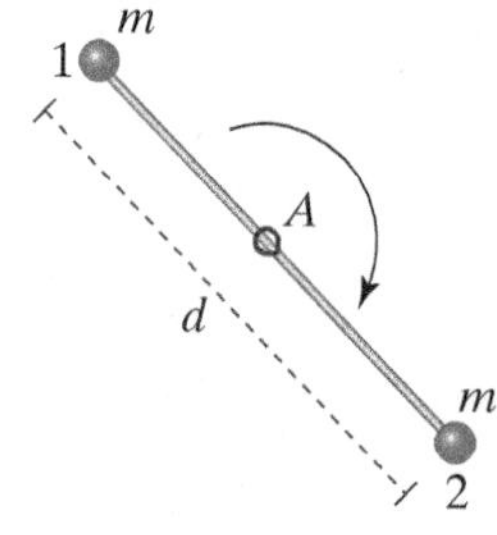

Figure 11.79

We can calculate the angular momentum and kinetic energy of this object in two different ways, by treating the object as two separate balls or as one barbell. Use the usual coordinate system, with x to the right, y toward the top of the page, and z out of the page, toward you.

I: Treat the object as two separate balls. Calculate the following quantities:

(a) The speed of ball 1, **(b)** $\vec{L}_{\text{trans},1,A}$ of ball 1, **(c)** $\vec{L}_{\text{trans},2,A}$ of ball 2, **(d)** $\vec{L}_{\text{tot},A}$, **(e)** the translational kinetic energy of ball 1, **(f)** the translational kinetic energy of ball 2, **(g)** the total kinetic energy of the barbell.

II: Treat the object as one barbell. Calculate the following quantities:

(h) The moment of inertia I of the barbell, **(i)** $\vec{\omega}$, expressed as a vector, **(j)** $\vec{L}_{\text{rot}}$ of the barbell, **(k)** K_{rot}.

III: Compare the two approaches:

1. Compare your result for $\vec{L}_{\text{tot},A}$ in part I to your result for $\vec{L}_{\text{rot}}$ in part II. Should these quantities be the same, or different?
2. Compare your result for K_{total} in part I to your result for K_{rot} in part II. Should these quantities be the same, or different?

•P25 A low-mass rod of length 0.30 m has a metal ball of mass 1.7 kg at each end. The center of the rod is located at the origin, and the rod rotates in the yz plane about its center. The rod rotates clockwise around its axis when viewed from a point on the $+x$ axis, looking toward the origin. The rod makes one complete rotation every 0.5 s. **(a)** What is the moment of inertia of the object (rod plus two balls)? **(b)** What is the rotational angular momentum of the object? **(c)** What is the rotational kinetic energy of the object?

••P26 Calculate the angular momentum of the Earth: **(a)** Calculate the magnitude of the translational angular momentum of the Earth relative to the center of the Sun. See the data on inside back cover. **(b)** Calculate the magnitude of the rotational angular momentum of the Earth. How does this compare to your result in part (a)?
(The angular momentum of the Earth relative to the center of the Sun is the sum of the translational and rotational angular momenta. The rotational axis of the Earth is tipped 23.5° away from a perpendicular to the plane of its orbit.)

••P27 In Figure 11.80 two small objects each of mass $m = 0.3$ kg are connected by a lightweight rod of length $d = 1.5$ m. At a particular instant they have velocities whose magnitudes are $v_1 = 38$ m/s and $v_2 = 60$ m/s and are subjected to external forces whose magnitudes are $F_1 = 41$ N and $F_2 = 26$ N. The distance $h = 0.3$ m, and the distance $w = 0.7$ m. The system is moving in outer space. Assuming the usual coordinate system with $+x$ to the right, $+y$ toward the top of the page, and $+z$ out of the page toward you, calculate these quantities for this system:

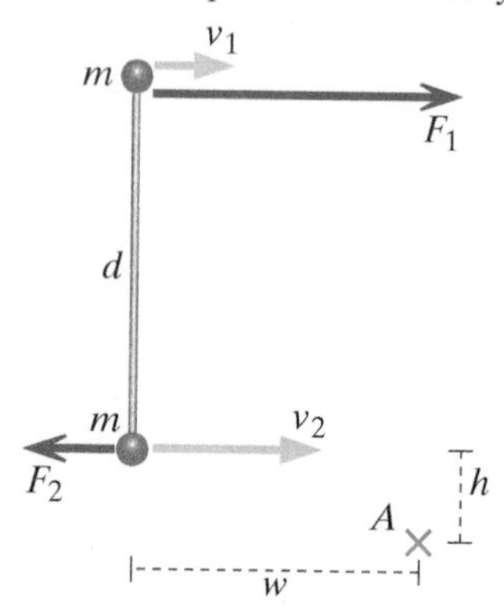

Figure 11.80

(a) $\vec{p}_{\text{total}}$, **(b)** $\vec{v}_{\text{CM}}$, **(c)** $\vec{L}_{\text{tot},A}$, **(d)** $\vec{L}_{\text{rot}}$, **(e)** $\vec{L}_{\text{trans},A}$, **(f)** $\vec{p}_{\text{total}}$ at a time 0.23 s after the initial time.

Section 11.3

•P28 A barbell consists of two small balls, each with mass $m = 0.4$ kg, at the ends of a very low mass rod of length $d = 0.6$ m. It is mounted on the end of a low-mass rigid rod of length $b = 0.9$ m (Figure 11.81). The apparatus is set in motion in such a way that although the rod rotates clockwise with angular speed $\omega_1 = 15$ rad/s, the barbell maintains its vertical orientation. Calculate these vector quantities: **(a)** $\vec{L}_{\text{rot}}$, **(b)** $\vec{L}_{\text{trans},B}$, **(c)** $\vec{L}_{\text{tot},B}$.

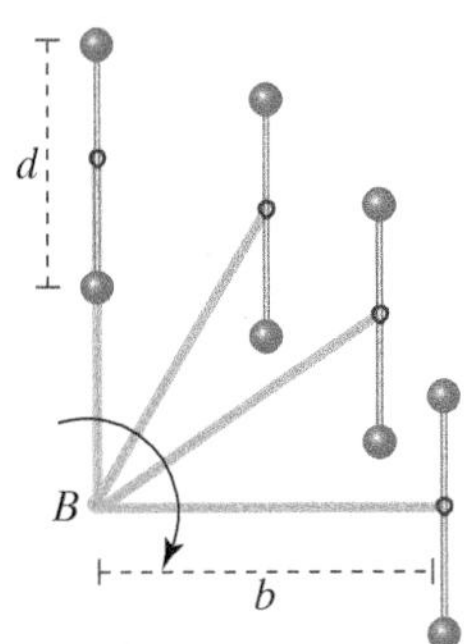

Figure 11.81

•P29 A barbell consists of two small balls, each with mass $m = 0.4$ kg, at the ends of a very low mass rod of length $d = 0.6$ m. It is mounted on the end of a low-mass rigid rod of length $b = 0.9$ m. The apparatus is set in motion in such a way that it again rotates clockwise with angular speed $\omega_1 = 15$ rad/s, but in addition, the barbell rotates clockwise about its center, with an angular speed $\omega_2 = 20$ rad/s (Figure 11.82). Calculate these vector quantities: **(a)** $\vec{L}_{\text{rot}}$, **(b)** $\vec{L}_{\text{trans},B}$, **(c)** $\vec{L}_{\text{tot},B}$.

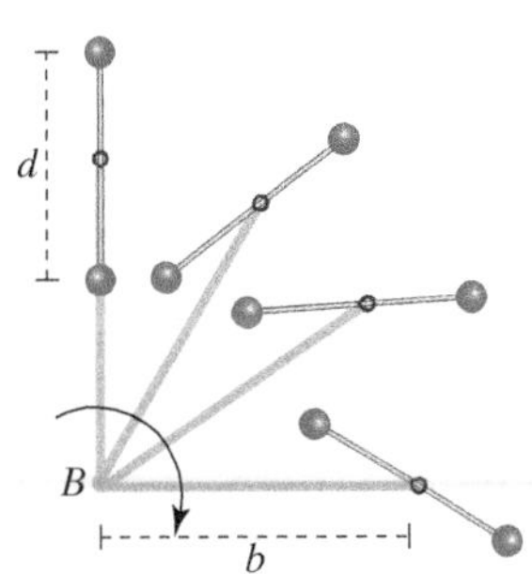

Figure 11.82

Section 11.5

•P30 As shown in Figure 11.83, seven forces all with magnitude $|\vec{F}| = 25$ N are applied to an irregularly shaped object. Each force is applied at a different location on the object, indicated by the tail of the arrow; the directions of the forces differ. The distances shown in the diagram have these values: $w = 9$ m, $h = 14$ m, and $d = 13$ m. For each force, calculate the z component of the torque due to that force, relative to location A (x to the right, y up, z out of the page). Make sure you give the correct sign. Relative to location A, what is the z component of the net torque acting on this object?

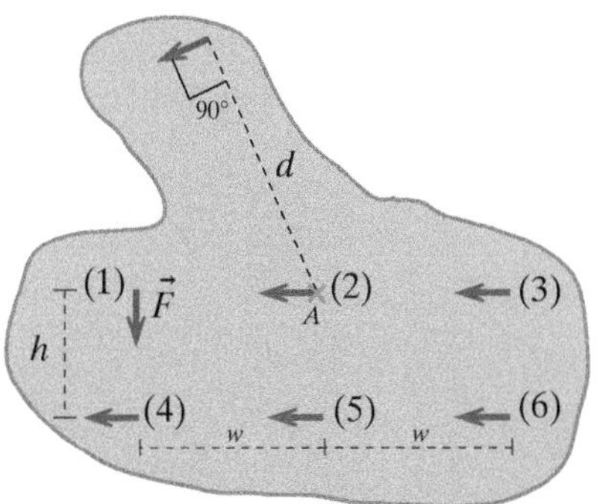

Figure 11.83

•P31 At $t = 15$ s, a particle has angular momentum $\langle 3,5,-2\rangle$ kg · m²/s relative to location A. A constant torque $\langle 10,-12,20\rangle$ N · m relative to location A acts on the particle. At $t = 15.1$ s, what is the angular momentum of the particle?

•P32 Calculating torque in Figure 11.84: **(a)** If $r_A = 3$ m, $F = 8$ N, and $\theta = 51°$, what is the magnitude of the torque about location A, including units? **(b)** If the force were perpendicular to $\vec{r}_A$ but gave the same torque as in the preceding question, what would be its magnitude?

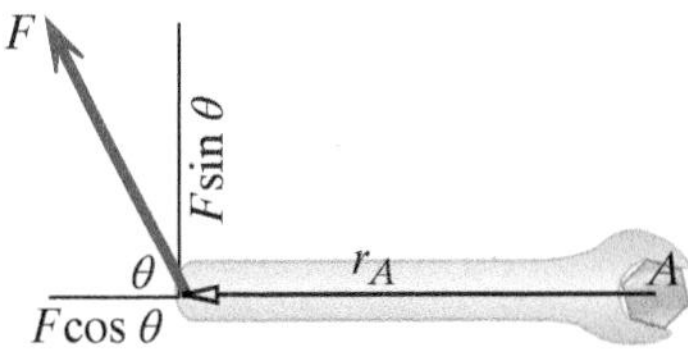

Figure 11.84

••P33 Let's compare the Momentum Principle and the Angular Momentum Principle in a simple situation. Consider a mass m falling near the Earth (Figure 11.85). Neglecting air resistance, the Momentum Principle gives $dp_y/dt = -mg$, yielding $dv_y/dt = -g$ (nonrelativistic). Choose a location A off to the side, on the ground. Apply the Angular Momentum Principle to find an algebraic expression for the rate of change of angular momentum of the mass about location A.

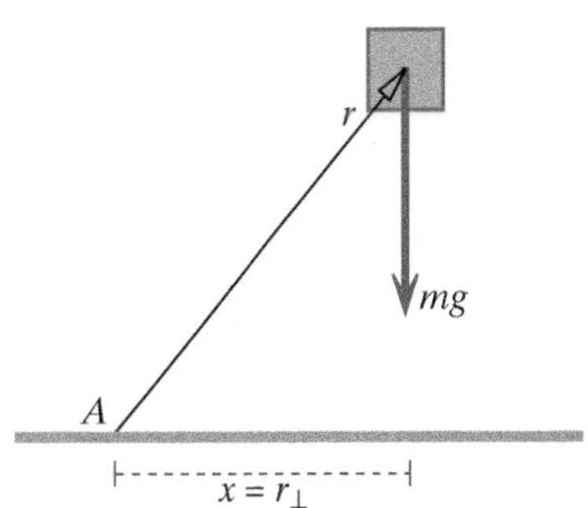

Figure 11.85

••P34 A small rock passes a massive star, following the path shown in red on the diagram. When the rock is a distance 4.5×10^{13} m (indicated as d_1 in Figure 11.86) from the center of the star, the magnitude p_1 of its momentum is 1.35×10^{17} kg·m/s, and the angle is 126°. At a later time, when the rock is a distance $d_2 = 1.3 \times 10^{13}$ m from the center of the star, it is heading in the $-y$ direction. There are no other massive objects nearby. What is the magnitude p_2 of the final momentum?

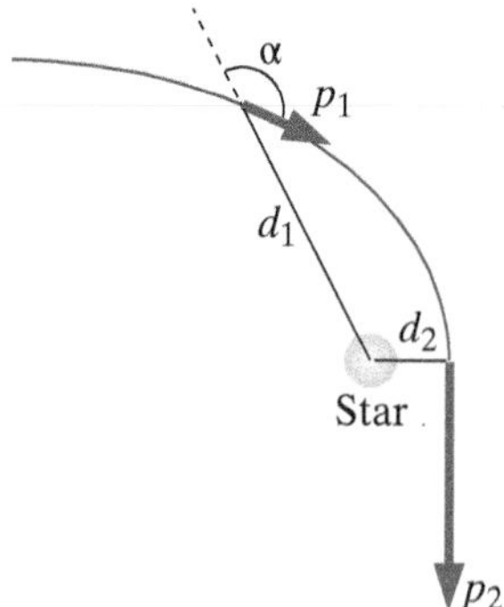

Figure 11.86

Section 11.7

•**P35** A stationary bicycle wheel of radius 0.9 m is mounted in the vertical plane (Figure 11.87). The axle is held up by supports that are not shown, and the wheel is free to rotate on the nearly frictionless axle. The wheel has mass 4.8 kg, all concentrated in the rim (the spokes have negligible mass). A lump of clay with mass 0.5 kg falls and sticks to the outer edge of the wheel at the location shown. Just before the impact the clay has speed 5 m/s, and the wheel is rotating clockwise with angular speed 0.33 rad/s.

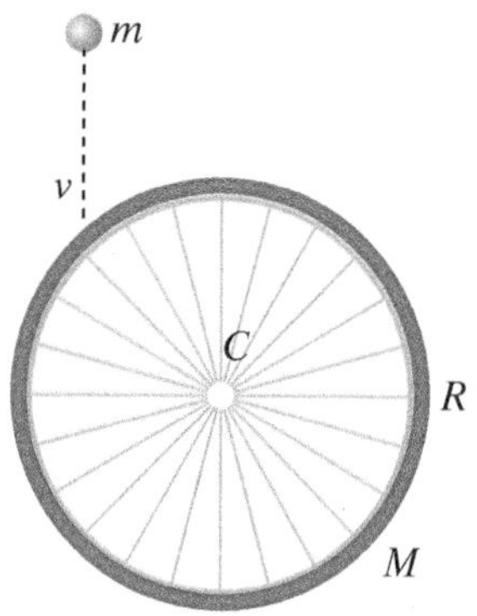

Figure 11.87

(a) Just before the impact, what is the angular momentum (magnitude and direction) of the combined system of wheel plus clay about the center C? (As usual, x is to the right, y is up, and z is out of the screen, toward you.) **(b)** Just after the impact, what is the angular momentum (magnitude and direction) of the combined system of wheel plus clay about the center C? **(c)** Just after the impact, what is the angular velocity (magnitude and direction) of the wheel? **(d)** Qualitatively, what happens to the linear momentum of the combined system? Why? (1) The downward linear momentum decreases because the axle exerts an upward force. (2) Some of the linear momentum is changed into angular momentum. (3) Some of the linear momentum is changed into energy. (4) There is no change because linear momentum is always conserved.

•**P36** A rotating uniform-density disk of radius 0.6 m is mounted in the vertical plane, as shown in Figure 11.88. The axle is held up by supports that are not shown, and the disk is free to rotate on the nearly frictionless axle. The disk has mass 5 kg. A lump of clay with mass 0.4 kg falls and sticks to the outer edge of the wheel at the location $\langle -0.36, 0.480, 0 \rangle$ m, relative to an origin at the center of the axle. Just before the impact the clay has speed 8 m/s, and the disk is rotating clockwise with angular speed 0.51 radians/s.

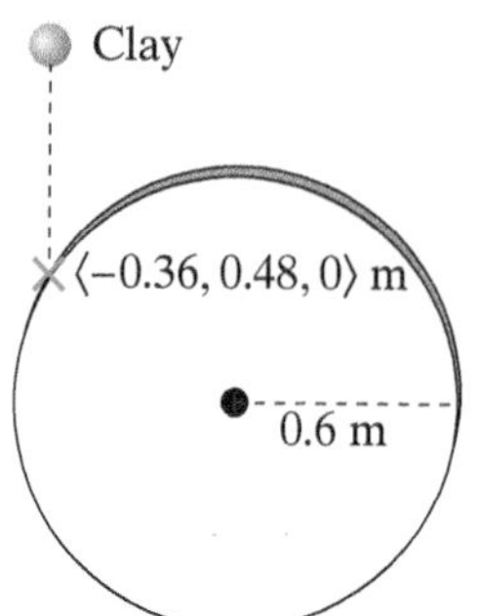

Figure 11.88

(a) Just before the impact, what is the angular momentum (magnitude and direction) of the combined system of wheel plus clay about the center C? (As usual, x is to the right, y is up, and z is out of the screen, toward you.) **(b)** Just after the impact, what is the angular momentum (magnitude and direction) of the combined system of wheel plus clay about the center C? **(c)** Just after the impact, what is the angular velocity (magnitude and direction) of the wheel? **(d)** Qualitatively, what happens to the linear momentum of the combined system? Why? (A) There is no change because linear momentum is always conserved. (B) Some of the linear momentum is changed into angular momentum. (C) Some of the linear momentum is changed into energy. (D) The downward linear momentum decreases because the axle exerts an upward force.

•**P37** Figure 11.89 depicts a device that can rotate freely with little friction with the axle. The radius is 0.4 m, and each of the eight balls has a mass of 0.3 kg. The device is initially not rotating. A piece of clay falls and sticks to one of the balls as shown in the figure. The mass of the clay is 0.066 kg and its speed just before the collision is 10 m/s.

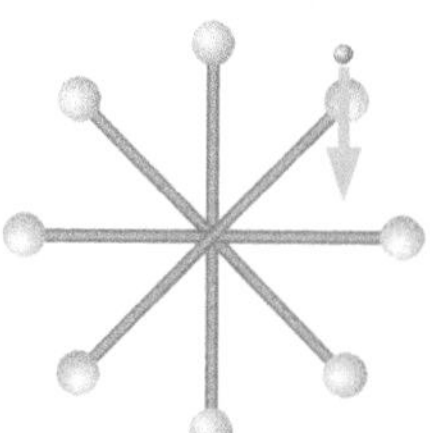

Figure 11.89

(a) Which of the following statements are true, for angular momentum relative to the the axle of the wheel? (1) Just before the collision, $r_\perp = 0.4\sqrt{2}/2 = 0.4\cos(45°)$ (for the clay). (2) The angular momentum of the wheel is the same before and after the collision. (3) Just before the collision, the angular momentum of the wheel is 0. (4) The angular momentum of the wheel is the sum of the angular momenta of all eight balls. (5) The angular momentum of the wheel + clay after the collision is equal to the initial angular momentum of the clay. (6) The angular momentum of the falling clay is zero because the clay is moving in a straight line. **(b)** Just after the collision, what is the speed of one of the balls?

••**P38** A stick of length L and mass M hangs from a low-friction axle (Figure 11.90). A bullet of mass m traveling at a high speed v strikes near the bottom of the stick and quickly buries itself in the stick.

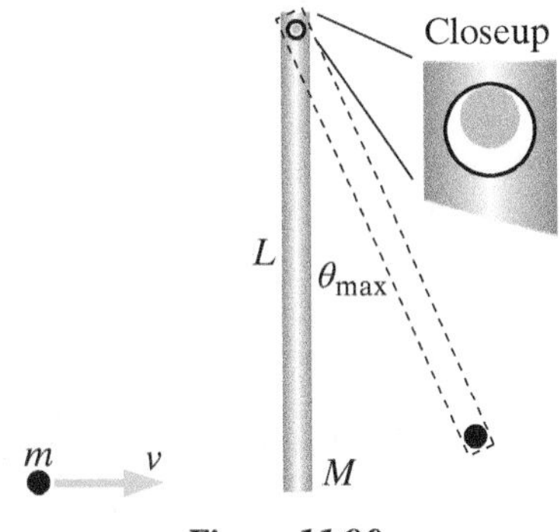

Figure 11.90

(a) During the brief impact, is the linear momentum of the stick + bullet system constant? Explain why or why not. Include in your explanation a sketch of how the stick shifts on the axle during the impact. **(b)** During the brief impact, around what point does the angular momentum of the stick + bullet system remain constant? **(c)** Just after the impact, what is the angular speed ω of the stick (with the bullet embedded in it)? (Note that the center of mass of the stick has a speed $\omega L/2$. The moment of inertia of a uniform rod about its center of mass is $\frac{1}{12}ML^2$.) **(d)** Calculate the change in kinetic energy from just before to just after the impact. Where has this energy gone? **(e)** The stick (with the bullet embedded in it) swings through a maximum angle θ_{max} after the impact, then swings back. Calculate θ_{max}.

•**P39** **(a)** What is the period of small-angle oscillations of a simple pendulum with a mass of 0.1 kg at the end of a string of length 1 m? **(b)** What is the period of small-angle oscillations of a meter stick suspended from one end, whose mass is 0.1 kg?

••**P40** A disk of mass 3 kg and radius 0.15 m hangs in the xy plane from a horizontal low-friction axle. The axle is 0.09 m from the center of the disk. What is the frequency f of small-angle oscillations of the disk? What is the period?

••**P41** Design a decorative "mobile" to consist of a low-mass rod of length 0.49 m suspended from a string so that the rod is horizontal, with two balls hanging from the ends of the rod. At the left end of the rod hangs a ball with mass 0.484 kg. At the right end of the rod hangs a ball with mass 0.273 kg. You need to decide how far from the left end of the rod you should attach the string that will hold up the mobile, so that the mobile hangs motionless with the rod horizontal ("equilibrium"). You also need to determine the tension in the string supporting the mobile. **(a)** What is the tension in the string that supports the mobile? **(b)** How far from the left end of the rod should you attach the support string?

••**P42** A space station has the form of a hoop of radius $R = 14$ m, with mass $M = 6250$ kg (Figure 11.91). Initially its center of mass is not moving, but it is spinning with angular speed $\omega_0 = 0.0013$ rad/s. Then a small package of mass $m = 6$ kg is thrown by a spring-loaded gun toward a nearby spacecraft as shown; the package has a speed $v = 40$ m/s after launch. Calculate the center-of-mass velocity of the space station and its rotational speed ω after the launch.

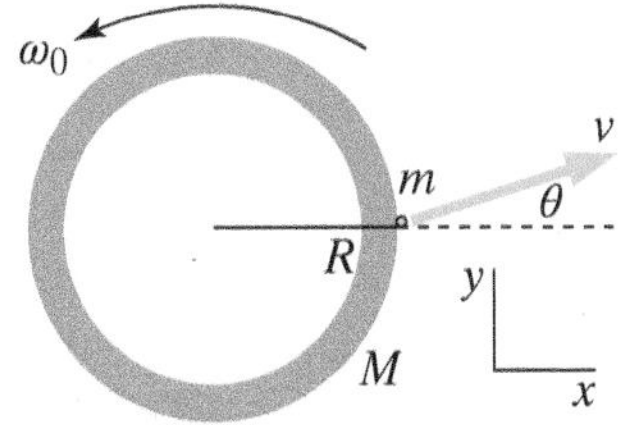

Figure 11.91

••**P43** You sit on a rotating stool and hold barbells in both hands with your arms fully extended horizontally. You make one complete turn in 2 s. You then pull the barbells in close to your body. **(a)** Estimate how long it now takes you to make one complete turn. Be clear and explicit about the principles you apply and about your assumptions and approximations. **(b)** About how much energy did you expend?

••**P44** A certain comet of mass m at its closest approach to the Sun is observed to be at a distance r_1 from the center of the Sun, moving with speed v_1 (Figure 11.92). At a later time the comet is observed to be at a distance r_2 from the center of the Sun, and the angle between $\vec{r}_2$ and the velocity vector is measured to be θ. What is v_2? Explain briefly.

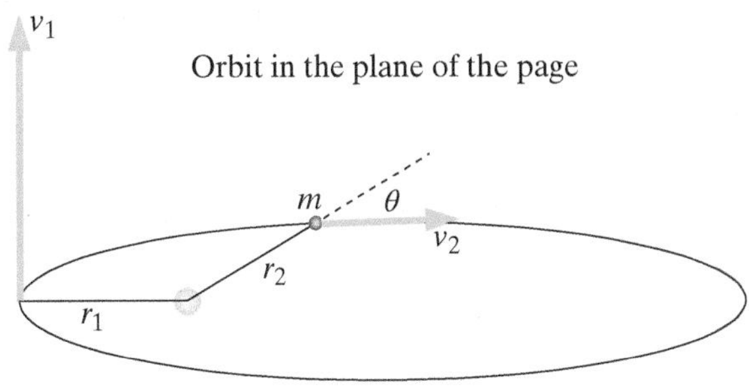

Figure 11.92

••**P45** An ice skater whirls with her arms and one leg stuck out as shown on the left in Figure 11.93, making one complete turn in 1 s. Then she quickly moves her arms up above her head and pulls her leg in as shown at the right in Figure 11.93.

Figure 11.93

(a) Estimate how long it now takes for her to make one complete turn. Explain your calculations, and state clearly what approximations and estimates you make. **(b)** Estimate the minimum amount of chemical energy she must expend to change her configuration.

••P46 A playground ride consists of a disk of mass $M = 43$ kg and radius $R = 1.7$ m mounted on a low-friction axle (Figure 11.94). A child of mass $m = 25$ kg runs at speed $v = 2.3$ m/s on a line tangential to the disk and jumps onto the outer edge of the disk.

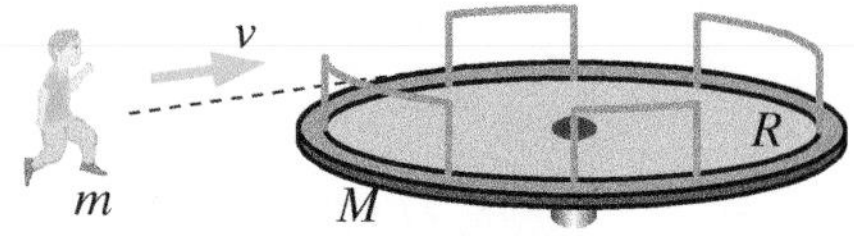

Figure 11.94

(a) If the disk was initially at rest, now how fast is it rotating? **(b)** What is the change in the kinetic energy of the child plus the disk? **(c)** Where has most of this kinetic energy gone? **(d)** Calculate the change in *linear* momentum of the system consisting of the child plus the disk (but not including the axle), from just before to just after impact. What caused this change in the linear momentum? **(e)** The child on the disk walks inward on the disk and ends up standing at a new location a distance 0.85 m from the axle. Now what is the angular speed? **(f)** What is the change in the kinetic energy of the child plus the disk, from the beginning to the end of the walk on the disk? **(g)** What was the source of this increased kinetic energy?

••P47 In Figure 11.95 two small objects each of mass m_1 are connected by a lightweight rod of length L. At a particular instant the center of mass speed is v_1 as shown, and the object is rotating counterclockwise with angular speed ω_1. A small object of mass m_2 traveling with speed v_2 collides with the rod at an angle θ_2 as shown, at a distance b from the center of the rod. After being struck, the mass m_2 is observed to move with speed v_4, at angle θ_4. All the quantities are positive magnitudes. This all takes place in outer space.

For the object consisting of the rod with the two masses, write equations that, in principle, could be solved for the center of mass speed v_3, direction θ_3, and angular speed ω_3 in terms of the given quantities. State clearly what physical principles you use to obtain your equations.

Don't attempt to solve the equations; just set them up.

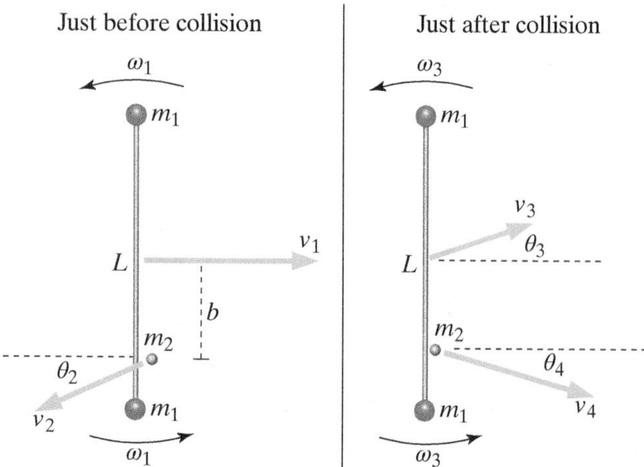

Figure 11.95

••P48 In Figure 11.96 a spherical nonspinning asteroid of mass M and radius R moving with speed v_1 to the right collides with a similar nonspinning asteroid moving with speed v_2 to the left, and they stick together. The impact parameter is d. Note that $I_{\text{sphere}} = \frac{2}{5}MR^2$.

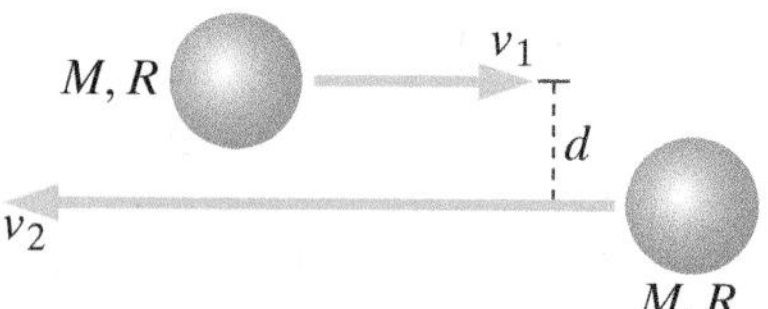

Figure 11.96

After the collision, what is the velocity v_{CM} of the center of mass and the angular velocity ω about the center of mass? (Note that each asteroid rotates about its own center with this same ω.)

••P49 A spherical satellite of approximately uniform density with radius 4.8 m and mass 205 kg is originally moving with velocity $\langle 2600, 0, 0\rangle$ m/s, and is originally rotating with an angular speed 2 rad/s, in the direction shown in the diagram. A small piece of space junk of mass 4.1 kg is initially moving toward the satellite with velocity $\langle -2200, 0, 0\rangle$ m/s. The space junk hits the edge of the satellite at location C as shown in Figure 11.97, and moves off with a new velocity $\langle -1300, 480, 0\rangle$ m/s. Both before and after the collision, the rotation of the space junk is negligible.

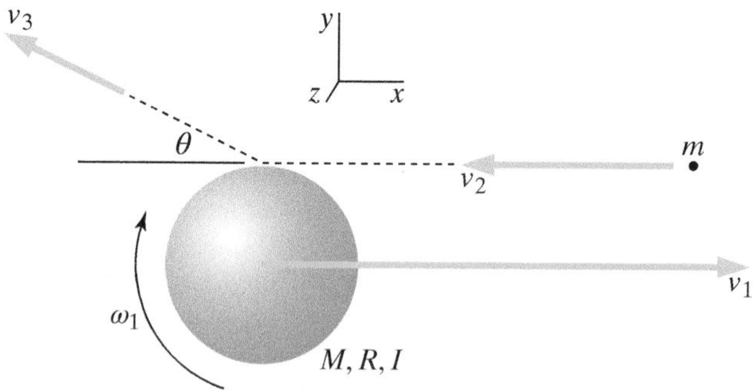

Figure 11.97

(a) Just after the collision, what are the components of the center-of-mass velocity of the satellite (v_x and v_y) and its rotational speed ω? **(b)** Calculate the rise in the internal energy of the satellite and space junk combined.

••P50 A device consisting of four heavy balls connected by low-mass rods is free to rotate about an axle, as shown in Figure 11.98. It is initially not spinning. A small bullet traveling very fast buries itself in one of the balls. $m = 0.002$ kg, $v = 550$ m/s, $M_1 = 1.2$ kg, $M_2 = 0.4$ kg, $R_1 = 0.6$ m, and $R_2 = 0.2$ m. The axle of the device is at the origin $\langle 0, 0, 0\rangle$, and the bullet strikes at location $\langle 0.155, 0.580, 0\rangle$ m. Just after the impact, what is the angular speed of the device? Note that this is an inelastic collision; the system's temperature increases.

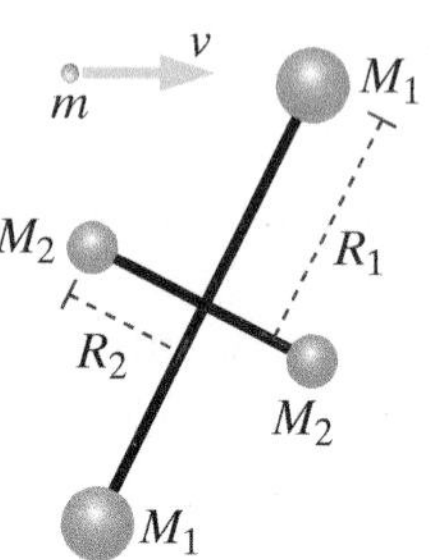

Figure 11.98

••**P51** A thin metal rod of mass 1.3 kg and length 0.4 m is at rest in outer space, near a space station (Figure 11.99). A tiny meteorite with mass 0.06 kg traveling at a high speed of 200 m/s strikes the rod a distance 0.2 m from the center and bounces off with speed 60 m/s as shown in the diagram. The magnitudes of the initial and final angles to the x axis of the small mass's velocity are $\theta_i = 26°$ and $\theta_f = 82°$. **(a)** Afterward, what is the velocity of the center of the rod? **(b)** Afterward, what is the angular velocity $\vec{\omega}$ of the rod? **(c)** What is the increase in internal energy of the objects?

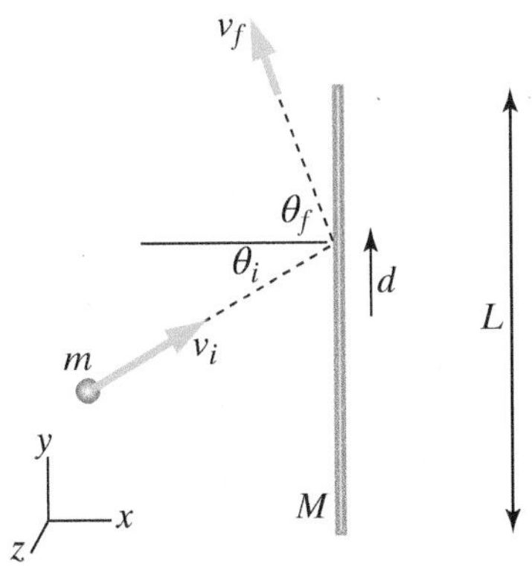

Figure 11.99

••**P52** A diver dives from a high platform (Figure 11.100). When he leaves the platform, he tucks tightly and performs three complete revolutions in the air, then straightens out with his body fully extended before entering the water. He is in the air for a total time of 1.4 s. What is his angular speed ω just as he enters the water? Give a numerical answer. Be explicit about details of your model, and include (brief) explanations. You will need to estimate some quantities.

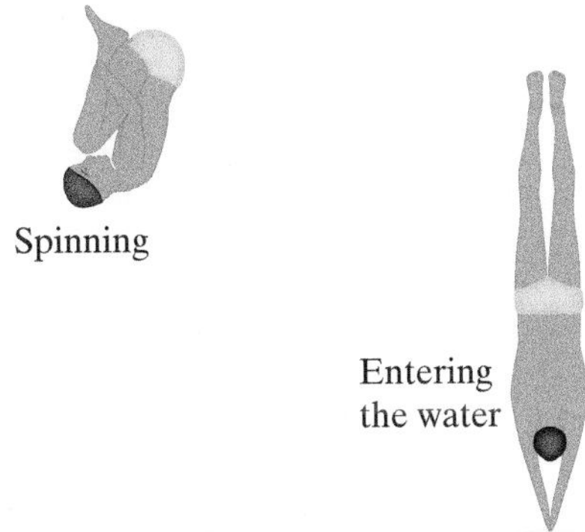

Figure 11.100

•••**P53** Suppose that an asteroid of mass 2×10^{21} kg is nearly at rest outside the solar system, far beyond Pluto. It falls toward the Sun and crashes into the Earth at the equator, coming in at an angle of 30° to the vertical as shown, against the direction of rotation of the Earth (Figure 11.101; not to scale). It is so large that its motion is barely affected by the atmosphere.

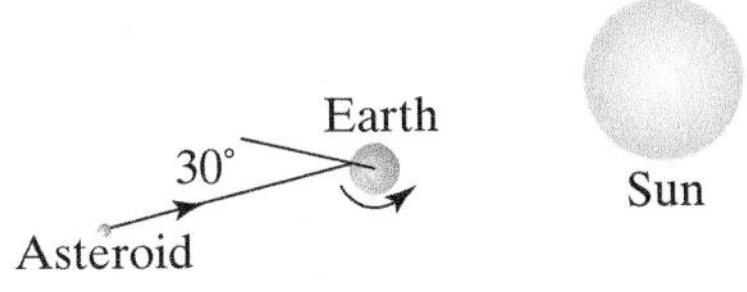

Figure 11.101

(a) Calculate the impact speed. **(b)** Calculate the change in the length of a day due to the impact.

Section 11.8

•**P54** A disk of radius 8 cm is pulled along a frictionless surface with a force of 10 N by a string wrapped around the edge (Figure 11.102). 24 cm of string has unwound off the disk. What are the magnitude and direction of the torque exerted about the center of the disk at this instant?

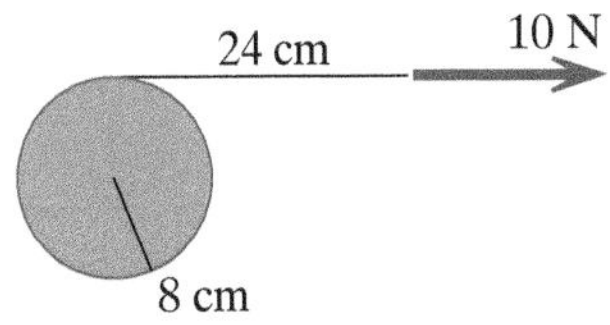

Figure 11.102

•**P55** In Figure 11.102, the uniform solid disk has mass 0.4 kg (moment of inertia $I = \frac{1}{2}MR^2$). At the instant shown, the angular velocity is 20 rad/s into the page. **(a)** At this instant, what are the magnitude and direction of the angular momentum about the center of the disk? **(b)** At a time 0.2 s later, what are the magnitude and direction of the angular momentum about the center of the disk? **(c)** At this later time, what are the magnitude and direction of the angular velocity?

•**P56** Two people of different masses sit on a seesaw (Figure 11.103). M_1, the mass of person 1, is 90 kg, M_2 is 42 kg, $d_1 = 0.8$ m, and $d_2 = 1.3$ m. The people are initially at rest. The mass of the board is negligible.

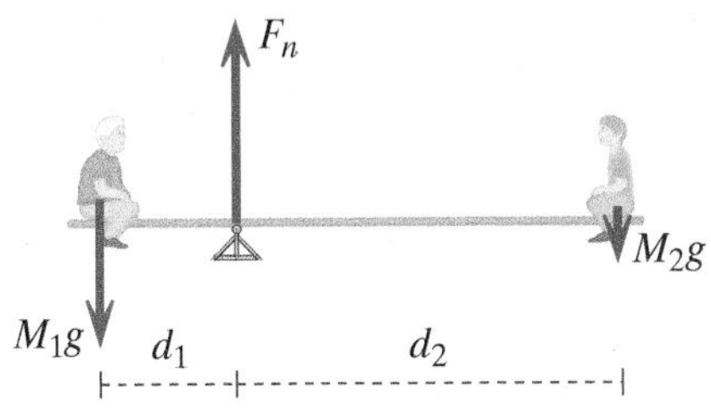

Figure 11.103

(a) What are the magnitude and direction of the torque about the pivot due to the gravitational force on person 1? **(b)** What are the magnitude and direction of the torque about the pivot due to the gravitational force on person 2? **(c)** Since at this instant the linear momentum of the system may be changing, we don't know the magnitude of the "normal" force exerted by the pivot. Nonetheless, it is possible to calculate the torque due to this force. What are the magnitude and direction of the torque about the pivot due to the force exerted by the pivot on the board? **(d)** What are the magnitude and direction of the net torque on the system (board + people)? **(e)** Because of this net torque, what will happen? (A) The seesaw will begin to rotate clockwise. (B) The seesaw will begin to rotate counterclockwise. (C) The seesaw will not move. **(f)** Person 2 moves to a new position, in which the magnitude of the net torque about the pivot is now 0, and the seesaw is balanced. What is the new value of d_2 in this situation?

•**P57** A board of length $2d = 6$ m rests on a cylinder (the "pivot"). A ball of mass 5 kg is placed on the end of the board. Figure 11.104 shows the objects at a particular instant. **(a)** On a free-body diagram, show the forces acting on the ball + board system, in their correct locations. **(b)** Take the point at which the board touches the cylinder as location A. What is the magnitude of the torque on the system of (ball + board) about location A? **(c)** Which of the following statements are correct? (1) Because there is a torque, the angular momentum of the system will change in the next tenth of a second. (2) The forces balance, so the angular momentum of the system about location A will not change. (3) The force by the cylinder on the board contributes nothing to the torque about location A.

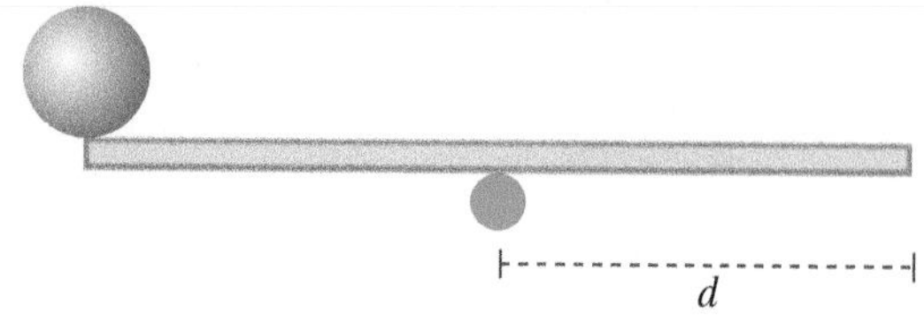

Figure 11.104

•P58 A barbell is mounted on a nearly frictionless axle through its center (Figure 11.105). At this instant, there are two forces of equal magnitude applied to the system as shown, with the directions indicated, and at this instant the angular velocity is 60 rad/s, counterclockwise. In the next 0.001 s, the angular momentum relative to the center increases by an amount 2.5×10^{-4} kg · m^2/s. What is the magnitude of each force? What is the net force?

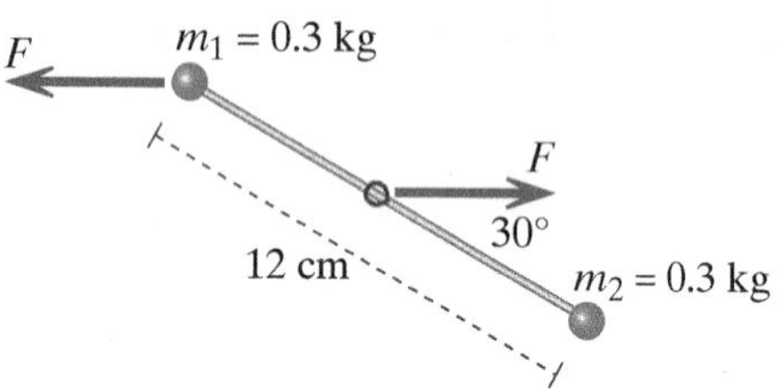

Figure 11.105

••P59 A disk of radius 0.2 m and moment of inertia 1.5 kg · m^2 is mounted on a nearly frictionless axle (Figure 11.106). A string is wrapped tightly around the disk, and you pull on the string with a constant force of 25 N. After a while the disk has reached an angular speed of 2 rad/s. What is its angular speed 0.1 seconds later? Explain briefly.

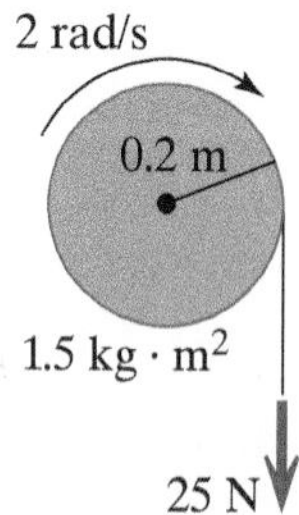

Figure 11.106

••P60 String is wrapped around an object of mass 1.2 kg, radius 0.06 m, and moment of inertia 0.0015 kg · m^2 (the density of the object is not uniform). With your hand you pull the string straight up with some constant force F such that the center of the object does not move up or down, but the object spins faster and faster (Figure 11.107). This is like a yo-yo; nothing but the vertical string touches the object.

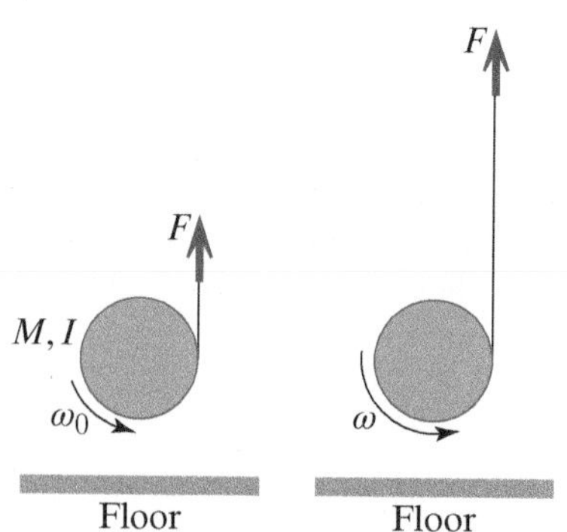

Figure 11.107

(a) Initially the object has an angular speed $\omega_0 = 12$ rad/s. After a time interval of 0.1 s, what is the angular speed ω of the object? **(b)** Your answer must be numeric and not contain the symbol F. Explain the physics principles you are using.

••P61 A solid object of uniform density with mass M, radius R, and moment of inertia I rolls without slipping down a ramp at an angle θ to the horizontal. The object could be a hoop, a disk, a sphere, etc. **(a)** Carefully follow the complete analysis procedure explained in earlier chapters, but with the addition of the Angular Momentum Principle about the center of mass. Note that in your force diagram you must include a small frictional force f that points *up* the ramp. Without that force the object will slip. Also note that the condition of nonslipping implies that the instantaneous velocity of the atoms of the object that are momentarily in contact with the ramp is zero, so $f < \mu F_N$ (no slipping). This zero-velocity condition also implies that $v_{CM} = \omega R$, where ω is the angular speed of the object, since the instantaneous speed of the contact point is $v_{CM} - \omega R$. **(b)** The moment of inertia about the center of mass of a uniform hoop is MR^2, for a uniform disk it is $(1/2)MR^2$, and for a uniform sphere it is $(2/5)MR^2$. Calculate the acceleration dv_{CM}/dt for each of these objects. **(c)** If two hoops of different mass are started from rest at the same time and the same height on a ramp, which will reach the bottom first? If a hoop, a disk, and a sphere of the same mass are started from rest at the same time and the same height on a ramp, which will reach the bottom first? **(d)** Write the energy equation for the object rolling down the ramp, and for the point-particle system. Show that the time derivatives of these equations are compatible with the force and torque analyses.

••P62 A yo-yo is constructed of three disks: two outer disks of mass M, radius R, and thickness d, and an inner disk (around which the string is wrapped) of mass m, radius r, and thickness d. The yo-yo is suspended from the ceiling and then released with the string vertical (Figure 11.108).

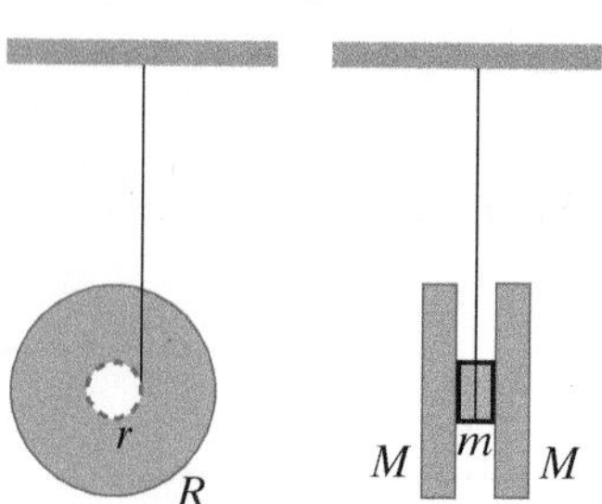

Figure 11.108

Calculate the tension in the string as the yo-yo falls. Note that when the center of the yo-yo moves down a distance y, the yo-yo turns through an angle y/r, which in turn means that the angular speed ω is equal to v_{CM}/r. The moment of inertia of a uniform disk is $\frac{1}{2}MR^2$.

••**P63** A string is wrapped around a uniform disk of mass $M = 1.2$ kg and radius $R = 0.11$ m (Figure 11.109). Attached to the disk are four low-mass rods of radius $b = 0.14$ m, each with a small mass $m = 0.4$ kg at the end. The device is initially at rest on a nearly frictionless surface. Then you pull the string with a constant force $F = 21$ N for a time of 0.2 s. Now what is the angular speed of the apparatus?

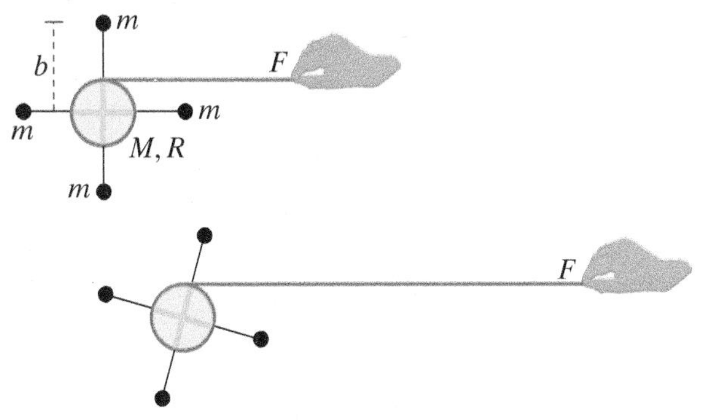

Figure 11.109

•••**P64** A small solid rubber ball of radius r hits a rough horizontal floor such that its speed v just before striking the floor at location A makes an angle of 60° with the horizontal and also has back spin with angular speed ω. It is observed that the ball repeatedly bounces from A to B, then from B back to A, etc. Assuming perfectly elastic impact determine **(a)** the required magnitude of ω of the back spin in terms of v and r, and **(b)** the minimum magnitude of co-efficient of static friction μ_s to enable this motion. *Hint:* Notice that the direction of $\vec{\omega}$ flips in each collision.

Section 11.9

•**P65** A wheel rotates at a rate of 200 rpm (revolutions per minute). **(a)** What is the angular speed in radians per second? **(b)** What is the angle in radians through which the wheel rotates in 5 s? **(c)** What is this angle in degrees?

•**P66** A wheel is mounted on a stationary axle that lies along the z axis. The center of the wheel is at $\langle 0,0,0 \rangle$ m. It is rotating with a constant angular speed of 30 rad/s. At time $t = 4.20$ s a red dot painted on the rim of the wheel is located at $\langle 0.25,0,0 \rangle$ m. At time $t = 4.26$ s, what is the location of the red dot?

•**P67** A rotating wheel accelerates at a constant rate from an angular speed of 24 rad/s to 36 rad/s in a time interval of 3 s. **(a)** What is the angular acceleration in rad/s/s? **(b)** What is the average angular speed? **(c)** What is the angle in radians through which the wheel rotates? **(d)** What is this angle in degrees?

•**P68** A string is wrapped around a wheel of radius 25 cm mounted on a stationary axle. The wheel is initially not rotating. You pull the string with a constant force through a distance of 40 cm. What is the angle in radians and degrees through which the wheel rotates?

••**P69** A wheel with radius 30 cm is rotating at a rate of 10 rev/s. **(a)** What is the angular speed in radians per second? **(b)** In a time interval of 6 s, what is the angle in radians through which the wheel rotates? **(c)** At $t = 10$ s the angular speed begins to increase at a rate of 1.4 rad/s/s. At $t = 15$ s, what is the angular speed in radians per second? **(d)** Through what angle in radians did the wheel rotate during the time between $t = 10$ s and $t = 15$ s? **(e)** If the wheel rolls along the ground without slipping, the instantaneous velocity of the atoms of the object that are momentarily in contact with the ground is zero. This zero-velocity condition implies that $v_{CM} = \omega R$, where ω is the angular speed of the object, since the instantaneous speed of the contact point is $v_{CM} - \omega R$. During the time between $t =$ 10 s and $t = 15$ s, how far did the center of the wheel move, in meters?

••**P70** If you did not already do Problem P59, do it now. Also calculate the angle through which the disk turns, in radians and degrees.

••**P71** If you did not already do Problem P60, do it now. Also calculate numerically the angle through which the yo-yo turns, in radians and degrees.

••**P72** If you did not already do Problem P63, do it now. Also calculate numerically the angle through which the apparatus turns, in radians and degrees.

•••**P73** Two balls of equal radius and mass, free to roll on a horizontal plane, are separated by a distance L large compared to their radius. One ball is solid, the other hollow with a thickness small compared to its radius. They are attracted by an electric force. How far will the solid ball roll before it collides with the hollow ball?

•••**P74** An amusing trick is to press a finger down on a marble on a horizontal table top, in such a way that the marble is projected along the table with an initial linear speed v and an initial backward rotational speed ω about a horizontal axis perpendicular to v. The coefficient of sliding friction between marble and top is constant. The marble has radius R. **(a)** If the marble slides to a complete stop, what was ω in terms of v and R? **(b)** If the marble skids to a stop and then starts returning toward its initial position, with a final constant speed of $(3/7)v$, what was ω in terms of v and R? *Hint for part (b):* When the marble rolls without slipping, the relationship between speed and angular speed is $v = \omega R$.

Section 11.11

•**P75** According to the Bohr model of the hydrogen atom, what is the magnitude of the translational angular momentum of the electron (relative to the location of the proton) when the atom is in the 2nd excited state above the ground state ($N = 3$)?

••**P76** Calculate rotational energy levels and photon emission energies for a carbon monoxide molecule (CO). **(a)** Calculate the energies of the quantized rotational energy levels for the CO molecule. Estimate any quantities you need. See the discussion of diatomic molecules in the section on quantized angular momentum; the parameter l has values $0, 1, 2, 3 \ldots$. **(b)** Describe the emission spectrum for electromagnetic radiation emitted in transitions among the rotational CO energy levels. Include a calculation of the lowest-energy emission in electron volts (1.6×10^{-19} J). **(c)** It is transitions among "electronic" states of atoms that produce visible light, with photon energies on the order of a couple of electron-volts. Each electronic energy level has quantized rotational and vibrational (harmonic oscillator) energy sublevels. Explain why this leads to a visible spectrum that contains "bands" rather than individual energies.

••P77 Review the derivation of the Bohr model of the hydrogen atom and apply this reasoning to predict the energy levels of ionized helium He^+ (a helium atom with only one electron, and a nucleus containing two protons and two neutrons). What are the energies in eV of the ground state and the first excited state? What is the energy of a photon emitted in a transition from the first excited state to the ground state? How do these results differ from those for a hydrogen atom?

••P78 The Bohr model correctly predicts the main energy levels not only for atomic hydrogen but also for other "one-electron" atoms where all but one of the atomic electrons has been removed, such as in He^+ (one electron removed) or Li^{++} (two electrons removed). **(a)** Predict the energy levels in eV for a system consisting of a nucleus containing Z protons and just one electron. You need not recapitulate the entire derivation for the Bohr model, but do explain the changes you have to make to take into account the factor Z. **(b)** The negative muon (μ^-) behaves like a heavy electron, with the same charge as the electron but with a mass 207 times as large as the electron mass. As a moving μ^- comes to rest in matter, it tends to knock electrons out of atoms and settle down onto a nucleus to form a "one-muon" atom. For a system consisting of a lead nucleus (Pb^{208} has 82 protons and 126 neutrons) and just one negative muon, predict the energy in eV of a photon emitted in a transition from the first excited state to the ground state. The high-energy photons emitted by transitions between energy levels in such "muonic atoms" are easily observed in experiments with muons. **(c)** Calculate the radius of the smallest Bohr orbit for a μ^- bound to a lead nucleus (Pb^{208} has 82 protons and 126 neutrons). Compare with the approximate radius of the lead nucleus (remember that the radius of a proton or neutron is about 1×10^{-15} m, and the nucleons are packed closely together in the nucleus).

Comments: This analysis in terms of the simple Bohr model hints at the result of a full quantum-mechanical analysis, which shows that in the ground state of the lead–muon system there is a rather high probability for finding the muon *inside* the lead nucleus. Nothing in quantum mechanics forbids this penetration, especially since the muon does not participate in the strong interaction. Electrons in an atom can also be found inside the nucleus, but the probability is very low, because on average the electrons are very far from the nucleus, unlike the muon.

The eventual fate of the μ^- in a muonic atom is that it either decays into an electron, neutrino, and antineutrino, or it reacts through the weak interaction with a proton in the nucleus to produce a neutron and a neutrino. This "muon capture" reaction is more likely if the probability is high for the muon to be found inside the nucleus, as is the case with heavy nuclei such as lead.

••P79 The nucleus dysprosium-160 (containing 160 nucleons) acts like a spinning object with quantized angular momentum, $L^2 = l(l+1)\hbar^2$, and for this nucleus it turns out that l must be an even integer (0,2,4...). When a Dy-160 nucleus drops from the $l = 2$ state to the $l = 0$ state, it emits an 87 keV photon (87×10^3 eV). **(a)** What is the moment of inertia of the Dy-160 nucleus? **(b)** Given your result from part (a), find the approximate radius of the Dy-160 nucleus, assuming it is spherical. (In fact, these and similar experimental observations have shown that some nuclei are not quite spherical.) **(c)** The radius of a (spherical) nucleus is given approximately by $(1.3 \times 10^{-15}\,\text{m})A^{1/3}$, where A is the total number of protons and neutrons. Compare this prediction with your result in part (b).

Section 11.12

•P80 Two gyroscopes are made exactly alike except that the spinning disk in one is made of low-density aluminum, whereas the disk in the other is made of high-density lead. If they have the same spin angular speeds and the same torque is applied to both, which gyroscope precesses faster?

••P81 This problem requires that you have a toy gyroscope available. The purpose of this problem is to make as concrete as possible the unusual motions of a gyroscope and their analysis in terms of fundamental principles. In all of the following studies, the effects are most dramatic if you give the gyroscope as large a spin angular speed as possible. **(a)** Hold the spinning gyroscope firmly in your hand, and try to rotate the spin axis quickly to point in a new direction. Explain qualitatively why this feels "funny." Also explain why you *don't* feel anything odd when you move the spinning gyroscope in any direction *without* changing the direction of the spin axis. **(b)** Support one end of the spinning gyroscope (on a pedestal or in an open loop of the string) so that the gyroscope precesses *counterclockwise* as seen from above. Explain this counterclockwise precession direction; include sketches of top and side views of the gyroscope. **(c)** Again support one end of the spinning gyroscope so that the gyroscope precesses *clockwise* as seen from above. Explain this clockwise precession direction; include sketches of top and side views of the gyroscope.

••P82 This problem requires that you have a toy gyroscope available. The purpose of this problem is to make as concrete as possible the unusual motions of a gyroscope and their analysis in terms of fundamental principles. In all of the following studies, the effects are most dramatic if you give the gyroscope as large a spin angular speed as possible. **(a)** If you knew the spin angular speed of your gyroscope, you could predict the precession rate. Invent an appropriate experimental technique and determine the spin angular speed approximately. Explain your experimental method and your calculations. Then predict the corresponding precession rate, and compare with your measurement of the precession rate. You will have to measure and estimate some properties of the gyroscope and how it is constructed. **(b)** Make a *quick* measurement of the precession rate with the spin axis horizontal, then make another quick measurement of the precession rate with the spin axis nearly vertical. (If you make quick measurements, friction on the spin axis doesn't have much time to change the spin angular speed.) Repeat, this time with the spin axis initially nearly vertical, then horizontal. Making all four of these measurements gives you some indication of how much the spin unavoidably changes due to friction while you are quickly changing the angle. What do you conclude about the dependence of the precession rate on the angle, assuming the same spin rate at these different angles? What is the theoretical prediction for the dependence of the precession rate on angle (for the same spin rate)?

••P83 The axis of a gyroscope is tilted at an angle of 30° to the vertical (Figure 11.110). The rotor has a radius of 15 cm, mass 3 kg, moment of inertia $0.06\,\text{kg}\cdot\text{m}^2$, and spins on its axis at 30 rad/s. It is supported in a cage (not shown) in such a way that without an added weight it does not precess. Then a mass of 0.2 kg is hung from the axis at a distance of 18 cm from the center of the rotor.

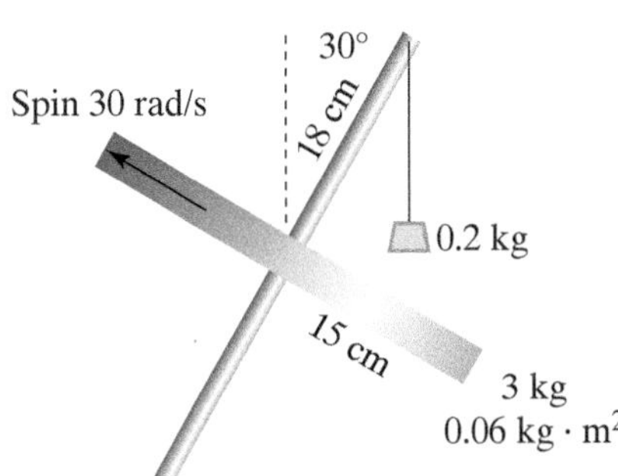

Figure 11.110

(a) Viewed from above, does the gyroscope precess in a (1) clockwise or a (2) counterclockwise direction? That is, does the top end of the axis move (1) out of the page or (2) into the page in the next instant? Explain your reasoning. **(b)** How long does it take for the gyroscope to make one complete precession?

••**P84** A bicycle wheel with a heavy rim is mounted on a lightweight axle, and one end of the axle rests on top of a post. The wheel is observed to precess in the horizontal plane. With the spin direction shown in Figure 11.111, does the wheel precess clockwise or counterclockwise? Explain in detail, including appropriate diagrams.

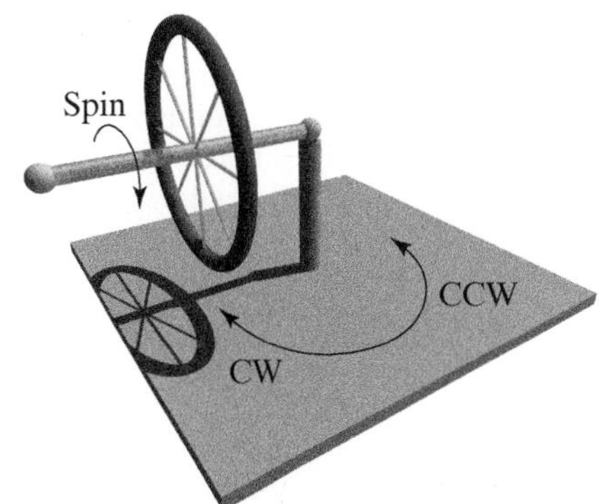

Figure 11.111

••**P85** A solid wood top spins at high speed on the floor, with a spin direction shown in Figure 11.112.

Figure 11.112

(a) Using appropriately labeled diagrams, explain the direction of motion of the top (you do not need to explain the magnitude). **(b)** How would the motion change if the top had a higher spin rate? Explain briefly. **(c)** If the top were made of solid steel instead of wood, explain how this would affect the motion (for the same spin rate).

COMPUTATIONAL PROBLEMS

•**P86** A rod on a low-friction axle is initially at rest when a constant torque begins to be applied. **(a)** Complete the skeleton program shown below and make sure that it runs properly. See Section 11.10 for suggestions. After exiting the loop, print the values of the final angle `theta` and the final angular speed `mag(omega)`. Given the fact that the torque is constant, what should be the value of the ratio `mag(omega)/theta`? Is this true for your program? **(b)** Make graphs of `theta` vs. time and `mag(omega)` vs. t. What should these graphs look like? Do they? **(c)** Change the torque to oscillate: $\tau_{\text{net},z} = 3\cos(5t)$. Graph `theta` vs. time and measure the period. Is it consistent with the angular frequency of 5 rad/s of the torque? **(d)** Plot on the same graph the rotational kinetic energy of the rod vs. time and the magnitude of the torque as a function of time. Is the rotational kinetic energy large when the torque is large?

```
from visual import *
from visual.graph import *

M = 2
Lrod = 1
R = 0.1
Laxle = 4*R
I = (1/12)*M*Lrod**2 + (1/4)*M*R**2

rod = cylinder(pos=vector(-1,0,0),
          radius=R, color=color.orange,
          axis=vector(Lrod,0,0))
axle = cylinder(pos=vector(-1+Lrod/2,0,-Laxle/2),
          radius=R/6, color=color.red,
          axis=vector(0,0,4*R))

L = vector(0,0,0) # angular momentum
deltat = 0.0001   # for accuracy in later parts
t = 0
theta = 0
dtheta = 0

while t < 2:
    rate(10000)
    torque = vector(0,0,2) # constant torque
    # Apply Angular Momentum Principle
    # Update angle and rod position
    t = t + deltat
```

•**P87** Start with the program for part **(c)** of Problem P86. Add a constant force $\vec{F} = \langle 0.1, 0, 0\rangle$ N applied to the axle. Let the while loop run to 7 s. Describe what you see.

••**P88** Start with the program from Chapter 3 Problem P68 to model the motion of a planet going around a fixed star. In this problem you will build on that program. **(a)** Use initial conditions that produce an elliptical orbit. At each step calculate the translational angular momentum $\vec{L}_{\text{trans},A}$ of the planet with respect to a location A chosen to be in the orbital plane but outside the orbit. Display this in two ways (i and ii below), and briefly describe in words what you observe: (i) Display $\vec{L}_{\text{trans},A}$ as an arrow with its tail at location A, throughout the orbit. Since the magnitude of $\vec{L}_{\text{trans},A}$ is quite different from the magnitudes of the distances involved, you will need to scale the arrow by some factor to fit it on the screen. (ii) Graph the component of $\vec{L}_{\text{trans},A}$ perpendicular to the orbital plane as a function of time. **(b)** Repeat part (a), but this time choose a different location B at the center of the fixed star, and calculate and display $\vec{L}_{\text{trans},B}$ relative to that location B. As in part (a), display $\vec{L}_{\text{trans},B}$ as an arrow (scaled appropriately), and also graph the component of $\vec{L}_{\text{trans},B}$ perpendicular to the orbital plane, as a function of time, and briefly describe in words what you observe. **(c)** Choose a location C which is not in the orbital plane, and calculate and display $\vec{L}_{\text{trans},C}$ as an arrow throughout the orbit. (You do not need to make a graph.) Briefly describe in words what you observe.

••P89 Figure 11.113 shows a glass disk mounted on a low-friction axle and held stationary. The center of the disk is at the origin, $\langle 0,0,0\rangle$ m. A peg in the disk is connected by two identical stretched springs to a wall. The disk is released from rest. **(a)** Run the skeleton program shown below and rotate the camera to understand the apparatus. Then complete the program so that the disk rotates. **(b)** Make a graph of the angle `theta` and determine the period. Is this system a harmonic oscillator? Try varying the initial angle and/or the relaxed length of the spring. **(c)** Show on a second graph the kinetic energy of the disk, the spring energy, and their sum, as a function of time. What should the sum look like? Try a smaller time step. (The setting `material=materials.rough` for the disk makes it easier to see its rotation; this may not have a visible effect if your computer graphics card is not up to date.)

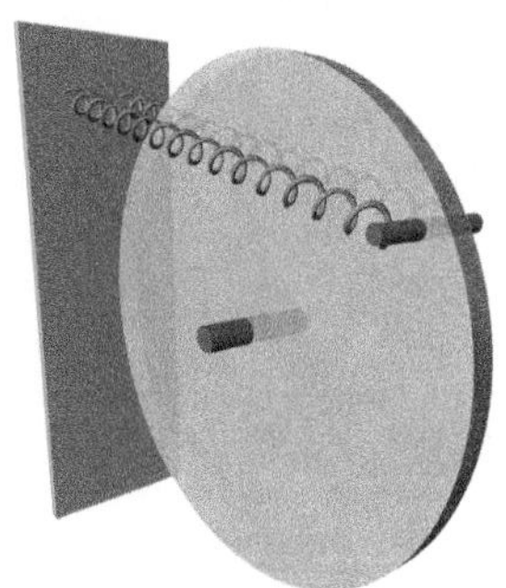

Figure 11.113

```
from visual import *
from visual.graph import *
M = 2             # mass of uniform-density disk
R = 0.2           # radius of disk
thick = 0.02      # thickness of disk
I = M*R**2        # moment of inertia of disk
r = 0.9*R         # peg in disk at this radius
Lpeg = 5*thick # length of peg
wall = box(pos=vector(-1.2*R,0,0),
           size=(0.01,2.4*R,0.8*R),
           color=color.green)
# Center of disk is at <0,0,0>:
disk = cylinder(pos=vector(0,0,-thick/2),
           radius=R, axis=vector(0,0,thick),
           color=color.white, opacity=0.7,
           material=materials.rough)
axle = cylinder(pos=vector(0,0,-Lpeg/2),
           radius=0.05*R,color=color.red,
           axis=vector(0,0,Lpeg))
# Place peg in the disk on x axis:
peg = cylinder(pos=(r,0,-Lpeg/2),
          radius=0.03*R, color=color.red,
          axis=vector(0,0,Lpeg))
# Rotate to initial position:
theta = pi/6      # radians; CCW from x axis
peg.rotate(angle=theta, axis=axle.axis,
           origin=axle.pos)
rspring = 0.05*R  # spring radius
# Front spring:
springF = helix(pos=(wall.x,r,1.5*thick),
           radius=0.05*R, color=color.orange,
           coils=15, thickness=0.4*rspring)
# Back spring:
springB = helix(pos=(wall.x,r,-1.5*thick),
           radius=0.05*R, color=color.orange,
           coils=15, thickness=0.4*rspring)
# Attach springs to peg:
end = peg.pos+vector(0,0,Lpeg/2+springF.pos.z)
springB.axis = springF.axis = end - springF.pos


t = 0
deltat = 0.01
dtheta = 0
ks = 1.5           # stiffness of each spring
L0 = 0.26          # relaxed length of each spring
L = vector(0,0,0) # initial angular momentum


while True:
    rate(100)
    # Calculate spring force F acting on peg
    # Calculate torque due to springs
    # Update angular momentum L
    # Calculate angular velocity omega
    # Calculate dtheta and theta
    # Update disk and peg positions:
    disk.rotate(angle=dtheta, axis=axle.axis,
              origin=axle.pos)
    peg.rotate(angle=dtheta, axis=axle.axis,
              origin=axle.pos)
    # Update spring lengths:
    end = peg.pos+vector(0,0,Lpeg/2+springF.pos.z)
    springF.axis = end - springF.pos
    springB.axis = springF.axis
```

•••P90 An electric "dipole" consists of an object that is positively charged at one end and negatively charged at the other. A molecular example is an HCl molecule in which the H end is positive and the Cl end is negative. In Figure 11.114 is a snapshot of a positively charged sphere on the left interacting with a dipole on the right in outer space; the two charges on the right are stuck together. The spheres are uniformly charged spheres of radius $R = 1$ mm which interact electrically as though they were point charges concentrated at the centers. Each sphere has uniform density with a mass of 1×10^{-6} kg. Each of the red spheres has an electric charge of $+2\times10^{-10}$ C and the negative sphere's charge is -2×10^{-10} C.

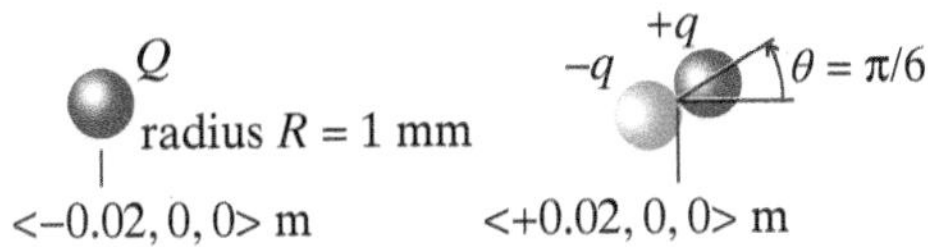

Figure 11.114

The charge on the left is fixed in position at $\langle -0.02, 0, 0\rangle$ m. The center of mass of the dipole is initially located at $\vec{r}_{CM} = \langle 0.02, 0, 0\rangle$ m, with the positive charge at $\vec{r}_{CM} + \vec{r}$ and the negative charge at $\vec{r}_{CM} - \vec{r}$, where $\vec{r} = \langle R\cos\theta, R\sin\theta, 0\rangle$ and $\theta = \pi/6$ rad. The dipole is free to rotate about an axle at its center of mass, but the axle is fixed and cannot move. **(a)** Create a computational model of the situation. Note in calculating the moment of inertia of the rotating dipole that each of the dipole charges has both translational and rotational angular momenta. We suggest using a time step of 0.001 s for accuracy, with a large value in the rate statement for display speed. Use the Angular Momentum Principle to predict and display the motion of the dipole, starting from rest (no rotation). Include a graph of θ vs. time. What is the period of the oscillation? **(b)** Determine the period for starting values of θ of 0.01π and 0.02π. Is this oscillator a harmonic oscillator for small oscillations (that is, is the period independent of the amplitude)? **(c)** Is the period the same if you increase the amplitude to $\theta = \pi/2$? **(d)** Measure the period for a small amplitude with quadruple the charge on the left. What is the ratio of this period to the small-amplitude period you measured in part (b)? How does this effect compare with quadrupling the spring stiffness for a spring–mass oscillator?

•••P91 This is a continuation of problem P90, in which we allow the center of mass of the dipole to move freely. Restore the charge on the left to $Q = +2\times10^{-10}$ C. **(a)** Use the Momentum Principle and the Angular Momentum Principle to predict and display the motion of the translating, rotating dipole. Stop the animation when the distance between the charge on the left and the center of mass of the dipole is less than 0.02 m. **(b)** Make a graph that shows as a function of time K_{trans}, K_{rot}, U_{elec}, and the total $E = K_{trans} + K_{rot} + U_{elec}$. (We can ignore the constant electric potential energy term associated with the dipole pair of charges; the distance between them does not change.) What does the Energy Principle predict for the graph of E vs. t? Is your graph consistent with the Energy Principle? **(c)** Let the charge on the left move freely. Make sure your graph of E vs. t still makes sense.

In this problem we've seen a situation where we needed to use all three fundamental mechanics principles: The Momentum Principle, the Energy Principle, and the Angular Momentum Principle.

•••P92 Model the motion of a meter stick suspended from one end on a low-friction axle. Do not make the small-angle approximation but allow the meter stick to swing with large angles. Plot on the same graph both θ and the z component of $\vec{\omega}$ vs. time. Try starting from rest at various initial angles, including nearly straight up (which would be $\theta_i = \pi$ radians). Is this a harmonic oscillator? Is it a harmonic oscillator for small angles?

ANSWERS TO CHECKPOINTS

1 $-y$
2 D: 50 kg · m²/s out of page; E: 50 kg · m²/s out of page; F: 50 kg · m²/s out of page; G: 0; H: 30 kg · m²/s into page
3 (a) $L_1 = 0.72$ kg · m²/s and $L_2 = 0.72$ kg · m²/s, both into page; **(b)** $L_{rot} = 1.44$ kg · m²/s into page; **(c)** $I = 0.072$ kg · m²; **(d)** $\vec{\omega}_0$ into page; **(e)** $I\omega_0 = 1.44$ kg · m²/s into page; **(f)** They're the same; **(g)** $K_{rot} = 14.4$ J
4 Nonzero; up (toward the sky); nonzero; up
5 6 N · m; 2 N
6 (a) $R\cos(45°)mv$ out of page; **(b)** $(M+m)R^2\omega$ out of page; **(c)** $R\cos(45°)mv/[(M+m)R^2]$ out of page; **(d)** The linear momentum of the clay decreases because the axle exerts an impulsive force upward (the wheel always has zero linear momentum because its center of mass doesn't move).
7 No torques around center of mass means no change in rotational angular momentum, so rotational angular momentum stays constant in magnitude (which determines length of day) and direction (which determines what "North Star" the axis points at). Doesn't matter that Earth is going around Sun; rotational angular momentum is affected solely by torque around center of mass.
8 $dL_z/dt = \frac{d}{dt}[RMv_{CM} - \frac{1}{2}MR^2\omega] = 0$, since there is no torque about a point under the string. Differentiating, we get $d\omega/dt = (2/R)(dv_{CM}/dt) = 2F_T/(MR)$, as before.
9 $dL_z/dt = d_1F_N - (d_1+d_2)M_2g = 0$ ($+z$ out of page), and since $F_N = M_1g + M_2g$, this is equivalent to $d_1M_1g - d_2M_2g$, as before.
10 (a) 10 rad/s; **(b)** 5 rad/s; **(c)** 3 rad; **(d)** 0.9 m
12 Same as in horizontal case: $\Omega = RMg/(I\omega)$

CHAPTER

12

Entropy: Limits on the Possible

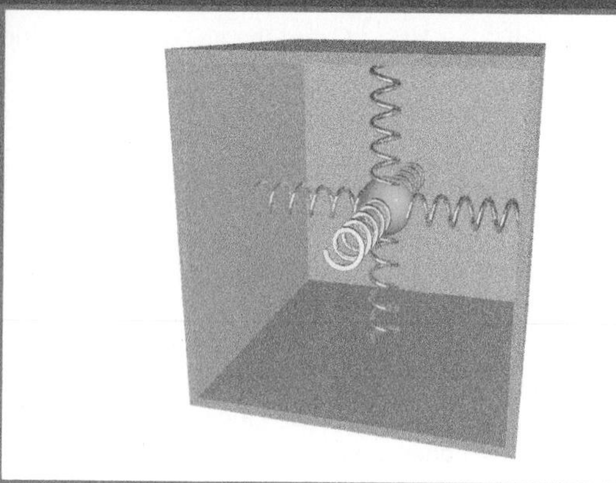

OBJECTIVES

After studying this chapter, you should be able to

- Calculate the entropy of a solid object, based on the Einstein model of a solid
- Calculate the absolute temperature of a solid object as a function of its energy
- Calculate the specific heat of a solid object as a function of temperature
- Calculate the probability of finding a microscopic system in a particular excited state

12.1 IRREVERSIBILITY

A pen lying on a table doesn't suddenly jump upward, despite the fact that there is plenty of energy in the table in the form of microscopic kinetic and potential energy. Why doesn't this happen? It wouldn't be a violation of the Energy Principle for energy to flow from the table into the pen. In fact, if instead of a pen you place a single atom on a table at room temperature, on average it will get so much energy from the table that it could jump thousands of meters upward (if there were no air or other obstacles in the way). This chapter deals with a statistical analysis of microscopic energy that puts limits on what is possible.

Much of this chapter is based on an article by Thomas A. Moore and Daniel V. Schroeder, "A different approach to introducing statistical mechanics," *American Journal of Physics*, vol. 65, pp. 26–36 (January 1997).

What Is Temperature?

We have repeatedly encountered a connection between the temperature of an object and the average motion of the atoms that make up the object, but our understanding of the meaning and role of temperature is incomplete. What is the quantitative relationship between temperature and microscopic energy?

High temperature

Low temperature

Q

Figure 12.1 We observe that energy flows from a hotter object to a cooler one.

Why Does Energy Flow from Hot to Cold?

When two blocks of different temperatures are placed in contact with each other, we observe that energy flows from the hotter block to the colder block (Figure 12.1). We can understand this in a rough way. The average energy of atoms in the hotter block is greater than the average energy of atoms in the colder block. In the interface region, where atoms of the two blocks are in contact with each other, it seems more likely that an atom in the hotter block

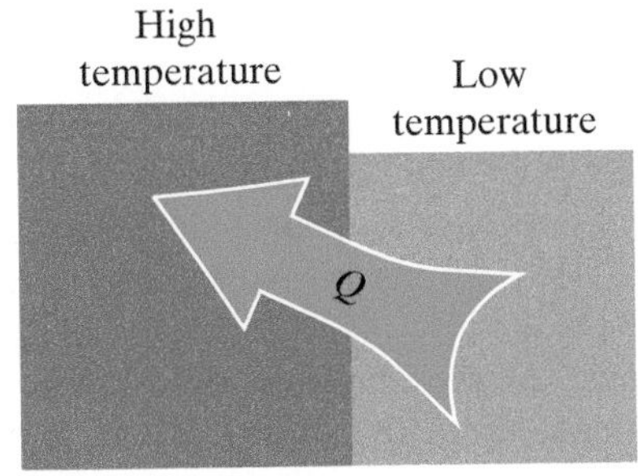

Figure 12.2 Could energy be transferred from the cooler object to the hotter one?

will lose energy rather than gain energy when it collides with an atom in the colder block.

Couldn't energy flow the other way? After all, sometimes an individual atom in the hotter block might happen to have a lot less than its average energy when it collides with an individual atom in the colder block that happens to have a lot more than its average energy. In that case energy would be transferred from the colder block to the hotter block (Figure 12.2).

This "uphill" movement of energy seems unlikely, but how unlikely is it? Should we occasionally observe a hot block get hotter when we put it in contact with a cold block? If you put an ice cube in your drink, could your drink get warmer and the ice cube get colder? This would not violate conservation of energy, because the total energy of the two objects would still be the same. So what physical principle would be violated?

Why Do Things Run Down?

A related question, and a very deep one, has to do with the "reversibility" of processes. The fundamental physical interactions seem to be completely "reversible" in the following technical, physics sense: Make a movie of an alpha particle scattering off a gold nucleus, then run the movie forward or reversed. Both views look entirely reasonable. In contrast, make a movie of a ball bouncing on the floor with decreasing height on each bounce, then run the movie forward or reversed. The reversed movie looks silly: you see a ball bouncing higher and higher with each bounce! Would such a motion violate conservation of energy? Not necessarily. It could be that the atoms in the floor happen on average to give energy to the ball, and the floor is getting colder as a result of this continuing loss of energy. So what physical principle is violated by a process represented by the reversed movie?

Statistical Models

This chapter addresses these questions by applying probabilistic and statistical ideas to the behavior of systems. This subject is called "statistical mechanics." We will first apply statistical mechanics to a simple model of a solid, and we'll find that we can go surprisingly far with just a few new concepts. The main concepts that we will develop for the specific case of a solid apply in general to a wide variety of systems.

How will we know whether our statistical model explains anything? Our criterion for understanding is whether the predictions of our microscopic model agree with measurements of macroscopic systems, such as measurements of heat capacity (the amount of energy required to increase the temperature of an object by one degree).

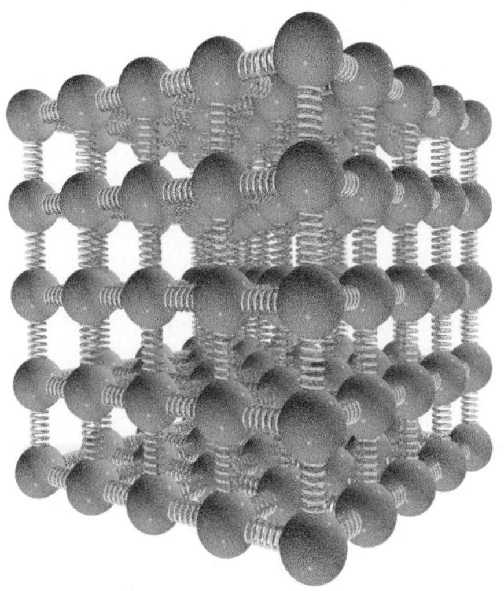

Figure 12.3 The familiar model of a solid as massive balls connected by springs.

12.2 THE EINSTEIN MODEL OF A SOLID

We have noted in previous chapters that the interactions between atoms in a solid are electric in nature, but that a detailed model of them involves quantum mechanics. The interatomic potential energy function encountered in Chapter 7 provides a reasonably accurate description of these interactions. Because of the similarity of this function to the potential energy curve for a harmonic oscillator (a mass on a spring), in previous chapters we have modeled a solid as a large number of tiny masses (the atoms) connected to their neighbors by springs (the interatomic bonds), as shown in Figure 12.3. This model has allowed us to understand qualitatively how solids interact with other objects. We would now like to use a modified version of this model to ask detailed quantitative questions about the distribution of energy in a solid.

In this chapter we focus on calculating how probable a particular distribution of speeds or energies in a solid would be. We do this for a solid because it happens to be easier to calculate probabilities for a simple model of a solid than it is for a gas or a liquid. The atoms in a solid are nearly fixed in position, so we don't have to consider how likely different spatial arrangements might be (unlike the situation for a gas). Reasoning about the energy distribution in a solid will also enable us to draw conclusions about the transfer of thermal energy from one solid object to another.

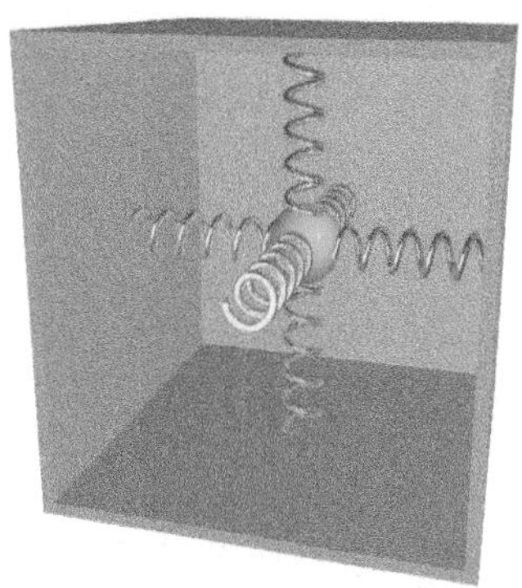

Figure 12.4 A single atom can oscillate in three dimensions. We simplify our model of a solid by assuming that each atom moves independently of the surrounding atoms—in effect, that it is connected to rigid walls instead of moving atoms.

Assume that Atoms Move Independently

Because our goal is to calculate the probability of particular distributions of energy among atoms, we can simplify our model of a solid even further. We can think of each atom as moving independently of its neighbors, as though it were connected to rigid walls rather than to other atoms (Figure 12.4). Of course in a real solid energy is exchanged with neighboring atoms. However, to address our current questions we do not need to worry about the mechanism of energy exchange between atoms, since we will focus on calculating probabilities for various distributions of energy among the atoms of the solid.

This is a different kind of model from those we have so far constructed, because this very simple model would not allow us to predict in detail the dynamic motion of each atom in the solid, nor to ask how long it would take for energy to flow from one end of a solid to the other. In fact, there could not be propagation of sound in such a solid, with its "rigid walls." The model will not shed light on the details of a process, but it does allow us to ask questions about initial and final states. Since at the moment the questions we want to address involve initial and final states, this model is useful, and it is mathematically much simpler than the connected-atoms model.

Einstein proposed this simple model in 1907, and he found that some basic properties of solids such as specific heat could in fact be understood using this model. This model also allows us to understand in detail the statistical nature of energy transfer between a hot object and a cold object, and why two objects come to "thermal equilibrium" (the same final temperature). This simple model of a solid will help us gain a more sophisticated and powerful understanding of the meaning of temperature.

A Three-Dimensional Oscillator—Three One-Dimensional Oscillators

We will consider each atom in a solid to be connected by springs to immovable walls. We will model each isolated atom as a three-dimensional spring–mass system, with $\vec{s}$ representing the three-dimensional vector displacement away from a fixed equilibrium position. Since $p^2 = p_x^2 + p_y^2 + p_z^2$ and $s^2 = s_x^2 + s_y^2 + s_z^2$, we can write the energy of a three-dimensional classical oscillator as the sum of three parts, corresponding to the x, y, and z oscillations:

$$K_{\text{vib}} + U_s = \left(\frac{p_x^2}{2m} + \frac{1}{2}k_s s_x^2\right) + \left(\frac{p_y^2}{2m} + \frac{1}{2}k_s s_y^2\right) + \left(\frac{p_z^2}{2m} + \frac{1}{2}k_s s_z^2\right)$$

Recall from Chapter 8 that a quantum oscillator (a quantum mechanical "ball and spring") has evenly spaced energy levels. A complete quantum mechanical analysis of an oscillator that is free to oscillate in three dimensions rather than just one dimension leads to the conclusion that the motion of a three-dimensional oscillator can be separated into x, y, and z

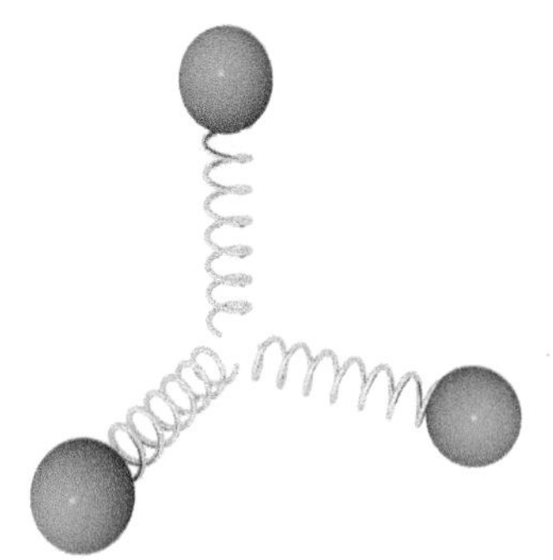

Figure 12.5 Since we are considering the atoms to be independent, in our model we can replace a single three-dimensional oscillator (one atom) by three independent one-dimensional oscillators.

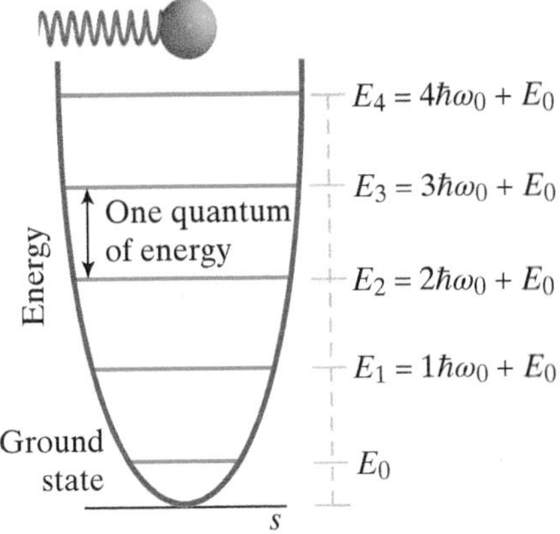

Figure 12.6 The energy levels of a one-dimensional atomic oscillator.

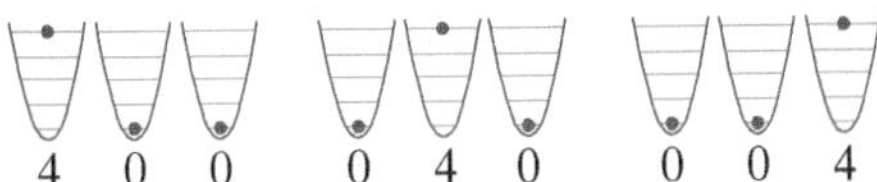

Figure 12.7 Three ways of distributing four quanta of vibrational energy among three one-dimensional oscillators.

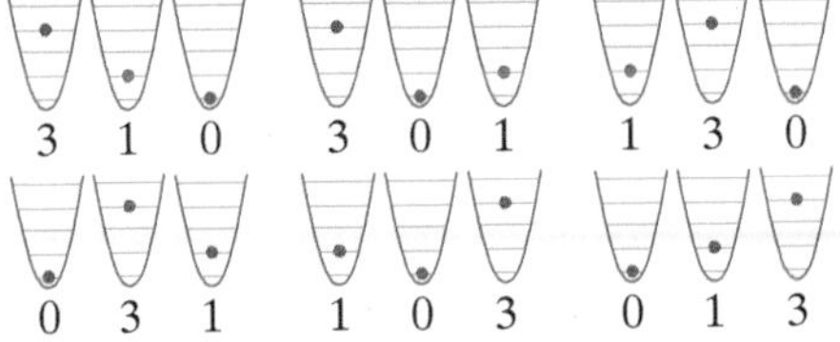

Figure 12.8 Six more ways of distributing four quanta of vibrational energy among three one-dimensional oscillators.

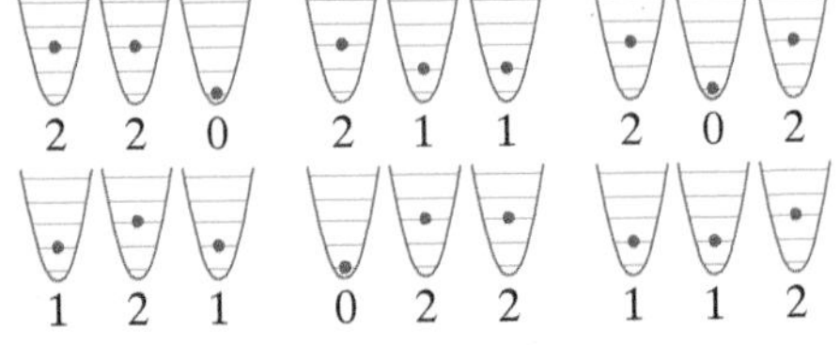

Figure 12.9 Still more ways of distributing four quanta of vibrational energy among three one-dimensional oscillators.

components, each of which has the same energy level structure as the familiar one-dimensional oscillator. This is mathematically equivalent to replacing each three-dimensional oscillator (which we're using to model an atom) with three ordinary one-dimensional oscillators, which we will do to simplify our model. We will think of a block as containing N one-dimensional oscillators, corresponding to $N/3$ atoms (Figure 12.5).

In Chapter 8 we discussed energy levels in atomic spring–mass systems. As predicted by quantum mechanics, the stable energy levels of a one-dimensional quantum oscillator are integer multiples of one "quantum" of energy $\hbar\omega_0$, where $\omega_0 = \sqrt{k_{s,i}/m_a}$ above the ground state (Figure 12.6). Here $\hbar = h/(2\pi) = 1.05 \times 10^{-34}$ J · s, $k_{s,i}$ is the interatomic spring stiffness, and m_a is the mass of the atom. The ground state of the quantum oscillator is not at the bottom of the potential energy well; $E_0 = \frac{1}{2}\hbar\omega_0$. This offset doesn't matter for what we are going to do; it just sets the baseline energy level, and we will measure energies starting from the ground state.

Distributing Energy Among Objects

When two identical blocks are brought into contact, their total energy is shared among all the oscillators (atoms) in both blocks. It seems plausible that the most likely outcome would be that the total thermal energy be shared equally between identical blocks. However, might there be some probability that the first block would have more energy than the other, or even have all of it? With a large number of atoms the probability of even a small deviation from the most probable distribution turns out to be hugely unlikely. In order to look at this question in detail, we need to find a way to calculate the probability of various possible distributions of the energy. If we can find the most probable energy distribution, we can predict the eventual equilibrium distribution of energy between two objects brought into thermal contact with each other.

Distributing Energy: A Single Atom Inside a Solid

Consider a single atom inside a solid, whose energy we model in terms of the energy of three one-dimensional oscillators (x, y, z), neglecting interactions with the neighbors. Each of these three oscillators can have 0, 1, 2, or more "quanta" of vibrational energy ($\hbar\omega_0$) added to its ground state.

Suppose that the total vibrational energy added to the atom is four quanta, and we ask how we might distribute this energy among the three oscillators. We will soon see that enumerating all the possible ways of distributing the energy among the oscillators leads to a deeper understanding of the statistical behavior of a solid. We could give all the energy to the first one-dimensional oscillator and none to the others, or all to the second, or all to the third, as shown in Figure 12.7.

We could also give three quanta to one oscillator, and give the remaining quantum to one of the others, as shown in Figure 12.8.

Or we could give two quanta to one oscillator, and distribute the other two to the others, as shown in Figure 12.9.

That's it—there aren't any more ways to distribute the energy. (Check to make sure that you can't think of any other arrangements.) By explicitly listing all the possible arrangements, we see that there are 15 different ways that the four energy quanta could be distributed among the three one-dimensional oscillators.

In each of these 15 cases the total energy of the three-oscillator atom is exactly the same. When we make macroscopic measurements of the energy of a block, we don't know and we usually don't care exactly how the energy is distributed among the atoms that make up the block, because the internal energy of the block is the same for all of these distributions. However, the number of different arrangements influences interactions with other objects, as we will see.

Microstates and Macrostates

In the context of the Einstein model of a solid, the term "macrostate" refers to the total amount of energy in the system. For every macrostate (for example, a total of four quanta of energy in the system), there are a number of possible "microstates." Each microstate reflects a different way of distributing the total energy among all the oscillators in the system. Figure 12.10 illustrates the 15 different microstates (possible ways of arranging energy) corresponding to a macrostate of a system of three oscillators with a total of four quanta of energy. These same microstates are shown pictorially in Figures 12.7–12.9.

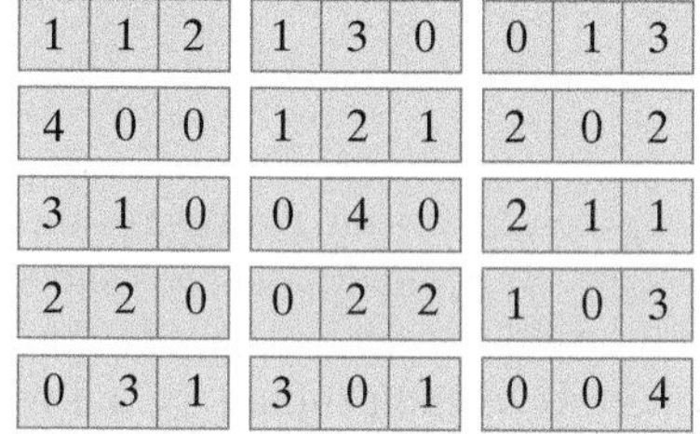

1	1	2	1	3	0	0	1	3
4	0	0	1	2	1	2	0	2
3	1	0	0	4	0	2	1	1
2	2	0	0	2	2	1	0	3
0	3	1	3	0	1	0	0	4

Figure 12.10 The 15 microstates corresponding to a macrostate of a system of three oscillators with four quanta of energy. According to the fundamental assumption of statistical mechanics, each of these microstates is equally probable.

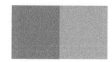

THE FUNDAMENTAL ASSUMPTION OF STATISTICAL MECHANICS

The fundamental assumption of statistical mechanics is that, over time, an isolated system in a given macrostate is equally likely to be found in any of its possible microstates.

For example, we isolate three oscillators (corresponding to one atom), containing a total energy of four quanta, which is a macrostate (we haven't said which oscillators have how many quanta). The fundamental assumption of statistical mechanics implies that if we repeatedly observe the detailed arrangement of the four quanta among the three oscillators (the microstate), over time we would find on the average that each of these microstates would occur 1/15th of the times that we looked. This fundamental assumption is plausible, but ultimately it is justified by the fact that deductions based on this assumption do agree with experimental observations.

In principle you could prepare a system in one particular microstate and hope it would stay that way, but if there is the slightest bit of interaction among the pieces of the system, or with the surroundings, over time the fundamental assumption of statistical mechanics claims that all microstates will occur with equal probability. The Einstein model is simplified in that it doesn't provide a mechanism for energy to transfer from one atom to another (a change of microstate) or even from one oscillator to another within the same atom. The Einstein model is useful for counting microstates, but for other purposes it needs to be supplemented by the notion that some slight interaction does occur.

Figure 12.11 Two small blocks in thermal contact.

Distributing Energy: Two Interacting Atoms

We can now consider the case of two blocks in thermal contact (Figure 12.11). The importance of counting arrangements ("microstates") can be seen when we have systems of oscillators interacting with each other. Consider the smallest possible "blocks," two neighboring atoms (a total of six one-dimensional oscillators), and suppose four quanta of vibrational energy, $4\hbar\omega_0$, are distributed among these six oscillators. We already know that there are 15 ways to distribute the four quanta among the three one-dimensional oscillators of the first atom, and in that case there is only one way to distribute zero quanta to the second atom. Similarly, in Figure 12.12 we see that there are 15 ways to distribute the four quanta among the three one-dimensional oscillators of the second atom, and in that case there is only one way to distribute zero quanta to the first atom.

Atom 1	Atom 2	# ways
4	0	$15 \cdot 1$
0	4	$1 \cdot 15$

Figure 12.12 If we give all four quanta to one atom or the other, there are 30 ways of distributing four quanta of energy between the two atoms (that is, among six independent oscillators).

However, since the two atoms are in contact with each other, the four quanta could also be shared between the atoms. For example, the first atom might have only three quanta, with the second atom having the other quantum.

QUESTION How many ways are there to distribute three quanta among three one-dimensional oscillators? Try to list all the possibilities, as we did for the case of four quanta.

You should have found 10 ways, which we could list as 300 (that is, three quanta on the first oscillator, zero on the second, and zero on the third), 030, 003, 201, 210, 021, 120, 012, 102, and 111.

QUESTION For each of these arrangements, how many ways are there to distribute the one remaining quantum among the other three one-dimensional oscillators?

There are just three ways to arrange one quantum among the other three oscillators: 100, 010, and 001.

The product of these two numbers is the total number of ways of distributing the four quanta in such a way that there are three quanta on the first atom and one on the other. This product is $10 \cdot 3 = 30$ ways. These results are summarized in Figure 12.13.

Atom 1	Atom 2	# ways
3	1	$10 \cdot 3$
1	3	$3 \cdot 10$

Figure 12.13 We find 60 more ways to distribute four quanta among two atoms (six oscillators) if we give three quanta to one atom and one quantum to the other.

Now let's give just two quanta to the first atom (leaving two quanta for the other atom).

Atom 1	Atom 2	# ways
2	2	$6 \cdot 6$

Figure 12.14 We find 36 ways of distributing four quanta of energy by giving two quanta to each atom.

QUESTION How many ways are there to distribute two quanta among three one-dimensional oscillators?

You should find that there are six ways: 200, 020, 002, 101, 110, and 011.

QUESTION For each of these arrangements, how many ways are there to distribute the two remaining quanta among the other three one-dimensional oscillators?

Clearly there are also six ways to arrange the remaining two quanta among the other three oscillators. The total number of ways of distributing the four quanta in such a way that there are two quanta on the first atom and two on the other is $6 \cdot 6 = 36$ (Figure 12.14).

Now we have carried out enough calculations to be able to make a table of all the ways of distributing four quanta between two atoms (that is, among six one-dimensional oscillators). In this table, q_1 is the number of quanta (0 to 4) assigned to the first atom, and q_2 is the number of quanta assigned to the second (with $q_1 + q_2$ necessarily equal to 4). In Figure 12.15 we show the number of ways to arrange the quanta among the first three one-dimensional oscillators, among the second three oscillators, and the product, which is the total number of ways of distributing q_1 and q_2 quanta.

q_1	q_2	# ways 1	# ways 2	# ways 1 · # ways 2
0	4	1	15	15
1	3	3	10	30
2	2	6	6	36
3	1	10	3	30
4	0	15	1	15

Figure 12.15 Summary of the 126 different ways to distribute four quanta of vibrational energy between two atoms (each consisting of three independent oscillators).

Figure 12.16 is a graph of the total number of ways of distributing the quanta vs. q_1. Remember that the basic assumption of statistical mechanics is that all of these 126 different arrangements $(15 + 30 + 36 + 30 + 15 = 126$ microstates) with a total energy of four quanta are equally probable.

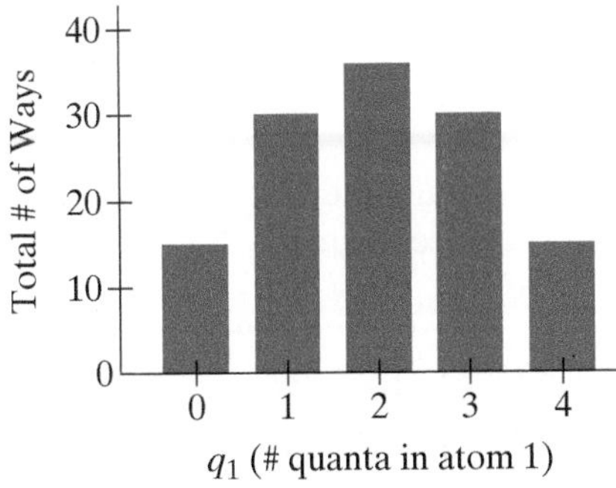

Figure 12.16 Bar graph showing the total number of ways of distributing four quanta of vibrational energy between two atoms (six one-dimensional oscillators).

QUESTION Therefore, if two atoms share four quanta, what is the most likely division of the energy between the two? What is the probability that at some instant the energy is equally divided? What is the probability that at some instant the first atom has all the energy?

Evidently the most probable division is 2 and 2, since there are 36 ways for this to be done, whereas all other divisions have fewer ways for this to happen. The probability for a 2-2 division is 36/126, or 29%, so even though this is the most probable division, it happens less than one-third of the time. The probability that the first atom has all the energy is the number of ways for this to happen (15) divided by 126: 15/126 = 12%.

This is a microcosm of what happens when two identical blocks are brought into contact: the most probable division of the thermal energy is that it is shared equally. In order to understand in detail how this works in a real macroscopic system, we need to consider large numbers of atoms and large numbers of quanta. For numbers of atoms or quanta even slightly larger than the few we've been considering, it becomes practically impossible to figure out the number of arrangements of quanta by explicitly listing all the possibilities as we did up to now. We need an equation for calculating the number of arrangements.

Checkpoint 1 For practice in counting microstates, determine how many ways there are to arrange three quanta among four one-dimensional oscillators. (This would represent one-and-a-third atoms, so this doesn't make physical sense.)

One System or Many?

The results for distributing four quanta among two atoms (six oscillators) can be thought of in two complementary ways. You can say, "I will make frequent observations of my isolated two-atom system, and I expect that in 29% of these observations I'll find the energy split evenly (2-2)." Alternatively, you can set up 100 of these isolated two-atom systems and say, "Whenever I look at my 100 systems, I expect that 29 of them will have the energy split evenly (2-2)." Sometimes the "one system, many observations" view is particularly helpful, and sometimes the "many systems, one observation" view is the more useful way to think about the statistical nature of a phenomenon.

An Equation for the Number of Arrangements of Quanta

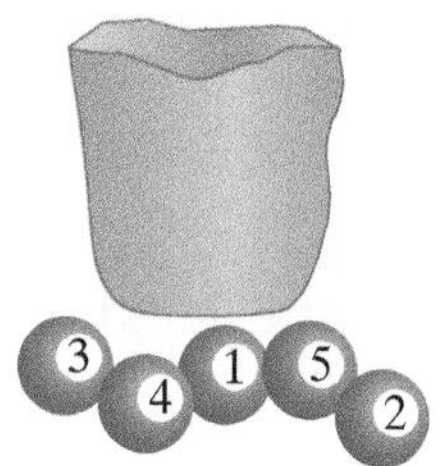

Figure 12.17 Five numbered billiard balls are taken one at a time from a bag. How many different number sequences are possible?

Clearly it would be not only tedious but also impractical to keep on counting states as we did above for systems involving very large numbers of atoms. We need an equation for calculating how many different ways a set of objects can be arranged. We can develop this equation in a general way, using concepts of probability, and then apply it to solids made up of atomic oscillators.

Suppose you have five billiard balls in a bag, numbered 1 through 5. You draw them out of the bag, one at a time, and record the sequence of numbers, such as 34152 (Figure 12.17). Then you put them back in the bag and repeat. How many different number sequences are possible?

- The first ball might be any one of the five balls.
- For each of these five possibilities, there are four possibilities for the second ball, or $5 \cdot 4 = 20$ choices so far.
- For each of the $5 \cdot 4 = 20$ choices of the first two balls, there are three different possible choices for the third ball, or $5 \cdot 4 \cdot 3$ choices so far.
- For each of the $5 \cdot 4 \cdot 3 = 60$ choices of the first three balls, there are two possible choices for the fourth ball, for a total of $5 \cdot 4 \cdot 3 \cdot 2 \cdot 1 = 120$ possibilities, since there is only one remaining ball to choose.

Evidently there are 120 different "permutations" of the five integers. It would be exceedingly tedious to list all these different arrangements, but we have a simple equation to calculate how many there are: 5!, which is the standard notation for "5 factorial," meaning $5 \cdot 4 \cdot 3 \cdot 2 \cdot 1$. Fortunately, for analyzing a solid all we care about is how many arrangements (microstates) there are for a particular total energy, not the details about how much of the energy is assigned to which oscillators. So all we need from an equation is the number of arrangements.

Checkpoint 2 We can easily check this factorial equation for the case of three balls. Explicitly list all possible permutations of the numbers 1, 2, and 3, and verify that there are indeed $3! = 3 \cdot 2 \cdot 1 = 6$ possible arrangements.

Making Fewer Distinctions

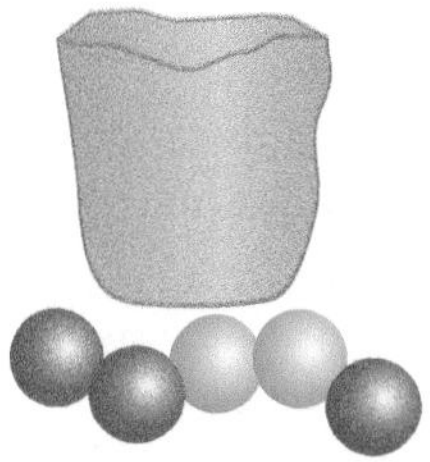

Figure 12.18 Five billiard balls are taken one at a time from a bag. Three are red and two are green. How many different color sequences are possible?

Now suppose that of the five balls, three are red (R) and two are green (G), and we ask how many different arrangements of the colors are possible, such as RRGGR shown in Figure 12.18. We know that there are $5! = 120$ different arrangements of the numbered balls, but if we're only interested in the color sequence, the numerical order of the red balls is irrelevant, as is the numerical order of the green balls.

There are $3! = 6$ permutations of the red balls among each other, and $2! = 2$ permutations of the green balls among each other. Therefore there are many fewer than 120 distinctively different color sequences, and we need to correct for this by dividing by the extra permutations:

$$\text{number of color sequences} = \frac{5!}{3!2!} = \frac{120}{(6)(2)} = 10$$

Checkpoint 3 Check this result by listing all the different ways of ordering three R's and two G's.

Arranging Quanta Among Oscillators

Figure 12.19 A representation of a situation where the first oscillator has 2 quanta, the second oscillator has 1 quantum, and the third oscillator has 1 quantum.

By an appropriate choice of visual representation, we can convert our problem of calculating the number of arrangements of q quanta among N one-dimensional oscillators ($N/3$ atoms) into the problem we just solved, for which we have an equation. Consider again the specific case of $q = 4$ quanta distributed among $N = 3$ one-dimensional oscillators. We'll represent a quantum of energy by the symbol •, and a (fictitious) boundary between oscillators by a vertical bar **|**. Figure 12.19 is a picture of a particular situation where the first oscillator has 2 quanta, the second oscillator has 1 quantum, and the third oscillator has 1 quantum; together with the boundaries between oscillators we have 6 objects arranged in a particular sequence.

Figure 12.20 Representations of two situations for the number of quanta in 3 oscillators.

This sequence represents a total energy of 4 quanta and 2 boundaries (a total of 6 things, if we consider a boundary to be a "thing"). Figure 12.20 illustrates two such sequences. Notice that we need $N - 1 = 2$ vertical bars to be able to indicate which oscillators have how much energy. How many such sequences like this are there? Rearranging quanta and boundaries is equivalent to moving quanta between oscillators. In this pictorial form, the problem is like having 4 red balls and 2 green ones. Therefore we can write an equation for the number of different arrangements (number of different microstates):

$$\frac{6!}{4!2!} = 15 \text{ ways of arranging 4 quanta among 3 oscillators}$$

This agrees with our earlier calculations. Generalizing to arbitrary numbers of quanta distributed among arbitrary numbers of oscillators, we have this important result for the number of microstates, written as Ω (Greek capital omega):

NUMBER OF WAYS Ω TO ARRANGE q QUANTA AMONG N ONE-DIMENSIONAL OSCILLATORS

$$\Omega = \frac{(q+N-1)!}{q!(N-1)!}$$

Checkpoint 4 Verify that this equation gives the correct number of ways to arrange 0, 1, 2, 3, or 4 quanta among 3 one-dimensional oscillators, given in earlier tables (1, 3, 6, 10, 15).

Very Big Numbers

As you increase q and/or N, this expression gets very big very fast. For example, the number of ways to distribute 100 quanta among 300 oscillators (100 atoms) is about 1.7×10^{96}, which is 17 followed by 95 zeros!

In striking contrast, there is only one way to arrange to have all 100 energy quanta be placed on just one particular oscillator out of all the 300 oscillators. Although this arrangement would satisfy the requirement that the total energy of the system be 100 quanta, this is extremely improbable. The fundamental assumption of statistical mechanics is that in an isolated system, over time, all microstates are equally probable, so the odds that the actual microstate is the one with all the energy given to just one particular oscillator, of your choice, is 1 in 1.7×10^{96}, which makes this unlikely event essentially impossible.

A typical macroscopic object such as a block of ordinary size contains 1×10^{23} or more atoms, not a mere hundred, and the number of quanta is even larger. How likely is it that all the energy will be found concentrated on one particular atom? The mind boggles at the astronomically huge odds against this ever happening. Could it happen? Yes. Is any human ever likely to observe it? No.

12.3 THERMAL EQUILIBRIUM OF BLOCKS IN CONTACT

We now have the tools necessary for analyzing in some detail what will happen when two blocks are brought into contact and approach thermal equilibrium. We'll choose two blocks made of the same material, so a quantum of energy is the same for the atomic oscillators in both blocks (same atomic mass m; same interatomic forces, so same "spring stiffness" k_s). We'll make the analysis somewhat general by choosing blocks of different sizes. One block contains N_1 one-dimensional oscillators ($N_1/3$ atoms) and initially contains q_1 quanta of energy, and the other block contains N_2 one-dimensional oscillators ($N_2/3$ atoms) and initially contains q_2 quanta of energy.

QUESTION We want to treat the simple case where the total energy $q_1 + q_2$ of the two-block system remains fixed at all times. What simplifying assumption should we make about the situation?

During the entire process we assume that there is little energy transferred into or out of the surrounding air or the supports for the blocks, so that the total energy of the two blocks doesn't change. However, the number of quanta in each block, q_1 or q_2, need not stay fixed, since energy can flow back and forth between the two blocks.

Consider a very concrete example. Suppose that $N_1 = 300$ (100 atoms), $N_2 = 200$ (about 67 atoms; or choose 201 oscillators if you wish to be exact), and the total energy distributed throughout the two blocks corresponds to $q_1 + q_2 = 100$ quanta (Figure 12.21).

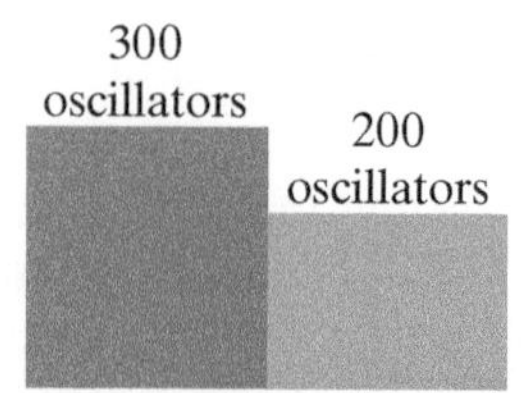

Figure 12.21 A total of 100 quanta of vibrational energy are available to be distributed between two systems, one consisting of 300 oscillators (100 atoms), the other consisting of 200 oscillators (about 67 atoms).

We use a computer to calculate the number of ways that q_1 quanta can be distributed among the $N_1 = 300$ oscillators of the first block (using the equation we developed earlier), and we multiply this number times the number of ways that $q_2 = (100 - q_1)$ quanta can be distributed among the $N_2 = 200$ oscillators of the second block.

The product of these two calculations is the number of ways that we can arrange the 100 quanta so that the first block has q_1 quanta and the second block has $q_2 = (100 - q_1)$ quanta, for a total energy shared between the two blocks of 100 quanta. We have the computer do this calculation for $q_1 = 0$, 1, 2, 3,…99, 100 quanta, which corresponds to $q_2 = 100$, 99, 98,…1, 0 quanta. The following table shows the first few results, where the number of microstates is denoted by Ω (Greek uppercase omega):

q_1	$q_2 = (100-q_1)$	$\Omega_1 = \frac{(q_1+300-1)!}{q_1!(300-1)!}$	$\Omega_2 = \frac{(q_2+200-1)!}{q_2!(200-1)!}$	**Total # of Ways** $\Omega_1\Omega_2$
0	100	1	2.772 E+81	2.772 E+81
1	99	300	9.271 E+80	2.781 E+83
2	98	4.515 E+04	3.080 E+80	1.391 E+85
3	97	4.545 E+06	1.016 E+80	4.619 E+86
4	96	3.443 E+08	3.331 E+79	1.147 E+88
	…	…	…	…

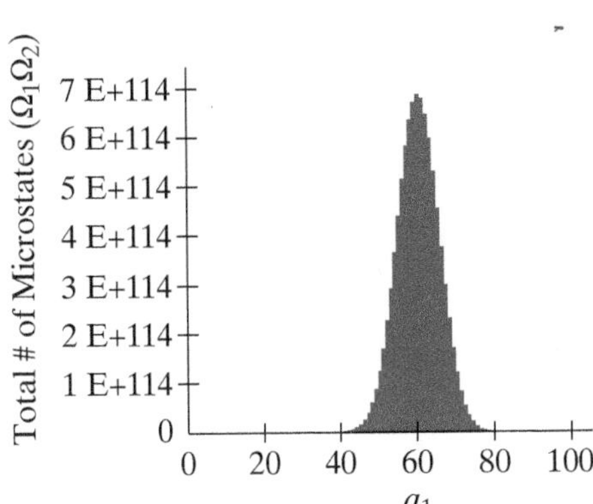

Figure 12.22 The number of ways of distributing 100 quanta of vibrational energy between two blocks having 300 and 200 oscillators, respectively. q_1 is the number of quanta in the first (larger) block.

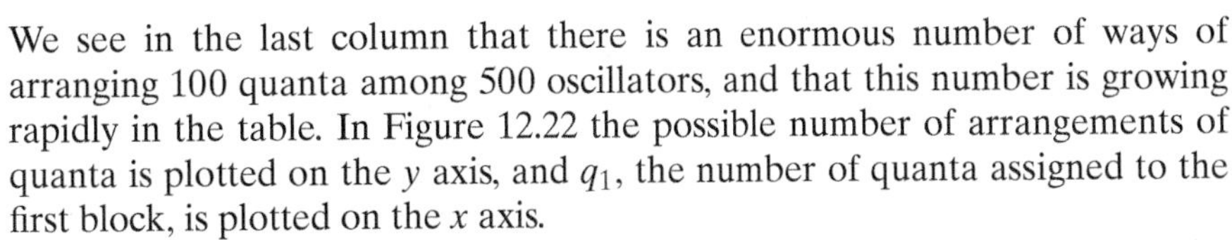

We see in the last column that there is an enormous number of ways of arranging 100 quanta among 500 oscillators, and that this number is growing rapidly in the table. In Figure 12.22 the possible number of arrangements of quanta is plotted on the *y* axis, and q_1, the number of quanta assigned to the first block, is plotted on the *x* axis.

The most probable arrangement (indicated by the highest point of the peak on the graph) is that 60 of the 100 quanta will be found in the first block, which contains 300 oscillators out of the total of 500 oscillators. This seems reasonable, since it does seem most probable that the energy would be distributed uniformly throughout the two blocks (since they're made of the same material), and that would give 3/5 (60%) of the energy to the first block. It is gratifying that our statistical analysis leads to this plausible result.

It appears from the width of the peak shown in the graph that we shouldn't be too surprised if occasionally we would find that the first block contains anywhere between 40 to 80 of the 100 quanta, but it appears that it is very unlikely to find fewer than 40 or more than 80 of the quanta in the first block.

Relatively speaking, how likely is it for *none* of the energy to be in the first block? In that case, $q_1 = 0$, and from the table we see that the number of ways to arrange the 100 quanta this way is 2.772 E+81 (2.772×10^{81}). That's an awfully big number, but how big is it compared to the most probable arrangement, where $q_1 = 60$, for which there are about 7×10^{114} ways according to the graph? Evidently it is less likely by about a factor of 10^{33} (!) to find the energy split 0-100 rather than 60-40. Is it possible according to the laws of physics for none of the energy to be in the first block? Yes. Is it likely that we would ever observe such an unusual distribution? Most emphatically not!

Note that the number of ways to arrange the quanta 0-100 (2.772×10^{81}) isn't actually visible on the graph, because it is 1×10^{33} times smaller than the peak of the graph. Values of q_1 outside the 40 to 80 range are invisible on the graph, because they are relatively so very small compared to the peak.

At this point it is very useful to do Problem P61, in which these ideas are made very concrete.

Width of the Distribution

Compared to the graph in Figure 12.16 of our earliest calculation (6 oscillators and 4 quanta), the peak shown in Figure 12.22 occupies a much narrower range

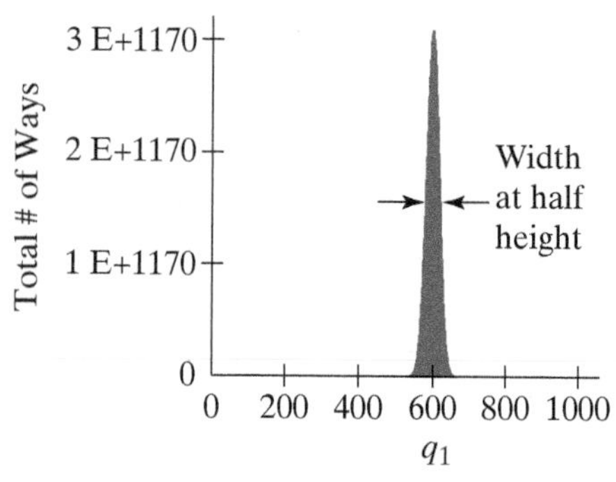

Figure 12.23 The number of ways of distributing 1000 quanta between two blocks containing 3000 and 2000 oscillators, respectively.

of the graph, reflecting the sharply decreased relative probability of observing extreme distributions of the energy.

For further comparison, Figure 12.23 is a graph of the number of ways of distributing 1000 quanta among two blocks containing 3000 and 2000 oscillators. This peak is much narrower than the peak in the previous graph (where there were only 100 quanta distributed among only 300 and 200 oscillators). This is a general trend. The larger the number of quanta and oscillators, the narrower is the peak around the most probable distribution.

The fractional width of the peak is the width of the peak at half height divided by the value of q_1 that gives the maximum probability (in the present case, the width divided by 600). It can be shown that the fractional width of the peak is proportional to $1/\sqrt{q}$ or to $1/\sqrt{N}$, whichever is larger. That is, if you quadruple the number of quanta or the number of oscillators, the fractional width of the peak decreases by a factor of 2.

> QUESTION Consider two blocks that are of ordinary macroscopic size, containing 3×10^{23} atoms and 2×10^{23} atoms, and many quanta per atom. Qualitatively, what would you expect about the width of the peak, if you could calculate it? How likely is it that you would ever observe a significant fluctuation away from the most probable 60-40 split?

You would expect the peak to be extremely narrow, and hence the probability of significant fluctuations would be very low.

The Most Probable Is the Only Real Possibility

These considerations show that in the world of macroscopic objects such as ordinary-sized blocks, the most probable arrangement is essentially the *only* arrangement that is ever observed. That is why you do not see a block suddenly leap up from the table when all of the thermal energy of the table floods into the block. On the other hand, at the microscopic level we should not be surprised if the energy of one of the atoms in a block varies a lot, since the most probable distribution includes many different arrangements of the quanta, with varying numbers of quanta on one particular atom.

Entropy and Equilibrium

We now have a good statistical description of the thermal equilibrium of two blocks in contact. At equilibrium, energy is distributed between the two blocks in the most probable manner, based on having the largest number of ways of achieving this distribution (largest number of microstates for the macrostate of given total energy). We have been studying very small systems consisting of a hundred or more atoms, and we found that in small systems there is some significant probability of finding the energy distributed somewhat differently than the most probable 60-40 division. (Objects consisting of only a few hundred atoms are called "nanoparticles" and are currently the subject of intense research, because many of their properties and behaviors are intermediate between those of atoms and those of large-scale objects.) For large macroscopic objects containing 10^{20} atoms or more, the most probable distribution of the energy is essentially the only energy distribution we will ever observe, because the probability of distributions that are only slightly different is very small.

Next we will study more deeply the details of why a particular thermal equilibrium becomes established. Suppose for example that the first block (a nanoparticle with 300 oscillators, or 100 atoms) starts out with 90 quanta and the second (200 oscillators, or about 67 atoms) with only 10 quanta (Figure 12.24). When we put them together, we expect this to shift toward a

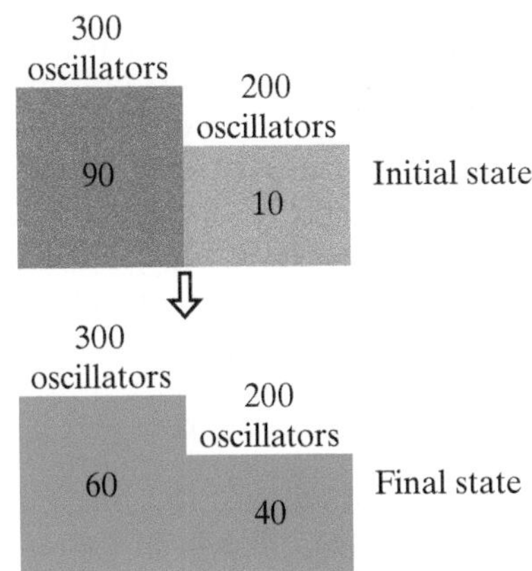

Figure 12.24 If the initial energy distribution between two systems in thermal contact is not the most probable energy distribution, energy will be exchanged until the most probable distribution is reached.

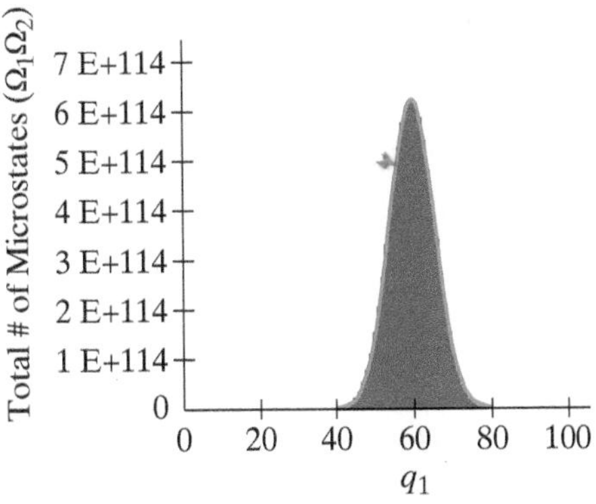

Figure 12.25 Ways of distributing 100 quanta of vibrational energy between a system of 300 oscillators and a system of 200 oscillators.

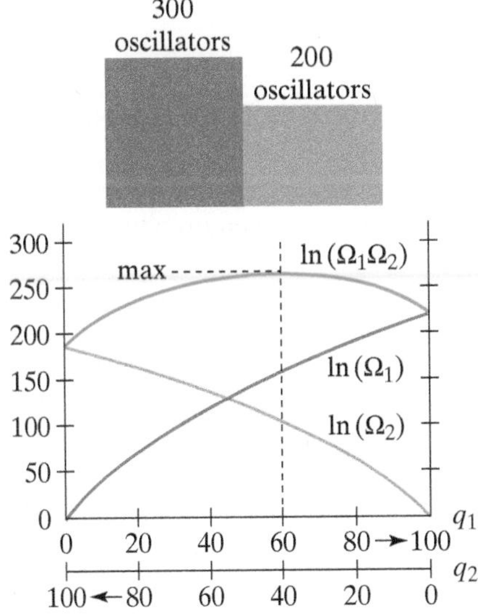

Figure 12.26 A logarithmic plot of the ways of distributing 100 quanta of vibrational energy between two systems, plotted against q_1, the number of quanta in object 1.

60-40 distribution (Figure 12.25). Studying the details of why this shift occurs will lead us to a deep understanding of the concepts of temperature and something called "entropy."

Since we start from a nonequilibrium energy distribution, it would be nice if we could make some kind of graph where we could see the total number of ways to arrange the quanta even when we are far from equilibrium. As we have seen, the graph of the total number of microstates is so strongly peaked that we don't see anything outside the peak. For example, on the graph in Figure 12.25 where we can see 1×10^{114} we can't see the relatively much smaller value of 1×10^{81}. A way to get around this problem is to plot the logarithm of the data, which makes the data visible across the whole range of energy distributions.

In statistical mechanics it is standard practice to use the base-*e* or "natural" logarithm ("ln"). For the case of distributing 100 quanta among 300 and 200 one-dimensional oscillators, which we studied before, Figure 12.26 shows a plot of the natural logarithms of the number of ways Ω_1 to arrange q_1 quanta in block 1, the number of ways Ω_2 to arrange $q_2 = (100 - q_1)$ quanta in block 2, and the total number of ways $\Omega_1\Omega_2$ to arrange the 100 quanta among the 500 oscillators. These quantities are plotted against q_1, of course running left to right, 0 to 100. You can think of q_2 as running right to left.

Because we are considering discrete microstates, each of these curves is really a set of 100 dots, but we connect the dots and make continuous curves. This is particularly appropriate when we deal with macroscopic objects, where an increment of one quantum along the *x* axis is practically an infinitesimal fraction of the axis.

By taking (natural) logarithms, we have converted a highly-peaked graph into a slowly varying one. To make sure you understand how this graph is related to the peaked graph, do the following:

> **Checkpoint 5** For this situation, we calculated earlier that there is just one way to arrange 0 quanta in block 1, and 2.772×10^{81} ways to arrange 100 quanta in block 2. Take the natural logarithm of this number and verify that all three curves on the graph make sense to you at $q_1 = 0$ ($q_2 = 100$). Note that at the right end of the graph, at $q_1 = 100$, the curve for $\ln(\Omega_1)$ goes higher than $\ln(2.772 \times 10^{81})$, because block 1 has a larger number of oscillators among which to distribute the 100 quanta (300 oscillators compared with 200).

Definition of Entropy

We will deal repeatedly with the natural logarithm of the number of ways to arrange energy among a group of atoms (the number of microstates corresponding to a particular macrostate of specified energy). This quantity, $\ln(\Omega)$, when multiplied by the Boltzmann constant k_B, is called the "entropy" of the object and is denoted by the letter "S" (the triple equal sign means "is defined as"):

DEFINITION OF ENTROPY *S*

$$S = k_B \ln \Omega$$

The Boltzmann constant is $k_B = 1.38 \times 10^{-23}$ J/K. (This is a simplified definition that is adequate for our purposes.)

Because the Boltzmann constant is $k_B = 1.38 \times 10^{-23}$ J/K, and the logarithm has no units, entropy has units of joules per kelvin. The Boltzmann constant is included in the definition for consistency with an older, macroscopically based definition of entropy.

Since $k_B \ln(\Omega_1\Omega_2) = k_B \ln(\Omega_1) + k_B \ln(\Omega_2)$ (this is a property of logarithms), the entropy S of the two-block system is equal to the entropy S_1

of the first block plus the entropy S_2 of the second block. A consequence of defining entropy in terms of a logarithm is that we can consider entropy as describing a property of a system, and when there is more than one object in a system we get the total entropy simply by adding up the individual entropies, $S_{tot} = S_1 + S_2$, as is the case with energy ($E_{tot} = E_1 + E_2$).

Figure 12.27 is a plot of the entropies of each block, and their sum. Since both $k_B \ln(\Omega_1\Omega_2)$ and $\Omega_1\Omega_2$ go through a maximum at the same value of q_1, we can state the condition for thermal equilibrium of the two blocks, in terms of entropy:

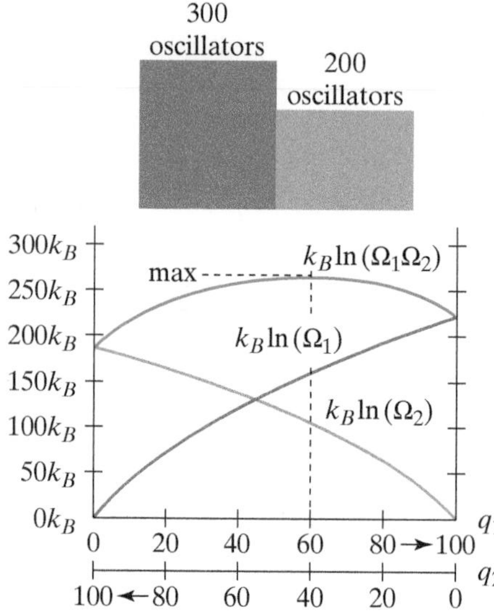

Figure 12.27 A plot of the entropies S_1, S_2, and $S_1 + S_2$ plotted against q_1, the number of quanta in object 1. Entropy is $k_B \ln \Omega$.

CONDITION FOR THERMAL EQUILIBRIUM

In equilibrium the most probable energy distribution is the one that maximizes the total entropy $S_{tot} = S_1 + S_2$ of two objects in thermal contact.

QUESTION In Figure 12.27, what is the physical significance of the fact that the top curve goes through a maximum for $q_1 = 60$?

The most probable energy distribution is a 60/40 sharing, corresponding to the 300/200 sizes of the two objects.

QUESTION In Figure 12.27, what is the physical significance of the point where the two lower curves cross each other?

Nothing! This point isn't anything special. It is the point where the entropy in one object is equal to the entropy in the other object (and equal to half the entropy of the combined system). It turns out that this doesn't have any real physical significance.

At this point it is very useful to do Problem P62, to see in detail how to produce Figure 12.26 or Figure 12.27.

12.4 THE SECOND LAW OF THERMODYNAMICS

The "first law of thermodynamics" is another name for the Energy Principle $\Delta E = W + Q$, which by now should be very familiar to you. The "second law of thermodynamics" is, however, something new. It can be stated in a number of equivalent forms, but there is a particularly useful formulation in terms of entropy:

THE SECOND LAW OF THERMODYNAMICS

If a closed system is not in equilibrium, the most probable consequence is that the entropy of the system will increase.

In other words, a closed system will tend toward maximum entropy.

As a specific example of the second law of thermodynamics, consider our two blocks. The most probable energy distribution is the one for which the total entropy is a maximum, and if initially the energy distribution is something else, it is highly likely that the entropy will increase. For nanoparticles there can be significant fluctuations away from this state of maximum entropy (with accompanying decrease in total entropy), but for ordinary-sized systems these fluctuations are extremely small, and for practical purposes the entropy of a closed macroscopic system never decreases.

Loosely, one can restate the second law of thermodynamics like this: "A closed system tends toward increasing disorder." This is overly vague without precise definitions of "order" and "disorder," and it is the definition of entropy as $k_B \ln \Omega$ that provides the needed precision. A closed system tends toward increasing entropy, which is a measure of the number of microstates corresponding to a particular macrostate of given energy.

Irreversibility

Any process in which the entropy of the Universe increases is "irreversible" in the technical, physics sense: a reversed movie of the process looks odd. However, any process in which the entropy of the Universe doesn't change is in principle "reversible," and there do exist processes that are approximately or nearly reversible. For example, a steel ball bearing that bounces vertically on a steel plate rebounds almost to the same height from which it was dropped, and this represents a (nearly) reversible process, because the system returns (nearly) to its original state.

However, if we watch for a while, the ball bearing returns to lower and lower heights and eventually settles down to rest on the plate. If the ball bearing kept returning to its original height, this would not be a violation of energy conservation. It is the second law of thermodynamics that says that we cannot expect the process to be completely reversible. There are a larger number of ways to share the energy between the ball and the plate than the number of ways for the ball to keep all the energy.

Suppose we make a movie of the ball bearing bouncing lower and lower and coming to rest on the plate. Then we run it backwards, and our friends see the ball starting to bounce, and bouncing higher and higher (Figure 12.28). This needn't violate energy conservation: energy in the plate could be flowing into the ball. This reversed movie certainly looks odd, though, presumably because we have an instinctive sense, based on lots of experience, that the entropy of the Universe increases rather than decreases.

Time running forward

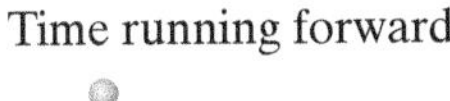

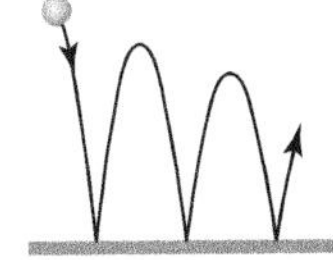

Time running in reverse

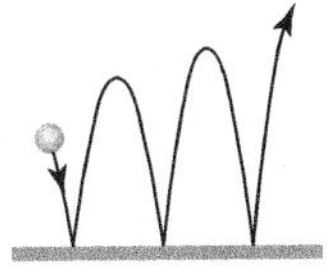

Figure 12.28 If the height of a bouncing ball increased with time instead of decreasing with time, this would not necessarily violate conservation of energy.

REVERSIBLE AND IRREVERSIBLE PROCESSES

Reversible process: $\Delta S_{\text{sys}} + \Delta S_{\text{surroundings}} = 0$

Irreversible process: $\Delta S_{\text{sys}} + \Delta S_{\text{surroundings}} > 0$

Checkpoint 6 You see a movie in which a shallow puddle of water coalesces into a perfectly cubical ice cube. How do you know the movie is being played backwards? Otherwise, what physical principle would be violated?

12.5 WHAT IS TEMPERATURE?

Until now, we have loosely associated temperature with the average energy of a molecule. The concept of entropy makes possible a deeper connection between our macroscopic measurements of temperature and a fundamental, atomic, statistical view of matter and energy. We will develop a statistically based definition of temperature.

Consider again our two blocks. On the graph in Figure 12.29 we plot entropy $S = k_B \ln \Omega$. Remember that we are approximating a set of dots by a smooth curve, since the dots are very close together (tiny energy spacing). At the maximum of the total entropy $S = S_1 + S_2$, the total entropy curve is horizontal (slope = 0), and the following relationship is true at equilibrium:

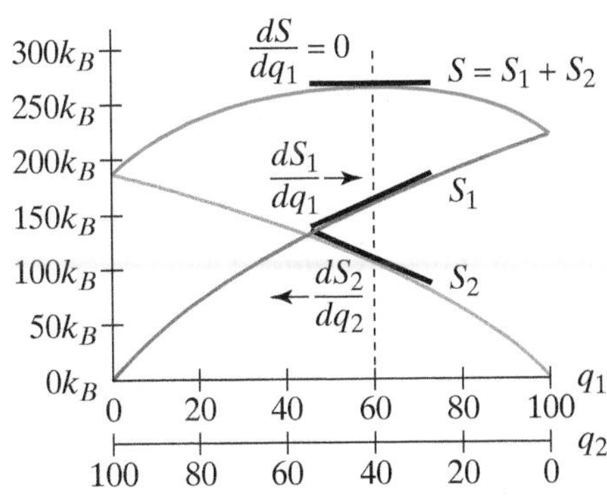

Figure 12.29 Entropy vs. number of quanta of energy in system 1.

$$\frac{dS}{dq_1} = \frac{dS_1}{dq_1} + \frac{dS_2}{dq_1} = 0$$

Since $q_2 = (100 - q_1)$, this leads to

$$\frac{dS_1}{dq_1} - \frac{dS_2}{dq_2} = 0 \quad \text{(at equilibrium)}$$

dS_2/dq_2 is the slope moving from right to left and is positive, since q_2 and S_2 increase from right to left. Therefore we see that these slopes are the same at equilibrium:

$$\frac{dS_1}{dq_1} = \frac{dS_2}{dq_2} \quad \text{(at equilibrium; maximum total entropy)}$$

dS_1/dq_1 is a measure of the state of the first block, and dS_2/dq_2 is a measure of the state of the second block.

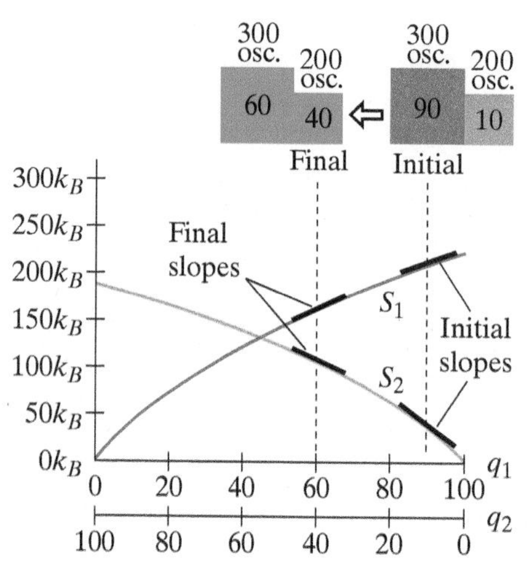

Figure 12.30 Location of the initial and final states of the two-block system on a plot of entropy vs. number of energy quanta in system 1.

QUESTION What physical property can these derivatives represent? What physical property of the blocks is the same when the two blocks reach thermal equilibrium?

The blocks reach the same temperature. It must be that the derivative of the entropy with respect to the energy is somehow related to temperature.

To see what this relationship is, suppose that when the two blocks were initially brought into contact, the first block contained $q_1 = 90$ quanta and the second block only $q_2 = 100 - 90 = 10$ quanta, as illustrated in Figure 12.30. Note that in Figure 12.30 the initial slope of the entropy curve for block 2 is steeper than the initial slope of the entropy curve for block 1. (Remember: dS_2/dq_2 is the slope moving from right to left and is positive, since q_2 and S_2 increase from right to left.) Therefore, if we remove one quantum of energy from block 1 and give it to block 2, we'll increase the entropy in block 2 more than we'll decrease the entropy in block 1.

QUESTION Will this result in a net increase or a net decrease in the total entropy S? (See Figure 12.30.)

Since the entropy of block 2 increases more than the entropy of block 1 decreases, there is a net increase of the entropy of the two blocks together. We already established that the state of the two blocks will evolve toward greater total entropy, not less total entropy.

QUESTION Consider which block gives up energy, and which block takes it up. Which block is initially at a higher temperature?

Evidently block 1 must be at the higher temperature, because on average it will give energy to block 2 in order that the total entropy increase.

QUESTION Does a steeper or a less steep slope of entropy vs. energy correspond to higher temperature?

Initially, block 1 has the higher temperature, and S_1 vs. q_1 has the smaller slope. Initially, block 2 has the lower temperature, and S_2 vs. q_2 has the larger slope.

QUESTION Therefore, which of the following looks like a better guess for relating temperature to entropy: $T_1 \propto dS_1/dq_1$ or $1/T_1 \propto dS_1/dq_1$?

Since block 1 has the higher temperature and the smaller slope (smaller dS/dq), it seems possible that $1/T_1 \propto dS_1/dq_1$.

Definition of Temperature

For these reasons, we *define* temperature in terms of dS/dq. However, the magnitude of one quantum of energy varies for different systems, so we can't compare different systems if we use q. So instead of q we use E_{int}, the internal energy above the ground state of a system measured in joules. For a group of oscillators, E_{int} is the number of quanta q times the energy per quantum: $E_{\text{int}} = q\hbar\sqrt{k_s/m}$ for a system of harmonic oscillators, where k_s is the spring stiffness.

DEFINITION OF TEMPERATURE T

$$\frac{1}{T} = \frac{dS}{dE_{int}}$$

The Boltzmann constant k_B in the definition $S = k_B \ln \Omega$ makes the units come out right. With the internal energy E_{int} of a system measured in joules (J), and entropy S measured in joules per kelvin (J/K), temperature T is measured in kelvins (K). (Remember that we're approximating a set of closely spaced dots by a curve, so taking a derivative makes sense.)

This is a highly sophisticated and abstract way of defining temperature. How does this relate to the temperature that is measured by an ordinary thermometer? In Supplement 1 we show that the temperature defined by $1/T = dS/dE_{int}$ is the same as the "absolute" temperature, which appears for example in the ideal gas law $PV = N_{moles}RT$, which you have probably encountered in previous chemistry or physics studies. On this temperature scale, ice melts at a temperature of +273.15 K (0 °C), and water boils at +373.15 K (100 °C). To put it another way, absolute zero Kelvin is at −273.15 °C. For our purposes, it will be adequate to say ice melts at 273 K. Room temperature is about 20 °C or about 293 K.

To recapitulate, the greater the dependence of the entropy on the energy for an object (the steeper the slope of S vs. E_{int}), the more "eager" the object is to take in energy to contribute to increasing the total entropy of the total system. It will do this if it can take energy from another object that has a smaller dependence of entropy on energy (smaller slope), since the second object's entropy will decrease less than the first object's entropy will increase. An object with a large dS/dE_{int} has a low temperature, since it is low-temperature objects that take in energy from high-temperature objects.

Conversely, the smaller the dependence of the entropy on the energy, the less reluctant the object is to give up energy and decrease its own entropy, if it can give the energy to an object with a greater dependence of entropy on energy. An object with a small dS/dE_{int} has a high temperature, since it is high-temperature objects that donate energy to low-temperature objects.

At this point it is useful to do Problem P63, in which you calculate the temperature of an object as a function of the amount of energy in the object.

W, Q, and Entropy

More formally, we should write $1/T = \partial S/\partial E_{int}$, involving a "partial derivative," which means that we hold everything but S and E constant when we take the derivative. In particular, we hold the volume constant, which means that we do no work on the system to compress it or stretch it, so we should write $1/T = (\partial S/\partial E_{int})_V$, where the subscript V means "take the (partial) derivative holding the volume constant."

One might ask whether there is any entropy change of the surroundings associated with energy exchange in the form of work. The answer turns out to be no for processes that are reversible (there do exist irreversible processes in which work can be associated with an increase in entropy). Energy transfer caused by a temperature difference between objects in contact, Q, is "disorganized" energy transfer and is associated with entropy change. Work is "organized" energy transfer and in many situations doesn't affect the entropy. It is beyond the scope of this textbook to prove this rigorously, but we can illustrate the basic issues in terms of an Einstein solid.

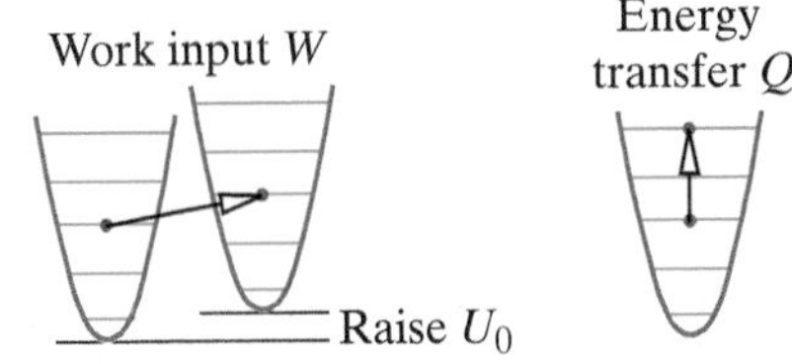

Figure 12.31 Mechanical work (compression or stretching of an oscillator system) compared to energy transfer due to a temperature difference to the same system.

If you mechanically squeeze or compress a block, doing work on it to raise its energy, it can be shown that you shift the energy levels upward without changing which state an oscillator is in, as shown in Figure 12.31. The "spring" stiffnesses don't change, but there is more energy stored in the "springs." On the other hand, if there is transfer of energy into the system due to a temperature difference, Q, there is no change in the energy level, but an oscillator will jump to a higher level. This jump from one level to another affects the probability calculations and the entropy, which is the natural log of the number of ways to arrange the energy quanta.

Roughly speaking, transfer of energy Q due to a temperature difference alters which state the system is in and affects the entropy. Work W alters the energy levels without changing which level the system is in, and doesn't affect the entropy.

Transfer of energy into a system due to a temperature difference, of amount Q, with $\Delta E_{\text{int}} = Q$, leads to an entropy change of that system. Since $1/T = \Delta S/\Delta E_{\text{int}} = \Delta S/Q$, we conclude that the entropy change associated with an input Q is $\Delta S = Q/T$ in the case that T doesn't change very much. If Q is large, causing a large temperature change in a system, we must add up (integrate) the contributions to the entropy change.

To repeat, there do exist irreversible processes in which work can be associated with an increase in entropy (for example, in rubbing two blocks together, with friction), so entropy change due to Q is only part of the story.

Checkpoint 7 There was transfer of energy of 5000 J into a system due to a temperature difference, and the entropy increased by 10 J/K. What was the approximate temperature of the system, assuming that the temperature didn't change very much?

Can the Entropy of an Object Decrease?

QUESTION We analyzed two blocks that initially had a 90-10 energy distribution. In the process of changing to a 60-40 distribution, was it possible for the entropy of *one* of the blocks to decrease significantly? Is this a violation of the second law of thermodynamics?

The entropy of block 1 decreased, but the entropy of block 2 increased even more. We believe that the Universe is a closed system, and that the total entropy of the Universe continually increases and never decreases. However, some portions of the Universe may experience a decrease of entropy, as long as there is at least as much increase elsewhere.

Irreversibility Again

We started with a 90-10 distribution of the energy but found the most probable final distribution to be 60-40. This change is effectively irreversible. The 60-40 distribution is so enormously more probable than the 90-10 arrangement that an observer will essentially *never* see a 90-10 distribution spontaneously recur. Closed macroscopic systems (such as two blocks inside an insulating box) always evolve toward the most probable arrangement, and stay there (or so close that you can hardly tell the difference).

12.6 SPECIFIC HEAT OF A SOLID

In Problem P63 you calculated the relationship between the temperature of a block and its energy. How can we compare these calculations with experiment, when we don't have a good way to measure the total energy in a solid block? It is, however, possible to measure *changes* in energy and temperature. Suppose that we add a known amount of energy ΔE to a system, and we measure the resulting rise in temperature ΔT. The ratio of the internal energy change to the temperature rise of an object is called the heat capacity of that object. If we express this on a per-gram or per-mole or per-atom basis, it is called "specific heat." We will work with specific heat on a per-atom basis:

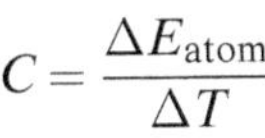

$$C = \frac{\Delta E_{atom}}{\Delta T}$$

where

$$\Delta E_{atom} = \frac{\Delta E_{system}}{N_{atoms}}$$

Different materials have different specific heats. For example, at room temperature the specific heat of water (on a per-gram basis) is 4.2 J/K/g, while the specific heat of iron at room temperature is 0.84 J/K/g. Specific heat is an important property of a material for two reasons. It has practical consequences in science and engineering because it, along with the thermal conductivity, determines the thermal interactions that the material has with other objects. Also, specific heat is an experimentally measurable quantity that we can compare with theoretical calculations, to test the validity of our models.

How would we measure the heat capacity of a block? One way is to enclose the block inside an insulating box and use an electric heater inside the box to raise the temperature of the block (Figure 12.32). The electric power (in watts) times the amount of time the current runs (in seconds) gives us the input energy ΔE (in joules), which is equal to the internal energy change ΔE_{int}, and we can use a thermometer to measure the temperature rise ΔT of the block. The specific heat (on a per-atom basis) is $\Delta E_{atom}/\Delta T = (\Delta E_{block}/\Delta T)/N_{atoms}$. The heater itself should have a small mass compared to the sample of material whose heat capacity we're measuring, so that the heater isn't a significant part of the material that is being warmed up. Problem P32 illustrates another way to measure specific heat, on a per-gram basis.

Figure 12.32 Schematic diagram of an apparatus for measuring the heat capacity of a solid.

Figure 12.33 Measured heat capacities of lead and aluminum, shown on a per-atom basis.

Experimental Results for Specific Heat

Measurements of the specific heat of aluminum (circles) and lead (squares) are displayed in Figure 12.33 (the numerical data are given in a table accompanying Problem P64). They are plotted as heat capacity (J/K) per atom to facilitate comparison. Note that at low temperatures the specific heat of both substances changes dramatically with temperature. It is not initially obvious why this should be the case, and in fact this temperature dependence was not predicted by classical theory developed in the 1800s. Note also that at room temperature and above, the specific heat (on a per-atom basis) of both materials is about the same.

Will our simple statistical model of a solid be able to predict the temperature dependence of the specific heat of a solid? In Problem P63 you determined the relationship between energy and temperature for a block (in fact, you did the calculation for two different blocks). This calculation can easily be extended to predict the heat capacity per atom, which is $\Delta E_{atom}/\Delta T$, where E_{atom} is the average energy per atom.

Effective Spring Stiffness

In Chapter 4 we determined the interatomic spring stiffness $k_{s,i}$ of aluminum and lead by relating it to the macroscopic stress–strain relationship (Young's modulus). In Problem P64 you will use the Einstein independent-oscillator model of a solid to predict the specific heat of aluminum and lead. We need to use a value for $k_{s,i}$ in this model, in order to convert the number of quanta to a value for energy in joules. Presumably the spring stiffness $k_{s,i}$ that we use in $\hbar\sqrt{k_{s,i}/m_a}$ should be related to the spring stiffness $k_{s,i}$ determined from Young's modulus. However, we shouldn't expect our

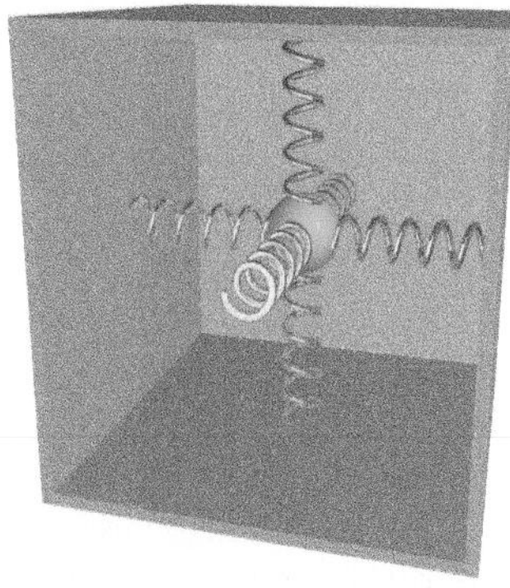

Figure 12.34 In our simple model of a solid, each atom experiences forces from two interatomic "springs." Each of these springs is half as long as an interatomic bond.

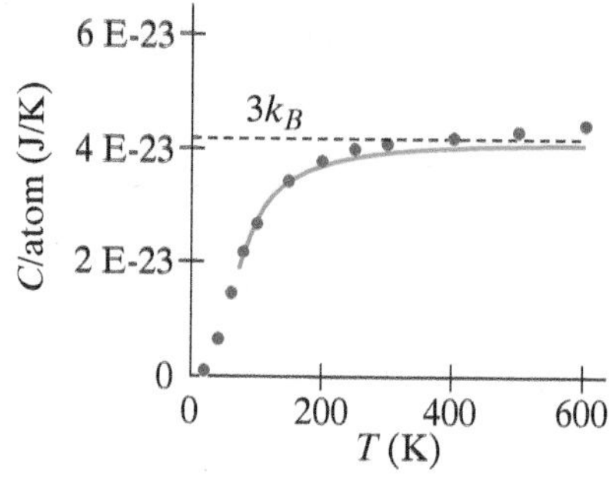

Figure 12.35 Computed specific heat of copper, compared with experimental values.

predictions to agree exactly with experimental data, because our model ignores the fact that the atoms are not actually isolated from each other.

Moreover, the effective spring stiffness for oscillations in the x, y, or z directions can be expected to be larger than the value of $k_{s,i}$ estimated in Chapter 4 for two reasons.

First, each atom is attached to two "springs" (Figure 12.34). When an atomic core is displaced a distance s_x from its equilibrium position, the spring to the left and the spring to the right each exert a force of $k_{s,i}s_x$, so the combined force is $2k_{s,i}s_x$. This implies that the "effective spring stiffness" for oscillations would be $2k_{s,i}$.

Second, in this simple model we divide the solid into "cubes" surrounding each independent atomic oscillator (Figure 12.34), so each of the two springs to the left and right of an atomic core is effectively half the length of a full interatomic spring, and each half-spring would have a spring stiffness of $2k_{s,i}$.

These considerations suggest that the effective spring stiffness for thermal oscillations in the context of the Einstein model might be a factor of 4 larger than the spring stiffness we determined in Chapter 4 from stretching a wire.

We emphasize that the Einstein model is clearly deficient, since an atom is not actually enclosed in rigid walls. However, having chosen this model we follow through in a consistent way, including using a spring stiffness k_s appropriate to the model. There are more sophisticated models for determining specific heat that fit experimental data even better than the Einstein model does, but the Einstein model captures the main features.

Specific Heat as a Function of Temperature

Figure 12.35 displays the result of a calculation of the specific heat of copper on a per-atom basis. The colored dots represent actual experimental data, and the solid line represents the computed values. An outline of the calculation that produced this curve is as follows:

- Use the mass of a copper atom and the effective interatomic spring stiffness for copper, calculated in Chapter 4 and modified by the factor discussed above, to find the value of one quantum of vibrational energy for this system.
- For a small particle of solid copper containing a given number of atoms, calculate the entropy of the system for increasing values of energy added to the system, in amounts of one quantum (as in Problem P62).
- Calculate the change in entropy due to the addition of each quantum of energy, and use the definition of temperature ($1/T \approx \Delta S/\Delta E$) to find the temperature of the particle for each value of energy (as in Problem P63).
- Calculate the change in temperature of the particle due to the addition of each quantum of energy, to find the specific heat ($C \approx \Delta E/\Delta T$), and plot values of C (per atom) versus T (in kelvins) as in Problem P64.

Our simple model of a solid as a collection of independent harmonic oscillators does a surprisingly good job of fitting experimental data over a wide range of temperatures. If we were able to include more oscillators in our calculations (by using double-precision arithmetic to handle larger numbers, or by using mathematical techniques such as Stirling's approximation for factorials), we could extend our predictions to lower temperatures.

The deviation of the experimental data from the prediction at very high temperatures suggests that at these temperatures a simple harmonic oscillator is not a good model of the atoms in this solid. At high quantum levels, the harmonic oscillator potential energy curve is a poor approximation to the actual potential energy curve describing interatomic interactions (see Chapter 7).

Note an interesting aspect of the graph: at room temperatures, the predicted specific heat on a per-atom basis approaches a constant value of $3k_B$ (three times the Boltzmann constant, 1.38×10^{-23} J/K). This value agrees quite well with the measured specific heat of a variety of substances at ordinary temperatures.

At this point it is important to do Problem P64, in which you compute the specific heat of aluminum and lead, and compare with actual experimental data.

Energy Quantization and Specific Heat

The key difference between our model of solids (the Einstein model) and earlier classical models that did not predict a temperature dependence of specific heat lies in the discrete energy levels of atomic oscillators. Statistical mechanics was originally developed in the 1800s, before the beginning of quantum theory. The classical theory did predict that at high temperatures the energy of each atom in a solid would be $3k_BT$ (k_BT per nonquantized one-dimensional oscillator), so that the heat capacity per atom would be $3k_B$, and this is discussed later in this chapter.

If you examine your calculations, you'll find that the high-temperature limit (heat capacity per atom $= 3k_B$) corresponds to temperatures high enough that k_BT is significantly larger than one quantum of energy. In a situation where the average energy per one-dimensional oscillator is about k_BT and is large compared to one quantum of energy, the quantization of the energy doesn't make much difference in the analysis. That is, if energy quanta are small compared to the energies of interest, mathematical analysis can be carried out adequately by considering energy to be continuous rather than discrete. In such cases pre-quantum and quantum calculations will give the same results, as is the case here at high temperatures.

On the other hand, at low temperatures the average energy per one-dimensional oscillator is comparable to or smaller than one quantum of energy, and the continuous, nonquantum calculations are not valid. The classical theory provided no explanation for the discovery that as materials were cooled down to very low temperatures, the specific heat decreased with decreasing temperature.

In 1907 Einstein carried through the analysis we have just done and predicted the curve we have just plotted. The good agreement with both low-temperature *and* high-temperature measurements of specific heat was strong additional evidence for the hypothesis that the energy levels of oscillators are indeed discrete. (At extremely low temperatures the model of a solid must be refined, taking into account the electrons in the metal, among other things, to achieve full agreement between theory and experiment.)

The fact that the specific heat for all materials decreases at low temperatures has practical consequences. For example, it makes it difficult to cool a sample to a very low temperature. Cooling a sample depends essentially on putting the sample in contact with a "sink," a large object that is already at a lower temperature, so that there is transfer of energy Q due to the temperature difference out of the sample into the sink, lowering the temperature of the sample and not raising the temperature of the sink very much. At very low temperatures, however, the sink has a low specific heat, so this is difficult to achieve.

Which of Our Results Are General?

We have analyzed simple models of solid matter. Nevertheless, the basic conclusions are quite general. For example, if our two model blocks were made of different materials, so that the energy quanta were of different size in the two blocks, this would complicate the procedures for evaluating the number of ways Ω to arrange the energy, but the basic conclusion would remain—that the entropy will increase to a maximum.

EXAMPLE **A Lead Nanoparticle**

In Chapter 4, from Young's modulus for lead we found that the effective interatomic "spring" stiffness was about 5 N/m. **(a)** For a nanoparticle consisting of 3 lead atoms, what is the approximate temperature when there are 5 quanta of energy in the nanoparticle? **(b)** What is the approximate specific heat (per atom) at this temperature? Compare with the approximate high-temperature specific heat (per atom) for lead.

Solution We'll use the Einstein model of a solid, though with only three atoms our conclusions will be very approximate.

(a) We model the three atoms as 9 independent quantized oscillators. One quantum of energy in one of these oscillators is this many joules:

$$\hbar\omega_0 = \hbar\sqrt{k_s/m} = (1.05 \times 10^{-34}\,\mathrm{J\cdot s})\sqrt{\frac{4(5\,\mathrm{N/m})}{(207 \times 1.7 \times 10^{-27}\,\mathrm{kg})}}$$

$$= 7.92 \times 10^{-22}\,\mathrm{J}$$

q	Ω	$\ln\Omega$
4	$\frac{12!}{4!8!} = 495$	6.20
5	$\frac{13!}{5!8!} = 1287$	7.16
6	$\frac{14!}{6!8!} = 3003$	8.01

Figure 12.36 Calculation of entropy ($S = k_B \ln\Omega$).

We need to calculate the number of ways to arrange q quanta in the neighborhood of $q = 5$, and the associated entropy (Figure 12.36).

Since $1/T = \partial S/\partial E$, $T \approx \Delta E/\Delta S$. Take differences from $q = 4$ to $q = 6$, in order to approximate the slope at $q = 5$; energy increase is 2 quanta:

$$T \approx \frac{N\hbar\omega_0}{k_B\Delta(\ln\Omega)} = \frac{2(7.92 \times 10^{-22}\,\mathrm{J})}{(1.38 \times 10^{-23}\,\mathrm{J/K})(8.01 - 6.20)} \approx 63.7\,\mathrm{K}$$

(b) Find T_1 at the midpoint of the 4 to 5 interval, T_2 at the midpoint of the 5 to 6 interval; energy increase from T_1 to T_2 is one quantum:

$$T_1 \approx \frac{(7.92 \times 10^{-22}\,\mathrm{J})}{(1.38 \times 10^{-23}\,\mathrm{J/K})(7.16 - 6.20)} \approx 60.1\,\mathrm{K}$$

$$T_2 \approx \frac{(7.92 \times 10^{-22}\,\mathrm{J})}{(1.38 \times 10^{-23}\,\mathrm{J/K})(8.01 - 7.16)} \approx 67.7\,\mathrm{K}$$

As expected, these two temperatures bracket the temperature of 63.7 K at $q = 5$. We want the specific heat on a per-atom basis, and there are 3 atoms:

$$C_{\text{per atom}} = \frac{1}{3}\frac{\Delta E}{\Delta T} \approx \frac{1}{3}\frac{(7.92 \times 10^{-22}\,\mathrm{J})}{(67.7 - 60.1)\,\mathrm{K}} \approx 3.44 \times 10^{-23}\,\mathrm{J/K}$$

The approximate high-temperature specific heat for a solid is $3k_B$ per atom, which is $3(1.38 \times 10^{-23}\,\mathrm{J/K}) = 4.14 \times 10^{-23}\,\mathrm{J/K}$. This suggests that at a temperature of about 63.7 K, lead is not quite at the high-temperature limit.

The actual experimental value for lead at 63.7 K (interpolation of the data accompanying Problem P64) is 22.6 J/K/mole, which on a per-atom basis is

$$(22.6\,\mathrm{J/K/mole})\left(\frac{1\,\mathrm{mole}}{6 \times 10^{23}\,\mathrm{atoms}}\right) = 3.8 \times 10^{-23}\,\mathrm{J/K}$$

Our three-atom calculation is not a very accurate model of a macroscopic block of lead, in part because with only three atoms it is not a very good model to say that each atom is connected by spring-like bonds to six neighboring atoms. Despite the failings of our simple model, our result is rather close to the actual value.

12.7 COMPUTATIONAL MODELS

Calculating entropy, energy, and specific heat by hand is possible but tedious. In order to be able to vary the number of atoms in the objects, their elemental composition, and the amount of energy available to the system, it is easier to use a computer.

The structure of the programs we will write to do statistical calculations is very similar to that of the programs we wrote to predict motion. Constants and initial values are set at the beginning of the program, and then calculations are done iteratively inside a while loop. The two most important differences are these:

- In models of motion, time is the variable that increases each time the loop repeats. In the statistical calculations you will do, the amount of energy in the system (or in a particular part of the system) is the variable that increases.
- In models of motion, the calculations done inside the loop involve forces and fundamental principles, and predict motion. In statistical models, calculations inside the loop involve calculating the number of ways of arranging energy among a specific number of independent quantum oscillators (three for each atom).

Combinations, Logs, and Bar Graphs in VPython

To get the `factorial` and `combin` functions, and be able to graph results, insert the following statements at the beginning of your program. These statements *must* be entered in the order shown:

```
from visual import*
from visual.graph import*
from visual.factorial import*
```

You will need to use the following functions:

- `combin()`. This function takes two arguments:

  $$\texttt{combin(a,b)} = \frac{a!}{b!(a-b)!}$$

  If we want to calculate the number of ways to arrange q quanta of energy among N oscillators, we can let $a = q+N-1$ and $b = q$. Then $(a-b) = (N-1)$, and

  $$\texttt{combin(q+N-1,q)} = \frac{(q+N-1)!}{q!(N-1)!}$$

- `log()`. In Python, the natural logarithm $ln(x)$ is written `log(x)`. The function to find the base 10 logarithm is `log10(x)`.

- `gvbars()`: To make a vertical bar graph, create a `gvbars` object. The parameter `delta` determines the width of a single bar:

  ```
  mybars = gvbars(color=color.red, delta=0.3)
  ```

  This code displays a vertical bar at location x:

  ```
  mybars.plot(pos=(x,height))
  ```

In addition, VPython does have a `factorial()` function, but the largest number whose factorial it can compute is 170. It is safer simply to use the `combin()` function, which does not have this limitation. `b = factorial(9)` will set b to $9! = 9 \cdot 8 \cdot 7 \cdot 6 \cdot 5 \cdot 4 \cdot 3 \cdot 2 \cdot 1 = 362880$. As is standard, `factorial(0)` is 1.

Computational Explorations

In Problems P61–P64 at the end of this chapter, you will apply the ideas discussed in this chapter to create computational models of nanoparticles—solid particles containing up to a few hundred atoms. Problems P61–P63 involve modeling two nanoparticles in thermal contact with each other.

The main part of such a program, referring to two nanoparticles labeled 1 and 2, might look like this:

```
Ntotal = 6 # total number of oscillators
N1 = 3 # number of oscillators in object 1
N2 = Ntotal-N1 # number of oscillators in object 2
qtotal = 4 # total quanta of energy available
q1 = 0 # initially 0 quanta in object 1

while q1 <= qtotal: # for each possible value of energy in 1
    q2 = ?? # number of quanta of energy in 2
    # Calculate number of ways to arrange q1 quanta in 1:
    ways1 = ?
    # Calculate number of ways to arrange q2 quanta in 2:
    ways2 = ?
    # Plot number of ways to arrange energy in both:
    waygraph.plot( pos=(q1,??) )
    q1 = q1+1
```

Since $\Delta S/\Delta E \approx 1/T$, to calculate one temperature we need two values of S. One way to do this inside a calculation loop is to "look ahead," like this:

```
while q1 < qtotal:
    # For each delta_S you need two values of S
    # so use q1, and q1a=(q1+1), etc.
    q2 = qtotal-q1
    q1a = q1+1
    q2a = ?
    ways1 = ?
    ways2 = ?
    ways1a = ?
    ways2a = ?
    # etc.
```

In Problem P64, you will modify your model to consider a single nanoparticle and will use the definitions of entropy, temperature, and specific heat to predict the specific heat of two different real substances (aluminum and lead) as a function of temperature. You'll compare the predictions of your model with published experimental data for these quantities.

12.8 THE BOLTZMANN DISTRIBUTION

How does the density of Earth's atmosphere change as altitude increases? How does the distribution of the speeds of molecules in a gas change as the temperature of the gas increases? So far we have mainly been concerned with the thermal equilibrium of two blocks, and how thermal equilibrium arises as that particular distribution of energy between the two blocks that has (by far) the largest number of ways to arrange the quanta. What can we say about the probability of observing a particular amount of energy associated with one particular atom or oscillator? Addressing this question will lead us to the "Boltzmann distribution," which provides insight into the behavior of a very wide variety of physical, chemical, and biological phenomena.

A Constant-Temperature Reservoir

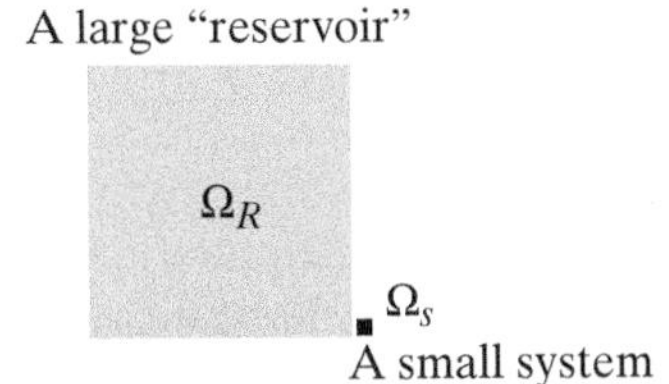

Figure 12.37 A small system in contact with a very large system (a "reservoir").

Consider a large system in contact with a small system, as shown in Figure 12.37. The two systems are isolated from their surroundings and share a fixed amount of energy E_{tot}. The large system's energy and temperature cannot change very much, so we call the large system a nearly constant-temperature "reservoir," which keeps the small system always at nearly the same temperature.

The entropy of the total system (reservoir plus small system) will increase rapidly if energy is transferred from the small system to the reservoir, because in the large system that energy can be distributed among a very much larger number of microstates than were available in the small system. Since we expect the total entropy to increase, we expect that most of the time, most of the total energy E_{tot} will be found in the reservoir, and little energy in the small system. We will show how to make this idea quantitative.

Let $\Omega_{res}(E_{res})$ be the number of microstates in the reservoir when the amount of energy in the reservoir is E_{res}. Similarly, let $\Omega(E)$ be the number of microstates in the small system when it has an amount of energy E, where E is much less than E_{res}, because the reservoir is so big. The total energy of the combined system is $E_{tot} = E_{res} + E$, which is a fixed number because the two systems are isolated from their surroundings. The total number of microstates for the combined system corresponding to E_{tot} is $\Omega_{tot}(E_{tot})$, which is also a fixed number.

Probability of a Particular Division of Energy

The number of ways of arranging E_{res} in the reservoir and E in the small system is $\Omega_{res}(E_{res})\Omega(E)$, while the total number of ways of arranging E_{tot} in the combined system is $\Omega_{tot}(E_{tot})$. Therefore the probability $P(E)$ of finding the energy split between the reservoir and the small system so that there is energy E in the small system is this:

$$P(E) = \frac{\Omega_{res}(E_{res})\Omega(E)}{\Omega_{tot}(E_{tot})}$$

The most probable value of the energy E to be found in the small system is zero, because the more energy E that is taken away from the big system and put into the small system, the fewer the microstates in the big system (without a comparable increase in the small system), which would mean a decrease in the total entropy.

We're interested in how fast $P(E)$ decreases as we move more and more of the total energy into the small system. We take logarithms and multiply by k_B to express everything in terms of entropy, then see how the expression varies with E:

$$k_B \ln P(E) = k_B \ln(\Omega_{res}(E_{res})) + k_B \ln(\Omega(E)) - k_B \ln(\Omega_{tot}(E_{tot}))$$

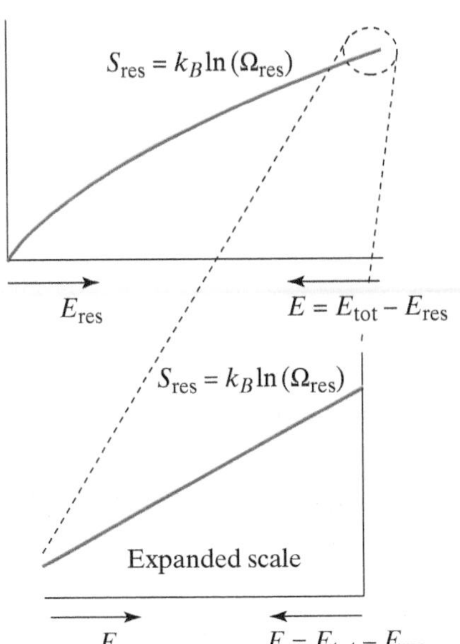

Figure 12.38 Entropy (in units of k_B) as a function of E_{res}, for the larger object.

Consider the term $k_B \ln(\Omega_{res}(E_{res}))$, which is the entropy $S_{res}(E_{res})$ of the reservoir when the energy in the reservoir is $E_{res} = E_{tot} - E$ due to some energy E having gone into the small system. In Figure 12.38 we show the familiar calculation of the entropy of the reservoir as a function of the energy in the reservoir. On an expanded scale we show a portion of a plot of the entropy vs. E_{res} near the maximum possible value of E_{res}, which is the total energy E_{tot} of the combined system.

We need to evaluate the entropy of the reservoir for small values of $E = (E_{tot} - E_{res})$. We can obtain this by the following argument. The nearly straight line on the graph has a slope dS_{res}/dE_{res} and goes through $S_{res}(E_{tot})$, where the energy in the reservoir is E_{tot}. Therefore in this region $S_{res}(E_{res})$ can be represented by

$$S_{res}(E_{res}) = S_{res}(E_{tot}) - \frac{dS_{res}}{dE_{res}}E$$

The entropy of the reservoir decreases as more energy E is shifted into the small system.

However, $dS_{res}/dE_{res} = 1/T$, where T is the temperature of the reservoir, and the fact that the slope is nearly constant in this region reflects the fact that removing a small amount of energy from this large object hardly changes its temperature. The large object is a nearly constant-temperature reservoir, which keeps the small object at that temperature.

Making the substitution $dS_{res}/dE_{res} = 1/T$, we have

$$S_{res}(E_{res}) = k_B \ln(\Omega_{res}(E_{res})) = S_{res}(E_{tot}) - \frac{E}{T}$$

Substitute this into the equation for $k_B \ln P(E)$ and then divide by k_B:

$$\ln P(E) = \frac{S_{res}(E_{tot})}{k_B} - \frac{E}{k_B T} + \ln(\Omega(E)) - \ln(\Omega_{tot}(E_{tot}))$$

Note that the terms that don't involve E are all constants because E_{tot} is constant. Take the exponential of this equation—that is, raise e to both sides of the equation:

$$e^{\ln P(E)} = e^{\text{constant}} e^{\ln(\Omega(E))} e^{-E/k_B T}$$

Since e raised to the natural logarithm of some quantity is that quantity, we have the following, where A is some constant:

$$P(E) = A\Omega(E)e^{-E/k_B T}$$

We have calculated the probability of finding a small amount of energy E in a small system that is in contact with a large reservoir. This is called the Boltzmann distribution in honor of the Austrian physicist who developed statistical mechanics in the 19th century:

THE BOLTZMANN DISTRIBUTION

The probability of finding energy E in a small system in contact with a large reservoir is proportional to

$$\Omega(E)e^{-E/k_B T}$$

The exponential part, $e^{-E/k_B T}$, is called the "Boltzmann factor." $\Omega(E)$ is the number of microstates corresponding to energy E.

This result is very general and applies to a very wide variety of phenomena. In many situations the number of microstates $\Omega(E)$ changes much more slowly with energy than does the Boltzmann factor, in which case the qualitative behavior of the system can be determined just from the exponential.

QUESTION What is the most likely value of E, the energy to be found in a microscopic system that is in thermal equilibrium with a large system, if the Boltzmann factor is the important factor? Are you likely to find an energy that is much larger than $k_B T$?

Since the exponent is negative, the exponential factor is largest for $E = 0$, so you're most likely to find the system in its ground state. This agrees with our expectation that taking energy out of the reservoir and putting it in the small system will reduce the number of total microstates and represent a decrease in entropy for the combined system, which is unlikely.

Since e^{-1} is 0.37, the probability of the energy being much larger than $k_B T$ is rather small.

If the temperature is so low that $k_B T$ is small compared to the energy E_1 of the first excited state, you are unlikely ever to find the system in one of its excited states. The system is thermally inert, because the probability of taking in *any* amount of energy is very small. In contrast, if the temperature is high enough that $k_B T > E_1$, then you will sometimes find the system in one of its excited states, although it is still true that the system is most likely to be found in its ground state (Figure 12.39).

The Boltzmann distribution has far-reaching consequences. For example, chemical and biochemical reaction rates typically depend strongly on temperature because with higher temperature the reactants are moving faster and may be found in excited states. Physical reaction rates are also affected. At the very high temperatures found in the interior of our Sun, kinetic energies are high enough to overcome the electric repulsion between nuclei and to allow the nuclei to come in contact, so that they can undergo thermonuclear fusion reactions. A gas becomes a plasma if the temperature is so high that $k_B T$ is comparable to the ionization energy.

High temperature

$k_B T$

E_1

$k_B T > E_1$
May find in
excited state

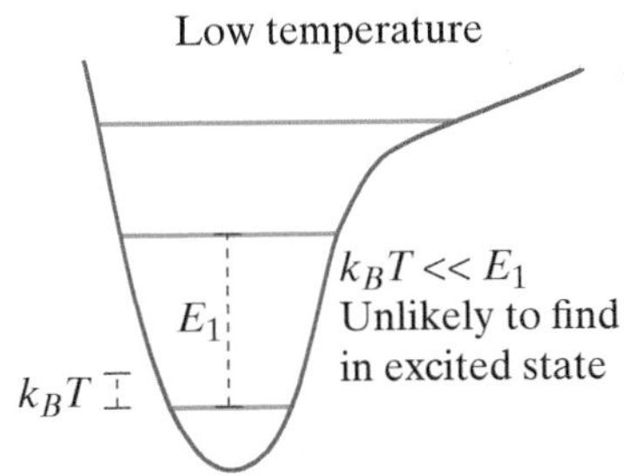

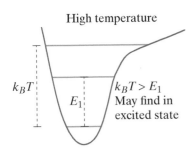

Figure 12.39 At a low temperature it is unlikely to find a system in an excited state, but at a high temperature the likelihood increases.

Checkpoint 8 At room temperature, show that $k_B T \approx \frac{1}{40}$ eV. It is useful to memorize this result, because it tells a lot about what phenomena are likely to occur at room temperature.

Many Observations vs. Many Systems

If a microscopic system (such as a single oscillator) is in contact with a large system, the probability of finding a particular amount of energy in the microscopic system is governed by the Boltzmann distribution.

One can think about this in two ways. The first is to imagine measuring the energy in this particular system repeatedly, over time, and recording the results. The fraction of the results that indicate a particular energy is predicted by the Boltzmann distribution.

A second way to think about this is to imagine assembling a large number of identical systems, each in contact with a large system, and taking one "snapshot" in which the energy of each system is recorded simultaneously. The fraction of systems with a particular energy is predicted by the Boltzmann distribution.

The second approach also gives us a way to predict the distribution of energy in a single large system. For example, consider a system consisting of all the air molecules in a room. Imagine that each molecule is identified by a letter: A, B, C ... One could consider molecule A to be a microscopic system in contact with the large system consisting of all other molecules (B, C, ...). However, one could also consider B to be the microscopic system, in contact with all other molecules (A, C, ...). We can use the Boltzmann distribution to predict the fraction of all molecules in the room that have a particular energy.

These two views of the Boltzmann distribution, one microscopic system observed repeatedly or a large number of microscopic systems observed once, complement each other. Sometimes one view is more helpful, sometimes the other.

12.9 THE BOLTZMANN DISTRIBUTION IN A GAS

The Boltzmann distribution applies to any kind of system—not just a solid. As a major application of the Boltzmann distribution, we will study a gas consisting of molecules that don't interact much with each other. Examples are the so-called "ideal gas" (with no interactions at all), and any real gas at sufficiently low density that the molecules seldom come near each other.

In order to apply the Boltzmann distribution, we need an equation for the energy of a molecule in the gas. We will omit rest energy, and we will also omit nuclear energy and electronic energy from our total, because at ordinary temperatures there is not enough energy available in the surroundings to raise the molecule to an excited nuclear state or an excited electronic state. Also, for simplicity, instead of writing ΔE_{vib} to represent an amount of vibrational energy above the ground vibrational state, we will simply write E_{vib}.

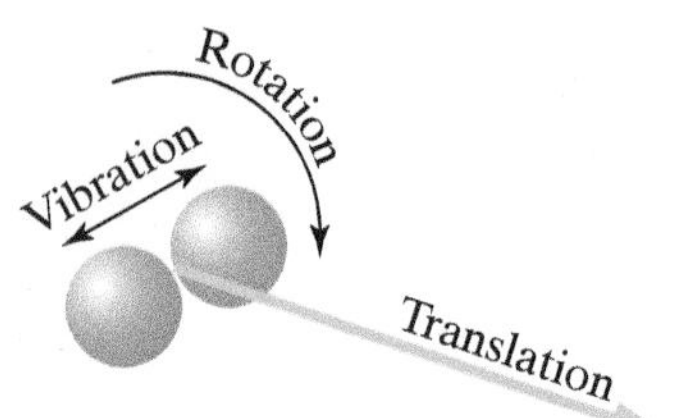

Figure 12.40 Energy of an oxygen molecule.

The energy of a single gas molecule in the gravitational field near the Earth's surface (excluding rest energy, nuclear energy, and electronic energy) is this:

$$K_{\text{trans}} + E_{\text{vib}} + E_{\text{rot}} + Mgy_{\text{CM}}$$

where E_{vib} and E_{rot} are the vibrational and rotational energies relative to the center of mass (if the molecule contains more than one atom, as in Figure 12.40) and y_{CM} is the height of the molecule's center of mass above the Earth's surface. The mass of the molecule is M. For brevity we will speak of "the energy of the gas molecule," but of course the gravitational potential energy really applies to the system of gas molecule plus Earth.

> QUESTION For this expression for the energy to be accurate, why must the gas be an "ideal" gas (or a real gas at low density)?

In a real gas at somewhat high density we may not be able to neglect the potential energy associated with the intermolecular forces, in which case this expression for the energy would not be adequate.

If the temperature T is the same everywhere in the (ideal) gas, the probability that a particular molecule will have a certain amount of energy is proportional to

$$\Omega(E)e^{-(K_{\text{trans}}+E_{\text{vib}}+E_{\text{rot}}+Mgy_{\text{CM}})/k_BT}$$

We mentally divide the gas into two systems: one particular molecule of interest and all the rest of the molecules. These two systems are in thermal equilibrium with each other, because the gas molecules are continually colliding with each other and can share energy. The energy for the one particular molecule is then expected to follow the Boltzmann distribution.

To avoid excess subscripts, in the following discussion we will simply write v for the center-of-mass speed v_{CM} and y for the center-of-mass height y_{CM}.

Separating the Various Factors

It is useful to group the terms of the Boltzmann factor like this:

$$\left[e^{-K_{\text{trans}}/k_BT}\right]\left[e^{-Mgy/k_BT}\right]\left[e^{-E_{\text{vib}}/k_BT}\right]\left[e^{-E_{\text{rot}}/k_BT}\right]$$

The first bracket is associated with the distribution of velocities, the second with the distribution of positions, the third with the distribution of vibrational energy, and the fourth with the distribution of rotational energy. We will discuss each of these individually.

Height Distribution in a Gas

A striking property of Earth's atmosphere is that in high mountains the air density and pressure are significantly lower than at sea level. The air density is so low on top of Mount Everest that most climbers carry oxygen tanks. Can we explain this? In order to get at the main issue, the variation of density with height, we make the rough approximation that the temperature is constant—the same in the mountains as at sea level.

> QUESTION How bad an approximation is it to consider the temperature to be the same at all altitudes?

Even in high mountains the temperature is typically above −29 °C (−20 °F, which is 244 K, and this is only 17% lower than room temperature of 293 K (20 °C or 68 °F). So maybe this approximation isn't too bad.

We speak of the probability that the x component of the gas molecule's position lies between some x and $x + dx$, and similarly for y and z. Here dx is considered to be a very short distance, small compared to the size of the container but large compared to the size of a molecule.

Focus just on that part of the Boltzmann distribution that deals with position, where the distribution is proportional to the probability of finding one particular molecule between x and $x + dx$, y and $y + dy$, and z and $z + dz$:

$$e^{-Mgy/k_BT} dx\,dy\,dz$$

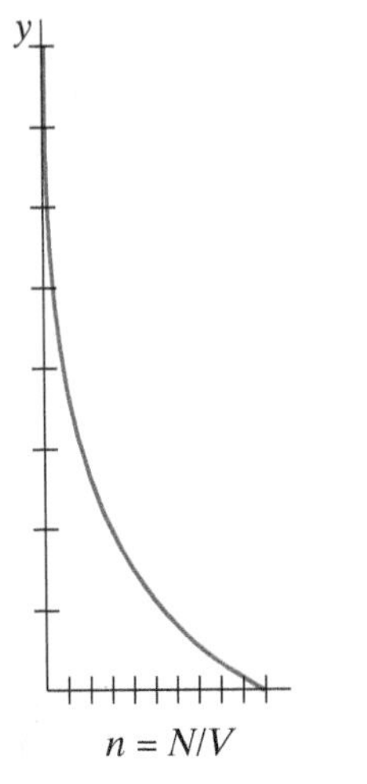

Figure 12.41 Number density vs. height in a constant-temperature atmosphere.

Evidently there's nothing very interesting about x and z (directions parallel to the ground). However, in the vertical direction there is an exponential fall-off with increasing height for the probability of finding a particular molecule at height y (Figure 12.41).

Looked at another way, our exponential equation tells us how the number density of the atmosphere depends on height, because in telling us about the behavior of one representative molecule, the equation also tells us something about all the molecules.

We shifted gears from thinking about one molecule to thinking about many. Think again about one single air molecule. Suppose that you place it on a table, which is at room temperature. It is not a very unusual event for a thermally agitated atom in the table to hit our air molecule hard enough to send it kilometers high into the air! Actually, our particular molecule will very soon run into another air molecule and not make it very high in one great leap, but on average we do find lots of air molecules very high above sea level rather than finding them all lying on the ground.

Notice again that k_BT is the important factor in understanding the statistical behavior of matter. Here it sets the scale for the variation with height of the atmosphere's number density.

At a height where the gravitational potential energy $U_g = mgy$ is equal to k_BT, the number density has dropped by a factor of $e^{-1} = 1/e = 0.37$. Dry air at sea level is 78% nitrogen (one mole $N_2 = 28$ grams), 21% oxygen (one mole $O_2 = 32$ grams), about 1% argon, and 0.03% CO_2. An average mass of 29 grams per mole is good enough for most calculational purposes.

Distribution of Velocities in a Gas

Next we look at the distribution of molecular velocities. When the gas is confined inside a finite container the energy levels in terms of momentum and height are discrete, but under almost all conditions the size of the energy quantum is so small compared to k_BT that it is appropriate to take a nonquantum approach for these variables.

We speak of the probability that the gas molecule has an x component of velocity within the range between some v_x and $v_x + dv_x$, and similarly for v_y

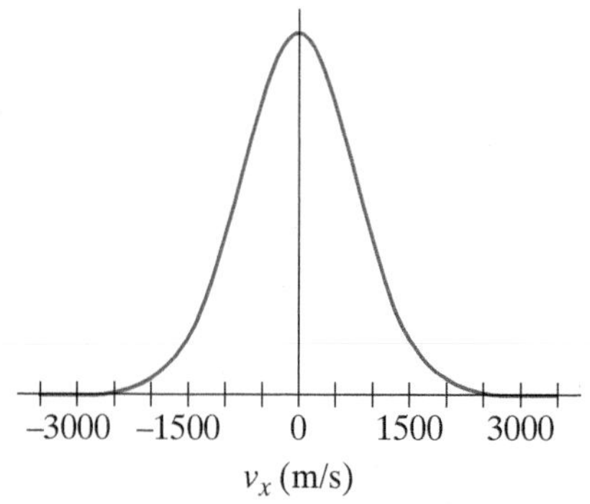

Figure 12.42 Distribution of the x component of velocity for helium at room temperature.

and v_z. Here dv_x is considered to be a very small amount, small compared to the average speed of the molecules.

Since we're explicitly interested in v_{CM}, it will be useful to express translational kinetic energy as $\frac{1}{2}Mv_{CM}^2$ rather than $p^2/(2M)$ in this discussion. Since $v^2 = v_x^2 + v_y^2 + v_z^2$, we can write the equation for the distribution of velocity in a gas as follows (remember that we are simply writing v for the center-of-mass speed v_{CM}):

$$\left[e^{-\frac{1}{2}Mv_x^2/k_BT}dv_x\right]\left[e^{-\frac{1}{2}Mv_y^2/k_BT}dv_y\right]\left[e^{-\frac{1}{2}Mv_z^2/k_BT}dv_z\right]$$

The distribution for each velocity component is a bell-shaped curve, called a "Gaussian." In Figure 12.42 we show the distribution of the x component of velocity for helium at room temperature.

QUESTION Judging from this graph, what is the average value for v_x? What are the average values of v_y and v_z? Why are these results plausible?

Evidently the average value for the x component of the velocity is zero. This is reassuring, because a molecule is as likely to be headed to the right as it is to be headed to the left. Similarly, the average values for the y and z components of velocity are zero.

It would not be particularly surprising to find a gas molecule with an x component of velocity such that $\frac{1}{2}Mv_x^2 \approx k_BT$. On the other hand, it would be very surprising to find a gas molecule with a value of v_x many times larger. As usual, k_BT sets the scale for thermal phenomena.

Distribution of Speeds in a Gas

By converting from rectangular coordinates to spherical coordinates, the velocity-component distribution can be converted into a speed distribution. We won't go into the details, but the main idea is that the volume element in rectangular coordinates in "velocity space" $dv_xdv_ydv_z$ turns into the volume of a spherical shell with surface area $4\pi v^2$, and thickness dv, as shown in Figure 12.43.

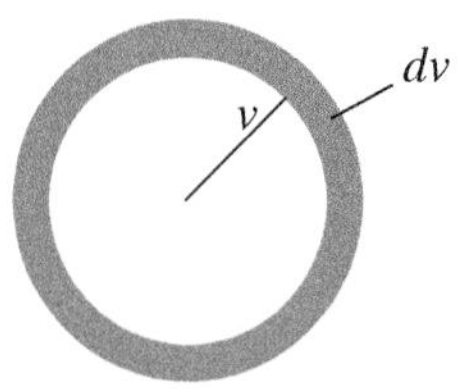

Figure 12.43 The volume of a shell with radius v and thickness dv is $4\pi v^2 dv$.

Transferring to spherical coordinates, we have the following:

$$e^{-\frac{1}{2}Mv^2/k_BT}dv_xdv_ydv_z = e^{-\frac{1}{2}Mv^2/k_BT}4\pi v^2 dv$$

The only thing missing is a "normalization" factor in front to make the integral over all speeds from 0 to infinity be equal to 1.0 (since our one molecule must have a speed somewhere in that range). Here is the Maxwell–Boltzmann distribution for a low-density gas:

MAXWELL–BOLTZMANN SPEED DISTRIBUTION (LOW-DENSITY GAS)

$$P(v) = 4\pi\left(\frac{M}{2\pi k_BT}\right)^{3/2} v^2 e^{-\frac{1}{2}Mv^2/k_BT}$$

The probability that a molecule of a gas has a center-of-mass speed within the range v to $v + dv$ is given by $P(v)dv$.

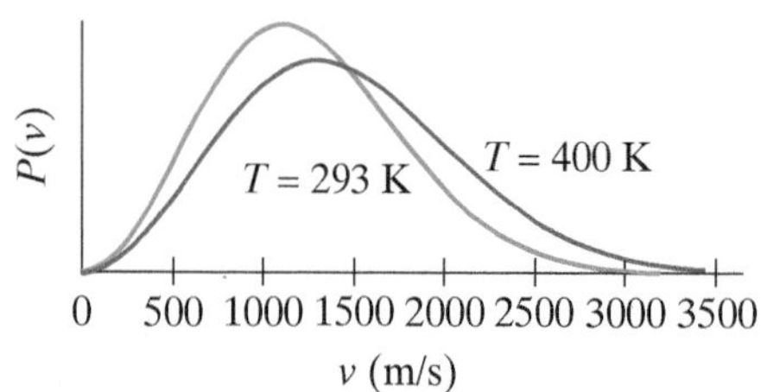

Figure 12.44 Distribution of speeds of helium atoms at two temperatures.

The distribution function for helium is shown for two different temperatures in Figure 12.44. You can see that the average speed of a helium atom is predicted to be about 1200 m/s at room temperature (20 °C, which is 293 K). The prediction was first made by Maxwell, in the mid-1800s, long before the development of quantum mechanics. In retrospect, this worked because the quantum levels of the translational kinetic energy are so close together as to be almost a continuum, so the classical model is a good approximation.

The average number of helium atoms in a container that have speeds in the range of 415.4 m/s to 415.7 m/s can be calculated by evaluating Maxwell's equation with $v = 415.4$ m/s and multiplying by $dv = 0.3$ m/s, which gives the probability of finding one molecule in this speed range. If there are N atoms in the container, multiplying by N gives the total number of molecules likely to be in that speed range at any given instant. The area of the vertical slice shown in Figure 12.45 represents the fraction of helium atoms that are likely to have speeds between 500 m/s and 600 m/s. At higher temperatures the distribution shifts to higher speeds (Figure 12.44).

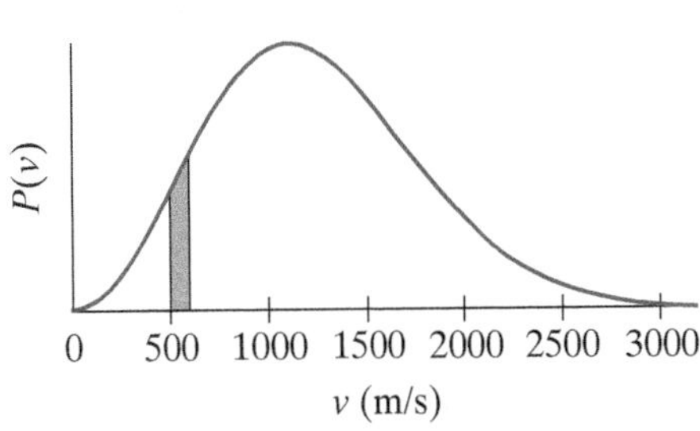

Figure 12.45 Fraction of helium atoms at 293 K with speeds between 500 and 600 m/s.

Measuring the Distribution of Speeds in a Gas

The actual distribution of speeds of molecules in any gas can be measured by an ingenious experiment. Make a tiny hole in a container of gas and let the molecules leak out into a vacuum (that is, a region from which the air is continually pumped out). The moving molecules run through collimating holes and then through a slot in a rapidly rotating drum and strike a row of devices at the other side of the drum that can detect gas molecules (Figure 12.46). Such measurements confirm the Maxwell prediction.

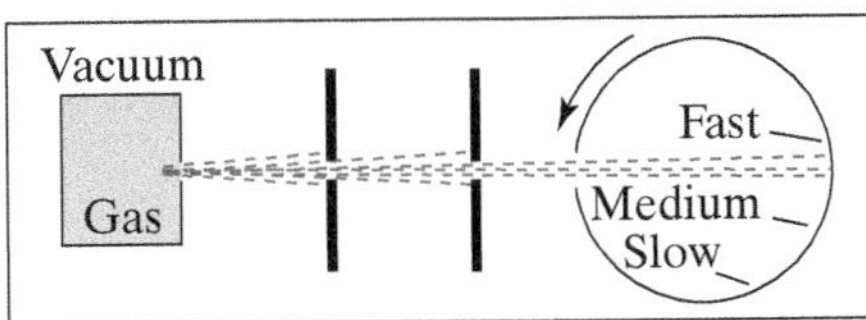

Figure 12.46 Apparatus for measuring the distribution of speeds of gas molecules.

Fast molecules strike soonest, followed by medium-speed molecules, and finally slow molecules, so molecules with different speeds strike at different locations along the rotating drum. The number of molecules striking various locations along the drum is a direct indication of the distribution of speeds of molecules in the gas. (A correction has to be applied to these data to obtain the speed distribution inside the gas, because high-speed molecules emerge through the hole more frequently than low-speed molecules do, even if their numbers are the same in the interior of the gas. Also, the apparatus measures the distribution of v_x rather than v.)

How might we display the results of this experiment? We could present the data as a graph (Figure 12.47) of the fraction $\Delta N/N$ of the N helium atoms that have speeds between 0 and 500 m/s, between 500 and 1000 m/s, between 1000 and 1500 m/s, and so on.

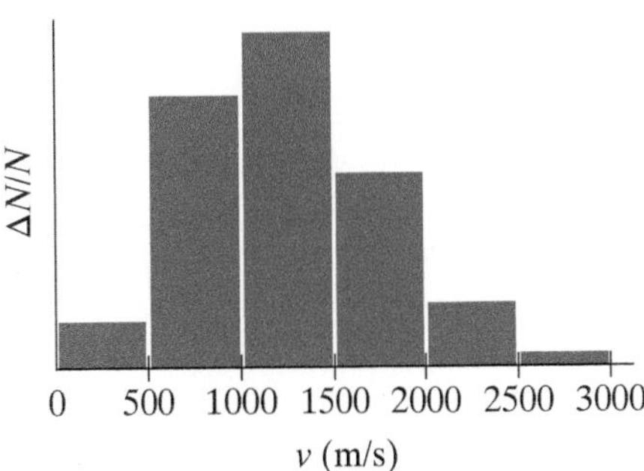

Figure 12.47 Distribution showing fraction of atoms with speeds in a given range.

Average Translational Kinetic Energy in a Gas

For a gas confined inside a container, quantum mechanics predicts that the kinetic energy $\frac{1}{2}Mv^2$ is quantized, but if the container is of ordinary macroscopic size, the energy quanta are extremely small compared to the average kinetic energy, even at very low temperatures. Most gases liquefy before the temperature drops so low as to invalidate the pre-quantum analysis for the velocity distribution.

Whenever the energy has a term containing a square of a position or momentum component, such as x^2 or p_x^2, pre-quantum theory predicts an associated average energy of $\frac{1}{2}k_BT$. This result is valid in quantum mechanics for high temperatures, where k_BT is large compared to the quantum energy spacing. This result follows from using an integral to take the average value of a quadratic term when the distribution is a Gaussian involving that quadratic term. That is, using integral tables (or integrating by parts), you can show that the average value of a quadratic term is $\frac{1}{2}k_BT$ (a bar over a quantity means "average value"):

$$\overline{w^2} = \frac{\displaystyle\int_0^\infty w^2 e^{-w^2/k_BT}dw}{\displaystyle\int_0^\infty e^{-w^2/k_BT}dw} = \frac{1}{2}k_BT$$

The average value is obtained by weighting the values of w^2 by the probability of finding that value; the denominator takes care of normalizing the distribution properly.

As an example, when we did quantum-based calculations for the Einstein solid, we found that at high temperatures the specific heat on a per-atom basis was $3k_B$, which implies an average energy of $3k_BT$. We modeled an atomic oscillator as three one-dimensional oscillators, each having an average energy of $\frac{1}{2}k_BT$ corresponding to kinetic energy $p^2/(2M)$ and another $\frac{1}{2}k_BT$ for the potential energy $\frac{1}{2}k_sx^2$. This is a total of six quadratic terms, implying an average energy per atom of $6(\frac{1}{2}k_BT)$, or $3k_BT$. We therefore expect $3k_BT$ the heat capacity per atom in a solid at high temperature to be $3k_B$, which is indeed what is observed.

HIGH-TEMPERATURE AVERAGE ENERGY

If $k_BT \gg$ the energy quantum, the average energy

associated with a quadratic energy term is $\dfrac{1}{2}k_BT$.

The number of quadratic terms in the expression for the energy is often called the "degrees of freedom."

QUESTION What is the average translational kinetic energy $\frac{1}{2}M\overline{v^2}$ in terms of k_BT? What is the associated contribution to the specific heat of the gas?

Since there are three quadratic terms in the translational kinetic energy, $K_{\text{trans}} = \frac{1}{2}mv_x^2 + \frac{1}{2}mv_y^2 + \frac{1}{2}mv_z^2$, the average value is three times $\frac{1}{2}k_BT$. We have the following important result:

AVERAGE K_{trans} FOR AN IDEAL GAS

$$\overline{K}_{\text{trans}} = \frac{1}{2}M\overline{v^2} = \frac{3}{2}k_BT$$

The contribution to the specific heat for a gas is $\frac{3}{2}k_B$.

Root-Mean-Square Speed

The square root of $\overline{v^2}$ is called the "root-mean-square" or "rms" speed:

$$v_{\text{rms}} = \sqrt{\overline{v^2}}$$

With this definition of v_{rms}, we write $\frac{1}{2}m\overline{v^2} = \frac{1}{2}mv_{\text{rms}}^2 = \frac{3}{2}k_BT$.

Average Speed vs. RMS Speed

The way you find the average speed of molecules in a gas is to weight the speed by the number of molecules that have that speed, and divide by the total number of molecules:

$$\overline{v} = \frac{N_1v_1 + N_2v_2 + \text{etc.}}{N} = \frac{N_1}{N}v_1 + \frac{N_2}{N}v_2 + \cdots$$

For a continuous distribution of speeds this turns into an integral, where the weighting factors are given by the Maxwell speed distribution, which gives the probability that a molecule has a center-of-mass speed in the range v to $v + dv$. Using integral tables one finds this:

$$\overline{v} = \int_0^\infty 4\pi\left(\frac{m}{2\pi k_BT}\right)^{3/2} e^{-\frac{1}{2}mv^2/k_BT}(v)v^2dv = \sqrt{\frac{8}{3\pi}}\sqrt{\frac{3k_BT}{m}} = 0.92v_{\text{rms}}$$

This shows that the average speed is smaller than the rms speed ($0.92v_{rms}$). This same calculational scheme can be used to determine any average. For example, if you calculate the average value of v^2 by this method, you find $\overline{v^2} = 3k_BT/m$, as expected.

Earlier we commented that the average speed $\overline{v}$ of helium atoms at room temperature is about 1200 m/s, but you can calculate that v_{rms} at room temperature is 1350 m/s. This is an example of the fact that the rms speed is higher than the average speed.

Checkpoint 9 The rms speed is somewhat higher than the average speed due to the averaging of squared speeds. Calculate the average of the numbers 1, 2, 3, and 4, then calculate the rms average (the square root of the average of their squares), and show that the rms average is larger than the simple average. Squaring gives extra weight to larger contributions.

Application: Retaining a Gas in the Atmosphere

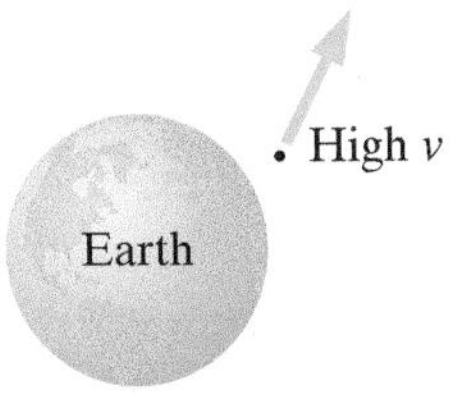

Since $v_{rms} = \sqrt{3k_B(T/m)}$, helium atoms typically travel much faster than nitrogen molecules in our atmosphere, due to the small mass of the helium atoms. Some few helium atoms will be going much faster than the average and may attain a high enough speed to escape from the Earth entirely (escape velocity from the Earth is about 1.1×10^4 m/s). This leads to a continuous leakage of high-speed helium atoms and other low-mass species such as hydrogen molecules (Figure 12.48). Other processes may also contribute to the flow of helium away from the Earth.

High v

Earth

Figure 12.48 Low-mass atoms or molecules in our atmosphere may have speeds high enough to escape from the Earth entirely.

Where do we obtain helium for party balloons and low-temperature refrigeration and scientific experiments? It is extracted from natural gas when the gas is pumped out of the ground. Heavy radioactive elements such as uranium in the Earth's crust emit alpha particles (helium nuclei), which capture electrons and become helium atoms. These atoms are trapped with natural gas in underground cavities. When the helium-bearing natural gas is pumped to the surface, the helium is extracted (at some cost).

QUESTION Why doesn't the Moon have any atmosphere at all?

The Moon's gravitational field is so weak that escape speed is quite small, and all common gases can escape, not just hydrogen and helium.

Application: Speed of Sound

Sound waves in a gas consist of propagation of variations in density (Figure 12.49), and the fundamental mechanism for this kind of wave propagation involves collisions between neighboring molecules, whose speeds are proportional to v_{rms} (and roughly comparable to the speed of sound). For example, compare the v_{rms} for nitrogen, 510 m/s, with the speed of sound in air (which is mostly nitrogen) at 293 K, which is measured to be 344 m/s. (You may know the approximate rule that a 1-s delay between lightning and thunder indicates a distance of about 1000 ft, which is about 300 m.)

v

High gas density

Low gas density

Figure 12.49 Periodic variations in gas density make a sound wave, which travels through a gas with speed v (the speed of sound).

Checkpoint 10 Should the speed of sound in air increase or decrease with increasing temperature? What percentage change would result from doubling the absolute temperature? (This effect is readily observed by measuring the speed of sound in a gas as a function of absolute temperature. The excellent agreement between theory and experiment provides additional evidence for our understanding of gases.)

Vibrational Energy in a Diatomic Gas Molecule

We have treated the distribution of velocity and position. Next we discuss the distribution of vibrational energy, with the Boltzmann factor

$$e^{-E_{\text{vib}}/k_B T}$$

For a monatomic gas such as helium, there is no vibrational energy term. However, for a diatomic molecule such as N_2 or HCl, the vibrational energy is

$$E_{\text{vib}} = \frac{p_1^2}{2m_1} + \frac{p_2^2}{2m_2} + \frac{1}{2}k_s s^2$$

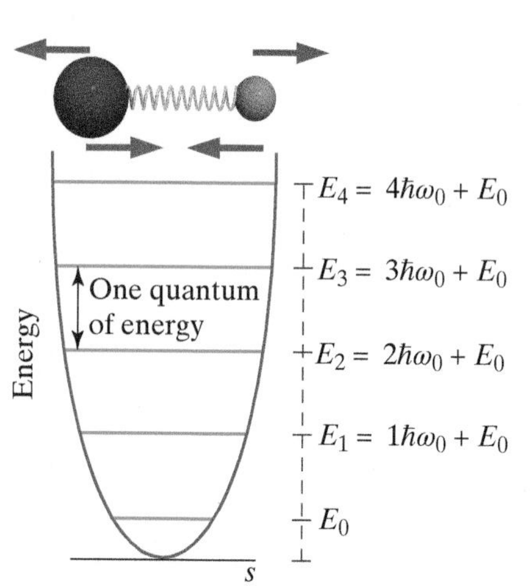

Figure 12.50 The quantized vibrational energy levels of a diatomic oscillator.

where k_s is the effective spring stiffness corresponding to the interatomic electric force (not to be confused with k_B, the Boltzmann constant), and s is the stretch of the interatomic bond. This equation for the vibrational energy is essentially the same as the equation for a one-dimensional spring–mass oscillator, which we have studied in detail (Figure 12.50).

Since $p_1 = p_2$ for momenta p_1 and p_2 relative to the center of mass, we can write

$$\frac{p_2^2}{2m_2} = \frac{p_1^2}{2m_2} = \left(\frac{m_1}{m_2}\right)\frac{p_1^2}{2m_1}$$

$$\frac{p_1^2}{2m_2} + \left(\frac{m_1}{m_2}\right)\frac{p_1^2}{2m_1} + \frac{1}{2}k_s s^2 = \left(1 + \frac{m_1}{m_2}\right)\frac{p_1^2}{2m_1} + \frac{1}{2}k_s s^2$$

This equation has a form exactly like that for a spring–mass oscillator, with a different mass.

Earlier we modeled a solid as a large number of isolated three-dimensional atomic oscillators (each corresponding to three one-dimensional oscillators, because there are spring-like interatomic forces on an atom from neighboring atoms in all directions). This is an overly simplified model of a solid, because in a solid the atoms interact with each other. For example, if an atom moves to the left this affects the atoms to the right and to the left.

In a low-density gas, however, the vibrational oscillators really are nearly independent of each other, because the gas molecules aren't even in contact with each other except when they happen to collide. So the analysis we carried out for the Einstein model of a solid applies even better to the vibrational portion of the energy in a real gas than it does to a real solid.

Among the results that apply immediately are that the specific heat associated with the vibrational energy of one oscillator is k_B at high temperatures. The specific heat decreases at very low temperatures, where $k_B T$ is small compared to the energy spacing of the quantized oscillator energies. Just as it was a surprise when the specific heat of metals was found to decrease at low temperatures, there was a similar surprise in the measurements of the specific heat of diatomic gases at low temperatures, because the contribution of the vibrational motion vanished.

QUESTION There are two quadratic terms in the vibrational energy (kinetic and spring). Therefore, at high temperatures what is the average vibrational energy in terms of $k_B T$? What is the associated contribution to the specific heat of the gas?

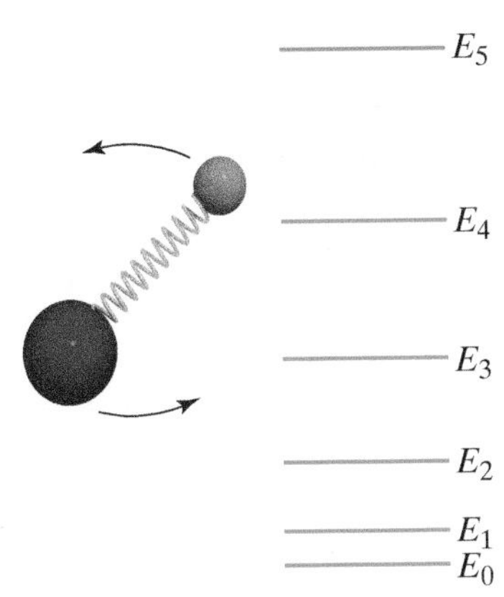

Figure 12.51 Rotational kinetic energy levels of a diatomic molecule.

At high temperatures the vibrational motion of a diatomic molecule (a one-dimensional oscillator) is expected to have an average energy that is approximately $2(\frac{1}{2}k_BT) = k_BT$, and contributes k_B to the specific heat.

Rotational Energy in a Diatomic Gas Molecule

Finally we consider the rotational-energy portion of the distribution, for which the Boltzmann factor is

$$e^{-E_{\text{rot}}/k_BT}$$

For a monatomic gas there is no rotational term, but for a diatomic gas there is rotational kinetic energy associated with rotation of the "dumbbell" consisting of the two nuclei. The rotational kinetic energy is this (Figure 12.51):

$$E_{\text{rot}} = \frac{1}{2}I\omega_x^2 + \frac{1}{2}I\omega_y^2 = \frac{L_{\text{rot},x}^2}{2I} + \frac{L_{\text{rot},y}^2}{2I}$$

This corresponds to rotational angular momenta $L_{\text{rot},x}$ and $L_{\text{rot},y}$ about the x axis and about the y axis. Note that only the nuclei contribute significantly to the rotational kinetic energy, because the electrons have much less mass.

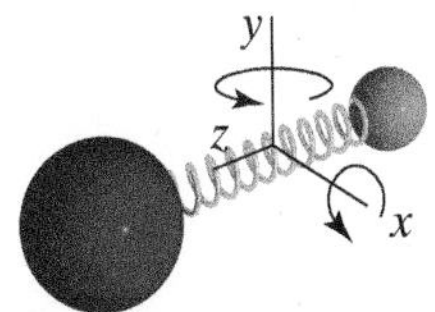

Figure 12.52 A diatomic molecule can rotate around the x or the y axis. We say there are two rotational degrees of freedom.

Rotation about the z axis connecting the two nuclei is irrelevant, for a somewhat subtle reason. The angular momentum is quantized, which leads to energy quantization. For rotations around the z axis (Figure 12.52) the energy is $L_{\text{rot},z}^2/(2I)$, where $L_{\text{rot},z}$ is the z component of the rotational angular momentum. Since I about the z axis for the tiny nuclei is extremely small compared with the moment of inertia about the x and y axes of the diatomic molecule, the rotational energy of the associated first excited state is enormous, and this state is not excited at ordinary temperatures.

Since there are two quadratic energy terms associated with rotation, we conclude that at high temperatures the rotational motion of a diatomic molecule has an average energy that is approximately $2(\frac{1}{2}k_BT) = k_BT$, and contributes k_B to the specific heat.

The spacing between quantized rotational energies for a diatomic molecule is even smaller than the vibrational energy quantum. As a result, the gas must be cooled to a very low temperature before the pre-quantum results become invalid. Many gases liquefy before this low-temperature regime is reached.

Specific Heat of a Gas

We are now in a position to discuss the specific heat of a gas as a function of temperature. This property is important in calculating the thermal interactions of a gas. Historically, measurements of the specific heat of gases were also important in testing theories of statistical mechanics. We'll concentrate on calculating C_V, the specific heat at constant volume (meaning no mechanical work is done on the gas). The associated experiment would be to add a known amount of energy to a gas in a rigid container and measure the temperature rise of the gas.

Consider the average energy of a diatomic molecule such as N_2 or HCl, consisting of the translational kinetic energy associated with the motion of the center of mass, plus the vibrational energy, plus the rotational energy, plus the gravitational energy of the molecule and the Earth:

$$\left[\frac{1}{2}m\overline{v_x^2} + \frac{1}{2}m\overline{v_y^2} + \frac{1}{2}m\overline{v_z^2}\right] + \left[\left(1+\frac{m_1}{m_2}\right)\frac{\overline{p_1^2}}{2m_1} + \frac{1}{2}k_s\overline{s^2}\right] + \left[\frac{\overline{L_{\text{rot},x}^2}}{2I} + \frac{\overline{L_{\text{rot},y}^2}}{2I}\right] + Mg\overline{y}_{\text{CM}}$$

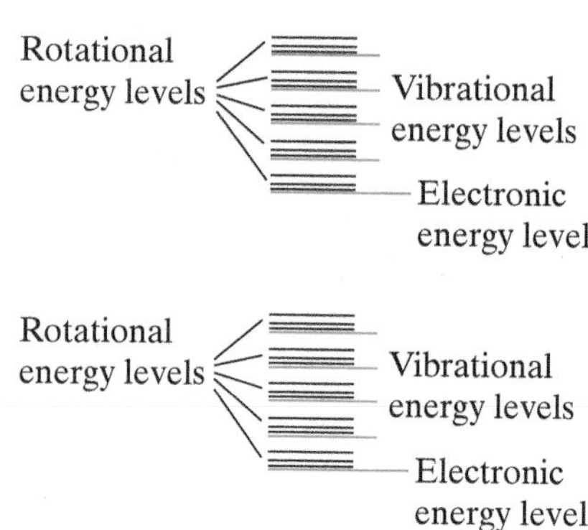

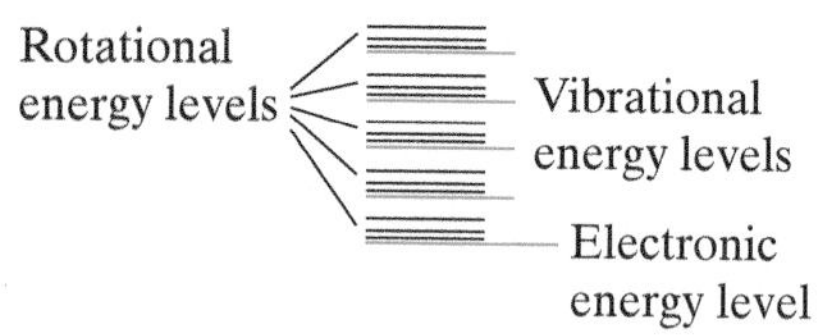

Figure 12.53 Electronic, vibrational, and rotational energy levels for a diatomic molecule.

QUESTION At high temperature, what should be the specific heat at constant volume?

There are seven quadratic terms in the expression for the energy, so if the temperature is very high, we expect an average energy of $7(\frac{1}{2}k_BT)$ and a specific heat at constant volume $C_V = \frac{7}{2}k_B$.

The discrete energy levels for a diatomic molecule include electronic, vibrational, and rotational energy levels (Figure 12.53). The electronic energy levels correspond to particular configurations of the electron clouds, and the spacing between these levels is typically 1 eV or more. Since at room temperature k_BT is about $\frac{1}{40}$ eV, the electronic levels do not contribute to the specific heat at ordinary temperatures. The vibrational energy spacing is much smaller, and the rotational energy spacing is smaller still. A diatomic gas at high temperature has a band-like spectrum, since transitions between electronic levels may be accompanied by one or more vibrational or rotational quanta.

What About Mgy?

What about the Mgy_{CM} term in the energy? Remember that we are trying to predict the specific heat of a gas at constant volume (so that no mechanical work is done on the system). We could measure C_V by having a known amount of energy transfer Q due to a temperature difference into a closed container of gas and observing the temperature rise of the gas. During this process there is an extremely small increase in y_{CM} of the gas in the container due to the temperature dependence of the height distribution, so the associated Mgy contribution to the specific heat is negligible.

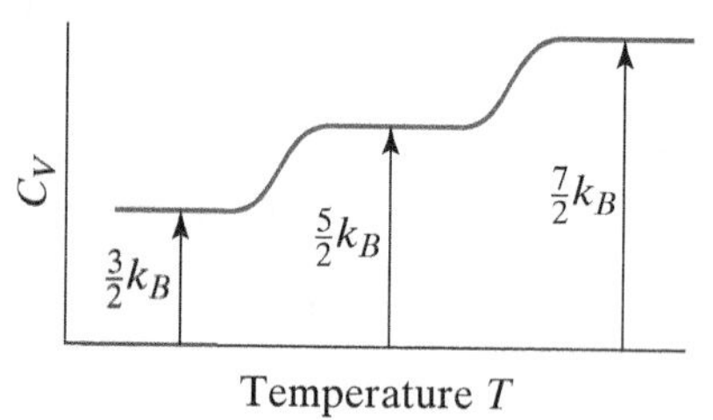

Figure 12.54 Specific heat C_V of a diatomic gas vs. temperature.

Specific Heat vs. Temperature

Figure 12.54 is a graph of the specific heat (at constant volume) of a diatomic gas as a function of temperature. In the following Checkpoint, see whether you can explain these values of the specific heat in terms of our analyses of the various contributions to the energy of a diatomic gas molecule.

Remember that k_BT is about $\frac{1}{40}$ eV at room temperature.

Checkpoint 11 What is the specific heat, on a per-molecule basis, at constant volume, of a diatomic gas at the following temperatures? **(a)** At high temperatures? **(b)** At a temperature where k_BT is small compared to the energy of the first excited vibrational state but larger than the energy of the first excited rotational state? **(c)** At a temperature low enough that k_BT is small compared to the energy of the first excited rotational state? (Some gases liquify above this temperature.) **(d)** For which gas, H_2 or D_2, does the specific heat become $\frac{3}{2}k_B$ at a higher temperature?

SUMMARY

The fundamental assumption of statistical mechanics is that, over time, an isolated system in a given macrostate is equally likely to be found in any of its possible microstates.

Two blocks in thermal contact evolve to that division of energy that has associated with it the largest number of ways $\Omega = \Omega_1\Omega_2$ of arranging the total energy among the atoms (largest number of microstates for a macrostate of given energy).

Second law of thermodynamics: The entropy of a closed system never decreases. Only in a reversible process does the entropy of a closed system stay constant.

Einstein solid: Each atom is modeled as three one-dimensional quantized oscillators.

Number of ways to arrange q quanta of energy among N one-dimensional oscillators:

$$\Omega = \frac{(q+N-1)!}{q!(N-1)!} \quad \text{(number of microstates)}$$

Entropy: $S \equiv k_B \ln \Omega$, where $k_B = 1.38 \times 10^{-23}$ J/K

Temperature: $\frac{1}{T} = \frac{\partial S}{\partial E}$

A small flow of energy Q due to a temperature difference into a system raises entropy by $\Delta S = \frac{Q}{T}$.

Specific heat per atom: $C = \frac{\Delta E_{\text{atom}}}{\Delta T}$, where

$$\Delta E_{\text{atom}} = \frac{\Delta E_{\text{system}}}{N_{\text{atoms}}}$$

The probability of finding energy E in a small system in contact with a large reservoir is proportional to

$$\Omega(E)e^{-E/k_BT}$$

$\Omega(E)$ is the number of microstates corresponding to energy E.

For each energy term involving a quadratic term such as x^2 or p_x^2 ("degree of freedom"), if the average energy is large compared to k_BT the average energy is $\frac{1}{2}k_BT$ and the contribution to the specific heat is $\frac{1}{2}k_B$. This contribution decreases at low temperatures. At high temperatures, specific heat on a per-atom basis in a solid $\approx 3k_B$.

Speed distribution for a gas:

$$P(v) = 4\pi \left(\frac{M}{2\pi k_BT}\right)^{3/2} e^{-\frac{1}{2}mv^2/k_BT}v^2$$

QUESTIONS

Q1 Energy conservation for two blocks in contact with each other is satisfied if all the energy is in one block and none in the other. Would you expect to observe this distribution in practice? Why or why not?

Q2 What is the advantage of plotting the (natural) logarithm of the number of ways of arranging the energy among the many atoms (natural logarithm of the number of microstates)?

Q3 Which has a higher temperature: a system whose entropy changes rapidly with increasing energy or one whose entropy changes slowly with increasing energy?

Q4 Explain why it is a disadvantage for some purposes that the specific heat of all materials decreases at low temperatures.

Q5 Two blocks with different temperatures had entropies of 10 J/K and 35 J/K before they were brought in contact. What can you say about the entropy of the combined system after the two came in contact with each other?

Q6 Can the entropy of a system decrease? What must be true about the entropy of the surroundings in this case?

Q7 Consider two blocks of copper. Block A contains 600 atoms and initially has a total of 20 quanta of energy. Block B contains 400 atoms and initially has 80 quanta of energy. The two blocks are placed in contact with each other, inside an insulated container (so no thermal energy can be exchanged with the surroundings). After waiting for a long time (for example, an hour), which of the following would you expect to be true? (1) Approximately 60 quanta of energy are in block A, and approximately 40 quanta of energy are in block B. (2) Approximately 50 quanta of energy are in block A, and approximately 50 quanta of energy are in block B. (3) The entropy of block A is equal to the entropy of block B. (4) The temperature of block A and the temperature of block B are equal.

Q8 Since $\Delta T = \Delta E/C$, what will happen at low temperatures to the temperature of a sink when some energy ΔE is transferred to it from the sample? Why is this unfortunate?

Q9 Which has more internal energy at room temperature—a mole of helium or a mole of air?

Q10 Which of the following are true about the Boltzmann factor $e^{-\Delta E/k_BT}$, where ΔE is the energy above the ground state? (1) $e^{-\Delta E/k_BT}$ is small when ΔE is large. (2) $e^{-\Delta E/k_BT}$ is small at low temperature T. (3) $e^{-\Delta E/k_BT}$ tends to zero as T gets larger and larger. (4) Even at high temperature, when many energy levels are excited, the ground state ($\Delta E = 0$) is the most populated state.

Q11 Explain qualitatively the basis for the Boltzmann distribution. Never mind the details of the math for the moment. Focus on the trade-offs involved with giving energy to a single oscillator vs. giving that energy to a large object.

Q12 Many chemical reactions proceed at rates that depend on the temperature. Discuss this from the point of view of the Boltzmann distribution.

Q13 A gas is made up of diatomic molecules. At temperature T_1, the ratio of the number of molecules in vibrational energy state 2 to the number of molecules in the ground state is measured, and found to be 0.35. The difference in energy between state 2 and the ground state is ΔE. **(a)** Which of the following conclusions is correct? (1) $\Delta E \approx k_BT_1$, (2) $\Delta E \ll k_BT_1$, (3) $\Delta E \gg k_BT_1$ **(b)** At a different temperature T_2, the ratio is found to be 8×10^{-5}. Which of the following is true? (1) $\Delta E \approx k_BT_2$, (2) $\Delta E \ll k_BT_2$, (3) $\Delta E \gg k_BT_2$.

Q14 How does the speed of sound in a gas change when you raise the temperature from 0 °C to 20 °C? Explain briefly.

Q15 This question follows the entire chain of reasoning involved in determining the specific heat of an Einstein solid. Start with two metal blocks, one consisting of one mole of aluminum (27 g) and the other of one mole of lead (207 g), both initially at a temperature very near absolute zero (0 K). From measurements of Young's modulus one finds that the effective stiffness of the interatomic bond modeled as a spring is 16 N/m for aluminum and 5 N/m for lead. **(a)** Is the number of quantized oscillators in the aluminum block greater, smaller, or the same as the number in the lead block? **(b)** What is the initial entropy of each block? **(c)** In which metal is the energy spacing of the quantized harmonic oscillators larger? **(d)** If we add 1 J of energy to each block, which metal now has the larger number of energy quanta? **(e)** In which block is the number of possible ways of arranging this 1 J of energy greater? **(f)** Which block now has the larger entropy? **(g)** Which block experienced a greater entropy change? **(h)** Which block experienced the larger temperature change? **(i)** Which metal has the larger specific heat at low temperatures? **(j)** Does your conclusion agree with the actual data given in Figure 12.33? (The numerical data are given in a table accompanying Problem P64.)

PROBLEMS

Section 12.2

•P16 List explicitly all the ways to arrange 2 quanta among 4 one-dimensional oscillators.

•P17 How many different ways are there to get 5 heads in 10 throws of a true coin? How many different ways are there to get no heads in 10 throws of a true coin?

•P18 How many different ways are there to arrange 4 quanta among 3 atoms in a solid?

•P19 A carbon nanoparticle (very small particle) contains 6000 carbon atoms. According to the Einstein model of a solid, how many oscillators are in this block?

•P20 In order to calculate the number of ways of arranging a given amount of energy in a tiny block of copper, the block is modeled as containing 8.7×10^5 independent oscillators. How many atoms are in the copper block?

•P21 In Chapter 4 you determined the stiffness of the interatomic "spring" (chemical bond) between atoms in a block of lead to be 5 N/m, based on the value of Young's modulus for lead. Since in our model each atom is connected to two springs, each half the length of the interatomic bond, the effective "interatomic spring stiffness" for an oscillator is 4×5 N/m = 20 N/m. The mass of one mole of lead is 207 g (0.207 kg). What is the energy, in joules, of one quantum of energy for an atomic oscillator in a block of lead?

•P22 Consider an object containing 6 one-dimensional oscillators (this object could represent a model of 2 atoms in an Einstein solid). There are 4 quanta of vibrational energy in the object. **(a)** How many microstates are there, all with the same energy? **(b)** If you examined a collection of 48,000 objects of this kind, each containing 4 quanta of energy, about how many of these objects would you expect to find in the microstate 000004?

•P23 Suppose that you look once every second at a system with 300 oscillators and 100 energy quanta, to see whether your favorite oscillator happens to have all the energy (all 100 quanta) at the instant you look. You expect that just once out of 1.7×10^{96} times you will find all of the energy concentrated on your favorite oscillator. On the average, about how many years will you have to wait? Compare this to the age of the Universe, which is thought to be about 1×10^{10} years. ($1\text{ y} \approx \pi \times 10^7$ s.)

••P24 The reasoning developed for counting microstates applies to many other situations involving probability. For example, if you flip a coin 5 times, how many different sequences of 3 heads and 2 tails are possible? Answer: 10 different sequences, such as HTHHT or TTHHH. In contrast, how many different sequences of 5 heads and 0 tails are possible? Obviously only one, HHHHH, and our equation gives $5!/[5!0!] = 1$, using the standard definition that 0! is defined to equal 1.

If the coin is equally likely on a single throw to come up heads or tails, any specific sequence like HTHHT or HHHHH is equally likely. However, there is only one way to get HHHHH, while there are 10 ways to get 3 heads and 2 tails, so this is 10 times more probable than getting all heads.

Use the expression $5!/[N!(5-N)!]$ to calculate the number of ways to get 0 heads, 1 head, 2 heads, 3 heads, 4 heads, or 5 heads in a sequence of 5 coin tosses. Make a graph of the number of ways vs. the number of heads.

Section 12.3

•P25 Object A and object B are two identical microscopic objects. Figure 12.55 below shows the number of ways to arrange energy in one of these objects, as a function of the amount of energy in the object.

E, J	0.8 E–20	1.0 E–20	1.2 E–20	1.4 E–20
# ways	37	60	90	122

Figure 12.55

(a) When there are 1.0×10^{-20} J of energy in object A, what is the entropy of this object? **(b)** When there are 1.4×10^{-20} J of energy in object B, what is the entropy of this object? **(c)** Now the two objects are placed in contact with each other. At this moment, before there is time for any energy flow between the objects, what is the entropy of the combined system of objects A and B?

••P26 For a certain metal the stiffness of the interatomic bond and the mass of one atom are such that the spacing of the quantum oscillator energy levels is 1.5×10^{-23} J. A nanoparticle of this metal consisting of 10 atoms has a total thermal energy of 18×10^{-23} J. **(a)** What is the entropy of this nanoparticle? **(b)** The temperature of the nanoparticle is 87 K. Next we add 18×10^{-23} J to the nanoparticle. By how much does the entropy increase?

Section 12.5

•P27 A block of copper at a temperature of 50 °C is placed in contact with a block of aluminum at a temperature of 45 °C in an insulated container. As a result of a transfer of 2500 J of energy from the copper to the aluminum, the final equilibrium temperature of the two blocks is 48 °C. **(a)** What is the approximate change in the entropy of the aluminum block? **(b)** What is the approximate change in the entropy of the copper block? **(c)** What is the approximate change in the entropy of the Universe? **(d)** What is the change in the energy of the Universe?

•P28 It takes about 335 J to melt one gram of ice. During the melting, the temperature stays constant. Which has higher entropy, a gram of liquid water at 0 °C or a gram of ice at 0 °C? Does this make sense? How large is the entropy difference?

•**P29** Suppose that the entropy of a certain substance (not an Einstein solid) is given by $S = a\sqrt{E}$, where a is a constant. What is the energy E as a function of the temperature T?

••**P30** A nanoparticle consisting of four iron atoms (object 1) initially has 1 quantum of energy. It is brought into contact with a nanoparticle consisting of two iron atoms (object 2), which initially has 2 quanta of energy. The mass of one mole of iron is 56 g. **(a)** Using the Einstein model of a solid, calculate and plot $\ln\Omega_1$ vs. q_1 (the number of quanta in object 1), $\ln\Omega_2$ vs. q_1, and $\ln\Omega_{\text{total}}$ vs. q_1 (put all three plots on the same graph). Show your work and explain briefly. **(b)** Calculate the approximate temperature of the objects at equilibrium. State what assumptions or approximations you made.

Section 12.6

•**P31** Suppose that the entropy of a certain substance (not an Einstein solid) is given by $S = a\sqrt{E}$, where a is a constant. What is the specific heat C as a function of the temperature T?

••**P32** A 100-g block of metal at a temperature of 20 °C is placed into an insulated container with 400 g of water at a temperature of 0 °C. The temperature of the metal and water ends up at 2 °C. What is the specific heat of this metal, per gram? Start from the Energy Principle. The specific heat of water is 4.2 J/K/g.

••**P33** A nanoparticle containing 6 atoms can be modeled approximately as an Einstein solid of 18 independent oscillators. The evenly spaced energy levels of each oscillator are 4×10^{-21} J apart. **(a)** When the nanoparticle's energy is in the range $5 \times 4 \times 10^{-21}$ J to $6 \times 4 \times 10^{-21}$ J, what is the approximate temperature? (In order to keep precision for calculating the specific heat, give the result to the nearest tenth of a kelvin.) **(b)** When the nanoparticle's energy is in the range $8 \times 4 \times 10^{-21}$ J to $9 \times 4 \times 10^{-21}$ J, what is the approximate temperature? (In order to keep precision for calculating the specific heat, give the result to the nearest tenth of a degree.) **(c)** When the nanoparticle's energy is in the range $5 \times 4 \times 10^{-21}$ J to $9 \times 4 \times 10^{-21}$ J, what is the approximate heat capacity per atom? Note that between parts (a) and (b) the average energy increased from 5.5 quanta to 8.5 quanta. As a check, compare your result with the high temperature limit of $3k_B$.

••**P34** The entropy S of a certain object (not an Einstein solid) is the following function of the internal energy E: $S = bE^{1/2}$, where b is a constant. **(a)** Determine the internal energy of this object as a function of the temperature. **(b)** What is the specific heat of this object as a function of the temperature?

••**P35** The interatomic spring stiffness for tungsten is determined from Young's modulus measurements to be 90 N/m. The mass of one mole of tungsten is 0.185 kg. If we model a block of tungsten as a collection of atomic "oscillators" (masses on springs), note that since each oscillator is attached to two "springs," and each "spring" is half the length of the interatomic bond, the effective interatomic spring stiffness for one of these oscillators is 4 times the calculated value given above.

Use these precise values for the constants: $\hbar = 1.0546 \times 10^{-34}$ J·s (Planck's constant divided by 2π), Avogadro's number $= 6.0221 \times 10^{23}$ molecules/mole, $k_B = 1.3807 \times 10^{-23}$ J/K (the Boltzmann constant). **(a)** What is one quantum of energy for one of these atomic oscillators? **(b)** Figure 12.56 contains the number of ways to arrange a given number of quanta of energy in a particular block of tungsten. Fill in the blanks to complete the table, including calculating the temperature of the block. The energy E is measured from the ground state. Nothing goes in the shaded boxes. Be sure to give the temperature to the nearest 0.1 kelvin. **(c)** There are about 60 atoms in this object. What is the heat capacity on a per-atom basis? (Note that at high temperatures the heat capacity on a per-atom basis approaches the classical limit of $3k_B = 4.2 \times 10^{-23}$ J/K/atom.)

q	# ways	E, J	S, J/K	ΔE, J	ΔS, J/K	T, K
20	4.91 E26					
21	4.44 E27					
22	3.85 E28					

Figure 12.56

••**P36** A 50-g block of copper (one mole has a mass of 63.5 g) at a temperature of 35 °C is put in contact with a 100-g block of aluminum (molar mass 27 g) at a temperature of 20 °C. The blocks are inside an insulated enclosure, with little contact with the walls. At these temperatures, the high-temperature limit is valid for the specific heat. Calculate the final temperature of the two blocks. Do NOT look up the specific heats of aluminum and copper; you should be able to figure them out on your own.

••**P37** Young's modulus for copper is measured by stretching a copper wire to be about 1.2×10^{11} N/m^2. The density of copper is about 9 g/cm^3, and the mass of a mole is 63.5 g. Starting from a very low temperature, use these data to estimate roughly the temperature T at which we expect the specific heat for copper to approach $3k_B$. Compare your estimate with the data shown on a graph in this chapter.

••**P38** Figure 12.57 shows a one-dimensional row of 5 microscopic objects each of mass 4×10^{-26} kg, connected by forces that can be modeled by springs of stiffness 15 N/m. These objects can move only along the x axis.

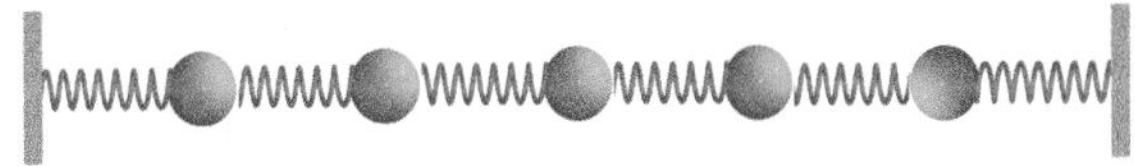

Figure 12.57

(a) Using the Einstein model, calculate the approximate entropy of this system for total energy of 0, 1, 2, 3, 4, and 5 quanta. Think carefully about what the Einstein model is, and apply those concepts to this one-dimensional situation. **(b)** Calculate the approximate temperature of the system when the total energy is 4 quanta. **(c)** Calculate the approximate specific heat on a per-object basis when the total energy is 4 quanta. **(d)** If the temperature is raised very high, what is the approximate specific heat on a per-object basis? Give a numerical value and compare with your result in part (c).

••**P39** In an insulated container a 100-W electric heating element of small mass warms up a 300-g sample of copper for 6 s. The initial temperature of the copper was 20 °C (room temperature). Predict the final temperature of the copper, using the $3k_B$ specific heat per atom.

••P40 The goal of this experiment is to understand, in a concrete way, what specific heat is and how it can be measured. You will need a microwave oven, a styrofoam coffee cup, and a clock or watch.

In the range of temperature where water is a liquid (0 C to 100 °C), it is approximately true that it takes 4.2 J of energy (1 calorie) to raise the temperature of 1 g of water through 1 Kelvin. To measure this specific heat of water, we need some way to raise a known mass of water from a known initial temperature to a final temperature that can also be measured, while we keep track of the energy supplied to the water. One way to do this, as discussed in the text, is to put water in a well-insulated container within which a heater, whose power output is known, warms up the water. In this experiment, instead of a well-insulated box with a heater, we will use microwave power, which preferentially warms up water by exciting rotational modes of the water molecules, as opposed to burners or heaters that warm up water in a pan by first warming up the pan.

Use a styrofoam coffee cup of known volume in which water can be warmed up. The density of water is 1 gram/cm^3.

> It is a good idea not to fill the cup completely full, because this makes it more likely to spill.

One method of recording the initial temperature of the water is to get water from the faucet and wait for it to equilibrate with room temperature (which can either be read off a thermostat or estimated based on past experience). After waiting about a half hour for this to happen, place the cup in the microwave oven and turn on the oven at maximum power. The cup needs to be watched as it warms up, so that when the water starts to boil, the elapsed time can be noted accurately.

> BE CAREFUL! A styrofoam cup full of hot liquid can buckle if you hold it near the rim. Hold the cup near the bottom. If the cup is full, do not attempt to move the cup while the water is hot. A spill can cause a painful burn.

On the back of the microwave oven (or inside the front door), there is usually a sticker with specifications that says "Output Power= ...Watts" which can be used to calculate the energy supplied. If there is no indication, use a typical value of 600 W for a standard microwave oven. Using all the quantities measured above and knowing the temperature interval over which you have warmed up the water, you can calculate the specific heat of water. **(a)** Show and explain all your data and calculations, and compare with the accepted value for water (4.2 J/K/g). **(b)** Discuss why your result might be expected to differ from the accepted value. For each effect that you consider, state whether this effect would lead to a result that is larger or smaller than the accepted value.

••P41 A box contains a uniform disk of mass M and radius R that is pivoted on a low-friction axle through its center (Figure 12.58). A block of mass m is pressed against the disk by a spring, so that the block acts like a brake, making the disk hard to turn. The box and the spring have negligible mass. A string is wrapped around the disk (out of the way of the brake) and passes through a hole in the box. A force of constant magnitude F acts on the end of the string. The motion takes place in outer space. At time t_i the speed of the box is v_i, and the rotational speed of the disk is ω_i. At time t_f the box has moved a distance x, and the end of the string has moved a longer distance d, as shown.

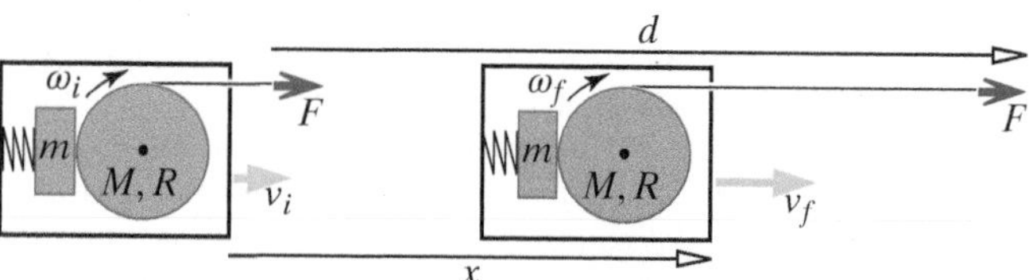

Figure 12.58

(a) At time t_f, what is the speed v_f of the box? **(b)** During this process, the brake exerts a tangential friction force of magnitude f. At time t_f, what is the angular speed ω_f of the disk? **(c)** At time t_f, assume that you know (from part b) the rotational speed ω_f of the disk. From time t_i to time t_f, what is the increase in thermal energy of the apparatus? **(d)** Suppose that the increase in thermal energy in part (c) is 8×10^4 J. The disk and brake are made of iron, and their total mass is 1.2 kg. At time t_i their temperature was 350 K. At time t_f, what is their approximate temperature?

Section 12.8

•P42 Consider the exponential function e^{-x}. Evaluate this function for $x = 1$, 10,000, and 0.01.

•P43 At room temperature (293 K), calculate $k_B T$ in joules and eV.

•P44 A microscopic oscillator has its first and second excited states 0.05 eV and 0.10 eV above the ground-state energy. Calculate the Boltzmann factor for the ground state, first excited state, and second excited state, at room temperature.

Section 12.9

•P45 Suppose that you put one air molecule on your desk, so it is in thermal equilibrium with the desk at room temperature. Suppose that there is no atmosphere to get in the way of this one molecule bouncing up and down on the desk. Calculate the typical height that the air molecule will be above your desk, so that $Mgy \approx k_B T$.

•P46 Approximately what fraction of the sea-level air density is found at the top of Mount Everest, a height of 8848 m above sea level?

•P47 Calculate v_{rms} for a helium atom and for a nitrogen molecule (N_2; molecular mass 28 g per mole) in the room you're in (whose temperature is probably about 293 K).

•P48 Calculate the escape speed from the Moon and compare with typical speeds of gas molecules. The mass of the Moon is 7×10^{22} kg, and its radius is 1.75×10^6 m.

•P49 Sketch and label graphs of specific heat vs. temperature for hydrogen gas (H_2) and oxygen gas (O_2), using the same temperature scale. Explain briefly.

•P50 Marbles of mass $M = 10$ g are lying on the floor. They are of course in thermal equilibrium with their surroundings. What is a typical height above the floor for one of these marbles? That is, for what value of y is $Mgy \approx k_B T$?

•P51 Viruses of mass $M = 2 \times 10^{-20}$ kg are lying on the floor at room temperature (about 20 °C = 293 K). They are of course in thermal equilibrium with their surroundings. What is a typical height above the floor for one of these viruses? That is, for what value of y is $Mgy \approx k_B T$?

•**P52** The temperature of the surface of a certain star is 7000 K. Most hydrogen atoms at the surface of the star are in the electronic ground state. What is the approximate fraction of the hydrogen atoms that are in the first excited state (and therefore could emit a photon)? The energy of the first excited state above the ground state is $(-13.6/2^2 eV) - (-13.6 eV) = 10.2 eV = 1.632 \times 10^{-18}$ J.

(In this estimate we are ignoring the fact that there may be several excited states with the same energies—for example, the 2s and 2p states in hydrogen—because this makes only a small difference in the answer.)

•**P53** Figure 12.59 shows the distribution of speeds of atoms in a particular gas at a particular temperature. Approximately what is the average speed? Is the rms (root-mean-square) speed bigger or smaller than this? Approximately what fraction of the molecules have speeds greater than 1000 m/s?

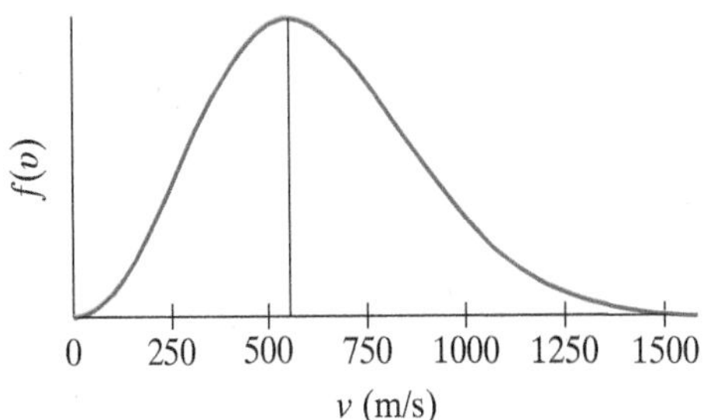

Figure 12.59

••**P54** Calculate the temperature rise of a gas. **(a)** You have a bottle containing a mole of a monatomic gas such as helium or neon. You warm up this monatomic gas with an electrical heater, which inputs $Q = 580$ J of energy. How much does the temperature of the gas increase? **(b)** You have a bottle containing a mole of a diatomic gas such as nitrogen (N_2) or oxygen (O_2). The initial temperature is in the range where many rotational energy levels are excited but essentially no vibrational energy levels are excited. You warm up this diatomic gas with an electrical heater, which inputs $Q = 580$ J of energy. How much does the temperature of the gas increase? **(c)** You have a bottle containing a mole of a diatomic gas such as nitrogen (N_2) or oxygen (O_2). The initial temperature is in the range where not only many rotational energy levels are excited but also many vibrational energy levels are excited. You warm up this diatomic gas with an electrical heater, which inputs $Q = 580$ J of energy. How much does the temperature of the gas increase?

••**P55** In studying a voyage to the Moon in Chapter 3 we somewhat arbitrarily started at a height of 50 km above the surface of the Earth. **(a)** At this altitude, what is the density of the air as a fraction of the density at sea level? **(b)** Approximately how many air molecules are there in one cubic centimeter at this altitude? **(c)** At what altitude is air density one-millionth (1×10^{-6}) that at sea level?

••**P56** At sufficiently high temperatures, the thermal speeds of gas molecules may be high enough that collisions may ionize a molecule (that is, remove an outer electron). An ionized gas in which each molecule has lost an electron is called a "plasma." Determine approximately the temperature at which air becomes a plasma.

••**P57** 100 J of energy transfer due to a temperature difference are given to air in a 50-L rigid container and to helium in a 50-L rigid container, both initially at STP (standard temperature and pressure). **(a)** Which gas experiences a greater temperature rise? **(b)** What is the temperature rise of the helium gas?

•••**P58** It is possible to estimate some properties of a diatomic molecule from the temperature dependence of the specific heat. **(a)** Below about 80 K the specific heat at constant volume for hydrogen gas (H_2) is $\frac{3}{2}k_B$ per molecule, but at higher temperatures the specific heat increases to $\frac{5}{2}k_B$ per molecule due to contributions from rotational energy states. Use these observations to estimate the distance between the hydrogen nuclei in an H_2 molecule. **(b)** At about 2000 K the specific heat at constant volume for hydrogen gas (H_2) increases to $\frac{7}{2}k_B$ per molecule due to contributions from vibrational energy states. Use these observations to estimate the stiffness of the "spring" that approximately represents the interatomic force.

•••**P59** In 1988, telescopes viewed Pluto as it crossed in front of a distant star. As the star emerged from behind the planet, light from the star was slightly dimmed as it went through Pluto's atmosphere. The observations indicated that the atmospheric density at a height of 50 km above the surface of Pluto is about one-third the density at the surface. The mass of Pluto is known to be about 1.5×10^{22} kg and its radius is about 1200 km. Spectroscopic data indicate that the atmosphere is mostly nitrogen (N_2). Estimate the temperature of Pluto's atmosphere. State what approximations and/or simplifying assumptions you made.

•••**P60** Buckminsterfullerene, C_{60}, is a large molecule consisting of 60 carbon atoms connected to form a hollow sphere. The diameter of a C_{60} molecule is about 7×10^{-10} m. It has been hypothesized that C_{60} molecules might be found in clouds of interstellar dust, which often contain interesting chemical compounds. The temperature of an interstellar dust cloud may be very low, around 3 K. Suppose you are planning to try to detect the presence of C_{60} in such a cold dust cloud by detecting photons emitted when molecules undergo transitions from one rotational energy state to another.

Approximately, what is the highest-numbered rotational level from which you would expect to observe emissions? Rotational levels are $l = 0, 1, 2, 3, \ldots$.

COMPUTATIONAL PROBLEMS

More detailed and extended versions of some computational modeling problems may be found in the lab activities included in the *Matter & Interactions, 4th Edition*, resources for instructors.

Working through Problems P61–P64 can be very informative and can help make the models discussed in this chapter concrete and clear.

••**P61** Write a program to calculate the number of ways to arrange energy in an Einstein solid. **(a)** Model a system consisting of two atoms (three oscillators each), among which 4 quanta of energy are to be distributed. Write a program to display a histogram showing the total number of possible microstates of the two-atom system vs. the number of quanta assigned to atom

1. Compare your histogram to the one shown in Figure 12.16 (you should get the same distribution). **(b)** Model a system consisting of two solid blocks, block 1 containing 300 oscillators and block 2 containing 200 oscillators. Find the possible distributions of 100 quanta among these blocks, and plot number of microstates vs. number of quanta assigned to block 1. Compare your histogram to the one shown in Figure 12.22. Determine the distribution of quanta for which the probability is half as large as the most probable 60-40 distribution. **(c)** Do a series of calculations distributing 100 quanta between two blocks whose total number of oscillators is 500, but whose relative number of atoms varies. For example, consider equal numbers of oscillators, and ratios of 2:1, 5:1, and so on. Describe your observations.

••P62 Start with your solution to Problem P61. For the same system of two blocks, with $N_1 = 300$ oscillators and $N_2 = 200$ oscillators, plot $\ln(\Omega_1)$, $\ln(\Omega_2)$, and $\ln(\Omega_1\Omega_2)$, for q_1 running from 0 to 100 quanta. Your graph should look like the one in Figure 12.26. Determine the maximum value of $\ln(\Omega_1\Omega_2)$ and the value of q_1 where this maximum occurs. What is the significance of this value of q_1?

••P63 Modify your program from Problem P62 to plot the temperature of block 1 in kelvins as a function of the number of quanta q_1 present in the first block. On the same graph, plot the temperature of block 2 in kelvins as a function of q_1 (of course, $q_2 = q_{\text{tot}} - q_1$).

In order to plot the temperature in kelvins, you must determine the values of ΔE and ΔS that correspond to a one-quantum change in energy. Consider the model we are using. The energy of one quantum, in joules, is $\Delta E = \hbar\sqrt{k_{s,i}/m_a}$. The increment in entropy corresponding to this increment in energy is $\Delta S = k_B\Delta(\ln\Omega)$. Assume that the blocks are made of aluminum. Based on Young's modulus measurements the interatomic spring constant $k_{s,i}$ for Al is approximately 16 N/m. In Section 12.6 we show that the effective $k_{s,i}$ for oscillations in the Einstein solid is expected to be about 4 times the value obtained from measuring Young's modulus.

What is the significance of the value of q_1 (and of q_2) where the temperature curves for the two blocks cross (the temperatures are equal)?

••P64 Modify your analysis of Problem P63 to determine the specific heat as a function of temperature for a single block of metal. In order to see all of the important effects, consider a single block of 35 atoms (105 oscillators) with up to 300 quanta of energy. Note that in this analysis you are calculating quantities for a *single* block, not two blocks in contact.

To make specific comparison with experimental data, consider the cases of aluminum (Al) and lead (Pb). For each metal, plot the theoretical specific heat C per atom vs. $T(K)$, with dots showing the actual experimental data given in the following table, which you should convert to the same per-atom basis as your theoretical calculations. Adjust the interatomic spring stiffness k_s until your calculations approximately fit the experimental data.

T	C, Al	C, Pb	T	C, Al	C, Pb
20	0.23	11.01	150	18.52	25.27
40	2.09	19.57	200	21.58	25.87
60	5.77	22.43	250	23.25	26.36
80	9.65	23.69	300	24.32	26.82
100	13.04	24.43	400	25.61	27.45

T is in kelvin; C is J/K/mole.

What value of k_s gives a good fit? Based on Young's modulus, estimated values of the interatomic spring stiffnesses are 16 N/m for Al and 5 N/m for lead. However, recall that these values may need modification since the "springs" in the Einstein model are half as long, and there are two of them per oscillator.

Show from your graph that the high-temperature limit of the specific heat is about $3k_B$ per atom. The rise above this limit at high temperatures may be due to the fact that the assumed uniform spacing of the quantized oscillator energy levels isn't a good approximation for highly excited states. See Chapter 8.

••P65 For some examples of your choice, demonstrate by carrying out actual computer calculations that the "square-root" rule holds true for the fractional width of the peak representing the most probable arrangements of the energy. *Warning:* Check to see what is the largest number you can use in your computations; some programs or programming environments won't handle numbers bigger than about 1×10^{307}, for example, and larger numbers are treated as "infinite."

••P66 Create a computational model of one atom of the Einstein solid as shown in Figure 12.4. Let the atom move under the influence of the six spring-like forces that act on the atom. Use the classical Momentum Principle, although the actual behavior is governed by quantum mechanics, with quantized energies. **(a)** Experiment with different initial displacements of the atom away from the equilibrium position. **(b)** In many materials the effective stiffness of the interatomic bonds is different in the x, y, and z directions; see how making the stiffnesses different affects the motion.

ANSWERS TO CHECKPOINTS

1 20
2 123, 132, 213, 231, 312, 321
3 RRRGG, RRGRG, RRGGR, RGRRG, RGRGR, RGGRR, GGRRR, GRGRR, GRRGR, GRRRG
4 See Figure 12.15.
6 This looks *very* improbable; the second law of thermodynamics would be violated by this process—the entropy of the Universe is decreasing.
7 500 K
9 2.50; 2.74
10 Increase, because the average speed of the air molecules increases; 40% (factor of $\sqrt{2}$)
11 (a) $\frac{7}{2}k_B$; (b) $\frac{5}{2}k_B$; (c) $\frac{3}{2}k_B$; (d) transition to $\frac{3}{2}k_B$ occurs at a lower temperature for deuterium because the rotational energy level spacing $L^2/(2I)$ is smaller for deuterium (larger nuclear mass means larger moment of inertia, since the internuclear distance is about the same; $L^2 = l(l+1)\hbar^2$ is the same for the rotational energy levels of hydrogen and deuterium molecules)

Answers to Odd-Numbered Problems

Chapter 1

P13 2.15×10^7 m/s **P15 (a)** $\vec{a} = \langle -4, -3, 0 \rangle$ **(b)** $\vec{b} = \langle -4, -3, 0 \rangle$ **(c)** True **(d)** $\vec{c} = \langle 4, 3, 0 \rangle$ **(e)** True **(f)** False **P17** $\langle 0.04, -3.4, 60.0 \rangle$ **P19** $\langle 0.58, 0.58, 0.58 \rangle$; $\langle 0.58, 0.58, 0.58 \rangle$ **P21** 458.26 m/s^2 $\langle 0.872872, 0.436436, -0.218218 \rangle$ **P23 (a)** $\langle 3 \times 10^{-10}, -3 \times 10^{-10}, 8 \times 10^{-10} \rangle$ m **(b)** 9.1×10^{-10} m **(c)** $\langle 0.33, -0.33, 0.88 \rangle$ **P25 (a)** $\langle -5.5, -20, 0 \rangle$ m **(b)** 20.74 m **P27 (a)** $\langle -10 \times 10^{10}, -17 \times 10^{10}, 0 \rangle$ m **(b)** $\langle 10 \times 10^{10}, 17 \times 10^{10}, 0 \rangle$ m **P29** $\langle x_p - x_e, y_p - y_e, z_p - z_e \rangle$; $\langle x_e - x_p, y_e - y_p, z_e - z_p \rangle$ **P31 (a)** $\langle -2.01 \times 10^5, 5.2 \times 10^4, -1 \times 10^3 \rangle$ m/s **(b)** 2.08×10^5 m/s **P33 (a)** $\langle 974, 0, 684 \rangle$ m **(b)** $\langle 1.50, 0, 1.05 \rangle$ m/s **P35** 3.0 s; $\langle 5.33 \times 10^2, -7.33 \times 10^2, 5.67 \times 10^2 \rangle$ m **P37 (a)** $\langle 4.4, -6.4, 0 \rangle$ m/s **(b)** $\langle 4, -7.7, 0 \rangle$ m/s **(c)** The time interval from $t = 6.3$ s to 6.8 s **(d)** $\langle 0.132, -0.192, 0 \rangle$ m **P39 (a)** $\langle 0, 9 \times 10^5, -4 \times 10^5 \rangle$ m/s **(b)** $\langle 0.02, 6.34, -2.86 \rangle$ m **P41** 147 m **P43 (a)** $\langle 22.3, 26.1, 0 \rangle$ m/s **(b)** $\langle 44.6, 52.2, 0 \rangle$ m **(c)** $\Delta t = 1.0$ s was too big. **P45** 6.2 kg · m/s **P47** 2.24×10^5 kg · m/s **P49** 6.7 kg · m/s **P51 (a)** $\langle -2mv_x, 0, 0 \rangle$ **(b)** 0 **P53** $\langle 0, 4.1, 0 \rangle$ kg · m/s **P55** 0; 500 kg · m/s **P57** $\langle 20, 50, -10 \rangle$ m/s **P59** $\langle 0, 0, -11 \rangle$ m **P61** 7.35×10^{-19} kg · m/s **P63** 1.93×10^{-20} kg · m/s **P65** 1250 **P67** $0.65c$

Chapter 2

P9 $\langle -9510, 0, 0 \rangle$ N **P11** $\langle 180, -180, 700 \rangle$ N **P13 (a)** $\langle 0.011, -0.005, -0.003 \rangle$ kg · m/s **(b)** $\langle 0.003, -0.049, -0.003 \rangle$ kg · m/s **(c)** $\langle -0.004, -0.022, 0 \rangle$ N **P15** $\langle 91, 106, 80 \rangle$ m/s **P17 (a)** 12 m/s **(b)** 0.06 s **(c)** 1.0×10^6 N **(d)** 41 **(e)** $\vec{v}_{avg} = (\vec{v}_i + \vec{v}_f)/2$ was valid **P19** Depends on what approximations you make. **P21** $\langle 0.15, 2.102, 0 \rangle$ m, $\langle 3, 2.04, 0 \rangle$ m/s; $\langle 0.3, 2.106, 0 \rangle$ m, $\langle 3, 0.08, 0 \rangle$ m/s; $\langle 0.45, 2.012, 0 \rangle$ m, $\langle 3, -1.88, 0 \rangle$ m/s **P23** $\langle -3.83, 0, 2.79 \rangle$ m **P25 (a)** 4 m/s **(b)** 12 m **P27 (a)** 90 mi **(b)** 37.5 mi/h **(c)** 45 mi/h **P29 (a)** none **(b)** 6 **(c)** 3 **(d)** 2 **(e)** none **(f)** none **(g)** 5 **(h)** 1 **(i)** 4 **P31 (a)** 6 **(b)** 7 **(c)** 5 **P33** assuming the flower pot fell from rest, it started from 5.5 m above the top of the window **P35 (a)** $\langle -11, 11.1, -6 \rangle$ m/s **(b)** arithmetic **(c)** $\langle -11, 13.6, -6 \rangle$ m/s **(d)** $\langle 3.5, 6.8, -9 \rangle$ m **(e)** 0 **(f)** $0 = 16\ \text{m/s} + (-9.8\ \text{N/kg})\Delta t$ **(g)** 1.63 s **(h)** 13 m **P37 (a)** $\Delta t = \sqrt{2h/g}$ **(b)** $v_{fy} = -g\sqrt{2h/g}$ **P39 (a)** 1 **(b)** 9/8 **(c)** 3/4 **P41** 23 cm **P43** $\langle 0, 0.2807, 0 \rangle$ m **P45** At $t = 0.04$ s, $s = 0.0157$ m and $L = 0.0843$ m. At $t = 0.08$ s, $s = 0.0466$ m and $L = 0.0554$ m **P47 (a)** $\Delta t \approx 6.15 \times 10^{-10}$ s **(b)** 18 cm

Chapter 3

P11 3/16 **P13 (a)** $\langle 2.8 \times 10^8, 0, -2.8 \times 10^8 \rangle$ m **(b)** 4.0×10^8 m **(c)** $\langle 0.7, 0, -0.7 \rangle$ **(d)** $\langle -4.9 \times 10^{28}, 0, 4.9 \times 10^{28} \rangle$ N **P15** 1.12×10^{24} N **P17 (a)** 2×10^{22} N **(b)** 2×10^{22} N **P19 (a)** $\langle -7 \times 10^{11}, 5 \times 10^{11}, 0 \rangle$ m **(b)** 8.6×10^{11} m **(c)** $\langle -0.81, 0.58, 0 \rangle$ **(d)** 1.8×10^{21} N **(e)** 1.8×10^{21} N **(f)** $\langle -1.5 \times 10^{21}, 1.0 \times 10^{21}, 0 \rangle$ N **(g)** $\langle 1.5 \times 10^{21}, -1.0 \times 10^{21}, 0 \rangle$ N **P21** about 1×10^{-6} N and about 10 N **P23** 3.2×10^4 m **P25** mg **P27 (a)** $\langle 1.39 \times 10^{-4}, -2.32 \times 10^{-4}, -1.86 \times 10^{-4} \rangle$ N **(b)** $\langle 8.4 \times 10^{-4}, 1.40 \times 10^{-3}, 1.12 \times 10^{-3} \rangle$ kg · m/s **P29 (a)** $\langle 3.04 \times 10^3, 1.496 \times 10^4, 0 \rangle$ m/s **(b)** $\langle 3.004 \times 10^{12}, 4.01 \times 10^{12}, 0 \rangle$ m **(c)** Force and velocity are not constant, so a large time interval will give inaccurate results. **P31 (b)** about 5 kg · m/s **(c)** about 500 m **P33** To the right **P35** c **P37 (a)** $\langle -0.9, 0.6, 0 \rangle$ m **(b)** 1.08 m **(c)** $\langle -0.833, 0.556, 0 \rangle$ **(d)** 2.29×10^{-16} N **(e)** $\langle 1.91 \times 10^{-16}, 1.27 \times 10^{-16}, 0 \rangle$ N **(f)** 6.17×10^{-8} N **(g)** $\langle -5.14 \times 10^{-8}, 3.43 \times 10^{-8}, 0 \rangle$ N **(h)** 2.69×10^8 **(i)** 2.69×10^8 **P39** 2.3×10^{-8} N, 5.4×10^{-51} N; will repel **P41 (a)** $\langle 2.2, -0.4, 2.6 \rangle$ kg · m/s **(b)** $\langle 0, -7.84, 0 \rangle$ N **(c)** $\langle 2.2, -1.2, 2.6 \rangle$ kg · m/s **P43** 4.6×10^6 m, so inside the Earth **P45 (a)** $\langle 1.875, 0.5, 0 \rangle$ kg · m/s **(b)** $\langle 5.36, 1.43, 0 \rangle$ m/s **P47** $\langle -2, 14, 7 \rangle$ kg · m/s **P49 (a)** $\langle 0, 0.09, -0.18 \rangle$ kg · m/s **(b)** $\langle 0, 0.09, -0.18 \rangle$ kg · m/s **(c)** $\langle 0, 1.5, -3 \rangle$ m/s **P51 (a)** $m\,|\vec{v}|$ **(b)** $\left|\vec{F}\right|$ **(c)** $m\,|\vec{v}|$ **(d)** $\left|\vec{v}_{\text{mosquito}}\right| \approx (M/m)\,|\vec{v}_{\text{car}}|$ **P53 (a)** $\langle 25, 1, 0 \rangle$ kg · m/s **(b)** approximately zero **(c)** $\langle 25, 1, 0 \rangle$ kg · m/s **(d)** $\langle 7, -4, 0 \rangle$ kg · m/s **P55** $\langle 3138, -1750, 4200 \rangle$ m/s **P57** $\langle 13.4, -1.4, 0 \rangle$ m/s **P59** $v_{fx} = [m(v_{1x} - v_{2x}) + Mv]/M$; $v_{fy} = [m(v_{1y} - v_{2y})]/M$ **P61** $-(m/M)\,\langle v\cos\theta, v\sin\theta, 0 \rangle$ **P63** about 1×10^{-13} m/s

Chapter 4

P21 1.7×10^4 kg/m^3 **P23 (a)** 1.06×10^{-25} kg/atom **(b)** 2.02×10^8 atoms **(c)** 0.870 kg **P25** 1.35×10^4 N/m **P27** 450 N/m **P29** 1950 N/m **P31** 27 N/m **P33** 46 N/m **P35 (a)** 5×10^{-5} m^2 **(b)** 7.2×10^{-5} m^2 **(c)** 3.6×10^{-5} m^2 **P37** 0.21 mm **P39** 2.0×10^{11} N/m^2 **P41 (a)** $-F/(m_1 + m_2 + m_3)$ **(b)** $-m_3F/(m_1 + m_2 + m_3)$; $-(m_2 + m_3)F/(m_1 + m_2 + m_3)$ **(c)** $\langle -m_3F/(m_1 + m_2 + m_3), 0, 0 \rangle$ **P43** 3 N **P45** $\langle 10.2, 0, 0 \rangle$ m; $\langle 4.23, 0, 0 \rangle$ m/s **P47** depends on your data **P49** $3Mg$ **P51** -0.105 m **P53** 1.6 s **P55** 0.4 s **P57 (a)** 0.94 m/s **(b)** 15 m/s^2 **P61 (a)** $dp/dt = -mg\sin(s/L)$ **(b)** $d^2s/dt^2 + (g/L)s = 0$ **(c)** $2\pi\sqrt{(L/g)}$ **P63** 2710 m/s **P65** about 1×10^{19} molecules **P67** 20.4 m; 19.8 m

Chapter 5

P7 **(a)** $\vec{0}$ **(b)** 3 **(c)** −392 N **(d)** 392 N **(e)** 497.4 N **(f)** 306.2 N **(g)** −306.2 N **P9** **(a)** $\vec{0}$ **(b)** $\vec{0}$ **(c)** rope, floor **(d)** 195.3 N **(e)** −195.3 N **(f)** Earth, floor, rope **(g)** 163.9 N **(h)** 294 N **(i)** 130 N **P11** **(a)** 6098 N, 3980 N, 7840 N **(b)** 0.049, 0.032, 0.063 **P13** **(a)** 1040 N, 1420 N **(b)** 528 N, 726 N **P15** **(a)** 2.95 m/s^2 **(b)** 39.3 N **P17** 2.24 m/s, tangential **P19** **(a)** $(1.8 \times 10^{23}\ N)\hat{p}$ **(b)** 3.13×10^{29} kg · m/s **P21** $\langle -4.97 \times 1022, 1.88 \times 10^{22}, 0\rangle$ N, $\langle 6.47 \times 10^{22}, 1.71 \times 10^{23}, 0\rangle$ N **P23** **(a)** 3.67 m/s **(b)** tangent to path **(c)** inward, toward center **P25** **(a)** d **(b)** b **(c)** 8.74×10^{-15} N **P27** $\vec{0}$, 0.349 m/s, 14.0 kg · m/s, 0.977 N **P29** 1.76 m/s, 66.3 N **P31** 8.78 N **P33** **(a)** a **(b)** $\vec{0}$ **(c)** 492 N **(d)** $+x$ **(e)** $\langle 0, 492, 0\rangle$ N **(f)** $\langle 296, 369, 0\rangle$ N **P35** $\sqrt{Rg}$ **P37** **(a)** $\vec{0}$ **(b)** 8.71 N **(c)** 871 N **(d)** $\langle 0, 1130, 0\rangle$ N **(e)** 1.4×10^4 N/m **P39** 6.65×10^{12} kg **P41** 1.46 m/s **P43** **(a)** 13.7 m/s **(b)** direction changes **(c)** spring exerts a force **(d)** 300 N **(e)** 32.8 kg **P45** $\sqrt{4\pi^2 mR/(k_s(R-L))}$ **P47** 3.0×10^4 m/s **P49** **(a)** $\langle 0, 200, 0\rangle$ kg · m/s^2 **(b)** $\langle 0, -549, 0\rangle$ N **(c)** $\langle 0, 749, 0\rangle$ N **(d)** $\langle 0, -200, 0\rangle$ kg · m/s^2 **(e)** $\langle 0, -549, 0\rangle$ N **(f)** $\langle 0, 349, 0\rangle$ N **(g)** heavier **(h)** lighter **P51** 2.69 m/s; 1.21 s **P53** **(a)** 4.24×10^7 m **(b)** 0,48 s **(c)** 85 min = 1.4 hr **(d)** 7.9×10^3 m/s **(e)** 3250 s = 0.9 hr **P55** **(a)** 1.60×10^6 m/s$(0.005c)$ **(b)** yes **(c)** 1.11×10^{37} kg **(d)** 5.6 million solar masses

Chapter 6

P9 6 **P11** 0.99999985c; 1.5×10^{-10} J **P13** **(a)** 5.76×10^{-10} J **(b)** no **(c)** 1.48×10^{-10} J **(d)** C **P15** 81.6 J **P17** 0 **P19** 104 J **P21** 585 J **P23** $\langle 0, -4.9, 0\rangle$ N; 4.9 J **P25** −8.6 J **P27** 2.5×10^{-19} J **P29** 189 J **P31** 2.58×10^{-9} J; 4.59×10^{-10} J; 2.12×10^{-9} J; 7.28×10^{-9} J; 4.59×10^{-10} J; 6.82×10^{-9} J **P33** 10.2 m/s **P35** 4.5 m/s **P37** 9.81 m/s **P39** 1×10^{-13} N **P41** a, b, e, and f; 0.99 **P43** 0.5 MeV **P45** **(a)** 3.499767×10^{-8} J **(b)** 3.499708×10^{-8} J **(c)** decreased **(d)** 3.72 MeV **P47** **(b)** 1.194×10^{-12} J **(c)** 2.2×10^{-14} J **(d)** $\gamma = 1.002$ **P49** 19.8 m/s **P51** −2200 J **P53** 6.3 m/s **P55** **(a)** 9.8 J **(b)** 6.3 m/s **P57** 102 m **P59** **(a)** 2970 m/s **(b)** 2835 m/s **P61** 25.5 m/s **P63** 4.26×10^3 m/s **P65** 0.99c **P67** $mgR/2$, half **P69** **(a)** $mgL(1-\cos\theta)$ **(c)** $-mg\sin\theta$ **(d)** $\sqrt{4gL}$ **P71** **(b)** $\sqrt{1-m^2/M^2}c$, high speed **(c)** 0.99c **(d)** 8.4×10^{-18} m **P73** **(a)** $\langle 0, 13, 0\rangle$ m**(b)** $\langle 0, -784, 0\rangle$ N **(c)** -1.02×10^4 J **(d)** 0 **(e)** -1.02×10^4 J **(f)** 0 **(g)** -1.02×10^4 J **(h)** 0 **(i)** -1.02×10^4 J **(j)** -1.02×10^4 J **(k)** all the same **P75** **(a)** 1.26×10^7 m/s **(b)** 1.57×10^{-14} m **(c)** 6.38×10^{-15} m; gap is only 3×10^{-15} m **(d)** 1.9×10^{13} J

Chapter 7

P17 0.5 m/s **P19** **(a)** Mg/k_s; $-\frac{1}{2}(Mg)^2/k_s$ **(b)** $-2Mg/k_s$ **(c)** $\sqrt{k_s s^2/M - 2g(L-s)}$ **P21** 2.37 m **P23** 0.0746 m **P25** **(a)** −0.2 eV **(b)** 0.2 eV **(c)** $6a/r^7$ **P27** 80.5°C **P29** −516 J |31| **(a)** 58.9°C **(b)** energy transfer between system and surroundings is negligible; heat capacities independent of temperature **(c)** 1.81×10^4 J **P33** 490 J; 2040 kg/s **P35** 3 W **P37** **(a)** 9 J **(b)** 9 J **(c)** 3 m/s **(d)** 0.2 s **(e)** 0 **(f)** −9 J **(g)** 15 m/s **(h)** 18 m/s **(i)** 225 J **(j)** 325 J **(k)** 99 J **(l)** 3.3 m **(m)** 99 J **(n)** 3 m **(o)** 90 J **(p)** −9 J **P39** 31.9 m; too large **P41** **(a)** 353 N **(b)** 696 N **P43** **(a)** same as hand **(b)** small **(c)** large **(d)** reduced

Chapter 8

P9 13.6 eV **P11** 6190 atoms **P13** 1.4 eV **P15** **(a)** excited states of 1.91 eV and 2.48 eV above the ground state; excited states of 0.57 eV and 2.48 eV above the ground state **(b)** Bombard with electrons whose kinetic energy is greater than 0.57 eV and less than 1.91 eV and see whether there is a dark line. **P17** **(a)** 0.8, 1.9, 1.1, 3.8, 3.0, and 1.9 eV (note two transitions of 1.9 eV) **(b)** 1.9, 3.0, and 3.8 eV **P19** 11, 5, 2, 9, 3, and 6 eV; 6, 9, and 11 eV **P21** **(a)** 3, 2.5, 1.9, 0.5, 1.1, and 0.6 eV **(b)** 1.9, 2.5, and 3.0 eV **P23** **(a)** one possible scheme: ground state, and 0.3, 0.8, and 2.8 eV above the ground state **(b)** no **(c)** 0.3, 0.8, and 2.8 eV **P25** **(a)** 0.023 eV **(b)** 4.1×10^{-38} kg **(c)** 0.023, 0.046, and 0.069 eV **P27** **(a)** 0.015 eV **(b)** 1.8 eV **(c)** 120 **(d)** 0.03 and 0.045 eV

Chapter 9

P9 4.6×10^6 m from center of Earth **P11** $\langle L/4, L/4, 0\rangle$ from lower left; $\langle -3L/4, L/4, 0\rangle$ from lower right **P13** 4.7×10^4 J **P15** **(a)** $\langle 25, 97, 0\rangle$ kg · m/s **(b)** $\langle 3.125, 12.125, 0\rangle$ m/s **(c)** 809 J **(d)** 627.125 J **(e)** 181.875 J **(f)** 68.203 J and 113.672 J, which add to 181.875 J **P17** **(a)** center of mass moves with constant velocity **(b)** the centers of the two orbits coincide at the center of mass; M_1 must be greater than M_2 **(c)** 7.5×10^{29} kg and 5×10^{29} kg **P19** 14.3 J **P21** 395 J **P23** 10.5 rad/s **P25** **(a)** 2.99×10^4 m/s **(b)** 2.686×10^{33} J **(c)** 7.27×10^{-5} rad/s **(d)** 2.60×10^{29} J **(e)** 2.69×10^{33} J **P27** **(a)** 0.007 J **(b)** 0.90 m/s **(c)** 0.207 J **(d)** 313 rad/s **P29** **(a)** 6.75 m/s **(b)** 225 J **(c)** 65 J **P31** typical results are 1100 N, −350 J, 0.2 s **P33** **(a)** $\sqrt{gh + v_i^2}$ **(b)** $\left((M+m)gh + (M+\frac{1}{2}m)v_i^2)/(M+\frac{1}{2}m)\right)^{\frac{1}{2}}$ **P35** **(a)** $\sqrt{Fb/M}$ **(b)** $F(d-b)$ **P37** **(a)** $\left(2Fw/M + v_i^2\right)^{\frac{1}{2}}$ **(b)** $\left((2Fd)/(3mr^2) + \omega_i^2\right)^{\frac{1}{2}}$ **P39** **(a)** 24.2 J **(b)** 25.0 J **P41** $\sqrt{2Mg(y-y_0)/I + \omega_i^2}$ **P43** **(a)** $\sqrt{2Fd/(M+4m)}$ **(b)** $\left(Fw/(2mb^2 + \frac{1}{4}MR^2)\right)^{\frac{1}{2}}$ **P45** **(a)** $\sqrt{2Fd/(M+2m)}$ **(b)** $Fs - \frac{1}{4}\left(MR^2 + mL^2\right)\omega^2$

Chapter 10

P15 $v_{1,f} = v_{1,i}(m_1 - m_2)/(m_1 + m_2)$; $v_{2,f} = 2m_1 v_{1,i}/(m_1 + m_2)$ **P17** **(a)** $\langle 312, -398, 185\rangle$ m/s **(b)** 4.3×10^6 J **(c)** 1 **P19** **(a)** $\langle 2500, -3500, -1500\rangle$ kg · m/s **(b)** 8.02×10^7 J **(c)** 0 **(d)** 6.82×10^7 J **(e)** 1.20×10^7 J **(f)** 4.8×10^7 J **(g)** 0 **P21** **(a)** car + truck; Momentum Principle **(b)** 0 **(c)** 2.73×10^6 J **(d)** inelastic **P23** **(a)** $\langle -1.40 \times 10^{-19}, 0, 0\rangle$ kg · m/s **(b)** $\langle 2.86 \times 10^{-19}, 0, 0\rangle$ kg · m/s **(c)** 9.17 MeV **(d)** 0.78 MeV **(e)** 2.28×10^{-14} m **P25** $|\vec{p}_n| = |\vec{p}_{\pi^-}|$; $E_n + E_{\pi^-} = 1196$ MeV; $E_n^2 - (p_n c)^2 = (939\ \text{MeV})^2$; $E_{\pi^-} - (p_{\pi^-}c)^2 = (140\ \text{MeV})^2$ **P27** **(a)** 0.5c **(b)** 1.87 m **P29** **(a)** 3 m/s **(b)** 4.8×10^{-8} eV; negligible compared to 10.2 eV **P31** 259 MeV; 8×10^7 m/s **P33** (answers given for x components) **(a)** $mv_1/(m+M)$ **(b)** $Mv_1/(m+M)$; $-mv_1/(m+M)$ **(c)** The momenta have equal magnitudes before the collision, and after the collision they also have to have equal magnitudes in order for the total momentum to remain zero. If a magnitude changes, that would mean the energy changes, so the Energy Principle would be violated. **(d)** $-(M-m)v/(M+m)$; $2mv/(M+m)$

Chapter 11

P13 $22\,\hat{k}$ **P15** $\langle 0,0,144\rangle$ kg·m²/s **P17 (a)** out of page **(b)** $d_1 p_1 \sin\alpha$ **(c)** out of page **(d)** $d_2 p_2$ **P19** 7.1×10^{33} kg·m²/s **P21** 34.1 J **P23 (a)** 0.18 kg·m² **(b)** 14.2 J **(c)** 2.26 kg·m²/s **P25 (a)** 0.0765 kg·m² **(b)** 0.91 kg·m²/s **(c)** 6.28 J **P27 (a)** $\langle 29.4,0,0\rangle$ kg·m/s **(b)** $\langle 49,0,0\rangle$ m/s **(c)** $\langle 0,0,-25.9\rangle$ kg·m²/s **(d)** $\langle 0,0,4.95\rangle$ kg·m²/s **(e)** $\langle 0,0,-30.9\rangle$ kg·m²/s **(f)** $\langle 32.9,0,0\rangle$ kg·m/s **P29 (a)** 1.44 kg·m²/s **(b)** 9.72 kg·m²/s, into page **(c)** 11.16 kg·m²/s, into page **P31** $\langle 4,3.8,0\rangle$ kg·m²/s **P33** $d/dt\langle 0,0,xmv_y\rangle=\langle 0,0,-xmg\rangle$ **P35 (a)** −0.323 kg·m²/s **(b)** −0.323 kg·m²/s **(c)** −0.138 kg·m²/s **(d)** 1 **P37 (a)** 1, 3, 4, and 5 **(b)** 0.189 m/s **P39 (a)** 2.01 s **(b)** 1.64 s **P41 (a)** 7.42 N **(b)** 0.177 m **P43 (a)** about 0.5 s **(b)** about 100 J **P45 (a)** about 0.2 s **(b)** about 800 J **P47 (a)** $2m_1v_1+(-m_2v_2\cos\theta_2)=m_1v_3\cos\theta_3+m_2v_4\cos\theta_4$ **(b)** $-m_2v_2\sin\theta_2=2m_1v_3\sin\theta_3-m_2v_4\sin\theta_4$ **(c)** $2m_1(L/2)^2\omega_1+(-bm_2v_2\cos\theta_2)=$ $2m_1(L/2)^2\omega_2+bm_2v_4\cos\theta_4$ **P49 (a)** $\langle 2582,-9.6,0\rangle$ m/s, 7.375 rad/s out of page **(b)** 1.55×10^7 J **P51 (a)** $\langle 8.68,1.3,0\rangle$ m/s **(b)** 166 rad/s **(c)** 803 J **P53 (a)** 4.37×10^4 m/s **(b)** increases 1 hour **P55 (a)** $\langle 0,0,-0.0256\rangle$ kg·m²/s **(b)** $L_z=-0.186$ kg·m²/s **(c)** $\omega_z=-145$ rad/s **P57 (b)** 147 N·m **(c)** 1 and 3 **P59** 3.67 rad/s **P61 (b)** $\frac{1}{2}g\sin\theta$; $\frac{2}{3}g\sin\theta$; $\frac{5}{7}g\sin\theta$ **(c)** same time; sphere **(d)** $\frac{1}{2}\left(M+I_{CM}/R^2\right)v_{CM}^2=Mg\sin(\theta)\Delta x$; $\frac{1}{2}Mv_{CM}^2=\left(Mg\sin\theta-I_{CM}/R^2\,(dv_{CM}/dt)\right)\Delta x$; time derivatives agree with force and torque analyses **P63** 12 rad/s **P65 (a)** 20.9 rad/s **(b)** 104.7 rad **(c)** 6000° **P67 (a)** 4 rad/s² **(b)** 30 rad/s **(c)** 90 rad **(d)** 5160° **P69 (a)** 62.8 rad/s **(b)** 377 rad **(c)** 69.8 rad/s **(d)** 332 rad **(e)** 99.5 m **P71** 3.55 rad, 203° **P73** $0.543L$ **P75** 3.15×10^{-34} J·s **P77** −54.4 eV, −13.6 eV, 40.8 eV, all 4 times the values for hydrogen **P79 (a)** 2.38×10^{-54} kg·m² **(b)** 4.7×10^{-15} m **(c)** 7×10^{-15} m; nucleus may not be spherical **P83 (a)** clockwise **(b)** 32 s **P85 (a)** clockwise as seen from above **(b)** slower precession **(c)** same

Chapter 12

P17 252; 1 **P19** 18000 **P21** 8.04×10^{-22} J **P23** 5.4×10^{88} year, or about 5×10^{78} times the age of the Universe **P25 (a)** 5.65×10^{-23} J/K **(b)** 6.62×10^{-23} J/K**(c)** 1.23×10^{-23} J/K **P27 (a)** 7.86 J/K **(b)** −7.74 J/K **(c)** 0.12 J/K **(d)** 0 **P29** $E=\frac{1}{4}a^2T^2$ **P31** $C=\frac{1}{2}a^2T/N_{atoms}$ (J/K/atom) **P33 (a)** 215.55 K **(b)** 273.02 K **(c)** 3.48×10^{-23} J/K/atom **P35 (a)** 3.61006×10^{-21} J **(b)** temperatures 118.7 K and 121.0 K **(c)** 2.62×10^{-23} J/K/atom **P37** about 250 K **P39** 25.1°C **P41 (a)** $v_i+F\Delta t/(M+m)$ **(b)** $\omega_i+2(F+f)\Delta t/(MR)$ **(c)** $F(d-x)-\frac{1}{4}MR^2(\omega_f^2-\omega_i^2)$ **(d)** 500 K **P43** 4.04×10^{-21} J, 1/40 eV **P45** 8300 m **P47** 1350 m/s; 510 m/s **P49** similar to Figure 12.54, but with different break points **P51** about 2 cm **P53** about 600 m/s; rms speed is larger; roughly one-tenth **P55 (a)** about 0.004 **(b)** 1×10^{17} **(c)** 120 km **P57 (a)** helium **(b)** 8 K **P59** 108 K

Chapter 13

P17 $\langle 0,-4.48\times 10^{-17},0\rangle$ N **P19** $\langle 3.2\times 10^{-15},3.2\times 10^{-15},0\rangle$ N **P21 (a)** negative **(b)** down and to the left **(c)** $\langle 0.707,0.707,0\rangle$ **(d)** $\langle -1.4\times 10^{-5},-1.4\times 10^{-5},0\rangle$ N **(e)** $\langle -0.707,-0.707,0\rangle$ **P23** $\langle 0,0,-9.13\times 10^4\rangle$ N/C **P25** 9.38×10^3 N/C **P27 (a)** $\langle -0.6,-0.7,-0.2\rangle$ m **(b)** $\langle 0.5,-0.1,-0.5\rangle$ m **(c)** $\langle 1.1,0.6,-0.3\rangle$ m **(d)** 1.29 m **(e)** $\langle 0.854,0.466,-0.233\rangle$ **(f)** 48.8 N/C **(g)** $\langle 41.7,22.7,-11.4\rangle$ N/C **P29** $\langle -2550,-2550,0\rangle$ N/C **P31** 1.13×10^4 N/C **P33** $\langle -225,0,0\rangle$ N/C **P35 (a)** $\langle 0.8,0.7,-0.8\rangle$ m **(b)** $\langle 0.5,1,-0.5\rangle$ m **(c)** $\langle -0.3,0.3,0.3\rangle$ m **(d)** 0.520 m **(e)** $\langle -0.577,0.577,0.577\rangle$ **(f)** -5.33×10^{-9} N/C **(g)** $\langle 3.08\times 10^{-9},-3.08\times 10^{-9},-3.08\times 10^{-9}\rangle$ N/C **P37** $\langle 3,0,0\rangle$ m **P39 (a)** $\langle 0,-5.92\times 10^{-7},0\rangle$ m **(b)** $\langle 0,5.92\times 10^{-7},0\rangle$ m **P41** $\langle -960,-960,0\rangle$ N/C **P43** $\langle 0,3\times 10^{-6},0\rangle$ m **P45 (a)** $+x$ **(b)** 7.31×10^5 N/C **(c)** -7.69×10^{-8} m to the left **P47 (a)** $\langle -1.15\times 10^7,8.64\times 10^6,0\rangle$ N/C **(b)** $\langle 0,-3\times 10^7,0\rangle$ N/C **(c)** $\langle -1.15\times 10^7,-2.13\times 10^7,0\rangle$ N/C **(d)** $\langle 23.0,42.6,0\rangle$ N/C **(e)** $\langle 0,4\times 10^7,0\rangle$ N/C **(f)** $\langle -8.64\times 10^6,-6.48\times 10^6,0\rangle$ N/C **(g)** $\langle 1.125\times 10^7,0,0\rangle$ N/C **(h)** $\langle 2.61\times 10^6,3.35\times 10^7,0\rangle$ N/C **(i)** $\langle -7.83\times 10^{-3},-0.101,0\rangle$ N/C **P49 (a)** $\langle 1.92\times 10^5,0,0\rangle$ N/C **(b)** $\langle -1.27\times 10^5,0,0\rangle$ N/C **(c)** $\langle -3.07\times 10^{-14},0,0\rangle$ N **P51 (a)** $\vec{0}$ **(b)** 3.38×10^4 N/C **(c)** 2.22×10^3 N/C **P53** 1.07×10^4 N/C **P55** $\langle 0,-4.43\times 10^{-15},0\rangle$ N **P57 (a)** 64.8 N/C **(b)** 32.4 N/C **P59** $\langle 0,105,0\rangle$ N/C **P61 (a)** right end **(b)** 8.3×10^{-8} C

Chapter 14

P27 4×10^{-10} C **P29** 2 and 3 **P31** 1 **P33** 1/243 **P35 (a)** 2 **(b)** g **(c)** c **(d)** g **P37 (a)** 894 m/s² **(b)** 1/32 **P41** c **P43** 1.25×10^{-4} m/s **P45** 3 **P47** b **P49** j **P51 (a)** 1 **(b)** 3 **(c)** 2 **P53 (a)** B **(b)** 3 **P55** 4.08×10^{-4} m/s **P57 (a)** 675 N/C **(b)** 9.1×10^{-3} N/C **(c)** increase **(d)** no **P59 (a)** $(1/(4\pi\epsilon_0)\langle 2p/(L+R+b)^3-Q/(R+b)^2,-p/(L+R+b)^3,0\rangle$ **(b)** $L\gg s$ **(c)** $\vec{E}_4$ points to the right and downward; $\vec{E}_{ball}$ points to the left and upward **(d)** positive surface charge on the lower right, negative surface charge on the upper left **(e)** $\vec{0}$ **P61** 1, 2, and 4 **P63 (a)** 2.5 nC **(b)** 3

Chapter 15

P21 (a) $-Q/(2A)$ **(b)** $-Qdx/(2A)$ **(c)** $\langle -x,y,0\rangle$ **(d)** $\sqrt{x^2+y^2}$ **(e)** x **P23 (a)** 0.024 m **(b)** -4×10^{-10} C **(c)** Q/N **(d)** L/dL **(e)** $(Q/L)dL$ **P25 (a)** $-(1/(4\pi\epsilon_0))2Q\,|\vec{p}|\,/(Lx^2)$ **(b)** 2×10^4 m/s² **P27 (a)** each piece is an arc of radius R and angle $\Delta\theta$ **(b)** $(1/(4\pi\epsilon_0))(|Q|/(\alpha R^2))\Delta\theta\,\langle\cos\theta,\sin\theta,0\rangle$ **(c)** $(1/(4\pi\epsilon_0))(|Q|/(\alpha R^2))\,(1-\cos\alpha)$ **(d)** units correct; field points toward rod midpoint (symmetry); for small α, field approaches that of a point charge **P29 (a)** small-radius rings of charge; contribution of each ring is to the right **(b)** $(1/(4\pi\epsilon_0))(Q/L)\Delta x/\,(L+d-x)^2\,\langle 1,0,0\rangle$ **(c)** $(1/(4\pi\epsilon_0))Q/(d(L+d))$ **(d)** for $d\gg L$, field approaches that of a point charge **P31 (a)** 3.048×10^4 N/C toward the negative ring **(b)** 2.743×10^{-4} N toward the positive ring **P33** 0 **P35 (a)** 2.031×10^6 N/C **(b)** 2.012×10^6 N/C **(c)** differs by less than 1% **P37 (a)** 2.795×10^6 N/C **(b)** 2.760×10^6 N/C **(c)** differs by only 1.3% **P39 (a)** negative on left face, positive on right face **(b)** 0 **(c)** 2.66×10^{-9} C **P41** 1.84×10^{-5} C **P43** 1.01×10^{-4} C **P45 (a)** curves downward between plates, then moves in a straight line down and to the right **(b)** 1.8×10^{16} m/s² **(c)** 3.18×10^{-9} C; negative **P47** $Q(s/R)/(2-s/R-2t/R)$ **P49 (a)** metal ball polarizes with right side more negative than left side; molecules in plastic polarize with positive ends pointing toward metal ball **(b)** 7780 N/C to the left, ignoring the field produced by the polarized molecules in the plastic **(c)** 0 **(d)** 7780 N/C to the right **P51** 4.59×10^4 N/C inward; 3.51×10^4 N/C outward **P53 (a)** $\langle -5 \times 10^{-5}, 0, 0\rangle$ N/C **(b)** 0 **(c)** left side more positive than right side **P55 (a)** dipoles are radial with negative ends pointing toward center **(b)** no polarization **(c)** same **(d)** negative on inner surface, positive on outer surface **(e)** $-Q$ on inner surface, $+Q$ on outer surface **(f)** same **P57 (a)** 249 N/C; 0 **(b)** 364 N/C; −182 N/C **(c)** -5.82×10^{-17} N; 2.91×10^{-17} N **P59 (a)** $(1/(4\pi\epsilon_0))qs/L^3$ to the right **(b)** 0; right side of ball more positive than left side; dipole field points to the right and field of charges on ball points to left **P61** $(1/(4\pi\epsilon_0))^2(4\alpha Q^2/L)/|\vec{d}|^4$

Chapter 16

P21 3.11×10^6 m/s **P23 (a)** 1.64×10^{-25} J **(b)** 3.01×10^{-20} J **P25** -7.52×10^{-23} J **P27** 3.6×10^4 V/m **P29 (a)** $\langle -1, 0, 0\rangle$ m **(b)** 750 V **(c)** 1.2×10^{-16} J = 750 eV **(d)** -1.2×10^{-16} J = −750 eV **P31 (a)** $\langle 0.7, 0, 0\rangle$ m **(b)** 595 V **(c)** 9.52×10^{-17} J = 595 eV **(d)** -9.52×10^{-17} J = −595 eV **P33** 7860 V/m **P35** 9×10^3 V **P37** 1500 V **P39** B; 150 V/m to the left **P41 (a)** to the right **(b)** for example, $R = 0.1$ m, $Q_A = 2.78 \times 10^{-8}$ C and $Q_B = -2.78 \times 10^{-8}$ C **P43** −1200 V **P45** 1.92×10^{-19} J **P47 (a)** −205 V **(b)** 205 V **(c)** -3.28×10^{-17} J **(d)** 4.64×10^{-17} J **P49 (a)** $(q^2/(4\pi\epsilon_0))(1/((x-s/2)^2+y^2)^{1/2} - 1/((x+s/2)^2+y^2)^{1/2})$ **(b)** $(1/(4\pi\epsilon_0))px/r^3$ **(c)** $(1/(4\pi\epsilon_0))2p/x^3$ **(d)** E is in x direction **P51** $(Q_2b - Q_1a)/(A\epsilon_0)$ **P53** $(1/(4\pi\epsilon_0))2(Q/L)b/r$ **P55** $(1/(4\pi\epsilon_0))Qe(1/R - 1/(R+h))$ **P57** **(a)** $(Q_1/(4\pi\epsilon_0))(1/L - 1/(L+R_2))$ **(b)** $(Q_2/(4\pi\epsilon_0))(1/(R_2+d) - 1/R_2) + (Q_1/(4\pi\epsilon_0))(1/(L+R_2) - 1/(L+R_2+d)$ **(c)** decrease **P59 (a)** 37 V **(b)** 3214 V **P61** $(2Q/(4\pi\epsilon_0))(1/(R+d+L) - 1/(R+d))$ **P63** $(1/(4\pi\epsilon_0))(Q/R)\ln(d/(d-h)) + (Q/(2\epsilon_0))(h/A)$ **P65** 8 V **P67 (a)** 0 **(b)** 0 **P69** $(1/(4\pi\epsilon_0))(q_1/r_{1,A} - q_2/r_{2,A})$ **P71** −450 V **P73** 608 V/m **P75** $-(1/(4\pi\epsilon_0))Q/(a^2+h^2)^{1/2}$ **P77** $(11/8)(1/(4\pi\epsilon_0))(Q/R)$ **P79 (a)** $Q_2 = -Q$; $Q_3 = 5Q$ **(b)** $(1/(4\pi\epsilon_0))5Q/r_3$; $(1/(4\pi\epsilon_0))5Q/r_3$; $(1/(4\pi\epsilon_0))5Q/r_3$; $(1/(4\pi\epsilon_0))(5Q/r_3 - Q/r_2 + Q/r_1)$; $(1/(4\pi\epsilon_0))(5Q/r_3 - Q/r_2 + Q/r_1)$ **P81 (a)** no polarization **(b)** $-(1/(4\pi\epsilon_0))Q/R$ **P83** 250 V; 100 V; 250 V; 600 V **P85** $(1/(4\pi\epsilon_0))Qz/(z^2+R^2)^{3/2}$ **P87 (a)** 0 **(b)** $-(1/(4\pi\epsilon_0))qs(1/a^2 - 1/b^2)$ **(c)** $(1/(4\pi\epsilon_0))eqs(1/a^2 - 1/b^2)$

Chapter 17

P15 $\langle 0, 0, 7.5\rangle$; $\langle 0, 0, -7.5\rangle$ **P17** $\langle 15, 15, 6\rangle$ **P19** 0 at 2 and 5; $\langle 0, 0, -1.28 \times 10^{-17}\rangle$ T at 1 and 6; $\langle 0, 0, 1.28 \times 10^{-17}\rangle$ T at 3 and 4 **P21** 0 at 2 and 5; $\langle 0, 0, -5.28 \times 10^{-18}\rangle$ T at 3 and 4; $\langle 0, 0, 5.28 \times 10^{-18}\rangle$ T at 1 and 6 **P23 (a)** electric $\langle -1, 0, 0\rangle$; magnetic $\langle 0, 0, 1\rangle$ **(b)** electric 5.76×10^9 N/C; magnetic 0.66 T **P25 (a)** 7.5×10^{19} electrons/s **(b)** 3.15×10^{-5} m/s **P27** 8.98×10^{28} m^{-3} **P29 (a)** 3.5×10^{-6} T **(b)** north **P31** 5.14×10^{-6} T **P33 (a)** $+x$ **(b)** 0.1625 m **(c)** 0.1625 m **(d)** $\langle 0.1625, 0, 0\rangle$ m **(e)** $\langle -0.08125, 0, 0\rangle$ m **(f)** $\langle 0.16225, 0.178, 0\rangle$ m **(g)** $\langle 0.674, 0.739, 0\rangle$ **(h)** $\langle 0, 0, 0.120\rangle$ m **(i)** $\langle 0, 0, 2.95 \times 10^{-6}\rangle$ T **P35 (a)** conventional current down and to the left; magnetic field down and to the right **(b)** 0.088 A **P37 (a)** 0 **(b)** $\langle 0, 0, hdx\rangle$ **(c)** $+x$ **P39 (a)** 0.0917 A **(b)** $\langle 0, 0, -1.63 \times 10^{-6}\rangle$ T **(c)** assumed $r \ll L$; neglected curved part of wire **P41 (a)** 1.11×10^{-7} T **(b)** 1.11×10^{-5} T **P43** 1.5 A **P45** $(\mu_0/(4\pi))I\pi(1/R_1 + 1/R_2)$; direction $\langle 0, 0, -1\rangle$ **P49 (a)** into the page **(b)** $(\mu_0/(4\pi))I(\pi/R + 2/h)$ **P51** $(\mu_0/(4\pi))I\pi(1/R + 3/r)$; $\langle 0, 0, -1\rangle$ **P53** 4.3 A·m^2 **P55** 0.5 A·m^2 **P57 (a)** S end **(b)** 1 and 5 **(c)** 12

Chapter 18

P21 1, 3, 6, 8, and 9 **P23** $i_A = i_B + i_C$, $i_B + i_C = i_D$, $i_A = i_D$; i_D **P25 (a)** 11 A **(b)** yes **(c)** 5 A **(d)** no **P27** 0.016 V/m **P29** 2, 5, 8, and 9 **P31** 6.24×10^{19} electrons/s **P33** 1; 6×10^{18} electrons/sl 0.26 V/m **P35 (a)** downward **(b)** upward **(c)** 19.8 V/m; to the left **P37** 5 V/m; same **P39** 3.33 V/m; same **P41 (a)** 3.5 V **(b)** C **P43 (a)** 2 **(b)** $1.3\text{ V} + V_E - V_G + 1.3\text{ V} + V_B - V_D = 0$ **(c)** 5 V/m **(d)** 6.73×10^{18} s^{-1} **(e)** 5 V/m **(f)** (c) **P45 (a)** 1, 2, and 4 **(b)** $1.3\text{ V} + 2(V_A - V_C) + V_C - V_E = 0$ **(c)** $i_B = i_D = i_F$ **(d)** 21.9 V/m; 0.925 V/m **P47 (a)** 1.2×10^{18} s^{-1} **(b)** neglect connection wires; filaments same lengths; ideal battery (no internal resistance) **(c)** high positive density near + terminal of battery; small gradient along wires; large gradient across bulbs (largest along thin-filament bulb) **P49 (a)** 3×10^{17} s^{-1} **(b)** 4.5×10^{17} s^{-1} **(c)** do the experiment **(d)** 6×10^{17} s^{-1} **P51 (a)** 2.27 V/m **(b)** 2.27 V/m **(c)** 3 **P53** 18.7°

Chapter 19

P29 d **P31** c **P33** d **P35** 1, 3, and 4 **P37 (a)** 1, 2, 3, and 5 **(b)** 1, 2, 3, and 5 **(c)** c **(d)** a **P39** 1.85×10^{-8} F **P41** 8×10^{-10} C **P43** battery: $-2E$, 0; bulb: 0, 0; capacitor: $+E$, $-E$; surroundings: $+E$, $+E$ **P45** 4.06×10^7 $(\Omega\cdot\text{m})^{-1}$ **P47** 40 Ω **P49 (a)** 0.83 Ω, 3.6 A **(b)** 4.2×10^{-4} Ω, 15.0015 V **P51** 240 Ω; 280 Ω **P53** bulb resistance increases with temperature **P55** 3 C; 1.9×10^{19} electrons **P57** 5×10^{-3} V/m **P59 (a)** field is small in thick wires, large in thin wires; field goes counterclockwise **(b)** roughly, a small gradient along the thick wires, large along the thin wires **(c)** $n(1/4)\pi d_2^2 uK/\left((d_2/d_1)^2(2L_1 + L_3) + 2L_2\right)$ **(d)** $n(1/4)\pi d_2^2 uK/(2L_2)$ **(e)** $K/4$ **P61 (a)** E decreases with x **(b)** mobility increases with x (lower temperature) **(c)** $I/(ne(wh)(u_0 + kx))$ **(d)** negative **(e)** $(I/(ne(wh)k))\ln((u_0 + kd)/u_0)$ **(f)** $(1/(ne(wh)k))\ln((u_0 + kd)/u_0)$ **P63 (a)** $3.0\text{ V} - I_4(15\,\Omega) - I_3(30\,\Omega + 20\,\Omega) = 0$; $3.0\text{ V} - I_4(15\,\Omega) - I_1(20\,\Omega) = 0$; $I_4 = I_3 + I_1$ **(b)** compass 1 deflects 3° toward NE; compass 2 deflects 4.2° toward NW **(c)** 126 V/m **(d)** $E_1 > E_2$ **(e)** 4.6 mm **(f)** −2.4 V **(g)** 4.8×10^{-19} J **(h)** 0.153 W **P67** 4 A **P69 (a)** 34.3 Ω **(b)** 0 **(c)** 0.26 A **P71 (a)** 15 A **(b)** 0 **P73** 10 V **P75 (a)** $C\times$emf **(b)** field: polarized molecules reduce fringe field, so current runs again; potential: gap voltage reduced, so current runs again **(c)** $(\text{emf}/R)(1 - 1/K)$ **(d)** $KC\times$emf **P77 (a)** $E = 3600$ V/m; too small to make a spark **(b)** 3.83×10^{-7} C **(c)** 6.12 V **(d)** 3.02 V **P79 (a)** $ABCEFGA$: $20 - 10I_1 - 15I_4 - 12I_6 - 20I_1 = 0$; $DECD$: $5 + 15I_4 - 20I_2 = 0$; $DFED$: $-30I_3 + 12I_6 - 5 = 0$; C: $I_1 = I_2 + I_4$; D: $I_2 = I_5 + I_3$; F: $I_3 + I_6 = I_1$ **(b)** do the check **(c)** 33.2 V **(d)** 1.63 W **(e)** 1730 V/m

Chapter 20

P25 $\langle 0, 0, 1.92 \times 10^{-14}\rangle$ N **P27** $\vec{0}$ **P29** 1.64 T **P31 (a)** left accelerating plate is positive; top deflection plate is positive; B into page **(b)** 1600 V; 320 V **(c)** 1371 V; 274 V **P33 (a)** 8.9° **(b)** 4×10^{-17} N **P35 (a)** CCW **(b)** 9.2 T **P37** $\langle 0, 0, 0.243\rangle$ N

P39 **(a)** 5.52 A **(b)** $-y$ **P41** **(a)** $\frac{1}{4}(\mu_0/(4\pi))2\pi I(1/a - 1/b)$; $\langle 0, 0, 1\rangle$ **(b)** $\frac{1}{4}(\mu_0/(4\pi))2\pi Iev(1/a - 1/b)$; $\langle 0, 1, 0\rangle$ **P43** 1.14 T into the page **P45** 9500 m/s **P47** $\langle 0, 0, -1863\rangle$ V/m **P49** **(a)** $-z$ **(b)** $-y$ **(c)** 1.08×10^{-4} T **(d)** 61.6 V/m **P51** **(a)** B **(b)** upward **(c)** upward **(d)** 3.30×10^7 m/s; 2.40×10^{-3} T **(e)** 2, 5, 6 **(f)** 650 V **P53** $(5/4)(\mu_0/(4\pi))2\pi IV/R, -y$ **P55** **(a)** positive **(b)** 4.8×10^{-3} m/s **(c)** 9.8×10^{-4} (m/s)/(V/m) **(d)** 4×10^{23} m^{-3} **(e)** 2.4 Ω **P57** **(a)** If you connect the + lead to the bottom of the bar (with the – lead straight above it) and the voltmeter reads positive, conventional current in the long wire goes to the right. **(b)** $nedh\Delta V/((\mu_0/(4\pi))2I_{\text{ammeter}})$ **P59** 1.77×10^{-5} V; 0.022 A **P61** 2 **P63** **(a)** 1 **(b)** 0 **(c)** slightly less than 4.48×10^{-19} N **(d)** 4.48×10^{-19} N **(e)** 1.26 V **(f)** to the left **P65** **(a)** 0; 0 **(b)** vBh/R, vB^2h^2/R **(c)** 0, 0 **(d)** vBh/R, vB^2h^2/R **(e)** 0, 0 **P67** west end has excess electrons; about 2500 electrons **P69** **(a)** negative below, positive above **(b)** 0.96A, CCW **(c)** 2 **(d)** 0.035 N to the right **P71** $\langle 0, 1.2, 0\rangle$ N **P73** 120 J **P75** **(a)** 24 V **(b)** 11.5 W **P77** **(a)** $+1.6 \times 10^{-19}$ C **(b)** 14.4 V **(c)** 0 **(d)** –14.4 V **(e)** 14.4 eV **(f)** 14.4 eV **P79** 7.5×10^{-6} C

Chapter 21

P7 left –0.24 V · m; right 0.60 V · m; all other faces 0; total 0.36 V · m; 3.19×10^{-12} C **P9** contains -5×10^{-11} C **P11** contains 1.8×10^{-12} C **P13** total flux 0.0672 V · m; contains 5.95×10^{-13} C **P15** –227 V · m **P17** **(a)** $+Q$ on surface of inner wire; $-Q$ on inner surface of outer tube **(b)** $(Q/L)/(2\pi\epsilon_0 r)$ **(c)** 0 **P19** **(a)** $(Q/A)/\epsilon_0$ **(b)** assume E varies and show this is inconsistent with Gauss's law **(c)** $(Q/A)/\epsilon_0$ **(d)** $Qs/(2R)$ **P21** **(a)** S/ϵ_0 **(b)** the field of a large sheet would be $(1/2)S/\epsilon_0$ **P23** 125 A, out of page **P25** **(a)** left above, right below **(b)** $\mu_0 NI/(2L)$ **(c)** $\mu_0 NI/L$ **(d)** 0 **P27** $(\mu_0/(4\pi))(I_1 - I_2)/R$, out of page **P29** 1×10^{36} N/C/m

Chapter 22

P13 1.5 V; CCW **P15** CCW; $|dB/dt|\pi r_1^2/R$; tangent, CCW; $|dB/dt|r_1^2/(2r_2)$ **P17** **(a)** 2.6×10^{-6} T · m^2 **(b)** treat coil as magnetic dipole **(c)** there is no electric field **(d)** $-y$ **(e)** 2.16×10^{-7} V **(f)** 2.16×10^{-7} V **(g)** 8.59×10^{-7} V/m **(h)** 8.59×10^{-7} V/m **P19** **(a)** $(\mu_0/(4\pi))(2N_1N_2\pi^2 r_1^2 r_2^2/x^3)(b + 2ct)$ **(b)** downward **(c)** $(\mu_0/(4\pi))(N_1N_2\pi r_1^2 r_2/x^3)(b + 2ct)$ **P21** **(a)** upward at 1, downward at 2 **(b)** 0.404 mV **(c)** B uniform across coil; used average dB/dt **P23** **(a)** sine wave in coil 1, –cosine wave in coil 2 **(b)** 0.233 V; assume $z >> r_1$ and coils are thin **P25** **(a)** $+z$ **(b)** $+x$ **(c)** 2.33×10^{-5} T · m^2 **(d)** coils are thin; B uniform across coil 2 **(e)** 7.76×10^{-6} T · m^2 **(f)** 3.88×10^{-5} V **(g)** 10.7 mV **(h)** 2.06×10^{-4} V/m **(i)** 1 and 4 **P27** **(a)** CCW **(b)** 13.6 mA **P29** **(a)** $(\pi r_1^2/R)\,|(\mu_0 N(-k))/d|$; to the right **(b)** much larger current in ring **P31** **(a)** eBR **(b)** $\pi r^2 be$ **P33** 1.13×10^{-6} V/m **P35** **(a)** CCW **(b)** $(e/m)(\mu_0/(4\pi))2I_0Ar/R^3$ **(c)** tangent to orbit, CW **(d)** slows down **P37** $(\mu_0/(4\pi))3\pi R^2 Iyv/x^4$ **P39** 420 V, 1.14 A **P41** **(a)** radius = 0.01 m, length = 0.04 m, 1600 turns **(b)** 18.8 A

Chapter 23

P17 $dE/dt = 1.3 \times 10^{18}$ V/m/s inside the loop **P19** 1×10^{-5} T in $-y$ direction **P21** 0.024 T in $-y$ direction **P23** **(a)** 1.16×10^{17} m/s^2 **(b)** 1.69×10^{-7} V/m **(c)** 4 **(d)** 5 **(e)** 8 **P25** **(a)** 5×10^{-10} s **(b)** $+y$ **(c)** $-y$ **P27** **(a)** electrons have low mass so experience larger acceleration than the ions; electric field at A is in $+y$ direction **(b)** electric field at B is in $+y$ direction; electric field at C is up and to the right **(c)** 1×10^{-8} s; into page **(d)** proportional to $1/r^2$ instead of $1/r$ **P29** **(a)** 6.4×10^{-12} V/m to the left **(b)** 50 ns **(c)** upward **(d)** 1.07×10^{-9} V/m; much larger **(e)** to the right **P31** **(a)** to the left **(b)** 4×10^{-8} s; 5×10^{-10} s **(c)** E out of page, B up **P33** 60 nM **P35** 300 m; 3 m; 698 nm; 400 nm **P37** 550 V/m **P39** 2.66×10^8 m/s **P41** **(a)** 3 **(b)** 65.6° **P43** **(a)** 63.4° **(b)** 42.3° **(c)** 4.2 m **P45** 0.23 m to the left of the lens; virtual; no image **P47** $\langle 100, -4, 0\rangle$ cm; real **P49** 0.302 m; 1.5 cm; inverted

Supplement S1

P3 2.4×10^{22} molecules/s **P5** **(b)** down to 1 percent in 4.6 min **P7** about 2 hours **P9** **(a)** 1.0×10^5 N/m^2 **(b)** 1300 m/s **(c)** 4×10^{20} atoms **(d)** helium leaves faster **(e)** decrease **P11** 3.17 **P13** **(a)** 2910 J **(b)** –2910 J **(c)** 0 J **P15** **(a)** 214 K **(b)** 313.6 J **(c)** 2784 N **(d)** 2652 N **(e)** 2514 N **P17** 25 s; increase **P19** 0.05 J **P21** **(a)** $NkT_H \ln(V_2/V_1)$ **(b)** $(T_H/T_L)^{3/2} = V_3/V_2$ **(c)** $NkT_L \ln(V_4/V_3)$ **(d)** $(T_H/T_L)^{3/2} = V_3/V_2 = V_4/V_1$; putting the pieces together, one finds that Q_H/T_H does indeed equal Q_L/T_L **P23** 5.9 mm

Supplement S2

P1 5×10^{-8} m

Supplement S3

P13 7.5×10^{14} Hz; 1.3×10^{-15} s **P15** 1.33 m **P17** **(a)** reradiation; electric field in and out of page, magnetic field in the xy plane, perpendicular to the electric field **(b)** 23.6°; 11.5° **P19** 23.6°; 53.1° **P21** 1.2×10^{-10} m; 1.6×10^{-10} m **P23** 212 nm **P25** **(a)** 28 cm **(b)** wider **(c)** narrower **P27** 9.4 N **P29** 0.02 m; 0.6 m; 7.5 ms; 1.67 m^{-1}; 838 rad/s; 133 Hz **P31** $0.04 \cos((2\pi/0.2)(x + 45t))$ **P33** 8.2×10^{-14} m; extremely small compared to 1×10^{-10} m **P35** **(a)** 89.6 m/s **(b)** 128 Hz, 70 cm; 192 Hz, 46.7 cm; 256 Hz, 35 cm; 320 Hz, 28 cm **P37** 365 nm **P39** 774 V/m; 2.8×10^{20} photons/s

Index

The text of Supplements 1–3 can be found at www.wiley.com/college/chabay

T